MODERN ELECTROMAGNETIC SCATTERING THEORY WITH APPLICATIONS

MODERN ELECTROMAGNETIC SCATTERING THEORY WITH APPLICATIONS

Andrey V. Osipov
Microwaves and Radar Institute
German Aerospace Center (DLR)
Germany

Sergei A. Tretyakov
School of Electrical Engineering
Aalto University
Finland

WILEY

This edition first published 2017

Registered office
John Wiley & Sons, Ltd, The Atrium, Southern Gate, Chichester, West Sussex, PO19 8SQ, United Kingdom

For details of our global editorial offices, for customer services and for information about how to apply for permission to reuse the copyright material in this book please see our website at www.wiley.com.

Library of Congress Cataloging-in-Publication Data

Names: Osipov, Andrey (Andrey V.) | Tretyakov, Sergei.
Title: Modern electromagnetic scattering theory with applications / Andrey Osipov, Sergei Tretyakov.
Description: Chichester, West Sussex : John Wiley & Sons, Inc., 2017. |
Includes bibliographical references and index.
Identifiers: LCCN 2016028695 | ISBN 9780470512388 (cloth) | ISBN 9781119293293 (epub)
Subjects: LCSH: Electromagnetic waves–Scattering. | Electromagnetic fields. | Radar cross sections.
Classification: LCC QC665.S3 .O85 2017 | DDC 539.2–dc23 LC record available at https://lccn.loc.gov/2016028695

A catalogue record for this book is available from the British Library.

Cover image: Art Phaneuf Photography/Shutterstock

Set in 9/11pt Times by Aptara Inc., New Delhi, India

10 9 8 7 6 5 4 3 2 1

MIX
Paper from responsible sources
FSC www.fsc.org FSC® C013604

Contents

Preface

Besides classical applications (radar and stealth technologies, antennas, microwave engineering, optics), scattering and diffraction are enabling phenomena for some emerging research fields (artificial electromagnetic materials or metamaterials, terahertz technologies, electromagnetic aspects of nanoscience, cloaking from electromagnetic fields). Although an overwhelming amount of material has been published in the literature over the years, addressing the mathematical formulation of scattering problems, solution methods, analysis and simulation techniques, and the electromagnetic properties of particular geometries and materials, there is still no textbook suitable for a graduate or post-graduate course in modern electromagnetic scattering theory. In the course of our teaching and research activities we have realized that the available books are either too general or too specialized or simply very old, and there is an urgent need to present the theory in a systematic and uniform way, addressing the latest developments and topical applications, and packed with new examples and numerical visualizations. We hope that this book will fill the gap between the general electromagnetic theory courses and collections of engineering formulas by connecting the theory to working examples of solutions for practical scattering and diffraction problems.

It is of course impossible to cover in a single book the whole spectrum of topics related to diffraction and scattering. When selecting the material we have limited the scope to those chapters of the theory that are relevant to deterministic electromagnetics and primarily focused on analytical methods and solutions. It is assumed that all relevant geometric and material parameters of the problem are known, and most of the solutions are derived for far fields in the low- or high-frequency approximations upon exact formulation of the scattering problem. The book addresses the following topics:

1. Fundamental definitions, equations, and concepts (Maxwell's equations; constitutive relations; exact and approximate boundary conditions; electromagnetic Huygens' principle; scattering matrix and cross sections; geometrical theory of diffraction and physical optics).
2. Solutions for basic canonical geometries (flat material interfaces; circular cylinders; spheres; wedges; polygonal and circular plates; conical points) with an emphasis on impedance and magneto-dielectric bodies.
3. Scattering and diffraction phenomena (specular reflection from flat or smoothly curved, PEC or non-metallic surfaces; edge-diffracted, surface, and creeping waves; resonant and Rayleigh scattering; superlensing and cloaking).
4. Solution and analysis methods (separation of variables; Sommerfeld–Maliuzhinets, Wiener–Hopf, and Meixner's methods; complex integration and Watson's transformation; asymptotic steepest descent and stationary phase methods).

To make the book more innovative and appealing to the reader, we put an emphasis on the following features:

1. A challenging and, in part, new content, targeted on electromagnetic scattering from non-metallic bodies and interfaces; in almost every chapter we attempt to extend and deepen the available literature. This concerns, for example, newly derived exact solutions for scattering from impedance wedges as well as the logarithmic terms in Meixner's expansions of fields at the edge of non-metallic wedge-shaped configurations (Chapter 5), low-frequency resonant scattering from magneto-dielectric spheres and high-frequency backscattering from low-absorption spheres (Chapter 7), a truly electromagnetic version of the physical optics method (Chapters 8 and 9), and Debye's asymptotics of Bessel functions of complex argument and order (Appendix E).

2. Coverage of the whole span, from rigorous mathematical formulation of scattering problems and their exact or approximate solution to simple approximations and physical interpretation of the obtained solutions. The results can always be back-traced to Maxwell's equations as we have attempted to document every essential step in the constructions. Furthermore, extended and self-contained mathematical appendices describe all essential mathematical tools used in the book. This approach may have increased the volume of the book but it has also made the book self-contained, suitable as a tutorial for those learning solution methods and curious to know how engineering concepts and techniques relate to the foundations of electromagnetics.
3. A large number of original figures, which, we think, is important to make the book more appealing not only aesthetically but also technically as diagrams can present the most characteristic features of the derived solutions in a comprehensible form, eliminating at the same time the danger of a too formal presentation of the material.
4. Extended problem sections supplementing each chapter with a number of control questions and problems (sometimes quite challenging) for self-study. We think that this is important as mastering the tough theory needs exercise and active examination of examples and special cases.

The book can be used as an essential reading for courses like advanced electromagnetics, scattering and diffraction, radar techniques, and fields and waves taught at graduate and post-graduate level in electrical engineering departments. We hope that the book will also be of interest to university professors and teachers, graduate and post-graduate students of electrical engineering and physics (mainly optics, but also acoustics) departments, as well as researchers at MSc and PhD levels in industrial and government laboratories concerned with studies of electromagnetic scattering for telecommunications, aerospace, automotive, and naval applications. The principal author of the book is Andrey Osipov. Sergei Tretyakov wrote several sections and contributed by discussing and developing the book material. The whole book was edited by both authors together.

Andrey Osipov and Sergei Tretyakov
January 2016

Acknowledgements

Some sections of this book use lecture notes written by the authors for the post-graduate course in radio engineering given at the former Radio Laboratory of the Helsinki University of Technology in the spring semester of 2005. We gratefully acknowledge the permission of the laboratory director to use the material in this book. We are indebted to Dr. Igor Nefedov for his contribution to section 6.3.

We acknowledge the use of Wolfram Mathematica (Wolfram Research, Inc.) to perform symbolic and numeric computations, generate and plot numerical data, the use of Origin (OriginLab Corporation) to create presentation-quality scientific diagrams, the use of CorelDRAW (Corel Corporation) for editing vector graphics, and the use of MiKTeX distribution of the typesetting system LaTeX in conjunction with TeXnicCenter, a free open source integrated development environment.

We are grateful to John Wiley & Sons Ltd for the support of this project, and to the many editorial staff members who encouraged and helped us on the long road to completion of the project.

Last but not least, we thank our families for their absolute understanding and unconditional patience.

List of Abbreviations

Vectors are denoted by bold letters ($\mathbf{a}$). The scalar product of two vectors is denoted by dot ($\mathbf{a} \cdot \mathbf{b}$), and the vector product is denoted by cross ($\mathbf{a} \times \mathbf{b}$). Unit vectors are denoted as $\hat{x}$, $\hat{n}$, etc. The double bar over a character denotes a 2 by 2 matrix or a second-order tensor ($\overline{\overline{M}}$). Transposition of a vector or a matrix is denoted by the superscript T, e.g. $\overline{\overline{M}}^{\mathrm{T}}$. Time-harmonic dependence is in the form $\exp(j\omega t)$.

Values related to *incident* wave are denoted using index "inc" and those related to *scattered* waves are marked by index "sc".

Asymptotic behaviour at large values of arguments (p real valued):
$o(k^p)$ means that $\lim_{k\to\infty} |o(k^p)k^{-p}| = 0$;
$O(k^p)$ means that there exists a finite constant C such that $\lim_{k\to\infty} |O(k^p)k^{-p}| \le C$.

X (dB) is equal to 20 $\log_{10} X$ if X is amplitude-related and to 10 $\log_{10} X$ if X is power-related.

$\overline{\overline{A}}$, A_{mn}	scattering matrix and its components
a	radius of a sphere or cylinder
$\mathrm{Ai}(x)$	Airy function
$\mathbf{B}$	magnetic induction
BEM	boundary element method
$c_0 = 1/\sqrt{\epsilon_0\mu_0}$	the speed of light in vacuum (free space)
C	capacitance, integration path
$\mathrm{d}l$	infinitesimal arc length of a curve in 2D or 3D space
$\mathrm{d}V = \mathrm{d}x\mathrm{d}y\mathrm{d}z$	infinitesimal volume element
$\mathrm{d}\Omega$	infinitesimal element of a surface in 3D space
D, $D_\perp$	scatterer size (diameter) in 3D space and in the cross section perpendicular to the axis of a cylindrical scatterer
$\mathbf{D}$	electric displacement vector
$\overline{\overline{D}}$, D_{ee}, D_{eh}, D_{he}, D_{hh}	tensor diffraction coefficient and its elements
DNG	double-negative, a medium with $\epsilon < 0$ and $\mu < 0$
$\mathbf{E}$	electric field
E_n	normal component of the electric field
$\mathbf{E}_t$	tangential electric field component
$\mathcal{E}$	eikonal
$\mathbf{F}$	far-field coefficient, scattering amplitude
FDTD	finite-difference time-domain method
FEM	finite element method
$G_0(\mathbf{r}, \mathbf{r}')$	free-space Green's function
$G_2(\boldsymbol{\rho}, \boldsymbol{\rho}', k)$	two-dimensional Green's function
GO	geometrical optics

GTD	geometrical theory of diffraction
$\mathbf{H}$	magnetic field
H_n	normal component of the magnetic field
$H_n^{(1,2)}(z)$	Bessel functions of the third kind or Hankel functions
$\mathbf{H}_t$	tangential magnetic field component
$\hat{i}$	incidence direction
j	imaginary unit
$J_n(z)$	Bessel function of the first kind
$\mathbf{k}$, k	wave vector in medium and its amplitude (wavenumber)
$k_0 = \omega\sqrt{\epsilon_0\mu_0}$	free-space wavenumber
$\mathbf{K}_e$, $\mathbf{K}_m$	equivalent electric and magnetic surface currents
l	length
L	inductance
$m = (ka/2)^{1/3}$	asymptotic parameter for propagation on curved surfaces
$\overline{\overline{M}}$	magnetic polarizability tensor
M_{ij}	elements of magnetic polarizability tensor $(i, j = 1, 2, 3)$
MoM	method of moments
N	refractive index, total number of material layers in a planar stratified structure, total number of wedges in a multi-wedge configuration, second fundamental form of a surface
$\hat{n}$	unit vector normal to a surface or an interface
$\frac{\partial}{\partial n} = \hat{n} \cdot \nabla$	normal derivative
$\mathbf{p}_e$	electric dipole moment
$\mathbf{P}_e$	electric polarization vector
$\mathbf{p}_m$	magnetic dipole moment
$\mathbf{P}_m$	magnetic polarization vector
$\overline{\overline{P}}$	electric polarizability tensor
P_{ij}	elements of electric polarizability tensor $(i, j = 1, 2, 3)$
P_{abs}	power absorbed by a scatterer
P_{inc}	power in the incident wave
P_{sc}	power in the scattered wave
PEC	perfect electric conductor
PMC	perfect magnetic conductor
PO	physical optics
PTD	physical theory of diffraction
$\mathbf{r}$	position vector in space
r_n^u, r_n^v	electric and magnetic multipole coefficients
$\tilde{r}_n^u$, $\tilde{r}_n^v$	electric and magnetic spherical reflection coefficients, reflection coefficients for spiral waves
R	reflection coefficient, distance, resistance
$\mathcal{R}$	reflectance, reflectivity
$R_\parallel$, $R_\perp$	planar reflection coefficients for parallel and perpendicular polarization
R_{TE}, R_{TM}	planar reflection coefficients for TE and TM polarization
R_E, R_H	planar reflection coefficients for electric and magnetic field components perpendicular to the plane of incidence
R_e, R_m	sheet resistivity and sheet magnetic conductivity
RCS	radar cross section
SDP	steepest descent path
$\mathbf{S}$	complex Poynting vector
$\hat{s}$	unit vector in the scattering direction
$\mathbf{s}$	2D vector built from Sommerfeld's spectral amplitudes

S_e, S_h	Sommerfeld's spectral amplitudes
$S_\nu^{(1)}(z)$, $S_\nu^{(2)}(z)$	base functions from Debye's asymptotics of Bessel functions
t	time
$\mathcal{T}$	transmittance, transmissivity
T, $T_\parallel$, $T_\perp$, T_{TE}, T_{TM}, T_E, T_H	transmission coefficients in various reference frames
u	electric Debye's potential
UTD	uniform theory of diffraction
v	magnetic Debye's potential
$\mathbf{w}$, w	mean intensity of energy flow in a time-harmonic wave and its amplitude
W	width of a scatterer
$\hat{x}$	unit vector along Cartesian axis x
$Y = \sqrt{\epsilon/\mu} = 1/Z$	intrinsic or wave admittance
$\hat{y}$	unit vector along Cartesian axis y
y_n	modal admittance at the surface of the sphere
$\tilde{y}_n$	normalized modal admittance at the surface of the sphere
$\hat{z}$	unit vector along Cartesian axis z
z_n	modal impedance at the surface of the sphere
$\tilde{z}_n$	normalized modal impedance at the surface of the sphere
$Z = \sqrt{\mu/\epsilon} = 1/Y$	intrinsic or wave impedance
$Z_0 = \sqrt{\mu_0/\epsilon_0}$	wave impedance of plane waves in vacuum, free-space impedance
Z_s	surface impedance
$\tilde{Z}_s$	equivalent surface impedance
δ	skin depth
$\delta(\mathbf{r})$	Dirac delta function
$\delta_n^u(r)$, $\delta_n^v(r)$	spherical modal impedance and admittance
ϵ	permittivity
ϵ_0	permittivity of vacuum
$\epsilon' = \mathrm{Re}\,\epsilon$	the real part of the permittivity
$\epsilon'' = -\mathrm{Im}\,\epsilon$	the losses-related part of the permittivity
$\zeta_n^{(1,2)}(z) = \sqrt{\pi z/2}\,H_{n+1/2}^{(1,2)}(z)$	Riccati–Bessel functions of the third kind or Riccati–Hankel functions
η	surface impedance normalized to free-space impedance
$\hat{\eta}$	unit vector perpendicular to the plane of incidence
θ	zenith angle in the spherical coordinates
$\Theta(x)$	unit step function equal to unity with $x \geq 0$ and zero with $x < 0$
λ	wavelength
μ	permeability
μ_0	permeability of vacuum
ν	$\nu = \pi/(2\Phi)$ (Chapter 5); integration variable in Watson's transformation (Chapters 6 and 7)
ν_n	Neumann symbol ($\nu_0 = 1$; $\nu_n = 2$ with $n = 1, 2, \ldots$)
$\hat{\xi}$	unit vector parallel to the plane of incidence
$\mathbf{\Pi}_e$, $\mathbf{\Pi}_m$	electric and magnetic Hertz vectors
$\boldsymbol{\rho}$	position vector on a plane
σ	conductivity, mono- and bistatic scattering cross section
σ_{abs}, σ_{ext}, σ_T	absorption, extinction, total scattering cross sections
σ_{back}, σ_{fwd}	backscattering and forward scattering cross sections
σ_{2D}	mono- and bistatic scattering width
σ_{abs}^{2D}, σ_{ext}^{2D}, σ_T^{2D}	absorption, extinction, total scattering widths
σ_E, σ_H, σ_{EE}, σ_{EH}, σ_{HE}, σ_{HH}	polarization-related bistatic scattering widths

φ	azimuthal angle in the circular cylinder coordinates
ϕ	azimuthal angle in the spherical coordinates
Φ	half-width of the region exterior to a wedge; scalar potential
$\Phi(\nu, z)$	phase function from Debye's asymptotics for cylindrical functions
$\psi_n(z) = \sqrt{\pi z/2}\, J_{n+1/2}(z)$	Riccati–Bessel function of the first kind
ω	angular frequency

1

Introduction

1.1 Scattering and Diffraction Theory

Defined broadly, scattering of electromagnetic waves is the process of re-radiation of an incident wave by a material body. In the course of scattering, the incident wave may change its type (e.g., from plane to spherical or cylindrical), the direction of propagation, amplitude, phase, and polarization state. The other name of scattering is *diffraction*. In this book we use these terms as synonyms, although in the optical literature a distinction between the two names is sometimes made. Understanding of the scattering of electromagnetic fields by material objects is essential to the design of optical, microwave, and radio devices and systems.

Electromagnetic fields scattered by an obstacle include versatile information about geometric and material properties of the scatterer, its position in space, orientation, speed, etc. This fact is the basis for remote sensing technologies that use electromagnetic waves to survey, probe, and study remotely and non-destructively the space and material objects surrounding an observer. To extract the information encoded in the scattered field, it is important to understand the physics of scattering (how electromagnetic waves interact with the target and get scattered from the target) and be able to efficiently simulate the scattered fields. Modern radar remote sensing systems predominantly operate at microwave frequencies between 100 MHz and 100 GHz, which corresponds to the wavelength from 3 m to 3 mm. Such wavelengths are comparable, smaller or larger, to the size of a typical scatterer in our everyday life. Even in the visible spectrum, with frequencies between 430 and 770 THz and corresponding wavelengths between 390 and 700 nm, the wave effects cannot be neglected since from the viewpoint of applications of nanoscience or even in atmospheric optics these wavelengths may be smaller, comparable, or larger than the size of a nanoparticle or a water droplet. Under these circumstances the wave nature of electromagnetic fields becomes noticeable via various diffraction effects such as penetration of electromagnetic energy into regions shadowed by obstacles, blurred shadow boundaries and focal points, internal resonances, etc. These wave phenomena cannot be adequately described and understood in the framework of simple ray optical constructions. Similar observations can be made with respect to radio wave propagation, where the effects of the ground, buildings, trees (or furniture in the microcell mobile communications scenario) cannot be understood without solid knowledge of scattering phenomena.

Electromagnetic scattering theory is a branch of electromagnetics that describes, explains, and predicts electromagnetic field behavior in the presence of material obstacles by fully accounting for the wave nature of the electromagnetic field.

Electromagnetic scattering theory lies on the intersection of physics (particularly, electromagnetics and optics) and mathematics. During several centuries of its history, many outstanding researchers, including René Descartes (1596–1650), Francesco Maria Grimaldi (1618–1663), Christiaan Huygens (1629–1695), Thomas Young (1773–1829), François Arago (1786–1853), Joseph von Fraunhofer (1787–1826), Augustin-Jean Fresnel (1788–1827), Hermann von Helmholtz (1821–1894), Gustav Kirchhoff (1824–1887), James Clerk Maxwell (1831–1879), Heinrich Hertz (1857–1894), Hector Macdonald (1865–1935), and Arnold Sommerfeld (1868–1951), have contributed to the establishment of the wave theory of electromagnetic diffraction. The originators of the diffraction theory were inspired by remarkable optical phenomena, like blurring shadow boundaries, lit spots behind impenetrable obstacles (Poisson's spot or the spot of

Modern Electromagnetic Scattering Theory with Applications, First Edition. Andrey V. Osipov and Sergei A. Tretyakov.

Arago), atmospheric rainbows, etc. The inability of simple ray-based concepts to explain these phenomena has led to the creation of electromagnetic scattering and diffraction theory. The need to solve the wave equations stimulated the development of various areas of mathematics (boundary value problems, special functions, analytical, asymptotic, and numerical methods of solution of integral and differential equations). The theory is still in the process of development motivated by new application fields, from stealthy aircraft and ships to metamaterials in photonics and nanoscience.

The physics of electromagnetic scattering can be explained in terms of a few basic scattering mechanisms. They can be studied by looking at solutions of Maxwell's equation for a limited number of canonical shapes, which are a planar interface (reflected and transmitted waves, surface waves), a circular cylinder and a sphere (reflection from a curved surface, creeping waves, interior resonances), a half-plane and a wedge (edge-diffracted waves), and a circular disc (multiply diffracted waves). The respective boundary value problems can be rigorously solved in a symbolic form, typically in terms of elementary or special functions, integrals or infinite sums, and the basic wave types and the associated scattering mechanisms are extracted from the solutions by asymptotic analysis.

Exact symbolic solutions of Maxwell's equations are rare. A solution is only possible for a simply shaped scatterer (planar infinite interfaces of homogeneous materials, wedges, circular cylinders, spheres, cones) with a simple material constitution (perfect electric conductor, homogeneous or layered materials) illuminated by a canonical source (plane, cylindrical, spherical waves). However, such solutions are extremely important for the modern system of knowledge about radiation, propagation, diffraction, and scattering of electromagnetic waves, which is explained by the following circumstances:

- In many cases, even solutions of these idealized problems adequately describe solutions of real-world problems, e.g. canonical targets, like metal spheres, are used for calibration of measurement facilities.
- Exact symbolic solutions can be typically calculated with a high numerical precision, which allows their use for validation of numerical solutions and procedures. Furthermore, solutions for idealized configurations may be used in the follow-up numerical iteration or optimization procedures.
- Symbolic solutions can be often evaluated approximately or asymptotically to give simple analytic or even closed-form expressions for the fields. Such expressions provide a physical insight and can spot simple qualitative and quantitative relations between various parameters of the problem, which enables explaining and forecasting the behavior of fields in more general and complex situations.
- Asymptotic solutions for idealized geometries serve as building blocks in engineering techniques (GTD/UTD, PO, PTD) for simulation of high-frequency electromagnetic fields in realistic environment.

Solutions of electromagnetic scattering and diffraction problems even for simple geometries and even when an exact solution is available can be very complex. Recent advances in computation technologies have enabled direct numerical simulation of electromagnetic phenomena for objects of moderate size compared with the wavelength, and computer-based visualization tools bring new ways of representing the behavior of scattered fields. However, it is important to understand that pure numerical approaches are subject to severe fundamental limitations. First, a direct application of discretization-based methods (BEM, FEM, FDTD) to objects larger than a dozen wavelengths leads to systems of equations with millions of unknowns, which are hardly amenable to a numerical solution because of unrealistic execution times and heavy memory requirements. Second, a numerical solution cannot provide a qualitative, physical insight into the basic mechanisms of scattering and diffraction.

Hence, there is a continuing need for the development of approximate and physically justified approaches that supplement the direct numerical methods by providing fast and acceptably accurate results for objects that are much larger than the wavelength. Several approximations are available for describing the scattering and diffraction of high-frequency electromagnetic fields.

- Geometrical optics (GO) is a classical ray-based technique for describing optical and microwave fields. The ray fields can be calculated by using simple laws of their propagation in free space and their reflection and transmission at material interfaces. However, the applicability of GO is limited to very short waves. The ray optics does not describe the penetration of fields in the regions shadowed by the scattering body and the excitation of many types of waves at geometric and material singularities on the scattering surface. Furthermore, GO fails at caustics, which is a major deficiency since the field intensity is typically at maximum there and many important phenomena (e.g., focusing of electromagnetic energy by dielectric lenses and parabolic reflectors, atmospheric glories and rainbows) take place at caustics.

- The geometrical theory of diffraction (GTD) extends GO through the addition of new types of rays which are generated at edges (edge-diffracted waves), conical points (tip-diffracted waves), and shadow boundaries on the scattering surface (creeping waves). However, GTD is a ray-based technique which fails at caustics. Extension of GTD to caustic regions is the subject of the uniform geometrical theory of diffraction (UTD).
- Physical optics (PO) and its extension, the physical theory of diffraction (PTD), are current-based approaches. The field scattered by a body is considered as radiated by secondary currents induced by the incident wave on the scatterer. Every PO/PTD solution uses Huygens' principle, which is a rigorous integral relation between the fields at every point in a volume and their values on a given (physical or mathematical) surface enclosing the volume. Since the surface fields are not known in advance, PO substitutes them with their GO approximation, thus reducing the problem to evaluation of a surface integral. When the observation point is located far from the body, the kernel of the integral can be simplified, leading to a fundamental formula known as the radiation integral. This formula is widely used in a great many of applications, from antenna design and prediction of radio wave propagation to simulation of scattering cross sections of radar targets. When an exact solution is impractical or not available, the PO method is unique in its ability to provide accurate numerical estimations of scattered fields in the situation (electrically large scatterers, observer in the far field) when neither discretization-based numerical techniques (BEM, FDTD, FEM) nor ray optical approximations (GO, GTD, UTD) apply.
- The GO approximation for the surface currents may be insufficiently accurate when the surface has geometric and/or material singularities (edges, tips, jumps in the values of material parameters). PTD corrects the GO surface currents in the vicinity of those singularities by using solutions of canonical problems for locally conformal geometries. This procedure may significantly improve the accuracy of the current-based approximations.

These approximate solution methods need justification. GO and PO were known well before Maxwell's theory of electromagnetism as empirical techniques, based on pre-Maxwell theories of light. GTD was also formulated as a set of postulates. It is therefore important to relate these methods and concepts to Maxwell's equations, validate their correctness, associated assumptions and simplifications, and estimate their accuracy. The only way to do this is to examine exact solutions to several characteristic scattering problems, that is, to look at solutions of canonical scattering problems.

1.2 Books on Related Subjects

The amount of literature on diffraction and scattering theory is overwhelming as the history of the theory, which can be counted from Fresnel's *Mémoire sur la diffraction de la lumière*, is almost two centuries old. It is therefore simply impossible to give a complete bibliography of all the books and papers that have been published over the years around the world, and we have not even attempted that. Every chapter in the main body of the book is accompanied with its own list of references, and we did our best to reference the most relevant publications. If the reader feels that we have overlooked an important contribution, please let us know!

Here, we briefly review books related to scattering and diffraction theory. They fall into the following categories:

- Historical texts (Descartes 1637; Drude 1902; Fresnel 1866; Green 1828; Grimaldi 1665; Heaviside 1893; Huygens 1690; Kirchhoff 1891; Macdonald 1902; Maxwell 1873; Sommerfeld 1927; Thomson 1893; Helmholtz 1897; Young 1845), see also English translations of classical texts (Crew 1900; Sommerfeld 2004). These and many other historical texts are now freely available on the Internet, thanks to ongoing digitizing efforts. The works by Descartes, Grimaldi, Huygens, Young, Fresnel, and Green appeared before the formulation of Maxwell's theory and present physical (Descartes, Grimaldi, Huygens, Young) and mathematical (Fresnel, Green) concepts that are used in the modern scattering theory. Descartes (1637) described the laws of reflection and refraction, and used them to explain atmospheric rainbows. Grimaldi (1665) observed the fringes in the shadow of an object placed in a beam of light and named the deviation of light from straight propagation path *diffraction*. Huygens (1690) was a proponent of the wave theory of light and formulated the principle of propagation of wavefronts (Huygens' principle). Young (1845) presented a number of theoretical reasons supporting the wave theory of light, and performed and analyzed demonstrations to support this viewpoint (interference in a double-slit experiment and upon reflection from thin films). Fresnel (1866) developed a first quantitative wave theory of light by building on experimental work by Young, and showed by mathematical methods that polarization could be explained only if light was entirely transverse. Green (1828) derived the integral relation that relates the value of a wave function in a volume to its values on a surface enclosing the volume (Green's formula) and applied it to electro- and magnetostatic problems.

The electromagnetic theory by Maxwell (1873) established the connection between light and electromagnetism. In its original formulation, the theory assumed the existence of aether and used quaternion algebra. Kirchhoff (1891) described optics based on the aether theory of light. The modern form of Maxwell's equations, in terms of the vectors of electric and magnetic fields, is attributed to Heaviside (1893). Experimental verification of Maxwell's theory and a first theoretical solution of Maxwell's equations (for the field excited by an oscillating electric dipole) are attributed to Hertz (1894). First descriptions of electromagnetic scattering, diffraction, and propagation entirely based on Maxwell's theory can be found in Drude (1902), Helmholtz (1897), and Thomson (1893). The beginning of rigorous electromagnetic scattering theory is associated with the works by Sommerfeld (1927) (formulation of scattering and diffraction problems as boundary value problems and their solution with rigorous mathematical methods, exact solutions for a half-plane and a wedge, for a dipole source over a homogeneous half-space) and Macdonald (1902) (electromagnetic formulation of Huygens' principle and conception of electromagnetic version of the PO method, solution for wedge diffraction).

- General courses in electromagnetics and electromagnetic waves (Balanis 1989; Borgnis and Papas 1955; Chew 1990; Felsen and Marcuvitz 1973; Franz 1957; Harrington 1961; Hönl *et al.* 1961; Ishimaru 1991; Jackson 1998; Jones 1964, 1979, 1989; Katsenelenbaum 2006; King and Wu 1959; Kong 1986; Plonus 1978; Schelkunoff 1943; Stratton 1941; Van Bladel 1964). These books give general theoretical knowledge about electromagnetic scattering and diffraction, and contain material on the most fundamental scattering problems, like scattering from wedges, cylinders, spheres, and apertures, mostly assuming PEC boundaries and often in scalar PO approximation.
- Handbooks, collections of formulas of diffraction and scattering theory (Bouman *et al.* 1987; Ruck *et al.* 1970), overviews of the state of the art (e.g., Langer (1962), Pike and Sabatier (2002), Ufimtsev (1999), and Uslenghi (1978)). These books offer a broad, eventually encyclopedic coverage but omit derivations of the presented results.
- Textbooks and monographs on computational methods of electromagnetics (e.g., Chew (1990), Chew *et al.* (2001), Harrington (1993), Jin (1993), Kunz and Luebbers (1993), Peterson *et al.* (1998), Silvester and Ferrari (1990), Taflove (1995), Tsang *et al.* (2001), and Volakis *et al.* (1998)). These books deal with various discretization-based methods for numerical solution of Maxwell's equations.
- Books on the wave theory of light (e.g., Akhmanov and Nikitin (1997), Born and Wolf (1959), Longhurst (1973), Newton (1966), Nieto-Vesperinas (2006), Petykiewicz (1992), Solimeno *et al.* (1986), and Sommerfeld (1964)). These books include chapters on diffraction of electromagnetic waves and address applications in optics.
- Books aimed at specific configurations or aspects of electromagnetic diffraction and scattering theory, e.g. dyadic methods and analytical modeling (Lindell 1992; Tai 1971; Tretyakov 2003), field singularities (Van Bladel 1991), stratified media (Brekhovskikh 1960; Wait 1962), wedges (Bobrovnikov and Fisanov 1988; Budaev 1995; Lyalinov and Zhu 2013), spheres, cylinders, and small particles (Bohren and Huffman 1983; Grandy 2000; Kerker 1969; Van de Hulst 1957; Wait 1959), cavities (Vinogradov *et al.* 2002), curved structures (Lewin *et al.* 1977), and rough surfaces and random media (Ishimaru 1978; Rytov *et al.* 1989; Tsang and Kong 2001).
- Books on the mathematical theory of diffraction (Babič and Buldyrev 1991; Baker and Copson 1953; Borovikov 1966; Colton and Kress 1983, 1998; Fock 1965; Kline and Kay 1965; Müller 1957; Northover 1971; Ramm 1986). Scattering and diffraction theory is presented as a branch of mathematics, mostly using the formal mathematical language and concentrating on methods rather than applications.
- Books on specific mathematical methods of diffraction and scattering theory (Babič and Kirpičnikova 1979; Babich *et al.* 2007; Bouche *et al.* 1997; Daniele and Zich 2014; Mittra and Lee 1971; Molinet *et al.* 2005; Noble 1958; Weinstein 1969).
- Books on engineering techniques such as GTD/UTD (Borovikov and Kinber 1994; James 1976; McNamara *et al.* 1990) and PO/PTD (Diaz and Milligan 1996; Rubinowicz 1957; Senior and Volakis 1995; Ufimtsev 1962, 2003, 2014).

- Application of scattering and diffraction theory to problems of RCS engineering (Crispin and Siegel 1968; Jenn 2005; Knott *et al.* 1993; Mentzer 1955; Saez de Adana *et al.* 2011; Tretyakov and Osipov 2006), remote sensing (Beckmann and Spizzichino 1963; Ulaby *et al.* 1982), radio wave propagation and wireless communication (Bertoni 2000; Bremmer 1949; Makarov *et al.* 1991; Wait 1981, 1962), and to closely related radiation problems, specifically to reflector and aperture antennas (Balanis 1997; Collin and Zucker 1969; Fradin 1961; Silver 1984; Wait, 1959).

1.3 Concept and Outline of the Book

The book describes deterministic scattering of electromagnetic waves from material bodies with full consideration of the polarization. Although many qualitative features of scattering can be described by using solutions of scalar problems, for example in the framework of the Fresnel–Kirchhoff approach, in practice vector Maxwell's equations always have to be solved, and formulation of a truly scalar scattering problem is possible only in exceptional cases, for example for PEC bodies with translational or spherical symmetry or for infinite planar material interfaces illuminated by plane waves. The components of the electric and magnetic fields are typically coupled through boundary conditions on non-metallic interfaces. Thus, throughout the book we study truly electromagnetic problems, which makes the analysis more involved, but it is these solutions that are really needed in practice. For the same reason the book does not address problems of acoustic scattering since acoustic fields are scalar fields, for which polarization coupling is irrelevant.

This book is intended to serve as a bridge between textbooks and handbooks. Textbooks describe the basic facts and ideas, simplifying the material as much as possible for better understanding, presenting the results that can rarely be used in applications. Handbooks provide broad collections of ready-to-use results but often do not explain how and under which assumptions these results have been obtained. In this book we formulate the rigorous boundary value problems, explain solution methods, and let the reader follow the derivations, showing how to arrive at the handbook formulas. In particular, through the high-frequency analysis of solutions for canonical bodies, the ray structure of fields is extracted, thus validating the postulates of GO and GTD.

An important feature of the book is its focus on scattering from non-metallic (imperfectly conducting) and impedance bodies. This is important because of the increasing use of advanced materials in electromagnetic engineering; also natural surfaces (ground, see, buildings, etc.) are non-metallic. Scattering from non-metallic scatterers is addressed in the book by using either exact solutions (for simply shaped scatterers, like wedges, cylinders, spheres) or approximate PO solutions (for generally shaped bodies). Reflection and scattering from impedance-matched bodies (when the surface impedance is equal to the intrinsic impedance of the surrounding medium) show a number of extremal features, so special attention is devoted to the impedance-matched plane (section 4.6.3), wedge (section 5.4.3), and sphere (section 7.5.4). We also discuss metamaterial scatterers, which includes the superlens (section 4.5.3) and the perfect cloak (section 7.6.2).

The book presents novel and less known results. We believe that the complete theory of low-frequency scattering from spheres (sections 7.4.2, 7.5.2, and 7.5.3) and a theory of backscattering from electrically large low-absorption spheres in terms of interior creeping waves (section 7.5.5) have not been described in the literature so far.

The less-known results include a simplified derivation of the electromagnetic Huygens principle in two and three dimensions (section 2.6), formulation of the forward-scattering theorem for cylindrical objects under oblique illumination (section 3.4.4), solutions for planar material layers excited by dipoles arbitrarily oriented with respect to the interface of the media (sections 2.5.4 and 4.4.2), a proof of the minimum reflection property for the planar impedance-matched boundary (section 4.6.3), a complete collection of the available exact solutions for impedance wedges (section 5.4), a discussion of the analytical structure of Meixner's series, including logarithmic terms (section 5.6.2), an exact solution for an obliquely illuminated impedance cylinder (section 6.5), a systematic exposition of a fully electromagnetic version of the PO method which is applicable to non-metallic bodies and describes fields not only in the far-field region but also at finite distances from the scatterer, up to the scattering surface, including caustics (Chapter 8), PO solutions for a broad variety of non-metallic simply shaped scatterers and a new derivation of Gordon's formulas by integrating by parts in the PO solution for plates (Chapter 9), and complete specification of Debye's asymptotic approximations of Bessel functions on the complex plane of the order and derivation of the Debye approximations by the method of phase integrals (Appendix E).

Solution and analysis of solutions of scattering problems require the use of various quite sophisticated mathematical methods. We present the methods not in an abstract way but rather by showing how they work in solving and analyzing the scattering problems. The book describes a variety of methods, including

- exact solution methods: separation of variables (Chapters 5, 6, and 7), the Wiener–Hopf (factorization) method (section 5.3), and the Sommerfeld–Maliuzhinets method (section 5.4);
- methods of approximate solution of scattering problems: GO, GTD, UTD (Chapters 5, 6, and 7), and PO (Chapters 8 and 9);
- methods of asymptotic analysis of solutions: the steepest descent method (section C.3), the stationary point method in one and two dimensions (Appendix D), and the WKB method (Appendix E).

Special functions play a key role in solutions of canonical scattering problems. Addressed are Legendre functions (section 7.2.2), Bessel and Riccati–Bessel functions (sections 7.2.2 and E.1.1), Maliuzhinets and Bobrovnikov–Fisanov functions (sections 5.4.2 and 5.4.3), Fresnel integrals and related functions (Appendix B), and Airy and Fock functions (sections D.2.3 and E.3).

In order to keep the size of the book reasonable, we had to restrict our attention to the time-harmonic deterministic linear direct electromagnetic scattering theory in open regions. Electromagnetics of waveguides, resonators, and periodic arrays is not addressed (e.g., see Collin (1960)), nor is the theory of inverse scattering (e.g., see Colton and Kress (1998)). A number of geometries, solutions, and solution methods are not included, for example the rigorous solutions for PEC circular discs and cones (Bouman *et al.* 1987), mainly because of missing generalizations to imperfectly reflecting boundaries (scattering from a conical point on an impedance boundary is addressed in the PO approximation in section 9.2). Such methods as PTD, the parabolic equation method, the boundary-layer method, Maslov's method, etc. are also not discussed, and the reader is referred to the many good sources on the approximate methods of scattering, diffraction, and propagation theory, for example Babič and Buldyrev (1991), Bouche *et al.* (1997), Fock (1965), Ufimtsev (2014), Ziolkowski and Deschamps (1984), Babič and Kirpičnikova (1979), and Levy (2000). No anisotropic materials are addressed in the book, except for the case of the spherical perfect cloak.

Organization of the book is seen from the table of contents, so here we only briefly outline the content of the chapters. Chapter 2 presents the fundamentals of the scattering theory. This includes Maxwell's equations and constitutive relations, exact and approximate boundary conditions, vector potentials and Green's functions, and basic solutions of Maxwell's equations (plane, cylindrical, and spherical waves). The fundamental properties of the electromagnetic fields are also addressed, including energy conservation, reciprocity, and the electromagnetic version of Huygens' principle for compact and cylindrical scatterers.

Chapter 3 describes scattered fields in the far (Fraunhofer) zone where they approach a spherical wave (cylindrical or conical for cylindrical scatterers), which is particularly important for radar and optical applications. We define the relevant quantities, including the far-field coefficient, the scattering matrix, and various scattering cross sections and scattering widths. The optical theorem is addressed both in two and three dimensions, and classification of far-field scattering according to the scattering regimes (low-frequency, resonant, and high-frequency) is given. The chapter concludes with a theory of Rayleigh scattering for simply shaped PEC or magneto-dielectric scatterers smaller than the wavelength.

Chapter 4 addresses reflection and transmission of electromagnetic waves at planar interfaces. We define reflection and transmission coefficients for various polarization frames (parallel–perpendicular, TE-TM, *E*-*H*) and present a vector solution for reflection and transmission of a plane wave at the interface between two homogeneous magneto-dielectric half-spaces (section 4.2). A general solution for multilayered slabs is derived in section 4.3. Fields excited by localized sources are superpositions of plane and inhomogeneous waves, which complicates the problem by requiring complex integration of reflected/transmitted components (section 4.4). In section 4.5, the general solution for the multilayered slab is specialized to the important particular case of a homogeneous layer between two homogeneous eventually different half-spaces. Finally, the accuracy of impedance formulations for planar interfaces is studied, and the extremal properties of the impedance-matched interface are addressed (section 4.6).

A wedge is a canonical body to model the presence of an edge in a material boundary. In Chapter 5 we derive and study solutions for various wedge-shaped configurations with PEC and impedance boundary conditions. A variety of solution methods is used, including separation of variables (section 5.2), and the Wiener–Hopf (section 5.3) and Maliuzhinets (section 5.4) methods. The general solution for an arbitrarily angled obliquely illuminated impedance wedge is not available, so section 5.4 addresses all the special cases that have an exact solution. In section 5.5, the solutions are analyzed far from the edge, and the GO, surface, and edge-diffracted waves are extracted, showing that the result has the structure postulated by GO/GTD. The high-frequency approximations have to be corrected at the shadow boundaries, and we show how to do that, presenting uniform approximations in terms of Fresnel integrals and UTD transition functions. The presence of an edge leads to singularities of field components at the edge, and Meixner's method permits determining the order of singularity of the fields. In section 5.6 we show how this method works and discuss the problem of construction of higher-order terms in the Meixner expansions.

Similarly to the wedge, an infinite circular cylinder is a configuration with translational symmetry. On the other hand, its boundary is smooth and of finite diameter in the plane perpendicular to the cylinder axis, which leads to a different scattering picture. In Chapter 6 we solve (by separation of variables) scattering problems and study solutions for PEC (section 6.2), magneto-dielectric homogeneous (section 6.3), coated (section 6.5.1), and impedance (section 6.5.2) circular cylinders. To construct high-frequency approximations for the fields, Watson's transformation is introduced and applied in section 6.4. In addition to the GO reflected wave, the scattered field includes creeping waves, gliding along the surface on their way from the source to the observer, which justifies the GO/GTD postulates about the ray structure in the presence of a smoothly curved boundary. In the transition zone across the shadow boundary (penumbra region) on the surface of the cylinder, GTD is not applicable, and a special treatment in terms of Fock functions is necessary (section 6.4.4). Finally, the high-frequency solution for the circular cylinder is extended to electrically large generally shaped convex impedance bodies (section 6.6).

A sphere is one of the most important canonical bodies. In contrast to infinite cylinders, which are idealizations, the sphere is compact and serves as a model of a basic scatterer. Chapter 7 is devoted to scattering from spheres of various material compositions (PEC, magneto-dielectric, multilayered). An exact solution is derived by separation of variables, and the high-frequency approximations for homogeneous spheres are obtained by Watson's transformation. The homogeneous sphere allows new types of waves, not only the interior resonances and GO rays multiply reflected in the interior of the sphere but also waves propagating on the concave side of the spherical boundary. We present the full GO solution for the homogeneous magneto-dielectric sphere and apply it to the theory of atmospheric rainbows. It is shown that the backscattering from electrically large low-absorption spheres is dominated by the interior creeping waves, which leads to a new explanation of the atmospheric glory. Electrically small spheres are also an extremely important canonical case, which has gained importance because of the developments in nanoscience, photonics, and plasmonics. The classical theory of Rayleigh scattering is insufficient when multipole and radiation corrections are required. Furthermore, in spheres with large values of permittivity/permeability internal resonances are possible, which is not covered by the Rayleigh theory. In the chapter we derive a complete theory of low-frequency scattering that closes this gap. The chapter concludes with the solution for the spherical perfect cloak and with a discussion of new applications of small spheres in the design of metamaterials.

Chapter 8 describes the PO method. We derive a fully electromagnetic version of the method by combining the electromagnetic version of Huygens' principle with the vector solution for reflection of a plane wave at material interfaces (section 8.2). In section 8.3 the PO solution is applied to the case of an aperture in an impenetrable screen, and it is shown that high-frequency analysis extracts a wave diffracted at the rim of the opening. The wave has the form of the edge-diffracted wave studied in the chapter on wedges, generated at the points on the rim where the incidence and scattering directions make the same angle with the edge, which is in agreement with the GTD concept of the edge-diffracted wave and the Keller cone. Next, in section 8.4, reflection from curved surfaces is studied in the PO approximation, and we arrive at the classical Fock reflection formula. Finally, the PO solution for an imperfectly reflecting surface with an edge is studied, and the edge-diffracted wave component is extracted (section 8.4.4). The PO approximation leads to the edge-diffracted wave with the correct phase but an inaccurate amplitude, the latter being determined by the PO diffraction coefficient. Advantages and limitations of the PO method are discussed in section 8.5.

The PO method permits an immediate solution, in the form of a surface integral, for the field scattered by an arbitrarily shaped metallic or non-metallic scatterer. In Chapter 9, PO solutions are constructed for polygonal plates, cones, and bodies of revolution of various cross sections.

Appendices A to E importantly supplement the main body of the book by providing the background mathematical information about the special functions and methods that are used. Appendix A is a short collection of necessary formulas from vector analysis. Appendix B describes the Fresnel integral, error function for complex argument, and UTD transition function. The principles of the complex analysis and complex integration, including the steepest descent method, are addressed in Appendix C, while Appendix D describes in a relatively detailed manner the stationary phase method in one and two dimensions. Appendix E contains a collection of formulas for Bessel (and related) functions and a study on Debye's asymptotic approximations, performed with a new method (WKB method on the complex plane of the order of the Bessel functions). Explicit Debye's approximations in every part of the complex plane of the order are provided.

Bibliographic references are collected at the end of every chapter. A significant part of the bibliography comes from German and Russian technical literature, less known to the English-speaking community. Problem sections at the end of every chapter are rich collections of some 250 original problems for practice, most with detailed solutions. Some of the problems are easy, but some are difficult. Many are of a technical nature, which is due to the fact that electromagnetic theory in general and scattering theory in particular use a lot of mathematics as a language to formulate the results and as a tool to derive solutions from Maxwell's equations.

References

Akhmanov A and Nikitin SY 1997 *Physical Optics*, Oxford University Press.

Babič VM and Kirpičnikova NY 1979 *The Boundary-Layer Method in Diffraction Problems*, Springer-Verlag.

Babič VM and Buldyrev VS 1991 *Short-Wavelength Diffraction Theory*, Springer-Verlag.

Babich VM, Lyalinov MA, and Grikurov VG 2007 *Sommerfeld-Malyuzhinets Technique in Diffraction Theory.* Alpha Science International Ltd.

Baker BB and Copson ET 1953 *The Mathematical Theory of Huygens' Principle*, Clarendon Press.

Balanis CA 1989 *Advanced Engineering Electromagnetics*, John Wiley & Sons.

Balanis CA 1997 *Antenna Theory: Analysis and Design*, Second edition, John Wiley & Sons.

Beckmann P and Spizzichino A 1963 *The Scattering of Electromagnetic Waves from Rough Surfaces*, Pergamon.

Bertoni HL 2000 *Radio Propagation for Modern Wireless Systems*, Prentice Hall.

Bobrovnikov MS and Fisanov VV 1988 *Diffraction of Waves in Angular Regions*, Tomsk University Press, Tomsk (in Russian).

Bohren CF and Huffman DR 1983 *Absorption and Scattering of Light by Small Particles*, Wiley-Interscience.

Borgnis FE and Papas CH 1955 *Randwertprobleme der Mikrowellenphysik*, Springer.

Born M and Wolf E (eds) 1959 *Principles of Optics: Electromagnetic Theory of Propagation, Interference and Diffraction of Light*, Pergamon Press.

Borovikov VA 1966 *Diffraction by Polygons and Polyhedra*, Nauka, Moscow (in Russian).

Borovikov VA and Kinber BY 1994 *Geometrical Theory of Diffraction*, IEE Electromagnetic Waves Series 37, The Institution of Electrical Engineers, London.

Bouche D, Molinet F, and Mittra R 1997 *Asymptotic Methods in Electromagnetics*, Springer-Verlag.

Bouman JJ, Senior TBA, and Uslenghi PLE (eds) 1987 *Electromagnetic and Acoustic Scattering by Simple Shapes*, Hemisphere Publishing Corporation.

Brekhovskikh LM 1960 *Waves in Layered Media*, Academic Press.

Bremmer H 1949 *Terrestrial Radio Waves*, Elsevier.

Budaev BV 1995 *Diffraction by Wedges*, Longman Scientific & Technical.

Chew WC 1990 *Waves and Fields in Inhomogeneous Media*, Van Nostrand Reinhold.

Chew WC, Jin JM, Michielssen E, and Song J (eds) 2001 *Fast and Efficient Algorithms in Computational Electromagnetics*, Artech House.

Collin RE 1960 *Field Theory of Guided Waves*, McGraw-Hill.

Collin RE and Zucker FJ 1969 *Antenna Theory: Part 2*, McGraw-Hill.

Colton D and Kress R 1983 *Integral Equation Methods in Scattering Theory*, John Wiley & Sons.

Colton D and Kress R 1998 *Inverse Acoustic and Electromagnetic Scattering Theory*, Springer.

Crew H (ed) 1900 *The Wave Theory of Light. Memoirs by Huygens, Young and Fresnel*, American Book Company.

Crispin JW and Siegel KM (eds) 1968 *Methods of Radar Cross-Section Analysis*, Academic Press.

Daniele V and Zich R 2014 *The Wiener–Hopf Method in Electromagnetics*, SciTech Publishing, Rayleigh, NC.

Descartes R 1637 *Discours de la Méthode, Pour Conduire la Raison et Chercher la Vérité des Sciences, Plus la Dioptrique et les Météores Qui Sont des Essais sur Cette Méthode*, Leiden.

Diaz L and Milligan T 1996 *Antenna Engineering Using Physical Optics: Practical CAD Techniques and Software*, Artech House.

Drude P 1902 *The Theory of Optics*, Longmans Green and Co.

Felsen LB and Marcuvitz N 1973 *Radiation and Scattering of Waves*, Prentice-Hall.

Fock VA 1965 *Electromagnetic Diffraction and Propagation Problems*, Pergamon Press.

Fradin AZ 1961 *Microwave Antennas*, Pergamon Press.

Franz W 1957 *Theorie der Beugung elektromagnetischer Wellen*, Springer-Verlag.
Fresnel A 1866 *Œuvres Complètes d'Augustin Fresnel, Tome Premier: Théorie de la Lumière*. De Sénarmont H, Verdet É, and Fresnel L (eds), Imprimerie impériale, Paris.
Grandy WT 2000 *Scattering of Waves from Large Spheres*, Cambridge University Press.
Green G 1828 *An Essay on the Application of Mathematical Analysis to the Theories of Electricity and Magnetism*, printed for the author by T. Wheelhouse, Nottingham.
Grimaldi FM 1665 *Physico-Mathesis de Lumine, Coloribus, et Iride*, Bologna.
Harrington RF 1961 *Time-Harmonic Electromagnetic Fields*, McGraw-Hill.
Harrington RF 1993 *Field Computation by Moment Methods*, IEEE Press.
Heaviside O 1893 *Electromagnetic Theory. Volume 1*, The Electrician Printing and Publishing Co, London.
Hertz H 1894 *Untersuchungen über die Ausbreitung der elektrischen Kraft*, Johann Ambrosius Barth, Leipzig.
Hönl H, Maue AW, and Westpfahl K 1961 *Theorie der Beugung*, in *Encyclopedia of Physics, vol. XXV/1*, Springer-Verlag.
Huygens C 1690 *Traite de la Lumiere*, Leyden.
Ishimaru A 1978 *Wave Propagation and Scattering in Random Media*, Academic Press.
Ishimaru A 1991 *Electromagnetic Wave Propagation, Radiation, and Scattering*, Prentice Hall.
Jackson JD 1998 *Classical Electrodynamics*, Third edition, Wiley.
James GL 1976 *Geometrical Theory of Diffraction for Electromagnetic Waves*, Peter Peregrinus, London.
Jenn DC 2005 *Radar and Laser Cross Section Engineering*, American Institute of Aeronautics and Astronautics, Reston.
Jin JM 1993 *The Finite Element Method in Electromagnetics*, John Wiley & Sons.
Jones DS 1964 *The Theory of Electromagnetism*, Pergamon Press.
Jones DS 1979 *Methods in Electromagnetic Wave Propagation*, Clarendon Press.
Jones DS 1989 *Acoustic and Electromagnetic Waves*, Clarendon Press.
Katsenelenbaum BZ 2006 *High-frequency Electrodynamics*, Wiley-VCH.
Kerker M 1969 *The Scattering of Light and Other Electromagnetic Radiation*, Academic Press.
King RWP and Wu TT 1959 *The Scattering and Diffraction of Waves*, Harvard University Press, Cambridge, MA.
Kirchhoff G 1891 *Mathematische Optik (Vorlesungen über mathematische Physik, 2. Band)*, Teubner Verlag, Leipzig.
Kline M and Kay IW 1965 *Electromagnetic Theory and Geometrical Optics*, John Wiley & Sons.
Knott EF, Shaeffer JF, and Tuley MT 1993 *Radar Cross Section*, Artech House.
Kong JA 1986 *Electromagnetic Wave Theory*, John Wiley & Sons.
Kouyoumjian RG and Pathak PH 1974 A uniform geometrical theory of diffraction for an edge in a perfectly conducting surface. *Proc. IEEE*, **62**, 1448–1461.
Kunz KS and Luebbers RJ 1993 *The Finite Difference Time Domain Method for Electromagnetics*, CRC Press.
Langer RE (ed.) 1962 *Electromagnetic Waves*, The University of Wisconsin Press.
Levy M 2000 *Parabolic Equation Methods for Electromagnetic Wave Propagation*, The Institution of Electrical Engineers, London.
Lewin L, Chang DC, and Kuester EF 1977 *Electromagnetics Waves and Curved Structures*, Peter Peregrinus.
Lindell IV 1992 *Methods for Electromagnetic Field Analysis*, Oxford Science Publications.
Longhurst RS 1973 *Geometrical and Physical Optics*, Longman.
Lyalinov MA and Zhu NY 2013 *Scattering of Waves by Wedges and Cones with Impedance Boundary Conditions*, SciTech Publishing.
Macdonald HM 1902 *Electric Waves*, Cambridge University Press.
Makarov GI, Novikov VV, and Rybachek ST 1991 *Electromagnetic Waves Propagation over Earth's Surface*, Nauka, Moscow (in Russian).
Maxwell JC 1873 *A Treatise on Electricity and Magnetism*, Clarendon Press, Oxford.
McNamara DA, Pistorius CWI, and Malherbe JAG 1990 *Introduction to the Uniform Geometrical Theory of Diffraction*, Artech House.
Mentzer JR 1955 *Scattering and Diffraction of Radio Waves*, Pergamon Press.
Mittra R and Lee SW 1971 *Analytical Techniques in the Theory of Guided Waves*, The Macmillan Company.
Molinet F, Andronov I, and Bouche D 2005 *Asymptotic and Hybrid Methods in Electromagnetics*, IEE Electromagnetic Waves Series 51, The Institution of Electrical Engineers, London.
Müller C 1957 *Grundprobleme der mathematischen Theorie elektromagnetischer Schwingungen*, Springer-Verlag.
Newton RG 1966 *Scattering Theory of Waves and Particles*, McGraw-Hill.
Nieto-Vesperinas M 2006 *Scattering and Diffraction in Physical Optics*, World Scientific.
Noble B 1958 *Methods Based on the Wiener–Hopf Technique*, Pergamon Press.
Northover FH 1971 *Applied Diffraction Theory*, American Elsevier.
Peterson AF, Ray SL, and Mittra R 1998 *Computational Methods for Electromagnetics*, IEEE Press.
Petykiewicz J 1992 *Wave Optics*, Kluwer.
Pike R and Sabatier P (eds) 2002 *Scattering: Scattering and Inverse Scattering in Pure and Applied Science*, Academic Press.
Plonus MA 1978 *Applied Electromagnetics*, McGraw-Hill.

Ramm AG 1986 *Scattering by Obstacles*, D. Reidel Publishing Company.
Rubinowicz A 1957 *Die Beugungswelle in der Kirchhoffschen Theorie der Beugung*, Panstwowe Wydawnictwo Naukowe, Warsaw.
Ruck GT, Barrick DE, Stuart WD, and Krichbaum CK 1970 *Radar Cross Section Handbook*, Plenum Press.
Rytov SM, Kravtsov YA, and Tatarskii VI 1989 *Principles of Statistical Radiophysics*, Springer.
Saez de Adana F, Gutierrez O, Gonzalez I, Catedra MF, and Lozano L 2011 *Practical Applications of Asymptotic Techniques in Electromagnetics*, Artech House.
Schelkunoff SA 1943 *Electromagnetic Waves*, D. Van Nostrand Company.
Senior TBA and Volakis JL 1995 *Approximate Boundary Conditions in Electromagnetics*, IEE Electromagnetic Waves Series 41, The Institution of Electrical Engineers, London.
Silver S (ed) 1984 *Microwave Antenna Theory and Design*, IEE Electromagnetic Waves Series 19, The Institution of Electrical Engineers, London.
Silvester PP and Ferrari RL 1990 *Finite Elements for Electrical Engineers*, Second edition, Cambridge University Press, Cambridge.
Solimeno S, Crosignani B, and DiPorto P 1986 *Guiding, Diffraction, and Confinement of Optical Radiation*, Academic Press.
Sommerfeld A 1927 Elektromagnetische Schwingungen, in *Die Differential- und Integralgleichungen der Mechanik und Physik. 2. Physikalischer Teil*, Frank P and von Mises R (eds), chapters 19–23, Vieweg & Sohn, Braunschweig.
Sommerfeld A 1964 *Lectures on Theoretical Physics: Optics*, Academic Press.
Sommerfeld A 2004 *Mathematical Theory of Diffraction*, Birkhäuser.
Stratton JA 1941 *Electromagnetic Theory*, McGraw-Hill.
Taflove A 1995 *Computational Electrodynamics: the Finite-difference Time-domain Method*, Artech House.
Tai C-T 1971 *Dyadic Green's Functions in Electromagnetic Theory*, Intext Educational Publishers.
Thomson JJ 1893 *Notes on Recent Researches in Electricity and Magnetism*, Clarendon Press, Oxford.
Tretyakov SA 2003 *Analytical Modeling in Applied Electromagnetics*, Artech House.
Tretyakov SA and Osipov AV 2006 *Applied Theory of Electromagnetic Scattering and Diffraction*, Helsinki University of Technology Radio Laboratory Publications, Espoo.
Tsang L and Kong JA 2001 *Scattering of Electromagnetic Waves: Advanced Topics*, John Wiley & Sons.
Tsang L, Kong JA, Ding K-H, and Ao CO 2001 *Scattering of Electromagnetic Waves: Numerical Simulations*, John Wiley & Sons.
Ufimtsev PY 1962 *Method of Edge Waves in the Physical Theory of Diffraction*, Sovetskoe Radio, Moscow (in Russian).
Ufimtsev PY 1999 Backscatter. In *Wiley Encyclopedia of Electrical and Electronics Engineering, vol. 2* (ed. Webster JG), pp. 175–187, John Wiley & Sons.
Ufimtsev PY 2003 *Theory of Edge Diffraction in Electromagnetics*, Tech Science Press.
Ufimtsev PY 2014 *Fundamentals of the Physical Theory of Diffraction*, Wiley.
Ulaby FT, Moore RK, and Fung AK 1982 *Microwave Remote Sensing: Active and Passive, vol.2: Radar Remote Sensing and Surface Scattering and Emission Theory*, Addison-Wesley.
Uslenghi PLE (ed) 1978 *Electromagnetic Scattering*, Academic Press.
van Bladel J 1964 *Electromagnetic Fields*, McGraw-Hill.
van Bladel J 1991 *Singular Electromagnetic Fields and Sources*, Oxford University Press, New York.
van de Hulst HC 1957 *Light Scattering by Small Particles*, John Wiley & Sons.
Vinogradov SS, Smith PD, and Vinogradova ED 2002 *Canonical Problems in Scattering and Potential Theory. Part II: Acoustic and Electromagnetic Diffraction by Canonical Structures*, CRC Press.
Volakis JL, Chatterjee A, and Kempel LC 1998 *Finite Element Method for Electromagnetics: Antennas, Microwave Circuits, and Scattering Applications*, IEEE Press.
von Helmholtz H 1897 *Vorlesungen über elektromagnetische Theorie des Lichtes*, Voss Verlag, Hamburg und Leipzig.
Wait JR 1959 *Electromagnetic Radiation from Cylindrical Structures*, Pergamon Press.
Wait JR 1962 *Electromagnetic Waves in Stratified Media*, Pergamon Press.
Wait JR 1981 *Lectures on Wave Propagation Theory*, Pergamon Press.
Weinstein LA 1969 *The Theory of Diffraction and the Factorization Method (Generalized Wiener-Hopf Technique)*, Golem Press.
Weston VH 1963 Theory of absorbers in scattering. *IEEE Trans. Antennas Propag.*, **11**, 578–584.
Young Th 1845 *A Course of Lectures on Natural Philosophy and Mechanical Arts*, Second edition, Taylor and Walton, London, lecture XXXIX.
Ziolkowski RW and Deschamps GA 1984 Asymptotic evaluation of high-frequency fields near a caustic: An introduction to Maslov's method. *Radio Sci.*, **19**, 1001–1025.

2

Fundamentals of Electromagnetic Scattering

2.1 Introduction

From the physical point of view, the electromagnetic scattering process means excitation of electric currents and electric and magnetic polarizations by incident electromagnetic fields, which in turn excite secondary, or *scattered*, fields in surrounding space. In classical electromagnetics, we deal with macroscopic bodies that contain huge amounts of molecules (individual scatterers). Thus, to study scattering we have to introduce models that allow us to treat materials as effectively homogeneous media. These models (constitutive relations) describe materials in terms of some effective materials parameters, usually permittivity and permeability. Furthermore, in the majority of practical situations interfaces between different materials are very sharp, and appear as discontinuities in the values of material parameters. To properly model such jumps in permittivity and permeability we need boundary conditions on interfaces.

From the mathematical point of view, solutions of scattering problems mean solutions of differential or integral equations for electric and magnetic field vectors subject to certain boundary conditions and constitutive relations. In this chapter we briefly review the fundamental field equations and provide necessary information on boundary conditions and material relations that are used to model electromagnetic properties of materials. Here we also describe the basic field solutions in the form of plane, cylindrical, and spherical waves, the tensor Green function of the vector wave equation, separation of fields in E and H modes, and introduce the electromagnetic form of the Huygens principle.

2.2 Fundamental Equations and Conditions

Fundamental equations of classical electromagnetics relate the four field vectors: electric field **E**, magnetic induction **B**, electric displacement vector **D**, and magnetic field **H**. Usually the first two vectors are considered as the fundamental field vectors because they determine the force acting on a moving charge

$$\mathbf{F} = q(\mathbf{E} + \mathbf{v} \times \mathbf{B}). \tag{2.1}$$

Here q is the particle charge and $\mathbf{v}$ is its velocity. In macroscopic electromagnetics, **E** and **B** are usually defined as the volume-averaged values of the microscopic electric and magnetic fields (e.g., Landau *et al.* (1984)). The other two vectors include the effects of material polarization, which are modeled by constitutive relations.

Modern Electromagnetic Scattering Theory with Applications, First Edition. Andrey V. Osipov and Sergei A. Tretyakov.

2.2.1 *Maxwell's Equations*

Exact solutions in the diffraction theory are solutions of Maxwell's equations for the four field vectors:

$$\nabla \times \mathbf{H} = \frac{\partial}{\partial t}\mathbf{D} + \mathbf{J}_e \tag{2.2}$$

$$\nabla \times \mathbf{E} = -\frac{\partial}{\partial t}\mathbf{B}. \tag{2.3}$$

Here, $\mathbf{J}_e$ stands for the volume density of electric source currents. The term is usually associated with the source currents but may include (some part of) induced currents. The divergence equations

$$\nabla \cdot \mathbf{D} = \rho_e \tag{2.4}$$

$$\nabla \cdot \mathbf{B} = 0 \tag{2.5}$$

follow from the curl equations and the law of charge conservation

$$\nabla \cdot \mathbf{J}_e = -\frac{\partial \rho_e}{\partial t}, \tag{2.6}$$

where ρ_e is the electric charge density.

In the theory of diffraction and scattering most often we deal with time-harmonic fields (in other words, we work with Fourier harmonics of the fields). For time-harmonic fields we write

$$\nabla \times \mathbf{H} = j\omega\mathbf{D} + \mathbf{J}_e \tag{2.7}$$

$$\nabla \times \mathbf{E} = -j\omega\mathbf{B} \tag{2.8}$$

(the time dependence is in the form $e^{j\omega t}$). In this case, vectors **E**, **H**, **D**, **B**, and $\mathbf{J}_e$ have the meaning of complex amplitudes (or phasors) and the physical, real-valued fields result from them as $\mathrm{Re}(\mathbf{E}e^{j\omega t})$, $\mathrm{Re}(\mathbf{H}e^{j\omega t})$, etc.

2.2.2 *Constitutive Relations*

Complete solutions for electromagnetic fields in the presence of material bodies mean solving diffraction problems for systems containing enormously large numbers of scatterers (atoms or molecules). As a result, one would know electromagnetic field distribution inside the material body with all tiny details and variations caused by individual molecules. Solving such problems is practically impossible, but fortunately for the vast majority of applications detailed knowledge of the microscopic field structure is not needed. Most often we are actually interested in averaged properties of the fields, in other words in the collective response of large ensembles of molecules. For instance, in the calculation of the reflection coefficient from a metal surface, the amplitude of the averaged surface current density is the relevant quantity, and not the microscopic current distribution in the crystal lattice of the material.

The averaging is made by introducing averaged fields **E** and **B** where the averaging volume is small compared to the wavelength and to the sizes of the body but contains many molecules. Furthermore, the averaged polarization densities **P** and **M** define the induction fields **D** and **H**. The material properties are modeled by constitutive relations between the fields and induction vectors. In the most general linear material one can write

$$\mathbf{D} = \overline{\overline{\epsilon}} \cdot \mathbf{E} + \overline{\overline{A}} \cdot \mathbf{H} = \epsilon_0 \overline{\overline{\epsilon}}_r \cdot \mathbf{E} + \sqrt{\epsilon_0\mu_0}\,\overline{\overline{a}} \cdot \mathbf{H} \tag{2.9}$$

$$\mathbf{B} = \overline{\overline{\mu}} \cdot \mathbf{H} + \overline{\overline{B}} \cdot \mathbf{E} = \mu_0 \overline{\overline{\mu}}_r \cdot \mathbf{H} + \sqrt{\epsilon_0\mu_0}\,\overline{\overline{b}} \cdot \mathbf{E}. \tag{2.10}$$

Tensorial relative permittivity $\overline{\overline{\epsilon}}_r$ and permeability $\overline{\overline{\mu}}_r$ measure the strength of electric and magnetic response in the medium, while the magneto-electric coefficients $\overline{\overline{a}}$ and $\overline{\overline{b}}$ measure the amplitudes of magneto-electric effects.

The free-space permittivity ϵ_0 and permeability μ_0 are fundamental constants which are related to the speed of light in free space (vacuum) by $c_0 = (\epsilon_0\mu_0)^{-1/2}$. In the international system of units SI, by definition, $\mu_0 = 4\pi \times 10^{-7}$ (H/m) and $c_0 = 299\,792\,458$ (m/s), so that the value $\epsilon_0 = 8.854187817\ldots \times 10^{-12}$ (F/m) can be calculated with any desired precision.

In reciprocal materials $\overline{\overline{\epsilon}} = \overline{\overline{\epsilon}}^T$, $\overline{\overline{\mu}} = \overline{\overline{\mu}}^T$ and $\overline{\overline{b}} = -\overline{\overline{a}}^T$, where T denotes the transpose operation (e.g., Kong (1986) and Serdyukov *et al.* (2001)). With this in mind, for reciprocal materials the constitutive equations are often written in form

$$\mathbf{D} = \epsilon_0\overline{\overline{\epsilon}}_r \cdot \mathbf{E} - j\sqrt{\epsilon_0\mu_0}\,\overline{\overline{\kappa}} \cdot \mathbf{H} \tag{2.11}$$

$$\mathbf{B} = \mu_0\overline{\overline{\mu}}_r \cdot \mathbf{H} + j\sqrt{\epsilon_0\mu_0}\,\overline{\overline{\kappa}}^T \cdot \mathbf{E}. \tag{2.12}$$

Here the reciprocal magneto-electric parameter $\overline{\overline{\kappa}}$ measures the effects of weak spatial dispersion, that is, dependence of the material response on the wavelength of the propagating wave. The character of this effect depends on the microstructure geometry (Serdyukov *et al.* 2001). In particular, the trace of $\overline{\overline{\kappa}}$, called the *chirality parameter*, is nonzero only if the microstructure of the material is chiral, that is, it is not symmetric with respect to mirror reflection.

The imaginary parts of the material parameters in the above relations measure absorptive properties of the material. In lossless reciprocal media all three parameters ($\overline{\overline{\epsilon}}$, $\overline{\overline{\mu}}$ and $\overline{\overline{\kappa}}$) are real functions of the frequency.

If the material structure is such that no preferred direction in space can be distinguished, the material is called *isotropic*. In isotropic media all the material parameters are scalars or pseudo-scalars. In this book, we will mostly deal with reciprocal isotropic materials with negligible spatial dispersion, modeled by the simple constitutive relations

$$\mathbf{D} = \epsilon\mathbf{E} = \epsilon_0\epsilon_r\mathbf{E}, \quad \mathbf{B} = \mu\mathbf{H} = \mu_0\mu_r\mathbf{H}, \tag{2.13}$$

where the parameters ϵ_r and μ_r are complex scalars, called the relative permittivity and the relative permeability.

Conducting bodies can be modeled by permittivity and conductivity, writing Maxwell's equation (2.7) as

$$\nabla \times \mathbf{H} = j\omega\epsilon\mathbf{E} + \sigma\mathbf{E} + \mathbf{J}_\mathrm{e}. \tag{2.14}$$

Here σ is the conductivity of the medium and $\mathbf{J}_\mathrm{e}$ is the current density of external (source) electric currents. In this model the imaginary part of ϵ describes losses occurring in the process of inducing electric polarization due to oscillations of bound charges, while σ models losses and electric polarization due to movements of free electrons. However, it is often convenient to introduce complex permittivity modeling losses of any nature by one single parameter. This is done by writing

$$j\omega\epsilon\mathbf{E} + \sigma\mathbf{E} = j\omega\left(\epsilon - j\frac{\sigma}{\omega}\right)\mathbf{E} = j\omega\hat{\epsilon}\mathbf{E}. \tag{2.15}$$

In this book we will often use this model denoting $\hat{\epsilon}$ simply as ϵ and, when necessary, separating the real and imaginary parts as

$$\epsilon_r = \epsilon/\epsilon_0 = \epsilon_r' - j\epsilon_r'', \tag{2.16}$$

where for media in which conducting losses prevail

$$\epsilon_r'' = \frac{\sigma}{\omega\epsilon_0}. \tag{2.17}$$

Common classification of materials is based on the relation between ϵ_r' and ϵ_r''. A conductor is a material with high conductivity such that $\epsilon_r' \ll \epsilon_r''$. An insulator has low conductivity and, thus, $\epsilon_r' \gg \epsilon_r''$. Semiconductors are materials with comparable ϵ_r' and ϵ_r''.

The permeability in (2.13) can also be complex-valued

$$\mu_r = \mu/\mu_0 = \mu_r' - j\mu_r'', \tag{2.18}$$

where the imaginary part accounts for losses associated with interaction of magnetic fields with the material. Magnetization rotation within the domains is the principal loss mechanism at microwave frequencies (Knott *et al.* 1993).

Permittivity ϵ and permeability μ are often referred to as dielectric and magnetic constants, and their imaginary parts as dielectric and magnetic loss factors. As with all complex numbers, ϵ and μ can be represented in polar notation

$$\epsilon_r = |\epsilon_r| e^{-j\delta_\epsilon}, \quad \mu_r = |\mu_r| e^{-j\delta_\mu}, \tag{2.19}$$

in which case δ_ϵ and δ_μ are called the dielectric and magnetic loss angles. The tangents of these angles relate the real and imaginary parts of ϵ and μ according to

$$\epsilon_r'' = \epsilon_r' \tan\delta_\epsilon, \quad \mu_r'' = \mu_r' \tan\delta_\mu, \tag{2.20}$$

and this is why the values of $\tan\delta_\epsilon$ and $\tan\delta_\mu$ (called loss tangents) often replace ϵ_r'' and μ_r'' in tables of material parameters.

Refractive index or index of refraction $N = \sqrt{\epsilon_r \mu_r}$ is another useful material parameter. In general it is complex-valued, $N = N' - jN''$, and N'' must be non-negative for any passive medium; the real part, however, can be positive or negative. In any lossless non-dispersive medium, the refractive index is equal to the ratio of the wave speeds in vacuum and in the medium and appears in the refraction (Snell's) law relating the directions of incidence and refraction at material interfaces (see (4.20) in Chapter 4), hence its name. In non-magnetic media, the refractive index and the relative permittivity are directly related by $\epsilon_r = N^2$ and are therefore equivalent parameters.

Tables 2.1 and 2.2, compiled by using mostly data from Hippel (1954) and Weast (1988), present the values of ϵ, μ, and σ for a selection of materials that are often encountered in problems of microwave scattering. Table 2.1 lists the real and imaginary parts of the relative dielectric and magnetic constants of a variety of natural and artificial non-metallic materials. At microwave frequencies, that is, from 1 GHz to 100 GHz, permittivity and permeability of most materials only weakly depend on frequency, so the values in Table 2.1, which are measured at 10 GHz, are good estimations for the whole microwave frequency range. The situation is different at optical frequencies as there exists strong frequency dependence of ϵ (dispersion) in most materials, and that is addressed in more detail later in this section.

Most of the materials, natural and artificial, are nearly non-magnetic with the permeability very close to unity (as it is shown for the air), so it is sufficient for practical application to assume $\mu_r = 1$. The deviation of μ from μ_0 in the case of the building wall (Table 2.1) is due to the experimental nature of the data, which are in fact effective values obtained by measurements on the real-world building facades. However, for ferrites, which are substances composed of iron oxides and other metallic oxides, the values of μ_r are significantly greater than unity, and their electric and magnetic constants are strongly dependent on the frequency, as seen from Table 2.1. The relative permeability of pure iron may arrive at values of several thousands. The highest reported values of μ_r are about 10^6 (magnetic alloys at frequencies lower than 100 kHz).

Table 2.2 lists the values of σ for several pure and alloy metals. Once σ is known, the dielectric loss factor is calculated from (2.17). The magnetic permeability of most metals is close to that of vacuum, that is, $\mu_r = 1$. Exceptions are ferromagnetic materials (cobalt, iron, nickel) where μ_r at radio frequencies can be several orders of magnitude greater than μ_0.

The precision of the data should not be overestimated, rather the given values should be considered as benchmarks or typical values. Electromagnetic properties of materials, whether it is a conductor or isolator, a natural or an artificial material, vary and are affected by differences in chemical composition, measurement conditions (frequency, humidity, temperature), production process, etc. For example, even for such a relatively simple material as stainless steel the conductivity σ may vary by more than an order of magnitude and the magnetic permeability μ_r can be between 1 and 2000. Furthermore, many materials have a fine but macroscopic internal structure (composites, metamaterials) or a considerable surface roughness (sea surface, building wall) and in many situations can be only approximately considered as homogeneous and isotropic; in such cases the values of the parameters found in tables are in fact effective or statistically averaged.

Frequency Dispersion in Metals

Let us next discuss the frequency dispersion of material parameters. We begin with metals. The electromagnetic response of metals is usually dominated by the response of electrons that can more or less freely move under the influence of applied electric fields. Frequency dispersion of electron plasmas can be modeled by the Drude model

$$\epsilon = \epsilon_0 \left(1 - \frac{\omega_\mathrm{p}^2}{\omega^2 - j\omega\Gamma} \right), \tag{2.21}$$

Table 2.1 Material parameters of a selection of natural and artificial non-metallic materials at 10 GHz and room temperature unless otherwise noted.

Material	ϵ_r'	ϵ_r''	μ_r'	μ_r''	Notes
Air dry, CO_2 free	1.005366	0	1.000001	0	20° C, 1 atm
Construction					
Asphalt	2.5	0.002	1	0	99.9 % bitumen
Concrete blocks	2.4	0.0016	1	0	
Brick building wall[a]	4.4	0.04	0.99	0	4 GHz
Limestone wall[a]	7.5	0.13	0.95	0	4 GHz
Water-related					
Fresh water	59	34	1	0	
Salted, sea water	50	32	1	0	0.6 molal NaCl
Snow (freshly fallen)	1.3	0.0005	1	0	−6° C
Freshwater ice	3.17	0.002	1	0	−12° C
Glasses					
Standard	6.7	0.1	1	0	Soda-lime glass
Flint	8.05	0.04	1	0	Lead glass
Pyrex	4.8	0.05	1	0	Soda-borosilic.
Quartz	3.78	0.0004	1	0	Pure fused SiO_2
Plastics					
Epoxy resins	3.1	0.07	1	0	Averaged values
FR-4[b]	4.4	0.1	1	0	Fiberglass-reinforced epoxy laminate
Plexiglas®	2.59	0.02	1	0	Acrylic glass
Rohacell®31 HF[c]	1.04	0.0018	1	0	Acrylic foam
Teflon®	2.08	0.0008	1	0	
Soils					
Sandy dry	2.5	0.009	1	0	
Sandy wet	13	3.8	1	0	16.8 % moisture
Loamy dry	2.4	0.001	1	0	
Loamy wet	14	2.5	1	0	13.8 % moisture
Clay dry	2.2	0.03	1	0	
Clay wet	11	2.8	1	0	20 % moisture, 3 GHz
Wood					
Balsa	1.2	0.01	1	0	
Birch	1.9	0.05	1	0	
Fir	1.8	0.05	1	0	
Mahogany	1.7	0.04	1	0	
Poplar	1.4	0.03	1	0	
Radar absorbers					
Nickel-zinc ferrite[d]	27	54	15	45	0.1 GHz
The same material[d]	20	9.0	1.2	12	1 GHz
The same material[d]	15	6.3	0.1	0.32	10 GHz

[a]From Landron *et al.* (1996). [b]From Djordjevic *et al.* (2001). [c]According to the Evonik's data sheet. [d]From Knott *et al.* (1993).

Table 2.2 Electrical conductivity of metals at 20° C (Weast 1988).

Metal	$\sigma \times 10^{-6}$ S/m
Silver	63
Copper	60
Gold	45
Aluminum	38
Iron	10
Plain carbon steel	10
Lead	4.8
Stainless steel	1.4

derived from the equation of movement of a free electron with a phenomenological loss factor Γ, which has the dimension of the frequency (e.g., Landau *et al.* (1984, section 78)). In the classical-physics model, the so-called *plasma frequency* ω_{p} is given by the formula

$$\omega_{\mathrm{p}}^2 = \frac{N_{\mathrm{e}} e^2}{\epsilon_0 m} \tag{2.22}$$

where N_{e} is the number of electrons per unit volume, e is the electron charge, and m is the electron mass.

For most metals at optical frequencies, the free-electron approximation is insufficiently accurate to model their properties in the visible or infrared range due to significant effects of interband transitions. However, for noble metals it is possible to find a set of Drude parameters which provide a good fit to the experimental data at least in the infrared frequency range. This model can be written as

$$\epsilon = \epsilon_0 \left(\epsilon_\infty - \frac{\omega_{\mathrm{p}}^2}{\omega^2 - j\omega\Gamma} \right). \tag{2.23}$$

Here the effective parameter ϵ_∞ models the interband transitions effect. The generalized Drude model (2.23) is known to be quite accurate well below the plasma frequency in the infrared region, while in the visible range and above the frequency dispersion is more complicated.

Among the currently known metals and alloys, silver has the smallest losses in the visible. The effective Drude parameters for bulk silver which provide a good fit to the experimental data in the infrared and far infrared (Ordal *et al.* 1985) are $\omega_{\mathrm{p}}/(2\pi) = 2175$ THz, $\Gamma/(2\pi) = 4.35$ THz, and $\epsilon_\infty = 5$ (El-Kady *et al.* 2000; Ordal *et al.* 1985). The frequency dependence of the bulk permittivity of silver is illustrated in Figure 2.1. Note that the Drude model considerably underestimates losses in the visible, while it is quite accurate in the infrared.

Measured values of the real and imaginary parts of the relative permittivity of silver for frequencies ranging from the near infrared over the visible ($f = 430 - 750$ THz) to the near ultraviolet are presented in Table 2.3. In addition to the pronounced frequency dependence (dispersion) the permittivity of the noble metals at optical frequencies features the negative values of ϵ_r'. Tables of the optical constants for various solid materials, including metals, semiconductors and isolators, can be found in Palik (1997) and for noble metals also in Johnson and Christy (1972). Systematic parametrizations of the complex dielectric function $\epsilon_r(\omega) = \epsilon_r'(\omega) - j\epsilon_r''(\omega)$ for 11 metals in a frequency range from 24.1 THz to 1446 THz can be found in Rakic *et al.* (1998).

At microwave frequencies for metals $\omega \ll \Gamma$, so that we can neglect ω^2 in the denominator of (2.21) and write

$$\epsilon \approx -j\frac{\sigma}{\omega}, \tag{2.24}$$

where

$$\sigma = \frac{\omega_{\mathrm{p}}^2 \epsilon_0}{\Gamma} \tag{2.25}$$

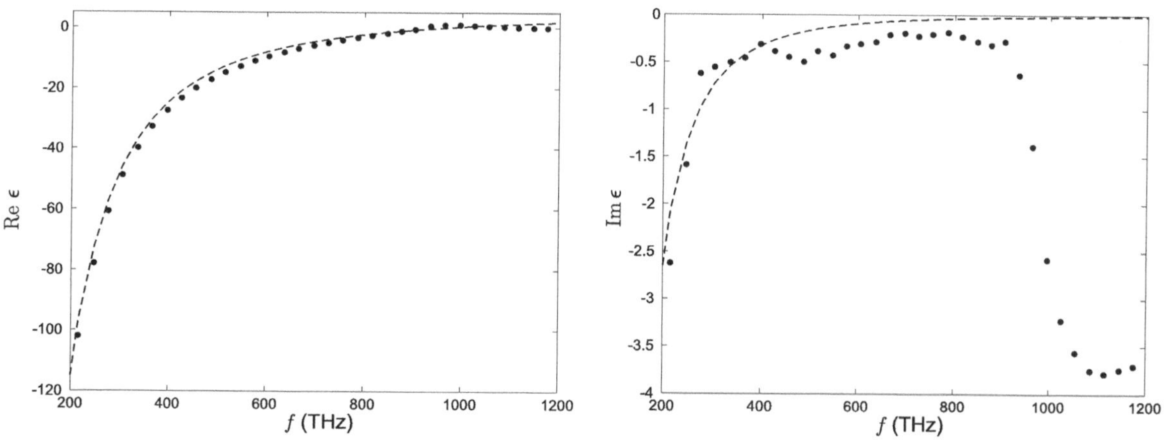

Figure 2.1 Real and imaginary parts of the permittivity of bulk silver from infrared to ultraviolet frequencies. Dots show measured data from Johnson and Christy (1972), and the dashed curves correspond to the generalized Drude model (2.23) with the parameters $\omega_p/(2\pi) = 2175$ THz, $\Gamma/(2\pi) = 4.35$ THz, and $\epsilon_\infty = 5$.

is the conductivity. For good conductors the real part of the permittivity is negligible as compared with the imaginary part, and for this reason it is typically sufficient to give only the value of conductivity σ and use the approximation $\epsilon_r \approx -j\epsilon_r''$, with ϵ_r'' given by (2.17).

It is seen from (2.17) and the values of σ from Table 2.2 that at microwave frequencies $|\epsilon_r''| = O(10^7)$ and metal surfaces can be regarded as nearly perfectly electrically conducting. At optical frequencies, however, since ϵ_r'' is inversely proportional to ω, ϵ_r'' becomes comparable to ϵ_r' and the approximation of a perfect electric conductor does not apply any more. For metals like silver or gold the real and imaginary parts of the permittivity in the visible are of the same order of magnitude.

Frequency Dispersion in Water and Biological Tissues

Electromagnetic properties of polar dielectrics can be modeled by the Debye model, which is based on the classical physics analysis of electric dipole orientation in the applied time-harmonic electric field. This consideration leads to the following dispersion rule (e.g., Sihvola (1999)):

$$\epsilon_r(\omega) = \epsilon_\infty + \frac{\epsilon_s - \epsilon_\infty}{1 + j\omega/\Gamma}. \tag{2.26}$$

A typical example of a polar dielectric is water, whose Debye parameters at 20° C are (Chang and Wilheit 1979) $\epsilon_\infty \approx 4.90$, $\epsilon_s \approx 80.1$, and $\Gamma \approx 9.9 \times 10^{10}$ 1/s. From here we see that water has strong dispersion in the microwave frequency range. Note that the Debye model usually fails at frequencies much larger than Γ. At infrared and optical frequencies, say between 200 THz and 1500 THz, ϵ_r' varies slowly around 1.8, and water has extremely low absorption (see Table 2.3).

Electromagnetic properties of biological materials are closely related to those of water, but the complexity of biological materials is such that each dispersion region may be broadened by multiple contributions to it. The broadening could be empirically accounted for by introducing an additional parameter in (2.26), thus giving a dispersion approximation known as the Cole–Cole equation,

$$\epsilon_r(\omega) = \epsilon_\infty + \frac{\Delta\epsilon}{1 + (j\omega\tau)^{1-\alpha}}, \tag{2.27}$$

where the parameter α is a measure of the broadening of the dispersion region. With $\alpha = 0$ the Cole–Cole formula reduces to the Debye equations with $\Delta\epsilon = \epsilon_s - \epsilon_\infty$ and $\tau = 1/\Gamma$. Table 2.4 presents the values of parameters from (2.27) for a selection of human tissues. The approximation is a simplified version of a more general one that is valid over a broader

Table 2.3 Relative permittivity of silver and water at optical frequencies.

Silver				Water			
f (THz)	λ (nm)	ϵ_r'	ϵ_r''	f (THz)	λ (nm)	ϵ_r'	ϵ_r''
154.751	1937.25	−198	6.76	99.9308	3000	1.805	0.745
186.185	1610.18	−140	3.55	101.625	2950	1.580	0.770
215.201	1393.08	−102	2.63	103.377	2900	1.370	0.643
246.635	1215.53	−77.9	1.59	105.19	2850	1.285	0.425
275.651	1087.58	−60.8	0.624	107.069	2800	1.290	0.262
304.667	984.001	−48.9	0.559	109.015	2750	1.335	0.136
336.101	891.973	−39.8	0.505	111.034	2700	1.410	0.0451
365.116	821.087	−32.8	0.458	113.129	2650	1.485	0.0163
396.55	756.001	−27.5	0.315	115.305	2600	1.542	0.00787
425.566	704.456	−23.4	0.387	124.914	2400	1.635	0.00244
454.582	659.49	−20.1	0.448	136.269	2200	1.679	0.000749
486.016	616.837	−17.2	0.498	149.896	2000	1.705	0.00287
515.032	582.085	−14.9	0.386	166.551	1800	1.721	0.000301
546.466	548.603	−12.9	0.430	187.37	1600	1.734	0.000225
575.481	520.942	−11.0	0.332	214.137	1400	1.745	0.000364
604.497	495.937	−9.56	0.309	249.827	1200	1.752	0.0000261
635.931	471.423	−8.23	0.287	299.792	1000	1.760	7.67×10^{-6}
664.947	450.852	−7.06	0.213	307.479	975	1.760	9.23×10^{-6}
696.381	430.501	−6.06	0.197	315.571	950	1.760	7.77×10^{-6}
725.397	413.281	−5.17	0.227	324.1	925	1.763	2.81×10^{-6}
754.413	397.385	−4.28	0.207	333.103	900	1.763	1.29×10^{-6}
785.847	381.49	−3.47	0.186	342.62	875	1.763	1.03×10^{-6}
814.862	367.906	−2.74	0.232	352.697	850	1.766	7.78×10^{-7}
846.296	354.241	−2.00	0.284	363.385	825	1.766	4.83×10^{-7}
875.312	342.498	−1.28	0.320	374.741	800	1.766	3.32×10^{-7}
904.328	331.509	−0.658	0.282	386.829	775	1.768	3.93×10^{-7}
935.762	320.373	0.502	0.635	399.723	750	1.768	4.14×10^{-7}
964.778	310.737	0.897	1.39	413.507	725	1.768	2.43×10^{-7}
996.212	300.932	0.866	2.58	428.275	700	1.771	8.91×10^{-8}
1025.23	292.416	0.584	3.23	444.137	675	1.771	5.93×10^{-8}
1054.24	284.367	0.390	3.56	461.219	650	1.771	4.36×10^{-8}
1085.68	276.134	0.216	3.75	479.668	625	1.774	3.70×10^{-8}
1114.69	268.946	0.022	3.79	499.654	600	1.774	2.90×10^{-8}
1146.13	261.57	−0.101	3.74	521.378	575	1.776	9.59×10^{-9}
1175.14	255.111	−0.172	3.71	545.077	550	1.776	5.22×10^{-9}
1204.16	248.964	−0.213	3.64	571.033	525	1.779	3.52×10^{-9}
1235.59	242.631	−0.209	3.58	599.585	500	1.782	2.67×10^{-9}
1264.61	237.063	−0.230	3.50	631.142	475	1.784	2.49×10^{-9}
1296.04	231.314	−0.203	3.47	666.205	450	1.787	2.72×10^{-9}
1325.06	226.249	−0.219	3.39	705.394	425	1.790	3.47×10^{-9}
1354.07	221.4	−0.238	3.36	749.481	400	1.792	4.98×10^{-9}
1385.51	216.377	−0.296	3.26	799.447	375	1.798	9.38×10^{-9}
1414.52	211.939	−0.316	3.18	856.55	350	1.803	1.74×10^{-8}
1445.96	207.331	−0.329	3.10	922.438	325	1.811	2.90×10^{-8}
1474.97	203.253	−0.357	2.98	999.308	300	1.819	4.31×10^{-8}
1503.99	199.331	−0.331	2.91	1090.15	275	1.833	6.36×10^{-8}
1535.42	195.251	−0.321	2.81	1199.17	250	1.855	9.12×10^{-8}
1564.44	191.629	−0.308	2.71	1332.41	225	1.885	1.34×10^{-7}
1595.87	187.855	−0.324	2.59	1498.96	200	1.948	3.07×10^{-7}

Based on Johnson and Christy (1972) and Hale and Querry (1973).

Table 2.4 Parameters of a Cole–Cole approximation for human tissues at 37° C and frequencies from 1 GHz to 100 GHz.

Tissue	ϵ_∞	$\Delta\epsilon$	τ (nm)	α	Notes
Blood	4.0	56	8.38	0.10	
Bone	2.5	18.	13.26	0.22	Cancellous
Fat	2.5	3.0	7.96	0.20	Not infiltrated
Muscle	4.0	50	7.23	0.10	
Dry skin	4.0	32	7.23	0.	
Water	4.9	75	10.1	0.	

Based on data from Gabriel *et al.* (1996).

range of frequencies (Gabriel *et al.* 1996), obtained by specializing the latter to microwave frequencies. The parameters for water are given for reference. Figure 2.2 shows the frequency dependence of ϵ_r' and ϵ_r'' of several human tissues in the microwave range calculated from the Cole–Cole approximation (2.27) with the parameters from Table 2.4. The curves indicate the significant role of dispersion in biological materials.

Effective properties of dielectric materials with nonpolar molecules can be modeled by the Lorentz model, which follows from consideration of molecules as linear oscillators excited by external electric fields. In the vicinity of each resonance frequency ω_0 we can write

$$\epsilon_r(\omega) = \epsilon_\infty + \frac{\omega_p^2}{\omega_0^2 - \omega^2 + j\omega\Gamma}. \tag{2.28}$$

Models for a wide frequency range can be constructed by linear combinations of several such terms. This model can also be used in a limited frequency range. In particular, it fails at very high frequencies, if $\epsilon_\infty \neq 1$. Many low-loss dielectric materials used in microwave engineering have weak dispersion in the microwave range.

To model the frequency response of materials in a wide frequency ranges, dispersion terms of various types need to be combined. For instance, molecules of polar dielectrics may have resonances at frequencies much higher than Γ, and they are modeled by Lorentz terms in the dispersion relation. Note that the Drude model (2.21) is a limiting case of the Lorentz formula (2.28) at $\omega_0 \to 0$.

2.2.3 Time-harmonic Scattering Problems

In this section we formulate the scattering problem for time-harmonic fields, which includes field equations supplemented with boundary conditions on material boundaries and on the spherical surface with the radius going to infinity. The whole

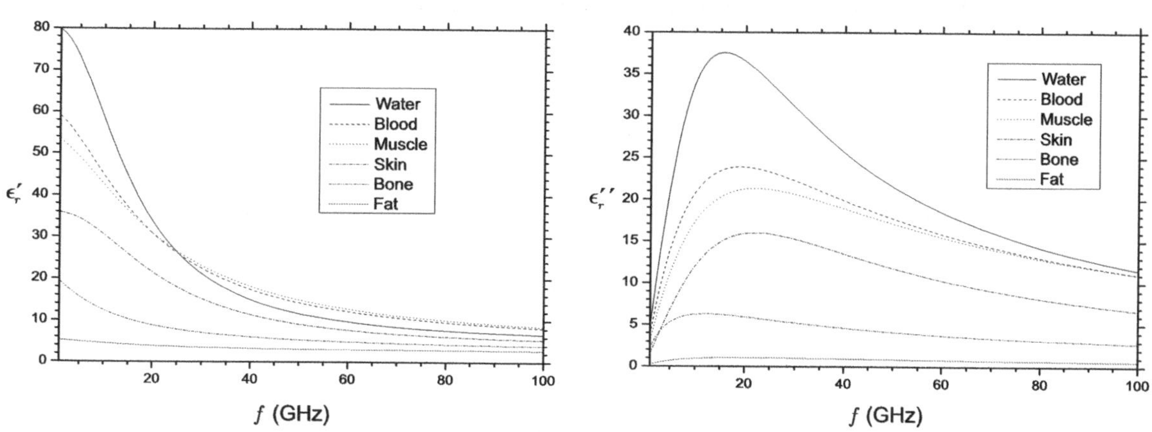

Figure 2.2 Real (ϵ_r') and imaginary (ϵ_r'') parts of the relative permittivity of human tissues at microwave frequencies.

book, with the exception of Chapter 4 where we treat the problem of reflection from infinite planar material boundaries, is based on the formulation of the scattering problem given in this section.

Consider a body (scatterer) immersed in a homogeneous infinite background medium, typically free space or vacuum. The body is illuminated by an incident field, $\mathbf{E}_{\text{inc}}$ and $\mathbf{H}_{\text{inc}}$, from a given source which can be located at a finite distance from the body but most typically will be assumed infinitely distant. The source and therefore the fields are time-harmonic with the time dependence $\exp(j\omega t)$. The incident field arrives at the scatterer and since material composition of the scatterer is different from the background medium, the scattered field, $\mathbf{E}_{\text{sc}}$ and $\mathbf{H}_{\text{sc}}$, are generated. The total field, $\mathbf{E}$ and $\mathbf{H}$, is the sum of the incident and scattered fields,

$$\mathbf{E} = \mathbf{E}_{\text{inc}} + \mathbf{E}_{\text{sc}}, \quad \mathbf{H} = \mathbf{H}_{\text{inc}} + \mathbf{H}_{\text{sc}}. \tag{2.29}$$

In scattering problems, we most often deal with piece-wise homogeneous scattering bodies. Inside the scatterer the material parameters are different from those in the background medium, but the scatterer is still homogeneous in each of its distinctive regions. For example, within this piece-wise homogeneous model we can study metal bodies covered by absorbing layers.

The assumption of the piece-wise homogeneous body does not limit the generality of our analysis. The scattering problems can also be formulated for smoothly inhomogeneous bodies, that is, when permittivity and/or permeability in the interior of the scatterer are smooth functions of the position. Approximating the material composition by a piece-wise constant media reduces the problem to the case of the piece-wise homogeneous body. Moreover, in practice many inhomogeneous artificial materials are assembled as a set of homogeneous layers.

The problem of scattering from the piece-wise homogeneous body can be expressed in terms of field equations in each homogeneous region and boundary conditions on the interfaces. If the boundaries include some geometric singularities, like edges or conical points, then the fields may become infinite there, and care should be taken to ensure that the infinities are physically acceptable. Finally, it is necessary to ensure that the scattered fields are outgoing waves. The suitable conditions follow from the physically apparent requirement that there is no energy supply to the system other than from the given excitation.

We begin by specializing Maxwell's equations to the case of homogeneous isotropic source-free medium and noting that their solutions are waves with different propagation directions. The connection of the fields at the material interfaces within the piece-wise homogeneous scatterer is achieved by imposing the continuity conditions on the field components tangential to the interfaces. Conditions at geometric singularities of the boundaries can be formulated as the requirement that the singularities are not sources of energy (Meixner's conditions). The correct solution for the scattered field is selected by demanding that the scattered energy be transported away from the scatterer (conditions at infinity). In section 2.4.1 we will show that the combination of Maxwell's equations with all these conditions ensures the uniqueness of the solution of the time-harmonic scattering problems.

Time-harmonic Maxwell's Equations in Homogeneous Isotropic Media

Writing down Maxwell's equations in homogeneous and isotropic media, and using constitutive relations (2.13) in equations (2.7) and (2.8), we obtain

$$\nabla \times \mathbf{H} = j\omega\epsilon\mathbf{E} + \mathbf{J}_{\text{e}} \tag{2.30}$$

$$\nabla \times \mathbf{E} = -j\omega\mu\mathbf{H}. \tag{2.31}$$

Sometimes, fictitious magnetic current sources are introduced, which do not exist but may serve as a convenient tool to describe fields of real sources. Substitution

$$\mathbf{E} = \tilde{\mathbf{E}} + \frac{j}{\omega\epsilon}\mathbf{J}_{\text{e}} \tag{2.32}$$

excludes the electric current from (2.30) and transforms (2.30) and (2.31) to

$$\nabla \times \mathbf{H} = j\omega\epsilon\tilde{\mathbf{E}} \tag{2.33}$$

$$\nabla \times \tilde{\mathbf{E}} = -j\omega\mu\mathbf{H} - \mathbf{J}_{\text{m}}, \tag{2.34}$$

where

$$\mathbf{J}_{\mathrm{m}} = \frac{j}{\omega\epsilon}\nabla \times \mathbf{J}_{\mathrm{e}}. \tag{2.35}$$

The quantity $\mathbf{J}_{\mathrm{m}}$ can be referred to as the magnetic source current density. According to equations (2.33) and (2.34), in the exterior of the source region the currents $\mathbf{J}_{\mathrm{e}}$ and $\mathbf{J}_{\mathrm{m}}$ excite the same electric and magnetic fields. Within the source region, the magnetic fields coincide, whereas the electric fields are related by (2.32).

The divergence equations (2.4) and (2.5), which for time-harmonic fields in homogeneous isotropic materials have the form

$$\epsilon\nabla \cdot \mathbf{E} = \rho_{\mathrm{e}}, \quad \nabla \cdot \mathbf{H} = 0, \tag{2.36}$$

follow from (2.30) and (2.31), and the relation

$$\nabla \cdot \mathbf{J}_{\mathrm{e}} = -j\omega\rho_{\mathrm{e}}, \tag{2.37}$$

which is the Fourier-transformed charge conservation law (2.6). Current densities of time-harmonic dipole sources can be expressed as

$$\mathbf{J}_{\mathrm{e}} = j\omega\mathbf{P}_{\mathrm{e}}, \quad \mathbf{J}_{\mathrm{m}} = j\omega\mu\mathbf{P}_{\mathrm{m}}, \tag{2.38}$$

where $\mathbf{P}_{\mathrm{e}}$ and $\mathbf{P}_{\mathrm{m}}$ are the densities of electric and magnetic dipole moments, respectively (e.g., Stratton (1941)). Note that $\nabla \cdot \mathbf{P}_{\mathrm{e}} = -\rho_{\mathrm{e}}$ because of (2.37) and $\nabla \cdot \mathbf{P}_{\mathrm{m}} = 0$ because of (2.35).

In source-free regions, Maxwell's equations have the form

$$\nabla \times \mathbf{H} = j\omega\epsilon\mathbf{E} \tag{2.39}$$

$$\nabla \times \mathbf{E} = -j\omega\mu\mathbf{H}, \tag{2.40}$$

or

$$\nabla \times \mathbf{H} = jkY\mathbf{E} \tag{2.41}$$

$$\nabla \times \mathbf{E} = -jkZ\mathbf{H}, \tag{2.42}$$

where $k = \omega\sqrt{\epsilon\mu}$ is called the wavenumber or propagation factor (and we will see why a bit later), $Z = \sqrt{\mu/\epsilon}$ is the *intrinsic impedance*, and $Y = 1/Z$ is the *intrinsic admittance* of the medium. The impedance Z is in ohms and is equal to the ratio of the complex amplitudes of the electric and magnetic fields in a plane wave propagating in the medium (section 2.5.1). That is why Z is often called the wave impedance of the medium.

The intrinsic impedance of free space (vacuum), often refereed to as impedance of free space, can be expressed though the fundamental constants as

$$Z_0 = \sqrt{\frac{\mu_0}{\epsilon_0}} = \mu_0 c_0, \tag{2.43}$$

and since the International System of Units (SI) assigns exact numerical values to the permeability of vacuum μ_0 and the speed of light in vacuum c_0, the impedance of free space has the *exact* numerical value $Z_0 = 119.9169832\pi$ ohm. Many textbooks give an approximate value of 120π ohm, which would result from (2.43) with c_0 substituted with 3×10^8 (m/s).

Equations (2.39)–(2.42) are invariant under the transformation

$$\mathbf{E} \rightarrow \mathbf{H}, \quad \mathbf{H} \rightarrow -\mathbf{E}, \quad \epsilon \leftrightarrow \mu, \quad Z \leftrightarrow Y \tag{2.44}$$

which is called the duality transformation. Application of (2.44) automatically gives a solution for, let us say, the magnetic field if a solution for the electric field is available, and vice versa. This property is also useful in checking the correctness of solutions of Maxwell's equations, as expressions for electric and magnetic fields must comply with the duality property.

Maxwell's equations are a system of two vector differential equations of first order for the fields **E** and **H**. By applying the curl operator to any of the equations and using another equation it is always possible to exclude one of the fields and obtain a single vector differential equation for either **E** or **H**. For example, this procedure transforms (2.30) and (2.31) to the equations:

$$\nabla \times \nabla \times \mathbf{H} - k^2\mathbf{H} = \nabla \times \mathbf{J}_e \tag{2.45}$$

$$\nabla \times \nabla \times \mathbf{E} - k^2\mathbf{E} = -j\omega\mu\mathbf{J}_e. \tag{2.46}$$

Every equation is for a single field but, in contrast to the original equations (2.30) and (2.31), involves differential operators of second order.

Equations (2.45) and (2.46) can be written as Helmholtz equations. For example, in source-free regions, by expressing the double curl operators in terms of the vector Laplacian (A.24) as

$$\nabla \times (\nabla \times \mathbf{E}) = \nabla(\nabla \cdot \mathbf{E}) - \nabla^2\mathbf{E} \tag{2.47}$$

$$\nabla \times (\nabla \times \mathbf{H}) = \nabla(\nabla \cdot \mathbf{H}) - \nabla^2\mathbf{H}, \tag{2.48}$$

and using the divergence equations (2.36) in the source-free case, equations (2.45) and (2.46) reduce to the Helmholtz equations (wave equations) for the electric and magnetic fields

$$\nabla^2\mathbf{E} + k^2\mathbf{E} = 0 \tag{2.49}$$

$$\nabla^2\mathbf{H} + k^2\mathbf{H} = 0. \tag{2.50}$$

In Cartesian coordinates, because of the property (A.25), the vector equations (2.49) and (2.50) split into separate scalar Helmholtz equations for every Cartesian component of the fields. This is the reason for the similarity between theories of electromagnetic and acoustic fields, and for the success of the scalar theories of diffraction and scattering of light developed in the nineteenth century.

The basic solutions of the Helmholtz equation are *waves*, which are described by exponential functions of the form

$$e^{-jk\hat{\imath}\cdot\mathbf{r}}, \tag{2.51}$$

where **r** is the position vector of the observer and $\hat{\imath}$ is a vector, possibly complex-valued, with Cartesian components that satisfy the relation

$$i_x^2 + i_y^2 + i_z^2 = 1. \tag{2.52}$$

A real-valued vector $\hat{\imath}$ is a unit vector that can be interpreted as the direction of propagation of the wave. In this case the wave (2.51) describes a periodic process in space with the period

$$\lambda = \frac{2\pi}{k}, \tag{2.53}$$

which is called the *wavelength*. Vector $\mathbf{k} = k\hat{\imath}$ is referred to as the *wave vector*. Every solution of the Helmholtz equation can be represented as a linear superposition of waves (2.51) with various vectors $\hat{\imath}$, in the most general case as a two-dimensional continuum of complex-valued propagation factors. Simple examples of the wave solutions of Maxwell's equations are given in section 2.5.1 and problem 13.

However, Maxwell's equations cannot be fully replaced with a single Helmholtz equation. Electric and magnetic time-variable fields are coupled by Maxwell's equations and only exist together. Even a single-component linear current excites in general all three Cartesian components of both fields (section 2.5.3), and individual components of the field vectors are typically coupled at material boundaries. Furthermore, it is important to understand that although equations (2.47) and (2.48) have been derived from Maxwell's equations, they are not equivalent to them. Equation

$$\nabla \cdot \mathbf{E} = 0 \tag{2.54}$$

follows from (2.39), but it does not follow from (2.49) and (2.50). Moreover, supplementing (2.49) and (2.50) with (2.54) does not make the system equivalent to (2.39) and (2.40). For example, non-trivial solutions where $\mathbf{H} \equiv 0$ are allowed by (2.49), (2.50), and (2.54), but such fields do not satisfy Maxwell's equations. The correct way to use the Helmholtz equations is to solve one of them, for instance (2.49), check that (2.54) is satisfied, and then find the other field from Maxwell's equation (2.40). An example of a difficulty in applying the scalar formulation to the individual components to the electromagnetic fields is described in Wait (1970).

It is therefore apparent that integration of Maxwell's equations requires special care to fully account for the vector nature of electromagnetic fields. Truly electromagnetic solutions are typically obtained by the introduction of various potential functions (section (2.5)), or by the use of a vector version of Huygens' principle, section (2.6).

Interface and Boundary Conditions

If the material parameters (permittivity and permeability) of the scatterer experience jumps (sharp interfaces between different media), the field equations have to be complemented with appropriate interface or boundary conditions.

On interface surfaces between two media the tangential to the interface components of **E** and **H** (marked by index t) must be continuous[1]

$$\mathbf{E}_{t1}|_S = \mathbf{E}_{t2}|_S, \quad \mathbf{H}_{t1}|_S = \mathbf{H}_{t2}|_S. \tag{2.55}$$

On ideally conducting (PEC) surfaces

$$\mathbf{E}_t|_S = 0. \tag{2.56}$$

From here it follows that on PEC surfaces

$$\hat{n} \cdot \mathbf{H}|_S = 0, \tag{2.57}$$

where $\hat{n}$ is the unit vector normal to the surface. Also,

$$\left.\frac{\partial \mathbf{H}_t}{\partial n}\right|_S = 0. \tag{2.58}$$

Conditions at Singular Points and Lines

Solutions of Maxwell's equations can be infinite at points and lines in space. Such points and lines are called singular, and they include (a) sources and (b) edges and tips on material interfaces. Assuming that the electromagnetic field energy must be finite in any finite volume V_0, we demand that the following integrals converge:

$$\int_{V_0} |\mathbf{E}|^2 \, \mathrm{d}V < \infty, \quad \int_{V_0} |\mathbf{H}|^2 \, \mathrm{d}V < \infty, \tag{2.59}$$

[1] Physically this is equivalent to the requirement that the induced volume current density is finite everywhere and cannot have a delta function singularity on the interface. In other words, this means that there is no surface current on the interface.

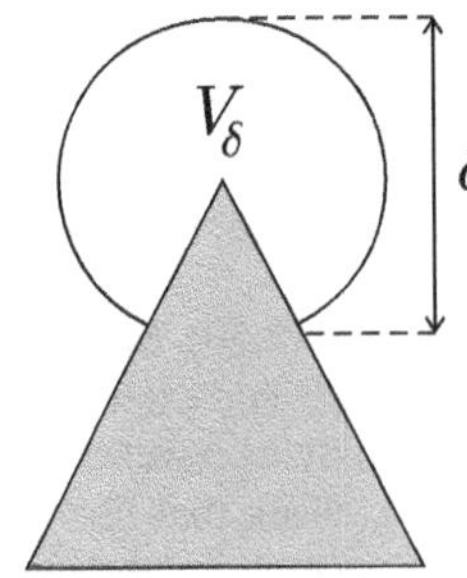

Figure 2.3 Integration volume enclosing a singularity.

where the amplitudes of complex-valued vectors are defined according to (A.3) as $|\mathbf{E}|^2 = |E_x|^2 + |E_y|^2 + |E_z|^2$ and $|\mathbf{H}|^2 = |H_x|^2 + |H_y|^2 + |H_z|^2$. From the convergence of the integrals it follows that if V_δ is a small volume enclosing a singularity (the interior of a spherical or cylindrical segment for point and edge singularities, respectively, Figure 2.3), they have to vanish as the diameter δ of the integration region goes to zero, that is,

$$\lim_{\delta \to 0} \int_{V_\delta} |\mathbf{E}|^2 \, \mathrm{d}V \to 0, \quad \lim_{\delta \to 0} \int_{V_\delta} |\mathbf{H}|^2 \, \mathrm{d}V \to 0. \tag{2.60}$$

The field of a point charge or any collection of (moving) point charges violates this condition. Thus, (2.60) assumes that at any singular point of a material boundary (like an edge) charges never form point dipoles or higher multipoles or infinitely thin lines: charges are always distributed over surfaces or volumes. However, field sources are often modeled as point dipoles, although conditions (2.60) are not satisfied near such singularities.

When applied to edge singularities, condition (2.59) is called the *Meixner condition*. It restricts the field behavior near edges. In general, the fields are allowed to be infinite at edges but in such a way that the integrals (2.59) converge. The same applies to the tip singularities. It is easily seen that the singularities in the field components cannot be stronger than $\rho^{\tau-1}$ with $\tau > 0$ for edges and $\tau > -1/2$ for tips, where ρ is the distance to the edge or to the tip, respectively. The exact value of τ depends on the shape and material properties of the scatterer's surface near the singularity. For example, near an edge of an ideally conducting 'surface' (infinitely thin sheet of an ideally conducting metal) the fields behave as $|\mathbf{E}|, |\mathbf{H}| \sim 1/\sqrt{\rho}$. The behavior of electromagnetic fields at edges is studied in section 5.6 in more detail.

Conditions at Infinity

The scattering theory is concerned with exterior problems for Maxwell's equations, that is, the fields are sought in an infinite region exterior to the scattering body. It is therefore necessary to be able to select solutions with correct behavior at infinity. This is achieved by means of conditions at infinity. Supplementing Maxwell's equations with the conditions at infinity is necessary to ensure a unique solution of a scattering problem.

There are various forms of the conditions, but all of them result from the simple and physically apparent requirement that the field scattered by an arbitrary body of a finite size at very large distances from the body must have the form of a spherical wave[2] traveling *from* the scatterer to infinity

$$E_{\mathrm{sc}} \sim \frac{\mathrm{e}^{-jkr}}{r}, \tag{2.61}$$

[2] Properties of spherical, cylindrical, and plane electromagnetic waves are described in section 2.5.1.

where r is the distance from a fixed origin. Here we assume that the scatterer is located in a medium where $\mathrm{Re}\, k > 0$ (in applications it is usually free space). This means that the solution of the form

$$E_{sc} \sim \frac{e^{+jkr}}{r} \tag{2.62}$$

is unphysical, although it may satisfy the field equations and the boundary conditions.

Note that the wave should propagate from the scattering body to infinity in the sense of the energy flow direction. The phase velocity may have an arbitrary direction. The above formulation assumes that in the medium where the body is located the energy flow direction is the same as the phase velocity (which is determined by the sign of the real part of k). This is of course true for vacuum or air, but there exist materials where the phase velocity and the energy flow directions are opposite (they are called *backward-wave media* or *double-negative materials*), for example materials with negative permittivity and permeability. In these situations the correct radiation condition reads (2.62).

A condition that ensures uniqueness is called the *radiation condition* or the *Sommerfeld condition* (e.g., Jones (1964), Colton and Kress (1983), Müller (1957), Wilcox (1959)). Demanding that the far fields behave as in (2.61), the radiation condition can be written in the form

$$\lim_{r\to\infty} r\left(\frac{\partial E^{sc}_{\theta,\phi}}{\partial r} + jkE^{sc}_{\theta,\phi}\right) = 0, \tag{2.63}$$

where E^{sc}_{θ} and E^{sc}_{ϕ} are the components of the scattered electric field in spherical coordinates (r, θ, ϕ). Sometimes, a coordinate-independent form of the radiation condition is more convenient,

$$\lim_{r\to\infty} r[\hat{r} \times (\nabla \times \mathbf{E}_{sc}) - jk\mathbf{E}_{sc}] = 0, \tag{2.64}$$

where $\hat{r} = \mathbf{r}/r$ is the unit vector pointing from the origin to the observer. The scattered magnetic field is subject to the same condition as the electric field,

$$\lim_{r\to\infty} r[\hat{r} \times (\nabla \times \mathbf{H}_{sc}) - jk\mathbf{H}_{sc}] = 0. \tag{2.65}$$

Conditions (2.64) and (2.65) are called the Silver–Müller conditions. The radiation conditions have to be satisfied uniformly in $\hat{r}$, that is, for all values of the angles θ and ϕ.

Yet another form of radiation conditions is obtained by using Maxwell's equations (2.41) and (2.42) in (2.64) and (2.65) to exclude the curls, which gives

$$\lim_{r\to\infty} r(Y\mathbf{E}_{sc} + \hat{r} \times \mathbf{H}_{sc}) = 0 \tag{2.66}$$

$$\lim_{r\to\infty} r(Z\mathbf{H}_{sc} - \hat{r} \times \mathbf{E}_{sc}) = 0. \tag{2.67}$$

One recognizes in (2.66) and (2.67) the relations between the electric and magnetic fields in a spherical outgoing wave (see section 2.5.1).

An alternative form of the conditions at infinity, which is valid independently of whether the phase velocity and the energy flow direction are co- or reversely directed, is possible if the medium is lossy ($\mathrm{Im}\, k < 0$). In this case it is sufficient to require that all components of the scattered field vanish at infinity, that is,

$$\lim_{r\to\infty} |\mathbf{E}_{sc}| = 0, \quad \lim_{r\to\infty} |\mathbf{H}_{sc}| = 0. \tag{2.68}$$

This approach to the formulation of the conditions at infinity can be extended to lossless media through a temporary assumption of small losses which, upon construction of the solution, are set to zero. This latter method is referred to as the principle of limiting absorption.

In many instances we will be concerned with scattering of a plane wave, the source of which is located at infinity. For plane-wave incidence it is necessary to separate the incident and scattered field, and only the scattered fields are required to satisfy the conditions at infinity.

Care should be taken if the scatterer is of infinite extent, like an infinite cylinder. The infinite size of the scatterer changes the behavior of outgoing waves at infinity, which become cylindrical waves. The corresponding conditions are presented in (2.322) and (2.323). Further complications arise when the cylinder has an infinite geometric cross section, like a wedge built from two intersecting half-planes. In this case, in addition to the incident plane wave, there can be more plane waves resulting from the reflection of the incident wave at the boundary of the scatterer. The scattered field, which satisfies the conditions (2.64) or (2.68), is then obtained by subtraction of all plane-wave fields from the total field.

2.3 Approximate Boundary Conditions

Solutions of many important scattering problems can be dramatically simplified by the use of approximate boundary conditions on the surface of the scattering body. When applicable, this technique eliminates the necessity to determine fields in the interior of the scatterer. This section describes several types of such conditions, starting with the impedance boundary conditions (Leontovich's, standard) which model interfaces of media with strong losses (e.g., metals at microwave frequencies) but can also account for the presence of thin layers of dielectric, magnetic, or absorbing materials on a conducting substrate (section 2.3.1). The accuracy of the impedance boundary conditions can be improved through the inclusion of tangential field derivatives in the boundary conditions, which leads to the higher-order or generalized impedance boundary conditions (section 2.3.2). An adequate simulation of a thin layer immersed in a homogeneous medium is achieved with the sheet transition conditions, which represent the layer as an infinitely thin sheet of equivalent surface currents (section 2.3.3).

2.3.1 Impedance Boundary Conditions

Impedance boundary conditions represent a material interface as a surface with an equivalent surface impedance. The true surface impedance of every material interface depends on the type of the incident wave and on the material composition of the body, but under specific circumstances this dependence may become weak and the approximation through an equivalent impedance remarkably accurate.

The most common situation when the approach is applicable is when the problem involves highly conducting materials or materials with very large values of the refractive index. So, we start with the simplest model for the interface between two media with extremely high contrast of refraction indices, which leads us to the condition traditionally referred to as the Leontovich boundary condition. Then we note that the condition has a broader field of applications (e.g., thin coatings) and introduce the standard impedance boundary conditions (SIBCs).

Surface Impedance

A general definition of the surface impedance reads

$$Z_s = \frac{E_t}{H_t}, \tag{2.69}$$

where E_t and H_t are components of the electric and magnetic fields tangential to a surface. The surface can be differently chosen and be, for example, a phase front of a wave,[3] a material boundary or any other mathematical or physical surface.

[3] In this case, Z is sometimes called characteristic impedance. The impedance can be defined, for example for plane, cylindrical, or spherical waves (section 2.5.1). In the two latter cases, the impedance is a function of the position of the observation point and approaches the characteristic impedance of the plane wave as the observation point goes away from the phase centers of the waves.

Thus, many various definitions are possible, and that is why one can speak of "the impedance offered by a given medium to a wave of given type" (Stratton 1941, p. 282).

Assuming that E_t and H_t are the components of the total fields tangential to a material interface between two different media, (2.69) defines the surface impedance of the interface. An important feature of the surface impedance is its continuity at material interfaces as the tangential components of fields are continuous there.

The use of total fields in the definition of the surface impedance makes the surface impedance of an interface dependent on the material constitution of the structure behind the interface (section 4.3). Waves transmitted through the interface into the adjacent medium can be reflected or scattered back from the scatterers or material boundaries there and, in the case of low absorption, arrive at the interface only slightly attenuated, thus influencing the value of the surface impedance. Furthermore, the surface impedance depends on such parameters as polarization and direction of incidence of the incoming wave.

In some cases, however, it is possible to assign a constant value to the impedance, without making an unacceptable error in the modeling of the interface. This value is denoted by $\tilde{Z}_s$ and called the *equivalent surface impedance*. Inserting the equivalent impedance in (2.69) leads to an approximate boundary condition, which is referred to as the impedance boundary condition.

Leontovich Boundary Condition

Consider a boundary between free space (vacuum or air) and a lossy or high refractive index medium (e.g., metal at radio or microwave frequency) with the parameters ϵ and μ. In metals at radio and microwave frequencies, $\epsilon = \epsilon' - j\sigma/\omega$ (see (2.17)) and $\sigma/\omega \gg \epsilon'$ because the conductivity σ is very large and the frequency is moderate (Table 2.2). In non-ferromagnetic metals $\mu \approx \mu_0$. In that case the propagation constant is given by

$$k = \omega\sqrt{-j\frac{\sigma}{\omega}\mu} = \frac{1-j}{\sqrt{2}}\sqrt{\omega\sigma\mu} \tag{2.70}$$

and has an imaginary part which implies exponentially decaying waves (see (2.51)). The waves get attenuated by the factor e upon traveling the distance

$$\delta = \frac{1}{|\mathrm{Im}\,k|} = \sqrt{\frac{2}{\omega\sigma\mu}}. \tag{2.71}$$

The penetration depth δ is called the skin depth. A slab of absorbing material thicker than a couple of skin depths reflects almost as a semi-infinite slab of the same material since the waves reflected back from the rear interface of the slab arrive at the front face strongly attenuated. In metals at microwave frequencies, the distance δ is about 1 μm $= 10^{-6}$ m and, therefore, much smaller than the wavelength λ_0 in the free space. For example, at 10 GHz, $\lambda_0 = 3 \times 10^{-2}$ m and $\delta = 0.63$ μm in silver and 4.3 μm in stainless steel. In such cases, that is, when δ is much smaller than λ_0 and the thickness of the body, we speak about *strong skin effect*. It should be noted, however, that at lower frequencies and when thin strips of metals are involved, for example in printed circuit boards with copper sheets laminated onto a non-conductive substrate, the skin depth can be comparable with the thickness of the metal (see Table 2.5), implying partial transparency of the material layer.

Consider now the tangential field components at a material boundary in the case when $|k| \gg k_0$ (the refractive index is very large or the material is very lossy) (Figure 2.4). If the interface is excited by propagating waves in free space, the

Table 2.5 Skin depth in copper at several microwave frequencies (conductivity of copper from Table 2.2).

f (GHz)	0.5	1	2	5	10	20
δ (μm)	2.9	2.1	1.5	0.92	0.65	0.46

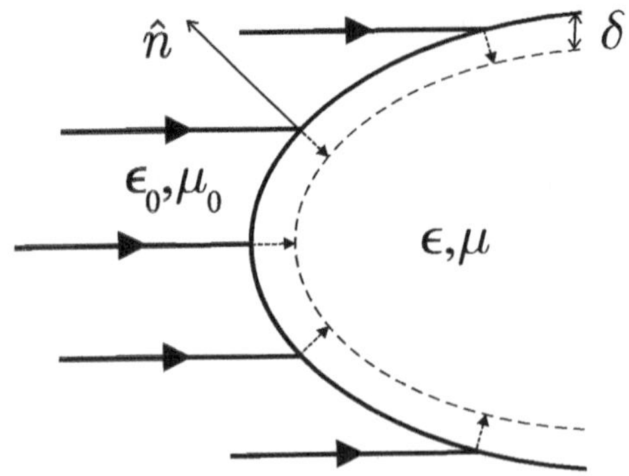

Figure 2.4 Skin effect at an air–metal boundary: waves penetrated into the metal body propagate almost normally to the boundary, being attenuated within a thin layer of the thickness δ.

tangential component of the wave vector is always smaller than the wavenumber in free space. This means that inside the body the normal to the interface component of the wave vector is very large compared to the tangential part, which is continuous across the interface because of Snell's law (Chapter 4). Thus, waves in the body propagate nearly normally to the interface, implying that the electric and magnetic fields are almost tangential to the interface, $\mathbf{E}_t \approx \mathbf{E}$ and $\mathbf{H}_t \approx \mathbf{H}$, and are related as in a plane wave propagating toward the interior of the body (see (2.158)),

$$\mathbf{E}_t = Z_s \hat{n} \times \mathbf{H}_t, \tag{2.72}$$

where $\hat{n}$ is the unit vector normal to the interface and directed into free space. If, in addition, the skin depth is so small that the waves in the body completely decay before reaching any inhomogeneity, then $Z_s = Z$ where Z is the intrinsic impedance of the material (most commonly, metal)

$$Z = \sqrt{\frac{\mu}{\epsilon}} \approx \sqrt{\frac{j\omega\mu}{\sigma}} = \frac{1+j}{\sqrt{2}}\sqrt{\frac{\omega\mu}{\sigma}}. \tag{2.73}$$

Because of the continuity of tangential fields at material interfaces, (2.72) is approximately correct also for the fields on the free-space side of the interface, so this relation can be used as a boundary condition to simulate the presence of the body if the equivalent surface impedance is chosen as $\tilde{Z}_s = Z$. The impedance boundary condition is often written as

$$\hat{n} \times \mathbf{E} = \tilde{Z}_s \hat{n} \times (\hat{n} \times \mathbf{H}), \tag{2.74}$$

where $\mathbf{E}$ and $\mathbf{H}$ are the interface values of the fields from the free space. Relation (2.74) with $\tilde{Z}_s$ given by (2.73) is called the *Leontovich boundary condition*.[4]

According to Levin *et al.* (1983), M.A. Leontovich proposed the impedance boundary conditions in 1938–1939 but published his results 10 years later (Leontovich 1948). The theory of the boundary conditions has been developed by Rytov (1940). The boundary conditions have been mostly applied to problems of propagation of radio waves over Earth's surface (e.g., Feinberg 1944; Grinberg 1942; Schukin 1940). First application of the conditions to electromagnetic scattering as opposed to propagation is attributed to V.A. Fock, who used the conditions to derive an approximate solution to the problem of the scattering of a plane electromagnetic wave by a large conducting convex body (Fock 1965, Chapter 5).

[4] Sometimes we use the plural because in the coordinate form (2.74) is written as a set of equations for the field components.

Standard Impedance Boundary Condition

A boundary condition of the form (2.74) can be used to model other types of material boundaries, most notably PEC-backed thin layers and metamaterial absorbers. We therefore call the boundary condition (2.74) the standard impedance boundary condition (SIBC) and consider the Leontovich boundary conditions as a specialization of the SIBC for the case of a material with significant loss.

The exact surface impedance Z_s (2.69) of a general boundary can be derived from the solution of the reflection and transmission problem for a plane electromagnetic wave at the interface of a locally conformal planar configuration of corresponding material composition. In Chapter 4 there are many examples, including Z_s of a PEC-backed homogeneous layer or of a homogeneous layer immersed in homogeneous medium, equations (4.341) and (4.349). The exact value of Z_s depends not only on the material composition of the interface but also on other factors, for example the propagation direction and polarization of the incident wave.

On the contrary, the SIBC includes a single excitation-independent parameter $\tilde{Z}_s$. A simple but successful choice is to take the value of the surface impedance at the incidence perpendicular to the interface,

$$\tilde{Z}_s = Z_s|_{\theta=0} \tag{2.75}$$

(where θ is the incidence angle), which leads to polarization-independent values of $\tilde{Z}_s$ for isotropic boundaries. Since every field can be expanded in a superposition of plane waves coming from various directions (including complex incidence angles in the most general case), SIBCs with the equivalent surface impedance (2.75) can be expected to work fine if the dependence of the exact surface impedance Z_s of the interface on the incidence angle and polarization of the incident field is weak and can be neglected. The choice of $\tilde{Z}_s$ according to (2.75) is most suitable for problems in which the incidence direction is not confined to a specific direction.[5] For many configurations involving planar material layers and boundaries the value of $Z_s(0)$ can be obtained in a closed form (see Chapter 4). In the case of more complicated structures, when a symbolic solution is not available, $Z_s(0)$ can be determined from direct numerical simulations of reflection or even from measurements of reflection (e.g., Culhaoglu *et al.* 2013).

The practical value of impedance boundary conditions derives from the fact that they exclude the interior of the scatterer from analysis and replace the scatterer with a boundary condition on its surface. This can significantly simplify both numerical simulations (as it eliminates the need to sample within the body, reducing the number of unknowns and the complexity of the problem) and theoretical studies (as it may lead to a simpler boundary value problem with a simpler solution). Furthermore, the impedance boundary conditions can be used for a wide variety of material boundaries, flat or curved, homogeneous or inhomogeneous, and for various types of the incidence field (e.g., plane, cylindrical or spherical waves). Finally, they can be improved and generalized (sections 2.3.2 and 2.3.3). SIBCs with tensorial $\tilde{Z}_s$ can successfully model anisotropic surfaces, for example with electrically small, periodic, or stochastic corrugations (see Senior and Volakis 1995, section 2.6).

Conditions of Applicability

The assumptions made in the process of formulation of SIBCs (2.74) imply several limitations on the applicability of the boundary conditions. First, the condition of strong refraction

$$|k| \gg k_0 = \omega\sqrt{\epsilon_0\mu_0} \tag{2.76}$$

(equivalent to $|N| \gg 1$, where N is the refractive index) must be satisfied to ensure that the waves transmitted through the boundary propagate almost at the right angle to the boundary. For conductors, condition (2.76) can be rewritten in terms

[5] For example, in problems of radio wave propagation over the Earth surface the fields propagate mostly along the material interface (Earth's surface). Formulating the impedance boundary conditions in the coordinate form (e.g. for the components normal to the interface), one can define $\tilde{Z}_s$ as Z_s at the *grazing* incidence and use polarization-dependent values of $\tilde{Z}_s$ for the alternative polarizations since polarizations are well separated in this class of problems.

of the skin depth δ. From (2.71) we have

$$\delta \ll \frac{\lambda_0}{2\pi}, \tag{2.77}$$

where $\lambda_0 = 2\pi/k_0$ is the free-space wavelength.

In case of excitation by propagating waves the error can be estimated by comparing solutions of plane-wave reflection problems for planar interfaces (section 4.6.1). The results look similar: in the impedance approximation we get N in place of $\sqrt{N^2 - \sin^2\theta}$ (θ is the incidence angle). Thus, the error is of the order of $1/|N|^2$. In configurations involving thin material layers, there can be further restrictions (section 4.6.1).

If the interface is excited by evanescent waves (e.g., if there is an electrically small source or a small scattering particle very close to the interface), the applicability condition (2.76) should be replaced by $|k| \gg k_t$, where k_t is the tangential to the interface component of the wavenumber in the incident evanescent waves.

An inherent feature of impedance boundary conditions is their local character, that is, the fact that they connect the electric and magnetic fields at the same point of the boundary. Physically, this is not a point but rather a small region of the size δ. In order for impedance boundary conditions to apply, the size of this region must be smaller than every relevant geometric parameter in the incident field and in the boundary constitution. More specifically this means the following.

If the scatterer is a partially transparent body of the diameter d, then we have the restriction

$$\delta \ll d, \tag{2.78}$$

which ensures that the waves transmitted through the body and those multiply reflected in the interior of the body can be neglected.

If the interface is illuminated by a wave with a curved wavefront, then there is an additional condition

$$\delta \ll R, \tag{2.79}$$

where R is the curvature radius of the wavefront. This ensures that the front of the incident wave is almost flat on the scale of δ and the curvature of the wavefront does not compromise the locality assumption.

If the surface of a lossy body is curved with the curvature radius a, we also demand that

$$\delta \ll a. \tag{2.80}$$

A too small value of a would mean the presence of an edge or a conical point on the boundary, in the vicinity of which the condition (2.78) is violated.

In the case of a geometrically smooth boundary but with varying material parameters the equivalent surface impedance is position-dependent. In order for the previous derivations of the boundary conditions to apply, the variation of the material properties should be sufficiently slow to ensure that on the scale of δ the boundary can be regarded as approximately homogeneous,

$$\delta \ll v, \tag{2.81}$$

where v is the scale of the spacial variation in the material parameters of the boundary.

SIBCs can be quite accurate when materials with high values of refractive index or significant loss are involved. Experience shows that SIBCs may lead to acceptable results even in circumstances where it is difficult to justify their use. In several chapters of the book we study the accuracy of SIBCs by using exact solutions as a reference. In section 4.6.1 the accuracy is estimated by looking at predictions of SIBCs for reflection coefficients of several planar configurations. In chapters on scattering from cylinders (section 6.5.2) and spheres (Chapter 7) we study the influence of the curvature of the scattering surface on the accuracy of the impedance formulation. Further discussions of the accuracy of the impedance boundary conditions can be found in the literature, for example Alexopoulos and Tadler (1975), Lee and Gee (1987), Wang (1987), and Senior and Volakis (1995, section 2.5).

2.3.2 Generalized (Higher-order) Impedance Boundary Conditions

The Leontovich conditions (2.74) can be improved for different particular situations, for example we can use the known exact solution for plane-wave excitation of planar interfaces. The idea is the following: the exact surface impedance in the Fourier space (for plane waves) is expanded in Taylor series with respect to the ratio of the tangential wavenumber to the wavenumber in metal. Usually only the two first terms are kept. Transforming the result back to the physical space, so-called *higher-order* boundary conditions follow. In that case the surface impedance is a differential operator which contains tangential derivatives of the magnetic field and sometimes also of the electric field.

Let us consider a planar interface between two isotropic materials. One of the materials (which we want to model by impedance boundary conditions) is characterized by its complex permittivity ϵ and permeability μ (conductivity is included in the imaginary part of permittivity as in (2.15)). The other material is most often free space, but it can be any other linear material. Assuming propagation (or attenuation) of plane waves in the material, we can consider waves of two orthogonal linear polarizations separately. For our problem of a planar interface it is convenient to define TM and TE polarizations (also called H and E polarizations, respectively, see section 2.5.2) with respect to the normal at the interface, as shown in Figure 2.5. Here, $\hat{n}$ is the unit normal vector and $\mathbf{k}$ is the wave vector in the medium ($k = \omega\sqrt{\epsilon\mu}$).

Simple geometric considerations (Figure 2.5) lead to the following relations between the field components orthogonal to $\hat{n}$:

$$\mathbf{E}_t = Z\frac{\sqrt{k^2 - k_t^2}}{k}\hat{n}\times\mathbf{H}_t, \quad \text{TM modes} \tag{2.82}$$

$$\mathbf{E}_t = Z\frac{k}{\sqrt{k^2 - k_t^2}}\hat{n}\times\mathbf{H}_t, \quad \text{TE modes,} \tag{2.83}$$

where $Z = \sqrt{\mu/\epsilon}$ is the wave impedance in the medium. In dyadic notations, these relations can be written as a single dyadic relation (Lindell 1992; Tretyakov 2003)

$$\mathbf{E}_t = \overline{\overline{Z}}\cdot\hat{n}\times\mathbf{H}_t \quad \text{with} \quad \overline{\overline{Z}} = Z\frac{k\left(\overline{\overline{I}}_t - \frac{\mathbf{k}_t\mathbf{k}_t}{k^2}\right)}{\sqrt{k^2 - k_t^2}}, \tag{2.84}$$

where $\overline{\overline{I}}_t$ is the transverse unit dyadic. Because at the interface between the medium and free space (or another material) the tangential field components are continuous, the same relations hold at the interface surface.

For individual plane-wave components of the fields, the above relations are exact boundary conditions which model electromagnetic properties of the material half-space. But in order to use them as tools to solve general scattering problems for arbitrary fields and not just plane waves, we must transform the relations into the physical domain. This can be done by replacing the tangential wave vector by the transverse nabla operator: $\mathbf{k}_t \to -\nabla_t$, where $\nabla_t = \nabla - \frac{\partial}{\partial z}\hat{n}$ (z is the axis

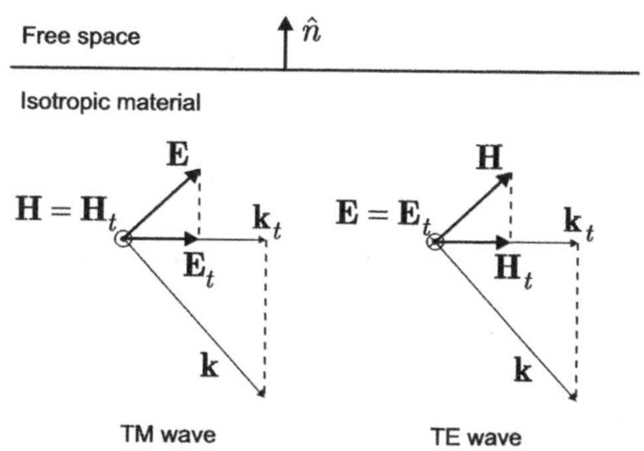

Figure 2.5 Interface between two isotropic media and plane waves in one medium.

along $\hat{n}$). This way we arrive at exact boundary conditions which contain pseudo-differential operators:

$$\mathbf{E}_t = Z\frac{\sqrt{k^2+\nabla_t^2}}{k}\,\hat{n}\times\mathbf{H}_t, \quad \text{TM modes} \tag{2.85}$$

$$\mathbf{E}_t = Z\frac{k}{\sqrt{k^2+\nabla_t^2}}\,\hat{n}\times\mathbf{H}_t, \quad \text{TE modes.} \tag{2.86}$$

To arrive at practically useful tools, we need to approximate these operators by some rational expressions containing only a small number of low-order differential operators.

Usually we are interested in situations when the medium is electrically very dense and the excitation field is a propagating wave. In this case $|k_t| \ll |k|$. Totally neglecting the tangential derivatives, we naturally arrive at the Leontovich boundary conditions (2.74). Using the Taylor expansion of the square root and keeping only the second-order terms, we obtain higher-order boundary conditions in the form

$$\mathbf{E}_t = Z\left(1+\frac{1}{2k^2}\nabla_t^2\right)\hat{n}\times\mathbf{H}_t, \quad \text{TM modes} \tag{2.87}$$

$$\mathbf{E}_t = Z\left(1-\frac{1}{2k^2}\nabla_t^2\right)\hat{n}\times\mathbf{H}_t, \quad \text{TE modes.} \tag{2.88}$$

If the medium is not very dense electrically or a more uniform (with respect to the incidence angle) approximation is desirable, the Padé approximation can be used for the square root

$$\sqrt{1-\frac{k_t^2}{k^2}} \approx \frac{1-\frac{3}{4}\frac{k_t^2}{k^2}}{1-\frac{1}{4}\frac{k_t^2}{k^2}}. \tag{2.89}$$

In this approximation the higher-order boundary conditions contain second-order tangential derivatives of both electric and magnetic fields (Tretyakov 2003):

$$\left(1+\frac{1}{4k^2}\nabla_t^2\right)\mathbf{E}_t = Z\left(1+\frac{3}{4k^2}\nabla_t^2\right)\hat{n}\times\mathbf{H}_t, \quad \text{TM modes} \tag{2.90}$$

$$\left(1+\frac{3}{4k^2}\nabla_t^2\right)\mathbf{E}_t = Z\left(1+\frac{1}{4k^2}\nabla_t^2\right)\hat{n}\times\mathbf{H}_t, \quad \text{TE modes.} \tag{2.91}$$

Higher-order impedance boundary conditions for metal surfaces coated by thin layers can be found, for example, in Tretyakov (2003). A similar approach to the derivation of higher-order impedance boundary conditions is used also in Hoppe and Rahmat-Samii (1995). Alternative derivations and examples of applications of higher-order conditions in solutions of diffraction problems can be found in Senior and Volakis (1995).

2.3.3 Sheet Transition Conditions

The sheet transition conditions model thin material layers immersed in a homogeneous medium. The approach consists of replacing the layer with an equivalent infinitely thin current sheet. Let us derive these conditions.

Consider a layer of thickness d_1 with permittivity ϵ_1 and permeability μ_1 immersed in a homogeneous medium (not necessarily vacuum) with material constants ϵ_0 and μ_0 (Figure 2.6). If the layer is sufficiently thin ($|k_1|d_1 \ll 1$) and electrically dense ($|\epsilon_1\mu_1| \gg |\epsilon_0\mu_0|$), we can attempt to replace it with an infinitely thin sheet of equivalent electric and magnetic currents. Because of the assumption $|\epsilon_1\mu_1| \gg |\epsilon_0\mu_0|$, inside the layer the normal components of the fields and induced polarizations are much weaker than the tangential components, and we can neglect them. Sheets of tangential currents induced in the layer correspond to jumps of the tangential components of the fields

$$\hat{n}\times\mathbf{H}_+ - \hat{n}\times\mathbf{H}_- = \mathbf{j}_\mathrm{e} \tag{2.92}$$

$$\hat{n}\times\mathbf{E}_+ - \hat{n}\times\mathbf{E}_- = -\mathbf{j}_\mathrm{m}, \tag{2.93}$$

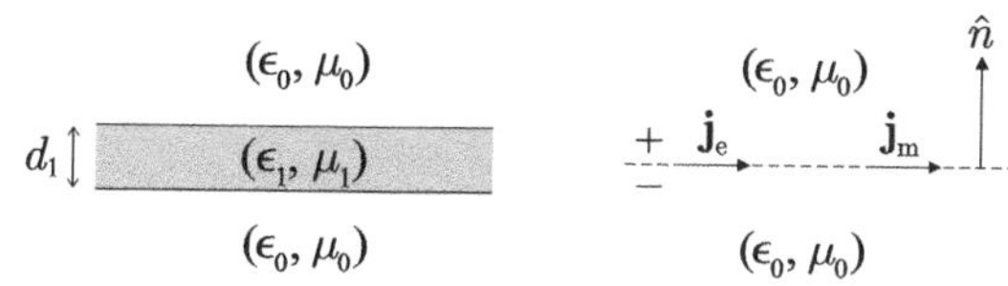

Figure 2.6 Thin layer (to the left) modeled as a current sheet (to the right). The unit vector $\hat{n}$ is perpendicular to the sheet and points to the medium on the + side of the sheet.

where $\mathbf{j}_e$ and $\mathbf{j}_m$ are the surface densities of the sheet currents and the $\pm$ signs denote the values of the fields at the upper and lower sides of the sheet, respectively. The sheet currents are obtained by integration of the volume densities of the induced polarization or conduction currents $\mathbf{J}_e$ and $\mathbf{J}_m$ over the thickness of the layer, and since the layer is thin,

$$\mathbf{j}_e \approx d_1 \mathbf{J}_e, \quad \mathbf{j}_m \approx d_1 \mathbf{J}_m. \tag{2.94}$$

The volume densities can be deduced from Maxwell's equations in the slab by rewriting them as

$$\nabla \times \mathbf{H}' = j\omega\epsilon_1 \mathbf{E}' = j\omega\epsilon_0 \mathbf{E}' + \mathbf{J}_e \tag{2.95}$$
$$\nabla \times \mathbf{E}' = -j\omega\mu_1 \mathbf{H}' = -j\omega\mu_0 \mathbf{H}' - \mathbf{J}_m \tag{2.96}$$

with $\mathbf{E}'$ and $\mathbf{H}'$ being the fields in the layer, and it follows from (2.95) and (2.96) that

$$\mathbf{J}_e = j\omega(\epsilon_1 - \epsilon_0)\mathbf{E}' \tag{2.97}$$
$$\mathbf{J}_m = j\omega(\mu_1 - \mu_0)\mathbf{H}'. \tag{2.98}$$

Since the layer is thin and the tangential fields are continuous on both interfaces, the averaged tangential fields in the layer can be approximated through the boundary values on the sides of the layer as

$$\mathbf{E}'_t \approx \frac{1}{2}(\mathbf{E}_{t+} + \mathbf{E}_{t-}) \tag{2.99}$$
$$\mathbf{H}'_t \approx \frac{1}{2}(\mathbf{H}_{t+} + \mathbf{H}_{t-}), \tag{2.100}$$

where t denotes the tangential components. This leads to the approximations

$$\mathbf{J}_e \approx \frac{j\omega}{2}(\epsilon_1 - \epsilon_0)(\mathbf{E}_{t+} + \mathbf{E}_{t-}) \tag{2.101}$$
$$\mathbf{J}_m \approx \frac{j\omega}{2}(\mu_1 - \mu_0)(\mathbf{H}_{t+} + \mathbf{H}_{t-}). \tag{2.102}$$

The jump conditions (2.92) and (2.93) can therefore be represented as

$$\hat{n} \times \mathbf{H}_+ - \hat{n} \times \mathbf{H}_- = \frac{j}{2}\omega d_1(\epsilon_1 - \epsilon_0)(\mathbf{E}_{t+} + \mathbf{E}_{t-}) \tag{2.103}$$
$$\hat{n} \times \mathbf{E}_+ - \hat{n} \times \mathbf{E}_- = -\frac{j}{2}\omega d_1(\mu_1 - \mu_0)(\mathbf{H}_{t+} + \mathbf{H}_{t-}), \tag{2.104}$$

and building a cross product of (2.103) and (2.104) with the vector $\hat{n}$ to select the tangential component of the equations, we arrive at the relations:

$$2R_e\, \hat{n} \times \hat{n} \times (\mathbf{H}_+ - \mathbf{H}_-) = \hat{n} \times (\mathbf{E}_+ + \mathbf{E}_-) \tag{2.105}$$
$$2R_m\, \hat{n} \times \hat{n} \times (\mathbf{E}_+ - \mathbf{E}_-) = -\hat{n} \times (\mathbf{H}_+ + \mathbf{H}_-), \tag{2.106}$$

where

$$R_e = \frac{-j}{\omega d_1(\epsilon_1 - \epsilon_0)} \tag{2.107}$$

$$R_m = \frac{-j}{\omega d_1(\mu_1 - \mu_0)}. \tag{2.108}$$

Conditions (2.105) and (2.106) relate the tangential components of the fields on the opposite sides of the thin layer and can serve as approximate boundary conditions, called the *sheet transition conditions*. Physically, they describe thin partially transparent material layers.

Parameters R_e and R_m are referred to as the *sheet resistivity* and the *sheet magnetic conductivity*, respectively, since they relate the sheet currents to the fields according to

$$R_e \mathbf{j}_e = \mathbf{E}', \quad R_m \mathbf{j}_m = \mathbf{H}'. \tag{2.109}$$

Their units are ohms and siemens, respectively. In a strongly conducting medium with conductivity σ_1, $\epsilon_1 = \epsilon_1' - j\sigma_1/\omega \approx -j\sigma_1/\omega$ and

$$R_e \approx \frac{1}{\sigma_1 d_1}. \tag{2.110}$$

The sheet resistivity R_e is a directly measurable quantity and sometimes its values are given in ohms per square (ohm/square). This refers to the measurement procedure, in the framework of which one measures the electrical resistance between the opposite edges of a square sample of the thin material. Indeed, the electrical resistance of a block of material with conductivity σ_1, thickness d_1, length l, and width w is equal to $l/(\sigma_1 d_1 w)$ and numerically coincides with (2.110) when $l = w$, that is, for any square sample.

If the material in the layer is nonmagnetic ($\mu_1 = \mu_0$), then $R_m = \infty$ and (2.105) and (2.106) reduce to the *resistive sheet transition conditions*:

$$R_e \hat{n} \times \hat{n} \times (\mathbf{H}_+ - \mathbf{H}_-) = \hat{n} \times \mathbf{E}_+ \tag{2.111}$$

$$\hat{n} \times (\mathbf{E}_+ - \mathbf{E}_-) = 0, \tag{2.112}$$

which describe a sheet supporting only an electric current. With $R_e = 0$, the resistive sheet becomes perfectly conducting and (2.111) and (2.112) reduce to the usual PEC boundary conditions $\hat{n} \times \mathbf{E}_+ = \hat{n} \times \mathbf{E}_- = 0$ (see (2.56)). Reflection and transmission coefficients of the resistive sheet are studied in section 4.5.3 (see also problem 31 from section 4.7).

The electromagnetically dual counterpart to the resistive sheet is the so-called *conductive sheet*, which is a thin layer of material with $\epsilon_1 = \epsilon_0$ and $\mu_1 \neq \mu_0$. The corresponding transition conditions result from (2.105) and (2.106) with $R_e = \infty$,

$$\hat{n} \times (\mathbf{H}_+ - \mathbf{H}_-) = 0 \tag{2.113}$$

$$R_m \hat{n} \times \hat{n} \times (\mathbf{E}_+ - \mathbf{E}_-) = -\hat{n} \times \mathbf{H}_+. \tag{2.114}$$

Finally, in the limit of a perfectly transparent thin layer ($\epsilon_1 = \epsilon_0$, $\mu_1 = \mu_0$), $R_e, R_m = \infty$, and the sheet transition conditions reduce to the conditions of the continuity of the tangential components of the fields: $\hat{n} \times \mathbf{E}_+ = \hat{n} \times \mathbf{E}_-$ and $\hat{n} \times \mathbf{H}_+ = \hat{n} \times \mathbf{H}_-$.

If the layer is electrically thin but the material contrast is not high (the condition $|\epsilon_1\mu_1| \gg |\epsilon_0\mu_0|$ is not satisfied), then the sheet conditions (2.103) and (2.104) can be improved by adding terms with second-order derivatives along the tangential directions. These conditions can be found in Tretyakov (2003, Eqs. (2.52), (2.53)). On the other hand, when the contrast is high but the layer is not electrically thin (condition $|k_1|d_1 \ll 1$ is not satisfied), the accuracy can be improved by using in the above derivation the dynamic formulas for the averaged tangential fields in the layer (Tretyakov 2003, Eq. (2.104)),

$$\mathbf{E}_t' \approx \frac{\tan(k_1 d_1/2)}{k_1 d_1}(\mathbf{E}_{t+} + \mathbf{E}_{t-}) \tag{2.115}$$

$$\mathbf{H}_t' \approx \frac{\tan(k_1 d_1/2)}{k_1 d_1}(\mathbf{H}_{t+} + \mathbf{H}_{t-}), \tag{2.116}$$

instead of the quasi-static relations (2.99) and (2.100). More about generalized sheet conditions can be found in Tretyakov (2003) and Senior and Volakis (1995). Similarly to SIBCs, the sheet transition conditions can be generalized to anisotropic materials by introducing the sheet resistivity and conductivity tensors.

2.4 Fundamental Properties of Time-harmonic Electromagnetic Fields

Time-harmonic electromagnetic fields are solutions of time-harmonic Maxwell's equations subject to a set of boundary conditions. These solutions possess a number of fundamental properties which can be used in the formulation of scattering problems and in the derivation and verification of the obtained solutions. This section is concerned with the properties of energy conservation and the uniqueness of the solutions (subsection 2.4.1) and reciprocity (subsection 2.4.2). The property of the equivalence of sources is addressed in section 2.6, where it is used in formulation of the electromagnetic version of the Huygens principle.

All these properties are derived from Maxwell's equations by applying the divergence theorem (A.28) and the vector identity

$$\nabla \cdot (\mathbf{A} \times \mathbf{B}) = \mathbf{B} \cdot (\nabla \times \mathbf{A}) - \mathbf{A} \cdot (\nabla \times \mathbf{B}) \tag{2.117}$$

with suitably chosen vectors **A** and **B**.

2.4.1 Energy Conservation and Uniqueness

Determination of the field scattered by an obstacle requires solving Maxwell's equations supplemented with a set of suitable boundary conditions, and it is therefore important to ensure that the boundary value problem has a unique solution.[6] In physical terms, uniqueness means that a problem without external excitation has only a trivial solution, that is, $\mathbf{E} = 0$ and $\mathbf{H} = 0$. It is therefore logical to expect that a proof of uniqueness should be somehow related to the property of energy conservation, which we consider next.

Energy Conservation

To formulate the energy conservation law we start with defining the corresponding physical quantities. Let

$$\mathbf{S} = \frac{1}{2}\mathbf{E} \times \mathbf{H}^*, \tag{2.118}$$

where the asterisk denotes the complex conjugate, be the complex Poynting vector. Then in a time-harmonic electromagnetic wave the mean (or time-averaged) intensity **w** of the energy flow at a point is given by the real part of this vector (Stratton 1941, sections 2.19, 2.20)

$$\mathbf{w} = \operatorname{Re} \mathbf{S}. \tag{2.119}$$

[6] The question of existence of a solution will not concern us in this book. The problems that are studied in this book always allow explicit solutions, either exact or approximate, so that we will always have one. The interested reader is referred to Colton and Kress (1983).

The time-average power flow across a surface Ω is given by the surface integral of $\mathbf{w}$ over that surface,

$$P = \int_{\Omega} \mathbf{w} \cdot \hat{n}\, \mathrm{d}\Omega, \tag{2.120}$$

where $\hat{n}$ is the unit vector normal to Ω. For closed surfaces we will always define $\hat{n}$ as the vector pointing to the exterior of the enclosed volume, which implies that for a closed surface Ω, P is positive if on the average (per period of oscillations and over the whole surface) more energy leaves the volume enclosed by Ω than enters the volume. Another feature of the integral (2.120) is that the value of the integral is completely determined by the tangential components of the fields on the integration surface, which follows from the property of the cyclic invariance of the scalar triple product (A.11),

$$(\mathbf{E} \times \mathbf{H}^*) \cdot \hat{n} = \mathbf{E} \cdot (\mathbf{H}^* \times \hat{n}) = \mathbf{H}^* \cdot (\hat{n} \times \mathbf{E}). \tag{2.121}$$

Now let V be the volume enclosed by Ω, free of sources and filled with a homogeneous medium. Integrating $\mathbf{w}$ over Ω leads to a result which can be interpreted as the law of energy conservation. We start with the vector identity (2.117) with $\mathbf{A} = \mathbf{E}$ and $\mathbf{B} = \mathbf{H}^*$,

$$\nabla \cdot (\mathbf{E} \times \mathbf{H}^*) = \mathbf{H}^* \cdot (\nabla \times \mathbf{E}) - \mathbf{E} \cdot (\nabla \times \mathbf{H}^*), \tag{2.122}$$

and rewrite the right-hand side as

$$-j\omega\mu|\mathbf{H}|^2 + j\omega\epsilon^*|\mathbf{E}|^2 \tag{2.123}$$

by using Maxwell's equation to replace the curls of $\mathbf{E}$ and $\mathbf{H}^*$. Integrating the identity over V and applying the divergence theorem (A.28) to the integral in the left-hand side of (2.122) yields

$$\int_{\Omega} (\mathbf{E} \times \mathbf{H}^*) \cdot \hat{n}\, \mathrm{d}\Omega = j\omega \int_{V} (\epsilon^*|\mathbf{E}|^2 - \mu|\mathbf{H}|^2)\, \mathrm{d}V, \tag{2.124}$$

which can be rewritten as

$$\int_{\Omega} \mathbf{S} \cdot \hat{n}\, \mathrm{d}\Omega = -\frac{\omega}{2}\left(\epsilon'' \int_{V} |\mathbf{E}|^2\, \mathrm{d}V + \mu'' \int_{V} |\mathbf{H}|^2\, \mathrm{d}V\right) + j\frac{\omega}{2}\left(\epsilon' \int_{V} |\mathbf{E}|^2\, \mathrm{d}V - \mu' \int_{V} |\mathbf{H}|^2\, \mathrm{d}V\right), \tag{2.125}$$

where $\epsilon = \epsilon' - j\epsilon''$ and $\mu = \mu' - j\mu''$. The real part of (2.125) describes the power flow across Ω as

$$P = -\frac{\omega}{2}\left(\epsilon'' \int_{V} |\mathbf{E}|^2\, \mathrm{d}V + \mu'' \int_{V} |\mathbf{H}|^2\, \mathrm{d}V\right). \tag{2.126}$$

If the material parameters do not depend on the frequency, then the imaginary part of the right-hand side of (2.125) is proportional to the difference of the time-average values of electric and magnetic energies stored in the volume V,

$$W_{\mathrm{e}} = \frac{1}{4}\mathrm{Re} \int_{V} \mathbf{D} \cdot \mathbf{E}^*\, \mathrm{d}V, \quad W_{\mathrm{m}} = \frac{1}{4}\mathrm{Re} \int_{V} \mathbf{B} \cdot \mathbf{H}^*\, \mathrm{d}V \tag{2.127}$$

(e.g., see Stratton 1941, sections 2.19, 2.20). Thus, equation (2.126) relates the power flow across the boundary of a volume to the energy stored and dissipated in the volume and represents, therefore, the energy balance in the volume, in other words, the energy conservation law.

A source-free volume filled with a passive medium can only absorb energy, which means that the power flow P cannot be positive. It follows therefore from (2.126) that ϵ and μ of passive media have negative imaginary parts, that is, $\epsilon'' \geq 0$ and $\mu'' \geq 0$. In real materials there are always at least small losses, implying more restrictive inequalities $\epsilon'' > 0$ and $\mu'' > 0$.

The property of conservation of electromagnetic energy in a volume, derived from Maxwell's equations and formulated in terms of energy flows through the surface of the volume, is due to Poynting (1884).

Uniqueness

Now we are ready to apply the energy conservation law to proving uniqueness of solutions of scattering problems for electromagnetic fields. Consider two fields $(\mathbf{E}_1, \mathbf{H}_1)$ and $(\mathbf{E}_2, \mathbf{H}_2)$ in a volume V which is free of sources and filled with a homogeneous medium. Since Maxwell's equations are linear, the difference field $(\mathbf{E}_1 - \mathbf{E}_2, \mathbf{H}_1 - \mathbf{H}_2)$ is also a solution of Maxwell's equations, which is subject to the energy conservation

$$\mathrm{Re} \int_{\Omega} (\mathbf{E}_1 - \mathbf{E}_2) \times (\mathbf{H}_1^* - \mathbf{H}_2^*) \cdot \hat{n} \, \mathrm{d}\Omega = -\omega \left(\epsilon'' \int_V |\mathbf{E}_1 - \mathbf{E}_2|^2 \, \mathrm{d}V + \mu'' \int_V |\mathbf{H}_1 - \mathbf{H}_2|^2 \, \mathrm{d}V \right). \tag{2.128}$$

Because of (2.121) the integral in the left-hand side of (2.128) depends entirely on the components of the field $\mathbf{E}_1 - \mathbf{E}_2$ and $\mathbf{H}_1 - \mathbf{H}_2$ that are tangential to the integration surface Ω. Thus, if the tangential components of the electric fields $\mathbf{E}_1$ and $\mathbf{E}_2$, and/or of the magnetic fields $\mathbf{H}_1$ and $\mathbf{H}_2$ are equal to each other on Ω, then the integral in the left-hand side of (2.128) vanishes, implying that the right-hand side of (2.128) must vanish as well, which (provided that $\epsilon'' > 0$, $\mu'' > 0$ and $\omega \neq 0$) is only possible when $\mathbf{E}_1 = \mathbf{E}_2$ and $\mathbf{H}_1 = \mathbf{H}_2$ throughout the whole volume V. Thus, specification of either the tangential components of $\mathbf{E}$ or the tangential components of $\mathbf{H}$ on a closed surface uniquely specifies the electromagnetic field in the interior of the surface.

The proof can be extended to the case when either $\epsilon'' = 0$ or $\mu'' = 0$ using Maxwell's equations. If two sets of, say, electric fields are the same at every point, then their derivatives are also equal, which ensures that the corresponding magnetic fields are also the same. When both $\epsilon'' = 0$ and $\mu'' = 0$ or $\omega = 0$ the proof does not apply and different approaches are necessary. The last case is an idealization as all media exhibit at least small absorption and thus the difficulty can be circumvented by simply assuming infinitesimal losses in the medium. The case of $\omega = 0$ relates to the static limit of Maxwell's equations and is therefore irrelevant to studies of scattering phenomena.

Scattering problems are typically formulated in an infinite region. The corresponding equation of the energy balance can be obtained by a slight modification of the approach employed in the derivation of (2.126). Consider a volume V, free of sources and filled with a homogeneous medium, which is bounded by a closed surface Ω and a spherical surface Ω_∞ with a large but finite radius R_∞. Then the divergence theorem is directly applicable to such a volume, and integrating (2.122) over V gives the power balance equation

$$P_\Omega + P_\infty + \frac{\omega}{2} \left(\epsilon'' \int_V |\mathbf{E}|^2 \, \mathrm{d}V + \mu'' \int_V |\mathbf{H}|^2 \, \mathrm{d}V \right) = 0, \tag{2.129}$$

where P_Ω and P_∞ are the time-average power flows across Ω and Ω_∞, respectively. In the absence of sources also in the interior of Ω and in the exterior of Ω_∞ (i.e., if there are no sources at all), no energy is supplied to the volume V and thus $P_\Omega + P_\infty = 0$. The integrals over V in (2.129) are strictly positive for any field that is not identically zero everywhere in V, implying that (2.129) can only be satisfied by the trivial field, $\mathbf{E} = 0$ and $\mathbf{H} = 0$, in which case all the terms in the left-hand side of (2.129) vanish. For fields compliant with the conditions at infinity, the surface Ω_∞ can be expanded to infinity, and we arrive at the conclusion that the source-free problem has only the trivial solution in the exterior of Ω. Non-trivial solutions in the absence of sources (eigenmodes) can exist only in lossless systems: in that case $\epsilon'' = 0$ and $\mu'' = 0$, and (2.129) can hold even if fields are not identically zero.

Boundary Conditions and Energy Balance

The integration surfaces Ω and Ω_∞ in (2.129) can be freely chosen as long as the volume between them remains free of sources and filled with a homogeneous medium. In principle the surface Ω can be contracted to coincide with the surface of a scattering body and the surface Ω_∞ expanded up to infinity. So the boundary conditions that are imposed on solutions of Maxwell's equations must ensure the correct power balance on the material interfaces and at infinity. Let us show that the conditions described in sections 2.2.3 and 2.3 do comply with the energy conservation.

Let us start with the conditions at infinity. Choosing the radius R_∞ of Ω_∞ to be sufficiently large such that the relation (2.67) applies, and since $\hat{n} = \hat{r}$, we can write for the fields at Ω_∞

$$(\mathbf{E} \times \mathbf{H}^*) \cdot \hat{n} = \mathbf{E} \cdot (\mathbf{H}^* \times \hat{r}) = Z|\mathbf{H}|^2 + o\left(\frac{1}{R_\infty^2}\right) \tag{2.130}$$

and thus

$$P_\infty = \frac{1}{2}\mathrm{Re}\int_{\Omega_\infty} (\mathbf{E} \times \mathbf{H}^*) \cdot \hat{n}\, \mathrm{d}\Omega = \frac{1}{2}\mathrm{Re}\, ZR_\infty^2 \int_0^\pi \int_0^{2\pi} |\mathbf{H}|^2 \sin\theta\, \mathrm{d}\theta \mathrm{d}\phi. \tag{2.131}$$

As $R_\infty \to \infty$, $|\mathbf{H}|$ decays either as $\exp(R_\infty \mathrm{Im}\, k)$ when $\mathrm{Im}\, k < 0$, or as R_∞^{-1} when $\mathrm{Im}\, k = 0$, and since $\mathrm{Re}\, Z \geq 0$ in any passive medium, equation (2.131) gives $P_\infty \geq 0$. The latter inequality means that the use of conditions at infinity (in any form) excludes energy supply from infinity.

The boundary conditions on material interfaces must ensure that the mean power flow integrated over the surface of the scatterer should be directed into the interior of the scatterer or vanish, provided that there are no sources inside the scatterer. It is easy to see that the exact boundary condition (2.56) on a PEC surface or the approximate impedance boundary conditions (2.72) and (2.74) are compliant with this physical requirement. For example, because of the property (2.121) the normal component of the power flow vanishes at a PEC surface as the tangential components of $\mathbf{E}$ are zero on the surface. If the surface is modeled with the impedance boundary conditions, say in the form (2.74), then accounting for the direction of the normal vector in (2.74),

$$(\mathbf{E} \times \mathbf{H}^*) \cdot \hat{n} = \tilde{Z}_s[|\mathbf{H}|^2 - (\mathbf{H}^* \cdot \hat{n})(\mathbf{H} \cdot \hat{n})] = \tilde{Z}_s|\mathbf{H}_t|^2, \tag{2.132}$$

and the mean power flow across the impedance boundary is always non-negative

$$P = \mathrm{Re}\left(\frac{\tilde{Z}_s}{2}\right) \int_\Omega |\mathbf{H}_t|^2\, \mathrm{d}\Omega \geq 0 \tag{2.133}$$

when

$$\mathrm{Re}\, \tilde{Z}_s \geq 0. \tag{2.134}$$

This latter condition ensures that a surface modeled by impedance boundary conditions is physically passive, that is, can only absorb the incident energy, and this is an important restriction on the values of the equivalent surface impedance $\tilde{Z}_s$ in the impedance boundary conditions.

Estimation (2.133) involves the total energy flow across the whole scattering surface. Boundaries modeled with the PEC or impedance boundary conditions are physically passive also locally since $\hat{n} \cdot \mathbf{w} = 0$ (PEC surface) or $\hat{n} \cdot \mathbf{w} \geq 0$ (impedance surface) at every point of the boundary.

Property (2.121) and the continuity of the tangential components of $\mathbf{E}$ and $\mathbf{H}$ at interfaces of two different media (2.55) mean that the normal component of the power flow vector is continuous across media interfaces. From this it follows that the mean power flow across the surface of an arbitrary lossless body is zero (Jones 1964, section 9.1). Indeed, if the scatterer is lossless, then the continuity of the tangential components enables us to move the surface of integration in (2.120) into the interior of the object and then contract it to a point.

If the scattering surface has edges or tips, then the edges can be surrounded by small cylindrical surfaces and the tips with small spherical segments (let us denote such a surface by Ω_0), and the surface integral over the scattering surface can be reduced to an integral over the surface and the remainder of the surface in the vicinities of the geometric singularities. Compliance with the energy conservation on such surfaces apparently requires that

$$\mathrm{Re} \int_{\Omega_0} (\mathbf{E} \times \mathbf{H}^*) \cdot \hat{n} \, d\Omega \to 0 \tag{2.135}$$

as Ω_0 contracts to the edge or to the tip. Physically, this means that there are no point- or edge-localized energy sinks (non-zero power loss requires non-zero volume). The expression on the left-hand side of (2.135) is proportional to the power absorbed in the volume enclosed by the surface (see (2.126)), and for materials with non-zero losses (when $\epsilon'' > 0$ and $\mu'' > 0$) we recover the conditions at the geometric singularities (see (2.60)).

2.4.2 *Reciprocity*

Another fundamental property of electromagnetic fields is the property of reciprocity. Roughly speaking, reciprocity means that interchanging source and receiver does not change the result of the measurement. The presumably first formulation of the principle for scalar fields was suggested by Helmholtz in his treatment of sound waves in organ pipes (Helmholtz 1860, p. 29). The electromagnetic version is attributed to Lorentz (1895). In this section we derive the property from Maxwell's equation for the general scattering geometry and general external currents and apply the resulting relation to the case of point dipoles.

Consider a volume V which is filled with an isotropic medium and contains all the sources (Figure 2.7). The boundary Ω of the volume consists of a surface Ω_{int}, which encloses all the scatterers and material boundaries, and a surface Ω_{ext}, which can be expanded to infinity without crossing any source or material boundary. Assume that we have two sets of time-harmonic sources in V, each consisting of electric and magnetic currents, $(\mathbf{J}_1^{\mathrm{e}}, \mathbf{J}_1^{\mathrm{m}})$ and $(\mathbf{J}_2^{\mathrm{e}}, \mathbf{J}_2^{\mathrm{m}})$. The currents excite the fields $(\mathbf{E}_1, \mathbf{H}_1)$ and $(\mathbf{E}_2, \mathbf{H}_2)$, respectively, and the reciprocity relation we are looking for is a relation between the fields due to one set of sources at the location of another set of sources.

If ϵ and μ are the electric and magnetic constants of the medium in V, which can depend on the spatial coordinates, then the fields are solutions of the two sets of Maxwell's equations (section 2.2.3):

$$\nabla \times \mathbf{H}_1 = j\omega\epsilon \mathbf{E}_1 + \mathbf{J}_1^{\mathrm{e}} \tag{2.136}$$

$$\nabla \times \mathbf{E}_1 = -j\omega\mu \mathbf{H}_1 - \mathbf{J}_1^{\mathrm{m}} \tag{2.137}$$

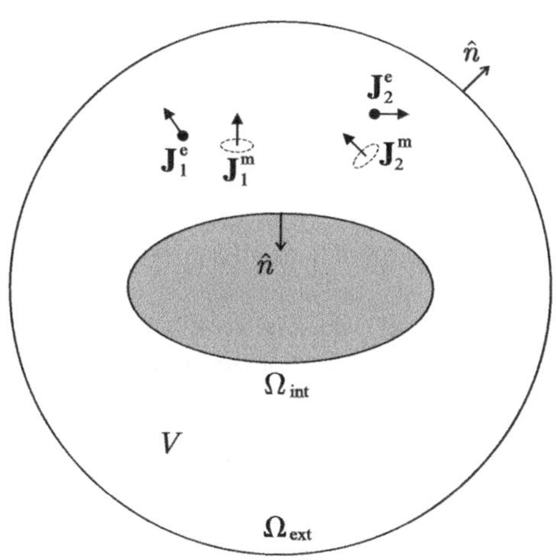

Figure 2.7 Geometry assumed in the derivation of reciprocity relations.

and

$$\nabla \times \mathbf{H}_2 = j\omega\epsilon\mathbf{E}_2 + \mathbf{J}_2^{\mathrm{e}} \tag{2.138}$$

$$\nabla \times \mathbf{E}_2 = -j\omega\mu\mathbf{H}_2 - \mathbf{J}_2^{\mathrm{m}}. \tag{2.139}$$

Building a scalar product of (2.136) with the vector $\mathbf{E}_2$ and of (2.139) with $\mathbf{H}_1$, subtracting the results, and using (2.117) with $\mathbf{A} = \mathbf{H}_1$ and $\mathbf{B} = \mathbf{E}_2$ in the left-hand side gives the relation

$$\nabla \cdot (\mathbf{H}_1 \times \mathbf{E}_2) = j\omega\epsilon\mathbf{E}_1 \cdot \mathbf{E}_2 + j\omega\mu\mathbf{H}_1 \cdot \mathbf{H}_2 + \mathbf{E}_2 \cdot \mathbf{J}_1^{\mathrm{e}} + \mathbf{H}_1 \cdot \mathbf{J}_2^{\mathrm{m}}. \tag{2.140}$$

In a similar way, multiplying (2.138) with $\mathbf{E}_1$ and (2.137) with $\mathbf{H}_2$ and subtracting leads to a symmetric relation:

$$\nabla \cdot (\mathbf{H}_2 \times \mathbf{E}_1) = j\omega\epsilon\mathbf{E}_1 \cdot \mathbf{E}_2 + j\omega\mu\mathbf{H}_1 \cdot \mathbf{H}_2 + \mathbf{E}_1 \cdot \mathbf{J}_2^{\mathrm{e}} + \mathbf{H}_2 \cdot \mathbf{J}_1^{\mathrm{m}}, \tag{2.141}$$

and the difference between (2.140) and (2.141) can be therefore written as

$$\nabla \cdot (\mathbf{H}_1 \times \mathbf{E}_2 - \mathbf{H}_2 \times \mathbf{E}_1) = \mathbf{E}_2 \cdot \mathbf{J}_1^{\mathrm{e}} + \mathbf{H}_1 \cdot \mathbf{J}_2^{\mathrm{m}} - \mathbf{E}_1 \cdot \mathbf{J}_2^{\mathrm{e}} - \mathbf{H}_2 \cdot \mathbf{J}_1^{\mathrm{m}}. \tag{2.142}$$

The left-hand side of (2.142) is a divergence of a vector, so integrating (2.142) over the volume V and using the divergence relation (A.28) we obtain

$$\int_{\Omega} (\mathbf{H}_1 \times \mathbf{E}_2 - \mathbf{H}_2 \times \mathbf{E}_1) \cdot \hat{n}\, \mathrm{d}\Omega = \int_{V} \left(\mathbf{E}_2 \cdot \mathbf{J}_1^{\mathrm{e}} + \mathbf{H}_1 \cdot \mathbf{J}_2^{\mathrm{m}} - \mathbf{E}_1 \cdot \mathbf{J}_2^{\mathrm{e}} - \mathbf{H}_2 \cdot \mathbf{J}_1^{\mathrm{m}}\right) \mathrm{d}V, \tag{2.143}$$

where $\hat{n}$ is the unit normal vector pointing to the exterior of V. In order to apply the divergence theorem we have to assume that the fields are continuous everywhere in V, which implies that the external currents as well as the material parameters ϵ and μ are smooth and bounded functions of coordinates. This limitation does not restrict the generality of the analysis as singular sources, for example a point source, can be approximated by a sequence of smooth and bounded source functions that approach the given singular source in the limit.

In the absence of sources in V, the right-hand side of (2.143) vanishes, implying that the surface integral in the left-hand side is equal to zero as well. The surface integral is a sum of contributions from Ω_{int} and Ω_{ext}, so that we have

$$\int_{\Omega_{\mathrm{ext}}} (\mathbf{H}_1 \times \mathbf{E}_2 - \mathbf{H}_2 \times \mathbf{E}_1) \cdot \hat{n}\, \mathrm{d}\Omega = \int_{\Omega_{\mathrm{int}}} (\mathbf{H}_1 \times \mathbf{E}_2 - \mathbf{H}_2 \times \mathbf{E}_1) \cdot \hat{n}'\, \mathrm{d}\Omega, \tag{2.144}$$

where $\hat{n}' = -\hat{n}$ is the interior unit normal. The equation can be interpreted as an identity allowing a free deformation of the integration surface in source-free regions without changing the value of the integral.

Now consider the surface integral in the left-hand side of (2.143) in the case when some sources are present in V. By construction, there are no sources and material boundaries beyond Ω_{ext}, so the property (2.144) permits us to extend Ω_{ext} to infinity, where the conditions at infinity (section 2.2.3) make the integral over Ω_{ext} vanish.

The surface Ω_{int} does not enclose sources but may enclose scattering bodies, that is, material boundaries. In order to arrive at the reciprocity relation, the integral over Ω_{int} must be zero. Let us show that this is always the case as long as media inside V are isotropic. First, we note that for the case of isotropic media relation (2.142) holds everywhere inside V because the terms which depend on the permittivity and permeability cancel out for any volumetric distributions of the material parameters, even if these functions are not continuous. If there are no material boundaries inside V, and V is filled with a medium with uniform or smoothly varying ϵ and μ, the surface can be contracted to a point by using (2.144) as the fields are continuous and the divergence theorem applies throughout the volume V. In the case of sharp material boundaries inside V, the integral is also zero since the integrand depends on the tangential components of the electric and magnetic fields only (see (2.121)) and is therefore continuous across any material boundary between isotropic media, implying that the integration surface can still be contracted to a point. In the case of localized scatterers inside V described by PEC or impedance boundary conditions, the integral vanishes because the fields $(\mathbf{E}_1, \mathbf{H}_1)$ and $(\mathbf{E}_2, \mathbf{H}_2)$ satisfy the same boundary conditions. It is important to note that in the case of anisotropic or bianisotropic media relation (2.142) is not necessarily valid and the reciprocity relation may be not satisfied (see problem 38).

Thus, for all isotropic scatterers the left-hand side of (2.143) is zero, and the relation (2.143) simplifies to

$$\int_V \left(\mathbf{E}_2 \cdot \mathbf{J}_1^{\mathrm{e}} + \mathbf{H}_1 \cdot \mathbf{J}_2^{\mathrm{m}} - \mathbf{E}_1 \cdot \mathbf{J}_2^{\mathrm{e}} - \mathbf{H}_2 \cdot \mathbf{J}_1^{\mathrm{m}}\right) \mathrm{d}V = 0, \tag{2.145}$$

which is the integral form of the reciprocity theorem.

The reciprocity relation takes a particularly simple form if the sources are point dipoles. For example, if the sources are electric dipoles (see (2.203)), with the moments $\mathbf{p}_{1,2}^{\mathrm{e}}$ located at the points $\mathbf{r} = \mathbf{r}_{1,2}$, then insertion of the currents

$$\mathbf{J}_{1,2}^{\mathrm{e}} = j\omega \mathbf{p}_{1,2}^{\mathrm{e}} \delta(\mathbf{r} - \mathbf{r}_{1,2}), \quad \mathbf{J}_{1,2}^{\mathrm{m}} = 0 \tag{2.146}$$

in (2.145) gives the formula

$$\mathbf{E}_2(\mathbf{r}_1) \cdot \mathbf{p}_1^{\mathrm{e}} = \mathbf{E}_1(\mathbf{r}_2) \cdot \mathbf{p}_2^{\mathrm{e}} \tag{2.147}$$

which relates the electric field created by one source at the position of a second source to the electric field created by the second source at the position of first source. If the sources are magnetic dipoles (see (2.209)) with the moments $\mathbf{p}_{1,2}^{\mathrm{m}}$,

$$\mathbf{J}_{1,2}^{\mathrm{e}} = 0, \quad \mathbf{J}_{1,2}^{\mathrm{m}} = j\omega\mu \mathbf{p}_{1,2}^{\mathrm{m}} \delta(\mathbf{r} - \mathbf{r}_{1,2}), \tag{2.148}$$

then the reciprocity reads

$$\mathbf{H}_2(\mathbf{r}_1) \cdot \mathbf{p}_1^{\mathrm{m}} = \mathbf{H}_1(\mathbf{r}_2) \cdot \mathbf{p}_2^{\mathrm{m}} \tag{2.149}$$

and relates the magnetic fields. In the case of mixed excitation such that

$$\mathbf{J}_1^{\mathrm{e}} = j\omega \mathbf{p}_1^{\mathrm{e}} \delta(\mathbf{r} - \mathbf{r}_1), \quad \mathbf{J}_2^{\mathrm{e}} = 0, \quad \mathbf{J}_1^{\mathrm{m}} = 0, \quad \mathbf{J}_2^{\mathrm{m}} = j\omega\mu \mathbf{p}_2^{\mathrm{m}} \delta(\mathbf{r} - \mathbf{r}_2), \tag{2.150}$$

the reciprocity takes the form

$$\mathbf{E}_2(\mathbf{r}_1) \cdot \mathbf{p}_1^{\mathrm{e}} = -\mu \mathbf{H}_1(\mathbf{r}_2) \cdot \mathbf{p}_2^{\mathrm{m}}. \tag{2.151}$$

As discussed before, equation (2.145) is derived for smooth and bounded currents and, strictly speaking, cannot be applied to the currents of the point dipoles described by delta functions. However, this limitation can be circumvented if we recall that delta functions can be approximated by sequences of continuous functions. For example, the sequence

$$\delta_n(\mathbf{r} - \mathbf{r}_0) = \frac{n^3}{\pi^{3/2}} \mathrm{e}^{-n^2 |r - r_0|^2} \tag{2.152}$$

consists of smooth functions for every finite value of the parameter n, but it approaches $\delta(\mathbf{r} - \mathbf{r}_0)$ in the limit $n \to \infty$ in the sense that

$$\lim_{n \to \infty} \int_V f(\mathbf{r}) \delta_n(\mathbf{r} - \mathbf{r}_0) \, \mathrm{d}V = f(\mathbf{r}_0) \tag{2.153}$$

for every volume V containing the point $\mathbf{r} = \mathbf{r}_0$ and every function $f(\mathbf{r})$ continuous at $\mathbf{r} = \mathbf{r}_0$. Thus, equations (2.147), (2.149), and (2.151) can be rigorously derived by approximating the delta functions in (2.146), (2.148), and (2.150) with δ_n, inserting these in the formula (2.145) and using the limiting formula (2.153).

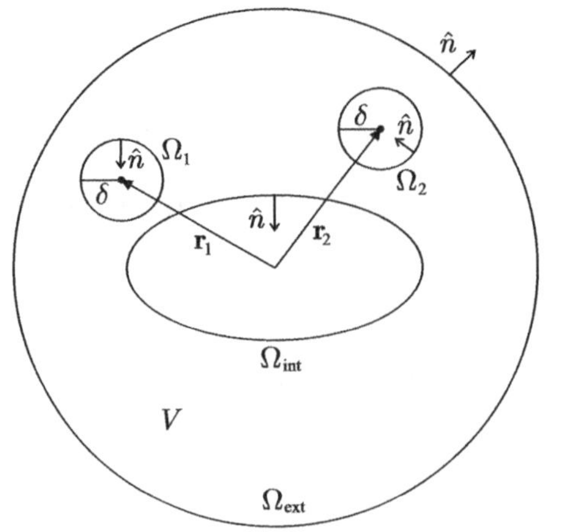

Figure 2.8 Small vicinities of point sources at $\mathbf{r} = \mathbf{r}_{1,2}$ excluded from integration region V.

Another approach to deriving the reciprocity relations for point sources consists in integrating (2.142) over the volume V with excluded vicinities of the sources (Figure 2.8). The volume integral in (2.143) is then equal to zero, but the surface integrals over surfaces $\Omega_{1,2}$ enclosing the points $\mathbf{r} = \mathbf{r}_{1,2}$ must be added to the equation, and since integrals over Ω_{int} and Ω_{ext} are equal to zero, (2.143) is replaced with the formula

$$\int_{\Omega_1} (\mathbf{H}_1 \times \mathbf{E}_2 - \mathbf{H}_2 \times \mathbf{E}_1) \cdot \hat{n}\, \mathrm{d}\Omega + \int_{\Omega_2} (\mathbf{H}_1 \times \mathbf{E}_2 - \mathbf{H}_2 \times \mathbf{E}_1) \cdot \hat{n}\, \mathrm{d}\Omega = 0. \tag{2.154}$$

For the point sources, the surfaces $\Omega_{1,2}$ can be chosen to be spheres of a small radius δ around the points $\mathbf{r}_{1,2}$, and letting δ go to zero and using the explicit expressions for the fields of point dipoles (see (2.207), (2.208), (2.211), and (2.212)) can be shown to recover the formulas (2.147), (2.149), and (2.151) (see Problems section).

2.5 Basic Solutions of Maxwell's Equations in Homogeneous Isotropic Media

In the previous section we formulated the field equations and all additional conditions that are necessary to ensure a unique solution of the equations. Derivation of an exact solution which is compliant with these equations and conditions is typically a difficult task and is only possible in a limited number of simple cases. This section presents some of the basic solutions and solution methods that are of particular importance for the electromagnetic scattering theory.

The section begins with a description of the simplest solutions of Maxwell's equations, namely, plane, cylindrical, and spherical waves. They are at the same time the most basic solutions as solutions of general scattering and radiation problems can be described in terms of these basic wave forms. Derivation of symbolic solutions of Maxwell's equations can be greatly simplified through the introduction of electromagnetic potentials, which may help to reduce the vector boundary value problem to a scalar and therefore much simpler problem. By using the Hertz potentials we describe the fundamental division of solutions of Maxwell's equations in E and H modes and present explicit solutions for the fields excited by arbitrary electric and magnetic currents, including the special case of elementary electric and magnetic dipoles. A useful idealization is the assumption of infinitely long line sources or cylindrical scatterers in which case the three-dimensional scattering problem can be reduced to a two-dimensional one. The section concludes with a discussion of the reduction of Maxwell's equations to the scalar two-dimensional Helmholtz equations.

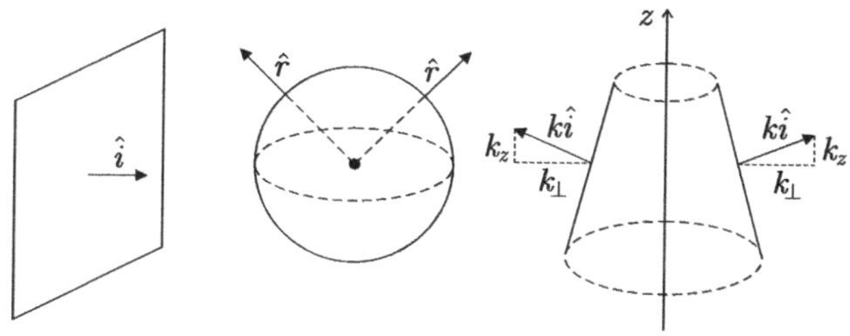

Figure 2.9 Wavefronts and propagation directions of the plane, spherical, and cylindrical (conical) waves.

2.5.1 Plane, Spherical, and Cylindrical Waves

The plane wave, the spherical wave, and the cylindrical wave are the three types of solutions of Maxwell's equations in free space that are particularly important in problems of electromagnetic scattering. This section is a summary of their characteristic features, including their amplitudes and phase factors, polarization, shapes of their wavefronts, propagation directions, corresponding power flux densities, and relations between electric and magnetic fields.

The Plane Wave

The plane wave is the simplest wave solution of Maxwell's equations; every solution of Maxwell's equations can be represented as a superposition (sum or integral) of plane waves using Fourier transformation. In a plane wave propagating in a homogeneous medium with constitutive parameters ϵ and μ in the direction $\hat{i}$ ($\hat{i}$ is a real-valued unit vector), the electric field is given by

$$\mathbf{E}_{\mathrm{pw}} = \mathbf{E}_0 e^{-j\mathbf{k}\cdot\mathbf{r}}, \tag{2.155}$$

where $\mathbf{k} = \hat{i}k$ with $k = \omega\sqrt{\epsilon\mu} = 2\pi/\lambda$ is the wave vector. In a lossy medium, the wavenumber k has a nonzero imaginary part ($\mathrm{Im}\, k < 0$) which results in an exponential decay of the wave (2.155) as the wave propagates in the medium. The factor $\mathbf{E}_0$ is a constant, real- or complex-valued vector that describes the amplitude, initial phase, and polarization of the wave. The wave is linearly polarized if $\mathbf{E}_0 = E_0\hat{p}$ where E_0 is a real- or complex-valued scalar and a real-valued unit vector $\hat{p}$ defines the orientation of the electric field vector in the wave. Otherwise, the wave (2.155) is elliptically polarized (e.g., see Stratton (1941) and Born and Wolf (1959)). The amplitude and phase of (2.155) are constant on every plane surface described by the equation $\hat{i}\cdot\mathbf{r} = \mathrm{const}$, and this is why these solutions are referred to as plane waves.

Insertion of (2.155) in Maxwell's equation (2.42) relates the magnetic and electric fields in the plane wave as

$$Z\mathbf{H}_{\mathrm{pw}} = \hat{i}\times\mathbf{E}_{\mathrm{pw}}, \tag{2.156}$$

where $Z = \sqrt{\mu/\epsilon}$ is the wave impedance. Thus, the magnetic field is also a plane wave of the form

$$\mathbf{H}_{\mathrm{pw}} = \mathbf{H}_0 e^{-j\mathbf{k}\cdot\mathbf{r}} \tag{2.157}$$

with $\mathbf{H}_0 = Y\hat{i}\times\mathbf{E}_0$, where $Y = 1/Z$ is the wave admittance of the medium. Insertion of (2.157) into the Maxwell equation (2.41) leads to a relation dual to (2.156),

$$Y\mathbf{E}_{\mathrm{pw}} = -\hat{i}\times\mathbf{H}_{\mathrm{pw}}. \tag{2.158}$$

Equations (2.156) and (2.158) make obvious the orthogonality of the fields in a plane electromagnetic wave to the direction of propagation, in the sense that

$$\hat{i}\cdot\mathbf{H}_{\mathrm{pw}} = 0, \quad \hat{i}\cdot\mathbf{E}_{\mathrm{pw}} = 0. \tag{2.159}$$

According to (2.119) the mean (or time-averaged) intensity $\mathbf{w}$ of the energy flow in the plane waves (2.155) and (2.157) is given by

$$\mathbf{w}_{\text{pw}} = \frac{\hat{\imath}}{2}|\mathbf{E}_0|^2 e^{2(\hat{\imath}\cdot\mathbf{r})\operatorname{Im}k}\operatorname{Re} Y \tag{2.160}$$

or, in terms of the amplitude of the magnetic field,

$$\mathbf{w}_{\text{pw}} = \frac{\hat{\imath}}{2}|\mathbf{H}_0|^2 e^{2(\hat{\imath}\cdot\mathbf{r})\operatorname{Im}k}\operatorname{Re} Z \tag{2.161}$$

with $|\mathbf{E}_0| = |E_0|$ and $|\mathbf{H}_0| = |H_0| = |YE_0|$ in a linearly polarized wave. The exponential factors in (2.160) and (2.161) describe the dissipation of the energy transported by the plane wave due to absorption in the medium. In a lossless medium, (2.160) and (2.161) become

$$\mathbf{w}_{\text{pw}} = \frac{\hat{\imath}}{2}Y|\mathbf{E}_0|^2 = \frac{\hat{\imath}}{2}Z|\mathbf{H}_0|^2. \tag{2.162}$$

Note, that if both ϵ and μ are real but negative, plane waves can propagate ($\mathbf{k}$ is a real vector), but, according to (2.162), the direction of the power flow is opposite to the direction of the wave vector, $\mathbf{k} = -|k|\hat{\imath}$. Such waves are called *backward waves.*[7]

The Spherical Wave

The spherical wave can be defined by the formula

$$\mathbf{E}_{\text{sph}} = \mathbf{F}\frac{e^{-jkr}}{kr}, \tag{2.163}$$

where r is the distance from the origin of the wave at $r = 0$ to the observation point at $\mathbf{r}$ and $\mathbf{F}$ is a constant complex-valued amplitude vector. The importance of this kind of solution of Maxwell's equations is due to the fact that electromagnetic fields radiated from compact sources or scattered by compact obstacles become spherical waves once the observation point is located sufficiently far from the object.

In the spherical coordinate system (r, θ, ϕ) the complex-valued vector $\mathbf{F}$ has the form

$$\mathbf{F} = \hat{\theta}F_0^{\theta}(\theta, \phi) + \hat{\phi}F_0^{\phi}(\theta, \phi) \tag{2.164}$$

and thus is tangential to the wavefront. For sufficiently large values of kr, the front of the wave is locally almost planar and the relation between the magnetic and electric fields is, therefore, similar to (2.156) and (2.158)

$$Z\mathbf{H}_{\text{sph}} = \hat{r}\times\mathbf{E}_{\text{sph}}, \quad Y\mathbf{E}_{\text{sph}} = -\hat{r}\times\mathbf{H}_{\text{sph}}, \tag{2.165}$$

where $\hat{r} = \mathbf{r}/r$. The mean density of the energy flow transported by the wave in a lossless medium is apparently given by

$$\mathbf{w}_{\text{sph}} = \frac{\hat{r}}{2}Y|\mathbf{E}_{\text{sph}}|^2 = \frac{\hat{r}Y}{2k^2r^2}|\mathbf{F}|^2, \tag{2.166}$$

implying that the energy flow in the spherical wave decreases as r^{-2} with growing distance from the origin, which is the consequence of the broadening of the wavefront as the wave propagates away from the source.

[7] For real negative ϵ and μ the values of the square roots defining the wavenumber $k = \omega\sqrt{\epsilon\mu}$ and the impedance $Z = \sqrt{\mu/\epsilon}$ are obtained by analytic continuation such that $\arg\epsilon$ and $\arg\mu$ vary from 0 ($\epsilon > 0$, $\mu > 0$) to $-\pi$ ($\epsilon < 0$, $\mu < 0$), which leads to the limiting values $k = -\omega|\sqrt{\epsilon\mu}|$ and $Z = |\sqrt{\mu/\epsilon}|$.

In contrast to the plane wave, the spherical wave (2.163) and (2.164) is not an exact solution of Maxwell's equations. It satisfies the field equations asymptotically as $kr \to \infty$. In fact, (2.164) is only the leading term of the expansion

$$\mathbf{F} = \mathbf{F}_0 + \sum_{n=1}^{+\infty} (kr)^{-n} \mathbf{F}_n(\theta, \phi), \tag{2.167}$$

where $\mathbf{F}_0$ is given by (2.164). Expansion coefficients $\mathbf{F}_n(\theta, \phi)$ with $n \geq 1$ may have nonzero radial components,

$$\mathbf{F}_n = \hat{\theta} F_n^{\theta}(\theta, \phi) + \hat{\phi} F_n^{\phi}(\theta, \phi) + \hat{r} F_n^{r}(\theta, \phi). \tag{2.168}$$

Inserting (2.163) together with (2.167) into Maxwell's equations and equating the like powers of kr leads to a recurrent chain of equations that allow the derivation of all the coefficients $\mathbf{F}_n(\theta, \phi)$ with $n \geq 1$ from the zero-order term $\mathbf{F}_0$ (Borovikov and Kinber 1994; Wilcox 1956).

The simplest example of a spherical electromagnetic wave is the field radiated by elementary dipole sources (section 2.5.2). For example, in the case of a magnetic dipole oriented along the z axis, the exact expressions for the field are given by (2.211) and (2.212), implying that the only nonvanishing terms in the expansion (2.167) are

$$F_0^{\phi} = Z \frac{k^3 p_{\mathrm{m}}}{4\pi} \sin\theta, \quad F_1^{\phi} = -j F_0^{\phi}. \tag{2.169}$$

In the case of an electric z-oriented dipole, these are the terms

$$\begin{aligned} F_0^{\theta} &= -\frac{k^3 p_{\mathrm{e}}}{4\pi\epsilon} \sin\theta, \quad F_1^{\theta} = -j F_0^{\theta}, \quad F_1^{r} = -2j \cot\theta F_0^{\theta}, \\ F_2^{\theta} &= -F_0^{\theta}, \quad F_2^{r} = -2 \cot\theta F_0^{\theta}, \end{aligned} \tag{2.170}$$

which follow from (2.207) and (2.208).

The Cylindrical Wave

If a plane wave gets scattered by a long, eventually infinitely long, cylindrical body with generators parallel to a straight line, at distances greater than several wavelengths from the body the scattered field approaches a cylindrical wave. Choosing the axis of the body to be the z-axis of a cylindrical coordinate system (ρ, φ, z), the electric field in the wave can be described by

$$\mathbf{E}_{\mathrm{cyl}} = \mathbf{C}(\varphi) \frac{\mathrm{e}^{-jk_z z - jk_\perp \rho}}{\sqrt{k_\perp \rho}}, \tag{2.171}$$

where k_z and $k_\perp$ are the components of the wave vector parallel and perpendicular to the z axis so that

$$k^2 = k_\perp^2 + k_z^2 \tag{2.172}$$

with k being the wavenumber of the homogeneous medium in which the scatterer is immersed. Complex-valued amplitude vector $\mathbf{C}(\varphi)$ defines the amplitude and polarization of the wave.

A characteristic feature of the cylindrical wave is that its electric and magnetic field vectors are completely determined by the z components of the fields,

$$\mathbf{E}_{\mathrm{cyl}} = -\frac{k_z}{k_\perp} E_z \hat{\rho} + \frac{k}{k_\perp} Z H_z \hat{\varphi} + E_z \hat{z} \tag{2.173}$$

$$\mathbf{H}_{\mathrm{cyl}} = -\frac{k_z}{k_\perp} H_z \hat{\rho} - \frac{k}{k_\perp Z} E_z \hat{\varphi} + H_z \hat{z}. \tag{2.174}$$

The latter can be represented as

$$E_z = C_E(\varphi)\frac{e^{-jk_z z - jk_\perp \rho}}{\sqrt{k_\perp \rho}}, \quad ZH_z = C_H(\varphi)\frac{e^{-jk_z z - jk_\perp \rho}}{\sqrt{k_\perp \rho}}, \tag{2.175}$$

implying that

$$\mathbf{C}(\varphi) = -\frac{k_z}{k_\perp}C_E(\varphi)\hat{\rho} + \frac{k}{k_\perp}C_H(\varphi)\hat{\varphi} + C_E(\varphi)\hat{z}. \tag{2.176}$$

The far field coefficients $C_E(\varphi)$, $C_H(\varphi)$, and $\mathbf{C}(\varphi)$ are functions of φ but not of z and ρ. Similarly to the case of spherical wave, the cylindrical wave is an asymptotic solution of Maxwell's equations, which satisfies the equations asymptotically when $k_\perp \rho$ is large (e.g., see (5.424) and (5.427)). The far field coefficients can be expanded into a series in inverse powers of $k_\perp \rho$, in which the leading terms are independent of ρ.

The normal to the wavefront, or the direction of propagation of the wave, $\hat{s}$, is determined by taking the phase function $\Phi = k_z z + k_\perp \rho$ and evaluating the gradient $\nabla\Phi = \hat{\rho}k_\perp + \hat{z}k_z$ and the absolute value $|\nabla\Phi| = k$, which gives

$$\hat{s} = \frac{\nabla\Phi}{|\nabla\Phi|} = \hat{\rho}\frac{k_\perp}{k} + \hat{z}\frac{k_z}{k}. \tag{2.177}$$

The vector $\hat{s}$ depends only on φ (through $\hat{\rho}$) and makes the same angle with the z axis in every azimuth plane with $\varphi = \text{const}$, which implies that the wavefront is a circular conical surface. The conical surface reduces to the surface of a circular cylinder when $k_z = 0$. So, strictly speaking, referring to the wave (2.171) as a cylindrical wave is inaccurate, as the wave is in fact a conical wave. This terminology is, however, widespread in the technical literature, so we refer to (2.171) as a cylindrical wave regardless of the value of k_z.

Electric and magnetic field vectors in the cylindrical wave are related by similar relations as in the plane and spherical waves,

$$Z\mathbf{H}_{\text{cyl}} = \hat{s} \times \mathbf{E}_{\text{cyl}}, \quad Y\mathbf{E}_{\text{cyl}} = -\hat{s} \times \mathbf{H}_{\text{cyl}}, \tag{2.178}$$

which is seen by direct inspection of (2.173) and (2.174). Finally, the time-averaged energy flow density is obtained from (2.119) as

$$\mathbf{w}_{\text{cyl}} = \frac{\hat{s}}{2}\text{Re}\, Y|\mathbf{E}_{\text{cyl}}|^2, \tag{2.179}$$

in which the intensity of electric field can be expressed through either E_z and H_z or the far field coefficients as follows:

$$|\mathbf{E}_{\text{cyl}}|^2 = \frac{k^2}{k_\perp^2}\left(|E_z|^2 + |ZH_z|^2\right) = \frac{k^2}{\rho k_\perp^3}\left(|C_E|^2 + |C_H|^2\right). \tag{2.180}$$

Equations (2.179) and (2.180) make it apparent that in the cylindrical wave the energy flows in the direction of $\hat{s}$ and the density of the energy flow decays as $1/\rho$.

2.5.2 *Electromagnetic Potentials and Fields of External Currents*

Maxwell's equations are a system of six first-order partial differential equations for six unknown functions, the components of vectors **E** and **H**. Transformation of Maxwell's equations to the Helmholtz equation, as shown in section 2.2.3, leads to a system of three second-order partial differential equations for three unknowns, the components of either vector **E** or vector **H**. By introducing new unknown functions instead of the electric and magnetic fields it is sometimes possible to further reduce the number of unknown functions and the corresponding equations. Electromagnetic potentials serve as such new functions, and the *vector and scalar potentials* $\mathbf{A}$ and Φ or the *electric and magnetic Hertz vectors* $\mathbf{\Pi}_{e,m}$ are the best known examples of these (Born and Wolf 1959; Jones 1964; Stratton 1941; Van Bladel 1964). For a variety of simple field sources, distributions of material (homogeneous, layered) and geometric shapes of the boundaries (flat, cylindrical, spherical, etc.) the original problem can be reduced to a boundary value problem with a single unknown function, which can then be solved analytically (Bouman *et al.* 1987).

For time-harmonic fields, the potentials $\mathbf{A}$ and Φ and the Hertz vectors are related to each other according to

$$\Phi = -\nabla \cdot \mathbf{\Pi}_e, \quad \mathbf{A} = j\epsilon\mu\omega\mathbf{\Pi}_e \tag{2.181}$$

or

$$\Phi = 0, \quad \mathbf{A} = \mu\nabla \times \mathbf{\Pi}_m, \tag{2.182}$$

and it is only a matter of convenience which pair of potentials is used. In this book the electric and magnetic Hertz vectors are used since they permit symmetric formulations with respect to electric and magnetic sources. This section describes the basic theory of the vector potentials, including reduction of Maxwell's equations to Helmholtz equations or to equations reducible to the Helmholtz equation, and derivation of solutions for dipole sources. It also provides a foundation for sections 2.5.3 and 2.5.4. A general theory of the Hertz vectors can be found in Nisbet (1955) and McCrea (1957).

Hertz Vectors

Consider a homogeneous isotropic medium and suppose that the external excitation consists of electric and magnetic currents. Because of the linearity of Maxwell's equations, the field can be divided into two parts due to the electric and magnetic currents,

$$\mathbf{E} = \mathbf{E}_e + \mathbf{E}_m, \quad \mathbf{H} = \mathbf{H}_e + \mathbf{H}_m, \tag{2.183}$$

and both parts can be described independently. The electric Hertz vector $\mathbf{\Pi}_e$ defines the electric part of the solution, and the magnetic Hertz vector $\mathbf{\Pi}_m$ the magnetic part. By definition, the electric part satisfies equations (2.30) and (2.31). Since $\nabla \cdot \mathbf{H} = 0$, it is natural to represent the magnetic field as a curl of a new vector function

$$\mathbf{H}_e = j\omega\epsilon\nabla \times \mathbf{\Pi}_e. \tag{2.184}$$

Representation for the electric field then results from (2.30) as

$$\mathbf{E}_e = \nabla \times \nabla \times \mathbf{\Pi}_e - \frac{\mathbf{J}_e}{j\omega\epsilon}. \tag{2.185}$$

Inserting (2.184) and (2.185) in (2.31) leads to the equation

$$\nabla \times \left(\nabla \times \nabla \times \mathbf{\Pi}_e - k^2\mathbf{\Pi}_e - \frac{\mathbf{J}_e}{j\omega\epsilon}\right) = 0, \tag{2.186}$$

which implies that

$$\nabla \times \nabla \times \mathbf{\Pi}_e - k^2\mathbf{\Pi}_e - \frac{\mathbf{J}_e}{j\omega\epsilon} = \nabla f_e, \tag{2.187}$$

where f_e is an arbitrary differentiable scalar function.

The magnetic Hertz vector $\mathbf{\Pi}_m$ is responsible for the portion of the field due to the external magnetic currents. The corresponding equations are obtained from (2.33) and (2.34) along similar guidelines, with the following result:

$$\mathbf{E}_m = -j\omega\mu\nabla \times \mathbf{\Pi}_m \tag{2.188}$$

$$\mathbf{H}_m = \nabla \times \nabla \times \mathbf{\Pi}_m - \frac{\mathbf{J}_m}{j\omega\mu} \tag{2.189}$$

$$\nabla \times \nabla \times \mathbf{\Pi}_m - k^2\mathbf{\Pi}_m - \frac{\mathbf{J}_m}{j\omega\mu} = \nabla f_m, \tag{2.190}$$

where f_m is an arbitrary differentiable function. Note that in the source region, due to the definition of magnetic currents, function $\mathbf{E}_m$ should be corrected according to (2.32) to obtain the physical electric field.

The general solution of Maxwell's equations is then the sum of the fields due to the electric and magnetic currents, and outside the sources we have

$$\mathbf{E} = \nabla \times \nabla \times \mathbf{\Pi}_e - j\omega\mu\nabla \times \mathbf{\Pi}_m \tag{2.191}$$

$$\mathbf{H} = j\omega\epsilon\nabla \times \mathbf{\Pi}_e + \nabla \times \nabla \times \mathbf{\Pi}_m. \tag{2.192}$$

It should be pointed out that the vector potentials $\mathbf{\Pi}_{e,m}$ are to satisfy (2.187) and (2.190) up to a gradient of an arbitrary scalar function. The choice of functions f_e and f_m is a matter of convenience since it may influence the values of the potentials but not the fields, which are uniquely defined by (2.191) and (2.192). For example, by choosing $f_e = \nabla \cdot \mathbf{\Pi}_e$ and $f_m = \nabla \cdot \mathbf{\Pi}_m$ equations (2.187) and (2.190) are reduced to the vector Helmholtz equations

$$\nabla^2\mathbf{\Pi}_e + k^2\mathbf{\Pi}_e = -\frac{\mathbf{J}_e}{j\omega\epsilon} \tag{2.193}$$

$$\nabla^2\mathbf{\Pi}_m + k^2\mathbf{\Pi}_m = -\frac{\mathbf{J}_m}{j\omega\mu}, \tag{2.194}$$

where the vector Laplace operator is defined by the relation

$$\nabla^2\mathbf{\Pi} = \nabla\nabla \cdot \mathbf{\Pi} - \nabla \times \nabla \times \mathbf{\Pi}. \tag{2.195}$$

Using (2.193), (2.194), and (2.195), we can rewrite expressions (2.191) and (2.192) for the fields in the exterior of the sources as

$$\mathbf{E} = \nabla\nabla \cdot \mathbf{\Pi}_e + k^2\mathbf{\Pi}_e - j\omega\mu\nabla \times \mathbf{\Pi}_m \tag{2.196}$$

$$\mathbf{H} = j\omega\epsilon\nabla \times \mathbf{\Pi}_e + \nabla\nabla \cdot \mathbf{\Pi}_m + k^2\mathbf{\Pi}_m. \tag{2.197}$$

In Cartesian coordinates (x, y, z), the vector Laplacian (2.195) is reduced to the scalar Laplacian applied to the corresponding components of the vector function $\mathbf{\Pi}$,

$$\nabla^2\mathbf{\Pi} = \hat{x}\nabla^2\Pi_x + \hat{y}\nabla^2\Pi_y + \hat{z}\nabla^2\Pi_z, \tag{2.198}$$

implying that equations (2.193) and (2.194) split into independent scalar Helmholtz equations for Cartesian components of the vector potentials excited by corresponding Cartesian components of the currents. These equations can be solved by using the Green function[8]

$$G_0(\mathbf{r}, \mathbf{r}') = \frac{e^{-jk|\mathbf{r}-\mathbf{r}'|}}{4\pi|\mathbf{r} - \mathbf{r}'|}, \tag{2.199}$$

[8] Green's function is differently defined in the literature by using different coefficients in front of the delta function in the right-hand side of (2.200), which can be -1 as in (2.200) or $+1$ or -4π, as in Morse and Feshbach (1953). This results in corresponding changes in the explicit expressions for the Green functions. The convention accepted in this book eliminates the factor $1/(4\pi)$ from expressions for the fields, which involve the Green function, by absorbing the factor in G_0.

which satisfies the equation

$$\nabla^2 G_0(\mathbf{r},\mathbf{r}') + k^2 G_0(\mathbf{r},\mathbf{r}') = -\delta(\mathbf{r}-\mathbf{r}') \tag{2.200}$$

with $\delta(\mathbf{r}-\mathbf{r}') = \delta(x-x')\delta(y-y')\delta(z-z')$ being the delta function in three-dimensional space (e.g., Morse and Feshbach (1953, p. 891)). The result is given by the formulas

$$\mathbf{\Pi}_\mathrm{e}(\mathbf{r}) = \frac{1}{j\omega\epsilon}\int_{V_0} \mathbf{J}_\mathrm{e}(\mathbf{r}')G_0(\mathbf{r},\mathbf{r}')\,\mathrm{d}V' \tag{2.201}$$

$$\mathbf{\Pi}_\mathrm{m}(\mathbf{r}) = \frac{1}{j\omega\mu}\int_{V_0} \mathbf{J}_\mathrm{m}(\mathbf{r}')G_0(\mathbf{r},\mathbf{r}')\,\mathrm{d}V', \tag{2.202}$$

where V_0 denotes the volume that contains all sources. Equations (2.201) and (2.202) represent the Hertz vectors of the fields excited by external currents $\mathbf{J}_\mathrm{e}$ and $\mathbf{J}_\mathrm{m}$ in infinite free space. The relations are not tensorial as, for example, the x components of the Hertz vectors are entirely determined by integration of the x components of the corresponding currents. On the contrary, relations between electromagnetic fields and the currents, which are obtained by inserting (2.201) and (2.202) in (2.191) and (2.192) or in (2.196) and (2.197), are fully tensorial as, say, a linear electric current produces both electric and magnetic fields not collinear with the direction of the current. This topic is addressed next.

Fields of Elementary Dipoles

The Hertz vectors are particularly useful when fields are excited by linear currents. In this case, representations (2.201) and (2.202) suggest that the electric and magnetic fields can be expressed through a single scalar function – the projection of the respective Hertz vector on the direction of the current (Hertz 1889).[9]

For example, if the source is an elementary (i.e. infinitely small) electric dipole with the moment $\mathbf{p}_\mathrm{e} = p_\mathrm{e}\hat{z}$ at point $\mathbf{r} = 0$, then $\mathbf{P}_\mathrm{e} = p_\mathrm{e}\delta(\mathbf{r})\hat{z}$, $\mathbf{P}_\mathrm{m} = 0$, and it follows from (2.38) that

$$\mathbf{J}_\mathrm{e} = j\omega p_\mathrm{e}\hat{z}\delta(\mathbf{r}), \quad \mathbf{J}_\mathrm{m} = 0. \tag{2.203}$$

Equations (2.201) and (2.202) give for the Hertz vectors

$$\mathbf{\Pi}_\mathrm{e} = \Pi_\mathrm{e}\hat{z}, \quad \mathbf{\Pi}_\mathrm{m} = 0 \tag{2.204}$$

with

$$\Pi_\mathrm{e} = \frac{p_\mathrm{e}}{4\pi\epsilon}\frac{\mathrm{e}^{-jkr}}{r}. \tag{2.205}$$

The field of a dipole is conveniently described in spherical coordinates (r,θ,ϕ) with the origin at the dipole position and the polar axis $\hat{z}$ along the dipole direction. Representing the electric Hertz vector in the spherical coordinates

$$\mathbf{\Pi}_\mathrm{e} = \Pi_\mathrm{e}(\hat{r}\cos\theta - \hat{\theta}\sin\theta) \tag{2.206}$$

and using the relations of the Hertz vectors to the fields lead to the expressions:

$$\mathbf{E} = -\frac{k^2 p_\mathrm{e}}{4\pi\epsilon r}\mathrm{e}^{-jkr}\left\{\frac{2\cos\theta}{jkr}\left[1-\frac{j}{kr}\right]\hat{r} + \sin\theta\left[1-\frac{j}{kr}-\frac{1}{(kr)^2}\right]\hat{\theta}\right\} \tag{2.207}$$

$$\mathbf{H} = -Y\frac{k^2 p_\mathrm{e}}{4\pi\epsilon r}\mathrm{e}^{-jkr}\sin\theta\left(1-\frac{j}{kr}\right)\hat{\phi}. \tag{2.208}$$

[9] Hertz noticed that the electromagnetic field of an oscillating electric dipole can be described in terms of a single scalar potential function. Introduction of the magnetic Hertz vector is attributed to Righi (1901).

Similarly, the electromagnetic field of an elementary magnetic dipole ($\mathbf{J}_e = 0$) with the moment $\mathbf{p}_m = p_m\hat{z}$ and the corresponding magnetic current density

$$\mathbf{J}_m = j\omega p_m \mu \hat{z}\delta(\mathbf{r}) \tag{2.209}$$

is described in terms of the one-component magnetic Hertz vector $\mathbf{\Pi}_m = \Pi_m\hat{z}$ with

$$\Pi_m = \frac{p_m}{4\pi}\frac{e^{-jkr}}{r}, \tag{2.210}$$

which gives for the electromagnetic field of the magnetic dipole

$$\mathbf{E} = Z\frac{k^2 p_m}{4\pi r}e^{-jkr}\sin\theta\left(1 - \frac{j}{kr}\right)\hat{\phi} \tag{2.211}$$

$$\mathbf{H} = -\frac{k^2 p_m}{4\pi r}e^{-jkr}\left\{\frac{2\cos\theta}{jkr}\left[1 - \frac{j}{kr}\right]\hat{r} + \sin\theta\left[1 - \frac{j}{kr} - \frac{1}{(kr)^2}\right]\hat{\theta}\right\}. \tag{2.212}$$

The fields radiated by elementary dipoles are examples of spherical waves, which are addressed in section 2.5.1. At distances such that $kr \gg 1$, the far field radiated by the electric dipole tends to

$$\mathbf{E} = -\frac{k^2 p_e}{4\pi\epsilon r}e^{-jkr}\hat{\theta}\sin\theta, \quad \mathbf{H} = -Y\frac{k^2 p_e}{4\pi\epsilon r}e^{-jkr}\hat{\phi}\sin\theta, \tag{2.213}$$

and since $Yk = \omega\epsilon$, $\hat{r}\times z = -\hat{\phi}\sin\theta$, and $\hat{\theta} = -\hat{r}\times\hat{\phi}$, these can be expressed as

$$\mathbf{H} = \hat{r}\times\mathbf{p}_e\frac{k\omega}{4\pi r}e^{-jkr}, \quad \mathbf{E} = -Z\hat{r}\times\mathbf{H}. \tag{2.214}$$

Correspondingly, the far field radiated by an elementary magnetic dipole can be written as

$$\mathbf{E} = -\hat{r}\times\mathbf{p}_m\frac{k\omega\mu}{4\pi r}e^{-jkr}, \quad \mathbf{H} = Y\hat{r}\times\mathbf{E}. \tag{2.215}$$

2.5.3 Tensor Green's Function

We are now ready to obtain expressions for the electromagnetic field excited by given currents in the infinite and empty space filled with a homogeneous material. These expressions, which are among the most fundamental solutions of Maxwell's equations, permit determining the fields by convolution of the currents with a solution of Maxwell's equations for an elementary source, with the so-called tensor Green's function.

Expressions (2.201) and (2.202) give in explicit form the electric and magnetic Hertz vectors $\mathbf{\Pi}_e$ and $\mathbf{\Pi}_m$ due to given distributions of external currents $\mathbf{J}_e$ and $\mathbf{J}_m$. These can be inserted in the relations (2.196) and (2.197) between Hertz vectors and the electric and magnetic fields $\mathbf{E}$ and $\mathbf{H}$ to express the fields excited by the currents. The result has the form of integral linear relations between the currents and the fields, and since these are vectors, the coefficients are in general tensors (dyadics),

$$\mathbf{E}(\mathbf{r}) = \int_{V_0}\left[\overline{\overline{G}}_{ee}(\mathbf{r},\mathbf{r}')\cdot\mathbf{J}_e(\mathbf{r}') + \overline{\overline{G}}_{em}(\mathbf{r},\mathbf{r}')\cdot\mathbf{J}_m(\mathbf{r}')\right]\mathrm{d}V' \tag{2.216}$$

$$\mathbf{H}(\mathbf{r}) = \int_{V_0}\left[\overline{\overline{G}}_{me}(\mathbf{r},\mathbf{r}')\cdot\mathbf{J}_e(\mathbf{r}') + \overline{\overline{G}}_{mm}(\mathbf{r},\mathbf{r}')\cdot\mathbf{J}_m(\mathbf{r}')\right]\mathrm{d}V', \tag{2.217}$$

where V_0 is the volume that includes all the currents. The definition of the integral terms in the right-hand sides of (2.216) and (2.217) follow from comparing (2.216) and (2.217) with (2.196) and (2.197),

$$\int_{V_0} \overline{\overline{G}}_{\rm ee}(\mathbf{r},\mathbf{r}') \cdot \mathbf{J}_{\rm e}(\mathbf{r}')\,{\rm d}V' = k^2\mathbf{\Pi}_{\rm e}(\mathbf{r}) + \nabla[\nabla \cdot \mathbf{\Pi}_{\rm e}(\mathbf{r})] \tag{2.218}$$

$$\int_{V_0} \overline{\overline{G}}_{\rm em}(\mathbf{r},\mathbf{r}') \cdot \mathbf{J}_{\rm m}(\mathbf{r}')\,{\rm d}V' = -j\omega\mu\nabla \times \mathbf{\Pi}_{\rm m}(\mathbf{r}) \tag{2.219}$$

$$\int_{V_0} \overline{\overline{G}}_{\rm me}(\mathbf{r},\mathbf{r}') \cdot \mathbf{J}_{\rm e}(\mathbf{r}')\,{\rm d}V' = j\omega\epsilon\nabla \times \mathbf{\Pi}_{\rm e}(\mathbf{r}) \tag{2.220}$$

$$\int_{V_0} \overline{\overline{G}}_{\rm mm}(\mathbf{r},\mathbf{r}') \cdot \mathbf{J}_{\rm m}(\mathbf{r}')\,{\rm d}V' = k^2\mathbf{\Pi}_{\rm m}(\mathbf{r}) + \nabla[\nabla \cdot \mathbf{\Pi}_{\rm m}(\mathbf{r})]. \tag{2.221}$$

Explicit forms of the tensors $\overline{\overline{G}}_{\rm ee}$, $\overline{\overline{G}}_{\rm em}$, $\overline{\overline{G}}_{\rm me}$, and $\overline{\overline{G}}_{\rm mm}$ follow from (2.201) and (2.202). For example, using (2.201) in (2.218) gives

$$\int_{V_0} \overline{\overline{G}}_{\rm ee}(\mathbf{r},\mathbf{r}') \cdot \mathbf{J}_{\rm e}(\mathbf{r}')\,{\rm d}V' = -j\omega\mu \int_{V_0} \left\{\mathbf{J}_{\rm e}(\mathbf{r}')G_0(\mathbf{r},\mathbf{r}') + \frac{1}{k^2}\nabla\nabla \cdot \left[\mathbf{J}_{\rm e}(\mathbf{r}')G_0(\mathbf{r},\mathbf{r}')\right]\right\} {\rm d}V', \tag{2.222}$$

and since the operator ∇ acts on variable $\mathbf{r}$,

$$\nabla \cdot \left[\mathbf{J}_{\rm e}(\mathbf{r}')G_0(\mathbf{r},\mathbf{r}')\right] = \mathbf{J}_{\rm e}(\mathbf{r}') \cdot \nabla G_0(\mathbf{r},\mathbf{r}'), \tag{2.223}$$

the result can be expressed as

$$\overline{\overline{G}}_{\rm ee}(\mathbf{r},\mathbf{r}') \cdot \mathbf{J}_{\rm e}(\mathbf{r}') = -j\omega\mu \left[\mathbf{G}_x(\mathbf{r},\mathbf{r}')J_x^{\rm e}(\mathbf{r}') + \mathbf{G}_y(\mathbf{r},\mathbf{r}')J_y^{\rm e}(\mathbf{r}') + \mathbf{G}_z(\mathbf{r},\mathbf{r}')J_z^{\rm e}(\mathbf{r}')\right], \tag{2.224}$$

where $J_x^{\rm e}$, $J_y^{\rm e}$, and $J_z^{\rm e}$ are the Cartesian coordinates of $\mathbf{J}_{\rm e}$ and vectors $\mathbf{G}_x$, $\mathbf{G}_y$, and $\mathbf{G}_z$ are given by

$$\mathbf{G}_x(\mathbf{r},\mathbf{r}') = \hat{x}G_0(\mathbf{r},\mathbf{r}') + \frac{1}{k^2}\frac{\partial}{\partial x}\nabla G_0(\mathbf{r},\mathbf{r}') \tag{2.225}$$

$$\mathbf{G}_y(\mathbf{r},\mathbf{r}') = \hat{y}G_0(\mathbf{r},\mathbf{r}') + \frac{1}{k^2}\frac{\partial}{\partial y}\nabla G_0(\mathbf{r},\mathbf{r}') \tag{2.226}$$

$$\mathbf{G}_z(\mathbf{r},\mathbf{r}') = \hat{z}G_0(\mathbf{r},\mathbf{r}') + \frac{1}{k^2}\frac{\partial}{\partial z}\nabla G_0(\mathbf{r},\mathbf{r}'). \tag{2.227}$$

Similarly, using (2.201) in (2.220) leads to

$$\int_{V_0} \overline{\overline{G}}_{\rm me}(\mathbf{r},\mathbf{r}') \cdot \mathbf{J}_{\rm e}(\mathbf{r}')\,{\rm d}V' = \int_{V_0} \nabla G_0(\mathbf{r},\mathbf{r}') \times \mathbf{J}_{\rm e}(\mathbf{r}')\,{\rm d}V' \tag{2.228}$$

and, therefore, to

$$\overline{\overline{G}}_{\rm me}(\mathbf{r},\mathbf{r}') \cdot \mathbf{J}_{\rm e}(\mathbf{r}') = \mathbf{G}_x'(\mathbf{r},\mathbf{r}')J_x^{\rm e}(\mathbf{r}') + \mathbf{G}_y'(\mathbf{r},\mathbf{r}')J_y^{\rm e}(\mathbf{r}') + \mathbf{G}_z'(\mathbf{r},\mathbf{r}')J_z^{\rm e}(\mathbf{r}'), \tag{2.229}$$

where

$$\mathbf{G}'_x(\mathbf{r},\mathbf{r}') = \hat{y}\frac{\partial}{\partial z}G_0(\mathbf{r},\mathbf{r}') - \hat{z}\frac{\partial}{\partial y}G_0(\mathbf{r},\mathbf{r}') \tag{2.230}$$

$$\mathbf{G}'_y(\mathbf{r},\mathbf{r}') = \hat{z}\frac{\partial}{\partial x}G_0(\mathbf{r},\mathbf{r}') - \hat{x}\frac{\partial}{\partial z}G_0(\mathbf{r},\mathbf{r}') \tag{2.231}$$

$$\mathbf{G}'_z(\mathbf{r},\mathbf{r}') = \hat{x}\frac{\partial}{\partial y}G_0(\mathbf{r},\mathbf{r}') - \hat{y}\frac{\partial}{\partial x}G_0(\mathbf{r},\mathbf{r}'). \tag{2.232}$$

The remaining tensors look similar to (2.224) and (2.229),

$$\overline{\overline{G}}_{\text{em}}(\mathbf{r},\mathbf{r}')\cdot\mathbf{J}_{\text{m}}(\mathbf{r}') = -\left[\mathbf{G}'_x(\mathbf{r},\mathbf{r}')J_x^{\text{m}}(\mathbf{r}') + \mathbf{G}'_y(\mathbf{r},\mathbf{r}')J_y^{\text{m}}(\mathbf{r}') + \mathbf{G}'_z(\mathbf{r},\mathbf{r}')J_z^{\text{m}}(\mathbf{r}')\right] \tag{2.233}$$

$$\overline{\overline{G}}_{\text{mm}}(\mathbf{r},\mathbf{r}')\cdot\mathbf{J}_{\text{m}}(\mathbf{r}') = -j\omega\epsilon\left[\mathbf{G}_x(\mathbf{r},\mathbf{r}')J_x^{\text{m}}(\mathbf{r}') + \mathbf{G}_y(\mathbf{r},\mathbf{r}')J_y^{\text{m}}(\mathbf{r}') + \mathbf{G}_z(\mathbf{r},\mathbf{r}')J_z^{\text{m}}(\mathbf{r}')\right], \tag{2.234}$$

and it is seen that the four tensors $\overline{\overline{G}}_{\text{ee}}$, $\overline{\overline{G}}_{\text{em}}$, $\overline{\overline{G}}_{\text{me}}$, and $\overline{\overline{G}}_{\text{mm}}$ can be expressed in terms of just two tensors $\overline{\overline{G}}$ and $\overline{\overline{G}}'$ as

$$\overline{\overline{G}}_{\text{ee}} = -j\omega\mu\overline{\overline{G}} \tag{2.235}$$

$$\overline{\overline{G}}_{\text{me}} = -\overline{\overline{G}}_{\text{em}} = \overline{\overline{G}}' \tag{2.236}$$

$$\overline{\overline{G}}_{\text{mm}} = -j\omega\epsilon\overline{\overline{G}}. \tag{2.237}$$

Tensors $\overline{\overline{G}}$ and $\overline{\overline{G}}'$ can be written as dyadics, that is, as sums of vector pairs,

$$\overline{\overline{G}}(\mathbf{r},\mathbf{r}') = \mathbf{G}_x(\mathbf{r},\mathbf{r}')\hat{x} + \mathbf{G}_y(\mathbf{r},\mathbf{r}')\hat{y} + \mathbf{G}_z(\mathbf{r},\mathbf{r}')\hat{z} \tag{2.238}$$

$$\overline{\overline{G}}'(\mathbf{r},\mathbf{r}') = \mathbf{G}'_x(\mathbf{r},\mathbf{r}')\hat{x} + \mathbf{G}'_y(\mathbf{r},\mathbf{r}')\hat{y} + \mathbf{G}'_z(\mathbf{r},\mathbf{r}')\hat{z} \tag{2.239}$$

or, expanding the vectors (2.225)–(2.227) and (2.230)–(2.232) into the components,

$$\overline{\overline{G}} = \hat{x}G_{xx}\hat{x} + \hat{y}G_{yx}\hat{x} + \hat{z}G_{zx}\hat{x} + \hat{x}G_{xy}\hat{y} + \hat{y}G_{yy}\hat{y} + \hat{z}G_{zy}\hat{y} + \hat{x}G_{xz}\hat{z} + \hat{y}G_{yz}\hat{z} + \hat{z}G_{zz}\hat{z} \tag{2.240}$$

$$\overline{\overline{G}}' = \hat{x}G'_{xx}\hat{x} + \hat{y}G'_{yx}\hat{x} + \hat{z}G'_{zx}\hat{x} + \hat{x}G'_{xy}\hat{y} + \hat{y}G'_{yy}\hat{y} + \hat{z}G'_{zy}\hat{y} + \hat{x}G'_{xz}\hat{z} + \hat{y}G'_{yz}\hat{z} + \hat{z}G'_{zz}\hat{z}, \tag{2.241}$$

where

$$G_{xx}(\mathbf{r},\mathbf{r}') = \left(1 + \frac{1}{k^2}\frac{\partial^2}{\partial x^2}\right)G_0(\mathbf{r},\mathbf{r}') \tag{2.242}$$

$$G_{xy}(\mathbf{r},\mathbf{r}') = G_{yx}(\mathbf{r},\mathbf{r}') = \frac{1}{k^2}\frac{\partial^2}{\partial x\partial y}G_0(\mathbf{r},\mathbf{r}') \tag{2.243}$$

$$G_{xz}(\mathbf{r},\mathbf{r}') = G_{zx}(\mathbf{r},\mathbf{r}') = \frac{1}{k^2}\frac{\partial^2}{\partial x\partial z}G_0(\mathbf{r},\mathbf{r}') \tag{2.244}$$

$$G_{yy}(\mathbf{r},\mathbf{r}') = \left(1 + \frac{1}{k^2}\frac{\partial^2}{\partial y^2}\right)G_0(\mathbf{r},\mathbf{r}') \tag{2.245}$$

$$G_{yz}(\mathbf{r},\mathbf{r}') = G_{zy}(\mathbf{r},\mathbf{r}') = \frac{1}{k^2}\frac{\partial^2}{\partial y\partial z}G_0(\mathbf{r},\mathbf{r}') \tag{2.246}$$

$$G_{zz}(\mathbf{r},\mathbf{r}') = \left(1 + \frac{1}{k^2}\frac{\partial^2}{\partial z^2}\right)G_0(\mathbf{r},\mathbf{r}') \tag{2.247}$$

and

$$G'_{xx}(\mathbf{r},\mathbf{r}') = G'_{yy}(\mathbf{r},\mathbf{r}') = G'_{zz}(\mathbf{r},\mathbf{r}') = 0 \tag{2.248}$$

$$G'_{xy}(\mathbf{r},\mathbf{r}') = -G'_{yx}(\mathbf{r},\mathbf{r}') = -\frac{\partial}{\partial z}G_0(\mathbf{r},\mathbf{r}') \tag{2.249}$$

$$G'_{xz}(\mathbf{r},\mathbf{r}') = -G'_{zx}(\mathbf{r},\mathbf{r}') = \frac{\partial}{\partial y}G_0(\mathbf{r},\mathbf{r}') \tag{2.250}$$

$$G'_{yz}(\mathbf{r},\mathbf{r}') = -G'_{zy}(\mathbf{r},\mathbf{r}') = -\frac{\partial}{\partial x}G_0(\mathbf{r},\mathbf{r}'). \tag{2.251}$$

The vector pairs in dyadic representations (2.238)–(2.241) are neither scalar nor vector products of the vectors; the scalar product of a dyadic **ab** with a vector **c** gives a vector $\mathbf{a}(\mathbf{b}\cdot\mathbf{c})$ if multiplied from the right or a vector $(\mathbf{c}\cdot\mathbf{a})\mathbf{b}$ if multiplied from the left. Vector products of dyadics and vectors, application of ∇ operators to dyadics and a number of further operations on dyadics can also be defined (e.g., see Morse and Feshbach (1953, section 1.6) and Lindell (1992)).

Vectors $\mathbf{G}_x(\mathbf{r},\mathbf{r}')$, $\mathbf{G}_y(\mathbf{r},\mathbf{r}')$, and $\mathbf{G}_z(\mathbf{r},\mathbf{r}')$ have the meaning of fundamental solutions (Green's functions) of the vector equations (2.45) and (2.46) satisfied by **E** and **H**. It is easy to see that

$$\nabla\times\nabla\times\mathbf{G}_x - k^2\mathbf{G}_x = \hat{x}\delta(\mathbf{r}-\mathbf{r}') \tag{2.252}$$

$$\nabla\times\nabla\times\mathbf{G}_y - k^2\mathbf{G}_y = \hat{y}\delta(\mathbf{r}-\mathbf{r}') \tag{2.253}$$

$$\nabla\times\nabla\times\mathbf{G}_z - k^2\mathbf{G}_z = \hat{z}\delta(\mathbf{r}-\mathbf{r}'). \tag{2.254}$$

The corresponding dyadic $\overline{\overline{G}}(\mathbf{r},\mathbf{r}')$ can be seen as the fundamental solution of the dyadic equation

$$\nabla\times\nabla\times\overline{\overline{G}} - k^2\overline{\overline{G}} = \overline{\overline{I}}\delta(\mathbf{r}-\mathbf{r}'), \tag{2.255}$$

where $\overline{\overline{I}} = \hat{x}\hat{x} + \hat{y}\hat{y} + \hat{z}\hat{z}$ is the unit dyadic. The dyadic is referred to as the tensor Green function of an infinite free space and is often written in the compact form

$$\overline{\overline{G}}(\mathbf{r},\mathbf{r}') = \left(\overline{\overline{I}} + \frac{1}{k^2}\nabla\nabla\right)G_0(\mathbf{r},\mathbf{r}'). \tag{2.256}$$

The second tensor $\overline{\overline{G}}'$ is written in the dyadic notations as

$$\overline{\overline{G}}'(\mathbf{r},\mathbf{r}') = -\nabla\times\left[G_0(\mathbf{r},\mathbf{r}')\overline{\overline{I}}\right]. \tag{2.257}$$

The integral relations (2.216) and (2.217) with the tensor kernels defined by (2.235)–(2.237) allow determination of the fields excited by arbitrary given external currents in the infinite free space. The relations are truly tensorial as any single current component excites in general all components of the fields. For example, an elementary arbitrarily directed electric dipole at a point $\mathbf{r}_0$ with the dipole moment $\mathbf{p}_e = p_e\hat{c}$ ($\hat{c}$ is the direction of the dipole moment) corresponds to the current densities

$$\mathbf{J}_e = j\omega\mathbf{p}_e\delta(\mathbf{r}-\mathbf{r}_0), \quad \mathbf{J}_m = 0, \tag{2.258}$$

which excite the fields

$$\mathbf{E} = \omega^2\mu\left\{\mathbf{p}_e G_0(\mathbf{r},\mathbf{r}_0) + \frac{1}{k^2}\nabla[\mathbf{p}_e\cdot\nabla G_0(\mathbf{r},\mathbf{r}_0)]\right\} \tag{2.259}$$

$$\mathbf{H} = -j\omega\mathbf{p}_e\times\nabla G_0(\mathbf{r},\mathbf{r}_0). \tag{2.260}$$

An elementary magnetic dipole with the dipole moment $\mathbf{p}_m = p_m\hat{c}$ is described by the current densities

$$\mathbf{J}_e = 0, \quad \mathbf{J}_m = j\omega\mu\mathbf{p}_m\delta(\mathbf{r}-\mathbf{r}_0), \tag{2.261}$$

and the excited fields are given by

$$\mathbf{E} = j\omega\mu\mathbf{p}_{\mathrm{m}} \times \nabla G_0(\mathbf{r},\mathbf{r}_0) \tag{2.262}$$

$$\mathbf{H} = k^2\left\{\mathbf{p}_{\mathrm{m}}G_0(\mathbf{r},\mathbf{r}_0) + \frac{1}{k^2}\nabla[\mathbf{p}_{\mathrm{m}}\cdot\nabla G_0(\mathbf{r},\mathbf{r}_0)]\right\}. \tag{2.263}$$

With $\hat{c} = \hat{z}$ these expressions recover the formulas (2.207), (2.208), (2.211), and (2.212).

2.5.4 E and H Modes

A fundamental property of Maxwell's equations in homogeneous and isotropic media is that given an arbitrary direction $\hat{\zeta}$ in space, the fields excited by arbitrary sources can be represented as a sum of two subfields, or modes, with every mode described by a single scalar function, the ζ component of the electric or magnetic Hertz vectors (section 2.5.2),

$$\mathbf{\Pi}_{\mathrm{e}} = \Pi_{\mathrm{e}}\hat{\zeta} \tag{2.264}$$

$$\mathbf{\Pi}_{\mathrm{m}} = \Pi_{\mathrm{m}}\hat{\zeta}. \tag{2.265}$$

Functions Π_{e} and Π_{m} are referred to as the electric and magnetic Hertz potentials, and each of them satisfies a scalar Helmholtz equation and describes the field with a specific polarization structure. Namely, in the electromagnetic field represented by the potential Π_{e} the electric field has a ζ component, but the magnetic field does not, and this kind of the electromagnetic field is called *E* mode or TM (transverse magnetic) mode. The potential Π_{m} describes the complimentary field, *H* mode or TE (transverse electric) mode, in which the magnetic field has a ζ component and the electric field does not.

Splitting up in the two modes is possible not only in Cartesian coordinates but also in a number of other coordinate systems. In what follows we present expressions for the field components in a general system of orthogonal curvilinear coordinates (ξ, η, ζ) in which the scale factors h_ξ, h_η, and h_ζ satisfy the conditions

$$h_\zeta = 1, \quad \frac{\partial}{\partial\zeta}\left(\frac{h_\xi}{h_\eta}\right) = 0. \tag{2.266}$$

Later in this section we give examples of the coordinate systems which comply with these requirements. The present analysis is based on the analysis from Makarov *et al.* (1991, Chapter 1), simplified to the case of homogeneous media. In this book we use the concept of *E* and *H* modes to represent the field of arbitrarily oriented dipoles in the presence of flat material boundaries (section 4.4) and to solve the problem of scattering of a plane wave by a sphere by using Debye's potentials in Chapter 7. Applications of the concept to the theory of wave propagation in waveguides are addressed in Collin (1960). A summary of curvilinear orthogonal coordinates and vector differential operators in such coordinates can be found in Appendix A.3.

The analysis is based on the material presented in section 2.5.2. Let us start with the fields in a source-free region, that is, $\mathbf{J}_{\mathrm{e,m}} = 0$. It can be shown that the potential Π_ζ^{e} describes a partial solution of Maxwell's equations with the components

$$E_\xi^{\mathrm{e}} = \frac{1}{h_\xi}\frac{\partial^2\Pi_{\mathrm{e}}}{\partial\xi\partial\zeta}, \quad E_\eta^{\mathrm{e}} = \frac{1}{h_\eta}\frac{\partial^2\Pi_{\mathrm{e}}}{\partial\eta\partial\zeta}, \quad E_\zeta^{\mathrm{e}} = \frac{\partial^2\Pi_{\mathrm{e}}}{\partial\zeta^2} + k^2\Pi_{\mathrm{e}}, \tag{2.267}$$

$$H_\xi^{\mathrm{e}} = j\frac{\omega\epsilon}{h_\eta}\frac{\partial\Pi_{\mathrm{e}}}{\partial\eta}, \quad H_\eta^{\mathrm{e}} = -j\frac{\omega\epsilon}{h_\xi}\frac{\partial\Pi_{\mathrm{e}}}{\partial\xi}, \quad H_\zeta^{\mathrm{e}} = 0 \tag{2.268}$$

provided that it satisfies the Helmholtz equation of the form

$$\nabla_{\xi\eta}^2\Pi_{\mathrm{e}} + \frac{\partial^2\Pi_{\mathrm{e}}}{\partial\zeta^2} + k^2\Pi_{\mathrm{e}} = 0, \tag{2.269}$$

where $\nabla^2_{\xi\eta}$ is the two-dimensional Laplacian[10] defined by

$$\nabla^2_{\xi\eta}\Pi_e = \frac{1}{h_\xi h_\eta}\left[\frac{\partial}{\partial\xi}\left(\frac{h_\eta}{h_\xi}\frac{\partial\Pi_e}{\partial\xi}\right) + \frac{\partial}{\partial\eta}\left(\frac{h_\xi}{h_\eta}\frac{\partial\Pi_e}{\partial\eta}\right)\right]. \tag{2.270}$$

Function Π_m satisfies the same equation (2.269) and defines the field with the components

$$E^m_\xi = -j\frac{\omega\mu}{h_\eta}\frac{\partial\Pi_m}{\partial\eta}, \quad E^m_\eta = j\frac{\omega\mu}{h_\xi}\frac{\partial\Pi_m}{\partial\xi}, \quad E^m_\zeta = 0, \tag{2.271}$$

$$H^m_\xi = \frac{1}{h_\xi}\frac{\partial^2\Pi_m}{\partial\xi\partial\zeta}, \quad H^m_\eta = \frac{1}{h_\eta}\frac{\partial^2\Pi_m}{\partial\eta\partial\zeta}, \quad H^m_\zeta = \frac{\partial^2\Pi_m}{\partial\zeta^2} + k^2\Pi_m. \tag{2.272}$$

Expressions (2.268) follow from (2.184) upon representing the curl in curvilinear orthogonal coordinates with the scale factors (2.266) as

$$\nabla\times\mathbf{\Pi}_e = \frac{\hat{\xi}}{h_\eta h_\zeta}\left[\frac{\partial}{\partial\eta}\left(h_\zeta\Pi^e_\zeta\right) - \frac{\partial}{\partial\zeta}\left(h_\eta\Pi^e_\eta\right)\right] + \frac{\hat{\eta}}{h_\xi h_\zeta}\left[\frac{\partial}{\partial\zeta}\left(h_\xi\Pi^e_\xi\right) - \frac{\partial}{\partial\xi}\left(h_\zeta\Pi^e_\zeta\right)\right] + \frac{\hat{\zeta}}{h_\xi h_\eta}\left[\frac{\partial}{\partial\xi}\left(h_\eta\Pi^e_\eta\right) - \frac{\partial}{\partial\eta}\left(h_\xi\Pi^e_\xi\right)\right] \tag{2.273}$$

and accounting for the fact that $\mathbf{\Pi}_e$ has only a ζ component, which gives

$$\nabla\times(\Pi_e\hat{\zeta}) = \frac{\hat{\xi}}{h_\eta}\frac{\partial\Pi_e}{\partial\eta} - \frac{\hat{\eta}}{h_\xi}\frac{\partial\Pi_e}{\partial\xi}. \tag{2.274}$$

Here $\hat{\xi}$, $\hat{\eta}$, and $\hat{\zeta}$ are the orthogonal unit vectors defining the coordinate frame at every point (ξ,η,ζ). The Helmholtz equation (2.269) is simply the ζ component of (2.187) with the arbitrary function chosen as $f_e = \partial\Pi_e/\partial\zeta$ and the double-curl term evaluated by application of (2.273) to (2.274),

$$\nabla\times\nabla\times\mathbf{\Pi}_e = \frac{\hat{\xi}}{h_\xi}\frac{\partial^2\Pi_e}{\partial\zeta\partial\xi} + \frac{\hat{\eta}}{h_\eta}\frac{\partial^2\Pi_e}{\partial\zeta\partial\eta} - \hat{\zeta}\nabla^2_{\xi\eta}\Pi_e. \tag{2.275}$$

Expressions (2.267) for the electric field components result from (2.185) upon accounting for (2.269) and (2.275). Formulas (2.269), (2.271), and (2.272) for the field described by the magnetic Hertz potential (2.265) follow from (2.188), (2.189), and (2.190) with $\mathbf{J}_m = 0$ and $f_m = \partial\Pi_m/\partial\zeta$.

Let us now consider excitation of E and H modes. General current sources excite both modes simultaneously. A separate excitation is only possible with ζ-directed dipoles or linear currents, in which case $\mathbf{J}_e$ excites the E mode and $\mathbf{J}_m$ the H mode. The corresponding Hertz potentials are then determined from (2.269) with the source term $J^e_\zeta/(j\omega\epsilon)$ in the right-hand side for Π_e and with $J^m_\zeta/(j\omega\mu)$ for Π_m. In general, however, the currents $\mathbf{J}_e$ and $\mathbf{J}_m$ may have ξ and η components, and one has to satisfy vector equations (2.187) and (2.190) in the source region. The problem is therefore to split up the physical external current into two parts

$$\mathbf{J} = \mathbf{J}_E + \mathbf{J}_H \tag{2.276}$$

responsible for the excitation of the E and H modes, respectively.

[10] In curvilinear coordinates, $\nabla^2_{\xi\eta} + \partial^2/\partial\zeta^2$ is not equal to the three-dimensional Laplacian $\nabla^2_{\xi\eta\zeta}$ unless the product $h_\xi h_\eta$ is independent of ζ.

Currents $\mathbf{J}_E$ and $\mathbf{J}_H$ can be determined from the condition of consistence of the vector equations (2.187) and (2.190). The two equations resulting from ξ and η components of (2.187) are only consistent when the current $\mathbf{J}_E$ is such that

$$\mathbf{J}_E = \frac{1}{h_\xi}\frac{\partial Q}{\partial \xi}\hat{\xi} + \frac{1}{h_\eta}\frac{\partial Q}{\partial \eta}\hat{\eta} + J_\zeta\hat{\zeta}, \tag{2.277}$$

where J_ζ is the ζ component of $\mathbf{J}$ and Q is an arbitrary function which will be specified later. Then choosing

$$f_e = \frac{\partial \Pi_e}{\partial \zeta} + \frac{j}{\omega\epsilon}Q \tag{2.278}$$

makes these two equations satisfied, and writing down the ζ component of (2.187) gives the equation

$$\nabla^2_{\xi\eta}\Pi_e + \frac{\partial^2 \Pi_e}{\partial \zeta^2} + k^2\Pi_e = \frac{j}{\omega\epsilon}\left(J_\zeta - \frac{\partial Q}{\partial \zeta}\right) \tag{2.279}$$

for the potential Π_e.

The H mode is due to the magnetic current $\mathbf{J}_m$, which is related to the physical electric current $\mathbf{J}_H$ by the formula (2.35). In terms of the components of the physical current $\mathbf{J}_H$ the conditions of consistency of the equations resulting from projecting (2.190) on the ξ and η directions can be expressed as

$$\mathbf{J}_H = \nabla \times (M\hat{\zeta}) = \frac{\hat{\xi}}{h_\eta}\frac{\partial M}{\partial \eta} - \frac{\hat{\eta}}{h_\xi}\frac{\partial M}{\partial \xi}, \tag{2.280}$$

where M is an arbitrary function to be specified later, and

$$f_m = \frac{\partial}{\partial \zeta}\left(\Pi_m + \frac{1}{k^2}M\right). \tag{2.281}$$

Projecting (2.190) on the ζ direction and using (2.35), (2.280), and (2.281) leads to the equation

$$\nabla^2_{\xi\eta}\Pi_m + \frac{\partial^2 \Pi_m}{\partial \zeta^2} + k^2\Pi_m = \frac{1}{k^2}\left(\nabla^2_{\xi\eta}M + \frac{\partial^2 M}{\partial \zeta^2}\right) \tag{2.282}$$

for the potential Π_m, which is reduced to the form

$$\nabla^2_{\xi\eta}\widetilde{\Pi}_m + \frac{\partial^2 \widetilde{\Pi}_m}{\partial \zeta^2} + k^2\widetilde{\Pi}_m = -M \tag{2.283}$$

upon the substitution

$$\Pi_m = \widetilde{\Pi}_m + \frac{1}{k^2}M. \tag{2.284}$$

The remaining task is to determine the functions Q and M since once they are available, the potentials are obtained as solutions of the inhomogeneous differential equations (2.279) and (2.283). We now show that Q and M are solutions of Poisson's equations. Indeed, it follows from (2.276), (2.277), and (2.280) that

$$J_\xi = \frac{1}{h_\xi}\frac{\partial Q}{\partial \xi} + \frac{1}{h_\eta}\frac{\partial M}{\partial \eta} \tag{2.285}$$

$$J_\eta = \frac{1}{h_\eta}\frac{\partial Q}{\partial \eta} - \frac{1}{h_\xi}\frac{\partial M}{\partial \xi}, \tag{2.286}$$

which is a system of two equations of the first order for Q and M, and excluding one of the functions gives Poisson's equations

$$\nabla^2_{\xi\eta} Q = \frac{1}{h_\xi h_\eta}\left[\frac{\partial}{\partial \xi}(h_\eta J_\xi) + \frac{\partial}{\partial \eta}(h_\xi J_\eta)\right] \tag{2.287}$$

$$\nabla^2_{\xi\eta} M = \frac{1}{h_\xi h_\eta}\left[\frac{\partial}{\partial \eta}(h_\xi J_\xi) - \frac{\partial}{\partial \xi}(h_\eta J_\eta)\right]. \tag{2.288}$$

Here the right-hand sides are expressed through the components of the known excitation current. The solutions Q and M should vanish at infinity.

Once Q and M are constructed, the potentials are determined from the differential equations (2.279) and (2.282) (or (2.283) with (2.284)). The electromagnetic field is the sum of the fields of E and H modes

$$\mathbf{E} = \mathbf{E}_E + \mathbf{E}_H, \quad \mathbf{H} = \mathbf{H}_E + \mathbf{H}_H, \tag{2.289}$$

and the field in the E mode results from (2.184) and (2.185) as

$$\mathbf{E}_E = \nabla \times \nabla \times \mathbf{\Pi}_\mathrm{e} + \frac{j}{\omega\epsilon}\mathbf{J}_E \tag{2.290}$$

$$\mathbf{H}_E = j\omega\epsilon \nabla \times \mathbf{\Pi}_\mathrm{e}. \tag{2.291}$$

The electromagnetic field in the H mode follows from (2.188) and (2.189) as

$$\mathbf{E}_H = -j\omega\mu \nabla \times \mathbf{\Pi}_\mathrm{m} + \frac{j}{\omega\epsilon}\mathbf{J}_H \tag{2.292}$$

$$\mathbf{H}_H = \nabla \times \nabla \times \mathbf{\Pi}_\mathrm{e} + \frac{j}{\omega\mu}\mathbf{J}_\mathrm{m}, \tag{2.293}$$

where the expression (2.292) accounts for the presence of sources according to (2.32). The equivalent magnetic current $\mathbf{J}_\mathrm{m}$ can be related using (2.35) to the physical electric current $\mathbf{J}_H$ responsible for the excitation of the H mode,

$$\mathbf{J}_\mathrm{m} = \frac{j}{\omega\epsilon}\nabla \times \mathbf{J}_H. \tag{2.294}$$

By using (2.280), (2.284), and (2.294), expressions (2.292) and (2.293) can be rewritten in terms of $\widetilde{\Pi}_\mathrm{m}$ as

$$\mathbf{E}_H = -j\omega\mu \nabla \times (\widetilde{\Pi}_\mathrm{m}\hat{\zeta}) \tag{2.295}$$

$$\mathbf{H}_H = \nabla \times \nabla \times (\widetilde{\Pi}_\mathrm{m}\hat{\zeta}). \tag{2.296}$$

Combining (2.264), (2.289), (2.290), (2.291), (2.295), and (2.296) gives for the total field

$$\mathbf{E} = \nabla \times \nabla \times (\Pi_\mathrm{e}\hat{\zeta}) - j\omega\mu \nabla \times (\widetilde{\Pi}_\mathrm{m}\hat{\zeta}) + \frac{j}{\omega\epsilon}\mathbf{J}_E \tag{2.297}$$

$$\mathbf{H} = j\omega\epsilon \nabla \times (\Pi_\mathrm{e}\hat{\zeta}) + \nabla \times \nabla \times (\widetilde{\Pi}_\mathrm{m}\hat{\zeta}), \tag{2.298}$$

where the current $\mathbf{J}_E$ is given in (2.277).

There are many coordinate systems that comply with the requirements (2.266). In Cartesian coordinates (x, y, z), all scale factors are equal to unity and every coordinate can be chosen as ζ. The next group of coordinate systems are cylinder coordinates with $\zeta = z$, if z is the distance between the planes $z = \mathrm{const}$ in which orthogonal curvilinear coordinates (ξ, η) with h_ξ and h_η independent of z (e.g., circular, elliptic, parabolic, etc.) are used. Such coordinates are useful for description of electromagnetic fields in cylindrical waveguides with the z axis being the axis of the guide and a cross section uniform with respect to the z coordinate. The fields in such waveguides consist of E and H modes, which have no longitudinal component in the magnetic and electric field, respectively (e.g., Collin (1960, Chapter 5)).

Another important coordinate system compliant with (2.266) is the spherical coordinate system, in which $\xi = \theta, \eta = \phi$ and $\zeta = r$. The scale factors are given by $h_\theta = r, h_\eta = r \sin\theta$ and $h_r = 1$ and satisfy (2.266). Fields in spherical coordinates can therefore be described by one-component Hertz vectors $\mathbf{\Pi}_e = \Pi_e \hat{r}$ and $\mathbf{\Pi}_m = \Pi_m \hat{r}$. The potentials Π_e and Π_m are solutions of equation (2.269) which is transformed to the Helmholtz equation by substituting $\Pi_e = ru$ and $\Pi_m = rv$. The new functions u and v are called Debye's potentials, and they provide a theoretical framework for solving problems of electromagnetic scattering from spheres (Chapter 7).

The modes propagate independently in homogeneous isotropic media, but if material interfaces are present (e.g., a scattering body or walls of a waveguide), then the modes are in general coupled through the boundary conditions on the interfaces. If, however, the shape of the boundary is conformal with a coordinate surface in an orthogonal curvilinear coordinate system compliant with (2.266) and the analytical form of the boundary conditions is such that the potentials Π_e and Π_e are not coupled (see below), then the E and H modes can be excited independently from each other and the solution of Maxwell's equations is reduced to the solution of a scalar Helmholtz equation for a single unknown function.

The Hertz potentials can be used to represent fields in inhomogeneous media with ϵ and μ being functions of ζ. If ϵ and μ are piecewise constant functions of ζ (layered media), then in every layer the fields can be described by equations (2.267)–(2.272) and represented by a superposition of E and H modes. On the material interfaces $\zeta = \text{const}$, boundary conditions have to be imposed to ensure continuity of the tangential field components E_ξ, E_η, H_ξ, and H_η. It follows from expressions (2.267), (2.268), (2.271), and (2.272) that the continuity is guaranteed if the quantities

$$\epsilon\Pi_e, \quad \frac{\partial}{\partial\zeta}\Pi_e, \quad \mu\Pi_m, \quad \frac{\partial}{\partial\zeta}\Pi_m \tag{2.299}$$

are continuous at every material interface. Note that the potentials are fully decoupled in these boundary conditions. The Hertz potentials are then determined independently by solving the differential equation (2.269) with the boundary conditions (2.299). This approach to the solution of Maxwell's equation can be extended to media with continuous dependencies of ϵ and μ on ζ, for example by approximating these dependencies by suitably chosen step-functions.

It should be pointed out that in a homogeneous isotropic medium the direction $\hat{\zeta}$ can be chosen at will, and the electromagnetic field can be expressed in terms of an electric Hertz vector and a magnetic Hertz vector both along this arbitrary direction. If, however, material boundaries coincide with coordinate surfaces $\zeta = \text{const}$, then it is particularly beneficial to choose $\hat{\zeta}$ to be orthogonal to the material boundaries since this simplifies the boundary conditions for potentials Π_e and Π_m. Determining the portions of the external current responsible for the excitation of the E and H modes and, therefore, determining the right-hand sides in the inhomogeneous Helmholtz equations satisfied by the potentials requires in general the solution of two Poisson's equations.

2.5.5 *Fields with Translational Symmetry*

Many theoretical problems involve infinite bodies whose shape and material properties do not change under translation in one specific direction. Examples are cylinders and wedges with geometric cross sections invariant with respect to a Cartesian coordinate measured along the axis of the cylinder or the edge of the wedge. Let this direction be the z axis of a Cartesian or a cylindrical coordinate system. If, furthermore, the surrounding medium is homogeneous (or inhomogeneous but with material properties independent of the z coordinate), then the fields can be described in terms of two scalar functions E_z and H_z – the z components of electric and magnetic fields, respectively. These functions satisfy uncoupled Helmholtz equations, and the rest of the field components can be calculated from them. In this sense, E_z and H_z serve as potentials that help reduce vector Maxwell's equations to scalar Helmholtz equations. This approach to simplification of Maxwell's equations is more direct than the separation of fields into E and H modes described in the previous section. A further advantage is that in contrast to the Hertz potentials the functions E_z and H_z have immediate physical meaning.

Without loss of generality, as every field can be expanded in a superposition of plane waves, the incident field can be assumed to be a plane wave with dependence on the z coordinate as $\exp(-jk_z z)$. Because of the translational symmetry of the configurations under study, the scattered field must depend on z in exactly the same way, which can be used in Maxwell's equations to replace differentiation with respect to z by multiplication according to the rule $\partial/(\partial z) \rightarrow -jk_z$.

We will now simplify Maxwell's equations in the homogeneous source-free region, starting with the case of Cartesian coordinates (x, y, z). Writing (2.41) and (2.42) in the coordinate form and differentiating with respect to z give the set of equations:

$$jkYE_x = \frac{\partial}{\partial y}H_z + jk_zH_y \tag{2.300}$$

$$jkYE_y = -jk_zH_x - \frac{\partial}{\partial x}H_z \tag{2.301}$$

$$jkYE_z = \frac{\partial}{\partial x}H_y - \frac{\partial}{\partial y}H_x \tag{2.302}$$

$$-jkZH_x = \frac{\partial}{\partial y}E_z + jk_zE_y \tag{2.303}$$

$$-jkZH_y = -jk_zE_x - \frac{\partial}{\partial x}E_z \tag{2.304}$$

$$-jkZH_z = \frac{\partial}{\partial x}E_y - \frac{\partial}{\partial y}E_x. \tag{2.305}$$

Assume for a while that functions E_z and H_z are known, then E_x and H_y can be determined from equations (2.300) and (2.304), and E_y and H_x from (2.301) and (2.303). These are linear algebraic equations which are easily solved to give

$$E_x = \frac{1}{jk_\perp^2}\left(k_z\frac{\partial}{\partial x}E_z + kZ\frac{\partial}{\partial y}H_z\right) \tag{2.306}$$

$$E_y = \frac{1}{jk_\perp^2}\left(k_z\frac{\partial}{\partial y}E_z - kZ\frac{\partial}{\partial x}H_z\right) \tag{2.307}$$

$$H_x = \frac{1}{jk_\perp^2}\left(k_z\frac{\partial}{\partial x}H_z - kY\frac{\partial}{\partial y}E_z\right) \tag{2.308}$$

$$H_y = \frac{1}{jk_\perp^2}\left(k_z\frac{\partial}{\partial y}H_z + kY\frac{\partial}{\partial x}E_z\right), \tag{2.309}$$

where

$$k_\perp^2 = k^2 - k_z^2. \tag{2.310}$$

Using (2.306)–(2.309) to exclude E_x, E_y, H_x, and H_y from the remaining equations (2.302) and (2.305) results in the Helmholtz equations for E_z and H_z,

$$\frac{\partial^2}{\partial x^2}E_z + \frac{\partial^2}{\partial y^2}E_z + k_\perp^2 E_z = 0 \tag{2.311}$$

$$\frac{\partial^2}{\partial x^2}H_z + \frac{\partial^2}{\partial y^2}H_z + k_\perp^2 H_z = 0. \tag{2.312}$$

The equations have to be supplemented with boundary conditions on the surface of the scattering body, and depending on whether E_z and H_z are coupled or not in the boundary conditions, one ends up with either a boundary value problem for two unknown functions or with two separate boundary value problems for a single unknown function each. The latter situation takes place for PEC scatterers and the former is typical for impedance boundary conditions.

Once the Helmholtz equations with the boundary conditions are solved, the rest of the field components can be determined from (2.306)–(2.309). This approach ensures equivalence of the formulation with the original problem for

Maxwell's equations since it can be easily seen that the divergence equations $\nabla \cdot \mathbf{E} = 0$ and $\nabla \cdot \mathbf{H} = 0$ are satisfied and solutions like $\mathbf{E} \equiv 0$ or $\mathbf{H} \equiv 0$ are excluded (see discussion in section 2.2.3).

A similar procedure is applicable in cylindrical coordinate systems. For example, in the circular cylinder coordinates (ρ, φ, z), one arrives at the equations

$$E_\rho = \frac{1}{jk_\perp^2}\left(k_z \frac{\partial}{\partial \rho} E_z + \frac{kZ}{\rho}\frac{\partial}{\partial \varphi} H_z\right) \tag{2.313}$$

$$E_\varphi = \frac{1}{jk_\perp^2}\left(\frac{k_z}{\rho}\frac{\partial}{\partial \varphi} E_z - kZ\frac{\partial}{\partial \rho} H_z\right) \tag{2.314}$$

$$H_\rho = \frac{1}{jk_\perp^2}\left(k_z \frac{\partial}{\partial \rho} H_z - \frac{kY}{\rho}\frac{\partial}{\partial \varphi} E_z\right) \tag{2.315}$$

$$H_\varphi = \frac{1}{jk_\perp^2}\left(\frac{k_z}{\rho}\frac{\partial}{\partial \varphi} H_z + kY\frac{\partial}{\partial \rho} E_z\right) \tag{2.316}$$

instead of (2.306)–(2.309), and

$$\frac{1}{\rho}\frac{\partial}{\partial \rho}\left(\rho\frac{\partial}{\partial \rho}E_z\right) + \frac{1}{\rho^2}\frac{\partial^2}{\partial \varphi^2}E_z + k_\perp^2 E_z = 0 \tag{2.317}$$

$$\frac{1}{\rho}\frac{\partial}{\partial \rho}\left(\rho\frac{\partial}{\partial \rho}H_z\right) + \frac{1}{\rho^2}\frac{\partial^2}{\partial \varphi^2}H_z + k_\perp^2 H_z = 0 \tag{2.318}$$

instead of (2.311) and (2.312).

Helmholtz equations (2.311) and (2.312), and (2.317) and (2.318) can be written in a unified form,

$$\nabla_\perp^2 E_z + k_\perp^2 E_z = 0 \tag{2.319}$$

$$\nabla_\perp^2 H_z + k_\perp^2 H_z = 0, \tag{2.320}$$

where $\nabla_\perp^2$ is the two-dimensional Laplacian. Since the fields depends on the z coordinate as

$$\begin{pmatrix} E_z \\ H_z \end{pmatrix} = e^{-jk_z z}\begin{pmatrix} U_E \\ U_H \end{pmatrix}, \tag{2.321}$$

where U_E and U_H are independent of z, the z dependence can be factored out from all equations and the analysis can be made completely two dimensional with functions U_E and U_H satisfying the two-dimensional Helmholtz equations (2.319) and (2.320) and two-dimensional boundary conditions. The conditions at infinity $\rho \to \infty$ for outgoing fields take the form

$$|U_E| \to 0, \quad |U_H| \to 0 \tag{2.322}$$

when there are losses in the surrounding medium ($\operatorname{Im} k_\perp < 0$), or

$$\sqrt{\rho}\left(\frac{\partial U_E}{\partial \rho} + jk_\perp U_E\right) \to 0, \quad \sqrt{\rho}\left(\frac{\partial U_H}{\partial \rho} + jk_\perp U_H\right) \to 0 \tag{2.323}$$

with bounded values of $\sqrt{\rho}|U_E|$ and $\sqrt{\rho}|U_H|$ when $\operatorname{Im} k_\perp = 0$.

To solve the two-dimensional Helmholtz equation with a source term

$$\nabla_\perp^2 U + k_\perp^2 U = -f, \tag{2.324}$$

we will need the two-dimensional Green function, for example (Morse and Feshbach 1953, p. 891),

$$G_2(x, y, x', y', k) = \frac{1}{4j} H_0^{(2)} \left[k\sqrt{(x - x')^2 + (y - y')^2} \right], \tag{2.325}$$

where $H_0^{(2)}(\ldots)$ is the Hankel function of the second kind. Sometimes, for brevity, we will write the function as

$$G_2(\boldsymbol{\rho}, \boldsymbol{\rho}', k) = \frac{1}{4j} H_0^{(2)}(k|\boldsymbol{\rho} - \boldsymbol{\rho}'|), \tag{2.326}$$

where $\boldsymbol{\rho}$ and $\boldsymbol{\rho}'$ are the position vectors on the plane (x, y).

The Green function is a solution of the equation

$$\nabla_\perp^2 G_2 + k^2 G_2 = -\delta(x - x')\delta(y - y'), \tag{2.327}$$

compliant with conditions (2.322) or (2.323) at $\rho \to \infty$. It is easy to see that the solution of (2.324) compliant with the conditions at infinity can be expressed as

$$U(x, y) = \int_{-\infty}^{\infty}\int_{-\infty}^{\infty} G_2(x, y, x', y', k_\perp) f(x', y')\, \mathrm{d}x' \mathrm{d}y'. \tag{2.328}$$

This section forms the basis for the analysis of scattering from circular cylinders in Chapter 6 and from wedges in Chapter 5.

2.6 Electromagnetic Formulation of Huygens' Principle

An important property of electromagnetic fields is that the fields in the interior or exterior of every bounded region can be obtained by integrating their values over the boundary of the region, provided that the boundary completely encloses either the source(s) or the observer. This property goes back to Huygens' concept of propagation of light (Huygens 1690), according to which the propagation can be seen as the result of superposition of elementary waves from secondary sources (Huygens' sources) excited by the primary wave and distributed over the wavefront of the primary wave at a preceding instant of time, traveling only in the forward direction (Figure 2.10). The mathematical formulation of this property requires a coherent summation of the secondary waves (Fresnel 1818) and surface integration (Helmholtz 1860; Kirchhoff 1883, 1891). The classical Fresnel and Helmholtz–Kirchhoff theories assume scalar fields and can be only conditionally applied to the vector electromagnetic fields (see discussion in section 2.2.3).

The rigorous formulation of Huygens' principle for electromagnetic fields is based on the equivalence principle. The principle means that the actual source(s) can be replaced by equivalent currents distributed over a closed surface in space, which can be a mathematical surface or a physical boundary of a scatterer. The contribution from the equivalent sources is a sum of elementary waves radiated at every point on the boundary in the forward direction but vanishes in the backward direction (null-field property), which fully corresponds to Huygens' principle.

The result of integration of the equivalent currents is determined by whether the source and observer are located in the interior or exterior of a bounded region V, implying four possible cases (Figure 2.11). In the context of scattering problems, the observer is located in an infinite region V_{ext} (Figure 2.11b) exterior to a bounded volume V containing all scattering bodies which are considered as sources of the scattered field. In the case of a single scatterer the integration surface may even coincide with the boundary of the scatterer.

The integral representations in terms of equivalent currents are important from both theoretical and practical points of view since on the one hand they can be seen as a rigorous mathematical formulation of the Huygens principle and on the other hand permit determination of fields in the whole region once the fields are available on the boundary of the region. By letting the observation point approach the surface, these representations are transformed into integral equations, useful for numerical solution of Maxwell's equation as the dimensionality of the equations is less than that of the original problem. If the boundary field is determined approximately, which may be simpler or more accurate than approximating

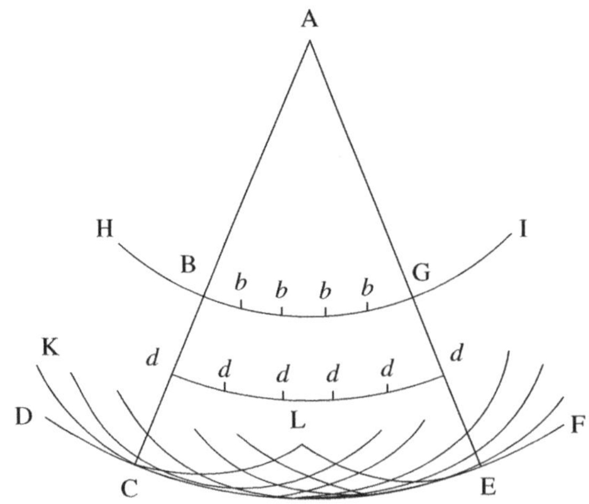

Figure 2.10 Propagation of light according to Huygens (1690, p. 22). DCEF is the wavefront of the principal wave emanating from the luminous point A. The aether particle B, one of those within the sphere DCEF, is a source of a secondary elementary wave which touches DCEF at C at the same moment that the principal wave has arrived at DCEF. Similarly, the other particles of the sphere DCEF, such as bb, dd, etc., will each make its own wave. Each of these waves contributes to the composition of DCEF by the part of its surface which is most distant from the center A.

the field in the whole region, then the integral representations give approximate representations of the fields in the region, leading, for example, to PO approximations when the boundary field is determined with the help of GO.

In what follows we derive these representations for compact (section 2.6.1) and cylindrical (section 2.6.2) scatterers. Our goal is to derive boundary integral representations in terms of the field components tangential to the integration surface. Such representations form the foundation of PO and PTD (Chapters 8 and 9).

2.6.1 Compact Scatterers

Let us derive the boundary integral representations of the fields **E** and **H** for compact regions in three dimensions. The simplest way might be to use the fact that every rectangular component of the fields satisfies a scalar Helmholtz equation (see section 2.2.3). Boundary integral relations for the scalar Helmholtz equation, derived with the help of an integral formula from the famous manuscript attributed to Green (1828), have been known since the work by Helmholtz (1860).

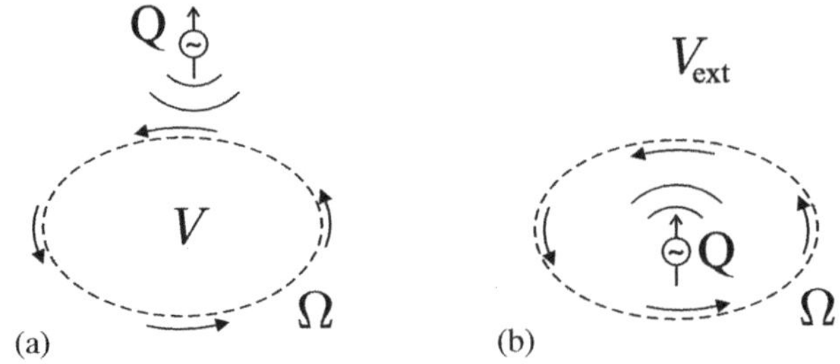

Figure 2.11 The equivalence principle. A closed surface Ω separates a bounded volume V and the rest of the space, the volume V_{ext}. (a) Source Q in V_{ext} can be replaced by equivalent currents on Ω to produce the same field in V; the currents produce null field in V_{ext}. (b) The source Q in V and equivalent boundary currents on Ω produce the same field in V_{ext}; the surface currents produce null field in V.

The Helmholtz integral formula for monochromatic fields together with Kirchhoff's extension to the time-variable fields (Kirchhoff 1883) laid the mathematical foundation for the scalar Fresnel–Kirchhoff theory of diffraction (Born and Wolf 1959; Kirchhoff 1891).

Provided that the exact boundary values of the field components are known, these representations would indeed describe solutions of Maxwell's equations in three dimensions. There are, however, significant disadvantages associated with such an approach. First, it requires the knowledge of six field components (three for **E** and three for **H** fields) over an integration surface, which is apparently superficial since according to the uniqueness theorem (section 2.4.1), the knowledge of the tangential components of the fields over a closed surface is sufficient to determine the electromagnetic field in the volume enclosed by the surface. Furthermore, the fields are typically available only approximately (from some analytical or numerical approximation), and it is therefore important to ensure that the integral representations, even with approximate boundary values of the fields, satisfy Maxwell's equations. As we discussed in section 2.2.3, the six scalar Helmholtz equations (2.49) and (2.50) are not necessarily equivalent to the original Maxwell equations.

So, it is desirable to derive boundary integral representations that are free of these disadvantages. The representations that use only the tangential components of **E** and **H** on the integration surface and automatically satisfy Maxwell's equations even with arbitrary boundary values of the fields were originally described by Love (1901) and Macdonald (1902, pp. 16, 17) and put in a modern form by Franz (1948, 1952, 1957). Following Tai (1972) we will refer to these representations as Franz' representations. An additional advantage of the formulation is that in some applications the integration surface can be chosen to be the physical surface of the scatterer and since the boundary conditions on material interfaces are formulated in terms of the tangential field components, the boundary conditions can be conveniently incorporated into the integral representations.

Another, perhaps better known, form of the boundary integral representations is due to Stratton and Chu (1939). They are in fact closely related to the Franz representations but have the disadvantages of involving the normal components of the fields under the integration sign and of changing their analytical form for surfaces with edges. We will discuss these representations a bit later, once the Franz formulas are derived. Both formulations, Franz' and Stratton-Chu's, may serve as the mathematical foundation for the electromagnetic, fully vectorial theory of scattering.

We will now derive the Franz formulas. First, following an elegant approach suggested by Lindell (1992, 1996), a mathematical formulation of the equivalence principle for Maxwell's equations in arbitrary media will be given, with equivalent currents expressed in terms of the field components tangential to the surface separating the sources and the observer. In contrast to classical formulations (Love 1901; Macdonald 1902, p. 16, 17; see also Macdonald 1911; Schelkunoff 1936; Collin 1960, section 1.7), the derivation is not limited to homogeneous isotropic media. Then, by assuming that the observer is located in a homogeneous isotropic space the fields due to the equivalent sources are obtained with the help of the free-space solutions of Maxwell's equations either in terms of the Hertz potentials and the scalar Green function (section 2.5.2) or by applying the tensor Green function (section 2.5.3), which leads to Franz' boundary integral representations.

Equivalence Principle and Huygens' Sources

Let **r** be the position vector of a point in three-dimensional space, V be a bounded region fully enclosed by a surface Ω and $\hat{n}$ be the unit vector normal to Ω and pointing to the exterior of V (Figure 2.12). The electromagnetic fields satisfy Maxwell's equations, which we write in the most general form

$$\nabla \times \mathbf{H} - j\omega \mathbf{D} = \mathbf{J}_{\mathrm{e}}, \quad \nabla \times \mathbf{E} + j\omega \mathbf{B} = -\mathbf{J}_{\mathrm{m}} \tag{2.329}$$

to account for every kind of medium, possibly inhomogeneous and/or anisotropic, and for arbitrary source current densities.

First we consider the case when the sources are located only outside of volume V (Figure 2.12, left). Let us introduce a scalar function of the position vector **r**

$$P(\mathbf{r}) = 1 \text{ if } \mathbf{r} \in V, \quad P(\mathbf{r}) = 0 \text{ if } \mathbf{r} \notin V \tag{2.330}$$

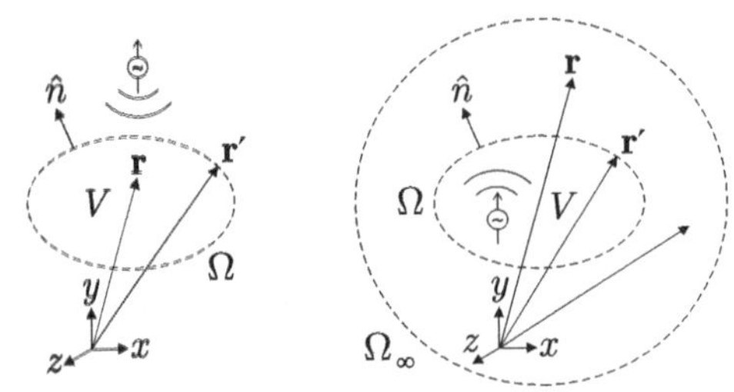

Figure 2.12 Geometries assumed in the derivation of three-dimensional boundary integral representations for the incident field (left) and the scattered or radiated field (right).

and multiply Maxwell's equations (2.329) by this function. Multiplication of the fields and sources truncates them to zero outside V and does not change them inside the volume. Multiplication of the curl terms can be made using the identity

$$P(\mathbf{r})\nabla \times \mathbf{F}(\mathbf{r}) = \nabla \times [P(\mathbf{r})\mathbf{F}(\mathbf{r})] - [\nabla P(\mathbf{r})] \times \mathbf{F}(\mathbf{r}) \tag{2.331}$$

which holds for any differentiable vector function $\mathbf{F}(\mathbf{r})$.

The gradient of the step-function can be expressed as

$$\nabla P(\mathbf{r}) = -\hat{n}(\mathbf{r})\delta_\Omega(\mathbf{r}), \tag{2.332}$$

where $\hat{n}(\mathbf{r})$ is an arbitrary smooth vector function coinciding with the unit normal $\hat{n}$ on Ω and $\delta_\Omega(\mathbf{r})$ is the Dirac function, which transforms every integral over a volume containing V to a surface integral over Ω, according to the chain of equations

$$\int f\nabla P\,\mathrm{d}V = \int \nabla(fP)\,\mathrm{d}V - \int P\nabla f\,\mathrm{d}V = -\int_V \nabla f\,\mathrm{d}V = -\int_\Omega f\hat{n}\,\mathrm{d}\Omega, \tag{2.333}$$

where $f(\mathbf{r})$ is an arbitrary differentiable function. In (2.333), Gauss' theorem (A.31) is applied twice,[11] first to evaluate the integral of $\nabla(fP)$ and then to integrate ∇f over V.

Thus, the equations for the truncated fields (marked by index V) read

$$\nabla \times \mathbf{H}_V - j\omega\mathbf{D}_V = -\delta_\Omega(\mathbf{r})\hat{n} \times \mathbf{H} \tag{2.334}$$

$$\nabla \times \mathbf{E}_V + j\omega\mathbf{B}_V = -\delta_\Omega(\mathbf{r})\hat{n} \times \mathbf{E}. \tag{2.335}$$

By construction, the solutions of these equations are equal to zero outside volume V and are equal to the solutions of the original Maxwell's equations (2.329) inside this volume. We see that the equivalent surface currents

$$\mathbf{K}_\mathrm{e}(\mathbf{r}) = -\hat{n}(\mathbf{r}) \times \mathbf{H}(\mathbf{r}), \quad \mathbf{K}_\mathrm{m}(\mathbf{r}) = -\mathbf{E}(\mathbf{r}) \times \hat{n}(\mathbf{r}) \quad (\mathbf{r} \in \Omega) \tag{2.336}$$

flowing within a sheet of infinitesimal thickness produce the same fields inside V as the original volumetric currents $\mathbf{J}_\mathrm{e}$ and $\mathbf{J}_\mathrm{m}$. In the space outside V they do not generate any fields, the property being referred to as the null-field property.

[11]The theorem assumes continuously differentiable vector fields, so strictly speaking, one should first approximate the step function P by a continuously differentiable function $\tilde{P}$ approaching P in a limit, apply the Gauss' theorem and only afterward go to the limit $\tilde{P} \to P$.

Similar representations can be derived for the exterior problem (Figure 2.12, right), in which the sources of the fields are located entirely in V and we look for solutions of Maxwell's equations in the exterior of V where the fields have to comply with conditions at infinity (2.68). For this case we define the truncation function $\mathbf{P}(\mathbf{r})$ in the complimentary way,

$$P(\mathbf{r}) = 0 \text{ if } \mathbf{r} \in V, \quad P(\mathbf{r}) = 1 \text{ if } \mathbf{r} \notin V \tag{2.337}$$

and repeat the same derivation with two modifications. The first one is due to the fact that the exterior of V is an infinite region and in order to apply Gauss' theorem one confines the integration region to a sphere Ω_∞ with a large but finite radius, and since P is not zero in the exterior of V one has to require that

$$\int_{\Omega_\infty} f \, \mathrm{d}\Omega \to 0, \tag{2.338}$$

as the radius of Ω_∞ tends to infinity. The second modification is the alternated direction of $\hat{n}$ in the formulas since the normal vector in Gauss' theorem is directed to the exterior of the integration volume. In contrast to (2.332) and (2.336), $\nabla P(\mathbf{r}) = \hat{n}\delta_\Omega(\mathbf{r})$ and

$$\mathbf{K}_\mathrm{e}(\mathbf{r}) = \hat{n}(\mathbf{r}) \times \mathbf{H}(\mathbf{r}), \quad \mathbf{K}_\mathrm{m}(\mathbf{r}) = \mathbf{E}(\mathbf{r}) \times \hat{n}(\mathbf{r}) \quad (\mathbf{r} \in \Omega). \tag{2.339}$$

Thus, rewriting Maxwell's equations for truncated fields has resulted in the field equivalence theorem, which states that the actual sources can be replaced by a system of virtual sources distributed on a closed surface separating the sources from the observer and fully enclosing either the sources or the observer. When the observer and the original sources are located in the same region, the equivalent sources produce the null field and have therefore the properties of Huygens' sources. Furthermore, if the fields produced by the original sources are $\mathbf{E}$ and $\mathbf{H}$, then the equivalent sources are electric and magnetic current sheets with the densities $\hat{n} \times \mathbf{H}$ and $\mathbf{E} \times \hat{n}$, where $\hat{n}$ is the unit normal vector directed into the region where the observer is located.

Boundary Integral Representations of Fields

If volume V is filled with an isotropic magnetodielectric (or free space), then we can make use of the free-space Green function and find the fields inside V by integrating the equivalent currents over Ω. We use the expressions for the Hertz vectors (2.201) and (2.202) where the source volumetric current densities $\mathbf{J}_\mathrm{e}$ and $\mathbf{J}_\mathrm{m}$ are replaced by the equivalent current densities $\delta_\Omega(\mathbf{r})\mathbf{K}_\mathrm{e}$ and $\delta_\Omega(\mathbf{r})\mathbf{K}_\mathrm{m}$, respectively. Substituting the Hertz vectors into the formulas (2.196) and (2.197) for the fields gives

$$\frac{j}{\omega\epsilon}[k^2 + \nabla(\nabla\cdot)] \int_\Omega G_0(\mathbf{r},\mathbf{r}')\hat{n}' \times \mathbf{H}' \mathrm{d}\Omega' - \nabla \times \int_\Omega G_0(\mathbf{r},\mathbf{r}')\hat{n}' \times \mathbf{E}' \mathrm{d}\Omega' = \begin{cases} \mathbf{E}(\mathbf{r}), & \mathbf{r} \in V \\ 0, & \mathbf{r} \notin V \end{cases}. \tag{2.340}$$

Here, $G_0(\mathbf{r},\mathbf{r}')$ is the Green function (2.199), and we have denoted for brevity $\mathbf{H}' = \mathbf{H}(\mathbf{r}')$, $\mathbf{E}' = \mathbf{E}(\mathbf{r}')$ and $\hat{n}' = \hat{n}(\mathbf{r}')$. For the magnetic field we get

$$\frac{1}{j\omega\mu}[k^2 + \nabla(\nabla\cdot)] \int_\Omega G_0(\mathbf{r},\mathbf{r}')\hat{n}' \times \mathbf{E}' \mathrm{d}\Omega' - \nabla \times \int_\Omega G_0(\mathbf{r},\mathbf{r}')\hat{n}' \times \mathbf{H}' \mathrm{d}\Omega' = \begin{cases} \mathbf{H}(\mathbf{r}), & \mathbf{r} \in V \\ 0, & \mathbf{r} \notin V \end{cases}. \tag{2.341}$$

With $\mathbf{r} \in V$, (2.340) and (2.341) represent the fields $\mathbf{E}$ and $\mathbf{H}$ in the region V which is free of field sources and filled with a homogeneous isotropic medium. Note that the shape of the boundary can be arbitrary as long as it encloses a homogeneous medium free of field sources. By construction, (2.340) and (2.341) represent exact solutions of Maxwell's equation in V, regardless of the choice of the boundary values $\hat{n}' \times \mathbf{E}'$ and $\hat{n}' \times \mathbf{H}'$ under the integration sign. This can be also directly proved by applying the curl operator (see problem 40).

In the alternative case when the sources are located entirely in V, and provided that in the exterior of V the medium is homogeneous and isotropic and the fields satisfy the conditions at infinity (2.68), integration of the equivalent current densities results in the formulas:

$$\frac{1}{j\omega\epsilon}[k^2 + \nabla(\nabla\cdot)] \int_\Omega G_0(\mathbf{r},\mathbf{r}')\hat{n}' \times \mathbf{H}' d\Omega' + \nabla \times \int_\Omega G_0(\mathbf{r},\mathbf{r}')\hat{n}' \times \mathbf{E}' d\Omega' = \begin{cases} 0, & \mathbf{r} \in V \\ \mathbf{E}(\mathbf{r}), & \mathbf{r} \notin V \end{cases} \tag{2.342}$$

$$\frac{j}{\omega\mu}[k^2 + \nabla(\nabla\cdot)] \int_\Omega G_0(\mathbf{r},\mathbf{r}')\hat{n}' \times \mathbf{E}' d\Omega' + \nabla \times \int_\Omega G_0(\mathbf{r},\mathbf{r}')\hat{n}' \times \mathbf{H}' d\Omega' = \begin{cases} 0, & \mathbf{r} \in V \\ \mathbf{H}(\mathbf{r}), & \mathbf{r} \notin V \end{cases}. \tag{2.343}$$

With $\mathbf{r} \notin V$, these formulas describe the electromagnetic fields due to sources in V and thus either the radiated fields if the sources are primary, or the scattered fields if the sources are secondary, that is, induced by incident fields on a scatterer or scatterers in V. Note that the scatterers can be arbitrary (PEC, homogeneous, inhomogeneous, bianisotropic, nonreciprocal or even nonlinear) since the only assumption made about the fields inside V is that they satisfy homogeneous Maxwell's equations (2.329). Representations (2.342) and (2.343) describe exact solutions of Maxwell's equations in the exterior of V. Their compliance with the conditions at infinity follows from the corresponding behavior of the Green function $G_0(\mathbf{r},\mathbf{r}')$ for $r \to \infty$ (this limit is studied in more detail in section 3.4.1).

Expressions (2.340), (2.341), (2.342), and (2.343) can be seen as a rigorous electromagnetic formulation of the Huygens principle: each point on a primary wavefront can be considered as a source of a secondary spherical wave and the field at any later instant is the result of superposition of the secondary waves. In this framework, the null-field property when the observation point is located in the same region as the field sources is interpreted as that the secondary sources do not radiate backwards.

The theory of electromagnetic scattering is concerned with the solutions of Maxwell's equations in the infinite region exterior to scatterers (section 2.2.3) and therefore we are mostly interested in representations of scattered fields $\mathbf{E}_{\mathrm{sc}}$ and $\mathbf{H}_{\mathrm{sc}}$. These are already given by (2.342) and (2.343) with $\mathbf{r} \notin V$. However, it is often more convenient to have under the integration sign the total fields (2.29) rather than the scattered fields since it is the total field that satisfies boundary conditions on scattering surfaces and is most often determined by measurement, numerical simulation or analytical approximation. In scattering problems, the incident fields $\mathbf{E}_{\mathrm{inc}}$ and $\mathbf{H}_{\mathrm{inc}}$ are excited by primary sources located in the exterior of V, so they are described by (2.340) and (2.341). When $\mathbf{r} \notin V$, the boundary integrals (2.340) and (2.341) for the incident fields vanish, implying that the scattered fields in the integrands of (2.342) and (2.343) can be replaced with the total fields (2.29). Thus, we can write

$$\mathbf{E}_{\mathrm{sc}}(\mathbf{r}) = \frac{1}{j\omega\epsilon}[k^2 + \nabla(\nabla\cdot)] \int_\Omega G_0(\mathbf{r},\mathbf{r}')\hat{n}' \times \mathbf{H}' d\Omega' + \nabla \times \int_\Omega G_0(\mathbf{r},\mathbf{r}')\hat{n}' \times \mathbf{E}' d\Omega' \quad (\mathbf{r} \notin V) \tag{2.344}$$

$$\mathbf{H}_{\mathrm{sc}}(\mathbf{r}) = \frac{j}{\omega\mu}[k^2 + \nabla(\nabla\cdot)] \int_\Omega G_0(\mathbf{r},\mathbf{r}')\hat{n}' \times \mathbf{E}' d\Omega' + \nabla \times \int_\Omega G_0(\mathbf{r},\mathbf{r}')\hat{n}' \times \mathbf{H}' d\Omega' \quad (\mathbf{r} \notin V), \tag{2.345}$$

where $\mathbf{E}$ and $\mathbf{H}$ under the integration sign have now the meaning of total fields.

Recalling the definition of equivalent surface currents (2.339) we rewrite (2.344) and (2.345) in a more compact form,

$$\mathbf{E}_{\mathrm{sc}}(\mathbf{r}) = \frac{1}{j\omega\epsilon}[k^2 + \nabla(\nabla\cdot)] \int_\Omega \mathbf{K}_{\mathrm{e}}(\mathbf{r}')G_0(\mathbf{r},\mathbf{r}') d\Omega' - \nabla \times \int_\Omega \mathbf{K}_{\mathrm{m}}(\mathbf{r}')G_0(\mathbf{r},\mathbf{r}') d\Omega' \quad (\mathbf{r} \notin V) \tag{2.346}$$

$$\mathbf{H}_{\mathrm{sc}}(\mathbf{r}) = \frac{1}{j\omega\mu}[k^2 + \nabla(\nabla\cdot)] \int_\Omega \mathbf{K}_{\mathrm{m}}(\mathbf{r}')G_0(\mathbf{r},\mathbf{r}') d\Omega' + \nabla \times \int_\Omega \mathbf{K}_{\mathrm{e}}(\mathbf{r}')G_0(\mathbf{r},\mathbf{r}') d\Omega' \quad (\mathbf{r} \notin V), \tag{2.347}$$

where

$$\mathbf{K}_{\mathrm{e}} = \hat{n} \times (\mathbf{H}_{\mathrm{inc}} + \mathbf{H}_{\mathrm{sc}}), \quad \mathbf{K}_{\mathrm{m}} = (\mathbf{E}_{\mathrm{inc}} + \mathbf{E}_{\mathrm{sc}}) \times \hat{n}. \tag{2.348}$$

Relations (2.344) and (2.345) or (2.346) and (2.347) are sometimes referred to as Franz' representations. They explicitly show that solutions of Maxwell's equations are determined by specification of the tangential electric and magnetic fields on a closed surface, which is in accord with the uniqueness theorem for Maxwell's equations.[12] The formulas are not limited to mathematical surfaces in space as they are equally applicable to physical boundaries of scattering bodies. In many cases, the latter choice may be the most advantageous because of the possibility of using the boundary conditions to simplify the integrals. For example, on the surface of a PEC scatterer $\mathbf{K}_{\mathrm{m}} = 0$, and (2.346) and (2.347) simplify to

$$\mathbf{E}_{\mathrm{sc}}(\mathbf{r}) = \frac{1}{j\omega\epsilon}[k^2 + \nabla(\nabla\cdot)] \int_{\Omega} \mathbf{K}_{\mathrm{e}}(\mathbf{r}')G_0(\mathbf{r},\mathbf{r}')\mathrm{d}\Omega' \tag{2.349}$$

$$\mathbf{H}_{\mathrm{sc}}(\mathbf{r}) = \nabla \times \int_{\Omega} \mathbf{K}_{\mathrm{e}}(\mathbf{r}')G_0(\mathbf{r},\mathbf{r}')\mathrm{d}\Omega'. \tag{2.350}$$

It should be noted that alternative forms of the boundary integral representations are available, most notably the Stratton–Chu integral representations (Stratton and Chu 1939). For example, for scattered fields they have the form

$$\mathbf{E}_{\mathrm{sc}}(\mathbf{r}) = \int_{\Omega} [-j\omega\mu G_0 \hat{n}' \times \mathbf{H}' + (\hat{n}' \times \mathbf{E}') \times \nabla' G_0 + (\hat{n}' \cdot \mathbf{E}')\nabla' G_0]\, \mathrm{d}\Omega' \tag{2.351}$$

$$\mathbf{H}_{\mathrm{sc}}(\mathbf{r}) = \int_{\Omega} [j\omega\epsilon G_0 \hat{n}' \times \mathbf{E}' + (\hat{n}' \times \mathbf{H}') \times \nabla' G_0 + (\hat{n}' \cdot \mathbf{H}')\nabla' G_0]\, \mathrm{d}\Omega', \tag{2.352}$$

which in addition to the tangential components includes the normal components of the fields on the integration surface. Moreover, if the scattering surface has edges, then (2.351) and (2.352) have to be modified by adding line integrals over the edges in order to make the fields $\mathbf{E}_{\mathrm{sc}}$ and $\mathbf{H}_{\mathrm{sc}}$ represented by the integrals divergence-free (Kottler 1923; Stratton and Chu 1939).

It can be checked by direct inspection that taking the curl transforms (2.351) and (2.352) to (2.344) and (2.345) (Tai 1972). An advantage of (2.344) and (2.345) is that they retain their analytical form even for surfaces with edges. For example, they can be used to formulate integral equations on the screens with apertures (Levine and Schwinger 1950) or serve as a basis for PTD, which extends PO to scatterers with edges. On the other hand, in contrast to Franz' representations, the Stratton–Chu formulas involve only first-order derivatives of the Green function, which may be of advantage in numerical implementations.

The integral formulas (2.346) and (2.347) do not solve any scattering problem, they simply replace the problem of finding the fields in the exterior of V with another problem of constructing the equivalent surface currents (2.348) on a physical or mathematical surface bounding the volume V. However, in some cases and particularly when the wavelength is short, a reasonable approximation to the surface currents can often be obtained by application of the ray optical laws, which leads to an approximate representation of the scattered fields everywhere in the exterior of V. This is the basic idea of the PO method (see Chapter 8).

2.6.2 *Cylindrical Scatterers*

The boundary integral representations for compact scatterers derived in section 2.6.1 cannot be directly applied to scatterers of infinite extent, for example to cylindrical scatterers, since the divergence theorem assumes bounded regions. For cylindrical scatterers (section 2.5.5) it is possible to develop a similar theory which is, however, in several aspects different from the theory for compact scatterers. In place of three-dimensional Maxwell's equations we have to solve two-dimensional Helmholtz equations for two unknown functions, the components E_z or H_z, if z is the Cartesian coordinate along the cylinder axis. So, we need to formulate a two-dimensional analogue of the equivalence theorem and supplement it with suitable integral representations. Historically, representations of solutions of the two-dimensional Helmholtz

[12] Strictly speaking, the knowledge of the tangential component of either electric or magnetic field is sufficient (see section 2.4.1).

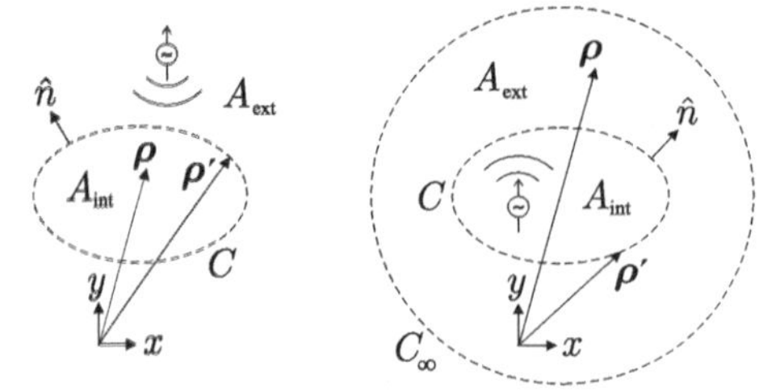

Figure 2.13 Geometries assumed in the derivation of two-dimensional equivalence theorem and boundary integral representations.

equation through line integrals are due to Weber (1869), who relied on Riemann's results on the Laplace equation derived in the framework of Riemann's theory of functions of a complex variable. The approach that follows is a modification of the approach by Lindell (1996, 1992).

As we have seen in section 2.5.5, scattering problems involving infinite scatterers with translational symmetry, like wedges and cylinders, can be reduced to two-dimensional Helmholtz equations (2.319) and (2.320). The following exposition is applicable to both polarization cases, in which we do not differentiate between the polarization cases and simply start with the equation

$$\nabla_\perp^2 U + k_\perp^2 U = 0, \tag{2.353}$$

where U represents either U_E or U_H. This function is independent of the z coordinate and is therefore a function of the position vector

$$\boldsymbol{\rho} = \hat{x}x + \hat{y}y \tag{2.354}$$

on the plane (x, y). The wave number and the two-dimensional gradient are defined by $k_\perp = \sqrt{k^2 - k_z^2}$ ($\mathrm{Im}\, k_\perp \leq 0$) and

$$\nabla_\perp = \hat{x}\frac{\partial}{\partial x} + \hat{y}\frac{\partial}{\partial y}. \tag{2.355}$$

Let A_{int} be a compact area on the (x, y) plane enclosed by a smooth or piecewise continuous contour C and A_{ext} be the area exterior to A_{int} (Figure 2.13). We start with the derivation of an integral representation for the field in A_{int} due to sources in A_{ext}. Introducing the step function

$$P(\boldsymbol{\rho}) = 1 \text{ if } \boldsymbol{\rho} \in A_{\mathrm{int}}, \quad P(\boldsymbol{\rho}) = 0 \text{ if } \boldsymbol{\rho} \notin A_{\mathrm{int}} \tag{2.356}$$

and multiplying (2.353) with $P(\boldsymbol{\rho})$ we see that the truncated function $U_P(\boldsymbol{\rho}) = P(\boldsymbol{\rho})U(\boldsymbol{\rho})$ satisfies the equation

$$\nabla_\perp^2 U_P + k_\perp^2 U_P = -f, \tag{2.357}$$

where

$$f = -2\nabla_\perp U \cdot \nabla_\perp P - U\nabla_\perp^2 P. \tag{2.358}$$

The source term f in (2.357) is everywhere zero except for the boundary of A_{int}, where it is singular. By construction this source leads to the solution of the Helmholtz equation that is equal to the original field U when $\boldsymbol{\rho} \in A_{\mathrm{int}}$ but vanishes when $\boldsymbol{\rho} \notin A_{\mathrm{int}}$, implying that f is an equivalent (Huygens') source. Thus, we have formulated a two-dimensional analogue of the equivalence principle. Note that the wave number $k_\perp$ in (2.357) can be an arbitrary function of $\boldsymbol{\rho}$.

In contrast to the case of compact scatterers, the source function includes not only the gradient of the step function but also its Laplacian. By referring to (2.332) we can formally write

$$\nabla_\perp P = -\hat{n}(\boldsymbol{\rho})\delta_C(\boldsymbol{\rho}), \quad \nabla_\perp^2 P = -\nabla_\perp \cdot [\hat{n}(\boldsymbol{\rho})\delta_C(\boldsymbol{\rho})], \tag{2.359}$$

where $\hat{n}(\boldsymbol{\rho})$ is an arbitrary smooth vector function of $\boldsymbol{\rho}$ that coincides with the outward normal at C and the delta function $\delta_C(\boldsymbol{\rho})$ acts on an arbitrary function $g(\boldsymbol{\rho})$ according to the rule

$$\int g\delta_C \, \mathrm{d}\sigma = \int_C g \, \mathrm{d}l, \tag{2.360}$$

so that

$$\int g\nabla_\perp \cdot (\delta_C\hat{n}) \, \mathrm{d}\sigma = \int [\nabla_\perp \cdot (\delta_C\hat{n}g) - \delta_C\hat{n} \cdot \nabla_\perp g] \, \mathrm{d}\sigma = -\int_{A_{\mathrm{int}}} \delta_C\hat{n} \cdot \nabla_\perp g \, \mathrm{d}\sigma = -\int_C \frac{\partial g}{\partial n} \, \mathrm{d}l \tag{2.361}$$

where $\partial g/(\partial n) = \hat{n} \cdot \nabla_\perp g$ and $\mathrm{d}l$ is the infinitesimal arc length. Thus, the source

$$f = 2\delta_C\hat{n} \cdot \nabla_\perp U + U\nabla_\perp \cdot (\hat{n}\delta_C) \tag{2.362}$$

distributed on C and the original sources located in A_{ext} produce the same field in A_{int}.

If the wave number $k_\perp$ is constant in A_{int},[13] then one obtains the integral representation of U_P by using the Green function (2.325) and integrating over the (x, y) plane

$$U_P(\boldsymbol{\rho}) = \int G_2(\boldsymbol{\rho}, \boldsymbol{\rho}', k_\perp) f(\boldsymbol{\rho}') \, \mathrm{d}\sigma' = \int_C \left\{ 2G_2(\boldsymbol{\rho}, \boldsymbol{\rho}', k_\perp) \frac{\partial}{\partial n'} U(\boldsymbol{\rho}') - \frac{\partial}{\partial n'} [G_2(\boldsymbol{\rho}, \boldsymbol{\rho}', k_\perp) U'(\boldsymbol{\rho}')] \right\} \mathrm{d}l' \tag{2.363}$$

or

$$\int_C \left[G_2(\boldsymbol{\rho}, \boldsymbol{\rho}', k_\perp) \frac{\partial}{\partial n'} U(\boldsymbol{\rho}') - U(\boldsymbol{\rho}') \frac{\partial}{\partial n'} G_2(\boldsymbol{\rho}, \boldsymbol{\rho}', k_\perp) \right] \mathrm{d}l' = \begin{cases} U(\boldsymbol{\rho}) & \boldsymbol{\rho} \in A_{\mathrm{int}} \\ 0 & \boldsymbol{\rho} \in A_{\mathrm{ext}} \end{cases}, \tag{2.364}$$

where $\hat{n}' = \hat{n}(\boldsymbol{\rho}')$ for brevity.

The formula represents the field due to secondary sources (2.362) distributed over the boundary curve C which encloses the observation point, separating it from the original sources. The superposition recovers the field of the original sources in the source-free region A_{int} but vanishes in the region A_{ext} where the original sources are located.

Similar representation of solutions of the Helmholtz equation (2.353) is possible in the complimentary case, in which all sources are located in A_{int} and the solution is sought in A_{ext} (Figure 2.13, right) where it satisfies conditions at infinity in the form of (2.322) or (2.323). Replacing $\hat{n}$ with $-\hat{n}$ in (2.362) to comply with the normal vector convention in Gauss' theorem, we obtain

$$f = -2\delta_C\hat{n} \cdot \nabla_\perp U - U\nabla_\perp \cdot (\hat{n}\delta_C) \tag{2.365}$$

for the equivalent source distribution. If the wave number $k_\perp$ is position independent, then the source distribution (2.365) produces the field

$$-\int_C \left[G_2(\boldsymbol{\rho}, \boldsymbol{\rho}', k_\perp) \frac{\partial}{\partial n'} U(\boldsymbol{\rho}') - U(\boldsymbol{\rho}') \frac{\partial}{\partial n'} G_2(\boldsymbol{\rho}, \boldsymbol{\rho}', k_\perp) \right] \mathrm{d}l' = \begin{cases} 0 & \boldsymbol{\rho} \in A_{\mathrm{int}} \\ U(\boldsymbol{\rho}) & \boldsymbol{\rho} \in A_{\mathrm{ext}} \end{cases}. \tag{2.366}$$

[13] The value of $k_\perp$ in A_{ext} does not play any role since the term $k_\perp^2 U_P$ in (2.357) is zero in A_{ext}.

To correctly apply Gauss' theorem to functions which are not identically zero in the unbounded region A_{ext}, we first integrate over a domain bounded by a circle C_∞ of a large but finite radius and then note that the contribution from the arc of C_∞ vanishes as its radius tends to infinity, provided that the fields satisfy the conditions at infinity.

With $\boldsymbol{\rho} \in A_{\text{ext}}$, (2.366) represents solutions of the homogeneous Helmholtz equation in the infinite source-free region, which comply with the conditions at infinity and are, therefore, outgoing waves as $\rho \to \infty$. The shape of the boundary C can be arbitrary as long as it encloses all the sources. From the physical viewpoint this case corresponds to the field either scattered from obstacles located in A_{int} or radiated by sources in A_{int}.

In scattering problems we are mostly interested in the field $U_{\text{sc}}(\boldsymbol{\rho})$ scattered by a scatterer or a group of scatterers located in a bounded region A_{int}. The scatterers are illuminated by an incident field $U_{\text{inc}}(\boldsymbol{\rho})$ from some given sources in A_{ext}. Thus, the incident field can be represented by the formula (2.364) and the scattered field by (2.366).

When $\boldsymbol{\rho} \in A_{\text{ext}}$, equation (2.366) represents $U_{\text{sc}}(\boldsymbol{\rho})$ in terms of the boundary values of $U_{\text{sc}}(\boldsymbol{\rho})$ and $\partial U_{\text{sc}}(\boldsymbol{\rho})/(\partial n)$. Subtracting (2.364) from (2.366) gives

$$U_{\text{sc}}(\boldsymbol{\rho}) = -\int_C \left[G_2(\boldsymbol{\rho}, \boldsymbol{\rho}', k_\perp)\frac{\partial}{\partial n'}U(\boldsymbol{\rho}') - U(\boldsymbol{\rho}')\frac{\partial}{\partial n'}G_2(\boldsymbol{\rho}, \boldsymbol{\rho}', k_\perp)\right] \mathrm{d}l', \tag{2.367}$$

where U is the total field

$$U = U_{\text{inc}} + U_{\text{sc}}. \tag{2.368}$$

The z components of the scattered fields $E_z^{\text{sc}}(x, y, z)$ and $H_z^{\text{sc}}(x, y, z)$ are obtained by multiplication of (2.367) with $\exp(-jk_z z)$, the latter being the assumed dependence of the fields on the coordinate along the axis of the infinite cylindrical body (section 2.5.5).

2.7 Problems

1. The free space impedance is defined by the relation $Z_0 = \sqrt{\mu_0/\epsilon_0}$. In many textbooks one finds that $Z_0 = 120\pi$ ohm. Is this the exact value?

   Answer: No. The exact value $Z_0 = 119.916983\pi$ ohm follows from the relation $Z_0 = \mu_0 c_0$ where c_0 is the speed of light in free space (see (2.43)).

2. Show that from the boundary condition on ideally conducting surfaces ($\mathbf{E}_t|_S = 0$) it follows that

$$\hat{n} \cdot \mathbf{H}|_S = 0 \tag{2.369}$$

   (formula (2.57)).
   Hint: Express the magnetic field vector $\mathbf{H}$ via the spatial derivatives of electric field

$$\mathbf{H} = \frac{1}{j\omega\mu_0}\nabla \times \mathbf{E} \tag{2.370}$$

   and check that the normal to the interface component of this expression contains only tangential derivatives of the tangential component of $\mathbf{E}$.

3. Prove that

$$\left.\frac{\partial \mathbf{H}_t}{\partial n}\right|_S = 0 \tag{2.371}$$

   holds on boundaries of ideally conducting bodies (formula (2.58)).
   Hint: Consider $\nabla \times \mathbf{H}$ and require that this vector has no component along the boundary surface because on the surface $\mathbf{E}_t|_S = 0$. Make use of formula (2.57) (see problem 2).

4. Show that the orthogonality condition $\mathbf{E} \cdot \mathbf{H} = 0$ holds if the fields depend only on a single spatial coordinate, let us say, on z, so that $\mathbf{E} = \mathbf{E}(z, t)$ and $\mathbf{H} = \mathbf{H}(z, t)$. Are the vectors of electric and magnetic fields $\mathbf{E}$ and $\mathbf{H}$ always orthogonal to each other?

   Answer: In general, the electric and magnetic fields are not orthogonal.

5. The Poynting vector $\mathbf{S} = \mathbf{e} \times \mathbf{h}$, where $\mathbf{e} = \mathbf{e}(\mathbf{r}, t)$ and $\mathbf{h} = \mathbf{h}(\mathbf{r}, t)$ are the time-dependent electric and magnetic field vectors, describes the intensity of energy flow at a point $\mathbf{r}$ at the instant t. In the time-harmonic case,

$$\mathbf{e}(\mathbf{r}, t) = \mathrm{Re}(\mathbf{E}e^{j\omega t}), \quad \mathbf{h}(\mathbf{r}, t) = \mathrm{Re}(\mathbf{H}e^{j\omega t}). \tag{2.372}$$

   Show that the time-average intensity of the energy flow can be expressed though the complex amplitudes $\mathbf{E}(\mathbf{r})$ and $\mathbf{H}(\mathbf{r})$ according to the formula (2.119).

   Hint: Define the time-average of a function $f(t)$ with a period $T = 2\pi/\omega$ by the formula

$$\frac{\omega}{2\pi} \int_0^{2\pi/\omega} f(t)\, dt$$

   and evaluate the average for every component in the vector product $\mathbf{e} \times \mathbf{h}$.

6. The impedance boundary condition (2.74) is sometimes written in an alternative form

$$\hat{n} \times \hat{n} \times \mathbf{E} = -\tilde{Z}_s \hat{n} \times \mathbf{H}. \tag{2.373}$$

   Show that this is an equivalent representation of the boundary conditions.

   Hint: Build a cross product of (2.74) with $\hat{n}$ and use vector relations from Appendix A.

7. Show that the impedance boundary conditions (2.74) are compliant with the duality of Maxwell's equations.

   Hint: Duality of Maxwell's equations means that these equations are invariant under the transformation (2.44).

   Answer: Representations (2.74) and (2.373) are related by the duality relations (2.44).

8. Let $\mathbf{E}$ and $\mathbf{H}$ be the electromagnetic field in the presence of a body whose surface is described by the impedance boundary condition (2.74) with the normalized equivalent surface impedance $\eta = \tilde{Z}_s/Z$ (Z is the intrinsic impedance of the medium surrounding the scatterer). Show that the field $\mathbf{E}' = Z\mathbf{H}$ and $\mathbf{H}' = -Y\mathbf{E}$ satisfies Maxwell's equations and the impedance boundary conditions with the inverse value of the normalized impedance $\eta' = 1/\eta$ (Senior 1962).

   Solution: That $\mathbf{E}'$ and $\mathbf{H}'$ are indeed solutions of Maxwell's equations can be checked by the direct substitution:

$$\nabla \times \mathbf{H}' = -Y\nabla \times \mathbf{E} = Yj\omega\mu\mathbf{H} = Y^2 j\omega\mu\mathbf{E}' = j\omega\epsilon\mathbf{E}' \tag{2.374}$$
$$\nabla \times \mathbf{E}' = Z\nabla \times \mathbf{H} = Zj\omega\epsilon\mathbf{E} = -Z^2 j\omega\epsilon\mathbf{H}' = -j\omega\mu\mathbf{H}'. \tag{2.375}$$

   In terms of the relative impedance η, the impedance boundary conditions (2.74) are expressed as

$$\hat{n} \times \mathbf{E} = \eta Z \hat{n} \times \hat{n} \times \mathbf{H}, \tag{2.376}$$

   and replacing $\mathbf{E}$ and $\mathbf{H}$ with $\mathbf{E} = -Z\mathbf{H}'$ and $\mathbf{H} = Y\mathbf{E}'$ gives

$$-Z\hat{n} \times \mathbf{H}' = \eta \hat{n} \times \hat{n} \times \mathbf{E}' \tag{2.377}$$

   which is the impedance boundary condition of the form (2.373) with the normalized equivalent surface impedance $1/\eta$.

A consequence of the property is that if $\mathbf{E}_{sc}$ and $\mathbf{H}_{sc}$ are the fields due to the incident fields $\mathbf{E}_{inc}$ and $\mathbf{H}_{inc}$ scattered by an impedance body with η, then $\mathbf{E}'_{sc} = Z\mathbf{H}_{sc}$ and $\mathbf{H}'_{sc} = -Y\mathbf{E}_{sc}$ are the fields due to the incident fields $\mathbf{E}'_{inc} = Z\mathbf{H}_{inc}$ and $\mathbf{H}'_{inc} = -Y\mathbf{E}_{inc}$ scattered by the body of the same shape but with $1/\eta$.

9. Let $\mathbf{E}_1(\mathbf{r}, \omega)$ be a solution of Maxwell's equations for a harmonic linearly polarized plane wave (2.155). Assuming that $\mathbf{E}_1(\mathbf{r}, \omega)$ is known, find a time domain solution $\mathbf{E}(\mathbf{r}, t)$ for excitation of the form

$$\mathbf{E}_{inc}(\mathbf{r}, t) = \hat{p} f(t - \hat{\imath} \cdot \mathbf{r}/c), \tag{2.378}$$

where $f(\tau)$ is an arbitrary smooth and integrable function and c is the speed of light.

Hint: Use the Fourier transform to relate time domain and frequency domain solutions of Maxwell's equations.

Answer:

$$\mathbf{E}(\mathbf{r}, t) = \int_{-\infty}^{+\infty} e^{j\omega t} \frac{F(\omega)}{E_0} \mathbf{E}_1(\mathbf{r}, \omega) d\omega \tag{2.379}$$

with

$$F(\omega) = \frac{1}{2\pi} \int_{-\infty}^{+\infty} e^{-j\omega t} f(t) dt. \tag{2.380}$$

10. Show that the orthogonality of vectors $\mathbf{E}_{pw}$ and $\hat{\imath}$ in a plane wave (2.155) and $\mathbf{H}_{pw}$ and $\hat{\imath}$ in (2.156) results from the fact that the fields are divergence free in a source-free medium.

Hint: Evaluate the divergence of $\mathbf{E}_{pw}$ and $\mathbf{H}_{pw}$ by using the relation (A.20) with $\mathbf{A}$ being a constant vector.

11. Prove by direct substitution that the plane wave (2.155) is indeed a solution of the vector Helmholtz equation (2.49).

12. A plane wave (2.155) with complex-valued amplitude $\mathbf{E}_0$ is referred to as elliptically polarized. Show that in any plane perpendicular to the direction of propagation the end point of the electric vector lies on an ellipse.

Hint: Write down expressions for the Cartesian components of the physical field $\mathbf{e}(\mathbf{r}, t)$ from (2.372) in the plane perpendicular to $\hat{\imath}$ and exclude the parameter $\tau = \omega t - k\hat{\imath} \cdot \mathbf{r}$ from these two equations.

Solution: See, for example, Born and Wolf (1959, section 1.4.2).

13. An important class of solutions of Maxwell's equations which is closely related to the plane waves are *inhomogeneous* waves, given by the relations (2.155) and (2.157) where the wave vector is complex-valued ($\mathbf{k} = \mathbf{k}' - j\mathbf{k}''$). Examples are surface waves over material interfaces, waves refracted in a lossy medium and waves transmitted through a material interface under the conditions of total reflection. In contrast to plane waves, in inhomogeneous waves the surfaces of constant phase ($\mathbf{k}' \cdot \mathbf{r} = \text{const}$) and of constant amplitude ($\mathbf{k}'' \cdot \mathbf{r} = \text{const}$), though still being planar, in general do not coincide.

Prove that the inhomogeneous wave satisfies Maxwell's equations if the wave vector is such that

$$\mathbf{k} \cdot \mathbf{k} = \omega^2 \epsilon \mu \tag{2.381}$$

$$\mathbf{k} \cdot \mathbf{E}_{iw} = 0 \tag{2.382}$$

$$\mathbf{k} \cdot \mathbf{H}_{iw} = 0. \tag{2.383}$$

14. Find the relations between the electric and magnetic fields in an inhomogeneous plane wave.

Answer: They have the form

$$\omega \mu \mathbf{H}_{iw} = \mathbf{k} \times \mathbf{E}_{iw} \tag{2.384}$$

$$-\omega \epsilon \mathbf{E}_{iw} = \mathbf{k} \times \mathbf{H}_{iw}. \tag{2.385}$$

15. Are equations (2.160) and (2.161) for the mean intensity $\mathbf{w}$ of the energy flow applicable to the case of an inhomogeneous plane wave?

Answer: In general, no. One has to use the general definition (2.119). However, if at least one of the following conditions is satisfied (a) $\mathbf{k} = k\hat{\imath}$ with $\hat{\imath}$ being a real-valued vector or (b) the wave is linearly polarized ($\mathbf{E}_0 = E_0\hat{p}$ with real-valued vector $\hat{p}$), then simplifications are possible and the following formulas can be derived:

$$\mathbf{w}_{\mathrm{iw}} = |\mathbf{E}_{\mathrm{iw}}|^2 \mathrm{Re}\left(\frac{\mathbf{k}}{2\omega\mu}\right) = |\mathbf{H}_{\mathrm{iw}}|^2 \mathrm{Re}\left(\frac{\mathbf{k}}{2\omega\epsilon}\right). \tag{2.386}$$

Solution: By using (2.384) and the formula (A.12) for the vector triple product, (2.119) can be expressed as

$$\mathbf{w}_{\mathrm{iw}} = \frac{1}{2}\mathrm{Re}\left\{\frac{1}{\omega\mu^*}\left[\mathbf{k}^*|\mathbf{E}_{\mathrm{iw}}|^2 - \mathbf{E}^*_{\mathrm{iw}}\left(\mathbf{k}^*\cdot\mathbf{E}_{\mathrm{iw}}\right)\right]\right\}. \tag{2.387}$$

The property (2.382) does not imply in general that $\mathbf{k}^*\cdot\mathbf{E}_{\mathrm{iw}} = 0$. However, when the conditions (a) and/or (b) are true, the latter scalar product vanishes because of (2.382) and (2.387) simplifies to (2.386). In case (a), the formulas in (2.386) reduce to (2.160) and (2.161).

16. Prove that the spherical wave (2.163) satisfies the radiation condition (2.64).

Solution: Using the relations of vector algebra (Appendix A), one finds that

$$\nabla\times\mathbf{E}_{\mathrm{sph}} = \left(\nabla\frac{\mathrm{e}^{-jkr}}{kr}\right)\times\mathbf{F} + \frac{\mathrm{e}^{-jkr}}{kr}\nabla\times\mathbf{F} \tag{2.388}$$

which, upon retaining the leading term with respect to r, gives

$$\nabla\times\mathbf{E}_{\mathrm{sph}} \approx -jk\hat{r}\times\mathbf{E}_{\mathrm{sph}}. \tag{2.389}$$

Next, using the formula for the double vector product and going to the limit $r\to\infty$ we obtain

$$\hat{r}\times(\nabla\times\mathbf{E}_{\mathrm{sph}}) \approx -jk[\hat{r}(\hat{r}\cdot\mathbf{E}_{\mathrm{sph}}) - (\hat{r}\cdot\hat{r})\mathbf{E}_{\mathrm{sph}}] \to jk\mathbf{E}_{\mathrm{sph}}, \tag{2.390}$$

since $\hat{r}\cdot\hat{r} = 1$ and $\hat{r}\cdot\mathbf{E}_{\mathrm{sph}} \to 0$. The correction terms that we have omitted in (2.390) decay as r^{-2} or faster. Thus, the condition (2.64) is satisfied.

17. Assume that $\mathbf{F}_0(\theta,\phi) = \hat{\phi}\sin\theta$ in the expansion (2.167). Derive $\mathbf{F}_1(\theta,\phi)$ from $\mathbf{F}_0(\theta,\phi)$.

Hint: Exclude the magnetic field from Maxwell's equations, use the expression for the curl operator in spherical coordinates from Appendix A.3, and write down the equations for the electric field and consider them when $kr\to\infty$.

18. Calculate the energy flows transported by the fields radiated from elementary electric and magnetic dipoles. Show that the terms $O(r^{-2})$ and $O(r^{-3})$ in expressions (2.207), (2.208), (2.211), and (2.212) do not contribute to the energy flow.

19. Calculate the total power radiated by an elementary electric dipole p_{e}.

Hint: In view of the solution to the previous problem, it is enough to calculate the integral (2.120) keeping only the far-field terms $O(r^{-1})$ in expressions (2.207) and (2.208).

Answer:

$$P = \frac{\mu\omega^4}{12\pi c}|p_{\mathrm{e}}|^2.$$

20. Derive expressions (2.207) and (2.208) for the components of the field of an electric dipole by inserting the Hertz vectors, which are given by (2.204) and (2.205), into the general formulas (2.191) and (2.192).

21. Consider an electric dipole located at the origin and directed along the z axis of the spherical coordinate system. Starting from (2.259) and (2.260), derive expressions for the field components in the spherical coordinates and compare the result with (2.207) and (2.208).

Hint: Use $\hat{c} = \hat{z}$, $G_0 = \exp(-jkr)/(4\pi r)$ and the relations (A.50)–(A.53).

22. Show that the two-dimensional Green function (2.325) can be represented by the Fourier expansion

$$G_2(x, x', y, y', k) = \frac{1}{4\pi j} \int_{-\infty}^{+\infty} e^{-j\xi(x-x')-j\gamma(\xi)|y-y'|} \frac{d\xi}{\gamma(\xi)} \tag{2.391}$$

with $\gamma(\xi) = \sqrt{k^2 - \xi^2}$ and the integration contour located entirely on the sheet of Riemann surface where $\mathrm{Im}\,\gamma(\xi) < 0$.

Solution: Function $G_2(x, x', y, y', k)$ is defined for $-\infty < x < +\infty$, exponentially decaying as $x \to \pm\infty$ if $\mathrm{Im}\,k < 0$ (which is always the case in any physical medium). So it can represented by the Fourier integral

$$G_2(x, x', y, y', k) = \int_{-\infty}^{+\infty} e^{-j\xi x} \tilde{G}_2(\xi, x', y, y', k)\, d\xi, \tag{2.392}$$

and inserting this in the Helmholtz equation (2.327) gives

$$\int_{-\infty}^{+\infty} e^{-j\xi x} \left[\frac{\partial^2}{\partial y^2} + k^2 - \xi^2\right] \tilde{G}_2(\xi, x', y, y', k)\, d\xi = -\delta(x - x')\delta(y - y'). \tag{2.393}$$

Next we multiply (2.393) with $\exp(j\xi' x)$ and integrate over the range $-\infty < x < +\infty$, and using the orthogonality relation

$$\int_{-\infty}^{+\infty} e^{jx(\xi'-\xi)}\, dx = 2\pi\delta(\xi - \xi') \tag{2.394}$$

arrive at the differential equation for the the spectral amplitude

$$2\pi \left[\frac{\partial^2}{\partial y^2} + k^2 - \xi'^2\right] \tilde{G}_2(\xi', x', y, y', k) = -e^{j\xi' x'}\delta(y - y'), \tag{2.395}$$

which we rewrite in the form

$$\left[\frac{\partial^2}{\partial y^2} + \gamma(\xi)^2\right] \tilde{G}_2(\xi, x', y, y', k) = -\frac{1}{2\pi} e^{j\xi x'}\delta(y - y'). \tag{2.396}$$

The solution of this equation, which vanishes at $y \to \pm\infty$ and has the discontinuity at $y = y'$ required by the presence of the delta function in the right-hand side of (2.396),

$$\tilde{G}_2(\xi, x', y' + 0, y', k) - \tilde{G}_2(\xi, x', y' - 0, y', k) = 0 \tag{2.397}$$

$$\frac{\partial \tilde{G}_2}{\partial y}(\xi, x', y' + 0, y', k) - \frac{\partial \tilde{G}_2}{\partial y}(\xi, x', y' - 0, y', k) = -\frac{1}{2\pi} e^{j\xi x'}, \tag{2.398}$$

can be represented as

$$\tilde{G}_2(\xi, x', y, y', k) = -\frac{1}{2\pi C} e^{j\xi x'} \begin{cases} \exp[j\gamma(\xi)(y' - y)], & y \ge y' \\ \exp[j\gamma(\xi)(y - y')], & y \le y' \end{cases}. \tag{2.399}$$

With $\mathrm{Im}\,\gamma(\xi) < 0$ and arbitrary coefficient C, (2.399) vanishes at $y \to \pm\infty$ and satisfies (2.397). Choosing $C = -2j\gamma(\xi)$ satisfies the condition (2.398), and (2.399) leads to the expression

$$\tilde{G}_2(\xi, x', y, y', k) = \frac{1}{4\pi j\gamma(\xi)} e^{j\xi x' - j\gamma(\xi)|y - y'|}. \tag{2.400}$$

23. Show that function

$$\psi = \frac{e^{-jkr}}{r} \tag{2.401}$$

is the solution to the scalar three-dimensional Helmholtz equation

$$\nabla^2\psi + k^2\psi = -4\pi\delta(\mathbf{r}), \tag{2.402}$$

where $\delta(\mathbf{r})$ is the three-dimensional delta function.

Hint: First, prove that $\psi = C\exp(-jkr)/r$ with an arbitrary constant C satisfies the homogeneous equation, evaluating $\nabla^2\psi$ in the spherical coordinate system (see (A.54)). Next, integrating over a small sphere around the origin, prove that the constant $C = 1$.

24. Referring to problem 23, prove the integral representation

$$\psi = \frac{1}{2\pi j} \int_{-\infty}^{+\infty}\int_{-\infty}^{+\infty} \frac{1}{\gamma} e^{-jk_x x - jk_y y - j\gamma|z|} dk_x dk_y, \tag{2.403}$$

where $\gamma = \sqrt{k^2 - k_x^2 - k_y^2}$ with $\mathrm{Im}\,\gamma \le 0$.

Hint: Apply Fourier transforms with respect to x and y to equation (2.402). The condition $\mathrm{Im}\,\gamma \le 0$ is necessary to ensure compliance of the solution with the conditions at infinity.

25. Prove the integral representation

$$\psi = -j \int_{0}^{+\infty} \frac{1}{\gamma} J_0(k_t\rho) e^{-j\gamma|z|} k_t dk_t, \tag{2.404}$$

where $J_0(k_t\rho)$ is the Bessel function and $\gamma = \sqrt{k^2 - k_t^2}$ with $\mathrm{Im}\,\gamma \le 0$.

Solution: Changing the integration variables in (2.403) according to $k_x = k_t \cos\chi$ and $k_y = k_t \sin\chi$ with $0 \le k_t < +\infty$ and $0 \le \chi < 2\pi$ gives

$$\psi = \frac{1}{2\pi j} \int_{0}^{+\infty}\int_{0}^{2\pi} \frac{1}{\gamma} e^{-jk_t\rho\cos(\chi - \phi) - j\gamma|z|} k_t\, dk_t d\chi, \tag{2.405}$$

where $x = \rho\cos\phi$ and $y = \rho\sin\phi$. The integration with respect to χ is performed in an explicit form by using the relation

$$\int_{0}^{2\pi} \mathrm{e}^{-jk_t\rho\cos(\chi-\phi)}\,\mathrm{d}\chi = 2\pi J_0(k_t\rho), \tag{2.406}$$

which transforms (2.405) to (2.404).

26. Prove the integral representation

$$\psi = \frac{1}{2j}\int_{\infty e^{-j\pi}}^{+\infty} \frac{1}{\gamma} H_0^{(2)}(k_t\rho)\mathrm{e}^{-j\gamma|z|}k_t\mathrm{d}k_t, \tag{2.407}$$

where $H_0^{(2)}(k_t\rho)$ is the Hankel function and $\gamma = \sqrt{k^2 - k_t^2}$ with $\mathrm{Im}\,\gamma \le 0$. The integration is performed in the complex k_t plane and the integration path goes from $-\infty$ to $+\infty$ below the branch cut $-\infty < k_t \le 0$ of the Hankel function.

Solution: Start with (2.404) and express the Bessel function of the first kind in terms of Hankel functions as

$$2J_0(k_t\rho) = H_0^{(1)}(k_t\rho) + H_0^{(2)}(k_t\rho). \tag{2.408}$$

Using the property

$$H_0^{(1)}(k_t\rho) = -H_0^{(2)}(k_t\rho\mathrm{e}^{-j\pi}), \tag{2.409}$$

rewrite the formula (2.404) as

$$\psi = \frac{1}{2j}\int_{0}^{+\infty} \frac{1}{\gamma} H_0^{(2)}(k_t\rho)\mathrm{e}^{-j\gamma|z|}k_t\mathrm{d}k_t - \frac{1}{2j}\int_{0}^{+\infty} \frac{1}{\gamma} H_0^{(2)}(k_t\rho\mathrm{e}^{-j\pi})\mathrm{e}^{-j\gamma|z|}k_t\mathrm{d}k_t. \tag{2.410}$$

Now choosing $k_t\mathrm{e}^{-j\pi}$ as the new integration variable in the second integral in the right-hand side of (2.410) transforms (2.410) to the form (2.407).

27. Express the Green function $G_0(\mathbf{r},\mathbf{r}_0)$ from (2.199) as the Fourier integral with respect to coordinates x and y.

Hint: Compare $G_0(\mathbf{r},\mathbf{r}_0)$ with ψ from (2.401).

Answer:

$$G_0(\mathbf{r},\mathbf{r}_0) = \frac{1}{8\pi^2 j}\int_{-\infty}^{+\infty}\int_{-\infty}^{+\infty} \frac{1}{\gamma}\mathrm{e}^{-jk_x(x-x_0)-jk_y(y-y_0)-j\gamma|z-z_0|}\mathrm{d}k_x\mathrm{d}k_y, \tag{2.411}$$

where $\mathbf{r} = (x,y,z)$, $\mathbf{r}_0 = (x_0,y_0,z_0)$, and $\gamma = \sqrt{k^2 - k_x^2 - k_y^2}$ with $\mathrm{Im}\,\gamma \le 0$. The expansion (2.411) can be interpreted as a superposition of plane waves. This kind of representation is useful in problems with material boundaries as the corresponding Green function can be obtained from (2.411) by simply multiplying the plane waves with corresponding reflection or transmission coefficients of the boundaries.

28. Prove the relation

$$\int_{-\infty}^{+\infty} \mathrm{e}^{jk_z z}G_0(\mathbf{r},\mathbf{r}')\,\mathrm{d}z = \mathrm{e}^{jk_z z'}G_2(x,x',y,y',k_\perp), \tag{2.412}$$

where k_z is a real-valued parameter ($-\infty < k_z < \infty$), $G_0(\mathbf{r}, \mathbf{r}')$ is the three-dimensional Green function (2.199) and $G_2(x, x', y, y', k_\perp)$ is the two-dimensional Green function (2.325) with the wavenumber $k_\perp = \sqrt{k^2 - k_z^2}$ ($\mathrm{Im}\, k_\perp \leq 0$).

Hint: The integral in the left-hand side of (2.412) can be evaluated directly by substituting $z - z' = q \sinh \alpha$, where $q = \sqrt{(x - x')^2 + (y - y')^2}$ and α is the new integration variable, and reducing the integral to an integral representation of the Hankel function (Ryshik and Gradstein 1957, section 6.431)

$$H_0^{(2)}(w) = \frac{j}{\pi} \int_{-\infty}^{+\infty} \mathrm{e}^{-jw \cosh \alpha}\, \mathrm{d}\alpha \quad (w > 0). \tag{2.413}$$

However, a more elegant way of proving (2.412) consists of multiplying (2.200) with $\exp(jk_z z)$ and integrating with respect to z.

Solution: Denote the integral on the left hand-side of (2.412) by I for brevity. Multiplying (2.200) with $\exp(jk_z z)$ and integrating with respect to z gives

$$\left(\nabla_{xy}^2 + k^2\right) I + \int_{-\infty}^{+\infty} \mathrm{e}^{jk_z z} \frac{\partial^2}{\partial z^2} G_0(\mathbf{r}, \mathbf{r}')\, \mathrm{d}z = -\mathrm{e}^{jk_z z'} \delta(x - x') \delta(y - y'), \tag{2.414}$$

and integrating by parts twice in the integral in (2.414) leads to the formula

$$\left(\nabla_{xy}^2 + k^2 - k_z^2\right) I = -\mathrm{e}^{jk_z z'} \delta(x - x') \delta(y - y'). \tag{2.415}$$

Apart from the exponential factor on the right-hand side, this is equation (2.327) for the the two-dimensional Green function $G_2(x, x', y, y', k_\perp)$, and since I decays at $x \to \pm\infty$ and $y \to \pm\infty$ (as an integral of a function decaying at $x \to \pm\infty$ and $y \to \pm\infty$), we obtain

$$I = \mathrm{e}^{jk_z z'} G_2(x, x', y, y', k_\perp). \tag{2.416}$$

29. Prove the integral representation

$$G_0(\mathbf{r}, \mathbf{r}') = \frac{1}{2\pi} \int_{-\infty}^{+\infty} \mathrm{e}^{jk_z(z' - z)} G_2(x, x', y, y', k_\perp)\, \mathrm{d}k_z, \tag{2.417}$$

where $G_2(x, x', y, y', k_\perp)$ is defined in problem 28.

Answer: Representation (2.417) can be seen as a Fourier expansion with respect to z, and the Fourier transform of $G_0(\mathbf{r}, \mathbf{r}')$ with respect to z is given by (2.412).

30. Let (ρ, φ, z) be circular cylinder coordinates. Is it possible to express the fields in free space through the Hertz vectors of the form $\mathbf{\Pi}_\mathrm{e} = \Pi_\rho^\mathrm{e} \hat{\rho}$ and $\mathbf{\Pi}_\mathrm{m} = \Pi_\rho^\mathrm{m} \hat{\rho}$?

Answer: No, since the scale factors (see (A.39)) do not comply with the requirements in (2.266).

31. A rectangular waveguide with perfectly conducting walls at $x = 0, a$ and $y = 0, b$ is aligned with the z axis of the Cartesian coordinate system (x, y, z). Show that the Hertz potential of the form

$$\Pi_z^\mathrm{e} = \mathrm{e}^{\pm \Gamma_{nm} z} \sin\left(n \frac{\pi}{a} x\right) \sin\left(m \frac{\pi}{b} y\right) \tag{2.418}$$

describes an E mode and

$$\Pi_z^\mathrm{m} = \mathrm{e}^{\pm \Gamma_{nm} z} \cos\left(n \frac{\pi}{a} x\right) \cos\left(m \frac{\pi}{b} y\right) \tag{2.419}$$

an H mode of the waveguide if

$$\Gamma_{nm}^2 = \left(n\frac{\pi}{a}\right)^2 + \left(m\frac{\pi}{b}\right)^2 + k^2 \tag{2.420}$$

with n and m being integer numbers.

Hint: Specialize equations (2.267)–(2.272) to the waveguide geometry and check compliance of the potentials with the differential equation and the boundary conditions.

32. Consider a circular cylindrical waveguide of radius a with a perfectly conducting wall. Derive boundary conditions satisfied by the electric and magnetic Hertz potentials on the wall of the waveguide.

Hint: Using the circular cylinder coordinates (ρ, ϕ, z) with the z axis as the axis of the waveguide, equations (2.267) and (2.271) for the electric fields, and (A.39) for the scale factors, write down expressions for the ϕ and z components of the electric field and impose the boundary condition on the wall of the waveguide.

Answer: The electric Hertz potential and the normal derivative of the magnetic Hertz potentials vanish on the wall.

33. Prove the relations

$$\int_0^{2\pi} e^{-jw\cos(\chi-\phi)} \cos\chi \, d\chi = -2\pi j \cos\phi J_1(w) \tag{2.421}$$

$$\int_0^{2\pi} e^{-jw\cos(\chi-\phi)} \sin\chi \, d\chi = -2\pi j \sin\phi J_1(w), \tag{2.422}$$

where $J_1(w)$ is the Bessel function.

Hint: By changing the integration variable, reduce the integrals to the integral representation of the Bessel function

$$J_1(w) = \frac{j}{2\pi} \int_0^{2\pi} e^{-jw\cos\chi} \cos\chi \, d\chi. \tag{2.423}$$

34. Let an elementary electric dipole with the current density

$$\mathbf{J}_e = j\omega p_e(\hat{x}\cos\phi_0 + \hat{y}\sin\phi_0)\delta(\mathbf{r} - \mathbf{r}_0) \tag{2.424}$$

be located in a homogeneous isotropic medium at point $\mathbf{r}_0 = (x_0, y_0, z_0)$. The angle ϕ_0 $(0 \le \phi_0 < 2\pi)$ defines the orientation of the dipole in the xOy plane. Determine the sources of the E and H modes defined with respect to the z axis in Cartesian coordinates (x, y, z).

Hint: Determine functions Q and M by solving Poisson's equations (2.287) and (2.288).

Solution: In Cartesian coordinates, equations (2.287) and (2.288) take the form

$$\nabla_{xy}^2 Q = \frac{\partial}{\partial x} J_x + \frac{\partial}{\partial y} J_y \tag{2.425}$$

$$\nabla_{xy}^2 M = \frac{\partial}{\partial y} J_x - \frac{\partial}{\partial x} J_y, \tag{2.426}$$

implying that Q and M can be represented as

$$Q = j\omega p_e \delta(z - z_0)(\psi_1 \cos\phi_0 + \psi_2 \sin\phi_0) \tag{2.427}$$

$$M = j\omega p_e \delta(z - z_0)(-\psi_1 \sin\phi_0 + \psi_2 \cos\phi_0), \tag{2.428}$$

where functions ψ_1 and ψ_2 are solutions of the equations:

$$\nabla^2_{xy}\psi_1 = \delta(y - y_0)\frac{\partial}{\partial x}\delta(x - x_0) \tag{2.429}$$

$$\nabla^2_{xy}\psi_2 = \delta(x - x_0)\frac{\partial}{\partial y}\delta(y - y_0). \tag{2.430}$$

The function

$$g_2(x, y, x_0, y_0) = -\frac{1}{2\pi}\ln\sqrt{(x - x_0)^2 + (y - y_0)^2} \tag{2.431}$$

is the Green function of the two-dimensional Laplace equation

$$\nabla^2_{xy}g_2 = -\delta(x - x_0)\delta(y - y_0) \tag{2.432}$$

(e.g., see Morse and Feshbach (1953, p. 891)), and functions ψ_1 and ψ_2 are obtained by convolving the Green function with the right-hand sides of equations (2.429) and (2.430), which gives

$$\psi_1 = -\int_{-\infty}^{+\infty}\int_{-\infty}^{+\infty} g_2(x, y, x', y')\delta(y' - y_0)\frac{\partial}{\partial x'}\delta(x' - x_0)\,\mathrm{d}x'\mathrm{d}y' = \frac{1}{2\pi}\frac{x - x_0}{(x - x_0)^2 + (y - y_0)^2} \tag{2.433}$$

$$\psi_2 = -\int_{-\infty}^{+\infty}\int_{-\infty}^{+\infty} g_2(x, y, x', y')\delta(x' - x_0)\frac{\partial}{\partial y'}\delta(y' - y_0)\,\mathrm{d}x'\mathrm{d}y' = \frac{1}{2\pi}\frac{y - y_0}{(x - x_0)^2 + (y - y_0)^2}. \tag{2.434}$$

The portions of the current $\mathbf{J}_e$ which are responsible for the excitation of the E and H modes follow from (2.277) and (2.280) as

$$\mathbf{J}_E = \frac{\partial Q}{\partial x}\hat{x} + \frac{\partial Q}{\partial y}\hat{y} \tag{2.435}$$

$$\mathbf{J}_H = \frac{\partial M}{\partial y}\hat{x} - \frac{\partial M}{\partial x}\hat{y}. \tag{2.436}$$

35. Construct the Hertz potentials $\mathbf{\Pi}_e = \Pi_e\hat{z}$ and $\mathbf{\Pi}_m = \Pi_m\hat{z}$ for the elementary electric dipole (2.424).
Hint: Work in Cartesian coordinates (x, y, z) and use the results from problem 34.
Solution: Differential equations (2.279) and (2.283) satisfied by Π_e and $\widetilde{\Pi}_m$ in Cartesian coordinates and for the source (2.424) have the form

$$\nabla^2\Pi_e + k^2\Pi_e = -\frac{j}{\omega\epsilon}\frac{\partial Q}{\partial z} \tag{2.437}$$

$$\nabla^2\widetilde{\Pi}_m + k^2\widetilde{\Pi}_m = -M, \tag{2.438}$$

and their solutions can be expressed as the convolutions with the Green function of the Helmholtz equation (see (2.199)),

$$\Pi_e(\mathbf{r}, \mathbf{r}_0) = \frac{j}{\omega\epsilon}\int_{-\infty}^{+\infty}\int_{-\infty}^{+\infty}\int_{-\infty}^{+\infty} G_0(\mathbf{r}, \mathbf{r}')\frac{\partial}{\partial z'}Q(\mathbf{r}', \mathbf{r}_0)\,\mathrm{d}x'\mathrm{d}y'\mathrm{d}z' \tag{2.439}$$

$$\widetilde{\Pi}_m(\mathbf{r}, \mathbf{r}_0) = \int_{-\infty}^{+\infty}\int_{-\infty}^{+\infty}\int_{-\infty}^{+\infty} G_0(\mathbf{r}, \mathbf{r}')M(\mathbf{r}', \mathbf{r}_0)\,\mathrm{d}x'\mathrm{d}y'\mathrm{d}z'. \tag{2.440}$$

Functions Q and M are determined in problem 34, and inserting them in (2.439) and (2.440) leads to the formulas:

$$\Pi_e(\mathbf{r},\mathbf{r}_0) = \frac{p_e}{2\pi\epsilon}\frac{\partial}{\partial z_0}\int_{-\infty}^{+\infty}\int_{-\infty}^{+\infty}\int_{-\infty}^{+\infty} G_0(\mathbf{r},\mathbf{r}')\delta(z'-z_0)\frac{(x'-x_0)\cos\phi_0 + (y'-y_0)\sin\phi_0}{(x'-x_0)^2+(y'-y_0)^2}\,\mathrm{d}x'\mathrm{d}y'\mathrm{d}z' \tag{2.441}$$

$$\widetilde{\Pi}_m(\mathbf{r},\mathbf{r}_0) = \frac{j\omega p_e}{2\pi}\int_{-\infty}^{+\infty}\int_{-\infty}^{+\infty}\int_{-\infty}^{+\infty} G_0(\mathbf{r},\mathbf{r}')\delta(z'-z_0)\frac{-(x'-x_0)\sin\phi_0 + (y'-y_0)\cos\phi_0}{(x'-x_0)^2+(y'-y_0)^2}\,\mathrm{d}x'\mathrm{d}y'\mathrm{d}z'. \tag{2.442}$$

In obtaining (2.441), the relation

$$\frac{\partial}{\partial z'}\delta(z'-z_0) = -\frac{\partial}{\partial z_0}\delta(z'-z_0) \tag{2.443}$$

has been used.

Using the plane wave expansion (2.411) of the Green function and integrating in (2.441) and (2.442) with respect to z' yield the representations:

$$\Pi_e(\mathbf{r},\mathbf{r}_0) = -\frac{jp_e}{16\pi^3\epsilon}\frac{\partial}{\partial z_0}(I_x\cos\phi_0 + I_y\sin\phi_0) \tag{2.444}$$

$$\widetilde{\Pi}_m(\mathbf{r},\mathbf{r}_0) = \frac{\omega p_e}{16\pi^3}(-I_x\sin\phi_0 + I_y\cos\phi_0), \tag{2.445}$$

where

$$I_x = \int_{-\infty}^{+\infty}\cdots\int_{-\infty}^{+\infty}\frac{(x'-x_0)\mathrm{e}^{-jk_x(x-x')-jk_y(y-y')-j\gamma|z-z_0|}}{\gamma[(x'-x_0)^2+(y'-y_0)^2]}\,\mathrm{d}k_x\mathrm{d}k_y\mathrm{d}x'\mathrm{d}y' \tag{2.446}$$

$$I_y = \int_{-\infty}^{+\infty}\cdots\int_{-\infty}^{+\infty}\frac{(y'-y_0)\mathrm{e}^{-jk_x(x-x')-jk_y(y-y')-j\gamma|z-z_0|}}{\gamma[(x'-x_0)^2+(y'-y_0)^2]}\,\mathrm{d}k_x\mathrm{d}k_y\mathrm{d}x'\mathrm{d}y'. \tag{2.447}$$

Expressions for I_x and I_y can be simplified by integrating with respect to x' and y' in succession. For example, using the result

$$\int_{-\infty}^{+\infty}\frac{\mathrm{e}^{jk_y y'}}{(x'-x_0)^2+(y'-y_0)^2}\,\mathrm{d}y' = \frac{\pi}{|x'-x_0|}\mathrm{e}^{jk_y y_0 - |k_y||x'-x_0|}, \tag{2.448}$$

which is obtained by integration in the complex y' plane as a residue at a pole $y' = y_0 \pm j|x'-x_0|$, followed by integration with respect to x', one obtains

$$I_x = 2\pi j\int_{-\infty}^{+\infty}\int_{-\infty}^{+\infty}\mathrm{e}^{-jk_x(x-x_0)-jk_y(y-y_0)-j\gamma|z-z_0|}\frac{k_x}{\gamma(k_x^2+k_y^2)}\,\mathrm{d}k_x\mathrm{d}k_y \tag{2.449}$$

and using the apparent symmetry between I_x and I_y,

$$I_y = 2\pi j\int_{-\infty}^{+\infty}\int_{-\infty}^{+\infty}\mathrm{e}^{-jk_x(x-x_0)-jk_y(y-y_0)-j\gamma|z-z_0|}\frac{k_y}{\gamma(k_x^2+k_y^2)}\,\mathrm{d}k_x\mathrm{d}k_y. \tag{2.450}$$

Now introducing the new integration variables $k_x = k_t \cos\chi$ and $k_y = k_t \sin\chi$ transform (2.449) and (2.450) to

$$I_x = 2\pi j \int_0^{+\infty} \mathrm{d}k_t \frac{\mathrm{e}^{-j\gamma|z-z_0|}}{\gamma} \int_0^{2\pi} \mathrm{d}\chi \cos\chi \mathrm{e}^{-jk_t\cos\chi(x-x_0)-jk_t\sin\chi(y-y_0)} \tag{2.451}$$

$$I_y = 2\pi j \int_0^{+\infty} \mathrm{d}k_t \frac{\mathrm{e}^{-j\gamma|z-z_0|}}{\gamma} \int_0^{2\pi} \mathrm{d}\chi \sin\chi \mathrm{e}^{-jk_t\cos\chi(x-x_0)-jk_t\sin\chi(y-y_0)}. \tag{2.452}$$

The coordinate system can be always chosen such that the dipole is located at the z axis, $x_0 = 0$ and $y_0 = 0$. Then using the polar coordinates $x = \rho\cos\phi$ and $y = \rho\sin\phi$ in (2.451) and (2.452) reduces the integrals with respect to χ to the integrals (2.421) and (2.422), and we arrive at the formulas

$$I_x = 4\pi^2 I_1(\rho, z, z_0)\cos\phi, \quad I_y = 4\pi^2 I_1(\rho, z, z_0)\sin\phi \tag{2.453}$$

with

$$I_1(\rho, z, z_0) = \int_0^{+\infty} J_1(k_t\rho)\mathrm{e}^{-j\gamma|z-z_0|}\frac{\mathrm{d}k_t}{\gamma} \tag{2.454}$$

and $\gamma = \sqrt{k^2 - k_t^2}$, $\operatorname{Im}\gamma \le 0$. Function $I_1(\rho, z, z_0)$ can be alternatively represented as

$$I_1(\rho, z, z_0) = \frac{1}{2}\int_{\infty\mathrm{e}^{-j\pi}}^{+\infty} H_1^{(2)}(k_t\rho)\mathrm{e}^{-j\gamma|z-z_0|}\frac{\mathrm{d}k_t}{\gamma} \tag{2.455}$$

by using analytical relations between Bessel functions:

$$2J_1(k_t\rho) = H_1^{(1)}(k_t\rho) + H_1^{(2)}(k_t\rho) = H_1^{(2)}(k_t\rho\mathrm{e}^{-j\pi}) + H_1^{(2)}(k_t\rho). \tag{2.456}$$

In (2.455), the integration contour goes below the branch cut $-\infty < k_t \le 0$ of the Hankel function.

Using (2.453) in (2.444) and (2.445) results in the expressions:

$$\Pi_\mathrm{e}(\mathbf{r}, \mathbf{r}_0) = \frac{p_\mathrm{e}}{4\pi j\epsilon}\cos(\phi - \phi_0)\frac{\partial}{\partial z_0}I_1(\rho, z, z_0) \tag{2.457}$$

$$\widetilde{\Pi}_\mathrm{m}(\mathbf{r}, \mathbf{r}_0) = \frac{\omega p_\mathrm{e}}{4\pi}\sin(\phi - \phi_0)I_1(\rho, z, z_0), \tag{2.458}$$

which describe the electric and magnetic Hertz vectors defined with respect to the direction perpendicular to the orientation of an electric dipole. In the case of free space it would be of course much simpler to define the potentials with respect to the direction of the dipole. Representations (2.457) and (2.458) are, however, necessary if a material interface is present and the dipole is not orthogonal to it because they can be simply extended to this case through a simple modification of the integrand in the plane wave expansion (2.455).

36. Assume that all materials, their boundaries and external fields are rotationally symmetric with respect to an axis z. Show that in the cylindrical coordinates (ρ, φ, z), all field components can be expressed in terms of the components E_φ and H_φ.

Hint: Expand the fields in the Fourier series

$$\mathbf{E} = \sum_{n=-\infty}^{+\infty} \mathrm{e}^{jn\varphi}\mathbf{E}_n, \quad \mathbf{H} = \sum_{n=-\infty}^{+\infty} \mathrm{e}^{jn\varphi}\mathbf{H}_n \tag{2.459}$$

and study equations satisfied by the harmonics $\mathbf{E}_n$ and $\mathbf{H}_n$.

Solution: Since the structure and the external fields are rotationally symmetric, the total fields must be periodic functions of φ and can be expanded as in (2.459). Writing Maxwell's equation (2.39) and (2.40) in the source-free region in the cylindrical coordinates (see Appendix A.3) and using the linearity of the equations, one arrives at the equations satisfied by the nth harmonics,

$$j\omega\epsilon E_{n\rho} = \frac{jn}{\rho} H_{nz} - \frac{\partial}{\partial z} H_{n\varphi} \tag{2.460}$$

$$j\omega\epsilon E_{n\varphi} = \frac{\partial}{\partial z} H_{n\rho} - \frac{\partial}{\partial \rho} H_{nz} \tag{2.461}$$

$$j\omega\epsilon E_{nz} = \frac{1}{\rho}\frac{\partial}{\partial \rho}(\rho H_{n\varphi}) - \frac{jn}{\rho} H_{n\rho} \tag{2.462}$$

$$-j\omega\mu H_{n\rho} = \frac{jn}{\rho} E_{nz} - \frac{\partial}{\partial z} E_{n\varphi} \tag{2.463}$$

$$-j\omega\mu H_{n\varphi} = \frac{\partial}{\partial z} E_{n\rho} - \frac{\partial}{\partial \rho} E_{nz} \tag{2.464}$$

$$-j\omega\mu H_{nz} = \frac{1}{\rho}\frac{\partial}{\partial \rho}(\rho E_{n\varphi}) - \frac{jn}{\rho} E_{n\rho}. \tag{2.465}$$

Equations (2.460), (2.462), (2.463), and (2.465) can be considered as a system of four linear equations for the components $E_{n\rho}$, E_{nz}, $H_{n\rho}$, and H_{nz}, and solving the system gives

$$E_{n\rho} = \frac{j}{(k\rho)^2 - n^2}\left[n\frac{\partial}{\partial \rho}(\rho E_{n\varphi}) + kZ\rho^2 \frac{\partial}{\partial z} H_{n\varphi}\right] \tag{2.466}$$

$$E_{nz} = \frac{j\rho}{(k\rho)^2 - n^2}\left[n\frac{\partial}{\partial z} E_{n\varphi} - kZ\frac{\partial}{\partial \rho}(\rho H_{n\varphi})\right] \tag{2.467}$$

$$H_{n\rho} = \frac{j}{(k\rho)^2 - n^2}\left[n\frac{\partial}{\partial \rho}(\rho H_{n\varphi}) - kY\rho^2 \frac{\partial}{\partial z} E_{n\varphi}\right] \tag{2.468}$$

$$H_{nz} = \frac{j\rho}{(k\rho)^2 - n^2}\left[n\frac{\partial}{\partial z} H_{n\varphi} + kY\frac{\partial}{\partial \rho}(\rho E_{n\varphi})\right], \tag{2.469}$$

where $Z = \sqrt{\mu/\epsilon} = 1/Y$. These relations show that all field components can be indeed expressed through the ρ and z derivatives of the two basic components $E_{n\varphi}$ and $H_{n\varphi}$. The latter satisfy a system of two coupled differential equations, which result from (2.461) and (2.464) upon the substitution of (2.466)–(2.469) as

$$\nabla^2 \tilde{E}_{n\varphi} + k^2 \tilde{E}_{n\varphi} = \frac{2k}{(k\rho)^2 - n^2}\left[k\rho\frac{\partial}{\partial \rho}\tilde{E}_{n\varphi} + nZ\frac{\partial}{\partial z}\tilde{H}_{n\varphi}\right] \tag{2.470}$$

$$\nabla^2 \tilde{H}_{n\varphi} + k^2 \tilde{H}_{n\varphi} = \frac{2k}{(k\rho)^2 - n^2}\left[k\rho\frac{\partial}{\partial \rho}\tilde{H}_{n\varphi} - nY\frac{\partial}{\partial z}\tilde{E}_{n\varphi}\right], \tag{2.471}$$

where $\tilde{E}_{n\varphi} = \rho E_{n\varphi}$, $\tilde{H}_{n\varphi} = \rho H_{n\varphi}$ and the operator ∇^2 is given by (A.33) with the scale factors (A.39).

37. By using the explicit solution (2.259) for the electric field of a point electric dipole, show the compliance of the solution with the reciprocity relation (2.147).

38. Generalize the derivation of the reciprocity theorem in Section 2.4.2 for arbitrary dyadic permittivities and permeabilities. Find under what conditions expression (2.142) is independent from the permitivity and permeability, and thus the reciprocity theorem (2.145) holds.

Hint: Make use of the relation $\mathbf{a} \cdot \overline{\overline{\epsilon}} \cdot \mathbf{b} = \mathbf{b} \cdot \overline{\overline{\epsilon}}^T \cdot \mathbf{a}$ valid for arbitrary vectors $\mathbf{a}$ and $\mathbf{b}$ and an arbitrary dyadic $\overline{\overline{\epsilon}}$.

Answer: The reciprocity relation (2.145) holds if the permittivity and permeability are symmetric dyadics, that is, $\overline{\overline{\epsilon}} = \overline{\overline{\epsilon}}^T$ and $\overline{\overline{\mu}} = \overline{\overline{\mu}}^T$.

39. Derive the reciprocity relations (2.147), (2.149), and (2.151) for point dipoles by excluding the small vicinities of the dipoles from integration region in (2.143) (see Figure 2.8).

Hint: Evaluate (2.154) in the limit $\delta \to 0$, where δ is the radius of the spherical regions excluded from the integration region.

Solution: Consider, for example, the case of point electric dipoles with the currents (2.146). Other cases, magnetic dipoles or electric and magnetic dipoles, can be studied in a similar way.

Denote the integral over the surface Ω_1 in (2.154) by I_1. Introducing spherical coordinates with the z axis along the direction of the dipole moment $\mathbf{p}_1^e$ and using the cyclic invariance of the triple scalar product (A.11) give

$$I_1 = \delta^2 \int_0^{2\pi} \int_0^{\pi} \sin\theta[\mathbf{E}_2 \cdot (\mathbf{H}_1 \times \hat{r}) + \mathbf{H}_2 \cdot (\mathbf{E}_1 \times \hat{r})]\, d\phi\, d\theta. \tag{2.472}$$

Since δ can be chosen arbitrarily small, the fields $\mathbf{E}_2(\mathbf{r})$ and $\mathbf{H}_2(\mathbf{r})$ can be replaced with $\mathbf{E}_2(\mathbf{r}_1)$ and $\mathbf{H}_2(\mathbf{r}_1)$. The fields $\mathbf{E}_1(\mathbf{r})$ and $\mathbf{H}_1(\mathbf{r})$ are due to the electric dipole at the center of the integration sphere, so using (2.207) and (2.208) with $r = \delta$ gives

$$\mathbf{H}_1 \times \hat{r} = -\frac{\omega k p_1^e}{4\pi\delta} e^{-jk\delta} \sin\theta \left(1 - \frac{j}{k\delta}\right)\hat{\theta}$$
$$\mathbf{E}_1 \times \hat{r} = \frac{k^2 p_1^e}{4\pi\epsilon\delta} e^{-jk\delta} \sin\theta \left[1 - \frac{j}{k\delta} - \frac{1}{(k\delta)^2}\right]\hat{\phi}.$$

Since $\hat{\phi} = -\sin\phi\hat{x} + \cos\phi\hat{y}$,

$$\int_0^{2\pi} \mathbf{H}_2 \cdot (\mathbf{E}_1 \times \hat{r})\, d\phi = 0. \tag{2.473}$$

The integral I_1 in the limit $\delta \to 0$ reduces therefore to

$$\lim_{\delta\to 0} I_1 = \frac{j\omega p_1^e}{4\pi} \int_0^{2\pi} \int_0^{\pi} \sin^2\theta[\mathbf{E}_2(\mathbf{r}_1) \cdot \hat{\theta}]\, d\phi\, d\theta \tag{2.474}$$

and since $\hat{\theta} = \cos\theta\cos\phi\hat{x} + \cos\theta\sin\phi\hat{y} - \sin\theta\hat{z}$, integration with respect to ϕ leaves only the z component, leading to

$$\lim_{\delta\to 0} I_1 = -\frac{j\omega p_1^e}{2}[\mathbf{E}_2(\mathbf{r}_1) \cdot \hat{z}] \int_0^{\pi} \sin^3\theta\, d\theta, \tag{2.475}$$

which is equal to

$$\lim_{\delta\to 0} I_1 = \frac{2\omega}{3j}[\mathbf{E}_2(\mathbf{r}_1) \cdot \mathbf{p}_1^e]. \tag{2.476}$$

The integral over Ω_2 in (2.154) can be formally obtained from (2.476) by simply interchanging the subscripts 1 and 2 and multiplying with -1,

$$\lim_{\delta\to 0} I_2 = -\frac{2\omega}{3j}[\mathbf{E}_1(\mathbf{r}_2) \cdot \mathbf{p}_2^e], \tag{2.477}$$

and (2.154) leads therefore to the relation

$$\mathbf{E}_2(\mathbf{r}_1) \cdot \mathbf{p}_1^e - \mathbf{E}_1(\mathbf{r}_2) \cdot \mathbf{p}_2^e = 0, \tag{2.478}$$

which is equivalent to the reciprocity relation (2.147).

40. Prove that the Franz representations (2.340), (2.341), (2.342), and (2.343) are compliant with Maxwell's equations.

Solution: In source-free space filled with homogeneous isotropic medium, Maxwell's equations are given by (2.39) and (2.40). Let us check that (2.340) and (2.341) describe solutions of the equations when $\mathbf{r} \in V$. A proof for (2.342) and (2.343) with $\mathbf{r} \notin V$ is similar.

Applying the curl operator to (2.340) and using the property (A.27) leads to the expression

$$\nabla \times \mathbf{E}(\mathbf{r}) = \frac{jk^2}{\omega\epsilon} \nabla \times \int_\Omega G_0(\mathbf{r}, \mathbf{r}')\hat{n}' \times \mathbf{H}' d\Omega' - \nabla \times \nabla \times \int_\Omega G_0(\mathbf{r}, \mathbf{r}')\hat{n}' \times \mathbf{E}' d\Omega'. \tag{2.479}$$

Dividing (2.479) by $-j\omega\mu$ and using the relation $k^2 = \omega^2\epsilon\mu$ results in the expression

$$\frac{j}{\omega\mu} \nabla \times \mathbf{E}(\mathbf{r}) = -\nabla \times \int_\Omega G_0(\mathbf{r}, \mathbf{r}')\hat{n}' \times \mathbf{H}' d\Omega' + \frac{1}{j\omega\mu} \nabla \times \nabla \times \int_\Omega G_0(\mathbf{r}, \mathbf{r}')\hat{n}' \times \mathbf{E}' d\Omega'. \tag{2.480}$$

Now, using the identity (A.24) for the double curl operator, the relation (A.76) for the vector Laplacian and the fact that when $\mathbf{r} \notin \Omega$ the function $G_0(\mathbf{r}, \mathbf{r}')$ satisfies the homogeneous Helmholtz equation, we arrive at the formula

$$\frac{j}{\omega\mu} \nabla \times \mathbf{E}(\mathbf{r}) = -\nabla \times \int_\Omega G_0(\mathbf{r}, \mathbf{r}')\hat{n}' \times \mathbf{H}' d\Omega' + \frac{1}{j\omega\mu}[k^2 + \nabla(\nabla \cdot)] \int_\Omega G_0(\mathbf{r}, \mathbf{r}')\hat{n}' \times \mathbf{E}' d\Omega'. \tag{2.481}$$

The right-hand side of (2.481) coincides with the right-hand side of (2.341) and is therefore equal to $\mathbf{H}$, which transforms (2.481) to the Maxwell equation (2.40).

In a similar fashion, application of the curl operator transforms (2.341) to (2.340), provided that the Maxwell equation (2.39) is satisfied.

References

Alexopoulos NG and Tadler GA 1975 Accuracy of the Leontovich boundary condition for continuous and discontinuous surface impedance. *J. Appl. Phys.*, **46**, 3326–3332.

Born M and Wolf E (eds) 1959 *Principles of Optics: Electromagnetic Theory of Propagation, Interference and Diffraction of Light*, Pergamon Press.

Borovikov VA and Kinber BY 1994 *Geometrical Theory of Diffraction*, The Institution of Electrical Engineers, London.

Bouman JJ, Senior TBA, and Uslenghi PLE (eds) 1987 *Electromagnetic and Acoustic Scattering by Simple Shapes*, Hemisphere Publishing Corporation.

Chang ATC and Wilheit TT 1979 Remote sensing of atmospheric water vapor, liquid water, and wind speed at the ocean surface by passive microwave technique from the Nimbus 5 satellite. *Radio Sci.*, **14**, 793–802.

Collin RE 1960 *Field Theory of Guided Waves*, McGraw-Hill.

Colton D and Kress R 1983 *Integral Equation Methods in Scattering Theory*, John Wiley & Sons.

Culhaoglu AE, Osipov AV, and Russer P 2013 Mono- and bistatic scattering reduction by a metamaterial low reflection coating. *IEEE Trans. Antennas Propag.*, **61**, 462–466.

Djordjevic AR, Biljic RM, Likar-Smiljanic VD, and Sarkar TK 2001 Wideband frequency-domain characterization of FR-4 and time-domain causality. *IEEE Trans. Electromagn. Compat.*, **43**, 662–667.

El-Kady I, Sigalas MM, Biswas R, Ho KM, and Soukoulis CM 2000 Metallic photonic crystals at optical wavelengths. *Phys. Rev. B*, **62**, 15299.

Feinberg EL 1944 On the propagation of radio waves along an imperfect surface. *J. Phys. USSR*, **8**, 317–330.

Fock VA 1965 *Electromagnetic Diffraction and Propagation Problems*, Pergamon Press.

Franz W 1948 Zur Formulierung des Huygenschen Prinzips. *Z. Naturforsch. A*, 500–506.

Franz W 1952 Einfache Herleitung der allgemeinen Kirchhoffschen Beugungsformel und ihres elektromagnetischen Analogons. *Z. Angew. Math. Mech.*, **32**, 26–27.

Franz W 1957 *Theorie der Beugung elektromagnetischer Wellen*, Springer-Verlag.

Fresnel A 1818 Mémoire sur la diffraction de la lumière. *Œuvres completes*, 247–382.

Gabriel S, Lau RW, and Gabriel C 1996 The dielectric properties of biological tissues: III. Parametric models for the dielectric spectrum of tissues. *Phys. Med. Biol.*, **41**, 2271–2293.

Green G 1828 *An Essay on the Application of Mathematical Analysis to the Theories of Electricity and Magnetism*, printed for the author by T. Wheelhouse, Nottingham.

Grinberg GA 1942 Theory of the coastal refraction of electromagnetic waves. *J. Phys. USSR*, **6**, 185–209.

Hale GM and Querry MR 1973 Optical constants of water in the 200-nm to 200-μm wavelength region. *Appl. Opt.*, **12**, 555–563.

Helmholtz H von 1860 Theorie der Luftschwingungen in Röhren mit offenen Enden. *J. Reine Angew. Math.*, **57**, 1–72.

Hertz H 1889 Die Kräfte electrischer Schwingungen, behandelt nach der Maxwellschen Theorie. *Ann. d. Phys. u. Chem.*, **36**, 1–22.

Hoppe DJ and Rahmat-Samii Y 1995 *Impedance Boundary Conditions in Electromagnetics*, Taylor and Francis.

Huygens C 1690 *Traite de la Lumiere*, Leyden.

Johnson PB and Christy RW 1972 Optical constants of noble metals. *Phys. Rev. B*, **6**, 4370–4379.

Jones DS 1964 *The Theory of Electromagnetism*, Pergamon Press.

Kirchhoff G 1883 Zur Theorie der Lichtstrahlen. *Ann. Phys.* (Berlin), **254**, 663–695.

Kirchhoff G 1891 *Mathematische Optik (Vorlesungen über mathematische Physik, 2. Band)*, Teubner Verlag, Leipzig.

Knott EF, Shaeffer JF, and Tuley MT 1993 *Radar Cross Section*, Artech House.

Kong JA 1986 *Electromagnetic Wave Theory*, John Wiley & Sons. EMW Publishing.

Kottler F 1923 Elektromagnetiche Theorie der Beugung an schwarzen Schirmen. *Ann. Phys.* (Berlin), **376**, 457–508.

Landau LD, Pitaevskii LP, and Lifshitz EM 1984 *Electrodynamics of Continuous Media (Course of Theoretical Physics, Volume 8)*, Second edition, Butterworth-Heinemann.

Landron O, Feuerstein MJ, and Rappaport TS 1996 A comparison of theoretical and empirical reflection coefficients for typical exterior wall surfaces in a mobile radio environment. *IEEE Trans. Antennas Propag.*, **44**, 341–351.

Lee SW and Gee W 1987 How good is the impedance boundary condition? *IEEE Trans. Antennas Propag.*, **35**, 1313–1315.

Leontovich MA 1948 On approximate boundary conditions for the electromagnetic field on the surface of a good conductor, in *Investigations on Radiowave Propagation, Part. II*, USSR Academy of Sciences, 5–12 (in Russian).

Levin ML, Rytov SM, and Shafranov VD 1983 Leontovich's researches in electrodynamics. *Sov. Phys. Usp.*, **26**, 353–355.

Levine H and Schwinger J 1950 On the theory of electromagnetic wave diffraction by an aperture in an infinite plane conducting screen. *Comm. Pure Appl. Math.*, **3**, 355–391.

Lindell IV 1992 *Methods for Electromagnetic Field Analysis*, Oxford Science Publications.

Lindell IV 1996 Huygens principle in electromagnetics. *IEE Proc-Sci. Meas. Technol.*, **143**, 103–105.

Lorentz HA 1895 The theorem of Poynting concerning the energy in the electromagnetic field and two general propositions concerning the propagation of light. *Versl. Kon. Ak. Wet.*, **4**, 176–187 (in Dutch).

Love AEH 1901 The integration of the equations of propagation of electric waves. *Phil. Trans. R. Soc. Lond. A*, **197**, 1–45.

Macdonald HM 1902 *Electric Waves*, Cambridge University Press.

Macdonald HM 1911 The integration of the equations of propagation of electric waves. *Proc. London Math. Soc.*, **10**, 91–95.

Makarov GI, Novikov VV, and Rybachek ST 1991 *Electromagnetic Waves Propagation over Earth's Surface*, Moscow: Nauka (in Russian).

McCrea WH 1957 Hertzian electromagnetic potentials. *Proc. R. Soc. Lond. A*, **240**, 447-457.

Morse PM and Feshbach H 1953 *Methods of Theoretical Physics*, McGraw-Hill.

Müller C 1957 *Grundprobleme der mathematischen Theorie elektromagnetischer Schwingungen*, Springer-Verlag.

Nisbet A 1955 Hertzian electromagnetic potentials and associated gauge transformations. *Proc. R. Soc. Lond. A*, **231**, 250–263.

Ordal MA, Bell RJ, Alexander RW, Jr, Long LL, and Querry MR 1985 Optical properties of fourteen metals in the infrared and far infrared: Al, Co, Cu, Au, Fe, Pb, Mo, Ni, Pd, Pt, Ag, Ti, V, and W. *Appl. Opt.*, **24**, 4493–4499.

Palik ED (ed.) 1997 *Handbook of Optical Constants of Solids*, Academic Press.

Poynting JH 1884 On the transfer of energy in the electromagnetic field. *Phil. Trans. R. Soc. Lond.*, **175**, 343–361.

Rakic AD, Djurisic AB, Elazar JM, and Majewski ML 1998 Optical properties of metallic films for vertical-cavity optoelectronic devices. *Appl. Opt.*, **37**, 5271–5283.

Righi A 1901 Sui campi elettromagnetici e particolarmente su quelli creati, da cariche elettriche o da poli magnetici in movimento. *Nuovo Cimento*, **2**, 104–121.

Ryshik IM and Gradstein IS 1957 *Tables of Series, Products, and Integrals*, VEB, Berlin.

Rytov SM 1940 Calcul du skin-effet par la méthode des perturbations. *J. Phys. USSR*, **2**, 233–242.

Schelkunoff SA 1936 Some equivalence theorems of electromagnetics and their application to radiation problems. *Bell Syst. Tech. J.*, **15**, 92–112.

Schukin AN 1940 *Propagation of Radio Waves*, Svyazizdat, Moscow (in Russian).

Senior TBA 1962 A note on impedance boundary conditions. *Can. J. Phys.*, **40**, 663–665.

Senior TBA and Volakis JL 1995 *Approximate Boundary Conditions in Electromagnetics*, The Institution of Electrical Engineers, London.

Serdyukov A, Semchenko I, Tretyakov S, and Sihvola A 2001 *Electromagnetics of Bi-anisotropic Materials: Theory and Applications*, Electrocomponent Science Monographs, Gordon and Breach Science Publishers.

Sihvola A 1999 *Electromagnetic Mixing Formulas and Applications*, The Institution of Electrical Engineers, London.

Stratton JA 1941 *Electromagnetic Theory*, McGraw-Hill.

Stratton JA and Chu LJ 1939 Diffraction of electromagnetic waves. *Phys. Rev.*, **56**, 99–107.

Tai C-T 1972 Kirchhoff Theory: Scalar, Vector, or Dyadic? *IEEE Trans. Antennas Propag.*, **20**, 114–115.

Tretyakov SA 2003 *Analytical Modeling in Applied Electromagnetics*, Artech House.

van Bladel J. 1964 *Electromagnetic Fields*, McGraw-Hill.

von Hippel AR (ed.) 1954 *Dielectric Materials and Applications*, MIT Press, Cambridge.

Wait JR 1970 A pitfall in the scalar electromagnetic formulation of Kirchhoff theory. *IEEE Trans. Educ.*, **13**, 98–99.

Wang DS 1987 Limits and validity of the impedance boundary condition on penetrable surfaces. *IEEE Trans. Antennas Propag.*, **35**, 453–457.

Weast RC (ed.) 1988 *CRC Handbook of Chemistry and Physics*, 69th edition, CRC Press.

Weber H 1869 Über die Integration der partiellen Differentialgleichung $\partial^2 u/\partial x^2 + \partial^2 u/\partial y^2 + k^2 u = 0$. *Math. Ann.*, **1**, no. 1, 1–36.

Wilcox CH 1956 An expansion theorem for electromagnetic fields. *Comm. Pure Appl. Math.*, **9**, 115–134.

Wilcox CH 1959 Spherical means and radiation conditions. *Arch. Rational Mech. Anal.*, **3**, 133–148.

3

Far-field Scattering

3.1 Introduction

Most practical applications of electromagnetic scattering (e.g., radar or remote sensing) involve targets that are located very far from the transmitter and receiver. At such distances the target appears to be a point scatterer and the scattered field at the receiver position looks like a spherical wave emanating from the scatterer. The location of the far-field region, the behavior of the scattered field, and the choice of analysis methods depend on the relation between the wavelength λ, the characteristic size D of the scatterer, and the distance R between the scatterer and the observation point.

This chapter describes parameters (various scattering cross sections, the scattering matrix, the far-field coefficient) that characterize scattering from distant targets. We also present mathematical conditions for the distance to the far-field region and derive integral representations of the far field in terms of equivalent currents. Furthermore, the forward-scattering (optical) theorem which helps us to determine the losses of energy in the incident wave due to scattering and absorption is derived. The chapter concludes with an overview of different scattering regimes (low-frequency, resonant, high-frequency) observed in the far-field region of scatterers of various electrical sizes.

3.2 Scattering Cross Section

Scattering cross section is a parameter that is widely used in practical applications to describe electromagnetic scattering in the far-field regime. In fact, this is not a single parameter but rather a whole family of related parameters which describe various aspects of the scattering process. In this section we introduce the various kinds of cross sections, including the special case of scattering from infinitely long scatterers.

3.2.1 Monostatic and Bistatic, Backscattering and Forward-scattering Cross Sections, Differential, Total, Absorption, and Extinction Cross Sections

The scattering cross section tells us how strongly an object scatters electromagnetic fields in a certain direction if it is excited by a given incident field (Figure 3.1). Obviously, this is the main parameter that measures the radar "visibility" of an object. The *radar cross section* (or *RCS* or *backscatter cross section* or *backscattering cross section* or *monostatic cross section*) is a special case of this parameter for the situation when the receiver is located at the same position as the transmitter[1] (Figure 3.2). We use the term "radar cross section" for the backscattering situation throughout the book. The scattering cross section applied to the situation of spatially separated transmitter and receiver will be referred to as

[1] Sometimes, "radar cross section" is used as a generic term which is then specialized to the backscatter cross section, monostatic cross section, bistatic cross section, etc. Since the word "radar" usually implies a receiver and transmitter at the same location, we prefer the term "scattering cross section" to address a general situation with unspecified source–receiver configuration.

Modern Electromagnetic Scattering Theory with Applications, First Edition. Andrey V. Osipov and Sergei A. Tretyakov.

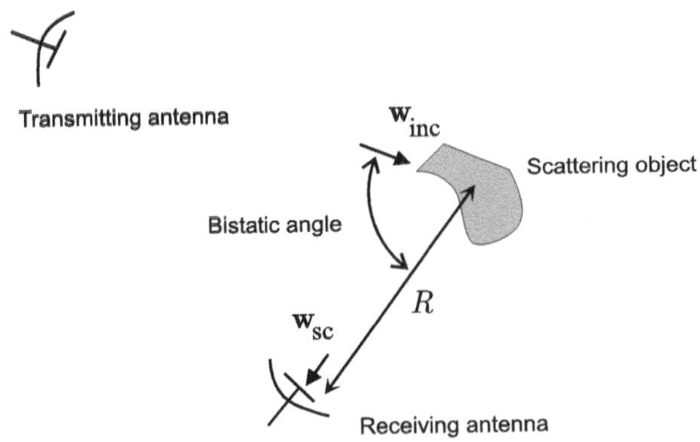

Figure 3.1 Definition of scattering cross section.

bistatic cross section. The angle between the directions from the scatterer to the transmitter and to the receiver is called the bistatic angle. The bistatic angle is equal to zero in monostatic configurations and to 180° in the case of forward scattering. The *forward-scatter(ing) cross section* is therefore defined as the bistatic scattering cross section at the separation angle of 180°.

How can we define the scattering cross section? First, the goal is to measure how much power scattered from an object reaches the receiving antenna, which means that our parameter should be proportional to the scattered power flow density *at the position of the receiving antenna* and should be normalized to the power flow density that excites the body, that is, to the incident power flow *at the position of the scattering body*:

$$\sigma \sim \frac{w_{\rm sc}}{w_{\rm inc}}, \tag{3.1}$$

where $w_{\rm inc,sc} = |\mathbf{w}_{\rm inc,sc}|$ and the vector $\mathbf{w}_{\rm inc,sc}$ is the mean intensity of the energy flow in the incident and scattered waves defined in (2.119). The set-up is shown in Figure 3.1. A scattering object is illuminated by an antenna, and the scattered field is measured by another antenna, whose location is in general different from that of the transmitting antenna. The distances between the antennas and the object are large compared to the wavelength so that the incident field can be considered as a spherical wave, and also the scattered field at the receiver is approximately a spherical wave. Next, we should agree on the proportionality coefficient in formula (3.1). It is clear that any reasonable proportionality coefficient must depend on the distance between the object and the receiving antenna as R^2. This is because otherwise parameter σ would depend on R (the scattered power density $w_{\rm sc} \sim 1/R^2$). Thus, we can refine the definition as

$$\sigma \sim R^2 \frac{w_{\rm sc}}{w_{\rm inc}}. \tag{3.2}$$

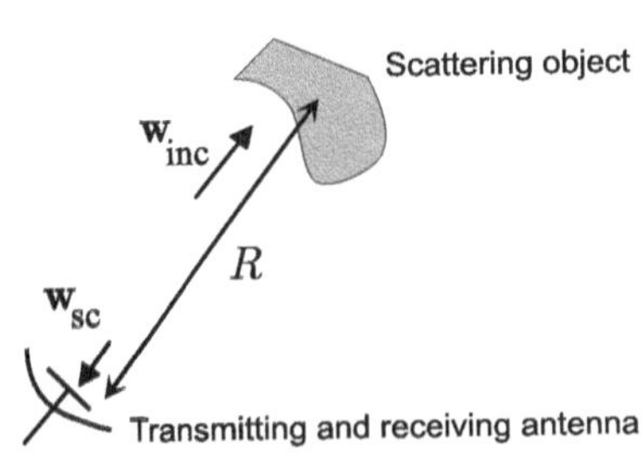

Figure 3.2 Definition of radar cross section.

To decide on an appropriate numerical proportionality coefficient, we need to choose a certain known scatterer for comparison. The commonly agreed "standard object" is a fictitious isotropic scatterer that scatters power uniformly in all directions in space and creates the same scattered field at the receiver as the actual body. Since the "standard scatterer" creates the field with the energy flow density $\mathbf{w}_{sc}$ at the receiver and is an isotropic scatterer, the total scattered power is simply

$$P_{sc} = 4\pi R^2 w_{sc}, \tag{3.3}$$

where R is the distance between the scatterer and the receiver. This power is in the first place created by the transmitting antenna. The scatterer first works as a "receiving antenna" with a certain effective area σ, so that the received power is $P_{rec} = \sigma w_{inc}$. If the scatterer does not absorb any power (e.g., as an ideally conducting body), all this received power is scattered back into space, so we have

$$P_{rec} = P_{sc} \quad \Rightarrow \quad 4\pi R^2 w_{sc} = \sigma w_{inc}. \tag{3.4}$$

We see that the effective area σ of this standard scatterer can be used as a good measure of radar visibility because it has the proper dependence on the incident and scattered power densities. To ensure that we deal with a spherical wave in the far field, we should take the limit for large distances R:

$$\sigma = \lim_{R\to\infty} 4\pi R^2 \frac{w_{sc}}{w_{inc}} \tag{3.5}$$

(compare with (3.2)), and indeed relation (3.5) has been agreed to define the *scattering cross section* of an object.

In terms of electric or magnetic fields (see (2.162)) we can write

$$\sigma = \lim_{R\to\infty} 4\pi R^2 \frac{w_{sc}}{w_{inc}} = \lim_{R\to\infty} 4\pi R^2 \frac{|\mathbf{E}_{sc}|^2}{|\mathbf{E}_{inc}|^2} = \lim_{R\to\infty} 4\pi R^2 \frac{|\mathbf{H}_{sc}|^2}{|\mathbf{H}_{inc}|^2}. \tag{3.6}$$

In words, the *scattering cross section* is the effective area that would intercept the amount of incident power necessary to produce the observed scattered power at the receiver, if the object scatters uniformly in all directions.

In general, the scattering cross section σ is not equal to the area of the geometric cross section of the target. For example, the scattering cross section of an electrically large plate can be much greater than the area of the plate (see Table 3.1). A notable exception is a perfectly electrically conducting sphere, whose RCS approaches the value πa^2 (a is the radius of the sphere) as the frequency goes to infinity (e.g., Figure 3.16).

The energy flow transported by an ideal plane wave in a lossless medium is the same at every point in the medium. In practical applications, however, fields are excited by localized sources and absorption in media may be significant, which implies that in reality the energy flow from the source is a function of coordinates. Therefore it is important to remember that the definition (3.6) assumes that over the whole extension of the scatterer the incident field must be close to a plane wave and the magnitudes of the incident field, $|\mathbf{E}_{inc}|$ or $|\mathbf{H}_{inc}|$, are to be taken at the scatterer.

The scattered field and, therefore, the scattering cross section depend on the polarization of the incident wave. Furthermore, the receiving antenna can also be polarized, in the sense that it registers not the complete magnitude $|\mathbf{E}_{sc}|$ of the arriving wave but rather the magnitude of a particular component of the scattered wave, $\mathbf{E}_{sc} \cdot \hat{p}$, where $\hat{p}$ is a unit vector defining the polarization of the antenna. It is therefore sometimes convenient to work with a different definition of the scattering cross section

$$\sigma(\hat{p}) = \lim_{R\to\infty} 4\pi R^2 \frac{|\mathbf{E}_{sc} \cdot \hat{p}|^2}{|\mathbf{E}_{inc}|^2}. \tag{3.7}$$

It is important to understand that (3.5) is only a formal definition. Isotropic scatterers do not exist, and sometimes this definition is not very convenient. Because of this, sometimes we use another parameter called *differential scattering cross section*, which is defined as

$$\sigma_{dif} = \lim_{R\to\infty} R^2 \frac{w_{sc}}{w_{inc}}. \tag{3.8}$$

In words, the differential cross section is the ratio of the power scattered into a unit solid angle along a certain direction to the incident power density. Finally, we define the *total scattering cross section* or, shorter, *total cross section* as the total scattered power normalized to the incident power density:

$$\sigma_{\mathrm{T}} = \int \sigma_{\mathrm{dif}}\, \mathrm{d}\Omega = \frac{1}{4\pi}\int \sigma\, \mathrm{d}\Omega = \frac{P_{\mathrm{sc}}}{w_{\mathrm{inc}}}, \tag{3.9}$$

where the integration is performed over the unit sphere and P_{sc} is the total scattered power. A somewhat arbitrary definition of the scattering cross section (3.5) may sometimes lead to "strange" results. For example, RCS of a small object may happen to be larger than the total cross section of the same object.

The energy incident on an obstacle is not only scattered but is also absorbed if the obstacle is lossy. To characterize the absorption losses, one may relate the rate P_{abs} at which the energy is being absorbed by the scatterer to the density w_{inc} of the energy flow in the incident wave and define the *absorption cross section* by the ratio

$$\sigma_{\mathrm{abs}} = \frac{P_{\mathrm{abs}}}{w_{\mathrm{inc}}}. \tag{3.10}$$

The total energy losses in the incident wave due to both scattering and absorption are described by the so-called *extinction cross section*

$$\sigma_{\mathrm{ext}} = \sigma_{\mathrm{T}} + \sigma_{\mathrm{abs}}. \tag{3.11}$$

In the *optical limit* ($\lambda \to 0$), the extinction cross section of an opaque scatterer made of ordinary materials has the limit

$$\sigma_{\mathrm{ext}} \to 2S_0, \tag{3.12}$$

where S_0 is the area of the geometric cross section of the shadow region behind the scatterer; this property is derived in section 3.4.3 (see section 3.4.4 for a similar property for cylindrical scatterers).

The cross sections σ, σ_{dif}, σ_{T}, σ_{ext}, and σ_{abs} depend on many factors:

- object size
- object shape
- object orientation with respect to the transmitter and receiver
- material of the object
- frequency
- polarization of the transmitting and receiving antennas.

Depending on these factors, the numerical values of a scattering cross section may differ by many orders of magnitude and that is why they are typically given on a logarithmic scale according to

$$10 \log_{10} \sigma. \tag{3.13}$$

The corresponding units are referred to as dBsm (decibel referenced to one square meter) since σ is defined in the units of area.

We continue with a note on the scattering cross section of objects placed above a ground plane (Figure 3.3). Such configurations are often encountered in practical applications, and examples are a car on a road or a ship on the sea surface. In the presence of the ground plane the cross section will differ from the scattered cross section of the same scatterer in free space. Indeed, the definition (3.5) assumes illumination by a single plane wave 1 and does not account for a wave 2 reflected from the ground, which may additionally illuminate the target. Furthermore, the field scattered by the target arrives at the receiver not only directly over path 3 but also upon reflection from the ground through path 4.

Numerical evaluation of the cross sections is typically based on the use of the far-field coefficient **F** (see section 3.4, equations (3.86) and (3.110)). The total scattering cross section σ_{T} is then obtained by integration of σ over all scattering directions according to (3.9). The extinction cross section σ_{ext} can be also expressed through the far-field coefficient in the forward direction, which is the direction of the incident wave, by using the so-called forward-scattering (or optical) theorem (see (3.147)). Once σ_{ext} and σ_{T} are known, the absorption cross section σ_{abs} results from the definition (3.11). For lossless scatterers (purely dielectric, PEC), $\sigma_{\mathrm{ext}} = \sigma_{\mathrm{T}}$ and the total scattering cross section can be found directly from the

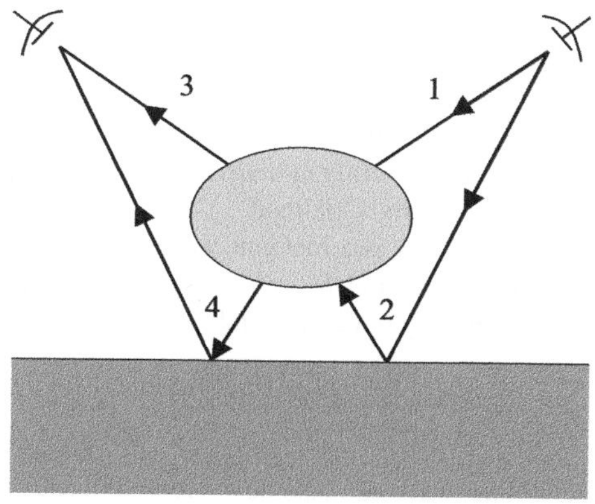

Figure 3.3 Scattering cross section of an obstacle placed above a ground plane differs from scattering cross section of the same objects in free space because of the additional illumination and scattering channels 2 and 4.

forward-scattering relation (3.147), that is, without integrating the bistatic scattering cross section. The forward-scattering relation is derived in section 3.4.3.

3.2.2 Scattering Width

Scattering width is a parameter that characterizes scattering from infinite scatterers, like cylinders and wedges, which are invariant with respect to translations in a specified direction. This kind of geometry requires definitions different from (3.5)–(3.11) since the scattered fields are no longer spherical waves but rather cylindrical waves which transport infinite power through any surface enclosing the scatterer because of the infinite size of the scatterer. Application of definitions (3.5)–(3.11) derived for compact objects to infinite cylinders or wedges would result in infinite values of scattering cross sections.

The scattering parameters that are finite can be defined by considering energy scattered per unit length of the scatterer. They can be seen as two-dimensional versions of scattering cross sections for compact objects and obtained by excluding the dependence of the fields on the coordinate along which the scatterer is invariant. Parameter σ_{2D}, a two-dimensional version of the bistatic scattering cross section σ, is measured in units of length and will be therefore referred to as bistatic scattering width. The parameter is defined as the ratio of the scattered to incident energy flow densities times the length of a circle with the radius going to infinity (Ruck *et al.* 1970),

$$\sigma_{2D} = \lim_{\rho\to\infty} \frac{2\pi\rho}{\sin\beta} \frac{w_{sc}}{w_{inc}}. \tag{3.14}$$

If z is the axis along which the scatterer shape is invariant, then ρ is the distance between the observer and the z axis, and β is the angle between the axis and the scattering direction $\hat{s}$ (Figure 3.4). The energy flow density of the incident wave is taken at the scatterer ($\rho = 0$) and that of the scattered wave at $\rho \to \infty$. The definition (3.14) accounts for the fact that the

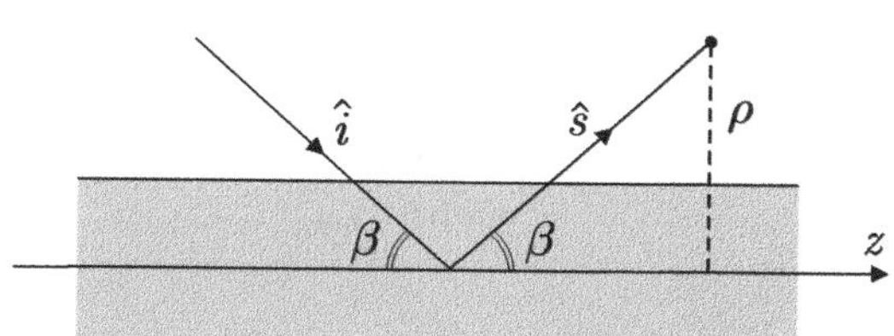

Figure 3.4 Geometry parameters used in the definition of scattering widths.

scattered wave is an outgoing conical wave (see section 2.5.1), and the distance $\rho/\sin\beta$, which appears in the definition, is the length of the propagation path of the scattered wave from the scatterer to the observer.

The angle β is at the same time the angle between the z axis and the incidence direction $\hat{\imath}$, which is a consequence of the translational invariance of the configuration. Note that the backscattering cross section can be only defined in the case of $\beta = \pi/2$, when the scattered waves degenerates to a truly cylindrical wave. In general, however, there is no backscatter power and, consequently, no backscattering width can be defined.

The densities of the power flows in the scattered and incident waves can be conveniently expressed through the z components of the electric and magnetic fields, E_z and H_z, which serve as potential functions in the problems with translational symmetry (see section 2.5.5). Referring to (2.179) and (2.180) we write

$$w_{\mathrm{sc}} = \frac{Y}{2\sin^2\beta}\left(\left|E_z^{\mathrm{sc}}\right|^2 + \left|ZH_z^{\mathrm{sc}}\right|^2\right). \tag{3.15}$$

The power density (2.162) in the incident plane wave can also be expressed in terms of the z components as

$$w_{\mathrm{inc}} = \frac{Y}{2\sin^2\beta}\left(\left|E_z^{\mathrm{inc}}\right|^2 + \left|ZH_z^{\mathrm{inc}}\right|^2\right) \tag{3.16}$$

so that (3.14) becomes

$$\sigma_{2\mathrm{D}} = \lim_{\rho\to\infty}\frac{2\pi\rho}{\sin\beta}\frac{\left|E_z^{\mathrm{sc}}\right|^2 + \left|ZH_z^{\mathrm{sc}}\right|^2}{\left|E_z^{\mathrm{inc}}\right|^2 + \left|ZH_z^{\mathrm{inc}}\right|^2}. \tag{3.17}$$

This general definition can be specialized to the two principal polarization cases. In the case of E polarization, $H_z^{\mathrm{inc}} = 0$ and, thus, (3.17) becomes[2]

$$\sigma_E = \lim_{\rho\to\infty}\frac{2\pi\rho}{\sin\beta}\frac{\left|E_z^{\mathrm{sc}}\right|^2 + \left|ZH_z^{\mathrm{sc}}\right|^2}{\left|E_z^{\mathrm{inc}}\right|^2}. \tag{3.18}$$

In the case of H polarization ($E_z^{\mathrm{inc}} = 0$),

$$\sigma_H = \lim_{\rho\to\infty}\frac{2\pi\rho}{\sin\beta}\frac{\left|E_z^{\mathrm{sc}}\right|^2 + \left|ZH_z^{\mathrm{sc}}\right|^2}{\left|ZH_z^{\mathrm{inc}}\right|^2}. \tag{3.19}$$

In general, the scattered field includes co- and cross-polar components so it is sometimes convenient to specialize the definitions (3.18) and (3.19) further for every individual component so that for the E-polarized incident wave, the co- and cross-polar scattering widths are

$$\sigma_{EE} = \lim_{\rho\to\infty}\frac{2\pi\rho}{\sin\beta}\frac{\left|E_z^{\mathrm{sc}}\right|^2}{\left|E_z^{\mathrm{inc}}\right|^2} \tag{3.20}$$

$$\sigma_{HE} = \lim_{\rho\to\infty}\frac{2\pi\rho}{\sin\beta}\frac{\left|ZH_z^{\mathrm{sc}}\right|^2}{\left|E_z^{\mathrm{inc}}\right|^2} \tag{3.21}$$

[2] We omit the label 2D from the polarization-specified definitions of the scattering width since the labels E and H assume the cylindrical shape of the scatterer.

and for the H-polarized incident wave,

$$\sigma_{EH} = \lim_{\rho\to\infty} \frac{2\pi\rho}{\sin\beta} \frac{\left|E_z^{\rm sc}\right|^2}{\left|ZH_z^{\rm inc}\right|^2} \tag{3.22}$$

$$\sigma_{HH} = \lim_{\rho\to\infty} \frac{2\pi\rho}{\sin\beta} \frac{\left|H_z^{\rm sc}\right|^2}{\left|H_z^{\rm inc}\right|^2}. \tag{3.23}$$

To define a two-dimensional version of the total scattering cross section $\sigma_{\rm T}^{\rm 2D}$, we need the total scattered energy per unit length of the scatterer, and this can be obtained by integration of the ρ component of the energy flow density over the circle of large radius ρ,

$$\rho \int_0^{2\pi} \hat{\mathbf{w}}_{\rm sc} \cdot \hat{\rho}\, {\rm d}\varphi. \tag{3.24}$$

It follows from (2.177) and (2.179) with $k_\perp = k \sin\beta$ that $\hat{\rho} \cdot \mathbf{w}_{\rm sc} = w_{\rm sc} \sin\beta$ and, normalizing (3.24) to the density of the energy flow of the incident wave at the scatterer, we arrive at the definition of the total scattering width

$$\sigma_{\rm T}^{\rm 2D} = \lim_{\rho\to\infty} \rho \frac{\sin\beta}{w_{\rm inc}} \int_0^{2\pi} w_{\rm sc}\, {\rm d}\varphi \tag{3.25}$$

or referring to (3.14),

$$\sigma_{\rm T}^{\rm 2D} = \frac{\sin^2\beta}{2\pi} \int_0^{2\pi} \sigma_{\rm 2D}\, {\rm d}\varphi. \tag{3.26}$$

Specialization of the total scattering width to particular polarization cases results from the use of the corresponding scattering widths (3.18)–(3.23) in the integrand in the right-hand side of (3.26).

The absorption width $\sigma_{\rm abs}^{\rm 2D}$ can be defined as the ratio of the power absorbed per unit length by the cylindrical scatterer, $p_{\rm abs}$, and the energy flow density of the incident wave,

$$\sigma_{\rm abs}^{\rm 2D} = \frac{p_{\rm abs}}{w_{\rm inc}}. \tag{3.27}$$

The extinction width $\sigma_{\rm ext}^{\rm 2D}$ is a measure of the total power losses due to scattering and absorption in the incident wave,

$$\sigma_{\rm ext}^{\rm 2D} = \sigma_{\rm abs}^{\rm 2D} + \sigma_{\rm T}^{\rm 2D}. \tag{3.28}$$

3.2.3 Backscattering from Impedance-matched Bodies

Scattering from impedance bodies with the surface impedance equal to the intrinsic impedance of the surrounding medium possess a number of remarkable features (Wagner and Lynch 1963; Osipov 2015; Weston 1963). The following general theorem was proved by Weston (1963). If a plane wave is incident upon a body, at the surface of which the total electromagnetic field satisfies the impedance boundary conditions (2.74), and the surface of the body is invariant under a 90° rotation about some axis, then the far-zone backscattering field is zero if the direction of incidence is along this axis and the surface impedance is equal to the wave impedance of the surrounding homogeneous medium. Examples of the shapes are shown in Figure 3.5.

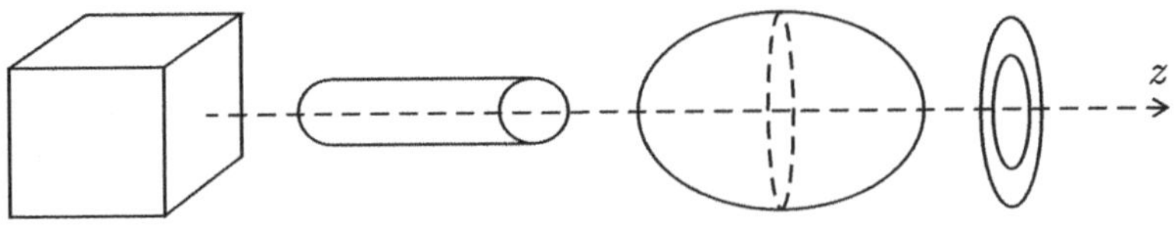

Figure 3.5 Examples of bodies compliant with the symmetry requirements of Weston's theorem: a cube, a piece of circular cylinder, a spheroid, and a ring. RCS of the bodies in the z direction vanishes when their surface impedance is equal to the intrinsic impedance of the background medium.

Let ϵ and μ be the permittivity and the permeability of the surrounding homogeneous medium. Let $Z = \sqrt{\mu/\epsilon}$ be the wave impedance of the surrounding medium and $\tilde{Z}_s$ be the equivalent surface impedance from impedance boundary conditions (2.74) imposed on the surface of the scatterer. Equations satisfied by the fields can be rewritten entirely in terms of either electric or magnetic field. The equations for the electric field can be written as

$$\nabla \times \nabla \times \mathbf{E} = k^2 \mathbf{E} \tag{3.29}$$

$$\nabla \cdot \mathbf{E} = 0 \tag{3.30}$$

$$\hat{n} \times (\hat{n} \times \mathbf{E}) = \frac{\tilde{Z}_s}{j\omega\mu} \hat{n} \times (\nabla \times \mathbf{E}), \tag{3.31}$$

where the first two equations are equivalent to Maxwell's equations and the last equation is the impedance boundary condition. The magnetic field satisfies the same equations

$$\nabla \times \nabla \times \mathbf{H} = k^2 \mathbf{H} \tag{3.32}$$

$$\nabla \cdot \mathbf{H} = 0 \tag{3.33}$$

$$\hat{n} \times (\hat{n} \times \mathbf{H}) = \frac{1}{j\omega\epsilon\tilde{Z}_s} \hat{n} \times (\nabla \times \mathbf{H}), \tag{3.34}$$

the only difference being in the coefficients of the impedance boundary conditions (3.31) and (3.34). The coefficients coincide, however, once $\tilde{Z}_s = Z$.

The fields must also satisfy the conditions at infinity and, when necessary, the proper conditions at geometric singularities of the scattering surface. These are available in various forms (sections 2.2.3 and 2.2.3), but they also are fully symmetric with respect to the electric and magnetic fields.

Assume that the z axis is the symmetry axis of the scatterer, the plane wave is incident along the z axis,

$$\mathbf{E}_{\text{inc}} = E_0 \hat{x}\, e^{jkz}, \quad \mathbf{H}_{\text{inc}} = -YE_0 \hat{y} e^{jkz}, \tag{3.35}$$

and the scattered electric field

$$\mathbf{E}_{\text{sc}}(x, y, z) = \hat{x}E_x^{\text{sc}}(x, y, z) + \hat{y}E_y^{\text{sc}}(x, y, z) + \hat{z}E_z^{\text{sc}}(x, y, z) \tag{3.36}$$

is known. Then it follows from the symmetry of the scatterer, the form of equations (3.31) and (3.34) and the relation between the electric and magnetic fields in the incident wave (3.35) that the magnetic field can be obtained from the electric field by multiplying with the wave admittance $Y = 1/Z$ and interchanging $x \to -y$, $y \to x$,

$$\mathbf{H}_{\text{sc}}(x, y, z) = \hat{x}YE_y^{\text{sc}}(-y, x, z) - \hat{y}YE_x^{\text{sc}}(-y, x, z) + \hat{z}YE_z^{\text{sc}}(-y, x, z). \tag{3.37}$$

At the symmetry axis $x = 0$ and $y = 0$,

$$\mathbf{H}_{\text{sc}}(0, 0, z) = \hat{x}YE_y^{\text{sc}}(0, 0, z) - \hat{y}YE_x^{\text{sc}}(0, 0, z) + \hat{z}YE_z^{\text{sc}}(0, 0, z), \tag{3.38}$$

which can be rewritten as

$$\mathbf{H}_{\text{sc}}(0, 0, z) = -Y\hat{z} \times \mathbf{E}_{\text{sc}}(0, 0, z) + Y\hat{z}E_z^{\text{sc}}(0, 0, z). \tag{3.39}$$

In the far zone in the backscattering direction ($z \to +\infty$), the z component of the scattered field is much smaller than the x and y components, so we can write

$$\mathbf{E}_{sc}(0,0,z) \approx \hat{x}E_x^{sc}(0,0,z) + \hat{y}E_y^{sc}(0,0,z) \tag{3.40}$$

$$\mathbf{H}_{sc}(0,0,z) \approx -Y\hat{z}\times\mathbf{E}_{sc}(0,0,z). \tag{3.41}$$

Now calculating the power flow $\mathbf{w}_{sc} = \frac{1}{2}\,\mathrm{Re}\,(\mathbf{E}_{sc}\times\mathbf{H}_{sc}^*)$ (section 2.4.1) in the backscattering direction ($x = 0$, $y = 0$, $z \to +\infty$) with the help of (3.40) and (3.41) gives

$$\mathbf{w}_{sc} = -\frac{Y}{2}\hat{z}|\mathbf{E}_{sc}(0,0,z)|^2, \tag{3.42}$$

which is a vector directed *toward* the scatterer. Hence, $\mathbf{E}_{sc}(0,0,z)$ must vanish in the backscattering direction as $z \to +\infty$, which proves the theorem.

A similar theorem can be formulated for homogeneous magneto-dielectric bodies with $\mu_1/\epsilon_1 = \mu/\epsilon$ where ϵ_1 and μ_1 are the permittivity and permeability inside the body (Weston 1963) (problem 13). For incidence directions off the symmetry axis or for non-symmetric bodies the theorem does not apply. However, it can be shown that the backscattering cross section of every impedance-matched convex body goes to zero as $k \to \infty$ (section 8.4.3). In section 4.6.3 we prove the minimum-reflection property for the infinite planar impedance-matched boundary. Polarization-averaged scattering cross sections (mono-, bistatic, total scattering, absorption, and extinction) of arbitrarily shaped compact bodies have an extremum as functions of $\eta = \tilde{Z}_s/Z$ at $\eta = 1$ (Osipov 2015).

Results regarding vanishing backscattering are also available for other bodies, for example bodies of revolution with anisotropic surface impedance (Yee and Chang 1991), spherical chiral structures (Uslenghi 1990), bodies of revolution composed of impenetrable impedance-matched bodies and penetrable bodies with tensor permittivity and permeability (Uslenghi 1996), and small particles (Vehmas *et al.* 2013). General conditions for zero backscattering from finite-size bodies are formulated in Lindell *et al.* (2009).

3.3 Scattering Matrix

3.3.1 Definition

Considering scattering from objects excited by transverse waves (objects located far from the illuminating antenna), we can write the incident electric field as a sum of two orthogonal components:

$$\mathbf{E}_{inc} = E_x^{inc}\hat{x}_{inc} + E_y^{inc}\hat{y}_{inc} \tag{3.43}$$

(x and y components, assuming that the incident wave propagates along z, see Figure 3.6). Likewise, we can decompose the scattered field at the position of the receiver:

$$\mathbf{E}_{sc} = E_x^{sc}\hat{x}_{sc} + E_y^{sc}\hat{y}_{sc}. \tag{3.44}$$

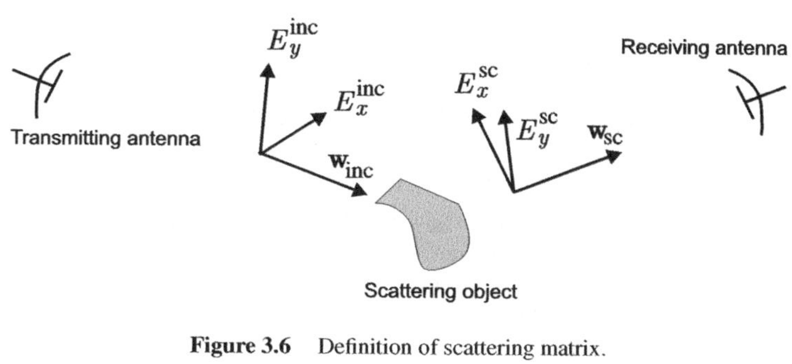

Figure 3.6 Definition of scattering matrix.

The unit vectors are in general different because two coordinate systems are bound to the positions of the transmitting and receiving antennas that may be arbitrary.

Assuming that the scattering is a linear process, we can write linear relations between the scattered and incident field components:

$$E_x^{\mathrm{sc}} = a_{11}E_x^{\mathrm{inc}} + a_{12}E_y^{\mathrm{inc}} \tag{3.45}$$

$$E_y^{\mathrm{sc}} = a_{21}E_x^{\mathrm{inc}} + a_{22}E_y^{\mathrm{inc}} \tag{3.46}$$

or, in the matrix form,

$$\begin{pmatrix} E_x^{\mathrm{sc}} \\ E_y^{\mathrm{sc}} \end{pmatrix} = \begin{pmatrix} a_{11} & a_{12} \\ a_{21} & a_{22} \end{pmatrix} \begin{pmatrix} E_x^{\mathrm{inc}} \\ E_y^{\mathrm{inc}} \end{pmatrix}. \tag{3.47}$$

Using the definition of the scattering cross section (3.6), the scattered field of an isotropic scatterer can be conveniently represented as

$$E_{\mathrm{sc}} = \frac{\mathrm{e}^{-jkR}}{\sqrt{4\pi}R} \sqrt{\sigma}\, \mathrm{e}^{j\Psi} E_{\mathrm{inc}}. \tag{3.48}$$

Here, k is the wave number in the surrounding space. The field at the receiver is a spherical wave, and parameter Ψ describes the phase change that takes place in the scattering process. Denoting $A = \sqrt{\sigma}\, \mathrm{e}^{j\Psi}$ and generalizing, we can write

$$\begin{pmatrix} E_x^{\mathrm{sc}} \\ E_y^{\mathrm{sc}} \end{pmatrix} = \frac{\mathrm{e}^{-jkR}}{\sqrt{4\pi}R} \begin{pmatrix} A_{11} & A_{12} \\ A_{21} & A_{22} \end{pmatrix} \begin{pmatrix} E_x^{\mathrm{inc}} \\ E_y^{\mathrm{inc}} \end{pmatrix}. \tag{3.49}$$

Matrix $\overline{\overline{A}}$ is called the scattering matrix. In terms of the scattering cross section,

$$\overline{\overline{A}} = \begin{pmatrix} \sqrt{\sigma_{11}}\, \mathrm{e}^{j\Psi_{11}} & \sqrt{\sigma_{12}}\, \mathrm{e}^{j\Psi_{12}} \\ \sqrt{\sigma_{21}}\, \mathrm{e}^{j\Psi_{21}} & \sqrt{\sigma_{22}}\, \mathrm{e}^{j\Psi_{22}} \end{pmatrix}. \tag{3.50}$$

Similarly to the definition of scattering cross sections, the scattering matrix is defined for an incident plane wave. In order for (3.49) to apply, the incident wave must be close to a plane wave over the whole region occupied by the scatterer.

Scattering matrix $\overline{\overline{A}}$ is more general than the scattering cross section σ since it additionally includes information about the phase of the scattered field or, more precisely, about the phase difference between the scattered and incident waves. In contrast to the amplitude, the phase of a plane wave is a function of spatial coordinates, and a unique definition of the phase of the scattered field requires a precise specification of the point at which the incident wave is taken and of the point from which the distance R to the observer is measured. Typically, the origin of a coordinate system located at a fixed but otherwise arbitrary distance from the scatterer is chosen as such a reference point. With this definition, a shift in the position of the origin results in a phase shift in the value of the scattering matrix.

The scattering matrix explicitly relates the polarization states of the incident and scattered waves. Diagonal elements A_{11} and A_{22} are often referred to as co-polar components and elements A_{12} and A_{21} as cross-polar components of the scattering matrix. This interpretation is apparent for monostatic configurations in which the transmitting and receiving antennas reside at the same place and are polarized either parallel or transversal. For bistatic configurations, with separated transmitter and receiver, the elements of the scattering matrix may not directly correspond to the co- and cross-polarization cases.

If the scatterer does not contain any nonreciprocal materials, interchanging the positions of the source and receiver leads to a transposed scattering matrix $\overline{\overline{A}}^T$ which has the same diagonal but interchanged off-diagonal elements. This property directly follows from reciprocity of electromagnetic phenomena (see (3.122)) and implies the symmetry of the scattering matrix ($A_{12} = A_{21}$) for monostatic configurations.

Traditionally, the scattering matrix is defined in terms of the electric field, but it can also be defined for the incident and scattered magnetic fields. This is not necessary, however, since the electric and magnetic fields in the incident plane wave and in the scattered spherical wave are uniquely related to each other according to (2.156) and (2.165), and thus the magnetic field scattering matrix can be directly calculated from the electric field scattering matrix (see problem 10).

One can choose the bases differently to represent the components of the incident and scattered waves. The z axis must coincide with the direction of propagation of the wave, but the other two axes, transversal to the z direction, can be chosen differently. An approach typically used in measurements is to choose one of the transverse axes to be parallel to the ground plane (horizontal axis). The remaining third axis is then uniquely defined since it must be orthogonal to the first two. This latter axis is then referred to as vertical, which is, strictly speaking, accurate only in the limit of grazing incidence and scattering. The polarization state of the waves are in this case described as horizontal or vertical.

From the theoretical point of view, two other ways of defining the polarization of the incident and scattered waves are more preferable. One way is to use the spherical coordinate system (r, θ, ϕ) as a global system of coordinates, in which case the scattering direction coincides with the unit vector $\hat{r}$ and the unit vectors $\hat{\theta}$ and $\hat{\phi}$ describe the polarization state. An alternative approach defines the polarization cases by using the so-called plane of scattering (PoS) as the reference plane. This latter approach preserves the physical meaning of the (off-) diagonal elements of the scattering matrix as the (cross-) co-polar components of the scattered field in the case of bistatic scattering. These two approaches are now discussed in succession.

3.3.2 Scattering Matrix in Spherical Coordinates

The spherical coordinate system is a convenient mathematical framework to describe the scattering matrix (Figure 3.7). A plane wave incident toward the origin of the coordinate system in the direction

$$\hat{i} = -\hat{x} \sin\theta_{\rm inc} \cos\phi_{\rm inc} - \hat{y} \sin\theta_{\rm inc} \sin\phi_{\rm inc} - \hat{z} \cos\theta_{\rm inc} \tag{3.51}$$

has no component in the direction of propagation and at every point in space can be represented as

$$\mathbf{E}_{\rm inc} = E^{\rm inc}_{\theta_{\rm inc}} \hat{\theta}_{\rm inc} + E^{\rm inc}_{\phi_{\rm inc}} \hat{\phi}_{\rm inc}, \tag{3.52}$$

where

$$E^{\rm inc}_{\theta_{\rm inc}} = \hat{\theta}_{\rm inc} \cdot \mathbf{E}_{\rm inc}, \quad E^{\rm inc}_{\phi_{\rm inc}} = \hat{\phi}_{\rm inc} \cdot \mathbf{E}_{\rm inc} \tag{3.53}$$

and

$$\hat{\theta}_{\rm inc} = \hat{x} \cos\theta_{\rm inc} \cos\phi_{\rm inc} + \hat{y} \cos\theta_{\rm inc} \sin\phi_{\rm inc} - \hat{z} \sin\theta_{\rm inc} \tag{3.54}$$

$$\hat{\phi}_{\rm inc} = -\hat{x} \sin\phi_{\rm inc} + \hat{y} \cos\phi_{\rm inc}. \tag{3.55}$$

Equations (3.51), (3.54), and (3.55) follow from (A.50)–(A.52) with $\hat{i} = -\hat{r}$, $\theta = \theta_{\rm inc}$, and $\phi = \phi_{\rm inc}$.

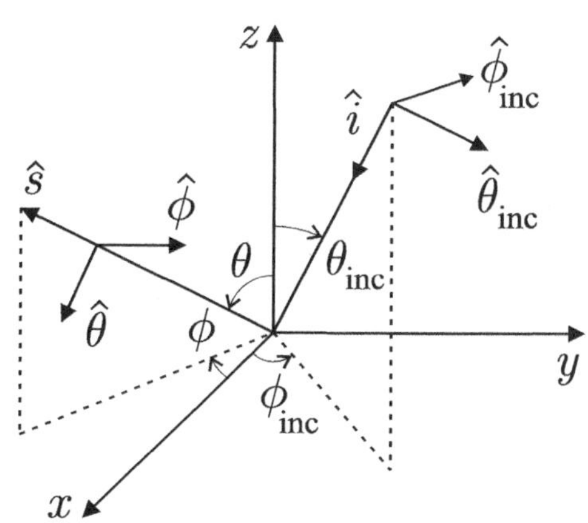

Figure 3.7 Bases of the incident and scattered fields in the spherical coordinate system.

Sufficiently far from the scatterer, at a point with coordinates r, θ, and ϕ, the scattered field is a spherical wave that can be considered as a locally plane wave propagating in the radial direction

$$\hat{s} = \hat{x}\sin\theta\cos\phi + \hat{y}\sin\theta\sin\phi + \hat{z}\cos\theta \tag{3.56}$$

so that

$$\mathbf{E}_{\text{sc}} = E_\theta^{\text{sc}}\hat{\theta} + E_\phi^{\text{sc}}\hat{\phi} \tag{3.57}$$

with

$$\hat{\theta} = \hat{x}\cos\theta\cos\phi + \hat{y}\cos\theta\sin\phi - \hat{z}\sin\theta \tag{3.58}$$

$$\hat{\phi} = -\hat{x}\sin\phi + \hat{y}\cos\phi \tag{3.59}$$

and

$$E_\theta^{\text{sc}} = \hat{\theta}\cdot\mathbf{E}_{\text{sc}}, \quad E_\phi^{\text{sc}} = \hat{\phi}\cdot\mathbf{E}_{\text{sc}}. \tag{3.60}$$

Equations (3.52), (3.53), (3.57), and (3.60) are representations of the incident and scattered waves in the bases ($\hat{i}$, $\hat{\theta}_{\text{inc}}$, $\hat{\phi}_{\text{inc}}$) and ($\hat{s}$, $\hat{\theta}$, $\hat{\phi}$), respectively. The corresponding scattering matrix relates the θ and ϕ components of the scattered field at the observation point (r, θ, ϕ) in the far field with the θ_{inc} and ϕ_{inc} components of the incident field,

$$\begin{pmatrix} E_\theta^{\text{sc}} \\ E_\phi^{\text{sc}} \end{pmatrix} = \frac{e^{-jkr}}{\sqrt{4\pi}r}\begin{pmatrix} A_{\theta\theta_{\text{inc}}} & A_{\theta\phi_{\text{inc}}} \\ A_{\phi\theta_{\text{inc}}} & A_{\phi\phi_{\text{inc}}} \end{pmatrix}\begin{pmatrix} E_{\theta_{\text{inc}}}^{\text{inc}} \\ E_{\phi_{\text{inc}}}^{\text{inc}} \end{pmatrix}. \tag{3.61}$$

The elements of the matrix can be determined from the fields by using the relations:

$$A_{\theta\theta_{\text{inc}}} = \lim_{r\to\infty}\sqrt{4\pi}\, r e^{jkr}\hat{\theta}\cdot\mathbf{E}_{\text{sc}} \tag{3.62}$$

$$A_{\phi\theta_{\text{inc}}} = \lim_{r\to\infty}\sqrt{4\pi}\, r e^{jkr}\hat{\phi}\cdot\mathbf{E}_{\text{sc}} \tag{3.63}$$

with $E_{\theta_{\text{inc}}}^{\text{inc}} = 1$ and $E_{\phi_{\text{inc}}}^{\text{inc}} = 0$ and

$$A_{\theta\phi_{\text{inc}}} = \lim_{r\to\infty}\sqrt{4\pi}\, r e^{jkr}\hat{\theta}\cdot\mathbf{E}_{\text{sc}} \tag{3.64}$$

$$A_{\phi\phi_{\text{inc}}} = \lim_{r\to\infty}\sqrt{4\pi}\, r e^{jkr}\hat{\phi}\cdot\mathbf{E}_{\text{sc}} \tag{3.65}$$

with $E_{\theta_{\text{inc}}}^{\text{inc}} = 0$ and $E_{\phi_{\text{inc}}}^{\text{inc}} = 1$. Depending on the polarization of the incident wave ($\hat{\theta}_{\text{inc}}$ or $\hat{\phi}_{\text{inc}}$) and the receiving antenna ($\hat{\theta}$ or $\hat{\phi}$), one has a set of four scattering cross sections

$$\sigma_{uv}(\theta,\phi,\theta_{\text{inc}},\phi_{\text{inc}}) = |A_{uv}(\theta,\phi,\theta_{\text{inc}},\phi_{\text{inc}})|^2, \tag{3.66}$$

where the subscripts u and v denote the polarization of the incident wave ($u = \theta$ or ϕ) and the receiver ($v = \theta_{\text{inc}}$ or ϕ_{inc}).

In the spherical coordinates, the general definition for the total scattering cross section (3.9) takes the form

$$\sigma_{\text{T}} = \lim_{r\to\infty} r^2 \int_0^{2\pi}\int_0^{\pi} \frac{\left|E_\theta^{\text{sc}}\right|^2 + \left|E_\phi^{\text{sc}}\right|^2}{\left|E_v^{\text{inc}}\right|^2}\sin\theta\, d\theta\, d\phi \tag{3.67}$$

with the result that

$$\sigma_{\text{T}}(\hat{\theta}_{\text{inc}}) = \frac{1}{4\pi}\int_0^{2\pi}\int_0^{\pi}\left(\left|A_{\theta\theta_{\text{inc}}}\right|^2 + \left|A_{\phi\theta_{\text{inc}}}\right|^2\right)\sin\theta\, d\theta\, d\phi \tag{3.68}$$

for the incident wave polarized in the $\hat{\theta}_{\mathrm{inc}}$ direction and

$$\sigma_{\mathrm{T}}(\hat{\phi}_{\mathrm{inc}}) = \frac{1}{4\pi}\int_0^{2\pi}\int_0^{\pi}\left(\left|A_{\theta\phi_{\mathrm{inc}}}\right|^2 + \left|A_{\phi\phi_{\mathrm{inc}}}\right|^2\right)\sin\theta\,\mathrm{d}\theta\,\mathrm{d}\phi \tag{3.69}$$

for the incident wave polarized in the $\hat{\phi}_{\mathrm{inc}}$ direction.

3.3.3 Scattering Matrix in the Plane of Scattering Coordinates

For bistatic arrangements, diagonal (off-diagonal) elements of scattered matrices do not necessarily represent the co- (cross-) polar components. For example, a bistatic scattering matrix of a sphere may include off-diagonal elements implying cross-polarization in a scattered field, which in fact is unphysical. When extending the concept of co- and cross-polarization to the bistatic case it is most natural to require that the field scattered by a sphere be fully co-polarized. A further requirement is that the definition should recover the usual definitions in the back or forward-scattering limit. Furthermore, to be sufficiently general, the definition must be independent of the particular target geometry. In this section we show that all these points are achieved by describing the scattering in the so-called *plane of scattering (PoS)* reference frame (Boerner *et al.* 1998; Van de Hulst 1957).

In a bistatic case, the vectors $\hat{i}$ and $\hat{s}$ are different and, thus, they uniquely define a plane of reference which is referred to as the PoS. To describe bistatic scattering we need two bases, a basis of the incident wave and a basis of the scattered wave. We take $\hat{i}$ as the first unit vector of the incident wave basis and supplement it with two further vectors

$$\hat{\eta}_{\mathrm{inc}} = \frac{\hat{s}\times\hat{i}}{|\hat{s}\times\hat{i}|},\qquad \hat{\xi}_{\mathrm{inc}} = \hat{i}\times\hat{\eta}_{\mathrm{inc}}, \tag{3.70}$$

the former being perpendicular to PoS and the latter parallel to PoS and perpendicular to both $\hat{i}$ and $\hat{\eta}_{\mathrm{inc}}$ (Figure 3.8). The basis of the scattered wave is defined analogously by the vector $\hat{s}$ and the unit vectors

$$\hat{\eta} = \hat{\eta}_{\mathrm{inc}},\qquad \hat{\xi} = \hat{\eta}\times\hat{s}. \tag{3.71}$$

The components of the fields in these bases can, therefore, be referred to as perpendicular and parallel, respectively. They are often denoted by the subscripts r and l, which stand for the last letters of the words perpendicular and parallel (Van de Hulst 1957).

In the monostatic case, when $\hat{s} = -\hat{i}$, or in the case of forward scattering when $\hat{s} = \hat{i}$, the vector product $\hat{s}\times\hat{i}$ vanishes and equations (3.70) and (3.71) do not apply, but the vectors $\hat{\xi}_{\mathrm{inc}}$, $\hat{\eta}_{\mathrm{inc}}$, $\hat{\xi}$, and $\hat{\eta}$ can still be defined as limiting values with the angle between the incidence and scattering directions approaching 0 (forward scattering) or π (backscattering). In the backscattering limit, the definitions (3.70) and (3.71) give $\hat{\xi}_{\mathrm{inc}} = \hat{\xi}$ and $\hat{\eta}_{\mathrm{inc}} = \hat{\eta}$, thus recovering a common reference frame for the incident and scattering waves and, therefore, the usual definitions of co- and cross-polarization in the monostatic case. In the forward-scattering case, (3.70) and (3.71) result in the limiting relations $\hat{\xi}_{\mathrm{inc}} = -\hat{\xi}$ and $\hat{\eta}_{\mathrm{inc}} = \hat{\eta}$. Changing the expression for ξ in (3.71) to $\hat{\xi} = \hat{s}\times\hat{\eta}$ leads to an alternative definition with the scattered wave basis coincident with

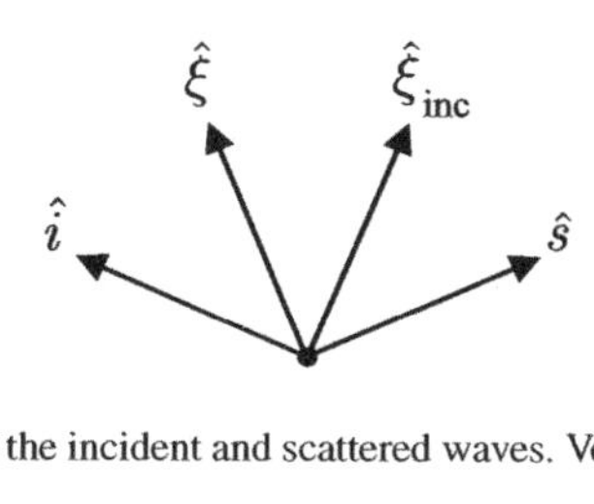

Figure 3.8 Plane of scattering and the bases of the incident and scattered waves. Vectors $\hat{\eta}$ and $\hat{\eta}_{\mathrm{inc}}$ are perpendicular to the plane and coincide with each other.

the basis of the incident wave in the limit of forward scattering, but in the backscattering limit this definition results in oppositely directed vectors ξ and ξ_{inc}. Both definitions are equivalent, and their choice is a matter of convenience.

In the PoS bases, the incident and scattered waves can be represented by

$$\mathbf{E}_{\text{inc}} = E_{\parallel}^{\text{inc}} \hat{\xi}_{\text{inc}} + E_{\perp}^{\text{inc}} \hat{\eta}_{\text{inc}} \tag{3.72}$$

$$\mathbf{E}_{\text{sc}} = E_{\parallel}^{\text{sc}} \hat{\xi} + E_{\perp}^{\text{sc}} \hat{\eta} \tag{3.73}$$

with the components

$$E_{\parallel}^{\text{inc}} = \hat{\xi}_{\text{inc}} \cdot \mathbf{E}_{\text{inc}}, \quad E_{\perp}^{\text{inc}} = \hat{\eta}_{\text{inc}} \cdot \mathbf{E}_{\text{inc}} \tag{3.74}$$

$$E_{\parallel}^{\text{sc}} = \hat{\xi} \cdot \mathbf{E}_{\text{sc}}, \quad E_{\perp}^{\text{sc}} = \hat{\eta} \cdot \mathbf{E}_{\text{sc}}. \tag{3.75}$$

The components are related by a corresponding scattering matrix,

$$\begin{pmatrix} E_{\parallel}^{\text{sc}} \\ E_{\perp}^{\text{sc}} \end{pmatrix} = \frac{\mathrm{e}^{-jkr}}{\sqrt{4\pi} r} \begin{pmatrix} A_{\parallel\parallel} & A_{\parallel\perp} \\ A_{\perp\parallel} & A_{\perp\perp} \end{pmatrix} \begin{pmatrix} E_{\parallel}^{\text{inc}} \\ E_{\perp}^{\text{inc}} \end{pmatrix} \tag{3.76}$$

which will be denoted $\overline{\overline{A}}_{\text{PoS}}$.

The scattering matrix $\overline{\overline{A}}_{\text{PoS}}$ can be obtained from the scattering matrix $\overline{\overline{A}}_{\text{sph}}$ in spherical coordinates (see (3.61)). By dot-multiplying (3.72) with the unit vectors $\hat{\theta}_{\text{inc}}$ and $\hat{\phi}_{\text{inc}}$, and (3.57) with $\hat{\xi}$ and $\hat{\eta}$ we obtain the relations

$$\begin{pmatrix} E_{\theta_{\text{inc}}}^{\text{inc}} \\ E_{\phi_{\text{inc}}}^{\text{inc}} \end{pmatrix} = \overline{\overline{V}} \cdot \begin{pmatrix} E_{\parallel}^{\text{inc}} \\ E_{\perp}^{\text{inc}} \end{pmatrix} \tag{3.77}$$

and

$$\begin{pmatrix} E_{\parallel}^{\text{sc}} \\ E_{\perp}^{\text{sc}} \end{pmatrix} = \overline{\overline{U}} \cdot \begin{pmatrix} E_{\theta}^{\text{sc}} \\ E_{\phi}^{\text{sc}} \end{pmatrix} \tag{3.78}$$

with

$$\overline{\overline{U}} = \begin{pmatrix} \hat{\xi} \cdot \hat{\theta} & \hat{\xi} \cdot \hat{\phi} \\ \hat{\eta} \cdot \hat{\theta} & \hat{\eta} \cdot \hat{\phi} \end{pmatrix} \tag{3.79}$$

$$\overline{\overline{V}} = \begin{pmatrix} \hat{\xi}_{\text{inc}} \cdot \hat{\theta}_{\text{inc}} & \hat{\eta}_{\text{inc}} \cdot \hat{\theta}_{\text{inc}} \\ \hat{\xi}_{\text{inc}} \cdot \hat{\phi}_{\text{inc}} & \hat{\eta}_{\text{inc}} \cdot \hat{\phi}_{\text{inc}} \end{pmatrix}. \tag{3.80}$$

Inserting (3.77) in (3.61), followed by multiplication of (3.61) with the matrix (3.80), leads, on comparison with (3.76), to the formula

$$\overline{\overline{A}}_{\text{PoS}} = \overline{\overline{U}} \cdot \overline{\overline{A}}_{\text{sph}} \cdot \overline{\overline{V}} \tag{3.81}$$

which relates the scattering matrix in spherical coordinates with the scattering matrix in the PoS bases.

In Chapter 7 we will show that the scattering matrix of a sphere in the PoS coordinates is diagonal for arbitrary values of the sphere radius, the wavelength, and the incidence and scattering aspects.

The concept of the PoS helps to identify the cases in which the scattering matrices of the same but differently positioned and oriented scatterers can be expressed through the same elements (Van de Hulst 1957). These configurations are obtained by mirroring an initial configuration with respect to two planes. To describe the transformations, one introduces a line halving the angle between the incidence and scattering directions (called *bisectrix*) in the PoS and a plane through this line and perpendicular to the plane of scattering (*bisectrix plane*) (Figure 3.9).

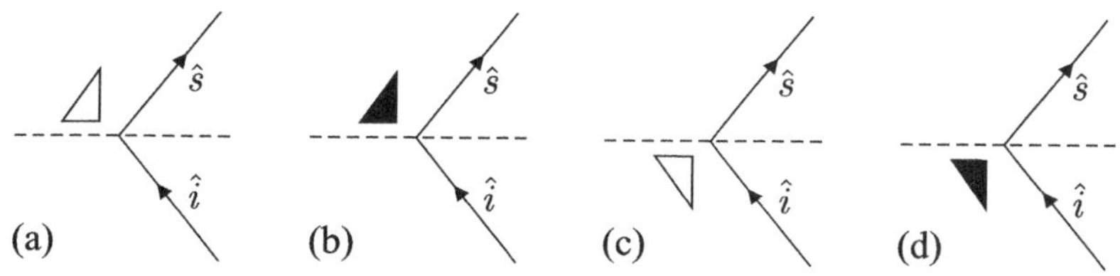

Figure 3.9 (a) The basic configuration and configurations mirrored with respect to (b) the PoS, (c) the bisectrix plane or (d) both. The dashed line is the bisectrix, which halves the angle between the incidence direction $\hat{\imath}$ and scattering direction $\hat{s}$. The scatterer is shown as a triangle with a white upper site and black lower side.

We now take an arbitrary scatter, arbitrarily oriented and positioned, and build the PoS matrix for the given incidence and scattering directions,

$$\overline{\overline{A}}_a = \begin{bmatrix} A_{11} & A_{12} \\ A_{21} & A_{22} \end{bmatrix}. \tag{3.82}$$

This is our initial configuration, which will be referred to as case a (Figure 3.9). Mirroring the scatterer with respect to the PoS gives a second configuration, referred to as configuration b (Figure 3.9). This operation is equivalent to alternating the sign of the perpendicular components in the incident and scattered field, implying the scattering matrix

$$\overline{\overline{A}}_b = \begin{bmatrix} A_{11} & -A_{12} \\ -A_{21} & A_{22} \end{bmatrix}. \tag{3.83}$$

Configuration c is obtained from configuration a by mirroring the scatterer with respect to the bisectrix plane (Figure 3.9). On interchanging the incidence and scattering directions and inverting the direction of propagation we obtain the same physical situation as in case a. Configuration c can be seen as reciprocal of a, and the reciprocity property for the scattered field (see section 3.4.2 and (3.122)) leads to the transposed scattering matrix

$$\overline{\overline{A}}_c = \begin{bmatrix} A_{11} & A_{21} \\ A_{12} & A_{22} \end{bmatrix}. \tag{3.84}$$

Finally, rotation of the scatterer in the basic configuration around the bisectrix by 180° is equivalent to mirroring either configuration c with respect to the PoS or configuration b with respect to the bisectrix plane. The resulting configuration is referred to as d, and the corresponding scattering matrix is obtained by transposing the matrix $\overline{\overline{A}}_b$ or by alternating the signs of the off-diagonal elements in $\overline{\overline{A}}_c$,

$$\overline{\overline{A}}_d = \begin{bmatrix} A_{11} & -A_{21} \\ -A_{12} & A_{22} \end{bmatrix}. \tag{3.85}$$

Thus, the concept of the PoS reference frame is a useful and physically meaningful way of describing bistatic scattering. Numerical illustrations of electromagnetic scattering from simple canonical targets in terms of the PoS can be found, for example, in Kemptner and Osipov (2007).

3.4 Far-field Coefficient

The field scattered by any compact object at sufficiently large distances from the object has the form of a spherical wave and can be represented in the form

$$\mathbf{E}_{sc}(\mathbf{r}) = \mathbf{F}\frac{e^{-jkr}}{kr}, \tag{3.86}$$

where the amplitude vector **F** will be referred to as the far-field coefficient. Another meaningful name, particularly in problems of electromagnetic scattering, is the scattering amplitude.[3] The far-field coefficient is in general a complex-valued vector proportional to the scattered electric field and perpendicular to the normal to the wavefront at the observation point. It also depends on the incident electric field at the location of the scatterer. The distance r is measured from the origin of a coordinate system co-located with the scattering body.

Far-field coefficient **F** in (3.86) is an important quantity that characterizes the electric field scattered by a compact scatterer sufficiently "far" from the scatterer. Although **F** is defined for the electric field, it also determines the scattered magnetic field since in a spherical wave the electric and magnetic fields are directly related (see (2.165)). Once **F** is known, it can be used to calculate the scattering cross section for a given polarization of an incident wave and given incidence and scattering directions. Two far-field coefficients for two linearly polarized incident waves with mutually orthogonal polarization directions provide the complete information for building the scattering matrix of the scatterer. Finally, the far-field coefficient for the scattering in the forward direction defines the extinction cross section of the scatterer by means of the optical theorem.

In this section, we formulate mathematical conditions for the far-field regime, under which the scattered field can be regarded as a spherical wave, and derive a useful integral representation for **F** which can be used when the electromagnetic field, exact or approximate, is available on a closed surface in the vicinity of the scatterer. The surface can be, for example, the surface of the scatterer itself, or a boundary of a simulation volume. This formula transforms the near field to the far scattered field and can be seen as a near-field-to-far-field transformation. We then derive a universal formula that describes the far-field coefficient for the forward scattering from impenetrable scatterers in the optical limit ($\lambda \to 0$). We present here a new derivation of the formula, which is fully electromagnetic and makes use of neither the scalar (Kirchhoff) version of the scattering theory nor Babinet's principle (which is limited to PEC scatterers). Finally, the section concludes with the derivation of the useful forward-scattering (optical) theorem.

3.4.1 *Integral Representations and Far-field Conditions*

The derivation of the far-field formula is based on the integral representation (2.346), in which we choose the integration surface Ω to be a closed surface enclosing a scatterer or a group of scatterers but not the sources of the field, so that the expression

$$\mathbf{E}_{sc}(\mathbf{r}) = \frac{1}{j\omega\epsilon}[\nabla(\nabla\cdot) + k^2] \int_{\Omega} \mathbf{K}_e(\mathbf{r}')G_0(\mathbf{r},\mathbf{r}')\,d\Omega' - \nabla \times \int_{\Omega} \mathbf{K}_m(\mathbf{r}')G_0(\mathbf{r},\mathbf{r}')\,d\Omega' \tag{3.87}$$

describes the scattered field. We suppose that the equivalent electric and magnetic currents $\mathbf{K}_e$ and $\mathbf{K}_m$ or, which is the same (see (2.348)), the tangential components of the fields are known on Ω. The geometric and material properties of the scatterers can be arbitrary. In the case of a single scatterer, Ω can coincide with the physical surface of the scatterer.

Representation (3.87) is applicable in the exterior of the integration surface at arbitrary distances from the surface, and we now start moving the observation point farther from the surface. We define the distance R between the observation point $\mathbf{r}$ and the scattering surface Ω by

$$R = \min_{\mathbf{r}' \in \Omega} |\mathbf{r} - \mathbf{r}'|, \tag{3.88}$$

and when R becomes greater than the wavelength

$$R \gg \lambda, \tag{3.89}$$

formula (3.87) permits significant simplifications, namely, the result of the application of the vector differential operators ∇, $\nabla\cdot$, and $\nabla\times$ to the Green function $G_0(\mathbf{r},\mathbf{r}')$, which is given by (2.199), can be evaluated in a simple form. For example,

[3] The scattering amplitude is often defined as $\mathbf{F}/k$.

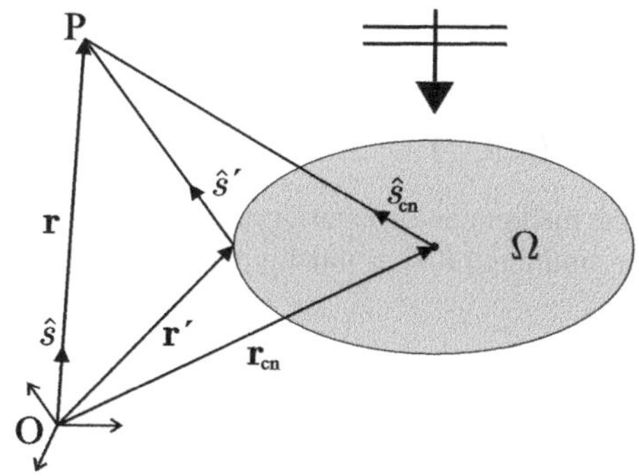

Figure 3.10 Notation used in the derivation of the far-field formula: O, the origin of a global coordinate system; P, the observation point; **r**, the position vector of P; $\hat{s}$, the unit vector pointing from O to P; Ω, the surface of integration; $\mathbf{r}'$, an integration point on Ω; $\hat{s}'$, the unit vector pointing from $\mathbf{r}'$ to $\mathbf{r}$; $\mathbf{r}_{cn}$, the position vector of the center of the scattering region; $\hat{s}_{cn}$, the unit vector pointing from $\mathbf{r}_{cn}$ to **r**.

applying the gradient operator to the Green function gives

$$\nabla G_0(\mathbf{r},\mathbf{r}') = \frac{e^{-jk|\mathbf{r}-\mathbf{r}'|}}{4\pi|\mathbf{r}-\mathbf{r}'|}\left(-jk - \frac{1}{|\mathbf{r}-\mathbf{r}'|}\right)\nabla|\mathbf{r}-\mathbf{r}'|. \tag{3.90}$$

Because of (3.89) the second term in the brackets on the right-hand side of (3.90) is small compared with the first one, and (3.90) simplifies to

$$\nabla G_0(\mathbf{r},\mathbf{r}') \approx -jk\hat{s}' G_0(\mathbf{r},\mathbf{r}'), \tag{3.91}$$

where

$$\hat{s}' = \frac{\mathbf{r}-\mathbf{r}'}{|\mathbf{r}-\mathbf{r}'|} \tag{3.92}$$

is the unit vector pointing from the integration point $\mathbf{r}'$ to the observation point **r** (Figure 3.10).

Applying the divergence operator to the integrand of the first integral in (3.87) and using (A.20), (3.91), and the fact that the current $\mathbf{K}_e$ depends on $\mathbf{r}'$ but not on **r**, we find

$$\nabla\cdot(G_0\mathbf{K}_e) = \mathbf{K}_e\cdot\nabla G_0 \approx -jkG_0\mathbf{K}_e\cdot\hat{s}'. \tag{3.93}$$

Next, using (3.91) and (A.19), and assuming $\hat{s}'$ to be slowly varying on the scale of the wavelength, we evaluate the gradient of the expression in (3.93)

$$\nabla[\nabla\cdot(G_0\mathbf{K}_e)] \approx -jk\nabla[G_0(\mathbf{K}_e\cdot\hat{s}')] \approx -k^2G_0(\mathbf{K}_e\cdot\hat{s}')\hat{s}' \tag{3.94}$$

and with the help of (A.12) arrive at the formula

$$[\nabla(\nabla\cdot)+k^2](G_0\mathbf{K}_e) \approx k^2G_0[\mathbf{K}_e-(\mathbf{K}_e\cdot\hat{s}')\hat{s}'] = k^2G_0\hat{s}'\times(\hat{s}'\times\mathbf{K}_e). \tag{3.95}$$

Finally, applying the curl operator to the integrand of the second integral in (3.87) and using (A.21) and (3.91) we find

$$\nabla\times(G_0\mathbf{K}_m) = \nabla G_0\times\mathbf{K}_m \approx -jkG_0\hat{s}'\times\mathbf{K}_m, \tag{3.96}$$

and inserting (3.95) and (3.96) in (3.87) we obtain

$$\mathbf{E}_{sc}(\mathbf{r}) \approx \frac{jk}{4\pi} \int_{\Omega} \frac{e^{-jk|\mathbf{r}-\mathbf{r}'|}}{|\mathbf{r}-\mathbf{r}'|} \hat{s}' \times [Z\hat{s}' \times \mathbf{K}_e(\mathbf{r}') + \mathbf{K}_m(\mathbf{r}')]\, d\Omega', \tag{3.97}$$

where $Z = \sqrt{\mu/\epsilon}$ is the impedance of the medium in the exterior of the surface Ω.

Representation (3.97) can be further simplified if R is much greater than the characteristic size D of the scattering region (which is the interior of Ω),

$$R \gg D. \tag{3.98}$$

Let $\mathbf{r}_{cn}$ be the position vector of the geometric center of the scattering region. The center of the region can be defined as the point for which the maximum distance to the boundary of the region is at minimum. Then for every point on the integration surface Ω,

$$|\mathbf{r}-\mathbf{r}'| = |\mathbf{r}-\mathbf{r}_{cn} - (\mathbf{r}'-\mathbf{r}_{cn})| = |\mathbf{r}-\mathbf{r}_{cn}| - \hat{s}_{cn} \cdot (\mathbf{r}'-\mathbf{r}_{cn}) + O(D^2/R) \tag{3.99}$$

with

$$\hat{s}_{cn} = \frac{\mathbf{r}-\mathbf{r}_{cn}}{|\mathbf{r}-\mathbf{r}_{cn}|} + O\left(\frac{D}{R}\right), \tag{3.100}$$

and equation (3.97) becomes

$$\mathbf{E}_{sc}(\mathbf{r}) \approx \frac{jk}{4\pi} \frac{e^{-jk|\mathbf{r}-\mathbf{r}_{cn}|}}{|\mathbf{r}-\mathbf{r}_{cn}|} \int_{\Omega} e^{jk\hat{s}_{cn}\cdot(\mathbf{r}'-\mathbf{r}_{cn})} \hat{s}_{cn} \times [Z\hat{s}_{cn} \times \mathbf{K}_e(\mathbf{r}') + \mathbf{K}_m(\mathbf{r}')]\, d\Omega'. \tag{3.101}$$

In order for the approximation (3.99) in the phase of the exponent to be valid, in addition to condition (3.98) we have to require that

$$\frac{D^2}{R} \ll \lambda. \tag{3.102}$$

In contrast to (3.87) and (3.97), the representation (3.101) has the form of a spherical wave coming from the point $\mathbf{r}_{cn}$ and polarized perpendicularly to the scattering direction $\hat{s}_{cn}$ ($\hat{s}_{cn} \cdot \mathbf{E}_{sc} = 0$).

Thus, we have just proved that field scattered from any compact scatterer or a group of scatterers approaches a spherical wave as the observation point goes sufficiently far away from the scatterer. The region in which the scattered field can be considered as a spherical wave is referred to as the far-field region. For scatterers that are larger than the wavelength, the far-field region is often referred to as the *Fraunhofer region* and the transition zone, where R is of the order of D^2/λ, as the *Fresnel region* (e.g., Born and Wolf (1959)). Figure 3.11 illustrates the changing behavior of the scattered field near an impenetrable scatterer (a PEC strip). The wavelength and the width of the strip are such that $D = 4\lambda$. An incident plane wave illuminates the strip from the right, and a shadow region with lower field values is seen directly adjacent to the left side of the strip. In this region the scattered field is close to a plane wave, compensating the incident wave. In front of the strip, in the immediate vicinity of it, the scattered field is close to the plane wave reflected from an infinite PEC plane. So, within the distance of about 2λ from the screen the total field is approximately equal to its GO values. As the observation point, however, goes farther away from the screen, the scattered component begins to decay, the shadow behind the strip gradually disappears, and the interference pattern in front of the strip weakens (Fresnel region). Finally, at distances comparable to and larger than $2D^2/\lambda = 3.2$ m (not shown in the plot) the scattered component approaches a wave with a circular wavefront and an amplitude decaying with the growing distance from the scatterer (Fraunhofer region).

Conditions (3.89), (3.98), and (3.102) that are used in the derivation of the far-field representation (3.101) can be summarized as

$$R \gg D_F = \max(\lambda, D, D^2/\lambda), \tag{3.103}$$

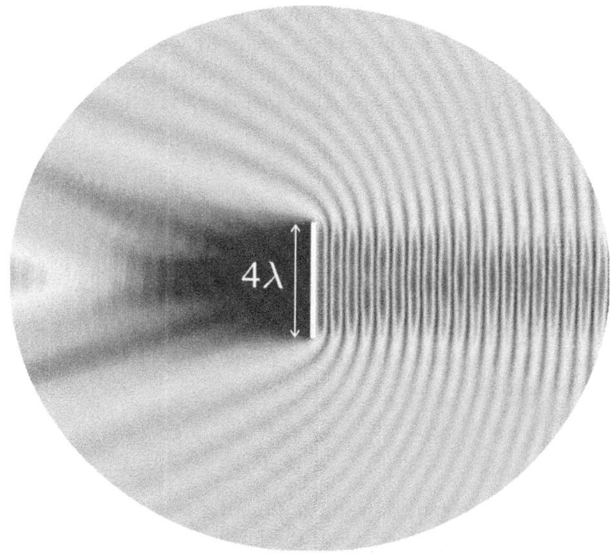

Figure 3.11 Distribution of the magnitude of the total electric field $|\mathbf{E}|$ around a perfectly conducting strip of width 4λ illuminated by a plane wave incident from the right with the wavelength λ. The wave is linearly polarized with the electric field vector perpendicular to the plane of the figure. Larger values of $|\mathbf{E}|$ are shown lighter.

implying that the position of the far-field region depends on the electrical size of the scatterer. For electrically large scatterers ($D \gg \lambda$), (3.103) gives

$$D_\mathrm{F} = \frac{D^2}{\lambda} \tag{3.104}$$

as the characteristic distance to the far-field region. Practitioners often regard $R = 2D_\mathrm{F}$ as the distance to the boundary of the far-field region. On the other hand, when the scatterer is electrically small ($D \ll \lambda$), it follows from (3.103) that $D_\mathrm{F} = \lambda$, and the far-field region begins at distances greater than a couple of wavelengths from the scatterer.

Finally, if the observation point is removed from the scatterer at a distance much greater than the distance between the origin of a global coordinate system (Figure 3.10) and the center of the scatterer, that is, $r \gg r_\mathrm{cn}$, then

$$|\mathbf{r} - \mathbf{r}_\mathrm{cn}| \approx r - \hat{s} \cdot \mathbf{r}_\mathrm{cn} \tag{3.105}$$

$$\hat{s}_\mathrm{cn} \approx \hat{s} = \frac{\mathbf{r}}{|\mathbf{r}|}, \tag{3.106}$$

and the formula (3.101) reduces to the expression

$$\mathbf{E}_\mathrm{sc}(\mathbf{r}) \approx jk\frac{\mathrm{e}^{-jkr}}{4\pi r} \int_\Omega \mathrm{e}^{jk\hat{s}\cdot\mathbf{r}'} \hat{s} \times [Z\hat{s} \times \mathbf{K}_\mathrm{e}(\mathbf{r}') + \mathbf{K}_\mathrm{m}(\mathbf{r}')]\, \mathrm{d}\Omega'. \tag{3.107}$$

By comparing (3.107) with (3.86) we obtain an integral representation of the far-field coefficient,

$$\mathbf{F} = \frac{jk^2}{4\pi} \int_\Omega \mathrm{e}^{jk\hat{s}\cdot\mathbf{r}'} \hat{s} \times [Z\hat{s} \times \mathbf{K}_\mathrm{e}(\mathbf{r}') + \mathbf{K}_\mathrm{m}(\mathbf{r}')]\, \mathrm{d}\Omega'. \tag{3.108}$$

As is apparent from (3.108), the far-field coefficient is a function of the scattering direction $\hat{s}$ and this function is determined by the frequency, material, and geometric properties of the scatterer, and by the incident field. In scattering problems, the incident field is typically assumed to be a plane wave.

The coordinate system in (3.108) can be freely chosen as long as its origin is located much closer to the scattering body than the observation point. In a coordinate system shifted by a vector $\delta\mathbf{r}$ with respect to the point O (Figure 3.10), such that $\mathbf{r} = \mathbf{r}_1 - \delta\mathbf{r}$, where $\mathbf{r}_1$ is the position vector of the observer in the shifted coordinate system, the far-field coefficient $\mathbf{F}_1$ is defined by (3.86) with r replaced with $r_1 = |\mathbf{r}_1|$ and since $|\delta\mathbf{r}| \ll r_1$, it follows from $r \approx r_1 - \hat{s} \cdot \delta\mathbf{r}$ that

$$\mathbf{F}_1 = \mathbf{F}\, e^{jk\hat{s}\cdot\delta\mathbf{r}}, \tag{3.109}$$

implying a phase shift in the far-field coefficient.

Relation to Scattering Matrix and Cross Sections

Once the far-field coefficient has been determined it can be used to calculate scattering cross sections and the scattering matrix. Using (3.86) in (3.6) and assuming illumination by a plane wave (2.155) with the amplitude $\mathbf{E}_0$, the bistatic scattering cross section can be expressed as

$$\sigma = \frac{4\pi}{k^2}\frac{|\mathbf{F}|^2}{|\mathbf{E}_0|^2}. \tag{3.110}$$

The total scattering cross section σ_T is then obtained by integration of σ over all scattering directions according to (3.9).

To obtain the scattering matrix of an obstacle, one needs the far-field coefficient for two alternative polarizations of the incident plane wave. For example, if $\mathbf{F}_{ix}$ and $\mathbf{F}_{iy}$ are the far-field coefficients that correspond to the incident waves $\mathbf{E}_{inc} = E_x^{inc}\hat{x}_{inc}$ and $\mathbf{E}_{inc} = E_y^{inc}\hat{y}_{inc}$, then the elements of the scattering matrix (3.49) can be found from the relations

$$A_{11} = \sqrt{4\pi}\frac{\hat{x}\cdot\mathbf{F}_{ix}}{kE_x^{inc}} \quad A_{12} = \sqrt{4\pi}\frac{\hat{x}\cdot\mathbf{F}_{iy}}{kE_y^{inc}}$$
$$A_{21} = \sqrt{4\pi}\frac{\hat{y}\cdot\mathbf{F}_{ix}}{kE_x^{inc}} \quad A_{22} = \sqrt{4\pi}\frac{\hat{y}\cdot\mathbf{F}_{iy}}{kE_y^{inc}}. \tag{3.111}$$

Expression (3.108) permits calculation of the far-field coefficient from tangential components of the total electric and magnetic fields on a surface enclosing the scatter(s). Integration of exact equivalent surface currents in (3.108) results in exact values of the scattering parameters. In many cases, the currents can be approximated by physically motivated values, which leads to approximated far-field parameters; this approach is followed in the framework of PO and PTD.

3.4.2 Reciprocity of Scattered Fields

The reciprocity property of electromagnetic fields (section 2.4.2) leads to a corresponding property of the scattered fields which relates the far-field coefficients on interchanging the incidence and scattering directions (de Hoop 1960; Saxon 1955).

Consider two elementary dipoles with electric moments $\mathbf{p}_1$ and $\mathbf{p}_2$ located at $\mathbf{r} = \mathbf{r}_1$ and $\mathbf{r}_2$ as the electromagnetic sources that illuminate a scattering body (Figure 3.12). Our goal is to relate the scattered field at the point $\mathbf{r}_2$ due to the first source to the scattered field at the point $\mathbf{r}_1$ due to the second source. The reciprocity relation is written for the total fields

$$\mathbf{p}_2 \cdot \mathbf{E}_1(\mathbf{r}_2) = \mathbf{p}_1 \cdot \mathbf{E}_2(\mathbf{r}_1), \tag{3.112}$$

and the total fields are the sums of the incident and scattered fields,

$$\mathbf{E}_n(\mathbf{r}) = \mathbf{E}_n^{inc}(\mathbf{r}) + \mathbf{E}_n^{sc}(\mathbf{r}) \tag{3.113}$$

with $n = 1, 2$. The field of an elementary dipole in free space satisfies the reciprocity, which can be checked by direct inspection of (2.259), implying that the scattered fields must also comply with the reciprocity relation

$$\mathbf{p}_2 \cdot \mathbf{E}_1^{sc}(\mathbf{r}_2) = \mathbf{p}_1 \cdot \mathbf{E}_2^{sc}(\mathbf{r}_1). \tag{3.114}$$

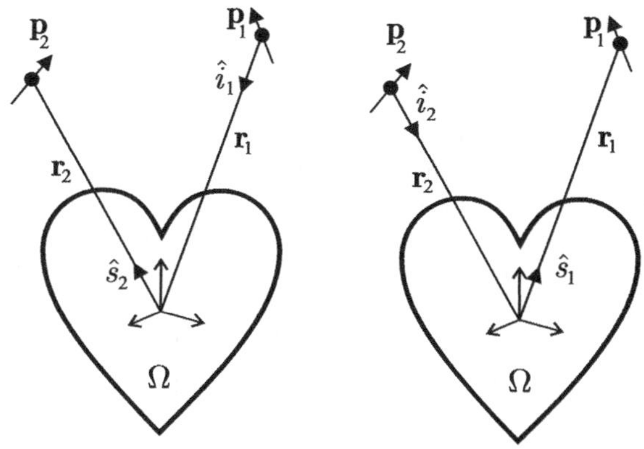

Figure 3.12 Reciprocal configurations. Left: a dipole with the electric moment $\mathbf{p}_1$ at a point $\mathbf{r}_1$ excites an incoming wave, the wave illuminates a scattering body Ω located in the direction $\hat{\imath}_1$, the body scatters the field in the direction $\hat{s}_2$, and the scattered field is measured at the point $\mathbf{r}_2$ with a dipole with the electric moment $\mathbf{p}_2$. Right: a reciprocal configuration, in which the dipole at $\mathbf{r}_2$ is transmitting and the dipole at $\mathbf{r}_1$ is receiving.

In terms of the far-field coefficients, the scattered fields are

$$\mathbf{E}_n^{\text{sc}}(\mathbf{r}) = \frac{\mathrm{e}^{-jkr}}{kr}\mathbf{F}_n(\hat{s}), \tag{3.115}$$

where $\hat{s} = \mathbf{r}/r$ is the unit vector pointing from the origin of a coordinate system, which is located at a finite distance from the scattering body, to the observation point, and it follows from (3.114) that

$$\frac{\mathrm{e}^{-jkr_2}}{kr_2}\mathbf{p}_2 \cdot \mathbf{F}_1(\hat{s}_2) = \frac{\mathrm{e}^{-jkr_1}}{kr_1}\mathbf{p}_1 \cdot \mathbf{F}_2(\hat{s}_1), \tag{3.116}$$

where $\hat{s}_n = \mathbf{r}_n/r_n$ with $n = 1, 2$. The scattered far field is always perpendicular to the scattering direction, $\hat{s} \cdot \mathbf{F}_n(\hat{s}) = 0$ (see (3.108)), so that we can write

$$\mathbf{p} \cdot \mathbf{F}_n(\hat{s}) = -(\hat{s} \times \hat{s} \times \mathbf{p}) \cdot \mathbf{F}_n(\hat{s}), \tag{3.117}$$

where $\mathbf{p}$ is an arbitrary vector, which transforms (3.116) to

$$\frac{\mathrm{e}^{-jkr_2}}{kr_2}(\hat{s}_2 \times \hat{s}_2 \times \mathbf{p}_2) \cdot \mathbf{F}_1(\hat{s}_2) = \frac{\mathrm{e}^{-jkr_1}}{kr_1}(\hat{s}_1 \times \hat{s}_1 \times \mathbf{p}_1) \cdot \mathbf{F}_2(\hat{s}_1). \tag{3.118}$$

If the dipoles are located far from the scattering body and from the origin of the coordinate system, then the fields of the sources in the free space are given by

$$\mathbf{E}_n^{\text{inc}}(\mathbf{r}) = -\frac{k^2\,\mathrm{e}^{-jk|\mathbf{r}-\mathbf{r}_n|}}{4\pi\epsilon|\mathbf{r}-\mathbf{r}_n|^3}(\mathbf{r}-\mathbf{r}_n) \times (\mathbf{r}-\mathbf{r}_n) \times \mathbf{p}_n, \tag{3.119}$$

where ϵ is the permittivity of the surrounding medium, $n = 1, 2$, and $(\mathbf{r}-\mathbf{r}_n)/|\mathbf{r} - r_n|$ is the unit vector pointing from the source at $\mathbf{r}_n$ to the observation point (see (2.214)). At observation points such that $r \ll r_n$, the wave (3.119) looks like the plane wave (2.155) with the incidence direction $\hat{\imath} = -\hat{s}_n$ and the amplitude

$$\mathbf{E}_n^{\text{inc}} = \mathbf{E}_n^{\text{inc}}(0) = -\frac{k^2\,\mathrm{e}^{-jkr_n}}{4\pi\epsilon r_n}\hat{s}_n \times \hat{s}_n \times \mathbf{p}_n, \tag{3.120}$$

implying that in the limit $r_n \to \infty$ the reciprocity relation (3.118) takes the form

$$\mathbf{E}_2^{\mathrm{inc}} \cdot \mathbf{F}_1(\hat{s}_2) = \mathbf{E}_1^{\mathrm{inc}} \cdot \mathbf{F}_2(\hat{s}_1). \tag{3.121}$$

The reciprocity relation (3.121) is formulated in terms of the far-field coefficients; it relates the field due to a plane wave with the incidence direction $\hat{\imath} = -\hat{s}_1$ and the amplitude $\mathbf{E}_1^{\mathrm{inc}}$ scattered in the direction $\hat{s}_2$ to the field due to a plane wave with $\hat{\imath} = -\hat{s}_2$ and the amplitude $\mathbf{E}_2^{\mathrm{inc}}$ scattered in the direction $\hat{s}_1$.

From the reciprocity relation (3.121) and the expressions (3.111) for the elements of scattering matrix it can be easily seen that interchanging the incident and scattering directions results in transposition of the scattering matrix,

$$\begin{bmatrix} A_{11} & A_{12} \\ A_{21} & A_{22} \end{bmatrix} \longrightarrow \begin{bmatrix} A_{11} & A_{21} \\ A_{12} & A_{22} \end{bmatrix}. \tag{3.122}$$

3.4.3 Forward Scattering

Scattering in the forward direction ($\hat{s} = \hat{\imath}$) is an important special case. The forward-scattering far-field coefficient of every scatterer is directly related to the extinction cross section.

Far-field Coefficient in Optical Limit

We start by deriving a useful and quite general relation for the far-field coefficient in the optical limit, that is, when $\lambda \to 0$ or, which is the same, $k \to \infty$. This limit is understood in the sense that the wavelength becomes arbitrarily small as compared with all characteristic dimensions of the scattering body (strictly speaking, in the limit of infinite frequency all materials become fully transparent because of the final mass of electrons). Consider a plane electromagnetic wave (2.155) with an amplitude $\mathbf{E}_0$ traveling in the direction $\hat{\imath}$ and illuminating an impenetrable but otherwise arbitrary (limitations are discussed later) scatterer. Then, the far-field coefficient $\mathbf{F}_{\mathrm{fwd}}$ for the field scattered by the scatterer in the forward direction, $\hat{s} = \hat{\imath}$, tends to the expression

$$\mathbf{F}_{\mathrm{fwd}} \to \frac{k^2 S_0}{2\pi j}\mathbf{E}_0, \tag{3.123}$$

where S_0 is the area of the geometric cross section of the shadow region behind the scatterer. This formula defines the optical limits for the extinction cross section (see (3.147)) and the bistatic scattering cross section (3.110) in the forward direction

$$\sigma_{\mathrm{fwd}} \to \frac{1}{\pi}k^2 S_0^2. \tag{3.124}$$

Equation (3.123) also shows that in the optical limit the forward-scattered field is polarized as the incident wave and has no cross-polarized components.

The proof is based on an integral representation (3.108) of the far-field coefficient. As long as exact equivalent currents $\mathbf{K}_{\mathrm{e,m}}$ are used, the integration surface in (3.108) can be an arbitrary surface enclosing the scatterer, and it is convenient to choose it as a smooth and convex surface Ω that encloses the scatterer (Figure 3.13). So we write for the far-field coefficient in the forward-scattering case

$$\mathbf{F}_{\mathrm{fwd}} = \frac{jk^2}{4\pi}\int_\Omega e^{jk\hat{\imath}\cdot\mathbf{r}'}\,\hat{\imath} \times [Z\hat{\imath} \times \mathbf{K}_{\mathrm{e}}(\mathbf{r}') + \mathbf{K}_{\mathrm{m}}(\mathbf{r}')]\,\mathrm{d}\Omega', \tag{3.125}$$

where the equivalent currents $\mathbf{K}_{\mathrm{e,m}}$ are related to the total electric and magnetic fields on the integration surface by the formulas (2.348).

Let us assume that when the frequency increases, not only the overall size of the body becomes very large as compared with the wavelength, but also that there are no inhomogeneities on the scale of the wavelength. In this situation none of the parts of the body can create resonant scattering response. If, in addition, we assume that the material properties are such that no resonant surface modes are possible (usual materials with positive real parts of the permittivity and permeability), the scattering obeys the PO equations. The shadow region behind the scatterer has the shape of a cylinder with generating

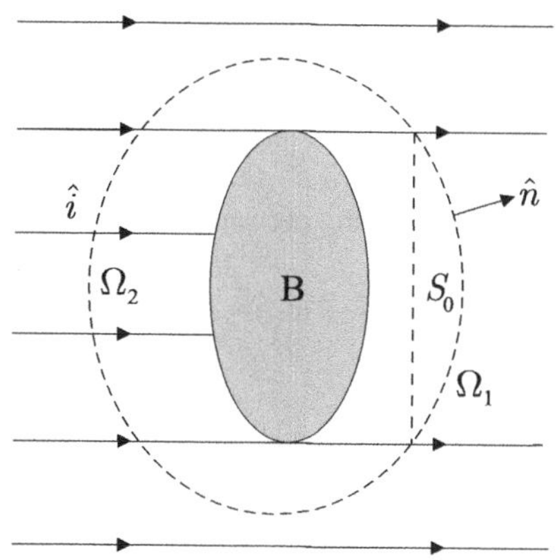

Figure 3.13 Geometry of integration: $\hat{\imath}$ is the incidence direction. The integration surface consists of two portions, Ω_1 and Ω_2, and encloses an arbitrarily shaped scatterer B. S_0 is the cross section of the shadow region behind the scatterer.

lines that are parallel to the incidence direction $\hat{\imath}$ and tangent to the surface of the scatterer. The intersection of the cross section S_0 of the shadow cylinder with the integration surface divides Ω into a shadowed portion Ω_1 and illuminated portion Ω_2. The total electric field on Ω can be represented as

$$\mathbf{E}(\mathbf{r}) = \begin{cases} 0, & \mathbf{r} \in \Omega_1 \\ \mathbf{E}_{\text{inc}}(\mathbf{r}) + \mathbf{E}_{\text{ref}}(\mathbf{r}), & \mathbf{r} \in \Omega_2 \end{cases}, \tag{3.126}$$

where $\mathbf{E}_{\text{ref}}(\mathbf{r})$ is the reflected field, which consists of all the waves reflected from the illuminated portion of the scatterer. The phase factors of the waves are given by $\exp(-jk\hat{s}\cdot\mathbf{r})$ with $\hat{s} \neq \hat{\imath}$, implying that in the limit $k \to \infty$ these waves do not contribute to the integral (3.125) because of the rapidly oscillating exponent $\exp[jk(\hat{\imath} - \hat{s})\cdot\mathbf{r}']$ in the integrand. Thus, retaining only the incident fields in the expressions for the equivalent currents and using (2.155) and (2.156) transforms (3.125) to

$$\mathbf{F}_{\text{fwd}} \to \frac{jk^2}{4\pi}\int_{\Omega_2} \hat{\imath}\times[\hat{\imath}\times\hat{n}\times\hat{\imath}\times\mathbf{E}_0 - \hat{n}\times\mathbf{E}_0]\,\mathrm{d}\Omega', \tag{3.127}$$

where the unit vector $\hat{n}$ points to the exterior of the integration surface and is a function of the position on the integration surface. The arrow in (3.127) means the optical limit ($k \to \infty$) of the expression.

The vector products in the integrand evaluate to

$$\hat{\imath}\times[\hat{\imath}\times\hat{n}\times\hat{\imath}\times\mathbf{E}_0 - \hat{n}\times\mathbf{E}_0] = 2\mathbf{E}_0\hat{\imath}\cdot\hat{n}, \tag{3.128}$$

and the expression for the far-field coefficient reduces to

$$\mathbf{F}_{\text{fwd}} \to \frac{jk^2}{2\pi}\mathbf{E}_0\int_{\Omega_2}\hat{\imath}\cdot\hat{n}\,\mathrm{d}\Omega'. \tag{3.129}$$

Because Ω is a closed surface and $\hat{\imath}$ is a constant vector with $\nabla\cdot\hat{\imath} = 0$, Gauss' theorem implies that

$$\int_{\Omega}\hat{\imath}\cdot\hat{n}\,\mathrm{d}\Omega' = 0, \tag{3.130}$$

and the formula (3.129) can be rewritten so as to integrate over the complimentary surface Ω_1,

$$\mathbf{F}_{\text{fwd}} \to -\frac{jk^2}{2\pi}\mathbf{E}_0\int_{\Omega_1}\hat{\imath}\cdot\hat{n}\,\mathrm{d}\Omega'. \tag{3.131}$$

Application of the Gauss theorem to the volume between the surfaces Ω_1 and S_0 suggests that

$$\int_{\Omega_1} \hat{\imath} \cdot \hat{n}\, \mathrm{d}\Omega' = -\int_{S_0} \hat{\imath} \cdot \hat{n}\, \mathrm{d}\Omega'. \tag{3.132}$$

Since $\hat{n} = -\hat{\imath}$ on S_0 ($\hat{n}$ is by definition the normal pointing outward from the volume),

$$\int_{S_0} \hat{\imath} \cdot \hat{n}\, \mathrm{d}\Omega' = -S_0, \tag{3.133}$$

and (3.131) results in the formula (3.123).

The derivations of this section have been inspired by (Van de Hulst 1949), where a formula similar to (3.123) is obtained by combining Kirchhoff's approximation with the Babinet principle (see also Born and Wolf (1959, p. 656)). Formula (3.123) seems to apply to penetrable scatterers as well, provided that they are non-planar. An example, a dielectric sphere, is given in section 7.5.5. On the other hand, it is important to note that the limit (3.123) is not general. For example, for large bodies made of double-negative materials the forward-scattering amplitude can be arbitrarily large because of excitation of resonant surface modes (Maslovski *et al.* 2016; Valagiannopoulos *et al.* 2015).

Forward-scattering (Optical) Theorem

The forward-scattering or optical theorem is a general law of wave-scattering theory, which relates the forward-scattering far-field coefficient to the extinction cross section of the scatterer. Its mathematical formulation is given by (3.147). The relation is a consequence of the conservation of energy and is therefore quite general. The physical argument behind this theorem is that the attenuation of the plane wave after it has passed the obstacle is a measure of the absorption and scattering of the obstacle. Since the scattered field and plane wave interfere at infinity only in the direction of the incident field, the attenuation can be estimated by looking at the scattered field in this direction. An implication of the optical theorem is that the far-field coefficient $\mathbf{F}$ of any physical scatterer cannot vanish in the forward-scattering direction. The relation was conjectured by Van de Hulst (1949), and Jones (1955) gave a mathematically rigorous proof for a linearly polarized incident wave. An extension to the case of an elliptically polarized wave was published by de Hoop (1958).

Consider a plane wave (2.155) incident on a scatterer of arbitrary form embedded in a dielectric medium with material parameters ϵ and μ (Figure 3.14). Our goal is to relate the energy absorbed and scattered by the obstacle to the far-field coefficient $\mathbf{F}$ with $\hat{s} = \hat{\imath}$ (forward-scattering case), which we denote by $\mathbf{F}_{\text{fwd}}$. The latter quantity is given by (3.125). The time-averaged power absorbed by the scatterer is equal to the time-averaged energy supply to the volume Ω

$$P_{\text{abs}} = -\frac{1}{2}\mathrm{Re} \int_{\Omega} (\mathbf{E} \times \mathbf{H}^*) \cdot \hat{n}\, \mathrm{d}\Omega', \tag{3.134}$$

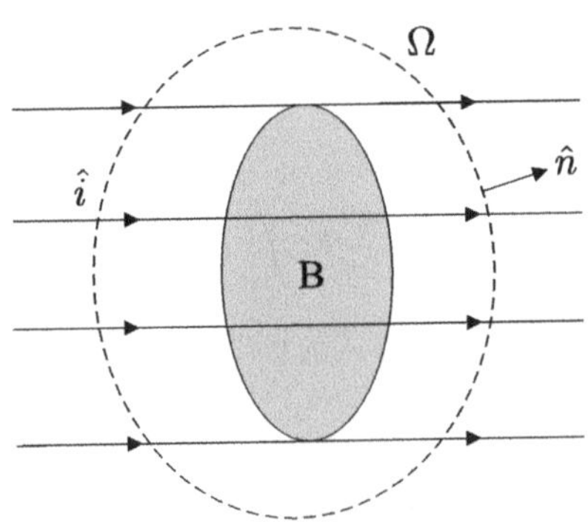

Figure 3.14 Geometry assumed in the derivation of the optical theorem: $\hat{\imath}$ is the incidence direction, Ω is an integration surface enclosing the scatterer B, and $\hat{n}$ is the outward unit vector normal to Ω.

where the minus sign in front of the integral is due to the choice of the direction of the unit normal. The integration surface Ω in (3.134) can be an arbitrary surface which encloses the scatterer. The formula (3.134) is simply an integral over the mean intensity of the energy flow (2.119) in the total field through Ω.

The total field is the sum of the incident and scattered fields,

$$\mathbf{E} = \mathbf{E}_{inc} + \mathbf{E}_{sc}, \quad \mathbf{H} = \mathbf{H}_{inc} + \mathbf{H}_{sc} \tag{3.135}$$

and inserting these in (3.134) leads to

$$P_{abs} = -P_{sc} - P_{inc} - \frac{1}{2}\mathrm{Re}\int_{\Omega} \left(\mathbf{E}_{inc} \times \mathbf{H}^*_{sc} + \mathbf{E}_{sc} \times \mathbf{H}^*_{inc}\right) \cdot \hat{n}\, \mathrm{d}\Omega', \tag{3.136}$$

where

$$P_{sc} = \frac{1}{2}\mathrm{Re}\int_{\Omega} \left(\mathbf{E}_{sc} \times \mathbf{H}^*_{sc}\right) \cdot \hat{n}\, \mathrm{d}\Omega' \tag{3.137}$$

is the time-averaged total power scattered by the obstacle and

$$P_{inc} = \frac{1}{2}\mathrm{Re}\int_{\Omega} \left(\mathbf{E}_{inc} \times \mathbf{H}^*_{inc}\right) \cdot \hat{n}\, \mathrm{d}\Omega' \tag{3.138}$$

is the time-averaged power carried through Ω by the incident wave in the absence of the scatterer. Since all field sources are assumed in the exterior of Ω, the latter integral vanishes, $P_{inc} = 0$, and (3.136) can be rewritten as

$$P_{sc} + P_{abs} = -\frac{1}{2}\mathrm{Re}\int_{\Omega} \left(\mathbf{E}^*_{inc} \times \mathbf{H}_{sc} + \mathbf{E}_{sc} \times \mathbf{H}^*_{inc}\right) \cdot \hat{n}\, \mathrm{d}\Omega' \tag{3.139}$$

in which the relation $\mathrm{Re}(\mathbf{E}_{inc} \times \mathbf{H}^*_{sc}) = \mathrm{Re}(\mathbf{E}^*_{inc} \times \mathbf{H}_{sc})$ has been used in order to relate the integral in the right-hand side of (3.139) to $\mathbf{F}_{fwd}$.

In fact, these two quantities are simply related to each other by the formula

$$\int_{\Omega} \left(\mathbf{E}^*_{inc} \times \mathbf{H}_{sc} + \mathbf{E}_{sc} \times \mathbf{H}^*_{inc}\right) \cdot \hat{n}\, \mathrm{d}\Omega' = \frac{4\pi}{jZk^2}\mathbf{E}^*_0 \cdot \mathbf{F}_{fwd}, \tag{3.140}$$

where $Z = \sqrt{\mu/\epsilon}$. To make (3.140) apparent, we represent the equivalent surface currents in (3.125) through tangential fields by using (2.348) and (3.135), and since the contribution from the incident field to (3.125) vanishes (see (3.130)) we can write

$$\mathbf{E}^*_0 \cdot \mathbf{F}_{fwd} = \frac{jk^2 Z}{4\pi}\int_{\Omega} e^{jk\hat{\imath}\cdot\mathbf{r}'}\mathbf{E}^*_0 \cdot \left[\hat{\imath} \times \hat{\imath} \times \hat{n} \times \mathbf{H}_{sc}(\mathbf{r}') - \frac{1}{Z}\hat{\imath} \times \hat{n} \times \mathbf{E}_{sc}(\mathbf{r}')\right] \mathrm{d}\Omega'. \tag{3.141}$$

By using the formulas (A.11) and (A.12) for the scalar and vector triple vector products and the transversality condition $\hat{\imath} \cdot \mathbf{E}^*_0 = 0$, one can derive the relations

$$\mathbf{E}^*_0 \cdot (\hat{\imath} \times \hat{\imath} \times \hat{n} \times \mathbf{H}_{sc}) = \left(\mathbf{E}^*_0 \times \mathbf{H}_{sc}\right) \cdot \hat{n} \tag{3.142}$$

$$\mathbf{E}^*_0 \cdot (\hat{\imath} \times \hat{n} \times \mathbf{E}_{sc}) = -\left(\mathbf{E}_{sc} \times \hat{\imath} \times \mathbf{E}^*_0\right) \cdot \hat{n} \tag{3.143}$$

and express the integrand in the right-hand side of (3.141) as

$$e^{jk\hat{\imath}\cdot\mathbf{r}}\left(\mathbf{E}^*_0 \times \mathbf{H}_{sc} + \frac{1}{Z}\mathbf{E}_{sc} \times \hat{\imath} \times \mathbf{E}^*_0\right) \cdot \hat{n} \tag{3.144}$$

which is equal to

$$\left(\mathbf{E}^*_{inc} \times \mathbf{H}_{sc} + \mathbf{E}_{sc} \times \mathbf{H}^*_{inc}\right) \cdot \hat{n} \tag{3.145}$$

because of (2.155) and (2.156). Inserting (3.145) in (3.141) proves the identity (3.140).

Using (3.140) in (3.139) we obtain

$$P_{sc} + P_{abs} = -\frac{2\pi}{Zk^2} \operatorname{Im}\left(\mathbf{E}_0^* \cdot \mathbf{F}_{fwd}\right). \tag{3.146}$$

Dividing (3.146) by the power density of the incident wave $w_{inc} = |\mathbf{w}_{inc}| = |\mathbf{E}_0|^2/(2Z)$ (see (2.162)) and using the definitions (3.9), (3.10), and (3.11) for the absorption, extinction, and total scattering cross sections result in the formula

$$\sigma_{ext} = -\frac{4\pi}{k^2} \operatorname{Im}\left(\frac{\mathbf{F}_{fwd} \cdot \mathbf{E}_0^*}{\mathbf{E}_0 \cdot \mathbf{E}_0^*}\right), \tag{3.147}$$

where $\mathbf{F}_{fwd}$ is the far-field coefficient in the forward direction, which is the direction of the incident wave.[4] The relation (3.147) is called the *forward-scattering* or *optical theorem*. For a linearly polarized incident wave of unit amplitude, $\mathbf{E}_0 = \hat{p}$ with $\hat{p}$ being the unit real-valued polarization vector, simplifies to

$$\sigma_{ext} = -\frac{4\pi}{k^2} \operatorname{Im}(\mathbf{F}_{fwd} \cdot \hat{p}). \tag{3.148}$$

The relation permits an easy evaluation of the extinction cross section. In the optical limit ($k \to \infty$) and for an impenetrable scatterer, the asymptotic formula (3.123) for $\mathbf{F}_{fwd}$ can be used in (3.147), which results in the high-frequency limit (3.12), according to which the extinction cross section is equal to twice the geometric cross section of the scatterer.

The forward-scattering relation (3.147) can be rewritten in terms of the diagonal elements of the scattering matrix. In the case of forward scattering, $\theta = \pi - \theta_{inc}$ and $\phi = \phi_{inc} \pm \pi$ and, thus, from (3.54), (3.55), (A.51), and (A.52) we have that

$$\hat{\theta} = \hat{\theta}_{inc}, \quad \hat{\phi} = -\hat{\phi}_{inc}. \tag{3.149}$$

If, for example, the incident linearly polarized wave is given by $\mathbf{E}_{inc} = E_{\theta_{inc}}^{inc} \hat{\theta}_{inc}$ with a real- or complex-valued amplitude $E_{\theta_{inc}}^{inc}$, then (3.147) becomes

$$\sigma_{ext}(\hat{\theta}_{inc}) = -\frac{4\pi}{k^2} \operatorname{Im}\left(\frac{E_{\theta_{inc}}^{inc\,*}}{|E_{\theta_{inc}}^{inc}|^2} \mathbf{F} \cdot \hat{\theta}_{inc}\right). \tag{3.150}$$

From the definition (3.86) of the far-field coefficient and from the relation (3.61) between the incident and scattered fields it follows that

$$\mathbf{F} = \frac{k}{\sqrt{4\pi}} E_{\theta_{inc}}^{inc} \left(A_{\theta\theta_{inc}} \hat{\theta} + A_{\phi\theta_{inc}} \hat{\phi}\right), \tag{3.151}$$

and inserting this result into (3.150) and using (3.149), we obtain the formula

$$\sigma_{ext}(\hat{\theta}_{inc}) = -\frac{\sqrt{4\pi}}{k} \operatorname{Im} A_{\theta\theta_{inc}}(\pi - \theta_{inc}, \phi_{inc} \pm \pi, \theta_{inc}, \phi_{inc}). \tag{3.152}$$

In a similar way, one shows that

$$\sigma_{ext}(\hat{\phi}_{inc}) = \frac{\sqrt{4\pi}}{k} \operatorname{Im} A_{\phi\phi_{inc}}(\pi - \theta_{inc}, \phi_{inc} \pm \pi, \theta_{inc}, \phi_{inc}). \tag{3.153}$$

The sign difference in the right-hand sides of (3.152) and (3.153) follows from (3.149) and is, in fact, the result of the employed definition of the bases of the incident and scattered waves.

[4] The minus sign in the right-hand side of (3.147) depends on the assumed time-dependence factor, which is $\exp(j\omega t)$ in this book. For the alternative time dependence $\exp(-i\omega t)$ the sign is plus.

The presented derivation of the optical theorem was suggested by de Hoop (1958). A historically first rigorous derivation of the electromagnetic version of the theorem (Jones 1955) is based on asymptotic evaluation of the integral in (3.139), in which the integration surface Ω is assumed to be a sphere of a large radius r going to infinity and the scattered fields under the integration sign are expressed through the far-field coefficient as in (3.86). An overview of the long history of the theorem in optics, electromagnetics, and quantum mechanics can be found in (Newton 1976).

3.4.4 Cylindrical Bodies

In this section we study the far-field behavior of fields scattered from cylindrical bodies infinite in one direction. Since the scattering body is infinite, the scattered field is no longer a spherical wave but rather a conical or cylindrical wave, and the definition of the far zone for compact scatterers does not apply. In this section we define the cylindrical counterparts of the scattering amplitude, derive integral representations for them, relate them to the scattering widths, and formulate the analogue of the optical theorem for cylindrical scatterers.

The scatterers are assumed to be infinite along the z axis. Their geometric cross section in the (x, y) plane can be arbitrary but still invariant in the z direction. It can be a single cylinder or a group of parallel cylinders with different cross sections in general. As we have seen in section 2.5.5, in this case Maxwell's equations are reduced to two-dimensional scalar Helmholtz equations (2.319) and (2.320) for the z components of the fields. The fields can be represented as in (2.321) with the dependence on the z coordinate given by $\exp(-jk_z z)$ where $k_z = k\cos\beta$ is the z component of the wave vector and β is the angle between the wave vector and the axis of the cylinder(s). For example, the scattered portion of the field can be represented as

$$E_z^{\rm sc}(\rho, \varphi, z) = U_E^{\rm sc}(\boldsymbol{\rho})\,{\rm e}^{-jk_z z}, \quad H_z^{\rm sc}(\rho, \varphi, z) = U_H^{\rm sc}(\boldsymbol{\rho})\,{\rm e}^{-jk_z z}, \tag{3.154}$$

where $\boldsymbol{\rho} = \hat{x}\rho\cos\varphi + \hat{y}\rho\sin\varphi$ is the position vector on the (x, y) plane and ρ and φ are the polar coordinates.

Scattering Amplitudes

We start by showing that at distances ρ such that

$$k_\perp \rho \gg 1, \quad \rho \gg k_\perp D_\perp^2, \tag{3.155}$$

where $k_\perp = k\sin\beta$ and $D_\perp$ is the diameter of the scattering region on the (x, y) plane, the scattering fields have the form of conical waves

$$E_z^{\rm sc}(\rho, \varphi, z) \approx -\sqrt{\frac{2}{\pi k_\perp \rho}} F_E(\varphi)\,{\rm e}^{-jk_z z - jk_\perp \rho + j\pi/4} \tag{3.156}$$

$$H_z^{\rm sc}(\rho, \varphi, z) \approx -\sqrt{\frac{2}{\pi k_\perp \rho}} F_H(\varphi)\,{\rm e}^{-jk_z z - jk_\perp \rho + j\pi/4}, \tag{3.157}$$

where $F_{E,H}(\varphi)$ are independent of ρ.[5] The quantities $F_{E,H}(\varphi)$ will be referred to as scattering amplitudes.

Equations (3.156) and (3.157) imply that

$$U_{E,H}^{\rm sc}(\boldsymbol{\rho}) \approx -\sqrt{\frac{2}{\pi k_\perp \rho}}\,{\rm e}^{-jk_\perp \rho + j\frac{\pi}{4}} F_{E,H}(\varphi). \tag{3.158}$$

To derive the representations (3.158), we use the integral formula (2.367)

$$U_{\rm sc}(\boldsymbol{\rho}) = \int\limits_C \left[U(\boldsymbol{\rho}')\frac{\partial}{\partial n'} G_2(\boldsymbol{\rho}, \boldsymbol{\rho}'; k_\perp) - G_2(\boldsymbol{\rho}, \boldsymbol{\rho}'; k_\perp)\frac{\partial}{\partial n'} U(\boldsymbol{\rho}') \right] {\rm d}l', \tag{3.159}$$

[5] Formulas (3.156) and (3.157) are in fact the leading terms of asymptotic expansions in inverse powers of ρ.

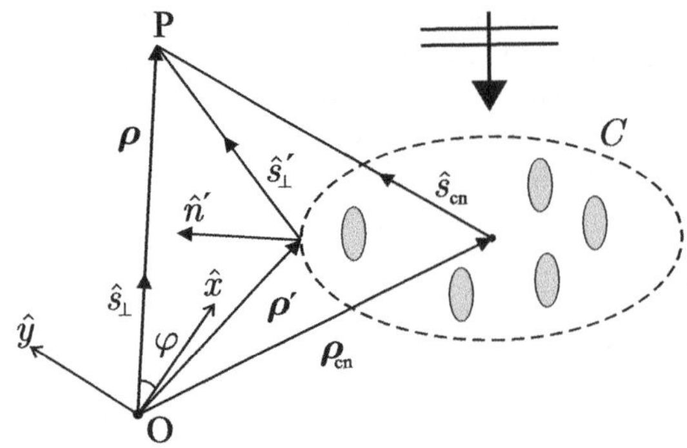

Figure 3.15 Coordinates and vectors used in the derivation of the integral representations for the scattering amplitude of cylindrical scatterers. The figure shows the case in which the integration contour C encloses several cylindrical scatterers.

where U can be either U_E or U_H. The relation (3.159) represents the amplitude of the z component of the scattered field in terms of the boundary values of the z components of the total field and its normal derivative on a contour C that encloses the cylinder(s) (Figure 3.15). The contour can be freely chosen and can be contracted to coincide with the physical boundaries of the scatterers. The two-dimensional unit vector $\hat{n}$ is normal to C and points to the exterior of the contour. The Green function has the form (2.325)

$$G_2(\boldsymbol{\rho}, \boldsymbol{\rho}'; k_\perp) = \frac{1}{4j} H_0^{(2)}(k_\perp |\boldsymbol{\rho} - \boldsymbol{\rho}'|), \tag{3.160}$$

where $\boldsymbol{\rho}$ and $\boldsymbol{\rho}'$ are the two-dimensional position vectors of the observation point $\boldsymbol{\rho} = \hat{x}x + \hat{y}y$ and integration point $\boldsymbol{\rho}' = \hat{x}x' + \hat{y}y'$ and $|\boldsymbol{\rho} - \boldsymbol{\rho}'| = \sqrt{(x-x')^2 + (y-y')^2}$. The prime in the quantities in the integrand of (3.159) means the dependence on the integration point.

The normal derivative of the Green function can be explicitly evaluated as

$$\frac{\partial}{\partial n'} G_2(\boldsymbol{\rho}, \boldsymbol{\rho}'; k_\perp) = \frac{1}{4j} k_\perp (\hat{n}' \cdot \hat{s}'_\perp) H_1^{(2)}(k_\perp |\boldsymbol{\rho} - \boldsymbol{\rho}'|), \tag{3.161}$$

where $\hat{s}'_\perp = (\boldsymbol{\rho} - \boldsymbol{\rho}')/|\boldsymbol{\rho} - \boldsymbol{\rho}'|$ is the two-dimensional unit vector pointing from $\boldsymbol{\rho}'$ to $\boldsymbol{\rho}$.

Assume that the distance between $\boldsymbol{\rho}$ and the integration contour C is large so that

$$k_\perp |\boldsymbol{\rho} - \boldsymbol{\rho}'| \gg 1 \tag{3.162}$$

for every $\boldsymbol{\rho}'$ from C. Then, the Hankel functions in (3.160) and (3.161) can be simplified by using their large-argument approximations over the whole integration contour in (3.159), which gives

$$U_{\rm sc}(\boldsymbol{\rho}) \approx \frac{e^{3j\pi/4}}{\sqrt{8\pi k_\perp}} \int_C \frac{e^{-jk_\perp|\boldsymbol{\rho}-\boldsymbol{\rho}'|}}{|\boldsymbol{\rho} - \boldsymbol{\rho}'|^{1/2}} \left[\frac{\partial}{\partial n'} U(\boldsymbol{\rho}') - jk_\perp (\hat{n}' \cdot \hat{s}'_\perp) U(\boldsymbol{\rho}')\right] dl'. \tag{3.163}$$

If, furthermore,

$$|\boldsymbol{\rho} - \boldsymbol{\rho}_{\rm cn}| \gg |\boldsymbol{\rho}' - \boldsymbol{\rho}_{\rm cn}|, \tag{3.164}$$

where $\boldsymbol{\rho}_{\rm cn}$ is the middle point of the geometric cross section of the scatterer (or a group of scatterers, as in Figure 3.15), then further simplifications are possible on using the Taylor approximation

$$|\boldsymbol{\rho} - \boldsymbol{\rho}'| \approx |\boldsymbol{\rho} - \boldsymbol{\rho}_{\rm cn}| - \hat{s}_{\rm cn} \cdot (\boldsymbol{\rho}' - \boldsymbol{\rho}_{\rm cn}) \tag{3.165}$$

and noticing that $\hat{s}'_\perp \approx \hat{s}_{\rm cn} = (\boldsymbol{\rho} - \boldsymbol{\rho}_{\rm cn})/|\boldsymbol{\rho} - \boldsymbol{\rho}_{\rm cn}|$, which transforms (3.163) to the formula

$$U_{\rm sc}(\boldsymbol{\rho}) \approx \frac{{\rm e}^{-jk_\perp|\boldsymbol{\rho}-\boldsymbol{\rho}_{\rm cn}|+j\frac{3}{4}\pi}}{\sqrt{8\pi k_\perp|\boldsymbol{\rho} - \boldsymbol{\rho}_{\rm cn}|}} \int_C {\rm e}^{jk_\perp\hat{s}_{\rm cn}\cdot(\boldsymbol{\rho}'-\boldsymbol{\rho}_{\rm cn})} \left[\frac{\partial}{\partial n'}U(\boldsymbol{\rho}') - jk_\perp(\hat{n}'\cdot\hat{s}_{\rm cn})U(\boldsymbol{\rho}')\right] {\rm d}l'. \tag{3.166}$$

In order to apply the approximation (3.165) to the exponent $\exp(-jk_\perp|\boldsymbol{\rho} - \boldsymbol{\rho}'|)$ in (3.163), it is necessary to require that

$$|\boldsymbol{\rho} - \boldsymbol{\rho}_{\rm cn}| \gg k_\perp|\boldsymbol{\rho}' - \boldsymbol{\rho}_{\rm cn}|^2 \tag{3.167}$$

for every $\boldsymbol{\rho}'$ from C. Note that the formula (3.166) is inapplicable to scatterers with infinite geometric cross section (e.g., an infinite wedge) since the condition (3.164) does not hold on distant portions of the integration contour C.

If, finally, the observation point $\boldsymbol{\rho}$ is located much farther from the origin of the coordinate system than the point $\boldsymbol{\rho}_{\rm cn}$, that is,

$$\rho \gg \rho_{\rm cn}, \tag{3.168}$$

where $\rho = |\boldsymbol{\rho}|$ and $\rho_{\rm cn} = |\boldsymbol{\rho}_{\rm cn}|$, then the simplifications

$$|\boldsymbol{\rho} - \boldsymbol{\rho}_{\rm cn}| \approx \rho - \hat{s}_\perp\cdot\boldsymbol{\rho}_{\rm cn}, \quad \hat{s}_{\rm cn} \approx \hat{s}_\perp = \frac{\boldsymbol{\rho}}{\rho} = \hat{x}\cos\varphi + \hat{y}\sin\varphi \tag{3.169}$$

can be made in (3.166), resulting in the formula

$$U_{\rm sc}(\boldsymbol{\rho}) \approx \frac{{\rm e}^{-jk_\perp\rho+j\frac{3}{4}\pi}}{\sqrt{8\pi k_\perp\rho}} \int_C {\rm e}^{jk_\perp\hat{s}_\perp\cdot\boldsymbol{\rho}'} \left[\frac{\partial}{\partial n'}U(\boldsymbol{\rho}') - jk_\perp(\hat{n}'\cdot\hat{s}_\perp)U(\boldsymbol{\rho}')\right] {\rm d}l' \tag{3.170}$$

which has the form of (3.158) with the scattering amplitude

$$F(\varphi) = \frac{1}{4j}\int_C {\rm e}^{jk_\perp\hat{s}_\perp\cdot\boldsymbol{\rho}'} \left[\frac{\partial}{\partial n'}U(\boldsymbol{\rho}') - jk_\perp(\hat{n}'\cdot\hat{s}_\perp)U(\boldsymbol{\rho}')\right] {\rm d}l'. \tag{3.171}$$

It can be seen that the assumptions (3.162), (3.164), (3.167), and (3.168), made in the derivation of (3.170), are equivalent to the conditions for (3.155). In particular, (3.170) and (3.171) are applicable to scatterers with compact cross sections only, otherwise (3.164) would be violated. The fields scattered by cylinders with geometric cross sections of infinite extent (e.g., a wedge, a half-plane) do not reduce to the cylindrical wave (3.158) even at infinitely distant observation points. Instead, the scattered field may include other components, for example plane waves reflected from the sides of the scatterer, and the formula (3.163) should be used to represent the scattered field.

In the general case, the fields scattered by cylinders are described by the two components E_z and H_z, and it follows from (3.171) that

$$F_E(\varphi) = \frac{1}{4j}\int_C {\rm e}^{jk_\perp\hat{s}_\perp\cdot\boldsymbol{\rho}'} \left[\frac{\partial}{\partial n'}U_E(\boldsymbol{\rho}') - jk_\perp(\hat{n}'\cdot\hat{s}_\perp)U_E(\boldsymbol{\rho}')\right] {\rm d}l' \tag{3.172}$$

$$F_H(\varphi) = \frac{1}{4j}\int_C {\rm e}^{jk_\perp\hat{s}_\perp\cdot\boldsymbol{\rho}'} \left[\frac{\partial}{\partial n'}U_H(\boldsymbol{\rho}') - jk_\perp(\hat{n}'\cdot\hat{s}_\perp)U_H(\boldsymbol{\rho}')\right] {\rm d}l'. \tag{3.173}$$

Equations (3.172) and (3.173) permit determination of the scattering amplitudes from the total fields given on a contour enclosing the scattering body (or bodies). The contour C can be chosen to be the boundary of the geometric cross section of the cylinder(s), which may be of advantage if the scattering boundary is PEC or impedance as the boundary value of the total field or the boundary value of the normal derivative of the total field can be excluded from the integral relations. For example, on a PEC scatterer,

$$U_E(\boldsymbol{\rho}') = 0, \quad \frac{\partial}{\partial n'}U_H(\boldsymbol{\rho}') = 0. \quad (\boldsymbol{\rho}' \in C) \tag{3.174}$$

Finally, if the incident field is a plane wave with the z components

$$E_z^{\mathrm{inc}} = E_{0z}\,\mathrm{e}^{-jk_z z - jk_\perp \rho\cos(\varphi-\varphi_0)}, \quad H_z^{\mathrm{inc}} = H_{0z}\,\mathrm{e}^{-jk_z z - jk_\perp \rho\cos(\varphi-\varphi_0)}, \tag{3.175}$$

and the scatterer is a single cylindrical scatterer impenetrable to the incident radiation, then as $k \to \infty$ (the optical limit) the scattering amplitudes in the direction $\varphi = \varphi_0$ behave as

$$F_E(\varphi_0) \to \frac{1}{2}k_\perp L E_{0z}, \quad F_H(\varphi_0) \to \frac{1}{2}k_\perp L H_{0z}, \tag{3.176}$$

where L is the width of the geometric shadow strip behind the scatterer. For example, $L = 2a$ for a circular cylinder of radius a. Relations (3.176) are the cylindrical counterparts of the relation (3.123). Derivation of (3.176) is similar to the three-dimensional case described in section 3.4.3 and is left to the problems section (problem 24).

Scattering Matrix

The scattering amplitudes can be particularized depending on the polarization of the incident field,

$$F_{EE} = F_E, \quad F_{HE} = F_H \tag{3.177}$$

when $H_z^{\mathrm{inc}} = 0$ and

$$F_{EH} = F_E, \quad F_{HH} = F_H \tag{3.178}$$

when $E_z^{\mathrm{inc}} = 0$. These quantities are useful to describe the scattering matrix of cylindrical scatterers. The matrix can be conveniently defined as the matrix coefficient relating the z components of the incident and scattered fields,

$$\begin{bmatrix} E_z^{\mathrm{sc}} \\ ZH_z^{\mathrm{sc}} \end{bmatrix} = \frac{\mathrm{e}^{-jk_z z - jk_\perp \rho - j\pi/4}}{\sqrt{2\pi k_\perp \rho}} \begin{bmatrix} D_{\mathrm{ee}} & D_{\mathrm{eh}} \\ D_{\mathrm{he}} & D_{\mathrm{hh}} \end{bmatrix} \begin{bmatrix} E_{0z} \\ ZH_{0z} \end{bmatrix}, \tag{3.179}$$

where E_{0z} and H_{0z} are equal to the values of E_z^{inc} and H_z^{inc} at the point $z = 0$, $\rho = 0$, respectively. The relations between the elements of the scattering matrix and the scattering amplitudes are as follows:

$$D_{\mathrm{ee}} = -2j\frac{F_{EE}}{E_{0z}}, \quad D_{\mathrm{eh}} = -2j\frac{F_{EH}}{ZH_{0z}}, \quad D_{\mathrm{he}} = -2j\frac{F_{HE}}{YE_{0z}}, \quad D_{\mathrm{hh}} = -2j\frac{F_{HH}}{H_{0z}}. \tag{3.180}$$

The scattering matrix is similar to the diffraction coefficient of an infinite wedge as both relate the amplitude of the scattered (edge-diffracted) wave with the amplitude of the incident wave at the scatterer. However, there are also significant differences, which are discussed in section 5.5.2.

Expressions for the Scattering Widths

The scattering widths defined in section 3.2.2 can be conveniently expressed through the scattering amplitudes $F_E(\varphi)$ and $F_H(\varphi)$ as this eliminates the need to evaluate the fields in the limiting case of $\rho \to \infty$, which can be a problem when the fields are determined numerically.

As ρ goes to infinity, the scattered fields behave as in (3.156) and (3.157), and these approximations can be used to determine the scattering widths. Thus, using (3.156) and (3.157) in the definition (3.17) of the bistatic scattering width gives

$$\sigma_{2\mathrm{D}} = \frac{4}{k\sin^2\beta}\frac{|F_E(\varphi)|^2 + |ZF_H(\varphi)|^2}{\left|E_z^{\mathrm{inc}}\right|^2 + \left|ZH_z^{\mathrm{inc}}\right|^2}. \tag{3.181}$$

where β is the angle between the axis of the cylinder(s) and the direction of propagation of the incoming wave (Figure 3.4). The scattered waves (3.156) and (3.157) are conical waves (section 2.5.1), and the angle β is at the same time the angle between the normal to the wavefront of the scattered conical wave and the axis of the cylinder.

The expression for the total scattering width follows on insertion of (3.181) in (3.26),

$$\sigma_{\mathrm{T}}^{2\mathrm{D}} = \frac{2}{\pi k} \frac{1}{\left|E_z^{\mathrm{inc}}\right|^2 + \left|ZH_z^{\mathrm{inc}}\right|^2} \int_0^{2\pi} \left[|F_E(\varphi)|^2 + |ZF_H(\varphi)|^2\right] \mathrm{d}\varphi. \tag{3.182}$$

Expressions (3.181) and (3.182) can be specialized to specific combinations of the components of the scattered and incident fields, as in (3.18)–(3.23). For example, in terms of the elements of the scattering matrix, such specializations of the scattering width read

$$\sigma_{EE} = \frac{|D_{\mathrm{ee}}|^2}{k \sin^2 \beta}, \quad \sigma_{EH} = \frac{|D_{\mathrm{eh}}|^2}{k \sin^2 \beta}, \quad \sigma_{HE} = \frac{|D_{\mathrm{he}}|^2}{k \sin^2 \beta}, \quad \sigma_{HH} = \frac{|D_{\mathrm{hh}}|^2}{k \sin^2 \beta}. \tag{3.183}$$

Optical Theorem for Cylinders

It is possible to derive an analogue of the optical theorem (section 3.4.3) for an infinite cylindrical scatterer (or a group of parallel cylindrical scatterers), which relates the total scattering width with the values of the scattering amplitudes in the direction of propagation of the incident wave. In this section we present the theorem in the general case of arbitrary incidence of the field with respect to the axis of the cylinder (arbitrary k_z or β). The case of normal incidence ($\beta = \pi/2$ or $k_z = 0$) is described in Jones (1964, section 8.6).

The approach is similar to that used in section 3.4.3 in the three-dimensional case. The surrounding medium is assumed homogeneous and lossless; the scatterer can have an arbitrary geometric cross section and must be passive. The starting point is the equation of the energy balance in a cylinder Z_ρ of a large radius ρ, which encloses the scatterer(s). In order to work with finite energies (the scatterer is of infinite extend and is illuminated by a plane wave), the energy balance is formulated per unit length along the axis of Z_ρ. The total energy p_{abs} absorbed per unit length by the scatterer(s) in the interior of Z_ρ is equal to the time-average energy flow per unit length from the exterior into the interior of Z_ρ, that is,

$$p_{\mathrm{abs}} = -\rho \int_0^{2\pi} \hat{\rho} \cdot \mathbf{w} \, \mathrm{d}\varphi, \tag{3.184}$$

where $\hat{\rho}$ is the outward unit vector normal to Z_ρ and $\mathbf{w}$ is the intensity of the energy flow (2.119). The intensity is independent of the z coordinate, and the integration in (3.184) is performed over a circle of radius ρ.

The radial component of the intensity is equal to

$$\hat{\rho} \cdot \mathbf{w} = w_\rho = \frac{1}{2} \mathrm{Re} \left(E_\varphi H_z^* - E_z H_\varphi^* \right) \tag{3.185}$$

or, in terms of the z components of the fields, see (2.314) and (2.316),

$$w_\rho = \frac{k}{2k_\perp^2} \mathrm{Im} \left[Y E_z \frac{\partial E_z^*}{\partial \rho} + Z H_z \frac{\partial H_z^*}{\partial \rho} + \frac{k_z}{k\rho} \left(E_z \frac{\partial H_z^*}{\partial \varphi} + H_z^* \frac{\partial E_z^*}{\partial \varphi} \right) \right]. \tag{3.186}$$

The components of the wave vector are related to the incidence angle β by $k_z = k \cos \beta$ and $k_\perp = k \sin \beta$.

The field consists of the incident and scattered waves,

$$E_z = E_z^{\mathrm{inc}} + E_z^{\mathrm{sc}}, \quad H_z = H_z^{\mathrm{inc}} + H_z^{\mathrm{sc}}, \tag{3.187}$$

and the incident wave is a plane electromagnetic wave, as in (3.175), which propagates in the direction

$$\hat{i} = \hat{x} \sin \beta \cos \varphi_0 + \hat{y} \sin \beta \sin \varphi_0 + \hat{z} \cos \beta. \tag{3.188}$$

The scattered fields have the simple form (3.156) and (3.157) as $\rho \to \infty$. So, let us consider the energy balance in the limit of $\rho \to \infty$. Inserting (3.187) in (3.186) and neglecting the terms that decay faster than $1/\rho$ (as they do not contribute to the integral (3.184) in the limit $\rho \to \infty$), we obtain

$$w_\rho = w_\rho^{\mathrm{inc}} + w_\rho^{\mathrm{sc}} + \widetilde{w}_\rho \tag{3.189}$$

with

$$w_\rho^{\text{inc}} = \frac{kY}{2k_\perp} \left(|E_{0z}|^2 + |ZH_{0z}|^2 \right) \cos(\varphi - \varphi_0) \tag{3.190}$$

$$w_\rho^{\text{sc}} = \frac{kY}{2k_\perp} \left(\left|E_z^{\text{sc}}\right|^2 + \left|ZH_z^{\text{sc}}\right|^2 \right) \tag{3.191}$$

$$\begin{aligned} \widetilde{w}_\rho = \text{Re} \Bigg\{ & \frac{kY}{2k_\perp} \left[E_z^{\text{inc}} E_z^{\text{sc}*} + \cos(\varphi - \varphi_0) E_z^{\text{inc}*} E_z^{\text{sc}} \right] + \frac{kZ}{2k_\perp} \left[H_z^{\text{inc}} H_z^{\text{sc}*} + \cos(\varphi - \varphi_0) H_z^{\text{inc}*} H_z^{\text{sc}} \right] \\ & + \frac{k_z}{2k_\perp} \sin(\varphi - \varphi_0) \left(E_z^{\text{inc}} H_z^{\text{sc}*} - H_z^{\text{inc}*} E_z^{\text{sc}} \right) \Bigg\}, \end{aligned} \tag{3.192}$$

where (3.190) and (3.191) are the ρ components of the energy flows transported by the incident and scattered waves, respectively.

Using (3.189) permits us to rewrite (3.184) as

$$p_{\text{abs}} = -p_{\text{inc}} - p_{\text{sc}} - \widetilde{p}. \tag{3.193}$$

The first term on the right-hand side

$$p_{\text{inc}} = \lim_{\rho \to \infty} \rho \int_0^{2\pi} w_\rho^{\text{inc}} \, d\varphi \tag{3.194}$$

vanishes for every finite value of ρ because of the relation

$$\int_0^{2\pi} w_\rho^{\text{inc}} \, d\varphi = \frac{k}{2Zk_\perp} \left(|E_{0z}|^2 + |ZH_{0z}|^2 \right) \int_0^{2\pi} \cos(\varphi - \varphi_0) \, d\varphi = 0, \tag{3.195}$$

which simply means that the incident field (3.175) has no sources in the interior of the cylinder Z_ρ.

The second term on the right-hand side (3.193) is proportional to the total scattering width of the scatterer(s) inside the cylinder of radius ρ, and it follows from (3.15)–(3.17), (3.26), and (3.191) that

$$p_{\text{sc}} = \lim_{\rho \to \infty} \rho \int_0^{2\pi} w_\rho^{\text{sc}} \, d\varphi = \sigma_{\text{T}}^{\text{2D}} w_{\text{inc}}. \tag{3.196}$$

The remaining term in (3.193)

$$\begin{aligned} \widetilde{p} = \frac{k}{2k_\perp} \text{Re} \lim_{\rho \to \infty} \rho \Bigg\{ & Y \int_0^{2\pi} \left[E_z^{\text{inc}} E_z^{\text{sc}*} + \cos(\varphi - \varphi_0) E_z^{\text{inc}*} E_z^{\text{sc}} \right] d\varphi + Z \int_0^{2\pi} \left[H_z^{\text{inc}} H_z^{\text{sc}*} + \cos(\varphi - \varphi_0) H_z^{\text{inc}*} H_z^{\text{sc}} \right] d\varphi \\ & + \frac{k_z}{k} \int_0^{2\pi} \sin(\varphi - \varphi_0) \left(E_z^{\text{inc}} H_z^{\text{sc}*} - H_z^{\text{inc}*} E_z^{\text{sc}} \right) d\varphi \Bigg\} \end{aligned} \tag{3.197}$$

can be evaluated asymptotically for $\rho \to \infty$ with the method of stationary phase (section D.2). Direct inspection shows that the integral in (3.197) can be rewritten as a linear combination of terms of the form

$$I_\pm(\rho) = \int_0^{2\pi} A(\varphi) \, e^{\pm j\rho S(\varphi)} \, d\varphi \tag{3.198}$$

with coefficients that may depend on ρ but are independent of φ. Here, $A(\varphi)$ is a slowly varying function of the integration variable φ and $S(\varphi) = k_\perp \cos(\varphi - \varphi_0)$, implying that $I_\pm$ are integrals of the type (D.1) with ρ serving as the large parameter. On the integration range there are two stationary (critical) points $\varphi_c = \varphi_0$ and $\varphi_c = \varphi_1 = \varphi_0 \pm \pi$ (in the latter expression the sign is chosen such that $0 \le \varphi_1 < 2\pi$). The integrals $I_\pm$ can therefore be asymptotically expressed as the sum of the contributions of the form (D.19) from the critical points,

$$I_\pm(\rho) = \sqrt{\frac{2\pi}{\rho k_\perp}}[A(\varphi_0)\,\mathrm{e}^{\pm jk_\perp\rho\mp j\pi/4} + A(\varphi_1)\,\mathrm{e}^{\mp jk_\perp\rho\pm j\pi/4}] + o\left(\frac{1}{\sqrt{\rho}}\right). \tag{3.199}$$

Now, using (3.156), (3.157), and (3.175) in (3.197) followed by application of (3.199) leads to

$$\begin{aligned}\widetilde{p} = -\frac{k}{k_\perp^2}\mathrm{Re}\,\{&YE_{0z}\left[F_E^*(\varphi_0) - jF_E^*(\varphi_1)\,\mathrm{e}^{2jk_\perp\rho}\right] + YE_{0z}^*[F_E(\varphi_0) - jF_E(\varphi_1)\,\mathrm{e}^{-2jk_\perp\rho}]\\ &+ ZH_{0z}\left[F_H^*(\varphi_0) - jF_H^*(\varphi_1)\,\mathrm{e}^{2jk_\perp\rho}\right] + ZH_{0z}^*[F_H(\varphi_0) - jF_H(\varphi_1)\,\mathrm{e}^{-2jk_\perp\rho}]\} + o(1),\end{aligned} \tag{3.200}$$

where $o(1)$ denotes terms which vanish in the limit $\rho \to \infty$ and can be neglected. Note that the third integral on the right-hand side of (3.197) does not contribute to the leading term of (3.200) since $\sin(\varphi - \varphi_0) = 0$ at both critical points $\varphi = \varphi_0$ and $\varphi = \varphi_1$. It is now seen that the terms in (3.200) resulting from contribution at the critical point φ_1 do not contribute to the real part of the expression on the right-hand side of (3.200), and we obtain

$$\widetilde{p} = \frac{2k}{k_\perp^2}\left\{Y\,\mathrm{Re}\left[E_{0z}^*F_E(\varphi_0)\right] + Z\,\mathrm{Re}\left[H_{0z}^*F_H(\varphi_0)\right]\right\}. \tag{3.201}$$

Accounting for (3.195) and (3.201) transforms the equation of energy balance (3.193) to the relation

$$p_{\mathrm{abs}} + p_{\mathrm{sc}} = \frac{2k}{k_\perp^2}\left\{Y\,\mathrm{Re}\left[E_{0z}^*F_E(\varphi_0)\right] + Z\,\mathrm{Re}\left[H_{0z}^*F_H(\varphi_0)\right]\right\}. \tag{3.202}$$

Normalizing (3.202) to the power density of the incident field w_{inc} from (3.16) and using (3.196) and the definitions of the absorption and extinction widths in (3.27) and (3.28), the relation (3.202) can be rewritten as

$$\sigma_{\mathrm{ext}}^{2\mathrm{D}} = \frac{4}{k}\frac{\mathrm{Re}\left[E_{0z}^*F_E(\varphi_0)\right] + Z^2\,\mathrm{Re}\left[H_{0z}^*F_H(\varphi_0)\right]}{|E_{0z}|^2 + |ZH_{0z}|^2}. \tag{3.203}$$

This formula expresses the extinction width of an arbitrary cylindrical body (or bodies) through the values of the scattering amplitudes $F_E(\varphi)$ and $F_H(\varphi)$ at $\varphi = \varphi_0$, that is, in the direction $\hat{\imath}$ of the incident wave or, in other words, in the direction of forward scattering (see (3.188)). The relation (3.203) can be therefore regarded as the forward-scattering or optical theorem for cylindrical scatterers.

The relation (3.203) is quite general as it applies to cylindrical scatterers with arbitrary material properties and arbitrarily shaped geometric cross sections illuminated by an arbitrarily polarized plane wave with the incidence direction making an arbitrary angle β with the axis of the cylinder(s). If the scatterer has no losses, then $\sigma_{\mathrm{abs}}^{2\mathrm{D}} = 0$ and the optical theorem permits determining the total scattering width $\sigma_{\mathrm{T}}^{2\mathrm{D}}$ of the scatterer(s) from the values of the scattering amplitudes in the direction of forward scattering.

In the case of illumination by a wave of E polarization ($H_{0z} = 0$) or H polarization ($E_{0z} = 0$), (3.203) reduces to

$$\sigma_{\mathrm{ext}}^{E} = \frac{4}{k}\mathrm{Re}\left[\frac{F_E(\varphi_0)}{E_{0z}}\right], \quad \sigma_{\mathrm{ext}}^{H} = \frac{4}{k}\mathrm{Re}\left[\frac{F_H(\varphi_0)}{H_{0z}}\right], \tag{3.204}$$

respectively. It is interesting to note that expressions in (3.204) formally coincide with the relations available for the normal incidence ($\beta = \pi/2$) (Jones 1964, section 8.6). However, it should be pointed out that the scattering amplitudes $F_E(\varphi)$ and $F_H(\varphi)$ are, in general, functions of β.

Finally, in the case of a single cylindrical scatterer impenetrable to the incident radiation, by using the asymptotic relations (3.176) for the scattering amplitudes for $k \to \infty$, one arrives at the limiting values of the extinction widths

$$\sigma_{\text{ext}}^{2\text{D}}, \sigma_{\text{ext}}^{E}, \sigma_{\text{ext}}^{H} \to 2L \sin\beta. \tag{3.205}$$

The limit (3.205) is valid under the same assumptions about the scattering body as the limit (3.124) for compact bodies. It is independent of polarization and depends on the width L of the strip of the geometric shadow behind the scatterer and on the angle β between the axis of the cylinder and the direction of incidence of the incoming plane waves. The extinction widths, therefore, vanish as $\beta \to 0$ or π, that is, as the incidence direction gets parallel to the axis of the cylinder.

3.5 Scattering Regimes

The behavior of the scattered field in the far zone of a scatterer is largely determined by the ratio between the characteristic size D of the scatterer and the wavelength λ. Depending on this ratio, one distinguishes the low-frequency ($D < \lambda$), resonant ($D \approx \lambda$), and high-frequency ($D > \lambda$) scattering regimes. A classical illustration of these is the dependence of RCS of a PEC sphere on $ka = 2\pi a/\lambda$, where a is the sphere radius (Figure 3.16). For small values of ka, when the wavelength is greater than the circumference of the sphere, the radar cross section monotonically increases with the growing value of ka. In the range $1 < ka < 10$, the curve changes its behavior and begins to oscillate. As the value of ka increases further, the amplitude of the oscillations gradually decreases and, when $ka > 10$, the radar cross section σ approaches a constant value πa^2. The following sections 3.5.1, 3.5.2, and 3.6 address in more detail the characteristic features of the scattering regimes and suitable simulation and analysis methods.

3.5.1 *Resonant-size Scatterers*

When the target dimensions are of the same order as the wavelength, the scattering cross section is a complicated function of the frequency, often with pronounced maxima and minima. The scattered field is the result of interference of many contributions from elementary pieces of the scattering surface and since the distances between these pieces are now

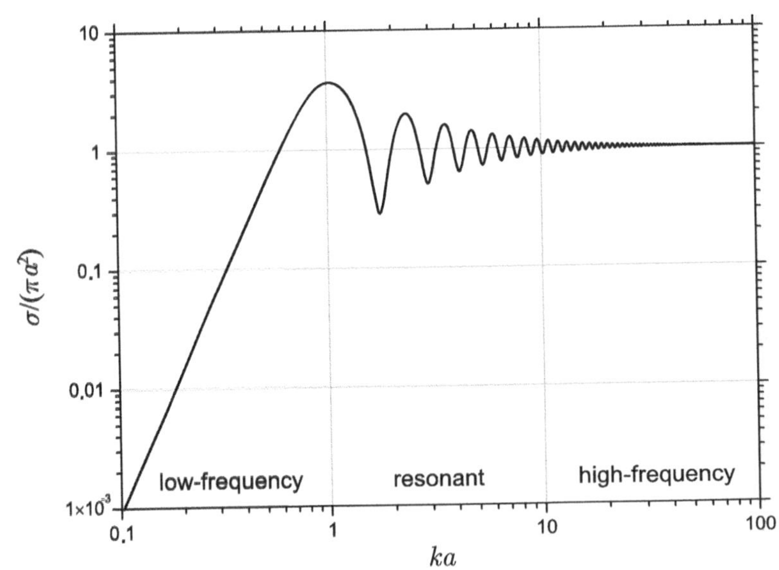

Figure 3.16 Radar cross section of an ideally conducting sphere normalized to πa^2, the area of the geometric cross section (radius a). The behavior of RCS corresponds to the low-frequency scattering regime when $0 < ka < 1$, to the resonant scattering regime when $1 < ka < 10$, and to the high-frequency scattering regime when $ka > 10$.

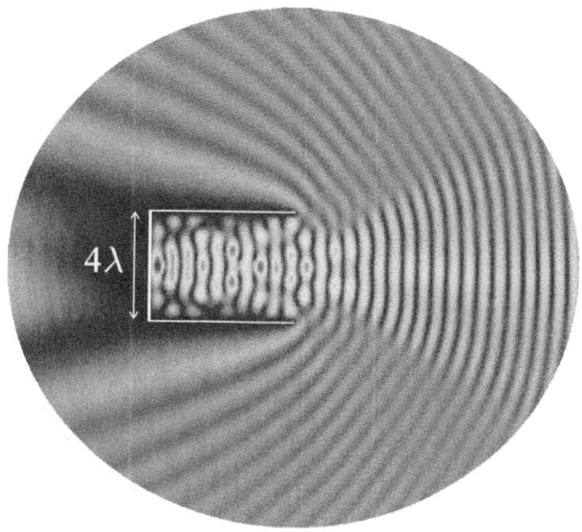

Figure 3.17 Distribution of the total electric field when a plane electromagnetic wave is scattered on a 4λ-wide open-end parallel-plate waveguide. The incident wave with the electric field vector perpendicular to the plane of the figure comes from the right-hand side along the axis of the waveguide.

comparable to the wavelength, the conditions for constructive or destructive interference can be satisfied, depending on the frequency and other factors, like incidence and scattering directions, polarization, etc. For example, in the case of a PEC sphere (Figure 3.16), the radar cross section has maxima when the circumference of the sphere $2\pi a$ is approximately equal to an integer number of λ, which can be interpreted as the result of the constructive interference of the wave reflected from the sphere with the waves circumventing the sphere one or more times before getting scattered toward the observer. Another mechanism that may lead to the resonant behavior of the scattered field is the excitation of resonant oscillations by the incident wave in scatterer's cavities. The fields scattered by such structures demonstrate strong resonance-type frequency dependence.

Apart from the special cases of a sphere and a PEC disc when exact solutions are available, an analytical study of scattering from resonant-size objects is not possible because of the absence of a small/large parameter. Numerical methods are used and a broad variety of them is applicable because the moderate size of the discretized problem permits computer-based simulations. Figure 3.17 is an illustrative result of such computer-based simulations of the scattering on a resonance-size scatterer. The scatterer is a rectangular resonator (waveguide) with perfectly conducting walls, which is open at the right-hand side. An incident electromagnetic wave comes from the right along the horizontal axis and gets scattered by the waveguide. The interference pattern inside the waveguide clearly suggests the excitation of resonance waves (waveguide modes).

3.5.2 Electrically Large Scatterers

In this frequency region the size of the scatterer is large compared with the wavelength of the incident field. This is the most typical situation for radiolocation. The scattering cross section very much depends on the shape of the scattering body and on the illumination and observation directions. A small change in the orientation of the target or in a position of the transmitter or receiver may lead to a dramatic change in the scattered field value. The characteristic scale of the angular variance is about λ/D where D is the characteristic size of the scatterer. Figure 3.18 shows a typical example: bistatic cross section of a generic aircraft as a function of the observation aspect at two different frequencies corresponding to $\lambda = 0.03$ m and 0.3 m. The aircraft is 17.34 m long, and the expected scale of the angular variance in the scattering diagram is about 0.1° and 1°, respectively. The plotted curves confirm this, and it is seen that the dotted curve is much smoother than the solid line. The sampling rates required to represent the diagrams correctly are 0.01° and 0.1°, respectively.

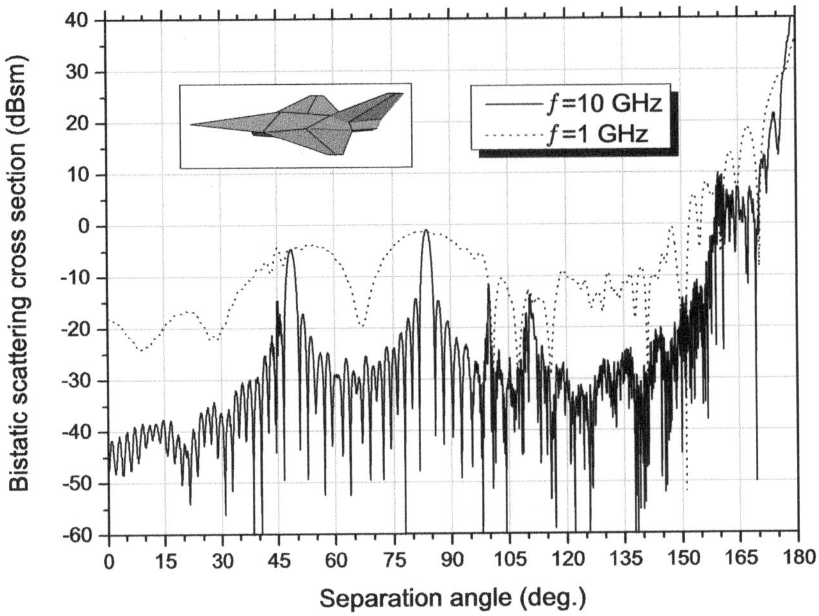

Figure 3.18 Bistatic scattering cross section of a generic aircraft DLR-F7 at two different frequencies. The aircraft lies in the plane $z = 0$ and is illuminated from the nose ($\theta_{\text{inc}} = 90°$, $\phi_{\text{inc}} = 180°$). The observation point moves around the aircraft in the plane $z = 0$ such that $\theta = 90°$ and $0 \le \phi \le 180°$. The separation angle is defined as $\phi_{\text{inc}} - \phi$ and is equal to zero in the backscattering case. Both the transmitting and receiving antennas are polarized in the z direction.

To illustrate these general statements about the properties of scattering in this regime, it is instructive to study scattering from a simple configuration which permits simple expressions for the scattered fields. Consider two electrically small PEC spheres of a radius a located at the points $\mathbf{r}_1$ and $\mathbf{r}_2$ (Figure 3.19). Without loss of generality we may assume that $\mathbf{r}_2 = -\mathbf{r}_1$ and write

$$\mathbf{r}_1 = \frac{D}{2}(\hat{x}\sin\theta_1\cos\phi_1 + \hat{y}\sin\theta_1\sin\phi_1 + \hat{z}\cos\theta_1). \tag{3.206}$$

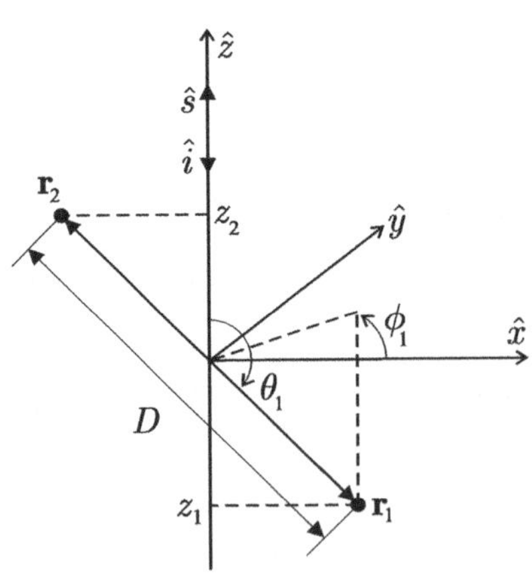

Figure 3.19 Geometry of the two-spheres configuration.

thus describing the configuration in terms of three parameters: zenith angle θ_1 $(0 \leq \theta_1 \leq \pi)$, azimuth angle ϕ_1 $(0 \leq \phi_1 \leq 2\pi)$, and the distance D between the spheres. Parameter D defines the size of the configuration, and we assume that $a \ll \lambda \ll D$ so that every individual sphere is electrically small, but the structure is electrically large. Under these limitations the mutual coupling between the spheres can be neglected, which allows a simple solution for the fields scattered from the configuration.

The configuration is illuminated by a plane wave

$$\mathbf{E}_{\text{inc}}(\mathbf{r}) = \hat{x}E_0 e^{-jk\hat{\imath}\cdot\mathbf{r}} \tag{3.207}$$

with $\hat{\imath} = -\hat{z}$, and we are going to derive a solution for the case of backscattering ($\hat{s} = -\hat{\imath}$). Using the results (3.324) and (3.325) from section 3.6 with $\phi = 0$ and $\theta = 0$ (backscattering) we write

$$\mathbf{E}_{\text{sc}}(\mathbf{r}) = \mathbf{E}_{\text{inc}}(\mathbf{r}_1)\frac{3}{2}(ka)^3\frac{e^{-jk|\mathbf{r}-\mathbf{r}_1|}}{k|\mathbf{r}-\mathbf{r}_1|} + \mathbf{E}_{\text{inc}}(\mathbf{r}_2)\frac{3}{2}(ka)^3\frac{e^{-jk|\mathbf{r}-\mathbf{r}_2|}}{k|\mathbf{r}-\mathbf{r}_2|}. \tag{3.208}$$

In the far field ($r \to \infty$),

$$|\mathbf{r}-\mathbf{r}_1| \approx r - \hat{s}\cdot\mathbf{r}_1 = r - z_1, \quad |\mathbf{r}-\mathbf{r}_2| \approx r - \hat{s}\cdot\mathbf{r}_2 = r - z_2, \tag{3.209}$$

and (3.208) becomes

$$\mathbf{E}_{\text{sc}}(\mathbf{r}) = \frac{e^{-jkr}}{kr}E_0\hat{x}\frac{3}{2}(ka)^3[e^{jk(\hat{s}-\hat{\imath})\cdot\mathbf{r}_1} + e^{jk(\hat{s}-\hat{\imath})\cdot\mathbf{r}_2}] \tag{3.210}$$

or

$$\mathbf{E}_{\text{sc}}(\mathbf{r}) = \frac{e^{-jkr}}{kr}E_0\hat{x}\frac{3}{2}(ka)^3(e^{2jkz_1} + e^{2jkz_2}). \tag{3.211}$$

Note that the backscattered field in (3.211) depends only on the z coordinates of the spheres, $z_1 = -z_2 = (D/2)\cos\phi_1$, that is, on the components of the position vectors $\mathbf{r}_1$ and $\mathbf{r}_2$ in the direction of incidence and scattering but not on transversal coordinates $x_{1,2}$ and $y_{1,2}$ and not on the angle ϕ_1 defining the rotation of the configuration around the incidence and scattering direction. This is a manifestation of a general property of bistatic scattering, which is seen from (3.210), that the field scattered by a collection of scatterers is independent of the distances between the scatterers and the PoS, the plane in which the vectors $\hat{\imath}$ and $\hat{s}$ lie (section 3.3.3).

The far-field coefficient follows from (3.211) as

$$\mathbf{F} = E_0\hat{x}\frac{3}{2}(ka)^3(e^{2jkz_1} + e^{2jkz_2}), \tag{3.212}$$

and (3.110) gives for the RCS

$$\sigma = \sigma_0\cos^2[kD(z_1 - z_2)] = \sigma_0\cos^2(kD\cos\theta_1) \tag{3.213}$$

with $\sigma_0 = 36\pi k^4 a^6$. By varying θ_1 in (3.213) we can study the dependence of RCS on the orientation of the configuration with respect to the incidence and scattering directions, and it is apparent that the angular scale of the variations is determined by the ratio λ/D, which is another important property of the scattering from electrically large configurations. Indeed, if θ_n and θ_{n+1} are the zenith angles of two subsequent minima or maxima in the dependence of σ on θ_1 (n is an integer number), then it follows from (3.213) that

$$\theta_n - \theta_{n+1} \approx \frac{\lambda}{2D\sin\theta_n} \quad (\theta_n \neq 0, \pi) \tag{3.214}$$

which makes the statement apparent. As to the frequency dependence, RCS oscillates (as a result of constructive or destructive interference of scattering contributions from individual scatterers) with an amplitude proportional to ω^4, which is a general property of scattering from configurations built from electrically small weakly interacting particles.

Table 3.1 Typical approximate frequency dependencies (on the left) and approximate values (on the right) of RCS for large metal bodies.

$\omega^2\ (\lambda^{-2})$	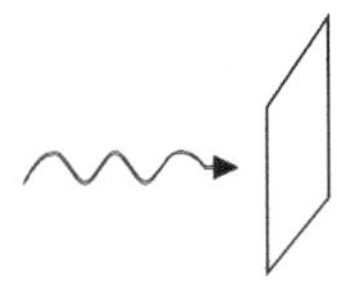	Metal plate (area S)	$\frac{4\pi S^2}{\lambda^2}$
$\omega\ (\lambda^{-1})$	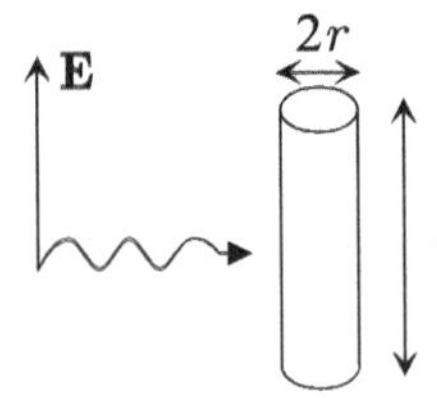	Metal cylinder	$\frac{2\pi r l^2}{\lambda}$
Constant	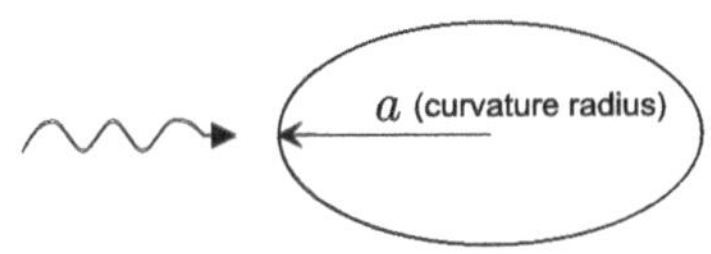	Sphere, spheroid	πa^2
$\omega^{-1}\ (\lambda)$	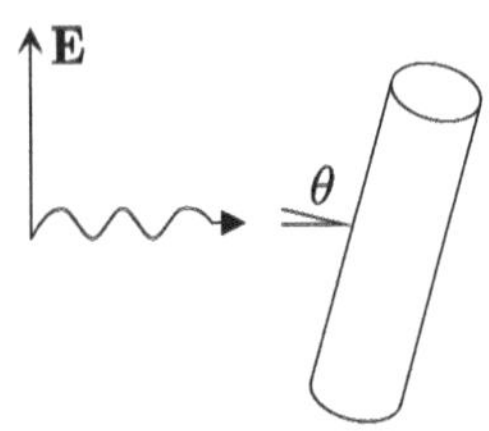	Metal cylinder	$\frac{r\lambda}{2\pi\theta^2}$
$\omega^{-2}\ (\lambda^2)$	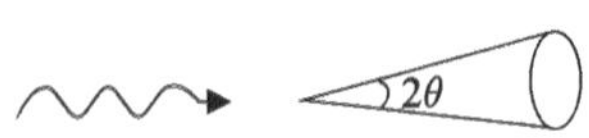	Metal cone	$\frac{\lambda^2}{16\pi}\tan^4\theta$

For different shapes, the cross section can increase or decrease with the frequency and the typical cases are listed in Table 3.1. Furthermore, the character of the frequency dependence of the scattering cross section may depend on the orientation of the scatterer with respect to the directions of incidence and scattering. For example, in the case of a flat scatterer (a plate) and scattering in either specular or forward direction, the typical dependence is $O(\omega^2)$, which is otherwise $O(\omega^{-1})$ for plates with a smoothly curved rim (e.g., a circular disc) or $O(\omega^{-2})$ for polygonal plates.

To simulate scattering from electrically large generally shaped scatterers, one needs suitable methods. The use of numerical techniques is typically difficult or even impractical because of a too high number of degrees of freedom in the problem. For example, application of the surface integral equation method in combination with the method of moments to a PEC sphere of about 100λ in diameter results in a problem with approximately 10 millions of unknowns. On the other hand, for electrically large scatterers the ratio λ/D is small and this opens up a possibility of analytical treatment by using various high-frequency methods, most notably GO, GTD, UTD, PO and PTD. All these methods provide approximate, physically meaningful solutions with an accuracy improving with growing frequency, for example Babič and Buldyrev (1991), Born and Wolf (1959), Borovikov and Kinber (1994), Bouche *et al.* (1997), Fock (1965), James (1976), Kline and Kay (1965), Kravtsov and Orlov (1990), McNamara *et al.* (1990), and Ufimtsev (2007).

According to Huygens' principle (section 2.6), the field scattered by a body can be seen as a result of radiation of the currents that are induced by the incident wave on the surface of the scatterer. When $k \to \infty$, the radiation integral can be evaluated asymptotically by the method of stationary phase (Appendix D) and the result is a sum of contributions from several critical points on the scattering surface, which are the points of specular reflection or the points at which the scattering surface is singular (edges, conical points, jumps in material properties). The scattered field gets split up,

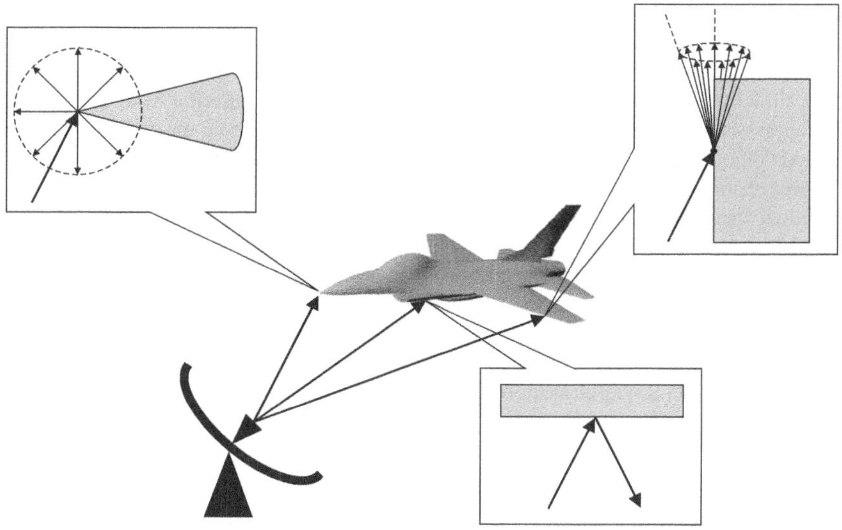

Figure 3.20 The locality principle.

therefore, into a limited number of contributions from isolated points or, more precisely, from regions around the points about a couple of wavelengths in size (locality principle, illustrated in Figure 3.20). Since these regions are small, the respective contributions can be modeled by solutions of canonical problems (reflection from a flat, cylindrical, or spherical surface, diffraction by an edge, a conical point, etc.) for bodies that conform with the scattering surface at and close to the critical point.

In this book we will study high-frequency solutions for wedges (Chapter 5), cylinders (Chapter 6), spheres (Chapter 7), and cones (Chapter 9). We will see that in the high-frequency limit, the exact solutions of these problems do reduce to sums of elementary contributions associated with specular reflection, creeping wave propagation, and edge- and tip diffraction, thus leading to GTD/UTD formulation of scattering. This formulation is best suited for observation points in the near or intermediate zone of the electrically large scatterer. For calculation of scattered fields in the far zone, the PO method is required, which uses the GO solutions for determination of equivalent currents induced on the scattering surface by the incident field. This method and its applications are described in Chapters 8 and 9.

3.6 Electrically Small Scatterers

If the characteristic size of a scatterer is much smaller than the wavelength in the surrounding space (such scatterers are referred to as electrically small scatterers), then the scattering problem can be often solved in the quasi-static (Rayleigh) approximation.[6] The scatterer is replaced by a pair of electric and magnetic dipoles with frequency-independent dipole moments, which are induced by the incident wave and serve as the sources of the scattered field. Note that in the low-frequency regime a non-zero frequency-independent magnetic moment can be induced by magnetic fields only in PEC bodies whose shape allows circulating currents (like spheres or closed loops). Indeed, in this situation the electromotive force which drives the current is proportional to the frequency (the time-derivative of the incident magnetic flux). Thus, the current amplitude will tend to a non-zero constant only if the equivalent impedance seen by the induced current is also proportional to the frequency (pure inductance). In the presence of losses, even very small ones, the current amplitude tends to zero when the frequency tends to zero. The foundation for this solution approach has been laid down by Strutt (Lord Rayleigh) first in the framework of ethereal (Lord Rayleigh 1871) and later Maxwell's theories of light (Lord Rayleigh 1881, 1897) and in the studies of Gans on the sub-wavelength resolution of small particles in optics (Gans 1920).

[6] The true static approximation means frequency equal to zero. In the regime that we study in this section, the frequency can be arbitrary, the only requirement being that the scatter is smaller than the wavelength and the induced dipole moments are frequency independent. The last assumption can be relaxed, still preserving the quasi-static nature of the solution.

This section is mostly devoted to Rayleigh scattering. There can be cases, however, in which this approximation is insufficient, for example when the dipole moments are more complex functions of frequency, resonant or including correction terms (radiation, dynamic corrections). Sometimes, higher-order multipole terms (quadrupole, octupole, etc.) are necessary. We use "low-frequency scattering" as a generic term to address scattering from electrically small scatterers, regardless of further approximation details. Approximations, which use only dipole moments, not necessarily frequency independent, are called "dipole approximations". The reader is referred to sections 7.4 and 7.5 where the whole variety of low-frequency approximations is presented for the particular case of spherical scatterers.

From the mathematical viewpoint, the quasi-static regime means that the spatial distribution of the fields in the vicinity of the scatterer can be assumed to be governed by the static equations (but, of course, the fields are time-varying with the same frequency at all points near and inside the body). Static problems involve the Laplace equation and are easier to solve than the original Maxwell equations, and this is used for determining the fields and the induced dipole moments. The fields are expanded in power series with respect to the wavenumber k in the surrounding space, and the expansion coefficients are found from a succession of static problems (Kleinman 1978; Stevenson 1953a).

The most characteristic feature of Rayleigh scattering is the dependence of the scattering cross sections on the frequency and the body dimensions as

$$\sigma = Ck^4 D^6, \tag{3.215}$$

where D is the characteristic size of the scatterer and the coefficient C depends on the body's shape and material, the incidence and scattering directions, and the polarization of the incident wave. Estimation (3.215) is applicable to arbitrary bodies, including those with vanishing volume, like a thin wire or a disc; for bodies with a finite volume V, quantity D^6 in (3.215) can be replaced by V^2.

Approximate expressions for scattering cross sections of a broad variety of electrically small bodies can be found in Ruck *et al.* (1970), Bouman *et al.* (1987), Kleinman and Senior (1975), and Siegel (1959). The available results are mostly limited to the bodies of revolution in the backscattering case with the incidence along the symmetry axis. Probably the most general result in the low-frequency scattering theory is that by Stevenson (1953a,b), who presented the first two terms in a power series expansion of the far zone scattered fields in powers of k for an arbitrary ellipsoid, dielectric or conducting, illuminated by an arbitrarily incident plane electromagnetic wave. The solution procedure appears to be quite cumbersome, but the results are elegant and physically understandable. Furthermore, in the special case of a spheroid (which is an ellipsoid with two equal semi-axes) the solution significantly simplifies and leads to simple expressions for the far scattered field in terms of elementary functions (Bouman *et al.* 1987; Collin 1960; Ruck *et al.* 1970; Van de Hulst 1957). It should be noted that in contrast to the acoustic case, the boundary value problem of electromagnetic scattering from ellipsoids, even in the simplest case of a perfectly reflecting spheroid, is not separable, so that no exact solution applicable at arbitrary frequencies and for arbitrary direction of incidence is available (Bouman *et al.* 1987, sections 11.4, 13.4).

We begin with a qualitative discussion of the dipole scattering from electrically small scatterers and then, once the physics is clarified, present a useful analysis method in terms of electric and magnetic polarizability tensors (Keller *et al.* 1972; Kleinman and Senior 1972). Stevenson's low-frequency solution for a magneto-dielectric ellipsoid is then presented, including the limiting cases of a PEC ellipsoid and a PEC elliptic plate. The analysis of low-frequency scattering significantly simplifies in the case of rotationally symmetric bodies. The low-frequency solutions for a broad variety of PEC and magneto-dielectric bodies of revolution are special cases of Stevenson's solution for an ellipsoid. We present explicit expressions for polarizability tensors, the scattering matrix, and various scattering cross sections of the magneto-dielectric and PEC spheroids, including the limits of a circular disc, a sphere, and a strongly elongated prolate spheroid. The fields scattered by electrically small PEC straight wire, circular loop, circular disc, and sphere are described in more detail. Estimations of the accuracy of the low-frequency solutions are given.

3.6.1 Physics of Dipole Scattering

A scatterer is placed in a homogeneous space with material parameters ϵ and μ, which is not necessarily vacuum. If the size D of the scatterer is much smaller than the wavelength λ in the surrounding space, the amplitude and the phase of an incident wave are almost constant over the whole volume of the scatterer. This field induces electric charges with alternative signs on the opposite ends of the scatterer and makes the scatterer act as an electric dipole. The scattered field should, therefore, be similar to the field radiated by a dipole, and it is natural to expect that dipole moments are the suitable means to describe the low-frequency scattering. The amplitude of the field scattered by the induced electric dipole with the dipole moment p in the direction orthogonal to the dipole reads (in the far zone)

$$E_{sc} = \frac{pk^2}{4\pi\epsilon R} e^{-jkR}. \tag{3.216}$$

The scattering cross section is, by definition (3.5),

$$\sigma = \frac{|p|^2 k^4}{4\pi\epsilon^2 |E_{inc}|^2}. \tag{3.217}$$

We can conclude that the scattering cross section of small objects depends on the frequency as $\sigma \sim \omega^4$, provided that p weakly depends on the frequency and tends to a nonzero constant in the limit of zero frequency. The scattering amplitude is determined by the quasi-static processes of particle polarization by uniform incident electric field.

For example, the electric dipole moment of an electrically small dielectric sphere with permittivity ϵ_1 and radius a excited by external electric field E_{inc} reads (e.g., Cheng (1992, p. 638), see also (7.497) in Chapter 7)

$$p = 4\pi a^3 \epsilon \frac{\epsilon_1 - \epsilon}{\epsilon_1 + 2\epsilon} E_{inc}. \tag{3.218}$$

The radar cross section is then

$$\sigma = 4\pi \left| \frac{\epsilon_1 - \epsilon}{\epsilon_1 + 2\epsilon} \right|^2 a^6 k^4. \tag{3.219}$$

As another example, let us consider scattering from a short, straight piece of thin metal wire. Let us denote the wire length by $2l$ ($l \ll \lambda$). We need to find the induced electric dipole in a given incident electric field. This can be done using the antenna theory because such a piece of an electric conductor can be considered as a short receiving antenna with short-circuited terminals. The amplitude of the induced current in the center of the wire reads (e.g., Schelkunoff and Friis (1952))

$$I = \frac{l_{eff} E_{inc}}{Z_{inp}}. \tag{3.220}$$

Here l_{eff} is the effective length of this dipole antenna, and for short dipoles $l_{eff} = l$. Z_{inp} is the input impedance of this antenna and is predominantly capacitive:

$$Z_{inp} \approx \frac{1}{j\omega C} + R_{rad} \tag{3.221}$$

with the capacitance (Schelkunoff and Friis 1952, p. 306)

$$C = \frac{\pi\epsilon l}{\ln(l/r_0) - 1} \tag{3.222}$$

(r_0 is the wire radius) and radiation resistance

$$R_{rad} = \frac{Z}{6\pi}(kl)^2 \tag{3.223}$$

($Z = \sqrt{\mu/\epsilon}$ is the impedance of the surrounding medium).

Now we can find the induced dipole moment

$$p = \int_{-l}^{l} q(x) x \, dx = -\frac{1}{j\omega} \int_{-l}^{l} \frac{dI(x)}{dx} x \, dx = \frac{1}{j\omega} \int_{-l}^{l} I(x) \, dx. \tag{3.224}$$

The current distribution of an electrically short receiving dipole antenna is approximately given by (King 1969)

$$I(x) = I\left(1 - \frac{x^2}{l^2}\right). \tag{3.225}$$

After a simple integration we find

$$p = \frac{4l^2}{3j\omega Z_{\text{inp}}} E_{\text{inc}} \approx \frac{4\pi l^3 \epsilon}{3[\ln(l/r_0) - 1]} E_{\text{inc}}. \tag{3.226}$$

In the last relation we have neglected the small radiation resistance R_{rad} in (3.221). The radar cross section (using (3.217)) reads

$$\sigma = \frac{4\pi k^4 l^6}{9[\ln(l/r_0) - 1]^2}. \tag{3.227}$$

Note the same dependence on the frequency (as k^4) and on the characteristic size (as a^6 or l^6) both for the small sphere and the short wire.

However, more complicated frequency dependence than that in (3.215) and stronger (resonant) scattering can happen even for very small scatterers because the induced current may have resonant behavior even if the particle is electrically small and the current distribution is quasi-static. For example, continuing the study of a short-wire scatterer, let us imagine that we cut the wire in the middle and insert a bulk inductance L there. In other words, the scatterer is now a short dipole antenna loaded by an inductive load. The induced current (3.220) reads

$$I = \frac{l_{\text{eff}} E_{\text{inc}}}{Z_{\text{inp}} + Z_{\text{load}}} \approx \frac{l_{\text{eff}} E_{\text{inc}}}{1/(j\omega C) + j\omega L + R_{\text{rad}}}. \tag{3.228}$$

Close to the resonant frequency the induced current is very strong, and the scattering cross section is very large. Let us calculate the scattering cross section at the resonant frequency, where the total reactance of particle is zero and, thus, $Z_{\text{inp}} + Z_{\text{load}} = R_{\text{rad}}$. If the particle is lossless, resistance R_{rad} equals to the radiation resistance given by (3.223). Substituting this value in place of Z_{inp} into (3.226) gives for the RCS of an electrically small dipole at its resonant frequency the following simple result:

$$\sigma = \frac{16\pi}{k^2} = \frac{4}{\pi}\lambda^2. \tag{3.229}$$

Notice that this value does not depend on the geometric parameters of the particle (its length and wire radius) but only on the wavelength. It is most important that the particle size can be very small as compared with the wavelength, but at the resonant frequency its scattering cross section is of the order of λ^2. It is interesting that while in radar and antenna engineering this property has been used for a long time, in optics it has been recognized only quite recently (Bohren 1983; Paul and Fischer 1983; Zumofen *et al.* 2008). More information about polarizabilities of loaded dipoles can be found in Tretyakov *et al.* (2003).

Let us next have a look at expression (3.219) for the radar cross section of a sphere filled with a material with permittivity ϵ_1. We see from this formula that if at some frequency the material permittivity happens to satisfy $\epsilon_1 \approx -2\epsilon$, the reflection is very strong, although the sphere radius is much smaller than the wavelength. Most of the materials which we use at radio frequencies and in the microwave region have positive permittivities (usually larger than the permittivity of vacuum ϵ_0), but materials with negative permittivity exist, and they are not exotic. The effective permittivity of metals like silver or gold have negative real part in the visible (see (2.21)). This means that at a certain frequency the real part of the denominator of (3.218) equals zero, and the absolute value of the sphere polarizability takes large values, although the electrical size of the sphere is small.

The corresponding resonance is usually called *plasmonic* resonance, and such small metal particles are called *plasmonic resonant particles.* With the recent advances in nanotechnologies, it is possible to manufacture high-quality metal spheres of nanometer dimensions. They exhibit strong resonant reflections in the visible. This is a very active area of current research (e.g., see Maier *et al.* (2001) and Schultz (2003)). One of the applications is the use of plasmonic spheres as small tags attached to biological molecules. These tags make individual molecules visible because of strong resonant reflections (Schultz 2003). Another important application is in the design of artificial electromagnetic materials (metamaterials). Here, electrically small but resonant plasmonic particles serve as artificial molecules in composite materials (see section 7.6).

Incident plane waves can induce in small conducting or dielectric particles not only electric dipole moments but also magnetic dipoles. In the limit of zero electrical size, the magnetic polarizability of any particle made of a non-magnetic material tends to zero as ω^2 (except the theoretical case of ideal PEC bodies), and eventually the electric-dipole scattering, which was discussed above, dominates. However, even very small particles can have resonant magnetic polarizability, and in the vicinity of such resonances magnetic-dipole scattering can be comparable to or even dominate the electric-dipole response. Examples of such scatterers are conductive split rings or small metal spirals (e.g., Tretyakov *et al.* (1996)).

3.6.2 Dipole Scattering in Terms of Polarizability Tensors

The electric and magnetic polarizability tensors $\overline{\overline{P}}$ and $\overline{\overline{M}}$ relate the incident electric and magnetic fields to the electric and magnetic dipole moments induced by an incident wave on an electrically small scatterer. If $\mathbf{E}_0$ and $\mathbf{H}_0$ are the electric and magnetic field vectors of the incident field at the scatterer, then the induced electric and magnetic dipole moments are given by

$$\frac{1}{\epsilon}\mathbf{p}_{\rm e} = \overline{\overline{P}} \cdot \mathbf{E}_0, \quad \mathbf{p}_{\rm m} = -\overline{\overline{M}} \cdot \mathbf{H}_0, \tag{3.230}$$

which is in fact the definition of the tensors. The polarizability tensors are independent of the incident field and are functions of the geometric and material properties of the scatterer.

The relations (3.230) can be expressed in the matrix form

$$\frac{1}{\epsilon}\begin{pmatrix} p_{\rm e1} \\ p_{\rm e2} \\ p_{\rm e3} \end{pmatrix} = \begin{pmatrix} P_{11} & P_{12} & P_{13} \\ P_{21} & P_{22} & P_{23} \\ P_{31} & P_{32} & P_{33} \end{pmatrix}\begin{pmatrix} E_{01} \\ E_{02} \\ E_{03} \end{pmatrix} \tag{3.231}$$

$$\begin{pmatrix} p_{\rm m1} \\ p_{\rm m2} \\ p_{\rm m3} \end{pmatrix} = -\begin{pmatrix} M_{11} & M_{12} & M_{13} \\ M_{21} & M_{22} & M_{23} \\ M_{31} & M_{32} & M_{33} \end{pmatrix}\begin{pmatrix} H_{01} \\ H_{02} \\ H_{03} \end{pmatrix}, \tag{3.232}$$

where $p_{{\rm e}i}, p_{{\rm m}i}, E_{0i}$, and E_{0i} with $i = 1, 2, 3$ are the components of the vectors $\mathbf{p}_{\rm e}$, $\mathbf{p}_{\rm m}$, $\mathbf{E}_0$, and $\mathbf{H}_0$ in a Cartesian coordinate system (x_1, x_2, x_3).

For generally shaped reciprocal scatterers the polarizability matrices are symmetric ($P_{in} = P_{ni}$, $M_{in} = M_{ni}$ with $i, n = 1, 2, 3$) as the symmetry ensures the reciprocity of the scattered field (see problem 39). Thus, only six elements per tensor are independent. The symmetry of the tensors can be expressed as the equality of their matrices to their transposed versions,

$$\overline{\overline{P}} = \overline{\overline{P}}^{\rm T}, \quad \overline{\overline{M}} = \overline{\overline{M}}^{\rm T}. \tag{3.233}$$

The polarizability tensors are also subject to a number of inequalities (Keller *et al.* 1972; Kleinman and Senior 1972).

The tensor or matrix character of the relations (3.230)–(3.232) represents the fundamental fact that the induced electric and magnetic moments are in general not collinear with the electric and magnetic fields. If P_i and M_i are eigenvalues and $\hat{\mathbf{e}}_i^{\rm e}$ and $\hat{\mathbf{e}}_i^{\rm m}$ the corresponding orthonormal eigenvectors[7] of the matrices $\overline{\overline{P}}$ and $\overline{\overline{M}}$,

$$\overline{\overline{P}} \cdot \hat{\mathbf{e}}_i^{\rm e} = P_i \hat{\mathbf{e}}_i^{\rm e}, \quad \overline{\overline{M}} \cdot \hat{\mathbf{e}}_i^{\rm m} = M_i \hat{\mathbf{e}}_i^{\rm m}, \quad (i = 1, 2, 3) \tag{3.234}$$

[7] The existence of a set of mutually orthogonal eigenvectors is guaranteed, for example, for every real symmetric matrix.

then the relations between the induced moments and the external fields can be expressed as

$$\frac{1}{\epsilon}\mathbf{p}_e = \sum_{i=1}^{3} P_i(\hat{\mathbf{e}}_i^e \cdot \mathbf{E}_0)\hat{\mathbf{e}}_i^e, \quad \mathbf{p}_m = -\sum_{i=1}^{3} M_i(\hat{\mathbf{e}}_i^m \cdot \mathbf{H}_0)\hat{\mathbf{e}}_i^m. \tag{3.235}$$

It is now seen that the moments and the external fields are collinear

$$\overline{\overline{P}} \cdot \mathbf{E}_0 = P_i\mathbf{E}_0, \quad \overline{\overline{M}} \cdot \mathbf{H}_0 = M_i\mathbf{H}_0 \tag{3.236}$$

only if $\mathbf{E}_0 = E_0\hat{\mathbf{e}}_i^e$ and $\mathbf{H}_0 = H_0\hat{\mathbf{e}}_i^m$.

It should be noted that elements of the polarizability tensors are complex-valued. For lossless scatterers (e.g., PEC or dielectric) the imaginary part of an element is usually small compared to the real part but is still finite, which can be understood as a consequence of radiation losses (e.g., Belov *et al.* (2003)). The polarizability tensors of lossless particles are therefore real-valued only in the leading terms of their expansions in the powers of k. We return to this question later in the section, once the scattering fields are expressed through the elements of the polarizability tensors (equations (3.300)–(3.305)).

Symmetries in the geometric and material composition of the particles may significantly simplify the structure of the polarizability matrices. For bodies which are axially symmetric about, let say, the x_3 axis of a Cartesian coordinate system (x_1, x_2, x_3),

$$P_{22} = P_{11}, \quad M_{22} = M_{11} \tag{3.237}$$

$$P_{jn} = 0 = M_{jn}, \quad j \neq n \tag{3.238}$$

implying that in these coordinates the polarizability matrices are diagonal and the number of independent non-vanishing elements of the tensors reduces to two per tensor, that is, to four in total. Relations (3.237) can be shown to follow from the condition that backscattering along the x_1 and x_2 axes is the same, and (3.238) are obtained by requiring that the bistatic scattering in the plane (x_1, x_2) is symmetric with respect to the incidence direction (see problem 40).

Furthermore, if the rotationally symmetric body is PEC and not ring-shaped, then

$$M_{33} = \frac{1}{2}P_{11} \tag{3.239}$$

(Keller *et al.* 1972), implying that the total number of non-vanishing independent elements in the polarizability matrices is equal to three.[8] Finally, spherically symmetric scatterers, not only PEC but also homogeneous or radially stratified magneto-dielectric spheres, are described by diagonal tensors with

$$P_{11} = P_{22} = P_{33}, \quad M_{11} = M_{22} = M_{33}, \tag{3.240}$$

reducing the total number of independent entrees in the polarizability matrices to two. In the case of a PEC sphere, just one element is independent because of the property (3.239).

The elements of the polarizability matrices can be deduced from solutions of potential problems that result from expansions of the fields in powers of k (Kleinman 1978), from numerical solutions of Maxwell's equations or from exact solutions of electromagnetic scattering problems when these latter are available. Once the tensors $\overline{\overline{P}}$ and $\overline{\overline{M}}$ are known, one can determine the dipole moments induced by an incident wave on the scatterer and, by considering them as secondary field sources, find the scattered field.

We assume that an electrically small scatterer is located in a homogeneous medium with material constants ϵ and μ. The electric field radiated by electric and magnetic dipoles can be expressed (section 2.5) as

$$\mathbf{E}(\mathbf{r}) = \nabla \times \nabla \times \mathbf{\Pi}_e - j\omega\mu\nabla \times \mathbf{\Pi}_m. \tag{3.241}$$

[8] The latter property may also hold for some ring-shaped bodies, like a small thin wire circular loop.

If we replace the scatterer by equivalent point dipoles with the moments $\mathbf{p}_e$ and $\mathbf{p}_m$, then the corresponding Hertz vectors $\mathbf{\Pi}_{e,m}$ (equations (2.205) and (2.210)) are

$$\mathbf{\Pi}_e = \mathbf{p}_e \frac{e^{-jkR}}{4\pi\epsilon R}, \quad \mathbf{\Pi}_m = \mathbf{p}_m \frac{e^{-jkR}}{4\pi R}, \tag{3.242}$$

where R is the distance from the scatterer to the observation point. In the far-field region, $|k|R \gg 1$ and therefore

$$\nabla \times \left(\mathbf{p}_{e,m} \frac{e^{-jkR}}{4\pi R} \right) \approx -jk\hat{s} \times \mathbf{p}_{e,m} \frac{e^{-jkR}}{4\pi R}, \tag{3.243}$$

where $\hat{s}$ is the unit vector pointing from the scatterer to the observer. The field radiated by the dipoles then reads

$$\mathbf{E}(\mathbf{r}) \approx -k^2 \frac{e^{-jkR}}{4\pi R} \left(\frac{1}{\epsilon} \hat{s} \times \hat{s} \times \mathbf{p}_e + Z\hat{s} \times \mathbf{p}_m \right) \tag{3.244}$$

with $Z = \omega\mu/k = \sqrt{\mu/\epsilon}$ being the surrounding space impedance. One relates the equivalent dipole moments to the polarizability tensors according to (3.230) and obtains for the scattered electric field

$$\mathbf{E}_{sc}(\mathbf{r}) \approx -k^2 \frac{e^{-jkR}}{4\pi R} [\hat{s} \times \hat{s} \times (\overline{\overline{P}} \cdot \mathbf{E}_0) - Z\hat{s} \times (\overline{\overline{M}} \cdot \mathbf{H}_0)]. \tag{3.245}$$

For a plane linearly polarized incident wave, $Z\mathbf{H}_0 = \hat{i} \times \mathbf{E}_0$, where $\hat{i}$ is the direction of incidence.

In a coordinate system collocated with the particle, (3.245) corresponds to the far-field coefficient

$$\mathbf{F}(\hat{i}, \hat{s}, \mathbf{E}_0, Z\mathbf{H}_0) = -\frac{k^3}{4\pi} [\hat{s} \times \hat{s} \times (\overline{\overline{P}} \cdot \mathbf{E}_0) - Z\hat{s} \times (\overline{\overline{M}} \cdot \mathbf{H}_0)]. \tag{3.246}$$

If the particle is located at a point $\mathbf{r} = \mathbf{r}_0$ shifted from the origin of the coordinate system, then $\mathbf{E}_0 = \mathbf{E}_{inc}(\mathbf{r}_0)$ and $\mathbf{H}_0 = \mathbf{H}_{inc}(\mathbf{r}_0)$, where $\mathbf{E}_{inc}(\mathbf{r})$ and $\mathbf{H}_{inc}(\mathbf{r})$ are the incident fields at the particle, and the right-hand side of (3.246) is to be multiplied with the factor $\exp(jk\hat{s} \cdot \mathbf{r}_0)$, according to (3.109).

The scattered magnetic field results from (3.245) and the far-field relation $Z\mathbf{H}_{sc}(\mathbf{r}) = \hat{s} \times \mathbf{E}_{sc}(\mathbf{r})$ as

$$\mathbf{H}_{sc}(\mathbf{r}) \approx k^2 \frac{e^{-jkR}}{4\pi R} [\hat{s} \times \hat{s} \times (\overline{\overline{M}} \cdot \mathbf{H}_0) + Y\hat{s} \times (\overline{\overline{P}} \cdot \mathbf{E}_0)]. \tag{3.247}$$

For magneto-dielectric particles, equations (3.245) and (3.247) must be related by the duality transformations (2.44), implying that

$$\overline{\overline{P}} \rightarrow -\overline{\overline{M}}, \quad \overline{\overline{M}} \rightarrow -\overline{\overline{P}} \tag{3.248}$$

on interchanging the dielectric and magnetic constants in all the involved media.

3.6.3 *Magneto-dielectric Ellipsoid*

The magneto-dielectric ellipsoid is the most general body for which polarizability tensors are available analytically. A full-wave solution is unavailable, but the low-frequency approximation can be constructed as an expansion in powers of k (Stevenson 1953a,b). Almost all specific cases (except for the loop) addressed later in this section follow from this solution.

An ellipsoid with permittivity ϵ_1 and permeability μ_1 is immersed in a homogeneous medium with permittivity ϵ and permeability μ. The incident electric and magnetic fields at the ellipsoid are $\mathbf{E}_0$ and $\mathbf{H}_0$; the fields in the interior of the ellipsoid are $\mathbf{E}_1$ and $\mathbf{H}_1$. The ellipsoid is centered at the origin of a Cartesian system ($\hat{x}_1 = \hat{x}$, $\hat{x}_2 = \hat{y}$, $\hat{x}_3 = \hat{z}$), and its principal axes are aligned with the axes of the coordinate system (Figure 3.21). This coordinate system will be referred to as principal coordinate system of the ellipsoid. In these coordinates, the shape of the ellipsoid is described by the equation

$$\left(\frac{x}{a}\right)^2 + \left(\frac{y}{b}\right)^2 + \left(\frac{z}{c}\right)^2 = 1, \tag{3.249}$$

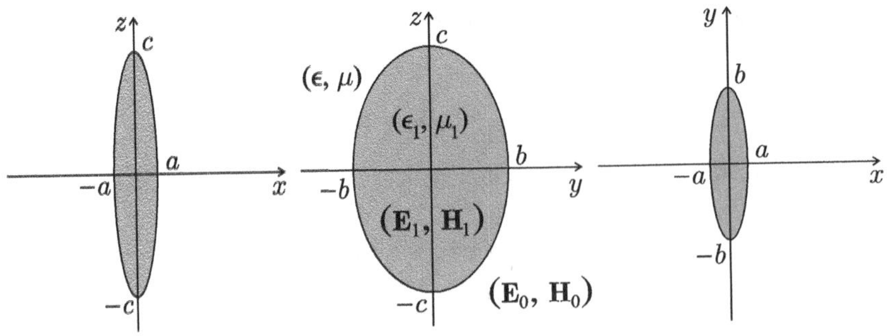

Figure 3.21 Ellipsoid with semi-principal axes of length a, b, and c aligned with the x, y, and z axes of a Cartesian coordinate system collocated with the center of the ellipsoid.

where a, b, and c are arbitrary positive parameters, which are referred to as semi-principal axes. The ellipsoid includes a wide variety of other shapes, for example spheroids (two equal semi-principal axes), circular and elliptic discs (one semi-principal axis equal to zero), and the sphere (all semi-principal axes equal).

In this section we construct the tensors of $\overline{\overline{P}}$ and $\overline{\overline{M}}$ of the ellipsoid by physical reasoning. For a rigorous derivation the reader is referred to Stevenson's works. In contrast to the analysis presented in Van de Hulst (1957), the particle may have magnetic properties as well. Once the polarizability tensors are known, explicit expressions for the scattering matrix of the electrically small magneto-dielectric ellipsoid are presented. Finally, the solution for a PEC ellipsoid is specialized to the case of a PEC elliptic plate.

Polarizability Tensors

From the symmetry reasons, the eigenvectors $\hat{\mathbf{e}}_i^{\mathrm{e}}$ and $\hat{\mathbf{e}}_i^{\mathrm{m}}$ of the polarizability matrices $\overline{\overline{P}}$ and $\overline{\overline{M}}$ (see (3.234)) should coincide with the Cartesian coordinate axes, so that we can write

$$\overline{\overline{P}} \cdot \mathbf{E}_0 = P_a E_{0x}\hat{x} + P_b E_{0y}\hat{y} + P_c E_{0z}\hat{z} \tag{3.250}$$

$$\overline{\overline{M}} \cdot \mathbf{H}_0 = M_a H_{0x}\hat{x} + M_b H_{0y}\hat{y} + M_c H_{0z}\hat{z}, \tag{3.251}$$

where P_i and M_i with $i = a, b, c$ are the eigenvalues that correspond to the eigenvectors $\hat{x}_a = \hat{x}$, $\hat{x}_b = \hat{y}$, and $\hat{x}_c = \hat{z}$.

We will now derive the eigenvalues. The external field vectors $\mathbf{E}_0$ and $\mathbf{H}_0$ can be expanded in the Cartesian components and the problem can be studied for every component separately. Assuming that $\mathbf{E}_0 = E_0\hat{x}_i$ or $\mathbf{H}_0 = H_0\hat{x}_i$, we have from (3.230), (3.250), and (3.251) that

$$\mathbf{p}_{\mathrm{e}} = \epsilon P_i \mathbf{E}_0, \quad \mathbf{p}_{\mathrm{m}} = -M_i \mathbf{H}_0. \tag{3.252}$$

The dipole moments of every magneto-dielectric body are equal to the electric and magnetic polarization vectors $\mathbf{P}_{\mathrm{e}}$ and $\mathbf{P}_{\mathrm{m}}$ integrated over the volume V of the body, and for electrically small bodies the integration can be approximately[9] replaced with multiplication by the volume,

$$\mathbf{p}_{\mathrm{e}} = \int \mathbf{P}_{\mathrm{e}}\, \mathrm{d}v \approx V\mathbf{P}_{\mathrm{e}}, \quad \mathbf{p}_{\mathrm{m}} = \int \mathbf{P}_{\mathrm{m}}\, \mathrm{d}v \approx V\mathbf{P}_{\mathrm{m}}. \tag{3.253}$$

In the case of ellipsoid, $V = 4\pi abc/3$.

---

[9] In fact, under the quasi-static conditions and uniform-field excitation, the polarization vector is constant inside the ellipsoid, and the equality is exact.

From (3.252) and (3.253),

$$V\mathbf{P}_e = \epsilon P_i \mathbf{E}_0, \quad V\mathbf{P}_m = -M_i \mathbf{H}_0. \tag{3.254}$$

The parameters P_i and M_i can be determined from (3.254) on expressing $\mathbf{P}_e$ and $\mathbf{P}_m$ through the exterior fields $\mathbf{E}_0$ and $\mathbf{H}_0$.

By definition (e.g., Stratton (1941, section 1.6), the polarization vectors are due to the difference of the electric and magnetic constants inside and outside the body,

$$\mathbf{P}_e = (\epsilon_1 - \epsilon)\mathbf{E}_1, \quad \mathbf{P}_m = \left(\frac{\mu_1}{\mu} - 1\right)\mathbf{H}_1, \tag{3.255}$$

where $\mathbf{E}_1$ and $\mathbf{H}_1$ are the fields in the interior of the magneto-dielectric body. Note that in many books polarization is defined relative to free space ($\mathbf{D} = \epsilon_0\mathbf{E} + \mathbf{P}$). Here we define it with respect to the environment, which is more convenient for studying scattering problems. For the external fields directed along the principal axes of the spheroid, the field and polarization vectors must be co-directed, and due to linearity we can write

$$\mathbf{E}_1 = \mathbf{E}_0 - \frac{1}{\epsilon} I_i^e \mathbf{P}_e, \quad \mathbf{H}_1 = \mathbf{H}_0 - I_i^m \mathbf{P}_m, \tag{3.256}$$

where I_i^e and I_i^m are some dimensionless scalar parameters which may depend on the geometry of the ellipsoid.

Eliminating $\mathbf{E}_1$ and $\mathbf{H}_1$ from (3.255) and (3.256) gives

$$\mathbf{P}_e = \frac{\epsilon(\epsilon_1 - \epsilon)}{\epsilon + I_i^e(\epsilon_1 - \epsilon)}\mathbf{E}_0, \quad \mathbf{P}_m = \frac{\mu_1 - \mu}{\mu + I_i^m(\mu_1 - \mu)}\mathbf{H}_0, \tag{3.257}$$

and using these in (3.254) results in the expressions for the eigenvalues

$$P_i = \frac{V(\epsilon_1 - \epsilon)}{\epsilon + I_i^e(\epsilon_1 - \epsilon)}, \quad M_i = -\frac{V(\mu_1 - \mu)}{\mu + I_i^m(\mu_1 - \mu)}. \tag{3.258}$$

So, the quasi-static electric and magnetic polarizabilities of the magneto-dielectric ellipsoid are proportional to the volume of the scatterer.

The parameters I_i^e and I_i^m are derived from the low-frequency expansions of the fields (Stevenson 1953a,b) as

$$I_a^e = I_a^m = I_a = \frac{abc}{2}\int_0^\infty \frac{ds}{(s+a^2)^{3/2}(s+b^2)^{1/2}(s+c^2)^{1/2}} \tag{3.259}$$

$$I_b^e = I_b^m = I_b = \frac{abc}{2}\int_0^\infty \frac{ds}{(s+a^2)^{1/2}(s+b^2)^{3/2}(s+c^2)^{1/2}} \tag{3.260}$$

$$I_c^e = I_c^m = I_c = \frac{abc}{2}\int_0^\infty \frac{ds}{(s+a^2)^{1/2}(s+b^2)^{1/2}(s+c^2)^{3/2}}. \tag{3.261}$$

They possess the property

$$I_a + I_b + I_c = 1 \tag{3.262}$$

which is derived by integration by parts in any of the equations (3.259)–(3.261). The integrals in (3.259)–(3.261) cannot be expressed in terms of elementary functions but can be easily evaluated by numerical integration. The parameters coincide with the depolarization factors appearing in static problems for conducting and dielectric ellipsoids in external electric fields and with demagnetization factors for magnetic ellipsoids in external magnetic fields (Stratton 1941, sections 3.25– 3.28, 4.18, 4.19).

The expressions (3.258) for the elements of polarizability matrices of the magneto-dielectric ellipsoid clearly comply with the duality relations (3.248). The limit of a PEC ellipsoid is obtained from (3.258) by letting $\epsilon_1 \to \infty$ and $\mu_1 \to 0$,

$$P_i = \frac{V}{I_i}, \quad M_i = \frac{V}{1 - I_i}. \tag{3.263}$$

Equations (3.258) and (3.263) with $i = a, b, c$ completely define the eigenvalues and, therefore, the polarizability matrices of the ellipsoid in the principal coordinate system of the ellipsoid,

$$\overline{\overline{P}} = \begin{pmatrix} P_a & 0 & 0 \\ 0 & P_b & 0 \\ 0 & 0 & P_c \end{pmatrix}, \quad \overline{\overline{M}} = \begin{pmatrix} M_a & 0 & 0 \\ 0 & M_b & 0 \\ 0 & 0 & M_c \end{pmatrix}. \tag{3.264}$$

In a rotated orthonormal coordinate frame $(\hat{x}', \hat{y}', \hat{z}')$ the polarizability matrices of the ellipsoid are no longer diagonal and are given by

$$\overline{\overline{P}}' = \overline{\overline{Q}}^{\mathrm{T}} \cdot \overline{\overline{P}} \cdot \overline{\overline{Q}}, \quad \overline{\overline{M}}' = \overline{\overline{Q}}^{\mathrm{T}} \cdot \overline{\overline{M}} \cdot \overline{\overline{Q}}, \tag{3.265}$$

where

$$\overline{\overline{Q}} = \begin{pmatrix} \hat{x}\cdot\hat{x}' & \hat{x}\cdot\hat{y}' & \hat{x}\cdot\hat{z}' \\ \hat{y}\cdot\hat{x}' & \hat{y}\cdot\hat{y}' & \hat{y}\cdot\hat{z}' \\ \hat{z}\cdot\hat{x}' & \hat{z}\cdot\hat{y}' & \hat{z}\cdot\hat{z}' \end{pmatrix}. \tag{3.266}$$

Elliptic Disc

An interesting and important special case of the ellipsoid is the elliptic disc, which is obtained in the limit of one vanishing principal semi-axis. Without loss of generality, we assume that $a \geq b$ and $c \to 0$. The resulting shape is a flat elliptic plate (Figure 3.22), which includes the special case of a circular disc ($a = b$) and a strip wire when the ratio of the remaining semi-axes approaches zero ($b/a \to 0$). These two shapes, the circular disc and the strip wire, are important elementary scatterers, and in this section their polarizability tensors are derived from the general expressions for the ellipsoid.

We start by specializing the general equations (3.258) and (3.263) to the special case $c \to 0$. The shape parameters I_a and I_b vanish because they are proportional to c,

$$I_a \approx \frac{abc}{2} i_a, \quad I_b \approx \frac{abc}{2} i_b \tag{3.267}$$

with

$$i_a = \int_0^\infty \frac{\mathrm{d}s}{\sqrt{s}(s+a^2)^{3/2}(s+b^2)^{1/2}} \tag{3.268}$$

$$i_b = \int_0^\infty \frac{\mathrm{d}s}{\sqrt{s}(s+a^2)^{1/2}(s+b^2)^{3/2}}, \tag{3.269}$$

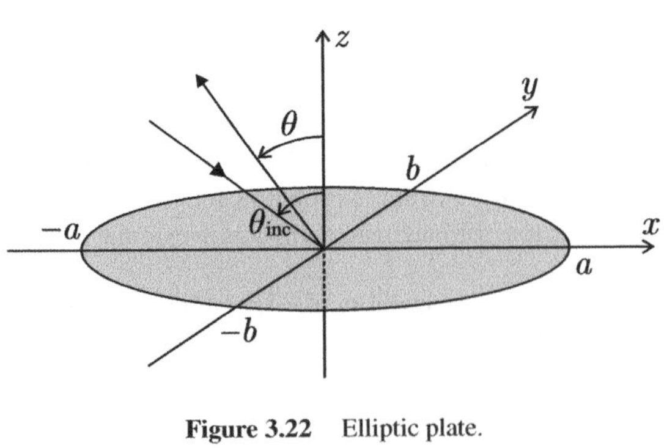

Figure 3.22 Elliptic plate.

whereas the parameter I_c approaches a finite limit

$$I_c \approx 1 - \frac{abc}{2}(i_a + i_b) \tag{3.270}$$

because of (3.267) and the identity (3.262). It is now easy to see that polarizabilities of the magneto-dielectric ellipsoid given by (3.258) vanish because of the vanishing volume of the particle.

The situation is different for PEC spheroids, with polarizabilities given by (3.263). Although P_c, M_a, and M_b still go to zero as $V \to 0$, the other parameters approach finite limits,

$$P_a \to \frac{8\pi}{3i_a}, \quad P_b \to \frac{8\pi}{3i_b}, \quad M_c \to \frac{8\pi}{3(i_a + i_b)}. \tag{3.271}$$

The integrals in (3.271) can be expressed in terms of the complete elliptic integrals

$$K(g) = \int_0^{\pi/2} \frac{\mathrm{d}\chi}{\sqrt{1 - g\sin^2\chi}}, \quad E(g) = \int_0^{\pi/2} \sqrt{1 - g\sin^2\chi}\,\mathrm{d}\chi \tag{3.272}$$

and the eccentricity of the ellipse

$$g = 1 - \frac{b^2}{a^2} \tag{3.273}$$

as follows:

$$i_a = \frac{2}{a^3 g}[K(g) - E(g)], \quad i_b = \frac{2}{a^3 g}\left[\frac{E(g)}{1-g} - K(g)\right] \tag{3.274}$$

so that we have

$$P_a = \frac{4\pi a^3 g}{3[K(g) - E(g)]} \tag{3.275}$$

$$P_b = \frac{4\pi a^3 g(1-g)}{3[E(g) - K(g)(1-g)]} \tag{3.276}$$

$$M_c = \frac{4\pi a^3 (1-g)}{3E(g)}. \tag{3.277}$$

When $b \to a$, the ellipse degenerates to the circular disc. The corresponding polarizabilities follow from (3.275)–(3.277) with $g \to 0$, and since

$$K(g) = \frac{\pi}{2} + \frac{\pi}{8}g + O(g^2), \quad E(g) = \frac{\pi}{2} - \frac{\pi}{8}g + O(g^2), \tag{3.278}$$

the result is given by the expressions:

$$P_a \to \frac{16a^3}{3}, \quad P_b \to \frac{16a^3}{3}, \quad M_c \to \frac{8a^3}{3}. \tag{3.279}$$

With $b \to 0$ we obtain an electrically small strongly elongated elliptic plate with a zero thickness ($c = 0$) and the maximum width $W = 2b$ much smaller than the length $L = 2a$. As $g \to 1$,

$$K(g) = 2\ln 2 - \frac{1}{2}\ln(1-g) + O[(1-g)\ln(1-g)] \tag{3.280}$$

$$E(g) = 1 + O[(1-g)\ln(1-g)], \tag{3.281}$$

implying that

$$P_a \to \frac{4\pi a^3}{3[2\ln 2 - 1 - (1/2)\ln(1-g)]}, \quad P_b \to 0, \quad M_c \to 0. \tag{3.282}$$

The shape, which looks like a strip wire, has no magnetic polarizability and the electric polarization is directed along its length,

$$P_a \to \frac{4\pi a^3}{3[\ln(4a/b) - 1]} = \frac{\pi L^3}{6[\ln(4L/W) - 1]}. \tag{3.283}$$

Comparing with the corresponding result for a round wire (3.226), we note that the cross section shape influences only the slow-varying log term.

Scattering Matrix

The scattering matrix provides a complete description of the far-field scattering from a given body (section 3.3). Let us derive explicit expressions for the scattering matrix of an electrically small ellipsoid in spherical coordinates associated with the principal coordinate system of the ellipsoid (with the z axis as the zenith axis). The origin of the coordinate systems is collocated with the ellipsoid. A more precise specification of the position of the origin in the interior of the ellipsoid is not necessary as the ellipsoid is electrically small and effectively appears as a point scatterer.

To express the elements of the scattering matrix we use the expression (3.246) for the far-field coefficient, and it follows from (3.245) and (3.62)–(3.65) that

$$A_{\theta\theta_{\text{inc}}} = \frac{\sqrt{4\pi}}{kE_0}\hat{\theta} \cdot \mathbf{F}(\hat{\imath}, \hat{s}, E_0\hat{\theta}_{\text{inc}}, -E_0\hat{\phi}_{\text{inc}}) \tag{3.284}$$

$$A_{\phi\theta_{\text{inc}}} = \frac{\sqrt{4\pi}}{kE_0}\hat{\phi} \cdot \mathbf{F}(\hat{\imath}, \hat{s}, E_0\hat{\theta}_{\text{inc}}, -E_0\hat{\phi}_{\text{inc}}) \tag{3.285}$$

$$A_{\theta\phi_{\text{inc}}} = \frac{\sqrt{4\pi}}{kE_0}\hat{\theta} \cdot \mathbf{F}(\hat{\imath}, \hat{s}, E_0\hat{\phi}_{\text{inc}}, E_0\hat{\theta}_{\text{inc}}) \tag{3.286}$$

$$A_{\phi\phi_{\text{inc}}} = \frac{\sqrt{4\pi}}{kE_0}\hat{\phi} \cdot \mathbf{F}(\hat{\imath}, \hat{s}, E_0\hat{\phi}_{\text{inc}}, E_0\hat{\theta}_{\text{inc}}). \tag{3.287}$$

In the chosen coordinate system, the matrices $\overline{\overline{P}}$ and $\overline{\overline{M}}$ are diagonal as in (3.264) with

$$P_{11} = P_a, \quad P_{22} = P_b, \quad P_{33} = P_c, \quad M_{11} = M_a, \quad M_{22} = M_b, \quad M_{33} = M_c, \tag{3.288}$$

so that equations (3.250) and (3.251) apply, and it follows from (3.51), (3.54), (3.55), (3.56), (3.58), and (3.59) that

$$\overline{\overline{P}} \cdot \mathbf{E}_0 = E_0(\hat{x}P_{11}\cos\theta_{\text{inc}}\cos\phi_{\text{inc}} + \hat{y}P_{22}\cos\theta_{\text{inc}}\sin\phi_{\text{inc}} - \hat{z}P_{33}\sin\theta_{\text{inc}}) \tag{3.289}$$

$$Z\overline{\overline{M}} \cdot \mathbf{H}_0 = E_0(\hat{x}M_{11}\sin\phi_{\text{inc}} - \hat{y}M_{22}\cos\phi_{\text{inc}}) \tag{3.290}$$

with $\mathbf{E}_0 = E_0\hat{\theta}_{\text{inc}}$ and $Z\mathbf{H}_0 = -E_0\hat{\phi}_{\text{inc}}$, and

$$\overline{\overline{P}} \cdot \mathbf{E}_0 = E_0(-\hat{x}P_{11}\sin\phi_{\text{inc}} + \hat{y}P_{22}\cos\phi_{\text{inc}}) \tag{3.291}$$

$$Z\overline{\overline{M}} \cdot \mathbf{H}_0 = E_0(\hat{x}M_{11}\cos\theta_{\text{inc}}\cos\phi_{\text{inc}} + \hat{y}M_{22}\cos\theta_{\text{inc}}\sin\phi_{\text{inc}} - \hat{z}M_{33}\sin\theta_{\text{inc}}) \tag{3.292}$$

with $\mathbf{E}_0 = E_0\hat{\phi}_{\text{inc}}$ and $Z\mathbf{H}_0 = E_0\hat{\theta}_{\text{inc}}$. Using these in (3.284)–(3.287) leads, after some vector algebra, to the explicit expressions for the elements of the scattering matrix of the ellipsoid,

$$\begin{aligned} A_{\theta\theta_{\text{inc}}} = \frac{k^2}{\sqrt{4\pi}}(&P_{11}\cos\theta_{\text{inc}}\cos\phi_{\text{inc}}\cos\theta\cos\phi + P_{22}\cos\theta_{\text{inc}}\sin\phi_{\text{inc}}\cos\theta\sin\phi \\ &+ P_{33}\sin\theta_{\text{inc}}\sin\theta + M_{11}\sin\phi_{\text{inc}}\sin\phi + M_{22}\cos\phi_{\text{inc}}\cos\phi) \end{aligned} \tag{3.293}$$

$$A_{\phi\theta_{\rm inc}} = -\frac{k^2}{\sqrt{4\pi}}(P_{11}\cos\theta_{\rm inc}\cos\phi_{\rm inc}\sin\phi - P_{22}\cos\theta_{\rm inc}\sin\phi_{\rm inc}\cos\phi$$
$$-M_{11}\sin\phi_{\rm inc}\cos\theta\cos\phi + M_{22}\cos\phi_{\rm inc}\cos\theta\sin\phi) \tag{3.294}$$

$$A_{\theta\phi_{\rm inc}} = -\frac{k^2}{\sqrt{4\pi}}(P_{11}\sin\phi_{\rm inc}\cos\theta\cos\phi - P_{22}\cos\phi_{\rm inc}\cos\theta\sin\phi$$
$$-M_{11}\cos\theta_{\rm inc}\cos\phi_{\rm inc}\sin\phi + M_{22}\cos\theta_{\rm inc}\sin\phi_{\rm inc}\cos\phi) \tag{3.295}$$

$$A_{\phi\phi_{\rm inc}} = \frac{k^2}{\sqrt{4\pi}}(P_{11}\sin\phi_{\rm inc}\sin\phi + P_{22}\cos\phi_{\rm inc}\cos\phi + M_{11}\cos\theta_{\rm inc}\cos\phi_{\rm inc}\cos\theta\cos\phi$$
$$+M_{22}\cos\theta_{\rm inc}\sin\phi_{\rm inc}\cos\theta\sin\phi + M_{33}\sin\theta_{\rm inc}\sin\theta). \tag{3.296}$$

The scattering matrix described by (3.293)–(3.296) is invariant with respect to the interchange of the incident and scattering directions, $\theta \leftrightarrow \theta_{\rm inc}$ and $\phi \leftrightarrow \phi_{\rm inc}$, as required by the reciprocity property of solutions of Maxwell's equations.

3.6.4 Rotationally Symmetric Particles

Scattering Matrix and Scattering Cross Sections

If the scatterer is a body of revolution, then the polarizability tensors have the simplified structures (3.237) and (3.238). Assuming that the z axis is the symmetry axis and using the relations

$$\begin{aligned}\overline{\overline{P}}\cdot\mathbf{E}_0 &= P_{11}\mathbf{E}_0 + (P_{33}-P_{11})(\mathbf{E}_0\cdot\hat{z})\hat{z}\\ \overline{\overline{M}}\cdot\mathbf{H}_0 &= M_{11}\mathbf{H}_0 + (M_{33}-M_{11})(\mathbf{H}_0\cdot\hat{z})\hat{z},\end{aligned} \tag{3.297}$$

the scattered field can be expressed as

$$\mathbf{E}_{\rm sc}(\mathbf{r}) \approx -k^2\frac{{\rm e}^{-jkR}}{4\pi R}\hat{s}\times[P_{11}\hat{s}\times\mathbf{E}_0 + (P_{33}-P_{11})(\mathbf{E}_0\cdot\hat{z})\hat{s}\times\hat{z} - ZM_{11}\mathbf{H}_0 - Z(M_{33}-M_{11})(\mathbf{H}_0\cdot\hat{z})\hat{z}] \tag{3.298}$$

which includes only four parameters: P_{11}, P_{33}, M_{11}, and M_{33}.

Equation (3.298) can be used to derive a generic form of the scattering matrix of electrically small rotationally symmetric scatterers. Using the spherical coordinate system with the zenith axis z (section 3.3.2), one expresses the incident field in (3.298) in the basis $\hat{\imath}$, $\hat{\theta}_{\rm inc}$, and $\hat{\phi}_{\rm inc}$ as

$$\begin{aligned}\mathbf{E}_0 &= E^{\rm inc}_{\theta_{\rm inc}}\hat{\theta}_{\rm inc} + E^{\rm inc}_{\phi_{\rm inc}}\hat{\phi}_{\rm inc}\\ Z\mathbf{H}_0 &= \hat{\imath}\times\mathbf{E}_0 = E^{\rm inc}_{\phi_{\rm inc}}\hat{\theta}_{\rm inc} - E^{\rm inc}_{\theta_{\rm inc}}\hat{\phi}_{\rm inc}\end{aligned} \tag{3.299}$$

and calculates the components $E^{\rm sc}_\theta(\mathbf{r})$ and $E^{\rm sc}_\phi(\mathbf{r})$ of the scattered field by multiplying (3.298) with $\hat{\theta}$ and $\hat{\phi}$, thus arriving at a representation of the form (3.61). The elements of the scattering matrix are as follows:

$$A_{\theta\theta_{\rm inc}} = \frac{k^2}{\sqrt{4\pi}}[(M_{11}+P_{11}\cos\theta\cos\theta_{\rm inc})\cos(\phi-\phi_{\rm inc}) + P_{33}\sin\theta\sin\theta_{\rm inc}] \tag{3.300}$$

$$A_{\theta\phi_{\rm inc}} = \frac{k^2}{\sqrt{4\pi}}(P_{11}\cos\theta + M_{11}\cos\theta_{\rm inc})\sin(\phi-\phi_{\rm inc}) \tag{3.301}$$

$$A_{\phi\theta_{\rm inc}} = -\frac{k^2}{\sqrt{4\pi}}(M_{11}\cos\theta + P_{11}\cos\theta_{\rm inc})\sin(\phi-\phi_{\rm inc}) \tag{3.302}$$

$$A_{\phi\phi_{\rm inc}} = \frac{k^2}{\sqrt{4\pi}}[(P_{11}+M_{11}\cos\theta\cos\theta_{\rm inc})\cos(\phi-\phi_{\rm inc}) + M_{33}\sin\theta\sin\theta_{\rm inc}]. \tag{3.303}$$

These expressions are in agreement with (3.293)–(3.296) specialized to a spheroid with $a = b$, for which $P_{11} = P_{22}$ and $M_{11} = M_{22}$.

Provided that the polarizability tensors are known, expressions (3.300)–(3.303) completely specify the dipole scattering from rotationally symmetric scatterers. The scattered field due to an arbitrarily polarized incident plane wave is obtained by multiplication of the θ_{inc} and ϕ_{inc} components of the incident wave with the scattering matrix as in (3.61).

The bistatic cross section immediately results from (3.66), and the total scattering cross section is obtained by integrating the bistatic cross section according to (3.68) and (3.69). For bodies of revolution, the integration can be performed in a closed form since the elements of the scattering matrix (3.300)–(3.303) are simple trigonometric functions, and the result is given by the formulas

$$\sigma_{\text{T}}(\hat{\theta}_{\text{inc}}) = \frac{k^4}{6\pi}\left(|M_{11}|^2 + |P_{11}|^2\cos^2\theta_{\text{inc}} + |P_{33}|^2\sin^2\theta_{\text{inc}}\right) \tag{3.304}$$

$$\sigma_{\text{T}}(\hat{\phi}_{\text{inc}}) = \frac{k^4}{6\pi}\left(|P_{11}|^2 + |M_{11}|^2\cos^2\theta_{\text{inc}} + |M_{33}|^2\sin^2\theta_{\text{inc}}\right) \tag{3.305}$$

which are independent of ϕ and ϕ_{inc} as it should be for rotationally symmetric scatterers.

The extinction cross section follows from (3.152), (3.153), (3.300), and (3.303) as

$$\sigma_{\text{ext}}(\hat{\theta}_{\text{inc}}) = -k\,\text{Im}\,(P_{11}\cos^2\theta_{\text{inc}} - M_{11} + P_{33}\sin^2\theta_{\text{inc}}) \tag{3.306}$$

$$\sigma_{\text{ext}}(\hat{\phi}_{\text{inc}}) = k\,\text{Im}\,(M_{11}\cos^2\theta_{\text{inc}} - P_{11} + M_{33}\sin^2\theta_{\text{inc}}). \tag{3.307}$$

According to (3.306) and (3.307), imaginary parts of the elements of polarizability tensors describe the losses in the energy of the incident wave, which are basically due to absorption in the scatterer and re-radiation of the electromagnetic energy in the surrounding medium. For lossless particles the former mechanism ceases to exist, but the latter mechanism remains, in which case $\sigma_{\text{ext}} = \sigma_{\text{T}}$. Thus, the polarizability tensors, even of lossless particles, are inherently complex-valued.

Compared to the real parts, the imaginary parts of polarizability tensors for low-loss particles are small, and their determination requires accounting for higher-order terms in the expansion of the scattered wave in powers of k. However, for a variety of simply shaped scatterers (see below) the number of independent elements in $\overline{\overline{P}}$ and $\overline{\overline{M}}$ can be one or two, in which case it can be shown by using the energy conservation principle (Belov *et al.* 2003) that if P and M are non-vanishing elements of tensors $\overline{\overline{P}}$ and $\overline{\overline{M}}$, then

$$\text{Im}\,\frac{1}{P} = -\text{Im}\,\frac{1}{M} = \frac{k^3}{6\pi}. \tag{3.308}$$

The validity of these relations is demonstrated in several problems in section 3.7 and in Chapter 7 for a variety of simple scatterers, including a lossless electric-dipole scatterer, a PEC disc, a thin wire loop, PEC, and homogeneous magneto-dielectric spheres. Application of (3.306) and (3.307) to particles with low losses requires therefore a particularly accurate knowledge of the elements of polarizability matrices, up to the terms of the order $O(k^3D^6)$ where D is the characteristic size of the particle.

In what follows, the general formulas (3.300)– (3.305) are applied to various bodies of revolution, including PEC and magneto-dielectric spheroids and spheres, PEC circular discs, short wires, and circular loops, which are examples of voluminous, flat (plate-like), and linear rotationally symmetric scatterers. All those bodies, except for the loop, are special cases of ellipsoids with $a = b$ and $c/a = 0, 1, \infty$. The presented approximations for the elements of polarizability tensors of PEC bodies are frequency independent and correspond therefore to the Rayleigh approximation.

PEC and Magneto-dielectric Spheroid

Every ellipsoid with two equal principal semi-axes degenerates to a spheroid, a body rotationally symmetric with respect to its third principal semi-axis. If we choose $a = b$, then the z axis is the axis of rotational symmetry. The spheroid becomes a circular disc when $c = 0$, a sphere when $a = c$ and a wire-like shape when $a \ll c$ (Figure 3.23).

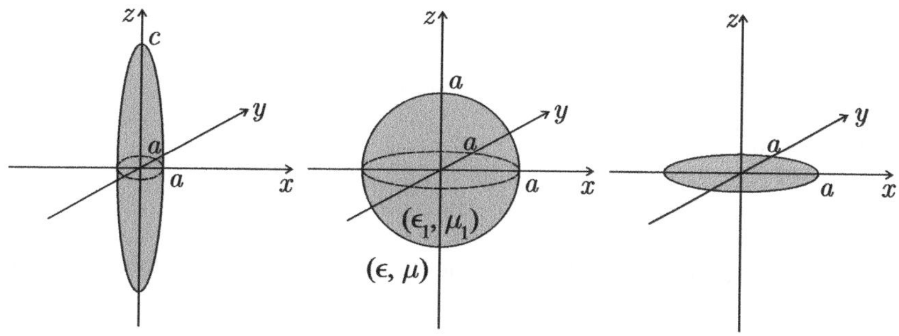

Figure 3.23 Limiting cases of a spheroid: a wire-like strongly prolate spheroid ($a \ll c$), a sphere ($a = b = c$), and a circular disc ($c = 0$).

Because of rotational symmetry, polarizability matrices of the spheroid must be of the same form in any Cartesian coordinate system rotated around the symmetry axis. So, specializing (3.258)–(3.261) to the case of the spheroid we obtain

$$I_a, I_b \to \frac{1}{2} I_1(c/a), \quad I_c \to \frac{1}{2} I_2(c/a), \tag{3.309}$$

where

$$I_1(t) = \int_0^\infty \frac{t^2\, d\tau}{\sqrt{1+\tau}(1+t^2\tau)^2} \tag{3.310}$$

$$I_2(t) = 2 - 2I_1(t) \tag{3.311}$$

and

$$P_{11} = P_{22} = \frac{2V(\epsilon_1 - \epsilon)}{(\epsilon_1 - \epsilon)I_1(c/a) + 2\epsilon}, \quad P_{33} = \frac{2V(\epsilon_1 - \epsilon)}{(\epsilon_1 - \epsilon)I_2(c/a) + 2\epsilon} \tag{3.312}$$

$$M_{11} = M_{22} = \frac{2V(\mu - \mu_1)}{(\mu_1 - \mu)I_1(c/a) + 2\mu}, \quad M_{33} = \frac{2V(\mu - \mu_1)}{(\mu_1 - \mu)I_2(c/a) + 2\mu} \tag{3.313}$$

with $V = 4\pi a^2 c/3$ being the volume of the spheroid. The off-diagonal elements of the matrices $\overline{\overline{P}}$ and $\overline{\overline{M}}$ are equal to zero, implying that the structure of the polarizability matrices agrees with the general properties (3.237) and (3.238).

The corresponding equations for PEC spheroids follow from (3.263), (3.309), and (3.310) as

$$P_{11} = P_{22} = \frac{2V}{I_1(c/a)}, \quad P_{33} = \frac{2V}{I_2(c/a)} \tag{3.314}$$

$$M_{11} = M_{22} = \frac{2V}{2 - I_1(c/a)}, \quad M_{33} = \frac{2V}{2 - I_2(c/a)}. \tag{3.315}$$

Because of (3.311), M_{33} and P_{11} comply with the general property (3.239) of polarizability tensors of rotationally symmetric non-toroidal PEC scatterers.

Integral $I_1(t)$ can be evaluated in a closed form,

$$I_1(t) = \begin{cases} \dfrac{t \arccos t}{(1-t^2)^{3/2}} - \dfrac{t^2}{1-t^2}, & t < 1 \\ \dfrac{t^2}{t^2-1} - \dfrac{t \operatorname{ArcCosh} t}{(t^2-1)^{3/2}}, & t > 1 \end{cases}. \tag{3.316}$$

When $t \to 0$,

$$I_1(t) \approx \frac{1}{2}\pi t - 2t^2 + O(t^3). \tag{3.317}$$

On the other hand, when $t \to \infty$,

$$I_1(t) \approx 1 + \frac{1}{t^2}[1 - \ln(2t)] + O\left(\frac{\ln t}{t^4}\right). \tag{3.318}$$

As apparent from its integral representation, $I_1(t)$ is analytic at $t = 1$; in the vicinity of this point, it is described by the Taylor series

$$I_1(t) = \frac{2}{3} + \frac{4}{15}(t-1) + O(|t-1|^2). \tag{3.319}$$

The continuity of $I_1(t)$ at $t = 1$ implies that expressions (3.312)–(3.315) are uniformly valid for $a < c$ (oblate spheroid), $a = c$ (sphere), and $a > c$ (prolate spheroid). Let us take a look at these limiting cases.

Sphere

A spheroid with $a = c$ reduces to a sphere, and since $I_1(1) = I_2(1) = 2/3$ (see (3.310) and (3.319)), equations (3.312) and (3.313) lead to the polarizabilities of a homogeneous magneto-dielectric sphere,

$$P_{11} = P_{22} = P_{33} = 4\pi a^3 \frac{\epsilon_1 - \epsilon}{\epsilon_1 + 2\epsilon} \tag{3.320}$$

$$M_{11} = M_{22} = M_{33} = 4\pi a^3 \frac{\mu - \mu_1}{\mu_1 + 2\mu}. \tag{3.321}$$

If the sphere is perfectly electrically conducting, then it follows from (3.314) and (3.315) that

$$P_{11} = P_{22} = P_{33} = 4\pi a^3, \quad M_{11} = M_{22} = M_{33} = \frac{1}{2}P_{11}. \tag{3.322}$$

Expressions (3.320)–(3.322) can also be derived from the low-frequency limit of the exact solution for the sphere. In section 7.4, equations (7.348), we derive polarizability tensors for a variety of magneto-dielectric spherical particles, both homogeneous and radially stratified, from the exact full-wave solution of Maxwell's equations.

The scattering matrix of the small sphere is given by (3.300)–(3.303), with (3.320) and (3.321) for the magneto-dielectric sphere and (3.322) for the PEC sphere. For the PEC sphere, the expressions for the scattering matrix, scattered fields, and scattering cross sections are particularly simple and useful, so we write these down. Inserting (3.322) in (3.300)–(3.303) and choosing, without loss of generality, $\theta_{\text{inc}} = 0$ and $\phi_{\text{inc}} = 0$ we obtain for the scattering matrix

$$\overline{\overline{A}} = \frac{\sqrt{4\pi}}{k}(ka)^3 \begin{bmatrix} \left(\frac{1}{2} + \cos\theta\right)\cos\phi & \left(\frac{1}{2} + \cos\theta\right)\sin\phi \\ -\left(1 + \frac{1}{2}\cos\theta\right)\sin\phi & \left(1 + \frac{1}{2}\cos\theta\right)\cos\phi \end{bmatrix}. \tag{3.323}$$

An incident wave of the form $\mathbf{E}_{\text{inc}} = E_0\hat{x}\,e^{jkz}$ has the components $E^{\text{inc}}_{\theta_{\text{inc}}} = E_0\,e^{jkz}$ and $E^{\text{inc}}_{\phi_{\text{inc}}} = 0$, and multiplying them with the scattering matrix (3.323) as in (3.61) gives the far scattered field

$$E^{\text{sc}}_\theta(\mathbf{r}) = E_0 \frac{e^{-jkR}}{kR}(ka)^3\left(\frac{1}{2} + \cos\theta\right)\cos\phi \tag{3.324}$$

$$E^{\text{sc}}_\phi(\mathbf{r}) = -E_0 \frac{e^{-jkR}}{kR}(ka)^3\left(1 + \frac{1}{2}\cos\theta\right)\sin\phi. \tag{3.325}$$

The formula (3.324) explains the presence of the minimum at $\theta = 120°$ in the E plane scattering diagram of every electrically small PEC sphere (see Figure 7.23).

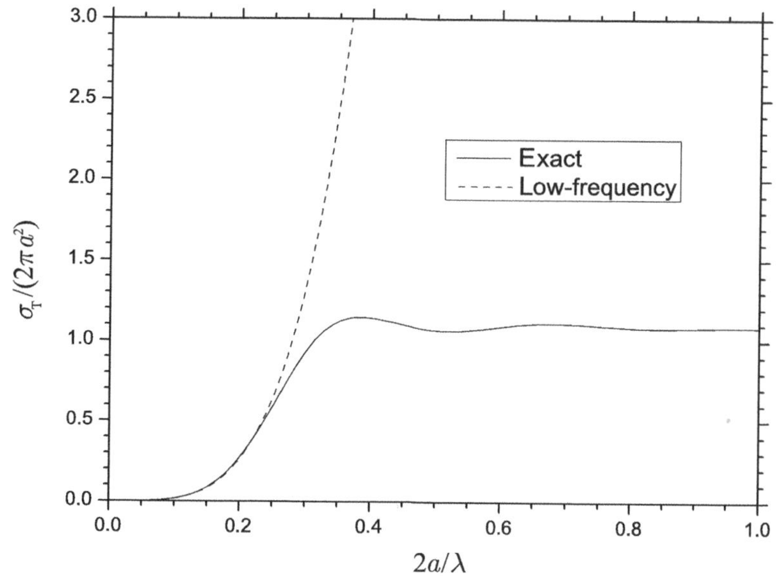

Figure 3.24 Normalized total scattering cross section of a PEC sphere as a function of the ratio between the sphere diameter $2a$ and the wavelength λ: exact vs. low-frequency approximation.

The total cross section of the PEC sphere results from (3.304) and (3.305) as

$$\frac{1}{2\pi a^2}\sigma_T(\hat{\theta}_{inc}) = \frac{1}{2\pi a^2}\sigma_T(\hat{\phi}_{inc}) = \frac{5}{3}(ka)^4 \tag{3.326}$$

and is independent of the incidence direction and polarization of the incident wave. The extinction cross section of an electrically small PEC sphere is equal to (3.326).

Figures 3.24 and 3.25 illustrate the accuracy of the low-frequency approximations. The approximate curves, which are calculated from (3.326) for the PEC sphere and from (3.304) and (3.305) with polarizabilities (3.320) and (3.321) for the dielectric sphere, demonstrate the simple behavior proportional to $(ka)^4$ and quickly grow as ka increases. The exact curves, however, follow this growing trend up to $2a/\lambda \approx 0.35$ ($ka \approx 1.1$) but change their behavior for larger values of ka, starting to approach their optical limits (see section 7.5.4).

This general feature that the low-frequency approximation works well when $D/\lambda < 0.3$, where D is the characteristic size of the body, is observed in other kinds of cross sections (e.g., backscattering, forward scattering) as well as other scatterer geometries (discs, short wires, spheroids, wire loops), studied later in this section. This is of course a manifestation of the key assumption of the low-frequency approach that the incident fields are almost uniform over the whole volume of the scatterer.

Disc

The disc is obtained from an oblate spheroid by letting the minor semi-axis go to zero (Figure 3.23). Polarizability tensors of a magneto-dielectric spheroid vanish, as apparent from (3.312) and (3.313), with $V \to 0$. For a PEC spheroid, however, the limit leads to finite polarizability tensors, similarly to the case of a PEC elliptic plate studied before.

The polarizability tensors of a PEC disc of radius a can be obtained from those of the PEC spheroid with $a = b$ and $c \to 0$ (see (3.314) and (3.315)), or from those of the elliptic plate ($c = 0$) in the limit $b \to a$ (see (3.279)). Alternatively, the tensors can be extracted from the exact solution of the electromagnetic scattering problem for a PEC disc (Meixner and Andrejewski 1950) with $ka \ll 1$ (Osipov 2010). Other approaches can also be used to describe the low-frequency

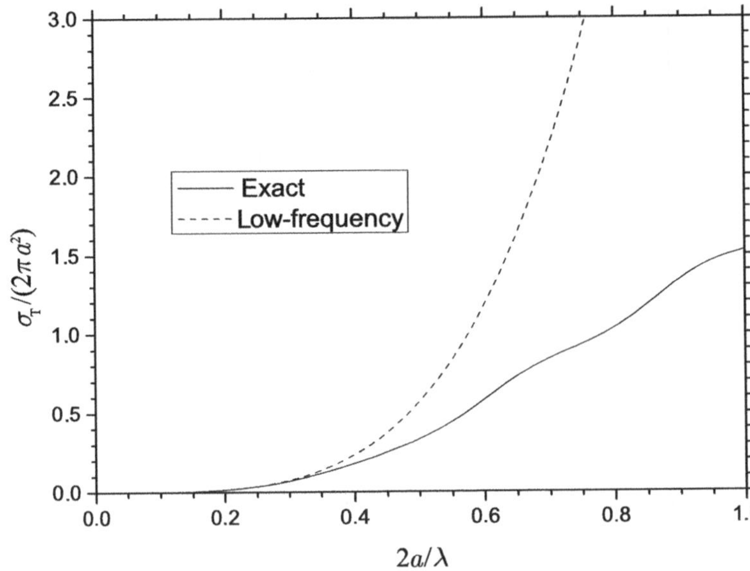

Figure 3.25 The same as in Figure 3.24 but for a sphere from teflon ($\epsilon_r = 2.08 - j0.0008$, $\mu_r = 1$).

scattering from the disc, most notably Eggimann (1961) and Kristensson and Waterman (1982). Regardless of the method, the result reads

$$P_{11} = P_{22} = \frac{16}{3}a^3, \quad P_{33} = 0, \quad M_{11} = M_{22} = 0, \quad M_{33} = \frac{1}{2}P_{11} \tag{3.327}$$

and leads to the scattering matrix

$$\overline{\overline{A}} = \frac{8(ka)^3}{3\sqrt{\pi k}} \cdot \begin{bmatrix} \cos\theta\cos\theta_{\text{inc}}\cos(\phi-\phi_{\text{inc}}) & \cos\theta\sin(\phi-\phi_{\text{inc}}) \\ -\cos\theta_{\text{inc}}\sin(\phi-\phi_{\text{inc}}) & \cos(\phi-\phi_{\text{inc}}) + \frac{1}{2}\sin\theta\sin\theta_{\text{inc}} \end{bmatrix}, \tag{3.328}$$

where the spherical coordinate system with its origin in the middle of the disc and the z axis perpendicular to the disc surface is assumed.

The total cross sections of the disc[10] follow from (3.304), (3.305), and (3.328) as

$$\frac{1}{2\pi a^2}\sigma_T(\hat{\theta}_{\text{inc}}) = \frac{64}{27\pi^2}(ka)^4\cos^2\theta_{\text{inc}} \tag{3.329}$$

$$\frac{1}{2\pi a^2}\sigma_T(\hat{\phi}_{\text{inc}}) = \frac{64}{27\pi^2}(ka)^4\left(1 + \frac{1}{4}\sin^2\theta_{\text{inc}}\right). \tag{3.330}$$

Let us write down the expressions for the bistatic scattered field for the PEC disc. It is necessary to distinguish between two different polarization cases, perpendicular ($\perp$) and parallel ($\|$), depending on whether the electric field vector in the incident wave is perpendicular or parallel to the incidence plane, which is the plane defined by the incidence direction $\hat{i}$ and the z axis. Because of the rotational symmetry of the scatterer, we can always choose the x–z plane to be the incidence plane ($\phi_{\text{inc}} = 0$). In a perpendicularly polarized wave, $\mathbf{E}_0 = E_0\hat{y}$ and, thus, $E^{\text{inc}}_{\theta_{\text{inc}}} = 0$ and $E^{\text{inc}}_{\phi_{\text{inc}}} = E_0$, and using

[10] The expression for the total cross section of a PEC disc that can be found in Eggimann (1961, eq. (48)) and also in Bouman *et al.* (1987, eq. (14.270)) gives an infinite value at $\theta_{\text{inc}} = \pi/2$ and is apparently an error.

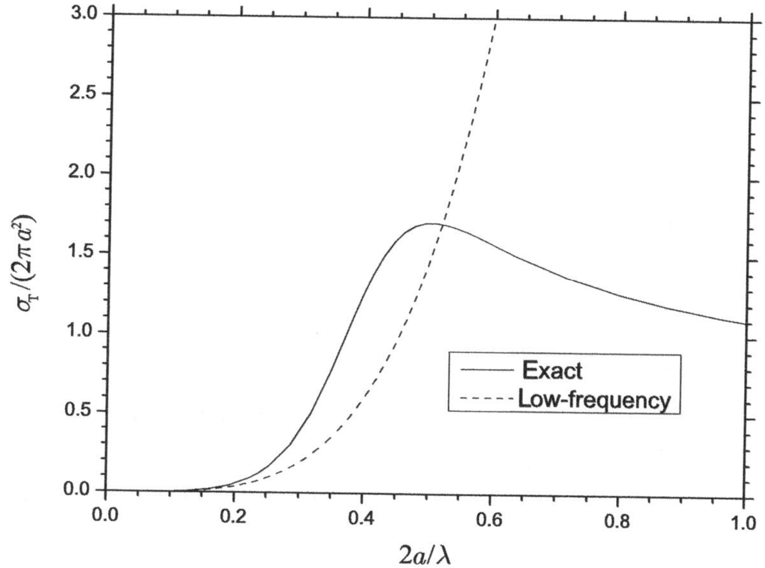

Figure 3.26 Normalized total scattering cross section of a PEC disc for normal illumination as a function of the ratio between the disc diameter $2a$ and the wavelength λ: exact vs. low-frequency approximation.

the definition (3.61) and the formula (3.328) we obtain the formulas:

$$E^{sc}_{\theta,\perp}(\mathbf{r}) = E_0 \frac{e^{-jkR}}{kR} \frac{4}{3\pi}(ka)^3 \cos\theta \sin\phi \tag{3.331}$$

$$E^{sc}_{\phi,\perp}(\mathbf{r}) = E_0 \frac{e^{-jkR}}{kR} \frac{4}{3\pi}(ka)^3 \left(\cos\phi + \frac{1}{2}\sin\theta \sin\theta_{inc}\right). \tag{3.332}$$

In the alternative case of parallel polarization, $Z\mathbf{H}_0 = E_0\hat{y}$ and $\mathbf{E}_0 = -E_0\hat{\theta}_{inc}$, leading to the relations[11]:

$$E^{sc}_{\theta,\parallel}(\mathbf{r}) = -E_0 \frac{e^{-jkR}}{kR} \frac{4}{3\pi}(ka)^3 \cos\theta \cos\theta_{inc} \cos\phi \tag{3.333}$$

$$E^{sc}_{\phi,\parallel}(\mathbf{r}) = E_0 \frac{e^{-jkR}}{kR} \frac{4}{3\pi}(ka)^3 \cos\theta_{inc} \sin\phi. \tag{3.334}$$

The accuracy of the low-frequency approximations (3.329) and (3.330) is illustrated in Figures 3.26 and 3.27 in the cases of normal ($\theta_{inc} = 0$) and grazing ($\theta_{inc} = \pi/2$) incidence on the disc, respectively. In both cases, the exact solution follows the law of $(ka)^4$ as long as $2a/\lambda < 0.4$ but deviates strongly for larger values of $2a/\lambda$. Note that at the grazing illumination, waves with the electric field vector perpendicular to the surface of the disc are not scattered at all, and the low-frequency formula (3.329) correctly predicts that.

Circular Loop

Another important configuration is an electrically small circular loop (torus) made of a thin PEC wire with circular cross section (Figure 3.28). The loop radius a (major radius), the wire radius r_0 (minor radius), and the wavelength λ are related

[11] In Ruck *et al.* (1970), formula (7.5-2), these equations are given with an incorrect sign.

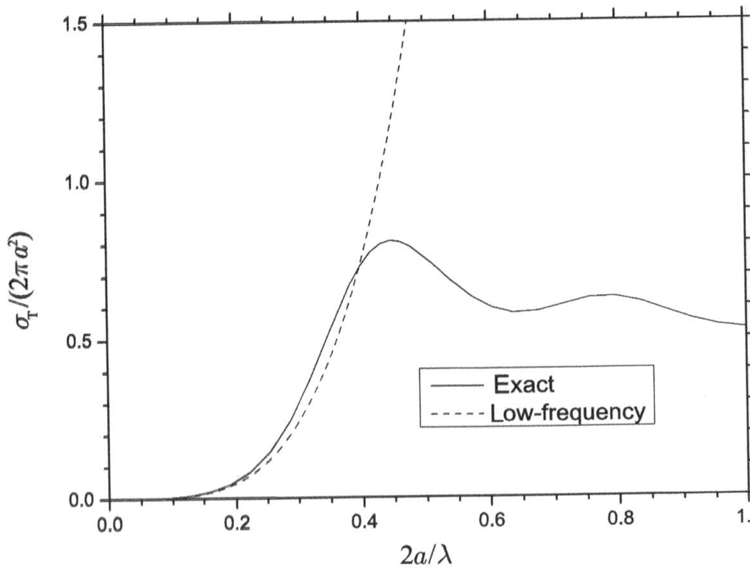

Figure 3.27 The same as in Figure 3.26 but for grazing illumination and an incident wave polarized parallel to the surface of the disc.

as $r_0 \ll a \ll \lambda$. The loop is centered at the origin of the coordinate system with the z axis oriented perpendicular to the plane of the loop. From an approximate solution for a small thin loop (Kouyoumjian 1956), elements of the tensors $\overline{\overline{P}}$ and $\overline{\overline{M}}$ can be obtained as

$$P_{11} = P_{22} = \frac{2\pi^2 a^3}{\ln(8a/r_0) - 2}, \quad P_{33} = 0, \quad M_{11} = M_{22} = 0, \quad M_{33} = \frac{1}{2}P_{11}. \tag{3.335}$$

The other, off-diagonal, elements of the tensors vanish because of the rotational symmetry of the loop (see (3.238)).

The structure of the polarizability tensors of the loop is the same as that of a PEC disc laying in the plane $z = 0$ (see (3.327)), the only difference being in the values of P_{11}. Given that the disc and the loop are of the same radius, the elements of the polarizability matrices differ by the factor

$$\chi = \frac{P_{11}^{\text{disc}}}{P_{11}^{\text{loop}}} = \frac{8}{3\pi^2}\left(\ln\frac{8a}{r_0} - 2\right) \approx \frac{8}{3\pi^2}\ln\frac{a}{r_0} \tag{3.336}$$

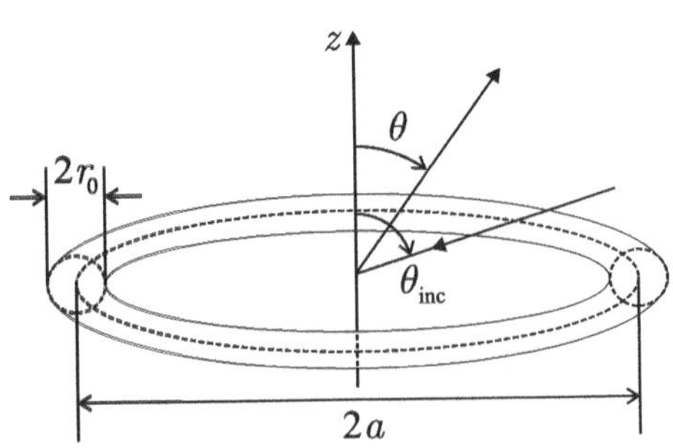

Figure 3.28 Circular loop.

which is greater than unity with $a/r_0 > 37.4$ and smaller than unity otherwise. So, a loop with $a/r_0 < 37.4$ scatters more strongly than a disc of the same radius, though the surface area of the disc ($2\pi a^2$) is larger than that of the wire loop ($4\pi^2 a r_0$) when $a/r_0 > 2\pi$.

Notice that the small loop does not represent a single magnetic dipole; rather it is equivalent to a magnetic dipole along the z axis and to an electric dipole in the plane of the loop. Furthermore, note that though the loop is an example of a toroidal scatterer, its polarizability matrices comply with the property (3.239) for non-toroidal scatterers.

The bistatic scattering matrix and the total cross sections are expressed as

$$\overline{\overline{A}} = \frac{\pi^{3/2}}{k}\frac{(ka)^3}{\ln(8a/r_0)-2}\cdot\begin{bmatrix}\cos\theta\cos\theta_{\text{inc}}\cos(\phi-\phi_{\text{inc}}) & \cos\theta\sin(\phi-\phi_{\text{inc}}) \\ -\cos\theta_{\text{inc}}\sin(\phi-\phi_{\text{inc}}) & \cos(\phi-\phi_{\text{inc}})+\frac{1}{2}\sin\theta\sin\theta_{\text{inc}}\end{bmatrix} \tag{3.337}$$

and

$$\frac{1}{\lambda^2}\sigma_{\text{T}}(\hat{\theta}_{\text{inc}}) = \frac{\pi(ka)^6\cos^2\theta_{\text{inc}}}{6[\ln(8a/r_0)-2]^2} \tag{3.338}$$

$$\frac{1}{\lambda^2}\sigma_{\text{T}}(\hat{\phi}_{\text{inc}}) = \frac{\pi(ka)^6}{6[\ln(8a/r_0)-2]^2}\left(1+\frac{1}{4}\sin^2\theta_{\text{inc}}\right). \tag{3.339}$$

The backscattering cross section at the broadside illumination ($\theta = \theta_{\text{inc}} = 0$) follows from (3.337) as

$$\frac{\sigma}{\lambda^2} = \frac{\pi(ka)^6}{4[\ln(8a/r_0)-2]^2}. \tag{3.340}$$

Figure 3.29 shows the frequency-dependent broadside backscattering cross sections of two loops with different major radii, calculated using the low-frequency approximation (3.340) and a numerical solution of exact integral equations. The low-frequency approximation is more accurate for the smaller loop, as expected. The electrical size $2a/\lambda$ of the loops at 20 GHz equals to 0.13 and 0.26, respectively, so that the low-frequency approximation for the smaller loop agrees well with the exact curve over all the plotted frequency range.

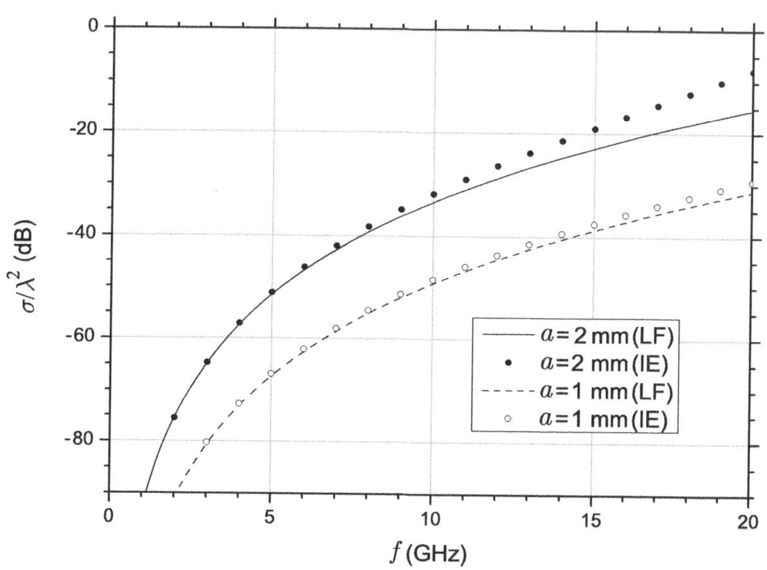

Figure 3.29 Normalized broadside backscattering cross sections of PEC circular loops with $r_0 = 0.1$ mm and $a = 1$ and 2 mm as functions of frequency.

Short Wire

A short wire, where the modifier "short" means that the length L of the body is much smaller than the wavelength and the word "wire" describes a strongly elongated body with a characteristic width W much smaller than L, is a useful model of an electrically small body, in which the currents induced by the incident wave can flow just in a single direction.

A prolate spheroid with the minor semi-axis a much shorter than the major semi-axis c (Figure 3.23) looks like a piece of a wire, and it is natural to expect that in the limit $a/c \to 0$, equations (3.314) and (3.315) describe polarizabilities of short wire structures. The model of the short wire is only meaningful for PEC or strongly conducting bodies with the skin depth much smaller than the diameter of the wire, since polarizabilities of small non-resonant non-PEC bodies are proportional to the volume of the body and quickly vanish in the wire limit. As we shall see, polarizabilties of short PEC wires vanish as well but at a much slower rate.

Indeed, using (3.311) and (3.318) in (3.314) and (3.315) let us conclude that

$$P_{11}, P_{22}, M_{11}, M_{22}, M_{33} = O(V), \tag{3.341}$$

where $V = 4\pi a^2 c/3$ is the volume of the spheroid, and

$$P_{33} = \frac{4\pi}{3} \frac{c^3}{\ln(2c/a) - 1}. \tag{3.342}$$

With $a \to 0$, the elements of the polarizability tensors in (3.341) vanish as $O(a^2)$, whereas P_{33} decays as $O(|\ln a|^{-1})$, that is, at a much slower logarithmic rate.

Formulas (3.341) and (3.342) are similar to those for an elongated elliptic plate; comparing (3.342) with (3.283), we conclude that the wire limits of the spheroid and of the elliptic plate lead to the same polarizabilities if the shapes have the same length $L = 2c$ and the maximum width W of the elliptic plate is such that $W = 4a$, where a is the minor radius of the spheroid.

The elliptic plate is not a rotationally symmetric body, but nevertheless, because of the electrically negligible width, its polarizability tensors coincide with those of the spheroid, implying that short wires with arbitrary cross sections can be described by the generic relations

$$P_{11}, P_{22}, M_{11}, M_{22}, M_{33} = 0 \tag{3.343}$$

$$P_{33} = \frac{\pi}{6} \frac{L^3}{\ln(L/W) + C}, \tag{3.344}$$

where L is the length of the body and W is the maximum width of the body in a plane perpendicular to its length, which are such that $W \ll L \ll \lambda$ (short wire). The parameter C is a shape-specific constant, for example $C = \ln 2 - 1$ for spheroids and $C = \ln 4 - 1$ for elliptic plates. The parameter serves as a correction in the denominator of (3.344), which is dominated by the log term.

A piece of a circular cylinder with its radius r_0 much smaller than its length $2l$ is a practically useful version of the short wire (Figure 3.30). The full electromagnetic scattering problem for the circular PEC cylinder of finite length has not been solved so far; approximate solutions for circular cylinders in the wire limit ($r_0 \ll l$) are available in the literature but mostly for the wires of resonant lengths ($2l = n\lambda/2$ with $n = 1, 2, \ldots$) (Bouman *et al.* 1987; Ruck *et al.* 1970; Schelkunoff and Friis 1952; Van Vleck *et al.* 1947). In section 3.6.1, an approximation for the electric polarizability is derived by modeling the short wire as a short dipole antenna, and extracting the polarizability from (3.226) we arrive at

$$P_{33} = \frac{4\pi}{3} \frac{l^3}{\ln(l/r_0) - 1}, \tag{3.345}$$

which corresponds to $C = -1$ in (3.344).[12]

[12]Ruck *et al.* (1970, section 4.3.1.1.1.2) suggest $C = \ln 4 - 1$, which does not comply with formula (3.226) and the results of our numerical simulations.

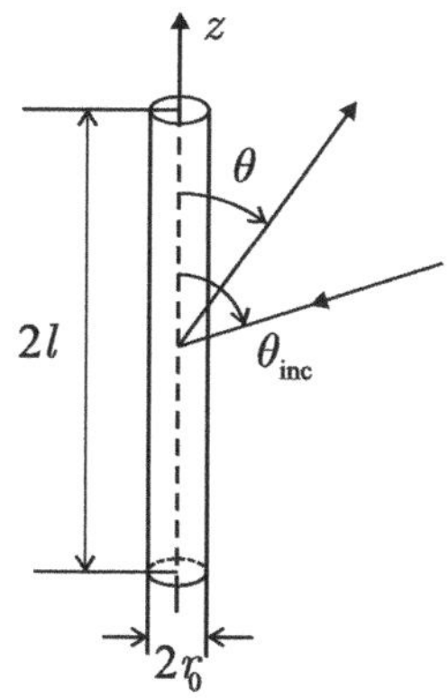

Figure 3.30 Cylindrical wire.

In the short-wire limit, the scattering from the truncated circular cylinder can be made equal to the scattering from cylinders with arbitrarily shaped cross sections by simply choosing a suitable radius r_0 (equivalent radius). Thus, expression (3.345) for the polarizability of a short PEC wire with circular cross section has a general meaning and can be applied to short wires with different geometric cross sections through the use of a suitable value of r_0, provided that the length is the same. For example, the truncated circular cylinder is equivalent to a prolate spheroid with $c = l$ and $a = 2r_0$, a right-angled strip wire of zero thickness with the length $2l$ and width $4r_0$, an elliptic strip wire with the major semi axis l and the minor semi axis $4r_0$, and truncated cylinders of height $2l$ with square and triangular cross sections with side lengths $1.69r_0$ and $2.38r_0$, respectively. Finally, if the cross section of the truncated cylinder is elliptic, with the semi axes a and b, then the radius of the equivalent cylinder is equal to $r_0 = (a + b)/2$ (Bouman *et al.* 1987, section 12.1). All those scatterers show the same scattering behavior in the short-wire limit ($W \ll L \ll \lambda$).

Using (3.345) in (3.300)–(3.303) and assuming that the wire is centered at the origin of a spherical coordinate system and aligned with the z axis leads to the scattering matrix

$$\bar{\bar{A}} = \frac{\sqrt{4\pi}}{3k} \frac{(kl)^3}{\ln(l/r_0) - 1} \begin{bmatrix} \sin\theta \sin\theta_{\text{inc}} & 0 \\ 0 & 0 \end{bmatrix}, \tag{3.346}$$

which implies that the polarization of transmitting and receiving antennas must have a finite projection on the wire to get a finite response from the wire. The short wire is not excited by a wave coming along it ($\theta_{\text{inc}} = 0, \pi$) and does not scatter in the axial direction ($\theta = 0, \pi$). The bistatic scattered field of the short wire can, therefore, be written as

$$\mathbf{E}_{\text{sc}}(\mathbf{r}) = \hat{\theta}(\mathbf{E}_0 \cdot \hat{\theta}_{\text{inc}}) \frac{e^{-jkR}}{kR} \frac{(kl)^3}{3} \frac{\sin\theta \sin\theta_{\text{inc}}}{\ln(l/r_0) - 1}. \tag{3.347}$$

The bistatic and total scattering cross sections of the wire results from (3.346) and (3.304) as

$$\frac{1}{\lambda^2}\sigma_{\theta\theta_{\text{inc}}} = \frac{(kl)^6 \sin^2\theta \sin^2\theta_{\text{inc}}}{9\pi[\ln(l/r_0) - 1]^2} \tag{3.348}$$

$$\frac{1}{\lambda^2}\sigma_{\text{T}}(\hat{\theta}_{\text{inc}}) = \frac{2(kl)^6 \sin^2\theta_{\text{inc}}}{27\pi[\ln(l/r_0) - 1]^2}, \tag{3.349}$$

where the cross sections are normalized to λ^2 to make the expressions dimensionless. The average backscattering cross section for an ensemble of randomly oriented short wires is the subject of problem 34.

Figure 3.31 compares backscattering cross sections of a cylindrical PEC wire ($l = 2$ mm, $r_0 = 0.1$ mm) and a strongly prolate spheroid ($c = 2$ mm, $a = 0.1$ mm) in the low-frequency case ($kl = kc \approx 0.84$ at $f = 20$ GHz). The broadside illumination ($\theta_{\text{inc}} = \pi/2$), a wave polarized along the bodies, and the backscattering case ($\theta = \pi/2$) are assumed. The

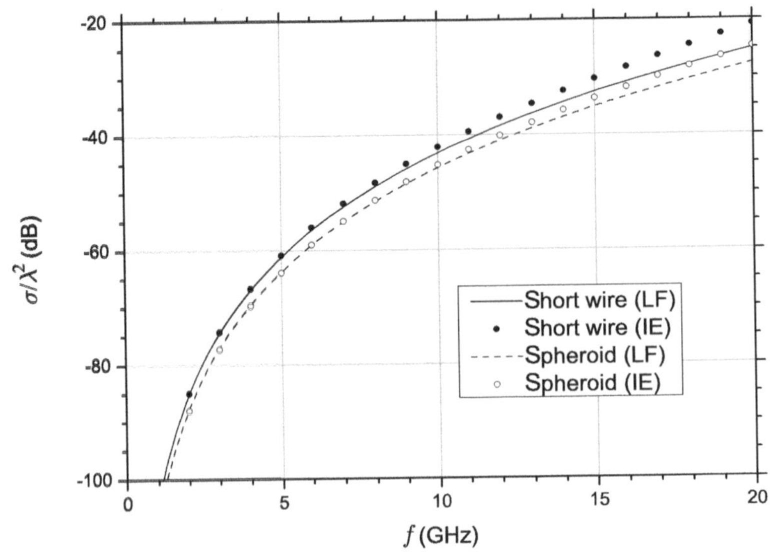

Figure 3.31 Normalized broadside backscattering cross sections of a PEC short wire (truncated circular cylinder with $l = 2$ mm, $r_0 = 0.1$ mm) and a PEC spheroid ($c = 2$ mm, $a = 0.1$ mm) as functions of frequency: integral equations solution vs. low-frequency approximation.

cylinder scatters stronger than the spheroid because r_0 is twice as large as the radius of the cylinder equivalent to the spheroid. It is also seen that the curves obtained from numerical solution of integral equations and from the low-frequency approximation (3.348) are hardly distinguishable when $f < 5$ GHz ($kl, kc < 0.2$).

3.7 Problems

1. Find the *total* scattering cross section of a large ideally conducting sphere in the GO approximation.

   Hint: GO approximation means that the field can be considered in terms of rays that hit the object and reflect as in a mirror. Note that in this approximation the total field in the shadow is zero, and remember that the total field is the sum of the incident field and the scattered field.

2. Find the total scattering cross section of a large ideally conducting disc of radius a in the GO approximation, assuming that an incoming wave propagates perpendicularly to the disc surface.

   Answer: $\sigma_T = 2\pi a^2$.

   Solution: The total cross section of the disc in the GO limit should be the same as that of a sphere of the same radius since both intercept the same amount of energy from the incident wave.

3. Given the unit vectors $\hat{i}$ and $\hat{s}$ that describe the directions of incidence and scattering, how is the bistatic angle calculated?

   Answer: If β denotes the bistatic angle, then $|\hat{i} \times \hat{s}| = \sin\beta$.

4. Assume that elements of the scattering matrix in (3.66) are known approximately with the relative error δ. Estimate an error in the scattering cross section in decibels.

   Solution: Let A_{uv}^{ex} and A_{uv}^{ap} denote the exact and an approximate value of an uv element of the scattering matrix, respectively. The relative error is then defined by

$$\delta = \left| \frac{A_{uv}^{\mathrm{ap}}}{A_{uv}^{\mathrm{ex}}} - 1 \right|. \tag{3.350}$$

The ratio of the corresponding cross sections $\sigma_{\rm ex} = |A_{uv}^{\rm ex}|^2$ and $\sigma_{\rm ap} = |A_{uv}^{\rm ap}|^2$ can be expressed as

$$\frac{\sigma_{\rm ap}}{\sigma_{\rm ex}} = \left|\frac{A_{uv}^{\rm ap}}{A_{uv}^{\rm ex}}\right|^2 = \left|\frac{A_{uv}^{\rm ap}}{A_{uv}^{\rm ex}} - 1 + 1\right|^2 \tag{3.351}$$

and if the relative error is small,

$$\frac{\sigma_{\rm ap}}{\sigma_{\rm ex}} \approx 1 + 2{\rm Re}\left(\frac{A_{uv}^{\rm ap}}{A_{uv}^{\rm ex}} - 1\right). \tag{3.352}$$

In decibels, using again the smallness of δ,

$$10\log_{10}\frac{\sigma_{\rm ap}}{\sigma_{\rm ex}} \approx \frac{20}{\ln 10}{\rm Re}\left(\frac{A_{uv}^{\rm ap}}{A_{uv}^{\rm ex}} - 1\right) \tag{3.353}$$

and since $20/\ln 10 \approx 8.7$ and

$$\left|{\rm Re}\left(\frac{A_{uv}^{\rm ap}}{A_{uv}^{\rm ex}} - 1\right)\right| \leq \left|\frac{A_{uv}^{\rm ap}}{A_{uv}^{\rm ex}} - 1\right|, \tag{3.354}$$

we finally arrive at the estimate

$$|\sigma_{\rm ap}({\rm dBsm}) - \sigma_{\rm ex}({\rm dBsm})| \approx 8.7\delta. \tag{3.355}$$

5. Let a scatterer with scattering matrix (3.61) be illuminated by a linearly polarized wave which at the origin of coordinate system is given by the relation

$$\mathbf{E}_{\rm inc}(0) = \hat{\theta}_{\rm inc}E_0\cos\psi + \hat{\phi}_{\rm inc}E_0\sin\psi, \tag{3.356}$$

where the angle ψ $(0 \leq \psi < 2\pi)$ defines the polarization of the incident wave. Express the bistatic scattering cross section σ_ψ through the elements of the scattering matrix $\overline{\overline{A}}$ and the polarization angle ψ.

Answer:

$$\sigma_\psi = \cos^2\psi\left(\left|A_{\theta\theta_{\rm inc}}\right|^2 + \left|A_{\phi\theta_{\rm inc}}\right|^2\right) + \sin^2\psi\left(\left|A_{\theta\phi_{\rm inc}}\right|^2 + \left|A_{\phi\phi_{\rm inc}}\right|^2\right) + 2\sin\psi\cos\psi\,{\rm Re}\left(A_{\theta\theta_{\rm inc}}A^*_{\theta\phi_{\rm inc}} + A_{\phi\theta_{\rm inc}}A^*_{\phi\phi_{\rm inc}}\right). \tag{3.357}$$

Solution: The electric field in the incident wave at the origin of coordinate system is given by (3.356), and thus $|\mathbf{E}_{\rm inc}(0)|^2 = |E_0|^2$ for arbitrary ψ. The scattered field is obtained by multiplication of (3.356) with the scattering matrix, which gives

$$\mathbf{E}_{\rm sc} = E_0\frac{{\rm e}^{-jkr}}{r\sqrt{4\pi}}\left[\left(A_{\theta\theta_{\rm inc}}\cos\psi + A_{\theta\phi_{\rm inc}}\sin\psi\right)\hat{\theta} + \left(A_{\phi\theta_{\rm inc}}\cos\psi + A_{\phi\phi_{\rm inc}}\sin\psi\right)\hat{\phi}\right]. \tag{3.358}$$

Evaluating $|\mathbf{E}_{\rm sc}|^2$ and using the definition of the bistatic scattering cross section in (3.6) results in the formula (3.357).

6. Let the incident field be a superposition of non-coherent, linearly polarized waves with randomly distributed orientations of polarization. Determine the average of the bistatic scattering cross section.

Hint: Use the result (3.357) and define the average over the polarization angle ψ by

$$\bar{\sigma}_\psi = \frac{1}{2\pi}\int_0^{2\pi}\sigma_\psi\,{\rm d}\psi. \tag{3.359}$$

Solution: Scattering matrices characterize scattering properties of scatterers and therefore do not depend on polarization of the incident wave. Thus, integration of (3.357) in (3.359) is straightforward and leads to the result

$$\bar{\sigma}_{\psi} = \frac{1}{2}\left(\sigma_{\theta\theta_{\rm inc}} + \sigma_{\theta\phi_{\rm inc}} + \sigma_{\phi\theta_{\rm inc}} + \sigma_{\phi\phi_{\rm inc}}\right), \tag{3.360}$$

in which relations (3.66) have been used.

7. Assume that a short linear antenna with polarization direction $\hat{p} = \hat{\theta}\cos\psi + \hat{\phi}\sin\psi$ serves as a transmitting and receiving antenna (monostatic configuration). Show that the backscattering cross section of a scatterer with scattering matrix (3.61) measured by the antenna can be expressed as

$$\sigma(\hat{p}) = \cos^4\psi\left|A_{\theta\theta_{\rm inc}}\right|^2 + 4\sin\psi\cos^3\psi\,{\rm Re}\left(A_{\theta\theta_{\rm inc}}A^*_{\theta\phi_{\rm inc}}\right) + 2\sin^2\psi\cos^2\psi\left[{\rm Re}\left(A_{\theta\theta_{\rm inc}}A^*_{\phi\phi_{\rm inc}}\right) + 2\left|A_{\theta\phi_{\rm inc}}\right|^2\right]$$
$$+4\sin^3\psi\cos\psi{\rm Re}\left(A_{\phi\phi_{\rm inc}}A^*_{\theta\phi_{\rm inc}}\right) + \sin^4\psi\left|A_{\phi\phi_{\rm inc}}\right|^2. \tag{3.361}$$

Hint: Use the definition (3.7) and account for the property $A_{\theta\phi_{\rm inc}} = A_{\phi\theta_{\rm inc}}$ of scattering matrices at backscattering.

Solution: The incident field at the origin of the coordinate system is given by (3.356) with $\hat{\theta}_{\rm inc} = \hat{\theta}$ and $\hat{\phi}_{\rm inc} = \hat{\phi}$, and the scattered field by (3.358). The antenna registers a component of the scattered field along $\hat{p}$,

$$\mathbf{E}_{\rm sc}\cdot\hat{p} = E_0\frac{{\rm e}^{-jkr}}{r\sqrt{4\pi}}\left[\left(A_{\theta\theta_{\rm inc}}\cos\psi + A_{\theta\phi_{\rm inc}}\sin\psi\right)\cos\psi + \left(A_{\phi\theta_{\rm inc}}\cos\psi + A_{\phi\phi_{\rm inc}}\sin\psi\right)\sin\psi\right]. \tag{3.362}$$

Evaluating the magnitude $|\mathbf{E}_{\rm sc}\cdot\hat{p}|^2$ and using $|\mathbf{E}_{\rm inc}(0)|^2 = |E_0|^2$, one derives from (3.7) the formula (3.361).

8. Find the average backscattering cross section which will be observed at a dipole antenna rotating with a constant speed around the direction to a scatterer in fixed position and orientation.
Solution: Describe the orientation and therefore the polarization of the antenna with the angle ψ $(0 \le \psi < 2\pi)$. At every instant value of ψ the antenna produces the field which can be represented by (3.356) at the scatterer. The instant radar cross section is then given by (3.361), and the average cross section can be obtained by averaging

$$\bar{\sigma} = \frac{1}{2\pi}\int_0^{2\pi}\sigma(\hat{p})\,{\rm d}\psi. \tag{3.363}$$

The result reads

$$\bar{\sigma} = \frac{3}{8}\sigma_{\theta\theta_{\rm inc}} + \frac{3}{8}\sigma_{\phi\phi_{\rm inc}} + \frac{1}{2}\sigma_{\theta\phi_{\rm inc}} + \frac{1}{4}{\rm Re}(A_{\theta\theta_{\rm inc}}A^*_{\phi\phi_{\rm inc}}). \tag{3.364}$$

9. Why is (3.364) different from (3.360)? Are there cases in which they coincide?
Answer: Expression (3.360) is derived for the situation in which the receiving antenna measures the total power of the scattered wave. Equation (3.364) is obtained for a dipole-type antenna that measures only a component of the scattered wave. They can, however, coincide when the received signal is polarization independent, which is a property of spherical scatterers. In this case, $A_{\theta\phi_{\rm inc}} = 0$, $A_{\theta\theta_{\rm inc}} = A_{\phi\phi_{\rm inc}}$, $\sigma_{\theta\theta_{\rm inc}} = \sigma_{\phi\phi_{\rm inc}} = \sigma$, and equations (3.360) and (3.364) give the same value σ.

10. The scattering matrix (3.61) is defined in terms of the components of the incident and scattered electric fields (3.52) and (3.57). One can also define the magnetic field scattering matrix by the formula

$$\begin{pmatrix} H^{\rm sc}_{\theta} \\ H^{\rm sc}_{\phi} \end{pmatrix} = \frac{{\rm e}^{-jkR}}{\sqrt{4\pi}R}\begin{pmatrix} A^H_{\theta\theta_{\rm inc}} & A^H_{\theta\phi_{\rm inc}} \\ A^H_{\phi\theta_{\rm inc}} & A^H_{\phi\phi_{\rm inc}} \end{pmatrix}\begin{pmatrix} H^{\rm inc}_{\theta_{\rm inc}} \\ H^{\rm inc}_{\phi_{\rm inc}} \end{pmatrix}. \tag{3.365}$$

How are the electric and magnetic field scattering matrices related to each other?

Hint: Use the relation between electric and magnetic fields in plane and spherical waves.

Answer: By using (2.156), (2.165), (3.52), (3.57), and the vector identities $\hat{i} \times \hat{\theta}_{\text{inc}} = -\hat{\phi}_{\text{inc}}$, $\hat{i} \times \hat{\phi}_{\text{inc}} = \hat{\theta}_{\text{inc}}$, $\hat{s} \times \hat{\theta} = \hat{\phi}$, and $\hat{s} \times \hat{\phi} = -\hat{\theta}$, one finds that

$$E^{\text{inc}}_{\theta_{\text{inc}}} = -ZH^{\text{inc}}_{\phi_{\text{inc}}}, \quad E^{\text{inc}}_{\phi_{\text{inc}}} = ZH^{\text{inc}}_{\theta_{\text{inc}}} \tag{3.366}$$

in the incident wave and

$$E^{\text{sc}}_{\theta} = ZH^{\text{sc}}_{\phi}, \quad E^{\text{sc}}_{\phi} = -ZH^{\text{sc}}_{\theta} \tag{3.367}$$

in the scattered wave. Using these in (3.61) and comparing with (3.365) gives the relations:

$$A^H_{\theta\theta_{\text{inc}}} = -A_{\phi\phi_{\text{inc}}}, \quad A^H_{\theta\phi_{\text{inc}}} = A_{\phi\theta_{\text{inc}}}, \tag{3.368}$$

$$A^H_{\phi\theta_{\text{inc}}} = A_{\theta\phi_{\text{inc}}}, \quad A^H_{\phi\phi_{\text{inc}}} = -A_{\theta\theta_{\text{inc}}}. \tag{3.369}$$

11. Let $\overline{\overline{A}}(\eta)$ be the scattering matrix of an impedance scatterer with the relative equivalent surface impedance η. How is this matrix related to the scattering matrix of the scatterer of the same shape but with the inverse value of the impedance, $1/\eta$?

Hint: Use the solution to problem 10 in this section and problem 8 in section 2.7.

Answer:

$$A_{\theta\theta_{\text{inc}}}(1/\eta) = -A_{\phi\phi_{\text{inc}}}(\eta), \quad A_{\theta\phi_{\text{inc}}}(1/\eta) = A_{\phi\theta_{\text{inc}}}(\eta), \tag{3.370}$$

$$A_{\phi\theta_{\text{inc}}}(1/\eta) = A_{\theta\phi_{\text{inc}}}(\eta), \quad A_{\phi\phi_{\text{inc}}}(1/\eta) = -A_{\theta\theta_{\text{inc}}}(\eta). \tag{3.371}$$

12. Let the equivalent surface impedance $\tilde{Z}_s$ of a scatterer be equal to the intrinsic impedance Z of the surrounding medium. Prove the symmetry relations:

$$A_{\theta\theta_{\text{inc}}} = -A_{\phi\phi_{\text{inc}}}, \quad A_{\theta\phi_{\text{inc}}} = A_{\phi\theta_{\text{inc}}}. \tag{3.372}$$

Hint: Use the result of problem 11 with $\eta = 1$.

13. Prove that the theorem from section 3.2.3 applies to scatterers filled with a material such that

$$\frac{\epsilon_1}{\epsilon} = \frac{\mu_1}{\mu}, \tag{3.373}$$

where the material parameters ϵ_1 and μ_1 can be functions of coordinates but in such a way so as to preserve the relation (3.373) and the rotational symmetry of the scatterer (Weston 1963).

Solution: Compared with the impedance scatterer, the impedance boundary conditions are to be replaced with the conditions of continuity of the tangential components of the total fields at the boundary of the body and by Maxwell's equations in the interior of the body. The latter can be written as

$$\nabla \times \left(\frac{1}{\mu_1}\nabla \times \mathbf{E}\right) = \omega^2 \epsilon_1 \mathbf{E} \qquad \nabla \cdot (\epsilon_1 \mathbf{E}) = 0 \tag{3.374}$$

for the **E** field and as

$$\nabla \times \left(\frac{1}{\epsilon_1}\nabla \times \mathbf{H}\right) = \omega^2 \mu_1 \mathbf{H} \qquad \nabla \cdot (\mu_1 \mathbf{H}) = 0 \tag{3.375}$$

for the **H** field. It is easy to see that (3.374) and (3.375) coincide once (3.373) is true.

The continuity conditions on the surface of the scattering body can be written as

$$\hat{n} \times \mathbf{E}|_{\text{int}} = \hat{n} \times \mathbf{E}|_{\text{ext}}, \quad \frac{1}{\mu_1}\hat{n} \times \nabla \times \mathbf{E}|_{\text{int}} = \frac{1}{\mu}\hat{n} \times \nabla \times \mathbf{E}|_{\text{ext}} \tag{3.376}$$

for the **E** field and as

$$\hat{n} \times \mathbf{H}|_{\text{int}} = \hat{n} \times \mathbf{H}|_{\text{ext}}, \quad \frac{1}{\epsilon_1}\hat{n} \times \nabla \times \mathbf{H}|_{\text{int}} = \frac{1}{\epsilon}\hat{n} \times \nabla \times \mathbf{H}|_{\text{ext}} \tag{3.377}$$

for the **H** field. Under the condition (3.373), equations (3.376) and (3.377) coincide.

Thus, the fields **E** and **H** satisfy the same equations. The only difference lies in the direction of the incident fields $\mathbf{E}_{\text{inc}}$ and $\mathbf{H}_{\text{inc}}$ given by (3.35). The next steps in the proof are the same as in the theorem for impedance-matched bodies.

14. Derive the following matrix relation between the scattering matrices in the spherical and PoS coordinate systems:

$$\overline{\overline{A}}_{\text{sph}} = \overline{\overline{U}}_1 \cdot \overline{\overline{A}}_{\text{PoS}} \cdot \overline{\overline{V}}_1, \tag{3.378}$$

where

$$\overline{\overline{U}}_1 = \begin{pmatrix} \hat{\theta} \cdot \hat{\xi} & \hat{\theta} \cdot \hat{\eta} \\ \hat{\phi} \cdot \hat{\xi} & \hat{\phi} \cdot \hat{\eta} \end{pmatrix} \tag{3.379}$$

$$\overline{\overline{V}}_1 = \begin{pmatrix} \hat{\xi}_{\text{inc}} \cdot \hat{\theta}_{\text{inc}} & \hat{\xi}_{\text{inc}} \cdot \hat{\phi}_{\text{inc}} \\ \hat{\eta}_{\text{inc}} \cdot \hat{\theta}_{\text{inc}} & \hat{\eta}_{\text{inc}} \cdot \hat{\phi}_{\text{inc}} \end{pmatrix}. \tag{3.380}$$

Hint: The derivation parallels the derivation of the inverse relation (3.81).

15. Prove the relations:

$$\overline{\overline{V}} \cdot \overline{\overline{V}}_1 = \overline{\overline{I}}, \quad \overline{\overline{U}} \cdot \overline{\overline{U}}_1 = \overline{\overline{I}}, \tag{3.381}$$

where $\overline{\overline{I}}$ is the unit 2 by 2 matrix and the matrices $\overline{\overline{U}}, \overline{\overline{V}}, \overline{\overline{U}}_1$, and $\overline{\overline{V}}_1$ are defined in (3.79), (3.80), (3.379), and (3.380).
Hint: Multiply the matrices and use the fact that the bases are orthonormal.

16. Prove relations (3.105) and (3.106).
Hint: Use (A.15) with $\mathbf{C} = 0$.

17. Let $\Delta = |\mathbf{r} - \mathbf{r}'| - r + \hat{s} \cdot r'$ be the absolute error of (3.105). Show that $k\Delta \geq \pi D^2/(r\lambda)$. What is the maximum phase error of the far-field approximation (3.107) at the distance $d = 2D_{\text{F}}$ from a scatterer?

Answer: 90°.

Hint: Without loss of generality one can assume that $\mathbf{r}' = r'\hat{x}$ and if $\mathbf{r} = r(\hat{x}\cos\theta\cos\phi + \hat{y}\cos\theta\sin\phi + \hat{z}\sin\theta)$, then $\Delta = 0.5r'^2(1 - \cos^2\theta\cos^2\phi)/r \leq 0.5D^2/r$.

18. Estimate the distance to the far-field zone (parameter D_{F}) for targets of the size $D = 1000$ nm, 1 mm, 0.03 m, 0.3 m, 3 m, and 30 m at a microwave frequency with $\lambda = 0.03$ m.
Answer: $D_{\text{F}} = 0.03$ m, 0.03 m, 0.03 m, 3 m, 300 m, and 30 km.

19. The same as above but for a light wave with $\lambda = 500$ nm.
Answer: $D_{\text{F}} = 2000$ nm, 2 m, 1800 m, 180 km, 18 000 km, and 1.8×10^6 km.

20. Derive from (3.107) an integral expression for H_{sc} by using the duality property of Maxwell's equations. Show that this expression is in agreement with the general relation (2.165).
Hint: Use the vector identity (A.12).

21. Let $\overline{\overline{A}}_1$ be the scattering matrix of a scatterer Ω_1. What will be the scattering matrix $\overline{\overline{A}}_2$ of another scatterer Ω_2 that is obtained from the first scatterer by a $\delta\mathbf{r}$ shift: $\Omega_2 = \Omega_1 + \delta\mathbf{r}$?

Hint: Consider the integral representation (3.108) for the far-field coefficient and its relation (3.111) to the scattering matrix.

Answer: The matrices are related by the formula

$$\overline{\overline{A}}_2 = \mathrm{e}^{-jk\mathbf{q}\cdot\delta\mathbf{r}}\overline{\overline{A}}_1 \tag{3.382}$$

with $\mathbf{q} = \hat{\imath} - \hat{s}$. This relation follows from the integral representation (3.108) of the far-field coefficients. Consider, for example, the integral representing the far-field coefficient of the body Ω_2. The change in the integration variable $\mathbf{r}'' = \mathbf{r}' + \delta\mathbf{r}$ transforms this integral to an integral representing the far-field coefficient of the body Ω_1 multiplied by the factor $\exp(jk\hat{s}\cdot\delta\mathbf{r})$, which results from the exponential factor in the integrand of (3.108), and by the factor $\exp(-jk\hat{\imath}\cdot\delta\mathbf{r})$, which is due to the fact that the equivalent currents $\mathbf{K}_{\mathrm{e,m}}(\mathbf{r}')$ are proportional to the phase of the incident plane wave and, thus, $\mathbf{K}_{\mathrm{e,m}}(\mathbf{r}') = \exp(-jk\hat{\imath}\cdot\delta\mathbf{r})\mathbf{K}_{\mathrm{e,m}}(\mathbf{r}'')$. The scattering matrices are proportional to the far-field coefficients because of (3.111), which leads to the property (3.382).

22. Is the optical theorem in the form of (3.147) applicable to an infinitely long cylindrical scatterer?

Answer: No. The field scattered by an infinite cylindrical scatterer is not a spherical wave but rather a cylindrical wave. For example, in the case of the incidence normal to the axis of the cylinder the scattered field u_{sc}, which depending on the polarization is the component of either electrical or magnetic field parallel to the axis of the cylinder, has the form

$$u_{\mathrm{sc}} = -\sqrt{\frac{2}{\pi k\rho}}F(\varphi)\,\mathrm{e}^{-j(k\rho-\frac{\pi}{4})}, \tag{3.383}$$

where ρ is the distance from the axis of the cylinder, φ is the azimuth coordinate in the plane perpendicular to the cylinder, and $F(\varphi)$ is the two-dimensional far-field coefficient (see Chapter 6).

23. Define the extinction width $\sigma_{\mathrm{ext}}^{\mathrm{2D}}$ as the ratio of the total energy per unit length of cylinder scattered and absorbed by the scatterer to the energy density per unit length of cylinder incident on the cylinder. Prove the two-dimensional analog of the optical theorem (Jones 1989)

$$\sigma_{\mathrm{ext}}^{\mathrm{2D}} = \frac{4}{k}\mathrm{Re}\frac{U_0^*F_{\mathrm{fwd}}}{|U_0|^2}, \tag{3.384}$$

where U_0 is the amplitude of the incident plane wave and F_{fwd} is the far-field coefficient for the scattering in the direction in which the incident wave is traveling.

24. Derive the formulas (3.176) for the forward scattering from cylindrical impenetrable bodies in the optical limit $(k \to \infty)$.

Hint: The derivation is similar to the three-dimensional case (section 3.4.3).

Solution: Consider, for example, the scattering amplitude F_E. Represent the component of the incident electric field along the axis of the cylindrical scatterer as

$$U_E^{\mathrm{inc}}(\rho) = E_{0z}\,\mathrm{e}^{-jk_\perp\hat{\imath}_\perp\cdot\rho}, \tag{3.385}$$

where $\hat{\imath}_\perp = \hat{x}\cos\varphi_0 + \hat{y}\sin\varphi_0$ is the direction of propagation of the incident plane wave on the (x, y) plane. In the forward-scattering case, $\hat{s}_\perp = \hat{\imath}_\perp$ and the integral representation (3.172) takes the form

$$F_E(\varphi_0) = \frac{1}{4j}\int_C \mathrm{e}^{jk_\perp\hat{\imath}_\perp\cdot\rho'}\left[\frac{\partial}{\partial n'}U_E(\rho') - jk_\perp(\hat{n}'\cdot\hat{\imath}_\perp)U_E(\rho')\right]\mathrm{d}l'. \tag{3.386}$$

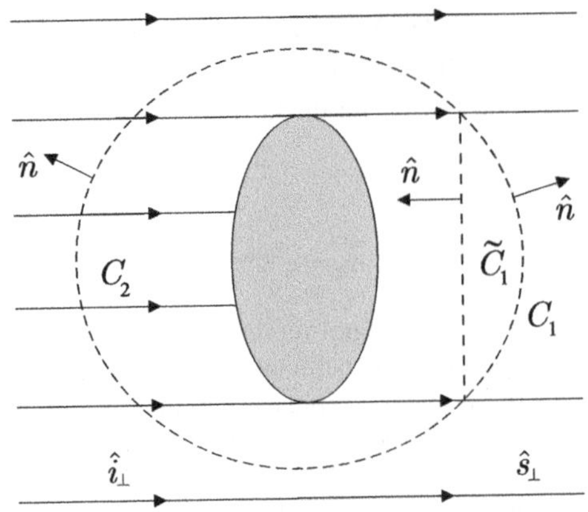

Figure 3.32 Geometry of integration: $\hat{\imath}_\perp$ and $\hat{s}_\perp$ are the incidence and scattering directions. The integration contour C consists of two portions C_1 and C_2 and encloses the cross section of an arbitrarily shaped cylindrical scatterer. C_1 is the portion of C shadowed by the scatterer, and $\tilde{C}_1$ is the straight line between the ends of C_1. $\hat{n}$ is the unit normal exterior to the regions enclosed by C and by $C_1 \cup \tilde{C}_1$.

Choose the integration contour C to enclose the boundary of the cylinder at some finite distance from it (Figure 3.32) such that in the optical limit the total field U_E on C approaches its GO limit

$$U_E(\rho) = \begin{cases} 0, & \rho \in C_1 \\ U_E^{\rm inc}(\rho) + U_E^{\rm ref}(\rho), & \rho \in C_2 \end{cases}. \tag{3.387}$$

So, the optical limit of $F_E(\varphi_0)$ can be obtained by integration in (3.386) over the illuminated portion of C. Furthermore, the reflected field does not contribute to the limit as the products of $e^{jk_\perp \hat{\imath}_\perp \cdot \rho'}$ with $U_E^{\rm ref}(\rho')$ and $\frac{\partial}{\partial n'} U_E^{\rm ref}(\rho')$ are rapidly oscillating functions.

Thus,

$$F_E(\varphi_0) \to \frac{1}{4j} \int_{C_2} e^{jk_\perp \hat{\imath}_\perp \cdot \rho'} \left[\frac{\partial}{\partial n'} U_E^{\rm inc}(\rho') - jk_\perp(\hat{n}' \cdot \hat{\imath}_\perp) U_E^{\rm inc}(\rho') \right] {\rm d}l'. \tag{3.388}$$

Using (3.385) and evaluating the normal derivative as

$$\frac{\partial}{\partial n} U_E^{\rm inc}(\rho) = \hat{n} \cdot \nabla_2 U_E^{\rm inc}(\rho) = -jk_\perp(\hat{n} \cdot \hat{\imath}_\perp) U_E^{\rm inc}(\rho), \tag{3.389}$$

we obtain

$$F_E(\varphi_0) \to -\frac{k_\perp}{2} E_{0z} \int_{C_2} (\hat{n}' \cdot \hat{\imath}_\perp) \, {\rm d}l'. \tag{3.390}$$

Since $\hat{\imath}_\perp$ is a constant vector, the divergence theorem (A.28) allows us to replace the integration over C_2 by integration over the complimentary part C_1 of the closed contour C,

$$\int_{C_2} (\hat{n}' \cdot \hat{\imath}_\perp) \, {\rm d}l' = -\int_{C_1} (\hat{n}' \cdot \hat{\imath}_\perp) \, {\rm d}l'. \tag{3.391}$$

Using again the divergence theorem, the integration over C_1 can be replaced by integration over the straight segment $\tilde{C}_1$, at which $\hat{n} = -\hat{\imath}_\perp$, so that we have

$$\int_{C_1} (\hat{n}' \cdot \hat{\imath}_\perp)\,\mathrm{d}l' = -\int_{\tilde{C}_1} (\hat{n}' \cdot \hat{\imath}_\perp)\,\mathrm{d}l' = \int_{\tilde{C}_1} \mathrm{d}l' = L, \tag{3.392}$$

where L is the arc length of $\tilde{C}_1$. Using (3.391) and (3.392) in (3.390) gives

$$F_E(\varphi_0) \to \frac{1}{2} k_\perp L E_{0z} \tag{3.393}$$

and recovers the corresponding formula from (3.176). The derivation of the formula for $F_H(\varphi_0)$ is similar.

25. Show that in the optical limit ($k \to \infty$) the scattering amplitudes of any cylindrical impenetrable scatterer taken in the direction of the incident plane wave behave as in (3.176).

Hint: The derivation is similar to that of section 3.4.3. Start with the integral representation (3.172) and (3.173) of the scattering amplitudes with $\varphi = \varphi_0$ and use GO approximate for the total field in the integrand.

26. Compare the scattering cross section of a small PEC sphere in the forward- ($\theta = \pi - \theta_{\text{inc}}, \phi = \phi_{\text{inc}} \pm \pi$) and backscattering ($\theta = \theta_{\text{inc}}, \phi = \phi_{\text{inc}}$) directions and show that the latter is greater than the former by a factor of nine.

Hint: Specialize (3.66) and (3.323) to the forward- and backscattering directions.

Answer: $\sigma_{\theta\theta_{\text{inc}}} = \sigma_{\phi\phi_{\text{inc}}} = 9\pi k^4 a^6$ in the backscattering case and $\pi k^4 a^6$ in the forward-scattering case.

27. Prove that the scattering cross section of a small PEC disc is the same in the forward- and specular scattering directions. The specular scattering direction is at $\theta = \theta_{\text{inc}}$ and $\phi = \phi_{\text{inc}} \pm \pi$ if the z axis is oriented perpendicularly to the plane of the disc.

Hint: Specialize (3.66) and (3.328) to the forward- and specular scattering directions.

Answer: For both directions, one finds that[13]

$$\sigma_{\theta\theta_{\text{inc}}} = \frac{64}{9\pi} k^4 a^6 \cos^4\theta, \quad \sigma_{\phi\phi_{\text{inc}}} = \frac{16}{9\pi} k^4 a^6 (2 - \sin^2\theta)^2. \tag{3.394}$$

28. Compare backscattering cross sections of an electrically small PEC sphere and disc (both of the same radius a). Which scatterer is stronger?

Answer: RCS of the sphere is greater than the RCS of an axially illuminated disc by a factor of $(9\pi/8)^2 \approx 12.5$ and greater than the RCS of a disc at grazing illumination by a factor $(3\pi/4)^2 \approx 5.55$ (polarization of the incidence wave in the plane of the disc).

Solution: The RCS of an electrically small PEC sphere is given by the formula $\sigma_{\text{sph}} = 9\pi k^4 a^6$ and the RCS of an electrically small PEC disc under axial illumination ($\theta_{\text{inc}} = 0$) by $\sigma_{\text{disc}}(0) = 64k^4a^6/(9\pi)$ (both radar cross sections are polarization independent), implying that $\sigma_{\text{sph}} = (9\pi/8)^2\sigma_{\text{disc}}(0)$. At the grazing illumination of the disc ($\theta_{\text{inc}} = \pi/2$), the RCS vanishes for an incident wave polarized perpendicularly to the plane of the disc and is given by $\sigma_{\text{disc}}(\pi/2) = 16k^4a^6/\pi$ for a wave polarized in the plane of the disc so that $\sigma_{\text{sph}} = (3\pi/4)^2\sigma_{\text{disc}}(\pi/2)$. Summarizing, we have the relations

$$\sigma_{\text{disc}}(0) < \sigma_{\text{disc}}(\pi/2) < \sigma_{\text{sph}}, \tag{3.395}$$

which can be explained by the fact that in contrast to a disc (which is assumed to be infinitely thin), a sphere has a finite volume. The same relations are held for total scattering cross sections.

It is interesting to note that in the high-frequency limit ($ka \to \infty$), total scattering cross sections of a PEC sphere and a PEC disc tend to each other (see problems 1 and 2); the backscattering cross section of the disc at axial

[13] In Ruck *et al.* (1970), equation (7.5-8), the low-frequency forward-scattering cross section $\sigma_{\phi\phi_{\text{inc}}}$ has an incorrect sign in front of $\sin^2\theta$.

illumination is greater than the RCS of the sphere by a factor $(ka)^2$ (see Chapter 8). Under grazing illumination, total scattering and backscattering cross sections of the disc vanish for both polarizations in the high-frequency limit.

29. Derive the condition under which the low-frequency scattering from a thin, circular, perfectly conducting loop with the loop radius a and wire radius r_0 is indistinguishable from the low-frequency scattering from a PEC disc of radius a.

Hint: Equate the scattering matrices (3.337) and (3.328).

Answer: The scattering matrices of the loop and the disc are equal if $\chi = 1$ where the factor χ is defined in (3.336). This condition is satisfied when the wire radius is chosen such that

$$\frac{r_0}{a} = 8\exp\left(-2 - \frac{3}{8}\pi^2\right) \approx 0.0267383. \tag{3.396}$$

30. Considering an electrically short PEC thin-wire scatterer, show that the imaginary part of its inverse polarizability is proportional to k^3 and does not depend on the wire length.

Solution: The relation between the induced electric dipole and applied electric field is given by (3.226), which defines the polarizability (only one component is needed, as the object is a straight thin wire)

$$P_{11} = \frac{4l^2}{3\epsilon_0 j\omega Z_{\text{inp}}}. \tag{3.397}$$

The imaginary part of $1/P_{11}$ depends only on the real part of Z_{inp}, the radiation resistance:

$$\text{Im}\left(\frac{1}{P_{11}}\right) = \frac{3\epsilon_0\omega R}{4l^2} \approx \frac{3}{4}\frac{k^3}{6\pi}. \tag{3.398}$$

Here we have used formula (3.223) for the radiation resistance. In fact, the exact result reads

$$\text{Im}\left(\frac{1}{P_{11}}\right) = \frac{k^3}{6\pi}. \tag{3.399}$$

The error in the numerical coefficient comes from approximate expressions for the current distribution along the wire. The exact result can be derived from the energy conservation principle (Belov *et al.* 2003), and it is valid for any lossless electric-dipole scatterer, if no cross-polarized scattered field is created.

31. By equating σ_{ext} and σ_{T} from (3.306), (3.307), (3.304), and (3.305) for an electrically small PEC disc, prove that

$$\text{Im}\left(\frac{1}{P_{11}}\right) = -\text{Im}\left(\frac{1}{M_{33}}\right) = \frac{k^3}{6\pi}. \tag{3.400}$$

Hint: Use the expressions (3.327) for the elements of the polarizability tensors of the disc, but note that in contrast to relations $P_{11} = P_{22}$, $P_{33} = 0$, $M_{11} = M_{22} = 0$, which are consequences of the scatterer's symmetries, relations $P_{11} = 2M_{33} = 16a^3/3$ are valid only for the lowest-order terms in the expansions of P_{11} and M_{33} in powers of k, implying that $\text{Im}\,P_{11}$ is not necessarily equal to $2\,\text{Im}\,M_{33}$ since the imaginary parts are quantities of a higher order.

Solution: Specializing the expressions for the total and extinction scattering cross sections to the case of the disc gives

$$\sigma_{\text{T}}(\hat{\theta}_{\text{inc}}) = \frac{k^4}{6\pi}|P_{11}|^2\cos^2\theta_{\text{inc}} \tag{3.401}$$

$$\sigma_{\text{T}}(\hat{\phi}_{\text{inc}}) = \frac{k^4}{6\pi}\left(|P_{11}|^2 + |M_{33}|^2\sin^2\theta_{\text{inc}}\right) \tag{3.402}$$

and

$$\sigma_{\text{ext}}(\hat{\theta}_{\text{inc}}) = -k\,\text{Im}\,P_{11}\cos^2\theta_{\text{inc}} \tag{3.403}$$

$$\sigma_{\text{ext}}(\hat{\phi}_{\text{inc}}) = k(-\text{Im}\,P_{11} + \text{Im}\,M_{33}\sin^2\theta_{\text{inc}}). \tag{3.404}$$

From (3.401) and (3.403) it follows that

$$\operatorname{Im} P_{11} = -\frac{k^3}{6\pi}|P_{11}|^2 \tag{3.405}$$

which can be rewritten as in (3.400) because of the relation

$$\operatorname{Im}\frac{1}{P_{11}} = -\frac{\operatorname{Im} P_{11}}{|P_{11}|^2}. \tag{3.406}$$

Equating (3.402) and (3.404) and accounting for (3.405) lead to the expression

$$\operatorname{Im}\frac{1}{M_{33}} = -\frac{k^3}{6\pi}. \tag{3.407}$$

32. Find the average backscattering cross section which will be observed at a dipole-type antenna when the signal is reflected from an ensemble of identically shaped but randomly oriented electrically small bodies of revolution.
Hint: Calculate the cross section by averaging

$$\bar{\sigma} = \frac{1}{4\pi}\int_0^{2\pi} \mathrm{d}\psi \int_0^{\pi} \mathrm{d}\theta \sin\theta\,\sigma(\psi,\theta), \tag{3.408}$$

where $\sigma(\psi,\theta)$ is given by (3.361) with

$$A_{\theta\theta_{\text{inc}}} = \frac{k^2}{\sqrt{4\pi}}(M_{11} + P_{11}\cos^2\theta + P_{33}\sin^2\theta) \tag{3.409}$$

$$A_{\theta\phi_{\text{inc}}} = 0 \tag{3.410}$$

$$A_{\phi\phi_{\text{inc}}} = \frac{k^2}{\sqrt{4\pi}}(P_{11} + M_{11}\cos^2\theta + M_{33}\sin^2\theta). \tag{3.411}$$

Expressions (3.409)–(3.411) result from (3.300)–(3.303) in the limit of backscattering ($\phi_{\text{inc}} = \phi$, $\theta_{\text{inc}} = \theta$).
Solution: Integration with respect to ψ in (3.408) can be performed by using formulas (3.363) and (3.364), which gives

$$\bar{\sigma} = \frac{1}{16}\int_0^{\pi}\left[3\left|A_{\theta\theta_{\text{inc}}}\right|^2 + 3\left|A_{\phi\phi_{\text{inc}}}\right|^2 + 2\operatorname{Re}\left(A_{\theta\theta_{\text{inc}}}A^*_{\phi\phi_{\text{inc}}}\right)\right]\sin\theta\,\mathrm{d}\theta. \tag{3.412}$$

Inserting (3.409)–(3.411) in (3.412) and integrating we obtain

$$\begin{aligned}\bar{\sigma} = \frac{k^4}{60\pi}\Big[&2M^*_{11}(4M_{11} + M_{33} + 3P_{11} + 2P_{33}) + M^*_{33}(2M_{11} + 3M_{33} + 4P_{11} + P_{33})\\ &+ P^*_{33}(2P_{11} + 3P_{33} + 4M_{11} + M_{33}) + 2P^*_{11}(4P_{11} + P_{33} + 3M_{11} + 2M_{33})\Big].\end{aligned} \tag{3.413}$$

33. Consider (3.413) in the limit of electrically small PEC spheres. Explain the result.
Answer: In the case of spherical particles, $P_{11} = P_{33}$ and $M_{11} = M_{33}$, and (3.413) reduces to the formula

$$\bar{\sigma} = \frac{k^4}{4\pi}|P_{11} + M_{11}|^2. \tag{3.414}$$

For a PEC sphere of radius a, the polarizabilities are given by (3.322), and using these in (3.414) we arrive at

$$\bar{\sigma} = 9\pi k^4 a^6, \tag{3.415}$$

which is equal to the RCS of an electrically small PEC sphere. The identity of the two cross sections is a consequence of the fact that scattering from a spherical particle is polarization independent.

34. Write down the average RCS of an ensemble of randomly oriented identical electrically small PEC (a) short wires and (b) discs.

Hint: Apply (3.413) to the respective bodies of revolution.

Solution: According to (3.345), except for P_{33} all other elements of polarizability tensors of a short wire vanish, and it follows from (3.413) that

$$\bar{\sigma} = \frac{k^4}{20\pi}|P_{33}|^2 = \frac{\lambda^2}{45\pi}\frac{(kl)^2}{[\ln(l/r_0) - 1]^2}. \tag{3.416}$$

35. Derive the average RCS of an ensemble of randomly oriented identical electrically small PEC discs.

Solution: In the case of a PEC disc, we can use (3.327) to express (3.413) entirely through P_{11} as

$$\bar{\sigma} = \frac{51k^4}{240\pi}|P_{11}|^2. \tag{3.417}$$

Inserting $P_{11} = 16a^3/3$ in (3.417) leads to the formula

$$\bar{\sigma} = \frac{272}{45\pi}k^4a^6. \tag{3.418}$$

The average cross section (3.418) is always smaller than the backscattering cross section of the disc at axial incidence,

$$\sigma = \frac{64}{9\pi}k^4a^6 = \frac{20}{17}\bar{\sigma}. \tag{3.419}$$

36. Determine if the total scattering cross section of electric dipole scatterers has an upper limit, and if it has, find the maximal possible cross section.

Solution: Arbitrary small particles which scatter as electric dipoles can be modeled by their polarizability P defined as in (3.230):

$$p_e = \epsilon P E_0. \tag{3.420}$$

Here we use scalar values of the dipole moment p_e and the incident field E_0 because to calculate the total scattered field we need only the amplitude of p_e. For the most general particle we can write for the inverse polarizability

$$\frac{1}{P} = \xi' + j\xi'' + j\frac{k^3}{6\pi}. \tag{3.421}$$

Here the last term is due to the scattering loss (see 3.308) and the complex value of $\xi = \xi' + j\xi''$ depends on the particle size, shape, material, and frequency. The absorption coefficient $\xi'' \geq 0$, and for lossless particles $\xi'' = 0$. Particle resonances correspond to the frequency points where $\xi' = 0$ because at these frequencies the absolute value of the polarizability has maxima, and the polarizability is an imaginary number.

The total power radiated by dipole p_e in far zone reads (see problem 19 in section 2.7)

$$P_{sc} = \frac{\mu\omega^4|p_e|^2}{12\pi c} = \frac{\mu\epsilon^2\omega^4|P|^2}{12\pi c}|E_0|^2 = \frac{\mu^2\epsilon^2\omega^4|P|^2}{6\pi}P_{inc}. \tag{3.422}$$

We have used the relation between the incident electric field and the incident power density $P_{inc} = |E_0|^2/(2Z)$. The scattered power increases with increasing absolute values of the polarizability $|P|$, but, according to (3.421), it cannot be made larger than $6\pi/k^3$. Thus, the total scattering cross section has a fundamental upper limit, and the ultimate value is

$$\sigma_{\mathrm{T\ max}} = \frac{\mu^2\epsilon^2\omega^4}{6\pi}\frac{36\pi^2}{k^6} = \frac{6\pi}{k^2} = \frac{3}{2\pi}\lambda^2. \tag{3.423}$$

Physically, this extreme value of the scattering cross section corresponds to resonant ($\xi' = 0$) lossless ($\xi'' = 0$) particles.

37. Determine if the absorption cross section of electric dipole scatterers has an upper limit, and if it has, find the maximal absorbed power.

Solution: Similarly to the previous problem, we model an arbitrary dipole particle by its polarizability P (3.420) and assume that it can be an arbitrary complex number, as in (3.421). The absorbed power is the difference between the power extracted by the particle from the incident field and the power scattered by the same particle into the surrounding space. The extracted power is given by (3.306) on multiplication by the incident power, and the radiated power reads (3.422). Thus,

$$P_{\mathrm{abs}} = -\epsilon\frac{\omega}{2}|E_0|^2\,\mathrm{Im}\,P - \frac{\mu\epsilon^2\omega^4|P|^2}{12\pi c}|E_0|^2 = \frac{\epsilon\frac{\omega}{2}\xi''}{\xi'^2 + \left(\xi'' + \frac{k^3}{6\pi}\right)^2}|E_0|^2. \tag{3.424}$$

Here we have substituted the general expression for the polarizability P from (3.421). Clearly, this value reaches its maximum value at $\xi' = 0$ and $\xi'' = k^3/(6\pi)$. The maximal possible value of absorbed power reads, on substitution of these values,

$$P_{\mathrm{abs\ max}} = \frac{3\pi}{2k^2}P_{\mathrm{inc}} = \frac{3}{8\pi}\lambda^2 P_{\mathrm{inc}} \tag{3.425}$$

and the ultimate value of the absorption cross section equals $3\lambda^2/(8\pi)$.

Physically, this corresponds to a particle at resonance ($\xi' = 0$) whose loss parameter is such that the amount of absorbed power is equal to the amount of scattered power. This extremum has the same meaning as the extremum of power delivered to a load from a generator when the load impedance is matched to the impedance of the source. Similarly, in that case half of the power is delivered to the load and half is dissipated in the internal resistance of the generator.

38. Determine if the scattering cross section of electric dipole scatterers has an upper limit, and if it has, find the maximal absorbed power.

Hint: Substitute the general expression for the particle polarizability (3.421) into the expression for the scattered power (3.422) and find its maximum for arbitrary ξ' and ξ''.

Answer: There is an upper limit and

$$\sigma_{\mathrm{sc\ max}} = \frac{3}{2\pi}\lambda^2. \tag{3.426}$$

This extreme value of the scattering cross section is the same as for the total cross section and corresponds to resonant ($\xi' = 0$) lossless ($\xi'' = 0$) particles. It is reached when particles are most strongly excited (resonance) and re-radiate all the received power back into space (no absorption).

39. Consider a generally shaped electrically small particle and show that the symmetry of polarizability tensors, that is, $\overline{\overline{P}} = \overline{\overline{P}}^T$ and $\overline{\overline{M}} = \overline{\overline{M}}^T$, ensures the compliance of the scattered field with the reciprocity.

Hint: Use the following relation

$$\mathbf{a}\cdot(\overline{\overline{L}}\cdot\mathbf{b}) = \mathbf{b}\cdot(\overline{\overline{L}}^T\cdot\mathbf{a}), \tag{3.427}$$

where $\mathbf{a}$ and $\mathbf{b}$ are arbitrary three-dimensional vectors, $\overline{\overline{L}}$ is an arbitrary 3 by 3 matrix, and $\overline{\overline{L}}^T$ is the transposed matrix.

Solution: The far-field coefficient of every electrically small particle is given by (3.246), and for illumination by a plane wave with the propagation direction $\hat{\imath}$ and the amplitude $\mathbf{E}_0$ it can be expressed as

$$\mathbf{F} = -\frac{k^3}{4\pi}\mathbf{f}(\hat{\imath}, \mathbf{E}_0, \hat{s}) \tag{3.428}$$

with

$$\mathbf{f}(\hat{\imath}, \mathbf{E}_0, \hat{s}) = \hat{s} \times \hat{s} \times (\overline{\overline{P}} \cdot \mathbf{E}_0) - \hat{s} \times [\overline{\overline{M}} \cdot (\hat{\imath} \times \mathbf{E}_0)], \tag{3.429}$$

where $\hat{s}$ is the scattering direction. The reciprocity relation (3.121) for the scattered field implies therefore the identity

$$\mathbf{E}_1 \cdot \mathbf{f}(\hat{\imath}, \mathbf{E}_0, \hat{s}) = \mathbf{E}_0 \cdot \mathbf{f}(-\hat{s}, \mathbf{E}_1, -\hat{\imath}), \tag{3.430}$$

where $\mathbf{E}_1$ is the amplitude of a plane wave incident in the direction $-\hat{s}$ and scattered in the direction $-\hat{\imath}$.
Using (3.429) in (3.430) gives

$$\mathbf{E}_1 \cdot [\hat{s} \times \hat{s} \times (\overline{\overline{P}} \cdot \mathbf{E}_0) - \hat{s} \times \overline{\overline{M}} \cdot (\hat{\imath} \times \mathbf{E}_0)] = \mathbf{E}_0 \cdot [\hat{\imath} \times \hat{\imath} \times (\overline{\overline{P}} \cdot \mathbf{E}_1) - \hat{\imath} \times \overline{\overline{M}} \cdot (\hat{s} \times \mathbf{E}_1)]. \tag{3.431}$$

Using the orthogonality of the electric field and the propagation direction in plane electromagnetic waves,

$$\hat{\imath} \cdot \mathbf{E}_0 = 0, \quad \hat{s} \cdot \mathbf{E}_1 = 0, \tag{3.432}$$

the cyclic invariance (A.11) of the scalar triple product and expanding the vector triple product according to (A.12), we obtain the relations

$$\mathbf{E}_1 \cdot [\hat{s} \times \hat{s} \times (\overline{\overline{P}} \cdot \mathbf{E}_0)] = -\mathbf{E}_1 \cdot (\overline{\overline{P}} \cdot \mathbf{E}_0) \tag{3.433}$$
$$\mathbf{E}_0 \cdot [\hat{\imath} \times \hat{\imath} \times (\overline{\overline{P}} \cdot \mathbf{E}_1)] = -\mathbf{E}_0 \cdot (\overline{\overline{P}} \cdot \mathbf{E}_1) \tag{3.434}$$
$$\mathbf{E}_1 \cdot \hat{s} \times [\overline{\overline{M}} \cdot (\hat{\imath} \times \mathbf{E}_0)] = -(\hat{s} \times \mathbf{E}_1) \cdot [\overline{\overline{M}} \cdot (\hat{\imath} \times \mathbf{E}_0)] \tag{3.435}$$
$$\mathbf{E}_0 \cdot \hat{\imath} \times [\overline{\overline{M}} \cdot (\hat{s} \times \mathbf{E}_1)] = -(\hat{\imath} \times \mathbf{E}_0) \cdot [\overline{\overline{M}} \cdot (\hat{s} \times \mathbf{E}_1)] \tag{3.436}$$

to express (3.431) as

$$-\mathbf{E}_1 \cdot (\overline{\overline{P}} \cdot \mathbf{E}_0) + (\hat{s} \times \mathbf{E}_1) \cdot [\overline{\overline{M}} \cdot (\hat{\imath} \times \mathbf{E}_0)] = -(\overline{\overline{P}} \cdot \mathbf{E}_1) \cdot \mathbf{E}_0 + [\overline{\overline{M}} \cdot (\hat{s} \times \mathbf{E}_1)] \cdot (\hat{\imath} \times \mathbf{E}_0). \tag{3.437}$$

It is now apparent that (3.437) is satisfied if the matrices $\overline{\overline{P}}$ and $\overline{\overline{M}}$ are such that

$$\mathbf{E}_1 \cdot (\overline{\overline{P}} \cdot \mathbf{E}_0) = (\overline{\overline{P}} \cdot \mathbf{E}_1) \cdot \mathbf{E}_0 \tag{3.438}$$
$$(\hat{s} \times \mathbf{E}_1) \cdot [\overline{\overline{M}} \cdot (\hat{\imath} \times \mathbf{E}_0)] = [\overline{\overline{M}} \cdot (\hat{s} \times \mathbf{E}_1)] \cdot (\hat{\imath} \times \mathbf{E}_0). \tag{3.439}$$

Because of (3.427), (3.438) is satisfied if $\overline{\overline{P}}$ is symmetric and (3.439) is satisfied if $\overline{\overline{M}}$ is symmetric.

40. Consider an electrically small particle rotationally symmetric with respect to the x_3 axis and show that $\overline{\overline{P}}$ and $\overline{\overline{M}}$ are diagonal

$$\overline{\overline{P}} = \begin{pmatrix} P_{11} & 0 & 0 \\ 0 & P_{22} & 0 \\ 0 & 0 & P_{33} \end{pmatrix}, \quad \overline{\overline{M}} = \begin{pmatrix} M_{11} & 0 & 0 \\ 0 & M_{22} & 0 \\ 0 & 0 & M_{33} \end{pmatrix} \tag{3.440}$$

with

$$P_{11} = P_{22}, \quad M_{11} = M_{22}. \tag{3.441}$$

Table 3.2 Monostatic far-field coefficients of rotationally symmetric electrically small particles in terms of the elements of polarizability matrices.

Case	$\hat{i}$	$\hat{e}_{\text{inc}}$	$\hat{s}$	$\hat{e}_{\text{sc}}$	$-(4\pi k^{-3})\hat{e}_{\text{sc}} \cdot \mathbf{F}(\hat{i}, \hat{e}_{\text{inc}}, \hat{s})$
1	$-\hat{x}$	$\hat{z}$	$\hat{x}$	$\hat{z}$	$-M_{22} - P_{33}$
2	$-\hat{y}$	$\hat{z}$	$\hat{y}$	$\hat{z}$	$-M_{11} - P_{33}$
3	$-\hat{x}$	$\hat{y}$	$\hat{x}$	$\hat{y}$	$-M_{33} - P_{22}$
4	$-\hat{y}$	$\hat{x}$	$\hat{y}$	$\hat{x}$	$-M_{33} - P_{11}$

Hint: Study scattering in the plane (x_1, x_2) perpendicular to the symmetry axis. The relations (3.441) follow from the condition that monostatic scattering is independent of the incidence direction in that plane. The diagonal structure of the polarizability matrices in (3.440) results from the condition that the bistatic scattering is symmetric with respect to the incidence direction.

Solution: Let $x_1 = x, x_2 = y$, and $x_3 = z$. The amplitude of the incident plane wave is assumed to be of unit amplitude polarized as the unit vector $\hat{e}_{\text{inc}}$, that is, $\mathbf{E}_0 = \hat{e}_{\text{inc}}$. The scattered field will be projected on the unit vector $\hat{e}_{\text{sc}}$.

Express the far-field coefficient in the monostatic case for the incidence/scattering directions $\hat{s} = -\hat{i} = \hat{x}$ and $\hat{s} = -\hat{i} = \hat{y}$ in terms of the elements of the polarizability matrices (see Table 3.2). Since the particle is symmetric, the scattered field must be the same. Thus, it follows from cases 1 and 2 that $M_{11} = M_{22}$ and from cases 3 and 4 that $P_{11} = P_{22}$.

To determine the off-diagonal elements, consider bistatic scattering in the plane (x, y) for the incidence along $-\hat{x}$ and $\hat{y}$ directions and for scattering in the directions $\pm\hat{y}$ and $\pm\hat{x}$, respectively. The results are summarized in Table 3.3. Because of the particle symmetry, for the incidence along the $-\hat{x}$ direction, the fields scattered in the directions $\pm\hat{y}$ must coincide (cases 1 through 8), from which we determine that

$$M_{12} = 0, \quad M_{32} = 0, \quad P_{32} = 0, \quad P_{12} = 0.$$

Table 3.3 Bistatic far-field coefficients of rotationally symmetric electrically small particles in terms of the elements of polarizability matrices.

Case	$\hat{i}$	$\hat{e}_{\text{inc}}$	$\hat{s}$	$\hat{e}_{\text{sc}}$	$-(4\pi k^{-3})\hat{e}_{\text{sc}} \cdot \mathbf{F}(\hat{i}, \hat{e}_{\text{inc}}, \hat{s})$
1	$-\hat{x}$	$\hat{z}$	$\hat{y}$	$\hat{z}$	$M_{12} - P_{33}$
2	$-\hat{x}$	$\hat{z}$	$-\hat{y}$	$\hat{z}$	$-M_{12} - P_{33}$
3	$-\hat{x}$	$\hat{z}$	$\hat{y}$	$\hat{x}$	$-M_{32} - P_{13}$
4	$-\hat{x}$	$\hat{z}$	$-\hat{y}$	$\hat{x}$	$M_{32} - P_{13}$
5	$-\hat{x}$	$\hat{y}$	$\hat{y}$	$\hat{z}$	$-M_{13} - P_{32}$
6	$-\hat{x}$	$-\hat{y}$	$-\hat{y}$	$\hat{z}$	$-M_{13} + P_{32}$
7	$-\hat{x}$	$\hat{y}$	$\hat{y}$	$\hat{x}$	$M_{33} - P_{12}$
8	$-\hat{x}$	$-\hat{y}$	$-\hat{y}$	$\hat{x}$	$M_{33} + P_{12}$
9	$\hat{y}$	$\hat{z}$	$\hat{x}$	$\hat{z}$	$-M_{21} - P_{33}$
10	$\hat{y}$	$\hat{z}$	$-\hat{x}$	$\hat{z}$	$M_{21} - P_{33}$
11	$\hat{y}$	$\hat{z}$	$\hat{x}$	$-\hat{y}$	$-M_{31} + P_{23}$
12	$\hat{y}$	$\hat{z}$	$-\hat{x}$	$-\hat{y}$	$M_{31} + P_{23}$
13	$\hat{y}$	$\hat{x}$	$\hat{x}$	$\hat{z}$	$M_{23} - P_{31}$
14	$\hat{y}$	$-\hat{x}$	$-\hat{x}$	$\hat{z}$	$M_{23} + P_{31}$
15	$\hat{y}$	$\hat{x}$	$\hat{x}$	$-\hat{y}$	$M_{33} + P_{21}$
16	$\hat{y}$	$-\hat{x}$	$-\hat{x}$	$-\hat{y}$	$M_{33} - P_{21}$

For the incidence in the $\hat{y}$ direction, the far field scattered in the directions $\pm\hat{x}$ must be the same (cases 9 through 16), from which we find that

$$M_{21} = 0, \quad M_{31} = 0, \quad P_{31} = 0, \quad P_{21} = 0.$$

These completely determine the off-diagonal elements of the polarizability matrices as the other elements follow from the symmetry of the matrices, that is, $P_{ik} = P_{ki}$ and $M_{ik} = M_{ki}$ with $i, k = 1, 2, 3$.

41. Prove the expression (3.265) for the polarizability matrices of an ellipsoid in the Cartesian coordinate frame not aligned with the principal axes of the ellipsoid.

Solution: Let $\hat{x}$, $\hat{y}$, and $\hat{z}$ be the unit vectors aligned with the principal axes of an ellipsoid. In this coordinate frame the coordinates of the vectors $\mathbf{p}_e$ and $\mathbf{E}_0$ are related by the diagonal polarizability matrix,

$$\frac{1}{\epsilon}\begin{pmatrix} p_{ex} \\ p_{ey} \\ p_{ez} \end{pmatrix} = \begin{pmatrix} P_a & 0 & 0 \\ 0 & P_b & 0 \\ 0 & 0 & P_c \end{pmatrix} \cdot \begin{pmatrix} E_{0x} \\ E_{0y} \\ E_{0z} \end{pmatrix}. \tag{3.442}$$

If $\hat{x}'$, $\hat{y}'$, and $\hat{z}'$ are the orthonormal vectors defining a rotated Cartesian frame, then

$$E_{0x}\hat{x} + E_{0y}\hat{y} + E_{0z}\hat{z} = E'_{0x}\hat{x}' + E'_{0y}\hat{y}' + E'_{0z}\hat{z}' \tag{3.443}$$

$$p_{0x}\hat{x} + p_{0y}\hat{y} + p_{0z}\hat{z} = p'_{0x}\hat{x}' + p'_{0y}\hat{y}' + p'_{0z}\hat{z}', \tag{3.444}$$

and projecting on the vectors $\hat{x}$, $\hat{y}$, and $\hat{z}$ gives the relation between coordinates in the two coordinate frames

$$\begin{pmatrix} E_{0x} \\ E_{0y} \\ E_{0z} \end{pmatrix} = \begin{pmatrix} \hat{x}\cdot\hat{x}' & \hat{x}\cdot\hat{y}' & \hat{x}\cdot\hat{z}' \\ \hat{y}\cdot\hat{x}' & \hat{y}\cdot\hat{y}' & \hat{y}\cdot\hat{z}' \\ \hat{z}\cdot\hat{x}' & \hat{z}\cdot\hat{y}' & \hat{z}\cdot\hat{z}' \end{pmatrix} \cdot \begin{pmatrix} E'_{0x} \\ E'_{0y} \\ E'_{0z} \end{pmatrix} \tag{3.445}$$

and similarly for the components of $\mathbf{p}_e$. The matrix on the right-hand side of (3.445) is the matrix Q defined in (3.266).

Inserting (3.445) and the similar relation for the components of $\mathbf{p}_e$ in (3.442) and multiplying from the left with the inverse matrix

$$\overline{\overline{Q}}^{-1} = \begin{pmatrix} \hat{x}\cdot\hat{x}' & \hat{y}\cdot\hat{x}' & \hat{z}\cdot\hat{x}' \\ \hat{x}\cdot\hat{y}' & \hat{y}\cdot\hat{y}' & \hat{z}\cdot\hat{y}' \\ \hat{x}\cdot\hat{z}' & \hat{y}\cdot\hat{z}' & \hat{z}\cdot\hat{z}' \end{pmatrix} \tag{3.446}$$

results in (3.265) since $\overline{\overline{Q}}^{-1} = \overline{\overline{Q}}^{T}$ for every orthogonal matrix. That (3.446) is indeed inverse to $\overline{\overline{Q}}$ can be checked by direct multiplication of the matrices using the fact that $\hat{x}'$, $\hat{y}'$ and $\hat{z}'$ are orthonormal vectors.

The matrix $\overline{\overline{M}}$ in the rotated coordinate system is expressed by the same matrix transformation.

References

Babič VM and Buldyrev VS 1991 *Short-Wavelength Diffraction Theory*, Springer-Verlag.

Belov PA, Malsovski SI, Simovski KR, and Tretyakov SA 2003 A condition imposed on the electromagnetic polarizability of a bianisotropic scatterer. *Tech. Phys. Lett.*, **29**, 718–720.

Boerner WM, Mott H, Lüneburg E et al. 1998 *Polarimetry in Radar Remote Sensing: Basic and Applied Concepts*. In *Manual of Remote Sensing*, Third edition, Ryerson RA (ed.), vol. 2, *Principles and Applications of Imaging Radar*, Henderson FM and Lewis AJ (eds), 271–357, John Wiley & Sons.

Bohren CF 1983 How can a particle absorb more than the light incident on it?. *Am. J. Phys.*, **51**, 323–327.

Born M and Wolf E (eds) 1959 *Principles of Optics: Electromagnetic Theory of Propagation, Interference and Diffraction of Light*, Pergamon Press.

Borovikov VA and Kinber BY 1994 *Geometrical Theory of Diffraction*, IEE Electromagnetic Waves Series 37, The Institution of Electrical Engineers, London.

Bouche D, Molinet F, and Mittra R 1997 *Asymptotic Methods in Electromagnetics*, Springer-Verlag.

Bouman JJ, Senior TBA, and Uslenghi PLE (eds) 1987 *Electromagnetic and Acoustic Scattering by Simple Shapes*, Hemisphere Publishing Corporation.

Cheng DK 1992 *Field and Wave Electromagnetics*, Second edition, Addison-Wesley.

Collin RE 1960 *Field Theory of Guided Waves*, McGraw-Hill.

de Hoop AT 1958 On the plane-wave extinction cross-section of an obstacle. *Appl. sci. Res. B*, **7**, 463–472.

de Hoop AT 1960 A reciprocity theorem for the electromagnetic field scattered by an obstacle. *Appl. sci. Res. B*, **8**, 135–140.

Eggimann WH 1961 Higher-order evaluation of electromagnetic diffraction by circular discs. *IRE Trans. Microw. Theory Techn.*, **9**, 408–418.

Fock VA 1965 *Electromagnetic Diffraction and Propagation Problems*, Pergamon Press.

Gans R 1920 Ultramikroskopische Studien. (Methoden zur Formbestimmung subultramikroskopischer Teilchen). *Ann. Phys. (Berlin)*, **367**, 331–357.

James GL 1976 *Geometrical Theory of Diffraction for Electromagnetic Waves*, Peter Peregrinus, London.

Jones DS 1955 On the scattering cross section of an obstacle. *Philos. Mag. Ser. 7*, **46**, 957–962.

Jones DS 1964 *The Theory of Electromagnetism*, Pergamon Press.

Jones DS 1989 *Acoustic and Electromagnetic Waves*, Clarendon Press.

Keller JB, Kleinman RE, and Senior TBA 1972 Dipole moments in Rayleigh scattering. *J. Inst. Math. Appl.*, **9**, 14–22.

Kemptner E and Osipov A 2007 Plane-of-scattering bistatic scattering matrix of simply shaped targets. In *Proc. 2nd European Conference on Antennas and Propagation (EuCAP 2007)*, Edinburgh, 11–16 Nov. 2007.

King RWP 1969 *Antennas and Waves: A Modern Approach*, The MIT Press.

Kleinman RE 1978 Low-frequency electromagnetic scattering. In *Electromagnetic Scattering* (ed. Uslenghi PLE), pp. 1–28. Academic Press.

Kleinman RE and Senior TBA 1972 Rayleigh scattering cross sections. *Radio Sci.*, **7**, 937–942.

Kleinman RE and Senior TBA 1975 Low-frequency scattering by space objects. *IEEE Trans. Aerosp. Electron. Syst.*, **11**, 672–675.

Kline M and Kay IW 1965 *Electromagnetic Theory and Geometrical Optics*, John Wiley & Sons.

Kouyoumjian RG 1956 The backscattering from a circular loop. *Appl. sci. Res. B*, **6**, 165–179.

Kravtsov YA and Orlov YI 1990 *Geometrical Optics of Inhomogeneous Media*, Springer.

Kristensson G and Waterman PC 1982 The T matrix for acoustic and electromagnetic scattering by circular discs. *J. Acoust. Soc. Am.*, **72**, 1612–1625.

Lindell IV, Sihvola A, Ylä-Oijala P, and Wallén H 2009 Zero backscattering from self-dual objects of finite size. *IEEE Trans. Antennas Propag.*, **57**, 2725–2731.

Maier SA, Brongersma ML, Kik PG, Meltzer S, Requicha A, and Atwater H 2001 Plasmonics – a route to nanoscale optical devices, *Adv. Mater.*, **13**, 1501–1505.

Maslovski SI, Simovski CR, and Tretyakov SA 2016 Overcoming black body radiation limit in free space: metamaterial superemitter. *New J. Phys.*, **18**, p. 013034.

McNamara DA, Pistorius CWI, and Malherbe JAG 1990 *Introduction to the Uniform Geometrical Theory of Diffraction*, Artech House.

Meixner J and Andrejewski W 1950 Strenge Theorie der Beugung ebener elektromagnetischer Wellen an der vollkommen leitenden Kreisscheibe und an der kreisförmigen Öffnung im vollkommen leitenden ebenen Schirm. *Ann. Phys. (Berlin)*, **442**, 157–168.

Newton RG 1976 Optical theorem and beyond. *Am. J. Phys.*, **44**, 639–642.

Osipov AV 2010 Low frequency electromagnetic scattering from metallic discs. *Proceedings of the URSI Commission B International Symposium on Electromagnetic Theory (EMTS 2010)*, Berlin, Germany, 612–615.

Osipov AV 2015 Scattering cross sections of impedance-matched bodies. *IEEE Trans. Antennas Propag.*, **63**, 3122–3126.

Paul H and Fischer R 1983 Light absorption by a dipole. *Sov. Phys. Usp.*, **26**, 923–926.

Ruck GT, Barrick DE, Stuart WD, and Krichbaum CK 1970 *Radar Cross Section Handbook*, Plenum Press.

Saxon DS 1955 Tensor scattering matrix for the electromagnetic field. *Phys. Rev.*, **100**, 1771–1775.

Schelkunoff SA and Friis HT 1952 *Antennas: Theory and Practice*, John Wiley & Sons.

Schultz DA 2003 Plasmon resonant particles for biological detection. *Curr. Opin. Biotechnol.*, **14**, 13–22.

Siegel KM 1959 Far-field scattering from bodies of revolution. *Appl. sci. Res. B*, **7**, 293–328.

Stevenson AF 1953a Solution of electromagnetic scattering problems as power series in the ratio (dimension of scatterer)/wavelength. *J. Appl. Phys.*, **24**, 1134–1142.

Stevenson AF 1953b Electromagnetic scattering by an ellipsoid in the third approximation. *J. Appl. Phys.*, **24**, 1143–1151.

Stratton JA 1941 *Electromagnetic Theory*, McGraw-Hill.

Strutt JW (Lord Rayleigh) 1871 On the scattering of light by small particles. *Philos. Mag. Ser. 4*, **41**, 447–454.

Strutt JW (Lord Rayleigh) 1881 On the electromagnetic theory of light. *Philos. Mag. Ser. 5*, **12**, 81–101.

Strutt JW (Lord Rayleigh) 1897 On the incidence of aerial and electric waves upon small obstacles in the form of ellipsoids or elliptic cylinders, and on the passage of electric waves through a circular aperture in a conducting screen. *Philos. Mag. Ser. 5*, **44**, 28–52.

Tretyakov SA, Mariotte F, Simovski CR, Kharina TG, and Heliot JP 1996 Analytical antenna model for chiral scatterers: Comparison with numerical and experimental data. *IEEE Trans. Antennas Propag.*, **44**, 1006–1014.

Tretyakov SA, Maslovski SI, and Belov PA 2003 An analytical model of metamaterials based on loaded wire dipoles. *IEEE Trans. Antennas Propag.*, **51**, 2652–2658.

Ufimtsev PY 2007 *Fundamentals of the Physical Theory of Diffraction*, Wiley-Interscience.

Uslenghi PLE 1990 Scattering by an impedance sphere coated with a chiral layer. *Electromagnetics*, **10**, 201–211.

Uslenghi PLE 1996 Three theorems on zero backscattering. *IEEE Trans. Antennas Propag.*, **44**, 269–270.

Valagiannopoulos CA, Vehmas J, Simovski CR, Tretyakov SA, and Maslovski SI 2015 Electromagnetic energy sink. *Phys. Rev. B*, **91**, p. 115305.

van de Hulst HC 1949 On the attenuation of plane waves by obstacles of arbitrary size and form. *Physica*, **15**, 740–746.

van de Hulst HC 1957 *Light Scattering by Small Particles*, John Wiley & Sons.

van Vleck JH, Bloch F, and Hamermesh M 1947 Theory of radar reflection from wires or thin metallic strips. *J. Appl. Phys.*, **18**, 274–294.

Vehmas J, Ra'di Y, Karilainen AO, and Tretyakov S 2013 Eliminating electromagnetic scattering from small particles. *IEEE Trans. Antennas Propag.*, **61**, 3747–3756.

Wagner RG and Lynch PJ 1963 Theorem on electromagnetic backscatter. *Phys. Rev.*, **131**, 21–23.

Weston VH 1963 Theory of absorbers in scattering. *IEEE Trans. Antennas Propag.*, **11**, 578–584.

Yee KS and Chang AH 1991 Scattering theorems with anisotropic surface boundary conditions for bodies of revolution. *IEEE Trans. Antennas Propag.*, **39**, 1041–1043.

Zumofen, G, Mojarad NM, Sandoghdar V, and Agio M 2008 Perfect reflection of light by an oscillating dipole. *Phys. Rev. Lett.*, **101**, 180404.

4

Planar Interfaces

4.1 Introduction

Reflection and transmission (or refraction) of waves at material interfaces is one of the most important mechanisms of electromagnetic scattering. This mechanism can be studied by looking at the canonical problem of reflection and transmission of a plane electromagnetic wave at a flat material interface. The assumption that the wavefront and the material interface are flat is justified even for curved surfaces if the curvature radii are greater than the wavelength.

Historically, the first solution for the problem of reflection and transmission of an electromagnetic wave at a boundary between two different media was obtained by Fresnel (1823) in the framework of a pre-Maxwellian theory of light, which described the wave properties of material media by a single parameter, the speed of waves.[1] The theory of reflection and transmission of light on the basis of Maxwell's equations has been developed by Drude (1902).

When a plane incident wave arrives at a material interface, a plane reflected wave and an in general inhomogeneous transmitted wave are produced. In each of these waves the field vectors are perpendicular to the respective wave vectors and can be completely represented by two components, implying that relations between the incident, reflected, and transmitted fields at a material interface can be described by 2 by 2 reflection and transmission matrices $\overline{\overline{R}}$ and $\overline{\overline{T}}$.

In isotropic media these matrices can be made diagonal by splitting up the fields into the components parallel and perpendicular to the so-called *plane of incidence*, which is the plane defined by the direction of incidence of the incoming wave and by the normal to the material interface (Figure 4.1). A fundamental feature of reflection and transmission of electromagnetic waves at isotropic and non-chiral material interfaces is that these components are completely decoupled in the sense that the parallel (perpendicular) component of an incident wave is related only to the parallel (perpendicular) components in the reflected and transmitted fields. If we denote the electric fields in the incident, reflected, and transmitted waves by $\mathbf{E}^{\mathrm{inc}}$, $\mathbf{E}^{\mathrm{ref}}$, and $\mathbf{E}^{\mathrm{tr}}$, and their parallel and perpendicular components by the symbols $\|$ and $\perp$, respectively, then the relations between the components can be expressed as

$$E_{\|}^{\mathrm{ref}} = R_{\|} E_{\|}^{\mathrm{inc}}, \quad E_{\perp}^{\mathrm{ref}} = R_{\perp} E_{\perp}^{\mathrm{inc}} \tag{4.1}$$

$$E_{\|}^{\mathrm{tr}} = T_{\|} E_{\|}^{\mathrm{inc}}, \quad E_{\perp}^{\mathrm{tr}} = T_{\perp} E_{\perp}^{\mathrm{inc}}, \tag{4.2}$$

[1] That is why the original Fresnel formulas do not involve ϵ and μ of the media but rather are written in terms of incidence and refraction angles and the velocities of waves in the respective media. Strictly speaking, these formulas only apply to interfaces between media with equal values of magnetic permeabilities (see section 4.2). However, since most of the natural and artificial media are non-magnetic, the Fresnel formulas, though based on a physically incorrect model of wave propagation in material media, correctly describe the great majority of practical situations. Many authors refer to virtually every set of equations representing the reflection and transmission coefficients of a material interface, even between magnetic materials, as Fresnel formulas.

Modern Electromagnetic Scattering Theory with Applications, First Edition. Andrey V. Osipov and Sergei A. Tretyakov.

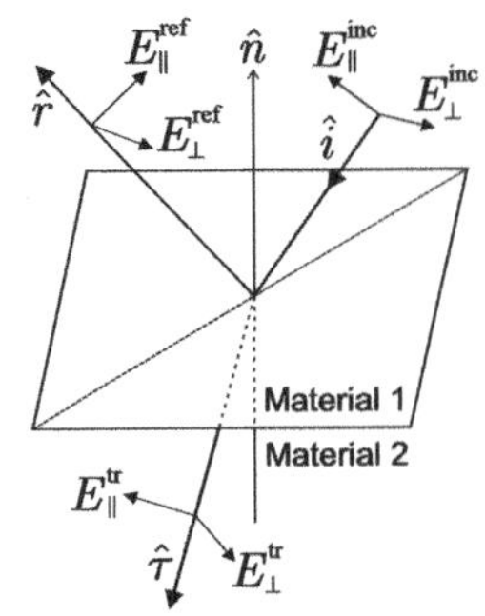

Figure 4.1 The plane of incidence is defined by the vectors $\hat{n}$ and $\hat{i}$. The parallel components of the electric fields in the incident, reflected, and transmitted waves lie in the plane of incidence. The perpendicular components are perpendicular to the plane of incidence.

implying that

$$\overline{\overline{R}} = \begin{pmatrix} R_{\perp} & 0 \\ 0 & R_{\parallel} \end{pmatrix}, \quad \overline{\overline{T}} = \begin{pmatrix} T_{\perp} & 0 \\ 0 & T_{\parallel} \end{pmatrix}. \tag{4.3}$$

The four factors $R_{\parallel}$, $R_{\perp}$, $T_{\parallel}$, and $T_{\perp}$, called the *reflection and transmission coefficients*, are sufficient to completely determine the relations between the complex amplitudes in the incident, reflected, and transmitted waves at a flat interface of isotropic materials.

The two fundamental polarization states of the electromagnetic fields in the problem are then the *perpendicular polarization*, when the wave is linearly polarized with its electric vector perpendicular to the plane of incidence, and the *parallel polarization*, when the the wave is linearly polarized with with its magnetic vector perpendicular to the plane of incidence (in which case the electric field vector lies in the plane of incidence). Alternatively, the perpendicular polarization is often called *transverse electric (TE)*, because in this case the electric field is purely transverse with respect to the plane of incidence, and the parallel polarization *transverse magnetic (TM)*, because the magnetic field is purely orthogonal to the plane of incidence. Still another name for the parallel (TM) polarization is *H polarization* because in this wave the only component of the magnetic field serves as a potential function from which all other components of the fields can be derived. Likewise, the perpendicular (TE) polarization is sometimes called *E polarization*.[2]

There are many possible definitions of reflection and transmission coefficients and great care should be taken when using results published by different authors. For example, the coefficients can be defined through different components of the electric fields or even through the magnetic fields, and this will result in different, but equivalent, definitions. Three definitions are most often encountered in the literature. The reflection and transmission coefficients $R_{\perp}$, $R_{\parallel}$, $T_{\perp}$, and $T_{\parallel}$ from (4.1) and (4.2) are ratios of the *electric* field components parallel or perpendicular to the plane of incidence; more formal definitions are given in section 4.2.6. Coefficients which will be denoted by R_{TE}, T_{TE}, R_{TM}, and T_{TM} relate the *tangential* to the interface components of the electric field in the incident, reflected and transmitted waves; these are defined in section 4.2.1. The quantities R_E, T_E, R_H, and T_H are used in section 4.3, and they relate the components of the electric fields perpendicular to the plane of incidence (R_E and T_E) in the case of a perpendicularly polarized incident wave and the components of the *magnetic* fields perpendicular to the plane of incidence (R_H and T_H) in the case of an incident wave with parallel polarization. All these definitions are equivalent; the relations between these variously defined reflection and transmission coefficients are given by (4.14), (4.15), (4.16), (4.86), (4.87), (4.88),

[2] It should be mentioned that in the theory of waveguides the polarization states are also referred to as TM/TE or *E/H*, but they are defined with respect to the axis of the waveguide and have, therefore, different meanings.

(4.89), (4.130), and (4.157). There are also further definitions possible and some of them are addressed in the problems section.

Reflection and transmission coefficients are, in general, complex-valued and describe amplitude and phase relations between the complex amplitudes of the incident, reflected, and transmitted waves at material interfaces. The energy relations between the energy flows transported by the waves to and from planar structures can be described by the closely related quantities of *reflectance* $\mathcal{R}$ and *transmittance* $\mathcal{T}$. In the particular case of a planar interface between two semi-infinite media, it is customary to call $\mathcal{R}$ and $\mathcal{T}$ *reflectivity* and *transmissivity*, in which case they describe the property of a material to reflect and transmit incident electromagnetic radiation (Born and Wolf 1959; International Commission on Illumination 2011).

In this book we consider solely isotropic materials, which permits description of reflection and transmission at the material interfaces by diagonal reflection and transmission matrices (4.3). The solution of the reflection and transmission problems consists then in determining the directions of reflection $\hat{r}$ and transmission $\hat{\tau}$[3] from the direction of incidence $\hat{i}$ and the normal $\hat{n}$ to the interface, splitting the vector $\mathbf{E}^{\text{inc}}$ into the components $E_{\parallel}^{\text{inc}}$ and $E_{\perp}^{\text{inc}}$, followed by multiplication of these with reflection and transmission coefficients as in (4.1) and (4.2).

In this chapter we derive solutions of the reflection and transmission problems, and present the reflection and transmission coefficients for various types of material interfaces, from the simplest case of an interface between two homogeneous half-spaces (section 4.2) to the most general case of a slab consisting of an arbitrary number of layers and separating two semi-infinite homogeneous media (section 4.3). Because of the constant progress in the development of artificial materials we do not limit our analysis to non-magnetic materials ($\mu = \mu_0$) or to lossless dielectrics, which are typical assumptions in many books on the subject, and derive expressions that allow arbitrary, in general, complex values of μ and ϵ, including artificial materials with negative values of μ and ϵ. We believe that the increased range of practical applications of these formulas compensates the slight increase in the complexity of the analysis. In section 4.2.6 we present a solution of the reflection and transmission problems at an interface of two homogeneous media in a coordinate-independent, vector form, which is well suited for use in ray-tracing algorithms. Material interfaces can support a specific type of inhomogeneous waves, surface waves, which are concentrated near the interface, and we discuss this mechanism of wave propagation with an emphasis on the case of metamaterial media in section 4.2.5. Reflection and transmission of cylindrical and spherical waves are studied in section 4.4. Section 4.5 is devoted to the case of a single homogeneous layer separating two semi-infinite materials, and we study the most important features of this practically important structure (section 4.5.1), including the special cases of a PEC-backed homogeneous layer (section 4.5.2) and a homogeneous layer in free space (section 4.5.3). With negative values of permittivity and permeability in the layer, a planar structure can act as a collecting lens (superlens), and we discuss this interesting configuration in section 4.5.3. The chapter concludes with an analysis of the accuracy of impedance boundary conditions (section 4.6) and a problems section.

4.2 Interface of Two Homogeneous Semi-infinite Media

4.2.1 Reflection and Transmission Coefficients

Let us consider a planar interface between two isotropic media characterized by scalar material parameters $\epsilon_{1,2}$ and $\mu_{1,2}$. The geometry of the problem of plane-wave incidence on the interface is illustrated in Figure 4.2.

Let us assume that the interface is excited from medium 1 by a plane wave, and find reflected and transmitted fields. As discussed above, if the incident wave is a linearly polarized wave of one of the two fundamental polarizations with respect to the incidence plane, the reflected and transmitted fields form plane waves of the same polarization as that of the incident plane wave. Because a plane wave of an arbitrary polarization can be presented as a linear combination of the TE- and TM-polarized components, it is enough to find the reflection and transmission coefficients $R_{\text{TE,TM}}$ and $T_{\text{TE,TM}}$ for these two basic polarizations.

[3] In contrast to $\hat{i}$ and $\hat{r}$, the vector $\hat{\tau}$ is complex-valued when the media are lossy or total internal reflection takes place, in which case it has no immediate geometric interpretation.

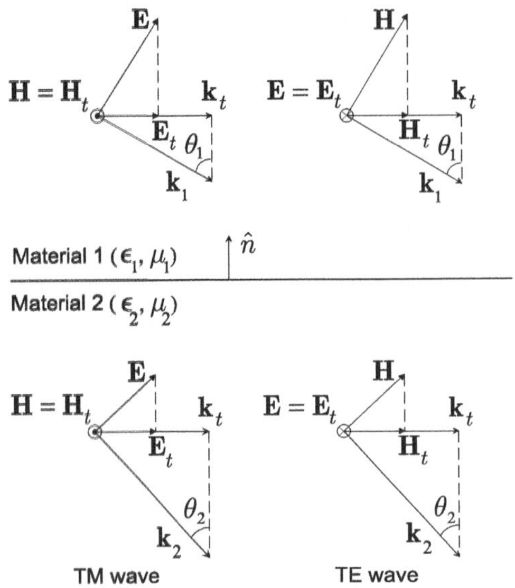

Figure 4.2 Interface between two isotropic media and plane waves in both media.

To do this, it is convenient to use the general relations (2.82) and (2.83) between the tangential to the interface field components of arbitrary plane waves in both media. The coefficients (plane-wave impedances) in (2.82) and (2.83) can be conveniently written in the following forms:

$$Z\frac{\sqrt{k^2-k_t^2}}{k} = \frac{\sqrt{k^2-k_t^2}}{\omega\epsilon} = \frac{\gamma}{\omega\epsilon} = \kappa^H, \qquad \text{TM modes} \tag{4.4}$$

$$Z\frac{k}{\sqrt{k^2-k_t^2}} = \frac{\omega\mu}{\sqrt{k^2-k_t^2}} = \frac{\omega\mu}{\gamma} = \frac{1}{\kappa^E}, \qquad \text{TE modes,} \tag{4.5}$$

where $\gamma = \sqrt{k^2 - k_t^2}$. These notations will also be useful in analysing properties of multilayered structures in the next section. The relations between the tangential fields (2.82) and (2.83) take the form

$$\mathbf{E}_t = \kappa^H\, \hat{n} \times \mathbf{H}_t, \qquad \text{TM modes} \tag{4.6}$$

$$\mathbf{E}_t = \frac{1}{\kappa^E}\, \hat{n} \times \mathbf{H}_t, \qquad \text{TE modes.} \tag{4.7}$$

To find the reflection and transmission coefficients we will next write these relations for the fields right at the interface, and make use of the continuity of the tangential field components. Assuming the unit amplitude of the tangential electric field component of the incident wave, and denoting the reflection coefficient as R, we have in the first medium $E_t = 1 + R$. In the second medium we have only one transmitted wave, whose amplitude we denote as $E_t = T$. Continuity of the tangential electric fields demands

$$1 + R = T. \tag{4.8}$$

To write the continuity condition for tangential magnetic fields we use (4.6) and (4.7). Writing equations for the TE polarization, and equating the amplitudes of tangential magnetic fields, we have

$$\kappa_1^E(1 - R) = \kappa_2^E T, \tag{4.9}$$

where indices 1 and 2 denote the first and the second medium ($\gamma_{1,2} = \sqrt{k_{1,2}^2 - k_t^2}$, etc.). For the reflected wave the unit vector has the opposite direction, so we have changed the sign of the second term on the left-hand side of the equation.

Solving (4.8) and (4.9) together, we find the reflection and transmission coefficients:

$$R_{\mathrm{TE}} = \frac{\kappa_1^E - \kappa_2^E}{\kappa_1^E + \kappa_2^E}, \quad T_{\mathrm{TE}} = \frac{2\kappa_1^E}{\kappa_1^E + \kappa_2^E}. \tag{4.10}$$

We can find the solution for the other (TM) polarization by replacing $\kappa^E \to 1/\kappa^H$. The result reads

$$R_{\mathrm{TM}} = \frac{\kappa_2^H - \kappa_1^H}{\kappa_1^H + \kappa_2^H}, \quad T_{\mathrm{TM}} = \frac{2\kappa_2^H}{\kappa_1^H + \kappa_2^H}. \tag{4.11}$$

The above formulas express the reflection and transmission coefficients in terms of the material parameters of the two media $\epsilon_{1,2}$ and $\mu_{1,2}$, the frequency, and the tangential component of the wavenumber k_t (see the definitions of parameters $\kappa^{E,H}$ in (4.4) and (4.5)).

The surface impedance of the interface, which is defined as the ratio of the tangential components of the electric and magnetic fields at the surface, can be expressed through the reflection coefficients. For the TE-polarized wave the tangential components on the upper side of the interface are given as in the left-hand sides of (4.8) and (4.9), leading to the expression

$$Z_E = \frac{1}{\kappa_1^E} \frac{1 + R_{\mathrm{TE}}}{1 - R_{\mathrm{TE}}} = \frac{1}{\kappa_2^E}. \tag{4.12}$$

For the TM polarization the impedance is obtained by using the symmetry $\kappa^E \to 1/\kappa^H$,

$$Z_H = \kappa_1^H \frac{1 + R_{\mathrm{TM}}}{1 - R_{\mathrm{TM}}} = \kappa_2^H. \tag{4.13}$$

Note that Z_E and Z_H are in general different (polarization dependent); the only exception is the case of normal incidence ($k_t = 0$), in which $Z_E = Z_H$ as $\kappa_2^H = 1/\kappa_2^E$.

The reflection and transmission coefficients in (4.10) and (4.11) have been defined as ratios of tangential components of the incident, reflected, and transmitted electric fields (E_t^{inc}, E_t^{ref}, and E_t^{tr}) at the material interface,

$$\frac{E_t^{\mathrm{ref}}}{E_t^{\mathrm{inc}}} = \begin{cases} R_{\mathrm{TE}}, & \text{TE case} \\ R_{\mathrm{TM}}, & \text{TM case} \end{cases}, \quad \frac{E_t^{\mathrm{tr}}}{E_t^{\mathrm{inc}}} = \begin{cases} T_{\mathrm{TE}}, & \text{TE case} \\ T_{\mathrm{TM}}, & \text{TM case} \end{cases}. \tag{4.14}$$

Depending on the polarization, these ratios take different values, which is indicated in (4.14) by the subscripts TE and TM.

An alternative definition of reflection and transmission coefficients uses ratios of the complex amplitudes of the full electric fields denoted by $E_\parallel^{\mathrm{inc}}$, $E_\parallel^{\mathrm{ref}}$, and $E_\parallel^{\mathrm{tr}}$ when the electric field lies in the plane of incidence (Figure 4.3), and by $E_\perp^{\mathrm{inc}}$, $E_\perp^{\mathrm{ref}}$, and $E_\perp^{\mathrm{tr}}$ when the electric field is perpendicular to the plane of incidence,

$$R_\parallel = \frac{E_\parallel^{\mathrm{ref}}}{E_\parallel^{\mathrm{inc}}}, \quad T_\parallel = \frac{E_\parallel^{\mathrm{tr}}}{E_\parallel^{\mathrm{inc}}} \tag{4.15}$$

$$R_\perp = \frac{E_\perp^{\mathrm{ref}}}{E_\perp^{\mathrm{inc}}}, \quad T_\perp = \frac{E_\perp^{\mathrm{tr}}}{E_\perp^{\mathrm{inc}}}. \tag{4.16}$$

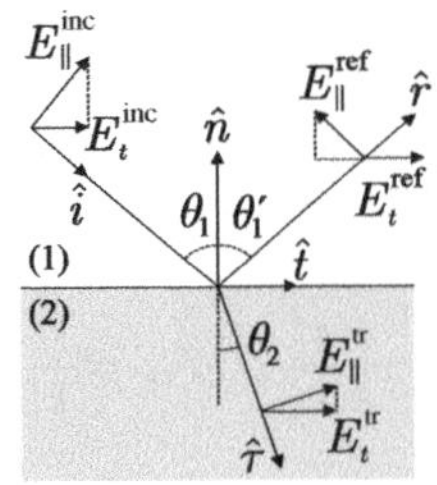

Figure 4.3 Definition of $E_{\parallel}^{\rm inc}$, $E_{\parallel}^{\rm ref}$, and $E_{\parallel}^{\rm tr}$ and their relation to $E_t^{\rm inc}$, $E_t^{\rm ref}$, and $E_t^{\rm tr}$.

In the case of perpendicular (TE) polarization, the electric fields are tangential to the interface and at the same time perpendicular to the plane of incidence, which implies that

$$R_{\perp} = R_{\rm TE}, \quad T_{\perp} = T_{\rm TE}. \tag{4.17}$$

In the case of parallel (TM) polarization, however, the electric fields have also a component normal to the interface, and if the fields are directed as shown in Figure 4.3, then it follows from simple geometric considerations that

$$R_{\parallel} = -R_{\rm TM}, \quad T_{\parallel} = T_{\rm TM}\frac{\cos\theta_1}{\cos\theta_2}, \tag{4.18}$$

where θ_1 is the angle of incidence and θ_2 is the angle of refraction. Both approaches to the definition of the reflection and transmission coefficients, as in (4.14) or as in (4.15) and (4.16), are apparently equivalent and the choice is mainly a matter of convenience.

The incidence, reflection and refraction angles θ_1, θ_1', and θ_2 shown in Figure 4.3 are related to each other by the laws of reflection

$$\theta_1' = \theta_1 \tag{4.19}$$

and refraction (Snell's law)

$$\theta_2 = \arcsin\left(\sqrt{\frac{\epsilon_1\mu_1}{\epsilon_2\mu_2}}\sin\theta_1\right). \tag{4.20}$$

These well-known relations follow from the condition that the tangential components of the wave vectors in the incident, reflected, and refracted waves must be the same in every wave, that is,

$$k_t = k_1\sin\theta_1 = k_1\sin\theta_1' = k_2\sin\theta_2, \tag{4.21}$$

to ensure continuity of the tangential field components at the material interface and, therefore, the fulfillment of the boundary conditions (2.55). The reflection and refraction laws are equally applicable to lossless and lossy media; in the later case, however, the angle θ_2 is complex-valued and no longer has the simple meaning of the angle of refraction, though it will still be referred to as such.

The angles define the unit vectors

$$\hat{i} = \hat{t}\sin\theta_1 - \hat{n}\cos\theta_1 \tag{4.22}$$
$$\hat{r} = \hat{t}\sin\theta_1 + \hat{n}\cos\theta_1 \tag{4.23}$$
$$\hat{\tau} = \hat{t}\sin\theta_2 - \hat{n}\cos\theta_2 \tag{4.24}$$

that describe the directions of incidence, reflection, and refraction (Figure 4.3). In the case where the lower medium is lossy, the vector τ is complex-valued[4] and the refracted wave is, therefore, an inhomogeneous wave (see problem 13 in

[4] Note that τ is a unit vector in the sense that $\hat{\tau}\cdot\hat{\tau} = 1$, but $|\hat{\tau}| \neq 1$.

section 2.7). The reflected wave is a plane wave, even when the upper medium is lossy, as long as the incidence angle θ_1 is real-valued and the incident wave is a plane wave. The reflected and refracted waves are linearly polarized provided that the incidence wave is polarized entirely parallel or perpendicular to the plane of incidence. We prove this general property in section 4.3 for more general material interfaces (see also problem 2).

It is often convenient to write formulas (4.10) and (4.11) in terms of the incidence and refraction angles. This is done by writing

$$\kappa^H_{1,2} = Z_{1,2}\frac{\sqrt{k^2_{1,2} - k^2_t}}{k_{1,2}} = Z_{1,2}\cos\theta_{1,2} \tag{4.25}$$

and, similarly,

$$\frac{1}{\kappa^E_{1,2}} = Z_{1,2}\frac{k_{1,2}}{\sqrt{k^2_{1,2} - k^2_t}} = Z_{1,2}\frac{1}{\cos\theta_{1,2}}. \tag{4.26}$$

Substituting into (4.10) and (4.11) gives

$$R_{\mathrm{TE}} = \frac{Z_2\cos\theta_1 - Z_1\cos\theta_2}{Z_2\cos\theta_1 + Z_1\cos\theta_2} \tag{4.27}$$

$$R_{\mathrm{TM}} = \frac{Z_2\cos\theta_2 - Z_1\cos\theta_1}{Z_2\cos\theta_2 + Z_1\cos\theta_1} \tag{4.28}$$

$$T_{\mathrm{TE}} = \frac{2Z_2\cos\theta_1}{Z_2\cos\theta_1 + Z_1\cos\theta_2} \tag{4.29}$$

$$T_{\mathrm{TM}} = \frac{2Z_2\cos\theta_2}{Z_1\cos\theta_1 + Z_2\cos\theta_2}. \tag{4.30}$$

Equations (4.10), (4.11), and (4.27)–(4.28) are applicable to media with arbitrary values of ϵ and μ. When $\mu_1 = \mu_2$ (e.g., both media are non-magnetic), the formulas can be reduced to the Fresnel formulas, obtained under the assumption that the speed of waves is a single parameter that characterizes the propagation of waves in a medium (Fresnel 1823). In this case, $Z_1/Z_2 = k_2/k_1$ and the law of refraction (4.20) helps to exclude the free space impedances from (4.27)–(4.30) and express the reflection and transmission coefficients entirely in terms of the incidence and reflection angles (see problem 1),

$$R_{\mathrm{TE}} = \frac{\sin(\theta_2 - \theta_1)}{\sin(\theta_1 + \theta_2)} \tag{4.31}$$

$$R_{\mathrm{TM}} = \frac{\tan(\theta_2 - \theta_1)}{\tan(\theta_1 + \theta_2)} \tag{4.32}$$

$$T_{\mathrm{TE}} = \frac{2\cos\theta_1\sin\theta_2}{\sin(\theta_1 + \theta_2)} \tag{4.33}$$

$$T_{\mathrm{TM}} = \frac{2\cos\theta_2\sin\theta_2}{\sin(\theta_1 + \theta_2)\cos(\theta_2 - \theta_1)}. \tag{4.34}$$

The important case of a PEC interface corresponds to the limit $|\epsilon_2| \to \infty$ with μ_2 finite, and because of the relations

$$Z_2 = \sqrt{\frac{\mu_2}{\epsilon_2}} \to 0, \quad \theta_2 \to 0, \tag{4.35}$$

(4.27)–(4.30) reduce to

$$R_{\mathrm{TE}} \to -1, \quad T_{\mathrm{TE}} \to 0, \quad R_{\mathrm{TM}} \to -1, \quad T_{\mathrm{TM}} \to 0. \tag{4.36}$$

At normal incidence ($\theta_1 = \theta_2 = 0$), equations (4.27) and (4.28) give equal values for the reflection coefficients

$$R_{\mathrm{TM}} = R_{\mathrm{TE}} = \frac{Z_2 - Z_1}{Z_2 + Z_1}, \tag{4.37}$$

which is in agreement with the simple consideration that at normal incidence, reflection from any flat and isotropic interface must be polarization independent. It is seen from (4.37) that the reflected field vanishes if the impedances of both media are matched, $Z_1 = Z_2$, implying perfect transmission into the lower semi-infinite medium, even when $k_2 \neq k_1$.

In contrast to reflection and transmission coefficients, which relate complex amplitudes of the incident, reflected, and refracted fields, the reflectivity $\mathcal{R}$ and transmissivity $\mathcal{T}$ describe relations between energy flows at a material interface. They are defined as ratios of the energy flows in the incident, reflected, and transmitted waves in the direction perpendicular to the interface (Born and Wolf 1959),

$$\mathcal{R} = \left| \frac{\mathbf{w}_{\mathrm{ref}} \cdot \hat{n}}{\mathbf{w}_{\mathrm{inc}} \cdot \hat{n}} \right| \tag{4.38}$$

$$\mathcal{T} = \left| \frac{\mathbf{w}_{\mathrm{tr}} \cdot \hat{n}}{\mathbf{w}_{\mathrm{inc}} \cdot \hat{n}} \right|. \tag{4.39}$$

For a linearly (either parallel or perpendicularly) polarized incident wave, the energy flows at a point of reflection result from (2.160) and (2.386) as

$$\mathbf{w}_{\mathrm{inc}} = \hat{\imath}\,\mathrm{Re}\left(Z_1^{-1}\right) |E_0|^2 \tag{4.40}$$

$$\mathbf{w}_{\mathrm{ref}} = \hat{r}\,\mathrm{Re}\left(Z_1^{-1}\right) |R|^2 |E_0|^2 \tag{4.41}$$

$$\mathbf{w}_{\mathrm{tr}} = \mathrm{Re}\left(\hat{t} Z_2^{-1}\right) |T|^2 |E_0|^2, \tag{4.42}$$

where E_0 is the amplitude of the incident wave at the point of reflection and R and T denote, depending on the polarization, the reflection and transmission coefficients (4.15) or (4.16). Using the relations

$$-\hat{\imath} \cdot \hat{n} = \hat{r} \cdot \hat{n} = \cos\theta_1, \quad -\hat{t} \cdot \hat{n} = \cos\theta_2, \tag{4.43}$$

which follow from (4.22)–(4.24), and the relation $\mathrm{Re}\,(Z_1^{-1}) = \mathrm{Re}\,Z_1/|Z_1|^2$, valid for every complex-valued number Z_1, we arrive at the formulas:

$$\mathcal{R} = |R|^2 \tag{4.44}$$

$$\mathcal{T} = \mathrm{Re}\left(\frac{\cos\theta_2}{Z_2}\right) \frac{|Z_1|^2}{\cos\theta_1 \mathrm{Re}\, Z_1} |T|^2. \tag{4.45}$$

In the case of lossless media, expression (4.45) simplifies to

$$\mathcal{T} = \frac{Z_1 \cos\theta_2}{Z_2 \cos\theta_1} |T|^2, \tag{4.46}$$

and since there is no energy dissipation in the media, we expect that

$$\mathcal{R} + \mathcal{T} = 1. \tag{4.47}$$

This is indeed so, and (4.47) can be derived by direct substitution of the reflection and transmission coefficients in (4.44) and (4.46) (see problem 3). In section 4.3 we show that (4.47) is valid for general lossless planar stratified media.

4.2.2 Brewster's Angle

When impedances Z_1 and Z_2 are different, it follows from (4.27) and (4.28) that reflection coefficients R_{TE} and R_{TM} may still vanish if

$$Z_2 \cos\theta_1 = Z_1 \cos\theta_2, \tag{4.48}$$

$$Z_2 \cos\theta_2 = Z_1 \cos\theta_1, \tag{4.49}$$

respectively, or in terms of the incidence angle θ_1 and the constitutive parameters,

$$\frac{\mu_2}{\mu_1}\cos\theta_1 = \sqrt{\frac{\epsilon_2\mu_2}{\epsilon_1\mu_1} - \sin^2\theta_1} \tag{4.50}$$

$$\frac{\epsilon_2}{\epsilon_1}\cos\theta_1 = \sqrt{\frac{\epsilon_2\mu_2}{\epsilon_1\mu_1} - \sin^2\theta_1}. \tag{4.51}$$

In lossless media, equations (4.50) and (4.51) may have real-valued solutions

$$\theta_{\mathrm{B}}^{\mathrm{TE}} = \arcsin\sqrt{\frac{(\mu_2/\mu_1)^2 - \epsilon_2\mu_2/(\epsilon_1\mu_1)}{(\mu_2/\mu_1)^2 - 1}}, \tag{4.52}$$

$$\theta_{\mathrm{B}}^{\mathrm{TM}} = \arcsin\sqrt{\frac{(\epsilon_2/\epsilon_1)^2 - \epsilon_2\mu_2/(\epsilon_1\mu_1)}{(\epsilon_2/\epsilon_1)^2 - 1}}, \tag{4.53}$$

respectively. The angles $\theta_{\mathrm{B}}^{\mathrm{TE}}$ and $\theta_{\mathrm{B}}^{\mathrm{TM}}$ are called *Brewster's angles* and these are the incidence angles at which no reflection of correspondingly polarized plane electromagnetic waves occurs from a material interface. Furthermore, under the conditions (4.48) and (4.49) equations (4.29) and (4.30) give $T_{\mathrm{TE}} = 1$ and $T_{\mathrm{TM}} = 1$, implying that a linearly (TE or TM) polarized plane wave incident at a corresponding Brewster's angle ($\theta_{\mathrm{B}}^{\mathrm{TE}}$ or $\theta_{\mathrm{B}}^{\mathrm{TM}}$) on a material interface is completely transmitted through the interface. It can be shown (problem 5) that for a given material interface there exists only one Brewster's angle, either $\theta_{\mathrm{B}}^{\mathrm{TE}}$ or $\theta_{\mathrm{B}}^{\mathrm{TM}}$, but not both simultaneously. The sole exception is the case of normal incidence on the impedance-matched interface discussed above.

Vanishing of reflection from a material interface at a specific angle is an inherent feature of materials with low losses. The Brewster effect is mostly observed for the TM (parallel) polarization and only seldom for the TE (perpendicular) polarization because most of the materials are non-magnetic (with μ equal to the permeability of vacuum), in which case $\mu_1 = \mu_2$ and equation (4.52) predicts an infinite value of $\theta_{\mathrm{B}}^{\mathrm{TE}}$.

Figure 4.4 shows the absolute values of the reflection and transmission coefficients of an interface between air and Plexiglas illuminated from the air as functions of the incidence angle θ_1 at 10 GHz. The dielectric and magnetic constants of the media are taken from Table 2.1. The curve for $|R_{\mathrm{TM}}|$ has a minimum near $\theta_1 = 58.1°$; without absorption in Plexiglas, this minimum would be exactly zero with the corresponding Brewster's angle given by (4.53). The corresponding transmission coefficient $|T_{\mathrm{TM}}|$ is unity at this point. It should be noted that in contrast to reflection coefficients, transmission coefficients may take values greater than unity. This does not violate energy conservation since the energy conservation condition involves not the transmission coefficient itself but rather a related quantity, transmissivity, defined in (4.45), which never exceeds unity.

The Brewster effect is also observed at optical frequencies. This is illustrated in Figure 4.5, which shows the reflection coefficients of an air/water interface at $\lambda = 0.55$ μm (green light). The TM reflection coefficient vanishes almost exactly at $\theta_1 = 53.1°$ since absorption in water is extremely low in the visible (see Table 2.1).

4.2.3 Total Internal Reflection

The phenomenon of total internal reflection occurs when both media are lossless and $\epsilon_1\mu_1 > \epsilon_2\mu_2$. Then for incidence angles such that $\theta_1 > \bar{\theta}$ with

$$\bar{\theta} = \arcsin\sqrt{\frac{\epsilon_2\mu_2}{\epsilon_1\mu_1}}, \tag{4.54}$$

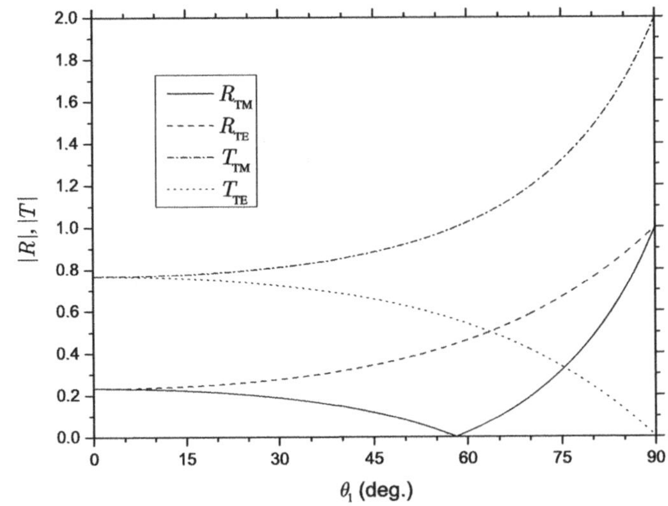

Figure 4.4 Reflection and transmission coefficients for an interface between air and Plexiglas at $f = 10$ GHz.

the refraction law (4.20) leads to a complex-valued refraction angle

$$\theta_2 = \frac{\pi}{2} \pm j\text{arccosh}\left(\sin\theta_1\sqrt{\frac{\epsilon_2\mu_2}{\epsilon_1\mu_1}}\right). \tag{4.55}$$

Under this condition, $\cos\theta_2$ is purely imaginary, and it follows from (4.27) and (4.28) that R_{TE} and R_{TM} are ratios of two complex-conjugated quantities and therefore

$$|R_{\text{TE}}| = |R_{\text{TM}}| = 1. \tag{4.56}$$

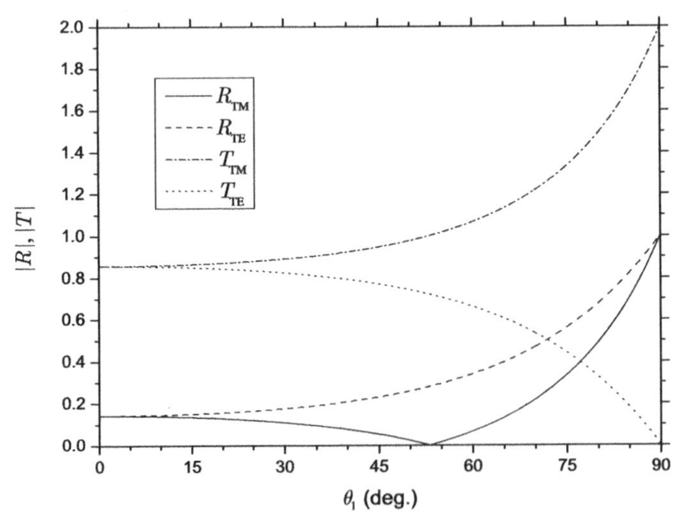

Figure 4.5 Reflection and transmission coefficients for an interface between air and water at an optical frequency ("green" light with $\lambda = 0.55\ \mu$m).

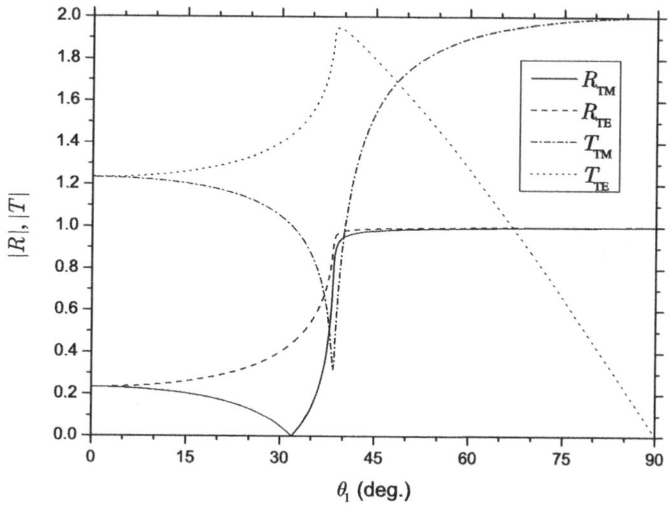

Figure 4.6 The same as Figure 4.4 but for incidence from Plexiglas.

Equations (4.56) in conjunction with (4.44) imply that

$$\mathcal{R}_{\rm TE} = \mathcal{R}_{\rm TM} = 1, \tag{4.57}$$

and the incident energy flow is completely reflected back from the interface (total reflection). In contrast to the Brewster effect, the total reflection is polarization independent and simultaneously involves both polarizations. This may lead to a confinement of electromagnetic energy flows within material layers, and the use of glass fibers for canalization and transmission of light is a prominent example of that.

It is worth pointing out that the transmission coefficients $T_{\rm TE}$ and $T_{\rm TM}$ do not vanish under the condition of total reflection and there exists a refracted wave with the complex-valued wave vector[5]

$$k_2 = k_1 \sin\theta_1 \hat{t} + jk_2 |\cos\theta_2| \hat{n}. \tag{4.58}$$

However, the energy flow

$$\mathbf{w}_{\rm tr} = \hat{t}\, |E_0 T|^2 \frac{\mu_1 \sin\theta_1}{2\mu_2 Z_1} \tag{4.59}$$

associated with this wave (see problems 6 and 7) is directed along the interface and, thus, does not take away the energy of the incident wave. Subsequently, in contrast to the reflection coefficient, the transmissivity $\mathcal{T}$ (which according to its definition (4.39) is proportional to the component of the energy flow normal to the interface) vanishes, and (4.57) represents the energy conservation at the interface in the case of total internal reflection.

Figures 4.6 and 4.7 show the behavior of the reflection and transmission coefficients for the same material interfaces as in Figures 4.4 and 4.5 but with the materials 1 and 2 interchanged so that now the plane wave is incident from the side of Plexiglas and water, respectively. The phenomenon of total reflection is clearly seen for $\theta_1 > 39°$ in Figure 4.6 and $\theta_1 > 48.6°$ in Figure 4.7. In addition to the total reflection, $R_{\rm TM}$ has also a Brewster angle at $\theta_1 \approx 31.8°$ (Plexiglass/air interface) and at $\theta_1 \approx 36.9°$ (water/air interface). The latter explains the pronounced polarization differences observed in rainbows, which we study in section 7.5.

In section 4.3.2 we study the phenomenon of the total internal reflection for more general planar interfaces.

---

[5] This wave can be classified as an inhomogeneous plane wave.

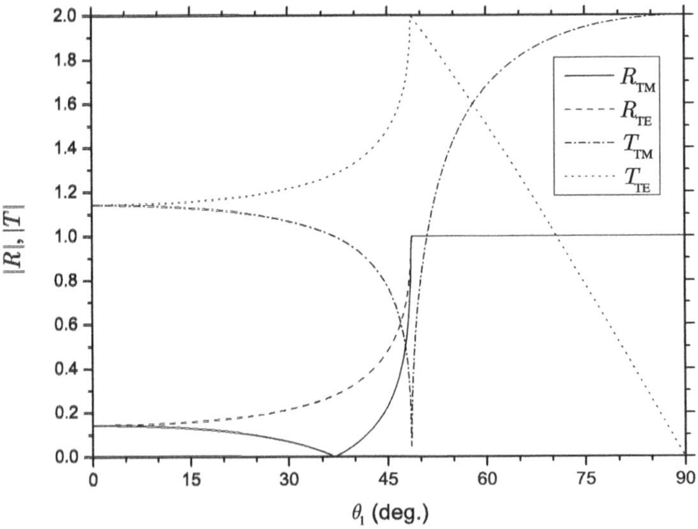

Figure 4.7 The same as Figure 4.5 but for incidence from water.

4.2.4 Interfaces with Double-negative Materials

Finally, we note interesting properties of planar interfaces between the usual isotropic materials and isotropic materials with negative real parts of permittivity and permeability. Such media, which belong to the class of metamaterials, are known as double-negative (DNG) or left-handed materials. They are not encountered in nature, but several types of artificial composite materials may be attributed effective values of ϵ and μ whose real parts are both negative in a specific, typically rather narrow, range of frequencies. They can be realized, for example, by including metal resonant elements in the form of split rings into electrically dense periodical lattice of metal wires embedded in a dielectric material or by using meshes of transmission lines with suitably chosen values of capacitance and inductance of the individual cells. For an overview of the electromagnetics of DNG media and further references see Eleftheriades and Balman (2005), Caloz and Itoh (2006), Engheta and Ziolkowski, eds. (2006), and Shalaev and Sarychev (2007).

In the following discussion of the interface properties we do not take dissipation into account and assume the values of the permittivity and permeability to be real. A DNG medium has a negative index of refraction $n = \sqrt{\epsilon\mu} = -\sqrt{|\epsilon\mu|}$, which leads to a negative transmission angle at an interface of a double-positive material and a double-negative material (Figure 4.8). The impedance $Z = \sqrt{\mu/\epsilon}$ of a DNG medium is, however, positive. The choice of the sign of the square-root functions follows from an analytic continuation of ϵ and μ in which arguments $\arg\epsilon$ and $\arg\mu$ are decreased from 0 to $-\pi$.

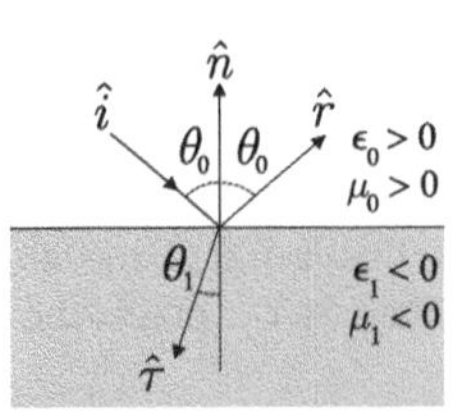

Figure 4.8 Transmission and reflection at an interface of a DNG medium. The arrows show the direction of the power flow (the Poynting vector).

From the physical point of view, the negative refraction follows from the fact that the DNG medium is a backward-wave medium: the directions of the Poynting vector and the wave vector in an isotropic DNG medium are antiparallel. This is directly seen from the expression for the Poynting vector of plane waves in isotropic media, (2.160) and (2.161).

Considering the reflection and refraction of propagating plane waves at an interface of free space and a DNG medium (Figure 4.8), we have to choose the eigensolution in the DNG medium which has the Poynting vector directed from the source. Because the second medium is a backward-wave medium, the normal component of the wavevector is directed towards the interface, along the unit vector $\hat{n}$. This defines the negative refraction angle as shown in Figure 4.8. To calculate the values of the reflection and transmission coefficients, we can use formulas (4.10) and (4.11). Considering, for instance, the TM-polarized case (conclusions will be the same for TE waves), we write

$$R_{\rm TM} = \frac{\kappa_1^H - \kappa_0^H}{\kappa_1^H + \kappa_0^H}, \qquad T_{\rm TM} = \frac{2\kappa_1^H}{\kappa_1^H + \kappa_0^H}. \tag{4.60}$$

Here, in the definition of κ^H (see (4.4))

$$\kappa^H = \frac{\sqrt{k^2 - k_t^2}}{\omega\epsilon} = \frac{\gamma}{\omega\epsilon} \tag{4.61}$$

the parameter γ is the normal component of the wave vector, and its sign should be chosen in such a way that the solution satisfies the physical conditions at infinity. Let us consider the special case when $\epsilon = -\epsilon_0$ and $\mu = -\mu_0$. For propagating modes, as we have just discussed, the DNG medium is a backward-wave material, thus, the sign of the normal component of the propagation constant γ changes sign, $\gamma = -\gamma_0$, because the sign of the normal component of the Poynting vector is determined only by the direction of the power flow from the source: it is the same in both media. Thus, because both γ and the permittivity change sign, $\kappa_1^H = \kappa_0^H$, and the reflection coefficient vanishes for all incidence angles. The transmission coefficient equals unity, also for all propagating waves.

The situation is completely different for evanescent incident fields. In this case, $k_t^2 > k^2$ and the normal component of the propagation constant γ is imaginary. Evanescent fields do not transport power, and the choice of the correct eigenwave in the second medium should be done so that the fields would decay away from the interface and vanish at infinity. This means that for evanescent fields γ does not change sign, and for the special case of $\epsilon = -\epsilon_0$ and $\mu = -\mu_0$ we have $\kappa_1^H = -\kappa_0^H$. The reflection and transmission coefficients tend to infinity for all transverse wave numbers corresponding to evanescent fields ($|k_t| > |k|$), which means that this interface supports surface modes with arbitrary wavenumbers (see the following section for more details).

4.2.5 *Surface Waves*

Surface waves are modes that propagate along surfaces so that their fields are localized near these surfaces. If the surface is an interface between two media, the fields of a surface wave in both contacting media exponentially decay away from the interface. Let us first study surface waves on planar interfaces between two different isotropic media characterized by material parameters $\epsilon_{1,2}$ and $\mu_{1,2}$, where the indices refer to two contacting media. To derive the dispersion equation for surface modes we make use of the general relations (4.6) and (4.7) between the tangential to the interface field components of arbitrary plane waves in both media. The tangential electric field components must be continuous on the interface. On the other hand, the fields of an eigenwave are nonzero, which brings us to the following simple eigenmode equations:

$$\kappa_1^H + \kappa_2^H = 0, \quad \kappa_1^E + \kappa_2^E = 0. \tag{4.62}$$

Note that the same result follows from equating the denominators of the reflection coefficients from the same interface (4.10) and (4.11) to zero. This is because the eigenmode propagating along the surface has nonzero fields in the absence of the incident field. Another conclusion from this result is that if the incident wave is in phase synchronism with a surface mode, the amplitude of the induced field tends to infinity (for an interface of infinite extent).

Writing (4.62) explicitly in terms of the material parameters of the two media we arrive at

$$\frac{\sqrt{\omega^2\epsilon_1\mu_1 - k_t^2}}{\epsilon_1} + \frac{\sqrt{\omega^2\epsilon_2\mu_2 - k_t^2}}{\epsilon_2} = 0 \tag{4.63}$$

and

$$\frac{\mu_1}{\sqrt{\omega^2\epsilon_1\mu_1 - k_t^2}} + \frac{\mu_2}{\sqrt{\omega^2\epsilon_2\mu_2 - k_t^2}} = 0. \tag{4.64}$$

Here the first equation corresponds to surface waves of TM polarization with respect to the propagation direction and to the normal to the interface. The only nonzero magnetic field component of this wave is parallel to the interface and orthogonal to the propagation direction. The second equation corresponds to TE modes.

The square roots in the above equations are equal to the normal to the interface components of the propagation factor: $k_{n1,2} = \gamma_{1,2} = \sqrt{\omega^2\epsilon_{1,2}\mu_{1,2} - k_t^2}$. If we are interested in surface-wave solutions that are bound to the interface, both these numbers should have nonzero imaginary parts $\text{Im}\, k_{n1,2} < 0$, so that the fields exponentially decay in both media.

Eigenwaves in the strict mathematical sense can exist only in lossless media, since otherwise fields decay along the propagation direction. Considering this case, we assume that the material parameters are real numbers. If there is a surface wave along the interface (the propagation factor k_t is a real number), both square roots in (4.63) and (4.64) are imaginary numbers of the same sign. Thus, non-trivial solutions of the eigenwave equations (4.63) and (4.64) can exist only if the material parameters of the two media have different signs, that is, if ϵ_1 is positive, then ϵ_2 should be negative, and the other way around. The same holds for μ in the case of TE-polarized waves.

Most natural materials have constitutive parameters with positive real parts. However, materials with negative parameters exist. In the microwave region, free-electron plasma of the ionosphere has permittivity whose real part is negative and losses can be small. Permeability tensor of magnetized ferrites has components that take negative values close to the ferrimagnetic resonant frequency. In the visible part of the spectrum, noble metals are characterized by permittivity that has a negative real part and moderate losses. Surface modes exist on interfaces of such media and free space. These modes at optical frequencies are often called plasmons.

On the other hand, surface waves on an interface between metal and free space in the microwave frequency range are very weakly bound to the surface. Indeed, metals at microwaves are highly conductive and can be modeled by complex permittivity which is very large in the absolute value. With $|\epsilon_2| \to \infty$ we have from (4.63) $\sqrt{\omega^2\epsilon_1\mu_1 - k_t^2}/\epsilon_1 \approx 0$, that is, $k_t \approx \omega\sqrt{\epsilon_1\mu_1}$. This solution corresponds to a usual plane wave in medium 1 propagating along the surface. Its electric field is normal to the interface with metal. If both material parameters, the permittivity and permeability, are negative in some frequency range, surface waves of both polarizations can travel along an interface of such a material and free space.

To study waves along an interface between an isotropic material (usually free space) and a surface characterized by an impedance boundary condition (2.72)

$$\mathbf{E}_t = Z_s\hat{n} \times \mathbf{H}, \tag{4.65}$$

where Z_s is the surface impedance, similar reasoning leads to the eigenwave equations

$$\frac{\sqrt{\omega^2\epsilon_1\mu_1 - k_t^2}}{\omega\epsilon_1} + Z_s = 0 \tag{4.66}$$

and

$$\frac{\omega\mu_1}{\sqrt{\omega^2\epsilon_1\mu_1 - k_t^2}} + Z_s = 0. \tag{4.67}$$

Because for surface-wave solutions the square roots in the above formulas take negative imaginary values (so that the wave decays away from the surface), we see that TM surface waves (see (4.66)) can propagate only over reactive surfaces with positive reactance (e.g., inductive impedance $Z_s = j\omega L$). In the dual case of TE modes (see (4.67)), solutions can exist only for surfaces characterized by negative reactance (e.g., capacitive reactance $Z_s = 1/(j\omega C)$). Here we assume that the material parameters of the isotropic medium ϵ_1 and μ_1 are positive. Note that in general the surface impedance Z_s can be different for the two polarizations and can also depend on the propagation factor k_t.

The eigenmode equations for surface waves along a multilayered structure formed by an arbitrary number of planar isotropic layers can be most conveniently derived by equating to zero the denominators of the general expressions for the reflection coefficients (4.129) and (4.155). In the most general case there can be as many solutions (for each polarization) as the number of interfaces.

4.2.6 Vector Solution of Reflection and Transmission Problems

This section describes a general procedure for simulation of reflection and transmission of a plane arbitrarily polarized wave at a planar interface of two semi-infinite media (Figure 4.9). The solution is given in a vector, coordinate-independent form, which is convenient for the use in the framework of ray-tracing algorithms when many subsequent reflections and transmissions at differently oriented material interfaces are to be taken into account. The presented procedure is applicable to more general incident fields and curved surfaces if these can be considered as approximately flat in a wavelength vicinity of the reflection/transmission point.

Let $\hat{i} = \mathbf{k}_1/k_1$ be a unit vector in the direction of incidence of a plane wave

$$\mathbf{E}_{\rm inc}(\mathbf{r}) = \mathbf{E}_{\rm inc}(\mathbf{r}_0)e^{-jk_1\hat{i}\cdot(\mathbf{r}-\mathbf{r}_0)} \tag{4.68}$$

and $\mathbf{r}_0$ be a position vector of a point of reflection/transmission on the interface in a global coordinate system (Figure 4.9). The following algorithm describes steps for calculation of the reflected and transmitted waves

$$\mathbf{E}_{\rm ref}(\mathbf{r}) = \mathbf{E}_{\rm ref}(\mathbf{r}_0)e^{-jk_1\hat{r}\cdot(\mathbf{r}-\mathbf{r}_0)} \tag{4.69}$$

$$\mathbf{E}_{\rm tr}(\mathbf{r}) = \mathbf{E}_{\rm tr}(\mathbf{r}_0)e^{-jk_2\hat{\tau}\cdot(\mathbf{r}-\mathbf{r}_0)}, \tag{4.70}$$

where $\hat{r}$ and $\hat{\tau}$ are the unit vectors describing the directions of reflection and transmission.

The first step of the procedure consists of splitting the incident wave into the components perpendicular and parallel to the plane of incidence. To this end, a local vector basis associated with the plane of incidence is introduced. If $\hat{n}$ is a unit vector normal to the interface at a point of reflection/transmission, then unit vectors

$$\hat{\alpha}_{\rm inc} = \frac{\hat{i}\times\hat{n}}{|\hat{i}\times\hat{n}|}, \quad \hat{\beta}_{\rm inc} = \hat{i}\times\hat{\alpha}_{\rm inc} \tag{4.71}$$

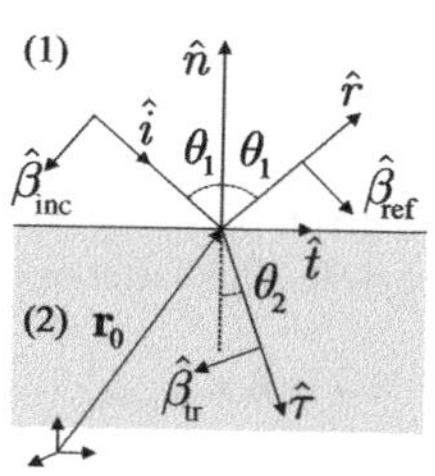

Figure 4.9 Bases used in the vector solution of the reflection and transmission problems. Vector $\hat{\alpha}_{\rm inc}$ is perpendicular to the plane of the figure and points toward the reader.

are perpendicular to the incidence direction $\hat{\imath}$ and to each other. The vector $\hat{\alpha}_{\text{inc}}$ is perpendicular to the plane of incidence, whereas the vector $\hat{\beta}_{\text{inc}}$ lies in the plane of incidence. These vectors allow us to decompose the incident electric field at a point $\mathbf{r}_0$ into a superposition of the perpendicular and parallel components

$$\mathbf{E}_{\text{inc}}(\mathbf{r}_0) = E^{\perp}_{\text{inc}}\hat{\alpha}_{\text{inc}} + E^{\parallel}_{\text{inc}}\hat{\beta}_{\text{inc}} \tag{4.72}$$

with

$$E^{\perp}_{\text{inc}} = \hat{\alpha}_{\text{inc}} \cdot \mathbf{E}_{\text{inc}}(\mathbf{r}_0), \quad E^{\parallel}_{\text{inc}} = \hat{\beta}_{\text{inc}} \cdot \mathbf{E}_{\text{inc}}(\mathbf{r}_0). \tag{4.73}$$

The representation (4.72) has no component along the incidence direction $\hat{\imath}$ since the incident wave is a plane wave.

The next step consists of finding the reflected field (4.69). The direction of propagation of the reflected wave $\hat{r}$ lies in the plane of incidence and thus can be sought as a linear combination of $\hat{\imath}$ and $\hat{n}$. The law of reflection (4.19) says that vectors $\hat{\imath}$ and $\hat{r}$ differ only in the sign of the component along $\hat{n}$, which leads to the expression

$$\hat{r} = \hat{\imath} - 2(\hat{n} \cdot \hat{\imath})\hat{n}. \tag{4.74}$$

Once $\hat{r}$ is known, we build the basis of the reflected wave by adding two unit vectors that are perpendicular to the direction of reflection and to each other,

$$\hat{\alpha}_{\text{ref}} = \hat{\alpha}_{\text{inc}}, \quad \hat{\beta}_{\text{ref}} = \hat{r} \times \hat{\alpha}_{\text{ref}}. \tag{4.75}$$

Now we can represent the reflected wave by

$$\mathbf{E}_{\text{ref}}(\mathbf{r}_0) = E^{\perp}_{\text{ref}}\hat{\alpha}_{\text{ref}} + E^{\parallel}_{\text{ref}}\hat{\beta}_{\text{ref}} \tag{4.76}$$

with

$$E^{\perp}_{\text{ref}} = \hat{\alpha}_{\text{ref}} \cdot \mathbf{E}_{\text{ref}}(\mathbf{r}_0), \quad E^{\parallel}_{\text{ref}} = \hat{\beta}_{\text{ref}} \cdot \mathbf{E}_{\text{ref}}(\mathbf{r}_0). \tag{4.77}$$

If the reflection coefficients relating the electric field components in the reflected and incident waves at the reflection point $\mathbf{r}_0$ are known,

$$R_{\perp} = \frac{E^{\perp}_{\text{ref}}}{E^{\perp}_{\text{inc}}}, \quad R_{\parallel} = \frac{E^{\parallel}_{\text{ref}}}{E^{\parallel}_{\text{inc}}}, \tag{4.78}$$

then we obtain the solution of the reflection problem as

$$\mathbf{E}_{\text{ref}}(\mathbf{r}_0) = R_{\perp}E^{\perp}_{\text{inc}}\hat{\alpha}_{\text{ref}} + R_{\parallel}E^{\parallel}_{\text{inc}}\hat{\beta}_{\text{ref}}. \tag{4.79}$$

The transmitted wave (4.70) can be determined along similar lines. One begins with building a basis of the transmitted wave by using vectors $\hat{\tau}$ and

$$\hat{\alpha}_{\text{tr}} = \hat{\alpha}_{\text{inc}}, \quad \hat{\beta}_{\text{tr}} = \hat{\tau} \times \hat{\alpha}_{\text{tr}}. \tag{4.80}$$

The vector $\hat{\tau}$ is defined by the Snell law (4.20) and given by the formula (4.24) as a linear combination of the vector $\hat{n}$ and the unit vector $\hat{t}$ tangent to the interface (Figure 4.9). This latter vector can be expressed as

$$\hat{t} = \hat{n} \times \hat{\alpha}_{\text{inc}}. \tag{4.81}$$

By using the basis (4.80) we calculate the parallel and perpendicular components of the transmitted wave (4.70)

$$E^{\perp}_{\text{tr}} = \hat{\alpha}_{\text{tr}} \cdot \mathbf{E}_{\text{tr}}(\mathbf{r}_0), \quad E^{\parallel}_{\text{tr}} = \hat{\beta}_{\text{tr}} \cdot \mathbf{E}_{\text{tr}}(\mathbf{r}_0) \tag{4.82}$$

and express the transmitted wave as

$$\mathbf{E}_{\rm tr}(\mathbf{r}_0) = E_{\rm tr}^{\perp}\hat{\alpha}_{\rm tr} + E_{\rm tr}^{\parallel}\hat{\beta}_{\rm tr}. \tag{4.83}$$

Transmission coefficients

$$T_{\perp} = \frac{E_{\rm tr}^{\perp}}{E_{\rm inc}^{\perp}}, \quad T_{\parallel} = \frac{E_{\rm tr}^{\parallel}}{E_{\rm inc}^{\parallel}} \tag{4.84}$$

relating the components of the incident and transmitted fields allow us to determine the transmitted wave

$$\mathbf{E}_{\rm tr}(\mathbf{r}_0) = T_{\perp}E_{\rm inc}^{\perp}\hat{\alpha}_{\rm tr} + T_{\parallel}E_{\rm inc}^{\parallel}\hat{\beta}_{\rm tr}. \tag{4.85}$$

Equations (4.79) and (4.85) solve the reflection/transmission problem when the reflection/transmission coefficients $R_{\perp,\parallel}$ and $T_{\perp,\parallel}$ defined with respect to the electric field components in the bases of the incident, reflected, and transmitted waves are available. These coefficients are simply related to the reflection and transmission coefficients $R_{\rm TE,TM}$ and $T_{\rm TE,TM}$ which, by definition, relate the electric field components tangential to the interface. Indeed, in the case of TE polarization, the electric field is proportional to the vector $\hat{\alpha}_{\rm inc}$ and is therefore tangential to the interface, implying that

$$R_{\rm TE} = \frac{\mathbf{E}_{\rm ref}\cdot\hat{\alpha}_{\rm inc}}{\mathbf{E}_{\rm inc}\cdot\hat{\alpha}_{\rm inc}} = \frac{E_{\rm ref}^{\perp}}{E_{\rm inc}^{\perp}} = R_{\perp} \tag{4.86}$$

$$T_{\rm TE} = \frac{\mathbf{E}_{\rm tr}\cdot\hat{\alpha}_{\rm inc}}{\mathbf{E}_{\rm inc}\cdot\hat{\alpha}_{\rm inc}} = \frac{E_{\rm tr}^{\perp}}{E_{\rm inc}^{\perp}} = T_{\perp}. \tag{4.87}$$

In the alternative case of TM polarization, the electric field in the incident, reflected, and transmitted waves is in general both normal and tangential to the interface components. The tangential components can be found by projecting the expressions (4.72), (4.76), and (4.83) onto the unit vector $\hat{t}$, leading to the expressions

$$R_{\rm TM} = \frac{\mathbf{E}_{\rm ref}\cdot\hat{t}}{\mathbf{E}_{\rm inc}\cdot\hat{t}} = \frac{E_{\rm ref}^{\parallel}\hat{\beta}_{\rm ref}\cdot\hat{t}}{E_{\rm inc}^{\parallel}\hat{\beta}_{\rm inc}\cdot\hat{t}} = -R_{\parallel} \tag{4.88}$$

$$T_{\rm TM} = \frac{\mathbf{E}_{\rm tr}\cdot\hat{t}}{\mathbf{E}_{\rm inc}\cdot\hat{t}} = \frac{E_{\rm tr}^{\parallel}\hat{\beta}_{\rm tr}\cdot\hat{t}}{E_{\rm inc}^{\parallel}\hat{\beta}_{\rm inc}\cdot\hat{t}} = T_{\parallel}\frac{\hat{\tau}\cdot\hat{n}}{\hat{\imath}\cdot\hat{n}} = T_{\parallel}\frac{\cos\theta_2}{\cos\theta_1}, \tag{4.89}$$

where we have used the relations:

$$\hat{t}\cdot\hat{\alpha}_{\rm inc} = 0, \quad \hat{t}\cdot\hat{\beta}_{\rm inc} = \hat{\imath}\cdot\hat{n}, \quad \hat{t}\cdot\hat{\beta}_{\rm ref} = -\hat{\imath}\cdot\hat{n}, \quad \hat{t}\cdot\hat{\beta}_{\rm tr} = \hat{\tau}\cdot\hat{n}. \tag{4.90}$$

By using the vector solution of the reflection problem we can now derive representations for the equivalent electric and magnetic surface currents $\mathbf{K}_{\rm e}$ and $\mathbf{K}_{\rm m}$ at the reflection point $\mathbf{r}_0$. The currents are defined in (2.348) and to find them one needs to determine the tangential components of the total electric and magnetic fields at the material interface. The total electric field is a sum of the incident and reflected waves, and multiplying this with $\hat{n}$ we obtain

$$\mathbf{K}_{\rm m} = -\hat{n}\times[\mathbf{E}_{\rm inc}(\mathbf{r}_0) + \mathbf{E}_{\rm ref}(\mathbf{r}_0)]. \tag{4.91}$$

Inserting (4.72) and (4.79) in (4.91) and using the vector relations (4.74), (4.75), and

$$\hat{n}\times\hat{\alpha}_{\rm inc} = \hat{t}, \quad \hat{n}\times\hat{\beta}_{\rm inc} = -\hat{\alpha}_{\rm inc}(\hat{\imath}\cdot\hat{n}), \quad \hat{n}\times\hat{\beta}_{\rm ref} = \hat{\alpha}_{\rm inc}(\hat{\imath}\cdot\hat{n}) \tag{4.92}$$

gives

$$\mathbf{K}_{\rm m} = (1 - R_{\parallel})(\hat{\imath}\cdot\hat{n})E_{\rm inc}^{\parallel}\hat{\alpha}_{\rm inc} - (1 + R_{\perp})E_{\rm inc}^{\perp}\hat{t}. \tag{4.93}$$

The equivalent electric surface current is determined through the total magnetic field as

$$\mathbf{K}_e = \hat{n} \times [\mathbf{H}_{inc}(\mathbf{r}_0) + \mathbf{H}_{ref}(\mathbf{r}_0)]. \tag{4.94}$$

According to the relation (2.156) the magnetic field in the incident and reflected waves is given by $Z_1\mathbf{H}_{inc} = \hat{\imath} \times \mathbf{E}_{inc}$ and $Z_1\mathbf{H}_{ref} = \hat{r} \times \mathbf{E}_{ref}$, and using representations (4.72) and (4.79) and the vector relations (4.92) and

$$\hat{n} \times (\hat{\imath} \times \hat{\beta}_{inc}) = -\hat{t}, \quad \hat{n} \times (\hat{r} \times \hat{\beta}_{ref}) = -\hat{t} \tag{4.95}$$

we arrive at the formula

$$Z_1\mathbf{K}_e = (R_\perp - 1)(\hat{\imath} \cdot \hat{n})E^\perp_{inc}\hat{\alpha}_{inc} - (1 + R_\parallel)E^\parallel_{inc}\hat{t}. \tag{4.96}$$

Equations similar to (4.93) and (4.96) can be found in Ulaby *et al.* (1982, Chapter 12).

As expected, the currents $\mathbf{K}_m$ and $\mathbf{K}_e$ are linear combinations of the vectors $\hat{\alpha}_{inc}$ and $\hat{t}$ tangential to the interface. In the limiting case of a perfectly electrically conducting interface, when $R_{TE,TM} = -1$ and thus $R_\perp = -1$ and $R_\parallel = 1$, it follows from (4.91) and (4.96) that

$$\mathbf{K}_m = 0, \quad Z_1\mathbf{K}_e = -2(\hat{\imath} \cdot \hat{n})E^\perp_{inc}\hat{\alpha}_{inc} - 2E^\parallel_{inc}\hat{t} = 2Z_1\hat{n} \times \mathbf{H}_{inc}. \tag{4.97}$$

Equations (4.91) and (4.96) are expressed in terms of the reflection coefficients of the material interface and are therefore not limited to the case of two semi-infinite half-spaces. They can be used to describe the equivalent currents on the boundary of the bodies of arbitrary material constitution as long as reflection from the boundary can be adequately described by reflection coefficients (e.g., impedance bodies). Equations (4.91) and (4.96) in conjunction with the integral representations (2.344), (2.345), (2.349), (2.350), (3.107), and (3.108) are an essential part of the PO approximation for the scattered fields (Chapter 8).

4.3 Arbitrary Number of Planar Layers

4.3.1 Solution by the Method of Characteristic Matrices

In this section we solve the problem of reflection and transmission of electromagnetic waves and derive reflection and transmission coefficients for planar structures consisting of an arbitrary number N of homogeneous isotropic layers surrounded by homogeneous isotropic semi-infinite media (Figure 4.10). The materials involved may be lossless or lossy.

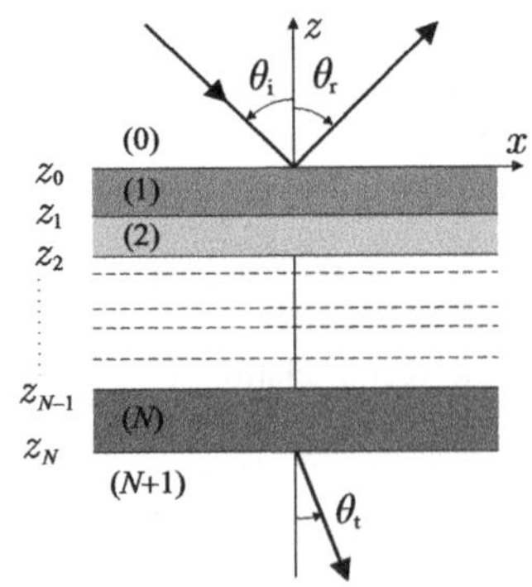

Figure 4.10 A set of planar layers.

An exact solution of this problem is constructed by using characteristic matrices which relate tangential components of the electric and magnetic fields at the opposite sides of homogeneous layers (Abelès 1950a,b; Born and Wolf 1959; Ruck *et al.* 1970; Tretyakov 2003). Note that in the microwave literature the characteristic matrix is usually called the *transmission matrix* or ABCD matrix. The characteristic matrix of the whole configuration, obtained by multiplication of characteristic matrices of individual layers, is well suited for computer implementation and applies to a great variety of practically interesting structures. Since any continuous dependence of ϵ and μ on z can be approximated by a set of planar layers with constant values of ϵ and μ, the approach provides a convenient way of simulating reflection and transmission in general stratified media.

Solution prior to Abelès was limited to a limited number of layers or to normal incidence or to the "optical" case in which the materials are characterized solely by the refraction index, implying that μ is the same throughout the structure (e.g. Crook (1948) and Weinstein (1947)). Some authors (e.g., Brekhovskikh (1960) and Wait (1962)) derive the input impedance of multilayered configurations, which determines the reflection coefficient but is insufficient to give the transmission coefficient. Equivalent solutions for the reflection and transmission of a plane electromagnetic wave incident at an arbitrary angle on a multilayered slab can be derived with other approaches, for example by using the analogy with transmission lines (e.g., Ishimaru (1991, section 3.7)), continuous fractions, and forward and backward propagation matrices (e.g., Ulaby *et al.* (1982, section 2.9) and Kong (1986, section 3.3)).

We use a Cartesian coordinate system (x, y, z), in which the interfaces of the media are located at $z = z_n$ with $n = 0, 1, \dots, N$ such that $z_0 > z_1 > \dots > z_n$. The upper ($z_0 < z < +\infty$) and lower ($-\infty < z < z_N$) half-spaces are filled with, in general, different materials with ϵ_0, μ_0, and ϵ_{N+1}, μ_{N+1}, respectively. Each layer is characterized by its permittivity ϵ_n, permeability μ_n, and thickness $d_n = z_{n-1} - z_n > 0$. The total thickness of the slab is, therefore, $d = z_0 - z_N = d_1 + d_2 + \dots + d_N$.

The slab is illuminated by a plane electromagnetic wave, and without loss of generality we can assume that the wave is incident from the upper half-space. If we choose the plane $X0Z$ to be the plane of incidence, then the incident field is independent of the y coordinate and since the configuration is also independent of this coordinate, the total electromagnetic field can be a function of the x and z coordinates only. Under such circumstances, Maxwell's equations reduce to the Helmholtz equations for the y components of the fields,

$$\left(\frac{\partial^2}{\partial x^2} + \frac{\partial^2}{\partial z^2} + k^2\right) E_y = 0 \tag{4.98}$$

$$\left(\frac{\partial^2}{\partial x^2} + \frac{\partial^2}{\partial z^2} + k^2\right) H_y = 0, \tag{4.99}$$

and to the four expressions for the other field components

$$H_x = \frac{-j}{\omega\mu}\frac{\partial}{\partial z}E_y, \quad H_z = \frac{j}{\omega\mu}\frac{\partial}{\partial x}E_y \tag{4.100}$$

$$E_x = \frac{j}{\omega\epsilon}\frac{\partial}{\partial z}H_y, \quad E_z = \frac{-j}{\omega\epsilon}\frac{\partial}{\partial x}H_y, \tag{4.101}$$

where we denote by k, ϵ, and μ the piecewise constant parameters which are equal to k_n, ϵ_n, and μ_n, respectively, in the nth material ($n = 0, 1, \dots, N+1$). These relations are supplemented with the requirement of continuity of tangential components of the fields, which are E_x, E_y, H_x, and H_y, at every material interface in the configuration.

It is now apparent that the electromagnetic field can be decomposed into a sum of two basic polarizations with $\mathbf{E} = (0, E_y, 0)$ and $\mathbf{H} = (H_x, 0, H_z)$ (TE polarization) and with $\mathbf{E} = (E_x, 0, E_z)$ and $\mathbf{H} = (0, H_y, 0)$ (TM polarization). The two polarization cases are fully decoupled and can be studied independently.

Before going to the analysis of the polarization cases, let us take a look at elementary solutions of (4.98) and (4.99) that can represent incident, reflected, and transmitted fields. These have the form

$$e^{-j\xi x \pm j\gamma z}, \tag{4.102}$$

where $\gamma = \sqrt{k^2 - \xi^2}$ and ξ is a parameter. A plane wave incident on the slab from the upper half-space at an angle θ_i $(-\pi/2 < \theta_i < \pi/2)$ is described by (4.102) with the upper sign in the exponent, $\xi = k_0 \sin\theta_i$, and $\gamma = \sqrt{k^2 - k_0 \sin^2\theta_i}$. In what follows we derive the expressions for the reflected and transmitted fields under a general assumption that ξ is an arbitrary, real- or complex-valued parameter. Such solutions include the plane wave incidence as a special case but are also applicable to incident fields of other kinds, like surface waves, in which case ξ is a complex-valued parameter, or cylindrical waves, in which case ξ is a spectral variable $(-\infty < \xi < \infty)$ from the Fourier expansion of the incident wave.

If k is real-valued and ξ is such that $-k \le \xi \le k$, then the expression (4.102) describes a plane wave; otherwise, the exponent in (4.102) is an inhomogeneous wave. Whether the wave (4.102) is incoming or outgoing depends on the sign in front of γ and on the selection of the branch of the square-root function in the complex plane ξ. If we define the branch by the conditions that $\mathrm{Im}\,\gamma \le 0$, which implies that $\gamma \ge 0$ when $-k \le \xi \le k$, then the exponent with the upper (lower) sign in (4.102) represents an inhomogeneous wave decaying (growing) as $z \to -\infty$ or a plane wave propagating toward the negative (positive) portion of the z axis depending on whether γ is complex- or real-valued.

TE Polarization

We begin with the TE case. The only component of the electric field in the incident wave in the upper half-space $(z > z_0)$ is given by

$$E_y^{\mathrm{inc}}(x,z) = E_0 e^{-j\xi x + j\gamma_0 z} = E_y^{\mathrm{inc}}(x,z_0) e^{j\gamma_0(z-z_0)}, \tag{4.103}$$

where $\gamma_0 = \sqrt{k_0^2 - \xi^2}$ and E_0 is the amplitude. At the interface with the slab $(z = z_0)$, the incident wave is partially reflected back in the upper half-space and partially transmitted in the slab. The reflected wave must be proportional to $E_y^{\mathrm{inc}}(x, z_0)$, the value of the incident wave at the interface, so we write

$$E_y^{\mathrm{ref}}(x,z) = R_E E_y^{\mathrm{inc}}(x,z_0) e^{-j\gamma_0(z-z_0)} \quad (z > z_0), \tag{4.104}$$

where R_E is a reflection coefficient defined by the relation

$$R_E = \frac{E_y^{\mathrm{ref}}(x,z_0)}{E_y^{\mathrm{inc}}(x,z_0)}. \tag{4.105}$$

To impose the continuity conditions on material interfaces we need expressions for the total fields E_y and H_x in every material. From (4.100), (4.103), and (4.104) it follows that in the upper half-space $(z > z_0)$ the total fields can be expressed as

$$E_y(x,z) = E_y^{\mathrm{inc}}(x,z_0)\left[e^{j\gamma_0(z-z_0)} + R_E e^{-j\gamma_0(z-z_0)}\right] \tag{4.106}$$

$$H_x(x,z) = \kappa_0^E E_y^{\mathrm{inc}}(x,z_0)\left[e^{j\gamma_0(z-z_0)} - R_E e^{-j\gamma_0(z-z_0)}\right], \tag{4.107}$$

where $\kappa_0^E = \kappa_n^E$ with $n = 0$ and

$$\kappa_n^E = \frac{\gamma_n}{\omega\mu_n} \tag{4.108}$$

$$\gamma_n = \sqrt{k_n^2 - \xi^2} \tag{4.109}$$

with $n = 0, 1, 2 \ldots, N+1$.

In the lower half-space ($z < z_N$), the total field coincides with the transmitted field. The latter must consist of a single outgoing wave and be proportional to the incident field on the upper side of the slab,

$$E_y(x,z) = E_y^{\text{tr}}(x,z) = E_y^{\text{inc}}(x,z_0) T_E e^{j\gamma_{N+1}(z-z_N)} \tag{4.110}$$

$$H_x(x,z) = H_x^{\text{tr}}(x,z) = \kappa_{N+1}^E E_y^{\text{inc}}(x,z_0) T_E e^{j\gamma_{N+1}(z-z_N)}. \tag{4.111}$$

The transmission coefficient T_E is thus defined by the relation

$$T_E = \frac{E_y^{\text{tr}}(x,z_N)}{E_y^{\text{inc}}(x,z_0)}. \tag{4.112}$$

In every individual layer inside the slab, the field may include both incoming and outgoing waves and can, therefore, be represented by the formulas

$$E_y(x,z) = E_y^{\text{inc}}(x,z_0)\{A_n \sin[\gamma_n(z - z_{n-1})] + B_n \cos[\gamma_n(z - z_{n-1})]\} \tag{4.113}$$

$$H_x(x,z) = -j\kappa_n^E E_y^{\text{inc}}(x,z_0)\{A_n \cos[\gamma_n(z - z_{n-1})] - B_n \sin[\gamma_n(z - z_{n-1})]\}, \tag{4.114}$$

where $z_n < z < z_{n-1}$ with $n = 1, 2, \ldots, N$. Coefficients A_n and B_n are directly related to the values of the tangential field components on the upper side of the nth layer,

$$A_n = \frac{j}{\kappa_n^E} \frac{H_x(x,z_{n-1})}{E_y^{\text{inc}}(x,z_0)} \tag{4.115}$$

$$B_n = \frac{E_y(x,z_{n-1})}{E_y^{\text{inc}}(x,z_0)}, \tag{4.116}$$

which is obtained by insertion of $z = z_{n-1}$ in equations (4.113) and (4.114) and leads to

$$E_y(x,z) = E_y(x,z_{n-1})\cos[\gamma_n(z - z_{n-1})] + \frac{j}{\kappa_n^E} H_x(x,z_{n-1}) \sin[\gamma_n(z - z_{n-1})] \tag{4.117}$$

$$H_x(x,z) = j\kappa_n^E E_y(x,z_{n-1}) \sin[\gamma_n(z - z_{n-1})] + H_x(x,z_{n-1}) \cos[\gamma_n(z - z_{n-1})]. \tag{4.118}$$

Setting $z = z_n$ in (4.117) and (4.118) allows us to express the limiting values of the tangential fields on the lower and upper sides of the layer as

$$\begin{bmatrix} E_y(x,z_n) \\ H_x(x,z_n) \end{bmatrix} = \overline{\overline{A}}_n^E \cdot \begin{bmatrix} E_y(x,z_{n-1}) \\ H_x(x,z_{n-1}) \end{bmatrix}, \tag{4.119}$$

where

$$\overline{\overline{A}}_n^E = \begin{bmatrix} \cos(\gamma_n d_n) & -j\sin(\gamma_n d_n)/\kappa_n^E \\ -j\kappa_n^E \sin(\gamma_n d_n) & \cos(\gamma_n d_n) \end{bmatrix} \tag{4.120}$$

and $d_n = z_{n-1} - z_n$. The matrix $\overline{\overline{A}}_n^E$ will be referred to as the characteristic matrix of an individual layer.

Since tangential field components are continuous across the material interfaces, we can use relation (4.119) in succession from $n = 1$ to $n = N$ to relate the components on the upper ($z = z_0$) and lower ($z = z_{N+1}$) sides of the slab,

$$\begin{bmatrix} E_y(x,z_n) \\ H_x(x,z_n) \end{bmatrix} = \overline{\overline{A}}_E \cdot \begin{bmatrix} E_y(x,z_0) \\ H_x(x,z_0) \end{bmatrix} \tag{4.121}$$

with the matrix

$$\overline{\overline{A}}_E = \begin{bmatrix} A_{11}^E & A_{12}^E \\ A_{21}^E & A_{22}^E \end{bmatrix} = \overline{\overline{A}}_N^E \cdot \overline{\overline{A}}_{N-1}^E \cdots \overline{\overline{A}}_1^E \tag{4.122}$$

being the characteristic matrix of the whole set of layers.

Inserting (4.106) and (4.107) with $z = z_0$ and (4.110) and (4.111) with $z = z_N$ in (4.121) gives a system of two algebraic relations for the two unknown coefficients R_E and T_E,

$$\begin{bmatrix} T_E \\ T_E \kappa_{N+1}^E \end{bmatrix} = \overline{\overline{A}}_E \cdot \begin{bmatrix} 1 + R_E \\ \kappa_0^E (1 - R_E) \end{bmatrix}, \tag{4.123}$$

and the solution of the system is

$$R_E = \frac{A_{21}^E + \kappa_0^E A_{22}^E - \kappa_{N+1}^E \left(A_{11}^E + \kappa_0^E A_{12}^E \right)}{-A_{21}^E + \kappa_0^E A_{22}^E + \kappa_{N+1}^E \left(A_{11}^E - \kappa_0^E A_{12}^E \right)} \tag{4.124}$$

$$T_E = \frac{2\kappa_0^E \det \overline{\overline{A}}_E}{-A_{21}^E + \kappa_0^E A_{22}^E + \kappa_{N+1}^E \left(A_{11}^E - \kappa_0^E A_{12}^E \right)} \tag{4.125}$$

with $\det \overline{\overline{A}}_E = A_{11}^E A_{22}^E - A_{12}^E A_{21}^E$. In (4.125) we can further simplify by using the property

$$\det \overline{\overline{A}}_E = 1 \tag{4.126}$$

which is a consequence of the fact that matrix $\overline{\overline{A}}_E$ is a product of characteristic matrices of individual layers $\overline{\overline{A}}_n^E$, and $\det \overline{\overline{A}}_n^E = 1$ as is seen from (4.120).

We define the input impedance of the structure (slab + lower half-space) as the ratio of the tangential electric and magnetic field components at the upper side of the slab. By using the boundary values of the fields in the upper half-space, the impedance can be expressed through the coefficient R_E as

$$Z_E = \left. \frac{E_y}{H_x} \right|_{z=z_0} = \frac{1}{\kappa_0^E} \frac{1 + R_E}{1 - R_E}, \tag{4.127}$$

and substituting R_E from (4.124) in (4.127) leads to the formula

$$Z_E = \frac{A_{22}^E - A_{12}^E \kappa_{N+1}^E}{-A_{21}^E + A_{11}^E \kappa_{N+1}^E}. \tag{4.128}$$

It follows from (4.127) that

$$R_E = \frac{\kappa_0^E Z_E - 1}{\kappa_0^E Z_E + 1}, \tag{4.129}$$

which permits determination of the reflection coefficient if the input impedance is known. In contrast, the transmission coefficient T_E cannot be expressed entirely through the input impedance of the structure.

According to their definitions (4.105) and (4.112), R_E and T_E relate the y components of the incident, reflected, and transmitted electric fields, and since the unit vector $\hat{y}$ is perpendicular to the incidence plane and, at the same time, tangential to the slab faces, they coincide with the differently defined reflection and transmission coefficients (see Figure 4.2 and section 4.2.6),

$$R_E = R_\perp = R_{\text{TE}}, \quad T_E = T_\perp = T_{\text{TE}}. \tag{4.130}$$

The solution for the perpendicular components of electric field can therefore be expressed as

$$E_{\perp}^{\mathrm{inc}}(x,z) = E_0^{\perp} \mathrm{e}^{-j\xi x + j\gamma_0 z} \quad (z \geq z_0) \tag{4.131}$$

$$E_{\perp}^{\mathrm{ref}}(x,z) = E_0^{\perp} R_{\perp} \mathrm{e}^{-j\xi x + j\gamma_0 (2z_0 - z)} \quad (z \geq z_0) \tag{4.132}$$

$$E_{\perp}^{\mathrm{tr}}(x,z) = E_0^{\perp} T_{\perp} \mathrm{e}^{-j\xi x + j\gamma_0 z_0 + j\gamma_{N+1}(z - z_N)} \quad (z \leq z_N), \tag{4.133}$$

where $E_0^{\perp} = E_{\perp}^{\mathrm{inc}}(0,0)$.

TM Polarization

In the alternative case of TM polarization, the magnetic field has a single component H_y, and it is therefore convenient to use this component to describe the solution of the problem. The derivation of the solution parallels that from the case of TE polarization and can be in principle derived by using the duality of Maxwell's equations. However, interchanging ϵ and μ in the expressions for the TE reflection and transmission coefficients would result in reflection and transmission coefficients that relate the incident, reflected, and transmitted *magnetic* fields. To obtain the coefficients for the electric fields, one should either express them through the magnetic reflection and transmission coefficients or repeat the derivations of the previous section for H_y instead of E_y. For the sake of completeness, we follow here the latter approach, leaving the former to the problems section.

Let us represent the incident wave by

$$H_y^{\mathrm{inc}}(x,z) = H_0 \mathrm{e}^{-j\xi x + j\gamma_0 z} = H_y^{\mathrm{inc}}(x,z_0) \mathrm{e}^{j\gamma_0 (z - z_0)} \tag{4.134}$$

with the amplitude H_0 and the reflected and transmitted waves by

$$H_y^{\mathrm{ref}}(x,z) = H_y^{\mathrm{inc}}(x,z_0) R_H \mathrm{e}^{-j\gamma_0 (z - z_0)} \quad (z > z_0) \tag{4.135}$$

$$H_y^{\mathrm{tr}}(x,z) = H_y^{\mathrm{inc}}(x,z_0) T_H \mathrm{e}^{j\gamma_{N+1}(z - z_N)} \quad (z < z_N) \tag{4.136}$$

with the reflection and transmission coefficients defined by the relations:

$$R_H = \frac{H_y^{\mathrm{ref}}(x,z_0)}{H_y^{\mathrm{inc}}(x,z_0)} \tag{4.137}$$

$$T_H = \frac{H_y^{\mathrm{tr}}(x,z_N)}{H_y^{\mathrm{inc}}(x,z_0)}. \tag{4.138}$$

The total magnetic field in the upper and lower half-spaces is then given by

$$H_y(x,z) = H_y^{\mathrm{inc}}(x,z) + H_y^{\mathrm{ref}}(x,z) \quad (z > z_0) \tag{4.139}$$

$$H_y(x,z) = H_y^{\mathrm{tr}}(x,z) \quad (z < z_N). \tag{4.140}$$

The tangential component of the electric field, E_x, results from (4.101), (4.134)–(4.136), (4.139), and (4.140) as

$$E_x(x,z) = E_x^{\mathrm{inc}}(x,z) + E_x^{\mathrm{ref}}(x,z) \quad (z > z_0) \tag{4.141}$$

$$E_x(x,z) = E_x^{\mathrm{tr}}(x,z) \quad (z < z_N), \tag{4.142}$$

where

$$E_x^{\mathrm{inc}}(x,z) = -\kappa_0^H H_y^{\mathrm{inc}}(x,z_0) \mathrm{e}^{j\gamma_0 (z - z_0)} \quad (z > z_0) \tag{4.143}$$

$$E_x^{\mathrm{ref}}(x,z) = \kappa_0^H H_y^{\mathrm{inc}}(x,z_0) R_H \mathrm{e}^{-j\gamma_0 (z - z_0)} \quad (z > z_0) \tag{4.144}$$

$$E_x^{\mathrm{tr}}(x,z) = -\kappa_{N+1}^H H_y^{\mathrm{inc}}(x,z_0) T_H \mathrm{e}^{j\gamma_{N+1}(z - z_N)} \quad (z < z_N) \tag{4.145}$$

and

$$\kappa_n^H = \frac{\gamma_n}{\omega\epsilon_n} \tag{4.146}$$

with $0 \leq n \leq N+1$.

Now we have to represent the fields H_y and E_x inside every layer and then impose the continuity condition on the material interfaces $z = z_n$ with $n = 0, 1, \ldots, N$. However, comparing (4.139), (4.140), (4.141), and (4.142) with (4.106), (4.107), (4.110), and (4.111) suggests that the solution for the TM case can be obtained from the solution for the TE case by simply substituting $E_y \to H_y$, $H_x \to -E_x$, and $\kappa_n^E \to \kappa_n^H$. Thus, we find that

$$\begin{bmatrix} H_y(x, z_n) \\ -E_x(x, z_n) \end{bmatrix} = \overline{\overline{A}}_n^H \cdot \begin{bmatrix} H_y(x, z_{n-1}) \\ -E_x(x, z_{n-1}) \end{bmatrix}, \tag{4.147}$$

and the characteristic matrix of an individual layer is equal to

$$\overline{\overline{A}}_n^H = \begin{bmatrix} \cos(\gamma_n d_n) & -j\sin(\gamma_n d_n)/\kappa_n^H \\ -j\kappa_n^H \sin(\gamma_n d_n) & \cos(\gamma_n d_n) \end{bmatrix}. \tag{4.148}$$

The characteristic matrix of the complete structure is a product of the characteristic matrices of individual layers,

$$\overline{\overline{A}}_H = \begin{bmatrix} A_{11}^H & A_{12}^H \\ A_{21}^H & A_{22}^H \end{bmatrix} = \overline{\overline{A}}_N^H \cdot \overline{\overline{A}}_{N-1}^H \cdots \overline{\overline{A}}_1^H, \tag{4.149}$$

and it relates the tangential field components on the opposite sides of the slab

$$\begin{bmatrix} H_y(x, z_n) \\ -E_x(x, z_n) \end{bmatrix} = \overline{\overline{A}}_H \cdot \begin{bmatrix} H_y(x, z_0) \\ -E_x(x, z_0) \end{bmatrix}. \tag{4.150}$$

Using (4.139)–(4.142) in (4.150) and solving the system for R_H and T_H gives the expressions:

$$R_H = \frac{A_{21}^H + \kappa_0^H A_{22}^H - \kappa_{N+1}^H \left(A_{11}^H + \kappa_0^H A_{12}^H\right)}{-A_{21}^H + \kappa_0^H A_{22}^H + \kappa_{N+1}^H \left(A_{11}^H - \kappa_0^H A_{12}^H\right)} \tag{4.151}$$

$$T_H = \frac{2\kappa_0^H}{-A_{21}^H + \kappa_0^H A_{22}^H + \kappa_{N+1}^H \left(A_{11}^H - \kappa_0^H A_{12}^H\right)}. \tag{4.152}$$

The input impedance of the configuration at the interface $z = z_0$ is given by

$$Z_H = \left.\frac{-E_x}{H_y}\right|_{z=z_0} = \kappa_0^H \frac{1 - R_H}{1 + R_H} \tag{4.153}$$

and, because of (4.151), can be expressed through the elements of the characteristic matrix as

$$Z_H = \frac{-A_{21}^H + A_{11}^H \kappa_{N+1}^H}{A_{22}^H - A_{12}^H \kappa_{N+1}^H}. \tag{4.154}$$

The relation to the reflection coefficient results from (4.153) as

$$R_H = \frac{\kappa_0^H - Z_H}{\kappa_0^H + Z_H}. \tag{4.155}$$

Similarly to the case of TE polarization, determination of the transmission coefficient T_H requires the knowledge of the characteristic matrix $\overline{\overline{A}}_H$ and cannot be reduced to finding the input impedance Z_H.

The reflection and transmission coefficients R_{TM} and T_{TM} are defined with respect to the tangential component of the electric field (see Figure 4.2), which is E_x in our case, so that we have

$$R_{\text{TM}} = \frac{E_x^{\text{ref}}(x, z_0)}{E_x^{\text{inc}}(x, z_0)}, \quad T_{\text{TM}} = \frac{E_x^{\text{tr}}(x, z_N)}{E_x^{\text{inc}}(x, z_0)}, \tag{4.156}$$

where E_x^{inc}, E_x^{ref}, and E_x^{tr} are the x components of the incident, reflected, and transmitted fields, and it follows from (4.156) and (4.143)–(4.145) that

$$R_{\text{TM}} = -R_H, \quad T_{\text{TM}} = \frac{\kappa_{N+1}^H}{\kappa_0^H} T_H. \tag{4.157}$$

To find the relations of R_H and T_H to $R_\parallel$ and $T_\parallel$ defined by

$$R_\parallel = \frac{E_\parallel^{\text{ref}}(x, z_0)}{E_\parallel^{\text{inc}}(x, z_0)}, \quad T_\parallel = \frac{E_\parallel^{\text{tr}}(x, z_N)}{E_\parallel^{\text{inc}}(x, z_0)}, \tag{4.158}$$

one has to express the parallel component of the electric field in the incident, reflected, and transmitted waves, which is the only component of the electric field in TM-polarized waves, through the y component of the magnetic field in the respective wave. Since the waves are plane, the electric and magnetic fields are related by the cross product (2.156) and $E_\parallel$ is equal to H_y multiplied by the wave impedance in the respective medium, leading to the relations:

$$R_\parallel = R_H, \quad T_\parallel = \frac{Z_{N+1}}{Z_0} T_H. \tag{4.159}$$

The solution for the parallel components of electric field is therefore given by

$$E_\parallel^{\text{inc}}(x, z) = E_0^\parallel \mathrm{e}^{-j\xi x + j\gamma_0 z} \quad (z \geq z_0) \tag{4.160}$$

$$E_\parallel^{\text{ref}}(x, z) = E_0^\parallel R_\parallel \mathrm{e}^{-j\xi x + j\gamma_0(2z_0 - z)} \quad (z \geq z_0) \tag{4.161}$$

$$E_\parallel^{\text{tr}}(x, z) = E_0^\perp T_\parallel \mathrm{e}^{-j\xi x + j\gamma_0 z_0 + j\gamma_{N+1}(z - z_N)} \quad (z \leq z_N) \tag{4.162}$$

with $E_0^\parallel = E_\parallel^{\text{inc}}(0, 0)$.

Finally, we note that R_H and T_H are related to R_E and T_E by the property of duality of Maxwell's equations, implying that they can be obtained from each other by interchanging the constitutive parameters ϵ and μ, which is equivalent to the substitutions $\kappa_n^E \leftrightarrow \kappa_n^H$ with $n = 0, 1, \ldots, N+1$.

4.3.2 Discussion and Limiting Cases

The expressions derived in the previous sections for the reflected and transmitted fields apply to planar structures with an arbitrary number of isotropic layers. The reflection and transmission coefficients of the structure are given by (4.124), (4.125), (4.151), and (4.152), and expressed through the elements of characteristic matrices (4.122) and (4.149), which are obtained by multiplication of the characteristic matrices (4.120) and (4.148) of individual layers.

The derived solutions include an arbitrary parameter ξ. When $\xi = k_0 \sin\theta_i$ with $\pi/2 \leq \theta_i \leq \pi/2$, we obtain the solution for a plane wave incident upon an angle θ_i on a multi-layered slab between two semi-infinite media. This case is of particular interest to the subject of this book, and in this section we study the properties of the reflection and transmission coefficients assuming illumination of the slab with a plane wave.

The case of an arbitrary, real- or complex-valued ξ is, however, also of significant importance. This parameter can be, for example, a spectral variable of a Fourier transform with respect to the x coordinate, in which case $-\infty < \xi < \infty$. With a complex-valued ξ, the incident, reflected, and transmitted waves become inhomogeneous waves that grow or decay in particular directions. An example of these is a surface wave that propagates along the interface, exponentially decaying in the direction perpendicular to the interface. If ξ is real-valued but such that $|\xi| > k_{N+1}$, then we have the situation of total internal reflection, and the transmitted field attains the character of a surface wave propagating along the lower side of the slab and concentrated close to the interface. The solution of the reflection/transmission problem with a complex-valued ξ is required when the incident field is not a single plane wave but rather a more general wave, for example cylindrical or spherical. In this case, the solution is found by expanding the incident field into a spectrum of plane or, in general, inhomogeneous plane waves, writing down the solution of the reflection/transmission problem for every spectral component, followed by integration of these over the whole spectrum. Excitation of multi-layered slabs with cylindrical and spherical waves is addressed in section 4.4.

Reflection and Transmission of Plane Waves

Consider now the derived solution of the reflection and transmission problem in the case of a plane wave. For a plane wave incident on the slab on angle θ_i, the solution results from the general solution on the substitutions

$$\xi = k_0 \sin\theta_i, \quad \gamma_n = \sqrt{k_n^2 - k_0^2 \sin^2\theta_i}, \tag{4.163}$$

where $0 \leq n \leq N+1$. For isotropic materials, a change in the sign of the incidence angle θ_i cannot change the values of the reflection and transmission coefficients, implying that these must be even functions of ξ or θ_i. This condition is indeed satisfied as the derived expressions depend on ξ^2 or $\cos\theta_i$ only. The incidence angle θ_i can therefore be always assumed to be positive, $0 \leq \theta_i \leq \pi/2$, which is the usual convention in books on the subject.

The law of reflection that has been previously discussed in the context of an interface between two semi-infinite media holds for interfaces of arbitrary planar layered media as well. It is seen from (4.103), (4.104), (4.134), and (4.135) that the direction of incidence and the direction of reflection have equal x components and opposite z components, which extends the law of reflection (4.19) and leads to the law of reflection for arbitrary planar layered media.

The Snell law for two semi-infinite media separated by a planar slab can be understood as a relation between the incidence angle θ_i and the transmission angle θ_t (which is the angle at which the wave transmitted through the slab emerges on the other side of the slab). This relation is a direct consequence of the fact that the dependence on the x coordinate in the incident and transmitted waves must be the same, otherwise the continuity conditions on the material boundaries in the configuration cannot be satisfied. The dependence of the fields on x is given by the exponent $\exp(-j\xi x)$ with $\xi = k_0 \sin\theta_i$ in the upper medium and $\xi = k_{N+1}\sin\theta_t$ in the lower medium, implying that

$$k_{N+1}\sin\theta_t = k_0 \sin\theta_i, \tag{4.164}$$

from which the transmission angle is determined as

$$\theta_t = \arcsin\left(\sqrt{\frac{\mu_0\epsilon_0}{\mu_{N+1}\epsilon_{N+1}}}\sin\theta_i\right). \tag{4.165}$$

The angle θ_t is complex-valued if the upper and/or lower media are lossy or if the lower medium is such that $\mu_{N+1}\epsilon_{N+1} < \mu_0\epsilon_0$ and the incidence angle exceeds the critical value

$$\bar{\theta}_i = \arcsin\left(\sqrt{\frac{\mu_{N+1}\epsilon_{N+1}}{\mu_0\epsilon_0}}\right). \tag{4.166}$$

In the latter case, total reflection occurs at the lower interface $z = z_N$ of the slab, and it follows from (4.110) and (4.136) that the wave transmitted in the medium $N+1$ propagates along the interface $z = z_N$, decaying exponentially as the observation point moves away from the interface toward $z = -\infty$.

It follows from (4.113) and (4.114) that the field in the nth layer ($n = 1, 2, \ldots = N$) can be always represented as a superposition of two waves (4.102) with $\gamma = \gamma_n$ propagating in the positive and negative directions of the z axis. This latter wave makes an angle θ_n with the z axis, which can be referred to as the refraction angle, and the condition of the invariance of the x dependence of the fields at every interface in the configuration leads to the generalized Snell's law

$$\xi = k_0 \sin\theta_{\rm i} = k_1 \sin\theta_1 = \ldots = k_N \sin\theta_N = k_{N+1} \sin\theta_{\rm t} \tag{4.167}$$

relating the incidence angle with the angles of refraction and transmission in planar multilayered structures.

Limiting Cases of Reflection and Transmission Coefficients

The derived expressions for the reflection and transmission coefficients are quite general and include a lot of important special cases. Let us consider some of these. When $\xi \to 0$ or in the limit of normal incidence $\theta_{\rm i} \to 0$, the general solution derived in this section should be compliant with the physically apparent requirements

$$R_{\rm TE} \to R_{\rm TM}, \quad T_{\rm TE} \to T_{\rm TM}, \tag{4.168}$$

which must be true for any planar isotropic configuration. Relations (4.168) are, indeed, recovered if one accounts for the relations $\kappa_n^E \to \sqrt{\epsilon_n/\mu_n}$, $\kappa_n^H \to \sqrt{\mu_n/\epsilon_n}$, and

$$A_{11}^E \to A_{22}^H, \quad A_{12}^E \to A_{21}^H, \quad A_{21}^E \to A_{12}^H, \quad A_{22}^E \to A_{11}^H. \tag{4.169}$$

A proof of (4.169) is left to the reader (problem 15).

At normal incidence, the reflection coefficients can therefore be written as

$$R_{\rm TE}(0) = R_{\rm TM}(0) = \frac{Z(0) - Z_0}{Z(0) + Z_0}, \tag{4.170}$$

where $Z_0 = \sqrt{\mu_0/\epsilon_0}$ is the free-space impedance and $Z(0) = Z_E(0) = Z_H(0)$ is the value of the input impedance of the interface at normal incidence, which is the same in both polarization cases for any isotropic planar configuration. A consequence of (4.170) is that reflection from any interface with $Z(0) = Z_0$ vanishes at normal illumination.

If the total thickness d of the slab vanishes, and we approach the case of an interface between two homogeneous half-spaces with $n = 0$ and $n = N+1$, then in every layer $|\gamma_n d_n| \to 0$ ($n = 1, 2, \ldots, N$) and it follows from (4.120) and (4.148) that all individual characteristic matrices $\overline{\overline{A}}_n^{E,H}$, and therefore the characteristic matrices $\overline{\overline{A}}_{E,H}$, tend to the unit matrix,

$$\overline{\overline{A}}_{E,H}, \overline{\overline{A}}_n^{E,H} \to \overline{\overline{I}} = \begin{bmatrix} 1 & 0 \\ 0 & 1 \end{bmatrix}, \tag{4.171}$$

and equations (4.124), (4.125), (4.130), (4.151), (4.152), and (4.157) give the formulas

$$R_{\rm TE} = \frac{\kappa_0^E - \kappa_{N+1}^E}{\kappa_0^E + \kappa_{N+1}^E} \tag{4.172}$$

$$T_{\rm TE} = \frac{2\kappa_0^E}{\kappa_0^E + \kappa_{N+1}^E} \tag{4.173}$$

$$R_{\mathrm{TM}} = -\frac{\kappa_0^H - \kappa_{N+1}^H}{\kappa_0^H + \kappa_{N+1}^H} \tag{4.174}$$

$$T_{\mathrm{TM}} = \frac{2\kappa_{N+1}^H}{\kappa_0^H + \kappa_{N+1}^H}, \tag{4.175}$$

which agree with (4.10) and (4.11).

A noteworthy property of characteristic matrices is that if two contiguous layers n and $n+1$ ($n = 1, 2, \ldots, N-1$) have the same constitutive parameters $\epsilon_{n+1} = \epsilon_n$ and $\mu_{n+1} = \mu_n$, then the product of the respective characteristic matrices is equal to the characteristic matrix of a layer with the aggregate thickness $d_n + d_{n+1}$,

$$\overline{\overline{A}}_{n+1}^{E,H}(d_{n+1}) \cdot \overline{\overline{A}}_{n}^{E,H}(d_n) = \overline{\overline{A}}_{n}^{E,H}(d_n + d_{n+1}), \tag{4.176}$$

which is obtained by direct multiplication of the matrices. This "fusion" property ensures that subdividing a homogeneous layer into several fictitious sub-layers does not change the characteristic matrices, and therefore the reflection and transmission coefficients of the whole structure.

In the limiting case that all materials are the same ($\epsilon_n = \epsilon_0$ and $\mu_n = \mu_0$ with $n = 1, 2, \ldots, N+1$), and a plane wave propagates in a homogeneous space, the characteristic matrices (4.122) and (4.149) become

$$\overline{\overline{A}}_{E,H} = \begin{bmatrix} \cos(\gamma_0 d) & -j\sin(\gamma_0 d)/\kappa_0^{E,H} \\ -j\kappa_0^{E,H}\sin(\gamma_0 d) & \cos(\gamma_0 d) \end{bmatrix}, \tag{4.177}$$

and it follows from (4.124), (4.125), (4.151), and (4.152) that

$$R_{E,H} = 0, \quad T_{E,H} = \mathrm{e}^{-j\gamma_0 d}. \tag{4.178}$$

As expected, there is no reflection and the transmitted waves (4.110) and (4.140) coincide with the incident waves (4.103) and (4.134).

Now consider what happens if the upper layer of a multi-layered slab is made of the same material as the upper half-space, $\mu_1 = \mu_0$, $\epsilon_1 = \epsilon_0$. Then, it can be shown that

$$R_{\mathrm{TE}} = \mathrm{e}^{-2j\gamma_0 d_1} R'_{\mathrm{TE}}, \quad T_{\mathrm{TE}} = \mathrm{e}^{-j\gamma_0 d_1} T'_{\mathrm{TE}} \tag{4.179}$$

$$R_{\mathrm{TM}} = \mathrm{e}^{-2j\gamma_0 d_1} R'_{\mathrm{TM}}, \quad T_{\mathrm{TM}} = \mathrm{e}^{-j\gamma_0 d_1} T'_{\mathrm{TM}}, \tag{4.180}$$

where $R'_{\mathrm{TE,TM}}$ and $T'_{\mathrm{TE,TM}}$ are the reflection and transmission coefficients of a slab consisting of the layers with $n = 2, 3, \ldots, N$. The phase-shifting factors in the reflection and transmission coefficients are apparent consequences of the displacement of the physical interface of the slab, which is now at $z = z_0 - d_1$, from the reference plane $z = z_0$, at which the relations between the incident, reflected, and transmitted waves are defined (see (4.105), (4.112), (4.137), and (4.138)).

If the lower layer of the slab with $n = N$ has the same material properties as the lower half-space ($\mu_N = \mu_{N+1}, \epsilon_N = \epsilon_{N+1}$), then

$$R_{\mathrm{TE}} = R''_{\mathrm{TE}}, \quad T_{\mathrm{TE}} = \mathrm{e}^{-j\gamma_{N+1} d_N} T''_{\mathrm{TE}} \tag{4.181}$$

$$R_{\mathrm{TM}} = R''_{\mathrm{TM}}, \quad T_{\mathrm{TM}} = \mathrm{e}^{-j\gamma_{N+1} d_N} T''_{\mathrm{TM}}, \tag{4.182}$$

where $R''_{\mathrm{TE,TM}}$ and $T''_{\mathrm{TE,TM}}$ are the reflection and transmission coefficients of a slab consisting of the layers with $n = 1, 2, \ldots, N-1$. So, in this case the reflection coefficients do not change (since the upper boundary of the slab coincides with the reference plane at $z = z_0$), but the transmission coefficients are multiplied with the phase-shifting factors which account for the change in the phase of the transmitted wave on the way from the lower boundary of the slab at $z = z_{N-1}$ to the plane $z = z_N$ with respect to which the transmission coefficients are defined.

In the limiting case of grazing incidence ($\theta_i \to \pm\pi/2$), $\gamma_0 \to 0$ and $\kappa_0^{E,H} \to 0$, implying that

$$R_{TE} \to -1, \quad R_{TM} \to 1 \tag{4.183}$$

as is easily seen from (4.124), (4.130), (4.151), and (4.157). The property (4.183) is valid as long as the upper layer of the slab is not a perfect conductor.

If the slab is backed by a PEC surface, which means that the lower half-space is a perfect electric conductor and, therefore, $|\epsilon_{N+1}| \to \infty$ and μ_{N+1} is finite, then $|\kappa_{N+1}^E| \to \infty$ and $|\kappa_{N+1}^H| \to 0$, which leads to the limiting values

$$R_{TE} = -\frac{A_{11}^E + \kappa_0^E A_{12}^E}{A_{11}^E - \kappa_0^E A_{12}^E}, \quad T_{TE} = 0, \quad R_{TM} = \frac{A_{21}^H + \kappa_0^H A_{22}^H}{A_{21}^H - \kappa_0^H A_{22}^H}, \quad T_{TM} = 0. \tag{4.184}$$

If, furthermore, we let the thickness of the slab go to zero ($d \to 0$), then using the property (4.171) in (4.184) we recover the reflection and transmission coefficients of a PEC ground plane (see (4.36)).

Consider again the general configuration shown in Figure 4.10. From reciprocity considerations it is natural to expect that transmission through the slab should be independent of the direction of propagation and the field of an electromagnetic wave incident from the upper half-space, transmitted through the slab and measured at a point in the lower half-space, should be in some way related to the field of an electromagnetic wave incident from the lower half-space and measured at a point in the upper half-space. This is indeed so (Abelès 1950a), and if we denote $T_E(\theta_i)$ and $T_H(\theta_i)$ by $T_E^{\downarrow}$ and $T_H^{\downarrow}$ and the transmission coefficients for a plane wave incident from the lower half-plane upon the angle θ_t on the slab by $T_E^{\uparrow}$ and $T_H^{\uparrow}$, then

$$\kappa_0^E T_E^{\uparrow} = \kappa_{N+1}^E T_E^{\downarrow}, \quad \kappa_0^H T_H^{\uparrow} = \kappa_{N+1}^H T_H^{\downarrow} \tag{4.185}$$

(the proof is left to problem 18). These can be rewritten as

$$T_{\perp}^{\uparrow} = \frac{Z_0 \cos\theta_t}{Z_{N+1}\cos\theta_i} T_{\perp}^{\downarrow}, \quad T_{\parallel}^{\uparrow} = \frac{Z_0 \cos\theta_t}{Z_{N+1}\cos\theta_i} T_{\parallel}^{\downarrow} \tag{4.186}$$

by taking into account the relations (4.130), (4.159), and

$$\kappa_0^E = \frac{\cos\theta_i}{Z_0}, \quad \kappa_{N+1}^E = \frac{\cos\theta_t}{Z_{N+1}}, \quad \kappa_0^H = Z_0\cos\theta_i, \quad \kappa_{N+1}^E = Z_{N+1}\cos\theta_t. \tag{4.187}$$

Energy Balance

Now let us take a look at the relations between energy flows of the incident, reflected, and transmitted waves. We will assume that the upper and lower half-spaces are lossless, but the materials in the slab can be lossy. The quantities that describe the power relations at planar structures are called reflectance and transmittance, and they can be defined by the same equations as the reflectivity $\mathcal{R}$ and transmissivity $\mathcal{T}$ in (4.38) and (4.39). Using the formulas (4.44) and (4.46) with the apparent change in the media labeling, we write

$$\mathcal{R}_{\perp} = |R_{\perp}|^2, \quad \mathcal{T}_{\perp} = \frac{Z_0\cos\theta_t}{Z_{N+1}\cos\theta_i}|T_{\perp}|^2 \tag{4.188}$$

$$\mathcal{R}_{\parallel} = |R_{\parallel}|^2, \quad \mathcal{T}_{\parallel} = \frac{Z_0\cos\theta_t}{Z_{N+1}\cos\theta_i}|T_{\parallel}|^2. \tag{4.189}$$

In terms of transmittances $\mathcal{T}_{\perp}$ and $\mathcal{T}_{\parallel}$ the reciprocity relations (4.186) take the simple form

$$\mathcal{T}_{\perp}^{\uparrow} = \mathcal{T}_{\perp}^{\downarrow}, \quad \mathcal{T}_{\parallel}^{\uparrow} = \mathcal{T}_{\parallel}^{\downarrow} \tag{4.190}$$

since

$$\mathcal{T}^{\downarrow}_{\perp,\parallel} = \frac{Z_0 \cos\theta_t}{Z_{N+1}\cos\theta_i}\left|T^{\downarrow}_{\perp,\parallel}\right|^2, \quad \mathcal{T}^{\uparrow}_{\perp,\parallel} = \frac{Z_{N+1}\cos\theta_i}{Z_0 \cos\theta_t}\left|T^{\uparrow}_{\perp,\parallel}\right|^2, \tag{4.191}$$

implying that the amount of energy transmitted through a planar arbitrarily layered slab is independent of the direction of propagation, even for slabs asymmetrically stratified with respect to their middle plane.

If all the media, including those in the slab, are lossless, then the reflectance and transmittance are related by the formulas

$$\mathcal{R}_\perp + \mathcal{T}_\perp = 1, \quad \mathcal{R}_\parallel + \mathcal{T}_\parallel = 1, \tag{4.192}$$

which are generalizations of the similar relations (4.47) for the interface of two semi-infinite media.

Relations (4.192) describe the division of the energy flow transported by the incident wave into the reflected and transmitted portions and clearly represent energy conservation. A proof immediately follows from the energy conservation law (see section 2.4.1) formulated for a surface Ω fully enclosing the slab,

$$\int_\Omega \mathbf{w}\cdot\hat{n}\,\mathrm{d}\Omega = \text{energy dissipated in the slab}, \tag{4.193}$$

where $\mathbf{w}$ is the density of the energy flow and $\hat{n}$ is the unit vector normal to the integration surface Ω. In our case the integration surface consists of two planes Ω_+ and Ω_- with $\hat{n} = \hat{z}$ on Ω_+ and $\hat{n} = -\hat{z}$ on Ω_- (Figure 4.11). The energy flow across Ω is determined by the incident, reflected, and transmitted waves

$$\mathbf{w} = \begin{cases} \mathbf{w}^{\mathrm{inc}} + \mathbf{w}^{\mathrm{ref}}, & \text{on } \Omega_+ \\ \mathbf{w}^{\mathrm{tr}}, & \text{on } \Omega_- \end{cases}, \tag{4.194}$$

and, in the absence of losses and total internal reflection, the associated energy flows can be expressed according to (2.162) as

$$\mathbf{w}^{\mathrm{inc}} = \frac{\hat{i}}{2Z_0}\,|E_0^{\mathrm{inc}}|^2 \tag{4.195}$$

$$\mathbf{w}^{\mathrm{ref}} = \frac{\hat{r}}{2Z_0}\,|E_0^{\mathrm{ref}}|^2 \tag{4.196}$$

$$\mathbf{w}^{\mathrm{tr}} = \frac{\hat{t}}{2Z_{N+1}}\,|E_0^{\mathrm{tr}}|^2, \tag{4.197}$$

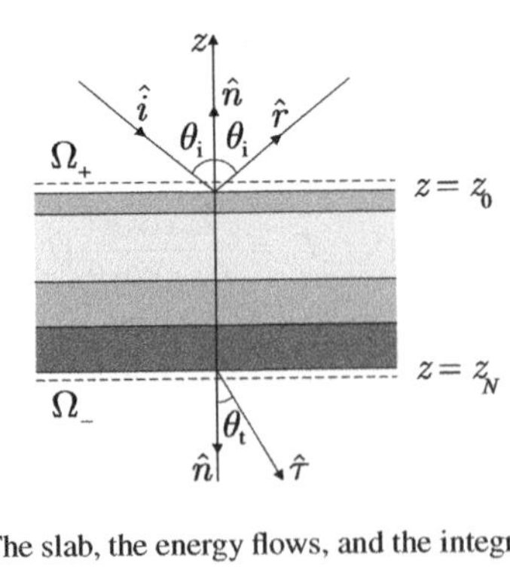

Figure 4.11 The slab, the energy flows, and the integration surfaces $\Omega_\pm$.

where $|E_0^{\rm inc}|$ is the amplitude of the electric field in the incident wave, $|E_0^{\rm ref}|$ in the reflected wave, and $|E_0^{\rm tr}|$ in the transmitted wave. Depending on the polarization case, these are related to each other by reflection and transmission coefficients, either $R_\perp$ and $T_\perp$ or $R_\parallel$ and $T_\parallel$.

Since there are no losses inside the volume enclosed by Ω, the right-hand side of (4.193) is zero, and energy conservation requires that

$$w_z^{\rm inc} + w_z^{\rm ref} - w_z^{\rm tr} = 0, \tag{4.198}$$

where we have taken into account that the integrand function on the left-hand side of (4.193) is constant over the integration surface. Now, dividing (4.198) by $w_z^{\rm inc}$, using (4.195)–(4.197) and the geometric relations

$$\hat{z}\cdot\hat{i} = -\cos\theta_{\rm i}, \quad \hat{z}\cdot\hat{r} = \cos\theta_{\rm i}, \quad \hat{z}\cdot\hat{t} = -\cos\theta_{\rm t}, \tag{4.199}$$

we arrive at relations (4.192) with $\mathcal{R}$ and $\mathcal{T}$ given by (4.188) or (4.189).

In the case of total internal reflection at the lower interface $z = z_N$ of the slab, when $\theta_{\rm i} > \bar{\theta}_{\rm i}$ with $\bar{\theta}_{\rm i}$ defined in (4.166), the wave transmitted in the medium $N+1$ is an inhomogeneous plane wave with the wave vector $\mathbf{k}_{N+1} = k_0 \sin\theta_{\rm i}\hat{x} + jk_{N+1}|\cos\theta_{\rm t}|\hat{z}$, and the expression for the energy flow (4.197) changes to

$$\mathbf{w}^{\rm tr} = \hat{x}\frac{\mu_0 \sin\theta_{\rm i}}{2Z_0\mu_{N+1}}|E_0^{\rm tr}|^2 \tag{4.200}$$

which has no component normal to the interface anymore (see section 4.2.3). Thus, the energy flow $w_z^{\rm tr}$ into the lower half-space vanishes, and the energy conservation relations (4.192) become simply

$$\mathcal{R}_\perp = |R_\perp| = 1, \quad \mathcal{R}_\parallel = |R_\parallel| = 1, \quad (\theta_{\rm i} > \bar{\theta}_{\rm i})$$

implying that the incident energy is completely reflected back into the upper half-space.

Note that in the case of total internal reflection, transmission coefficients $T_\perp$ and $T_\parallel$ do not vanish, but the transmitted energy flows along the interface $z = z_N$ and the transmitted fields decay exponentially as the observation point goes away from the interface into the lower medium (see (4.110) and (4.136)).

4.4 Reflection and Transmission of Cylindrical and Spherical Waves

Real sources excite not just a single plane wave but rather a continuum of plane waves, each of which is reflected and transmitted at material interfaces in accordance with the respective incidence angle and polarization. However, reflection and transmission of waves other than plane waves can still be described in terms of transmission and reflection coefficients for plane waves. In this section we consider the excitation of a multi-layered slab by a line electric current parallel to the slab surface and by an arbitrarily oriented electric dipole. Solutions for corresponding magnetic sources are obtained by using the duality property. Solutions of these problems are of fundamental importance for understanding and modeling of such problems as propagation of radio waves over the Earth surface, wireless communications, and radiation from antennas in the presence of supporting constructions. Detailed analysis of various problems arising when a non-plane wave impinges on a flat material interface can be found, for example, in Brekhovskikh (1960), Chew (1990), Felsen and Marcuvitz (1973), Makarov *et al.* (1991), Stratton (1941), and Wait (1962).

4.4.1 Excitation by a Linear Electric Current

Exact Solution

Consider an infinite linear electric current $\mathbf{J}_{\rm e} = \hat{y}I_0\delta(x - x_*)\delta(z - z_*)$ placed above a multi-layered slab (Figure 4.10) at a point $x = x_*$ and $z = z_*$ with $z_* > z_0$. Assuming for brevity that the amplitude I_0 of the current does not vary with y, it follows from Maxwell's equations (see (2.46) and section 2.5.5) that the electric field has only an y component, and this component satisfies the Helmholtz equation

$$\frac{\partial^2 E_y}{\partial x^2}+\frac{\partial^2 E_y}{\partial z^2}+k^2 E_y = j\omega\mu_0 I_0 \delta(x-x_*)\delta(z-z_*). \tag{4.201}$$

The magnetic field has x and z components which are expressed through E_y according to (4.100).

The incident field can be expressed through the two-dimensional Green function (2.325) as

$$E_y^{\rm inc} = -j\omega\mu_0 I_0 G_2(x,x_*,z,z_*,k_0), \tag{4.202}$$

and using the Fourier expansion (see (2.391))

$$G_2(x,x_*,z,z_*,k_0)=\frac{1}{4\pi j}\int_{-\infty}^{+\infty} e^{-j\xi(x-x_*)-j\gamma_0|z-z_*|}\frac{d\xi}{\gamma_0} \tag{4.203}$$

with $\gamma_0=\sqrt{k_0^2-\xi^2}$ ($\mathrm{Im}\,\gamma_0\le 0$), the electric field component in the incident cylindrical wave can be represented as

$$E_y^{\rm inc}(x,z)=\int_{-\infty}^{+\infty} e^{-j\xi x}\tilde{E}_y^{\rm inc}(z,\xi)\,d\xi, \tag{4.204}$$

where

$$\tilde{E}_y^{\rm inc}(z,\xi) = -\frac{\omega\mu_0 I_0}{4\pi\gamma_0}e^{j\xi x_*-j\gamma_0|z-z_*|}. \tag{4.205}$$

In the region between the source and the upper side of the slab, where $z_0\le z<z_*$, the Fourier transform of the incident wave can be written as

$$\tilde{E}_y^{\rm inc}(z,\xi)=\tilde{E}_y^{\rm inc}(z_0,\xi)e^{j\gamma_0(z-z_0)}. \tag{4.206}$$

The reflected and transmitted portions of the field, as well as the total field in every individual layer, can be sought in the form of a Fourier expansion similar to (4.204),

$$E_y(x,z)=\int_{-\infty}^{+\infty} e^{-j\xi x}\tilde{E}_y(z,\xi)\,d\xi, \tag{4.207}$$

and in order to represent solutions of the homogeneous Helmholtz equation the Fourier amplitude $\tilde{E}_y(z,\xi)$ in (4.207) has to satisfy the differential equation

$$\frac{\partial^2 \tilde{E}_y(z,\xi)}{\partial z^2}+\gamma^2\tilde{E}_y(z,\xi)=0 \tag{4.208}$$

with $\gamma=\sqrt{k^2-\xi^2}$ and $\mathrm{Im}\,\gamma\le 0$. Thus, in every medium $\tilde{E}_y(z,\xi)$ is a linear combination of the exponents $\exp(\pm j\gamma z)$, which implies that the integral in (4.207) is in fact a superposition of elementary waves of the form (4.102). Furthermore, the conditions of continuity of E_y and H_x at material interfaces are satisfied if Fourier transforms of E_y and H_x satisfy them.

Thus, the solution of the reflection and transmission problem for the cylindrical wave can be taken directly from the analysis of section 4.3.1,

$$\tilde{E}_y(z,\xi) = \tilde{E}_y(z_0,\xi)\cdot\begin{cases} e^{-j\gamma_0(z+z_0-2z_*)} + R_E(\xi)e^{-j\gamma_0(z-z_0)}, & z_* \le z < +\infty \\ e^{j\gamma_0(z-z_0)} + R_E(\xi)e^{-j\gamma_0(z-z_0)}, & z_0 \le z \le z_* \\ A_n(\xi)\cos[\gamma_n(z-z_{n-1})] + B_n(\xi)\sin[\gamma_n(z-z_{n-1})], & z_n \le z \le z_{n-1} \\ T_E(\xi)e^{j\gamma_{N+1}(z-z_N)}, & -\infty < z \le z_N \end{cases}, \quad (4.209)$$

where $R_E(\xi)$ and T_E are the reflection and transmission coefficients of the multi-layered slab described in section 4.3.1. The reflected and transmitted fields are given by

$$E_y^{\mathrm{ref}}(x,z) = \int_{-\infty}^{+\infty} \tilde{E}_y(z_0,\xi)R_E(\xi)e^{-j\xi x - j\gamma_0(z-z_0)}\,d\xi \quad (z \ge z_0) \quad (4.210)$$

$$E_y^{\mathrm{tr}}(x,z) = \int_{-\infty}^{+\infty} \tilde{E}_y(z_0,\xi)T_E(\xi)e^{-j\xi x + j\gamma_{N+1}(z-z_N)}\,d\xi \quad (z \le z_N). \quad (4.211)$$

The integrals in (4.210) and (4.211) are contour integrals in the complex plane of the spectral parameter ξ, and their numerical or analytical evaluation requires the knowledge of analytical properties of the reflection and transmission coefficients as functions of the complex variable ξ. These properties are seen from expressions (4.120) for the elements of the characteristic matrices $\overline{\overline{A}}_n^E$ $(n = 1, 2, \ldots, N)$, which include only $\cos(\gamma_n d_n)$, $\gamma_n \sin(\gamma_n d_n)$ or $\sin(\gamma_n d_n)/\gamma_n$ and are, therefore, even functions of ξ, free of pole and branch point singularities on the complex ξ plane. The elements of the characteristic matrix $\overline{\overline{A}}_E$ of the whole structure given by the products of the individual characteristic matrices have apparently the same analytical properties. Reflection and transmission coefficients are expressed in (4.124) and (4.125) as ratios of functions involving the elements of the characteristic matrix and the functions γ_0 and γ_{N+1}, implying that the reflection and transmission coefficients are even functions of ξ and have pole singularities at zeros of the denominators and the branch points at $\xi = \pm k_0$ and $\pm k_{N+1}$. The reflection and transmission coefficients R_E and T_E share the same denominator and have therefore pole singularities at the same locations. The singularities are, in general, complex-valued and located symmetrically with respect to the point $\xi = 0$ on the complex ξ plane (Figure 4.12).

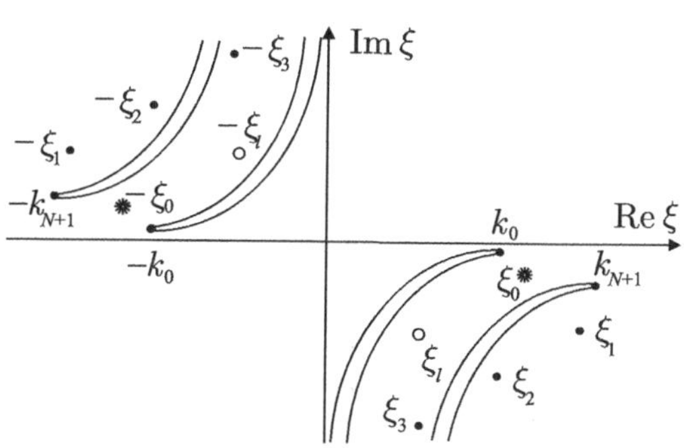

Figure 4.12 Singularities of reflection and transmission coefficients of a multilayered slab between two homogeneous semi-infinite media. The branch cuts defined by the conditions $\mathrm{Im}\,\gamma_0(\xi) = 0$ and $\mathrm{Im}\,\gamma_{N+1}(\xi) = 0$ start at the branch points $\xi = \pm k_0$ and $\xi = \pm k_{N+1}$. Small losses are assumed in the media surrounding the slab so that the branch points are shifted from the real axis. The poles ξ_1, ξ_2, and ξ_3 that correspond to trapped modes are shown by dots; a surface wave pole at ξ_0 by a star. The leaky wave poles at ξ_l (a small circle) lie on the lower sheet of the Riemann surface. Every pole has a counterpart symmetric with respect to the point $\xi = 0$.

For a general multilayered slab consisting of N layers there is an infinite number of poles which may be located at every sheet of the Riemann surface. In qualitative terms, there are N branches of poles residing along the lines $\operatorname{Im}\gamma_n(\xi) = 0$ with $n = 1, 2, \dots, N$ and the distance between the poles within the same branch is inversely proportional to the thickness of the nth layer so that the poles disappear if the thickness d_n of the layer vanishes. These poles, as we shall see later, are related to the modes that may propagate in every layer as in a waveguide. The poles on the upper sheet ($\operatorname{Im}\gamma_0(\xi) \le 0$ and $\operatorname{Im}\gamma_{N+1}(\xi) \le 0$) of the Riemann surface describe the so-called trapped waveguide modes, and the poles on the lower sheet, where $\operatorname{Im}\gamma_0(\xi) > 0$, are responsible for the so-called leaky waves. In addition to them there can be poles related to surface waves, which are waves that may exist on an interface between two homogeneous media (section 4.2.5). These waves do not need a second boundary and are not of the waveguide nature. In principle, every material interface of a multilayered slab may be represented by a respective surface-wave pole.

In the limit of a slab with vanishing thickness (two semi-infinite media with a flat interface between them), all the poles disappear except for one which is associated with a surface wave on the interface between the upper and lower semi-infinite media. Similar pole singularity is included in the reflection coefficient of a surface modeled with impedance boundary conditions. Finally, in the limit of a PEC interface the reflection coefficient becomes a constant ($R_E = -1$) and no pole singularities are present in the integrand of (4.210).

The position of the branch points $\xi = \pm k_0, \pm k_{N+1}$ with respect to the integration contour in (4.210) and (4.211) is determined by the requirement that the contour be entirely located on the sheet of the Riemann surface of the multi-valued functions γ_0 and γ_{N+1} where $\operatorname{Im}\gamma_0 \le 0$ and $\operatorname{Im}\gamma_{N+1} \le 0$, which is necessary to ensure that the reflected and transmitted waves decay as the observation points move away from the slab faces. The suitable Riemann surface has branch cuts along the lines at which $\operatorname{Im}\gamma_0 = 0$ and $\operatorname{Im}\gamma_{N+1} = 0$. It is convenient to assume a finite absorption in the media such that $\operatorname{Im} k_0 < 0$ and $\operatorname{Im} k_{N+1} < 0$, in which case the branch points and the branch cuts are moved off the real axis, making apparent the mutual location of these with respect to the integration contour (Figure 4.12).

Asymptotic Evaluation of the Reflected and Transmitted Fields

Once the position of singularities of the integrands in (4.210) and (4.211) is clarified, one can proceed with numerical or asymptotic integration. In both cases, by using the analyticity of the integrands, one can deform the integration contour onto the most suitable contour to apply numerical quadratures or the steepest descent method. It is instructive to take a look at the asymptotic evaluation of the integrals with the steepest descent method (section C.3) since it highlights the physics of the wave processes in the structure. The application of the method is possible when the distance between the source and observer is greater than several wavelengths. Let us start with the reflected field and rewrite (4.210) in a form suitable to application of the method,

$$E_y^{\mathrm{ref}}(x,z) = \int_{-\infty}^{+\infty} g(\xi) e^{iS(\xi,k)}\, \mathrm{d}\xi \tag{4.212}$$

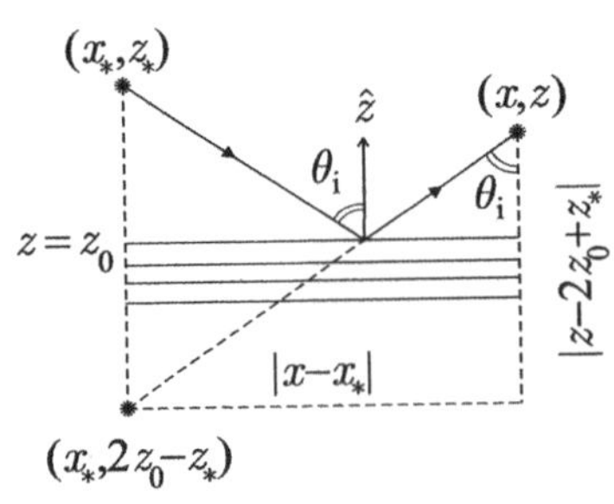

Figure 4.13 Reflection of a cylindrical wave from a slab: the source (x_*, z_*), the image of the source $(x_*, 2z_0 - z_*)$, the observation point (x, z), and the angle θ_i.

with

$$g(\xi) = -\frac{\omega\mu_0 I_0}{4\pi\gamma_0(\xi)} R_E(\xi) \tag{4.213}$$

$$S(\xi, k) = -\xi(x - x_*) - \gamma_0(\xi)(z - 2z_0 + z_*). \tag{4.214}$$

By using the evenness of $g(\xi)$ on the integration contour and the fact that $z - 2z_0 + z_* > 0$, it is convenient to rewrite (4.214) as

$$S(\xi, k) = -\xi|x - x_*| - \gamma_0(\xi)|z - 2z_0 + z_*| \tag{4.215}$$

so as to have non-negative coefficients in the expression for $S(\xi, k)$. The equation (C.12) for the saddle point ξ_c has the form

$$\frac{\partial S}{\partial \xi}(\xi_c, k) = -|x - x_*| + \frac{\xi_c}{\gamma_0(\xi_c)}|z - 2z_0 + z_*| = 0, \tag{4.216}$$

and its solution can be expressed as

$$\xi_c = k_0 \sin\theta_i, \quad \gamma_0(\xi_c) = k_0 \cos\theta_i, \tag{4.217}$$

where the angle θ_i ($0 \le \theta_i \le \pi/2$) is an angle in the triangle with vertices at the source (x_*, z_*), the observation point (x, z), and the image of the source (with respect to the upper side of the slab) at $(x_*, 2z_0 - z_*)$ (Figure 4.13). The angle has apparently the meaning of the incidence angle in a GO solution for the linear source.

Denoting the distance between the image of the source and the observation point by l so that $|x - x_*| = l \sin\theta_i$ and $|z - 2z_0 + z_*| = l \cos\theta_i$, we find that

$$S(\xi_c, k) = -k_0 l, \quad \frac{\partial^2 S}{\partial \xi^2}(\xi_c, k) = \frac{l}{k_0 \cos^2\theta_i}, \tag{4.218}$$

and using (C.24) gives the approximation

$$E_y^{\text{ref}}(x, z) \approx \frac{\omega\mu_0 I_0}{2\sqrt{2\pi k_0 l}} R_E(k_0 \sin\theta_i) \mathrm{e}^{-jk_0 l - j3\pi/4} \tag{4.219}$$

which describes the reflected wave as a wave from the image of the source, corrected by the reflection coefficient of the slab.

The transmitted wave (4.211) is evaluated along similar lines by rewriting it as in (4.212) with

$$g(\xi) = -\frac{\omega\mu_0 I_0}{4\pi\gamma_0(\xi)} T_E(\xi) \tag{4.220}$$

$$S(\xi, k) = -\xi|x - x_*| - \gamma_0(\xi)|z_* - z_0| - \gamma_{N+1}(\xi)|z_N - z|. \tag{4.221}$$

The saddle point equation

$$|x - x_*| = \frac{\xi_c}{\gamma_0(\xi_c)}|z_* - z_0| + \frac{\xi_c}{\gamma_{N+1}(\xi_c)}|z_N - z| \tag{4.222}$$

has the solution

$$\xi_c = k_0 \sin\theta_i = k_{N+1} \sin\theta_t \tag{4.223}$$

$$\gamma_0(\xi_c) = k_0 \cos\theta_i, \quad \gamma_{N+1}(\xi_c) = k_{N+1} \cos\theta_t, \tag{4.224}$$

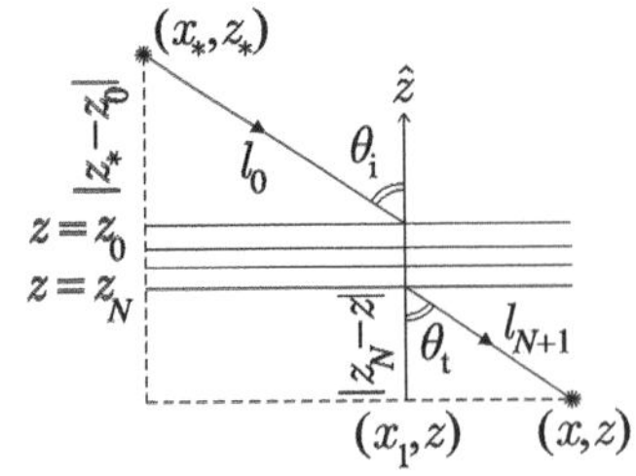

Figure 4.14 Transmission of a cylindrical wave through a slab: the source (x_*, z_*), the observation point (x, z), and the angles θ_i and θ_t.

where θ_i (or θ_t since both are related by (4.223)) is determined from the equation

$$|x - x_*| = |z_* - z_0| \tan\theta_i + |z_N - z| \tan\theta_t. \tag{4.225}$$

The angles θ_i and θ_t are such that

$$|z_* - z_0| = |x_* - x_1| \cot\theta_i, \quad |z_N - z| = |x_1 - x| \cot\theta_t, \tag{4.226}$$

where x_1 is an intermediate point between x_* and x. A geometric interpretation of θ_i and θ_t is given in Figure 4.14.

Defining l_0 as the distance between the source and the point (x_1, z_0) on the upper side of the slab and l_{N+1} as the distance between the point (x_1, z_N) on the lower side of the slab and the observer, we have

$$S(\xi_c, k) = -k_0 \sin\theta_i(|x_* - x_1| + |x_1 - x|) - k_0 \cos\theta_i |z_* - z_0| - k_{N+1} \cos\theta_t |z_N - z| = -k_0 l_0 - k_{N+1} l_{N+1} \tag{4.227}$$

and

$$\frac{\partial^2 S}{\partial \xi^2}(\xi_c, k) = \frac{|z_* - z_0|}{k_0 \cos^3\theta_i} + \frac{|z_N - z|}{k_{N+1} \cos^3\theta_t} = \frac{l_0}{k_0 \cos^2\theta_i} + \frac{l_{N+1}}{k_{N+1} \cos^2\theta_t}, \tag{4.228}$$

thus arriving at the formula

$$E_y^{\mathrm{tr}}(x, z) \approx \frac{\omega\mu_0 I_0 T_E(k_0 \sin\theta_i)}{2\sqrt{2\pi} \left| k_0 l_0 + k_{N+1} l_{N+1} \left(\frac{k_0 \cos\theta_i}{k_{N+1} \cos\theta_t} \right)^2 \right|^{\frac{1}{2}}} \mathrm{e}^{-jk_0 l_0 - jk_{N+1} l_{N+1} - j\frac{3}{4}\pi}. \tag{4.229}$$

Asymptotic formulas (4.219) and (4.229) include the case of a slab with vanishing thickness, that is, the limit of two semi-infinite homogeneous media. The formulas are also applicable to electrically thin slabs, but for slabs of finite thickness they are limited to almost normal incidence on the slab, otherwise the coefficients $R_E(\xi)$ and $T_E(\xi)$ and therefore functions $g(\alpha)$ in (4.213) or in (4.220) are no longer slowly varying. For the same reason, the approximations fail when the saddle point ξ_c approaches a pole or a branch point of $R_E(\xi)$ and $T_E(\xi)$.

The formulas (4.219) and (4.229) should be corrected if in the process of deforming the original integration contour C onto an SDP, poles or branch points of the integrands are crossed. The precise shape of SDP can be obtained from (C.14), which in the case of the phase functions (4.215) and (4.221) requires the solution of a quadratic or a quadric algebraic equation, respectively. Qualitatively, the SDP is a contour that starts at $\xi = \infty \exp[j(\pi + \theta_i)]$, goes through the saddle point $\xi = \xi_c$ at an angle $\chi = \pi/4$ (when k_0 and k_{N+1} are lossless), and ends at $\xi = \infty \exp(-j\theta_i)$. With $\theta_i = 0$ (normal incidence), the ends of the contour go parallel to the real axis in the first and third quadrants of the complex ξ plane. As $\theta_i \to \pi/2$, that is, $|x - x_*| \gg |z_* - z_0| + |z_N - z|$ (grazing incidence), the ends become parallel to the imaginary axis and go to infinity in the third and fourth quadrants. Portions of the contour may enter the lower sheets of the Riemann surface

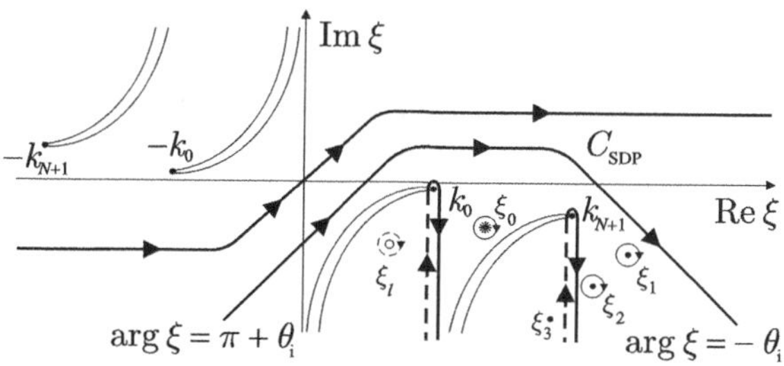

Figure 4.15 The shape of the SDP of the phase functions (4.215) and (4.221) for $\theta_i = 0$, an intermediate value of θ_i $(0 < \theta_i < \pi/2)$, and $\theta_i = \pi/2$. In the latter case, one half of the contour lies on the lower sheets of the Riemann surface (dashed lines). Deformation of the integration contour from the real axis on the SDP may lead to capturing various poles of the reflection or transmission coefficients as well the branch points at $\xi = k_0$ and k_{N+1}.

for the square root functions $\gamma_0(\xi)$ and $\gamma_{N+1}(\xi)$. In the limit $\theta_i = \pi/2$, the SDP reduces to two contours running around the branch points at $\xi = k_0$ and k_{N+1} on the opposite sheets of the Riemann surface (Figure 4.15).

In general, on deformation of the integration contour on SDP we have for the field

$$E_y(x, z) = E_y^{\text{SDP}}(x, z) + E_y^{\text{poles}}(x, z) + E_y^{\text{bc}}(x, z), \tag{4.230}$$

where E_y^{SDP} is the integral over SDP, E_y^{poles} is the sum of residues at the captured poles, and E_y^{bc} is an integral over the sides of a branch cut either from the point $\xi = k_{N+1}$ (reflected field) or $\xi = k_0$ (transmitted field). Let us discuss the physical meaning of the terms of representation (4.230). The integral over SDP can be evaluated at the saddle point and, as we have seen in (4.219) and (4.229), represents the GO rays, either specularly reflected or transmitted according to the Snell law.

The contribution from the captured poles, which we denote by ξ_p with $0 \leq p \leq P$, can be written as

$$E_y^{\text{poles}}(x, z) = -2\pi j \sum_{p=0}^{P} \text{res}\, g(\xi_p) e^{jS(\xi_p, k)}. \tag{4.231}$$

Note that despite the fact that $R_E(\xi)$ and $T_E(\xi)$ share the same pole singularities, the captured poles in (4.231) for the reflected and transmitted fields may be different as the functions $S(\xi, k)$ are different and therefore the SDP contours are different. Furthermore, the amplitude factors $g(\xi, k)$ are also different. Every pole from the upper sheet of the Riemann surface is such that $\text{Im}\,\xi_p < 0$, $\text{Im}\,\gamma_0(\xi_p) < 0$ and $\text{Im}\,\gamma_{N+1}(\xi_p) < 0$, and it follows from (4.215) and (4.221) that the corresponding term in the sum (4.231) decays with growing horizontal $|x - x_*|$ between the source and the observer as well as with growing distance between the source and the slab and between the observer and the slab in the direction perpendicular to the slab. These contributions are associated either with the modes trapped in the layers of the multilayered structure or with the surface waves localized at interfaces of the structure. In the limit of a homogeneous half-space, when, for example, the thickness of the slab goes to zero, just one of these poles, the one at ξ_0, remains on the complex ξ plane (Figure 4.12). This pole is called the *special pole* and the residue at ξ_0 describes a surface wave propagating along the material interface. The presence of this wave in the asymptotic estimation of the field depends on whether or not the pole is crossed in the process of deformation of the integration contour from the real axis onto the SDP (the pole is crossed in Figure 4.15).

Some poles lying on the lower sheet of the Riemann surface where $\text{Im}\,\gamma_0(\xi_p) > 0$ or $\text{Im}\,\gamma_{N+1}(\xi_p) > 0$ may also be crossed in the process of deformation of the original integration contour from the real axis onto the SDP as parts of the SDP may lie on the lower sheets (an example is shown in Figure 4.15). The corresponding residue contributions are interesting in the sense that though they still decay with growing horizontal distance between the source and the observer (as $\text{Im}\,\xi_p < 0$), in the direction perpendicular to the slab they grow as the observer or the source depart from the slab. Such modes are called *leaky modes* as these are modes propagating in the interior of the slab, loosing energy on their way. This behavior does not violate the radiation conditions, however, since as the distance from the slab grows, the shape of the SDP changes, the contour approaches the real axis on the complex ξ plane, and beyond some critical value of the vertical distance the leaky mode pole is not captured any more.

The term E_y^{bc} in (4.230) is an integral over a contour one half of which runs on a lower sheet of the Riemann surface and another half on the upper sheet around a branch point. In the case of reflected wave, this is the branch point $\xi = k_{N+1}$ (Figure 4.15). The integral of this type can also be asymptotically evaluated with the main contribution coming from a small vicinity of the branch point, which leads to a wave of the form

$$E_y^{\text{bc}}(x,z) \sim \mathrm{e}^{-jk_{N+1}|x-x_*|-j\sqrt{k_0^2-k_{N+1}^2}|z-2z_0+z_*|} \quad (z \geq z_0). \tag{4.232}$$

Judging from the phase factor, it is a wave that on its way from the source to the observer, both located above the slab, propagates in part in the lower medium along the lower side of the slab. This wave is termed the *lateral wave*. When the upper medium is optically more dense ($k_0 > k_{N+1}$), the wave in (4.232) is simply the GO ray that arrives at the lower interface of the slab at the angle of the full internal reflection, is refracted into the lower semi-infinite medium in such a way that it propagates along the interface, and then is refracted at the angle of the total internal reflection back into the upper medium toward the receiver. This contribution decays as $O(k^{-3/2})$, that is, faster than the saddle point contribution, which is $O(k^{-1/2})$. In the alternative case where $k_0 < k_{N+1}$, the wave is exponentially attenuating in the vertical direction and in general it is a negligible contribution compared to the saddle point contribution. However, when the upper medium has strong losses and the lower medium has not, the lateral wave may play a dominating role in the field at the observation point. A more detailed analysis of the lateral wave contribution can be found in Felsen and Marcuvitz (1973, Chapter 5) and in Brekhovskikh (1960, section 21).

In general, changing the position of the source and/or observer changes the shape of the SDP, which results in a different number of contributions from the non-ray-optical terms E_y^{poles} and E_y^{bc} to the field. Physically this means excitation of a smaller or larger number of eigenwaves associated with the layered structure. The mathematical condition for the excitation of these non-ray-optical waves due to singularity at a point $\xi = \xi_p$ can be formulated as the condition that

$$\operatorname{Re} S(\xi_p, k) = \operatorname{Re} S(\xi_c, k) \tag{4.233}$$

with $\operatorname{Im} S(\xi_p, k) \geq \operatorname{Im} S(\xi_c, k)$. In terms of the incidence angle θ_i, (4.233) gives the critical value θ_p^i such that the pth wave is excited when $\theta_p^i < \theta_i$. In this sense, non-ray-optical waves are not excited when $\theta_i = 0$.

The spectrum of non-ray-optical waves, even for simple configurations, can be extremely rich and many modes can be excited. Furthermore, the singularities associated with the modes may approach the saddle point, which significantly complicates the asymptotic evaluation of the fields reflected and transmitted at layered structures (see sections C.3.3 and C.3.4). Practical approaches are typically based on numerical integration over a contour which circumvents the singularities and is different from SDP, provided that the behavior of the integrand remains acceptable from the viewpoint of numerical integration.

4.4.2 Excitation by an Electric Dipole

This section describes the solution of the reflection and transmission problem for an elementary electric dipole. Without loss of generality we can assume that the dipole is located in the upper semi-infinite medium at a point $\mathbf{r} = \mathbf{r}_* = (0, 0, z_*)$ with $z_* > z_0$, where z_0 is the position of the upper material interface (Figure 4.10). The current density of the dipole

$$\mathbf{J}_\text{e} = j\omega p_\text{e} \hat{c} \delta(\mathbf{r} - \mathbf{r}_*) \tag{4.234}$$

includes the dipole moment p_e and the unit vector $\hat{c}$ defining the direction of the dipole. An arbitrarily oriented dipole can be represented as a superposition of two dipoles, one of which is perpendicular and another parallel to the interface of the configuration. The analytical form of the solution depends on the dipole orientation, which is due to the fact that a perpendicular electric dipole excites a TM field only and a parallel dipole excites both TE and TM modes. We consider these two types of excitation in succession, present exact integral representations for the reflected and transmitted fields, and show that the solution is expressed in terms of the plane wave reflection and transmission coefficients described in section 4.3.1. The problem is solved by introduction of the Hertz potentials $\mathbf{\Pi}_\text{e} = \Pi_\text{e}\hat{z}$ and $\mathbf{\Pi}_\text{m} = \Pi_\text{m}\hat{z}$ (section 2.5.4), which satisfy the Helmholtz equation and the conditions

$$[\epsilon\Pi_e] = 0, \quad \left[\frac{\partial}{\partial z}\Pi_e\right] = 0, \tag{4.235}$$

$$[\mu\Pi_m] = 0, \quad \left[\frac{\partial}{\partial z}\Pi_m\right] = 0 \tag{4.236}$$

at every material interface $z = z_n$ with $n = 0, 1, \ldots, z_N$. The square brackets in (4.235) and (4.236) denote the jump of the quantity across an interface, that is, $[f] = f(z+0) - f(z-0)$. The conditions at infinity require that the fields reflected from and transmitted through the layered structure are outgoing waves.

Dipole Perpendicular to the Interface

A dipole perpendicular to the interface ($\hat{c} = \hat{z}$) does not excite the H modes and since the boundary conditions for Π_m and Π_e are decoupled, $\Pi_m = 0$. The potential Π_e describes the E modes and is a solution of the equation

$$\nabla^2\Pi_e + k^2\Pi_e = -\frac{p_e}{\epsilon_0}\delta(\mathbf{r} - \mathbf{r}_*). \tag{4.237}$$

The field incident on the upper face of the slab is described by the potential

$$\Pi_e^{\mathrm{inc}} = \frac{p_e}{4\pi\epsilon_0|\mathbf{r} - \mathbf{r}_*|}\mathrm{e}^{-jk_0|\mathbf{r}-\mathbf{r}_*|}, \tag{4.238}$$

which is a solution of (4.237) in the infinite space filled with the medium as in the half-space $z > z_0$. Because of the problem geometry it is convenient to use cylindrical coordinates (ρ, φ, z) and represent the potential Π_e^{inc} as the spectral expansion (equation (2.404), section 2.7)

$$\Pi_e^{\mathrm{inc}}(\rho, z) = \int_0^{+\infty} J_0(k_t\rho)\tilde{\Pi}_e^{\mathrm{inc}}(z, k_t)\,\mathrm{d}k_t \tag{4.239}$$

with

$$\tilde{\Pi}_e^{\mathrm{inc}}(z, k_t) = \frac{p_e k_t}{4\pi j\epsilon_0\gamma_0(k_t)}\mathrm{e}^{-j\gamma_0(k_t)|z-z_*|}. \tag{4.240}$$

Expansion (4.239) is a particular case of integral representations used in the framework of the so-called *Hankel transformation*. The integrals can represent quite general functions and the expansion functions $J_\nu(k_t\rho)$ with $\nu \geq -1/2$ form a complete and orthogonal set (e.g. Watson (1962, section 14.3) and Ryshik and Gradstein (1957, section 5.3)). The integral in (4.239) belongs to the Hankel transform of order zero. Thus, we can represent the potential in every media in the form

$$\Pi_n^e(\rho, z) = \int_0^{+\infty} J_0(k_t\rho)\tilde{\Pi}_n^e(z, k_t)\,\mathrm{d}k_t, \tag{4.241}$$

where $n = 0, 1, 2, \ldots, N+1$. In order to comply with the Helmholtz equation (4.237), the spectral amplitudes $\tilde{\Pi}_n^e(z, k_t)$ have to be linear combinations of the functions $\exp[\pm jz\gamma_n(k_t)]$, where $\gamma_n(k_t) = \sqrt{k_n^2 - k_t^2}$. The branches of the the square root functions are defined by $\operatorname{Im}\gamma_n(k_t) \leq 0$, and the sheet of the Riemann surface with

$$\operatorname{Im}\gamma_0(k_t) \leq 0, \quad \operatorname{Im}\gamma_{N+1}(k_t) \leq 0$$

is referred to as the upper sheet.[6]

[6] Spectral amplitudes are single valued with respect to $\gamma_n(k_t)$ with $n = 1, 2, \ldots, N$.

The spectral function of the field reflected by the interface $z = z_0$ should describe an outgoing wave and can therefore be sought in the form

$$\tilde{\Pi}_{\mathrm{e}}^{\mathrm{ref}}(z, k_t) = \tilde{\Pi}_{\mathrm{e}}^{\mathrm{inc}}(z_0, k_t) R_{\mathrm{e}}(k_t) \mathrm{e}^{-j\gamma_0(k_t)(z-z_0)}, \tag{4.242}$$

where $R_{\mathrm{e}}(k_t)$ is the reflection coefficient that describes reflection of every spectral component in the Hankel expansion of the electric Hertz potential for the field reflected from the interface $z = z_0$. The total field in the medium $z > z_0$ is a sum

$$\tilde{\Pi}_0^{\mathrm{e}}(z, k_t) = \tilde{\Pi}_{\mathrm{e}}^{\mathrm{inc}}(z, k_t) + \tilde{\Pi}_{\mathrm{e}}^{\mathrm{ref}}(z, k_t), \tag{4.243}$$

and in the region between the source and the upper face of the slab ($z_0 < z < z_*$) we have

$$\tilde{\Pi}_0^{\mathrm{e}}(z, k_t) = \tilde{\Pi}_{\mathrm{e}}^{\mathrm{inc}}(z_0, k_t) \left[\mathrm{e}^{j\gamma_0(k_t)(z-z_0)} + R_{\mathrm{e}}(k_t) \mathrm{e}^{-j\gamma_0(k_t)(z-z_0)} \right]. \tag{4.244}$$

The spectral amplitudes of the fields in the layers may include contributions traveling in both directions with respect to the z coordinate so that

$$\tilde{\Pi}_n^{\mathrm{e}}(z, k_t) = \tilde{\Pi}_{\mathrm{e}}^{\mathrm{inc}}(z_0, k_t) \{ A_n(k_t) \cos[\gamma_n(k_t)(z - z_{n-1})] + B_n(k_t) \sin[\gamma_n(k_t)(z - z_{n-1})] \}, \tag{4.245}$$

where $z_n < z < z_{n-1}$ with $n = 1, 2, \ldots, N$. Finally, the wave transmitted to the lower semi-infinite medium ($-\infty < z < z_N$) can be expressed as

$$\tilde{\Pi}_{N+1}^{\mathrm{e}}(z, k_t) = \tilde{\Pi}_{\mathrm{e}}^{\mathrm{inc}}(z_0, k_t) T_{\mathrm{e}}(k_t) \mathrm{e}^{j\gamma_{N+1}(k_t)(z-z_N)} \tag{4.246}$$

with $T_{\mathrm{e}}(k_t)$ being the transmission coefficient for spectral amplitudes in the Hankel expansion of the electric field potential.

Coefficients $R_{\mathrm{e}}(k_t)$ and $T_{\mathrm{e}}(k_t)$ are determined from boundary conditions (4.235). Since these have to be satisfied for $0 < \rho < \infty$, they hold for spectral amplitudes in the Hankel expansions (4.241) for all k_t. The spectral amplitudes in (4.244)–(4.246) have exactly the same form as the corresponding representations (4.106), (4.110), and (4.113) for the TM plane waves but are subject to different boundary conditions on the material interfaces. If, however, we introduce new unknown functions $\pi_n^{\mathrm{e}}(z, k_t)$ by the substitutions $\tilde{\Pi}_n^{\mathrm{e}}(z, k_t) = \pi_n^{\mathrm{e}}(z, k_t)/\epsilon_n$, then they satisfy exactly the same boundary conditions as H_y, namely

$$\pi_n^{\mathrm{e}}(z_n, k_t) = \pi_{n+1}^{\mathrm{e}}(z_n, k_t), \qquad \frac{1}{\epsilon_n} \frac{\partial \pi_n^{\mathrm{e}}}{\partial z}(z_n, k_t) = \frac{1}{\epsilon_{n+1}} \frac{\partial \pi_{n+1}^{\mathrm{e}}}{\partial z}(z_n, k_t) \tag{4.247}$$

with $n = 0, 1, \ldots, N$, implying that the results of section 4.3.1 for TM-polarized plane waves can be directly used, to obtain

$$R_{\mathrm{e}}(k_t) = R_H(k_t), \qquad T_{\mathrm{e}}(k_t) = \frac{\epsilon_0}{\epsilon_{N+1}} T_H(k_t), \tag{4.248}$$

where $R_H(k_t)$ and $T_H(k_t)$ are given by (4.151) and (4.152), respectively.

The fields are obtained from the expressions for the potential by using general relations (2.290) and (2.291) for the E mode. The field components result from (2.267) and (2.268), where one can choose either Cartesian ($\xi = x, \eta = y, \zeta = z, h_\xi = h_\eta = h_\zeta = 1$) or cylindrical ($\xi = \rho, \eta = \varphi, \zeta = z, h_\xi = 1, h_\eta = \rho$ and $h_\zeta = 1$) coordinates. An application example is given in section 4.5.3, where transmission of the field due to a point dipole through a layer of double-negative material (superlens) is studied.

This completes the construction of the exact solution that describes reflection and transmission of a spherical wave from an electric dipole oriented perpendicularly to the interface of the general planar layered structure. To evaluate fields, one can integrate either the Hankel expansion or the equivalent representation (see problem 26)

$$\int_{0}^{+\infty} J_0(k_t\rho)\tilde{\Pi}(k_t)\,\mathrm{d}k_t = \frac{1}{2}\int_{\infty e^{-j\pi}}^{+\infty} H_0^{(2)}(k_t\rho)\tilde{\Pi}(k_t)\,\mathrm{d}k_t, \tag{4.249}$$

which results from relations (E.1) and (E.19) between Bessel functions when the integrand $\tilde{\Pi}(k_t)$ is an odd function of k_t on the upper sheet of the Riemann surface. The integration contour on the right-hand side of (4.249) goes along the real axis on the complex k_t plane, slightly below the negative part of the real axis, which is the branch cut of $H_0^{(2)}(k_t\rho)$. Apart from this branch cut, the other singularities of the integrand as functions of the complex variable k_t coincide with those of the Fourier transformed E_y studied in the previous section for the line source excitation. The contour can be deformed in the complex k_t plane on the steepest descent path (SDP) to apply the saddle point method. The shape of the SDP is close to that studied in the previous section (Figure 4.15) since the Hankel function with $|k_t\rho| > 1$ can be replaced with its large argument approximation, which gives the exponent $\exp(-jk_t\rho)$. All these considerations suggest that the point dipole may excite the same kinds of waves as the linear source, the only difference being the amplitudes of the waves.

Dipole Parallel to the Interface

The solution in the case of an electric dipole parallel to the interface, that is, with $\hat{c} = \hat{x}\cos\varphi_0 + \hat{y}\sin\varphi_0$, where φ_0 is a parameter that describes the orientation of the dipole, is derived along similar lines as for the case of a perpendicular dipole. The main difference is that both potentials, electric Π_e and magnetic Π_m, are necessary to describe the fields. One begins with integral representations of the potentials for the incident field (which is the field of the dipole in free space, problem 35) and, specializing (2.457) and (2.458) to the source at $\rho = 0$ and $z = z_*$ with $z_* > z_0$, arrives at

$$\Pi_e^{\mathrm{inc}}(\rho,\varphi,z) = \frac{p_e}{4\pi j\epsilon_0}\cos(\varphi-\varphi_0)\frac{\partial}{\partial z_*}I_1(\rho,z,z_*) \tag{4.250}$$

$$\tilde{\Pi}_m^{\mathrm{inc}}(\rho,\varphi,z) = \frac{\omega p_e}{4\pi}\sin(\varphi-\varphi_0)I_1(\rho,z,z_*) \tag{4.251}$$

with

$$I_1(\rho,z,z_0) = \int_{0}^{+\infty} J_1(k_t\rho)\mathrm{e}^{-j\gamma_0(k_t)|z-z_*|}\frac{\mathrm{d}k_t}{\gamma_0(k_t)}. \tag{4.252}$$

The potentials Π_m and $\tilde{\Pi}_m$ are related by (2.284) and coincide with each other everywhere except for the plane $z = z_*$. In what follows we will not distinguish between them and use $\tilde{\Pi}_m$ throughout.

Rewriting (4.250)–(4.252) as the Hankel integrals gives

$$\Pi_e^{\mathrm{inc}}(\rho,\varphi,z) = \cos(\varphi-\varphi_0)\int_{0}^{+\infty} J_1(k_t\rho)\tilde{\Pi}_e^{\mathrm{inc}}(z,k_t)\,\mathrm{d}k_t \tag{4.253}$$

$$\Pi_m^{\mathrm{inc}}(\rho,\varphi,z) = \sin(\varphi-\varphi_0)\int_{0}^{+\infty} J_1(k_t\rho)\tilde{\Pi}_m^{\mathrm{inc}}(z,k_t)\,\mathrm{d}k_t, \tag{4.254}$$

where

$$\tilde{\Pi}_e^{\mathrm{inc}}(z,k_t) = \mathrm{sgn}(z-z_*)\frac{p_e}{4\pi\epsilon_0}\mathrm{e}^{-j\gamma_0(k_t)|z-z_*|} \tag{4.255}$$

$$\tilde{\Pi}_m^{\mathrm{inc}}(z,k_t) = \frac{\omega p_e}{4\pi\gamma_0(k_t)}\mathrm{e}^{-j\gamma_0(k_t)|z-z_*|}. \tag{4.256}$$

These representations, which are categorized as the Hankel transforms of the first order, are superpositions of particular solutions of the Helmholtz equation

$$J_1(k_t\rho)e^{\pm j\varphi}e^{\pm j\gamma(k_t)z}. \tag{4.257}$$

Therefore the potentials in every medium can be sought in the form

$$\Pi_n^{e}(\rho,\varphi,z) = \cos(\varphi-\varphi_0)\int_0^{+\infty} J_1(k_t\rho)\tilde{\Pi}_n^{e}(z,k_t)\,dk_t \tag{4.258}$$

$$\Pi_n^{m}(\rho,\varphi,z) = \sin(\varphi-\varphi_0)\int_0^{+\infty} J_1(k_t\rho)\tilde{\Pi}_n^{m}(z,k_t)\,dk_t, \tag{4.259}$$

and the Helmholtz equation in every medium is satisfied if the spectral amplitudes $\tilde{\Pi}_n^{e}(z,k_t)$ and $\tilde{\Pi}_n^{m}(z,k_t)$ are linear combinations of the functions $\exp[\pm j\gamma_n(k_t)z]$.

The remaining task is to satisfy the conditions at $|z|\to\infty$ and the conditions (4.235) and (4.236) at material interfaces $z=z_n$ with $n=0,1,2,\dots,N$. These are satisfied if the spectral amplitudes satisfy the same conditions. Thus, we write

$$\tilde{\Pi}_0^{e}(z,k_t) = \tilde{\Pi}_e^{inc}(z_0,k_t)\left[e^{j\gamma_0(k_t)(z-z_0)} + R_e(k_t)e^{-j\gamma_0(k_t)(z-z_0)}\right] \tag{4.260}$$

$$\tilde{\Pi}_0^{m}(z,k_t) = \tilde{\Pi}_m^{inc}(z_0,k_t)\left[e^{j\gamma_0(k_t)(z-z_0)} + R_m(k_t)e^{-j\gamma_0(k_t)(z-z_0)}\right] \tag{4.261}$$

when $z_0 < z < z_*$,

$$\tilde{\Pi}_n^{e}(z,k_t) = \tilde{\Pi}_e^{inc}(z_0,k_t)\left\{A_n^{e}(k_t)\cos[\gamma_n(k_t)(z-z_{n-1})] + B_n^{e}(k_t)\sin[\gamma_n(k_t)(z-z_{n-1})]\right\} \tag{4.262}$$

$$\tilde{\Pi}_n^{m}(z,k_t) = \tilde{\Pi}_m^{inc}(z_0,k_t)\left\{A_n^{m}(k_t)\cos[\gamma_n(k_t)(z-z_{n-1})] + B_n^{m}(k_t)\sin[\gamma_n(k_t)(z-z_{n-1})]\right\} \tag{4.263}$$

when $z_n < z < z_{n-1}$ with $n=1,2,\dots,N$, and

$$\tilde{\Pi}_{N+1}^{e}(z,k_t) = \tilde{\Pi}_e^{inc}(z_0,k_t)T_e(k_t)e^{-j\gamma_{N+1}(k_t)(z-z_N)} \tag{4.264}$$

$$\tilde{\Pi}_{N+1}^{m}(z,k_t) = \tilde{\Pi}_m^{inc}(z_0,k_t)T_m(k_t)e^{-j\gamma_{N+1}(k_t)(z-z_N)} \tag{4.265}$$

when $-\infty < z < z_N$. The conditions at $z\to\pm\infty$ are satisfied if $\operatorname{Im}\gamma_0(k_t)\le 0$ and $\operatorname{Im}\gamma_{N+1}(k_t)\le 0$ for all k_t on the integration contour $0\le k_t < +\infty$, which specifies the sheet of the Riemann surface for the square root functions, on which the integration contours in (4.258) and (4.259) are located.

Coefficients R_e, R_m, A_n^{e}, B_n^{e}, A_n^{m}, B_n^{m}, T_e, and T_m are determined from the boundary conditions (4.235) and (4.236). Since the electric and magnetic potentials satisfy the conditions separately, R_e and T_e can be taken from the case of a perpendicular dipole (see (4.248)). The coefficients R_m and T_m satisfy equations that can be reduced to those solved in the case of TE-polarized plane waves, so that we have

$$R_m(k_t) = R_E(k_t), \quad T_m(k_t) = \frac{\mu_0}{\mu_{N+1}}T_E(k_t), \tag{4.266}$$

where $R_E(k_t)$ and $T_E(k_t)$ are given by (4.124) and (4.125), respectively.

Historical Notes

The description of electromagnetic fields excited by dipoles in the presence of a planar interface between two different media is a classical problem of wave propagation theory. This solution models the propagation of electromagnetic waves over the Earth surface at a range of distances over which the surface can be assumed planar. The first solution for an electric dipole perpendicular to the interface of two homogeneous media with different ϵ, μ, and σ in the form of Hankel's integrals is attributed to Sommerfeld (1909). The case of an electric or magnetic dipole parallel to the interface of two media with different ϵ was described by Sommerfeld (1926). The solution includes the special pole (ξ_0 in Figure 4.12), the residue at which describes a surface wave propagating along the interface $z = z_0$, the so-called Zenneck wave (Zenneck 1907). This wave cannot be excited by a plane wave but by localized sources. Excitation of such a wave at a material interface by a dipole source depends on the material and the position of the source and receiver, the mathematical criterion being whether the pole is crossed in the process of deformation of the integration contour from the real axis onto the SDP on the complex k_t plane (Weyl 1919). The format used by Sommerfeld (1961) to represent the solution for the dipole over a homogeneous half-space involves integration not along the SDP but rather along the sides of the branch cuts, in which case the special pole is always crossed and the Zenneck wave seems to be always present. However, when the SPD does not catch the pole, the pole contribution is effectively compensated by the branch cut integral and no separate Zenneck wave is observed. A further characteristic feature of the Zenneck wave is the proximity of the pole to the branch point, implying a strong interference of the surface and free-space waves.

A solution in terms of a double integral of plane waves and its asymptotic evaluation with the steepest descent method were presented by Weyl (1919). Other authors (e.g., Brekhovskikh (1960)) prefer to work not on the complex ξ or k_t planes but rather on the complex plane of a new variable θ defined by $k_x = k \sin\theta \cos\phi$, $k_y = k \sin\theta \sin\phi$, and $k_z = k\cos\theta$, which leads to Sommerfeld-type integrals on integration with respect to ϕ.

4.5 A Layer between Homogeneous Half-spaces

The solution presented in section 4.3 includes a great variety of specific configurations. A configuration perhaps most useful for practical applications is a single homogeneous layer separating two homogeneous semi-infinite media, which corresponds to $N = 1$ in Figure 4.10. The layer may serve as an adapter (matching layer) to adjust the reflection or transmission properties of the half-spaces and to arrive thereby at a desired reflection or transmission behavior of the whole configuration as in the design of anti-reflection coatings in optical devises by using thin dielectric films. Particular cases of the configuration are a PEC-backed dielectric layer and a layer in free space, which are useful models of a material coating on a metal vehicle, of a planar antenna, or of a dielectric radome. In this section we study these cases, starting from the more general case of two different and penetrable half-spaces (Figure 4.16).

4.5.1 Different Half-spaces

Specializing the general theory to the case $N = 1$, we obtain

$$R_{\mathrm{TE}} = \frac{\cos(\gamma_1 d_1)\kappa_1^E\left(\kappa_0^E - \kappa_2^E\right) + j\sin(\gamma_1 d_1)\left[\kappa_0^E\kappa_2^E - \left(\kappa_1^E\right)^2\right]}{\cos(\gamma_1 d_1)\kappa_1^E\left(\kappa_0^E + \kappa_2^E\right) + j\sin(\gamma_1 d_1)\left[\kappa_0^E\kappa_2^E + \left(\kappa_1^E\right)^2\right]} \tag{4.267}$$

$$T_{\mathrm{TE}} = \frac{2\kappa_0^E\kappa_1^E}{\cos(\gamma_1 d_1)\kappa_1^E\left(\kappa_0^E + \kappa_2^E\right) + j\sin(\gamma_1 d_1)\left[\kappa_0^E\kappa_2^E + \left(\kappa_1^E\right)^2\right]} \tag{4.268}$$

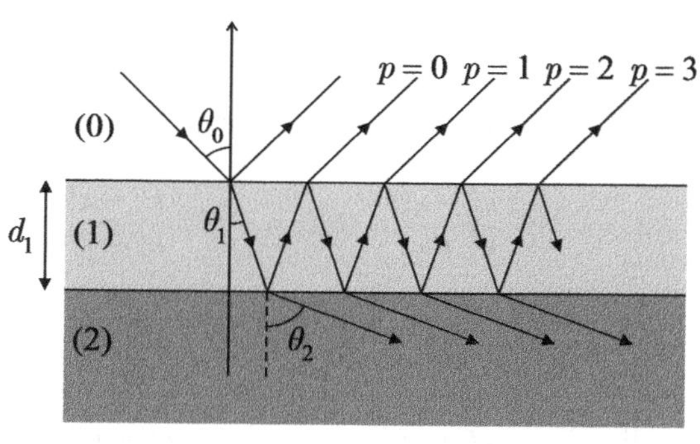

Figure 4.16 Multiple reflections in the interior of a layer separating two homogeneous half-spaces.

$$R_{\mathrm{TM}} = -\frac{\cos(\gamma_1 d_1)\kappa_1^H\left(\kappa_0^H - \kappa_2^H\right) + j\sin(\gamma_1 d_1)\left[\kappa_0^H\kappa_2^H - \left(\kappa_1^H\right)^2\right]}{\cos(\gamma_1 d_1)\kappa_1^H\left(\kappa_0^H + \kappa_2^H\right) + j\sin(\gamma_1 d_1)\left[\kappa_0^H\kappa_2^H + \left(\kappa_1^H\right)^2\right]} \tag{4.269}$$

$$T_{\mathrm{TM}} = \frac{2\kappa_1^H\kappa_2^H}{\cos(\gamma_1 d_1)\kappa_1^H\left(\kappa_0^H + \kappa_2^H\right) + j\sin(\gamma_1 d_1)\left[\kappa_0^H\kappa_2^H + \left(\kappa_1^H\right)^2\right]}, \tag{4.270}$$

where the parameters γ_1 and $\kappa_{0,1,2}^{E,H}$ are defined by (4.163), (4.108), and (4.146). In terms of the incidence angle $\theta_{\mathrm{i}} = \theta_0$, the refraction angle θ_1, and the transmission angle $\theta_{\mathrm{t}} = \theta_2$ (Figure 4.16), which are related by Snell's law (4.167), the parameters are uniformly expressed as

$$\gamma_1 = k_1 \cos\theta_1 \tag{4.271}$$

$$\kappa_n^E = Z_n^{-1}\cos\theta_n \tag{4.272}$$

$$\kappa_n^H = Z_n \cos\theta_n, \tag{4.273}$$

where $Z_n = \sqrt{\mu_n/\epsilon_n}$ is the intrinsic impedance of the nth medium and $n = 0, 1, 2$.

Reflection coefficients R_{TE} and R_{TM} permit alternative representations in terms of reflection and transmission coefficients of individual interfaces in the configuration. Namely, if

$$R_{\mathrm{TE}}^{01} = \frac{\kappa_0^E - \kappa_1^E}{\kappa_0^E + \kappa_1^E}, \quad R_{\mathrm{TM}}^{01} = \frac{\kappa_1^H - \kappa_0^H}{\kappa_0^H + \kappa_1^H} \tag{4.274}$$

and

$$T_{\mathrm{TE}}^{01} = \frac{2\kappa_0^E}{\kappa_0^E + \kappa_1^E}, \quad T_{\mathrm{TM}}^{01} = \frac{2\kappa_1^H}{\kappa_0^H + \kappa_1^H} \tag{4.275}$$

are the reflection and transmission coefficients of the interface separating media 0 and 1 and illuminated from the side of medium 0,

$$R_{\mathrm{TE}}^{10} = \frac{\kappa_1^E - \kappa_0^E}{\kappa_0^E + \kappa_1^E}, \quad R_{\mathrm{TM}}^{10} = \frac{\kappa_0^H - \kappa_1^H}{\kappa_0^H + \kappa_1^H} \tag{4.276}$$

and

$$T_{\mathrm{TE}}^{10} = \frac{2\kappa_1^E}{\kappa_0^E + \kappa_1^E}, \quad T_{\mathrm{TM}}^{10} = \frac{2\kappa_0^H}{\kappa_0^H + \kappa_1^H} \tag{4.277}$$

are the reflection and transmission coefficients of the same interface but illuminated from medium 1, and

$$R_{\mathrm{TE}}^{12} = \mathrm{e}^{-2j\gamma_1 d_1}\frac{\kappa_1^E - \kappa_2^E}{\kappa_1^E + \kappa_2^E}, \quad R_{\mathrm{TM}}^{12} = \mathrm{e}^{-2j\gamma_1 d_1}\frac{\kappa_2^H - \kappa_1^H}{\kappa_1^H + \kappa_2^H} \tag{4.278}$$

and

$$T_{\mathrm{TE}}^{12} = \mathrm{e}^{-j\gamma_1 d_1}\frac{2\kappa_1^E}{\kappa_1^E + \kappa_2^E}, \quad T_{\mathrm{TM}}^{12} = \mathrm{e}^{-j\gamma_1 d_1}\frac{2\kappa_2^H}{\kappa_1^H + \kappa_2^H} \tag{4.279}$$

are the reflection and transmission coefficients[7] of the interface separating media 1 and 2 and illuminated from the side of medium 1, then the expressions (4.267)–(4.270) can be rewritten (see problem 21) in a remarkably simple form:

$$R_{\mathrm{TE}} = \frac{R_{\mathrm{TE}}^{01} + R_{\mathrm{TE}}^{12}}{1 + R_{\mathrm{TE}}^{01} R_{\mathrm{TE}}^{12}} \tag{4.280}$$

$$T_{\mathrm{TE}} = \frac{T_{\mathrm{TE}}^{01} T_{\mathrm{TE}}^{12}}{1 + R_{\mathrm{TE}}^{01} R_{\mathrm{TE}}^{12}} \tag{4.281}$$

$$R_{\mathrm{TM}} = \frac{R_{\mathrm{TM}}^{01} + R_{\mathrm{TM}}^{12}}{1 + R_{\mathrm{TM}}^{01} R_{\mathrm{TM}}^{12}} \tag{4.282}$$

$$T_{\mathrm{TM}} = \frac{T_{\mathrm{TM}}^{01} T_{\mathrm{TM}}^{12}}{1 + R_{\mathrm{TM}}^{01} R_{\mathrm{TM}}^{12}}. \tag{4.283}$$

The physical meaning of the representations (4.280)–(4.283) become apparent upon noting that $R_{\mathrm{TE}}^{01} = -R_{\mathrm{TE}}^{10}$ and $R_{\mathrm{TM}}^{01} = -R_{\mathrm{TM}}^{10}$, followed by expansion of the denominators in a geometric series in powers of $R_{\mathrm{TE}}^{10} R_{\mathrm{TE}}^{12}$ or $R_{\mathrm{TM}}^{10} R_{\mathrm{TM}}^{12}$. For example, doing this in (4.280) and (4.282) results in the expressions (see problem 22)

$$R_{\mathrm{TE}} = R_{\mathrm{TE}}^{01} + T_{\mathrm{TE}}^{01} T_{\mathrm{TE}}^{10} R_{\mathrm{TE}}^{12} \sum_{p=0}^{+\infty} \left(R_{\mathrm{TE}}^{10} R_{\mathrm{TE}}^{12}\right)^p \tag{4.284}$$

$$R_{\mathrm{TM}} = R_{\mathrm{TM}}^{01} + T_{\mathrm{TM}}^{01} T_{\mathrm{TM}}^{10} R_{\mathrm{TM}}^{12} \sum_{p=0}^{+\infty} \left(R_{\mathrm{TM}}^{10} R_{\mathrm{TM}}^{12}\right)^p, \tag{4.285}$$

which represent the reflection coefficients as a superposition of all possible reflections of the incident wave. The first term on the right-hand sides of (4.284) and (4.285) is due to the direct reflection at the upper side of the layer, whereas the sum members describe contributions to the reflected field in the upper half-space upon p-fold reflection of the incident wave in the interior of the layer (Figure 4.16). This way of representing the reflection and transmission coefficients of a homogeneous layer situated between two homogeneous media is attributed to Airy, who used this approach to explain the intensities of Newton's rings (Airy 1833). These physically appealing representations are, however, impractical in calculations because of the need to sum up infinite series. The series diverge when $|R_{\mathrm{TE}}^{10} R_{\mathrm{TE}}^{12}| > 1$, which may take place if the incident field is an inhomogeneous wave. Furthermore, the approach is difficult to extend to structures with more layers because of the difficulty to track all the interior reflections and transmissions when the number of layers in the slab is greater then one.

An interesting feature of the configuration is the possibility of nullifying reflection by choosing the thickness d_1 and the material parameters μ_1 and ϵ_1 of the layer properly. This opens up a possibility of creating anti-reflection coatings to reduce the reflectivity of a given surface. Indeed, if

$$\gamma_1 d_1 = \pi \left(m + \frac{1}{2}\right) \tag{4.286}$$

with $m = 0, 1, \ldots$, then

$$R_{\mathrm{TE}} = \frac{\kappa_0^E \kappa_2^E - \left(\kappa_1^E\right)^2}{\kappa_0^E \kappa_2^E + \left(\kappa_1^E\right)^2} \tag{4.287}$$

$$R_{\mathrm{TM}} = \frac{\left(\kappa_1^H\right)^2 - \kappa_0^H \kappa_2^H}{\kappa_0^H \kappa_2^H + \left(\kappa_1^H\right)^2}, \tag{4.288}$$

[7] The phase shifts in (4.278) and (4.279) account for the displacement between the interface and the plane at which the reflected and transmitted waves are related to the incident wave in the definitions (4.105), (4.112), (4.137), and (4.138) (see also (4.179) and (4.180)).

and the reflection coefficients vanish when

$$\left(\kappa_1^E\right)^2 = \kappa_0^E \kappa_2^E \tag{4.289}$$

$$\left(\kappa_1^H\right)^2 = \kappa_0^H \kappa_2^H, \tag{4.290}$$

or, using (4.272) and (4.273),

$$Z_0 Z_2 \cos^2\theta_1 = Z_1^2 \cos\theta_0 \cos\theta_2 \tag{4.291}$$

$$Z_1^2 \cos^2\theta_1 = Z_0 Z_2 \cos\theta_0 \cos\theta_2. \tag{4.292}$$

The conditions (4.291) and (4.292) are in general different, implying separate nullification of reflection for perpendicular and parallel polarization. With the condition (4.291) satisfied,

$$T_{\text{TE}} = (-1)^{m-1} j \frac{\kappa_0^E}{\kappa_1^E} = (-1)^{m-1} j \frac{Z_1 \cos\theta_0}{Z_0 \cos\theta_1}, \tag{4.293}$$

and because of (4.188)

$$\mathcal{R}_\perp = 0, \quad \mathcal{T}_\perp = 1, \tag{4.294}$$

implying perfect transmission of the energy transported by an incident wave of perpendicular polarization through the layer into the lower medium.

For perfect transmission of the energy transported by a wave of parallel polarization, the condition (4.292) should be satisfied, in which case

$$T_{\text{TM}} = (-1)^{m-1} j \frac{\kappa_1^H}{\kappa_0^H} = (-1)^{m-1} j \frac{Z_1 \cos\theta_1}{Z_0 \cos\theta_0} \tag{4.295}$$

and, because of (4.189),

$$\mathcal{R}_\parallel = 0, \quad \mathcal{T}_\parallel = 1. \tag{4.296}$$

The polarization difference disappears at normal incidence when $\theta_0 = \theta_1 = \theta_2 = 0$, in which case equations (4.289) and (4.290) coincide to give

$$Z_1^2 = Z_0 Z_2. \tag{4.297}$$

Figure 4.17 illustrates the dependence of the reflection coefficient on the thickness of the layer in the case of normal incidence. The permeabilities are the same in all the media ($\mu_0 = \mu_1 = \mu_2$), and the permittivities of the semi-infinite media are such that $\epsilon_2/\epsilon_0 = 4$. The curves show the reflection coefficient for several values of ϵ_1/ϵ_0: 1 (dotted line), 2 (solid line), 4 (dotted line), and 8 (dashed line). When $\epsilon_1/\epsilon_0 = 1$ or 4 (dotted line), an interface between the layer and one of the half-spaces disappears, and the remaining interface between two homogeneous semi-infinite media with ϵ_0 or ϵ_2 leads to the same absolute values of the reflection coefficient in both cases. When $\epsilon_1 = 2\epsilon_0$ (solid line), the condition (4.297) is satisfied and the reflection coefficient vanishes at $d_1/\lambda_1 = 0.25$ and 0.75, which is in agreement with the formula (4.286) with $m = 0, 1$. With $\epsilon_1 > \epsilon_2$ (dashed line), the reflection coefficient has a maximum at the thicknesses defined by (4.286).

For oblique illumination, however, equations (4.289) and (4.290) are different, and to cancel the reflection for both polarizations the three equations (4.286), (4.289), and (4.290) have to be satisfied simultaneously. This can be done by

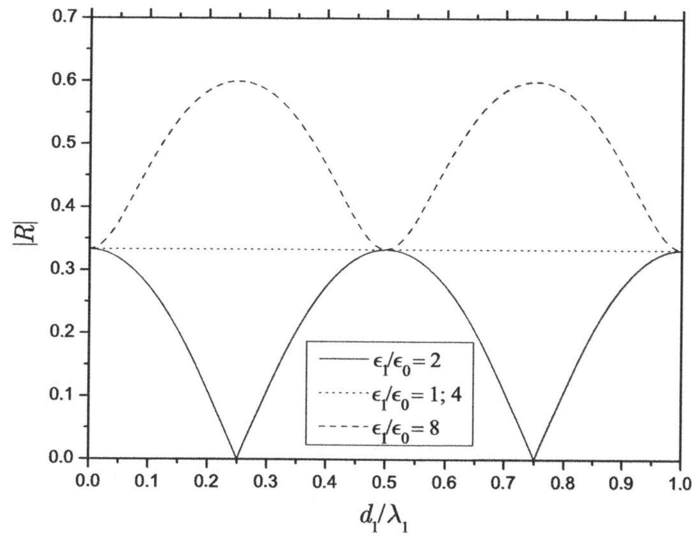

Figure 4.17 Reflection coefficient of a layer separating two homogeneous semi-infinite media as a function of the layer thickness d_1 normalized to the wavelength λ_1 in the material of the layer for several values of permittivity ϵ_1 in the case of normal incidence on the interface.

choosing the three parameters d_1, μ_1, and ϵ_1 according to the equations

$$\frac{\epsilon_1}{\epsilon_0} = \left(\frac{\epsilon_2\mu_0}{\epsilon_0\mu_2}q^2\right)^{1/4} \tag{4.298}$$

$$\frac{\mu_1}{\mu_0} = \left(\frac{\epsilon_0\mu_2}{\epsilon_2\mu_0}q^2\right)^{1/4} \tag{4.299}$$

$$d_1 = \frac{\lambda_0(m+1/2)}{2\sqrt{q-\sin^2\theta_i}}, \tag{4.300}$$

where θ_i is the angle of incidence, λ_0 is the wavelength in the upper semi-infinite medium, and

$$q = \frac{\epsilon_1\mu_1}{\epsilon_0\mu_0} = \frac{\sin^2\theta_i}{1-\cos\theta_i\sqrt{1-\frac{\epsilon_0\mu_0}{\epsilon_2\mu_2}\sin^2\theta_i}}. \tag{4.301}$$

A physical solution (with real-valued d_1) requires that the upper and lower media be lossless. Then, a solution exists for arbitrary θ_i when $\epsilon_2\mu_2 > \epsilon_0\mu_0$ and for $\theta_i < \bar{\theta}_i$ with $\bar{\theta}_i$ defined in (4.166) when $\epsilon_2\mu_2 < \epsilon_0\mu_0$. Note that the optimum permittivity and permeability are independent of the layer thickness, and the thickness divided by λ_0 does not depend on the frequency as long as the materials are non-dispersive.

Figure 4.18 presents a sample solution of equations (4.298)–(4.300) for the case that $\epsilon_2/\epsilon_0 = 4$ and $\mu_2/\mu_0 = 3$. It is seen that the solution exists for every value of θ_i. The thickness of the optimum layer, shown for $m = 0$, grows as θ_i approaches 90°.

Figure 4.19 shows a two-dimensional density distribution of the matrix norm $\|\overline{\overline{R}}\| = \sqrt{|R_{TE}|^2 + |R_{TM}|^2}$ of the reflection coefficient matrix for a hypothetical layer with optimum values of ϵ_1 and μ_1 at every value of θ_i. The matrix norm gives the upper-bound polarization-averaged estimate of the reflection coefficient and provides, therefore, the worst-case estimate of reflection from the layer. The parameters of the surrounding media are $\epsilon_2/\epsilon_0 = 4$

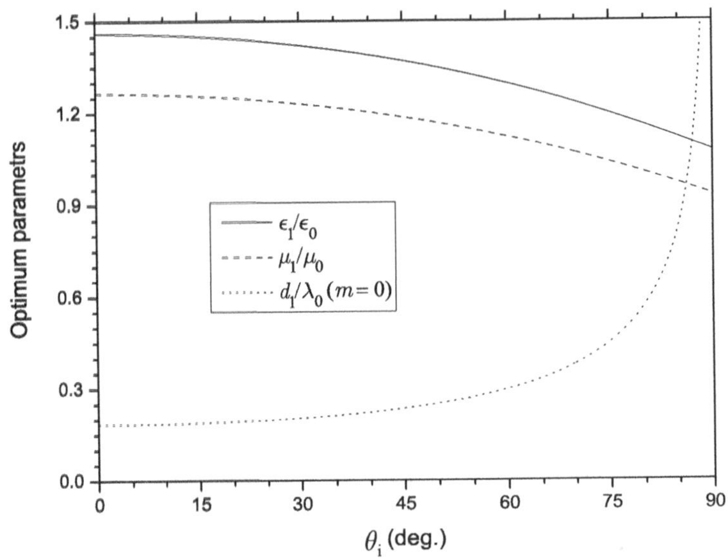

Figure 4.18 Optimum parameters of a layer separating two homogeneous semi-infinite media as a function of the illumination angle.

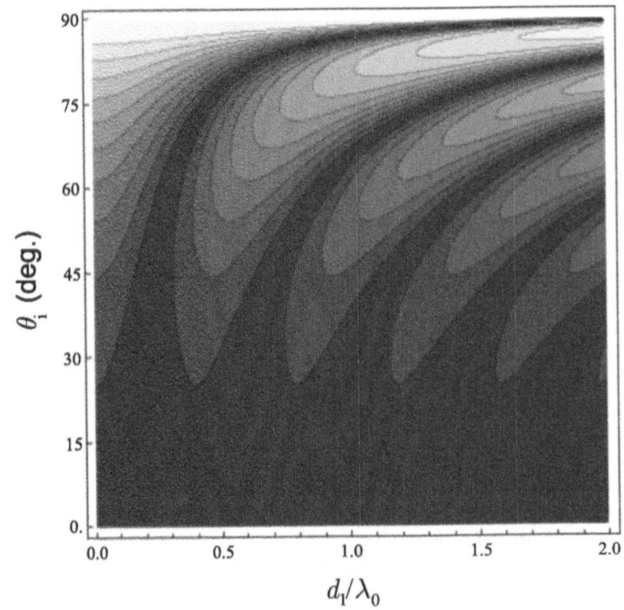

Figure 4.19 Upper-bound polarization-averaged estimate of the reflection coefficient of a layer with optimum values of ϵ_1 and μ_1 as a function of the incidence angle and the thickness of the layer normalized to the wavelength in the upper semi-infinite medium.

and $\mu_2/\mu_0 = 3$. The lighter color corresponds to greater values of $\|\bar{\bar{R}}\|$. The darker regions are regions where $\|\bar{\bar{R}}\|$ is small and the value of d_1 is close to the optimum thickness defined by equation (4.300). The shape of the darker regions indicates, therefore, the dependence of the optimum thickness on the angle of incidence. The five valleys that are visible on the plot correspond to $m = 0, 1, 2, 3, 4$. At $\theta_i = 0$ they start from the points where $d_1 = 0.5(m + 0.5)\lambda_1$, and then, as θ_i grows, follow a nearly parabolic-shaped path, bending constantly to the right, and approaching horizontal asymptotes as $\theta_i \to 90°$.

4.5.2 A PEC-backed Layer

A special case of the configuration with $N = 1$ is a homogeneous layer backed by a perfectly electrically conducting plate. This is a useful model of a metal scatterer coated with a dielectric material. If the layer is lossy, then the configuration models the Dallenbach layer – one of the oldest and simplest types of radar absorbers (Dällenbach and Kleinsteuber 1938; Knott *et al.* 1993).

If the lower half-space is a perfect conductor ($|\epsilon_2| = \infty$), then the transmission coefficients vanish and reflection coefficients (4.267) and (4.269) simplify to

$$R_{\text{TE}} = \frac{-\kappa_1^E \cos(\gamma_1 d_1) + j\kappa_0^E \sin(\gamma_1 d_1)}{\kappa_1^E \cos(\gamma_1 d_1) + j\kappa_0^E \sin(\gamma_1 d_1)} \tag{4.302}$$

$$R_{\text{TM}} = \frac{-\kappa_0^H \cos(\gamma_1 d_1) + j\kappa_1^H \sin(\gamma_1 d_1)}{\kappa_0^H \cos(\gamma_1 d_1) + j\kappa_1^H \sin(\gamma_1 d_1)}. \tag{4.303}$$

In lossless materials, when μ_1 and ϵ_1 are real-valued, reflection coefficients in (4.302) and (4.303) are ratios of complex conjugates and, therefore, $|R_{\text{TE}}| = |R_{\text{TM}}| = 1$. This property is a consequence of the energy conservation and guarantees the equality of the energy fluxes in the incoming and outgoing waves.

The surface impedances result from (4.127), (4.130), (4.153), and (4.157) as

$$Z_E = \frac{j}{\kappa_1^E} \tan(\gamma_1 d_1) \tag{4.304}$$

$$Z_H = j\kappa_1^H \tan(\gamma_1 d_1). \tag{4.305}$$

When $\theta_i = 0$,

$$Z_E = Z_H = j\sqrt{\frac{\mu_1}{\epsilon_1}} \tan(k_1 d_1) \tag{4.306}$$

and the dependence on polarization disappears, which is the general property of the surface impedance of any isotropic surface. Furthermore, it follows from (4.304) and (4.305) that in the limiting case of vanishing thickness of the layer $Z_E \to 0$ and $Z_H \to 0$, whereas in the alternative limit of $\operatorname{Im} k_1 d_1 \to \infty$ (thick layer), expressions (4.12) and (4.13) for the surface impedances of the material half-space are recovered.

Figure 4.20 shows the reflection coefficient of a Dallenbach-like structure for normal incidence $\theta_i = 0$ as a function of d_1/λ_1 where $\lambda_1 = \lambda_0 \operatorname{Re} \sqrt{\epsilon_0\mu_0/(\epsilon_1\mu_1)}$ is the wavelength in the material of the layer (Knott *et al.* 1993). The material parameters

$$\epsilon_1/\epsilon_0 = A_\epsilon \exp(-j\delta_\epsilon), \quad \mu_1/\mu_0 = A_\mu \exp(-j\delta_\mu)$$

correspond to a dominantly electric material $A_\epsilon = 16$, $\delta_\epsilon = 20°$, $A_\mu = 1$, $\delta_\mu = 0$ (dashed line), dominantly magnetic material $A_\epsilon = 1$, $\delta_\epsilon = 0$, $A_\mu = 16$, $\delta_\mu = 10°$ (dotted line), and a hypothetic material $A_\epsilon = A_\mu = 16$, $\delta_\epsilon = \delta_\mu = 15°$ (solid line). All these materials are lossy and therefore the absolute value of the reflection coefficient is smaller than unity. In the latter material $\mu_1/\mu_0 = \epsilon_1/\epsilon_0$ and thus $Z_1 = Z_0$, implying that reflection from the layer is entirely due to the PEC ground plane, which decays exponentially as the thickness of the layer and, therefore, the associated energy losses increase.

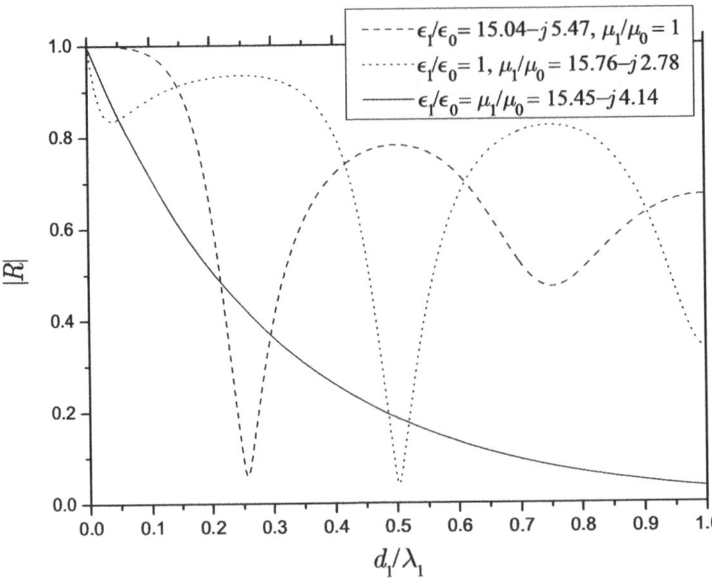

Figure 4.20 Reflection coefficient of a homogeneous absorbing layer backed by a PEC plate at normal illumination as a function of the layer thickness normalized to the wavelength in the material of the layer.

As the incidence angle θ_i deviates from the normal direction, the reflecting properties of the configuration change in that R_{TE} and R_{TM} are no longer equal, approaching the limiting values -1 and 1 as $\theta_i \to 90°$, respectively. Figure 4.21 is a contour plot showing the dependence of the matrix norm $\|\overline{\overline{R}}\| = \sqrt{|R_{TE}|^2 + |R_{TM}|^2}$ on θ_i and d_1/λ_1 for the configuration with $A_\epsilon = A_\mu = 16$, $\delta_\epsilon = \delta_\mu = 15°$. It is seen that with growing θ_i, $\|\overline{\overline{R}}\|$ constantly increases and the structure looses its absorbing performance. A physical explanation is that at oblique illumination the impedance of the upper interface of the

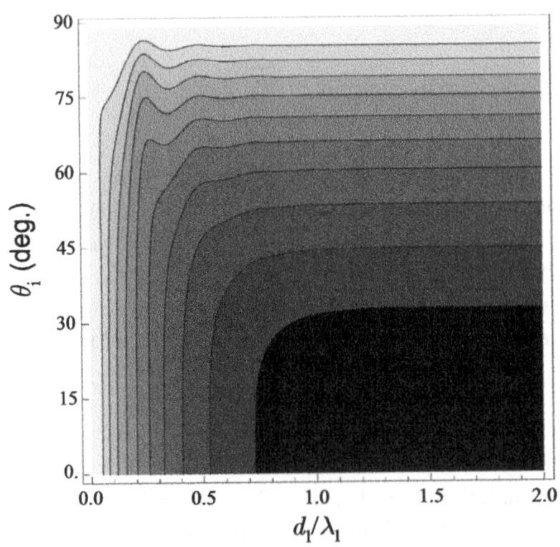

Figure 4.21 Upper-bound polarization-averaged estimate of the reflection coefficient of an absorbing layer with $\epsilon_1/\epsilon_0 = \mu_1/\mu_0 = 16\exp(-j\pi/12)$ as a function of the incidence angle and the layer thickness. Darker shading corresponds to lower reflection.

layer is no longer matched with the impedance on the upper semi-infinite medium. A detailed review of thin absorbing layers can be found in Ra'di *et al.* (2015).

4.5.3 Layer Immersed in a Homogeneous Space

A further important sub-type of the configuration with $N = 1$ is a single layer placed between two identical semi-infinite media. This can model, for example, a radome of an antenna, a window or a wall of a building. Reflection and transmission coefficients of the structure are obtained by setting $\epsilon_2 = \epsilon_0$ and $\mu_2 = \mu_0$ in (4.267)–(4.270),

$$R_{\mathrm{TE}} = \frac{j\sin(\gamma_1 d_1)\left[\left(\kappa_0^E\right)^2 - \left(\kappa_1^E\right)^2\right]}{2\kappa_0^E\kappa_1^E\cos(\gamma_1 d_1) + j\sin(\gamma_1 d_1)\left[\left(\kappa_0^E\right)^2 + \left(\kappa_1^E\right)^2\right]} \tag{4.307}$$

$$T_{\mathrm{TE}} = \frac{2\kappa_0^E\kappa_1^E}{2\kappa_0^E\kappa_1^E\cos(\gamma_1 d_1) + j\sin(\gamma_1 d_1)\left[\left(\kappa_0^E\right)^2 + \left(\kappa_1^E\right)^2\right]} \tag{4.308}$$

$$R_{\mathrm{TM}} = -\frac{j\sin(\gamma_1 d_1)\left[\left(\kappa_0^H\right)^2 - \left(\kappa_1^H\right)^2\right]}{2\kappa_0^H\kappa_1^H\cos(\gamma_1 d_1) + j\sin(\gamma_1 d_1)\left[\left(\kappa_0^H\right)^2 + \left(\kappa_1^H\right)^2\right]} \tag{4.309}$$

$$T_{\mathrm{TM}} = \frac{2\kappa_0^H\kappa_1^H}{2\kappa_0^H\kappa_1^H\cos(\gamma_1 d_1) + j\sin(\gamma_1 d_1)\left[\left(\kappa_0^H\right)^2 + \left(\kappa_1^H\right)^2\right]}. \tag{4.310}$$

It can be seen from (4.307)–(4.310) that when $\gamma_1 d_1 = \pi n$ with $n = 0, 1, 2\ldots$, the reflection coefficients vanish and the transmission coefficients become simply $(-1)^n$, implying perfect transmission through the layer. This effect is called *Fabry–Perot resonance*. The corresponding values of the thickness of the layer are given by the relation

$$d_1 = \frac{0.5\lambda_0 n}{\sqrt{\epsilon_1\mu_1/(\epsilon_0\mu_0) - \sin^2\theta_{\mathrm{i}}}}, \tag{4.311}$$

where λ_0 is the wavelength in the surrounding medium. Since d_1 must be real-valued, the perfect transmission requires lossless media.

R_{TE} and R_{TM} also vanish if $\kappa_0^E = \pm\kappa_1^E$ and $\kappa_0^H = \pm\kappa_1^H$, respectively, which is identical with the conditions (4.52) and (4.53) specifying Brewster's angles at an interface between two semi-infinite media with electric and magnetic constants ϵ_0 and μ_0 on one side and ϵ_1 and μ_1 on the other.

Figure 4.22 shows $|R_{\mathrm{TE}}|$ as a function of θ_{i} and d_1/λ_0 for a layer of Plexiglas in free space at 10 GHz (Table 2.1). The lighter shading corresponds to stronger reflection, and the general tendency is that $|R_{\mathrm{TE}}|$ increases as θ_{i} becomes more grazing. The darker regions correspond to lower reflection and, in addition to the vicinity of the vertical axis where $d_1 = 0$ (no layer), six further regions are seen, each, in accordance with (4.311), starting at $d_1/\lambda_0 = 0.31n$ with $n = 1, 2, \ldots 6$ on the horizontal axis ($\theta_{\mathrm{i}} = 0$) and then smoothly bending to the right as θ_{i} increases.

Figure 4.23 shows the corresponding behavior of the transmission coefficient. The distributions for $|R_{\mathrm{TE}}|$ and $|T_{\mathrm{TE}}|$ are nearly complementary to each other because the absorption in the material is small and in the lossless case the absolute values of the reflection and transmission coefficients are related by the energy conservation (4.188).

Figures 4.24 and 4.25 illustrate the behavior of the reflection and transmission coefficients in the alternative case of TM polarization. In addition to the regions of lower reflection/higher transmission similar to those in Figures 4.22 and 4.23, a horizontal region around $\theta_{\mathrm{i}} = 58°$ is clearly seen, which corresponds to the Brewster's angle θ_{B}^H, at which an incident TM-polarized plane wave is not reflected at the layer.

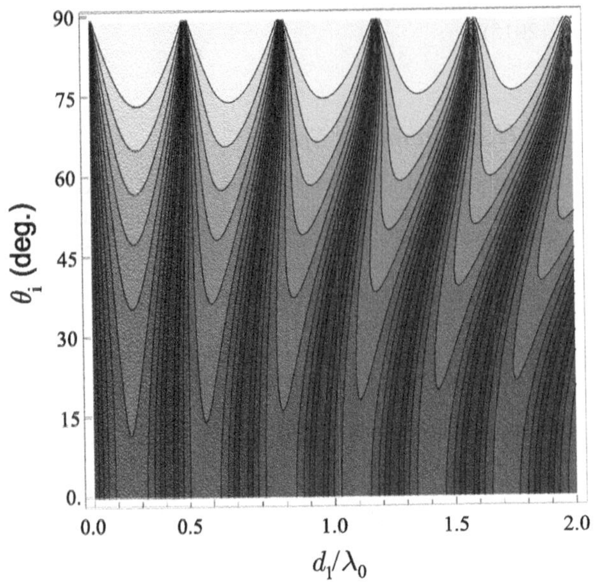

Figure 4.22 Absolute value of the TE reflection coefficient of a homogeneous Plexiglas layer in free space at 10 GHz ($\epsilon_r = 2.59 - j0.02$, $\mu_r = 1$) as a function of the incidence angle and the layer thickness relative to the wavelength in the free space. Darker shading corresponds to lower reflection.

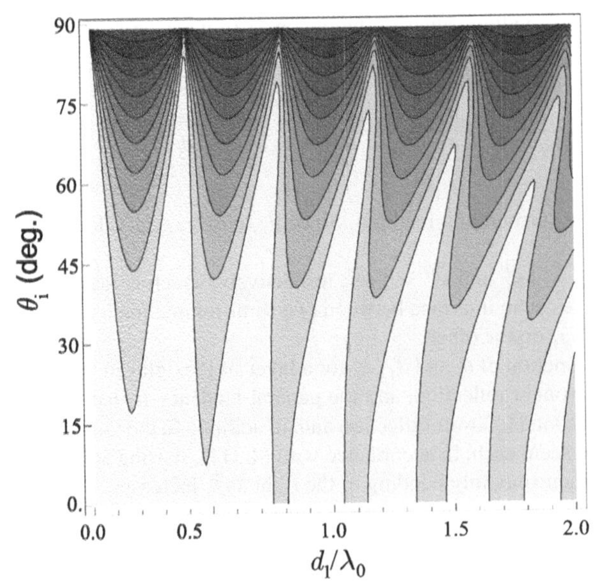

Figure 4.23 The same as Figure 4.22 but for $|T_{TE}|$. Lighter shading corresponds to higher transmission.

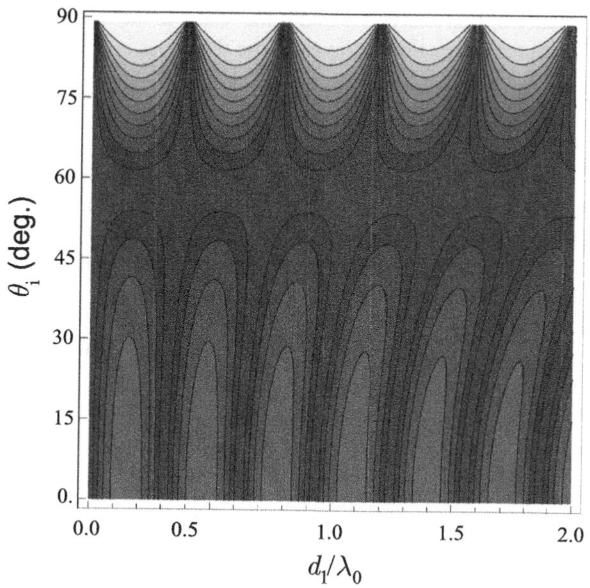

Figure 4.24 The same as Figure 4.22 but for $|R_{\text{TM}}|$.

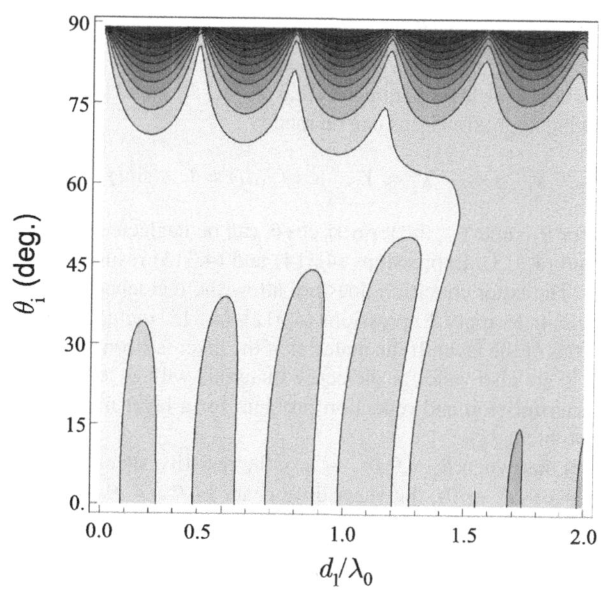

Figure 4.25 The same as Figure 4.23 but for $|T_{\text{TM}}|$.

An important practical consequence of this analysis is that if the losses in the layer are small, then a thinner layer may reflect more strongly and transmit worse than a thicker layer whose thickness is chosen in accordance with (4.311).

The Resistive Sheet

An interesting and practically important limiting case of the single layer configuration is the resistive sheet (e.g., Jenn (2005), Knott *et al.* (1993), and Senior and Volakis (1995)). Suppose that the layer is a good conductor such that $\epsilon_1 \approx -j\sigma_1/\omega$ and $|\epsilon_1| \gg \epsilon_0$, but its thickness d_1 is smaller than the skin depth $\delta = \sqrt{2/(\omega\sigma_1\mu_1)}$ so that the layer is partially transparent. Under these conditions the material layer can be modeled by the simple resistive sheet transition conditions (2.111) and (2.112), and its reflecting and transmitting properties can be entirely described by a single parameter, the resistivity $R_e = (\sigma_1 d_1)^{-1}$ (see (2.110)).

Let us start by showing that if $|k_1 d_1| \ll 1$, $|\epsilon_1/\epsilon_0| \gg 1$ and $|\mu_1/\mu_0|$ is neither small nor large, then the exact expressions (4.307)–(4.310) for the reflection and transmission coefficients simplify to

$$R_{\mathrm{TE}} = \frac{-Z_0}{2\tilde{R}_e \cos\theta_i + Z_0} \tag{4.312}$$

$$T_{\mathrm{TE}} = \frac{2\tilde{R}_e \cos\theta_i}{2\tilde{R}_e \cos\theta_i + Z_0} \tag{4.313}$$

$$R_{\mathrm{TM}} = \frac{-Z_0 \cos\theta_i}{2\tilde{R}_e + Z_0 \cos\theta_i} \tag{4.314}$$

$$T_{\mathrm{TM}} = \frac{2\tilde{R}_e}{2\tilde{R}_e + Z_0 \cos\theta_i} \tag{4.315}$$

with

$$\tilde{R}_e = \frac{Z_1}{jk_1 d_1} = \frac{1}{j\omega d_1 \epsilon_1}. \tag{4.316}$$

It follows from the assumptions about the layer parameters that $|k_0/k_1| \ll 1$ and $|Z_1/Z_0| = |Y_0/Y_1| \ll 1$, which implies that in (4.307)–(4.310) the following approximations can be made:

$$\gamma_1 \approx k_1, \quad \kappa_1^H \approx Z_1, \quad \kappa_1^E \approx Y_1, \quad \cos(\gamma_1 d_1) \approx 1, \quad \sin(\gamma_1 d_1) \approx k_1 d_1. \tag{4.317}$$

Furthermore, in the expressions for R_{TE} and T_{TE} the term $Y_0 \cos\theta_i$ can be neglected compared with Y_1, which transforms (4.307) and (4.308) to (4.312) and (4.313). Expressions (4.314) and (4.315) result from (4.309) and (4.310) under the assumption that $Z_0 \cos\theta_i \gg |Z_1|$. The latter condition does not allow the incidence angle to be too grazing, in order for the approximation for TM polarization to apply. Expressions (4.312)–(4.315) include just one parameter $\tilde{R}_e$ that describes the thickness and material properties of the layer. If the material of the layer is strongly conducting, then $\epsilon_1 \approx -j\sigma_1/\omega$ and $\tilde{R}_e \approx R_e$. Note that (4.312)–(4.315) are also valid for dielectric materials with $\epsilon_1 \gg \epsilon_0$. The expressions (4.312)–(4.315) can be obtained by solving the transmission and reflection problem for a layer modeled by the approximate transition conditions (2.111) and (2.112) (problem 31).

It follows from (4.312)–(4.315) that when $R_e = 0$ ($\sigma_1 \to \infty$), the resistive sheet is fully opaque since $T_{\mathrm{TE}} = T_{\mathrm{TM}} = 0$ and $R_{\mathrm{TE}} = R_{\mathrm{TM}} = -1$. When $R_e = \infty$ ($d_1 \to 0$), the sheet disappears as $R_{\mathrm{TE}} = R_{\mathrm{TM}} = 0$ and $T_{\mathrm{TE}} = T_{\mathrm{TM}} = 1$. For finite conductivity and thickness, it is the ratio R_e/Z_0 that determines the reflection and transmission at the sheet. In terms of a characteristic thickness $d_c = (\sigma_1 Z_0)^{-1}$, the sheet is penetrable when $d_1 < d_c$ and opaque when $d_1 > d_c$. Strictly speaking, in order for a thin layer to be penetrable, it is necessary that both conditions $d_1 < d_c$ and $d_1 < \delta$ are fulfilled (the latter condition ensures that $|k_1 d_1| \ll 1$), but for a good conductor d_c is much smaller than the skin depth δ as $d_c/\delta = \sqrt{\omega\mu_1\epsilon_0/(2\mu_0\sigma_1)} \ll 1$.

Because of the high conductivity values, purely metal films are practically impenetrable at microwave frequencies. For example, for copper $\sigma = 6 \times 10^7$ S/m and $d_c \approx 4 \times 10^{-5}$ μm = 0.4 Å which is comparable to the sizes of atoms; the skin depth is much greater, for example $\delta = 65$ mm at 1 Hz, 2 mm at 1 KHz, 65 μm at 1 MHz, 2 μm at 1 GHz, and 0.65 μm at

10 GHz (see Table 2.5). To arrive at resistivity values comparable to that of free space ($Z_0 \approx 377$ ohm), resistive sheets have to be artificially fabricated by placing thin metal deposits on low-permittivity dielectric films. Resistive sheets with surface resistance up to several thousand ohms are now available, and they are an effective means for controlling reflection from a surface. An example of such structure, the Salisbury screen, which involves a resistive sheet with $R_e = Z_0$, is addressed in problem 27.

The Superlens

Finally, we note interesting properties of parallel-sided slabs with negative values of permittivity and permeability (Veselago 1968). Negative refraction (see section 4.2) can make a parallel-sided slab of a DNG material act as a collecting lens (Pendry 2000; Veselago 1968) if the source is at a finite distance from the slab. A DNG material with $\epsilon = -\epsilon_0$ and $\mu = -\mu_0$, where ϵ_0 and μ_0 are the permittivity and permeability of the surrounding medium ($\epsilon_0 > 0$, $\mu_0 > 0$), is particularly advantageous because in this case the DNG layer, similarly to the single interface, is perfectly matched to the surrounding medium, implying no reflection from the slab at any incidence angle and for any polarization. Indeed, in this limit, $\kappa_1^E \to \kappa_0^E$ and $\kappa_1^H \to \kappa_0^H$ for propagating modes and $\kappa_1^E \to -\kappa_0^E$ and $\kappa_1^H \to -\kappa_0^H$ for evanescent waves (see the discussion in section 4.2), and reflection coefficients (4.307) and (4.309) vanish for arbitrary incidence directions and even for evanescent incident waves.

Figure 4.26 shows a configuration with a point source Q placed at a distance l in front of a perfectly matched DNG slab of thickness d. If $d > l$, then a simple ray-optics analysis predicts the appearance of two images F_{in} and F_{out} of the source, one inside the slab and another beyond the slab shifted by the distances $2l$ and $2d$ from the source, respectively. Note that the ray analysis does not account for evanescent waves that can be present in the spectrum of the incident field, and a more accurate, full-wave analysis is required to determine the quality of focusing of the slab (e.g., see Ziolkowski and Heyman (2001) and later in this section). It can be shown (Pendry 2000) that the image F_{out} is perfect in the sense that the fields in the semi-infinite half-space beyond the image point F_{out} coincide with those that would be excited by a source Q if it were placed at point F_{out}. If $d < l$, the images disappear.

In the ideal case of $\epsilon = -\epsilon_0$ and $\mu = -\mu_0$, the problem has no physically meaningful solution in the vicinity of the back interface of the lens (e.g., see Tretyakov (2003)). Due to the surface-wave resonance the fields here tend to infinity, as well as the field energy stored in the system. In addition, the Meixner condition (2.59) is not satisfied in the vicinity of the image point (the field behind the image is exactly the same as that of a point source, but there is no physical source at the image point). In reality, any backward-wave medium has losses, and introduction of even small losses removes the singularity of the solution (and destroys the perfectness of the image). The other fundamental limitation of this idealized model is that no medium can be considered as homogeneous when the amplitude of the transverse wave number increases with no limit. Wave components with extremely high values of $|k_t|$ do not create any resonant response, which also

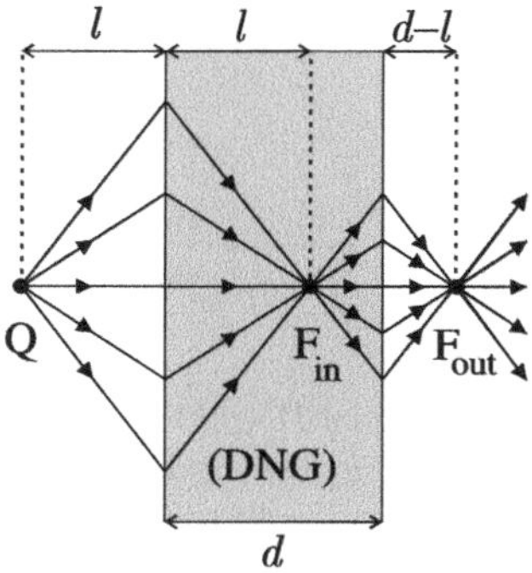

Figure 4.26 A planar DNG slab with $\epsilon_1 = -\epsilon_0$ and $\mu_1 = -\mu_0$ focuses the ray-optical portion of the field from a point source at Q into focal points F_{in} and F_{out}.

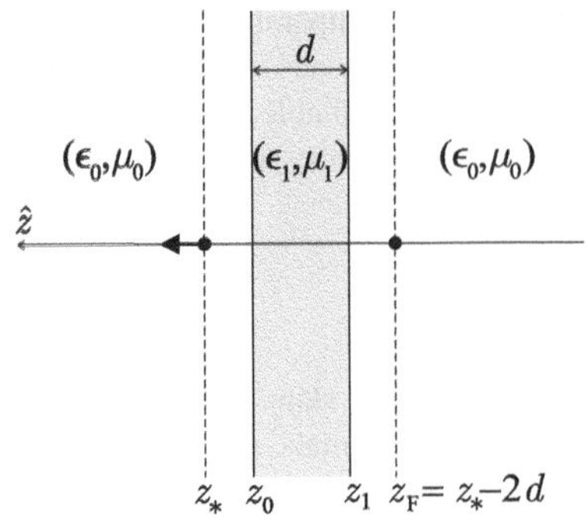

Figure 4.27 Electric dipole in front of a planar slab.

removes the singularity of the solution (Tretyakov 2003). Furthermore, every realistic lens is always bounded in space, which results in finite fields and further distortion of the image (e.g. Culhaoglu *et al.* (2008)).

To illustrate these physical considerations, let us study the full-wave solution for the field excited by an electric point dipole perpendicular to the interface of an infinite slab (Figure 4.27). The dipole is located at a point $\rho = 0$, $z = z_*$ and is directed along the z axis. The material parameters are ϵ_0 and μ_0 in the medium surrounding the slab, and ϵ_1 and μ_1 in the interior of the slab. The sides of the slab are at $z = z_0$ and $z = z_1$, and $d = z_0 - z_1$ is the thickness of the slab. We will study the field in the half-space $z < z_1$.

Adopting the general solution for the dipole source from section 4.4.2, equations (4.240), (4.241), (4.246), and (4.248), we obtain for the Hertz potential of the field transmitted through the slab

$$\Pi_e^{tr}(\rho, z) = \int_0^{+\infty} J_0(k_t\rho)\widetilde{\Pi}_e^{tr}(z, k_t)\, dk_t \quad (z < z_1) \tag{4.318}$$

with

$$\widetilde{\Pi}_e^{tr}(z, k_t) = \frac{p_e k_t}{4\pi j\epsilon_0\gamma_0} T_{TM}(k_t) e^{j\gamma_0(z - z_* + d)}, \tag{4.319}$$

where $\gamma_0 = \sqrt{k_0^2 - k_t^2}$ with $\mathrm{Im}\,\gamma_0 \leq 0$. The transmission coefficient of the homogeneous slab is given in (4.310), and it is convenient to express it as

$$T_{TM}(k_t) = \frac{1}{\cos(\gamma_1 d) + jq\sin(\gamma_1 d)} \tag{4.320}$$

with

$$q = \frac{1}{2}\left(\frac{\epsilon_1\gamma_0}{\epsilon_0\gamma_1} + \frac{\epsilon_0\gamma_1}{\epsilon_1\gamma_0}\right) \tag{4.321}$$

and $\gamma_1 = \sqrt{k_1^2 - k_t^2}$.

The field components are obtained from the potential $\Pi_e^{tr}(\rho, z)$ according to (2.267) and (2.268), where we choose cylindrical coordinates ($\xi = \rho$, $\eta = \varphi$, $\zeta = z$, $h_\xi = 1$, $h_\eta = \rho$, and $h_\zeta = 1$) because of the rotational symmetry of the configuration with respect to the z axis. The derivatives with respect to ρ and z in (2.267) and (2.268) are explicitly evaluated as

$$\frac{\partial}{\partial\rho}J_0(k_t\rho) = -k_t J_1(k_t\rho), \quad \frac{\partial}{\partial z}\widetilde{\Pi}_e^{tr}(z, k_t) = j\gamma_0\widetilde{\Pi}_e^{tr}(z, k_t),$$

and the result is given by

$$E_\rho^{\text{tr}} = -j \int\limits_0^{+\infty} k_t \gamma_0 J_1(k_t \rho) \widetilde{\Pi}_{\text{e}}^{\text{tr}}(z, k_t)\, \mathrm{d}k_t \tag{4.322}$$

$$E_z^{\text{tr}} = \int\limits_0^{+\infty} k_t^2 J_0(k_t \rho) \widetilde{\Pi}_{\text{e}}^{\text{tr}}(z, k_t)\, \mathrm{d}k_t \tag{4.323}$$

$$H_\varphi^{\text{tr}} = j \frac{k_0}{Z_0} \int\limits_0^{+\infty} k_t J_1(k_t \rho) \widetilde{\Pi}_{\text{e}}^{\text{tr}}(z, k_t)\, \mathrm{d}k_t. \tag{4.324}$$

The other field components E_φ^{tr}, H_ρ^{tr}, and H_z^{tr} are equal to zero. Finally, it is seen from the solution that the fields can be expressed entirely through the dimensionless parameters $k_0 d$, $k_0 \rho$, and $k_0(z - z_* + d)$.

Let us start by taking a look at the convergence of the integrals in (4.318) and (4.322)–(4.324) in the limit of the perfect lens when $\epsilon_1 \to -\epsilon_0$ and $\mu_1 \to -\mu_0$. Depending on whether $0 \le k_t < k_0$ or $k_0 < k_t < +\infty$, we have that $\gamma_1(k_t) \to \mp\gamma_0(k_t)$, $\kappa_1^H \to \pm\kappa_0^H$, and $q \to \pm 1$, but since $T_{\text{TM}}(k_t)$ is an even function of γ_1, the transmission coefficient tends to the limit

$$T_{\text{TM}}(k_t) \to \mathrm{e}^{j\gamma_0(k_t)d} \tag{4.325}$$

over the whole integration range $0 \le k_t < +\infty$, leading to

$$\widetilde{\Pi}_{\text{e}}^{\text{tr}}(z, k_t) \to \frac{p_{\text{e}} k_t}{4\pi j \epsilon_0 \gamma_0} \mathrm{e}^{j\gamma_0(z - z_* + 2d)}. \tag{4.326}$$

When $z < z_{\text{F}}$ with $z_{\text{F}} = z_* - 2d$, expression (4.326) coincides with the Hankel transform (4.240) of the Hertz potential for an electric dipole at a point $\rho = 0$ and $z = z_{\text{F}}$ since $z - z_* + 2d = -|z - z_{\text{F}}|$. The plane $z = z_{\text{F}}$ will be referred to as the image (or focal) plane. However, in the region $z_{\text{F}} < z < z_1$, that is, between the slab and the image plane, (4.326) does not reduce to (4.240) since $z - z_* + 2d = |z - z_{\text{F}}|$, and the estimation

$$|\widetilde{\Pi}_{\text{e}}^{\text{tr}}(z, k_t)| = O\left[\mathrm{e}^{k_t(z - z_F)}\right], \quad k_t \to +\infty \tag{4.327}$$

suggests that the integrals (4.318) and (4.322)–(4.324) diverge in the region $z_{\text{F}} < z < z_1$ because of the exponential growth of their integrands at $k_t \to +\infty$. So, in the limit of the perfect lens, the field approaches the field of an electric point dipole at the point $z = z_{\text{F}}$ and $\rho = 0$ in the half-space behind the image plane ($z < z_{\text{F}}$) but becomes infinite in the region between the slab and the image plane ($z_{\text{F}} < z < z_1$).

The divergence in the region $z_{\text{F}} < z < z_1$ and the pole singularity at the image point $\rho =$ and $z = z_{\text{F}}$ are consequences of the idealization of the problem (no absorption, perfectly homogeneous material of the slab, ideal values of ϵ_1 and μ_1 for all k_t). As soon as ϵ_1 and μ_1 deviate from the ideal values (even an arbitrarily small deviation suffices), $T_{\text{TM}}(k_t)$ dramatically changes its behavior at $k_t \to +\infty$ from the exponential growth $O[\exp(k_t d)]$ as in (4.325) to exponential decay as $O[\exp(-k_t d)]$, and this is the mechanism that ensures convergence of the integral representations (4.318), (4.322), (4.323), and (4.324) for realistic models of the DNG media.

Let us assume small losses in the slab ($\operatorname{Im}\epsilon_1 < 0$ and $\operatorname{Im}\mu_1 < 0$) and consider the integrals (4.318), (4.322), (4.323), and (4.324). The integrand functions have singularities at the branch point $k_t = k_0$ and at zeros of the denominator of $T_{\text{TM}}(k_t)$ in (4.320), which will be denoted k_{tm} with $m = 0, \pm 1, \pm 2, \ldots$. Because in any passive medium there are at least infinitesimal losses, $\operatorname{Im} k_0 < 0$ and the branch point lies below the integration contour (Figure 4.28). Location of the pole singularities can be determined from the approximate formula

$$k_{tm} \approx \pm\frac{1}{d}\left[-\ln\left(1 + \frac{\epsilon_1}{\epsilon_0}\right) + \ln 2 + j\pi m\right], \quad m = 0, \pm 1, \pm 2, \ldots \tag{4.328}$$

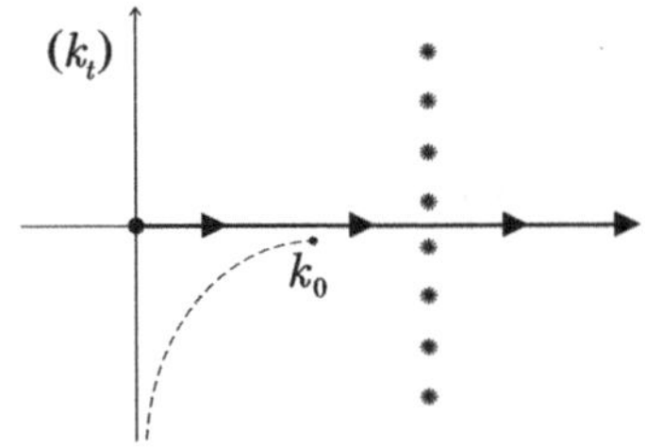

Figure 4.28 Integration contour, the branch point $k_t = k_0$, the associated branch cut (dotted line), and the poles k_{tm} in the right half-plane of the complex k_t plane.

which is valid for $\epsilon_1 \to -\epsilon_0$ and $\mu_1 \to -\mu_0$ (see problem 29). It follows from (4.328) that the poles are located symmetrically with respect to the point $k_t = 0$, along lines perpendicular to the real axis. Figure 4.28 shows the branch of the poles in the right half-plane of the complex k_t plane. A symmetric branch is located in the left half-plane. Denoting

$$\frac{\epsilon_1}{\epsilon_0} = -1 + \delta e^{j\chi}$$

with δ being a small positive parameter and χ being the phase factor ($-\pi < \chi < 0$ in any passive medium), k_{tm} can be split up in the real and imaginary parts as

$$k_{tm} \approx \pm\frac{1}{d}\left[\ln\frac{2}{\delta} + j(\pi m - \chi)\right], \tag{4.329}$$

implying that no poles are located on the real axis and therefore on the integration contour. In the limit $\epsilon_1 \to -\epsilon_0$ and $\mu_1 \to -\mu_0$, the poles disappear as $\mathrm{Re}\, k_{tm} \to \pm\infty$.

We are now ready to study the field behavior in the image region. The asymptotic steepest descent method (section C.3) is not suited for evaluation of fields in caustic regions where multiple rays intersect,[8] but the fields can be calculated by numerical integration. The form of Hankel's transform with the Bessel functions $J_{0,1}(k_t\rho)$ in the kernel as in (4.318), (4.322), (4.323), and (4.324) is convenient for numerical evaluation since the kernel is then bounded at the z axis where $\rho = 0$, which is in contrast to the alternative form with the Hankel function in the kernel as in (4.249). The integrands in (4.318), (4.322), (4.323), and (4.324) are bounded on the whole integration contour, including the branch point $k_t = k_0$ and the intersection of the integration contour with the line along which the pole singularities are located (poles k_{tm} are shifted from the real axis as long as $\mathrm{Im}\,(\epsilon_1/\epsilon_0) < 0$, see (4.328)).

Figures 4.29 and 4.30 illustrate the behavior of $|E_z|$ in the image plane $z = z_F$ and in a longitudinal plane that includes the z axis, respectively. Configuration parameters are as follows: $k_0 z_* = 0.5$, $z_0 = 0$, $z_1 = -d$, and $k_0 d = 1$. Furthermore, it is assumed that

$$\epsilon_1/\epsilon_0 = \mu_1/\mu_0 = -1 - j\delta$$

with a small and positive value of δ representing losses in the slab material. The field distribution in the image plane looks like a spot around the focal point at $\rho = 0$. As material parameters of a DNG layer approach the ideal values, the size of the spot goes to zero, implying that DNG lenses can overcome the diffraction limit. However, the distribution in the longitudinal plane is completely different: the field magnitude has a maximum at the back side of the slab ($z = z_1$) and gradually decreases as the observation point goes away from the slab, showing a pattern with the main lobe and side lobes. DNG lenses therefore do not create a point image of a point source, and it is more correct to speak of a focal region rather than of a focal point.

[8] From the mathematical viewpoint, the main trouble is that the transmission coefficient of the slab with $\epsilon_1 \to -\epsilon_0$ and $\mu_1 \to -\mu_0$ is not a slowly varying function.

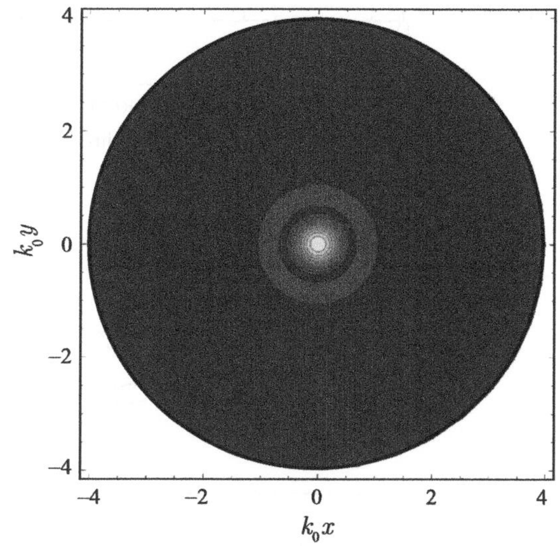

Figure 4.29 Distribution of $|E_z(\rho, z_F)|$ in the focal plane of a DNG lens with $\delta = 0.01$ as a function of dimensionless coordinates $k_0 x$ and $k_0 y$. The maximum value $|E_z^{\max}| = 58.1915$ is achieved at the point $\rho = 0$. The contour lines are at the levels of $p|E_z^{\max}|/10$ with $p = 1, 2, \ldots, 9$. Configuration parameters: $k_0 z_* = 0.5$, $z_0 = 0$, $z_1 = -d$, and $k_0 d = 1$.

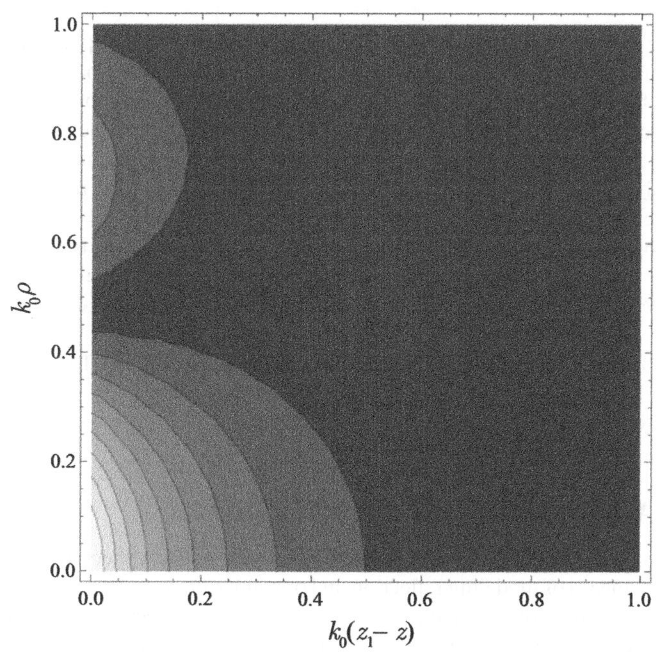

Figure 4.30 Distribution of $|E_z(\rho, z)|$ in the longitudinal plane as a function of dimensionless coordinates $k_0\rho$ and $k_0(z_1 - z)$. The focal plane is at $k_0(z_1 - z) = 0.5$. The maximum value $|E_z^{\max}| = 588.157$ is achieved at the side of the lens at $\rho = 0$ and $z = z_1$. The contour lines are at the levels of $p|E_z^{\max}|/10$ with $p = 1, 2, \ldots, 9$. Other parameters are as in Figure 4.29.

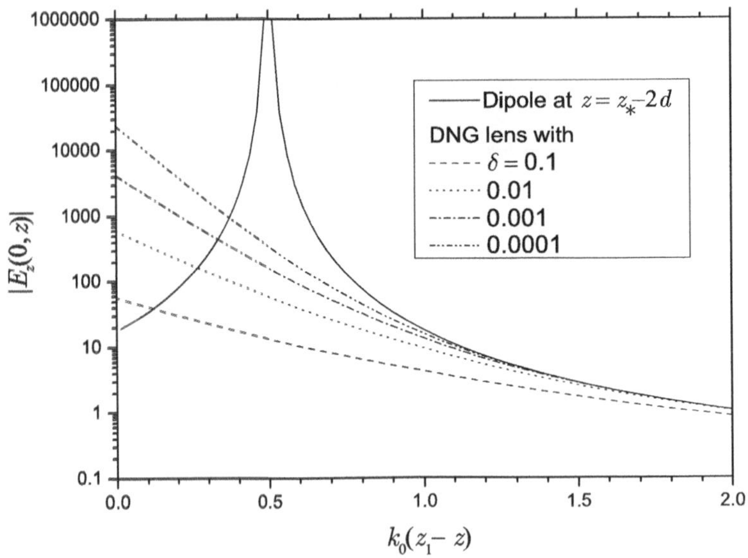

Figure 4.31 Field magnitude $|E_z(0, z)|$ on the z axis for a DNG lens with various values of δ compared with the field of an electric dipole at the image point. The configuration parameters are as in Figure 4.29.

Evanescent waves dominate over propagating waves in the region between the image plane and the back side of the lens, but their influence rapidly decays as the observation point goes farther away from the back side of the lens. In the region beyond the image plane, the field approaches the field of a point electric dipole at the point $z = z_F$, $\rho = 0$. The closer the parameters ϵ_1 and μ_1 to the ideal values $-\epsilon_0$ and $-\mu_0$, the greater the field magnitude in the region $z_F < z < z_1$ and the closer the field in the region $-\infty < z < z_F$ to that of the dipole. In the limit of the ideal DNG medium, the field becomes infinite in the region $z_F < z < z_1$ and equal to the field of the dipole at the image point in the region $-\infty < z < z_F$. Figure 4.31 illustrates these trends for the configuration with $k_0 z_* = 0.5$, $z_0 = 0$, $z_1 = -d$, and $k_0 d = 1$ by showing the dependence of $|E_z|$ at $\rho = 0$ on the dimensionless distance $k_0(z_1 - z)$ from the back side of the slab as the material parameters approach the ideal values. This picture actually shows that even with extremely small losses the image does not properly resolve sub-wavelength details, as the field is close to the dipole field only starting from the distance about $\lambda/(4\pi)$ from the image point.

To conclude, we note that a dipole with a component parallel to the slab excites both TM and TE modes (section 4.4.2). Transmission properties of a DNG slab with non-ideal parameters ($\epsilon_1 \neq -\epsilon_0$ and $\mu_1 \neq -\mu_0$) are slightly different for TE and TM modes which leads to an additional art of distortion of the image (Culhaoglu *et al.* 2014).

4.6 Modeling with Approximate Boundary Conditions

The approximate boundary conditions (Leontovich, standard impedance, resistive sheet transition conditions) introduced in section 2.3 provide a simple means of modeling reflection and transmission properties of material interfaces. For planar layered structures, as we have seen in this chapter, exact expressions for reflection and transmission coefficients are always available, so that they can be compared with predictions of the approximate boundary conditions, which allows derivation of justified expressions for the surface impedance and verification of the applicability of the approximate boundary conditions.

We start this section by looking at the exact and approximate reflection coefficients of the two configurations (interface of two semi-infinite homogeneous media and a homogeneous layer backed by a PEC surface), the two kinds of configurations that can be well modeled with Leontovich and standard impedance boundary conditions (SIBCs). Next we will take a look at transmission and reflection at a thin homogeneous layer immersed in a homogeneous space, and show that compared with SIBCs, the resistive sheet transition conditions are better suited for this kind of structure and lead to more accurate results in the limit of a very thin coating. Both are more accurate for the perpendicular (TE) polarization, and their accuracy for the parallel (TM) polarization, particularly for grazing incidence angles, may be insufficient. The section concludes with a study of an important special case of a surface with the surface impedance equal to that of the surrounding medium. Such a surface is characterized by a low level of reflection, and its reflection properties are addressed in this section.

4.6.1 Accuracy of Impedance Boundary Conditions

As we have seen in section 4.3.1, the reflection coefficients of an arbitrary planar material interface result from the general relations (4.129) and (4.155) as

$$R_{\perp}(\theta_\mathrm{i}) = R_{\mathrm{TE}}(\theta_\mathrm{i}) = R_E(\theta_\mathrm{i}) = \frac{Z_E(\theta_\mathrm{i})\cos\theta_\mathrm{i} - Z}{Z_E(\theta_\mathrm{i})\cos\theta_\mathrm{i} + Z} \tag{4.330}$$

$$R_{\parallel}(\theta_\mathrm{i}) = -R_{\mathrm{TM}}(\theta_\mathrm{i}) = R_H(\theta_\mathrm{i}) = \frac{Z\cos\theta_\mathrm{i} - Z_H(\theta_\mathrm{i})}{Z\cos\theta_\mathrm{i} + Z_H(\theta_\mathrm{i})}, \tag{4.331}$$

where Z is the wave impedance of the medium adjacent to the surface and $Z_E(\theta_\mathrm{i})$ and $Z_H(\theta_\mathrm{i})$ are the input impedances of the interface. On the other hand, the SIBCs (2.74) describe the surface in terms of a single parameter, the equivalent surface impedance $\tilde{Z}_\mathrm{s}$, which is independent of the incidence angle and polarization. So, the possibility of reducing the exact solution to the solution of a reflection problem with the impedance boundary conditions on the interface depends on whether the two impedances $Z_E(\theta_\mathrm{i})$ and $Z_H(\theta_\mathrm{i})$ in (4.330) and (4.331) can be replaced with a scalar constant $\tilde{Z}_\mathrm{s}$ without unacceptable losses in accuracy.

By using the fact that at normal incidence there is no difference between polarization cases, one can define the equivalent surface impedance (see (2.75)) by

$$\tilde{Z}_\mathrm{s} = Z_E(0) = Z_H(0) \tag{4.332}$$

and approximate the input impedances in (4.330) and (4.331) as

$$Z_E(\theta_\mathrm{i}) \approx Z_E(0) = \tilde{Z}_\mathrm{s}, \quad Z_H(\theta_\mathrm{i}) \approx Z_H(0) = \tilde{Z}_\mathrm{s}, \tag{4.333}$$

which gives the following approximation for the reflection coefficients:

$$\tilde{R}_{\perp}(\theta_\mathrm{i}) = \tilde{R}_{\mathrm{TE}}(\theta_\mathrm{i}) = \tilde{R}_E(\theta_\mathrm{i}) = \frac{\eta\cos\theta_\mathrm{i} - 1}{\eta\cos\theta_\mathrm{i} + 1} \tag{4.334}$$

$$\tilde{R}_{\parallel}(\theta_\mathrm{i}) = -\tilde{R}_{\mathrm{TM}}(\theta_\mathrm{i}) = \tilde{R}_H(\theta_\mathrm{i}) = \frac{\cos\theta_\mathrm{i} - \eta}{\cos\theta_\mathrm{i} + \eta}, \tag{4.335}$$

where $\eta = \tilde{Z}_\mathrm{s}/Z$ is the normalized surface impedance. Expressions (4.334) and (4.335) can be also derived by directly solving the problem of reflection of a plane wave from a flat interface with impedance boundary conditions (2.74) (problem 30).

Impedance boundary conditions are an approximation as the formulation neglects the polarization difference between the input impedances and the dependence of the impedances on the incidence angle. Relations (4.333) are the necessary conditions of their applicability. Now we use the solutions for the interface between two homogeneous half-spaces (section 4.2.1) and for that of a PEC-backed homogeneous layer (section 4.5.2) to derive the suitable values of $\tilde{Z}_\mathrm{s}$ and estimate the accuracy of the impedance approximation for these structures. In what follows the upper medium, from which the incident wave comes, is designated as medium "0".

Interface Between Two Semi-infinite Spaces

It can be checked that approximation (4.333) recovers the impedance (2.73) used in the Leontovich boundary conditions for strongly conducting bodies. Indeed, it follows from (4.12) and (4.13) that the exact surface impedances of the interface between two semi-infinite media are

$$Z_E = \frac{1}{\kappa_1^E} = \frac{\omega\mu_1}{\gamma_1}, \quad Z_H = \kappa_1^H = \frac{\gamma_1}{\omega\epsilon_1} \tag{4.336}$$

and (4.333) implies that

$$\tilde{Z}_s = \sqrt{\frac{\mu_1}{\epsilon_1}} = Z_1. \tag{4.337}$$

The dependence of Z_E and Z_H on θ_i is entirely due to $\gamma_1 = \sqrt{k_1^2 - k_0^2 \sin^2\theta_i}$ and if

$$\left|\frac{k_1}{k_0}\right|^2 = \left|\frac{\mu_1\epsilon_1}{\mu_0\epsilon_0}\right| \gg 1, \tag{4.338}$$

then

$$\gamma_1 \approx k_1, \quad \kappa_1^E \approx Y_1, \quad \kappa_1^H \approx Z_1, \quad Z_E \approx Z_H \approx Z_1, \tag{4.339}$$

and the dependence on θ_i in Z_E and Z_H can be neglected. Condition (4.338) is satisfied if the lower half-space is filled with a strongly conducting medium but also for dielectric materials with $\epsilon_{r1} = \epsilon_1/\epsilon_0 \gg 1$ or for ferromagnetic materials with $\mu_{r1} = \mu_1/\mu_0 \gg 1$. Figure 4.32 shows an example: despite the low absorption in the material and moderate value of $\mathrm{Re}\,\epsilon_{r1}$ the impedance formulation leads to a reasonable approximation of the reflection coefficients.

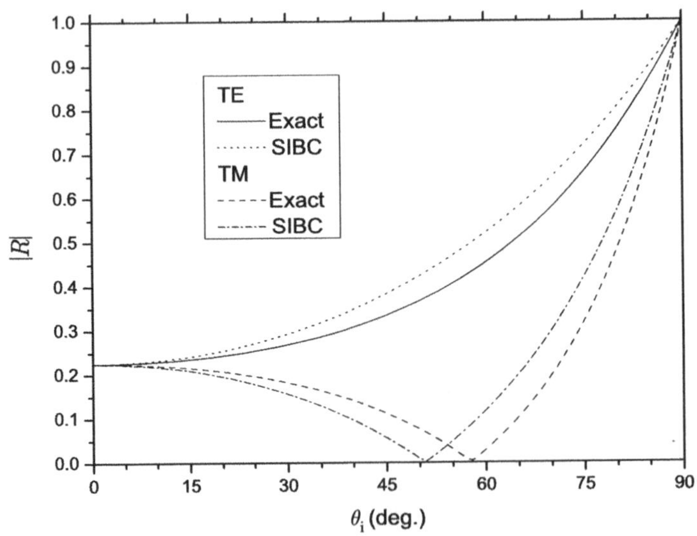

Figure 4.32 Reflection coefficient of the air–asphalt interface at 10 GHz ($\epsilon_1/\epsilon_0 = 2.5 - j0.003$, $\mu_1 = \mu_0$) as a function of the incidence angle θ_i.

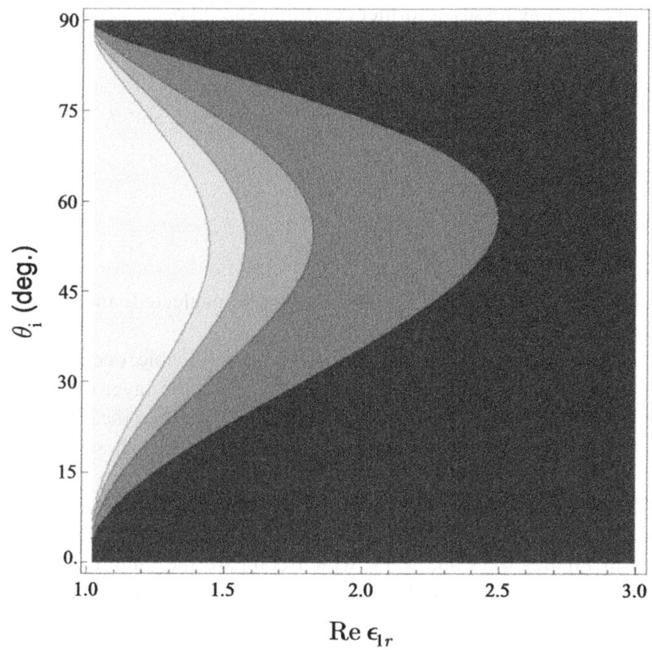

Figure 4.33 Relative error of impedance boundary conditions for the interface between two semi-infinite spaces as a function of the incidence angle θ_i and Re ϵ_{1r}. The contour lines are drown at the levels 0.25, 0.5, 0.75, and 1. In the darkest region the error is smaller than 25%.

The quality of the impedance approximation depends on the values of θ_i and the material parameters of the media on both sides of the interface. To describe the error it is convenient to define the relative error by

$$\delta = \sqrt{\frac{|R_{\mathrm{TM}} - \tilde{R}_{\mathrm{TM}}|^2 + |R_{\mathrm{TE}} - \tilde{R}_{\mathrm{TE}}|^2}{|R_{\mathrm{TM}}|^2 + |R_{\mathrm{TE}}|^2}}, \tag{4.340}$$

which accumulates errors in both polarization cases. Figure 4.33 shows the error as a contour plot in the case that $\mathrm{Im}\,\epsilon_{1r} = 0$ and $\mu_{1r} = 1$. The lighter the color the greater is the error δ. It is seen that with $\mathrm{Re}\,\epsilon_{1r} > 2.5$ the error of the impedance approximation is less than 25% uniformly in θ_i.

PEC-backed Layer

Another case in which the approximation (4.333) can be valid is that of a homogeneous layer backed by a metal (PEC) boundary. The surface impedances for this kind of structure are given in (4.304) and (4.305), and it follows from (4.306) and (4.333) that

$$\tilde{Z}_s = jZ_1 \tan(k_1 d_1). \tag{4.341}$$

For this structure the condition (4.338) is also necessary to make the impedances Z_E and Z_H compliant with (4.333). For example, for an electrically thin layer, $\tan(\gamma_1 d_1) \approx \gamma_1 d_1$ and (4.304) and (4.305) reduce to

$$Z_E \approx j\omega\mu_1 d_1, \quad Z_H \approx j\omega\mu_1 d_1 \left[1 - \left(\frac{k_0}{k_1}\right)^2 \sin^2\theta_i\right], \tag{4.342}$$

and if the condition (4.338) is satisfied, then $Z_E \approx Z_H \approx \tilde{Z}_s \approx j\omega\mu_1 d_1$.

For thicker layers of materials with low losses, a further condition should be accounted for. To ensure that $\tan(\gamma_1 d_1) \approx \tan(k_1 d_1)$, the absolute value of the correction term in the expansion

$$\gamma_1 d_1 = d_1 \sqrt{k_1^2 - k_0^2 \sin^2 \theta_i} \approx k_1 d_1 - \frac{k_0^2}{2k_1} d_1 \sin^2 \theta_i \tag{4.343}$$

should be small compared to π, which implies that

$$d_1 < |k_1|/k_0^2. \tag{4.344}$$

For greater values of d_1, the dependence of Z_E and Z_H on θ_i cannot be neglected, and the impedance formulation is no longer adequate.

Provided that (4.338) is true, the impedance approximation is always applicable, once the thickness of the layer exceeds the skin depth, that is, $d_1 > |\mathrm{Im}\, k_1|^{-1}$, since $\tan(\gamma_1 d_1) \approx -j$, $\tilde{Z}_s \approx Z_1$, and the layer can be regarded as a semi-infinite medium. Summarizing, if the losses in the layer are such that $|k_1 \mathrm{Im}\, k_1| > k_0^2$, then the impedance approximation with the surface impedance $\tilde{Z}_s$ from (4.341) is valid for arbitrary thicknesses, that is, $0 \le d_1 < \infty$. Otherwise, the approximation does not apply when $|k_1|/k_0^2 < d_1 < |\mathrm{Im}\, k_1|^{-1}$.

An important feature in the impedance approximations (4.334) and (4.335) for PEC-backed layers with small or moderate electrical thicknesses is that most of the error is in the phase, whereas the amplitude is modeled much more accurately. This is illustrated in Figures 4.34 and 4.35, which compare the amplitudes and phases of the exact and approximated reflection coefficients of a layer with the electrical thickness $k_1 d_1 = 1.01$.

Figure 4.36 shows the relative error defined by (4.340). The interval in θ_i, where the error is smaller than 25%, decreases with increasing electrical thickness d_1/λ. The accuracy starts improving again when the layer gets so thick that the response from the backing plane becomes negligible. In this particular case this corresponds to $d_1 > |\mathrm{Im}\, k_1|^{-1} \approx 0.77$ m or $d_1/\lambda > 25.6$ (this region is not shown in Figure 4.36).

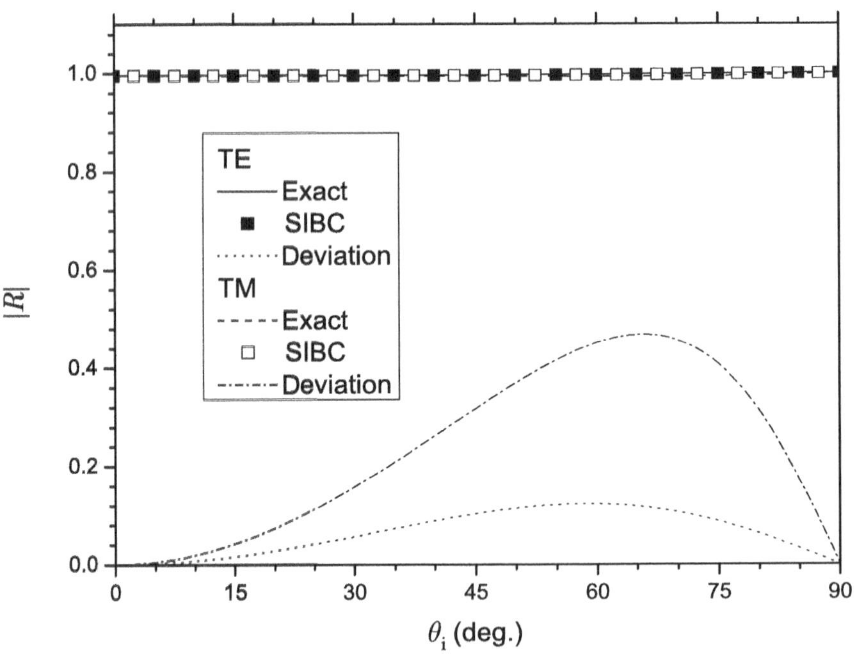

Figure 4.34 Exact and SIBC-based reflection coefficients of a 3-mm thick PEC-backed Plexiglas layer at 10 GHz as a function of the incidence angle. Amplitudes are close to unity and almost indistinguishable. The absolute error is, nevertheless, considerable as most of the error of the impedance approximation is in the phase.

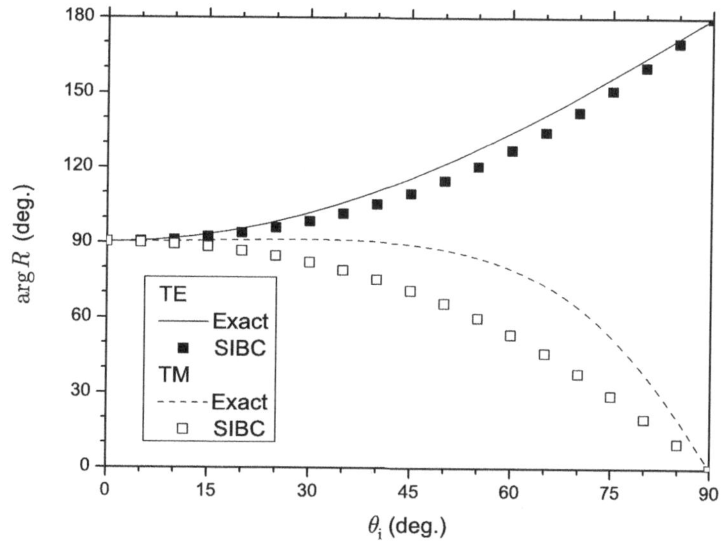

Figure 4.35 Phase of the exact and SIBC-based reflection coefficients of a 3-mm thick PEC-backed Plexiglas layer at 10 GHz as a function of the incidence angle.

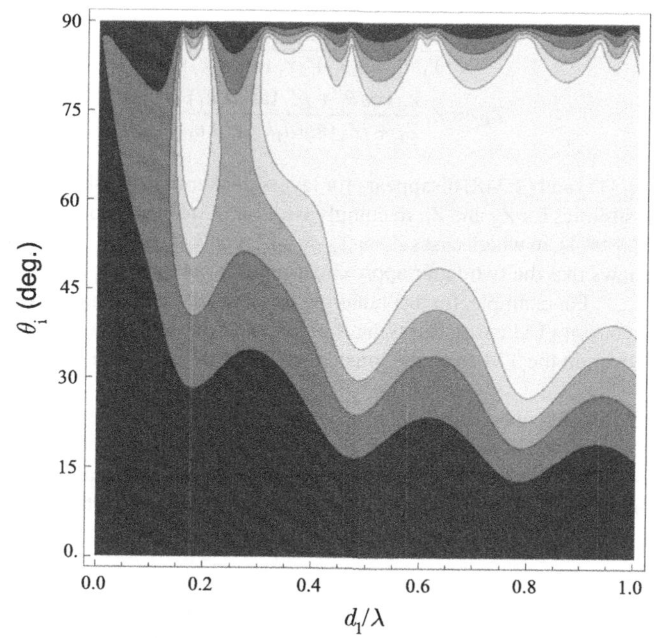

Figure 4.36 Relative error of impedance boundary conditions for the PEC-backed Plexiglas layer in the air ($\epsilon_{1r} = 2.59 - j0.02$, $\mu_{1r} = 1$) at 10 GHz as a function of θ_i and d_1/λ. Contour lines are drawn at the levels 0.25, 0.5, 0.75, and 1. In the black region the error is smaller than 25%.

4.6.2 Accuracy of Transition Boundary Conditions

Consider the approximate modeling of a homogeneous layer immersed in a homogeneous medium (section 4.5.3). A layer with thickness greater that the skin depth in the material of the layer is close to a semi-infinite space and, therefore, can be well described with the impedance boundary conditions. However, it will be shown in this section that a constant impedance $\tilde{Z}_s$ cannot be assigned to very thin layers, implying that such configurations cannot be adequately described by the impedance boundary conditions. The sheet transition conditions, say, in the form of the resistive sheet conditions (2.111) and (2.112), can still be applied, and they provide a more accurate description of reflection and transmission at a thin layer with an accuracy which is particularly good for perpendicular (TE) polarization. Approximation for parallel (TM) polarization is less accurate, and better modeling requires the use of higher-order transition boundary conditions.

We start by showing that the impedance approach does not work for thinner layers. The input impedances Z_E and Z_H follow from (4.127), (4.153), (4.307), and (4.309) as

$$Z_E = \frac{1}{\kappa_0^E}\frac{1+R_{\text{TE}}}{1-R_{\text{TE}}} = \frac{1}{\kappa_1^E}\frac{\kappa_1^E + j\tan(d_1\gamma_1)\kappa_0^E}{\kappa_0^E + j\tan(d_1\gamma_1)\kappa_1^E}, \tag{4.345}$$

$$Z_H = \kappa_0^H\frac{1+R_{\text{TM}}}{1-R_{\text{TM}}} = \kappa_1^H\frac{\kappa_0^H + j\tan(d_1\gamma_1)\kappa_1^H}{\kappa_1^H + j\tan(d_1\gamma_1)\kappa_0^H} \tag{4.346}$$

with $\kappa_0^E = Y_0\cos\theta_i$, $\kappa_0^H = Z_0\cos\theta_i$, and

$$\kappa_1^E = Y_1\sqrt{1-\left(\frac{k_0}{k_1}\right)^2\sin^2\theta_i}, \quad \kappa_1^H = Z_1\sqrt{1-\left(\frac{k_0}{k_1}\right)^2\sin^2\theta_i}.$$

If conditions (4.338) and (4.344) are true, then the dependence on θ_i in κ_1^E, κ_1^H, and $\tan(\gamma_1 d_1)$ can be neglected, which leads to the approximate expressions:

$$Z_E \approx \frac{1}{Y_1}\frac{Y_1 + jY_0\tan(d_1k_1)\cos\theta_i}{Y_0\cos\theta_i + jY_1\tan(d_1k_1)} \tag{4.347}$$

$$Z_H \approx Z_1\frac{Z_0\cos\theta_i + jZ_1\tan(d_1k_1)}{Z_1 + jZ_0\tan(d_1k_1)\cos\theta_i}. \tag{4.348}$$

The dependence on $\cos\theta_i$ in (4.347) and (4.348) disappears for layers thicker than the skin depth ($|\text{Im}\,k_1|d_1 > 1$), in which case $Z_E \approx Z_H \approx Z_1$. Other possibilities for Z_E and Z_H to comply with (4.333) are a conducting material ($|Z_1/Z_0| \ll 1$) or a ferromagnetic material ($|Z_1/Z_0| \gg 1$), in which cases $Z_E \approx Z_H \approx -jZ_1\cot(k_1d_1)$ or $Z_E \approx Z_H \approx jZ_1\tan(k_1d_1)$, respectively. However, closer inspection shows that the two latter approximations are not uniform with respect to k_1d_1 and fail when $k_1d_1 \approx m\pi/2$ with $m = 0, 1, 2, \ldots$. For example, for a conducting layer, the electrical thickness $|k_1d_1|$ must be greater than $|Z_1/Z_0|$ (TE case) or $|Z_1/(Z_0\cos\theta_i)|$ (TM case), implying the failure of the impedance formulation for very thin layers when $|k_1d_1| < |Z_1/Z_0|$. Moreover, in the TM case, the impedance formulation always fails for grazing incidence angles ($\theta_i \to \pi/2$).

A more accurate description of reflection and transmission at a thin layer can be achieved with the resistive sheet transition conditions (2.111) and (2.112). Their application results in approximations for the reflection and transmission coefficients, as in (4.312)–(4.315) (problem 31). Because for thin layers in this approximation $T_{\text{TE}} \approx 1 + R_{\text{TE}}$ and $T_{\text{TM}} \approx 1 + R_{\text{TM}}$, it is sufficient to study the accuracy of approximation for the reflection coefficients. Figures 4.37 and 4.38 compare the reflection coefficients of a thin Plexiglas layer in free space calculated by using the impedance formulation (2.74) with

$$\tilde{Z}_s = Z_E(0) = Z_H(0) = Z_1\frac{Z_0 + jZ_1\tan(d_1\gamma_1)}{Z_1 + jZ_0\tan(d_1\gamma_1)} \tag{4.349}$$

and the resistive sheet transition conditions (2.111) and (2.112). Two versions of the latter are used, one with R_e defined in (4.316) and another with R_e from (2.107). The definitions are close to each other when $|\epsilon_1/\epsilon_0| \gg 1$ but differ when $\epsilon_1 \approx \epsilon_0$. It is apparent from the figures that the transition conditions provide better accuracy than the impedance formulation; for a thinner layer the accuracy of the impedance formulation would be even worse. Furthermore, it is seen that the use of R_e from (2.107) significantly improves the accuracy of the transition conditions. In the case of TM polarization the accuracy is generally worse than for TE, particularly for greater values of the incidence angle.

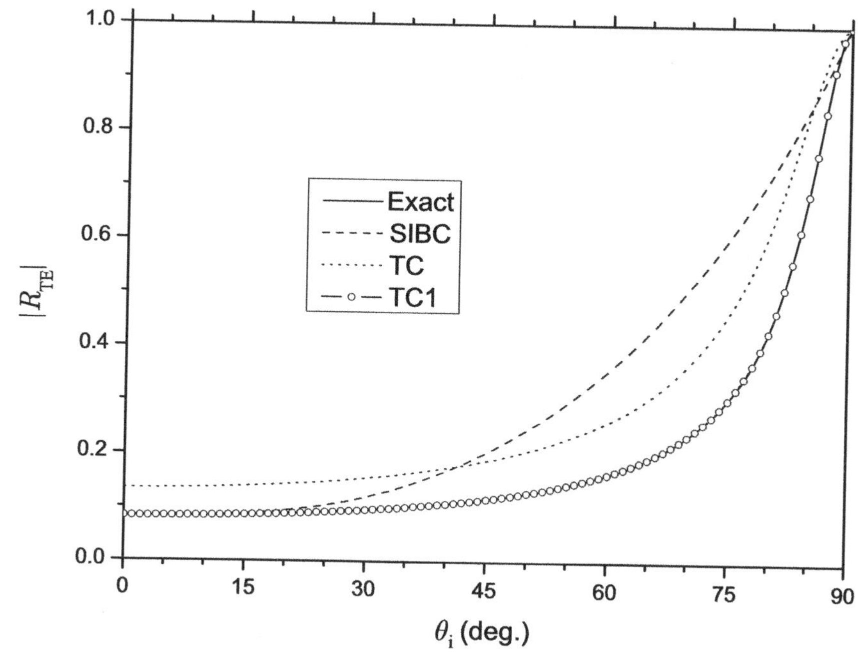

Figure 4.37 Amplitude of the reflection coefficient of a 0.5-mm thick Plexiglas layer in free space at 10 GHz for TE polarization as a function of the incidence angle: exact vs. approximate formulations. Curves TC and TC1 result from transition conditions with R_e from (4.316) and (2.107), respectively.

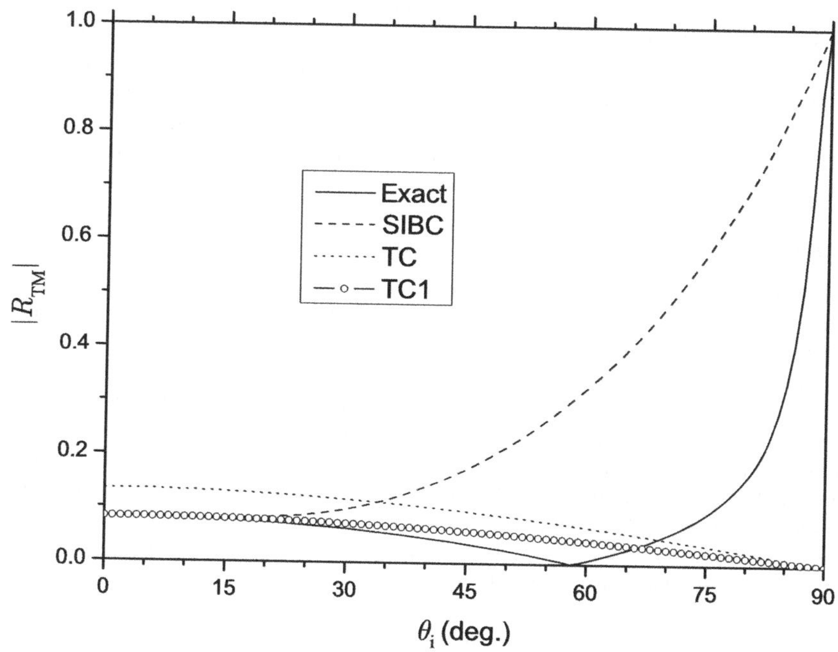

Figure 4.38 The same as Figure 4.37 but for TM polarization.

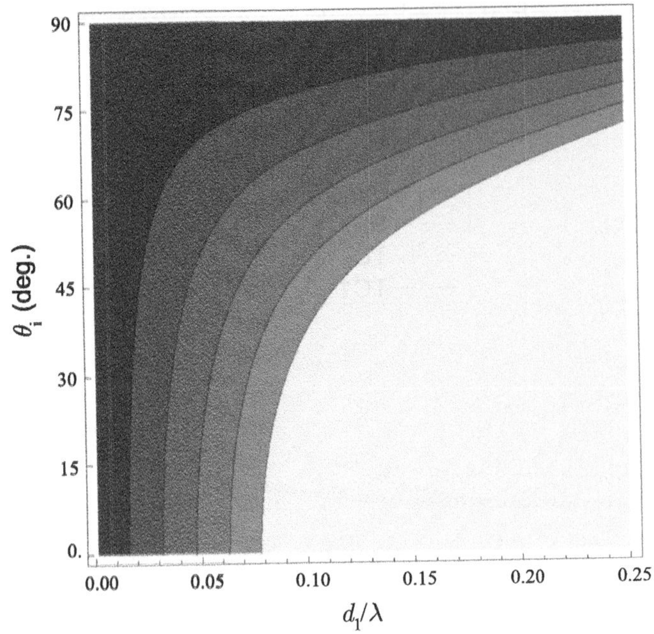

Figure 4.39 Relative error of the transition conditions for a thin Plexiglas layer in free space ($\epsilon_{1r} = 2.59 - j0.02$, $\mu_{1r} = 1$) at 10 GHz as a function of θ_i and d_1/λ for TE polarization. Contour lines are drawn at the levels 0.1, 0.2, 0.3, 0.4, and 0.5. In the darkest region the error is smaller than 10%.

The relative error of the transition conditions for reflection coefficients of the Plexiglas layer in dependence on θ_i and d_1/λ is shown in Figures 4.39 and 4.40. λ is the wavelength in free space and is equal to 0.03 m. Since ϵ_1 is relatively close to ϵ_0, parameter R_e is chosen according to (2.107). In the TE case (Figure 4.39), the error is less than 10% uniformly in θ_i when $d_1/\lambda < 0.015$. In the TM case (Figure 4.40), the error is greater than 50% for every θ_i when $d_1/\lambda > 0.08$, and even very thin layers, for example with $d_1/\lambda < 0.01$, are modeled inaccurately in the case of grazing ($\theta_i > 45°$) incidence. More accurate results for TM-polarized fields can be obtained by using second-order transition conditions (e.g. Senior and Volakis (1995, section 5.3.1).

4.6.3 *Impedance-matched Surface*

As we have seen in section 4.6.1, the interface of a half-space filled with a lossy material has a surface impedance that is almost polarization- and incidence-angle-independent, and close to the wave impedance of the medium (see (4.339)). If the material is such that $\mu_1/\epsilon_1 = \mu_0/\epsilon_0$, then $Z_1 = Z_0$ and the interface is impedance-matched with the medium from which the incident wave comes, in this case $\eta = 1$.

The matched surface impedance minimizes the reflection of the incident wave and permits a perfect transmission in the lower lossy medium where the transmitted power gets absorbed. This is the operation principle of a matched absorber, and a sufficiently thick layer of a lossy material is a way of practical realization. Furthermore, by using modern technologies, for example the concept of metamaterials, it is possible to realize thin artificial absorbers which, at least within a bounded range of frequencies and incidence angles, approach the desired value of the surface impedance (Ra'di *et al.* 2015).

The impedance-matched boundary only weakly reflects incident waves and possesses the property of minimum average reflectance (Osipov 2014). Let us take a closer look at this property. It follows from (4.330) and (4.331) that

$$R_{\perp}(\theta_i) = R_{\parallel}(\theta_i) = \frac{\cos\theta_i - 1}{\cos\theta_i + 1} \tag{4.350}$$

and, therefore, reflection becomes polarization-independent. Furthermore, at normal incidence $\theta_i = 0$, the reflection coefficients vanish. At other incidence angles, the reflection coefficients are not zero but still quite low, therefore we term

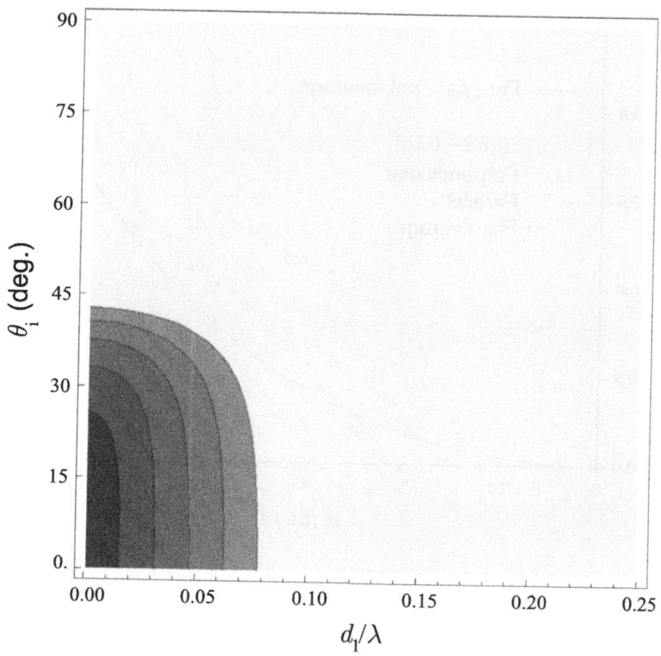

Figure 4.40 The same as Figure 4.39 but for TM polarization. The relative accuracy in the lighter region is greater than 50%. It is seen that the transition conditions fail to accurately describe the reflection coefficient when $\theta_i > 45°$, even for very small thicknesses.

such surfaces *low-reflection surfaces*. Figure 4.41 compares the reflection coefficients of the low-reflection surface with those of a test surface with $\eta = 0.53 - j0.08$. It is seen that in the range $0 \leq \theta_i < 40°$, reflection from the low-reflection surface is indeed much weaker. However, with $\theta_i > 45°$ this is no longer the case as $|R_{\parallel}(\theta_i)|$ of the test surface is smaller than that of the low-reflection surface. Clearly, a plane wave with parallel polarization incident on a material interface sufficiently close to the Brewster angle is reflected more weakly than from a surface with $\eta = 1$.

However, when averaged over the whole range of values of the incidence angle and over all possible polarization cases, reflection from the low-reflection surface achieves the lowest level compared with other impedance surfaces. We start with polarization averaging. If polarization of the incident wave can, with equal probability, be parallel or perpendicular, then it is convenient to describe reflection by the polarization-average reflectance

$$\mathcal{R}(\theta_i, \eta) = \frac{1}{2}\left(|R_{\parallel}(\theta_i, \eta)|^2 + |R_{\perp}(\theta_i, \eta)|^2\right) \tag{4.351}$$

which relates the averaged intensities of the reflected and incident waves. For the low-reflection surface,

$$\mathcal{R}(\theta_i, 1) = |R_{\parallel}(\theta_i, 1)|^2 = |R_{\perp}(\theta_i, 1)|^2. \tag{4.352}$$

Compared to boundaries with other values of the surface impedance, the low-reflection surface can be shown to ensure the lowest values of the polarization-averaged reflectance over the range of the incidence angles $0 \leq \theta_i < \theta_1$, where

$$\theta_1 = \arccos(2 - \sqrt{3}) \approx 1.2995 = 74.46°. \tag{4.353}$$

For larger angles of incidence (grazing illumination) with $\theta_1 < \theta_i \leq \pi/2$, surfaces with η other than unity can provide lower levels of $\mathcal{R}(\theta_i, \eta)$ (problem 33). This is seen, for example, from Figure 4.41, in which the solid line representing the function $\sqrt{\mathcal{R}(\theta_i, 1)}$ goes above the dot-dashed line for $\sqrt{\mathcal{R}(\theta_i, \eta)}$ of a surface with $\eta = 0.58 - j0.08$ when $\theta_i > 75°$.

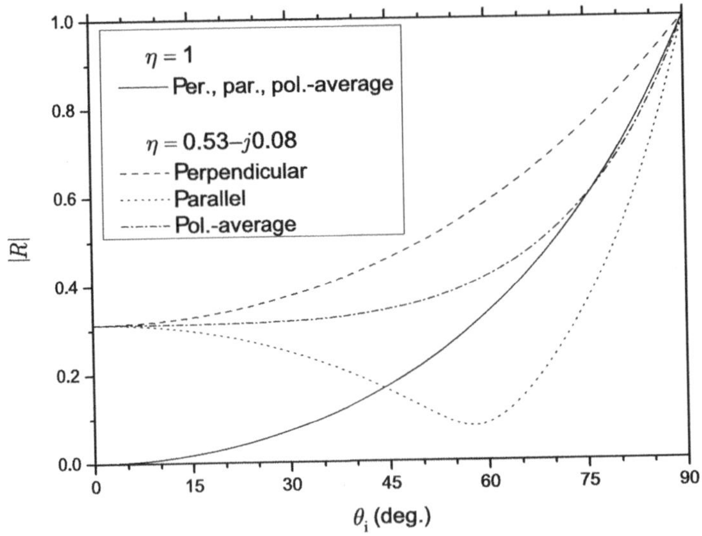

Figure 4.41 Reflection coefficient of the low-reflection surface ($\eta = 1$) compared with reflection coefficients of a surface with $\eta = 0.53 - j0.08$

One can further assume that the direction of incidence of the incoming wave can be arbitrary and average the reflectance (4.351) over the whole range of values of θ_i,

$$F(\eta) = \frac{2}{\pi} \int_0^{\pi/2} \mathcal{R}(\theta_i, \eta)\, d\theta_i. \tag{4.354}$$

Let us show that this mean reflectance is minimized when $\eta = 1$. To prove this statement, we consider $F(\eta)$ as a function of two arguments η' and η'' (the real and imaginary parts of the normalized impedance η) and begin by showing that $F(\eta)$, as a function of η'', is at minimum when $\eta'' = 0$. Evaluating the partial derivative of the reflectance gives the expression

$$\frac{\partial}{\partial \eta''}\mathcal{R} = 4\eta'\eta'' \cos\theta_i \left\{ \frac{1}{\left(\eta'^2 + \eta''^2 + 2\eta' \cos\theta_i + \cos^2\theta_i\right)^2} + \frac{\cos^2\theta_i}{\left[1 + 2\eta' \cos\theta_i + (\eta'^2 + \eta''^2)\cos^2\theta_i\right]^2} \right\}, \tag{4.355}$$

and it follows from (4.354) and (4.355) that with $\eta' \geq 0$, $\partial F(\eta)/(\partial \eta'')$ is negative when $\eta'' < 0$, vanishes at $\eta'' = 0$, and becomes positive when $\eta'' > 0$, implying that $F(\eta)$ has a minimum when $\eta'' = 0$.

The remaining task is to study the function $F(\eta)$ with purely real impedance $\eta = \eta'$ and show that $F(\eta')$ has a single minimum at $\eta = 1$ on the semi-infinite interval $\eta' \geq 0$. To this end, we evaluate the integral in (4.354) with $\eta = \eta'$ and obtain the expressions:

$$F(\eta) = 1 + \frac{8\eta}{\pi(1 - \eta^2)^{3/2}} \left(\eta \arctan\sqrt{\frac{1-\eta}{1+\eta}} - \operatorname{arctanh}\sqrt{\frac{1-\eta}{1+\eta}} \right) \tag{4.356}$$

when $0 \leq \eta < 1$ and

$$F(\eta) = 1 - \frac{8\eta}{\pi(\eta^2 - 1)^{3/2}} \left(\eta \operatorname{arctanh}\sqrt{\frac{\eta-1}{\eta+1}} - \arctan\sqrt{\frac{\eta-1}{\eta+1}} \right) \tag{4.357}$$

when $\eta > 1$. These describe a continuous function of η which is such that $F(0) = F(+\infty) = 1$ and has a minimum at the point $\eta = 1$ with the value $F(1) = 1 - 8/(3\pi) \approx 0.15117$. The function is shown in Figure 4.42 by the solid line. Further

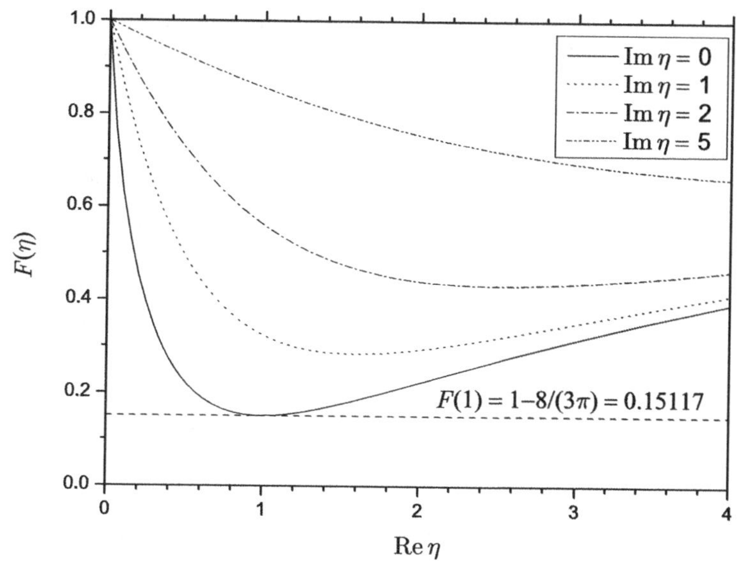

Figure 4.42 The mean reflectance (4.354) of an impedance plane as a function of Re η with various values of Im η. The dashed line shows the smallest possible level of the mean reflectance, which is achieved when Re $\eta = 1$ and Im $\eta = 0$.

curves in Figure 4.42 show the dependence of the mean reflectance (4.354) on η' for various values of η''. The curves are independent of the sign of η'' as $F(\eta' + j\eta'') = F(\eta' - j\eta'')$ by construction of $F(\eta)$.

The polarization-averaged reflectance (4.351) has the meaning of mathematical expectation of the reflectance of an impedance surface illuminated by plane waves with uniformly distributed random polarizations. The mean reflectance (4.354) can be interpreted as the expectation in the case when both polarization and incidence direction are uniformly distributed independent random variables.

4.7 Problems

1. Derive the Fresnel formulas (4.31)–(4.34) from the general formulas (4.27)–(4.28).
   Hint: Assume that $\mu_1 = \mu_2$.

2. A flat material interface is illuminated by a plane linearly polarized wave with polarization vector arbitrarily oriented with respect to the plane of incidence. Is the reflected wave linearly polarized?
   Answer: No, the reflected wave is in general elliptically polarized. In the incident wave, the ratio $E_{\parallel}^{\mathrm{inc}}/E_{\perp}^{\mathrm{inc}}$ is a real number. In the reflected wave, however, the ratio

$$\frac{E_{\parallel}^{\mathrm{ref}}}{E_{\perp}^{\mathrm{ref}}} = \frac{R_{\parallel} E_{\parallel}^{\mathrm{inc}}}{R_{\perp} E_{\perp}^{\mathrm{inc}}}$$

   is in general a complex number, which implies elliptic polarization of the reflected wave. The reflected wave is linearly polarized when either the involved media are lossless (then the reflection coefficients $R_{\parallel}$ and $R_{\perp}$ are real-valued) or the incident wave is polarized entirely parallel ($E_{\perp}^{\mathrm{inc}} = 0$) or perpendicular ($E_{\parallel}^{\mathrm{inc}} = 0$) to the plane of incidence.

3. Prove the energy conservation relation (4.47) for an interface of two lossless semi-infinite media by direct substitution of explicit formulas for reflection and transmission coefficients. Consider both polarization cases.

4. Consider an interface between two homogeneous semi-infinite media with ϵ_1, μ_1, ϵ_2, and μ_2, as in section 4.2. Let $R_{\text{TE}}(\theta_1)$ and $R_{\text{TM}}(\theta_1)$ be the TE and TM reflection coefficients, θ_1 the incidence angle, and $\theta_{\text{B}}^{\text{TE}}$ and $\theta_{\text{B}}^{\text{TM}}$ solutions of equations (4.52) and (4.53). Can it be that $\theta_{\text{B}}^{TE} = \theta_{\text{B}}^{\text{TM}}$ with $\theta_{\text{B}}^{\text{TE}}, \theta_{\text{B}}^{\text{TM}} > 0$?

   Answer: No.

   Solution: A simultaneous solution θ_1 of equations (4.50) and (4.51) can only exist if the equations are linearly dependent, which implies that $\epsilon_2/\epsilon_1 = \mu_2/\mu_1$ or $Z_1 = Z_2$. It follows then from (4.52) and (4.53) that $\theta_1 = 0$.

5. Do there exist interfaces for which both $R_{\text{TE}}(\theta_1)$ and $R_{\text{TM}}(\theta_1)$ have Brewster's angles (not necessarily equal)?

   Answer: No.

   Solution: In order for (4.52) and (4.53) to have real-valued solutions, it is necessary that $0 < \sin^2(\theta_{\text{B}}^{\text{TE}}) < 1$ and $0 < \sin^2(\theta_{\text{B}}^{\text{TM}}) < 1$ simultaneously. Rewrite (4.52) and (4.53) as $\sin^2(\theta_{\text{B}}^{\text{TE}}) = F(\epsilon_2/\epsilon_1, \mu_2/\mu_1)$ and $\sin^2(\theta_{\text{B}}^{\text{TM}}) = F(\mu_2/\mu_1, \epsilon_2/\epsilon_1)$ with

$$F(u, v) = v\frac{v - u}{v^2 - 1}.$$

   On the quadrant $0 < u < +\infty$ and $0 < v < +\infty$ find regions where $0 < F(u, v) < 1$ and $0 < F(v, u) < 1$. Show that the intersection of the regions is empty.

6. Derive the formula (4.59) for the energy flow in the case of total internal reflection at an interface of two media.

   Hint: The refracted wave is an inhomogeneous wave, but since it is linearly polarized, one can still use the simpler formula (2.386) from problem 15 of section 2.7.

   Solution: Using (4.58) in (2.386) leads to (4.59).

7. Show that under the condition of total internal reflection, when the first medium is lossless but the second medium is lossy, the energy flow (4.59) gets a component normal to the interface and directed in the second medium.

   Hint: Represent the wave vector in the second medium by

$$k_2 = \hat{t}k_1 \sin\theta_1 - \gamma_2\hat{n}, \tag{4.358}$$

   where $\gamma_2 = \sqrt{k_2^2 - k_1^2 \sin^2\theta_1}$, and insert this into (2.386).

   Solution: Representing $\mu_2 = \mu_2' - j\mu_2''$ and $\gamma_2 = \gamma_2' - j\gamma_2''$ and inserting (4.358) in (2.386) give for the normal component of the energy flow

$$\hat{w} \cdot \hat{n} = -\frac{1}{2\omega|\mu_2|^2}|E_0T|^2 \left(\gamma_2'\mu_2' + \gamma_2''\mu_2''\right).$$

   The component vanishes when $\mu_2'' = 0$ and $\gamma_2' = 0$ (total reflection, no losses), as expected. In the presence of losses in the second medium, when $\mu_2'' \le 0$ and $\operatorname{Im} k_2 < 0$, the branch of the square root function γ_2 has to be selected by the condition $\gamma_2'' < 0$ to ensure that the refracted wave decays in the second medium. The selection of the branch implies that $\gamma_2' > 0$ since $\operatorname{Im}(k_1 \sin\theta_1) = 0$, which leads to the conclusion that $\hat{w} \cdot \hat{n} < 0$.

8. In section 4.2.6 we have defined the reflection and transmission coefficients with respect to the electric field components in the bases of the incidence, reflected, and transmitted waves. Alternatively, definitions in terms of the components of the magnetic field,

$$\tilde{R}_\perp = \frac{H_{\text{ref}}^\perp}{H_{\text{inc}}^\perp}, \quad \tilde{R}_\| = \frac{H_{\text{ref}}^\|}{H_{\text{inc}}^\|}, \quad \tilde{T}_\perp = \frac{H_{\text{tr}}^\perp}{H_{\text{inc}}^\perp}, \quad \tilde{T}_\| = \frac{H_{\text{tr}}^\|}{H_{\text{inc}}^\|}$$

   are equally possible. Find the relations of thus defined $\tilde{R}_{\perp,\|}$ and $\tilde{T}_{\perp,\|}$ to the reflection and transmission coefficients $R_{\perp,\|}$, $T_{\perp,\|}$, $R_{\text{TE,TM}}$ and $T_{\text{TE,TM}}$.

   Hint: Use the relation between the electric and magnetic field vectors in a plane wave.

   Solution: For the incident, reflected, and transmitted waves, the relation (2.156) can be rewritten as $Z_1\mathbf{H}_{\text{inc}} = \hat{\imath} \times \mathbf{E}_{\text{inc}}$, $Z_1\mathbf{H}_{\text{ref}} = \hat{r} \times \mathbf{E}_{\text{ref}}$, and $Z_2\mathbf{H}_{\text{tr}} = \hat{t} \times \mathbf{E}_{\text{tr}}$. By using representations (4.72), (4.76), and (4.83) of the electric fields in the bases of the incident, reflected, and transmitted waves, and the vector relations $\hat{\imath} \times \hat{\alpha}_{\text{inc}} = \hat{\beta}_{\text{inc}}$ and $\hat{\imath} \times \hat{\beta}_{\text{inc}} = -\hat{\alpha}_{\text{inc}}$ (and similar relations for the bases of the reflected and transmitted waves), one finds the expressions $Z_1 H^{\perp}_{\text{inc}} = E^{\perp}_{\text{inc}}$, $Z_1 H^{\parallel}_{\text{inc}} = -E^{\parallel}_{\text{inc}}$, $Z_1 H^{\perp}_{\text{ref}} = E^{\perp}_{\text{ref}}$, $Z_1 H^{\parallel}_{\text{ref}} = -E^{\parallel}_{\text{ref}}$, $Z_2 H^{\perp}_{\text{tr}} = E^{\perp}_{\text{tr}}$, and $Z_2 H^{\parallel}_{\text{tr}} = -E^{\parallel}_{\text{tr}}$, implying that $\tilde{R}_{\perp} = R_{\perp}$, $\tilde{R}_{\parallel} = R_{\parallel}$, $Z_2 \tilde{T}_{\perp} = Z_1 T_{\perp}$ and $Z_2 \tilde{T}_{\parallel} = Z_1 T_{\parallel}$. Relations of $\tilde{R}_{\perp,\parallel}$ and $\tilde{T}_{\perp,\parallel}$ to $R_{\text{TE,TM}}$ and $T_{\text{TE,TM}}$ follow from equations (4.86)–(4.89).

9. We have formulated the reflection laws with respect to the components of the electromagnetic field, which are parallel and perpendicular to the plane of incidence. Sometimes, however, it is desirable to describe the reflection in terms of other components that are not necessarily parallel or perpendicular to the plane of incidence. This is the case, for example, when the problem geometry is translation-invariant with respect to a particular direction in space.

Prove the following relation. Let $\hat{\xi}$ be a unit vector defining an arbitrary direction in space and let $\mathbf{E}_{\text{inc}}$, $\mathbf{H}_{\text{inc}}$, $\mathbf{E}_{\text{ref}}$, and $\mathbf{H}_{\text{ref}}$ be the electric and magnetic field vectors in the plane waves incident on and reflected from a material interface. Then, the ξ components of the fields, which are defined by the scalar products $E^{\text{inc}}_{\xi} = \mathbf{E}_{\text{inc}} \cdot \hat{\xi}$, $E^{\text{ref}}_{\xi} = \mathbf{E}_{\text{ref}} \cdot \hat{\xi}$, $H^{\text{inc}}_{\xi} = \mathbf{H}_{\text{inc}} \cdot \hat{\xi}$, and $H^{\text{ref}}_{\xi} = \mathbf{H}_{\text{ref}} \cdot \hat{\xi}$, satisfy the relation

$$\begin{bmatrix} E^{\text{ref}}_{\xi} \\ Z_1 H^{\text{ref}}_{\xi} \end{bmatrix} = \overline{\overline{R}}_{\xi} \cdot \begin{bmatrix} E^{\text{inc}}_{\xi} \\ Z_1 H^{\text{inc}}_{\xi} \end{bmatrix}, \tag{4.359}$$

where Z_1 is the impedance of the space in which the waves propagate and the reflection coefficient is given by

$$\overline{\overline{R}}_{\xi} = \overline{\overline{Q}} \cdot \overline{\overline{P}}^{-1} \tag{4.360}$$

with

$$\overline{\overline{Q}} = \begin{bmatrix} R_{\perp}\hat{\xi} \cdot \hat{\alpha}_{\text{ref}} & R_{\parallel}\hat{\xi} \cdot \hat{\beta}_{\text{ref}} \\ R_{\perp}\hat{\xi} \cdot \hat{\beta}_{\text{ref}} & -R_{\parallel}\hat{\xi} \cdot \hat{\alpha}_{\text{ref}} \end{bmatrix}, \quad \overline{\overline{P}} = \begin{bmatrix} \hat{\xi} \cdot \hat{\alpha}_{\text{inc}} & \hat{\xi} \cdot \hat{\beta}_{\text{inc}} \\ \hat{\xi} \cdot \hat{\beta}_{\text{inc}} & -\hat{\xi} \cdot \hat{\alpha}_{\text{inc}} \end{bmatrix}. \tag{4.361}$$

Here, $\hat{\alpha}_{\text{inc,ref}}$ and $\hat{\beta}_{\text{inc,ref}}$ are the unit vectors defined in (4.71) and (4.75).

Solution: Represent the reflected electric field by the formula (4.79). The vector multiplication of this field with the unit vector $\hat{\imath}$ gives the magnetic field as

$$Z_1 \mathbf{H}_{\text{ref}} = R_{\perp} E^{\perp}_{\text{inc}} \hat{\beta}_{\text{ref}} - R_{\parallel} E^{\parallel}_{\text{inc}} \hat{\alpha}_{\text{ref}}.$$

By multiplying the reflected fields with $\hat{\xi}$ we obtain therefore the formula

$$\begin{bmatrix} E^{\text{ref}}_{\xi} \\ Z_1 H^{\text{ref}}_{\xi} \end{bmatrix} = \overline{\overline{Q}} \cdot \begin{bmatrix} E^{\perp}_{\text{inc}} \\ E^{\parallel}_{\text{inc}} \end{bmatrix}.$$

In a similar way we derive an analogous relation for the ξ components of the electric and magnetic fields in the incident wave,

$$\begin{bmatrix} E^{\text{inc}}_{\xi} \\ Z_1 H^{\text{inc}}_{\xi} \end{bmatrix} = \overline{\overline{P}} \cdot \begin{bmatrix} E^{\perp}_{\text{inc}} \\ E^{\parallel}_{\text{inc}} \end{bmatrix},$$

from which the quantities $E^{\perp}_{\text{inc}}$ and $E^{\parallel}_{\text{inc}}$ are obtained by inverting the matrix $\overline{\overline{P}}$,

$$\begin{bmatrix} E^{\perp}_{\text{inc}} \\ E^{\parallel}_{\text{inc}} \end{bmatrix} = \overline{\overline{P}}^{-1} \cdot \begin{bmatrix} E^{\text{inc}}_{\xi} \\ Z_1 H^{\text{inc}}_{\xi} \end{bmatrix}.$$

Combining the latter equation with the formula for the ξ components of the reflected field concludes the proof.

10. A procedure for the measurement of reflection coefficients may consist of the following steps. The field reflected from a reference object (typically a metal plate) is measured. Then the reference plate is replaced with a test plate (in such a way that no change in the position of the front interface occurs), and the field at the observation point is measured again. Given that E_0 and E_1 are the measured values of the electric field component *tangential* to the surface of the plates, and assuming that the incident field is a plane wave, calculate reflection coefficients $R_{\mathrm{TM,TE}}$.
Answer: For both polarizations and arbitrary incidence angles, $R_{\mathrm{TM,TE}} = -E_1/E_0$.

11. A typical procedure for measurements of transmission coefficients is that one measures the field of a transmitting antenna at a receiving antenna, inserts a test object (typically a rectangular plate) between the transmitter and receiver, and measures the field again. Given the results of the measurements, E_0 and E_1 respectively, and assuming that the incident field is a plane wave, calculate transmission coefficients $T_{\parallel,\perp}$.
Answer: One has to include a phase correction factor, $T_{\parallel,\perp} = \mathrm{e}^{-jk_0 d\cos\theta} E_1/E_0$, where θ is the angle between the normal to the plate and the propagation direction and d is the thickness of the plate.

12. In section 4.3.1 it has been shown that on the interface of an arbitrary multi-layered slab the tangential field components are related by the input impedances, Z_E or Z_H, depending on the polarization case. Represent relations (4.127) and (4.153) in the form of the impedance boundary conditions (see (2.74))

$$\hat{z} \times \mathbf{E} = \overline{\overline{Z}}_s \cdot \hat{z} \times (\hat{z} \times \mathbf{H}) \tag{4.362}$$

with the tensor (dyadic) impedance

$$\overline{\overline{Z}}_s = \hat{x}\hat{x}Z_{xx} + \hat{x}\hat{y}Z_{xy} + \hat{y}\hat{x}Z_{yx} + \hat{y}\hat{y}Z_{yy}. \tag{4.363}$$

Here, $\hat{z}$ is the unit normal to the interface, $\hat{x}$ is the unit vector tangential to the interface in the plane of incidence, and $\hat{y}$ is the unit vector perpendicular to the plane of incidence (see Figure 4.10).
Solution: Representing the fields in the coordinate frame (x, y, z) as

$$\mathbf{E} = \hat{x}E_x + \hat{y}E_y + \hat{z}E_z, \quad \mathbf{H} = \hat{x}H_x + \hat{y}H_y + \hat{z}H_z,$$

we find that

$$\hat{z} \times \mathbf{E} = -E_y\hat{x} + E_x\hat{y}, \quad \hat{z} \times \mathbf{H} = -H_y\hat{x} + H_x\hat{y}, \quad \hat{z} \times \hat{z} \times \mathbf{H} = -H_x\hat{x} - H_y\hat{y},$$

and using these and (4.363) in (4.362) gives the relation

$$E_y\hat{x} - E_x\hat{y} = \hat{x}(Z_{xx}H_x + Z_{xy}H_y) + \hat{y}(Z_{yx}H_x + Z_{yy}H_y),$$

which implies that

$$E_y = Z_{xx}H_x + Z_{xy}H_y \tag{4.364}$$

$$-E_x = Z_{yx}H_x + Z_{yy}H_y. \tag{4.365}$$

Comparing (4.364) with (4.127) and (4.365) with (4.153), we determine the elements of the dyadic input impedance

$$Z_{xx} = Z_E, \quad Z_{xy} = Z_{yx} = 0, \quad Z_{yy} = Z_H \tag{4.366}$$

and obtain

$$\overline{\overline{Z}}_s = \hat{x}\hat{x}Z_E + \hat{y}\hat{y}Z_H. \tag{4.367}$$

Note that though (4.362) has the form of *approximate* impedance boundary conditions (2.74), the use of the impedance (4.367) makes the relation (4.362) to be exact at the interface of every multi-layered structure (provided that there are no field sources below the interface). In contrast to the equivalent surface impedance $\tilde{Z}_s$ from (2.74), the impedance $\overline{\overline{Z}}_s$ is a tensor (dyadic) and its elements depend on the incidence angle. The relation (4.362) can be used as the exact boundary condition to model the interface of every isotropic multi-layered structure.

13. Expression (4.367) is formulated in the coordinate frame (x, y, z) aligned with the plane of incidence. Express the elements of the tensor impedance $\overline{\overline{Z}}_s$ in a Cartesian coordinate frame (ξ, η, z) obtained by rotation of the frame (x, y, z) around the z axis through an arbitrary angle χ.

Solution: The most general representation of a dyadic impedance is apparently given by

$$\overline{\overline{Z}}_s = \hat{\xi}\hat{\xi}Z_{\xi\xi} + \hat{\xi}\hat{\eta}Z_{\xi\eta} + \hat{\eta}\hat{\xi}Z_{\eta\xi} + \hat{\eta}\hat{\eta}Z_{\eta\eta}$$

with the relations between the tangential field components similar to (4.364) and (4.365),

$$E_\eta = Z_{\xi\xi}H_\xi + Z_{\xi\eta}H_\eta \tag{4.368}$$

$$-E_\xi = Z_{\eta\xi}H_\xi + Z_{\eta\eta}H_\eta. \tag{4.369}$$

On the other hand, inserting (4.367) in the right-hand side of (4.362) gives

$$\overline{\overline{Z}}_s \cdot \hat{z} \times (\hat{z} \times \mathbf{H}) = -\hat{x}Z_E[(\hat{x}\cdot\hat{\xi})H_\xi + (\hat{x}\cdot\hat{\eta})H_\eta] - \hat{y}Z_H[(\hat{y}\cdot\hat{\xi})H_\xi + (\hat{y}\cdot\hat{\eta})H_\eta] \tag{4.370}$$

and since $\hat{z} \times \mathbf{E} = -\hat{\xi}E_\eta + \hat{\eta}E_\xi$, it follows from (4.362) with (4.370) that

$$E_\eta = Z_E(\hat{x}\cdot\hat{\xi})[(\hat{x}\cdot\hat{\xi})H_\xi + (\hat{x}\cdot\hat{\eta})H_\eta] + Z_H(\hat{y}\cdot\hat{\xi})[(\hat{y}\cdot\hat{\xi})H_\xi + (\hat{y}\cdot\hat{\eta})H_\eta] \tag{4.371}$$

$$E_\xi = -Z_E(\hat{x}\cdot\hat{\eta})[(\hat{x}\cdot\hat{\xi})H_\xi + (\hat{x}\cdot\hat{\eta})H_\eta] - Z_H(\hat{y}\cdot\hat{\eta})[(\hat{y}\cdot\hat{\xi})H_\xi + (\hat{y}\cdot\hat{\eta})H_\eta]. \tag{4.372}$$

Comparing with (4.368) and (4.369) leads to

$$\begin{aligned}
Z_{\xi\xi} &= Z_E(\hat{x}\cdot\hat{\xi})^2 + Z_H(\hat{y}\cdot\hat{\xi})^2 \\
Z_{\xi\eta} &= Z_E(\hat{x}\cdot\hat{\xi})(\hat{x}\cdot\hat{\eta}) + Z_H(\hat{y}\cdot\hat{\xi})(\hat{y}\cdot\hat{\eta}) \\
Z_{\eta\xi} &= Z_{\xi\eta} \\
Z_{\eta\eta} &= Z_E(\hat{x}\cdot\hat{\eta})^2 + Z_H(\hat{y}\cdot\hat{\eta})^2
\end{aligned}$$

or, in terms of the rotation angle χ,

$$\begin{aligned}
Z_{\xi\xi} &= Z_E\cos^2\chi + Z_H\sin^2\chi \\
Z_{\xi\eta} &= Z_{\eta\xi} = (Z_H - Z_E)\sin\chi\cos\chi \\
Z_{\eta\eta} &= Z_E\sin^2\chi + Z_H\cos^2\chi.
\end{aligned}$$

14. Consider transmission through a multi-layered slab ($N \geq 1$), as shown in Figure 4.10. Does the angle θ_t of transmission into medium $N + 1$ depend on the properties of the adjacent layer N?

Answer: No, see equation (4.165). It only depends on electric and magnetic constants of the upper and lower semi-infinite media and the incidence angle θ_i.

15. Prove properties (4.169).

Hint: Use mathematical induction. The proof may consist of two steps. Begin by showing that (4.169) holds when $N = 1$. Then show that if (4.169) holds for some N, then it also holds when $N + 1$ is substituted for N.

16. Prove (4.179) and (4.180).

Hint: Split $\overline{\overline{A}}_{E,H}$ into $\overline{\overline{A}}'_{E,H}$ and $\overline{\overline{A}}_1^{E,H}$, and rewrite reflection and transmission coefficients in terms of elements of the primed matrices.

17. Prove (4.181) and (4.182).

Hint: Split $\overline{\overline{A}}_{E,H}$ into $\overline{\overline{A}}'_{E,H}$ and $\overline{\overline{A}}_N^{E,H}$, and rewrite reflection and transmission coefficients in terms of elements of the primed matrices.

18. Prove the reciprocity property (4.185) (Abelès 1950a).

Solution: Denote the characteristic matrices of a multi-layered slab illuminated from the upper and lower half-spaces by

$$\overline{\overline{A}}_\downarrow = \begin{bmatrix} A_{11}^\downarrow & A_{12}^\downarrow \\ A_{21}^\downarrow & A_{22}^\downarrow \end{bmatrix}$$

and

$$\overline{\overline{A}}_\uparrow = \begin{bmatrix} A_{11}^\uparrow & A_{12}^\uparrow \\ A_{21}^\uparrow & A_{22}^\uparrow \end{bmatrix},$$

respectively. The subscripts E and H, which indicate the polarization, will be omitted for brevity. The proof is essentially based on the following relations between the elements of the matrices:

$$A_{11}^\uparrow = A_{22}^\downarrow, \quad A_{12}^\uparrow = A_{12}^\downarrow, \quad A_{21}^\uparrow = A_{21}^\downarrow, \quad A_{22}^\uparrow = A_{11}^\downarrow. \tag{4.373}$$

These relations, which will be derived later, make the rest of the proof simple. Interchanging κ_0 and κ_{N+1} in (4.125) and (4.152), we express the transmission coefficient $T^\uparrow$ as

$$T^\uparrow = \frac{2\kappa_{N+1}}{-A_{21}^\uparrow + \kappa_{N+1}A_{22}^\uparrow + \kappa_0\left(A_{11}^\uparrow - \kappa_{N+1}A_{12}^\uparrow\right)} \tag{4.374}$$

and, using (4.373), obtain the reciprocity relation (4.185),

$$T^\uparrow = \frac{2\kappa_{N+1}}{-A_{21}^\downarrow + \kappa_0 A_{22}^\downarrow + \kappa_{N+1}\left(A_{11}^\downarrow - \kappa_0 A_{12}^\downarrow\right)} = \frac{\kappa_{N+1}}{\kappa_0}T^\downarrow. \tag{4.375}$$

Let us now prove (4.373). The matrix $\overline{\overline{A}}_\downarrow$ has been constructed in (4.122) or (4.149) as the product

$$\overline{\overline{A}}_\downarrow = \overline{\overline{A}}_N \ldots \overline{\overline{A}}_2 \cdot \overline{\overline{A}}_1 \tag{4.376}$$

of the matrices

$$\overline{\overline{A}}_n = \begin{bmatrix} \cos(\gamma_n d_n) & -j\sin(\gamma_n d_n)/\kappa_n \\ -j\kappa_n \sin(\gamma_n d_n) & \cos(\gamma_n d_n) \end{bmatrix} \tag{4.377}$$

of individual layers with $n = 1, 2, \ldots, N$, where N is the number of layers in the slab. The characteristic matrix of the same layer but illuminated from the lower half-space is obtained by reversing the order of multiplication,

$$\overline{\overline{A}}_\uparrow = \overline{\overline{A}}_1 \cdot \overline{\overline{A}}_2 \ldots \overline{\overline{A}}_N. \tag{4.378}$$

Multiplying (4.378) subsequently with the inverse matrices

$$\overline{\overline{A}}_n^{-1}(d_n) = \begin{bmatrix} \cos(\gamma_n d_n) & j\sin(\gamma_n d_n)/\kappa_n \\ j\kappa_n \sin(\gamma_n d_n) & \cos(\gamma_n d_n) \end{bmatrix} \tag{4.379}$$

transforms (4.378) to

$$\overline{\overline{A}}_N^{-1} \ldots \overline{\overline{A}}_2^{-1} \cdot \overline{\overline{A}}_1^{-1} \cdot \overline{\overline{A}}_\uparrow = \overline{\overline{I}}, \tag{4.380}$$

where $\overline{\overline{I}}$ is the unit matrix. Noting that $\overline{\overline{A}}_n^{-1}(d_n) = \overline{\overline{A}}_n(-d_n)$ and using (4.376) we rewrite (4.380) as

$$\overline{\overline{A}}_\uparrow = \overline{\overline{A}}_\downarrow^{-1}(-d_1, -d_2, \ldots, -d_N). \tag{4.381}$$

Next we use the property

$$\overline{\overline{A}}_\downarrow(-d_1, -d_2, \ldots, -d_N) = \begin{bmatrix} A_{11}^\downarrow(d_1, d_2, \ldots, d_N) & -A_{12}^\downarrow(d_1, d_2, \ldots, d_N) \\ -A_{21}^\downarrow(d_1, d_2, \ldots, d_N) & A_{22}^\downarrow(d_1, d_2, \ldots, d_N) \end{bmatrix} \tag{4.382}$$

which is derived by induction, starting from $\overline{\overline{A}}_1$, which is apparently compliant with (4.382), then checking that the product $\overline{\overline{A}}_2\overline{\overline{A}}_1$ is also compliant with (4.382) and so on.

Since $\det\overline{\overline{A}}_\downarrow = 1$ (see (4.126)), inversion of (4.382) gives

$$\overline{\overline{A}}_\downarrow^{-1}(-d_1, -d_2, \ldots, -d_N) = \begin{bmatrix} A_{22}^\downarrow(d_1, d_2, \ldots, d_N) & A_{12}^\downarrow(d_1, d_2, \ldots, d_N) \\ A_{21}^\downarrow(d_1, d_2, \ldots, d_N) & A_{22}^\downarrow(d_1, d_2, \ldots, d_N) \end{bmatrix}, \tag{4.383}$$

and using this result in (4.381) leads to the relations (4.373).

19. Is it possible that $|T| > 1$?

Hint: Consider energy fluxes in incident and transmitted fields.

Answer: Yes, but only when the upper and lower half-spaces are different. T is an amplitude transmission coefficient, whereas the energy conservation limits the value of the transmissivity $\mathcal{T}$, equations (4.192) and (4.39). From equations (4.188) and (4.189) relating the transmission coefficient and transmissivity, it is seen that $|T|$ can be greater then unity without violating the condition $\mathcal{T} < 1$ (e.g., if Z_0/Z_{N+1} is small). For example, the transmission coefficient of an interface between two semi-infinite media at normal incidence is

$$T = \frac{2Z_{N+1}}{Z_{N+1} + Z_0},$$

and this is greater than unity when $Z_{N+1} > Z_0$.

20. Is it possible that $|\mathcal{R}| > 1$ or $|\mathcal{T}| > 1$?

Hint: Consider situations when surface modes can be excited.

Answer: Yes, but only if the exciting wave is an evanescent wave. Reflection and transmission coefficients have poles at the eigenfrequencies of surface-wave modes, but these modes can be excited only by evanescent incident waves.

21. Derive (4.280)–(4.283) from (4.267)–(4.270).

Hint: Rewrite the numerators and denominators in (4.267)–(4.270) as linear functions of the quantity

$$E = \mathrm{e}^{-2j\gamma_1 d_1}. \tag{4.384}$$

Solution: Consider, for example, the reflection coefficient R_{TE}. The derivations of the formulas for T_{TE}, R_{TM}, and T_{TM} go along similar lines.

Expressing sine and cosine functions as

$$2\cos(\gamma_1 d_1) = e^{j\gamma_1 d_1}(1+E), \quad 2j\sin(\gamma_1 d_1) = e^{j\gamma_1 d_1}(1-E) \tag{4.385}$$

transforms (4.267) to the formula

$$R_{TE} = \frac{\kappa_1^E\left(\kappa_0^E - \kappa_2^E\right)(1+E) + \left[\kappa_0^E\kappa_2^E - \left(\kappa_1^E\right)^2\right](1-E)}{\kappa_1^E\left(\kappa_0^E + \kappa_2^E\right)(1+E) + \left[\kappa_0^E\kappa_2^E + \left(\kappa_1^E\right)^2\right](1-E)} \tag{4.386}$$

which, on combing terms with the like powers of E, takes the form

$$R_{TE} = \frac{\left(\kappa_0^E - \kappa_1^E\right)\left(\kappa_1^E + \kappa_2^E\right) + E\left(\kappa_1^E - \kappa_2^E\right)\left(\kappa_0^E + \kappa_1^E\right)}{\left(\kappa_0^E + \kappa_1^E\right)\left(\kappa_1^E + \kappa_2^E\right) + E\left(\kappa_1^E - \kappa_2^E\right)\left(\kappa_0^E - \kappa_1^E\right)}. \tag{4.387}$$

The latter expression can be rewritten as

$$R_{TE} = \left(\frac{\kappa_0^E - \kappa_1^E}{\kappa_0^E + \kappa_1^E} + E\frac{\kappa_1^E - \kappa_2^E}{\kappa_1^E + \kappa_2^E}\right)\left(1 + E\frac{\kappa_1^E - \kappa_2^E}{\kappa_1^E + \kappa_2^E}\frac{\kappa_0^E - \kappa_1^E}{\kappa_0^E + \kappa_1^E}\right)^{-1}, \tag{4.388}$$

which is equal to (4.280) because of (4.274) and (4.278).

22. Obtain expansion (4.284).

Hint: Replace R_{TE}^{01} with $-R_{TE}^{10}$ in the denominator of (4.280) and expand (4.280) in a geometric series in powers of $R_{TE}^{10}R_{TE}^{12}$.

Solution: Since $R_{TE}^{01} = -R_{TE}^{10}$ we can rewrite (4.280) as

$$R_{TE} = \frac{R_{TE}^{01} + R_{TE}^{12}}{1 - R_{TE}^{10}R_{TE}^{12}} \tag{4.389}$$

and assuming that $|R_{TE}^{10}R_{TE}^{12}| < 1$ expand (4.389) in the series

$$R_{TE} = \left(R_{TE}^{01} + R_{TE}^{12}\right)\sum_{p=0}^{+\infty}\left(R_{TE}^{10}R_{TE}^{12}\right)^p. \tag{4.390}$$

The latter can be expressed as

$$R_{TE} = R_{TE}^{01} + \left(R_{TE}^{01} + \frac{1}{R_{TE}^{10}}\right)\sum_{p=1}^{+\infty}\left(R_{TE}^{10}R_{TE}^{12}\right)^p \tag{4.391}$$

which is equal to

$$R_{TE} = R_{TE}^{01} + \left(1 + R_{TE}^{01}R_{TE}^{10}\right)R_{TE}^{12}\sum_{p=0}^{+\infty}\left(R_{TE}^{10}R_{TE}^{12}\right)^p. \tag{4.392}$$

Using expressions for R_{TE}^{01}, R_{TE}^{10}, T_{TE}^{01}, and T_{TE}^{10} from (4.274)–(4.277) leads to the relation

$$1 + R_{TE}^{01}R_{TE}^{10} = T_{TE}^{01}T_{TE}^{10} \tag{4.393}$$

which transforms (4.392) to the formula (4.284).

23. Will the solid line in Figure 4.20 change if we take a different value for A_ϵ and A_μ ($A_\epsilon = A_\mu$) while keeping δ_ϵ and δ_μ unchanged?

Answer: It does not change since with $\theta_i = 0$ and $\mu_1/\mu_0 = \epsilon_1/\epsilon_0$ equations (4.302) and (4.303) give

$$|R_{TE}| = |R_{TM}| = \exp\left[-4\pi \tan\left(\frac{\delta_\epsilon + \delta_\mu}{2}\right)\frac{d_1}{\lambda_1}\right],$$

which is independent of the absolute value of ϵ_1 and μ_1.

24. Suppose that R_∞ is the reflection coefficient (TE or TM) of a semi-infinite slab with μ_1 and ϵ_1, and R and T are the respective (TE or TM) reflection and transmission coefficients of a slab of the same material but with a finite thickness d_1. The slabs are immersed in a medium with μ_0 and ϵ_0. Prove the following relations:

$$R = R_\infty \frac{1 - E_-^2}{1 - (R_\infty E_-)^2}, \qquad T = E_- \frac{1 - R_\infty^2}{1 - (R_\infty E_-)^2},$$

where $E_- = \exp(-j\gamma_1 d_1)$.

Hint: Use equations (4.307)–(4.310).

25. The relations from the previous problem can be useful in determining the reflection coefficient of a bulk (i.e., of infinite extent) material from the experimentally available data for finite-thickness material samples (i.e., plates). Assuming that R and T are known, find R_∞.

Hint: Solve the above equations for R_∞ and E_-.

Answer:

$$R_\infty = \frac{1}{2R}\left[1 + R^2 - T^2 \pm \sqrt{(1 - R^2 + T^2)^2 - 4T^2}\right]$$
$$E_- = \frac{1}{2T}\left[1 - R^2 + T^2 \pm \sqrt{(1 - R^2 + T^2)^2 - 4T^2}\right],$$

where the sign in front of the square-root functions is to be chosen in such a way that $|E_-| \le 1$.

26. Study the reflection of a plane electromagnetic wave from a PEC-backed homogeneous layer (section 4.5.2). Determine the conditions under which the reflection coefficient of the configuration vanishes. Assume, for simplicity, the normal incidence ($\theta_i = 0$) and that the semi-infinite half-space is vacuum.

Answer: The reflection coefficient vanishes when

$$|\mathrm{Im}\,\mu_1| \gg \mathrm{Re}\,\mu_1, \quad |Z_1| \gg Z_0 \tag{4.394}$$

and

$$d_1 = \frac{Z_0}{\omega|\mathrm{Im}\,\mu_1|}. \tag{4.395}$$

Solution: For the case of normal incidence, $\gamma_1 = k_1$, $\kappa_0^H = 1/\kappa_0^E = Z_0$, $\kappa_1^H = 1/\kappa_1^E = Z_1$, reflection coefficients in (4.302) and (4.303) coincide, and the condition of their vanishing reads

$$Z_0 \cos(k_1 d_1) = jZ_1 \sin(k_1 d_1). \tag{4.396}$$

Assuming that $|k_1 d_1| \ll 1$ (a condition that will be checked later) so that $\cos(k_1 d_1) \approx 1$ and $\sin(k_1 d_1) \approx k_1 d_1$, we arrive at the relation $Z_0 \approx j\mu_1\omega d_1$, and in order for this to be satisfied, it is necessary to require that $\mu_1 \approx j\mathrm{Im}\,\mu_1$, which leads to the condition (4.395) since $\mathrm{Im}\,\mu_1 = -|\mathrm{Im}\,\mu_1|$ in any passive medium. From (4.395) it follows that $|k_1 d_1| \approx |Z_0/Z_1|$, which is small if $|Z_1| \gg Z_0$ or $|\mathrm{Im}\,\mu_1|/\mu_0 \gg |\epsilon_1|/\epsilon_0$.

Materials which comply with the requirements (4.394) are referred to as "magnetic absorbers" (Knott *et al.* 1993, section 8.6). The thickness (4.395) of the absorbing layer can be a small fraction of the wavelength. For example, for a nickel-zinc ferrite with $\epsilon_r' = 24$, $\epsilon_r'' = 24$, $\mu_r' = 9$, and $\mu_r'' = 45$ at frequency 0.5 GHz ($\lambda \approx 0.6$ m), $d_1 \approx 2$ mm.

27. The *Salisbury screen* is a resonant absorber created by placing a resistive sheet with the surface resistance equal to that of the surrounding medium at a distance of a quarter-wavelength in front of a metal plate. Assuming the case of normal incidence on the sheet, derive the conditions on the surface resistance of the sheet and on the distance between the sheet and a PEC backing plane from the rigorous theory of reflection from layered media.

Hint: Examine the general solution (4.184) with $N = 2$, $\epsilon_0 = \epsilon_2$, $\mu_0 = \mu_2$, and the thickness d_1 of the sheet as a small parameter. The permittivity of the sheet is given by $\epsilon_1 = \epsilon_0\epsilon'_{r1} - j\sigma_1/\omega \approx -j\sigma_1/\omega$, the permeability by $\mu_1 = \mu_0$, and the surface resistance by $R_e = (d_1\sigma_1)^{-1}$.

Solution: In the case of normal incidence ($\theta_i = 0$), reflection coefficients for alternative polarization coincide, and it is sufficient to consider any of the two polarization cases. From (4.184) it follows that the reflection coefficients vanish if

$$A_{12}^E + Z_0 A_{11}^E = 0. \tag{4.397}$$

Multiplication of characteristic matrices (4.120) of the individual layers according to (4.122) gives

$$A_{11}^E = C_1 C_2 - \frac{Z_0}{Z_1} S_1 S_2$$
$$A_{12}^E = -jZ_1 S_1 C_2 - jZ_0 C_1 S_2,$$

where $C_1 = \cos(k_1 d_1)$, $S_1 = \sin(k_1 d_1)$, $C_2 = \cos(k_0 d_2)$, $S_2 = \sin(k_0 d_2)$, and d_2 is the distance between the sheet and the backing plane. Inserting these in (4.397) leads to the equation

$$C_1 C_2 - \frac{Z_0}{Z_1} S_1 S_2 - jC_1 S_2 - j\frac{Z_1}{Z_0} S_1 C_2 = 0. \tag{4.398}$$

If d_1 as a small parameter, then $C_1 \approx 1$ and $S_1 \approx k_1 d_1$, and (4.398) becomes

$$C_2 - jS_2(1 - Z_0\sigma_1 d_1) \approx k_0 d_1 \left(\epsilon'_{r1} S_2 + jC_2\right), \tag{4.399}$$

in which the right-hand side is small and can be neglected. The left-hand side of (4.399) vanishes when $C_2 = 0$ and $Z_0\sigma_1 d_1 = 1$, which implies that $d_2 = \lambda/4 + n\lambda/2$ with $n = 0, 1, 2 \ldots$ and $R_e = Z_0$.

28. Does a linear electric current excite surface waves at the interface between two semi-infinite media when (a) $\mu_0 = \mu_1$ and (b) μ_1/μ_0 purely real and positive.

Hint: The reflection coefficient is given by

$$R_E(\xi) = \frac{\kappa_0^E - \kappa_1^E}{\kappa_0^E + \kappa_1^E} \tag{4.400}$$

with

$$\kappa_0^E = \frac{\gamma_0}{\omega\mu_0}, \quad \kappa_1^E = \frac{\gamma_1}{\omega\mu_1}.$$

The pole singularity when

$$\kappa_0^E + \kappa_1^E = 0$$

defines the surface wave contribution.

Solution: In (a) the pole goes to infinity on the complex ξ plane. In (b) there is no solution on the sheet of the Riemann surface where the integration contour lies ($\mathrm{Im}\,\gamma_0 \le 0$, $\mathrm{Im}\,\gamma_1 \le 0$). In both cases the surface wave cannot be excited.

29. Derive the approximate formula (4.328) for the poles of the transmission coefficient $T_{\mathrm{TM}}(k_t)$ of an almost perfect DNG lens ($\epsilon_1 \approx -\epsilon_0$ and $\mu_1 \approx -\mu_0$).

Solution: From (4.320) it follows that the poles of $T_{\mathrm{TM}}(k_t)$ are zeros of the function

$$\cos(\gamma_1 d) + jq\sin(\gamma_1 d) = \mathrm{e}^{j\gamma_1 d}\frac{1+q}{2} + \mathrm{e}^{-j\gamma_1 d}\frac{1-q}{2}, \tag{4.401}$$

which leads to the equation

$$\mathrm{e}^{2j\gamma_1 d} = \frac{q-1}{q+1}, \tag{4.402}$$

implying that

$$\gamma_1 d = \frac{1}{2j}\ln\frac{q-1}{q+1} + m\pi \tag{4.403}$$

with $m = 0, \pm 1, \pm 2, \ldots$. Denoting the zeros by k_{tm}, we obtain from (4.403)

$$k_{tm} = \pm\sqrt{\frac{1}{4d^2}\ln^2\left(\frac{q-1}{q+1}\right) + \frac{jm\pi}{d^2}\ln\frac{q-1}{q+1} + k_1^2 - \frac{m^2\pi^2}{d^2}}. \tag{4.404}$$

Parameter q is given by (4.321) and since $\gamma_0/\gamma_1 \to 1$ for $|k_t| \to \infty$,

$$q \approx \frac{1}{2}\left(\frac{\epsilon_1}{\epsilon_0} + \frac{\epsilon_0}{\epsilon_1}\right), \tag{4.405}$$

which is close to -1 for an almost perfect DNG lens. So, the logarithmic function in (4.404) is large, allowing the approximation

$$k_{tm} \approx \pm\frac{1}{2d}\left(\ln\frac{-2}{q+1} + 2jm\pi\right). \tag{4.406}$$

Because of (4.405)

$$q + 1 \approx \frac{\epsilon_0}{2\epsilon_1}\left(1 + \frac{\epsilon_1}{\epsilon_0}\right)^2, \tag{4.407}$$

and using this value in (4.406) we arrive at the formula

$$k_{tm} \approx \pm\frac{1}{d}\left[\ln(-2) - \ln\left(1 + \frac{\epsilon_1}{\epsilon_0}\right) + jm\pi\right], \tag{4.408}$$

which can be rewritten as in (4.328) since $\ln(-2) = \ln 2 \pm j\pi$.

30. A plane electromagnetic wave is incident on a flat surface (Figure 4.43), and the material properties of the surface are modeled by the impedance boundary conditions (2.74). Derive expressions (4.334) and (4.335) for the reflection coefficients of the impedance plane from the solution of the reflection problem in the vector form (section 4.2.6).

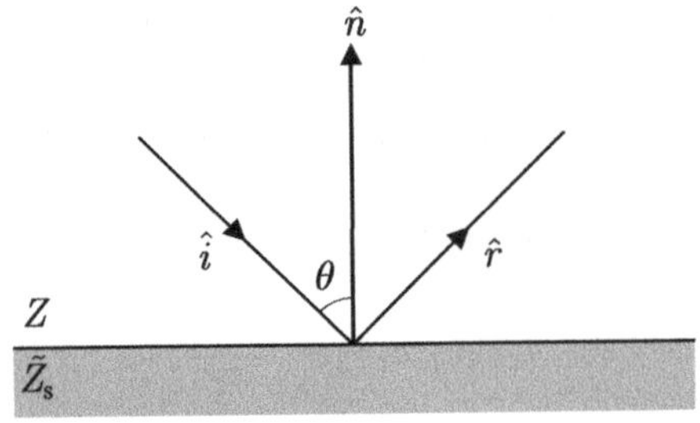

Figure 4.43 Reflection of a plane wave incident from a medium with the wave impedance Z at an impedance plane with the surface impedance $\tilde{Z}_s$.

Hint: Rewrite the impedance boundary conditions in terms of the equivalent currents $\mathbf{K}_m$ and $\mathbf{K}_e$, using the formulas (4.93) and (4.96).

Solution: In terms of the equivalent currents $\mathbf{K}_m$ and $\mathbf{K}_e$, the impedance boundary conditions take the form

$$\mathbf{K}_m = -\tilde{Z}_s \hat{n} \times \mathbf{K}_e. \tag{4.409}$$

Expressions (4.93) and (4.96) relate the equivalent currents to the reflection coefficients $R_\parallel$ and $R_\perp$. Evaluating the cross product $\hat{n} \times \mathbf{K}_e$ and using the mutual orthogonality of the vectors $\hat{\alpha}_{inc}$ and $\hat{\imath}$ permit rewriting of (4.409) as the scalar equations:

$$(R_\parallel - 1)\hat{n} \cdot \hat{\imath} = \eta(R_\parallel + 1) \tag{4.410}$$

$$R_\perp + 1 = \eta(R_\perp - 1)\hat{n} \cdot \hat{\imath}, \tag{4.411}$$

where $\eta = \tilde{Z}_s/Z$ is the normalized surface impedance. Since $\hat{n} \cdot \hat{\imath} = -\cos\theta$, where θ is the incidence angle (Figure 4.43), it follows from (4.410) and (4.411) that

$$R_\parallel = \frac{\cos\theta - \eta}{\cos\theta + \eta}, \quad R_\perp = \frac{\eta\cos\theta - 1}{\eta\cos\theta + 1}, \tag{4.412}$$

which recovers equations (4.334) and (4.335).

31. A plane electromagnetic wave is incident on a thin dielectric layer immersed in a homogeneous medium (Figure 4.44). Derive a vector solution for the reflection and transmission problems assuming that the layer is modeled by the resistive sheet transition conditions (2.111) and (2.112) with the sheet resistivity R_e.

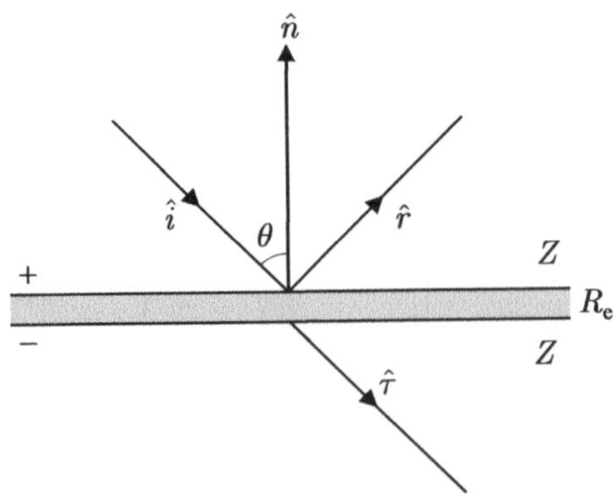

Figure 4.44 Reflection and transmission of a plane wave at a thin layer modeled as a resistive sheet.

Hint: Write down expressions for the tangential field components on the upper and lower sides of the sheet, insert these into the transition conditions, and obtain equations for the reflection and transmission coefficients (4.312)–(4.315) by projecting the conditions on the vectors $\hat{\alpha}_{\text{inc}}$ and $\hat{\imath}$.

Solution: Let the incident wave come from the upper half-plane so that above the sheet the fields consist of the incident and reflected components. On the upper side of the sheet they can therefore be expressed as in (4.72) and (4.76), and the vector products involved in the boundary conditions (2.111) and (2.112) result from (4.93) and (4.96) as

$$\hat{n}\times\mathbf{E}_+ = \hat{\imath}E^{\perp}_{\text{inc}}(1+R_\perp) + \hat{\alpha}_{\text{inc}}E^{\parallel}_{\text{inc}}(R_\parallel - 1)\hat{n}\cdot\hat{\imath} \tag{4.413}$$

$$R_e\hat{n}\times\hat{n}\times\mathbf{H}_+ = R_eY\left[\hat{\imath}E^{\perp}_{\text{inc}}(R_\perp - 1)\hat{n}\cdot\hat{\imath} + \hat{\alpha}_{\text{inc}}E^{\parallel}_{\text{inc}}(R_\parallel + 1)\right], \tag{4.414}$$

where Y is the intrinsic admittance of the surrounding medium. The fields behind the sheet have the form of a transmitted plane wave with the electric and magnetic field vectors

$$\mathbf{E}_{\text{tr}} = E^{\perp}_{\text{tr}}\hat{\alpha}_{\text{tr}} + E^{\parallel}_{\text{tr}}\hat{\beta}_{\text{tr}} \tag{4.415}$$

$$\mathbf{H}_{\text{tr}} = Y\hat{\tau}\times\mathbf{E}_{\text{tr}}, \tag{4.416}$$

where the unit vectors $\hat{\tau}$, $\hat{\alpha}_{\text{tr}}$, and $\hat{\beta}_{\text{tr}}$ define the basis of the transmitted wave (see (4.80)). Since the medium on both sides of the sheet is the same, $\hat{\alpha}_{\text{tr}} = \hat{\alpha}_{\text{inc}}$, $\hat{\tau} = \hat{\imath}$ and $\hat{\beta}_{\text{tr}} = \hat{\beta}_{\text{inc}}$. On the lower side of the sheet, the parallel and perpendicular components of the electric field are related by the transmission coefficients $T_\perp$ and $T_\parallel$ to the incident field as $E^{\perp}_{\text{tr}} = T_\perp E^{\perp}_{\text{inc}}$ and $E^{\parallel}_{\text{tr}} = T_\parallel E^{\parallel}_{\text{inc}}$, which leads to the formulas:

$$\hat{n}\times\mathbf{E}_- = \hat{\imath}E^{\perp}_{\text{inc}}T_\perp - \hat{\alpha}_{\text{inc}}E^{\parallel}_{\text{inc}}T_\parallel\hat{n}\cdot\hat{\imath} \tag{4.417}$$

$$R_e\hat{n}\times\hat{n}\times\mathbf{H}_- = R_eY\left[-\hat{\imath}E^{\perp}_{\text{inc}}T_\perp\hat{n}\cdot\hat{\imath} + \hat{\alpha}_{\text{inc}}E^{\parallel}_{\text{inc}}T_\parallel\right]. \tag{4.418}$$

Using (4.413), (4.414), (4.417), and (4.418) in the transition conditions (2.111) and (2.112), followed by projection of the conditions on the mutually orthogonal vectors $\hat{\imath}$ and $\hat{\alpha}_{\text{inc}}$, we arrive at two pairs of equations:

$$1 + R_\perp = T_\perp \tag{4.419}$$

$$1 + R_\perp = R_eY\hat{n}\cdot\hat{\imath}(R_\perp + T_\perp - 1) \tag{4.420}$$

and

$$1 - R_\parallel = T_\parallel \tag{4.421}$$

$$(R_\parallel - 1)\hat{n}\cdot\hat{\imath} = R_eY(R_\parallel - T_\parallel + 1), \tag{4.422}$$

the solution of which is given by (4.312)–(4.315) if the relation $\hat{n}\cdot\hat{\imath} = -\cos\theta$ is accounted for.

32. Derive the relations between reflection coefficients of impedance planes with the relative surface impedance η and $1/\eta$. Do these relations agree with the general relations between the fields scattered by impedance bodies with relative surface impedances η and $1/\eta$ (problem 8 in section 2.7)?

Solution: It follows from the explicit formulas (4.334) and (4.335) for the reflection coefficients of an impedance plane that

$$R_E(\eta) = R_H(\eta^{-1}), \quad R_H(\eta) = R_E(\eta^{-1})$$

$$R_\perp(\eta) = R_\parallel(\eta^{-1}), \quad R_\parallel(\eta) = R_\perp(\eta^{-1})$$

$$R_{\text{TE}}(\eta) = -R_{\text{TM}}(\eta^{-1}), \quad R_{\text{TM}}(\eta) = -R_{\text{TE}}(\eta^{-1}).$$

These relations make the solution of the problem of reflection from an impedance plane to comply with the general relations between solutions of Maxwell's equation for bodies with inverse values of the normalized surface impedance.

33. In (4.351), the polarization-averaged reflectance $\mathcal{R}(\theta_\mathrm{i}, \eta)$ of an impedance plane with the normalized impedance η has been defined. Show that $\mathcal{R}(\theta_\mathrm{i}, \eta)$ has a minimum at $\eta = 1$ for incidence angles θ_i from the interval $0 \leq \theta_\mathrm{i} < \theta_1$, where $\theta_1 = \arccos(2 - \sqrt{3})$.

Hint: Study $\mathcal{R}(\theta_\mathrm{i}, \eta)$ as a function of $\mathrm{Re}\,\eta$ and $\mathrm{Im}\,\eta$. Use the relation (4.355).

Solution: It follows from (4.355) that $\mathcal{R}(\theta_\mathrm{i}, \eta) \geq \mathcal{R}(\theta_\mathrm{i}, \eta')$ where $\eta' = \mathrm{Re}\,\eta$. Assuming that η is purely real, one obtains

$$\mathcal{R}(\theta_\mathrm{i}, \eta) = \frac{\eta^4 \cos^2\theta_\mathrm{i} + \eta^2(1 - 4\cos^2\theta_\mathrm{i} + \cos^4\theta_\mathrm{i}) + \cos^2\theta_\mathrm{i}}{(\eta\cos\theta_\mathrm{i} + 1)^2(\cos\theta_\mathrm{i} + \eta)^2}. \tag{4.423}$$

With $\theta_\mathrm{i} = \pi/2$, this expression reduces to unity, $\mathcal{R}(\pi/2, \eta) = 1$. For other angles of incidence, (4.423) is a function of η, such that

$$\mathcal{R}(\theta_\mathrm{i}, 0) = \mathcal{R}(\theta_\mathrm{i}, +\infty) = 1, \quad \mathcal{R}(\theta_\mathrm{i}, 1) = 2\left(\frac{1 - \cos\theta_\mathrm{i}}{1 + \cos\theta_\mathrm{i}}\right)^2. \tag{4.424}$$

The derivative of the function in (4.423) is equal to

$$\frac{\partial}{\partial\eta}\mathcal{R}(\theta_\mathrm{i}, \eta) = \frac{2\cos\theta_\mathrm{i}(\eta^2 - 1)f(\theta_\mathrm{i}, \eta)}{(\eta\cos\theta_\mathrm{i} + 1)^3(\cos\theta_\mathrm{i} + \eta)^3} \tag{4.425}$$

with

$$f(\theta_\mathrm{i}, \eta) = \eta^2(\cos\theta_\mathrm{i} + \cos^3\theta_\mathrm{i}) - \eta(1 - 6\cos^2\theta_\mathrm{i} + \cos^4\theta_\mathrm{i}) + \cos\theta_\mathrm{i} + \cos^3\theta_\mathrm{i}. \tag{4.426}$$

(4.425) vanishes at $\eta = \pm 1$ and at zeros of the function $f(\theta_\mathrm{i}, \eta)$. The latter is a quadratic function of η with zeros at

$$\eta = \frac{1 - 6\cos^2\theta_\mathrm{i} + \cos^4\theta_\mathrm{i} \pm \sqrt{1 - 14\cos^2\theta_\mathrm{i} + \cos^4\theta_\mathrm{i}}}{2\left(\cos\theta_\mathrm{i} + \cos^3\theta_\mathrm{i}\right)}. \tag{4.427}$$

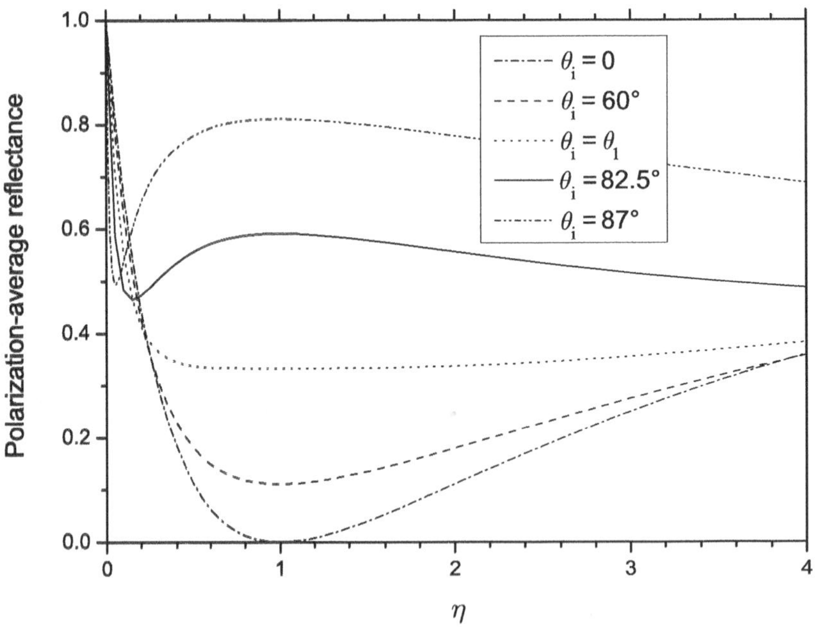

Figure 4.45 Polarization-averaged reflectance of impedance planes with real-valued impedances for various angles of incidence.

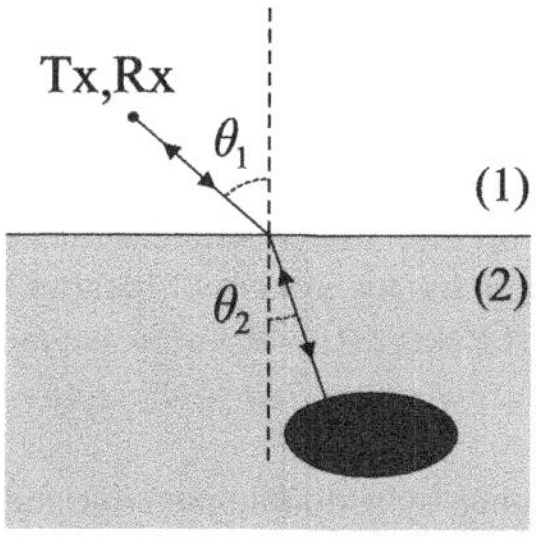

Figure 4.46 A scattering object is located in medium 2. The transmitter (Tx) and the receiver (Rx) are at the same location in medium 1.

When $\cos\theta_i = 2 - \sqrt{3}$, the square root function in (4.427) vanishes, implying that $f(\theta_i, \eta)$ has no real-valued zeros when $0 \le \theta_i < \theta_1$ and two real-valued zeros when $\theta_1 < \theta_i \le \pi/2$. Checking the sign of the second derivative $\frac{\partial^2}{\partial\eta^2}\mathcal{R}(\theta_i, \eta)$, one concludes that for $0 \le \eta < \infty$ (only positive values of $\mathrm{Re}\,\eta$ are physically meaningful) $\mathcal{R}(\theta_i, \eta)$ has either a single minimum at $\eta = 1$ or a maximum at $\eta = 1$ and two minima at the zeros (4.427). This behavior is illustrated in Figure 4.45.

34. According to the reciprocity property of electromagnetic fields, the monostatic scattering matrix of an arbitrary scatterer in the free space is symmetric (section 3.3). Will it remain symmetric if the scatterer is placed in a half-space filled with a different (isotropic) material?

Hint: Assume for simplicity the high-frequency case in which the incident and scattered waves propagate as nearly plane waves.

Answer: Yes.

Solution: The symmetry of the scattering matrix means that if a transmitting antenna sends a wave of one polarization (let say TE) and the receiving antenna is adjusted to receiving an alternative polarization (TM), then an interchange of the polarizations (transmitter sends TM polarization and the receiver receives TE polarization) does not change the received signal. Compared to the free space, we now have to account for the transmission of the incident and scattered waves through the material interface between the half-spaces. At the interface, according to (4.85), the electric field in the incident TE wave on its way to the scatterer is to be multiplied by the transmission coefficient $T_\perp^{12}$, where the subscript means the transmission from medium 1 into medium 2, and the scattered TM-polarized wave on its way back to the receiver by $T_\parallel^{21}$. By sending TM and receiving TE waves, the incident wave is multiplied by $T_\parallel^{12}$ and the scattered TE wave by $T_\perp^{21}$. The identity of the received signal in both cases follows then from the symmetry of the scattering matrix of the scatterer and the identity

$$T_\perp^{12} T_\parallel^{21} = T_\parallel^{12} T_\perp^{21},$$

which is seen from the expressions (4.29) and (4.30) for the transmission coefficients T_{TE} and T_{TM} of a material interface and the relations (4.87) and (4.89).

References

Abelès F 1950a Recherches sur la propagation des ondes électromagnétiques sinusoidales dans les milieux stratifiés. Application aux couches minces. Première partie. *Ann. Phys. Fr.*, **5**, 596–640.

Abelès F 1950b Recherches sur la propagation des ondes électromagnétiques sinusoidales dans les milieux stratifiés. Application aux couches minces. Deuxième partie. *Ann. Phys. Fr.*, **5**, 706–782.

Airy GB 1833 On the phenomena of Newton's rings when formed between two transparent substances of different refractive powers. *Philos. Mag., Ser. 3*, **2**, no. 7, 20–30.

Born M and Wolf E (eds) 1959 *Principles of Optics: Electromagnetic Theory of Propagation, Interference and Diffraction of Light*, Pergamon Press.

Brekhovskikh LM 1960 *Waves in Layered Media*, Academic Press.

Caloz C and Itoh T 2006 *Electromagnetic Metamaterials: Transmission Line Theory and Microwave Applications*, John Wiley & Sons.

Chew WC 1990 *Waves and Fields in Inhomogeneous Media*, Van Nostrand Reinhold.

Crook AW 1948 The reflection and transmission of light by any system of parallel isotropic films. *J. Opt. Soc. Am.*, **38**, 954–964.

Culhaoglu AE, Zedler M, Hoefer WJR, Osipov A, and Russer P 2008 Full wave numerical simulation of a finite 3D metamaterial lens. In *Proceedings of the 24th Annual Review of Progress in Applied Computational Electromagnetics (ACES 2008)*, March 30–April 4, 2008, Niagara Falls, Canada, 989–994.

Culhaoglu AE, Osipov AV, and Russer P 2014 Imaging by a double negative metamaterial slab excited with an arbitrarily oriented dipole. *Radio Sci.*, **49**, 68–79.

Dällenbach W and Kleinsteuber W 1938 Reflexion und Absorption von Dezimeterwellen an ebenen, dielektrischen Schichten. *Hochfrequenztechnik und Elektroakustik*, **51**, 152–156.

Drude P 1902 *The Theory of Optics*, Longmans Green and Co.

Eleftheriades GV and Balmain KG (eds) 2005 *Negative-Refraction Metamaterials: Fundamental Principles and Applications*, John Wiley & Sons.

Engheta N and Ziolkowski R (eds) 2006 *Electromagnetic Metamaterials: Physics and Engineering Explorations*, Wiley-IEEE Press.

Felsen LB and Marcuvitz N 1973 *Radiation and Scattering of Waves*, Prentice-Hall.

Fresnel A 1823 Mémoire sur la loi des modifications que la réflexion imprime á la lumière polarisée. *Œuvres completes*, 767–799.

International Commission on Illumination 2011 *ILV: International Lighting Vocabulary*, CIE S 017/E:2011, Vienna.

Ishimaru A 1991 *Electromagnetic Wave Propagation, Radiation, and Scattering*, Prentice Hall.

Jenn DC 2005 *Radar and Laser Cross Section Engineering*, American Institute of Aeronautics and Astronautics, Reston.

Knott EF, Shaeffer JF, and Tuley MT 1993 *Radar Cross Section*, Artech House.

Kong JA 1986 *Electromagnetic Wave Theory*, John Wiley & Sons.

Makarov GI, Novikov VV, and Rybachek ST 1991 *Electromagnetic Waves Propagation over Earth's Surface*, Moscow: Nauka (in Russian).

Osipov AV 2014 Minimum reflection properties of planar impedance-matched boundaries. *IEEE Trans. Antennas Propag.*, **62**, 5666–5670.

Pendry JB 2000 Negative refraction makes a perfect lens. *Phys. Rev. Lett.*, **85**, no. 18, 3966–3969.

Ra'di Y, Simovski CR, and Tretyakov SA 2015 Thin perfect absorbers for electromagnetic waves: Theory, design, and realizations, *Phys. Rev. Appl.*, **3**, p. 037001.

Ruck GT, Barrick DE, Stuart WD, and Krichbaum CK 1970 *Radar Cross Section Handbook*, Plenum Press.

Ryshik IM and Gradstein IS 1957 *Tables of Series, Products, and Integrals*, VEB, Berlin.

Shalaev VM and Sarychev AK 2007 *Electrodynamics of Metamaterials*, World Scientific Publishing Company.

Senior TBA and Volakis JL 1995 *Approximate Boundary Conditions in Electromagnetics*, IEE Electromagnetic Waves Series 41, The Institution of Electrical Engineers, London.

Sommerfeld A 1909 Über die Ausbreitung der Wellen in der drahtlosen Telegraphie. *Ann. Phys. (Berlin)*, **333**, 665–736.

Sommerfeld A 1926 Über die Ausbreitung der Wellen in der drahtlosen Telegraphie. *Ann. Phys. (Berlin)*, **386**, 1135–1153.

Sommerfeld A 1961 Drahtlose Telegraphie, in *Die Differential- und Integralgleichungen der Mechanik und Physik*, Riemann B, Weber HM, Frank P, and von Mises R (eds), chapter 23, Dover Publications.

Stratton JA 1941 *Electromagnetic Theory*, McGraw-Hill.

Tretyakov SA 2003 *Analytical Modeling in Applied Electromagnetics*, Artech House.

Ulaby FT, Moore RK, and Fung AK 1982 *Microwave Remote Sensing: Active and Passive*, Addison-Wesley.

Veselago VG 1968 The electrodynamics of substances with simultaneously negative values of ϵ and μ. *Sov. Phys. Usp.*, **10**, 509–514.

Wait JR 1962 *Electromagnetic Waves in Stratified Media*, Pergamon Press.

Watson GN 1962 *A Treatise on the Theory of Bessel Functions*, Cambridge University Press.

Weinstein W 1947 The reflectivity and transmissivity of multiple thin coatings. *J. Opt. Soc. Am.*, **37**, 576–581.

Weyl H 1919 Ausbreitung elektromagnetischer Wellen über einem ebenen Leiter. *Math. Ann.*, **365**, 481–500.

Zenneck J 1907 Über die Fortpflanzung ebener elektromagnetischer Wellen längs einer ebenen Leiterfläche und ihre Beziehung zur drahtlosen Telegraphie. *Ann. Phys. (Berlin)*, **328**, 846–866.

Ziolkowski RW and Heyman E 2001 Wave propagation in media having negative permittivity and permeability. *Phys. Rev. E*, **64**, 056625-1–056625-15.

5

Wedges

5.1 Introduction

This chapter is devoted to the scattering of time-harmonic electromagnetic fields by wedges with infinite flat faces. The configuration can be a set of contiguous, for example magneto-dielectric, wedges (Figure 5.1a) or just a single impenetrable, PEC or impedance, wedge in a homogeneous background medium (Figure 5.1b). Important special cases are the right-angled wedges with the included angles $\pi/2$ (exterior wedge, the field is sought in the wedge-shaped domain of the angular width $3\pi/2$) and $3\pi/2$ (interior wedge, like a corner reflector) (see Figure 5.2). With the included angle equal to zero, the impedance wedge shown in Figure 5.1b reduces to a half-plane (Figure 5.3a), and with the included angle π to an infinite flat impedance boundary, homogeneous or with a jump in the impedance value, depending on whether the face impedances $Z_{1,2}$ of the wedge are same or different (Figure 5.3b). A homogeneous wedge with the included angle $\pi/2$ recovers the homogeneous half-space studied in Chapter 4.

Such wedge-shaped configurations are among the most important canonical configurations of scattering and diffraction theory as they help model the influence of sharp edges on electromagnetic fields. In many cases exact solutions are available, allowing a physical insight and a meaningful description of the fields near the edge or far from the edge. The first exact treatments of the PEC half-plane problem were motivated by the need for more accuracy than in the framework of Kirchhoff's scalar diffraction theory, a truly electromagnetic description of the diffraction of light at the rim of an impenetrable screen (Poincaré 1892, 1897; Sommerfeld 1896). Exact solutions for an arbitrary-angled PEC wedge, illuminated by a plane wave incident perpendicularly to the edge or by a line source parallel to the edge, were presented in Macdonald (1902, Appendix D). Extensions on three-dimensional scalar problems for perfectly reflecting wedges described in Wiegrefe (1912) and Carslaw (1920) are applicable to problems of scattering of electromagnetic plane and cylindrical waves by PEC wedges.

As soon as the wedge becomes imperfectly reflecting, for example impedance or dielectric, the scattering problem can be solved neither by separation of variables nor by Sommerfeld's method of multivalued solutions. The problem gets even more difficult once an oblique incidence direction of plane electromagnetic waves is assumed, which leads to polarization coupling and truly vector boundary value problems. Exact solutions for wedges with impedance boundary conditions have been obtained in the scalar case (Maliuzhinets 1950, 1958a) (the solution is relevant to the case of normal incidence of a plane electromagnetic wave) and in the vector cases of oblique incidence on the full-plane impedance junction (Greenberg and Fock 1948; Vaccaro 1980), the impedance half-plane (Bucci and Franceschetti 1976), exterior and interior right-angle wedges with one face perfectly conducting (Senior 1978 1986; Vaccaro 1981), and the impedance-matched wedge (Bernard 1989; Lyalinov and Zhu 1999). The most general case of arbitrary impedance values and oblique incidence requires a numerical solution (Bilow 1991; Budaev and Bogy 2006; Daniele and Lombardi 2006; Lyalinov and Zhu 2006; Osipov and Senior 2008a; Pelosi *et al.* 1996).

Scattering problems involving penetrable wedges (see Figure 5.1a) have turned out to be hardly amenable to exact solution methods as different wave speeds in the media make the boundary value problem inseparable. The condition of separation is that the product $\epsilon\mu$ is the same in all involved media, which includes the cases of an isorefractive

Modern Electromagnetic Scattering Theory with Applications, First Edition. Andrey V. Osipov and Sergei A. Tretyakov.

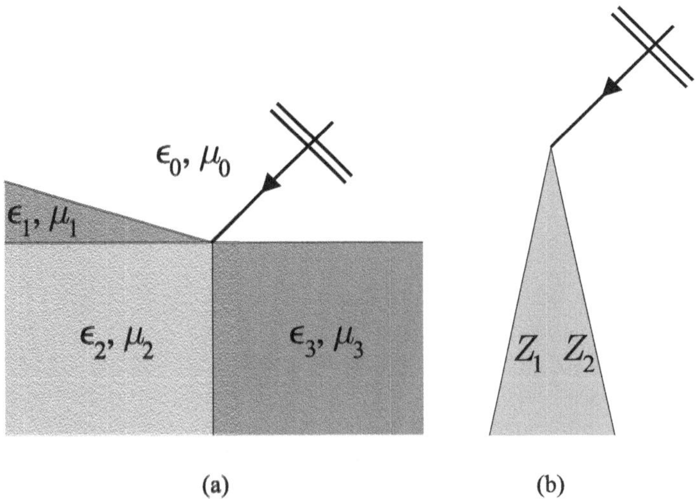

Figure 5.1 Examples of wedge-shaped configurations: (a) several contiguous magneto-dielectric wedges and (b) an impedance wedge.

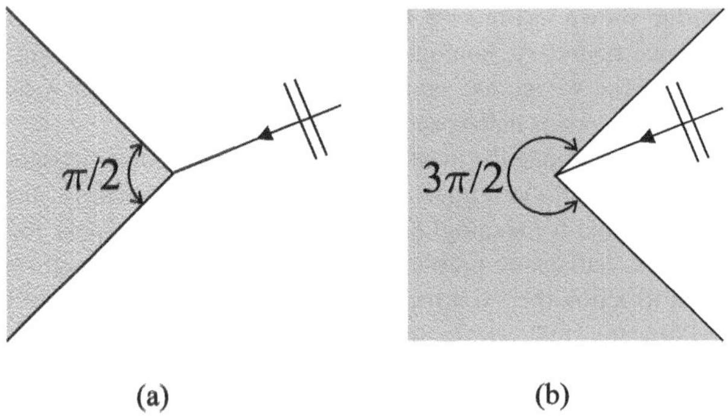

Figure 5.2 Examples of wedge-shaped configurations: (a) a right-angled wedge with included angle $\pi/2$ (exterior wedge) and (b) a right-angled wedge with included angle $3\pi/2$ (interior wedge).

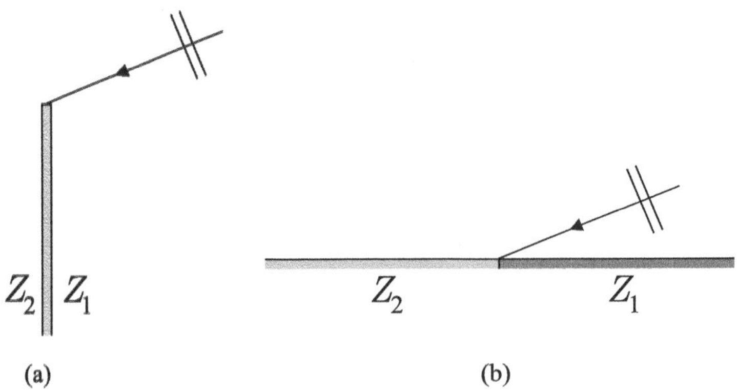

Figure 5.3 Examples of wedge-shaped configurations: (a) an impedance half-plane and (b) a full-plane impedance junction.

(diaphanous) wedge, that is, with equal wave speeds in all media (Knockaert *et al.* 1997; Osipov 1993b), and a DNG wedge, that is, with wave speeds with alternative sign (Monzon *et al.* 2005). The problem for an arbitrary number of contiguous wedges can also be solved in the static limit (corresponds to the case of infinite wave speeds in all involved media) (Greenberg 1948). A further explicitly solvable dynamic case involves a PEC half-plane placed on the interface between two homogeneous half-spaces (Ciarkowski 1987). Most of the available treatments of other penetrable wedge configurations reduce the boundary value problem to integral equations which are then solved numerically (e.g., Budaev (1995), Daniele (2010), Daniele (2011a,b), Daniele and Lombardi (2011), Kim *et al.* (1991a,b), Salem *et al.* (2006), Vasilév and Solodukhov (1976), Vasilév *et al.* (1991), and Wu and Tsai (1977)).

As far as the type of incident field is concerned, the analysis of this chapter is mostly limited to a plane wave, which is a justifiable simplification in scattering problems. In section 5.2, a solution for an infinite line source parallel to the edge of the PEC wedge is presented. For the Green function of an impedance wedge, see Osipov and Hongo (1998) and references therein. The reader interested in solutions for dipole sources in the presence of PEC wedges is referred to Bouman *et al.* (1987, section 6.5).

The chapter is organized as follows. In section 5.2 electromagnetic scattering by a PEC wedge is studied. The exact solution for a line source is derived by the method of separation of variables; the case of a plane wave follows from the solution in the limit of an infinitely distant source. The solution in the series form is employed to study the behavior of the electromagnetic field near the edge of the wedge. The behavior of the scattered field far from the edge is determined on transformation of the solution into an alternative, integral form which can be evaluated by the steepest descent method.

In section 5.3 we consider the limiting case of a PEC wedge, the perfectly conducting half-plane, to illustrate the basic features of the factorization method.

In section 5.4 electromagnetic scattering from wedges with impedance boundary conditions is studied using Maliuzhinets' method, which permits solutions for a number of important wedge-shaped impedance configurations in the form of the so-called Sommerfeld integrals. All exact solutions available in the literature for impedance wedges are presented.

Sommerfeld integrals are ideally suited for high-frequency analysis. As applied to wedges, the modifier "high-frequency" means that the distance between the observer and the edge of the wedge is greater than the wavelength. A generic high-frequency analysis of solutions for impenetrable wedges represented by Sommerfeld integrals is the topic of section 5.5. The results of the analysis are given in the format of GTD/UTD, which is the format most used in engineering applications.

Section 5.6 is devoted to the analysis of the field behavior near edges in the most general case of N contiguous magneto-dielectric wedges. A general technique for determining the degree of singularity of the fields at the edge is described. The analytical structure of the so-called Meixner series and the appearance of logarithmic terms in them are also addressed.

5.2 The Perfectly Conducting Wedge

A wedge with perfectly conducting faces is a simple configuration which permits an exact and relatively simple solution with the classical method of separation of variables. The solution can be studied in detail, in particular regarding its behavior at the edge and far from the edge. The model is important for many applications in wireless propagation, antenna design, and RCS prediction that involve edges in metal surfaces.

The presentation in section 5.2 partly follows Macdonald (1902, Appendix D) and Ufimtsev (1962). The solution is derived with the method of separation of variables, and to avoid convergence problems a linear current source parallel to the edge of the wedge is assumed. The solution for the plane wave excitation is obtained by letting the source go to infinity.

In section 5.2.1 Maxwell's equations are reduced to two independent boundary value problems for Green's functions of the Helmholtz equation in a wedge-shaped domain with Neumann and Dirichlet boundary conditions. We discuss the correct formulation of the boundary value problem, including the edge (Meixner's) conditions and the conditions at infinity.

In section 5.2.2 the Green functions are constructed in the form of eigenfunction expansions. This form of the solution permits us to investigate the field behavior near the edge of the wedge. We shall see that the components of the electromagnetic field that are transversal to the edge diverge at the edge (section 5.2.3).

The series form is unsuitable for the analysis of the solution when the observation point is located far (on the scale of the wavelength) from the edge. That is why our next step is the transformation of the solution into an alternative, integral form (section 5.2.4). The approach uses the technique of contour integration in a complex plane and is similar to the Watson transformation used in the theory of scattering by cylinders and spheres (see sections 6.4.1 and 7.3.4).

In section 5.2.5 we shall see that for several specific configurations, namely, internal and external right-angled wedges and a full plane, the solution can be expressed in a particularly simple form as a combination of exponential functions or Fresnel integrals.

For a general wedge angle, however, the solution is given by a contour integral, irreducible to known special functions. Fortunately, the integral is well suited for approximate (asymptotic) analysis. The asymptotic evaluation in section 5.2.6 shows that the field scattered by the wedge is split up into a sum of GO plane waves and an edge-diffracted component. The latter component is a cylindrical wave with a focus at the edge of the wedge and its presence is the characteristic phenomenon that explains the finite field levels in the regions shadowed by wedge-shaped impenetrable obstacles.

5.2.1 Formulation of Boundary Value Problem

We use a cylindrical coordinate system (ρ, φ, z), in which the wedge faces are at $\varphi = \pm\Phi$ and the fields are sought in the region defined by the relations $-\Phi \leq \varphi \leq \Phi$, $0 \leq \rho < \infty$, and $-\infty < z < \infty$ (Figure 5.4). The source is a combination of linear electric and magnetic currents

$$\mathbf{J}_{\mathrm{e}} = \hat{z} J_z(\rho, \varphi) \mathrm{e}^{-jk_z z}, \quad \mathbf{J}_{\mathrm{m}} = \hat{z} M_z(\rho, \varphi) \mathrm{e}^{-jk_z z}. \tag{5.1}$$

This latter equation defines the currents that flow in the z direction and have a purely exponential dependence on z with k_z being a real-valued constant parameter. The currents are located at $\rho = \rho_0$ and $\varphi = \varphi_0$, which implies that

$$\begin{bmatrix} J_z(\rho, \varphi) \\ M_z(\rho, \varphi) \end{bmatrix} = \begin{bmatrix} I_{\mathrm{e}} \\ I_{\mathrm{m}} \end{bmatrix} \frac{1}{\rho_0} \delta(\rho - \rho_0) \delta(\varphi - \varphi_0), \tag{5.2}$$

where $\delta(\ldots)$ is the delta function, so that

$$\int_{-\Phi}^{\Phi} \mathrm{d}\varphi \int_{0}^{\infty} \mathrm{d}\rho\rho \begin{bmatrix} J_z(\rho, \varphi) \\ M_z(\rho, \varphi) \end{bmatrix} = \begin{bmatrix} I_{\mathrm{e}} \\ I_{\mathrm{m}} \end{bmatrix} \tag{5.3}$$

with $I_{\mathrm{e,m}}$ defining the amplitudes of the currents.

The form of the z dependence of the currents together with the translation symmetry of the wedge in the z direction implies that the fields are proportional to $\exp(-jk_z z)$. Under such conditions, the derivatives with respect to z in the Maxwell

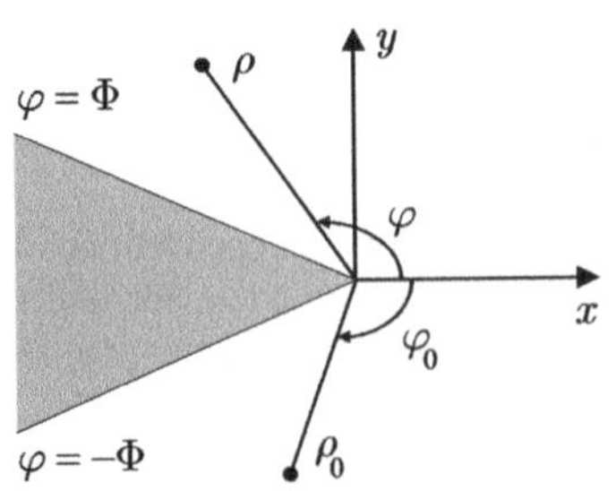

Figure 5.4 The wedge geometry and the coordinate system. A line current at ρ_0 and φ_0 is parallel to the edge of the wedge.

equations can be evaluated explicitly and the ρ and φ components of the fields expressed through the z components, as shown in section 2.5.5. The z components are to satisfy inhomogeneous Helmholtz equations

$$\left(\frac{\partial^2}{\partial\rho^2}+\frac{1}{\rho}\frac{\partial}{\partial\rho}+\frac{1}{\rho^2}\frac{\partial^2}{\partial\varphi^2}+k_\perp^2\right)E_z=j\frac{Z}{k}k_\perp^2 J_z(\rho,\varphi)\mathrm{e}^{-jk_z z} \tag{5.4}$$

$$\left(\frac{\partial^2}{\partial\rho^2}+\frac{1}{\rho}\frac{\partial}{\partial\rho}+\frac{1}{\rho^2}\frac{\partial^2}{\partial\varphi^2}+k_\perp^2\right)H_z=\frac{j}{kZ}k_\perp^2 M_z(\rho,\varphi)\mathrm{e}^{-jk_z z}, \tag{5.5}$$

where $k_\perp^2=k^2-k_z^2$ and k and Z are the wave number and the intrinsic impedance in the medium filling the wedge-shaped region. The field splits up into two independent polarization cases: one with $H_z\equiv 0$, excited by the electric current, and another with $E_z\equiv 0$, excited by the magnetic current. These two cases will be referred to as *E and H polarizations*. Notice that each has non-vanishing E_ρ, E_φ, H_ρ, and H_φ field components.

Introducing the new functions

$$E_z=E_0\mathrm{e}^{-jk_z z}G_D(\rho,\varphi) \tag{5.6}$$

$$H_z=H_0\mathrm{e}^{-jk_z z}G_N(\rho,\varphi) \tag{5.7}$$

with $E_0=jZI_e k_\perp^2/k$ and $H_0=jk_\perp^2 I_m/(kZ)$ transforms equations (5.4) and (5.5) into

$$\left(\frac{\partial^2}{\partial\rho^2}+\frac{1}{\rho}\frac{\partial}{\partial\rho}+\frac{1}{\rho^2}\frac{\partial^2}{\partial\varphi^2}+k_\perp^2\right)G_D=\frac{1}{\rho_0}\delta(\rho-\rho_0)\delta(\varphi-\varphi_0) \tag{5.8}$$

and

$$\left(\frac{\partial^2}{\partial\rho^2}+\frac{1}{\rho}\frac{\partial}{\partial\rho}+\frac{1}{\rho^2}\frac{\partial^2}{\partial\varphi^2}+k_\perp^2\right)G_N=\frac{1}{\rho_0}\delta(\rho-\rho_0)\delta(\varphi-\varphi_0). \tag{5.9}$$

We need to define the boundary conditions for these new functions. Since the wedge faces are perfectly conducting, the tangential electric field components must vanish on the faces,

$$E_z=0,\ E_\rho=0 \text{ when } \varphi=\pm\Phi. \tag{5.10}$$

This implies that G_D satisfies the Dirichlet boundary conditions

$$G_D=0 \text{ when } \varphi=\pm\Phi, \tag{5.11}$$

whereas G_N satisfies the Neumann boundary conditions

$$\frac{\partial}{\partial\varphi}G_N=0 \text{ when } \varphi=\pm\Phi. \tag{5.12}$$

The latter relation results from (2.313) and the fact that $\partial E_z/\partial\rho=0$ at $\varphi=\pm\Phi$ if $E_z=0$ at $\varphi=\pm\Phi$.

In order to complete the formulation of the boundary value problem, we should add conditions at infinity $\rho\to\infty$ and at the edge $\rho=0$ (section 2.2.3). These conditions exclude unphysical solutions from the analysis and make the solution unique. By requiring that

$$|G_D|,|G_N|\to 0 \quad \text{when } \rho\to\infty \quad \text{and} \quad \mathrm{Im}\,k<0, \tag{5.13}$$

we ensure that the solutions will have a form of outgoing waves for $\rho\to\infty$, in other words, there is no energy incoming in the system from infinity.

The edge of the wedge is a singularity of the geometry, and one has to impose an additional condition for $\rho \to 0$, the Meixner condition (2.59). This condition results from the requirement that the electromagnetic energy must be finite in any finite volume V_0 and restricts the behavior of the field components at $\rho = 0$. Indeed, if we assume that the volume V_0 is a cylinder of radius R enclosing the edge and the field components behave as $E_i, H_i = f_i(\rho)\psi_i(\varphi, z)$ for $\rho \to 0$, then (2.59) leads to

$$\int_0^R |f_i(\rho)|^2 \rho \, \mathrm{d}\rho < \infty. \tag{5.14}$$

For the integral in (5.14) to be finite, it is necessary that $\rho|f_i(\rho)| \to 0$, which implies that the most singular field components cannot diverge at the edge as ρ^{-1} or faster. By looking at the relations (2.313)–(2.316) we conclude that the edge condition (2.59) is satisfied if $E_z, H_z = O(1)$, while $E_\rho, E_\varphi, H_\rho, H_\varphi = O(\rho^{\tau-1})$ with $\tau > 0$. Since G_N and G_D are proportional to E_z and H_z (see (5.6) and (5.7)), we have to require that

$$G_D, G_N = O(1) \text{ when } \rho \to 0, \tag{5.15}$$

which completes the formulation of the boundary value problem.

5.2.2 *Solution by Separation of Variables*

Functions G_D and G_N are the Green functions of the two-dimensional Helmholtz equations (5.8) and (5.9) with Dirichlet and Neumann boundary conditions, respectively. They can be constructed by expanding in a series of eigenfunctions on the interval $-\Phi < \varphi < \Phi$.

We define the eigenfunctions $\psi_n(\varphi)$ as solutions of the ordinary differential equations

$$\psi_n'' = -\lambda_n^2 \psi_n \text{ when } -\Phi < \varphi < \Phi \tag{5.16}$$

with boundary conditions $\psi_n(\pm\Phi) = 0$ or $\psi_n'(\pm\Phi) = 0$, where the prime denotes the derivative with respect to the argument. A simple analysis shows that

$$\psi_n(\varphi) = \sin[\lambda_n(\varphi + \Phi)] \tag{5.17}$$

if the Dirichlet conditions are imposed, and

$$\psi_n(\varphi) = \cos[\lambda_n(\varphi + \Phi)] \tag{5.18}$$

if the Neumann conditions are imposed. The eigenvalues λ_n are the same for both types of boundary conditions,

$$\lambda_n = n\nu, \tag{5.19}$$

where $\nu = \pi/(2\Phi)$ and n is any integer number. Since the eigenfunctions with $n \geq 1$ and $n \leq -1$ are linearly dependent, only those with $n \geq 0$ will be retained in the analysis. It can be easily checked that the functions satisfy the orthogonality condition

$$\int_{-\Phi}^{\Phi} \psi_m(\varphi)\psi_n(\varphi)\mathrm{d}\varphi = \frac{\Phi}{q_n}\delta_{mn}, \tag{5.20}$$

where $\delta_{mn} = 0$ if $n \neq m$ and $\delta_{mn} = 1$ if $n = m$. Parameter q_n equals $1/2$ if $n = 0$ and 1 if $n \geq 1$.

Next, we represent the Green functions as the eigenfunction series

$$G(\rho, \varphi) = \sum_{n=0}^{\infty} R_n(\rho)\psi_n(\varphi), \tag{5.21}$$

insert this into the Helmholtz equation, and use (5.20) to obtain equations for the expansion coefficients,

$$R''_m + \frac{1}{\rho}R'_m + \left[k_\perp^2 - \left(\frac{\nu m}{\rho}\right)^2\right] R_m = \frac{q_m}{\Phi\rho_0}\psi_m(\varphi_0)\delta(\rho - \rho_0) \tag{5.22}$$

with $m = 0, 1, 2\ldots$ and $0 < \rho < \infty$. $R_m(\rho)$ is expected to be bounded at $\rho = 0$ and vanishing when $\rho \to \infty$ and $\operatorname{Im} k < 0$. To recover the delta-source singularity on the right-hand side of (5.22), $R_m(\rho)$ must satisfy the homogeneous equation (5.22) everywhere except for the point $\rho = \rho_0$, at which it must be continuous with a jump in the first derivative,

$$\begin{aligned} R_m(\rho_0 + 0) - R_m(\rho_0 - 0) &= 0 \\ R'_m(\rho_0 + 0) - R'_m(\rho_0 - 0) &= q_m\psi_m(\varphi_0)/(\Phi\rho_0). \end{aligned} \tag{5.23}$$

One may check by direct inspection that all these requirements are satisfied by the function

$$R_m(\rho) = jq_m\nu\psi_m(\varphi_0)J_{\nu m}(k_\perp\rho_<)H^{(2)}_{\nu m}(k_\perp\rho_>), \tag{5.24}$$

where $J_{\nu m}(k_\perp\rho_<)$ and $H^{(2)}_{\nu m}(k_\perp\rho_>)$ are the Bessel functions (Appendix E) with $\rho_< = \min(\rho, \rho_0)$ and $\rho_> = \max(\rho, \rho_0)$.

Thus, the expansions (5.21) for the Green functions G_D and G_N are constructed, and in terms of the field components (5.6) and (5.7), the solution is given by the relations:

$$E_z = j\nu E_0 e^{-jk_z z}\cdot\sum_{n=1}^{\infty}\sin[\nu n(\varphi + \Phi)]\sin[\nu n(\varphi_0 + \Phi)]J_{\nu n}(k_\perp\rho_<)H^{(2)}_{\nu n}(k_\perp\rho_>) \tag{5.25}$$

$$H_z = j\nu H_0 e^{-jk_z z}\cdot\sum_{n=0}^{\infty} q_n\cos[\nu n(\varphi + \Phi)]\cos[\nu n(\varphi_0 + \Phi)]J_{\nu n}(k_\perp\rho_<)H^{(2)}_{\nu n}(k_\perp\rho_>). \tag{5.26}$$

Note that (5.25) and (5.26) are invariant under the interchange of ρ and ρ_0, φ and φ_0, which corresponds to the reciprocity and the symmetry of the configuration.

When the source currents are located far from the edge ($k_\perp\rho_0 \gg 1$), the fields in (5.25) and (5.26) can be related to the amplitude of the incident field at $\rho = 0$. By using the asymptotic relation

$$H^{(2)}_{\nu n}(k_\perp\rho_0) \approx H^{(2)}_0(k_\perp\rho_0)e^{j\pi\nu n/2}, \tag{5.27}$$

which follows from the asymptotic formula (E.25), and the value of the two-dimensional free-space Green function $G_2(\boldsymbol{\rho}, \boldsymbol{\rho}_0, k_\perp)$ (2.326) at $\rho = 0$

$$G_2(0, \boldsymbol{\rho}_0, k_\perp) = \frac{1}{4j}H^{(2)}_0(k_\perp\rho_0), \tag{5.28}$$

we rewrite the solutions as

$$E_z \approx 4\nu E^{\text{inc}}_z(0)\sum_{n=1}^{\infty}\sin[\nu n(\varphi + \Phi)]\sin[\nu n(\varphi_0 + \Phi)]J_{\nu n}(k_\perp\rho)e^{j\pi\nu n/2} \tag{5.29}$$

$$H_z \approx 4\nu H^{\text{inc}}_z(0)\sum_{n=0}^{\infty} q_n\cos[\nu n(\varphi + \Phi)]\cos[\nu n(\varphi_0 + \Phi)]J_{\nu n}(k_\perp\rho)e^{j\pi\nu n/2}, \tag{5.30}$$

where $E^{\text{inc}}_z(0)$ and $H^{\text{inc}}_z(0)$ are the values of the incident fields

$$E^{\text{inc}}_z(\rho, \varphi, \rho_0, \varphi_0, z) = -E_0 G_2(\boldsymbol{\rho}, \boldsymbol{\rho}_0, k_\perp)e^{-jk_z z} \tag{5.31}$$

$$H^{\text{inc}}_z(\rho, \varphi, \rho_0, \varphi_0, z) = -H_0 G_2(\boldsymbol{\rho}, \boldsymbol{\rho}_0, k_\perp)e^{-jk_z z} \tag{5.32}$$

at $\rho = 0$.

An important particular case is that of excitation by plane waves with

$$E_z^{\mathrm{inc}}(\rho,\varphi,\rho_0,\varphi_0,z) = E_{0z}e^{jk_\perp\rho\cos(\varphi-\varphi_0)-jk_zz} \quad (5.33)$$
$$H_z^{\mathrm{inc}}(\rho,\varphi,\rho_0,\varphi_0,z) = H_{0z}e^{jk_\perp\rho\cos(\varphi-\varphi_0)-jk_zz}, \quad (5.34)$$

which corresponds to the current sources moved to infinity ($k_\perp\rho_0 \to \infty$). In this limit, approximation (5.27) is exact and the formulas (5.29) and (5.30) with

$$E_z^{\mathrm{inc}}(0) = E_{0z}e^{-jk_zz} \quad (5.35)$$
$$H_z^{\mathrm{inc}}(0) = H_{0z}e^{-jk_zz}, \quad (5.36)$$

which are the values of (5.33) and (5.34) at $\rho = 0$, describe exact solutions for E_z and H_z.

The solutions can be conveniently expressed in terms of a single function

$$u(\rho,\theta,\nu) = 2\nu\sum_{n=0}^{\infty} q_n e^{j\pi\nu n/2}\cos(\nu n\theta)J_{\nu n}(k_\perp\rho), \quad (5.37)$$

since the use of the trigonometric relations

$$2\sin[\nu n(\varphi+\Phi)]\sin[\nu n(\varphi_0+\Phi)] = \cos[\nu n(\varphi-\varphi_0)] - \cos[\nu n(\varphi+\varphi_0+2\Phi)] \quad (5.38)$$
$$2\cos[\nu n(\varphi+\Phi)]\cos[\nu n(\varphi_0+\Phi)] = \cos[\nu n(\varphi-\varphi_0)] + \cos[\nu n(\varphi+\varphi_0+2\Phi)] \quad (5.39)$$

reduces (5.29) and (5.30) to the formulas:

$$E_z(\rho,\varphi,z) = E_z^{\mathrm{inc}}(0)[u(\rho,\varphi-\varphi_0,\nu) - u(\rho,\varphi+\varphi_0+2\Phi,\nu)] \quad (5.40)$$
$$H_z(\rho,\varphi,z) = H_z^{\mathrm{inc}}(0)[u(\rho,\varphi-\varphi_0,\nu) + u(\rho,\varphi+\varphi_0+2\Phi,\nu)]. \quad (5.41)$$

For moderate values of $k_\perp\rho$, the series form of the solution permits efficient calculation of the fields scattered by wedges. The density plots in Figures 5.5 and 5.6 show the simulated distributions of the total fields within the circle of the 5λ radius around a right-angled wedge ($\Phi = 3\pi/4$) and a half-plane ($\Phi = \pi$) illuminated by a plane wave of unit amplitude. Both bodies create a clearly visible shadow region ($\varphi < -90°$) and a region where the incident wave interferes with the wave reflected from the upper face ($0 < \varphi < 135°$ and $90° < \varphi < 180°$, respectively). In the intermediate region ($-90° < \varphi < 0$ in Figure 5.5 and $-90° < \varphi < 90°$ in Figure 5.6) the incident wave is the only GO contribution and the interference pattern suggests overlapping with a non-GO contribution. This result is an indication for the presence of the edge-diffracted wave that will be studied later (section 5.2.6).

5.2.3 *Fields and Currents at the Edge*

The series form (5.29) and (5.30) of the solution is well suited to the analysis of the fields in an immediate proximity to the edge of the wedge. When $k_\perp\rho \to 0$, the small-argument approximation

$$J_p(w) \approx \left(\frac{w}{2}\right)^p \frac{1}{\Gamma(p+1)}, \quad (5.42)$$

where $\Gamma(p+1)$ is the Gamma function, which follows from (E.10), suggests that the terms with $n = 0, 1$ are dominant and describe the field components near the edge.

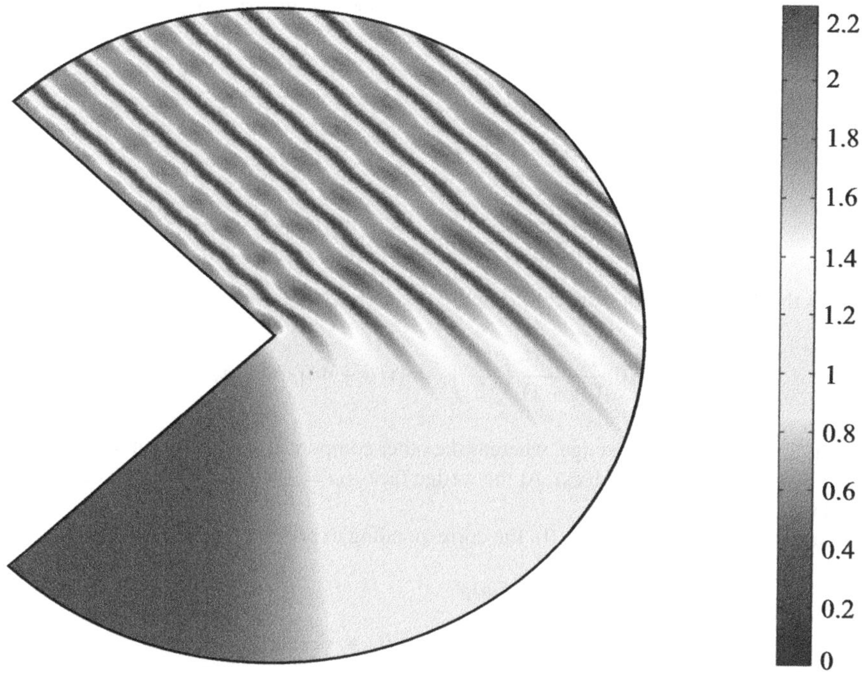

Figure 5.5 Field distribution around a right-angled PEC wedge. An H-polarized plane wave is incident from above ($\varphi_0 = \pi/2, k_z = 0$). The figure shows $|H_z|$ as a function of x and y.

Figure 5.6 The same as Figure 5.5 but for a half-plane and E polarization.

We begin with the case of E polarization, in which $H_z = 0$ everywhere in space. By using (5.42) in (5.29) and the relations (2.213)–(2.316) for the field components we obtain

$$E_z \approx a_0(\varphi)\rho^\nu \tag{5.43}$$

$$H_\varphi \approx -j\nu\frac{kY}{k_\perp^2}a_0(\varphi)\rho^{\nu-1}, \quad E_\rho = \frac{k_z}{kY}H_\varphi \tag{5.44}$$

$$H_\rho \approx j\frac{kY}{k_\perp^2}a_0'(\varphi)\rho^{\nu-1}, \quad E_\varphi = -\frac{k_z}{kY}H_\rho, \tag{5.45}$$

where the prime means the derivative with respect to φ and

$$a_0(\varphi) = E_z^{\text{inc}}(0)\frac{4\nu\, e^{j\pi\nu/2}}{\Gamma(\nu+1)}\left(\frac{k_\perp}{2}\right)^\nu \sin[\nu(\varphi+\Phi)]\sin[\nu(\varphi_0+\Phi)]. \tag{5.46}$$

Since $\nu = \pi/(2\Phi) > 1/2$, E_z vanishes at the edge, whereas the other components vanish when $\Phi < \pi/2$ (interior wedges) but are singular when $\Phi > \pi/2$ (exterior wedges). At the wedge faces ($\varphi = \pm\Phi$), E_ρ and H_φ are equal to zero for every ρ because of the boundary conditions.

In the alternative case of H polarization ($E_z = 0$), the corresponding expressions are as follows:

$$H_z \approx b_0 + c_0(\varphi)\rho^\nu \tag{5.47}$$

$$E_\varphi \approx j\nu\frac{kZ}{k_\perp^2}c_0(\varphi)\rho^{\nu-1}, \quad H_\rho = -\frac{k_z}{kZ}E_\varphi \tag{5.48}$$

$$E_\rho \approx -j\frac{kZ}{k_\perp^2}c_0'(\varphi)\rho^{\nu-1}, \quad H_\varphi = \frac{k_z}{kZ}E_\rho, \tag{5.49}$$

where $b_0 = 2\nu H_z^{\text{inc}}(0)$ and

$$c_0(\varphi) = H_z^{\text{inc}}(0)\frac{4\nu\, e^{j\pi\nu/2}}{\Gamma(\nu+1)}\left(\frac{k_\perp}{2}\right)^\nu \cos[\nu(\varphi+\Phi)]\cos[\nu(\varphi_0+\Phi)]. \tag{5.50}$$

As $\rho \to 0$, H_z tends to b_0, implying that $H_z(0) = 2\nu H_z^{\text{inc}}(0)$ where $H_z(0)$ and $H_z^{\text{inc}}(0)$ are the values of the total and incident fields at the edge. The other components are singular as $O(\rho^{\nu-1})$ at the edge when $\Phi > \pi/2$ but go to zero when $\Phi < \pi/2$. Components E_ρ and H_φ vanish when $\varphi = \pm\Phi$ because of the boundary conditions on the PEC faces.

In the general case, the fields may include both E- and H-polarized components, and as $\rho \to 0$, $E_z \to 0$ and $H_z = O(1)$ (the components parallel to the edge are always bounded), while E_φ, E_ρ, H_φ, and H_ρ either vanish when $\Phi < \pi/2$ (interior wedge) or are singular as $O(\rho^{\nu-1})$ when $\Phi > \pi/2$ (exterior wedge). The strongest field singularity with $\nu = 1/2$ occurs at the edge of a half-plane ($\Phi = \pi$). When $\Phi = \pi/2$, the wedge degenerates to a full plane and the edge disappears.

5.2.4 *Reduction to an Integral Form*

The members of series (5.37) do not decrease until the order νn of the Bessel functions becomes greater than the argument $k_\perp\rho$ (see section E.2.3). This makes the series form unsuitable for field calculations far from the edge ($\rho \to \infty$) or in the GO limit ($\lambda \to 0$). Transforming the solution into an alternative form can alleviate this difficulty. This approach uses the theory of integration in the complex plane and is similar to Watson's transformation, which is discussed in sections 6.4.1 and 7.3.4.

The idea of the transformation is based on the integral representation (E.38) of the Bessel function of the first kind,

$$J_{\nu n}(k_\perp\rho) = \frac{1}{2\pi}e^{-j\nu n\pi/2}\int_{C_3} e^{jk_\perp\rho\cos\alpha + j\nu n\alpha}\,d\alpha, \tag{5.51}$$

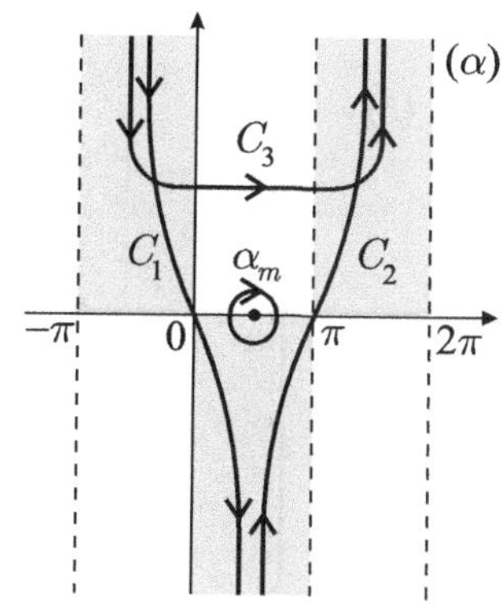

Figure 5.7 Integration contours on the complex plane α.

where the integration contour C_3 is shown in Figure 5.7. The ends of the integration contour reside in those parts of the complex plane where Im $\cos\alpha > 0$ (indicated by the filling in Figure 5.7). The series solution in (5.37) can then be recast in the following form:

$$u(\rho,\theta,\nu) = \frac{\nu}{2\pi}\int_{C_3} \mathrm{e}^{jk_\perp\rho\cos\alpha}\left[1+\sum_{n=1}^{\infty}\mathrm{e}^{j\nu n(\alpha+\theta)}+\sum_{n=1}^{\infty}\mathrm{e}^{j\nu n(\alpha-\theta)}\right]\mathrm{d}\alpha. \quad (5.52)$$

The sequences of the exponents in the integrand are geometric progressions that can be evaluated in a closed form, and one arrives at the following formula:

$$u(\rho,\theta,\nu) = \int_{C_3} \mathrm{e}^{jk_\perp\rho\cos\alpha}\,U(\alpha,\theta,\nu)\,\mathrm{d}\alpha \quad (5.53)$$

with

$$U(\alpha,\theta,\nu) = \frac{\nu}{2\pi}\left[\frac{1}{1-\mathrm{e}^{j\nu(\alpha+\theta)}}+\frac{1}{1-\mathrm{e}^{j\nu(\alpha-\theta)}}-1\right] = \frac{\nu}{2\pi j}\frac{\sin(\nu\alpha)}{\cos(\nu\alpha)-\cos(\nu\theta)}. \quad (5.54)$$

Integrals of this form, over an integration path as in the definition of the Bessel function of the first kind and with an integrand function odd with respect to the integration variable, are called Sommerfeld's integrals. Their properties are discussed in section 5.4.1.

The great advantage of Sommerfeld's integrals is that they are ideally suited for the asymptotic analysis of solutions when $k\rho \gg 1$. The analysis uses the steepest descent method, and in what follows we describe the typical steps. First we note that representation (5.53) can be further simplified by deforming the integration contour onto another one, along which the integrand function behaves particularly advantageously. This is achieved by deforming C_3 onto contours C_1 and C_2 shown in Figure 5.7. The integral over C_1 vanishes since the contour is symmetric with respect to $\alpha = 0$ and $U(\alpha,\theta,\nu)$ is an odd function of α. So, the result of the contour deformation is given by the formula

$$u(\rho,\theta,\nu) = u_{\text{res}}(\rho,\theta,\nu) + u_1(\rho,\theta,\nu), \quad (5.55)$$

where

$$u_1(\rho,\theta,\nu) = \int_{C_2} \mathrm{e}^{jk_\perp\rho\cos\alpha}\,U(\alpha,\theta,\nu)\,\mathrm{d}\alpha. \quad (5.56)$$

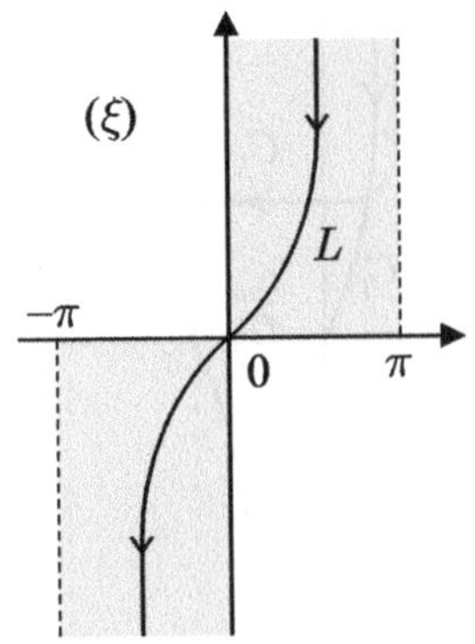

Figure 5.8 Integration contour in the complex plane ξ.

The integral in (5.56) can be further modified by changing the integration variable $\alpha = \pi - \xi$, which gives

$$u_1(\rho, \theta, \nu) = \int_L e^{-jk_\perp \rho \cos \xi} U(\xi - \pi, \theta, \nu)\, d\xi, \tag{5.57}$$

where the integration contour begins at $\xi = \pi/2 + j\infty$, goes through $\xi = 0$, and ends at $\xi = -\pi/2 - j\infty$ (Figure 5.8).

The term $u_{res}(\rho, \theta, \nu)$ describes contributions from the poles of $U(\alpha, \theta, \nu)$ that have been captured in the process of the contour deformation. As seen from (5.54), $U(\alpha, \theta, \nu)$ has poles at the points $\alpha = \alpha_m^\pm$, where

$$\alpha_m^\pm = \pm\theta + 4m\Phi \tag{5.58}$$

with $m = 0, \pm 1, \pm 2, \ldots$ The corresponding residues can be easily evaluated, and we obtain the formula

$$u_{res}(\rho, \theta, \nu) = \sum_m \omega_m^\pm e^{jk_\perp \rho \cos \alpha_m^\pm}, \tag{5.59}$$

in which $\omega_m^\pm = 1$ if a pole $\alpha = \alpha_m^\pm$ is captured and $\omega_m^\pm = 0$ otherwise. Since parameter θ is real-valued, the poles reside on the real axis in the complex α plane, and by looking at the contour topology in Figure 5.7 we conclude that only the poles with $0 < \alpha_m^\pm < \pi$ can be captured. A more detailed analysis shows that the poles with $m \le -1$ are never captured and can be omitted from the analysis.

Representation (5.55) is well suited for analytical evaluation since the sum (5.59) involves only a finite number of terms, whereas the integral (5.57) can be evaluated either exactly (for a set of special values of Φ) or asymptotically (for $k\rho \gg 1$). The special cases are addressed in section 5.2.5 and the asymptotic analysis in section 5.2.6. In what follows, the plane wave excitation as in (5.33) and (5.34) is assumed.

5.2.5 *Special Cases*

When $\Phi = \pi/(2M)$ with $M = 1, 2, 3, \ldots$, integral (5.57) vanishes since $U(\xi - \pi, \theta, \nu)$ with $\nu = M$ is an odd function of ξ. With $\Phi = \pi$ the integral does not vanish but can be reduced to a known special function. Let us take a closer look at the cases $M = 1$ (infinite plane), $M = 2$ (interior right-angle wedge) and $\Phi = \pi$ (half-plane).

The Infinite Plane

When $\Phi = \pi/2$, the wedge degenerates to an infinite plane (Figure 5.9). One finds from (5.54) that

$$U(\xi - \pi, \theta, 1) = \frac{-j}{2\pi} \frac{\sin \xi}{\cos \xi + \cos \theta}, \tag{5.60}$$

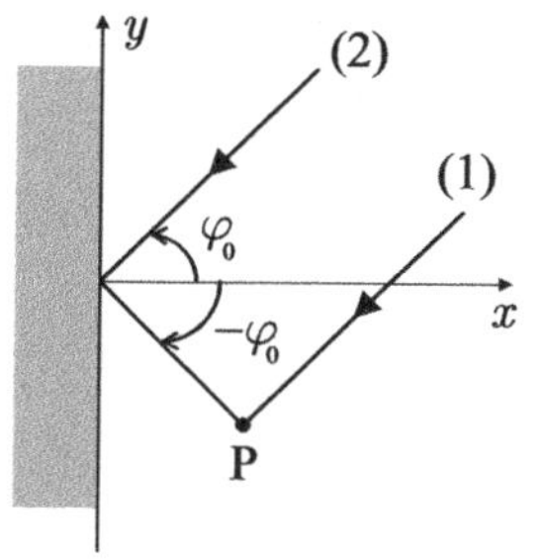

Figure 5.9 The full-plane case: (1) incident wave and (2) reflected wave.

which is an odd function of ξ, and since the integration contour in (5.57) is symmetric with respect to $\xi = 0$,

$$u_1(\rho, \theta, 1) = 0. \tag{5.61}$$

Thus, $u(\rho, \theta, 1) = u_{\text{res}}(\rho, \theta, 1)$. Since θ is equal to either $\varphi - \varphi_0$ or $\varphi + \varphi_0 + 2\Phi$, the value of θ belongs to the interval $-\pi < \theta < 2\pi$. A closer look at the poles shows that only three of them can contribute to (5.59), namely, $\alpha = \alpha_0^-$ (when $-\pi < \theta < 0$), $\alpha = \alpha_0^+$ (when $0 < \theta < \pi$), and $\alpha = \alpha_1^-$ (when $\pi < \theta < 2\pi$). The residues at those poles are equal to $\exp(jk_\perp \rho \cos\theta)$, which implies that

$$u(\rho, \theta, 1) = \mathrm{e}^{jk_\perp \rho \cos\theta}, \tag{5.62}$$

regardless of the value of θ. The fields are obtained by inserting (5.62) in (5.40) and (5.41). For plane wave incidence, the solution is a sum of an incident plane wave and a plane wave reflected from the boundary,

$$\begin{bmatrix} E_z \\ H_z \end{bmatrix} = \begin{bmatrix} E_{0z} \\ H_{0z} \end{bmatrix} \mathrm{e}^{-jk_z z}[\mathrm{e}^{jk_\perp \rho \cos(\varphi-\varphi_0)} \mp \mathrm{e}^{-jk_\perp \rho \cos(\varphi+\varphi_0)}], \tag{5.63}$$

where the upper sign corresponds to E_z and the lower one to H_z.

The Interior Right-angled Wedge

The case of $M = 2$ corresponds to an interior right-angled wedge ($\Phi = \pi/4$), which is an important canonical geometry (Figure 5.10). The integrand function in (5.57) is given by

$$U(\xi - \pi, \theta, 2) = -\frac{j}{\pi} \frac{\sin(2\xi)}{\cos(2\xi) - \cos(2\theta)} \tag{5.64}$$

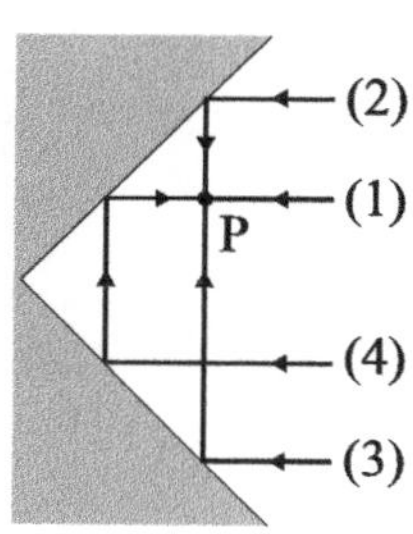

Figure 5.10 Reflections in a right-angled interior wedge: (1) incident wave, (2) wave singly reflected from the upper face, (3) wave singly reflected from the lower face, and (4) doubly reflected wave.

and $u_1(\rho, \theta, 2) = 0$. The range of variation of θ is $-\pi/2 < \theta < \pi$, and the captured poles are $\alpha = \alpha_0^-$ and α_1^+ when $-\pi/2 < \theta < 0$, and $\alpha = \alpha_0^+$ and α_1^- when $0 < \theta < \pi$. Thus,

$$u(\rho, \theta, \nu) = u_{\text{res}}(\rho, \theta, \nu) = e^{jk_\perp \rho \cos\theta} + e^{-jk_\perp \rho \cos\theta}, \tag{5.65}$$

which expresses the field as a combination of four plane waves,

$$\begin{bmatrix} E_z \\ H_z \end{bmatrix} = \begin{bmatrix} E_{0z} \\ H_{0z} \end{bmatrix} e^{-jk_z z} \cdot \left[e^{jk_\perp \rho \cos(\varphi - \varphi_0)} + e^{-jk_\perp \rho \cos(\varphi - \varphi_0)} \mp e^{-jk_\perp \rho \sin(\varphi + \varphi_0)} \mp e^{jk_\perp \rho \sin(\varphi + \varphi_0)} \right], \tag{5.66}$$

namely (from left to right in the square brackets), the incident wave, the doubly reflected wave, and the waves reflected from the upper ($\Phi = \pi/4$) and lower ($\Phi = -\pi/4$) faces of the wedge-shaped region (Figure 5.10).

The superposition of the four plane waves in (5.66) results in the periodic pattern with the periods

$$\lambda \frac{k}{k_\perp} \sec\left(\varphi_0 + \frac{\pi}{4}\right), \quad \lambda \frac{k}{k_\perp} \csc\left(\varphi_0 + \frac{\pi}{4}\right) \tag{5.67}$$

along the lower ($\varphi = -\pi/4$) and upper ($\varphi = \pi/4$) faces of the wedge, respectively. The magnitudes are periodic with the half of the periods in (5.67). For an incident wave of unit amplitude, the field magnitudes vary between the values 0 and 4, depending on whether the four plane waves in (5.66) superimpose constructively or destructively. Figures 5.11 and 5.12 show the distribution of the electric and magnetic fields due to a plane electromagnetic wave with $k_z = 0$ and $\varphi_0 = -\pi/6$ in an interior right-angled wedge with PEC boundaries.

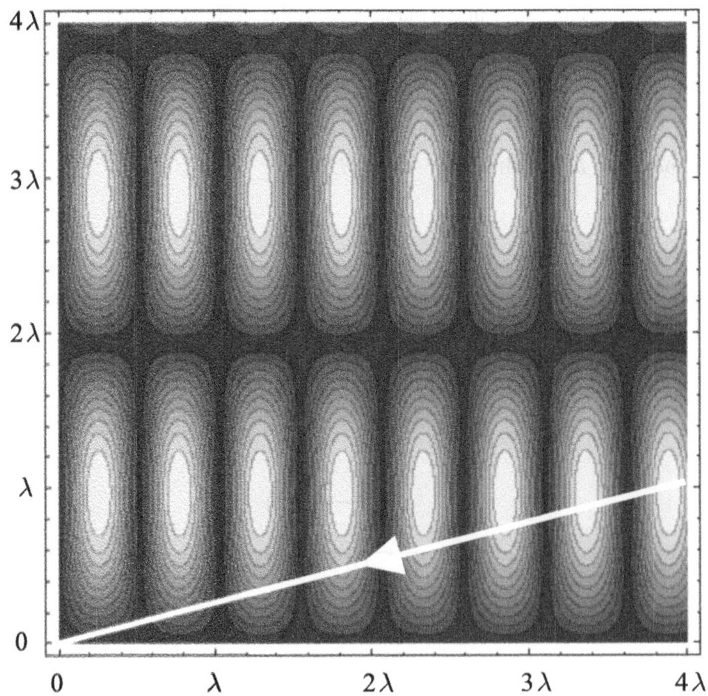

Figure 5.11 Distribution of $|E_z|$ in an interior right-angled wedge with PEC boundaries. An E-polarized plane wave of unit amplitude ($E_{0z} = 1$, $H_{0z} = 0$) is incident at a right angle to the edge of the wedge and makes a 15° angle with the lower (horizontal) face of the wedge. The field is plotted within a $4\lambda \times 4\lambda$ region. Lower field values are shown darker.

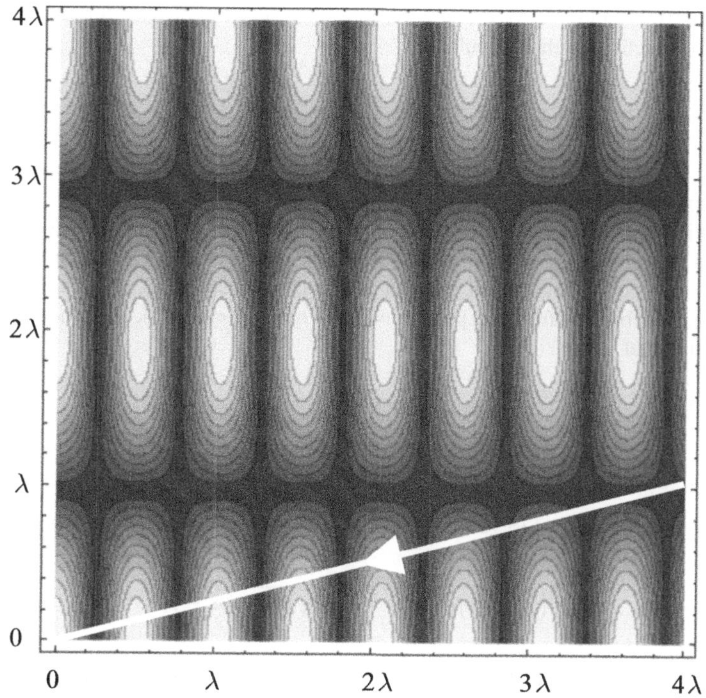

Figure 5.12 The same as Figure 5.11 but for $|H_z|$ and an H-polarized wave.

The Half-plane

Next, we consider the solution for the half-plane, which is an important special case of the wedge geometry, obtained when $\Phi = \pi$. In this case, the integral $u_1(\rho, \theta, \nu)$ does not vanish but can be expressed in terms of the Fresnel integral $F(X)$, the basic properties of which are described in Appendix B.

Indeed, when $\Phi = \pi$, the integrand function (5.54) becomes

$$U\left(\xi - \pi, \theta, \frac{1}{2}\right) = \frac{j}{4\pi} \frac{\cos(\xi/2)}{\sin(\xi/2) - \cos(\theta/2)}, \tag{5.68}$$

implying that integral (5.57) can be considerably simplified by using $\sin(\xi/2)$ as a new integration variable. The result of the substitution is particularly simple if we define the new integration variable t as

$$\sin\frac{\xi}{2} = \frac{t}{\sqrt{2k_\perp \rho}} e^{j\pi/4}, \tag{5.69}$$

which leads to the formula

$$u_1\left(\rho, \theta, \frac{1}{2}\right) = e^{-jk_\perp \rho} \frac{j}{2\pi} \int_{-\infty}^{\infty} \frac{e^{-t^2}}{t_0 - t} dt \tag{5.70}$$

with

$$t_0 = \sqrt{2k_\perp \rho} e^{-j\pi/4} \cos\frac{\theta}{2}. \tag{5.71}$$

According to the formulas (B.25) and (B.26), the integral in (5.70) can be expressed through the Fresnel integral $F(X)$ with $X = \sqrt{2k_\perp \rho}\cos(\theta/2)$, which gives the formula

$$u_1\left(\rho,\theta,\frac{1}{2}\right) = e^{jk_\perp\rho\cos\theta}\left[F\left(\sqrt{2k_\perp\rho}\cos\frac{\theta}{2}\right) - \Theta\left(\cos\frac{\theta}{2}\right)\right], \tag{5.72}$$

where $\Theta(x)$ is the unit step function defined by (B.12).

Now we turn our attention to the residue contribution $u_{\mathrm{res}}(\rho,\theta,1/2)$ (see (5.59)). Parameter θ, which is either $\varphi - \varphi_0$ or $\varphi + \varphi_0 + 2\pi$, varies in the range $-2\pi < \theta < 4\pi$. If $-2\pi < \theta < -\pi$ or $\pi < \theta < 3\pi$, then $u_{\mathrm{res}}(\rho,\theta,1/2) = 0$ since no poles (5.58) are captured when deforming the integration contour C_3 into C_1 and C_2 (Figure 5.7). On the other hand, one has to account for residue contributions from the poles $\alpha = \alpha_0^-$ (when $-\pi < \theta < 0$), α_0^+ (when $0 < \theta < \pi$) and α_1^- (when $3\pi < \theta < 4\pi$). Thus, we find that

$$u_{\mathrm{res}}\left(\rho,\theta,\frac{1}{2}\right) = \begin{cases} 0 & \text{if } \theta \in (-2\pi,-\pi)\cup(\pi,3\pi) \\ e^{jk_\perp\rho\cos\theta} & \text{if } \theta \in (-\pi,\pi)\cup(3\pi,4\pi) \end{cases}. \tag{5.73}$$

Regardless of the value of θ, the sum of (5.72) and (5.73) can always be rewritten in a surprisingly simple form

$$u\left(\rho,\theta,\frac{1}{2}\right) = e^{jk_\perp\rho\cos\theta}F\left(\sqrt{2k_\perp\rho}\cos\frac{\theta}{2}\right), \tag{5.74}$$

and the use of (5.40) and (5.41) leads to the following formulas for the total electromagnetic field of a plane wave diffracted by a perfectly conducting half-plane:

$$\begin{bmatrix} E_z \\ H_z \end{bmatrix} = \begin{bmatrix} E_{0z} \\ H_{0z} \end{bmatrix} e^{-jk_z z}\left\{e^{jk_\perp\rho\cos(\varphi-\varphi_0)}F\left[\sqrt{2k_\perp\rho}\cos\left(\frac{\varphi-\varphi_0}{2}\right)\right] \mp e^{jk_\perp\rho\cos(\varphi+\varphi_0)}F\left[-\sqrt{2k_\perp\rho}\cos\left(\frac{\varphi+\varphi_0}{2}\right)\right]\right\}. \tag{5.75}$$

Note that (5.75) is an exact solution of the problem of scattering of a plane electromagnetic wave from a PEC half-plane; it was obtained in the case of normal incidence ($k_z = 0$) by Macdonald (1902, Appendix D). The solution is valid for an arbitrary position of the observation point, including the edge of the configuration. By using the small-argument expansion (B.13) of the Fresnel integral, it can be shown that the behavior of the solution at the edge ($\rho \to 0$) is in agreement with the approximations (5.43)–(5.50).

Figures 5.13 and 5.14 illustrate the behavior of the tangential components of the magnetic field on the upper face of a PEC half-plane. When the face is illuminated, $\varphi_0 = 45°$ and $90°$, H_z is an oscillating function of ρ/λ. The oscillations disappear when the illumination is grazing ($\varphi_0 = 0$) or the face is in the shadow ($\varphi_0 = -90°$), in the latter case the field quickly vanishes with growing ρ/λ. Compared with H_z, the component H_ρ is singular at $\rho \to 0$, decays much faster with growing ρ/λ, and its oscillations are much less pronounced (Figure 5.14). Explanation for the different behavior of the components is the subject of problem 5.

5.2.6 *Edge-diffracted and GO Components. Diffraction Coefficient*

Oscillations of the surface current on the lit side of the PEC half-plane and finite values of the current on the shadowed side of the PEC half-plane, which are observed in Figures 5.13 and 5.14, cannot be explained by GO reasons. In this section we asymptotically evaluate the exact solution for an arbitrary-angle wedge for $k\rho \gg 1$ and extract a non-ray-optical component: the edge-diffracted wave. The wave is a cylindrical wave with the phase center at the edge of the wedge; it is omnipresent in space around the wedge, defining the field in the shadow and interfering with the ray optical components in the lit region.

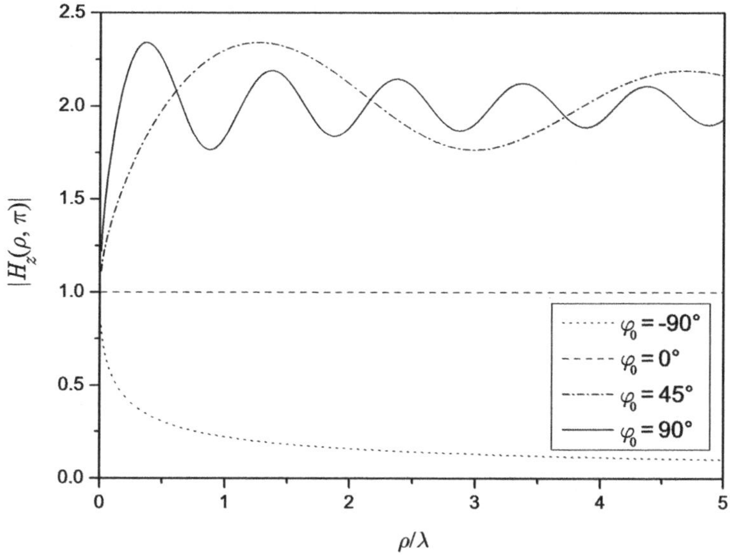

Figure 5.13 Amplitude of the tangential magnetic field component on the upper face ($\varphi = \pi$) of a PEC half-plane as a function of ρ/λ for an H-polarized plane wave of unit amplitude ($H_{0z} = 1$, $E_{0z} = 0$), incident normally to the edge of the half-plane at various incidence angles φ_0.

For a general value of Φ, function $u_1(\rho, \theta, \nu)$ cannot be reduced to known special functions. If, however, parameter $k_\perp \rho$ is large, we can perform integration in (5.57) approximately. By looking at the absolute value of the exponential factor in (5.57)

$$|e^{-jk_\perp \rho \cos \xi}| = \exp\left[-2k_\perp \rho \, \text{Im}\left(\sin^2 \frac{\xi}{2}\right)\right] \tag{5.76}$$

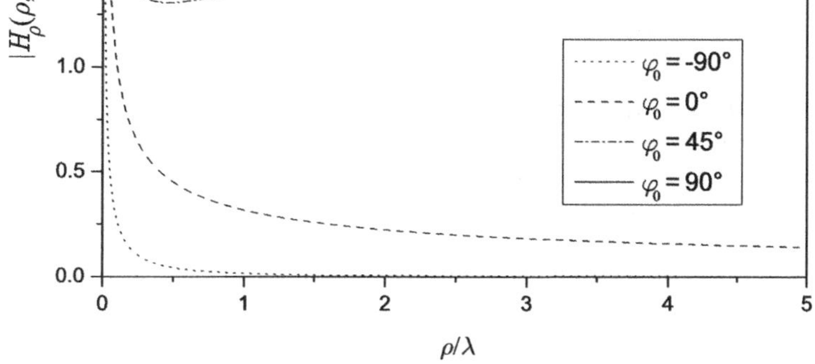

Figure 5.14 The same as Figure 5.13 but for the component $|H_\rho|$ and an E-polarized plane wave with $H_{0z} = 0$ and $YE_{0z} = 1$.

we note that on the integration contour L this function has a sharp maximum at $\xi = 0$ and rapidly decays as $\xi \to \pm(\pi/2 + j\infty)$. This suggests that the main contribution to the integral comes from a small vicinity of the point $\xi = 0$. If the integrand function $U(\xi - \pi, \theta, \nu)$ is sufficiently smooth in this vicinity, we can replace it with its value at $\xi = 0$ and write

$$u_1(\rho, \theta, \nu) \approx U(-\pi, \theta, \nu) \int_L e^{-jk_\perp \rho \cos\xi} d\xi. \tag{5.77}$$

The integral relates to the Hankel function

$$H_0^{(2)}(k_\perp \rho) = \frac{-1}{\pi} \int_L e^{-jk_\perp \rho \cos\xi} d\xi, \tag{5.78}$$

see (E.37), and since $k_\perp \rho \gg 1$, we can use a large-argument approximation (E.25) of the Hankel function to get

$$u_1(\rho, \theta, \nu) \approx \sqrt{\frac{2\pi}{k_\perp \rho}} U(-\pi, \theta, \nu) e^{-jk_\perp \rho - j3\pi/4}, \tag{5.79}$$

or, by evaluating $U(-\pi, \theta, \nu)$,

$$u_1(\rho, \theta, \nu) \approx \frac{e^{-jk_\perp \rho - j\pi/4}}{\sqrt{2\pi k_\perp \rho}} \frac{\nu \sin(\nu\pi)}{\cos(\nu\pi) - \cos(\nu\theta)}. \tag{5.80}$$

The latter expression describes a cylindrical wave outgoing from the edge of the wedge. On the other hand, the residue term (5.59) is a sum of plane waves, which suggests that $u_{res}(\rho, \theta, \nu)$ and $u_1(\rho, \theta, \nu)$ have different physical meaning. It is convenient, therefore, to separate these contributions in the total field by writing

$$\begin{bmatrix} E_z \\ H_z \end{bmatrix} = \begin{bmatrix} E_z^{go} \\ H_z^{go} \end{bmatrix} + \begin{bmatrix} E_z^{d} \\ H_z^{d} \end{bmatrix}, \tag{5.81}$$

where

$$\begin{bmatrix} E_z^{go} \\ H_z^{go} \end{bmatrix} = \begin{bmatrix} E_{0z} \\ H_{0z} \end{bmatrix} e^{-jk_z z}[u_{res}(\rho, \varphi - \varphi_0, \nu) \mp u_{res}(\rho, \varphi + \varphi_0 + 2\Phi, \nu)] \tag{5.82}$$

describes the contribution of the plane waves, and

$$\begin{bmatrix} E_z^{d} \\ H_z^{d} \end{bmatrix} = \begin{bmatrix} E_{0z} \\ H_{0z} \end{bmatrix} e^{-jk_z z}[u_1(\rho, \varphi - \varphi_0, \nu) \mp u_1(\rho, \varphi + \varphi_0 + 2\Phi, \nu)] \tag{5.83}$$

is the contribution of the cylindrical wave.

The GO part of the solution is, therefore, given by the residue contributions. The position of poles on the complex α plane depends on the values of φ, φ_0, and Φ, and evaluation of this part of the solution requires a careful analysis of whether a pole gets captured or not. For wedges with $\Phi > \pi/2$ the analysis is relatively simple, and the result is described below.

When $\pi - \Phi < \varphi_0 < \Phi$, the incident wave illuminates only the upper face of the wedge and the GO part of the solution is given by the formula

$$\begin{bmatrix} E_z^{go} \\ H_z^{go} \end{bmatrix} = \begin{bmatrix} E_{0z} \\ H_{0z} \end{bmatrix} e^{-jk_z z}[\Theta(\pi - |\varphi - \varphi_0|)e^{jk_\perp \rho \cos(\varphi - \varphi_0)} \mp \Theta(\pi - |\varphi + \varphi_0 - 2\Phi|)e^{jk_\perp \rho \cos(\varphi + \varphi_0 - 2\Phi)}], \tag{5.84}$$

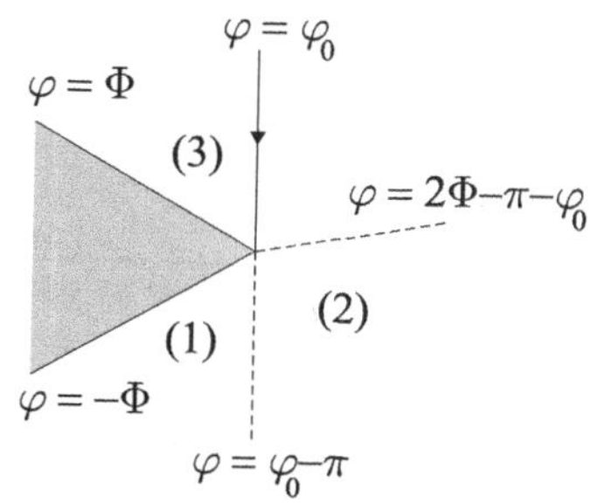

Figure 5.15 Structure of the GO field when the incident wave illuminates the upper face of the wedge only. Shadow boundaries are shown by the dashed lines. Region (1) $-\Phi < \varphi < \varphi_0 - \pi$, totally shadowed; region (2) $\varphi_0 - \pi < \varphi < 2\Phi - \pi - \varphi_0$, illuminated by the incident wave; region (3) $2\Phi - \pi - \varphi_0 < \varphi < \Phi$, illuminated by the incident wave and the wave reflected from the upper face.

where the unit step functions Θ define the lit and shadowed regions. The exponents in the square brackets represent the incident wave and the wave reflected from the upper face of the wedge. The shadow boundaries of these waves are at $\varphi = \varphi_0 - \pi$ and $\varphi = 2\Phi - \pi - \varphi_0$, respectively. This case is illustrated in Figure 5.15.

If the incident wave illuminates only the lower face of the wedge ($-\Phi < \varphi_0 < \Phi - \pi$), then

$$\begin{bmatrix} E_z^{\text{go}} \\ H_z^{\text{go}} \end{bmatrix} = \begin{bmatrix} E_{0z} \\ H_{0z} \end{bmatrix} \text{e}^{-jk_z z}[\Theta(\pi - |\varphi - \varphi_0|)\text{e}^{jk_\perp \rho \cos(\varphi - \varphi_0)} \mp \Theta(\pi - |\varphi + \varphi_0 + 2\Phi|)\text{e}^{jk_\perp \rho \cos(\varphi + \varphi_0 + 2\Phi)}]. \tag{5.85}$$

The two exponent terms in the brackets are the incident wave and the wave reflected from the lower face of the wedge. The shadow boundaries of the incident and reflected waves are at $\varphi = \varphi_0 + \pi$ and $\varphi = \pi - 2\Phi - \varphi_0$, respectively.

Both faces of the wedge are illuminated simultaneously if $\Phi - \pi < \varphi_0 < \pi - \Phi$ (Figure 5.16), in which case the GO field is a superposition of the incident and reflected waves,

$$\begin{bmatrix} E_z^{\text{go}} \\ H_z^{\text{go}} \end{bmatrix} = \begin{bmatrix} E_{0z} \\ H_{0z} \end{bmatrix} \text{e}^{-jk_z z}[\Theta(\pi - |\varphi - \varphi_0|)\text{e}^{jk_\perp \rho \cos(\varphi - \varphi_0)} \mp \Theta(\pi - |\varphi + \varphi_0 - 2\Phi|)\text{e}^{jk_\perp \rho \cos(\varphi + \varphi_0 - 2\Phi)}$$
$$\mp \Theta(\pi - |\varphi + \varphi_0 + 2\Phi|)\text{e}^{jk_\perp \rho \cos(\varphi + \varphi_0 + 2\Phi)}]. \tag{5.86}$$

The non-GO, or edge-diffracted, part of the solution, which is given by (5.83), can be rewritten as

$$\begin{bmatrix} E_z^{\text{d}} \\ H_z^{\text{d}} \end{bmatrix} = \frac{\exp(-jk_z z - jk_\perp \rho - j\pi/4)}{\sqrt{2\pi k_\perp \rho}} \begin{bmatrix} D_{\text{ee}} & 0 \\ 0 & D_{\text{hh}} \end{bmatrix} \begin{bmatrix} E_{0z} \\ H_{0z} \end{bmatrix}, \tag{5.87}$$

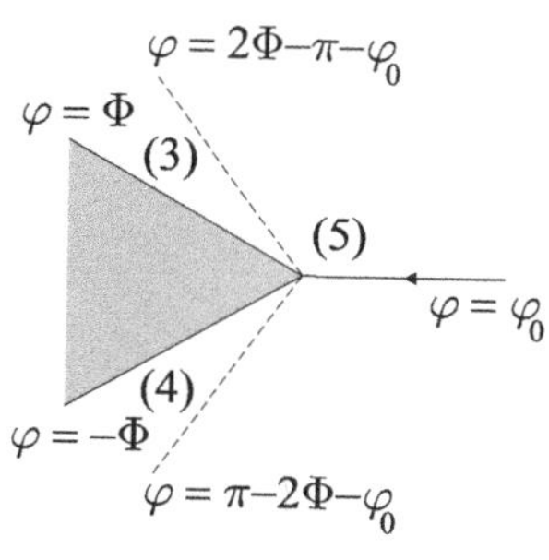

Figure 5.16 Shadow regions when both faces are illuminated. Region (3) $2\Phi - \pi - \varphi_0 < \varphi < \Phi$, illuminated by the incident wave and the wave reflected from the upper face; region (4) $-\Phi < \varphi < \pi - 2\Phi - \varphi_0$, illuminated by the incident wave and the wave reflected from the lower face; region (5) $\pi - 2\Phi - \varphi_0 < \varphi < 2\Phi - \pi - \varphi_0$, illuminated by the incident wave only.

where

$$D_{ee} = \frac{\nu \sin(\nu\pi)}{\cos(\nu\pi) - \cos[\nu(\varphi - \varphi_0)]} - \frac{\nu \sin(\nu\pi)}{\cos(\nu\pi) + \cos[\nu(\varphi + \varphi_0)]} \quad (5.88)$$

$$D_{hh} = \frac{\nu \sin(\nu\pi)}{\cos(\nu\pi) - \cos[\nu(\varphi - \varphi_0)]} + \frac{\nu \sin(\nu\pi)}{\cos(\nu\pi) + \cos[\nu(\varphi + \varphi_0)]}. \quad (5.89)$$

Quantities D_{ee} and D_{hh} are called *diffraction coefficients*. They relate the amplitude of the incident plane wave at the edge to the amplitude of the cylindrical wave scattered by the edge. The cross-polar components D_{eh} and D_{he} (the off-diagonal elements in the matrix in (5.87)) vanish for perfectly conducting wedges.

In contrast to the GO components, the edge-diffracted wave is omnipresent in space around the wedge, including the region of the optical shadow. The amplitude of the wave decays as either the observer moves away from the edge or the frequency grows. The presence of the edge-diffracted wave explains the fine features in the fields diffracted by wedges (see Figures 5.5 and 5.6), like the finite field level in the shadow and the interference pattern in the region illuminated by a single GO wave.

At the directions of the GO shadow boundaries, where

$$\cos(\nu\pi) = \mp\cos[\nu(\varphi \pm \varphi_0)], \quad (5.90)$$

diffraction coefficients (5.88) and (5.89) have pole singularities. These singularities are unphysical and result from the assumption we have made when approximating the integral $u_1(\rho, \theta, \nu)$. For these directions, $U(\xi - \pi, \theta, \nu)$ cannot be regarded as a slowly varying function in a vicinity of the point $\xi = 0$, and a more sophisticated approach is necessary to derive correct, finite approximations for the wave field in these regions (see section 5.5).

Figure 5.17 illustrates the dependence of the diffraction coefficients on the observation angle φ. The configuration corresponds to that shown in Figure 5.15. Both diffraction coefficients are singular at the shadow boundaries of the

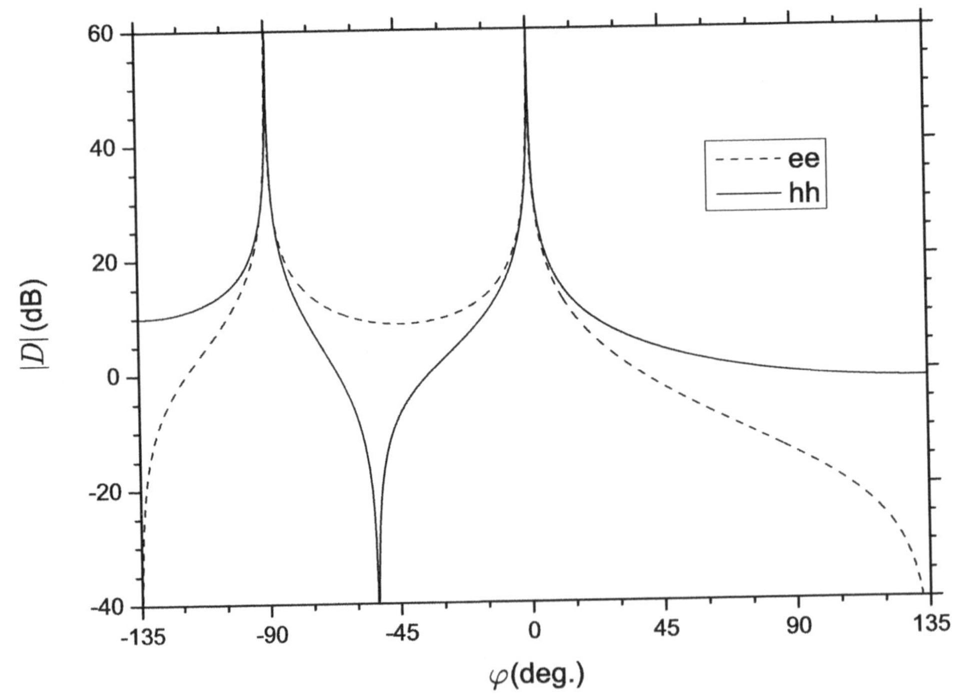

Figure 5.17 Diffraction coefficients of a PEC right-angled wedge illuminated at $\varphi_0 = 90°$ as a function of φ.

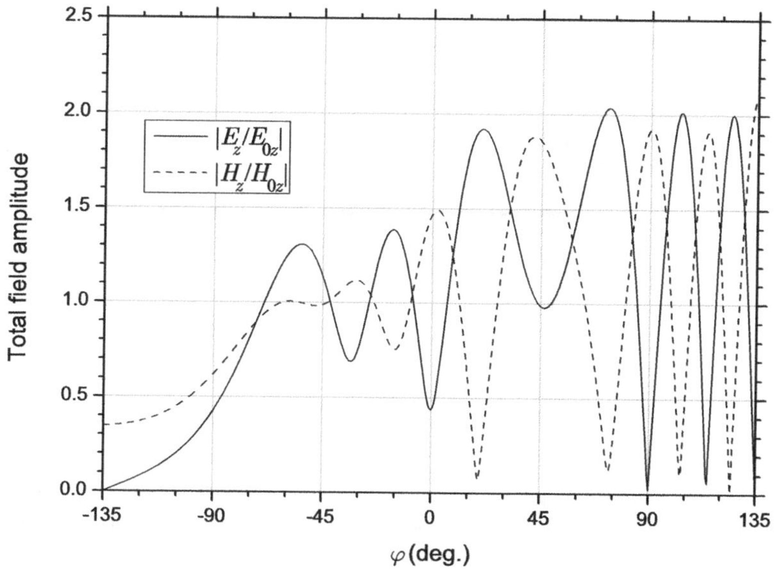

Figure 5.18 Total field at the distance $\rho = 2\lambda$ for a plane wave normally incident at a PEC right-angled wedge under $\varphi_0 = 90°$.

incident wave ($\varphi = -90°$) and the wave reflected from the upper face ($\varphi = 0$). Note the different behavior of D_{ee} and D_{hh} for $\varphi \to \pm\Phi$.

Exact fields are, of course, continuous everywhere but show the features predicted by the high-frequency asymptotics already at moderate distances from the edge. An example is given in Figure 5.18 which shows the distribution of the total electric and magnetic fields at the distance $\rho = 2\lambda$ around the edge of a PEC right-angled wedge in the case of normal illumination ($k_z = 0$). The pronounced oscillations in the region ($0 < \varphi \leq 135°$) are mostly due to the interference of the incident and reflected waves. In the portion of the lit region with $-90° < \varphi < 0$, oscillations are weaker and due to the incident and edge-diffracted waves. The field in the shadow region ($-135° \leq \varphi < -90°$) is formed by the edge-diffracted wave alone.

5.3 Scattering from a Half-plane (Solution by Factorization Method)

Here we will study the limiting case of a wedge: a semi-infinite ideally conducting plane of negligible thickness. The purpose of this section is to introduce a power technique for solving scattering problems for semi-infinite structures. The factorization method (sometimes called the *Wiener–Hopf technique*) allows exact solutions for many problems involving semi-infinite boundaries (Daniele and Zich 2014; Mittra and Lee 1971; Noble 1958; Weinstein 1969). As applied to electromagnetic scattering from wedges, the method leads to exact solutions for the half-plane ($\Phi = \pi$) and the full plane ($\Phi = \pi/2$) not only for PEC but also for impedance boundaries (see Senior and Volakis (1995, Chapter 3) and Rojas (1988a)).

5.3.1 Statement of the Problem

The geometry of the problem is shown in Figure 5.19. An ideally conducting half-plane is illuminated by a plane wave so that the wave vector of the incident wave **k** is orthogonal to the edge (axis z). During the derivations we will assume that the incidence angle is within the limits $\pi/2 < \varphi_0 < \pi$. The results can be analytically extended to arbitrary values of φ_0. We will distinguish between two orthogonal polarizations, E polarization:

$$\mathbf{E}_{\text{inc}} = \hat{z}E_0 e^{jk(x\cos\varphi_0 + y\sin\varphi_0)} \tag{5.91}$$

Figure 5.19 Geometry of the problem. The half-plane occupies the region $x > 0, y = 0$. The observation point has the coordinates (ρ, φ). The incidence angle is φ_0.

and H polarization:

$$\mathbf{H}_{\text{inc}} = \hat{z}H_0 e^{jk(x\cos\varphi_0 + y\sin\varphi_0)}. \tag{5.92}$$

Here, the notations E ("electric") and H ("magnetic") refer to the orientation of the fields vectors with respect to the edge (the z axis). We will write formulas for the E polarization only, as the solution technique is the same for both polarizations.

The half-plane is a geometry with translational symmetry so we can use the theory of sections (2.5.5) and (2.6.2). For normal incidence, the electric field has a single component $\mathbf{E}_z = \hat{z}E_z$ and is independent of z. According to the Huygens principle, the scattered field E_z^{sc} can be obtained by the integration of its boundary values at the half-plane, and since $E_z = 0$ on the half-plane, equation (2.367) gives

$$E_z^{\text{sc}}(x, y) = \frac{j}{4}\int_0^{+\infty} H_0^{(2)}[k\sqrt{y^2 + (x - x')^2}]\left[\frac{\partial E_z}{\partial y'}(x', +0) - \frac{\partial E_z}{\partial y'}(x', -0)\right] \mathrm{d}x', \tag{5.93}$$

where ± 0 denote the value of y' on the upper and lower sides of the half-plane. So, the scattered field is fully determined by the jump in the normal derivative of the total electric field across the half-plane, and since

$$\frac{\partial E_z}{\partial y}(x, \pm 0) = -jkZH_x(x, \pm 0), \tag{5.94}$$

see (2.308) with $k_z = 0$ and $k_\perp = k$, we can introduce a new unknown function $J(x)$, an equivalent electric surface current, as

$$J(x') = H_x(x', -0) - H_x(x', +0) \tag{5.95}$$

and express the scattered field as

$$E_z^{\text{sc}}(x, y) = -\frac{kZ}{4}\int_0^{\infty} J(x')H_0^{(2)}[k\sqrt{y^2 + (x - x')^2}]\,\mathrm{d}x'. \tag{5.96}$$

The tangential component of the total field should vanish on the PEC surface, but no restriction is imposed on this field at the other half of the xy plane. Thus, we write the boundary condition as

$$E_z^{\text{sc}}(x, 0) = \begin{cases} -E_z^{\text{inc}}(x, 0), & x > 0 \\ \mathcal{E}(x), & x < 0 \end{cases}, \tag{5.97}$$

where $\mathcal{E}(x)$ is an unknown function. Substituting from (5.91) and (5.96), we get the boundary condition in the following form:

$$-\frac{kZ}{4}\int_{0}^{\infty} J(x')H_0^{(2)}(k|x-x'|)\,\mathrm{d}x' = \begin{cases} -\mathrm{e}^{jkx\cos\varphi_0}, & x>0 \\ \mathcal{E}(x), & x<0 \end{cases}. \tag{5.98}$$

This is an integral equation with two unknown functions: $J(x)$ and $\mathcal{E}(x)$.

5.3.2 Functional Equation

To solve (5.98), we first Fourier transform both sides (multiply by e^{-jwx} and integrate over x). Calculating the left-hand side, we change the order of integration:

$$-\frac{kZ}{4}\int_{-\infty}^{\infty} \mathrm{e}^{-jwx}\int_{0}^{\infty} J(x')H_0^{(2)}(k|x-x'|)\,\mathrm{d}x'\mathrm{d}x = -\frac{kZ}{4}\int_{0}^{\infty} J(x')\int_{-\infty}^{\infty} \mathrm{e}^{-jwx}H_0^{(2)}(k|x-x'|)\,\mathrm{d}x\,\mathrm{d}x' \tag{5.99}$$

and change the variable as $x - x' = t$:

$$-\frac{kZ}{4}\int_{0}^{\infty} J(x')\int_{-\infty}^{\infty} \mathrm{e}^{-jwx}H_0^{(2)}(k|x-x'|)\,\mathrm{d}x\,\mathrm{d}x' = -\frac{kZ}{4}\int_{0}^{\infty} J(x')\mathrm{e}^{-jwx'}\int_{-\infty}^{\infty} \mathrm{e}^{-jwt}H_0^{(2)}(k|t|)\,\mathrm{d}t\,\mathrm{d}x'. \tag{5.100}$$

The internal integral can be evaluated using the known formula

$$\int_{-\infty}^{\infty} \mathrm{e}^{-jwt}H_0^{(2)}(k\sqrt{y^2+t^2})\,\mathrm{d}t = \frac{2\mathrm{e}^{-j\sqrt{k^2-w^2}|y|}}{\sqrt{k^2-w^2}}. \tag{5.101}$$

In our particular case,

$$\int_{-\infty}^{\infty} \mathrm{e}^{-jwt}H_0^{(2)}(k|t|)\,\mathrm{d}t = \frac{2}{\sqrt{k^2-w^2}}. \tag{5.102}$$

Let us denote $\gamma = \sqrt{k^2 - w^2}$ and

$$\tilde{J}\_(w) = \int_{0}^{\infty} J(x')\mathrm{e}^{-jwx'}\,\mathrm{d}x', \tag{5.103}$$

which is the Fourier transform of the induced surface current density function. Finally, the Fourier-transformed left-hand side of (5.98) becomes

$$-\frac{kZ}{2\gamma}\tilde{J}\_(w). \tag{5.104}$$

Note that using (5.101) we can express the Fourier-transformed scattered field in terms of the Fourier-transformed induced surface current density *for any value of the coordinate y*:

$$\tilde{E}_{\mathrm{sc}}(w,y) = -\frac{kZ}{2\gamma}\mathrm{e}^{-j\gamma|y|}\tilde{J}\_(w). \tag{5.105}$$

The right-hand side of (5.98) for $x > 0$ becomes, after Fourier transformation,

$$\int_{0}^{\infty} e^{jkx\cos\varphi_0} e^{-jwx}\, dx = -\frac{j}{w - k\cos\varphi_0}. \tag{5.106}$$

Here we have assumed that the integrand at $x \to \infty$ vanishes because there are at least very small losses in space (k has a small negative imaginary part). Integrating over negative x, we do not know the function that we integrate, so we simply write

$$\int_{-\infty}^{0} \mathcal{E}(x) e^{-jwx}\, dx = \tilde{\mathcal{E}}_+(w), \tag{5.107}$$

where $\tilde{\mathcal{E}}_+(w)$ is an unknown function. Finally, we have a functional equation to solve:

$$-\frac{kZ}{2\gamma}\tilde{J}_-(w) = \frac{j}{w - k\cos\varphi_0} + \tilde{\mathcal{E}}_+(w). \tag{5.108}$$

5.3.3 *Factorization and Solution*

To find the solution we will use the analytical properties of the unknown functions. More precisely, we want to rewrite equation (5.108) as the condition of an equality of two functions holomorphic (section C.2.1) either in the upper or in the lower half-planes of w. Function $\tilde{J}_-(w)$ is holomorphic in the lower half-plane of the complex variable w because the integral has absolute convergence there. For the same reason, $\tilde{\mathcal{E}}_+(w)$ is holomorphic in the upper half-plane. Function $\gamma = \sqrt{k^2 - w^2}$ has two branch points at $w = \pm k$ (see Figure 5.20). We will fix the branch of the square root by choosing the branch with $\operatorname{Im}\gamma < 0$.

Function $\gamma = \sqrt{k^2 - w^2}$ can be *factorized*, that is, written as a product of two functions

$$\gamma = \sqrt{k^2 - w^2} = \sqrt{k + w}\,\sqrt{k - w}, \tag{5.109}$$

so that one of the factors is holomorphic in the upper half-plane ($\sqrt{k - w}$) and the other one is holomorphic in the lower half-plane ($\sqrt{k + w}$). Now we can rewrite the functional equation (5.108) as

$$-\frac{kZ}{2\sqrt{k + w}}\tilde{J}_-(w) = \frac{j\sqrt{k - w}}{w - k\cos\varphi_0} + \sqrt{k - w}\,\tilde{\mathcal{E}}_+(w) = F(w) + \sqrt{k - w}\,\tilde{\mathcal{E}}_+(w). \tag{5.110}$$

The left-hand side is holomorphic in the lower half-plane. Function $\sqrt{k - w}\,\tilde{\mathcal{E}}_+(w)$ is holomorphic in the upper half-plane. Function $F(w)$ belongs to neither of these two classes, but we can write it as

$$F(w) = \frac{j\sqrt{k - w}}{w - k\cos\varphi_0} = j\left[\frac{\sqrt{k - k\cos\varphi_0}}{w - k\cos\varphi_0} + \frac{\sqrt{k - w} - \sqrt{k - k\cos\varphi_0}}{w - k\cos\varphi_0}\right] = F_-(w) + F_+(w). \tag{5.111}$$

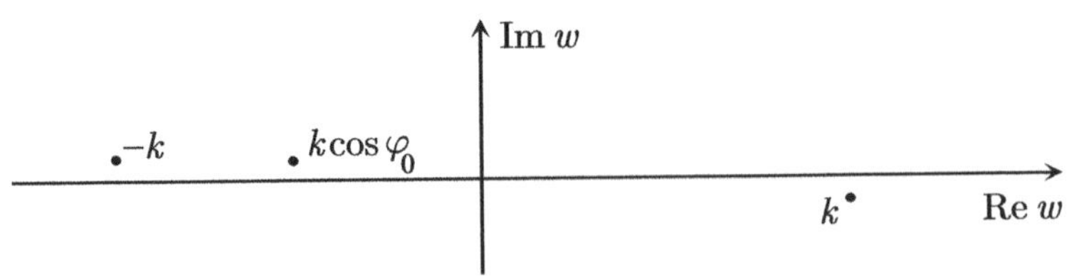

Figure 5.20 The branch points and the pole in the complex plane w.

Here, the first term is holomorphic in the lower half-plane. The second term has a finite value at $w \to k\cos\varphi_0$ and a branch point at $w = k$, that is, it is holomorphic in the upper half-plane. Finally, we write equation (5.110) as

$$-\frac{kZ}{2\sqrt{k+w}}\tilde{J}\_(w) - F\_(w) = F_+(w) + \sqrt{k-w}\,\tilde{\mathcal{E}}_+(w). \tag{5.112}$$

The left-hand side is holomorphic in the lower half-plane, the right-hand side is holomorphic in the upper half-plane, and these functions are continuous and finite at the boundary, where they actually coincide.[1] This means that both these expressions actually represent one holomorphic function in the complete complex plane of w, possibly except the infinite point (entire function, section C.2.1). To determine the behavior at infinity, we use the edge condition because the behavior of the transformed function at $w \to \infty$ is determined by the behavior of the original function at $x \to 0$. Indeed, considering the integral defining the Fourier-transformed induced surface current density (5.103)

$$\tilde{J}\_(w) = \int_0^\infty J(x)e^{-jwx}\,dx, \tag{5.113}$$

we can note that if w tends to infinity in the lower half-plane, the exponent quickly decays with increasing x. Thus, we can estimate the integral substituting the asymptotic expression for the surface current density $J(x)$ for small values of the argument (near the edge of the half-plane). The behavior near the edge is determined by the Meixner condition, which gives for $x \to 0$

$$J(x) \sim x^{\delta-1}, \tag{5.114}$$

where $\delta > 0$. For the half-plane, $\delta = 1/2$. The integral

$$\int_0^\infty x^{\delta-1}e^{-jwx}\,dx = \frac{\Gamma(\delta)}{(jw)^\delta} \tag{5.115}$$

is known (Ryshik and Gradstein 1957). The above formula is valid for $\operatorname{Im} w < 0$ and $\operatorname{Re}\delta > 0$.

The important conclusion is that this function tends to zero at infinity. This means that our entire function is holomorphic *everywhere*, so this is simply a constant (Liouville's theorem, section C.2.1). And this constant is zero because it is zero at infinity. Finally, we find that

$$-\frac{kZ}{2\sqrt{k+w}}\tilde{J}\_(w) - F\_(w) = 0. \tag{5.116}$$

We have solved our functional equation and have found the Fourier-transformed induced surface current density:

$$\tilde{J}\_(w) = -2\frac{\sqrt{k+w}}{kZ}F\_(w) = -2j\frac{\sqrt{k+w}}{kZ}\frac{\sqrt{k-k\cos\varphi_0}}{w-k\cos\varphi_0}. \tag{5.117}$$

The Fourier-transformed scattered field is given by (5.105):

$$\tilde{E}_{sc}(w,y) = -\frac{kZ}{2\gamma}e^{-j\gamma|y|}\tilde{J}\_(w) = j\frac{\sqrt{2k}\sin\frac{\varphi_0}{2}}{\sqrt{k-w}(w-k\cos\varphi_0)}e^{-j\sqrt{k^2-w^2}|y|}. \tag{5.118}$$

[1] More precisely, they coincide at the boundary and they are holomorphic in a common strip that contains the real axis.

(We have used the identity

$$\sqrt{k - k\cos\varphi_0} = \sqrt{2k}\left|\sin\frac{\varphi_0}{2}\right| \tag{5.119}$$

and the fact that in our case $\sin(\varphi_0/2) > 0$).

5.3.4 Scattered Field Far from the Edge

To find the scattered field, we have to calculate the inverse Fourier transform of (5.118):

$$E_z^{sc}(x, y) = j\frac{\sqrt{2k}}{2\pi}\sin\frac{\varphi_0}{2}\int_{-\infty}^{\infty}\frac{e^{jwx - j\sqrt{k^2-w^2}\,|y|}}{\sqrt{k - w}(w - k\cos\varphi_0)}\,dw. \tag{5.120}$$

Let us change the integration variable:

$$w = k\cos\tau, \tag{5.121}$$

where τ is the new complex integration variable. With this change

$$\gamma = \sqrt{k^2 - w^2} = k\sin\tau. \tag{5.122}$$

The coordinates of the observation point (x, y) we express in the polar coordinate system:

$$x = \rho\cos\varphi, \quad y = \rho\sin\varphi, \quad 0 \le \varphi \le 2\pi. \tag{5.123}$$

The exponent in (5.120) takes the simple form $e^{jk\rho\cos(\varphi+\tau)}$, and the function that we integrate transforms to

$$\frac{e^{jk\rho\cos(\varphi+\tau)}(-k\sin\tau)\,d\tau}{\sqrt{2k}\sin(\tau/2)k(\cos\tau - \cos\varphi_0)}. \tag{5.124}$$

The integral now reads

$$E_z^{sc}(x, y) = \frac{-j}{\pi}\sin\frac{\varphi_0}{2}\int_{\Gamma}\frac{e^{jk\rho\cos(\varphi+\tau)}\cos\frac{\tau}{2}}{\cos\tau - \cos\varphi_0}\,d\tau. \tag{5.125}$$

The integration path Γ is the result of transformation of the real axis w in $w = k\cos\tau$. Point $w = 0$ transforms into $\tau = \pi/2 + 2n\pi$, $w = \infty$ corresponds to $\tau = -j\infty + 2n\pi$, and $w = -\infty$ becomes $\tau = \pi + j\infty + 2n\pi$. Because $\sqrt{k^2 - w^2}$ transforms into $k\sin\tau$, the integrand is an entire function in the τ plane. The branch cuts in the plane of w transform into a set of curves, which pass through $\tau = 0, \pm\pi, \pm 2\pi, \ldots$. These curves are shown as dashed lines in Figure 5.21. We have to choose one of the strips between these lines. The integration path is shown in Figure 5.21.

For large $k\rho$ this integral can be approximately evaluated using the saddle point method. The exponential function in the integrand oscillates fast if $k\rho \gg 1$. The idea is to deform the integration path so that it goes through the point or points where the derivative of the phase (of the argument of the exponent) is zero. The main contribution in the integral is then from the vicinity of this point. The path should go via the stationary point (*the saddle point*) so that the exponent function decays as quickly as possible. This path is called the *steepest descent path* (SDP). For details of this method, see Appendix C.

In our case the equation for the saddle point is

$$\frac{d}{d\tau}\cos(\tau + \varphi) = -\sin(\tau + \varphi) = 0, \tag{5.126}$$

which gives for the saddle point

$$\tau_0 = \pi - \varphi \tag{5.127}$$

(we have taken the solution that belongs to the chosen strip, see Figure 5.21).

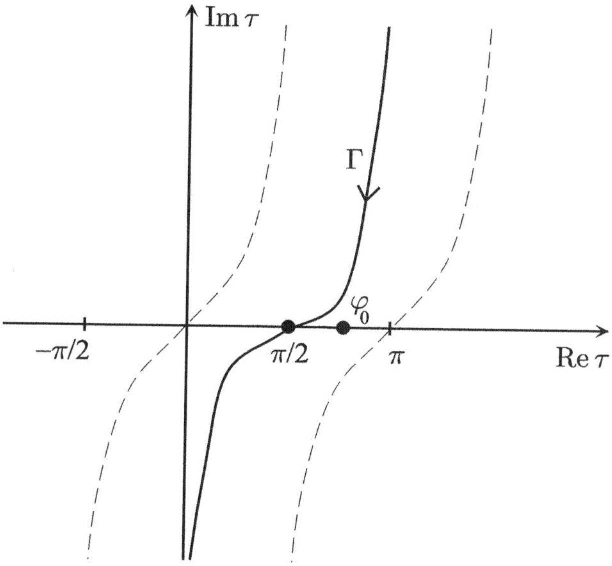

Figure 5.21 Integration path Γ in the complex plane τ. Branch cuts in the w plane transform into the curves shown by dashing. There is a pole at $\tau = \varphi_0$.

The new path is shown in Figure 5.22. The difference between the original integral over Γ and the integral over the deformed path depends on whether we cross the pole at φ_0 or not:

$$\int_\Gamma - \int_{\Gamma'} = \begin{cases} 2\pi j \operatorname{res}_{\varphi_0}, & \pi - \varphi > \varphi_0 \\ 0, & \pi - \varphi < \varphi_0 \end{cases}. \tag{5.128}$$

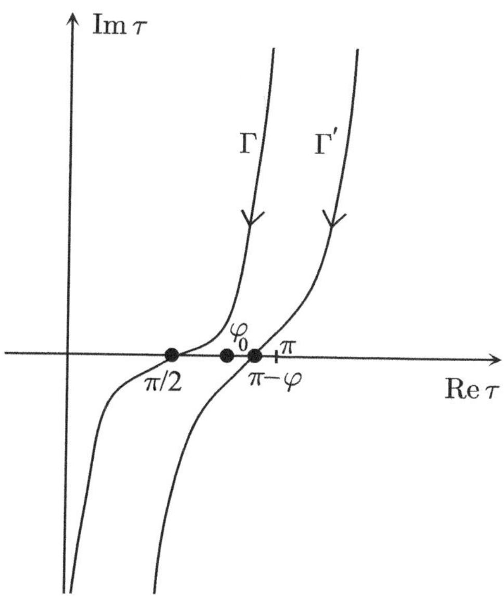

Figure 5.22 The steepest descent path Γ′ passes through $\pi - \varphi$, making a 45° angle with the real axis.

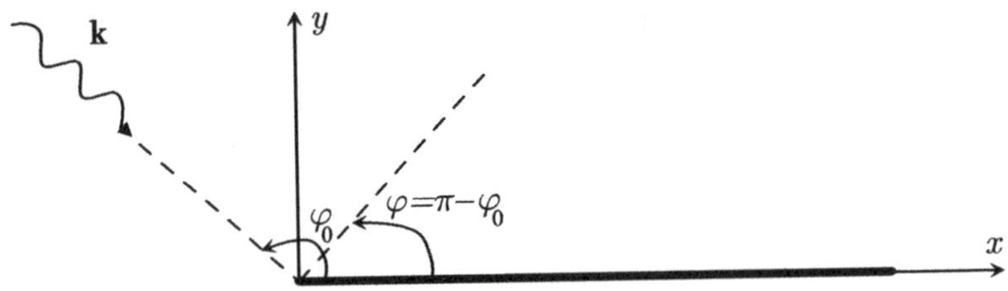

Figure 5.23 The reflected wave exists only if $\varphi < \pi - \varphi_0$.

The pole contribution

$$-2\pi j\frac{j}{\pi}\sin\frac{\varphi_0}{2}\frac{e^{jk\rho\cos(\varphi+\varphi_0)}\cos\frac{\varphi_0}{2}}{-\sin\varphi_0} = -e^{jk\rho\cos(\varphi+\varphi_0)} \quad (5.129)$$

is the reflected plane wave, which exists only at $\varphi < \pi - \varphi_0$ (see Figure 5.23).

The integral over Γ'

$$E_z^d(r,\varphi) = \frac{-j}{\pi}\sin\frac{\varphi_0}{2}\int_{\Gamma'}\frac{e^{jk\rho\cos(\varphi+\tau)}\cos\frac{\tau}{2}}{\cos\tau-\cos\varphi_0}d\tau \quad (5.130)$$

will be approximately calculated by changing the variable as

$$t = \sqrt{-2j}\cos\frac{\tau+\varphi}{2}. \quad (5.131)$$

This choice makes good sense because the argument of the exponent in (5.130) becomes

$$j\cos(\tau+\varphi) = j[-1+(1+\cos(\tau+\varphi))] = -j+2j\cos^2\frac{\tau+\varphi}{2} = -j-t^2. \quad (5.132)$$

The integral transforms into

$$E_z^d(r,\varphi) = j\frac{\sin\frac{\varphi_0}{2}}{\pi}\sqrt{\frac{2}{-j}}e^{-jk\rho}\int_{-\infty}^{\infty}\frac{e^{-k\rho t^2}\cos\frac{\tau}{2}}{(\cos\tau-\cos\varphi_0)}\frac{dt}{\sin\frac{\tau+\varphi}{2}}. \quad (5.133)$$

The main contribution into the integral comes from the vicinity of the saddle point. If the pole is not close to the saddle point, then all the terms except for the exponent are slowly varying functions near $t = 0$, so we can approximate them by their values at the saddle point $\tau = \pi - \varphi$. The remaining integral is known (Ryshik and Gradstein 1957, eq. 3.271):

$$\int_{-\infty}^{\infty} e^{-k\rho t^2}dt = \sqrt{\frac{\pi}{k\rho}}. \quad (5.134)$$

Finally, we have

$$E_z^d(r,\varphi) = \sqrt{\frac{2}{\pi k\rho}}e^{-j(k\rho+\frac{\pi}{4})}\frac{\sin\frac{\varphi}{2}\sin\frac{\varphi_0}{2}}{\cos\varphi+\cos\varphi_0}. \quad (5.135)$$

This is a cylindrical wave propagating from the edge of the half-plane. This asymptotic expression is not valid near $\varphi = \pi - \varphi_0$, which is the shadow boundary of the reflected wave. This is because at this point the integrand of (5.133) has a pole right at the saddle point, which means that it is not possible to assume that only the exponential term is a fast-varying function.

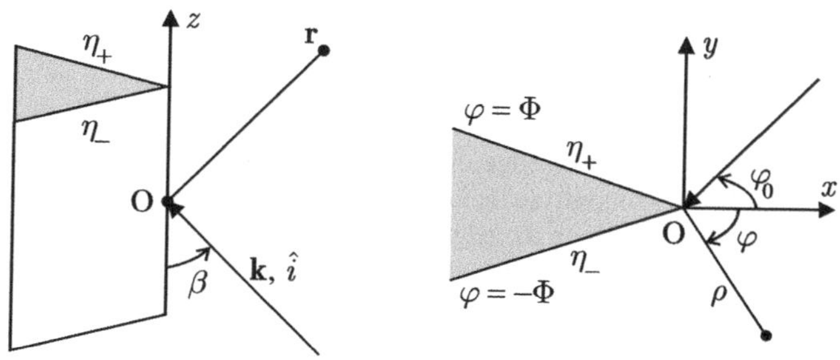

Figure 5.24 A plane wave with the wave vector $\mathbf{k} = k\hat{\imath}$ is incident on an infinite wedge with an edge along the z-coordinate axis. The incidence direction makes the angle β with the edge of the wedge. The faces of the wedge at $\varphi = \pm\Phi$ are impedance with $\eta_\pm$. ρ, φ, and z are cylindrical coordinates of the observer at the point $\mathbf{r}$. The x–y plane is orthogonal to the edge of the wedge. The angle φ_0 specifies the projection of the incidence direction on the x–y plane.

5.4 The Impedance Wedge

In this section we consider scattering of an electromagnetic plane wave by an infinite wedge with faces described by impedance boundary conditions (2.74), the face impedances being in general different (Figure 5.24). The boundary conditions model imperfectly reflecting surfaces which partly reflect and partly absorb the incident power. The geometry includes the case of a PEC wedge in the limit of vanishing equivalent surface impedance and the case of a half-plane (impedance or PEC) in the limit of the included wedge angle going to zero. However, in contrast to the PEC configurations studied in sections 5.2 and 5.3, the problem of scattering from an impedance wedge is mathematically much more difficult since the corresponding boundary value problem cannot be solved by separating variables. A further difficulty is the polarization coupling in the boundary conditions for general incidence directions.

As the problem cannot be solved by the separation of variables, another method is followed here, namely the method proposed by G.D. Maliuzhinets (also Malyuzhinets, Malyughinetz). The method was described in Maliuzhinets (1950, 1958a,b, 1960) and developed further in view of electromagnetic applications by a number of other authors, for example Bobrovnikov and Fisanov (1988), Osipov (1993a), Budaev (1995), Babich *et al.* (2007), Vaccaro (1980), and Senior (1986). The method uses representations of fields in terms of plane-wave expansions in circular cylinder coordinates (Sommerfeld 1896, 1927, 1964) followed by reduction of the problem to functional difference equations for the spectral amplitudes. In general, an explicit solution of the equations does not seem possible, but it is available in a number of special cases, which includes the impedance-matched wedge (Bernard 1989; Lyalinov and Zhu 1999), the impedance half-plane (Bucci and Franceschetti 1976), the full-plane impedance junction (Greenberg and Fock 1948; Vaccaro 1980), right-angled wedges with one face perfectly conducting and the included angle $\pi/2$ or $3\pi/2$ (Senior 1978, 1986; Vaccaro 1981), as well as the case of normal incidence on an arbitrarily angled wedge with arbitrary face impedances (Maliuzhinets 1958a).

To solve the problem for a general impedance wedge configuration it is necessary to resort to a numerical solution, and there are several techniques that are currently available (Bilow 1991; Budaev and Bogy 2006; Daniele and Lombardi 2006; Lyalinov and Zhu 2006, 2013; Osipov and Senior 2008a; Pelosi *et al.* 1996).

We start by formulating the scattering problem as a boundary value problem. Then the principles of the Maliuzhinets method are outlined, and the boundary value problem is reduced to functional equations (section 5.4.1). The section presents solutions for all known configurations that permit exact solutions (sections 5.4.2, 5.4.3 and 5.4.4). The high-frequency asymptotic analysis of the solutions is the subject of section 5.5.

5.4.1 Boundary Value Problem, Sommerfeld's Integrals, and Functional Equations

Boundary Value Problem

The wedge is immersed in a homogeneous medium with the wave impedance Z. The incident field is a plane wave

$$\mathbf{E}_{\text{inc}} = \mathbf{E}_0 e^{-j\mathbf{k}\cdot\mathbf{r}}, \quad Z\mathbf{H}_{\text{inc}} = \hat{\imath} \times \mathbf{E}_{\text{inc}} \tag{5.136}$$

with the wave vector $\mathbf{k} = k\hat{\imath}$ and

$$\hat{\imath} = -\hat{x}\cos\varphi_0 \sin\beta - \hat{y}\sin\varphi_0 \sin\beta + \hat{z}\cos\beta, \tag{5.137}$$

incident at an oblique angle β with respect to the edge of the wedge. Parameter β is a measure of obliqueness of the incident wave, for example $\beta = \pi/2$ corresponds to an incidence direction perpendicular to the edge of the wedge (normal incidence); with $\beta \to 0$ the wave propagates along the edge in the direction of the positive z axis and with $\beta \to \pi$ in the direction of the negative z axis. $\mathbf{E}_0$ is the amplitude of the electric field in the incident wave at the origin of the coordinate system.

A cylindrical coordinate system (ρ, φ, z) with $0 \le \rho < +\infty$, $-\Phi \le \varphi \le \Phi$ and $-\infty < z < +\infty$ is used. The parameter Φ $(0 < \Phi \le \pi)$ describes the angular width of the wedge so that $\Phi = \pi$ corresponds to a half-plane, $\Phi = 3\pi/4$ to a right-angled exterior wedge, $\Phi = \pi/2$ to a flat junction of two half-planes, and $\Phi = \pi/4$ to a right-angled interior wedge.

At the faces of the wedge the impedance boundary conditions (2.74) are imposed

$$\hat{n}_\pm \times \mathbf{E}_\pm = Z\eta_\pm \hat{n}_\pm \times (\hat{n}_\pm \times \mathbf{H}_\pm), \tag{5.138}$$

where $\pm$ signs in the subscripts relate the respective quantity to the upper $(\varphi = \Phi)$ or the lower $(\varphi = -\Phi)$ face of the wedge. The unit normal vectors $\hat{n}_\pm$, directed into the exterior of the wedge, are equal to $\pm\hat{\varphi}$. The scalar parameters $\eta_\pm$ will be referred to as normalized surface impedances. For every physically passive boundary, $\operatorname{Re}\eta_\pm \ge 0$ (see (2.134)).

Equations (5.138) relate tangential components of the fields, $E_\rho^\pm$, $E_z^\pm$, $H_\rho^\pm$, and $H_z^\pm$, on the faces of the wedge, and in the component form they give the four conditions:

$$E_\rho^\pm = \mp\eta_\pm Z H_z^\pm \tag{5.139}$$

$$E_z^\pm = \pm\eta_\pm Z H_\rho^\pm . \tag{5.140}$$

The wedge geometry is invariant with respect to the coordinate z, and the incident wave depends on z as $\exp(-jk\sin\beta z)$, implying the same dependence on z in the total field as well. This is therefore the case of the problem with translational symmetry studied in section 2.5.5 with $k_z = k\cos\beta$ and $k_\perp^2 = k^2\sin^2\beta$. The problem of electromagnetic scattering can be therefore formulated in two dimensions in terms of z components of the electric and magnetic fields, E_z and H_z. The components satisfy the Helmholtz equations (2.317) and (2.318),

$$\frac{\partial^2}{\partial\rho^2}E_z + \frac{1}{\rho}\frac{\partial}{\partial\rho}E_z + \frac{1}{\rho^2}\frac{\partial^2}{\partial\varphi^2}E_z + k^2\sin^2\beta E_z = 0 \tag{5.141}$$

$$\frac{\partial^2}{\partial\rho^2}H_z + \frac{1}{\rho}\frac{\partial}{\partial\rho}H_z + \frac{1}{\rho^2}\frac{\partial^2}{\partial\varphi^2}H_z + k^2\sin^2\beta H_z = 0, \tag{5.142}$$

and the incident field corresponds to

$$E_z^{\text{inc}}(\mathbf{r}) = E_{0z}\mathrm{e}^{-jkz\cos\beta + jk\rho\sin\beta\cos(\varphi-\varphi_0)} \tag{5.143}$$

$$H_z^{\text{inc}}(\mathbf{r}) = H_{0z}\mathrm{e}^{-jkz\cos\beta + jk\rho\sin\beta\cos(\varphi-\varphi_0)} \tag{5.144}$$

with

$$E_{0z} = \hat{z}\cdot\mathbf{E}_0, \quad ZH_{0z} = \hat{z}\cdot(\hat{\imath}\times\mathbf{E}_0). \tag{5.145}$$

The other components can be determined from z components as in (2.213)–(2.316),

$$E_\rho = \frac{1}{jk\sin^2\beta}\left[\cos\beta\frac{\partial}{\partial\rho}E_z + \frac{1}{\rho}\frac{\partial}{\partial\varphi}(ZH_z)\right] \tag{5.146}$$

$$E_\varphi = \frac{1}{jk\sin^2\beta}\left[\frac{\cos\beta}{\rho}\frac{\partial}{\partial\varphi}E_z - \frac{\partial}{\partial\rho}(ZH_z)\right] \tag{5.147}$$

$$H_\rho = \frac{1}{jk\sin^2\beta}\left[\cos\beta\frac{\partial}{\partial\rho}H_z - \frac{1}{\rho}\frac{\partial}{\partial\varphi}(YE_z)\right] \tag{5.148}$$

$$H_\varphi = \frac{1}{jk\sin^2\beta}\left[\frac{\cos\beta}{\rho}\frac{\partial}{\partial\varphi}H_z + \frac{\partial}{\partial\rho}(YE_z)\right], \tag{5.149}$$

where $Y = Z^{-1}$. Using (5.146)–(5.149) in (5.139) and (5.140) gives the boundary conditions

$$\frac{1}{\rho}\frac{\partial E_z}{\partial\varphi} \pm j\frac{k}{\eta_\pm}\sin^2\beta E_z = Z\cos\beta\frac{\partial H_z}{\partial\rho} \tag{5.150}$$

$$\frac{1}{\rho}\frac{\partial H_z}{\partial\varphi} \pm jk\eta_\pm\sin^2\beta H_z = -Y\cos\beta\frac{\partial E_z}{\partial\rho} \tag{5.151}$$

imposed at $\varphi = \pm\Phi$.

We look for solutions of Helmholtz equations (5.141) and (5.142) that satisfy the boundary conditions (5.150) and (5.151). Furthermore, the solution must be compliant with the conditions at the edge (section 5.2.1), which implies that independently of φ,

$$E_z \to E_z(0), \quad H_z \to H_z(0), \quad \rho|\nabla E_z| \to 0, \quad \rho|\nabla H_z| \to 0, \tag{5.152}$$

so that the z components of the total electric and magnetic fields are bounded at the edge, $|E_z(0)| < \infty$ and $|H_z(0)| < \infty$; their gradients may have a singularity but weaker than $O(\rho^{-1})$ as $\rho \to 0$.

Finally, because of the infinite extent of the wedge faces the conditions at infinity (2.322) and (2.323) are to be taken in a modified form. They can be formulated as the requirement that $E_z - E_z^{go}$ and $H_z - H_z^{go}$ have the form of an outgoing cylindrical wave as $\rho \to \infty$. Here, E_z^{go} and H_z^{go} are the GO approximation for the fields, which includes the incident wave and the plane waves reflected from the boundaries. It is insufficient to subtract only E_z^{inc} and H_z^{inc} from the total field since reflected waves are also plane waves, which depending on φ and φ_0 may be incoming or outgoing as functions of ρ. For wedges with infinite and flat faces, this formulation can be specialized to another form by requiring that E_z^{inc} and H_z^{inc} be the only waves with the *propagation direction within the sector* $-\Phi < \varphi < \Phi$. The latter condition excludes reflected waves as their propagation directions lie outside the sector.

The boundary value problem is therefore formulated for two unknown functions E_z and H_z that satisfy the Helmholtz equations (5.141) and (5.142), the boundary conditions (5.150) and (5.151), and the conditions at the edge (5.152) and at infinity. Except for the case $\beta = \pi/2$, E_z and H_z are coupled through the boundary conditions. A further problem is the presence of the factor ρ in the coefficients of (5.150) and (5.151), which makes the boundary value problem not amenable to the method of separation of variables.

Sommerfeld's Integrals

Plane waves are elementary solutions of the Helmholtz equation, and every solution of the equation can be expressed as a superposition of plane waves. In the general case, integration over a continuous spectrum of plane waves is required to represent the field. The plane-wave expansions written in circular-cylinder coordinates with an integration path in the form of two symmetric loops in a complex plane are called Sommerfeld's integrals (Sommerfeld (1896), see also Sommerfeld (1927, Chapter 20) and (Sommerfeld 1964, section 38)). For example, if $U(\rho, \varphi)$ is a solution of a two-dimensional Helmholtz equation, either (5.141) or (5.142), then

$$U(\rho,\varphi) = \frac{1}{2\pi j}\int_{\gamma'} S(\alpha')e^{j\kappa\rho\cos(\varphi-\alpha')}\,d\alpha' \tag{5.153}$$

$$= \frac{1}{2\pi j}\int_{\gamma} S(\alpha+\varphi)e^{j\kappa\rho\cos\alpha}\,d\alpha \tag{5.154}$$

$$= \frac{1}{2\pi j}\int_{\gamma_+} [S(\alpha+\varphi) - S(-\alpha+\varphi)]e^{j\kappa\rho\cos\alpha}\,d\alpha, \tag{5.155}$$

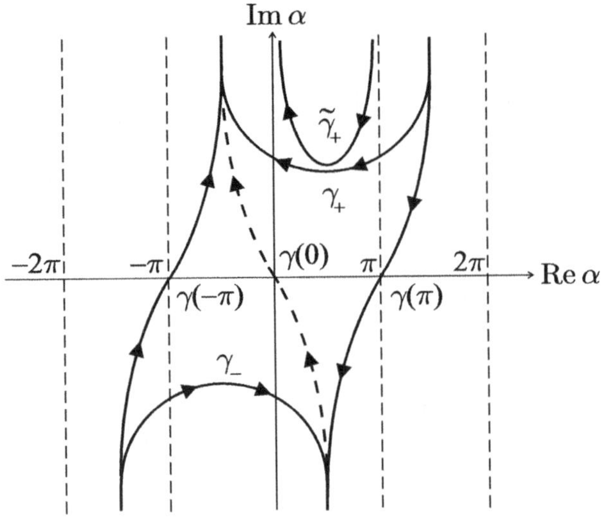

Figure 5.25 Integration paths relevant to Sommerfeld's integrals.

where $\kappa = k\sin\beta$, the integration path γ consists of two symmetric loops $\gamma_\pm$ (Figure 5.25), and the contour γ' is shifted with respect to γ by the value of φ along the real axis on the complex α plane. The ends of $\gamma_\pm$ are located in those regions of the complex α plane where $\operatorname{Im}\cos\alpha > 0$, so that the factor $\exp(j\kappa\rho\cos\alpha)$ decays as $|\operatorname{Im}\alpha| \to \infty$. These regions are strips arranged in a chessboard pattern on the complex α plane, $-\pi < \operatorname{Re}\alpha + 2m\pi < 0$ for $\operatorname{Im}\alpha > 0$ and $0 < \operatorname{Re}\alpha + 2m\pi < \pi$ for $\operatorname{Im}\alpha < 0$ where $m = 0, \pm1, \pm2, \ldots$ It is assumed that the ends of γ_+ are located at $-\pi/2 + j\infty$ and $3\pi/2 + j\infty$ and the ends of γ_- at $-3\pi/2 - j\infty$ and $\pi/2 - j\infty$ to ensure the fastest decay of the exponential factors with $\kappa > 0$ when $\operatorname{Im}\alpha \to \infty$. The loops are located sufficiently far from the real axis on the complex plane of the integration variable so that no singularities of the integrand functions are located in the regions bounded by $\gamma_\pm$ above γ_+ and below γ_-.

The integral (5.153) is a superposition of plane waves (including inhomogeneous ones when α' is complex-valued) with spectral amplitude $S(\alpha')$. Integrals in (5.154) and (5.153) are related by the change in the integration variable $\alpha' = \alpha + \varphi$, and the integral in (5.155) is a consequence of the symmetry of the integration contour γ in (5.154). All three forms are equivalent and will be referred to as Sommerfeld's integrals. It is seen from (5.155) that only the odd part of $S(\alpha + \varphi)$ contributes to the value of the Sommerfeld integral. For example, adding a constant to the spectral amplitude does not change the value of the integral.

The oddness of the integrand function with respect to α or, which is the same, the symmetry of the integration path γ is necessary for the compliance of Sommerfeld's integrals with conditions at infinity as $\rho \to +\infty$. Indeed, deforming γ_+ in (5.155) onto $\gamma(\pi)$ and $\gamma(0)$ (Figure 5.25) leads to integrals with integrand functions vanishing at every point on the integration paths in the limit $\rho \to \infty$, except for the point $\alpha = \pi$ in the integral over $\gamma(\pi)$ and $\alpha = 0$ in the integral over $\gamma(0)$. The contribution of the latter integral can be expected to be proportional to $\exp(j\kappa\rho)$, which is an incoming and therefore unphysical cylindrical wave. Such contribution is excluded, however, thanks to the oddness of the integrand function and the symmetry of $\gamma(0)$ with respect to $\alpha = 0$.

Plane wave components of the wave field, for example incident, reflected, and surface waves, are to be described, therefore, by some singularities of the spectral amplitudes that are crossed in the course of the contour deformation. Consider the Sommerfeld integral in the form (5.154) and deform the integration path γ onto contours $\gamma(\pm\pi)$ (Figure 5.25), which gives

$$U(\rho, \varphi) = \frac{1}{2\pi j} \int\limits_{\gamma(-\pi)\cup\gamma(\pi)} S(\alpha + \varphi) e^{j\kappa\rho\cos\alpha}\, d\alpha + \sum_m \operatorname{res} S(\alpha_m) e^{j\kappa\rho\cos\alpha_m}, \tag{5.156}$$

where α_m denotes the pole singularities of $S(\alpha + \varphi)$ that are located in the region enclosed by the contours $\gamma(\pm\pi)$ and $\gamma_\pm$. For example, an incoming plane wave $\exp[j\kappa\rho\cos(\varphi - \varphi_0)]$ corresponds to a pole singularity of $S(\alpha)$ at $\alpha = \varphi_0 - \varphi$ with a unit residue.

The integral in (5.156) is taken over the contours $\gamma(\pi)$ and $\gamma(-\pi)$, which are shifted by 2π with respect to each other and have opposite integration directions, so (5.156) can be rewritten as

$$U(\rho,\varphi) = \frac{1}{2\pi j}\int_{\gamma(\pi)} [S(\alpha+\varphi) - S(\alpha+\varphi-2\pi)]\mathrm{e}^{j\kappa\rho\cos\alpha}\,\mathrm{d}\alpha + \sum_m \operatorname{res} S(\alpha_m)\mathrm{e}^{j\kappa\rho\cos\alpha_m}. \tag{5.157}$$

This representation suggests a vanishing integral term if $S(\alpha)$ is 2π periodic. In such cases, the solution, the total field $U(\rho,\varphi)$, is expressed in a particularly simple form as a sum of several residue contributions. Examples can be found in sections 5.2.5 and 5.4.2.

Representations (5.153)–(5.155) are closely related to the inverse Laplace transform and their properties can be derived from those of the Laplace integral transformation (Osipov 1993a). We will now use this connection to explain the structure of (5.153)–(5.155) and derive useful properties of Sommerfeld integrals. Consider a function $F(\rho)$ which is the function $U(\rho,\varphi)$ at a fixed value of φ. The function $F(\rho)$ is a holomorphic function of ρ on the semi-axis $0 < \rho < +\infty$, bounded at $\rho = 0$ because of the edge conditions and $O[\exp(\rho|\operatorname{Im}\kappa|)]$ at $\rho \to +\infty$ due to the incident wave. Such a function can be represented by the inverse Laplace transform

$$F(\rho) = \frac{1}{2\pi j}\int_{b-j\infty}^{b+j\infty} Q(p)\mathrm{e}^{p\rho}\,\mathrm{d}p \tag{5.158}$$

with

$$Q(p) = \int_0^{+\infty} F(\rho)\mathrm{e}^{-p\rho}\,\mathrm{d}\rho \tag{5.159}$$

and $b > |\operatorname{Im}\kappa|$ (b is a parameter determined by the behavior of the original function $F(\rho)$ at $\rho \to \infty$) (e.g., Ryshik and Gradstein (1957, section 5.2), Abramowitz and Stegun (1965, Chapter 29), and Morse and Feshbach (1953, section 4.8)). Changing the integration variable in (5.158) as $p = j\kappa\cos\alpha$ gives

$$F(\rho) = \frac{1}{2\pi j}(-j\kappa)\int_{\tilde{\gamma}_+} Q(j\kappa\cos\alpha)\mathrm{e}^{j\kappa\rho\cos\alpha}\sin\alpha\,\mathrm{d}\alpha, \tag{5.160}$$

where the integration path $\tilde{\gamma}_+$ is shown in Figure 5.25.

The integrand function in (5.160) is an odd function of α and can therefore always be represented in the form

$$-j\kappa\sin\alpha\, Q(j\kappa\cos\alpha) = \tilde{S}(\alpha) - \tilde{S}(-\alpha), \tag{5.161}$$

where $\tilde{S}(\alpha)$ is a new function, which reduces (5.160) to the form similar to (5.155), the only difference being in the shape of the integration contours. For functions $F(\rho)$ that are holomorphic and compliant with the edge and infinity conditions not only on the semi-axis $0 < \rho < \infty$ but also within a sector $|\arg\rho| \le \delta$ with $\delta > 0$ on the complex ρ plane (Maliuzhinets 1958b; Osipov 1993a), the contour $\tilde{\gamma}_+$ can be deformed into the wider contour γ_+ and we obtain an integral representation of the form

$$F(\rho) = \frac{1}{2\pi j}\int_{\gamma_+} [\tilde{S}(\alpha) - \tilde{S}(-\alpha)]\mathrm{e}^{j\kappa\rho\cos\alpha}\,\mathrm{d}\alpha = \frac{1}{2\pi j}\int_{\gamma} \tilde{S}(\alpha)\mathrm{e}^{j\kappa\rho\cos\alpha}\,\mathrm{d}\alpha. \tag{5.162}$$

With $\tilde{S}(\alpha) = S(\alpha+\varphi)$ the representation describes a solution of the Helmholtz equation and coincides with the Sommerfeld representations (5.153)–(5.155).

So, we see that the Sommerfeld representation is in fact a variant of the Laplace inverse transform which expresses the original function $F(\rho)$ by integration of an image function. On the other hand, given the function $F(\rho)$, one can calculate the image function $\tilde{S}(\alpha) - \tilde{S}(-\alpha)$ by using (5.159) and (5.161); this relation is called the inversion formula for the Sommerfeld integral (Maliuzhinets 1958b).

Another useful property is formulated as the so-called nullification theorem (Maliuzhinets 1958b). The theorem describes solutions of the homogeneous integral equation

$$\int_{\gamma} \tilde{S}(\alpha) e^{j\kappa\rho\cos\alpha}\, d\alpha = 0, \quad 0 < \rho < \infty. \tag{5.163}$$

Apparently, $\tilde{S}(\alpha) = 0$ is always a solution of (5.163). However, there can be further solutions if $\tilde{S}(\alpha)$ growing at infinity is allowed. More precisely, if $|\tilde{S}(\alpha) - \tilde{S}(-\alpha)| = O[\exp(a \operatorname{Im}\alpha)]$ for $|\operatorname{Im}\alpha| \to \infty$, then

$$\tilde{S}(\alpha) - \tilde{S}(-\alpha) = \begin{cases} 0, & -\infty < a < 1 \\ \sin\alpha \sum_{m=1}^{[a]} c_m \cos^{m-1}\alpha, & a \geq 1 \end{cases}, \tag{5.164}$$

where $[a]$ denotes the integer part of a.

To see that, it is sufficient to rewrite (5.163) as

$$\int_{C} Q(p) e^{p\rho}\, dp = 0, \quad 0 < \rho < \infty, \tag{5.165}$$

where the integration contour C is the image of γ_+ on the complex p plane and it differs from $\operatorname{Re} p = b$, which is the image of $\tilde{\gamma}_+$, in that it goes to infinity in the left half-plane along the directions $\arg p = \pm(\delta + \pi/2)$ with $0 < \delta < \pi/2$. Function $Q(p)$ is holomorphic in the region to the right of C as a Laplace image function and is estimated by $O(|p|^{a-1})$ for $|p| \to \infty$, which follows from (5.161). On the other hand, equation (5.165) means that $Q(p)$ is holomorphic in the region to the *left* of C, implying that $Q(p)$ is holomorphic on the whole complex p plane and because of the extended Liouville theorem (section C.2.1), $Q(p)$ is either zero if $a < 1$ or a polynomial in p of the degree not exceeding $a - 1$. Formulated in terms of $\tilde{S}(\alpha)$ (see (5.161)), these statements lead to (5.164).

The behavior of the original function $F(\rho)$ at $\rho \to 0$ is related to the behavior of the spectral amplitude at $\operatorname{Im}\alpha \to \pm\infty$. This follows from the similarity of (5.162) with the integral representation (E.38) and from the small-argument approximation (5.42) of the Bessel function. For example, for spectral amplitudes bounded at $\operatorname{Im}\alpha \to \pm\infty$,

$$\frac{1}{2\pi j} \int_{\gamma_+} [\tilde{S}(\alpha) - \tilde{S}(-\alpha)] e^{j\kappa\rho\cos\alpha}\, d\alpha \to j[\tilde{S}(j\infty) - \tilde{S}(-j\infty)] J_0(\kappa\rho) \tag{5.166}$$

when $\rho \to 0$, and

$$F(0) = j[\tilde{S}(j\infty) - \tilde{S}(-j\infty)]. \tag{5.167}$$

Equations for Spectral Amplitudes

To solve the problem of scattering from impedance wedges, we follow a method which consists of representing the fields by Sommerfeld's integrals and inserting them into the boundary conditions, followed by obtaining functional equations for the spectral amplitudes and solving the equations. The method was proposed by Maliuzhinets (1958a), and its extension to vector problems was studied by Vaccaro (1980). The theory of associated functional difference equations was developed by Tuzhilin (1973).

The fields are expressed as

$$E_z(\rho,\varphi,z) = \frac{e^{-jkz\cos\beta}}{2\pi j}\int_\gamma S_e(\alpha+\varphi)e^{jk\rho\sin\beta\cos\alpha}\,d\alpha \tag{5.168}$$

$$ZH_z(\rho,\varphi,z) = \frac{e^{-jkz\cos\beta}}{2\pi j}\int_\gamma S_h(\alpha+\varphi)e^{jk\rho\sin\beta\cos\alpha}\,d\alpha. \tag{5.169}$$

For compactness sake, we will also use the vector notation

$$\mathbf{u}(\rho,\varphi,z) = \frac{e^{-jkz\cos\beta}}{2\pi j}\int_\gamma \mathbf{s}(\alpha+\varphi)e^{jk\rho\sin\beta\cos\alpha}\,d\alpha, \tag{5.170}$$

where

$$\mathbf{u} = \begin{bmatrix} E_z \\ ZH_z \end{bmatrix}, \quad \mathbf{s} = \begin{bmatrix} S_e \\ S_h \end{bmatrix}. \tag{5.171}$$

The conditions at $\rho \to \infty$ imply that the spectral amplitudes have the pole singularities

$$S_e(\alpha) \to \frac{E_{0z}}{\alpha - \varphi_0}, \quad S_h(\alpha) \to \frac{ZH_{0z}}{\alpha - \varphi_0} \tag{5.172}$$

as $\alpha \to \varphi_0$, and these are their only singularities in the strip

$$\Pi_0 = \{\alpha : \ |\mathrm{Re}\,\alpha| < \Phi, \ -\infty < \mathrm{Im}\,\alpha < +\infty\} \tag{5.173}$$

on the complex plane of their argument, as the incident plane wave, described by (5.143) and (5.144), is the only incoming wave with the propagation direction in the sector $-\Phi < \varphi < \Phi$. These conditions can be seen as an extension of radiation conditions on infinite wedge-shaped geometries.

The edge conditions (5.152) are satisfied by requiring that when $\mathrm{Im}\,\alpha \to \pm\infty$,

$$S_e(\alpha+\varphi) \to S_e(\pm j\infty), \quad S_h(\alpha+\varphi) \to S_h(\pm j\infty), \tag{5.174}$$

where $S_{e,h}(\pm j\infty)$ are bounded constants independent of φ.

The Sommerfeld representations (5.168) and (5.169) by construction satisfy the Helmholtz equations (5.141) and (5.142). The spectral amplitudes are determined from the boundary conditions (5.150) and (5.151). When an integral of the form (5.168) or (5.169) is inserted in the boundary conditions, the derivative $\partial/\partial\rho$ is replaced with the factor $jk\sin\beta\cos\alpha$ and the derivative $\partial/\partial\varphi$, on integration by parts, with the factor $jk\rho\sin\beta\sin\alpha$, and we obtain

$$\int_\gamma e^{jk\rho\sin\beta\cos\alpha}[(\sin\alpha \pm \frac{\sin\beta}{\eta_\pm})S_e(\alpha\pm\Phi) - \cos\alpha\cos\beta S_h(\alpha\pm\Phi)]\,d\alpha = 0 \tag{5.175}$$

$$\int_\gamma e^{jk\rho\sin\beta\cos\alpha}[\cos\alpha\cos\beta S_e(\alpha\pm\Phi) + (\sin\alpha \pm \eta_\pm\sin\beta)S_h(\alpha\pm\Phi)]\,d\alpha = 0, \tag{5.176}$$

or in vector notation (5.171),

$$\int_\gamma e^{jk\rho\sin\beta\cos\alpha}\,\overline{\overline{M}}_\pm(\alpha)\cdot\mathbf{s}(\alpha\pm\Phi)\,d\alpha = \begin{bmatrix} 0 \\ 0 \end{bmatrix}, \tag{5.177}$$

where

$$\overline{\overline{M}}_{\pm}(\alpha) = \begin{bmatrix} m_{11}^{\pm}(\alpha) & m_{12}^{\pm}(\alpha) \\ m_{21}^{\pm}(\alpha) & m_{22}^{\pm}(\alpha) \end{bmatrix} \tag{5.178}$$

with

$$m_{11}^{\pm}(\alpha) = \sin\alpha \pm \sin\beta/\eta_{\pm} \tag{5.179}$$

$$m_{12}^{\pm}(\alpha) = -\cos\alpha\cos\beta = -m_{21}^{\pm}(\alpha) \tag{5.180}$$

$$m_{22}^{\pm}(\alpha) = \sin\alpha \pm \eta_{\pm}\sin\beta. \tag{5.181}$$

If one of the wedge faces is perfectly conducting, for example $\eta_- = 0$, then m_{il}^- with $i, l = 1, 2$ should be replaced with

$$m_{11}^{-}(\alpha) = -\sin\beta \tag{5.182}$$

$$m_{12}^{-}(\alpha) = 0 \tag{5.183}$$

$$m_{21}^{-}(\alpha) = \cos\alpha\cos\beta \tag{5.184}$$

$$m_{22}^{-}(\alpha) = \sin\alpha, \tag{5.185}$$

whereas coefficients m_{il}^+ remain unchanged.

Accounting for the boundedness of spectral functions at $\operatorname{Im}\alpha \to \pm\infty$ (see (5.174)) and using nullification theorem (5.164) suggests that the solution of the homogeneous equation (5.177) satisfies the equation

$$\overline{\overline{M}}_{\pm}(\alpha)\cdot \mathrm{s}(\alpha \pm \Phi) - \overline{\overline{M}}_{\pm}(-\alpha)\cdot \mathrm{s}(-\alpha \pm \Phi) = \sin\alpha \begin{bmatrix} C_1^{\pm} \\ C_2^{\pm} \end{bmatrix}, \tag{5.186}$$

where $C_{1,2}^{\pm}$ are arbitrary parameters independent of α. Without loss of generality, the parameters can be set to zero. To see that, we let α in (5.186) go to infinity and, using (5.174), obtain

$$C_1^+ = C_1^- = S_{\mathrm{e}}(j\infty) + S_{\mathrm{e}}(-j\infty) + j\cos\beta[S_{\mathrm{h}}(j\infty) - S_{\mathrm{h}}(-j\infty)] \tag{5.187}$$

$$C_2^+ = C_2^- = -j\cos\beta[S_{\mathrm{e}}(j\infty) - S_{\mathrm{e}}(-j\infty)] + S_{\mathrm{h}}(j\infty) + S_{\mathrm{h}}(-j\infty), \tag{5.188}$$

implying that the relations $C_1^+ = C_1^-$ and $C_2^+ = C_2^-$ are necessary to ensure solvability of (5.186). Since Sommerfeld's integrals are invariant with respect to adding constants to their spectra, we can define new spectral amplitudes as

$$\tilde{S}_{\mathrm{e}}(\alpha) = S_{\mathrm{e}}(\alpha) - \frac{1}{2}C_1, \quad \tilde{S}_{\mathrm{h}}(\alpha) = S_{\mathrm{h}}(\alpha) - \frac{1}{2}C_2 \tag{5.189}$$

with $C_1 = C_1^{\pm}$ and $C_2 = C_2^{\pm}$ to see that they satisfy (5.186) with the right-hand side equal to zero. Thus, the different choice of constants in the right-hand side of (5.186) leads to solutions differing by constants, which does not influence the values of the integrals (5.168) and (5.169). So, in what follows we set $C_{1,2}^{\pm} = 0$ in (5.186) and study the homogeneous equation

$$\overline{\overline{M}}_{\pm}(\alpha)\cdot \mathrm{s}(\alpha \pm \Phi) = \overline{\overline{M}}_{\pm}(-\alpha)\cdot \mathrm{s}(-\alpha \pm \Phi). \tag{5.190}$$

The matrix equation (5.190) is a set of four functional difference equations with variable coefficients for two unknown functions $S_{\mathrm{e,h}}(\alpha)$. Although (5.190) is valid over the entire complex α plane, it is convenient to impose it only on the imaginary axis ($\operatorname{Re}\alpha = 0, -\infty < \operatorname{Im}\alpha < \infty$), thereby confining the analysis to the strip Π_0 (5.173). Once the spectra are determined here, they can be extended to the rest of the complex plane by using (5.190) as a functional relation. For example, when $\alpha \in \Pi_{\pm 1}$ where

$$\Pi_{\pm 1} = \{\alpha : -\Phi < \operatorname{Re}\alpha \mp 2\Phi < \Phi, -\infty < \operatorname{Im}\alpha < +\infty\}, \tag{5.191}$$

relations

$$\mathbf{s}(\alpha) = -\overline{\overline{R}}_{\pm}(\pm\alpha - \Phi) \cdot \mathbf{s}(\pm 2\Phi - \alpha) \tag{5.192}$$

with

$$\overline{\overline{R}}_{\pm}(\alpha) = -\overline{\overline{M}}_{\pm}^{-1}(\pm\alpha) \cdot \overline{\overline{M}}_{\pm}(\mp\alpha) \tag{5.193}$$

extend $\mathbf{s}(\alpha)$ from Π_0 into the neighboring strips $\Pi_{\pm 1}$. By performing matrix inversion and multiplication in (5.193), the matrices $\overline{\overline{R}}_{\pm}$ can be expressed as

$$\overline{\overline{R}}_{\pm}(\alpha) = \frac{1}{L_{\pm}(\alpha)} \begin{pmatrix} Q_{\text{ee}}^{\pm} & Q_{\text{eh}}^{\pm} \\ Q_{\text{he}}^{\pm} & Q_{\text{hh}}^{\pm} \end{pmatrix}, \tag{5.194}$$

where

$$L_{\pm}(\alpha) = \det \overline{\overline{M}}_{\pm}(\pm\alpha) = \sin^2\beta \left(\sin\alpha + \frac{1}{\eta_{\pm}\sin\beta}\right)\left(\sin\alpha + \frac{\eta_{\pm}}{\sin\beta}\right) \tag{5.195}$$

$$Q_{\text{ee}}^{\pm} = 2\sin\alpha(\eta_{\pm}\sin\beta + \sin\alpha) - L_{\pm}(\alpha) \tag{5.196}$$

$$Q_{\text{eh}}^{\pm} = -Q_{\text{he}}^{\pm} = \pm 2\sin\alpha\cos\alpha\cos\beta \tag{5.197}$$

$$Q_{\text{hh}}^{\pm} = 2\sin\alpha\left(\frac{\sin\beta}{\eta_{\pm}} + \sin\alpha\right) - L_{\pm}(\alpha). \tag{5.198}$$

An explicit solution of (5.190) is not possible for general values of Φ, $\eta_{\pm}$, and β, but it is possible in a broad variety of special cases. We begin with the case of a PEC wedge, a solution for which is derived in section 5.2 with a different method. Letting $\eta_{\pm} \to 0$ in (5.190) gives

$$S_{\text{e}}(\alpha \pm \Phi) = S_{\text{e}}(-\alpha \pm \Phi) \tag{5.199}$$

$$\cos\alpha\cos\beta S_{\text{e}}(\alpha \pm \Phi) + \sin\alpha S_{\text{h}}(\alpha \pm \Phi) = \cos\alpha\cos\beta S_{\text{e}}(-\alpha \pm \Phi) - \sin\alpha S_{\text{h}}(-\alpha \pm \Phi), \tag{5.200}$$

and using (5.199), equation (5.200) can be rewritten as

$$S_{\text{h}}(\alpha \pm \Phi) = -S_{\text{h}}(-\alpha \pm \Phi). \tag{5.201}$$

Solutions of functional equations (5.199) and (5.201) with the pole singularity at $\alpha = \varphi_0$ required by (5.172) can be easily constructed from $\sin(\nu\alpha)$ and $\cos(\nu\alpha)$ with

$$\nu = \frac{\pi}{2\Phi} \tag{5.202}$$

on noting that $\sin[\nu(\alpha \pm \Phi)] = \pm\cos(\nu\alpha)$ and $\cos[\nu(\alpha \pm \Phi)] = \mp\sin(\nu\alpha)$, and the result reads

$$S_{\text{e}}(\alpha) = \frac{E_{0z}\nu\cos(\nu\varphi_0)}{\sin(\nu\alpha) - \sin(\nu\varphi_0)} \tag{5.203}$$

$$S_{\text{h}}(\alpha) = \frac{ZH_{0z}\nu\cos(\nu\alpha)}{\sin(\nu\alpha) - \sin(\nu\varphi_0)}. \tag{5.204}$$

When inserted in (5.168) and (5.169), expressions (5.203) and (5.204) recover the solution derived in section 5.2, which can be checked by direct substitution of (5.53) with (5.54) in (5.40) and (5.41) (problem 11).

5.4.2 Normal Incidence (Maliuzhinets' Solution)

An exact solution is also possible for an impedance wedge illuminated by a plane wave with the incidence direction perpendicular to the edge of the wedge (normal incidence, $\beta = \pi/2$). In this case, S_e and S_h satisfy separate functional equations, and an explicit solution of the scattering problem is possible with the approach described in Maliuzhinets (1958a). The solution is practically important as the other parameters, Φ and $\eta_\pm$, can be arbitrary. It is also of mathematical importance as solutions for other explicitly solvable cases are obtained by reduction to the difference equations of the type solved by Maliuzhinets.

With $\beta = \pi/2$,

$$m_{11}^{\pm}(\alpha) = \sin\alpha \pm \frac{1}{\eta_\pm}, \quad m_{12}^{\pm}(\alpha) = m_{21}^{\pm}(\alpha) = 0, \quad m_{22}^{\pm}(\alpha) = \sin\alpha \pm \eta_\pm \tag{5.205}$$

and functional equations (5.190) split up into two separate pairs of equations:

$$(\sin\alpha \pm \sin\theta_e^{\pm})S_e(\alpha \pm \Phi) = (-\sin\alpha \pm \sin\theta_e^{\pm})S_e(-\alpha \pm \Phi) \tag{5.206}$$

$$(\sin\alpha \pm \sin\theta_h^{\pm})S_h(\alpha \pm \Phi) = (-\sin\alpha \pm \sin\theta_h^{\pm})S_h(-\alpha \pm \Phi), \tag{5.207}$$

where

$$\sin\theta_e^{\pm} = \frac{1}{\eta_\pm}, \quad \sin\theta_h^{\pm} = \eta_\pm. \tag{5.208}$$

The new parameters $\theta_{e,h}^{\pm}$ have the meaning of Brewster's angles of the respective impedance faces (measured from the face); their real parts are limited to the interval $0 < \operatorname{Re}\theta_{e,h}^{\pm} \leq \pi/2$ since $\operatorname{Re}\eta_\pm > 0$ for passive boundaries.

Equations (5.206) and (5.207) are of the same form

$$(\sin\alpha \pm \sin\theta_\pm)S(\alpha \pm \Phi) = (-\sin\alpha \pm \sin\theta_\pm)S(-\alpha \pm \Phi) \tag{5.209}$$

and can be solved by the same procedure. Let us seek the solution in the following form

$$S(\alpha) = \sigma(\alpha)\frac{\Psi(\alpha)}{\Psi(\varphi_0)}, \tag{5.210}$$

where $\sigma(\alpha)$ is such that

$$\sigma(\alpha \pm \Phi) = \sigma(-\alpha \pm \Phi) \tag{5.211}$$

and has the pole singularity required by (5.172), whereas $\Psi(\alpha)$ has neither poles nor zeros in Π_0 and is a solution of the equation

$$(\sin\alpha \pm \sin\theta_\pm)\Psi(\alpha \pm \Phi) = (-\sin\alpha \pm \sin\theta_\pm)\Psi(-\alpha \pm \Phi). \tag{5.212}$$

It is apparent that (5.210) satisfies (5.209) and has the required analytical properties in Π_0.

Equation (5.211) coincides with (5.199), and using (5.203) we have

$$\sigma(\alpha) = \frac{\nu\cos(\nu\varphi_0)}{\sin(\nu\alpha) - \sin(\nu\varphi_0)}. \tag{5.213}$$

The function $\Psi(\alpha)$ can be determined by reformulating (5.212) in terms of the logarithmic derivative of $\Psi(\alpha)$, which is achieved by writing

$$\frac{\Psi(\alpha \pm \Phi)}{\Psi(-\alpha \pm \Phi)} = \frac{-\sin\alpha \pm \sin\theta_\pm}{\sin\alpha \pm \sin\theta_\pm}, \tag{5.214}$$

followed by taking the logarithm of both sides and differentiation with respect to α. This can be done either directly (Osipov and Norris 1998) or in a less direct but more elegant way suggested by Maliuzhinets (1958a).

The Maliuzhinets Function

Maliuzhinets' approach consists in the introduction of a new function $\psi_\Phi(\alpha)$, named after him, and representing $\Psi(\alpha)$ as

$$\Psi(\alpha) = \psi_\Phi\left(\alpha + \Phi + \frac{\pi}{2} - \theta_+\right)\psi_\Phi\left(\alpha + \Phi - \frac{\pi}{2} + \theta_+\right)\cdot\psi_\Phi\left(\alpha - \Phi + \frac{\pi}{2} - \theta_-\right)\psi_\Phi\left(\alpha - \Phi - \frac{\pi}{2} + \theta_-\right). \quad (5.215)$$

By inserting (5.215) into (5.214), one may find that the difference equations are met if $\psi_\Phi(\alpha)$ satisfies two functional relations

$$\psi_\Phi(-\alpha) - \psi_\Phi(\alpha) = 0 \quad (5.216)$$

$$\frac{\psi_\Phi(\alpha + 2\Phi)}{\psi_\Phi(\alpha - 2\Phi)} = \cot\left(\frac{\alpha}{2} + \frac{\pi}{4}\right). \quad (5.217)$$

In order for $\Psi(\alpha)$ to have the required analytical properties, $\psi_\Phi(\alpha)$ must be holomorphic and free of poles and zeros within the strip $\Pi_{2\Phi+\pi/2} = \{\alpha : |\mathrm{Re}\,\alpha| \leq 2\Phi + \pi/2, -\infty < \mathrm{Im}\,\alpha < +\infty\}$.

In terms of the logarithmic derivative

$$\eta_\Phi(\alpha) = \frac{\partial}{\partial\alpha}\ln\psi_\Phi(\alpha), \quad (5.218)$$

functional equations (5.216) and (5.217) can be reduced to linear equations with constant coefficients of the form

$$\eta_\Phi(\alpha) + \eta_\Phi(-\alpha) = 0 \quad (5.219)$$

$$\eta_\Phi(\alpha + 2\Phi) - \eta_\Phi(\alpha - 2\Phi) = -\frac{1}{\cos\alpha}. \quad (5.220)$$

It follows from (5.218) and the analytical properties of $\psi_\Phi(\alpha)$ that $\eta_\Phi(\alpha)$ has no poles when $\alpha \in \Pi_{2\Phi+\pi/2}$. Once $\eta_\Phi(\alpha)$ is constructed, $\psi_\Phi(\alpha)$ is obtained from (5.218) as

$$\psi_\Phi(\alpha) = \exp\left[\int_0^\alpha \eta_\Phi(z)\,\mathrm{d}z\right]. \quad (5.221)$$

The integration range in (5.221) is chosen so as to comply with the normalization convention

$$\psi_\Phi(0) = 1 \quad (5.222)$$

that is usually assumed.

Equations (5.219) and (5.220) are imposed on the imaginary axis and can be solved by applying the modified Fourier transform

$$\eta_\Phi(\alpha) = \int_{-j\infty}^{+j\infty} \mathrm{e}^{-j\alpha t} G_\Phi(t)\,\mathrm{d}t \quad (5.223)$$

$$G_\Phi(t) = -\frac{1}{2\pi}\int_{-j\infty}^{+j\infty} \mathrm{e}^{j\alpha t}\eta_\Phi(\alpha)\,\mathrm{d}\alpha, \quad (5.224)$$

which differs from the standard one defined on the real axis by a corresponding change of the integration variables. The Fourier transform $G_\Phi(t)$ satisfies the equations

$$G_\Phi(t) + G_\Phi(-t) = 0 \quad (5.225)$$

$$G_\Phi(t)(\mathrm{e}^{-2j\Phi t} - \mathrm{e}^{2j\Phi t}) = \frac{j}{2\cos(\pi t/2)} \quad (5.226)$$

and is apparently given by the formula

$$G_\Phi(t) = \frac{-1}{4\sin(2\Phi t)\cos(\pi t/2)}. \tag{5.227}$$

Thus, from (5.223),

$$\eta_\Phi(\alpha) = -\int_{-j\infty}^{+j\infty} \frac{e^{-j\alpha t}\,dt}{4\sin(2\Phi t)\cos(\pi t/2)}, \tag{5.228}$$

where at the point $t = 0$ the integral is taken in the principal value sense to comply with (5.219). The integral in (5.228) is convergent for $|\mathrm{Re}\,\alpha| < 2\Phi + \pi/2$ and, therefore, defines $\eta_\Phi(\alpha)$ as a function free of poles in $\Pi_{2\Phi+\pi/2}$.

Changing the integration variable in (5.228) to integrate over the positive half-axis

$$\eta_\Phi(\alpha) = -\frac{1}{2}\int_0^{+\infty} \frac{\sinh(\alpha t)\,dt}{\sinh(2\Phi t)\cosh(\pi t/2)} \tag{5.229}$$

and using (5.221), the Maliuzhinets function is expressed as

$$\psi_\Phi(\alpha) = \exp\left[-\frac{1}{2}\int_0^{\alpha}\int_0^{+\infty} \frac{\sinh(\beta t)\,dt\,d\beta}{\sinh(2\Phi t)\cosh(\pi t/2)}\right] \tag{5.230}$$

or, after evaluation of the integral with respect to β,

$$\psi_\Phi(\alpha) = \exp\left[-\frac{1}{2}\int_0^{+\infty} \frac{\cosh(\alpha t) - 1}{t\sinh(2\Phi t)\cosh(\pi t/2)}\,dt\right] \tag{5.231}$$

(Zavadskii and Sakharova 1967). The integral in (5.231) is convergent for $|\mathrm{Re}\,\alpha| < 2\Phi + \pi/2$ and, therefore, defines a function $\psi_\Phi(\alpha)$ which is free of poles and zeros in $\Pi_{2\Phi+\pi/2}$.

The Maliuzhinets function has a number of further useful analytical properties, and here we just list the most relevant ones. For a complete theory of $\psi_\Phi(\alpha)$ see, for example, Maliuzhinets (1950) and Osipov and Stein (1999). For $\mathrm{Im}\,\alpha \to \pm\infty$, $\psi_\Phi(\alpha)$ grows according to the formula

$$\psi_\Phi(\alpha) \to \frac{1}{\sqrt{2}}\psi_\Phi\left(\frac{\pi}{2}\right)\exp\left[-j\frac{\pi\alpha}{8\Phi}\mathrm{sgn}(\mathrm{Im}\,\alpha)\right]. \tag{5.232}$$

In addition to (5.216) and (5.217), the function satisfies the relation

$$\psi_\Phi\left(\alpha + \frac{\pi}{2}\right)\psi_\Phi\left(\alpha - \frac{\pi}{2}\right) = \psi_\Phi^2\left(\frac{\pi}{2}\right)\cos\left(\frac{\pi\alpha}{4\Phi}\right). \tag{5.233}$$

The relation can be seen as a means for extending the function from the strip of width π over the whole complex α plane.

An important representation from the theoretical viewpoint is the double product representation

$$\psi_\Phi(\alpha) = \prod_{n=1}^{\infty}\prod_{m=1}^{\infty}\left[1 - \left(\frac{\alpha}{\alpha_{nm}}\right)^2\right]^{(-1)^{m+1}} \tag{5.234}$$

with

$$\alpha_{nm} = 2\Phi(2n-1) + \frac{\pi}{2}(2m-1). \tag{5.235}$$

This expansion clearly demonstrates that $\psi_\Phi(\alpha)$ is meromorphic on the complex α plane with zeros (m odd) and poles (m even) at $\alpha = \pm\alpha_{nm}$. The smallest zeros and poles are located at $\alpha = \pm(2\Phi + \pi/2)$ and $\alpha = \pm(2\Phi + 3\pi/2)$, respectively.

In the particular case that $\Phi = \pi N/(4M)$ with integer N and M, $\psi_\Phi(\alpha)$ simplifies to

$$\psi_\Phi(\alpha) = \prod_{m=1}^{M}\prod_{n=1}^{N}\left\{\frac{\cos[a(m,M,n,N)/2]}{\cos[\alpha/(2N) + a(m,M,n,N)/2]}\right\}^{(-1)^n} \quad (N = 1, 3, 5, \ldots) \tag{5.236}$$

and to

$$\psi_\Phi(\alpha) = \prod_{m=1}^{M}\prod_{n=1}^{N}\exp\left[\frac{(-1)^n}{\pi}\int_{a(m,M,n,N)}^{\alpha/N+a(m,M,n,N)} u\cot u\,\mathrm{d}u\right] \quad (N = 2, 4, 6\ldots), \tag{5.237}$$

where

$$a(m,M,n,N) = \frac{\pi}{2}\left(\frac{2n-1}{N} - \frac{2m-1}{M}\right). \tag{5.238}$$

So, for odd values of N the Maliuzhinets function is expressed as a finite product of trigonometric functions. On the other hand, when N is even, $\psi_{\pi N/(4M)}(\alpha)$ is given by a finite product of integrals over bounded integration ranges.

Specific examples of (5.236) and (5.237) with $M = 1$ are as follows:

$$\psi_{\pi/4}(\alpha) = \cos\left(\frac{\alpha}{2}\right) \tag{5.239}$$

$$\psi_{\pi/2}(\alpha) = \exp\left(\int_0^\alpha \frac{2t - \pi\sin t}{4\pi\cos t}\,\mathrm{d}t\right) \tag{5.240}$$

$$\psi_{3\pi/4}(\alpha) = \frac{4}{3}\cos\left(\frac{\alpha}{6}\right) - \frac{1}{3}\sec\left(\frac{\alpha}{6}\right) \tag{5.241}$$

$$\psi_\pi(\alpha) = \exp\left[\int_0^\alpha \frac{2\pi\sqrt{2}\sin(t/2) - 2t - \pi\sin t}{8\pi\cos t}\,\mathrm{d}t\right]. \tag{5.242}$$

These particular values of Φ are very important since they relate to important canonical configurations, namely impedance interior and exterior right-angle wedges, the full-plane impedance junction and the impedance half-plane. The formulas (5.239) through (5.242) allow one to express solutions of associated canonical diffraction and scattering problems in a particularly simple way.

For general values of the wedge angle, ψ_Φ can be accurately calculated by combining numerical integration in (5.231) in a strip of width 4Φ or π with the extension rules (5.217) or (5.233) (Osipov and Stein 1999). When $\pi/4 \le \Phi \le \pi$, the cosine approximation (Osipov 2005)

$$\psi_\Phi(\alpha) \approx \left[\cos\left(\frac{\nu\alpha}{2+\nu}\right)\right]^{(2+\nu)/4} \tag{5.243}$$

can be particularly useful for analytical studies and rapid calculations. When combined with the extension rule (5.233), its maximum relative error does not exceed 6% on the whole complex α plane. A wide variety of different approaches to calculating the Maliuzhinets function is available, and the reader is referred to Osipov and Stein (1999) and Osipov (2005) for further references.

Spectral Amplitudes and their Properties

Now we are ready to write down the solution for the spectra. Introducing

$$\Psi_{\mathrm{e,h}}(\alpha) = \Psi^{+}_{\mathrm{e,h}}(\alpha)\Psi^{-}_{\mathrm{e,h}}(\alpha) \tag{5.244}$$

with

$$\Psi^{\pm}_{\mathrm{e,h}}(\alpha) = \psi_{\Phi}\left(\alpha \pm \Phi + \frac{\pi}{2} - \theta^{\pm}_{\mathrm{e,h}}\right)\psi_{\Phi}\left(\alpha \pm \Phi - \frac{\pi}{2} + \theta^{\pm}_{\mathrm{e,h}}\right), \tag{5.245}$$

we obtain the solution in the following form:

$$S_{\mathrm{e}}(\alpha) = E_{0z}\sigma(\alpha)\frac{\Psi_{\mathrm{e}}(\alpha)}{\Psi_{\mathrm{e}}(\varphi_0)} \tag{5.246}$$

$$S_{\mathrm{h}}(\alpha) = ZH_{0z}\sigma(\alpha)\frac{\Psi_{\mathrm{h}}(\alpha)}{\Psi_{\mathrm{h}}(\varphi_0)} \tag{5.247}$$

with $\sigma(\alpha)$ defined in (5.213), where multiplication of $\sigma(\alpha)$ with E_{0z} and ZH_{0z} recovers the pole singularities required by (5.172).

If one of the faces is PEC, for example $\eta_{-} \to 0$, then $\operatorname{Im}\theta^{-}_{\mathrm{e}} \to \infty$, $\theta^{-}_{\mathrm{h}} \to 0$ and it follows from (5.232) that

$$\frac{\Psi_{\mathrm{e}}(\alpha)}{\Psi_{\mathrm{e}}(\varphi_0)} \to \frac{\Psi^{+}_{\mathrm{e}}(\alpha)}{\Psi^{+}_{\mathrm{e}}(\varphi_0)} \tag{5.248}$$

and from (5.233) that

$$\frac{\Psi_{\mathrm{h}}(\alpha)}{\Psi_{\mathrm{h}}(\varphi_0)} \to \frac{\Psi^{+}_{\mathrm{h}}(\alpha)}{\Psi^{+}_{\mathrm{h}}(\varphi_0)}\frac{\cos[\nu(\alpha - \Phi)/2]}{\cos[\nu(\varphi_0 - \Phi)/2]}, \tag{5.249}$$

and the solutions in (5.246) and (5.247) modify accordingly. If, furthermore, $\eta_{+} \to 0$, so that both faces become PEC, then

$$\frac{\Psi_{\mathrm{e}}(\alpha)}{\Psi_{\mathrm{e}}(\varphi_0)} \to 1, \tag{5.250}$$

$$\frac{\Psi_{\mathrm{h}}(\alpha)}{\Psi_{\mathrm{h}}(\varphi_0)} \to \frac{\cos[\nu(\alpha + \Phi)/2]}{\cos[\nu(\varphi_0 + \Phi)/2]}\frac{\cos[\nu(\alpha - \Phi)/2]}{\cos[\nu(\varphi_0 - \Phi)/2]} = \frac{\cos(\nu\alpha)}{\cos(\nu\varphi_0)}, \tag{5.251}$$

and (5.246) and (5.247) together with (5.168) and (5.169) recover the solution for the PEC wedge (see (5.203) and (5.204)).

Let us now take a look at properties of the obtained solution. The spectra are meromorphic functions on the complex α plane as their only singularities are poles. There is an infinite number of poles on the complex plane, but those located in the region between the contours $\gamma_{\pm}$ and $\gamma(\pm\pi)$ (Figure 5.25) are particularly important as they must be accounted for in the analysis of the solution when $k\rho \to \infty$. The analysis will be performed later in a unified manner for all configurations for which an exact solution is available; here we just summarize information about the poles of the spectra (5.246) and (5.247).

The poles belong to two families: one associated with the poles of $\sigma(\alpha + \varphi)$ and another with the poles of $\Psi_{\mathrm{e,h}}(\alpha + \varphi)$. The poles of $\sigma(\alpha)$ are zeros of the equation $\sin(\nu\alpha) = \sin(\nu\varphi_0)$, so they can be represented in the form

$$\alpha_n = 2n\Phi + (-1)^n\varphi_0, \tag{5.252}$$

where $n = 0, \pm1, \pm2, \ldots$ Since Φ, φ, and φ_0 are real-valued parameters, the poles reside on the real axis in the complex α plane and those for which $-\pi < \alpha_n - \varphi < \pi$ are enclosed by the contours $\gamma_{\pm}$ and $\gamma(\pm\pi)$. The residues of $S_{\mathrm{e,h}}(\alpha + \varphi)$ at $\alpha = \alpha_n - \varphi$ are given by

$$\operatorname{res} S_{\mathrm{e,h}}(\alpha_n) = (-1)^n\frac{\Psi_{\mathrm{e,h}}[2n\Phi + (-1)^n\varphi_0]}{\Psi_{\mathrm{e,h}}(\varphi_0)}. \tag{5.253}$$

As we shall see in section 5.5, these quantities describe amplitudes of GO waves, incident and reflected from the wedge faces. Here we just note a relation of the ratio of the Ψ functions in (5.253) to reflection coefficients of the wedge faces. Equations (5.214) for Ψ can be written as

$$\frac{\Psi_{\mathrm{e,h}}(\alpha+\Phi)}{\Psi_{\mathrm{e,h}}(-\alpha+\Phi)} = -R^{+}_{\mathrm{e,h}}(\alpha) \tag{5.254}$$

$$\frac{\Psi_{\mathrm{e,h}}(-\alpha-\Phi)}{\Psi_{\mathrm{e,h}}(\alpha-\Phi)} = -R^{-}_{\mathrm{e,h}}(\alpha), \tag{5.255}$$

where

$$R^{\pm}_{\mathrm{e,h}}(\alpha) = \frac{\sin\alpha - \sin\theta^{\pm}_{\mathrm{e,h}}}{\sin\alpha + \sin\theta^{\pm}_{\mathrm{e,h}}}. \tag{5.256}$$

Recalling the definition (5.208) of $\theta^{\pm}_{\mathrm{e,h}}$ and comparing with (4.334) and (4.335), we see that $R^{\pm}_{\mathrm{e,h}}(\alpha)$ coincides with reflection coefficients $\tilde{R}_{\perp,\parallel}(\alpha-\pi/2)$ of an impedance plane with $\eta=\eta_{\pm}$ provided that α is understood as the angle between the incidence direction and the tangent to the impedance plane.

Relations (5.254) and (5.255) can be used as functional relations to simplify the ratio of the Ψ functions in (5.253) (Griesser and Balanis 1989; Osipov 1990; Zavadskaya and Starovoitova 1988). The analytical form of the result depends on whether n is odd or even. If $n=2n'+1$ with $n'=0,\pm1,\pm2,\ldots$, then

$$\frac{\Psi_{\mathrm{e,h}}[2n\Phi+(-1)^n\varphi_0]}{\Psi_{\mathrm{e,h}}(\varphi_0)} = \frac{\Psi_{\mathrm{e,h}}(4n'\Phi+2\Phi-\varphi_0)}{\Psi_{\mathrm{e,h}}(\varphi_0)}$$
$$= \begin{cases} -R^{+}_{\mathrm{e,h}}(\Phi-\varphi_0)\prod\limits_{i=1}^{n'} R^{-}_{\mathrm{e,h}}(4i\Phi-\Phi-\varphi_0)R^{+}_{\mathrm{e,h}}(4i\Phi+\Phi-\varphi_0), & n'\ge 0 \\ -R^{-}_{\mathrm{e,h}}(\Phi+\varphi_0)\prod\limits_{i=1}^{-n'-1} R^{+}_{\mathrm{e,h}}(4i\Phi-\Phi+\varphi_0)R^{-}_{\mathrm{e,h}}(4i\Phi+\Phi+\varphi_0), & n'\le -1 \end{cases}. \tag{5.257}$$

Otherwise, when $n=2n'$ with $n'=0,\pm1,\pm2,\ldots$,

$$\frac{\Psi_{\mathrm{e,h}}[2n\Phi+(-1)^n\varphi_0]}{\Psi_{\mathrm{e,h}}(\varphi_0)} = \frac{\Psi_{\mathrm{e,h}}(4n'\Phi+\varphi_0)}{\Psi_{\mathrm{e,h}}(\varphi_0)} = \begin{cases} \prod\limits_{i=1}^{n'} R^{+}_{\mathrm{e,h}}(4i\Phi-\Phi+\varphi_0)R^{-}_{\mathrm{e,h}}(4i\Phi-3\Phi+\varphi_0), & n'\ge 0 \\ \prod\limits_{i=1}^{-n'} R^{-}_{\mathrm{e,h}}(4i\Phi-\Phi-\varphi_0)R^{+}_{\mathrm{e,h}}(4i\Phi-3\Phi-\varphi_0), & n'\le -1 \end{cases}. \tag{5.258}$$

In (5.257) and (5.258) we use the convention $\prod_{i=1}^{0}=1$.

The poles of the second family are associated with the poles of $\Psi_{\mathrm{e,h}}(\alpha)$, which will be denoted by $\alpha^{\mathrm{e,h}}_n$. Because of (5.244) and (5.245) they belong to four branches, corresponding to the poles of every function $\psi_{\Phi}[\alpha\pm\Phi\pm(\pi/2-\theta^{\pm}_{\mathrm{e,h}})]$ in (5.245). According to (5.234) and (5.235), the Maliuzhinets function has an infinite double-periodic set of poles when its argument is equal to $\pm\alpha_{mn}$ with $m=2,4,6,\ldots$ and $n=1,2,3\ldots$. Since the loops of the integration contour γ in the Sommerfeld integral are entirely located within the strip $|\mathrm{Re}\,\alpha|\le 3\pi/2$ (Figure 5.25), only the poles with $m=2$ and satisfying equations

$$\alpha^{\mathrm{e,h}}_n+\Phi+\frac{\pi}{2}-\theta^{+}_{\mathrm{e,h}}=\alpha_{2n} \tag{5.259}$$

$$\alpha^{\mathrm{e,h}}_n+\Phi-\frac{\pi}{2}+\theta^{+}_{\mathrm{e,h}}=-\alpha_{2n} \tag{5.260}$$

$$\alpha^{\mathrm{e,h}}_n-\Phi+\frac{\pi}{2}-\theta^{-}_{\mathrm{e,h}}=\alpha_{2n} \tag{5.261}$$

$$\alpha^{\mathrm{e,h}}_n-\Phi-\frac{\pi}{2}+\theta^{-}_{\mathrm{e,h}}=-\alpha_{2n} \tag{5.262}$$

are relevant. When $\Phi > \pi/4$, only those defined by (5.259) and (5.262) with $n = 1$ can reside within the strip bounded by $\gamma(\pm\pi)$. These poles are given by

$$\alpha^+_{\mathrm{e,h}} = \Phi + \pi + \theta^+_{\mathrm{e,h}}, \quad \alpha^-_{\mathrm{e,h}} = -\Phi - \pi - \theta^-_{\mathrm{e,h}}, \tag{5.263}$$

and the corresponding residues of $S_{\mathrm{e,h}}(\alpha + \varphi)$ at $\alpha = \alpha^\pm_{\mathrm{e,h}} - \varphi$ are

$$\operatorname{res} S_{\mathrm{e,h}}(\alpha^\pm_{\mathrm{e,h}}) = \mp 2\tan\theta^\pm_{\mathrm{e,h}}\,\sigma(\pm\theta^\pm_{\mathrm{e,h}} \pm \pi \pm \Phi)\frac{\Psi_{\mathrm{e,h}}(\mp\theta^\pm_{\mathrm{e,h}} \mp \pi \pm \Phi)}{\Psi_{\mathrm{e,h}}(\varphi_0)}. \tag{5.264}$$

The residues describe amplitudes of the surface waves excited by the incident plane wave at the edge of the wedge (section 5.5). In narrow wedge-shaped domains ($\Phi < \pi/4$), solutions of all equations (5.259)–(5.262) with $n \geq 1$ must be taken into account as the associated poles may also be captured in the process of the high-frequency evaluation of the Sommerfeld integrals, leading to contributions that can be interpreted as surface waves mirrored in the wedge faces (Osipov 1990).

An important property of the spectra is that they are 2π periodic under some specific conditions. In such cases, according to (5.157), the solution simplifies to a sum of several plane waves. Let us derive the conditions of 2π periodicity by looking at the difference of the spectra, either (5.246) or (5.247), which, for brevity, can be written as

$$S(\alpha - \pi) - S(\alpha + \pi) = \frac{\nu\cos(\nu\varphi_0)}{\Psi(\varphi_0)}\left\{\frac{\Psi(\alpha - \pi)}{\sin[\nu(\alpha - \pi)] - \sin(\nu\varphi_0)} - \frac{\Psi(\alpha + \pi)}{\sin[\nu(\alpha + \pi)] - \sin(\nu\varphi_0)}\right\}. \tag{5.265}$$

The result for S_{e} (S_{h}) can then be obtained by multiplication with E_{0z} (ZH_{0z}) and replacing $\theta_\pm$ with $\theta^\pm_{\mathrm{e}}$ ($\theta^\pm_{\mathrm{h}}$). Using the definitions (5.244) and (5.245) for the Ψ function and the property (5.233) of the Maliuzhinets function, we write

$$\begin{aligned}\Psi(\alpha \pm \pi) = {} & \frac{\psi^8_\Phi(\pi/2)}{\Psi(\alpha)}\cos\left[\frac{\nu}{2}(\alpha + \Phi \pm \theta_+)\right]\cos\left[\frac{\nu}{2}(\alpha + \Phi \pm \pi \mp \theta_+)\right] \\ & \cdot\cos\left[\frac{\nu}{2}(\alpha - \Phi \pm \theta_-)\right]\cos\left[\frac{\nu}{2}(\alpha - \Phi \pm \pi \mp \theta_-)\right]\end{aligned} \tag{5.266}$$

and substituting this in (5.265), after some further algebra, arrive at

$$S(\alpha - \pi) - S(\alpha + \pi) = \frac{\nu}{4}\tan\left(\frac{\nu\pi}{2}\right)\psi^8_\Phi\left(\frac{\pi}{2}\right)\frac{\cos(\nu\alpha)}{\Psi(\alpha)}\frac{\cos(\nu\varphi_0)}{\Psi(\varphi_0)}[1 + \Lambda_+(\alpha)\Lambda_-(\alpha)] \tag{5.267}$$

with

$$\Lambda_\pm(\alpha) = \frac{\sin(\nu\alpha) + \sin(\nu\varphi_0) \mp 2\cos(\nu\pi/2)\cos[\nu(\theta_\pm - \pi/2)]}{\cos[\nu(\alpha \pm \varphi_0)] \pm \cos(\nu\pi)}. \tag{5.268}$$

These formulas are significant in their being symmetric with respect to α and φ_0; since the diffraction coefficient of the wedge is given by $S(\varphi - \pi) - S(\varphi + \pi)$ (see (5.411)–(5.414)), formulas (5.267) and (5.268) with $\alpha = \varphi$ explicitly show the expected symmetry of the diffraction coefficient. If $\theta_+ = \theta_- = \theta$, formulas (5.267) and (5.268) further simplify to

$$\begin{aligned}S(\alpha - \pi) - S(\alpha + \pi) = {} & \frac{\nu}{2}\sin(\nu\pi)\psi^8_\Phi\left(\frac{\pi}{2}\right)\frac{\cos(\nu\alpha)}{\Psi(\alpha)}\frac{\cos(\nu\varphi_0)}{\Psi(\varphi_0)} \\ & \frac{\sin(\nu\alpha)\sin(\nu\varphi_0) - \cos(\nu\theta)\cos[\nu(\pi - \theta)]}{\{\cos[\nu(\alpha + \varphi_0)] + \cos(\nu\pi)\}\{\cos[\nu(\alpha - \varphi_0)] - \cos(\nu\pi)\}}.\end{aligned} \tag{5.269}$$

It is now seen that regardless of the incidence direction φ_0, the spectra $S_{\mathrm{e,h}}(\alpha)$ are 2π periodic if $\Phi = \pi/(4M)$ ($\eta_+ \neq \eta_-$) or $\Phi = \pi/(2M)$ ($\eta_+ = \eta_-$) with $M = 1, 2, 3, \ldots$ These cases correspond to interior wedges, in which the fields are exactly equal to a sum of the incident wave and several reflected plane waves, that is, to a sum of images of an incident wave with respect to the wedge faces. See section 5.2.5 for similar solutions for PEC wedges.

A further group of interesting cases follows from (5.269) with $\varphi_0 = 0$, in which case $S(\alpha)$ is 2π periodic if the Brewster angle θ of the impedance faces, which, depending on polarization, is equal to either $\arcsin \eta$ or $\arcsin(1/\eta)$, and the wedge angle Φ are such that

$$\cos(\nu\theta)\cos[\nu(\pi - \theta)] = 0. \tag{5.270}$$

Solutions of the equation are given by $\theta = (2m-1)\Phi$ and/or $\theta = \pi - (2m-1)\Phi$ with m an integer and chosen so that $0 < \theta \le \pi/2$. For example, for wedges with $\pi/2 < \Phi \le \pi$, a unique solution is given by $\theta = \pi - \Phi$. In this case, the faces of the wedge are illuminated on their Brewster angle, which fully eliminates the scattered field in the respective polarization (Norris and Osipov 1998).

Fields at the Edge

The solutions for the impedance wedge in the form of Sommerfeld's integrals with the spectra (5.246) and (5.247) permit detailed analysis of their behavior for small $k\rho$, that is, near the edge of the wedge. One should remember, however, that applicability of the results is limited by the applicability conditions of the impedance boundary conditions (section 2.3.1). For lossy wedges, one can expect an adequate description of the fields at distances from the edge greater than or comparable with the skin depth in the material of the wedge (see (2.71) and Table 2.5).

According to (5.167), the limiting values of E_z and H_z at $\rho = 0$ are expressed through the limiting values of $S_{\mathrm{e,h}}(\alpha)$ at $\operatorname{Im}\alpha \to \pm\infty$. It follows from (5.232), (5.244), and (5.245) that

$$\Psi_{\mathrm{e,h}}(\alpha) \to \frac{1}{4}\psi_\Phi^4\left(\frac{\pi}{2}\right)\mathrm{e}^{\mp j\nu\alpha} \tag{5.271}$$

and since

$$\sigma(\alpha) \to \mp 2j\nu\cos(\nu\varphi_0)\mathrm{e}^{\pm j\nu\alpha}, \tag{5.272}$$

we obtain

$$E_z(0) = E_{0z}\frac{\nu\cos(\nu\varphi_0)}{\Psi_{\mathrm{e}}(\varphi_0)}\psi_\Phi^4\left(\frac{\pi}{2}\right) \tag{5.273}$$

$$H_z(0) = H_{0z}\frac{\nu\cos(\nu\varphi_0)}{\Psi_{\mathrm{h}}(\varphi_0)}\psi_\Phi^4\left(\frac{\pi}{2}\right). \tag{5.274}$$

The equations relate the edge values $E_z(0)$ and $H_z(0)$ of the z components of total fields to the amplitudes of the incidence wave E_{0z} and H_{0z} at the edge. The ratios $E_z(0)/E_{0z}$ and $H_z(0)/H_{0z}$, which will be referred to as edge factors, depend on the incidence angle φ_0 and the face impedances $\eta_\pm$. Figure 5.26 illustrates the dependence of the factor $H_z(0)/H_{0z}$ on the angular width of the wedge. The incidence direction along the bisecting direction ($\varphi_0 = 0$) and equal face impedances are assumed. The smaller the face impedance, the closer the curves approach the edge factor in the PEC limit, which is given by π/Φ. With $\eta > 1$, $H_z(0)/H_{0z}$ becomes small and tends to zero with increasing η. The figure can also be used to illustrate the ratio $E_z(0)/E_{0z}$ since the edge factors for E and H polarization are related by the substitution $\eta_\pm \to \eta_\pm^{-1}$ (see (5.208)).

In the vicinity of the edge, E_z and H_z can be expanded in power series with respect to $k\rho$, in which $E_z(0)$ and $H_z(0)$ serve as leading terms. Such expansions are important as they permit analysis of edge singularities in the ρ and φ components of the fields (see (5.146), (5.147), (5.148), and (5.149)). The expansions are obtained by expanding $S_{\mathrm{e,h}}(\alpha)$ with $\operatorname{Im}\alpha \to \infty$ in series of $\exp[j(p\nu + q)\alpha\,\mathrm{sgn}(\operatorname{Im}\alpha)]$ where $p, q = 0, 1, 2, \ldots$, followed by the use of the integral representation (E.38) and the small-argument approximation (5.42) for the Bessel function (Osipov 1991). Retaining the most important, leading terms with $p = 0, 1$ and $q = 0, 1$ gives

$$E_z(\rho,\varphi) = E_z(0)\{1 + c_{10}^{\mathrm{e}}(\varphi)(k\rho)^\nu + c_{01}^{\mathrm{e}}(\varphi)k\rho + O[(k\rho)^{2\delta}]\} \tag{5.275}$$

$$H_z(\rho,\varphi) = H_z(0)\{1 + c_{10}^{\mathrm{h}}(\varphi)(k\rho)^\nu + c_{01}^{\mathrm{h}}(\varphi)k\rho + O[(k\rho)^{2\delta}]\}, \tag{5.276}$$

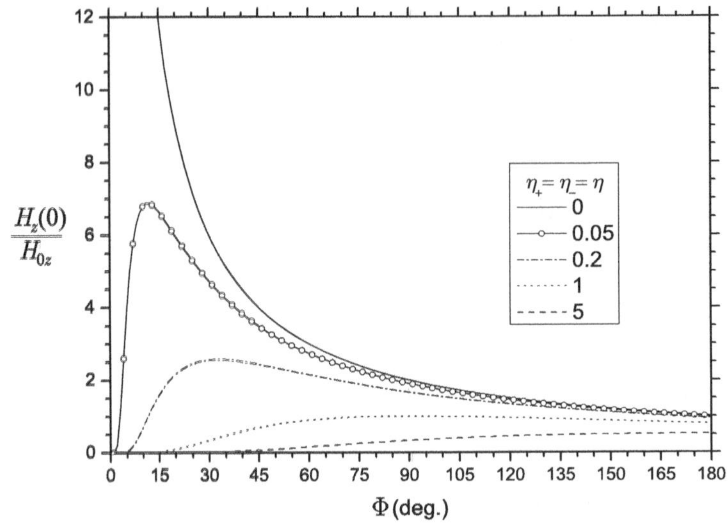

Figure 5.26 The edge factor for H polarization, equal face impedances, and $\varphi_0 = 0$.

where $\delta = \min(\nu, 1)$ and

$$c_{10}^{\mathrm{e,h}} = \left(\frac{j}{2}\right)^{\nu} \frac{\sin(\nu\varphi)}{\Gamma(\nu+1)} \left\{ 2\sin(\nu\varphi_0) - \frac{\cos\left[\nu(\theta_{\mathrm{e,h}}^{+} - \pi/2)\right] - \cos\left[\nu(\theta_{\mathrm{e,h}}^{-} - \pi/2)\right]}{\cos(\nu\pi/2)} \right\}$$

$$c_{01}^{\mathrm{e,h}} = \frac{j}{\sin(2\Phi)} \left[\sin\theta_{\mathrm{e,h}}^{+}\cos(\varphi+\Phi) + \sin\theta_{\mathrm{e,h}}^{-}\cos(\varphi-\Phi)\right]. \tag{5.277}$$

When one or both faces is/are PEC, corresponding representations for H_z are obtained from (5.274) and (5.276) by simply letting either θ_{h}^{+} or θ_{h}^{-} or both go to zero and using the formula

$$\Psi_{\mathrm{h}}^{\pm}(\varphi_0) \to \psi_{\Phi}^{2}\left(\frac{\pi}{2}\right)\cos\left[\frac{\nu}{2}(\varphi_0 \mp \Phi)\right], \tag{5.278}$$

the latter follows from (5.233) and (5.245). For example, with $\theta_{\mathrm{h}}^{\pm} \to 0$,

$$\Psi_{\mathrm{h}}(\varphi_0) \to \frac{1}{2}\psi_{\Phi}^{4}\left(\frac{\pi}{2}\right)\cos(\nu\varphi_0) \tag{5.279}$$

and equation (5.276) recovers the formula (5.47) for the PEC wedge.

The PEC limit for E_z is more complicated, however, since $E_z(0) \to 0$ and the expansion (5.275) does not apply anymore. Vanishing $E_z(0)$, which mathematically follows from the estimation $\Psi_{\mathrm{e}}(\varphi_0) \to \infty$ for $\mathrm{Im}\,\theta_{\mathrm{e}}^{\pm} \to \infty$, is a manifestation of the fact that the tangential components of the electric field vanish on every PEC surface. The correct result is obtained by expanding (5.246) with (5.248) or (5.250) for $\mathrm{Im}\,\alpha \to \infty$, which leads to an approximation of E_z different from (5.275). For example, in the case that both faces are PEC, the leading terms of the expansion are given by (5.43) and (5.46) (section 5.2.3). In the case that only one face is PEC, for example $\mathrm{Im}\,\theta_{\mathrm{e}}^{-} \to \infty$ and θ_{e}^{+} is finite, $S_{\mathrm{e}}(\alpha)$ expands in a series of $\exp\{j[p(\nu+1/2)+q]\alpha\,\mathrm{sgn}(\mathrm{Im}\,\alpha)\}$ with $p, q = 0, 1, 2, \ldots$ for $\mathrm{Im}\,\alpha \to \infty$, the main term of which leads to the approximation

$$E_z(\rho,\varphi) = E_z(0)\frac{4\psi_{\Phi}^{2}(\pi/2)\cos(\nu\varphi_0)}{\Gamma(\nu/2)\Psi_{\mathrm{e}}^{+}(\varphi_0)}\left(\frac{j}{2}k\rho\right)^{\nu/2}\sin\left[\frac{\nu}{2}(\Phi+\varphi)\right] + O[(k\rho)^{\delta+\nu/2}]. \tag{5.280}$$

The component E_z vanishes as $O[(k\rho)^{\nu/2}]$, implying that H_ρ and H_φ are $O[(k\rho)^{-1+\nu/2}]$ and singular at the edge if $\nu < 2$ or $\Phi > \pi/4$. This singularity is stronger than in the case of a wedge with both faces PEC or impedance, where H_ρ and H_φ are $O[(k\rho)^{\nu-1}]$, or in the case of H polarization, where E_ρ and E_φ are $O[(k\rho)^{\nu-1}]$. The stronger singularity observed in H_ρ and H_φ for wedges with one PEC face can be explained by a combined action of geometric (edge) and material (jump in the impedance value) singularities.

Even a flat surface (no geometric singularity) with a jump in the value of the surface impedance leads to a singularity in the ρ and φ components of the fields, but the singularity changes its character from power to logarithmic. Indeed, if $\Phi \to \pi/2$ and $\nu \to 1$, then coefficients $c_{10}^{\mathrm{e,h}}$ and $c_{01}^{\mathrm{e,h}}$ have pole singularities (see (5.277)), which cancel each other in (5.275) and (5.276), so that the correction term from the expansion

$$(k\rho)^{\nu-1} = 1 + (\nu - 1)\ln(k\rho) + \ldots \tag{5.281}$$

contributes to the limit, resulting in the appearance of logarithmic terms,

$$E_z(\rho, \varphi) = E_z(0)\left[1 + \frac{j}{\pi}\sin\varphi\left(\sin\theta_\mathrm{e}^+ - \sin\theta_\mathrm{e}^-\right) k\rho\ln(k\rho) + O(k\rho)\right] \tag{5.282}$$

$$H_z(\rho, \varphi) = H_z(0)\left[1 + \frac{j}{\pi}\sin\varphi\left(\sin\theta_\mathrm{h}^+ - \sin\theta_\mathrm{h}^-\right) k\rho\ln(k\rho) + O(k\rho)\right], \tag{5.283}$$

where $E_z(0)$ and $H_z(0)$ are taken at $\Phi = \pi/2$. Expansions (5.282) and (5.283) imply that the ρ and φ components of the electric and magnetic fields are $O[\ln(k\rho)]$ for $k\rho \to 0$. The logarithmic terms are present in expansions of the fields for other wedge angles as well, once Φ becomes a rational fraction of π, but only in the higher-order terms, and it is only for $\Phi = \pi/2$ that they determine the singularity of ρ and φ components at the edge (Osipov 1991).

5.4.3 Unit Surface Impedance

The problem of scattering by an impedance wedge permits an exact solution without limiting the incidence direction to the plane perpendicular to the edge if $\eta_\pm = 1$ (Bernard 1989; Lyalinov and Zhu 1999). The configuration is important since it models the interaction of electromagnetic waves with the edge in an impedance-matched boundary. The solution is derived in terms of a special function $\chi_\Phi(\alpha)$ introduced by Bobrovnikov and Fisanov (1972, 1988) in their studies of diffraction of normally incident plane waves by impedance wedges surrounded by anisotropic medium (gyrotropic plasma).

Functional Equations

By looking at (5.150) and (5.151) with $\eta_\pm = 1$, one may notice that combinations

$$u_1 = jE_z + ZH_z, \quad u_2 = E_z + jZH_z \tag{5.284}$$

satisfy the decoupled boundary conditions

$$\frac{1}{\rho}\frac{\partial u_1}{\partial\varphi} \pm jk\sin^2\beta u_1 = j\cos\beta\frac{\partial u_1}{\partial\rho} \tag{5.285}$$

$$\frac{1}{\rho}\frac{\partial u_2}{\partial\varphi} \pm jk\sin^2\beta u_2 = -j\cos\beta\frac{\partial u_2}{\partial\rho} \tag{5.286}$$

imposed at $\varphi = \pm\Phi$. Representing u_1 and u_2 by Sommerfeld's integrals of the form (5.168) and (5.169) with the spectra

$$S_1(\alpha) = jS_\mathrm{e}(\alpha) + S_\mathrm{h}(\alpha), \qquad S_2(\alpha) = S_\mathrm{e}(\alpha) + jS_\mathrm{h}(\alpha) \tag{5.287}$$

and inserting these integrals in (5.285) and (5.286) results in the functional equations which can be conveniently written as

$$\frac{S_1(\alpha \pm \Phi)}{S_1(-\alpha \pm \Phi)} = \frac{-\sin\alpha - j\cos\alpha\cos\beta \pm \sin\beta}{\sin\alpha - j\cos\alpha\cos\beta \pm \sin\beta} \tag{5.288}$$

$$\frac{S_2(\alpha \pm \Phi)}{S_2(-\alpha \pm \Phi)} = \frac{-\sin\alpha + j\cos\alpha\cos\beta \pm \sin\beta}{\sin\alpha + j\cos\alpha\cos\beta \pm \sin\beta}. \tag{5.289}$$

The equations can be also obtained from functional equations (5.190) by reformulating the latter in terms of S_1 and S_2. Provided that $\beta \neq \pi/2$, the equations are different from those studied in the case of normal incidence on an impedance wedge and cannot be solved in terms of the Maliuzhinets special function.

In solving these equations, let us follow, nevertheless, the approach used in the case of normal incidence on an impedance wedge. By representing

$$S_1(\alpha) = \Psi_1(\alpha)\sigma(\alpha)(jE_{0z} + ZH_{0z}), \quad S_2(\alpha) = \Psi_2(\alpha)\sigma(\alpha)(E_{0z} + jZH_{0z}), \tag{5.290}$$

where $\sigma(\alpha)$ is defined by (5.213) and ensures the required pole singularity of the spectra (see (5.172)), the problem is reduced to finding $\Psi_{1,2}(\alpha)$ that satisfy the same equations as $S_{1,2}(\alpha)$ but are holomorphic in the strip Π_0. Once $S_{1,2}(\alpha)$ are determined, the spectra of E_z and ZH_z follow from (5.287) as

$$S_e(\alpha) = \frac{1}{2}[-jS_1(\alpha) + S_2(\alpha)], \quad S_h(\alpha) = \frac{1}{2}[S_1(\alpha) - jS_2(\alpha)]. \tag{5.291}$$

Functions $\Psi_{1,2}(\alpha)$ can be expressed in terms of the special function $\chi_\Phi(\alpha)$ (Bobrovnikov and Fisanov 1972, 1988). Rewriting equations for $\Psi_{1,2}(\alpha)$ in the form

$$\frac{\Psi_1(\alpha \pm \Phi)}{\Psi_1(-\alpha \pm \Phi)} = \frac{-\sin\alpha\csc\beta - j\cos\alpha\cot\beta \pm 1}{\sin\alpha\csc\beta - j\cos\alpha\cot\beta \pm 1} \tag{5.292}$$

$$\frac{\Psi_2(\alpha \pm \Phi)}{\Psi_2(-\alpha \pm \Phi)} = \frac{-\sin\alpha\csc\beta + j\cos\alpha\cot\beta \pm 1}{\sin\alpha\csc\beta + j\cos\alpha\cot\beta \pm 1} \tag{5.293}$$

and introducing new parameters

$$\theta_1 = -\theta_2 = -j\ln\left(\cot\frac{\beta}{2}\right) \tag{5.294}$$

so that

$$\csc\beta = \cos\theta_{1,2}, \quad \cot\beta = j\sin\theta_1 = -j\sin\theta_2, \tag{5.295}$$

equations (5.292) and (5.293) are expressed as

$$\frac{\Psi_1(\alpha \pm \Phi)}{\Psi_1(-\alpha \pm \Phi)} = \frac{\sin(-\alpha + \theta_1) \pm 1}{\sin(\alpha + \theta_1) \pm 1} \tag{5.296}$$

$$\frac{\Psi_2(\alpha \pm \Phi)}{\Psi_2(-\alpha \pm \Phi)} = \frac{\sin(-\alpha + \theta_2) \pm 1}{\sin(\alpha + \theta_2) \pm 1}. \tag{5.297}$$

Equations (5.296) and (5.297) are particular cases of more general equations

$$\frac{\Psi(\alpha \pm \Phi)}{\Psi(-\alpha \pm \Phi)} = \frac{\sin(-\alpha + \theta) \pm \sin\theta_\pm}{\sin(\alpha + \theta) \pm \sin\theta_\pm} \tag{5.298}$$

encountered in the problem of diffraction of a normally incident plane wave by an impedance wedge in an anisotropic plasma where $\sin\theta_\pm$ describes the impedances of the wedge faces and θ is related to the degree of anisotropy of the plasma. The solution of (5.298), holomorphic in Π_0, can be expressed as

$$\Psi(\alpha) = \frac{\chi_\Phi(\alpha+\Phi+\pi-\theta-\theta_+)\chi_\Phi(\alpha+\Phi-\theta+\theta_+)}{\chi_\Phi(\alpha+\Phi-\pi+\theta+\theta_+)\chi_\Phi(\alpha+\Phi+\theta-\theta_+)} \cdot \frac{\chi_\Phi(\alpha-\Phi+\pi+\theta-\theta_-)\chi_\Phi(\alpha-\Phi+\theta+\theta_-)}{\chi_\Phi(\alpha-\Phi-\pi-\theta+\theta_-)\chi_\Phi(\alpha-\Phi-\theta-\theta_-)}, \tag{5.299}$$

where $\chi_\Phi(\alpha)$ is a new special function (Bobrovnikov and Fisanov 1972, 1988). Solutions of (5.296) and (5.297) are apparently given by (5.299) with $\theta_\pm = \pi/2$ and θ equal to either θ_1 or θ_2,

$$\Psi_1(\alpha) = \Psi(\alpha,\theta_1), \quad \Psi_2(\alpha) = \Psi(\alpha,\theta_2), \tag{5.300}$$

where

$$\Psi(\alpha,\theta) = \left[\frac{\chi_\Phi(\alpha+\Phi-\theta+\pi/2)\chi_\Phi(\alpha-\Phi+\theta+\pi/2)}{\chi_\Phi(\alpha+\Phi+\theta-\pi/2)\chi_\Phi(\alpha-\Phi-\theta-\pi/2)}\right]^2. \tag{5.301}$$

Function $\chi_\Phi(\alpha)$

$\chi_\Phi(\alpha)$ is a meromorphic function on the whole complex α plane with zeros and poles at $\alpha = \pi(2m+1)+2\Phi(2n+1)$ and $-\pi(2m+1)-2\Phi(2n+1)$, respectively $(m,n = 0,1,2\ldots)$. For arbitrary values of Φ it is represented by the formula (Avdeev 1994)

$$\chi_\Phi(\alpha) = \exp\left\{\frac{1}{2}\int_0^{+\infty} \frac{1}{s\sinh(\pi s)}\left[\frac{\alpha}{2\Phi} - \frac{\sinh(\alpha s)}{\sinh(2\Phi s)}\right] \mathrm{d}s\right\} \tag{5.302}$$

which is convergent when $|\mathrm{Re}\,\alpha| < \pi + 2\Phi$. When $\mathrm{Im}\,\alpha \to \pm\infty$,

$$\chi_\Phi(\alpha) = \exp\left[\mp\frac{j\alpha^2}{16\Phi} - \frac{\alpha\ln 2}{4\Phi} \pm \frac{j}{12}\left(\Phi + \frac{\pi^2}{4\Phi}\right) + o(1)\right] \tag{5.303}$$

(Avdeev 1994). It follows from these properties that $\Psi(\alpha)$ given by (5.299) has neither zeros nor poles in Π_0 when $0 < \mathrm{Re}\,\theta_\pm < \pi$ and $\mathrm{Re}\,\theta = 0$. When $\mathrm{Im}\,\alpha \to \pm\infty$, it behaves as

$$\Psi(\alpha) = \exp[\mp j\nu\alpha \pm j\theta - \nu\ln 4 + o(1)] \tag{5.304}$$

with $\nu = \pi/(2\Phi)$.

Further useful properties are as follows:

$$\chi_\Phi(-\alpha) = \frac{1}{\chi_\Phi(\alpha)} \tag{5.305}$$

$$\chi_\Phi(\alpha+2\Phi) = \chi_\Phi(\alpha-2\Phi)\cos\left(\frac{\alpha}{2}\right) \tag{5.306}$$

$$\chi_\Phi(\alpha+\pi) = 2^{1-\nu}\chi_\Phi(\alpha-\pi)\cos\left(\frac{\nu\alpha}{2}\right). \tag{5.307}$$

Functional relations (5.306) or (5.307) allow extension of $\chi_\Phi(\alpha)$ from the strip of convergence of the integral representation (5.302) into the whole complex α plane. In the particular case of $\Phi = \pi N/(2M)$ with $M,N = 1,2,\ldots$, the χ_Φ function can be represented by integrals over finite integration ranges,

$$\chi_\Phi(\alpha) = \exp\left[-\alpha\frac{M\ln 2}{2\pi N} + \frac{1}{\pi}\sum_{m=1}^{M}\sum_{n=1}^{N}\int_{a(m,M,n,N)}^{\alpha/(2N)+a(m,M,n,N)} \tau\cot\tau\,\mathrm{d}\tau\right] \tag{5.308}$$

with $a(m, M, n, N)$ defined in (5.238). The χ_Φ function relates to the Maliuzhinets function $\psi_\Phi(\alpha)$ according to

$$\psi_\Phi(\alpha) = \frac{1}{\chi_\Phi^2(\pi/2)} \frac{\chi_\Phi(\alpha + \pi/2)}{\chi_\Phi(\alpha - \pi/2)} \tag{5.309}$$

but in contrast to the latter never reduces to elementary functions.

Solution

Thus, we arrive at the solution

$$S_e(\alpha) = \frac{1}{2}\sigma(\alpha)\left[(E_{0z} - jZH_{0z})\frac{\Psi_1(\alpha)}{\Psi_1(\varphi_0)} + (E_{0z} + jZH_{0z})\frac{\Psi_2(\alpha)}{\Psi_2(\varphi_0)}\right] \tag{5.310}$$

$$S_h(\alpha) = \frac{1}{2}\sigma(\alpha)\left[(ZH_{0z} + jE_{0z})\frac{\Psi_1(\alpha)}{\Psi_1(\varphi_0)} + (ZH_{0z} - jE_{0z})\frac{\Psi_2(\alpha)}{\Psi_2(\varphi_0)}\right], \tag{5.311}$$

where $\Psi_{1,2}(\alpha)$ are defined in (5.300) and (5.301). In the limit of normal incidence ($\beta \to \pi/2$), equations (5.310) and (5.311) recover the Maliuzhinets solution (5.246) and (5.247) with $\eta_\pm = 1$ (problem 15).

The limiting values of E_z and H_z at the edge ($\rho = 0$) are obtained from the general formula (5.167). By letting $\operatorname{Im}\alpha \to \pm\infty$ in (5.310) and (5.311), using (5.272) and (5.295), and applying (5.304) to $\Psi_{1,2}(\alpha)$, we arrive at the formulas:

$$E_z(0) = 2\nu\cos(\nu\varphi_0)\csc\beta e^{-\nu\ln 4}\left[\frac{E_{0z} - jZH_{0z}}{\Psi_1(\varphi_0)} + \frac{E_{0z} + jZH_{0z}}{\Psi_2(\varphi_0)}\right] \tag{5.312}$$

$$ZH_z(0) = 2\nu\cos(\nu\varphi_0)\csc\beta e^{-\nu\ln 4}\left[\frac{ZH_{0z} + jE_{0z}}{\Psi_1(\varphi_0)} + \frac{ZH_{0z} - jE_{0z}}{\Psi_2(\varphi_0)}\right]. \tag{5.313}$$

In the limit of $\beta \to \pi/2$ these formulas agree with (5.274) because of the relation

$$\Psi(\varphi_0)\big|_{\theta_\pm=\pi/2} = \frac{1}{4}\psi_\Phi^4\left(\frac{\pi}{2}\right)e^{\nu\ln 4}\Psi(\varphi_0, 0) \tag{5.314}$$

between $\Psi(\varphi_0)$ defined in (5.215) and $\Psi(\varphi_0, 0)$ obtained from (5.301) with $\theta = 0$ (problem 16).

5.4.4 Further Exactly Solvable Cases

When no restrictions on the incidence direction and face impedances are imposed, exact solutions of functional equations (5.190) are still available for wedges with $\Phi = \pi$ (half-plane), $\pi/2$ (full-plane impedance junction), $3\pi/4$ (right-angle wedge with included angle $\pi/2$), and $\pi/4$ (right-angle wedge with included angle $3\pi/2$). In the latter two cases, however, one face of the wedge has to be PEC or PMC. Despite the fact that solutions of these problem have been derived by different authors and by partly different methods, it is nevertheless possible to describe the solutions in a unified manner (Rojas 1988b; Senior 1986; Senior and Volakis 1995). In this section we basically follow the presentation from Senior (1986) and Senior and Volakis (1995, section 4.3.1), extending the former to finite values of η_- and the latter to the cases of $\Phi = \pi/4$ and $3\pi/4$. In our analysis of wedges with $\Phi = \pi/4$ and $3\pi/4$ it is assumed that the lower face is PEC ($\eta_- = 0$). The case of a PMC face ($\eta_- = \infty$) is included through the duality relations: $\mathbf{E} \to \mathbf{H}$, $\mathbf{H} \to -\mathbf{E}$, $\eta_+ \to 1/\eta_+$, and $Z \to 1/Z$.

General Form of the Solutions

Regardless of the specific value of Φ, the spectra can be written in a unified form:

$$\mathbf{s}(\alpha) = \overline{\overline{R}}(\alpha) \cdot \mathbf{t}(\alpha), \tag{5.315}$$

where $\mathbf{s}(\alpha)$ is defined in (5.171),

$$\overline{\overline{R}}(\alpha) = \begin{bmatrix} r_1(\alpha) & r_2(\alpha) \\ r_2(\alpha) & -r_1(\alpha) \end{bmatrix} \tag{5.316}$$

$$r_1(\alpha) = \frac{\cos^2(\alpha - \Phi)}{1 - \sin^2\beta \sin^2(\alpha - \Phi)} \tag{5.317}$$

$$r_2(\alpha) = \cos\beta \frac{\cos(\alpha - \Phi)\sin(\alpha - \Phi)}{1 - \sin^2\beta \sin^2(\alpha - \Phi)} \tag{5.318}$$

$$\mathbf{t}(\alpha) = \begin{bmatrix} t_1(\alpha) \\ t_2(\alpha) \end{bmatrix}, \quad \begin{matrix} t_1(\alpha) = \Psi_{\mathrm{e}}(\alpha)\sigma_{\mathrm{e}}(\alpha) \\ t_2(\alpha) = \Psi_{\mathrm{h}}(\alpha)\sigma_{\mathrm{h}}(\alpha) \end{matrix}, \tag{5.319}$$

and $\sigma_{\mathrm{e,h}}(\alpha)$ are some solutions of equations (5.211) to be specified later.

Functions $\Psi_{\mathrm{e,h}}(\alpha)$ are given by (5.244) and (5.245) with

$$\sin\theta_{\mathrm{e}}^{\pm} = \frac{1}{\eta_\pm \sin\beta}, \quad \sin\theta_{\mathrm{h}}^{\pm} = \frac{\eta_\pm}{\sin\beta}, \tag{5.320}$$

where $0 \leq \operatorname{Re}\theta_{\mathrm{e,h}}^{\pm} \leq \pi/2$. They satisfy equations (5.214) which we rewrite for convenience as

$$\Psi_{\mathrm{e,h}}(-\alpha \pm \Phi) = \lambda_{\mathrm{e,h}}^{\pm}(\alpha)\Psi_{\mathrm{e,h}}(\alpha \pm \Phi) \tag{5.321}$$

with

$$\lambda_{\mathrm{e,h}}^{\pm}(\alpha) = \frac{\sin\alpha \pm \sin\theta_{\mathrm{e,h}}^{\pm}}{-\sin\alpha \pm \sin\theta_{\mathrm{e,h}}^{\pm}}. \tag{5.322}$$

In the case that $\eta_- = 0$, the Ψ functions are defined by

$$\Psi_{\mathrm{e}}(\alpha) = \Psi_{\mathrm{e}}^{+}(\alpha) \tag{5.323}$$

$$\Psi_{\mathrm{h}}(\alpha) = \Psi_{\mathrm{h}}^{+}(\alpha)\cos\left[\frac{\nu}{2}(\Phi - \alpha)\right] \tag{5.324}$$

with $\Psi_{\mathrm{e,h}}^{+}(\alpha)$ from (5.245) and $\nu = \pi/(2\Phi)$. They satisfy (5.321) with

$$\lambda_{\mathrm{e}}^{-}(\alpha) = 1, \quad \lambda_{\mathrm{h}}^{-}(\alpha) = -1. \tag{5.325}$$

Let us check by direct substitution that the spectra (5.315) are solutions of functional equations (5.190). Indeed, it follows from (5.211) and (5.321) that

$$\mathbf{s}(-\alpha \pm \Phi) = \overline{\overline{R}}(-\alpha \pm \Phi) \cdot \mathbf{t}(-\alpha \pm \Phi) = \overline{\overline{R}}(-\alpha \pm \Phi) \cdot \overline{\overline{L}}_{\pm}(\alpha) \cdot \mathbf{t}(\alpha \pm \Phi), \tag{5.326}$$

where

$$\overline{\overline{L}}_{\pm}(\alpha) = \begin{bmatrix} \lambda_{\mathrm{e}}^{\pm}(\alpha) & 0 \\ 0 & \lambda_{\mathrm{h}}^{\pm}(\alpha) \end{bmatrix}, \tag{5.327}$$

so that the difference between the right- and left-hand sides of (5.190) can be expressed as

$$\overline{\overline{M}}_{\pm}(\alpha) \cdot \mathbf{s}(\alpha \pm \Phi) - \overline{\overline{M}}_{\pm}(-\alpha) \cdot \mathbf{s}(-\alpha \pm \Phi) = \left[\overline{\overline{M}}_{\pm}(\alpha) \cdot \overline{\overline{R}}(\alpha \pm \Phi) - \overline{\overline{M}}_{\pm}(-\alpha) \cdot \overline{\overline{R}}(-\alpha \pm \Phi) \cdot \overline{\overline{L}}_{\pm}(\alpha)\right] \cdot \mathbf{t}(\alpha \pm \Phi), \tag{5.328}$$

where $\overline{\overline{M}}_{\pm}$ are given in (5.178)–(5.185). The matrix expression in the square brackets involves simple trigonometric matrices and can be explicitly evaluated. The definition of $\overline{\overline{R}}(\alpha)$ in (5.316), (5.317), and (5.318) ensures that

$$\overline{\overline{M}}_{+}(\alpha) \cdot \overline{\overline{R}}(\alpha + \Phi) = \overline{\overline{M}}_{+}(-\alpha) \cdot \overline{\overline{R}}(-\alpha + \Phi) \cdot \overline{\overline{L}}_{+}(\alpha) \tag{5.329}$$

with arbitrary values of Φ and η_+, so that the boundary conditions on the upper face of the wedge are always satisfied. The second identity

$$\overline{\overline{M}}_-(\alpha) \cdot \overline{\overline{R}}(\alpha - \Phi) = \overline{\overline{M}}_-(-\alpha) \cdot \overline{\overline{R}}(-\alpha - \Phi) \cdot \overline{\overline{L}}_-(\alpha), \tag{5.330}$$

which corresponds to the boundary conditions on the lower face of the wedge, is satisfied when $\Phi = \pi/2$ or π with arbitrary η_-. If $\eta_- = 0$, then (5.330) is true with $\Phi = \pi/4, \pi/2, 3\pi/2$ and π.

The role of functions $\sigma_{e,h}(\alpha)$ is to ensure the necessary analytical properties of the spectra, which includes the pole structure according to (5.172) (radiation conditions) and the behavior at infinity according to (5.174) (edge conditions). Suitable functions $\sigma_{e,h}(\alpha)$ can be constructed from particular solutions of functional equations (5.211). When $\operatorname{Im}\alpha \to \pm\infty$, $\Psi_{e,h}(\alpha) = O[\exp(v|\operatorname{Im}\alpha|)]$ with $v = \pi/(2\Phi)$ (see (5.271)), so $\sigma_{e,h}(\alpha)$ should decay as $O[\exp(-v|\operatorname{Im}\alpha|)]$ to ensure that $S_{e,h}(\alpha) = O(1)$. Such solutions of (5.211) have the form $[\sin(v\alpha) - \sin(v\tau)]^{-1}$ where the parameter τ defines the position of the pole singularity at $\alpha = \tau$. Since in the strip Π_0, defined by (5.173), $\Psi_{e,h}(\alpha)$ have neither zeros nor poles and $S_{e,h}(\alpha)$ are permitted to have a single pole singularity at $\alpha = \varphi_0$, the only acceptable values of τ are φ_0, $\alpha_1 = \Phi - \pi/2$, and $\alpha_2 = \alpha_1 - \pi$. The latter two values of τ do not lead to any singularity in the spectra because they coincide with zeros of $\cos(\alpha - \Phi)$ from coefficients $r_{1,2}(\alpha)$. Thus, the most general form of $\sigma_{e,h}(\alpha)$ is given by

$$\sigma_{e,h}(\alpha) = \frac{B_{e,h}}{\sin(v\alpha) - \sin(v\varphi_0)} + \frac{C_{e,h}}{\sin(v\alpha) - \sin(v\alpha_1)} + \frac{D_{e,h}}{\sin(v\alpha) - \sin(v\alpha_2)}, \tag{5.331}$$

where $B_{e,h}$, $C_{e,h}$, and $D_{e,h}$ are coefficients, independent of α, which are to be determined.

Coefficients $B_{e,h}$ in (5.331) result from (5.172) as

$$B_e = \frac{v\cos(v\varphi_0)}{\Psi_e(\varphi_0)}[E_{0z} + ZH_{0z}\cos\beta\tan(\varphi_0 - \Phi)] \tag{5.332}$$

$$B_h = \frac{v\cos(v\varphi_0)}{\Psi_h(\varphi_0)}[E_{0z}\cos\beta\tan(\varphi_0 - \Phi) - ZH_{0z}] \tag{5.333}$$

and ensure the correct residues of the spectra at the pole $\alpha = \varphi_0$. Coefficients $C_{e,h}$ and $D_{e,h}$ in (5.331) are chosen so as to compensate the poles of $r_{1,2}(\alpha)$ at zeros of the function $1 - \sin^2\beta\sin^2(\alpha - \Phi)$ in the denominator in (5.317) and (5.318). The zeros that may possibly reside in Π_0 are denoted by $\alpha = \omega_{1,2,3,4}$, where

$$\omega_1 = \Phi - 3\pi/2 - j\delta \tag{5.334}$$

$$\omega_2 = \Phi - 3\pi/2 + j\delta \tag{5.335}$$

$$\omega_3 = \Phi - \pi/2 - j\delta \tag{5.336}$$

$$\omega_4 = \Phi - \pi/2 + j\delta \tag{5.337}$$

with $\delta = \ln\tan(\beta/2)$. To comply with the radiation conditions (5.172), one has to require that

$$\operatorname{res} S_e(\omega_k) = 0, \quad \operatorname{res} S_h(\omega_k) = 0 \tag{5.338}$$

with $k = 1, 2, 3, 4$. A closer inspection shows, however, that for each value of k the two conditions in (5.338) are linearly dependent, which follows from the property

$$\frac{r_2(\omega_k)}{r_1(\omega_k)} = -\frac{r_1(\omega_k)}{r_2(\omega_k)}, \tag{5.339}$$

so that the eight conditions in (5.338) reduce to the four equations:

$$t_1(\omega_1) + jt_2(\omega_1) = 0 \tag{5.340}$$

$$t_1(\omega_2) - jt_2(\omega_2) = 0 \tag{5.341}$$

$$t_1(\omega_3) + jt_2(\omega_3) = 0 \tag{5.342}$$

$$t_1(\omega_4) - jt_2(\omega_4) = 0 \tag{5.343}$$

with $t_{1,2}(\alpha)$ defined in (5.319). Let us now specialize the solution to specific values of Φ.

The Half-plane

Here, $\Phi = \pi$ and the parameters from equations (5.331) and (5.340)–(5.343) are as follows: $\nu = 1/2$, $\alpha_1 = \pi/2$, $\alpha_2 = -\pi/2$, $\omega_1 = -\pi/2 - j\delta$, $\omega_2 = -\pi/2 + j\delta$, $\omega_3 = \pi/2 - j\delta$, and $\omega_4 = \pi/2 + j\delta$. The normalized face impedances $\eta_\pm$ are arbitrary, and the Maliuzhinets function $\psi_\Phi(\alpha)$ involved in $\Psi_{\rm e,h}$ is given by (5.242).

All ω_k from (5.334)–(5.337) lie inside Π_0, which is defined by $-\pi < \operatorname{Re}\alpha < \pi$ in this case, so all equations (5.340)–(5.343) have to be satisfied, for which we have four coefficients $C_{\rm e,h}$ and $D_{\rm e,h}$ in (5.331) at our disposal. Inserting (5.319) and (5.331) in (5.340)–(5.343) leads to four linearly independent inhomogeneous algebraic equations for four unknowns $C_{\rm e,h}$ and $D_{\rm e,h}$, from which $C_{\rm e,h}$ and $D_{\rm e,h}$ are uniquely and easily determined.

Consider the solution in the limit of the normal incidence. If $\beta \to \pi/2$, then $\delta \to 0, \omega_{1,2} \to \alpha_2, \omega_{3,4} \to \alpha_1$, and the system (5.340)–(5.343) tends to a homogeneous system with a trivial solution ($C_{\rm e,h}, D_{\rm e,h} \to 0$); the vanishing coefficients ensure compensation of the singular terms in (5.331) in this limit. Finally, by noting that $B_{\rm e} \to \nu\cos(\nu\varphi_0)E_{0z}/\Psi_{\rm e}(\varphi_0)$, $B_{\rm h} \to -\nu\cos(\nu\varphi_0)ZH_{0z}/\Psi_{\rm h}(\varphi_0)$, $r_1(\alpha) \to 1$, $r_2(\alpha) \to 0$, $\sin\theta_{\rm e}^\pm \to \eta_\pm^{-1}$, and $\sin\theta_{\rm h}^\pm \to \eta_\pm$, we arrive at the limiting expressions for $S_{\rm e,h}$ that exactly coincide with the Maliuzhinets solution (5.246) and (5.247) for the normal incidence.

The solution can be simplified by accounting for the specific values of the parameters Φ and ν. Rewriting $\sigma_{\rm e,h}$ as

$$\sigma_{\rm e,h}(\alpha) = \sec(\alpha - \Phi)\left[\frac{B_{\rm e,h}\cos(\alpha-\Phi)}{\sin(\nu\alpha) - \sin(\nu\varphi_0)} + \frac{C_{\rm e,h}\cos(\alpha-\Phi)}{\sin(\nu\alpha) - \sin(\nu\alpha_1)} + \frac{D_{\rm e,h}\cos(\alpha-\Phi)}{\sin(\nu\alpha) - \sin(\nu\alpha_2)}\right] \tag{5.344}$$

and evaluating the ratios in the square brackets as

$$\frac{\cos(\alpha-\Phi)}{\sin(\nu\alpha) - \sin(\nu\alpha_1)} = 2\sin\frac{\alpha}{2} + \sqrt{2} \tag{5.345}$$

$$\frac{\cos(\alpha-\Phi)}{\sin(\nu\alpha) - \sin(\nu\alpha_2)} = 2\sin\frac{\alpha}{2} - \sqrt{2} \tag{5.346}$$

$$\frac{\cos(\alpha-\Phi)}{\sin(\nu\alpha) - \sin(\nu\varphi_0)} = 2\sin\frac{\alpha}{2} + 2\sin\frac{\varphi_0}{2} - \frac{\cos\varphi_0}{\sin\frac{\alpha}{2} - \sin\frac{\varphi_0}{2}} \tag{5.347}$$

leads to the representation

$$\sigma_{\rm e,h}(\alpha) = \sec(\alpha-\Phi)\left(\frac{b_{\rm e,h}}{\sin\frac{\alpha}{2} - \sin\frac{\varphi_0}{2}} + c_{\rm e,h} + d_{\rm e,h}\sin\frac{\alpha}{2}\right), \tag{5.348}$$

where

$$b_{\rm e,h} = -B_{\rm e,h}\cos\varphi_0 \tag{5.349}$$

$$c_{\rm e,h} = 2B_{\rm e,h}\sin\frac{\varphi_0}{2} + \sqrt{2}(C_{\rm e,h} - D_{\rm e,h}) \tag{5.350}$$

$$d_{\rm e,h} = 2(B_{\rm e,h} + C_{\rm e,h} + D_{\rm e,h}). \tag{5.351}$$

This form of the solution is encountered in Bucci and Franceschetti (1976) and Vaccaro (1980).

The Wedge with Included Angle $\pi/2$

Here, $\Phi = 3\pi/4$ and $\eta_- = 0$. The Maliuzhinets function $\psi_\Phi(\alpha)$ is given by the simple formula (5.241). The other parameters are as follows: $\nu = 2/3$, $\alpha_1 = \pi/4$, $\alpha_2 = -3\pi/4$, $\omega_1 = -3\pi/4 - j\delta$, $\omega_2 = -3\pi/4 + j\delta$, $\omega_3 = \pi/4 - j\delta$, and $\omega_4 = \pi/4 + j\delta$. Despite the fact that the poles $\alpha = \omega_1$ and ω_2 lie on the boundary of the analyticity strip Π_0, they are still unphysical and must be excluded from the spectra.

In contrast to the half-plane case, function $[\sin(\nu\alpha) - \sin(\nu\alpha_2)]^{-1}$ in the right-hand side of (5.331) has a second-order pole when $\alpha \to \alpha_2$ since $\sin(\nu\alpha_2) = -1$. This implies that the corresponding term is not permitted because the singularity cannot be compensated by the factor $\cos(\alpha - \Phi)$ from $r_1(\alpha)$ and $r_2(\alpha)$. A more detailed analysis reveals, however, that such a term is allowed in $\sigma_{\rm h}(\alpha)$ because this latter function is always multiplied with $\Psi_{\rm h}(\alpha)$ which, when $\eta_- = 0$, includes the factor $\cos[(\alpha - \Phi)/3]$ vanishing at $\alpha = \alpha_2$ (see (5.324)). Thus, $D_{\rm e} = 0$ and the three remaining constants in (5.331) have to be chosen so as to satisfy four equations (5.340)–(5.343).

Fortunately, equations (5.340) and (5.341) turn out to be linearly dependent. This can be seen from the identities

$$t_1(\omega_2) = t_1(\omega_1), \quad t_2(\omega_2) = -t_2(\omega_1) \tag{5.352}$$

which result from functional equations (5.211) and (5.321) with (5.325). Thus, one has three linearly independent equations (5.342), (5.343), and, for example, (5.340) to determine the constants $C_{\rm e}$, $C_{\rm h}$, and $D_{\rm h}$. In the limit of $\beta \to \pi/2$, $C_{\rm e,h}$ and $D_{\rm h}$ vanish as expected, which recovers the Maliuzhinets solution for the normal incidence.

By using the values of Φ and ν, the spectra can be transformed to the form employed by Vaccaro (1981). Representing

$$\begin{aligned}\sigma_{\rm h}(\alpha) = \sec(\alpha - \Phi)\sec\frac{\alpha - \Phi}{3}\Bigg\{ & B_{\rm h}\frac{\cos(\alpha - \Phi)\cos[(\alpha - \Phi)/3]}{\sin(\nu\alpha) - \sin(\nu\varphi_0)} + C_{\rm h}\frac{\cos(\alpha - \Phi)\cos[(\alpha - \Phi)/3]}{\sin(\nu\alpha) - \sin(\nu\alpha_1)} \\ & + D_{\rm h}\frac{\cos(\alpha - \Phi)\cos[(\alpha - \Phi)/3]}{\sin(\nu\alpha) - \sin(\nu\alpha_2)}\Bigg\}\end{aligned} \tag{5.353}$$

and using the trigonometric identities

$$\frac{\cos(\alpha - \Phi)\cos[(\alpha - \Phi)/3]}{\sin(\nu\alpha) - \sin(\nu\alpha_1)} = \sin\frac{2\alpha}{3} + 1 \tag{5.354}$$

$$\frac{\cos(\alpha - \Phi)\cos[(\alpha - \Phi)/3]}{\sin(\nu\alpha) - \sin(\nu\alpha_2)} = \sin\frac{2\alpha}{3} - \frac{1}{2} \tag{5.355}$$

$$\frac{\cos(\alpha - \Phi)\cos[(\alpha - \Phi)/3]}{\sin(\nu\alpha) - \sin(\nu\varphi_0)} = \sin\frac{2\alpha}{3} + \sin\frac{2\varphi_0}{3} + \frac{1}{2} + \frac{\left(\sin\frac{2\varphi_0}{3} + 1\right)\left(\sin\frac{2\varphi_0}{3} - \frac{1}{2}\right)}{\sin\frac{2\alpha}{3} - \sin\frac{2\varphi_0}{3}} \tag{5.356}$$

transform $\sigma_{\rm h}(\alpha)$ into

$$\sigma_{\rm h}(\alpha) = \sec(\alpha - \Phi)\sec\frac{\alpha - \Phi}{3}\left(\frac{b_{\rm h}}{\sin\frac{2\alpha}{3} - \sin\frac{2\varphi_0}{3}} + c_{\rm h} + d_{\rm h}\sin\frac{2\alpha}{3}\right), \tag{5.357}$$

where

$$b_{\rm h} = B_{\rm h}\left(\sin\frac{2\varphi_0}{3} + 1\right)\left(\sin\frac{2\varphi_0}{3} - \frac{1}{2}\right) \tag{5.358}$$

$$c_{\rm h} = B_{\rm h}\left(\sin\frac{2\varphi_0}{3} + \frac{1}{2}\right) + C_{\rm h} - \frac{1}{2}D_{\rm h} \tag{5.359}$$

$$d_{\rm h} = B_{\rm h} + C_{\rm h} + D_{\rm h}. \tag{5.360}$$

Similarly, function $\sigma_e(\alpha)$ can be expressed as

$$\sigma_{\rm e}(\alpha) = \sec(\alpha - \Phi)\sec\frac{\alpha - \Phi}{3}\left(\frac{b_{\rm e}}{\sin\frac{2\alpha}{3} - \sin\frac{2\varphi_0}{3}} + c_{\rm e} + d_{\rm e}\sin\frac{2\alpha}{3}\right) \tag{5.361}$$

with

$$b_e = B_e\left(\sin\frac{2\varphi_0}{3}+1\right)\left(\sin\frac{2\varphi_0}{3}-\frac{1}{2}\right) \tag{5.362}$$

$$c_e = B_e\left(\sin\frac{2\varphi_0}{3}+\frac{1}{2}\right)+C_e \tag{5.363}$$

$$d_e = B_e + C_e. \tag{5.364}$$

The Full-plane Impedance Junction

With $\Phi = \pi/2$, the parameters are as follows: $\nu = 1$, $\alpha_1 = 0$, $\alpha_2 = -\pi$, $\omega_1 = -\pi - j\delta$, $\omega_2 = -\pi + j\delta$, $\omega_3 = -j\delta$, and $\omega_4 = j\delta$. The corresponding Maliuzhinets function is given by (5.240). Since $\sin(\nu\alpha_1) = \sin(\nu\alpha_2)$, the two last terms on the right-hand side of (5.331) are linearly dependent, implying that the representations of $\sigma_{e,h}$ have only two degrees of freedom. The terms proportional to $D_{e,h}$ are, therefore, unessential and can be excluded from the analysis by setting $D_{e,h} = 0$. The poles at $\alpha = \omega_1$ and ω_2 are outside the strip Π_0 and do not need to be compensated. Coefficients $C_{e,h}$ are therefore to be determined from equations (5.342) and (5.343), which are conditions of compensation of the poles at $\alpha = \omega_3$ and ω_4. In the limit of $\beta \to \pi/2$, $C_{e,h} \to 0$ and the Maliuzhinets solution for the normal incidence is recovered.

To reduce the spectra to the form employed in Vaccaro (1980), we write

$$\sigma_{e,h}(\alpha) = \sec(\alpha-\Phi)\left[\frac{B_{e,h}\cos(\alpha-\Phi)}{\sin(\nu\alpha)-\sin(\nu\varphi_0)} + \frac{C_{e,h}\cos(\alpha-\Phi)}{\sin(\nu\alpha)-\sin(\nu\alpha_1)}\right] \tag{5.365}$$

and, by using the identities

$$\frac{\cos(\alpha-\Phi)}{\sin(\nu\alpha)-\sin(\nu\alpha_1)} = 1 \tag{5.366}$$

$$\frac{\cos(\alpha-\Phi)}{\sin(\nu\alpha)-\sin(\nu\varphi_0)} = 1 + \frac{\sin\varphi_0}{\sin\alpha-\sin\varphi_0}, \tag{5.367}$$

arrive at

$$\sigma_{e,h}(\alpha) = \sec(\alpha-\Phi)\left(\frac{b_{e,h}}{\sin\alpha-\sin\varphi_0} + c_{e,h}\right), \tag{5.368}$$

where

$$b_{e,h} = B_{e,h}\sin\varphi_0 \tag{5.369}$$

$$c_{e,h} = B_{e,h} + C_{e,h}. \tag{5.370}$$

The Wedge with Included Angle $3\pi/2$

In this case, $\Phi = \pi/4$, $\eta_- = 0$, $\nu = 2$, $\alpha_1 = -\pi/4$, $\alpha_2 = -5\pi/4$, $\omega_1 = -5\pi/4 - j\delta$, $\omega_2 = -5\pi/4 + j\delta$, $\omega_3 = -\pi/4 - j\delta$, $\omega_4 = -\pi/4 + j\delta$, and $\psi_\Phi(\alpha)$ is given by (5.239). Since $\sin(\nu\alpha_1) = \sin(\nu\alpha_2)$, the terms with $D_{e,h}$ in (5.331) can be omitted. Furthermore, the coefficient C_e has to be set to zero since the function $\sin(\nu\alpha) - \sin(\nu\alpha_1)$ has a second-order zero at $\alpha = \alpha_1$, which cannot be compensated in $\sigma_e(\alpha)$. In $\sigma_h(\alpha)$, however, a similar term is permitted because of the presence of an additional compensating factor $\cos[\nu(\alpha-\Phi)/2]$ in $\Psi_h(\alpha)$ (see (5.324)).

So, we have one constant C_h to compensate the pole singularities of $S_{e,h}(\alpha)$ at $\alpha = \omega_3$ and ω_4 ($\alpha = \omega_{1,2}$ lie outside Π_0). The conditions of the compensation are given by equations (5.342) and (5.343). Similarly to (5.352), we note the relations

$$t_1(\omega_3) = t_1(\omega_4) \tag{5.371}$$

$$t_2(\omega_3) = -t_2(\omega_4) \tag{5.372}$$

to conclude that (5.342) and (5.343) are linearly dependent. Any of them can be used to determine C_h, and the result is

$$C_h = \frac{2\cot^2\beta}{\sin(2\omega_3) - \sin(2\varphi_0)} \left[B_h - jB_e \frac{\Psi_e(\omega_3)}{\Psi_h(\omega_3)} \right]. \tag{5.373}$$

It follows from (5.373) that $C_h = 0$ when $\beta = \pi/2$, which recovers the Maliuzhintes solution for the normal incidence.

Specializing $\sigma_e(\alpha)$ to the case $\Phi = \pi/4$ gives

$$\sigma_e(\alpha) = \frac{B_e}{\sin(2\alpha) - \sin(2\varphi_0)}. \tag{5.374}$$

As far as $\sigma_h(\alpha)$ is concerned, we rewrite it as

$$\sigma_h(\alpha) = \sec^2(\alpha - \Phi) \left[\frac{B_h \cos^2(\alpha - \Phi)}{\sin(\nu\alpha) - \sin(\nu\varphi_0)} + \frac{C_h \cos^2(\alpha - \Phi)}{\sin(\nu\alpha) - \sin(\nu\alpha_1)} \right] \tag{5.375}$$

and use the identity

$$2\cos^2(\alpha - \Phi) = 1 + \sin(2\alpha) \tag{5.376}$$

to obtain

$$\sigma_h(\alpha) = \sec^2(\alpha - \Phi) \left[\frac{b_h}{\sin(2\alpha) - \sin(2\varphi_0)} + c_h \right] \tag{5.377}$$

with

$$b_h = \frac{1}{2} B_h [\sin(2\varphi_0) + 1] \tag{5.378}$$

$$c_h = \frac{1}{2}(B_h + C_h). \tag{5.379}$$

It is evident from (5.239), (5.315), (5.317), (5.318), (5.323), (5.324), (5.374), and (5.377) that spectra $S_{e,h}(\alpha)$ are 2π periodic, so that the exact solution of the problem is equal to the sum of the incident wave and the reflected waves on single and double reflections (see (5.157)), that is, they can be obtained by simply reflecting the incident wave in the wedge faces (Senior 1986).

5.5 High-frequency Scattering from Impenetrable Wedges

Once the spectra are determined, the Sommerfeld integral (5.170) gives the solution for the total field in the entire space outside the wedge. In the high-frequency limit, when $k\rho \sin\beta \gg 1$, the integral can be evaluated asymptotically using the steepest descent method (Appendix C.3). The analysis of this section is applicable to all the solutions presented in section 5.4, as well as to solutions, exact or approximate, of any other problem for impenetrable wedges, provided that the solution is expressed as a Sommerfeld integral. The idea of the analysis consists of deforming the double-loop integration path γ onto SDP, which transforms the solution into a sum of residue contributions and an integral over the SDP (Maliuzhinets 1958a; Sommerfeld 1927).

We start with the residue components and their interpretation as GO and surface waves (section 5.5.1). Then, in section 5.5.2, the integral over SDP is asymptotically evaluated using $k\rho \sin\beta$ as a large parameter. The result describes the edge-diffracted field, which is a conical or cylindrical wave with an apparent focus at the edge of the wedge. The amplitude of the wave is determined by the *diffraction coefficient*. The diffraction coefficient is the key quantity that completely defines the high-frequency solution. We show how to derive the full high-frequency expansion for the edge-diffracted field from the diffraction coefficient and present a number of characteristic diagrams for scattering width of various wedges.

The diffraction coefficient is singular at shadow boundaries of GO waves. A technique for constructing asymptotic approximations valid at shadow boundaries, the so-called uniform approximations, is described in section 5.5.3. We conclude with a short outline of postulates of the geometrical and uniform theories of diffraction (GTD/UTD), followed by formulation of the results of the high-frequency analysis for impenetrable wedges in the framework of GTD/UTD (section 5.5.4).

5.5.1 GO Components and Surface Waves

The starting point of the analysis is the integral representation of the form (5.157), in which the contour $\gamma(\pi)$, shown in Figure 5.25, is SDP of the Sommerfeld integral. The path is described by the equation $\mathrm{Re}\,\alpha = \pi + \mathrm{gd}\,(\mathrm{Im}\,\alpha)$ where function gd(...) is defined in (C.53) and the direction of integration in $\gamma(\pi)$ corresponds to $\mathrm{Im}\,\alpha$ varying from ∞ to $-\infty$.

So, we write

$$\mathbf{u} = \mathbf{u}_{\mathrm{res}} + \mathbf{u}_{\mathrm{d}}, \tag{5.380}$$

where

$$\mathbf{u}_{\mathrm{res}} = \mathrm{e}^{-jkz\cos\beta} \sum_n \mathrm{res}\,\mathbf{s}(\alpha_n) \mathrm{e}^{jk\rho\sin\beta\cos(\alpha_n-\varphi)} \tag{5.381}$$

is the sum of residues at the poles $\alpha = \alpha_n - \varphi$ of $\mathbf{s}(\alpha + \varphi)$ located in the region bounded by $\gamma_\pm$ on its top and bottom and by $\gamma(\pm\pi)$ to its right and left (Figure 5.25). This region will be referred to as Π_{res}. The second term in (5.380) is given by the integral over the SDP,

$$\mathbf{u}_{\mathrm{d}} = \mathrm{e}^{-jkz\cos\beta} \frac{1}{2\pi j} \int_{\gamma(\pi)} [\mathbf{s}(\alpha + \varphi) - \mathbf{s}(\alpha + \varphi - 2\pi)] \mathrm{e}^{jk\rho\sin\beta\cos\alpha}\, d\alpha. \tag{5.382}$$

We will now subsequently write down the expressions for these components and describe their physical meaning, starting with the residue contributions. As shown in section 5.4.2, in the case of normal incidence the poles of the spectra can be classified in two branches, the poles associated with the incident and reflected waves (poles of $\sigma(\alpha)$, GO poles) and the poles related to material properties of the wedge faces (poles of $\Psi(\alpha)$, surface wave poles). As we shall now see, the pole structure remains the same in the case of oblique illumination of the wedge as well. According to this classification we subdivide the residue contributions into two groups:

$$\mathbf{u}_{\mathrm{res}} = \mathbf{u}_{\mathrm{go}} + \mathbf{u}_{\mathrm{sw}}. \tag{5.383}$$

The position of poles can be determined without really solving the functional equations (5.190), by simply using the extension formulas (5.192) and the analytical properties of the spectra in the basic strip Π_0, where $\mathbf{s}(\alpha)$ has a single pole at $\alpha = \varphi_0$ with the residue $\mathbf{U}_0 = (E_{0z}, ZH_{0z})^{\mathrm{T}}$ (see (5.172)).

For example, assuming that $\alpha \in \Pi_1$ (see (5.191)) and noting that the equation

$$\mathbf{s}(\alpha) = -\overline{\overline{R}}_+(\alpha - \Phi) \cdot \mathbf{s}(2\Phi - \alpha) \tag{5.384}$$

maps Π_0 onto Π_1, one concludes that singularities of $\mathbf{s}(\alpha)$ in Π_1 are due to the poles of either $\mathbf{s}(2\Phi - \alpha)$ at $2\Phi - \alpha = \varphi_0$ or $\overline{\overline{R}}_+(\alpha - \Phi)$ at $\alpha - \Phi = \chi_+$ where $L_+(\chi_+) = 0$ (see (5.195)). In the vicinities of these poles,

$$\mathbf{s}(\alpha) \approx \frac{\overline{\overline{R}}_+(\Phi - \varphi_0) \cdot \mathbf{U}_0}{\alpha - 2\Phi + \varphi_0} \tag{5.385}$$

and

$$\mathbf{s}(\alpha) \approx -\frac{\overline{\overline{Q}}_+(\chi_+) \cdot \mathbf{s}(\Phi - \chi_+)}{L_+'(\chi_+)(\alpha - \Phi - \chi_+)}, \tag{5.386}$$

where $L'_+(\chi_+)$ denotes the derivative of the function. Similarly, the second equation

$$\mathbf{s}(\alpha) = -\overline{\overline{R}}_-(-\alpha - \Phi) \cdot \mathbf{s}(-2\Phi - \alpha) \tag{5.387}$$

from (5.192) can be used to map Π_0 onto Π_{-1}, and we find that $\mathbf{s}(\alpha)$ has a pole in Π_{-1} when either $-2\Phi - \alpha = \varphi_0$ or $-\alpha - \Phi = \chi_-$ where $L_-(\chi_-) = 0$. Close to those poles,

$$\mathbf{s}(\alpha) \approx \frac{\overline{\overline{R}}_-(\Phi + \varphi_0) \cdot \mathbf{U}_0}{\alpha + 2\Phi + \varphi_0} \tag{5.388}$$

and

$$\mathbf{s}(\alpha) \approx \frac{\overline{\overline{Q}}_-(\chi_-) \cdot \mathbf{s}(-\Phi + \chi_-)}{L'_-(\chi_-)(\alpha + \Phi + \chi_-)}. \tag{5.389}$$

As easily seen from (5.195), $\chi_\pm$ can be equal to either $\pi + \theta_{\mathrm{e}}^\pm$ or $\pi + \theta_{\mathrm{h}}^\pm$, with $\theta_{\mathrm{e,h}}^\pm$ defined in (5.320).

Combining (5.384) and (5.387) one can determine the position of poles in farther strips on the complex α plane and, eventually, on the whole complex plane. For example, inserting (5.387) in (5.384) gives a relation

$$\mathbf{s}(\alpha) = \overline{\overline{R}}_+(\alpha - \Phi) \cdot \overline{\overline{R}}_-(\alpha - 3\Phi) \cdot \mathbf{s}(\alpha - 4\Phi) \tag{5.390}$$

which connects the spectra in Π_0 and Π_2, the latter being a strip defined by $3\Phi < \operatorname{Re}\alpha < 5\Phi$ and $-\infty < \operatorname{Im}\alpha < \infty$.

In the case of the GO poles, not only the position of the poles but also the residua at the poles can be explicitly determined (see (5.385) and (5.388)). The result, similar to (5.257) and (5.258) obtained in the case of normal incidence, is important as it permits a complete description of GO components in electromagnetic scattering by impenetrable wedges of arbitrary angles at oblique illumination. Summarizing the results of the extension procedure, one concludes that the poles of the GO branch are described by the formula $\alpha_n = 2n\Phi + (-1)^n\varphi_0$ with $n = 0, \pm 1, \pm 2, \ldots$, which coincides with (5.252), and the expressions for the corresponding residua depend on whether n is odd or even. If $n = 2n' + 1$ with $n' = 0, \pm 1, \pm 2, \ldots$, then

$$\operatorname{res}\mathbf{s}(\alpha_n) = \begin{cases} \overline{\overline{R}}_+(\Phi - \varphi_0) \cdot \prod_{i=1}^{n'} \overline{\overline{R}}_-(4i\Phi - \Phi - \varphi_0) \cdot \overline{\overline{R}}_+(4i\Phi + \Phi - \varphi_0) \cdot \mathbf{U}_0, & n' \geq 0 \\ \overline{\overline{R}}_-(\Phi + \varphi_0) \cdot \prod_{i=1}^{-n'-1} \overline{\overline{R}}_+(4i\Phi - \Phi + \varphi_0) \cdot \overline{\overline{R}}_-(4i\Phi + \Phi + \varphi_0) \cdot \mathbf{U}_0, & n' \leq -1 \end{cases}. \tag{5.391}$$

Otherwise, when $n = 2n'$ with $n' = 0, \pm 1, \pm 2, \ldots$,

$$\operatorname{res}\mathbf{s}(\alpha_n) = \begin{cases} \prod_{i=1}^{n'} \overline{\overline{R}}_+(4i\Phi - \Phi + \varphi_0) \cdot \overline{\overline{R}}_-(4i\Phi - 3\Phi + \varphi_0) \cdot \mathbf{U}_0, & n' \geq 0 \\ \prod_{i=1}^{-n'} \overline{\overline{R}}_-(4i\Phi - \Phi - \varphi_0) \cdot \overline{\overline{R}}_+(4i\Phi - 3\Phi - \varphi_0) \cdot \mathbf{U}_0, & n' \leq -1 \end{cases}. \tag{5.392}$$

In (5.391) and (5.392) we use the convention $\prod_{i=1}^{0} \ldots = \mathbf{U}_0$. In contrast to (5.257) and (5.258), these are matrix expressions, which is due to coupling between E_z and H_z on reflection from an impedance surface illuminated at an oblique angle.

Factors $\overline{\overline{R}}_\pm(\Phi \mp \varphi_0)$ can be expressed through reflection coefficients $R_\perp^\pm$ and $R_\parallel^\pm$ of the upper/lower faces (see (4.334) and (4.335)). For the incidence direction $\hat{\imath}$ defined in (5.137) and the unit normal vectors $\hat{n}_\pm = \hat{x}\sin\Phi \mp \hat{y}\cos\Phi$,

$$R_\perp^\pm = \frac{\eta_\pm \cos\theta_{\mathrm{i}}^\pm - 1}{\eta_\pm \cos\theta_{\mathrm{i}}^\pm + 1}, \quad R_\parallel^\pm = \frac{\cos\theta_{\mathrm{i}}^\pm - \eta_\pm}{\cos\theta_{\mathrm{i}}^\pm + \eta_\pm}, \tag{5.393}$$

where

$$\cos\theta_{\mathrm{i}}^\pm = -\hat{\imath} \cdot \hat{n}_\pm = \sin\beta \sin(\Phi \mp \varphi_0).$$

It can be shown by direct inspection that the representation

$$\overline{\overline{R}}_{\pm}(\Phi \mp \varphi_0) = \frac{1}{1 - \sin^2\beta \sin^2(\Phi \mp \varphi_0)} \begin{bmatrix} P^{\pm}_{\mathrm{ee}} & P^{\pm}_{\mathrm{eh}} \\ P^{\pm}_{\mathrm{he}} & P^{\pm}_{\mathrm{hh}} \end{bmatrix} \tag{5.394}$$

with

$$P^{\pm}_{\mathrm{ee}} = R^{\pm}_{\perp} \cos^2(\Phi \mp \varphi_0) - R^{\pm}_{\parallel} \cos^2\beta \sin^2(\Phi \mp \varphi_0) \tag{5.395}$$

$$P^{\pm}_{\mathrm{eh}} = \mp\left(R^{\pm}_{\perp} + R^{\pm}_{\parallel}\right) \cos\beta \cos(\Phi \mp \varphi_0) \sin(\Phi \mp \varphi_0) \tag{5.396}$$

$$P^{\pm}_{\mathrm{he}} = -P^{\pm}_{\mathrm{eh}} \tag{5.397}$$

$$P^{\pm}_{\mathrm{hh}} = R^{\pm}_{\parallel} \cos^2(\Phi \mp \varphi_0) - R^{\pm}_{\perp} \cos^2\beta \sin^2(\Phi \mp \varphi_0) \tag{5.398}$$

is equivalent to (5.194). This representation permits interpretation of the factors $\overline{\overline{R}}_{\pm}(\Phi \mp \varphi_0)$ as tensor reflection coefficients of the wedge faces. Indeed, in problem 9 of section 4.7 a tensor reflection coefficient $\overline{\overline{R}}_{\xi}(\hat{\imath}, \hat{n})$ of a planar surface with the unit normal $\hat{n}$ for a plane wave with the incidence direction $\hat{\imath}$ is constructed in terms of the ξ components of the electric and magnetic fields where $\hat{\xi}$ is not perpendicular to the plane of incidence. Specializing the general formula (4.359) to $\hat{\xi} = \hat{z}$, $\hat{\imath}$ from (5.137) and $\hat{n} = \hat{n}_{\pm}$ results in the important identities

$$\overline{\overline{R}}_{\pm}(\Phi \mp \varphi_0) = \overline{\overline{R}}_{z}(\hat{\imath}, \hat{n}_{\pm}). \tag{5.399}$$

This implies that the residue contributions (5.391) and (5.392) are expressed in terms of planar reflection coefficients of the respective impedance faces and exactly equal to the amplitudes of the GO waves.

Evaluating the contributions of the GO poles to the sum (5.381) gives

$$\mathbf{u}_{\mathrm{go}} = \sum_{n=-\infty}^{+\infty} \Theta(\pi - |\alpha_n - \varphi|) \operatorname{res} \mathbf{s}(\alpha_n) \mathrm{e}^{jk\rho \sin\beta \cos(\alpha_n - \varphi) - jkz\cos\beta}. \tag{5.400}$$

The sum has a limited number of terms because of the presence of the unit step function $\Theta(x)$. Only the poles α_n with $|\alpha_n - \varphi| < \pi$ belong to Π_{res} and contribute to the residue sum. In the important case that $\Phi > \pi/2$, $\mathbf{u}_{\mathrm{go}}$ consists of three terms with $n = 0, \pm 1$,

$$\mathbf{u}_{\mathrm{go}} = \mathbf{u}_0 + \mathbf{u}_1 + \mathbf{u}_{-1}, \tag{5.401}$$

where

$$\begin{aligned} \mathbf{u}_0 &= \Theta(\pi - |\varphi_0 - \varphi|) \mathbf{U}_0\, \mathrm{e}^{jk\rho \sin\beta \cos(\varphi_0 - \varphi) - jkz\cos\beta} \\ \mathbf{u}_1 &= \Theta(\pi - |2\Phi - \varphi_0 - \varphi|) \overline{\overline{R}}_{+}(\Phi - \varphi_0) \cdot \mathbf{U}_0\, \mathrm{e}^{jk\rho \sin\beta \cos(2\Phi - \varphi_0 - \varphi) - jkz\cos\beta} \\ \mathbf{u}_{-1} &= \Theta(\pi - |2\Phi + \varphi_0 + \varphi|) \overline{\overline{R}}_{-}(\Phi + \varphi_0) \cdot \mathbf{U}_0\, \mathrm{e}^{jk\rho \sin\beta \cos(2\Phi + \varphi_0 + \varphi) - jkz\cos\beta}. \end{aligned} \tag{5.402}$$

Obviously, the phase and amplitude factors correspond to those of the incident wave and the waves reflected from the upper and lower faces of the wedge. The unit step functions ensure vanishing of the waves in their respective shadow regions (see Figures 5.15 and 5.16).

In general, reflection matrices $\overline{\overline{R}}_{\pm}$ have finite off-diagonal elements, implying polarization coupling on reflection. However, in the limit of the perfectly conducting wedge ($R^{\pm}_{\perp} \to -1$, $R^{\pm}_{\parallel} \to 1$), it follows from (5.394) that

$$\overline{\overline{R}}_{\pm} = \begin{bmatrix} -1 & 0 \\ 0 & 1 \end{bmatrix} \tag{5.403}$$

and (5.401) with (5.402) recover formulas (5.84)–(5.86) for the GO part of the field scattered by a perfectly conducting wedge.

Contributions from the poles of reflection coefficients are collected in the term $\mathbf{u}_{sw}$. In the strips $\Pi_{\pm 1}$ these are the poles of $\mathbf{s}(\alpha)$ at $\alpha = \alpha^{\pm}_{e,h} = \mp(\Phi + \chi^{\pm}_{e,h})$ where $\chi^{\pm}_{e,h} = \pi + \theta^{\pm}_{e,h}$ (see (5.386) and (5.389)). In addition to the poles in $\Pi_{\pm 1}$, there is a multitude of further poles of this type in the neighboring strips. However, for wedges with $\Phi > \pi/2$ only four of these poles at $\alpha = \alpha^{\pm}_{e,h} - \varphi$ can reside in Π_{res} and contribute to the residue sum (5.381), which we write as

$$\mathbf{u}_{sw} = \mathbf{u}^{+}_{e} + \mathbf{u}^{+}_{h} + \mathbf{u}^{-}_{e} + \mathbf{u}^{-}_{h} \tag{5.404}$$

with

$$\mathbf{u}^{\pm}_{e,h} = \begin{cases} A^{\pm}_{e,h} e^{jk\rho \sin\beta \cos(\alpha^{\pm}_{e,h} - \varphi) - jkz\cos\beta}, & \alpha^{\pm}_{e,h} - \varphi \in \Pi_{res} \\ 0, & \alpha^{\pm}_{e,h} - \varphi \notin \Pi_{res} \end{cases} \tag{5.405}$$

and

$$A^{\pm}_{e,h} = \mp \frac{\overline{\overline{Q}}_{\pm}(\chi^{\pm}_{e,h}) \cdot \mathbf{s}(\pm\Phi \mp \chi^{\pm}_{e,h})}{L'_{\pm}(\chi^{\pm}_{e,h})}. \tag{5.406}$$

A necessary condition of capturing a pole at $\alpha^{\pm}_{e,h} - \varphi$ is that $\operatorname{Im}\theta^{\pm}_{e,h} > 0$ (see Figure 5.25). Because of the definition (5.320) of $\theta^{\pm}_{e,h}$, $\operatorname{Im}\theta^{\pm}_{e} < 0$ and $\operatorname{Im}\theta^{\pm}_{h} > 0$ when $\operatorname{Im}\eta_{\pm} > 0$ and $\operatorname{Im}\theta^{\pm}_{e} > 0$ and $\operatorname{Im}\theta^{\pm}_{h} < 0$ when $\operatorname{Im}\eta_{\pm} < 0$, implying that the poles $\alpha^{\pm}_{e} - \varphi$ and $\alpha^{\pm}_{h} - \varphi$ may never reside in Π_{res} simultaneously, so that either $\mathbf{u}^{\pm}_{e}$ or $\mathbf{u}^{\pm}_{h}$ or both are zero. It can be easily checked that $\mathbf{u}_{sw}$ describes surface waves, which are inhomogeneous waves concentrated near the faces of the wedge, propagating away from the edge (problem 17). In contrast to the GO poles, determining the residua at these poles requires finding the spectra, at least in the strip Π_0, that is, a complete solution of the problem. The surface waves do not contribute directly to the field scattered to the space surrounding the wedge (they may only take a part of power from the incident wave), so we do not investigate them any further. They can, however, be important in problems of radiation where the incident wave is a surface wave propagating along a face of the wedge; on hitting the edge, it generates reflected and transmitted surface waves and a scattered space wave. An interested reader is referred to Bobrovnikov and Fisanov (1988), Norris and Osipov (1998), and Rojas (1988b) for further references.

5.5.2 *Edge-diffracted Field, Diffraction Coefficient, and Scattering Widths*

Let us now take a closer look at the term $\mathbf{u}_d$ given by the integral over SDP, (5.382). Compared with the GO component, which is known in advance, and the surface wave component, which does not contribute to scattering in the exterior of the wedge, the term $\mathbf{u}_d$ plays a key role in the description of scattering by wedges, particularly in the regions of specular reflection, most notably in the shadow, but requires a full solution of the scattering problem. The good news is that once the spectra are known, the analysis of $\mathbf{u}_d$ is straightforward as the integral in (5.382) is ideally suited for the evaluation with the steepest descent method (Appendix C). When $k\rho\sin\beta \gg 1$, the main contribution to the integral comes from a small vicinity of the saddle point at $\alpha = \pi$, so factoring out the slowly varying factor gives

$$\mathbf{u}_d \approx e^{-jkz\cos\beta}[\mathbf{s}(\varphi + \pi) - \mathbf{s}(\varphi - \pi)]\frac{1}{2\pi j}\int_{\gamma(\pi)} e^{jk\rho\sin\beta\cos\alpha}\,d\alpha. \tag{5.407}$$

The integral is equal to $-\pi H^{(2)}_0(k\rho\sin\beta)$ (see (E.37)), and using the large-argument approximation (E.25) of the Hankel function we arrive at the expression

$$\mathbf{u}_d \approx [\mathbf{s}(\varphi - \pi) - \mathbf{s}(\varphi + \pi)]\frac{e^{-jk\rho\sin\beta - jkz\cos\beta - j\pi/4}}{\sqrt{2\pi k\rho\sin\beta}}. \tag{5.408}$$

The formula describes a conical (cylindrical when $\beta = \pi/2$, see section 2.5.1) wave, omnipresent in space around the wedge, with an apparent focal line at the edge of the wedge. The contribution is therefore referred to as the edge-diffracted wave. In section 5.2.6 the edge-diffracted component has already been extracted from the solution for a PEC wedge.

The edge-diffracted field must be a linear function of the amplitudes of the incident wave at the edge, so (5.408) can be expressed as

$$\mathbf{u}_{\mathrm{d}} \approx \frac{\mathrm{e}^{-jk\rho\sin\beta - jkz\cos\beta - j\pi/4}}{\sqrt{2\pi k\rho\sin\beta}}\overline{\overline{D}}\cdot\mathbf{U}_0, \tag{5.409}$$

where $\overline{\overline{D}}$ is the diffraction coefficient matrix

$$\overline{\overline{D}} = \begin{bmatrix} D_{\mathrm{ee}} & D_{\mathrm{eh}} \\ D_{\mathrm{he}} & D_{\mathrm{hh}} \end{bmatrix} \tag{5.410}$$

which relates the amplitude of the scattered wave to the amplitude of the incident wave at the point $\rho = 0$, $z = 0$. The relation (5.409) has the form of the scattering relation (3.179) for cylindrical bodies with a finite diameter $D_\perp$ of the geometric cross section. It should be pointed out, however, that (3.179) describes the complete scattered field in the far-field region, that is, at distances greater than $D_\perp^2/\lambda$, whereas (5.409) describes only a part of the scattered field as there are other contributions, namely the GO waves, that contribute to the scattered field and the far-field region cannot be defined for infinite wedges. However, aside from the directions of propagation of the GO waves, the field scattered by an infinite wedge has the form of a cylindrical wave and can therefore be described by various scattering widths directly related by (3.183) to the elements of $\overline{\overline{D}}$.

Since $S_{\mathrm{e,h}}(\alpha)$ are linear functions of E_{0z} and ZH_{0z}, the elements of $\overline{\overline{D}}$ can be expressed through the spectra as

$$D_{\mathrm{ee}}(\varphi) = S_{\mathrm{e}}(\varphi - \pi) - S_{\mathrm{e}}(\varphi + \pi) \tag{5.411}$$

$$D_{\mathrm{he}}(\varphi) = S_{\mathrm{h}}(\varphi - \pi) - S_{\mathrm{h}}(\varphi + \pi) \tag{5.412}$$

with $E_{0z} = 1$ and $ZH_{0z} = 0$ and

$$D_{\mathrm{eh}}(\varphi) = S_{\mathrm{e}}(\varphi - \pi) - S_{\mathrm{e}}(\varphi + \pi) \tag{5.413}$$

$$D_{\mathrm{hh}}(\varphi) = S_{\mathrm{h}}(\varphi - \pi) - S_{\mathrm{h}}(\varphi + \pi) \tag{5.414}$$

with $E_{0z} = 0$ and $ZH_{0z} = 1$.

Since $S_{\mathrm{e,h}}(\alpha)$ have poles on the real axis in the complex α plane, which are the GO poles α_n given by (5.252), the diffraction coefficient must also be singular at corresponding values of φ, namely when $\varphi = \tilde{\varphi}_n$ with $\tilde{\varphi}_n$ being equal to either $\alpha_n + \pi$ or $\alpha_n - \pi$ so that $|\tilde{\varphi}_n| \leq \Phi$. These values of φ correspond to the position of the shadow boundaries of the respective GO waves (see the argument of the unit step function in the GO solution (5.400)). The presence of pole singularities at the shadow boundaries of GO fields is a fundamental property of the diffraction coefficients of wedges. A closer inspection shows that when $\varphi \to \tilde{\varphi}_n$,

$$\overline{\overline{D}}(\varphi)\cdot\mathbf{U}_0 \approx \sigma_n \frac{\operatorname{res}\mathbf{s}(\alpha_n)}{|\varphi - \tilde{\varphi}_n|}, \tag{5.415}$$

where $\sigma_n = 1$ on the shadowed side of the shadow boundary and $\sigma_n = -1$ on the lit side of the shadow boundary. For example, for wedges with $\pi/2 < \Phi \leq \pi$,

$$\overline{\overline{D}}(\varphi) \approx \frac{\pm\overline{\overline{I}}}{\varphi - \varphi_0 \mp \pi} \tag{5.416}$$

close to the shadow boundaries associated with the incident wave, and

$$\overline{\overline{D}}(\varphi) \approx \mp\frac{\overline{\overline{R}}_{\pm}(\Phi \mp \varphi_0)}{\varphi + \varphi_0 \mp 2\Phi \pm \pi} \tag{5.417}$$

close to the shadow boundaries associated with the reflected waves.

Diffraction coefficients of impenetrable wedges have a number of symmetry properties. For example, reciprocity requirements (section 2.4.2), applied to wedges with isotropic face impedances, lead to the following symmetry relations (Osipov and Senior 2008a):

$$D_{\mathrm{ee,hh}}(\varphi, \varphi_0, \beta) = D_{\mathrm{ee,hh}}(\varphi_0, \varphi, \beta) \tag{5.418}$$

$$D_{\mathrm{he,eh}}(\varphi, \varphi_0, \beta) = D_{\mathrm{eh,he}}(\varphi_0, \varphi, \beta) \tag{5.419}$$

and

$$D_{\mathrm{ee,hh}}(\varphi, \varphi_0, \beta) = D_{\mathrm{ee,hh}}(\varphi, \varphi_0, \pi - \beta) \tag{5.420}$$

$$D_{\mathrm{he,eh}}(\varphi, \varphi_0, \beta) = -D_{\mathrm{he,eh}}(\varphi, \varphi_0, \pi - \beta). \tag{5.421}$$

A consequence of (5.419) is the equal cross-polar elements $D_{\mathrm{eh}}(\varphi, \varphi_0, \beta)$ and $D_{\mathrm{he}}(\varphi, \varphi_0, \beta)$ when $\varphi = \varphi_0$. Furthermore, because of (5.421) the cross-polar elements vanish when $\beta = \pi/2$, regardless of φ and φ_0. For example, the Maliuzhinets solution described in section 5.4.2 is in agreement with these properties: there are no cross-polar components ($S_{\mathrm{e}} = 0$ when $E_{0z} = 0$ and $S_{\mathrm{h}} = 0$ when $H_{0z} = 0$, see (5.246), (5.247), (5.412), and (5.413)) and the differences $S_{\mathrm{e,h}}(\varphi - \pi) - S_{\mathrm{e,h}}(\varphi + \pi)$ are explicitly symmetric with respect to φ and φ_0 (see (5.267) and (5.268)).

Inverting the values of the normalized face impedances ($\eta_+ \to 1/\eta_+, \eta_- \to 1/\eta_-$) and replacing β with $\pi - \beta$ transform the boundary conditions (5.150) into (5.151) and vice versa, thus leading to a problem which is equivalent to the original one with interchanged E_z and ZH_z. For the elements of $\overline{\overline{D}}$ this property, together with (5.420) and (5.421), means that the impedance inversion results in the following relations:

$$D_{\mathrm{ee}}(\varphi, \varphi_0, \beta) \to D_{\mathrm{hh}}(\varphi, \varphi_0, \beta), \quad D_{\mathrm{hh}}(\varphi, \varphi_0, \beta) \to D_{\mathrm{ee}}(\varphi, \varphi_0, \beta) \tag{5.422}$$

$$D_{\mathrm{eh}}(\varphi, \varphi_0, \beta) \to -D_{\mathrm{he}}(\varphi, \varphi_0, \beta), \quad D_{\mathrm{he}}(\varphi, \varphi_0, \beta) \to -D_{\mathrm{eh}}(\varphi, \varphi_0, \beta). \tag{5.423}$$

Formulas (5.408) and (5.409) correspond to the leading term in the asymptotic expansion of the integral in (5.382) in inverse powers of $k\rho \sin\beta$,

$$\mathbf{u}_{\mathrm{d}} = \frac{\mathrm{e}^{-jk\rho \sin\beta - jkz\cos\beta - j\pi/4}}{\sqrt{2\pi k\rho \sin\beta}} \sum_{n=0}^{+\infty} \frac{\mathbf{a}_n(\varphi)}{(k\rho \sin\beta)^n} \tag{5.424}$$

with

$$\mathbf{a}_0(\varphi) = \overline{\overline{D}} \cdot \mathbf{U}_0. \tag{5.425}$$

In most applications the leading term is sufficient, but there are cases in which higher-order terms are necessary. For example, diffraction coefficients of finitely conducting wedges vanish at the directions tangent to the wedge faces,

$$\overline{\overline{D}}(\pm\Phi) = \overline{\overline{0}}. \tag{5.426}$$

The statement (5.426) is sometimes referred to as the Karp–Karal lemma (Collin and Zucker 1969, p. 304; Rawlins 2005). Thus, it is the higher-order term with $n = 1$ in (5.425) that describes the edge-diffracted field on the face of an imperfectly reflecting wedge (Osipov *et al.* 2002).

The higher-order expansion coefficients can be easily obtained by noting that $\mathbf{u}_{\mathrm{d}}$ must satisfy the Helmholtz equation, (5.141) or (5.142), since the residue terms $\mathbf{u}_{\mathrm{res}}$ are solutions of the Helmholtz equation. By inserting (5.424) into the Helmholtz equation and equating terms with like powers of ρ, one can obtain the relation[2]

$$\mathbf{a}_n = \frac{j}{2n}\left[\frac{\partial^2}{\partial\varphi^2} + \left(n - \frac{1}{2}\right)^2\right]\mathbf{a}_{n-1} \tag{5.427}$$

which determines all $\mathbf{a}_n(\varphi)$ with $n \geq 1$ from the zero-order term $\mathbf{a}_0(\varphi)$, that is, from the diffraction coefficient.

[2] The full expansion can be derived also in the framework of the steepest descent method, see (C.22).

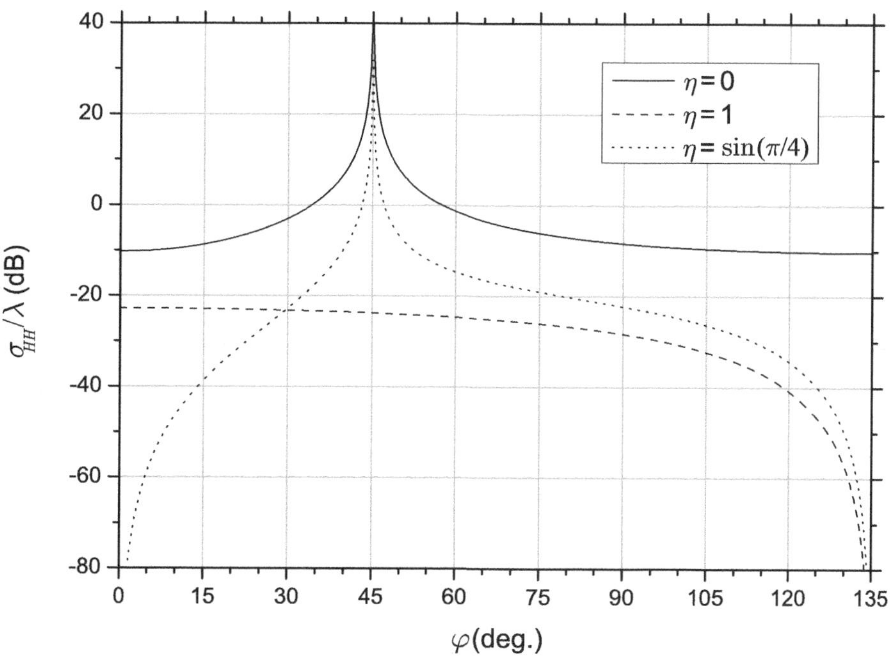

Figure 5.27 Monostatic scattering width of a right-angled wedge for several face impedances.

Let us take a look at several characteristic diagrams for the fields scattered by variously reflecting wedges. In all cases it will be assumed that $\varphi = \varphi_0$ (the observation point lies in the plane defined by the incidence direction and the edge of the wedge), so the scattered field is entirely formed by the edge-diffracted field and the scattering widths can be calculated from the elements of the diffraction coefficient $\overline{\overline{D}}$ according to (3.183). Figures 5.27 and 5.28 illustrate backscattering in the plane perpendicular ($\beta = \pi/2$) to the edge of a right-angled wedge ($\Phi = 3\pi/4$) for several values of the face impedance, equal on both faces ($\eta = \eta_+ = \eta_-$). Because of the configuration symmetry, the curves are symmetric with respect to $\varphi = 0$, so only the range $0 \leq \varphi \leq \Phi$ is shown. The solid curves correspond to the PEC wedge and, as expected, σ_{EE} vanishes as $\varphi \to \Phi$, whereas σ_{HH} tends to a finite limit. The dashed curves correspond to the impedance-matched ($\eta = 1$) wedge, in which case $\sigma_{EE} = \sigma_{HH}$. In contrast to other values of η, there are no singularities in the scattering widths at the direction of specular reflection ($\varphi = 45°$), which is explained by vanishing reflection from the impedance-matched plane under broadside illumination. The dotted curve in Figure 5.27 is calculated for η that satisfies the condition (5.270), and that is why σ_{HH} vanishes at $\varphi = 0$. The width σ_{EE} has a zero at $\varphi \approx 79.4°$, which is a solution of the equation

$$\sin^2(\nu\varphi) = \cos(\nu\theta_e)\cos[\nu(\pi - \theta_e)], \tag{5.428}$$

the latter being the condition under which diffraction coefficients of wedges with equal face impedance vanish (see (5.269)). Because of (5.422), the curves for σ_{HH} and σ_{EE} with $\eta = 0, 1$, and $1/\sqrt{2}$ in Figures 5.27 and 5.28 describe at the same time σ_{EE} and σ_{HH} for $\eta = \infty, 1$, and $\sqrt{2}$.

Figure 5.29 illustrates the behavior of σ_{EE} for PEC ($\eta_\pm = 0$), PMC ($\eta_\pm = \infty$), and impedance-matched ($\eta_\pm = 1$) half-planes ($\Phi = \pi$) and right-angled wedges ($\Phi = 3\pi/4$) when $\varphi = \varphi_0 = 0$ and β varies from 0 to 90°. These are bistatic configurations, in which the directions of incidence and scattering lie in the plane bisecting the space outside the wedge making the same angle β with the edge at the origin of the coordinate system. The scattered field is nevertheless fully determined by the edge-diffracted field. Only σ_{EE} is plotted since $\sigma_{EH} = \sigma_{HE} = 0$ and the results for σ_{HH} are equal to those for σ_{EE} with $\eta \to 1/\eta$. Since σ_{EE} is a symmetric function of β with respect to $\beta = \pi/2$, it is sufficient to plot the

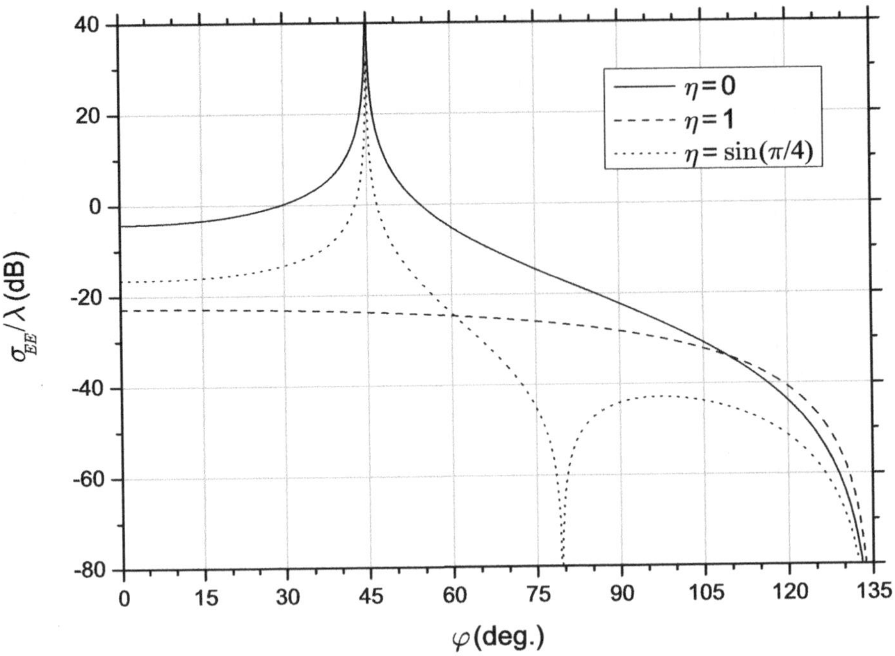

Figure 5.28 The same as Figure 5.27 but for the alternative polarization.

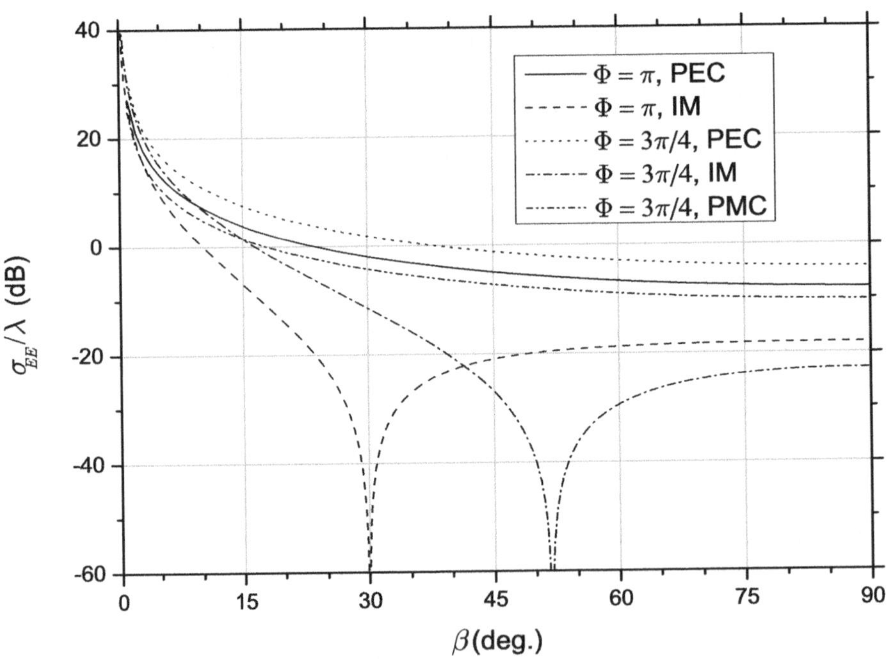

Figure 5.29 Normalized scattering width of a half-plane and a right-angled wedge with PEC, PMC, and impedance-matched (IM) boundary conditions as a function of β when $\varphi=\varphi_0=0$.

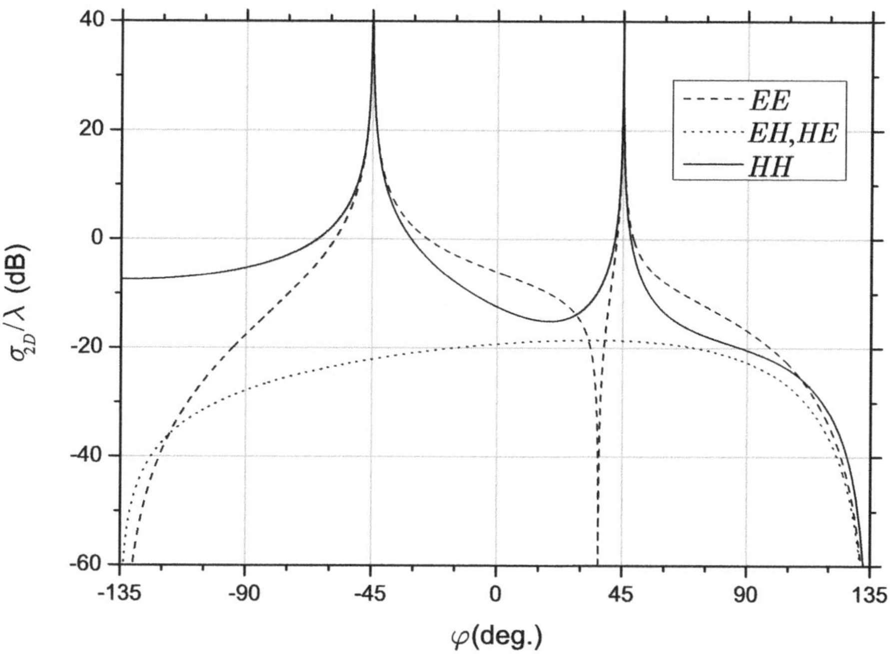

Figure 5.30 Normalized scattering width of a right-angled wedge with $\eta_+ = 1$ and $\eta_- = 0$ as a function of φ when $\varphi = \varphi_0$ and $\beta = \pi/4$.

curves over the range $0 < \beta \leq \pi/2$. The increasing values of σ_{EE} for $\beta \to 0$ are due to the factor $\csc\beta$ in the definition (3.183) of the scattering width. The permissible values of $\sin\beta$ are, however, bounded since asymptotic formulas are only applicable as long as $k\rho\sin\beta \gg 1$. The curve for a PMC half-plane is not shown in Figure 5.29 as $\sigma_{EE} = 0$ in this case. In all other cases, the level of scattering from impedance-matched configurations is typically lower than from PEC and PMC configurations. In the range $40° < \beta \leq 90°$ the impedance-matched right-angled wedge scatters less than the impedance-matched half-plane.

Scattering widths of a right-angled wedge with mixed boundary conditions ($\eta_+ = 1, \eta_- = 0$) as a function of φ when $\varphi = \varphi_0$ and $\beta = \pi/4$ are shown in Figure 5.30. Since $\varphi = \varphi_0$ and $\beta \neq \pi/2$, $\sigma_{EH} = \sigma_{HE} \neq 0$. In contrast to the case of normal incidence (see Figures 5.27 and 5.28), there are the specular reflection peaks at $\varphi = \pm 45°$ in the co-polar components. Note that the cross-polar components are everywhere finite. In agreement with the Karp–Karal lemma, all widths vanish on the finitely conducting face at $\varphi = \Phi$; the lower face is PEC, and that is why σ_{HH} goes to a finite limit at $\varphi = -\Phi$.

The reader interested in further simulation results is referred to (Osipov and Senior 2008b, 2009) where multiple plots illustrating the behavior of diffraction coefficients of right-angled impedance wedges (interior and exterior, isotropic and anisotropic) are presented.

Thus, we have seen in this section that the diffraction coefficient is the key quantity in the asymptotic analysis of scattering from wedges. The diffraction coefficient includes the complete information about the position of the shadow boundaries where it has pole singularities. The residues at these poles are associated with the amplitudes of the respective GO waves and can be expressed through reflection coefficients of the wedge faces. Once the diffraction coefficient is known, the complete asymptotic expansion of the edge-diffracted field for $k \to \infty$ can be constructed by applying (5.427). Finally, the diffraction coefficient is the basis for deriving improved approximations for the scattered fields, applicable at the shadow boundaries, which will be addressed next.

5.5.3 Uniform Asymptotic Approximations

Expression (5.409) for the edge-diffracted field is singular at every shadow boundary of the GO field (see (5.415)), which is an unphysical consequence of an insufficiently accurate evaluation of the integral in (5.382). With φ close to $\tilde{\varphi}_n$ a pole of the integrand function in (5.382) approaches the saddle point $\alpha = \pi$, making invalid the assumption about a slow variation of the integrand function. Corresponding modifications of the steepest descent method are available (see Appendix C) and result in modified asymptotic approximations in terms of the Fresnel integral $F(X)$ or related functions, most notably $F_{KP}(X^2)$ (see Appendix B), which are valid at the shadow boundaries while recovering the original asymptotic approximation far from the shadow boundary. In this section, we describe an algorithm for building high-frequency approximations, valid at shadow boundaries, by following an elegant approach described in Borovikov and Kinber (1994, Chapter 5) which permits the construction without resorting to the modified steepest descent method.

Let us pick up any term from the GO residue sum (5.400) and express it as

$$\mathbf{u}_n = \Theta(\pi - |\alpha_n - \varphi|) \operatorname{res} \mathbf{s}(\alpha_n) \mathrm{e}^{-jk\mathcal{E}_n - jkz\cos\beta}, \tag{5.429}$$

where α_n is given by (5.252), $\mathcal{E}_n = \rho \sin\beta \cos(\tilde{\varphi}_n - \varphi)$ is the eikonal of the respective wave, and $\tilde{\varphi}_n = \alpha_n \pm \pi$ ($|\tilde{\varphi}_n| < \Phi$). At the shadow boundary of the wave ($\varphi = \tilde{\varphi}_n$), the unit step function in (5.429) has a jump and the elements of $\overline{\overline{D}}$ in (5.409) have a pole singularity (5.415). Let us rewrite the formula (5.409) as

$$\mathbf{u}_\mathrm{d} \approx \frac{\mathrm{e}^{-jk\mathcal{E}_\mathrm{d} - jkz\cos\beta - j\pi/4}}{\sqrt{2\pi k\mathcal{E}_\mathrm{d}}} \overline{\overline{D}} \cdot \mathbf{U}_0, \tag{5.430}$$

where $\mathcal{E}_\mathrm{d} = \rho \sin\beta$ is the eikonal of the edge-diffracted wave. Both the jump in $\mathbf{u}_n$ and the pole singularity in $\mathbf{u}_\mathrm{d}$ are unphysical. Recalling the exact solution for the field scattered by a PEC half-plane (see (5.75)), we notice that the solution is described by Fresnel integrals, which are smooth functions at the shadow boundaries. So, let us replace the unit step function in (5.429) with a corresponding Fresnel integral and consider the modified function

$$\tilde{\mathbf{u}}_n = \operatorname{res} \mathbf{s}(\alpha_n) F\left[-\sigma_n \sqrt{k(\mathcal{E}_\mathrm{d} - \mathcal{E}_n)}\right] \mathrm{e}^{-jk\mathcal{E}_n - jkz\cos\beta}, \tag{5.431}$$

where the parameter σ_n, introduced in (5.415), is equal to 1 on the shadowed side of the shadow boundary and to -1 on the lit side of the shadow boundary. The argument of the Fresnel integral is nevertheless continuous at the shadow boundary since $\mathcal{E}_\mathrm{d} = \mathcal{E}_n$ when $\varphi = \tilde{\varphi}_n$. Off the shadow boundary, $\mathcal{E}_\mathrm{d} \geq \mathcal{E}_n$ and the square-root function in the argument of the Fresnel integral is understood in the arithmetic sense. The sign of the argument of the Fresnel integral is chosen so as to recover $\mathbf{u}_n$ from (5.429) far from the shadow boundary, when $k(\mathcal{E}_\mathrm{d} - \mathcal{E}_n) \gg 1$, by application of the asymptotic approximation (B.11) for the Fresnel integral. The field $\tilde{\mathbf{u}}_n$, called the penumbra field (Borovikov and Kinber 1994), is an exact solution of the Helmholtz equation, which can be checked by direct substitution.

In addition to the unit step function, the approximation (B.11) includes further terms with inverse powers of the argument of the Fresnel integral, which have the form of an edge-diffracted wave. In order to recover the original edge-diffracted wave (5.430) far from the shadow boundary, these additional terms have to be compensated by modifying (5.430) as

$$\tilde{\mathbf{u}}_\mathrm{d} = \frac{\mathrm{e}^{-jk\mathcal{E}_\mathrm{d} - jkz\cos\beta - j\pi/4}}{\sqrt{2\pi k\mathcal{E}_\mathrm{d}}} \overline{\overline{D}}_n \cdot \mathbf{U}_0, \tag{5.432}$$

where

$$\overline{\overline{D}}_n \cdot \mathbf{U}_0 = \overline{\overline{D}} \cdot \mathbf{U}_0 - \sigma_n \frac{\sqrt{\mathcal{E}_\mathrm{d}} \operatorname{res} \mathbf{s}(\alpha_n)}{\sqrt{2(\mathcal{E}_\mathrm{d} - \mathcal{E}_n)}}. \tag{5.433}$$

A direct inspection shows that

$$\tilde{\mathbf{u}}_n + \tilde{\mathbf{u}}_\mathrm{d} \to \mathbf{u}_n + \mathbf{u}_\mathrm{d} \tag{5.434}$$

when $k(\mathcal{E}_\mathrm{d} - \mathcal{E}_n) \to \infty$. At the shadow boundary, as $\varphi \to \tilde{\varphi}_n$,

$$\sqrt{\frac{\mathcal{E}_\mathrm{d}}{2(\mathcal{E}_\mathrm{d} - \mathcal{E}_n)}} \to \frac{1}{|\varphi - \tilde{\varphi}_n|} \tag{5.435}$$

and the singularity of $\overline{\overline{D}} \cdot \mathbf{U}_0$ on the right-hand side of (5.433) is compensated. Thus, the algorithm for obtaining smooth and bounded asymptotic approximations of the fields scattered by wedges consists of replacing the singular terms $\mathbf{u}_n + \mathbf{u}_\mathrm{d}$ in (5.380) with the modified terms $\tilde{\mathbf{u}}_n + \tilde{\mathbf{u}}_\mathrm{d}$ described in (5.431), (5.432), and (5.433). The modified approximation is called the *uniform asymptotic approximation*, in contrast to the original, non-uniform approximation, in which the analytical form of $\mathbf{u}_n$ is different on the different sides of the shadow boundary.

The uniform approximation can be expressed explicitly in terms of the parameters α_n, $\tilde{\varphi}_n$, and σ_n as

$$\tilde{\mathbf{u}}_n = \operatorname{res}\mathbf{s}(\alpha_n) F\left[-\sigma_n\sqrt{2k\rho\sin\beta}\left|\sin\left(\frac{\varphi - \tilde{\varphi}_n}{2}\right)\right|\right] \mathrm{e}^{-jk\rho\sin\beta\cos(\varphi - \tilde{\varphi}_n) - jkz\cos\beta} \tag{5.436}$$

$$\tilde{\mathbf{u}}_\mathrm{d} = \frac{\mathrm{e}^{-jk\rho\sin\beta - jkz\cos\beta - j\pi/4}}{\sqrt{2\pi k\rho\sin\beta}}\left[\overline{\overline{D}} \cdot \mathbf{U}_0 - \frac{1}{2}\sigma_n \operatorname{res}\mathbf{s}(\alpha_n)\left|\csc\left(\frac{\varphi - \tilde{\varphi}_n}{2}\right)\right|\right]. \tag{5.437}$$

Expression (5.437), or (5.432), describes a leading, zero-order term in the asymptotic expansion of $\tilde{\mathbf{u}}_\mathrm{d}$ in inverse powers of $k\rho\sin\beta$. Since $\tilde{\mathbf{u}}_\mathrm{d}$ must satisfy the Helmholtz equation, the complete asymptotic expansion has the form of (5.424) with coefficients related by (5.427) to the zero-order term.

In the uniform approximation, the transition between the lit and shadowed portions of space is described by the Fresnel integral, which can be seen as a smooth approximation of the unit step function (see Figure B.1). In contrast to transition regions of vanishing width predicted by GO, the real-world transition regions are of finite extent. Let us estimate the width of the transition regions.

Consider, for example, the shadow boundary $\varphi = \varphi_0 + \pi$ of the incident wave (Figure 5.31), where the transition is described by

$$F\left[\sqrt{2k_\perp\rho}\cos\left(\frac{\varphi - \varphi_0}{2}\right)\right] \tag{5.438}$$

with $k_\perp = k\sin\beta$ for brevity. At the shadow boundary, the argument of the Fresnel integral vanishes and the Fresnel integral is equal to 0.5. According to Figure B.1, $F(X)$ is close to unity when $X \geq 1$ and to zero when $X \leq -1$. It is therefore natural to define the transition region by the condition $-1 \leq X \leq 1$. Let δ_ess be the corresponding angular width

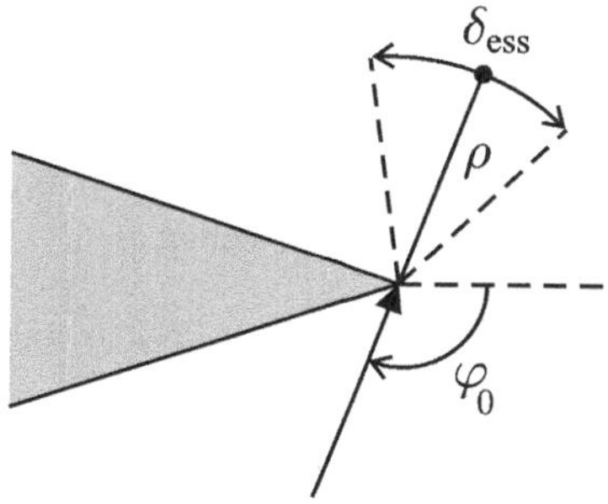

Figure 5.31 An impenetrable wedge is illuminated from the direction $\varphi = \varphi_0$. The transition zone around the shadow boundary at $\varphi = \varphi_0 + \pi$, within which the total field smoothly changes between the lit and shadowed portions of space, has the angular width δ_ess at distance ρ from the edge.

of the region. In the vicinity of the shadow boundary, $\varphi = \varphi_0 + \pi - \delta\varphi$ where $\delta\varphi$ is small, and the argument of the Fresnel integral in (5.438) can be expressed as

$$\sqrt{2k_\perp \rho}\cos\left(\frac{\varphi - \varphi_0}{2}\right) \approx \delta\varphi\sqrt{\frac{k_\perp \rho}{2}}. \tag{5.439}$$

The absolute value of this argument is equal to unity when

$$|\delta\varphi| = \sqrt{\frac{2}{k_\perp \rho}} \tag{5.440}$$

and since $\delta_{\mathrm{ess}} = 2|\delta\varphi|$, we find that

$$\delta_{\mathrm{ess}} = \sqrt{\frac{8}{k_\perp \rho}}. \tag{5.441}$$

In the case of normal incidence, $k_\perp = k = 2\pi/\lambda$, and we obtain

$$\delta_{\mathrm{ess}} = \sqrt{\frac{4\lambda}{\pi\rho}} \approx \sqrt{\frac{\lambda}{\rho}}, \tag{5.442}$$

which is a fundamental estimate of the angular width of the transition region associated with every shadow boundary behind a wedge with a sharp edge. The width is independent of the material properties of the wedge and orientation of the faces with respect to the incidence direction. For example, δ_{ess} is the same for a right-angled impedance wedge and a PEC half-plane. As the distance ρ grows, the angular width decreases, but the linear size of the transition region, given by the product $\rho\delta_{\mathrm{ess}}$, increases. In the GO limit ($\lambda \to 0$), the width of the region vanishes and the transition between the lit and shadowed portions of space becomes increasingly sharp.

Finally, we note an alternative, but equivalent, form of the uniform approximation. This is obtained by replacing the Fresnel integral in (5.436) with the function F_{KP} (see (B.29)). Using (B.32) in (5.436) gives

$$\tilde{\mathbf{u}}_n + \tilde{\mathbf{u}}_{\mathrm{d}} = \mathbf{u}_n + \tilde{\tilde{\mathbf{u}}}_{\mathrm{d}}, \tag{5.443}$$

where $\mathbf{u}_n$ is the GO term (5.429) and

$$\tilde{\tilde{\mathbf{u}}}_{\mathrm{d}} = \frac{\mathrm{e}^{-jk\rho\sin\beta - jkz\cos\beta - j\pi/4}}{\sqrt{2\pi k\rho\sin\beta}} \overline{\overline{D}}_n^{\mathrm{KP}} \cdot \mathbf{U}_0 \tag{5.444}$$

with

$$\overline{\overline{D}}_n^{\mathrm{KP}} \cdot \mathbf{U}_0 = \overline{\overline{D}} \cdot \mathbf{U}_0 + \frac{\sigma_n \operatorname{res} \mathbf{s}(\alpha_n)}{2\left|\sin\left(\frac{\varphi - \tilde{\varphi}_n}{2}\right)\right|}\left\{F_{\mathrm{KP}}\left[2k\rho\sin\beta\sin^2\left(\frac{\varphi - \tilde{\varphi}_n}{2}\right)\right] - 1\right\}. \tag{5.445}$$

It can be easily checked that the modified edge-diffracted field $\tilde{\tilde{\mathbf{u}}}_{\mathrm{d}}$ ensures the correct behavior of the approximation on the right-hand side of (5.443) at and far from the shadow boundary. Far from the shadow boundary, the argument of F_{KP} in (5.445) is large and because of the large-argument approximation (B.35), $F_{\mathrm{KP}} \to 1$ and $\overline{\overline{D}}_n^{\mathrm{KP}} \to \overline{\overline{D}}$, as expected. On the other hand, close to the shadow boundary (5.445) can be rewritten as

$$\overline{\overline{D}}_n^{\mathrm{KP}} \cdot \mathbf{U}_0 = \overline{\overline{D}}_n \cdot \mathbf{U}_0 + \frac{\sigma_n \operatorname{res} \mathbf{s}(\alpha_n)}{2\left|\sin\left(\frac{\varphi - \tilde{\varphi}_n}{2}\right)\right|} F_{\mathrm{KP}}\left[2k\rho\sin\beta\sin^2\left(\frac{\varphi - \tilde{\varphi}_n}{2}\right)\right], \tag{5.446}$$

where $\overline{\overline{D}}_n \cdot \mathbf{U}_0$, given by (5.433), is continuous at the shadow boundary by construction. According to the small-argument approximation (B.36), $F_{\mathrm{KP}}(X^2)$ can be replaced with $\sqrt{\pi}|X|\mathrm{e}^{j\pi/4}$, which leads to the appearance of the term

$$\frac{\sigma_n}{2} \operatorname{res} \mathbf{s}(\alpha_n) \mathrm{e}^{-jk\rho\sin\beta - jkz\cos\beta} \tag{5.447}$$

in $\tilde{\tilde{\mathbf{u}}}_{\mathrm{d}}$ that precisely compensates the discontinuity in the GO term $\mathbf{u}_n$ at the shadow boundary because of the corresponding jump in the parameter σ_n.

The advantage of the uniform representation on the right-hand side of (5.443) is that only the edge-diffracted field has to be modified in order to transform a non-uniform approximation into a uniform one. The use of this format is the basic feature of the uniform theory of diffraction (UTD).

5.5.4 GTD/UTD Formulation

The Concept of GTD/UTD

The importance of high-frequency ($k \to \infty$) solutions for canonical shapes is in their use as building blocks in the simulation of scattering from electrically large generally shaped bodies. Despite the fact that an infinite wedge with flat faces is a mathematical abstraction and such structures do not exist in nature, the high-frequency solution for the wedge can be used to simulate scattering from bounded and curved edges once the length of the edge and the curvature radii are large compared to the wavelength. This is a manifestation of the locality principle in high-frequency diffraction, according to which scattering from a complex shape in the high-frequency limit splits up in a sum of contributions from asymptotically small neighborhoods of several characteristic points, which are the points of specular reflection and the points of diffraction on the edges and tips of the scattering surface (Figure 3.20).

This asymptotic approach to describing electromagnetic scattering has been formulated in the framework of the geometrical theory of diffraction (GTD) (Borovikov and Kinber 1994; James 1976; Keller 1962). The theory compliments GO rays with additional types of rays: the edge-diffracted ray, the tip-diffracted ray, and the creeping ray (Figure 5.32). At the point where an incident ray hits the edge, a cone of edge-diffracted rays is generated in such a way that the angle between the scattered rays and the edge is equal to the angle between the incident ray and the edge (Keller's cone). Once the incident ray arrives at a tip (or a conical point) on the scattering surface, the GTD rays are scattered in all directions, and the tip appears as the source of a spherical wave. Creeping waves are excited at shadow boundaries on smooth surfaces, where the incident ray tangentially touches the surfaces. At that point (attachment point A) a diffracted ray is generated which propagates along the scattering surface in the shadow region up to a launching point L where it is launched tangentially to the surface toward the observer. These new types of rays are subject to an extended Fermat law, that is, they propagate along the shortest path from the source to the receiver under the condition that the scattering surface is touched at the edge, the tip or the shadow boundary, which is used for calculating the eikonal of the respective ray. To calculate their amplitudes, solutions of scattering problems for simply shaped bodies that coincide with the scattering surface in a vicinity of the diffraction points are used. In the case of an edge in the scattering surface, the canonical geometry is an infinite wedge with the locally conformal geometric and material constitution.

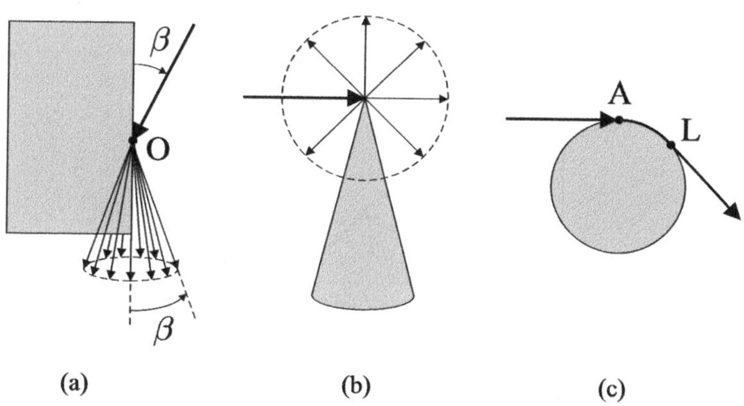

Figure 5.32 GTD diffraction rays: (a) edge rays, (b) tip rays, and (c) creeping rays.

GTD is based on the ray concept and, therefore, fails at caustics of rays, for example on shadow boundaries. This failure makes GTD more a theoretical concept rather than an engineering approach. Making the GTD formulas applicable at caustics is the subject of uniform theories of diffraction. There are various versions of these (Boersma and Rahmat-Samii 1980), most notably the uniform asymptotic theory of diffraction (UAT) (Ahluwalia *et al.* 1968; Borovikov and Kinber 1994) and the uniform geometrical theory of diffraction (UTD) (Kouyoumjian and Pathak 1974; McNamara *et al.* 1990), the latter being nowadays the standard technique implemented in the leading simulation software.

In this section we formulate the results for edge-diffracted rays in the format that is used in GTD/UTD. The creeping wave is extracted from exact solutions for circular cylinders in section 6.4, and its GTD representation for arbitrarily curved surfaces is described in section 6.6.3. Scattering at the tip is addressed in the framework of PO in section 9.2. In Chapters 8 and 9 we show how the various GTD rays can be extracted from asymptotic PO solutions for variously shaped bodies.

Edge-diffracted Rays in the Framework of GTD/UTD

A GTD interpretation of the expression (5.409) for the edge-diffracted wave is that an incident ray hits the edge at the diffraction point O, which is the origin of the chosen coordinate system, and gets scattered toward an observation point (Figure 5.33). At the diffraction point, the angle between the edge and the incidence direction $\hat{\imath}$, which is given by (5.137), is equal to the angle between the edge and the direction $\hat{s}$ to the observer. If (ρ, φ, z) are the coordinates of the observer in the associated circular cylinder coordinates and r is the distance between the observer and the diffraction point, then

$$\hat{s} = \hat{x}\cos\varphi\sin\beta + \hat{y}\sin\varphi\sin\beta + \hat{z}\cos\beta, \tag{5.448}$$

$\rho = r\sin\beta$, and $z = r\cos\beta$, implying that $r = \rho\sin\beta + z\cos\beta$.

The diffracted ray in (5.409) is a cylindrical (conical) wave with a wavefront that can be considered as locally plane as soon as $k\rho\sin\beta$ is sufficiently large, implying that the scattered field can be described by two components of the electric field orthogonal to the direction of diffraction $\hat{s}$. The incident plane wave can be also represented by two components of the electric field orthogonal to the incidence direction $\hat{\imath}$. It is therefore convenient to introduce the so-called "ray-fixed" coordinate frames (Kouyoumjian and Pathak 1974), one associated with the incident ray and another with the edge-diffracted ray, by building the unit vectors

$$\hat{\varphi}_{\rm inc} = \frac{\hat{\imath}\times\hat{z}}{\sin\beta}, \quad \hat{\beta}_{\rm inc} = \hat{\imath}\times\hat{\varphi}_{\rm inc} \tag{5.449}$$

$$\hat{\varphi} = \frac{\hat{s}\times\hat{z}}{\sin\beta}, \quad \hat{\beta} = \hat{s}\times\hat{\varphi} \tag{5.450}$$

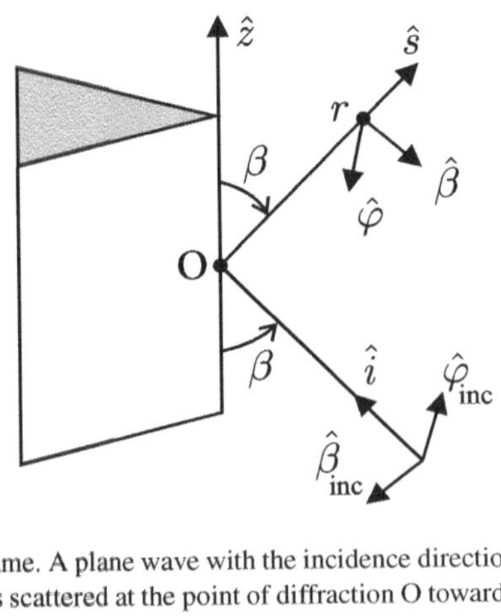

Figure 5.33 The GTD/UTD coordinate frame. A plane wave with the incidence direction $\hat{\imath}$ illuminates an impenetrable wedge with an edge along the z-coordinate axis. The wave is scattered at the point of diffraction O toward an observer at the point $\mathbf{r} = r\hat{s}$. The incidence and scattering directions make the same angle β with the edge of the wedge. The unit orthogonal vectors $\hat{\imath}$, $\hat{\beta}_{\rm inc}$, and $\hat{\varphi}_{\rm inc}$ ($\hat{s}$, $\hat{\beta}$, and $\hat{\varphi}$) define the ray-fixed coordinate frame of the incident (diffracted) ray.

which define the orthonormal bases of the incident and diffracted rays, $(\hat{\imath}, \hat{\varphi}_{\text{inc}}, \hat{\beta}_{\text{inc}})$ and $(\hat{s}, \hat{\varphi}, \hat{\beta})$, respectively. The electric fields in the incident and diffracted rays are then expressed as

$$\mathbf{E}_{\text{inc}} = \hat{\beta}_{\text{inc}} E^{\text{inc}}_{\beta_{\text{inc}}} + \hat{\varphi}_{\text{inc}} E^{\text{inc}}_{\varphi_{\text{inc}}} \tag{5.451}$$

$$\mathbf{E}_{\text{d}} = \hat{\beta} E^{\text{d}}_{\beta} + \hat{\varphi} E^{\text{d}}_{\varphi} \tag{5.452}$$

with

$$E^{\text{inc}}_{\beta_{\text{inc}}} = \hat{\beta}_{\text{inc}} \cdot \mathbf{E}_{\text{inc}}, \quad E^{\text{inc}}_{\varphi_{\text{inc}}} = \hat{\varphi}_{\text{inc}} \cdot \mathbf{E}_{\text{inc}} \tag{5.453}$$

$$E^{\text{d}}_{\beta} = \hat{\beta} \cdot \mathbf{E}_{\text{d}}, \quad E^{\text{d}}_{\varphi} = \hat{\varphi} \cdot \mathbf{E}_{\text{d}}, \tag{5.454}$$

which resolves the incident and diffracted rays into components parallel and perpendicular to the edge-fixed planes of incidence and diffraction.

The components E_z and ZH_z are simply related to the components in the ray-fixed coordinates by

$$E^{\text{inc}}_z = -\sin\beta E^{\text{inc}}_{\beta_{\text{inc}}} \tag{5.455}$$

$$ZH^{\text{inc}}_z = -\sin\beta E^{\text{inc}}_{\varphi_{\text{inc}}} \tag{5.456}$$

$$E^{\text{d}}_z = -\sin\beta E^{\text{d}}_{\beta} \tag{5.457}$$

$$ZH^{\text{d}}_z = -\sin\beta E^{\text{d}}_{\varphi} \tag{5.458}$$

(problem 21), and (5.409) for the edge-diffracted wave can be rewritten as

$$\begin{bmatrix} E^{\text{d}}_{\beta} \\ E^{\text{d}}_{\varphi} \end{bmatrix} = \frac{\exp(-jkr - j\pi/4)}{\sqrt{2\pi kr}\sin\beta} \begin{bmatrix} D_{\text{ee}} & D_{\text{eh}} \\ D_{\text{he}} & D_{\text{hh}} \end{bmatrix} \begin{bmatrix} E^{\text{inc}}_{\beta_{\text{inc}}}(\text{O}) \\ E^{\text{inc}}_{\varphi_{\text{inc}}}(\text{O}) \end{bmatrix} \tag{5.459}$$

since at the point of diffraction $E_{0z} = -\sin\beta E^{\text{inc}}_{\beta_{\text{inc}}}(\text{O})$ and $ZH_{0z} = -\sin\beta E^{\text{inc}}_{\varphi_{\text{inc}}}(\text{O})$.

The GTD format of writing the edge-diffracted ray is as follows:

$$\mathbf{E}_{\text{d}} = \overline{\overline{D}}_{\text{GTD}} \cdot \mathbf{E}_{\text{inc}}(\text{O}) \frac{\text{e}^{-jkr}}{\sqrt{r}}, \tag{5.460}$$

where $\mathbf{E}_{\text{inc}}(\text{O})$ is the incident electric field at the point of diffraction and $\overline{\overline{D}}_{\text{GTD}}$ is the dyadic diffraction coefficient

$$\overline{\overline{D}}_{\text{GTD}} = \hat{\beta} D_{\beta\beta_{\text{inc}}} \hat{\beta}_{\text{inc}} + \hat{\beta} D_{\beta\varphi_{\text{inc}}} \hat{\varphi}_{\text{inc}} + \hat{\varphi} D_{\varphi\beta_{\text{inc}}} \hat{\beta}_{\text{inc}} + \hat{\varphi} D_{\varphi\varphi_{\text{inc}}} \hat{\varphi}_{\text{inc}}. \tag{5.461}$$

Comparing (5.460) with (5.459) gives

$$D_{\beta\beta_{\text{inc}}} = \frac{\text{e}^{-j\pi/4}}{\sqrt{2\pi k}\sin\beta} D_{\text{ee}}, \quad D_{\beta\varphi_{\text{inc}}} = \frac{\text{e}^{-j\pi/4}}{\sqrt{2\pi k}\sin\beta} D_{\text{eh}} \tag{5.462}$$

$$D_{\varphi\beta_{\text{inc}}} = \frac{\text{e}^{-j\pi/4}}{\sqrt{2\pi k}\sin\beta} D_{\text{he}}, \quad D_{\varphi\varphi_{\text{inc}}} = \frac{\text{e}^{-j\pi/4}}{\sqrt{2\pi k}\sin\beta} D_{\text{hh}}. \tag{5.463}$$

According to UTD, the uniform representation of the edge-diffracted ray is obtained by replacing the non-uniform diffraction coefficient $\overline{\overline{D}}_{\text{GTD}}$ in (5.460) with a uniform diffraction coefficient $\overline{\overline{D}}_{\text{UTD}}$,

$$\mathbf{E}_{\text{d}} = \overline{\overline{D}}_{\text{UTD}} \cdot \mathbf{E}_{\text{inc}}(\text{O}) \frac{\text{e}^{-jkr}}{\sqrt{r}}. \tag{5.464}$$

Using (5.444) and (5.445) and introducing a diadic $\overline{\overline{\Omega}}_n$ with elements

$$\Omega_{\beta\beta_{\text{inc}}} = \text{res}\, S_{\text{e}}(\alpha_n), \quad \Omega_{\varphi\beta_{\text{inc}}} = \text{res}\, S_{\text{h}}(\alpha_n) \quad (E_{0z} = 1, ZH_{0z} = 0) \tag{5.465}$$

and

$$\Omega_{\beta\varphi_{\rm inc}} = \operatorname{res} S_{\rm e}(\alpha_n), \quad \Omega_{\varphi\varphi_{\rm inc}} = \operatorname{res} S_{\rm h}(\alpha_n) \quad (E_{0z} = 0, ZH_{0z} = 1) \tag{5.466}$$

we arrive at the expression

$$\overline{\overline{D}}_{\rm UTD} = \overline{\overline{D}}_{\rm GTD} + \frac{\sigma_n \overline{\overline{\Omega}}_n {\rm e}^{-j\pi/4}}{2\sqrt{2\pi k}\sin\beta}\left|\csc\left(\frac{\varphi - \tilde{\varphi}_n}{2}\right)\right| \cdot \left\{F_{\rm KP}\left[2k\rho\sin\beta\sin^2\left(\frac{\varphi-\tilde{\varphi}_n}{2}\right)\right] - 1\right\} \tag{5.467}$$

which corrects the singularity of $\overline{\overline{D}}_{\rm GTD}$ at the shadow boundary $\varphi = \tilde{\varphi}_n$ and compensates the jump in the GO ray $\mathbf{u}_n$ (see (5.429)). For example, at the shadow boundary of the incident wave, $\alpha_0 = \varphi_0$, $\tilde{\varphi}_0 = \varphi_0 \pm \pi$, $\sigma_0 = \pm{\rm sgn}(\varphi - \tilde{\varphi}_0)$, and $\overline{\overline{\Omega}}_0 = \overline{\overline{I}}$, so (5.467) reduces to

$$\overline{\overline{D}}_{\rm UTD} = \overline{\overline{D}}_{\rm GTD} - \frac{\overline{\overline{I}}{\rm e}^{-j\pi/4}}{2\sqrt{2\pi k}\sin\beta}\sec\left(\frac{\varphi - \varphi_0}{2}\right) \cdot \left\{F_{\rm KP}\left[2k\rho\sin\beta\cos^2\left(\frac{\varphi-\varphi_0}{2}\right)\right] - 1\right\}. \tag{5.468}$$

The UTD diffraction coefficients valid at the shadow boundaries of the rays reflected from the faces of the wedge follow from (5.467) with $n = \pm 1$, $\alpha_{\pm 1} = -\varphi_0 \pm 2\Phi$, $\tilde{\varphi}_{\pm 1} = \mp\pi \pm 2\Phi - \varphi_0$, and $\sigma_{\pm 1} = \mp{\rm sgn}\,(\varphi - \tilde{\varphi}_{\pm 1})$ as

$$\overline{\overline{D}}_{\rm UTD} = \overline{\overline{D}}_{\rm GTD} - \frac{\overline{\overline{\Omega}}_{\pm 1}{\rm e}^{-j\pi/4}}{2\sqrt{2\pi k}\sin\beta}\sec\left(\frac{\varphi + \varphi_0 \mp 2\Phi}{2}\right) \cdot \left\{F_{\rm KP}\left[2k\rho\sin\beta\cos^2\left(\frac{\varphi+\varphi_0 \mp 2\Phi}{2}\right)\right] - 1\right\}, \tag{5.469}$$

where the elements of $\overline{\overline{\Omega}}_{\pm 1}$ are given by

$$\Omega_{\beta\beta_{\rm inc}} = R^{\pm}_{\rm ee}(\Phi \mp \varphi_0), \quad \Omega_{\beta\varphi_{\rm inc}} = R^{\pm}_{\rm eh}(\Phi \mp \varphi_0) \tag{5.470}$$

$$\Omega_{\varphi\beta_{\rm inc}} = R^{\pm}_{\rm he}(\Phi \mp \varphi_0), \quad \Omega_{\varphi\varphi_{\rm inc}} = R^{\pm}_{\rm hh}(\Phi \mp \varphi_0) \tag{5.471}$$

with $R^{\pm}_{lm}$ $(l, m = {\rm e}, {\rm h})$ defined in (5.194).

The accuracy of the asymptotic approximations is illustrated in Figures 5.34 and 5.35. Non-uniform formulas are inadequate close to the shadow boundaries at $\varphi = \pm 45°$. The uniform formulas, by contrast, provide excellent accuracy even with the very moderate value of $k\rho$ (here 2π). A small jump at $\varphi = 0$ in the uniform approximation in Figure 5.35 is due to the switch between uniform approximations with $n = 0$ $(-180° \le \varphi < 0)$ and $n = 1$ $(0 < \varphi \le 180°)$. The accuracy of high-frequency approximations improves with growing $k\rho$.

5.6 Behavior of Electromagnetic Fields at Edges

5.6.1 Determining the Degree of Singularity

As shown in sections 5.2.3 and 5.4.2, the transversal components E_ρ, E_φ, H_ρ, and H_φ of the fields scattered by PEC and impedance wedges become infinite as $\rho \to 0$. In the studied cases, the singularity follows the power law $O(\rho^{\tau-1})$ with $\tau = \pi/(2\Phi)$ where 2Φ is the angular width of the wedge-shaped domain exterior to the wedge. The components parallel to the edge, E_z and H_z, are always bounded at $\rho = 0$, otherwise the singularity in the rest components would be too strong and violate the edge conditions (2.60).

The parameter τ defines the degree of singularity, which is a quantity useful for practice. The knowledge of τ is important, for example, in applications related to power transmission and insulation as the singular behavior must be known to estimate the danger of breakdown near the edge. The information about τ can be used to improve the efficiency of numerical procedures in the domains with edges or to check the validity of numerical and analytical solutions by verifying the field behavior in the vicinity of the edge (e.g. Mittra and Lee (1971), Mur (1981), Braver *et al.* (1988), Van Bladel (1991)).

The good news is that the degree of singularity can be found for every wedge-shaped configuration by a local analysis, that is, without solving the full boundary value problem. The idea of the approach was formulated by Meixner (1972) on

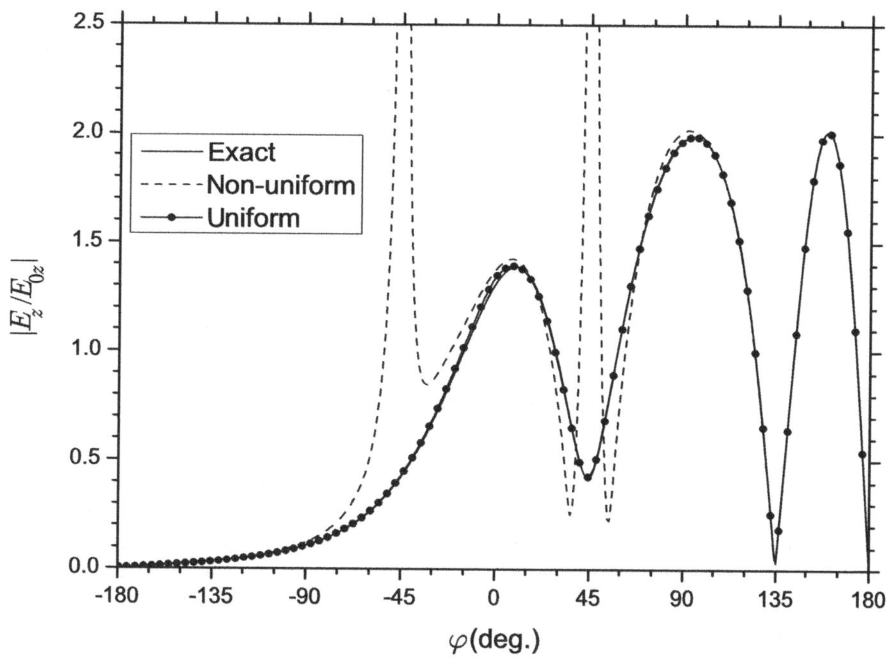

Figure 5.34 Total electric field at the distance $\rho = \lambda$ from the edge of a PEC half-plane illuminated by a plane wave with $\beta = \pi/2$ and $\varphi_0 = 3\pi/4$.

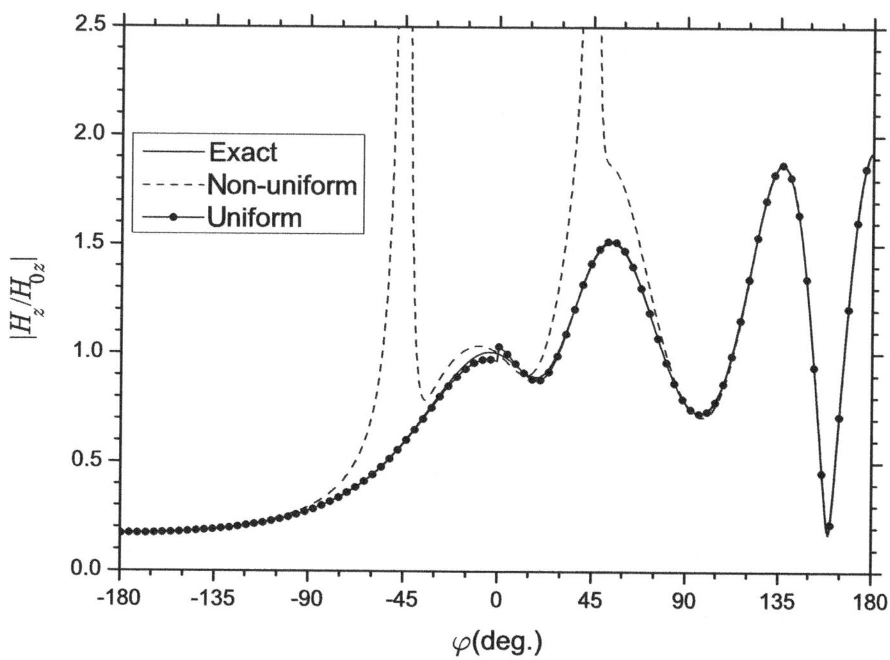

Figure 5.35 The same as Figure 5.34 but for the magnetic field.

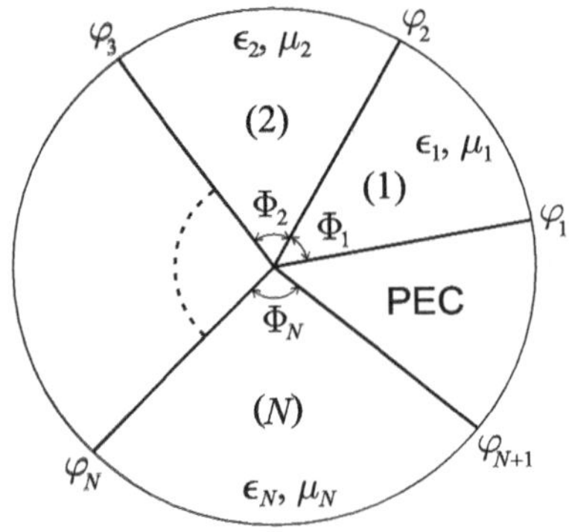

Figure 5.36 Generic wedge-shaped configuration with an optional PEC wedge.

the example of two magneto-dielectric wedges with an optional PEC wedge. Extensions to three isotropic dielectrics are described by Mittra and Lee (1971, section 1.3) and Brooke and Kharadly (1977), and a generalization to a PEC wedge with N adjacent dielectric wedges in (Hurd 1976). A lot of detailed results for configurations with multiple magneto-dielectric and anisotropic (e.g., magnetized ferrite or plasma) wedges in the two-dimensional case (normal incidence) can be found in (Bobrovnikov and Fisanov 1988, Chapter 5). In this section, a general procedure for determining the parameter τ in wedge-shaped configurations involving an arbitrary number of magneto-dielectric contiguous wedges with a possible inclusion of a PEC wedge is described in a three-dimensional formulation, that is, allowing for the dependence of fields on z.

Consider the general wedge-shaped configuration shown in Figure 5.36, which consists of N wedges of angular widths Φ_n with a common edge. The interfaces of the wedges are located at φ_n with $n = 1, 2, \ldots, N+1$, and the wedges are filled with, in general, different materials. The configuration may include a PEC wedge as well.

Let us search for the solutions of the form

$$E_z = \mathrm{e}^{-jk_z z}\rho^{\tau} E_0(\varphi), \quad H_z = \mathrm{e}^{-jk_z z}\rho^{\tau} H_0(\varphi), \tag{5.472}$$

where τ is a parameter with $\mathrm{Re}\,\tau > 0$. Other components follow from (2.313)–(2.316) and (5.472) as

$$E_\rho = \mathrm{e}^{-jk_z z}\frac{\rho^{\tau-1}}{jk_\perp^2}\left[\tau k_z E_0(\varphi) + \omega\mu H_0'(\varphi)\right] \tag{5.473}$$

$$E_\varphi = \mathrm{e}^{-jk_z z}\frac{\rho^{\tau-1}}{jk_\perp^2}\left[k_z E_0'(\varphi) - \omega\mu\tau H_0(\varphi)\right] \tag{5.474}$$

$$H_\rho = \mathrm{e}^{-jk_z z}\frac{\rho^{\tau-1}}{jk_\perp^2}\left[\tau k_z H_0(\varphi) - \omega\epsilon E_0'(\varphi)\right] \tag{5.475}$$

$$H_\varphi = \mathrm{e}^{-jk_z z}\frac{\rho^{\tau-1}}{jk_\perp^2}\left[k_z H_0'(\varphi) + \omega\epsilon\tau E_0(\varphi)\right], \tag{5.476}$$

where the prime denotes the derivative with respect to φ and $k_\perp^2 = \omega^2\epsilon\mu - k_z^2$. These components are singular at the edge if $\mathrm{Re}\,\tau < 1$.

Insertion of (5.472) in the Helmholtz equations (2.317) and (2.318) gives

$$\rho^{\tau-2}E_0''(\varphi) + \rho^{\tau-2}\tau^2 E_0(\varphi) + \rho^{\tau}k_\perp^2 E_0(\varphi) = 0 \tag{5.477}$$

$$\rho^{\tau-2}H_0''(\varphi) + \rho^{\tau-2}\tau^2 H_0(\varphi) + \rho^{\tau}k_\perp^2 H_0(\varphi) = 0, \tag{5.478}$$

which is satisfied in the limit $\rho \to 0$ if

$$E_0''(\varphi) + \tau^2 E_0(\varphi) = 0 \tag{5.479}$$

$$H_0''(\varphi) + \tau^2 H_0(\varphi) = 0. \tag{5.480}$$

The equations imply that E_0 and H_0 are linear combinations of $\sin(\tau\varphi)$ and $\cos(\tau\varphi)$ in every sector of the configuration (Figure 5.36), and it is convenient to represent the solutions as

$$E_0(\varphi) = a_n \cos[\tau(\varphi - \varphi_n)] + b_n \sin[\tau(\varphi - \varphi_n)] \tag{5.481}$$

$$H_0(\varphi) = c_n \cos[\tau(\varphi - \varphi_n)] + d_n \sin[\tau(\varphi - \varphi_n)], \tag{5.482}$$

where $\varphi_n \leq \varphi \leq \varphi_{n+1}$, $n = 1, 2, \ldots, N$.

Coefficients of the linear combinations must ensure meeting the conditions of continuity of the components E_z, E_ρ, H_z, and H_ρ at the material interfaces. For example, at the interface between the media n and $n+1$, the continuity conditions lead to the equations:

$$a_{n+1} = a_n C_n + b_n S_n \tag{5.483}$$

$$c_{n+1} = c_n C_n + d_n S_n \tag{5.484}$$

$$\frac{k_z a_{n+1} + \omega\mu_{n+1} d_{n+1}}{k^2_{\perp,n+1}} = \frac{k_z C_n a_n + k_z S_n b_n - \omega\mu_n S_n c_n + \omega\mu_n C_n d_n}{k^2_{\perp,n}} \tag{5.485}$$

$$\frac{-\omega\epsilon_{n+1} b_{n+1} + k_z c_{n+1}}{k^2_{\perp,n+1}} = \frac{\omega\epsilon_n S_n a_n - \omega\epsilon_n C_n b_n + k_z C_n c_n + k_z S_n d_n}{k^2_{\perp,n}}, \tag{5.486}$$

where $k^2_{\perp,n} = \omega^2\epsilon_n\mu_n - k_z^2$, $C_n = \cos(\tau\Phi_n)$, and $S_n = \cos(\tau\Phi_n)$. The system of equations can be solved for the coefficients in the $n+1$ medium,

$$\mathbf{F}_{n+1} = \overline{\overline{T}}_n(\tau) \cdot \mathbf{F}_n, \tag{5.487}$$

where $\mathbf{F}_n = (a_n, b_n, c_n, d_n)$ and

$$\overline{\overline{T}}_n(\tau) = \begin{bmatrix} C_n & S_n & 0 & 0 \\ -q_n \frac{\epsilon_n S_n}{\epsilon_{n+1}} & q_n \frac{\epsilon_n C_n}{\epsilon_{n+1}} & (1 - q_n)\frac{k_z C_n}{\omega\epsilon_{n+1}} & (1 - q_n)\frac{k_z S_n}{\omega\epsilon_{n+1}} \\ 0 & 0 & C_n & S_n \\ (q_n - 1)\frac{k_z C_n}{\omega\mu_{n+1}} & (q_n - 1)\frac{k_z S_n}{\omega\mu_{n+1}} & -q_n \frac{\mu_n S_n}{\mu_{n+1}} & q_n \frac{\mu_n C_n}{\mu_{n+1}} \end{bmatrix} \tag{5.488}$$

with $q_n = (k_{\perp,n+1}/k_{\perp,n})^2$, and imposing the continuity at material interfaces φ_n with $n = 1, 2, \ldots, N-1$ results in the chain of equations

$$\mathbf{F}_2 = \overline{\overline{T}}_1(\tau) \cdot \mathbf{F}_1, \quad \mathbf{F}_3 = \overline{\overline{T}}_2(\tau) \cdot \mathbf{F}_2, \ldots, \quad \mathbf{F}_N = \overline{\overline{T}}_{N-1}(\tau) \cdot \mathbf{F}_{N-1}. \tag{5.489}$$

If the configuration has no PEC inclusion, then applying the continuity conditions at the last interface $\varphi_{N+1} = \varphi_1 + 2\pi$ gives the equation

$$\mathbf{F}_1 = \overline{\overline{T}}_N(\tau) \cdot \mathbf{F}_N \tag{5.490}$$

which together with (5.489) leads to a system of N vector equations for N unknown vectors $\mathbf{F}_n$. Using the recursive structure of (5.489), $\mathbf{F}_N$ can be expressed through $\mathbf{F}_1$ and inserted in (5.490) to arrive at the equation

$$\mathbf{F}_1 = \overline{\overline{T}}_N(\tau) \cdot \overline{\overline{T}}_{N-1}(\tau) \ldots \overline{\overline{T}}_1(\tau) \cdot \mathbf{F}_1. \tag{5.491}$$

This is a homogeneous system of four linear algebraic equations for the four unknowns a_1, b_1, c_1, and d_1. A non-trivial solution (5.491) exists if and only if

$$\det\left[\overline{\overline{I}} - \overline{\overline{T}}_N(\tau)\cdot\overline{\overline{T}}_{N-1}(\tau)\ldots\overline{\overline{T}}_1(\tau)\right] = 0, \tag{5.492}$$

which is an equation on the parameter τ. The characteristic equation (5.492) involves the trigonometric functions C_n and S_n with $n = 1, 2, \ldots, N$ and has therefore infinite number of discrete solutions τ_s with $s = 1, 2, \ldots$ In the case of lossy media, the material parameters are complex-valued, implying complex-valued τ_s. The singularity of the electromagnetic field is determined by the solution with the smallest positive real part, which we denote by τ_1.

The procedure of determining τ is slightly different if the configuration includes a PEC wedge (Figure 5.36). Requiring that E_z and E_ρ vanish at φ_1 means that $a_1 = 0$ and $d_1 = 0$. The same condition at φ_{N+1} can be written as

$$\hat{e}_1\cdot\mathbf{F}_N = 0 \tag{5.493}$$

$$\hat{e}_4\cdot\mathbf{F}_N = 0, \tag{5.494}$$

where $\hat{e}_1$ and $\hat{e}_4$ are the unit vectors: $\hat{e}_1 = (C_N, S_N, 0, 0)$ and $\hat{e}_4 = (0, 0, -S_N, C_N)$. Again, from the chain of equations in (5.489),

$$\mathbf{F}_N = \overline{\overline{T}}\cdot\mathbf{F}_1 \tag{5.495}$$

with $\overline{\overline{T}} = \overline{\overline{T}}_{N-1}\cdot\ldots\cdot\overline{\overline{T}}_1$ (for brevity we omit the argument τ from the matrices), and since $\mathbf{F}_1 = (0, b_1, c_1, 0)$ equations (5.493) and (5.494) can be expressed as

$$m_{11}b_1 + m_{12}c_1 = 0 \tag{5.496}$$

$$m_{21}b_1 + m_{22}c_1 = 0 \tag{5.497}$$

with

$$m_{11} = \hat{e}_1\cdot\overline{\overline{T}}\cdot\hat{e}_2,\quad m_{12} = \hat{e}_1\cdot\overline{\overline{T}}\cdot\hat{e}_3,\quad m_{21} = \hat{e}_4\cdot\overline{\overline{T}}\cdot\hat{e}_2,\quad m_{22} = \hat{e}_4\cdot\overline{\overline{T}}\cdot\hat{e}_3, \tag{5.498}$$

where $\hat{e}_2 = (0, 1, 0, 0)$ and $\hat{e}_3 = (0, 0, 1, 0)$, leading to the characteristic equation

$$\det\overline{\overline{m}} = m_{11}m_{22} - m_{12}m_{21} = 0. \tag{5.499}$$

For example, when the configuration consists of two magneto-dielectric wedges and a PEC wedge, it follows from (5.488) with $n = 1$ and (5.498) that

$$m_{11} = S_1C_2 + C_1S_2\frac{\epsilon_1 k_{\perp,2}^2}{\epsilon_2 k_{\perp,1}^2} \tag{5.500}$$

$$m_{12} = C_1S_2(\epsilon_1\mu_1 - \epsilon_2\mu_2)\frac{k_z\omega}{\epsilon_2 k_{\perp,1}^2} \tag{5.501}$$

$$m_{21} = S_1C_2(\epsilon_2\mu_2 - \epsilon_1\mu_1)\frac{k_z\omega}{\mu_2 k_{\perp,1}^2} \tag{5.502}$$

$$m_{22} = -C_1S_2 - S_1C_2\frac{\mu_1 k_{\perp,2}^2}{\mu_2 k_{\perp,1}^2}, \tag{5.503}$$

and equation (5.499) reduces to

$$\det\overline{\overline{m}} = -\frac{k_{\perp,2}^2}{k_{\perp,1}^2}\left(S_1C_2 + \frac{\epsilon_1}{\epsilon_2}C_1S_2\right)\left(C_1S_2 + \frac{\mu_1}{\mu_2}S_1C_2\right). \tag{5.504}$$

Two factors in the brackets imply two branches of zeros, the equations for which can be represented in the form (Meixner 1972):

$$\frac{\epsilon_1 - \epsilon_2}{\epsilon_1 + \epsilon_2} = \frac{\sin[\tau(\Phi_1 + \Phi_2)]}{\sin[\tau(\Phi_1 - \Phi_2)]} \tag{5.505}$$

$$\frac{\mu_2 - \mu_1}{\mu_2 + \mu_1} = \frac{\sin[\tau(\Phi_1 + \Phi_2)]}{\sin[\tau(\Phi_1 - \Phi_2)]}. \tag{5.506}$$

In the limit $\epsilon_1 = \epsilon_2$ and $\mu_1 = \mu_2$, (5.505) and (5.506) reduce to $\sin[\tau(\Phi_1 + \Phi_2)] = 0$ with the solutions $\tau_s = \pi s/(\Phi_1 + \Phi_2)$, where $s = 1, 2, \ldots$, the smallest of which with $s = 1$ is exactly the degree of singularity that is observed for PEC wedges (see section 5.2.3). Finally, note that the characteristic exponents of symmetric configurations ($\Phi_1 = \Phi_2$) do not depend on material parameters and are equal to those for a PEC wedge in a homogeneous background medium (Meixner 1972).

The characteristic equation for the structure with $N = 2$ and without PEC inclusion (magneto-dielectric wedge in a homogeneous background medium) follows from (5.492) as

$$\det\left[\overline{\overline{I}} - \overline{\overline{T}}_2(\tau) \cdot \overline{\overline{T}}_1(\tau)\right] = Q\left(\tau, \frac{\epsilon_1}{\epsilon_2}, \Phi_1\right) Q\left(\tau, \frac{\mu_1}{\mu_2}, \Phi_1\right) = 0 \tag{5.507}$$

with

$$Q\left(\tau, \frac{\epsilon_1}{\epsilon_2}, \Phi_1\right) = 2 - 2C_1 \cos[\tau(2\pi - \Phi_1)] + \left(\frac{\epsilon_1}{\epsilon_2} + \frac{\epsilon_2}{\epsilon_1}\right) S_1 \sin[\tau(2\pi - \Phi_1)] \tag{5.508}$$

and similar for $Q(\tau, \mu_1/\mu_2, \Phi_1)$, which is equivalent to the four equations (Meixner 1972)

$$\frac{\epsilon_1 - \epsilon_2}{\epsilon_1 + \epsilon_2} = \pm\frac{\sin(\pi\tau)}{\sin[\tau(\pi - \Phi_1)]} \tag{5.509}$$

$$\frac{\mu_1 - \mu_2}{\mu_1 + \mu_2} = \pm\frac{\sin(\pi\tau)}{\sin[\tau(\pi - \Phi_1)]}. \tag{5.510}$$

Again, the characteristic equation (5.507) has an infinite number of various solutions τ_s ($s = 1, 2, \ldots$), and the one with the smallest positive real part determines the degree of singularity of the fields at the edge. In the limit of homogeneous medium ($\epsilon_1 = \epsilon_2$ and $\mu_1 = \mu_2$) equations (5.509) and (5.510) have only integer solutions $\tau_s = s$, and the fields are non-singular, as expected.

A special case of interest to plasmonics is when $\epsilon_2 \to -\epsilon_1$ and / or $\mu_2 \to -\mu_1$. Such structures are known for rapid oscillations and strong singularities of the fields at the edge (Wallén *et al.* 2008). Assume, for example, that $\epsilon_2 \to -\epsilon_1$. Then, the left-hand side of (5.509) goes to infinity, implying solutions with $\mathrm{Im}\,\tau \to \pm\infty$. Assuming that $\pi < \Phi_1 < 2\pi$ (Figure 5.37) and $\chi = \arg(\epsilon_2/\epsilon_1 + 1)$, where $-\pi < \chi < 0$ (condition of passivity of the media), and approximating the

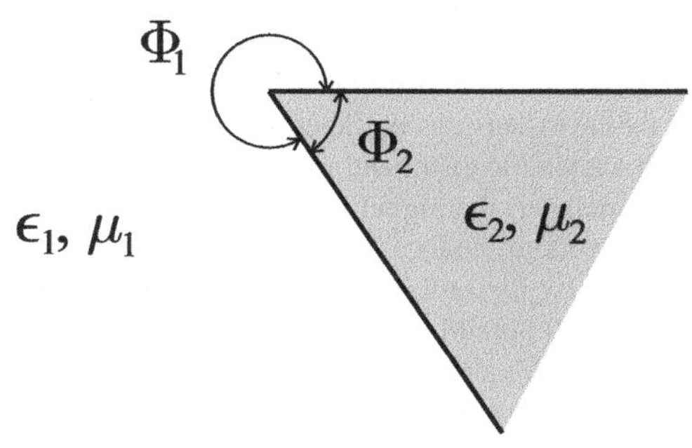

Figure 5.37 A wedge in a homogeneous space.

trigonometric functions on the right-hand side of (5.509) by growing exponent functions give for the characteristic exponents with the smallest positive real parts (problem 28)

$$\tau \approx \begin{cases} \dfrac{\pi + \chi}{\Phi_2} + \dfrac{j}{\Phi_2} \ln \left| \dfrac{\epsilon_2 - \epsilon_1}{\epsilon_2 + \epsilon_1} \right|, & \operatorname{Im} \tau \to +\infty \\ -\dfrac{\chi}{\Phi_2} - \dfrac{j}{\Phi_2} \ln \left| \dfrac{\epsilon_2 - \epsilon_1}{\epsilon_2 + \epsilon_1} \right|, & \operatorname{Im} \tau \to -\infty \end{cases}, \tag{5.511}$$

where $\Phi_2 = 2\pi - \Phi_1$ is the angular width of the second wedge. When $\epsilon_2 \to -\epsilon_1$, imaginary parts of characteristic exponents in (5.511) grow, implying rapid oscillations of the singular field components E_ρ, E_φ, H_ρ, and H_φ as $O[\exp(j \operatorname{Im} \tau \ln \rho)]$ at a rate increasing as $\rho \to 0$. Furthermore, for smaller losses, when $\chi \to 0$ or $-\pi$, the real part of a solution from (5.511) approaches zero, implying a strong field singularity at the edge. When ϵ_2 is exactly equal to $-\epsilon_1$, equations (5.509) degenerate to $\sin[\tau(\pi - \Phi_1)] = 0$, and the fields become regular since the smallest positive solution of (5.509) is equal to $\tau = \pi/|\pi - \Phi_1|$, which is always real-valued and greater than unity. Similar analysis applies to the case $\mu_2 \to -\mu_1$.

In both considered cases ($N = 2$, with and without the PEC inclusion), the characteristic equations, which are (5.505) and (5.506) or (5.509) and (5.510), respectively, are independent of k_z, implying the degree of singularity as in the case of normal illumination ($k_z = 0$), despite the presence of k_z in the continuity conditions (5.483)–(5.486). This result can be seen as a confirmation of the implicit assumption made by Meixner (1972) that the degree of singularity is independent of the incident field and is entirely determined by the geometry and the material constitution of the structure at the edge.

5.6.2 Analytical Structure of Meixner's Series

Since exact solutions of full scattering problems for wedge-shaped configurations with $N \geq 1$ are in general unavailable, one is tempted to try to construct local solutions by complementing the Meixner ansatz (5.472) with further terms. Referring to the structure of equations (5.477) and (5.478), one may search for the expansions of the form (Meixner's series)

$$E_z = e^{-jk_z z} \rho^\tau [E_0(\varphi) + \rho^2 E_1(\varphi) + \rho^4 E_2(\varphi) + \ldots] \tag{5.512}$$

$$H_z = e^{-jk_z z} \rho^\tau [H_0(\varphi) + \rho^2 H_1(\varphi) + \rho^4 H_2(\varphi) + \ldots]. \tag{5.513}$$

Insertion of (5.512) and (5.513) in the Helmholtz equations and equating like powers of ρ result in the recursive set of equations

$$E''_m(\varphi) + (\tau + 2m)^2 E_m(\varphi) = -k_\perp^2 E_{m-1}(\varphi) \tag{5.514}$$

$$H''_m(\varphi) + (\tau + 2m)^2 H_m(\varphi) = -k_\perp^2 H_{m-1}(\varphi), \tag{5.515}$$

where $m = 0, 1, 2, \ldots$ and $E_{-1}, H_{-1} = 0$. The conditions at material boundaries (5.483)–(5.486) do not depend on τ and do not couple coefficients with different values of n, so they remain the same for all terms in the expansions. The equations with $m = 0$ are homogeneous and together with the conditions at material boundaries define the values of τ_s ($s = 1, 2, \ldots$) for which the system with $m = 0$ has non-trivial solutions. The systems with $m \geq 1$ are inhomogeneous, with their right-hand sides determined by the expansion coefficients from the previous step of the recursion, and provided that $\tau + 2m$ does not coincide with a characteristic exponent, the systems can be solved and the coefficients E_m and H_m determined from E_{m-1} and H_{m-1}. In this manner, the higher-order expansion coefficients in (5.512) and (5.513) can be expressed through E_0 and H_0. In the framework of the local procedure, these latter, as solutions of homogeneous systems, remain undefined. They are required to satisfy the radiation conditions as the total field is a superposition of the local modes of the form (5.512) and (5.513) with all possible values of τ (Osipov 2002, 2004).

A special case occurs when $\tau_s + 2m$ with some s and m is coincidentally equal to another solution τ_q of the characteristic equation. This happens, for example, when the angular width of a wedge is a rational fraction of π (Bach Andersen and Solodukhov 1978). In such a case, E_m and H_m cannot be uniquely determined from (5.514) and (5.515) as the corresponding homogeneous system permits non-trivial solutions, and the construction of the expansions (5.512) and (5.513) breaks down

(Bach Andersen and Solodukhov 1978). This observation has questioned the whole Meixner procedure for determining the degrees of singularities at edges (Marx 1990).

The problem can be solved by noting that the situation corresponds to the case of external excitation of a resonant system, implying infinite solutions for E_m and H_m. It is therefore convenient to assume for a while that parameters of the configuration (e.g., angular width of a wedge) are slightly different from the degenerated case and consider the limit. The full representation of the total field is a superposition of the local solutions with all possible values of τ, including τ_s and τ_q. In the limit, the mth term of the expansion for $\tau = \tau_s$ tends to infinity and its power of ρ approaches that in the leading term of the expansions for $\tau = \tau_q$. Since physical solutions must be bounded for every parameter of the wedge-shaped configuration, the only way out is the compensation of singularities in the terms with equal powers of ρ. On cancellation of singularities the remaining term will include a derivative of ρ^τ with respect to τ, which is $\ln \rho$. This means that in the degenerated cases the assumption about the purely power structure of the expansions as in (5.512) and (5.513) does not apply and the Meixner series have to be complemented with logarithmic terms (Makarov and Osipov 1986). Such expansions are free of the resonance problem and can be constructed, which was demonstrated for a structure consisting of two right-angled dielectric wedges and a homogeneous half-space ($N = 3$, $\Phi_1 = \pi$, $\Phi_2 = \Phi_3 = \pi/2$) (Makarov and Osipov 1986) and for structures involving semi-transparent half-planes (Braver *et al.* 1988; Idemen 2000). The log terms can be explicitly derived from the exact Maliuzhinets solution for an impedance wedge in the limits of a half-plane and a full plane (Osipov 1991). In section 5.4.2, equations (5.282) and (5.283), the log term is extracted in the case of the full-plane impedance junction.

It should be pointed out that the log terms appear in the higher-order terms of Meixner's expansions, so they do not influence the lowest-order term, which determines the degree of singularity τ_1. The only exception is the case of the full-plane impedance junction, in which the transverse field components have the logarithmic singularity at the edge rather than a power singularity (see equations (5.282) and (5.283)). A deeper theoretical insight into the problem can be gained from the framework of a theory based on Kontorovich–Lebedev representations of the fields; the interested reader is referred to Osipov (1993b,c) and Salem *et al.* (2006).

5.7 Problems

1. Show that expression (5.75)
   (a) describes an exact solution of Maxwell's equations;
   (b) satisfies boundary conditions (5.10) on the sides of the half-plane;
   (c) behaves at the edge according to (5.43) and (5.47);
   (d) consists of GO and edge-diffracted components as in (5.85)–(5.87) when $k_\perp \rho \gg 1$.

   Hint: Use properties of the Fresnel integral (Appendix B) and relations (2.313)–(2.316) between the field components.

2. By using exact solutions (5.66) and (5.75) derive analytical expressions for the field components near the edge. Assume the H-polarized incident plane wave. Compare the results with approximations (5.47)–(5.50). Compare the field behavior at the edge of an interior right-angled wedge and at the edge of a half-plane.

3. Express the equivalent currents $\mathbf{K}_\mathrm{e}$ and $\mathbf{K}_\mathrm{m}$ on the faces of a PEC wedge through the field components.

   Hint: The currents are defined in (2.348).

   Answer: $\mathbf{K}_\mathrm{m} = 0$ and $\mathbf{K}_\mathrm{e} = \mp H_z^\pm \hat{\rho} \pm H_\rho^\pm \hat{z}$ where $H_z^\pm$ and $H_\rho^\pm$ denote the tangential components of the magnetic field at the faces $\varphi = \pm\Phi$, respectively.

4. Use the exact solution (5.75) for the PEC half-plane and specialize it to the tangential components of the magnetic field.

   Hint: Use the formula $F'(X) = \mathrm{e}^{-jX^2+j\pi/4}/\sqrt{\pi}$ which follows from the definition (B.1) of the Fresnel integral.

   Answer:

$$H_z(\rho, \pm\pi) = 2H_{0z} F\left(\pm\sqrt{2k_\perp \rho}\,\sin\frac{\varphi_0}{2}\right) \mathrm{e}^{-jk_z z - jk_\perp \rho\cos\varphi_0} \tag{5.516}$$

$$H_\rho(\rho, \pm\pi) = AF\left(\pm\sqrt{2k_\perp \rho}\,\sin\frac{\varphi_0}{2}\right) \mathrm{e}^{-jk_z z - jk_\perp \rho\cos\varphi_0} \pm \frac{B}{\sqrt{2\pi k_\perp \rho}} \mathrm{e}^{-jk_z z - jk_\perp \rho - j\pi/4}, \tag{5.517}$$

where

$$A = \frac{2k}{Zk_\perp}\left(E_{0z}\sin\varphi_0 - ZH_{0z}\frac{k_z}{k}\cos\varphi_0\right) \tag{5.518}$$

$$B = \frac{2k}{Zk_\perp}\left(E_{0z}\cos\frac{\varphi_0}{2} + ZH_{0z}\frac{k_z}{k}\sin\frac{\varphi_0}{2}\right). \tag{5.519}$$

5. Explain the difference in the behavior of the components H_z and H_ρ observed in Figures 5.13 and 5.14.
Hint: Specialize the previous result to normal incidence and use an asymptotic approximation for the Fresnel integral of large argument.
Answer: With $k_z = 0$, equations (5.516)–(5.519) give

$$H_z(\rho,\pi) = 2H_{0z}\mathrm{e}^{-jk\rho\cos\varphi_0}F\left(\sqrt{2k\rho}\sin\frac{\varphi_0}{2}\right) \tag{5.520}$$

$$H_\rho(\rho,\pi) = 2YE_{0z}\sin\varphi_0\mathrm{e}^{-jk\rho\cos\varphi_0}F\left(\sqrt{2k\rho}\sin\frac{\varphi_0}{2}\right) + 2YE_{0z}\cos\frac{\varphi_0}{2}\frac{\mathrm{e}^{-jk\rho-j\pi/4}}{\sqrt{2\pi k\rho}}. \tag{5.521}$$

At the edge ($\rho \to 0$), $H_z(\rho,\pi) \to H_{0z}$ whereas $H_\rho(\rho,\pi) = O(1/\sqrt{k\rho})$. Far from the edge, when $\sqrt{k\rho}|\sin(\varphi_0/2)| \gg 1$, using (B.11),

$$H_z(\rho,\pi) \approx H_{0z}\left[2\Theta(\varphi_0)\mathrm{e}^{-jk\rho\cos\varphi_0} + \csc\frac{\varphi_0}{2}\frac{\mathrm{e}^{-jk\rho+j3\pi/4}}{\sqrt{2\pi k\rho}}\right] \tag{5.522}$$

$$H_\rho(\rho,\pi) \approx YE_{0z}\left[2\Theta(\varphi_0)\sin\varphi_0\mathrm{e}^{-jk\rho\cos\varphi_0} + \csc\frac{\varphi_0}{2}\cot\frac{\varphi_0}{2}\frac{\mathrm{e}^{-jk\rho-j3\pi/4}}{\sqrt{\pi}(2k\rho)^{3/2}}\right]. \tag{5.523}$$

The unit step function $\Theta(\varphi_0)$ ensures the appearance/disappearance of the GO term when the face is illuminated ($\varphi_0 > 0$) or lies in the shadow ($\varphi_0 < 0$). The edge-diffracted term is $O[(k\rho)^{-1/2}]$ for H polarization and $O[(k\rho)^{-3/2}]$ for E polarization.

6. Describe the solution for a PEC half-plane when the incident wave comes from the direction $\varphi_0 = 0$ perpendicularly to the edge ($k_z = 0$). Is there a scattered field?
Hint: Use the exact solution (5.75) and the property (B.7) of the Fresnel integral.
Answer: The answer depends on the polarization. For H polarization,

$$H_z = H_{0z}\mathrm{e}^{jk\rho\cos\varphi} = H_z^{\mathrm{inc}} \tag{5.524}$$

and no scattered field occurs. For the E-polarized field,

$$E_z = E_{0z}\mathrm{e}^{jk\rho\cos\varphi}\left[2F\left(\sqrt{2k\rho}\cos\frac{\varphi}{2}\right) - 1\right], \tag{5.525}$$

and there is a scattered component due to the presence of the Fresnel integral.

7. A plane wave

$$\mathbf{E}^{\mathrm{inc}} = \hat{z}E_{0z}\mathrm{e}^{jk\rho\cos(\varphi-\varphi_0)} \tag{5.526}$$

with $\varphi_0 = 3\pi/4$ is incident on a PEC half-plane. Estimate numerically the accuracy of the high-frequency approximations (5.81), (5.85), and (5.87) for (a) $\rho = \lambda$ and (b) $\rho = 10\lambda$.

8. By using the solution (5.75), derive the time-domain solution for a half-plane illuminated by an impulse plane wave.

9. A wedge is illuminated by a line source.
(a) Generalize expression (5.87) to the case of illumination by cylindrical waves (5.31) and (5.32).
(b) Assume that the line source is located far from the edge of the wedge and simplify the obtained expressions.

10. Derive formula (5.135) as a limiting case of the solution for the ideally conducting wedge, when the wedge angle tends to zero.

11. Show that the solution for the PEC wedge problem obtained by Maliuzhinets' method in the form of Sommerfeld's integrals (5.168) and (5.169) with the spectra (5.203) and (5.204) recovers the solution of the same problem derived in section 5.2 with a different method in the form of expressions (5.40), (5.41), (5.53), and (5.54).
Hint: Rewrite (5.168) and (5.169) as integrals over the contour γ_+ and use the trigonometric relations

$$\frac{-\sin(\nu\alpha)}{\cos(\nu\alpha)-\cos[\nu(\varphi-\varphi_0)]} \pm \frac{\sin(\nu\alpha)}{\cos(\nu\alpha)-\cos[\nu(\varphi+\varphi_0+2\Phi)]} = \begin{cases} \dfrac{\cos(\nu\varphi_0)}{\sin[\nu(\alpha+\varphi)]-\sin(\nu\varphi_0)} - \dfrac{\cos(\nu\varphi_0)}{\sin[\nu(-\alpha+\varphi)]-\sin(\nu\varphi_0)} \\ \dfrac{\cos[\nu(\alpha+\varphi)]}{\sin[\nu(\alpha+\varphi)]-\sin(\nu\varphi_0)} - \dfrac{\cos[\nu(-\alpha+\varphi)]}{\sin[\nu(-\alpha+\varphi)]-\sin(\nu\varphi_0)} \end{cases}. \quad (5.527)$$

12. Check that the function $\Psi(\alpha)$ in the form (5.215) does satisfy the functional equations (5.214) if the function $\psi_\Phi(\alpha)$ has the properties (5.216) and (5.217).

13. Prove that $\Psi(\alpha)$ is 2π periodic when $\Phi = \pi/(4M)$ with $M = 1, 2, \ldots$ (Osipov and Norris 1998).
Hint: By repeatedly using (5.233), show that with arbitrary Φ

$$\frac{\Psi(\alpha+2\pi)}{\Psi(\alpha)} = \prod_{k=1}^{4} \frac{\cos\left[\frac{\pi}{4\Phi}(\alpha_k+3\pi/2)\right]}{\cos\left[\frac{\pi}{4\Phi}(\alpha_k+\pi/2)\right]}, \quad (5.528)$$

where $\alpha_1 = \alpha+\Phi+\theta_+-\pi/2$, $\alpha_2 = \alpha+\Phi-\theta_++\pi/2$, $\alpha_3 = \alpha-\Phi+\theta_--\pi/2$, and $\alpha_4 = \alpha-\Phi-\theta_-+\pi/2$; $\Psi(\alpha)$ and $\theta_\pm$ are either $\Psi_e(\alpha)$ and $\theta_e^\pm$ or $\Psi_h(\alpha)$ and $\theta_h^\pm$. Then, prove that

$$\cos\left[\frac{\pi}{4\Phi}\left(\alpha_k+3\frac{\pi}{2}\right)\right] = (-1)^M \cos\left[\frac{\pi}{4\Phi}\left(\alpha_k+\frac{\pi}{2}\right)\right] \quad (k=1,2,3,4) \quad (5.529)$$

when $\Phi = \pi/(4M)$ with $M = 1, 2, \ldots$

14. Prove that for impedance wedges with equal face impedances ($\eta_+ = \eta_-$), $\Psi(\alpha)$ is 2π periodic when $\Phi = \pi/(2M)$ with $M = 1, 2, \ldots$ (Osipov and Norris 1998).
Hint: Note that in this case

$$\frac{\Psi(\alpha+2\pi)}{\Psi(\alpha)} = \frac{\cos\left[\frac{\pi}{2\Phi}(\alpha+\theta+\pi)\right]\cos\left[\frac{\pi}{2\Phi}(\alpha-\theta+2\pi)\right]}{\cos\left[\frac{\pi}{2\Phi}(\alpha+\theta)\right]\cos\left[\frac{\pi}{2\Phi}(\alpha-\theta+\pi)\right]}, \quad (5.530)$$

which is unity when $\Phi = \pi/(2M)$ with $M = 1, 2, \ldots$. Here, $\Psi(\alpha)$ and θ are either $\Psi_e(\alpha)$ and $\theta_e^\pm$ or $\Psi_h(\alpha)$ and $\theta_h^\pm$.

15. Show that in the limit of normal incidence ($\beta \to \pi/2$), the solution for the impedance-matched wedge given by (5.310) and (5.311) recovers the Maliuzhinets solution (5.246) and (5.247) with $\eta_\pm = 1$.
Solution: If $\beta \to \pi/2$, then it follows from (5.294), (5.300), and (5.301) that $\theta_{1,2} \to 0$ and

$$\Psi_{1,2}(\alpha) \to \Psi(\alpha,0) = \left[\frac{\chi_\Phi(\alpha+\Phi+\pi/2)\chi_\Phi(\alpha-\Phi+\pi/2)}{\chi_\Phi(\alpha+\Phi-\pi/2)\chi_\Phi(\alpha-\Phi-\pi/2)}\right]^2, \quad (5.531)$$

so that (5.310) and (5.311) reduce to

$$S_e(\alpha) \to E_{0z}\sigma(\alpha)\frac{\Psi(\alpha,0)}{\Psi(\varphi_0,0)}, \quad S_h(\alpha) \to ZH_{0z}\sigma(\alpha)\frac{\Psi(\alpha,0)}{\Psi(\varphi_0,0)}. \tag{5.532}$$

Using the relation (5.309) between χ_Φ and ψ_Φ gives

$$\frac{\Psi(\alpha,0)}{\Psi(\varphi_0,0)} = \left[\frac{\psi_\Phi(\alpha+\Phi)\psi_\Phi(\alpha-\Phi)}{\psi_\Phi(\varphi_0+\Phi)\psi_\Phi(\varphi_0-\Phi)}\right]^2, \tag{5.533}$$

which coincides with $\Psi_{e,h}(\alpha)/\Psi_{e,h}(\varphi_0)$ from (5.244) and (5.245) with $\theta^{\pm}_{e,h} = \pi/2$, so that the expressions in (5.532) recover the Maliuzhinets solution (5.246) and (5.247).

16. Prove the relation (5.314).
Solution: From (5.309) and (5.531),

$$\Psi(\varphi_0,0) = \chi_\Phi^8\left(\frac{\pi}{2}\right)\Psi(\varphi_0)|_{\theta_\pm=\pi/2}, \tag{5.534}$$

and from (5.309) with $\alpha = \pi/2$,

$$\psi_\Phi\left(\frac{\pi}{2}\right) = \frac{\chi_\Phi(\pi)}{\chi_\Phi(0)\chi_\Phi^2(\pi/2)}. \tag{5.535}$$

By definition $\chi_\Phi(0) = 1$ (see (5.302)) and from (5.305) and (5.307)

$$\chi_\Phi(\pi) = 2^{(1-\nu)/2}, \tag{5.536}$$

so that (5.535) becomes

$$\psi_\Phi\left(\frac{\pi}{2}\right) = \frac{2^{(1-\nu)/2}}{\chi_\Phi^2(\pi/2)} \tag{5.537}$$

or

$$\frac{1}{\chi_\Phi^8(\pi/2)} = \frac{1}{4}\psi_\Phi^4\left(\frac{\pi}{2}\right)e^{\nu\ln 4}, \tag{5.538}$$

which transforms (5.534) to (5.314).

17. Show that the waves $\mathbf{u}_{sw}$ in (5.405) describe inhomogeneous waves concentrated near the faces of the wedge and propagating away from the edge.
Solution: The exponent in (5.405) can be rewritten as $e^{-jk\rho\sin\beta\cos(\Phi\mp\varphi+\theta^{\pm}_{e,h})}$. Introducing positive parameters $l_\pm = \rho\cos(\Phi\mp\varphi)$ and $h_\pm = \rho\sin(\Phi\mp\varphi)$, which have the meaning of the distances between the observation point and the edge along a face and between the observation point and the respective face, the exponent is rewritten as

$$\exp(-jkl_\pm\sin\beta\cos\theta^{\pm}_{e,h} + jkh_\pm\sin\beta\sin\theta^{\pm}_{e,h}). \tag{5.539}$$

The necessary condition of pole capturing is that $\operatorname{Im}\sin\theta^{\pm}_{e,h} > 0$, implying exponential decay of the exponent with growing $h_\pm$. The dependence on $l_\pm$ is best seen if we rewrite the corresponding exponent as

$$\exp\left[-jkl_\pm\sin\beta\cos(\operatorname{Re}\theta^{\pm}_{e,h})\cosh(\operatorname{Im}\theta^{\pm}_{e,h}) - kl_\pm\sin\beta\sin(\operatorname{Re}\theta^{\pm}_{e,h})\sinh(\operatorname{Im}\theta^{\pm}_{e,h})\right].$$

Since $0 < \operatorname{Re}\theta^{\pm}_{e,h} \le \pi/2$ and $\operatorname{Im}\theta^{\pm}_{e,h} > 0$, it is an outgoing wave, exponentially decaying with increasing $l_\pm$ ($\operatorname{Re}\theta^{\pm}_{e,h} > 0$ means absorption of power in the impedance boundary, see (2.134)).

18. Write down a high-frequency approximation of the total field diffracted by an impenetrable wedge with $\Phi > \pi/2$ applicable at the shadow boundaries of the incident wave at $\tilde{\varphi}_0 = \varphi_0 \pm \pi$.

Hint: Specialize (5.436) and (5.437) by noting that $n = 0$, $\alpha_0 = \varphi_0$, res $\mathbf{s}(\alpha_0) = \mathbf{u}_0$, and σ_0 is equal to either sgn $(\varphi - \tilde{\varphi}_0)$ when $-\Phi < \varphi_0 < \Phi - \pi$ or to $-$sgn $(\varphi - \tilde{\varphi}_0)$ when $\pi - \Phi < \varphi_0 < \Phi$.

Answer: In both cases ($\tilde{\varphi}_0 = \varphi_0 \pm \pi$) the approximation is given by the formulas:

$$\tilde{\mathbf{u}}_0 = \mathbf{U}_0 F\left[\sqrt{2k\rho\sin\beta}\cos\left(\frac{\varphi-\varphi_0}{2}\right)\right] e^{jk\rho\sin\beta\cos(\varphi-\varphi_0)-jkz\cos\beta} \tag{5.540}$$

$$\overline{\overline{D}}_0 = \overline{\overline{D}} + \frac{1}{2}\sec\left(\frac{\varphi-\varphi_0}{2}\right)\overline{\overline{I}}. \tag{5.541}$$

19. Repeat problem 18 for the shadow boundaries of the waves reflected from the faces of the wedge.

Answer: For the shadow boundaries at $\tilde{\varphi}_{\pm 1} = \pm 2\Phi \mp \pi - \varphi_0$,

$$\tilde{\mathbf{u}}_{\pm 1} = \overline{\overline{R}}_{\pm}(\Phi \mp \varphi_0)\cdot \mathbf{U}_0 F\left[\sqrt{2k\rho\sin\beta}\cos\left(\frac{\varphi+\varphi_0\mp 2\Phi}{2}\right)\right]\cdot e^{jk\rho\sin\beta\cos(\varphi+\varphi_0\mp 2\Phi)-jkz\cos\beta} \tag{5.542}$$

$$\overline{\overline{D}}_{\pm 1} = \overline{\overline{D}} + \frac{1}{2}\sec\left(\frac{\varphi+\varphi_0\mp 2\Phi}{2}\right)\overline{\overline{R}}_{\pm}(\Phi\mp\varphi_0). \tag{5.543}$$

20. Prove the formulas (5.394)–(5.398).

Hint: Specialize the solution of problem 9 in section 4.7 to the case that $\hat{\xi} = \hat{z}$, the normal vectors are $\hat{n}_{\pm} = \hat{x}\sin\Phi \mp \hat{y}\cos\Phi$, and the incidence direction is given by (5.137).

Solution: Elements of matrices $\overline{\overline{P}}$ and $\overline{\overline{Q}}$ for the upper face are expressed through the following scalar products:

$$\hat{z}\cdot\hat{\alpha}_{\text{inc}} = \hat{z}\cdot\hat{\alpha}_{\text{ref}} = \frac{\cos(\Phi-\varphi_0)\sin\beta}{\sqrt{1-\sin^2\beta\sin^2(\Phi-\varphi_0)}}$$

$$\hat{z}\cdot\hat{\beta}_{\text{inc}} = -\hat{z}\cdot\hat{\beta}_{\text{ref}} = -\frac{\sin(\Phi-\varphi_0)\sin\beta\cos\beta}{\sqrt{1-\sin^2\beta\sin^2(\Phi-\varphi_0)}}.$$

The formulas for the lower face immediately follow from those for the upper face on the substitution $\Phi \to \pi - \Phi$.

21. Prove the formulas (5.455)–(5.458).

Hint: Use the definitions (5.453) and (5.454) of the components in the ray-fixed coordinates and the relations (2.156) and (2.158) between the electric and magnetic field components in a plane wave.

Solution: Consider, for example, the incident wave. From (5.449) and (5.453),

$$E^{\text{inc}}_{\beta_{\text{inc}}} = (\hat{\imath}\times\hat{\varphi}_{\text{inc}})\cdot\mathbf{E}_{\text{inc}} = \frac{1}{\sin\beta}(\hat{\imath}\times\hat{\imath}\times\hat{z})\cdot\mathbf{E}_{\text{inc}}, \tag{5.544}$$

and since $\hat{\imath}\times\hat{\imath}\times\hat{z} = -\hat{z} + \hat{\imath}(\hat{\imath}\cdot\hat{z})$ (see the formula for the triple vector product (A.12)) and $\hat{\imath}\cdot\mathbf{E}_{\text{inc}} = 0$ (transversality of the electric field in the plane wave), we obtain the formula (5.455)

$$E^{\text{inc}}_{\beta_{\text{inc}}} = -\frac{1}{\sin\beta}(\hat{z}\cdot\mathbf{E}_{\text{inc}}). \tag{5.545}$$

The component of the incident wave, orthogonal to the edge-fixed plane of incidence, is expressed with the help of (2.158) and (5.449) as

$$E^{\text{inc}}_{\varphi_{\text{inc}}} = \hat{\varphi}_{\text{inc}}\cdot\mathbf{E}_{\text{inc}} = -\frac{Z}{\sin\beta}(\hat{\imath}\times\hat{z})\cdot(\hat{\imath}\times\mathbf{H}_{\text{inc}}), \tag{5.546}$$

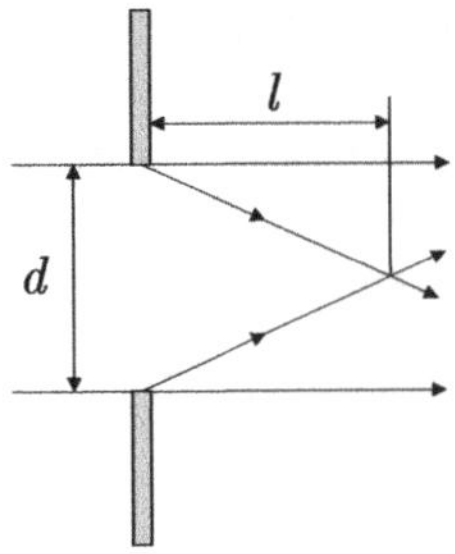

Figure 5.38 An opening in an impenetrable screen illuminated by a plane wave.

and using the formula (A.11) for the scalar triple product, the transversality of the magnetic field ($\hat{i} \cdot \mathbf{H}_{\text{inc}} = 0$) and the relation $(\hat{i} \times \hat{z}) \times \hat{i} = \hat{z} - \hat{i}(\hat{i} \cdot \hat{z})$ gives the formula (5.456)

$$E^{\text{inc}}_{\varphi_{\text{inc}}} = -\frac{Z}{\sin\beta}((\hat{i} \times \hat{z}) \times \hat{i}) \cdot \mathbf{H}_{\text{inc}} = -\frac{Z}{\sin\beta}\hat{z} \cdot \mathbf{H}_{\text{inc}}. \tag{5.547}$$

22. A plane wave with wavelength λ is incident on an opening of width d in an impenetrable screen, perpendicularly to the aperture. Estimate the distance l behind the aperture at which diffraction at the rim begins to influence the transmitted wave over the whole geometric cross section of the opening (Figure 5.38).

Hint: Use the formula (5.442) for the width δ_{ess} of the penumbra region behind an impenetrable wedge.

Answer: $l = d^2/\lambda$.

23. The tensor diffraction coefficient of a wedge in (5.459) relates the components of the edge-diffracted electric field to those of the incident electric field in the ray-fixed coordinates. What is the relation between the components of the magnetic fields?

Hint: Express the electric field components in the diffracted and incident waves through the magnetic field components.

Solution: For brevity, let us express (5.459) as

$$\begin{bmatrix} E^{\text{d}}_{\beta} \\ E^{\text{d}}_{\varphi} \end{bmatrix} = \frac{\exp(-jkr)}{\sqrt{r}} \begin{bmatrix} \mathcal{D}^E_{11} & \mathcal{D}^E_{12} \\ \mathcal{D}^E_{21} & \mathcal{D}^E_{22} \end{bmatrix} \begin{bmatrix} E^{\text{inc}}_{\beta_{\text{inc}}} \\ E^{\text{inc}}_{\varphi_{\text{inc}}} \end{bmatrix}. \tag{5.548}$$

In terms of magnetic fields we can expect a similar representation:

$$\begin{bmatrix} H^{\text{d}}_{\beta} \\ H^{\text{d}}_{\varphi} \end{bmatrix} = \frac{\exp(-jkr)}{\sqrt{r}} \begin{bmatrix} \mathcal{D}^H_{11} & \mathcal{D}^H_{12} \\ \mathcal{D}^H_{21} & \mathcal{D}^H_{22} \end{bmatrix} \begin{bmatrix} H^{\text{inc}}_{\beta_{\text{inc}}} \\ H^{\text{inc}}_{\varphi_{\text{inc}}} \end{bmatrix}. \tag{5.549}$$

The incident wave is a plane wave and the edge-diffracted wave is locally plane, implying that

$$\mathbf{E}_{\text{inc}} = -Z\hat{i} \times \mathbf{H}_{\text{inc}}, \quad \mathbf{E}_{\text{d}} = -Z\hat{s} \times \mathbf{H}_{\text{d}}. \tag{5.550}$$

Using the definition of the ray-fixed coordinate frames in (5.449) and (5.450), we find that

$$E^{\text{d}}_{\beta} = -ZH^{\text{d}}_{\varphi}, \quad E^{\text{d}}_{\varphi} = ZH^{\text{d}}_{\beta}, \quad E^{\text{inc}}_{\beta_{\text{inc}}} = -ZH^{\text{inc}}_{\varphi_{\text{inc}}}, \quad E^{\text{inc}}_{\varphi_{\text{inc}}} = ZH^{\text{inc}}_{\beta_{\text{inc}}}. \tag{5.551}$$

Inserting these in (5.548) gives

$$\begin{bmatrix} H^{\text{d}}_{\beta} \\ H^{\text{d}}_{\varphi} \end{bmatrix} = \frac{\exp(-jkr)}{\sqrt{r}} \begin{bmatrix} \mathcal{D}^E_{22} & -\mathcal{D}^E_{21} \\ -\mathcal{D}^E_{12} & \mathcal{D}^E_{11} \end{bmatrix} \begin{bmatrix} H^{\text{inc}}_{\beta_{\text{inc}}} \\ H^{\text{inc}}_{\varphi_{\text{inc}}} \end{bmatrix}, \tag{5.552}$$

and comparing (5.552) with (5.549) we obtain the relations:

$$\mathcal{D}^H_{11} = \mathcal{D}^E_{22}, \quad \mathcal{D}^H_{12} = -\mathcal{D}^E_{21}, \quad \mathcal{D}^H_{21} = -\mathcal{D}^E_{12}, \quad \mathcal{D}^H_{22} = \mathcal{D}^E_{11}. \tag{5.553}$$

24. Assume that the edge-diffraction coefficient $\overline{\overline{D}}$ of a wedge with the relative equivalent surface impedance $\eta = \eta_+ = \eta_-$ is known. Derive the edge-diffraction coefficient of a wedge with the impedance $1/\eta$ by using the result from problem 23 and the symmetry properties of the fields scattered by impedance scatterers with inverse values of η addressed in problem 8 in section 2.7.

Answer: The result reads

$$D_{\mathrm{ee}}(1/\eta) = D_{\mathrm{hh}}(\eta), \quad D_{\mathrm{eh}}(1/\eta) = -D_{\mathrm{he}}(\eta),$$
$$D_{\mathrm{he}}(1/\eta) = -D_{\mathrm{eh}}(\eta), \quad D_{\mathrm{hh}}(1/\eta) = D_{\mathrm{ee}}(\eta),$$

and is a special case of more general symmetry properties (5.422) and (5.423).

25. Derive (5.505) and (5.506) from (5.504).

Hint: Use the trigonometric relations

$$2S_1C_2 = \sin[\tau(\Phi_1 + \Phi_2)] + \sin[\tau(\Phi_1 - \Phi_2)] \tag{5.554}$$
$$2C_1S_2 = \sin[\tau(\Phi_1 + \Phi_2)] - \sin[\tau(\Phi_1 - \Phi_2)]. \tag{5.555}$$

26. Derive the characteristic equation for the degree of singularity τ in the wedge-shaped configuration as in Figure 5.36 but with a PMC wedge in place of the PEC wedge.

Hint: For the PMC wedge, $H_z = 0$ and $H_\rho = 0$ at $\varphi = \varphi_1$ and at $\varphi = \varphi_N$.

Answer:

$$m_{11} = \hat{e}_2 \cdot \overline{\overline{T}} \cdot \hat{e}_1, m_{12} = \hat{e}_2 \cdot \overline{\overline{T}} \cdot \hat{e}_4, m_{21} = \hat{e}_3 \cdot \overline{\overline{T}} \cdot \hat{e}_1, m_{22} = \hat{e}_3 \cdot \overline{\overline{T}} \cdot \hat{e}_4, \tag{5.556}$$

where $\overline{\overline{T}}$ is defined in (5.495) and $\hat{e}_1 = (1,0,0,0)$, $\hat{e}_2 = (0,0,C_N,S_N)$, $\hat{e}_3 = (-S_N, C_N, 0, 0)$, and $\hat{e}_4 = (0,0,0,1)$.

27. Show that $\tau_1 = \pi/(2\Phi)$ for oblique illumination of the impedance wedge shown in Figure 5.24.

Solution: Representing the z components as

$$E_z = \mathrm{e}^{-jk_z z}\rho^\tau[a\cos(\tau\varphi) + b\sin(\tau\varphi)] \tag{5.557}$$
$$H_z = \mathrm{e}^{-jk_z z}\rho^\tau[c\cos(\tau\varphi) + d\sin(\tau\varphi)] \tag{5.558}$$

and inserting in the boundary conditions (5.150) and (5.151) give the equations:

$$-aS + bC - Z\cos\beta(cC + dS) = 0 \tag{5.559}$$
$$Y\cos\beta(aC + bS) - cS + dC = 0 \tag{5.560}$$
$$aS + bC - Z\cos\beta(cC - dS) = 0 \tag{5.561}$$
$$Y\cos\beta(aC - bS) + cS + dC = 0, \tag{5.562}$$

where $C = \cos(\tau\Phi)$ and $S = \sin(\tau\Phi)$. The equations can be rewritten as

$$2C(b - Zc\cos\beta) = 0 \tag{5.563}$$
$$2S(a + Zd\cos\beta) = 0 \tag{5.564}$$
$$2C(aY\cos\beta + d) = 0 \tag{5.565}$$
$$2S(bY\cos\beta - c) = 0, \tag{5.566}$$

which has non-trivial solutions for a, b, c, and d when either $C = 0$ or $S = 0$. Thus, $\tau_s = s\pi/(2\Phi)$ with $s = 1, 2, \ldots$, and the degree of singularity is determined by τ_1.

28. Derive solutions (5.511) of equations (5.509) for $\epsilon_2 \to -\epsilon_1$.

Hint: Note that when $\mathrm{Im}\,\tau \to \pm\infty$, the right-hand side of (5.509) simplifies to $\exp[\mp j\tau(2\pi - \Phi_1)]$.

References

Abramowitz M and Stegun IA (eds) 1965 *Handbook of Mathematical Functions with Formulas, Graphs, and Mathematical Tables*, Dover Publications.

Ahluwalia DS, Lewis RM, and Boersma J 1968 Uniform asymptotic theory of diffraction by a plane screen. *SIAM J. Appl. Math.*, **16**, 788–807.

Avdeev AD 1994 On the special function of the problem of diffraction by a wedge in an anisotropic plasma. *J. Commun. Techn. Electr.*, **39**, no. 10, 70–78.

Babich VM, Lyalinov MA, and Grikurov VG 2007 *Sommerfeld-Malyuzhinets Technique in Diffraction Theory*, Alpha Science International Ltd.

Bach Andersen J and Solodukhov VV 1978 Field behavior near a dielectric wedge. *IEEE Trans. Antennas Propag.*, **26**, 598–602.

Bernard JML 1994 Exact analytical solution for the diffraction at skew incidence by a class of wedge with absorbing boundary. *Rev. Tech. Thomson-CSF*, **20–21**, 523–527.

Bilow HJ 1991 Scattering by an infinite wedge with tensor impedance boundary conditions – a moment method/physical optics solution for the currents. *IEEE Trans. Antennas Propag.*, **39**, 767–773.

Bobrovnikov MS and Fisanov VV 1972 Plane wave diffraction by an impedance wedge in an anisotropic plasma. *Russian Physics Journal*, **15**, 372–376.

Bobrovnikov MS and Fisanov VV 1988 *Diffraction of Waves in Angular Regions*, Tomsk University Press, Tomsk (in Russian).

Boersma J and Rahmat-Samii Y 1980 Comparison of two leading uniform theories of edge diffraction with the exact uniform asymptotic solution. *Radio Sci.*, **15**, 1179–1194.

Borovikov VA and Kinber BY 1994 *Geometrical Theory of Diffraction*, The Institution of Electrical Engineers, London.

Bouman JJ, Senior TBA, and Uslenghi PLE (eds) 1987 *Electromagnetic and Acoustic Scattering by Simple Shapes*, Hemisphere Publishing Corporation.

Braver IM, Fridberg PS, Garb KL, and Yakover IM 1988 The behavior of the electromagnetic field near the edge of resistive half-plane. *IEEE Trans. Antennas Propag.*, **36**, 1760–1768.

Brooke G and Kharadly MMZ 1977 Field behavior near anisotropic and multidielectric edges. *IEEE Trans. Antennas Propag.*, **25**, 571–575.

Bucci OM and Franceschetti G 1976 Electromagnetic scattering by a half plane with two face impedances. *Radio Sci.*, **11**, 49–59.

Budaev BV 1995 *Diffraction by Wedges*, Longman Scientific & Technical.

Budaev BV and Bogy DB 2006 Diffraction of a plane skew electromagnetic wave by a wedge with general anisotropic impedance boundary conditions. *IEEE Trans. Antennas Propag.*, **54**, 1559–1567.

Carslaw HS 1920 Diffraction of waves by a wedge of any angle. *Proc. London Math. Soc.*, **18**, 291–306.

Ciarkowski A 1987 Three-dimensional electromagnetic plane wave diffraction by a perfectly conducting half-plane placed at the interface of two dielectric media. Exact and uniform asymptotic solutions. *Radio Sci.*, **22**, 969–975.

Collin RE and Zucker FJ (eds) 1969 *Antenna Theory: Part 2*, McGraw-Hill.

Daniele V 2010 The Wiener–Hopf formulation of the dielectric wedge problem: Part I. *Electromagnetics*, **30**, 625–643.

Daniele V 2011a The Wiener–Hopf formulation of the dielectric wedge problem: Part II. *Electromagnetics*, **31**, 1–17.

Daniele V 2011b The Wiener–Hopf formulation of the dielectric wedge problem: Part III: The skew incidence case. *Electromagnetics*, **31**, 550–570.

Daniele VG and Lombardi G 2006 Wiener–Hopf solution for impenetrable wedges at skew incidence. *IEEE Trans. Antennas Propag.*, **54**, 2472–2485.

Daniele VG and Lombardi G 2011 The Wiener–Hopf solution of the isotropic penetrable wedge problem: diffraction and total field. *IEEE Trans. Antennas Propag.*, **59**, 3797–3818.

Daniele V and Zich R 2014 *The Wiener–Hopf Method in Electromagnetics*, SciTech Publishing.

Greenberg GA 1948 *Selected Topics in the Mathematical Theory of Electric and Magnetic Phenomena*, Academy of Sciences of USSR, Moscow (in Russian).

Greenberg GA and Fock VA 1948 On the theory of the coastal refraction of radiowaves. In *Researches on the Propagation of Radiowaves. Academy of Sciences of USSR, Moscow, USSR*, **2**, 69–96 (in Russian).

Griesser T and Balanis CA 1989 Reflections, diffractions, and surface waves for an interior impedance wedge of arbitrary angle. *IEEE Trans. Antennas Propag.*, **37**, 927–935.

Hurd RA 1976 The edge condition in electromagnetics. *IEEE Trans. Antennas Propag.*, **24**, 70–73.

Idemen M 2000 Confluent edge conditions for the electromagnetic wave at the edge of a wedge bounded by material sheets. *Wave Motion*, **32**, 37–55.

James GL 1976 *Geometrical Theory of Diffraction for Electromagnetic Waves*, Peter Peregrinus, London.

Keller JB 1962 Geometrical theory of diffraction. *J. Opt. Soc. Am.*, **52**, 116–130.

Kim SY, Ra JW, and Shin SY 1991a Diffraction by an arbitrary-angled dielectric wedge. I. Physical optics approximation. *IEEE Trans. Antennas Propag.*, **39**, 1272–1281.

Kim SY, Ra JW, and Shin SY 1991b Diffraction by an arbitrary-angled dielectric wedge. II. Correction to physical optics solution. *IEEE Trans. Antennas Propag.*, **39**, 1282–1292.

Knockaert L, Olyslager F, and De Zutter D 1997 The diaphanous wedge. *IEEE Trans. Antennas Propag.*, **45**, 1374–1381.

Kontorovich MI and Lebedev NN 1938 About a method of solution of some problems in diffraction theory and related problems. *J. Exper. Theor. Phys. USSR*, **8**, 1192–1206.

Kouyoumjian RG and Pathak PH 1974 A uniform geometrical theory of diffraction for an edge in a perfectly conducting surface. *Proc. IEEE*, **62**, 1448–1461.

Lyalinov MA and Zhu NY 1999 Diffraction of a skewly incident plane wave by an anisotropic impedance wedge – a class of exactly solvable cases. *Wave Motion*, **30**, 275–288.

Lyalinov MA and Zhu NY 2006 Diffraction of a skew incident plane electromagnetic wave by an impedance wedge. *Wave Motion*, **44**, 21–43.

Lyalinov MA and Zhu NY 2013 *Scattering of Waves by Wedges and Cones with Impedance Boundary Conditions*, SciTech Publishing.

Macdonald HM 1902 *Electric Waves*, Cambridge University Press.

Makarov GI and Osipov AV 1986 The structure of Meixner's series. *Radiophys. Quantum Electr.*, **29**, 544–549.

Maliuzhinets GD 1950 *Some Generalizations of the Method of Reflections in the Theory of Sinusoidal Wave Diffraction*. Doctor of Science Dissertation, FIAN, Moscow (in Russian).

Maliuzhinets GD 1958a Excitation, reflection and emission of surface waves from a wedge with given face impedances. *Sov. Phys. Doklady*, **3**, 752–755.

Maliuzhinets GD 1958b Inversion formula for the Sommerfeld integral. *Sov. Phys. Doklady*, **3**, 52–56.

Malyughinetz GD 1960 Das Sommerfeldsche Integral und die Lösung von Beugungsaufgaben in Winkelgebieten. *Ann. Phys. (Berlin)*, **461**, 107–112.

Marx E 1990 Computed fields near the edge of a dielectric wedge. *IEEE Trans. Antennas Propag.*, **38**, 1438–1442.

McNamara DA, Pistorius CWI, and Malherbe JAG 1990 *Introduction to the Uniform Geometrical Theory of Diffraction*, Artech House.

Meixner J 1972 The behavior of electromagnetic fields at edges. *IEEE Trans. Antennas Propag.*, **20**, 442–446.

Mittra R and Lee SW 1971 *Analytical Techniques in the Theory of Guided Waves*, The Macmillan Company.

Monzon C, Forester DW, and P Loschialpo 2005 Exact solution to line source scattering by an ideal left-handed wedge. *Phys. Rev. E*, **72**, 056606, 1–5.

Morse PM and Feshbach H 1953 *Methods of Theoretical Physics*, McGraw-Hill.

Mur G 1981 The modeling of singularities in the finite-difference approximation of the time-domain electromagnetic field equations. *IEEE Trans. Microw. Theory Tech.*, **29**, 1073–1077.

Noble B 1958 *Methods Based on the Wiener–Hopf Technique*, Pergamon Press.

Norris AN and Osipov AV 1998 Far-field analysis of the Malyuzhinets solution for plane and surface waves diffraction by an impedance wedge. *Wave Motion*, **30**, 69–89.

Oberhettinger F 1954 Diffraction of waves by a wedge. *Comm. Pure Appl. Math.*, **7**, 551–563.

Osipov AV 1990 Asymptotic representation of the sound field in a narrow angular region with impedance boundaries. *Sov. Phys. Acoust.*, **36**, 287–290.

Osipov AV 1991 Diffraction of a plane wave in a corner region with impedance boundary conditions. *Sov. Phys. Acoust.*, **37**, 381–385.

Osipov AV 1993a Malyuzhinets transform and the method of Sommerfeld integral in the theory of diffraction of waves in wedge-shaped domains. *Problems of Diffraction and Wave Propagation, St. Petersburg State University, St. Petersburg, Russia*, **25**, 148–173 (in Russian).

Osipov AV 1993b On the method of Kontorovich–Lebedev integrals in problems of wave diffraction in sectional media. *Problems of Diffraction and Wave Propagation, St. Petersburg State University, St. Petersburg, Russia*, **25**, 173–219 (in Russian).

Osipov AV 1993c Harmonic wave diffraction problems in sectored media. *Vestn. S-Peterb. Univ., Ser. 4: Fiz, Khim*, no. 2, 10–21 (in Russian).

Osipov AV 2002 Analysis of electromagnetic diffraction by wedges with the method of edge functions. In *Proc. 27th General Assembly of the International Union of Radio Science, August 17–24, 2002, Maastricht, the Netherlands*, paper no. 1235, 4 p.

Osipov AV 2004 A hybrid technique for the analysis of scattering by impedance wedges. In *Proc. URSI Int. Symposium on Electromagnetic Theory, May 23– 27, 2004, Pisa, Italy*, 1140–1142.

Osipov AV 2005 Simple approximation of the Maliuzhinets function for describing wedge diffraction. *IEEE Trans. Antennas Propag.*, **53**, 2773–2776.

Osipov AV and Hongo K 1998 Green's function of an impedance wedge. *Electromagnetics*, **18**, 135–165.

Osipov AV and Norris AN 1998 The Malyuzhinets theory for scattering from wedge boundaries: a review. *Wave Motion*, **29**, 313–340.

Osipov AV and Senior TBA 2008a Electromagnetic diffraction by arbitrary-angle impedance wedges. *Proc. R. Soc. Lond. A*, **464**, 177–195.

Osipov AV and Senior TBA 2008b Diffraction by a right-angled impedance wedge. *Radio Sci.*, **43**, RS4S02, 1–15.

Osipov AV and Senior TBA 2009 Diffraction and reflection of a plane electromagnetic wave by a right-angled impedance wedge. *IEEE Trans. Antennas Propag.*, **57**, 1789–1797.

Osipov AV and Stein V 1999 *The Theory and Numerical Computation of Maliuzhinets Special Function*, German Aerospace Center (DLR), Institut für Hochfrequenztechnik, Research Report DLR-IB 551-5/1999, 57p.

Osipov AV, Hongo K, and Kobayashi H 2002 High-frequency approximations for electromagnetic field near a face of an impedance wedge. *IEEE Trans. Antennas Propag.*, **50**, 930–940.

Pelosi G, Selleri S, and Graglia RD 1996 The parabolic equation model for the numerical analysis of the diffraction at an impedance wedge: skew incidence case. *IEEE Trans. Antennas Propag.*, **44**, 267–268.

Poincaré H 1892 Sur la polarisation par diffraction. *Acta Math.*, **16**, 297–339.

Poincaré H 1897 Sur la polarisation par diffraction (Seconde partie). *Acta Math.*, **20**, 313–355.

Rawlins AD 2005 The optimum orientation of an absorbing barrier. *Proc. R. Soc. Lond. A*, **461**, 2369–2383.

Rojas RG 1988a Wiener–Hopf analysis of the EM diffraction by an impedance discontinuity in a planar surface and by an impedance half-plane. *IEEE Trans. Antennas Propag.*, **36**, 71–83.

Rojas RG 1988b Electromagnetic diffraction of an obliquely incident plane wave field by a wedge with impedance faces. *IEEE Trans. Antennas Propag.*, **36**, 956–970.

Ryshik IM and Gradstein IS 1957 *Tables of Series, Products, and Integrals*, VEB, Berlin.

Salem MA, Kamel AH, and Osipov AV 2006 Electromagnetic fields in the presence of an infinite dielectric wedge. *Proc. R. Soc. Lond. A*, **462**, 2503–2522.

Senior TBA 1978 Skew incidence on a right-angled impedance wedge. *Radio Sci.*, **13**, 639–647.

Senior TBA 1986 Solution of a class of imperfect wedge problems for skew incidence. *Radio Sci.*, **21**, 185–191.

Senior TBA and Volakis JL 1995 *Approximate Boundary Conditions in Electromagnetics*, The Institution of Electrical Engineers, London.

Sommerfeld A 1896 Mathematische Theorie der Diffraktion. *Math. Ann.*, **47**, 317–374.

Sommerfeld A 1927 Elektromagnetische Schwingungen, in *Die Differential- und Integralgleichungen der Mechanik und Physik. 2. Physikalischer Teil*, Frank P and von Mises R (eds), chapters 19–23, Vieweg & Sohn, Braunschweig.

Sommerfeld A 1964 *Lectures on Theoretical Physics: Optics*, Academic Press.

Tuzhilin AA 1973 On the theory of Maliuzhinets' inhomogeneous functional equations. *Differentsial'nye Uravneniya*, **9**, 2058–2064 (in Russian).

Ufimtsev PY 1962 *Method of Edge Waves in the Physical Theory of Diffraction*, Sovetskoe Radio, Moscow (in Russian).

Ufimtsev PY 2007 *Fundamentals of the Physical Theory of Diffraction*, Wiley-Interscience.

Vaccaro VG 1980 The generalized reflection method in electromagnetism. *AEÜ*, **34**, 493–500.

Vaccaro VG 1981 Electromagnetic diffraction from a right-angled wedge with soft conditions on one face. *Optica Acta*, **28**, 293–311.

Van Bladel J 1991 *Singular Electromagnetic Fields and Sources*, IEEE Press.

Vasilév EN and Solodukhov VV 1976 Diffraction of electromagnetic waves by a dielectric wedge. *Radiophys. Quantum Electr.*, **17**, 1161–1169.

Vasilév EN, Solodukhov VV, and Fedorenko AI 1991 The integral equation method in the problem of electromagnetic waves diffraction by complex bodies. *Electromagnetics*, **11**, 161–182.

Wallén H, Kettunen H, and Sihvola A 2008 Surface modes of negative-parameter interfaces and the importance of rounding sharp corners. *Metamaterials*, **2**, 113–121.

Weinstein LA 1969 *The Theory of Diffraction and the Factorization Method (Generalized Wiener–Hopf Technique)*, Golem Press.

Wiegrefe A 1912 Über einige mehrwertige Lösungen der Wellengleichung $\Delta u + k^2 u = 0$ und ihre Anwendung in der Beugungstheorie. *Ann. Phys. (Berlin)*, **344**, 449–484.

Wu TK and Tsai LL, 1977 Scattering by a dielectric wedge: a numerical solution. *IEEE Trans. Antennas Propag.*, **25**, 570–571.

Zavadskaya MP and Starovoitova RP 1988 Edge waves in G.D. Malyuzhinets solution for an impedance wedge. *Sov. Phys. Acoust.*, **34** 678–680.

Zavadskii VY and Sakharova MP 1967 Application of the special function ψ_Φ in problems of wave diffraction in wedge-shaped regions. *Sov. Phys. Acoust.*, **13**, 48–54.

6

Circular Cylinders and Convex Bodies

6.1 Introduction

Most of our knowledge about the physics of electromagnetic scattering comes from solutions of Maxwell's equations for several simply shaped bodies, for which it is possible to obtain exact solutions of the scattering problems. In addition to planar media (Chapter 4), the number of solvable cases is limited to just three geometries: a wedge (Chapter 5), a circular cylinder and a sphere (Chapter 7).[1]

This chapter is devoted to the infinite circular cylinder, which, after planar media, is the geometry with the simplest solution. The solution involves only Bessel functions (Appendix E) and is therefore particularly amenable to a detailed analysis. By using the cylinder as an example, it is very simple to illustrate such analysis methods as separation of variables and Watson's transformation. Furthermore, solutions for material cylinders are relatively easy available and simple, which makes the cylinder particularly important for studies of the effect of finite conductivity of scattering bodies on the scattered field.

Despite its infinite (and thus unphysical) dimension, the cylinder is a useful model of mechanical and electrical connection elements (wires, cables, fibers) or mechanical supporting elements (poles, rods, tubes) on a platform or cylindrical portions of the platform itself (conformal antennas). This is due to the fact that the fields on the surface of a finite cylinder with length more than several wavelengths are similar to the fields on an infinite cylinder, provided that the incident field is the same and the observation point is shifted by a wavelength or more away from the ends of the finite cylinder.

Perhaps the most important fact about the circular cylinder is that solutions for electrically large cylinders have a general meaning since they can be extended to generally shaped smooth and convex bodies. The high-frequency solution for an infinite circular cylinder can be used to describe reflection of electromagnetic waves from a piece of a cylindrically curved surface or transmission of electromagnetic waves from the lit to the shadowed region in the presence of smooth-shaped convex opaque bodies.

The electromagnetic formulation for a PEC cylinder is simpler than for a material cylinder, and the solution for a wave incident at a right angle to the axis of the cylinder is simpler than that for oblique incidence. So, we begin with the solution for a PEC cylinder under normal illumination (section 6.2), continue with the case of a dielectric cylinder (section 6.3), and conclude with solutions for material cylinders under oblique illumination (section 6.5). We examine the scattering from electrically thin dielectric and metal cylinders, showing in particular the possibility of anomalously strong scattering from thin cylinders with negative permittivity (section 6.3.3). Furthermore, comparing the solutions for an arbitrarily-sized coated cylinder and an impedance cylinder helps us to establish the conditions of applicability of the impedance boundary conditions on curved surfaces (section 6.5.2).

By using Watson's transformation (section 6.4), the solution at an observation point far from the surface of the cylinder can be cast in a physically clear form in terms of incident, reflected, and creeping waves. This is a format used in the

[1] In the case of PEC boundary conditions, more geometries are solvable, e.g. elliptic cylinder, circular and elliptic cones, paraboloid of revolution, circular disc.

Modern Electromagnetic Scattering Theory with Applications, First Edition. Andrey V. Osipov and Sergei A. Tretyakov.

framework of the geometrical theory of diffraction (GTD). On the other hand, application of Watson's transformation to the exact solution on the surface of an impedance cylinder leads to an asymptotic solution which is valid in the transition region between the lit and shadowed portions of the surface (Fock-type representation). The GTD and Fock-type asymptotic solutions can be extended to generally shaped smooth and convex impedance bodies by placing them in the context of GTD or by using the principle of locality of the fields in the penumbra, respectively (section 6.6).

The treatment in this chapter is limited to the incident plane wave, which is the typical incidence assumed in problems of scattering. Furthermore, the phenomena associated with multiple scattering in the interior of the scatterer, which are present in dielectric cylinders, are not addressed in this chapter; instead, this subject will be studied in Chapter 7, where we show how multiple scattering results in the spectacular phenomenon of rainbows.

The first exact solutions for electromagnetic scattering from cylinders were derived as series expansions in Bessel functions in the pioneering works by Lord Rayleigh (1881) and Thomson (1893, pp. 427–437). An overview of historical works on scattering from cylinders can be found in Van de Hulst (1957, Chapter 15). A vast literature is devoted to solutions for other sources and analysis of other aspects of the scattering from cylinders, and the reader is referred to Bouman *et al.* (1987), Felsen and Marcuvitz (1973), King and Wu (1959), and Ruck *et al.* (1970) for further information and references.

6.2 Perfectly Conducting Cylinders: Separation of Variables and Series Solution

For some very simple geometries the field solutions can be found by separation of variables. This method can be used in cases when the scatterer shape can be described as a set of coordinate surfaces in a certain coordinate system. It is a simple but powerful method, allowing exact solutions. As a first illustrative example, we will solve the plane-wave diffraction problem for an infinite circular ideally conducting cylinder excited by a normally incident plane wave. Although this is a very idealized scattering problem, the solution will allow us to understand a number of important concepts, such as creeping waves and plasmonic resonance.

In view of the problem symmetry, it is convenient to introduce a polar coordinate system (ρ, φ, z), as shown in Figure 6.1. The circular cylinder is a body with translational symmetry (section 2.5.5), and Maxwell's equations for this case are given by (2.313)–(2.316), (2.317), and (2.318). For the normal incidence that we consider in this section, the fields do not depend on the z coordinate, which means that $k_z = 0$ and Maxwell's equations split into two independent sub-systems:

$$\frac{1}{\rho}\frac{\partial E_z}{\partial \varphi} + j\omega\mu H_\rho = 0, \qquad -\frac{\partial E_z}{\partial \rho} + j\omega\mu H_\varphi = 0 \tag{6.1}$$

$$\frac{1}{\rho}\frac{\partial}{\partial \rho}\left(\rho \frac{\partial}{\partial \rho}E_z\right) + \frac{1}{\rho^2}\frac{\partial^2}{\partial \varphi^2}E_z + k^2 E_z = 0 \tag{6.2}$$

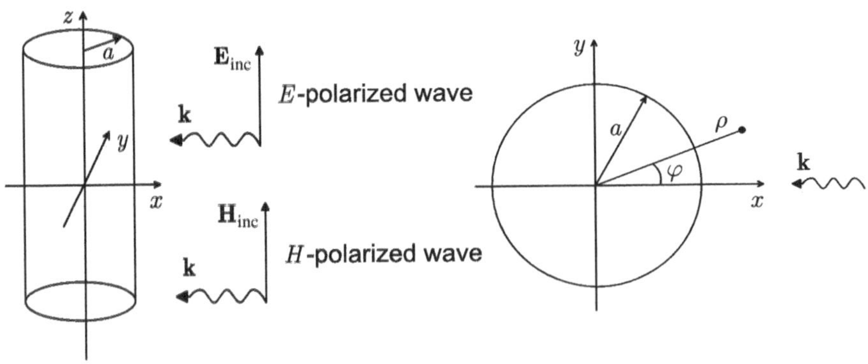

Figure 6.1 A circular cylinder excited by plane electromagnetic waves and the associated polar coordinate system. The scatterer surface is defined by the cylindrical coordinate surface $\rho = a$.

and

$$\frac{1}{\rho}\frac{\partial H_z}{\partial \varphi} - j\omega\epsilon E_\rho = 0, \qquad \frac{\partial H_z}{\partial \rho} + j\omega\epsilon E_\varphi = 0 \tag{6.3}$$

$$\frac{1}{\rho}\frac{\partial}{\partial \rho}\left(\rho\frac{\partial}{\partial \rho}H_z\right) + \frac{1}{\rho^2}\frac{\partial^2}{\partial \varphi^2}H_z + k^2 H_z = 0, \tag{6.4}$$

where $k = \omega\sqrt{\epsilon\mu}$ is the wavenumber in the exterior of the cylinder. The medium surrounding the cylinder will be assumed lossless so that $k > 0$.

The tangential components of the electric field must vanish on the PEC surface of the cylinder, which gives the boundary conditions

$$E_\varphi|_{\rho=a} = 0, \quad E_z|_{\rho=a} = 0 \tag{6.5}$$

or, using (6.3),

$$\left.\frac{\partial H_z}{\partial \rho}\right|_{\rho=a} = 0, \quad E_z|_{\rho=a} = 0. \tag{6.6}$$

It follows from (6.1), (6.2), (6.3), (6.4), and (6.6) that the fields are split up into two independent polarization cases: the E case with $H_z \equiv 0$, $E_\rho \equiv E_\varphi \equiv 0$ and the H case with $E_z \equiv 0$, $H_\rho \equiv H_\varphi \equiv 0$. This means that if $H_z \equiv 0$ in the incident field, then this property holds also for the total field H_z in the whole space. Equivalently, for the other polarization $E_z \equiv 0$ everywhere if $E_z \equiv 0$ in the incident field. In other words, the fields of these two polarizations are not coupled in scattering by a PEC cylinder.

Let the incident field be a plane wave

$$\mathbf{E}_{\text{inc}} = \mathbf{E}_0 e^{jkx} = \mathbf{E}_0 e^{jk\rho\cos\varphi}. \tag{6.7}$$

For E-polarized incident waves the electric field is directed along z at all points. The total electric field is the sum of the incident and scattered fields: $E_z = E_z^{\text{inc}} + E_z^{\text{sc}}$. The scattered field satisfies the Helmholtz equation

$$\nabla^2 E_z^{\text{sc}} + k^2 E_z^{\text{sc}} = 0, \tag{6.8}$$

the boundary condition

$$\left.\left(E_z^{\text{sc}} + E_z^{\text{inc}}\right)\right|_{\rho=a} = 0, \tag{6.9}$$

and the condition at infinity

$$\sqrt{\rho}\left(\frac{\partial E_z^{\text{sc}}}{\partial \rho} + jkE_z^{\text{sc}}\right) \to 0 \tag{6.10}$$

which follows from (2.323).

For H-polarized incident waves the magnetic field is directed along z at all points and has the amplitude H_0. Likewise, we write $H_z = H_z^{\text{inc}} + H_z^{\text{sc}}$,

$$\nabla^2 H_z^{\text{sc}} + k^2 H_z^{\text{sc}} = 0 \tag{6.11}$$

$$\left.\left(\frac{\partial H_z^{\text{inc}}}{\partial \rho} + \frac{\partial H_z^{\text{sc}}}{\partial \rho}\right)\right|_{\rho=a} = 0, \tag{6.12}$$

and

$$\sqrt{\rho}\left(\frac{\partial H_z^{\text{sc}}}{\partial \rho} + jkH_z^{\text{sc}}\right) \to 0. \tag{6.13}$$

The fields E_z^{sc} and H_z^{sc} satisfy similar equations, so in what follows for the sake of brevity we will use a scalar function u_{sc}, which is either E_z^{sc} or H_z^{sc}, to address both polarization cases simultaneously.

6.2.1 *Separation of Variables*

Looking for a solution of (6.8) and (6.9) for the scattered field in the form $u_{sc}(\rho, \varphi) = V(\rho)W(\varphi)$ we have (dividing by VW)

$$\frac{1}{\rho V(\rho)} \frac{d}{d\rho}\left(\rho \frac{dV(\rho)}{d\rho}\right) + \frac{1}{W(\varphi)\rho^2} \frac{d^2 W(\varphi)}{d\varphi^2} + k^2 = 0. \tag{6.14}$$

Multiplying by ρ^2, we get a sum of two functions depending only on ρ or φ. This means that

$$\frac{1}{W(\varphi)} \frac{d^2 W(\varphi)}{d\varphi^2} = \text{const} = -\lambda^2 \tag{6.15}$$

$$\frac{1}{\rho} \frac{d}{d\rho}\left(\rho \frac{dV(\rho)}{d\rho}\right) + \left(k^2 - \frac{\lambda^2}{\rho^2}\right) V(\rho) = 0. \tag{6.16}$$

The last equation is called the *Bessel* equation.

Particular solutions of the two main equations read

$$W(\varphi) = A_\lambda e^{-j\lambda\varphi} \tag{6.17}$$

$$V(\rho) = Z_\lambda(k\rho), \tag{6.18}$$

where Z_λ is a Bessel function and A_λ is a constant. From the radiation condition we choose the Hankel function

$$V(\rho) = H_\lambda^{(2)}(k\rho). \tag{6.19}$$

Function $W(\varphi)$ must be a periodical function with the period 2π, which means that $\lambda = n$, where $n = 0, \pm 1, \pm 2, \ldots$. Let us represent the unknown constant as[2]

$$A_n = C_n e^{jn\pi/2} \tag{6.20}$$

and write

$$W(\varphi) = C_n e^{-jn(\varphi - \pi/2)}. \tag{6.21}$$

Thus, the general solution can be written as

$$u_{sc}(\rho, \varphi) = \sum_{n=-\infty}^{\infty} C_n H_n^{(2)}(k\rho) e^{-jn(\varphi - \pi/2)}, \tag{6.22}$$

where C_n is a set of unknown coefficients.

6.2.2 *Satisfying the Boundary Conditions*

Next we must ensure that the field satisfies the boundary conditions. To do this, let us expand the incident plane wave into the Fourier series (Ryshik and Gradstein 1957):

$$e^{jk\rho\cos\varphi} = \sum_{n=-\infty}^{\infty} J_n(k\rho) e^{-jn(\varphi - \pi/2)}. \tag{6.23}$$

---

[2] Exponential factor is introduced simply for convenience in writing the following formulas.

The expansion coefficients are Bessel functions $J_n(x)$. Assuming that the incident field has the unit amplitude ($E_0 = 1$ for E polarization or $H_0 = 1$ for H polarization), the total field takes the form

$$u_{\rm tot} = \sum_{n=-\infty}^{\infty} \left[J_n(k\rho) + C_n H_n^{(2)}(k\rho)\right] {\rm e}^{-jn(\varphi-\pi/2)}. \tag{6.24}$$

For the E-polarized incidence this function, which represents the total electric field along the cylinder, should be zero at the cylinder surface $\rho = a$ (see 6.9). This determines the expansion coefficients:

$$C_n = -\frac{J_n(ka)}{H_n^{(2)}(ka)}. \tag{6.25}$$

For the H-polarized incidence this function represents the longitudinal magnetic field H_z. Its derivative with respect to ρ must vanish at the cylinder surface (see 6.12). Thus,

$$C_n = -\frac{J_n'(ka)}{H_n^{(2)\prime}(ka)}, \tag{6.26}$$

where the prime denotes the derivative with respect to the argument.

6.2.3 Scattered Fields

For compact writing, let us introduce an operator

$$\kappa = \begin{cases} 1, & \text{for } E \text{ polarization} \\ \left.\dfrac{\rm d}{{\rm d}(k\rho)}\right|_{\rho=a}, & \text{for } H \text{ polarization} \end{cases}. \tag{6.27}$$

With this notation, the z component of the scattered field for any polarization reads

$$u_{\rm sc}(\rho,\varphi) = -\sum_{n=-\infty}^{\infty} \frac{\kappa J_n(ka)}{\kappa H_n^{(2)}(ka)} H_n^{(2)}(k\rho){\rm e}^{-jn(\varphi-\pi/2)}. \tag{6.28}$$

This representation of the solution is called the Rayleigh series solution.[3]

The other field components are determined from Maxwell's equations according to (6.1) and (6.3). For the E-polarized incidence, E_z is given by (6.28), and

$$E_\rho = E_\varphi = 0, \quad H_z = 0, \quad H_\varphi = \frac{1}{j\omega\mu}\frac{\partial E_z}{\partial\rho}, \quad H_\rho = -\frac{1}{j\omega\mu}\frac{1}{\rho}\frac{\partial E_z}{\partial\varphi}. \tag{6.29}$$

The last component quickly decays with $k\rho \to \infty$.

For the H-polarized incidence, H_z is given by (6.28), and

$$H_\rho = H_\varphi = 0, \quad E_z = 0, \quad E_\varphi = -\frac{1}{j\omega\epsilon}\frac{\partial H_z}{\partial\rho}, \quad E_\rho = \frac{1}{j\omega\epsilon}\frac{1}{\rho}\frac{\partial H_z}{\partial\varphi}. \tag{6.30}$$

In this case, E_ρ quickly decays with $k\rho \to \infty$.

[3] Rayleigh in his pioneering work (Lord Rayleigh 1881) derived series solutions for a homogeneous circular cylinder with material constants differing from those in the surrounding medium. A solution specialized to the case of a PEC cylinder was published by Thomson (1893).

Far-zone Fields

For large distances from the cylinder (far zone, $k\rho \gg 1$) we may use the large-argument asymptotic approximation (E.25) for the Hankel function

$$H_n^{(2)}(k\rho) \sim \sqrt{\frac{2}{\pi k\rho}}\mathrm{e}^{-j\left(k\rho-\frac{\pi}{4}-n\frac{\pi}{2}\right)}. \tag{6.31}$$

In this approximation

$$u_{\mathrm{sc}}(\rho,\varphi) \approx -\sqrt{\frac{2}{\pi k\rho}}\mathrm{e}^{-j\left(k\rho-\frac{\pi}{4}\right)}\sum_{n=-\infty}^{\infty}\frac{\kappa J_n(ka)}{\kappa H_n^{(2)}(ka)}\mathrm{e}^{-jn(\varphi-\pi)}. \tag{6.32}$$

This is a cylindrical wave which decays as $1/\sqrt{k\rho}$. The angular dependence of the scattered amplitude is determined by the Fourier series in the above equation.

For such two-dimensional problems the usual definition of the scattering cross section cannot be used, since the scattering object has an infinite size. Instead, "scattering width", which refers to the scattered power *per unit length* of the cylinder, is used (section 3.2.2). When applied to the scattered field given by (6.29) and (6.30), the definitions (3.20)–(3.23) lead to

$$\sigma_{EE} = \lim_{\rho\to\infty} 2\pi\rho\frac{|E_z^{\mathrm{sc}}|^2}{|E_z^{\mathrm{inc}}|^2}, \quad \sigma_{HE} = 0, \quad \sigma_{EH} = 0, \quad \sigma_{HH} = \lim_{\rho\to\infty} 2\pi\rho\frac{|H_z^{\mathrm{sc}}|^2}{|H_z^{\mathrm{inc}}|^2}. \tag{6.33}$$

By using (6.32), the bistatic scattering width of a PEC circular cylinder under normal illumination can be written in the unified form

$$\sigma = \frac{4}{k}\left|\sum_{n=-\infty}^{\infty}\frac{\kappa J_n(ka)}{\kappa H_n^{(2)}(ka)}\mathrm{e}^{-jn(\varphi-\pi)}\right|^2, \tag{6.34}$$

where σ means either σ_{EE} or σ_{HH}.

The total scattering width σ_{T} of PEC cylinders can be obtained by either integration of σ over all scattering angles (section 3.2.2) or application of the forward-scattering theorem (section 3.4.4) since in the absence of absorption the extinction width σ_{ext} is equal to σ_{T}. For electrically large cylinders, the widths approach the limiting values:

$$\sigma_{E,H}(0) \to \pi a, \quad \sigma_{\mathrm{ext}}^{E,H} = \sigma_{\mathrm{T}}^{E,H} \to 4a, \tag{6.35}$$

where $\sigma_{E,H}(0)$ denote the backscattering widths. These are special cases of more general formulas derived for material cylinders and oblique incidence directions in section 6.5.1.

Electrically Thin Cylinder

If the radius of the cylinder is small compared to the wavelength, the main contribution to the series in (6.32) comes from the terms with $n = 0$ and $n = \pm 1$. In the expression for the E-polarized fields the terms with non-zero n can be neglected because all Bessel functions of the first kind with $n \neq 0$ tend to zero at small arguments, and the Hankel functions with nonzero indices and small arguments tend to infinity in the absolute value. Using the small-argument approximations of Bessel functions (see (E.10) and (E.11)),

$$J_0(ka) \approx 1, \qquad H_0^{(2)}(ka) = J_0(ka) - jY_0(ka) \approx 1 - j\frac{2}{\pi}\left(\gamma + \ln\frac{ka}{2}\right), \tag{6.36}$$

where $\gamma \approx 0.57721\ldots$ is the Euler constant, followed by their substitution into (6.32), gives for E polarization

$$u_{\mathrm{sc}} \approx -j\sqrt{\frac{\pi}{2k\rho}}\mathrm{e}^{-j\left(k\rho-\frac{\pi}{4}\right)}\left(\frac{1}{\ln\frac{ka}{2}+\gamma+j\frac{\pi}{2}}+\ldots\right). \tag{6.37}$$

For H polarization, we must take into account not only the terms with $n = 0$ but also with $n = \pm 1$ because they give contributions of the same order. We use the following approximations for derivatives of cylindrical functions:

$$J_0'(ka) = -J_1(ka) \approx -\frac{ka}{2} \tag{6.38}$$

$$H_0^{(2)'}(ka) = -J_1(ka) + jY_1(ka) \approx -\frac{2j}{\pi ka} \tag{6.39}$$

$$J_1'(ka) = \frac{1}{ka}J_1(ka) - J_2(ka) \approx \frac{1}{2} \tag{6.40}$$

(here we have taken into account that $Y_1(ka) \approx -2/(\pi ka)$, $J_1(ka) \approx ka/2$, and $J_2(ka) \approx (ka)^2/8$, so the contribution from J_2 can be neglected in the last equation)

$$H_1^{(2)'}(ka) = \frac{1}{ka}H_1^{(2)}(ka) - H_2^{(2)}(ka) \approx j\left(\frac{2}{\pi(ka)^2} - \frac{4}{\pi(ka)^2}\right) = -\frac{2j}{\pi(ka)^2}. \tag{6.41}$$

In the latter equation we have used the small-argument approximation

$$Y_2(ka) \approx -\frac{4}{\pi(ka)^2}. \tag{6.42}$$

For the term with $n = -1$ we get, in the same limit of small electrical sizes,

$$J_{-1}'(ka) = -\frac{1}{ka}J_{-1}(ka) - J_0(ka) = \frac{1}{ka}J_1(ka) - J_0(ka) \approx -\frac{1}{2} \tag{6.43}$$

$$H_{-1}^{(2)'}(ka) = -\frac{1}{ka}H_{-1}^{(2)}(ka) - H_0^{(2)}(ka) = \frac{1}{ka}H_1^{(2)}(ka) - H_0^{(2)}(ka) \approx \frac{2j}{\pi(ka)^2}. \tag{6.44}$$

Finally, substituting into (6.32) gives for H polarization

$$u_{\text{sc}} \approx j\sqrt{\frac{\pi}{2k\rho}}\mathrm{e}^{-j\left(k\rho-\frac{\pi}{4}\right)}\left[(ka)^2\left(\frac{1}{2} + \cos\varphi\right) + \ldots\right]. \tag{6.45}$$

It follows from (6.37) and (6.45) that electrically thin PEC cylinders scatter E-polarized waves much stronger than H-polarized waves. The apparent physical explanation is the much stronger interaction of E-polarized waves with the scatterer: the electric field in the incident wave, which is parallel to the axis of the cylinder, induces a much stronger current in the cylinder.

6.2.4 Numerical Examples

Let us express the final result for the scattered field in the far zone (6.32) in terms of the far-field scattering coefficient (see section 3.4.4) as

$$u_{\text{sc}}(\rho, \varphi) \approx -\sqrt{\frac{2}{\pi k\rho}}e^{-j\left(k\rho-\frac{\pi}{4}\right)}F(\varphi), \tag{6.46}$$

where the scattering coefficient reads

$$F(\varphi) = \sum_{n=-\infty}^{\infty} \frac{\kappa J_n(ka)}{\kappa H_n^{(2)}(ka)}\mathrm{e}^{-jn(\varphi-\pi)}. \tag{6.47}$$

$F(\varphi)$ depends on the angle between the incidence and scattering directions, on the cylinder radius a, and on the frequency via the wavenumber k. The cylindrical-wave factor in (6.46) is a function of the distance ρ to the observation point and also the frequency.

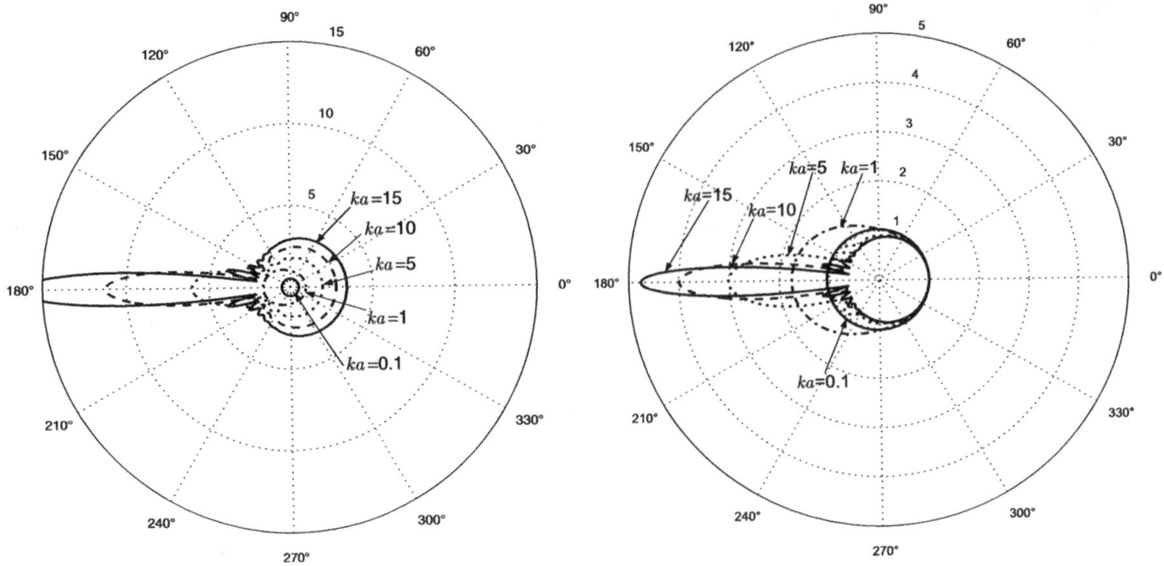

Figure 6.2 Angular dependence of the field scattered from an ideally conducting cylinder for several normalized thicknesses. E-polarized incidence (electric field is parallel to the cylinder axis). The cylinder is illuminated from the right. On the right-hand picture the values are normalized to $|F(0)|$.

To study the angular and size dependence of the scattered field we can, therefore, consider function $F(\varphi)$. For moderate diameters of the cylinder, the series in (6.47) can be calculated by direct summation since the number of terms to be accounted for is of the same order of magnitude as ka. Some computed results for function $|F(\varphi)|$ are shown in Figure 6.2 (E-polarized incidence) and Figure 6.3 (H-polarized incidence).[4] We see that the scattered field amplitude grows when the cylinder radius increases, both in the forward and back directions. To see in which direction this growth is stronger, we can plot the same results, normalizing them so that the backscattered value is the same for all curves. The result (for E polarization) is shown on the right in Figure 6.2: the forward-scattered field monotonically grows with increased radius. One can check that the scattered field in the forward direction does not become larger than the incident field (in the far zone) because the distance to the far zone also increases with the increased radius of the cylinder (as a^2).

The strong forward scattering from large cylinders means that there is a shadow behind the body. In the back direction, the cylinder scatters rather uniformly in all directions, while the scattering in the forward direction is "focused" into the shadow. This is illustrated by Figure 6.4, where we plot the amplitude of the *total* field in the far zone

$$u_{\text{tot}} = e^{jk\rho\cos\varphi} - \sqrt{\frac{2}{\pi k\rho}} e^{-j\left(k\rho - \frac{\pi}{4}\right)} F(\varphi) \tag{6.48}$$

for $k\rho = 2000$ and two normalized cylinder radii ($ka = 0.1$ and $ka = 15$). The shadow is seen in the field behind the thick cylinder, while the thin cylinder scatters weakly and rather uniformly in all directions. The oscillations result from interference of the incident and scattered fields.

Figure 6.5 shows the extinction width in dependence on ka. The extinction width for PEC cylinders is equal to the total scattering width; the absorption width is zero. It is seen that already with $ka = 10$, σ_{ext} for both polarizations is close to its asymptotic value $4a$. When $ka \to 0$, the dependence on polarization is much more apparent as E polarization (electric field vector parallel to the axis of the cylinder) is scattered much stronger than H polarization (magnetic field vector

[4] One can check that for a small cylinder the pattern (Figure 6.3) is very well modeled by function $0.5 + \cos\varphi$, see formula (6.45).

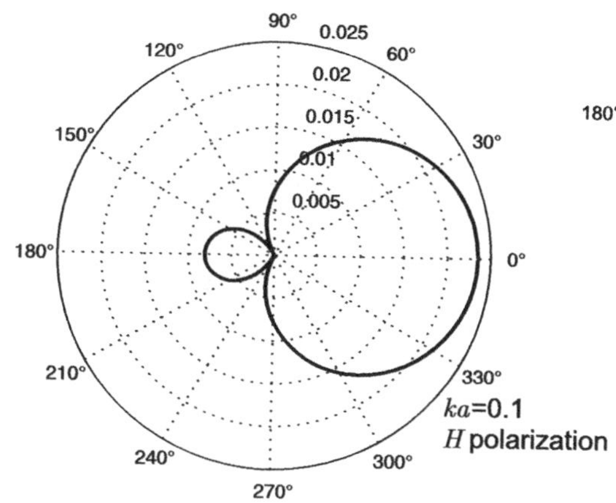

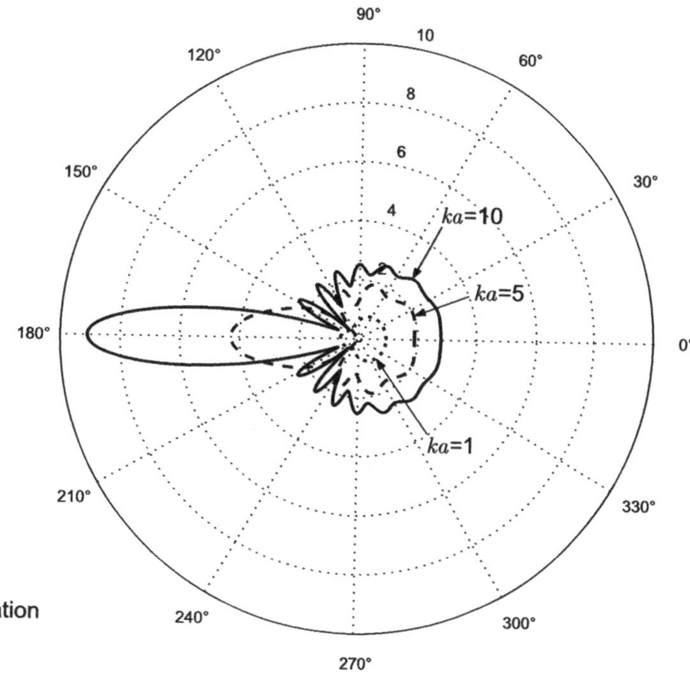

Figure 6.3 The same as Figure 6.2 (left) but for H-polarized incidence (magnetic field is parallel to the cylinder axis). The small cylinder case ($ka = 0.1$) is shown separately on the left.

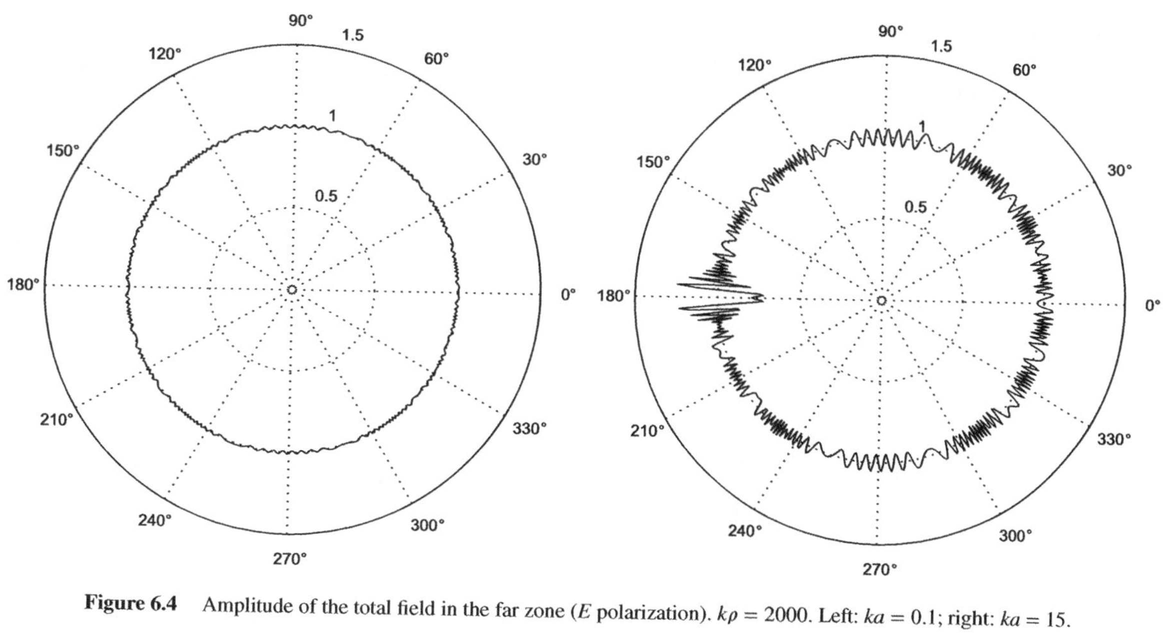

Figure 6.4 Amplitude of the total field in the far zone (E polarization). $k\rho = 2000$. Left: $ka = 0.1$; right: $ka = 15$.

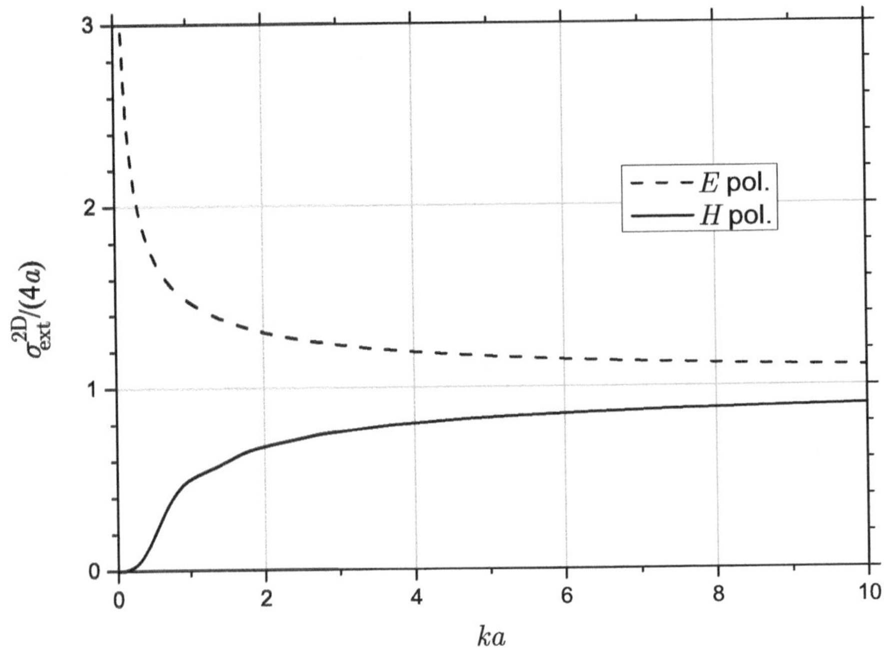

Figure 6.5 Extinction (total scattering) width of a PEC cylinder as a function of ka for E- and H-polarization cases. The width is normalized to its high-frequency limit $4a$.

parallel to the axis of the cylinder). Both widths vanish for $ka = 0$, the width for E polarization in the figure becomes infinite because of the normalization.

If we are interested in the frequency dependence of the scattered field amplitude, we should calculate the values of $|F(\varphi)|/\sqrt{ka}$ to account for the frequency dependence of the wave amplitude. The results for the forward and backscattering are shown in Figure 6.6.

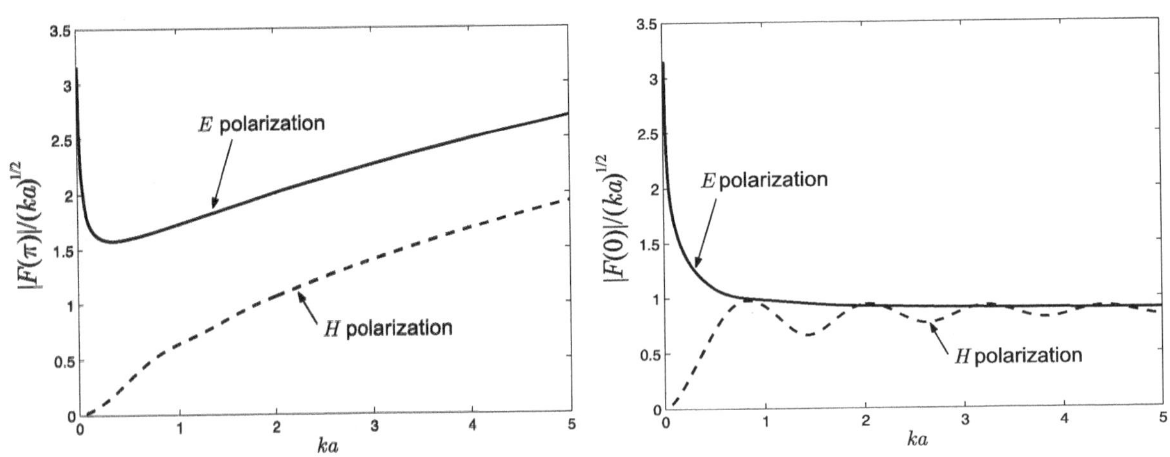

Figure 6.6 Frequency dependence of the forward-scattered (left) and backscattered field (right) in the far zone.

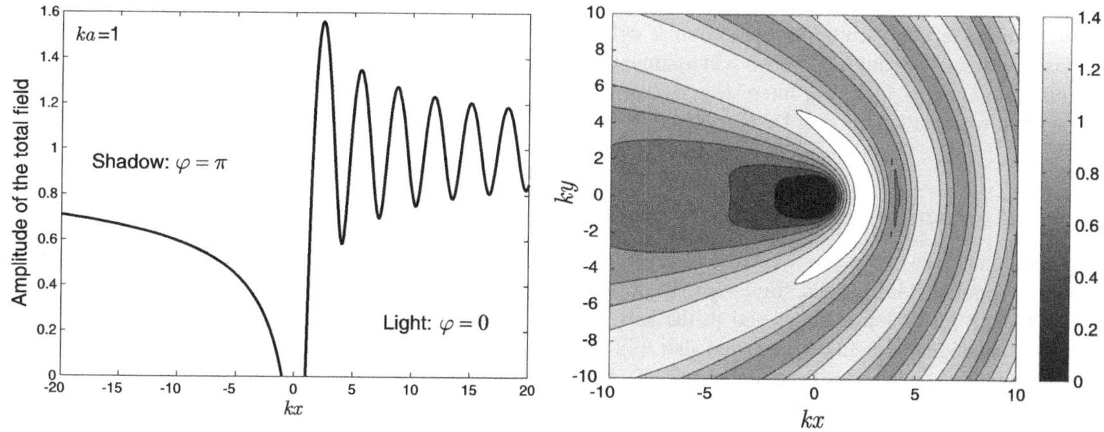

Figure 6.7 Amplitude of the total field on the x axis near the scatterer (left) and in the vicinity of the cylinder (right). E polarization.

Finally, let us have a look at the field distribution near the cylinder (Figure 6.7). We calculate the total field using formula (6.28):

$$u_{\text{tot}}(\rho,\varphi) = e^{jk\rho\cos\varphi} - \sum_{n=-\infty}^{\infty} \frac{\kappa J_n(ka)}{\kappa H_n^{(2)}(ka)} H_n^{(2)}(k\rho) e^{-jn(\varphi-\pi/2)} \tag{6.49}$$

for $\varphi = 0$ or $\varphi = \pi$. The normalized radius of the cylinder in this example equals $ka = 1$. In front of the cylinder, we see a standing wave formed by the sum of the incident and reflected waves. In the shadow, the field has no oscillations, since the field penetrates here only due to diffraction.

Figure 6.8 shows a comparison of the field distribution behind the cylinder on the x axis and on parallel lines shifted from the axis. The normalized cylinder radius $ka = 10$. Note that in the deep shadow the field at the axis is stronger than at off-axis points; this phenomenon can be explained by the constructive interference of creeping waves, which are studied in section 6.4.3. In the illuminated region the field oscillates, and far from the scatterer the amplitude tends to the amplitude of the incident field.

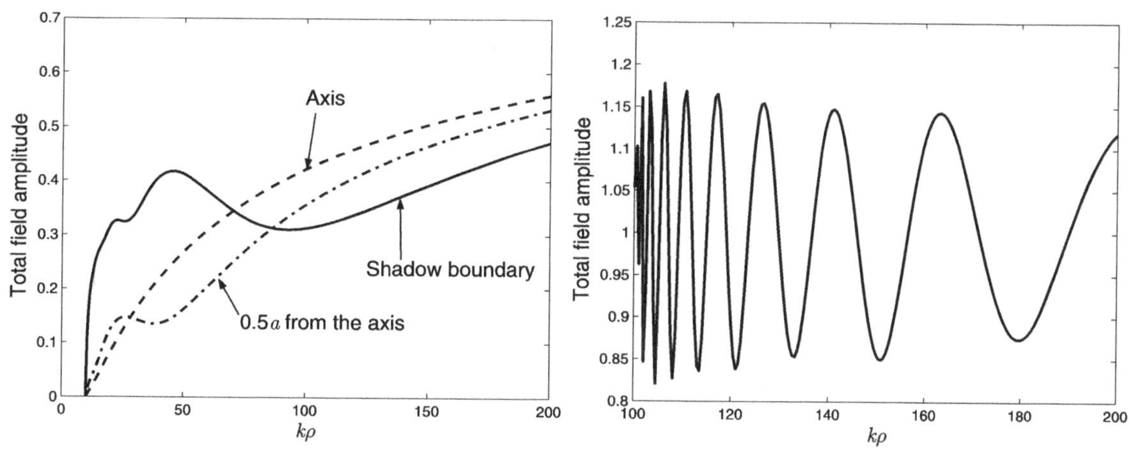

Figure 6.8 Amplitude of the total field in the shadow region (left) compared with the illuminated region (along a line parallel to x and shifted by $10a$) (right). E polarization.

6.3 Homogeneous Cylinders under Normal Illumination

The problem of plane-wave scattering by a circular cylinder made of an isotropic homogeneous material can also be solved by separation of variables. Here we will assume that the material filling the cylinder is characterized by arbitrary scalar parameters ϵ and μ, including interesting cases of metamaterials with negative material parameters. This section has been written with participation of Dr Igor Nefedov.

6.3.1 *Field Equations and Boundary Conditions*

For the normal incidence of a plane wave (the same geometry as shown in Figure 6.1), the field equations split into two sets for two orthogonal polarizations, similarly to the case of PEC cylinders. This is seen from writing the field equations as (6.1)–(6.4) both inside the cylinder and in the surrounding free space, and noticing that the boundary conditions (continuity of the tangential field components) also split into two independent sets for E and H polarizations.

Let us denote the parameters of the cylinder material by ϵ_2 and μ_2, and the parameters of the ambient space by ϵ_1 and μ_1. The scattered field satisfies the Helmholtz equation both in the surrounding space and inside the cylinder. For the E polarization, where the non-zero field components are E_z, H_φ, and H_ρ, we write

$$\nabla^2 E_{1z}^{\rm sc} + k_1^2 E_{1z}^{\rm sc} = 0, \quad \rho > a; \qquad \nabla^2 E_{2z}^{\rm sc} + k_2^2 E_{2z}^{\rm sc} = 0, \qquad \rho < a. \tag{6.50}$$

The boundary condition on the cylinder surface demands continuity of the total electric field

$$\left(E_{1z}^{\rm sc} + E_z^{\rm inc}\right)\Big|_{r=a} = E_{2z}^{\rm tot}\Big|_{\rho=a} \tag{6.51}$$

at the cylinder surface. Here the index "tot" refers to the total field: $E_z^{\rm tot} = E_z^{\rm inc} + E_z^{\rm sc}$. The other field component tangential to the surface is H_φ. This component is related to E_z as in (6.1)

$$H_\varphi = \frac{1}{j\omega\mu}\frac{\partial E_z}{\partial \rho}, \tag{6.52}$$

and the continuity of H_φ is ensured if[5]

$$\frac{1}{\mu_1}\frac{\partial}{\partial\rho}\left(E_z^{\rm inc} + E_{1z}^{\rm sc}\right)\Big|_{\rho=a} = \frac{1}{\mu_2}\frac{\partial}{\partial\rho}E_{2z}^{\rm tot}\Big|_{\rho=a}. \tag{6.53}$$

For H-polarized incident waves we write the Helmholtz equation for the longitudinal magnetic field component $H_z = H_z^{\rm inc} + H_z^{\rm sc}$,

$$\nabla^2 H_{1z}^{\rm sc} + k_1^2 H_{1z}^{\rm sc} = 0, \quad \rho > a; \qquad \nabla^2 H_{2z}^{\rm sc} + k_2^2 H_{2z}^{\rm sc} = 0, \quad \rho < a. \tag{6.54}$$

The boundary conditions demand the continuity of $H_z^{\rm tot}$ at $\rho = a$ and the continuity of E_φ, which is equivalent to

$$\frac{1}{\epsilon_1}\frac{\partial}{\partial\rho}H_{1z}^{\rm tot}\Big|_{\rho=a} = \frac{1}{\epsilon_2}\frac{\partial}{\partial\rho}H_{2z}^{\rm tot}\Big|_{\rho=a}. \tag{6.55}$$

The expressions for the two polarizations are dual in the sense that if the solution for one polarization is known, the solution for the other one can be obtained by the following replacements: $E_z \leftrightarrow H_z$, $\epsilon_{1,2} \leftrightarrow \mu_{1,2}$. The other field components can be expressed in terms of E_z and H_z using (6.29) and (6.30).

[5] Note that here we assume that $\mu_{1,2} \neq 0$. For the other polarization the following analysis is restricted to $\epsilon_{1,2} \neq 0$.

6.3.2 *Rayleigh Series Solution*

Series Expansion

Using the same generic notation for the axial field component u, we look for the general solution for the scattered field outside the cylinder in the form of the same series of Hankel functions as for the PEC cylinder (6.22):

$$u_{sc}(\rho, \varphi) = \sum_{n=-\infty}^{\infty} C_n H_n^{(2)}(k_1\rho) e^{-jn(\varphi-\pi/2)}, \qquad \rho \geq a. \tag{6.56}$$

To represent the field inside the cylinder, we retain only those cylindrical functions which have no singularities at $\rho = 0$, that is, the Bessel functions of the first kind:

$$u_{tot}(\rho, \varphi) = u_{inc}(\rho, \varphi) + u_{sc}(\rho, \varphi) = \sum_{n=-\infty}^{\infty} A_n J_n(k_2\rho) e^{-jn(\varphi-\pi/2)}, \qquad \rho \leq a. \tag{6.57}$$

With the same expansion of the incident field outside the cylinder (6.23)

$$e^{jk_1\rho\cos\varphi} = \sum_{n=-\infty}^{\infty} J_n(k_1\rho) e^{-jn(\varphi-\pi/2)}, \tag{6.58}$$

we write the boundary conditions on the cylinder surface for E-polarized incidence as

$$A_n J_n(k_2 a) = J_n(k_1 a) + C_n H_n^{(2)}(k_1 a) \tag{6.59}$$

$$\frac{k_1}{\mu_1}\left[J_n'(k_1 a) + C_n H_n^{(2)\prime}(k_1 a)\right] = \frac{k_2}{\mu_2} A_n J_n'(k_2 a). \tag{6.60}$$

The solution of these equations reads

$$A_n = \frac{J_n(k_1 a)}{J_n(k_2 a)} + C_n \frac{H_n^{(2)}(k_1 a)}{J_n(k_2 a)} \tag{6.61}$$

$$C_n = \frac{Z_2 J_n'(k_1 a) J_n(k_2 a) - Z_1 J_n'(k_2 a) J_n(k_1 a)}{Z_1 H_n^{(2)}(k_1 a) J_n'(k_2 a) - Z_2 H_n^{(2)\prime}(k_1 a) J_n(k_2 a)}, \tag{6.62}$$

where $Z_{1,2} = \sqrt{\mu_{1,2}/\epsilon_{1,2}}$.

Electrically Thin Cylinder

If the cross-section size of the cylinder is electrically small, formulas can be simplified assuming that $|k_{1,2}|a \ll 1$. Using the small-argument approximations of cylindrical functions (6.36) and (6.38)–(6.41), we get for the two lowest-order coefficients

$$C_0 \approx \frac{\pi a^2}{4j}\left(\frac{Z_1}{Z_2}k_1 k_2 - k_1^2\right) = \frac{\pi (k_1 a)^2}{4j}\left(\frac{\epsilon_2}{\epsilon_1} - 1\right) \tag{6.63}$$

$$C_1 = C_{-1} \approx \frac{\pi (k_1 a)^2}{4j}\frac{\mu_2 - \mu_1}{\mu_2 + \mu_1}. \tag{6.64}$$

These coefficients give the approximate solution for the E-polarized incidence. For the H-polarized case we replace $\epsilon_{1,2} \leftrightarrow \mu_{1,2}$ and arrive at

$$C_0 \approx \frac{\pi (k_1 a)^2}{4j}\left(\frac{\mu_2}{\mu_1} - 1\right) \tag{6.65}$$

$$C_1 = C_{-1} \approx \frac{\pi (k_1 a)^2}{4j}\frac{\epsilon_2 - \epsilon_1}{\epsilon_2 + \epsilon_1}. \tag{6.66}$$

Let us consider the last case in more detail because of the apparent possibility of resonant phenomena when $\epsilon_2 \approx -\epsilon_1$. Substituting the above approximate relations for $C_{\pm 1}$ and small-argument approximations of cylindrical functions in (6.61) we get

$$A_1 = A_{-1} \approx \frac{\sqrt{\epsilon_1 \mu_1}}{\sqrt{\epsilon_2 \mu_2}} \frac{2\epsilon_2}{\epsilon_2 + \epsilon_1}. \tag{6.67}$$

The expression (6.57) for the total field inside the cylinder describes the longitudinal magnetic field component $u_{\text{tot}} = H_z$ for the incident plane wave with $\mathbf{H}_{\text{inc}} = \hat{z}$ and $\mathbf{E}_{\text{inc}} = -\sqrt{\mu_1/\epsilon_1}\,\hat{y}$. Contributions from the leading resonant terms (proportional to $A_{\pm 1}$) to the total electric field inside the cylinder read

$$E_\varphi = -\frac{1}{j\omega\epsilon_2}\frac{\partial H_z}{\partial \rho} = -\frac{2\sqrt{\epsilon_1 \mu_1}}{\epsilon_2 + \epsilon_1}\cos\varphi. \tag{6.68}$$

Likewise,

$$E_\rho = \frac{1}{j\omega\epsilon_2\,\rho}\frac{\partial H_z}{\partial \varphi} = -\frac{2\sqrt{\epsilon_1 \mu_1}}{\epsilon_2 + \epsilon_1}\sin\varphi. \tag{6.69}$$

This means that in Cartesian coordinates $E_x = 0$ and

$$E_y = -\frac{2\sqrt{\epsilon_1 \mu_1}}{\epsilon_2 + \epsilon_1}. \tag{6.70}$$

In other words, the field inside is approximately uniform, and the following relation between the incident field and the field inside the cylinder holds:

$$\mathbf{E}_{\text{tot}} = \frac{2\epsilon_1}{\epsilon_2 + \epsilon_1}\mathbf{E}_{\text{inc}}. \tag{6.71}$$

This is the same as the quasi-static solution for a cylinder in uniform electric field directed orthogonally to the axis (Landau *et al.* 1984). The dipole moment induced in the cylinder (per unit length) is, from here,

$$\mathbf{P} = (\epsilon_2 - \epsilon_1)\frac{2\epsilon_1}{\epsilon_2 + \epsilon_1}\pi a^2 \mathbf{E}_{\text{inc}}. \tag{6.72}$$

Close to the resonant point where $\epsilon_2 \approx -\epsilon_1$, the field inside the cylinder as well as the induced electric dipole moments can attain high amplitudes.

6.3.3 Numerical Examples

A typical field distribution for diffraction on a cylinder made of a usual lossless dielectric is shown in Figure 6.9. One can see that inside the cylinder the waves propagate with a shorter wavelength. The field strength takes large values close to the cylinder center due to a kind of "focusing".

More interesting is the case of a cylinder made of a material with the permittivity having a negative real part. Such properties exhibit noble metals at optical frequencies (see section 2.2.2). Fields cannot propagate inside a medium with a negative permittivity, but an interface between media with positive and negative permittivities can support propagation of surface waves, which are called surface plasmons (or surface polaritons or surface plasmon polaritons). The surface waves have wavelengths much smaller than the wavelength in free space. Figure 6.10 illustrates diffraction on a silver cylinder having the same electrical size $k_1 a = 15$ as in the previous case. The complex permittivity corresponds to silver at $\lambda = 500$ nm (Johnson and Christy 1972). We can observe that the field weakly penetrates into the cylinder, where the waves are evanescent.

Let us consider the case of an electrically thin cylinder and H polarization. As was shown in section 6.3.2, we can find approximate expressions for coefficients C_n using the small-argument approximations of cylindrical functions.

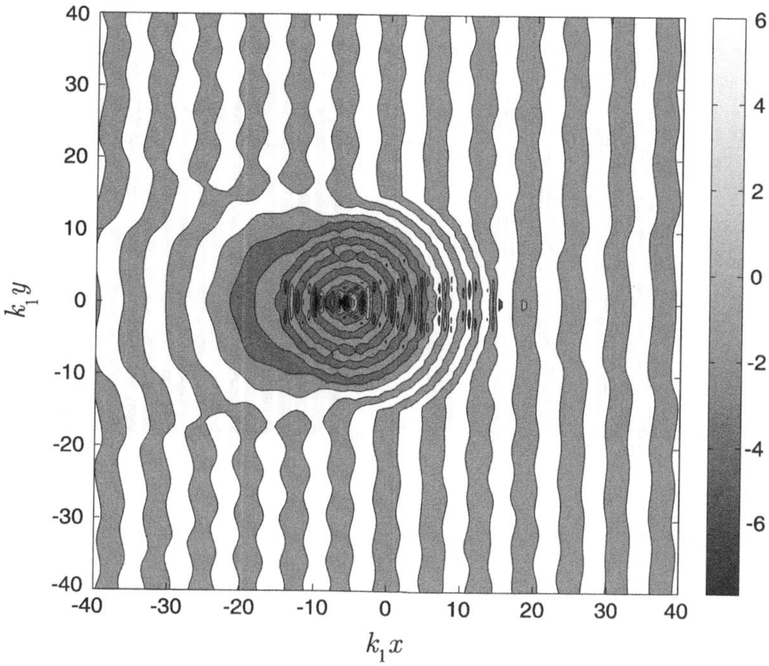

Figure 6.9 Magnetic field distribution calculated at $k_1 a = 15$, $\mu_2/\mu_1 = 1$, and $\epsilon_2/\epsilon_1 = 9$. H polarization of the incident plane wave. The cylinder is illuminated from the right.

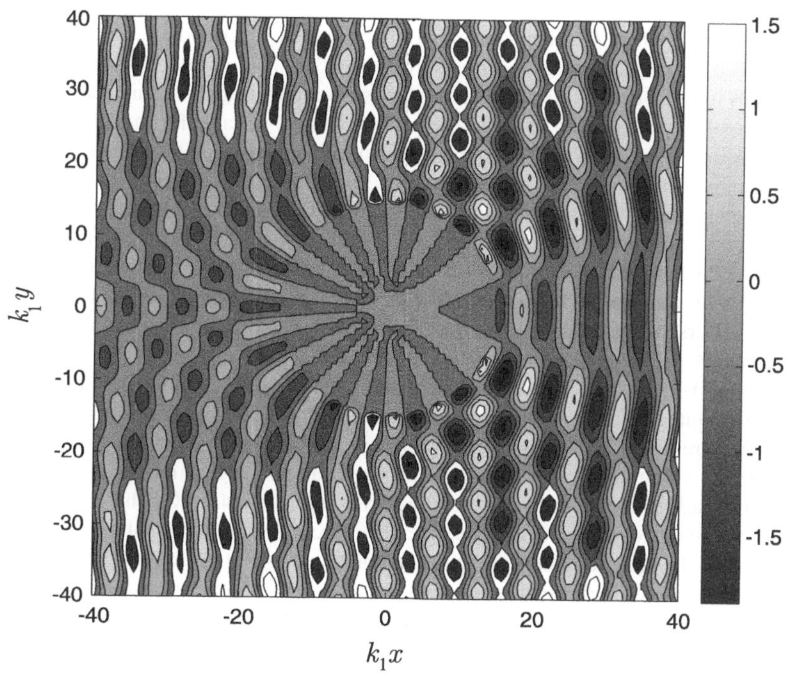

Figure 6.10 Magnetic field distribution calculated at $k_1 a = 15$, $\mu_2/\mu_1 = 1$, and $\epsilon_2/\epsilon_1 = -9.5456 - j0.309$. H polarization.

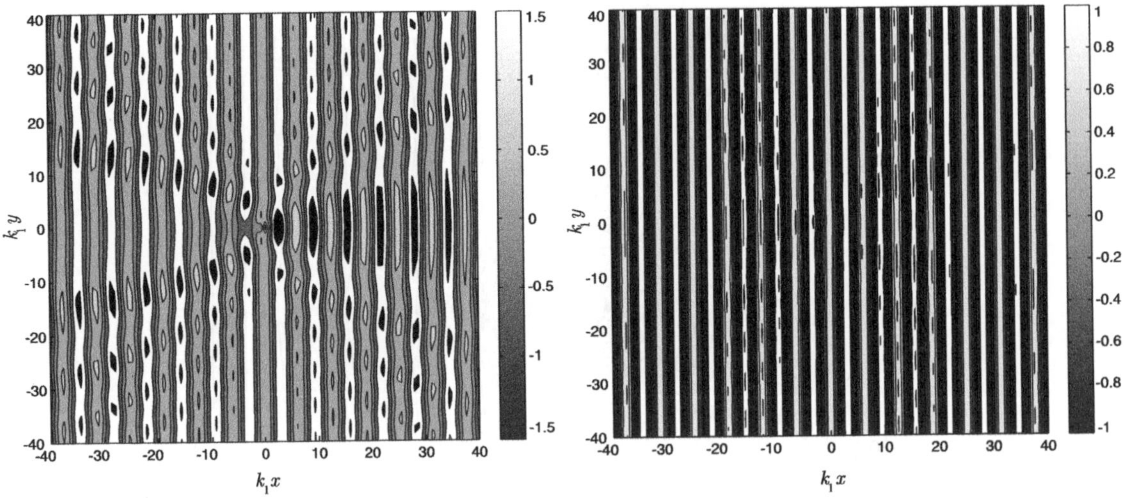

Figure 6.11 Left: Magnetic field distribution in the vicinity of a small *resonant* cylinder near the plasmon resonance frequency, calculated at $k_1a = 0.1$, $\mu_2/\mu_1 = 1$, and $\epsilon_2/\epsilon_1 = -1.01$. H polarization. Right: The same for a small *non-resonant* cylinder of the same size but with $\epsilon_2/\epsilon_1 = -1.2$.

Coefficients $C_{\pm 1}$ are of particular interest. As is seen from (6.66), $C_1 \to \infty$ if $\epsilon_2 \to -\epsilon_1$ ($\epsilon_1 > 0$). This resonant case is illustrated by Figure 6.11, in comparison with the negligibly small scattering from a cylinder with a slightly different but off-resonance permittivity.

6.4 Watson's Transformation and High-frequency Approximations

For large cylinder radii ($ka \gg 1$, where k is the wavenumber outside the cylinder) the series for the scattered field (6.28) converges very slowly. The convergence begins with the terms with $|n| > ka$, after the functions $J_n(ka)$ change their behavior from oscillations to rapid decay (Appendix E). A physical interpretation of this behavior is that the expansion (6.28) is basically a multipole representation of the scatterer, which is not suitable for modeling large bodies.

The Watson transformation allows us to transform the series into a form suitable for effective calculations and offering a clear physical insight into scattering from large bodies. The transformation consists of representing the series as a contour integral in a complex plane, which can be evaluated either as a sum of residues or by asymptotic methods, like the steepest descent method (Appendix C) and the method of stationary phase (Appendix D). The transformation was originally proposed in the context of radio propagation over the Earth surface to evaluate the exact series solution for the field in the presence of an electrically large sphere (Watson 1918). A similar idea can be applied to the cylindrical geometry; its realization differs, however, from the case of a sphere in some details as the solutions for the spherical and cylindrical geometries involve different special functions.

We start by describing the Watson transformation as applied to the series solution for the PEC cylinder (section 6.4.1). The series is expressed as an integral in the complex plane of the order of the Bessel functions. Evaluating the integral as a sum of residues leads to a representation that can be interpreted as an expansion over radial eigenfunctions (section 6.4.2), in contrast to the series (6.28) which can be seen as an expansion over angular eigenfunctions. This former series converges rapidly in the shadow and leads to the concept of creeping waves (section 6.4.3). On the other hand, application of the stationary point method to the contour integral recovers the GO representation of the solution in the form of the incident and reflected waves, which is the most suitable form in the illuminated region (section 6.4.3). To describe the fields in the transition region between the lit and shadowed portions of the scattering surface, a transformation of the contour integral solution to a pair of universal functions proposed by Fock can be used (section 6.4.4). This representation expresses the property of locality of high-frequency electromagnetic fields in the vicinity of the shadow boundary on the surface of any convex smooth body.

The reader can find further examples of the application of the Watson transformation in the sections on scattering from PEC wedges (section 5.2) and from electrically large spheres (section 7.3.4), where the solutions are obtained in a series form and then transformed to the contour integrals to evaluate the solutions in the high-frequency limit. An interesting overview of the Watson approach can be found in Bremmer (1949, Chapter 3).

6.4.1 Watson's Transformation

Let us introduce function

$$A(\nu) = \left[J_\nu(k\rho) - \frac{\kappa J_\nu(ka)}{\kappa H_\nu^{(2)}(ka)} H_\nu^{(2)}(k\rho) \right] e^{j\nu\pi/2} \tag{6.73}$$

of a complex variable ν, so that the series solution for the total field (6.24) can be expressed as

$$u_{\text{tot}}(\rho, \varphi) = \sum_{n=-\infty}^{+\infty} A(n) e^{-jn\varphi}. \tag{6.74}$$

Function $A(\nu)$ can be rewritten in terms of Hankel functions only. Using

$$J_\nu(x) = \frac{H_\nu^{(1)}(x) + H_\nu^{(2)}(x)}{2},$$

we have

$$A(\nu) = \frac{1}{2} \left[H_\nu^{(1)}(k\rho) - \frac{\kappa H_\nu^{(1)}(ka)}{\kappa H_\nu^{(2)}(ka)} H_\nu^{(2)}(k\rho) \right] e^{j\nu\pi/2}. \tag{6.75}$$

On the complex plane of order ν, cylindrical functions are holomorphic functions with an infinite number of zeros located along lines that start at $\nu = \pm x$ (x is the argument of the cylindrical function) and go to infinity (Appendix E). Because of the symmetry properties (E.17) and (E.18), every Hankel function has two families of zeros, and $A(\nu)$ is an even function of ν,

$$A(-\nu) = A(\nu), \tag{6.76}$$

with pole singularities at points ν_s, where

$$\kappa H_{\nu_s}^{(2)}(ka) = 0. \tag{6.77}$$

With $\text{Re}\, k > 0$, the points ν_s are located in the second and fourth quadrants of the complex ν plane, along the lines that start at $\nu = \pm ka$ and go to infinity in the respective quadrant (anti-Stokes lines $\gamma_\pm$ on Figure E.12).

Using the evenness of $A(\nu)$, the series (6.74) can be expressed as

$$u_{\text{tot}}(\rho, \varphi) = A(0) + 2 \sum_{n=1}^{+\infty} A(n) \cos(n\varphi). \tag{6.78}$$

This form makes it apparent that the solution depends on the absolute value $|\varphi|$, which means the symmetry with respect to the plane that includes the axis of the cylinder and the direction of incidence of the incoming wave.[6]

Now we are prepared for the first step of the Watson transformation, which consists in rewriting the series (6.78) as an integral

$$u_{\text{tot}}(\rho, \varphi) = -j \int_\Gamma \frac{\cos[\nu(\pi - |\varphi|)]}{\sin \nu\pi} A(\nu)\, d\nu, \tag{6.79}$$

[6] In the rest of the section we assume that $-\pi \le \varphi \le \pi$.

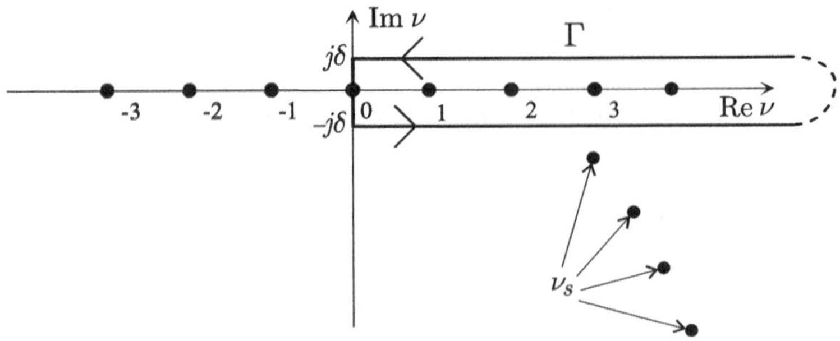

Figure 6.12 Integration path in formula (6.79). The portion of the integration path where the integrand vanishes is shown by a dashed line. The path goes between the positive real axis and the branch of zeros ν_s of (6.77).

where the integration path Γ (Figure 6.12) encloses the positive real axis of the complex ν plane and goes across the point $\nu = 0$, at which it is understood in the principal value sense. The positive parameter δ specifies the distance between the contour and the real axis, and can be an arbitrary number compliant with the requirement that $0 < \delta < |\mathrm{Im}\, \nu_1|$, where ν_1 is a zero of (6.77) with the smallest imaginary part.

Since function $A(\nu)$ is holomorphic on the real axis, the integrand has simple poles at $\nu = n = 0, 1, 2, \ldots$ with the residues

$$\mathrm{res}\left\{\frac{\cos\left[\nu(\pi - |\varphi|)\right]}{j \sin \nu\pi} A(\nu)\right\} = \frac{\cos\left[n(\pi - |\varphi|)\right]}{j\pi \cos n\pi} A(n) = \frac{1}{j\pi} \cos(n|\varphi|) A(n). \tag{6.80}$$

The integral is then

$$-j \int_\Gamma \frac{\cos\left[\nu(\pi - |\varphi|)\right]}{\sin \nu\pi} A(\nu)\, \mathrm{d}\nu = A(0) + 2 \sum_{n=1}^{+\infty} A(n) \cos(n|\varphi|) \tag{6.81}$$

(the half residue is taken at the pole $\nu = 0$), which shows that formula (6.79) represents the same function as (6.78).

The main idea behind the transformation of the series representation into an integral is the possibility of deforming the integration path, thus transforming the integral into a more appropriate form. The possibility of deforming the integration path is determined by the singularities of the integrand function on the complex ν plane and by the behavior of the integrand when $|\nu| \to \infty$. The integrand in (6.79) has poles on the real axis at $\nu = 0, \pm 1, \pm 2 \ldots$ and in the second and forth quadrants at the points ν_s defined by (6.77). The behavior at $|\nu| \to \infty$ is determined by the behavior of the Bessel functions. Applying the results of section E.2.3 to the case of the real-valued argument ($1 \ll x \ll |\nu|$), we have

$$J_\nu(x) \approx \frac{1}{\sqrt{2\pi\nu}} \left(\frac{2\nu}{ex}\right)^{-\nu}, \quad -\pi < \arg \nu < \pi \tag{6.82}$$

$$H_\nu^{(1)}(x) \approx \begin{cases} -j\sqrt{\dfrac{2}{\pi\nu}} \left(\dfrac{2\nu}{ex}\right)^{\nu}, & -\dfrac{\pi}{2} \le \arg \nu < \dfrac{\pi}{2} \\ \sqrt{\dfrac{2}{\pi\nu}} \left(\dfrac{2\nu}{ex}\right)^{-\nu}, & \dfrac{\pi}{2} \le \arg \nu < \pi \end{cases} \tag{6.83}$$

$$H_\nu^{(2)}(x) \approx \begin{cases} \sqrt{\dfrac{2}{\pi\nu}} \left(\dfrac{2\nu}{ex}\right)^{-\nu}, & -\pi < \arg \nu \le -\dfrac{\pi}{2} \\ j\sqrt{\dfrac{2}{\pi\nu}} \left(\dfrac{2\nu}{ex}\right)^{\nu}, & -\dfrac{\pi}{2} < \arg \nu \le \dfrac{\pi}{2} \end{cases} . \tag{6.84}$$

In particular, these approximations imply that in the first and fourth quadrants the Bessel function $J_\nu(x)$ decays as $O(\nu^{-\nu})$ and the Hankel functions $H_\nu^{(1)}(x)$ and $H_\nu^{(2)}(x)$ grow as $O(\nu^\nu)$, except for the imaginary axis ($\text{Re}\,\nu = 0$, $\text{Im}\,\nu \to \pm\infty$), at which the behavior is purely exponential,

$$|J_\nu(x)| \approx \frac{e^{|\nu|\pi/2}}{\sqrt{2\pi|\nu|}}, \quad |H_\nu^{(1)}(x)| \approx \sqrt{\frac{2}{\pi|\nu|}} e^{\pm|\nu|\pi/2}, \quad |H_\nu^{(2)}(x)| \approx \sqrt{\frac{2}{\pi|\nu|}} e^{\mp|\nu|\pi/2}. \tag{6.85}$$

Let us transform the integration path Γ into a contour that is entirely located in the forth quadrant of the complex ν plane. To this end, we begin by noting that in the first quadrant the integrand in (6.79) has no singularities and because of (6.82), (6.84), and the estimates

$$\left|\frac{\cos[\nu(\pi - |\varphi|)]}{j\sin\nu\pi}\right| \approx e^{-\text{Im}\,\nu|\varphi|} \tag{6.86}$$

$$|A(\nu)| \leq \left|e^{j\nu\frac{\pi}{2}} J_\nu(k\rho)\right| + \left|e^{j\nu\frac{\pi}{2}} \frac{\kappa J_\nu(ka)}{\kappa H_\nu^{(2)}(ka)} H_\nu^{(2)}(k\rho)\right| = O(1), \tag{6.87}$$

vanishes for $|\nu| \to \infty$ and $0 < \arg\nu \leq \pi/2$, that is, on the whole arc C_I, provided that $\varphi \neq 0$. Thus, we can deform the upper portion of Γ onto the positive imaginary axis (Figure 6.13),

$$\int_{+\infty+j\delta}^{j\delta} \frac{\cos[\nu(\pi - |\varphi|)]}{\sin\nu\pi} A(\nu)\,d\nu = \int_{j\infty}^{j\delta} \frac{\cos[\nu(\pi - |\varphi|)]}{\sin\nu\pi} A(\nu)\,d\nu \tag{6.88}$$

and noting that because of (6.76),

$$\int_{-j\delta}^{j\delta} \frac{\cos[\nu(\pi - |\varphi|)]}{\sin\nu\pi} A(\nu)\,d\nu = 0$$

$$\int_{j\infty}^{j\delta} \frac{\cos[\nu(\pi - |\varphi|)]}{\sin\nu\pi} A(\nu)\,d\nu = \int_{-j\infty}^{-j\delta} \frac{\cos[\nu(\pi - |\varphi|)]}{\sin\nu\pi} A(\nu)\,d\nu,$$

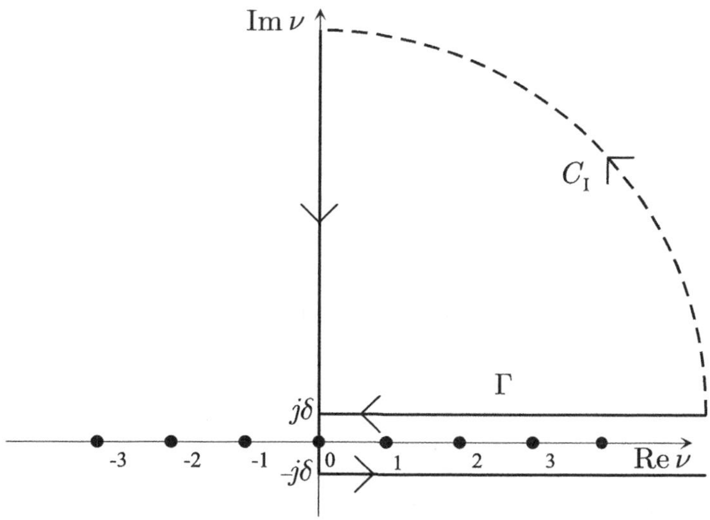

Figure 6.13 Deformation of the upper portion of Γ onto the positive imaginary axis.

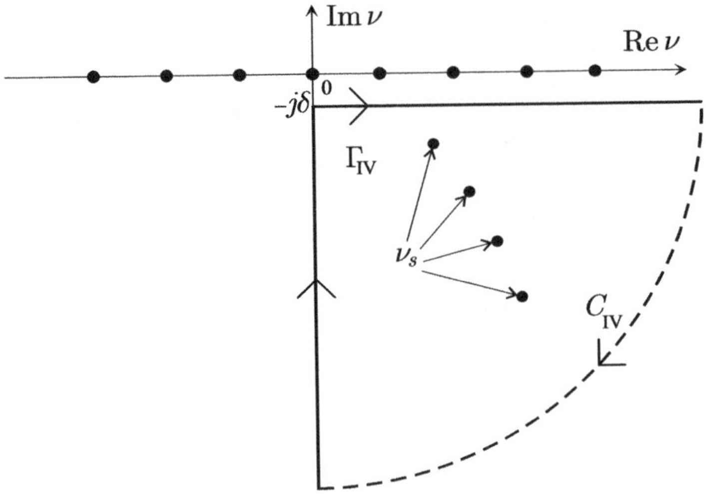

Figure 6.14 Integration path in formula (6.89).

obtain the formula

$$u_{\mathrm{tot}}(\rho, \varphi) = -j \int_{\Gamma_{\mathrm{IV}}} \frac{\cos\left[\nu(\pi - |\varphi|)\right]}{\sin \nu\pi} A(\nu)\,\mathrm{d}\nu, \tag{6.89}$$

where the integration path Γ_{IV} consists of two portions, one along the negative imaginary axis and another along the lower portion of the original contour Γ (Figure 6.14).

The contour Γ_{IV} can be freely deformed within the fourth quadrant as long as no pole singularities ν_s are crossed. As we shall see later, the representation (6.89) is very convenient and may serve as the basis for obtaining GO, shadow and semi-shadow high-frequency representations for the fields scattered by PEC cylinders.

When $|\varphi| > \pi/2$, the integral over the quarter-circle C_{IV} (Figure 6.14) is zero[7] for $|\nu| \to \infty$, and we can express (6.89) as the sum of residues

$$u_{\mathrm{tot}}(\rho, \varphi) = 2\pi \sum_s \mathrm{e}^{j\pi\nu_s/2} H_{\nu_s}^{(2)}(k\rho) \frac{\cos[\nu_s(\pi - |\varphi|)]}{\sin \pi\nu_s} \left\{ \frac{\kappa J_\nu(ka)}{\partial/\partial\nu\left[\kappa H_\nu^{(2)}(ka)\right]} \right\}_{\nu=\nu_s} \tag{6.90}$$

at points ν_s, which are the first-order poles of the integrand. This series is equivalent to the original form (6.24), but for a large cylinder radius it converges very fast and its members have a clear physical meaning, as will be seen from the following. Representation (6.90) is also important from the mathematical point of view, and we first consider this aspect before going to the physical interpretation of the representation.

6.4.2 *Alternative Solution by Separation of Variables*

The same representation as (6.90) can be obtained directly by separation of variables in the Helmholtz equation. In the previous derivation that led to a series over Hankel functions with integer orders and periodical exponentials for the angular dependence, the expansion functions (the periodic exponentials) were not subject to the boundary conditions on the surface of the cylinder. Alternatively, we can build a solution in terms of a different set of expansion functions, not with respect to the angular coordinate but with respect to the radial coordinate, demanding that *each* expansion function satisfy the boundary condition on the cylinder surface.

[7] The proof uses subtle estimations of the asymptotic behavior of Bessel functions and is addressed in problems 11 and 12.

Let us write for the total field:

$$u_{\text{tot}}(\rho, \varphi) = V(\rho)W(\varphi). \tag{6.91}$$

The radial function $V(\rho)$ should be a cylindrical function (to satisfy the Helmholtz equation in polar coordinates) and it should satisfy the radiation condition. The correct form is the Hankel function of an arbitrary index ν:

$$V(\rho) = H_\nu^{(2)}(k\rho). \tag{6.92}$$

Now, let us demand that each of the expansion terms satisfy the boundary condition on the cylinder surface, that is,

$$\kappa H_\nu^{(2)}(ka) = 0. \tag{6.93}$$

This is an equation for the unknown order ν. The roots of this equation have been denoted above as ν_s. Note that they are exactly the poles of function $A(\nu)$ (see Equation 6.77)). This way we arrive at the expansion

$$u_{\text{tot}} = \sum_s B_s W_s(\varphi) H_{\nu_s}^{(2)}(k\rho), \tag{6.94}$$

which is basically the same as (6.90). B_s are the expansion coefficients, and $W_s(\varphi)$ are solutions of the angular part of the field equation, where the separation constant is now ν_s^2, that is,

$$W_s(\varphi) = a\sin(\nu_s\varphi) + b\cos(\nu_s\varphi). \tag{6.95}$$

Let us assume that the field sources are localized in the plane $\varphi = 0$. For the E-polarized incidence, $u_{\text{tot}}(\rho, \varphi)$ equals E_z, and the sources are electric currents along z, which we denote by J_z. This means that the magnetic field component H_ρ is discontinuous at $\varphi = 0$:

$$\left.\frac{\partial u_{\text{tot}}(\rho, \varphi)}{\partial \varphi}\right|_{\varphi=+0} - \left.\frac{\partial u_{\text{tot}}(\rho, \varphi)}{\partial \varphi}\right|_{\varphi=-0} \sim J_z(\rho), \tag{6.96}$$

where ± 0 denote the values infinitesimally greater or smaller than 0. The electric field itself must be continuous:

$$u_{\text{tot}}(\rho, \varphi)|_{\varphi=+0} = u_{\text{tot}}(\rho, \varphi)|_{\varphi=-0}. \tag{6.97}$$

Conditions (6.96) and (6.97) can be satisfied for all ρ if we impose the following two conditions on $W_s(\varphi)$:

$$W_s(+0) = W_s(-0) \tag{6.98}$$

$$W_s'(+0) - W_s'(-0) = 1. \tag{6.99}$$

These conditions determine the coefficients a and b in (6.95), and we get

$$W_s(\varphi) = \frac{\cos[\nu_s(\pi - |\varphi|)]}{2\nu_s \sin \pi\nu_s}. \tag{6.100}$$

This result is, again, the same angular function as obtained by the Watson transform (see (6.90)).

Finally, we come to the equation

$$\left.\frac{\partial u_{\text{tot}}(\rho, \varphi)}{\partial \varphi}\right|_{\varphi=+0} - \left.\frac{\partial u_{\text{tot}}(\rho, \varphi)}{\partial \varphi}\right|_{\varphi=-0} = \sum_s B_s H_{\nu_s}^{(2)}(k\rho). \tag{6.101}$$

Relation (6.101) connects the known external currents at $\varphi = 0$ with the expansion coefficients B_s. From here, using the orthogonality of basis functions $H_{\nu_s}^{(2)}(k\rho)$, these coefficients can be finally found. For a line source at infinity (a plane wave), we again arrive at (6.90).

This possibility of using different expansions to represent the solution is a general feature of boundary value problems in separable coordinate systems. Since original partial differential equations can be split up in separate ordinary differential equations, each of these can be chosen to build a basis of expansion functions and expand the solution in a series. The expansion coefficients are functions of the other coordinates, and by using the orthogonality of the expansion functions one can reduce the problem to finding the coefficients from differential equations with respect to the other coordinate or coordinates, that is, from the boundary value problem of reduced dimensionality. The Watson transformation provides a way of connecting different representations of the solution. A systematic application of the approach to various problems in electromagnetics is described in Felsen and Marcuvitz (1973) and Makarov *et al.* (1991).

6.4.3 High-frequency Approximations

Let us find approximate expressions for the solution in the case of an electrically large cylinder ($ka \gg 1$) at distances considerably larger than the wavelength. We will use the representations obtained with the Watson transformation in section 6.4.1. The contour integral (6.89) will be asymptotically evaluated using the stationary phase method to obtain the GO approximation of the solution in the form of the incident wave and the wave reflected from the surface of the cylinder. Then we will study the series representation (6.90) to see that the terms have the meaning of waves that glide along the cylinder surface before arriving at the observer. These waves are termed creeping waves.

The mathematical basis of the analysis is Debye's approximations for the Bessel functions involved in the solution. A detailed description of the approximations can be found in Appendix E. Their use permits a straightforward physical interpretation of the obtained expressions. We will need the approximations of the Hankel functions of a real-valued argument in the fourth quadrant on the complex ν plane since both the integration contour Γ_{IV} in (6.89) and the points ν_s in the series (6.90) are located there. In this quadrant,

$$H_\nu^{(1)}(x) \approx \sqrt{\frac{2}{\pi x}} \frac{\exp[j\Phi(\nu,x) - j\pi/4]}{[1-(\nu/x)^2]^{1/4}} \tag{6.102}$$

with $\Phi(\nu,x)$ defined in (E.32) and $H_\nu^{(2)}(x) \approx -H_\nu^{(1)}(x)$ to the right and

$$H_\nu^{(2)}(x) \approx \sqrt{\frac{2}{\pi x}} \frac{\exp[-j\Phi(\nu,x) + j\pi/4]}{[1-(\nu/x)^2]^{1/4}} \tag{6.103}$$

to the left of the anti-Stokes line that starts at the point $\nu = x$ and goes to infinity in the fourth quadrant (line γ_+ in Figure E.12). The approximations are valid when $|\nu^2 - x^2| > x^{4/3}$ and $x \gg 1$, implying that in the fourth quadrant they are applicable everywhere except for $O(x^{1/3})$ in the vicinity of the point $\nu = x$. Multivalued functions in (6.102) and (6.103) are defined by conditions (E.33).

Since the contour Γ_{IV} and the points ν_s are located below the real axis ($\operatorname{Im}\nu < 0$), the trigonometric factor in (6.89) and (6.90) can be expanded in the geometric series

$$\frac{\cos[\nu(\pi-\varphi)]}{\sin\nu\pi} e^{j\nu\frac{\pi}{2}} = j\frac{e^{-j\nu(\varphi-\frac{\pi}{2})} + e^{-j\nu(\frac{3}{2}\pi-\varphi)}}{1-e^{-2j\nu\pi}} \tag{6.104}$$

$$= j\left[e^{-j\nu(\varphi-\frac{\pi}{2})} + e^{-j\nu(\frac{3}{2}\pi-\varphi)}\right]\sum_{q=0}^{+\infty} e^{-2jq\nu\pi}. \tag{6.105}$$

Because of the symmetry of the scattering problem, it is sufficient to assume that $0 \le \varphi \le \pi$. In this domain of φ the leading term of this expansion reads $j\exp[-j\nu(\varphi-\pi/2)]$.

GO Components

Let us show that the GO solution can be derived from the exact solution in the high-frequency limit ($k \to \infty$). We start with the integral (6.89), in which the trigonometric factor (6.104) is replaced just with the leading term $j\exp[-j\nu(\varphi-\pi/2)]$ in the geometric series (6.105), and the integrand is taken in the form (6.75),

$$u_{\mathrm{tot}}(\rho,\varphi) = \frac{1}{2}\int_{\Gamma_{\mathrm{IV}}} \left[H_\nu^{(1)}(k\rho) - \frac{\kappa H_\nu^{(1)}(ka)}{\kappa H_\nu^{(2)}(ka)} H_\nu^{(2)}(k\rho)\right] e^{j\nu(\frac{\pi}{2}-\varphi)}\, d\nu. \tag{6.106}$$

Note that the integrand is free of poles on the real axis in the complex ν plane and the contour can be deformed on the real axis, when necessary, without changing the value of the integral (the ends of the contour must be fixed in order to keep the integral convergent). The integral will be evaluated with the stationary phase method (Appendix D). As we shall see, the integrand has several critical points, so the contributions from these points are evaluated in succession.

Let us start with the second term in the integrand, in which the Hankel functions are approximated by (6.102) and (6.103) to give

$$\frac{\kappa H_\nu^{(1)}(ka)}{\kappa H_\nu^{(2)}(ka)} H_\nu^{(2)}(k\rho) \approx \pm\sqrt{\frac{2}{\pi k\rho}}\,\frac{\exp[2j\Phi(\nu,ka) - j\Phi(\nu,k\rho) - j\pi/4]}{[1-(\nu/k\rho)^2]^{1/4}}, \tag{6.107}$$

where the upper sign corresponds to $\kappa = 1$ (E polarization) and the lower sign to $\kappa = \mathrm{d}/\mathrm{d}(ka)$ (H polarization). The amplitude and the exponent in the right-hand side of (6.107) depend on k but in such a way that a change of integration variable by $\nu = kt$ would transform the integral to the form (D.1) amenable to the application of the stationary point method.

As stated in a note to (D.1), once the required behavior of the integrand function for $k \to +\infty$ is established, there is no need to explicitly factor out the large parameter in the phase function, and we can use the formula (D.21) for the stationary point contribution. So, we write

$$S(\nu) = \nu\left(\frac{\pi}{2} - \varphi\right) + 2\Phi(\nu,ka) - \Phi(\nu,k\rho), \tag{6.108}$$

and the equation for the critical point ν_c is

$$S'(\nu_c) = \frac{\pi}{2} - \varphi - 2\arccos\frac{\nu_c}{ka} + \arccos\frac{\nu_c}{k\rho} = 0. \tag{6.109}$$

The solution of (6.109) can be written as

$$\nu_c = ka\cos\chi_c = ka\sin\theta_c, \tag{6.110}$$

where $\theta_c = \pi/2 - \chi_c$ is the GO angle of reflection (Figure 6.15). This interpretation follows from the fact that equation (6.109) can be seen as the identity for the angles in the plane triangle OCP,

$$\angle\mathrm{O} + \angle\mathrm{C} + \angle\mathrm{P} = \pi. \tag{6.111}$$

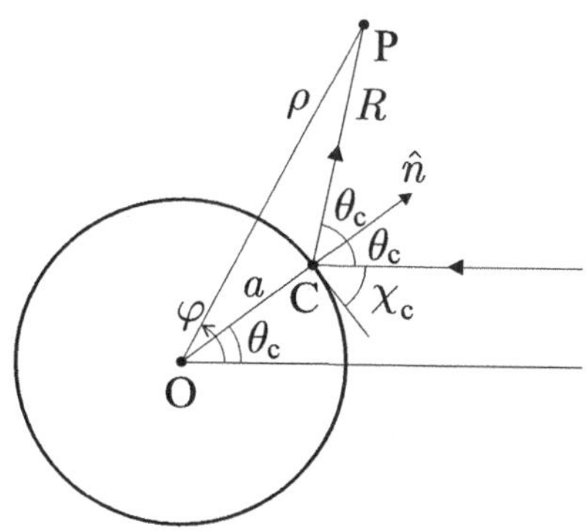

Figure 6.15 Reflection from a circular cylinder in the GO approximation. The GO reflection law is satisfied at the critical point C which is therefore the point of reflection. $\hat{n}$ is the normal to the surface at C.

Indeed, $\angle O = \varphi - \theta_c$, $\angle C = \pi - \theta_c$, and $\angle P$ is obtained as

$$\angle P = \frac{\pi}{2} - \arccos\left(\frac{a}{\rho}\sin\theta_c\right) \tag{6.112}$$

from the relation

$$\frac{a}{\sin(\angle P)} = \frac{\rho}{\sin(\angle C)}. \tag{6.113}$$

Since $0 < \nu_c < ka$, approximation (6.107) is applicable to calculating the contribution from the critical point ν_c. Using the formula (D.19) with

$$S''_c = \frac{2}{ka\cos\theta_c} - \frac{1}{k\sqrt{\rho^2 - a^2\sin^2\theta_c}} > 0, \tag{6.114}$$

we obtain

$$u_{\text{ref}} = \mp\left(\frac{a\cos\theta_c}{2\sqrt{\rho^2 - a^2\sin^2\theta_c} - a\cos\theta_c}\right)^{1/2} e^{2jka\cos\theta_c - jk\sqrt{\rho^2 - a^2\sin^2\theta_c}} \tag{6.115}$$

for the contribution of the critical point to the field.

Formula (6.115) has a clear physical interpretation: it describes a ray reflected from the surface of the cylinder with the angle of reflection equal to the angle of incidence θ_c. The formula can be expressed as

$$u_{\text{ref}} = \mp u_{\text{inc}}(a, \theta_c)\sqrt{\frac{\tilde{R}}{\tilde{R} + R}}e^{-jkR}, \tag{6.116}$$

where $\tilde{R} = (a/2)\cos\theta_c$, $u_{\text{inc}}(a, \theta_c) = \exp(jka\cos\theta_c)$ is the incident wave at the point of reflection, and R is the distance between the reflection point C and the observation point P. It follows from the relation

$$\rho = \sqrt{R^2 + a^2 - 2aR\cos(\angle C)} \tag{6.117}$$

between the sides of the triangle OCP that

$$R = \sqrt{\rho^2 - a^2\sin^2\theta_c} - a\cos\theta_c. \tag{6.118}$$

The ray is reflected from the curved surface and has therefore an apparent source (a focus) in the interior of the scatterer at the distance $\tilde{R}$ from the reflection point along the direction of reflection. The factor ∓ 1 on the right-hand side of (6.116) relates to the E and H polarization cases, respectively, and can be interpreted as the reflection coefficient of the PEC surface for the respective polarization.

When $\rho \to \infty$, we have that $\theta_c \to \varphi/2$ and

$$u_{\text{ref}} \to \mp\sqrt{\frac{a}{2\rho}\cos\frac{\varphi}{2}}\, e^{-jk\rho + 2jka\cos(\varphi/2)}. \tag{6.119}$$

The integral (6.106) has a further critical point which is responsible for the incident wave. The point is located at $\nu_c = k\rho\sin\varphi$ and occurs either in the first term of the integrand when $0 \le \varphi \le \pi/2$, or in the second term when $\pi/2 \le \varphi < \pi - \arcsin(a/\rho)$. Approximating $H^{(1)}_\nu(k\rho)$ by (6.102) with $x = k\rho$ leads to the phase function

$$S(\nu) = \nu\left(\frac{\pi}{2} - \varphi\right) + \Phi(\nu, k\rho), \tag{6.120}$$

and the equation for the critical point

$$\arccos\frac{\nu_c}{k\rho} = \frac{\pi}{2} - \varphi \tag{6.121}$$

has a solution only with $0 \le \varphi \le \pi/2$ since $\arccos(\nu/k\rho) \ge 0$ according to the choice of the branch in the Debye asymptotics. The contribution from the critical point is equal to $\exp(jk\rho\cos\varphi)$ when $0 \le \varphi \le \pi/2$, and zero otherwise.

When $\varphi > \pi/2$, it is the second term in the integrand of (6.106) that describes the incident wave. On the interval $ka < \nu < k\rho$, instead of (6.107) we have

$$\frac{\kappa H_\nu^{(1)}(ka)}{\kappa H_\nu^{(2)}(ka)} H_\nu^{(2)}(k\rho) \approx -\sqrt{\frac{2}{\pi k\rho}} \frac{\exp[-j\Phi(\nu, k\rho) + j\pi/4]}{[1-(\nu/k\rho)^2]^{1/4}} \tag{6.122}$$

since $H_\nu^{(1)}(ka) \approx -H_\nu^{(2)}(ka)$. This results in the phase function

$$S(\nu) = \nu\left(\frac{\pi}{2} - \varphi\right) - \Phi(\nu, k\rho) \tag{6.123}$$

and the equation for the critical point

$$\arccos\frac{\nu_c}{k\rho} = \varphi - \frac{\pi}{2} \tag{6.124}$$

which has a solution when $\varphi > \pi/2$ only. The contribution to the integral is $\exp(jk\rho\cos\varphi)$ when $\varphi > \pi/2$, and zero when $\varphi < \pi/2$. The value of φ has the upper boundary at $\varphi = \pi - \arcsin(a/\rho)$; otherwise, ν_c leaves the interval $ka < \nu_c < k\rho$ and (6.122) should be replaced with (6.107).

Summarizing, in the limit of $ka \gg 1$ the representation (6.106) recovers the GO solution

$$u_{\text{tot}}(\rho, \varphi) \approx u_{\text{inc}}(\rho, \varphi) + u_{\text{ref}}(\rho, \varphi) \tag{6.125}$$

with $u_{\text{ref}}(\rho, \varphi)$ given by (6.116) or (6.119). The approximation assumes that $0 \le \varphi < \pi - \arcsin(a/\rho)$, implying that the observation point is located in the exterior of the geometric shadow of the cylinder, since $\rho\sin\varphi = a$ on the shadow boundary.

Creeping Waves

When $ka \gg 1$, the representation (6.90) leads to a physically important interpretation of the solution. For the analysis that follows, the zeros of (6.77) with the smallest absolute values of the imaginary part of ν_s are particularly important. For electrically large cylinders it is possible to express these zeros in a simple form in terms of zeros of the Airy function

$$\text{Ai}(p) = \frac{1}{2\pi}\int_{-\infty}^{\infty} e^{j(p\tau+\tau^3/3)}\,d\tau = \frac{1}{\pi}\int_0^{\infty}\cos\left(p\tau + \frac{\tau^3}{3}\right)d\tau \tag{6.126}$$

and its derivative $\text{Ai}'(p)$. The zeros are located close to the point $\nu = ka$ in the sense that $|\nu_s - ka| \ll ka$. In this region of the complex ν plane, the argument and order of the Hankel function are almost equal and it can be approximated by the Airy function (see (E.162) in section E.3), which leads to

$$H_\nu^{(2)}(ka) \approx \frac{2}{m} e^{j\pi/3}\text{Ai}\left(\frac{\nu - ka}{m} e^{-2j\pi/3}\right) \tag{6.127}$$

$$H_\nu^{(2)\prime}(ka) \approx \frac{2}{m^2} e^{2j\pi/3}\text{Ai}'\left(\frac{\nu - ka}{m} e^{-2j\pi/3}\right) \tag{6.128}$$

with

$$m = \left(\frac{ka}{2}\right)^{1/3}. \tag{6.129}$$

Table 6.1 First nulls of the Airy function (p_s^E) and of its derivative (p_s^H) and the corresponding parameters from (6.132) and (6.133).

s	p_s^E	p_s^H	$-\left[\frac{3\pi}{2}\left(s-\frac{1}{4}\right)\right]^{2/3}$	$-\left[\frac{3\pi}{2}\left(s-\frac{3}{4}\right)\right]^{2/3}$
1	−2.33810	−1.01879	−2.32025	−1.11546
2	−4.08794	−3.24819	−4.08181	−3.26163
3	−5.52055	−4.82009	−5.51716	−4.82632

These approximations, known as transition region asymptotic approximations, are valid when ka and $|\nu|$ are large, but $|ka-\nu| \leq O(|\nu|^{1/3})$, and they therefore have the validity area complimentary to that of the Debye approximation (Appendix E).

It follows from (6.77) that ν_s are zeros of the function $H_\nu^{(2)}(ka)$ in the case of E polarization and of the function $H_\nu^{(2)\prime}(ka)$ in the case of H polarization, and approximations (6.127) and (6.128) imply that

$$\frac{\nu_s - ka}{m} e^{-2j\pi/3} = \begin{cases} p_s^E, & E \text{ polarization} \\ p_s^H, & H \text{ polarization} \end{cases}, \tag{6.130}$$

where p_s^E is the sth zero of $\mathrm{Ai}(p)$ and p_s^H is the sth zero of $\mathrm{Ai}'(p)$. Thus, the roots of equation (6.77) can be approximately written as

$$\nu_s \approx ka - me^{-j\pi/3} p_s = ka - \frac{1-j\sqrt{3}}{2} mp_s, \tag{6.131}$$

where p_s are the nulls of either the Airy function or its derivative. The first three p_s are given in Table 6.1. The real parts of the first zeros ν_s are of the order of ka, whereas their imaginary parts are of the order $(ka)^{1/3}$. Both the real and imaginary parts of the roots grow in the absolute value with increasing number s and normalized radius ka.

In the exterior of the $(ka)^{1/3}$ vicinity of the point $\nu = ka$, the simple formula (6.131) is inaccurate. In this region the Hankel functions are better described by the Debye approximations (Appendix E). In this approximation, the zeros of $H_\nu^{(2)}(ka)$ and $H_\nu^{(2)\prime}(ka)$ in the vicinity of $\nu = ka$ can be obtained from (E.106), (E.136), and (E.142) as

$$\nu_s \approx ka + me^{-j\pi/3}\left[\frac{3\pi}{2}\left(s-\frac{1}{4}\right)\right]^{2/3} \tag{6.132}$$

and

$$\nu_s \approx ka + me^{-j\pi/3}\left[\frac{3\pi}{2}\left(s-\frac{3}{4}\right)\right]^{2/3}, \tag{6.133}$$

respectively. In fact, already the zeros with $s \geq 2$ are quite accurately expressed through (6.132) and (6.133) (see Table 6.1). The accuracy of the Debye approximations for the zeros improves even further if the function $\Phi(\nu, ka)$ in (E.136) and (E.142) is calculated exactly, without being approximated by (E.106) (see Table E.2).

Now we are ready to start the asymptotic analysis of the solution in the form (6.90). Using (6.105), we rewrite (6.90) as

$$u_{\text{tot}} = \sum_{s=1}^{+\infty}\sum_{q=0}^{+\infty}\left(u_{sq}^{+} + u_{sq}^{-}\right), \tag{6.134}$$

where

$$u_{sq}^{+} = \pi j H_{\nu_s}^{(2)}(k\rho) Q_s e^{-j\nu_s(\varphi - \frac{\pi}{2} + 2\pi q)} \tag{6.135}$$

$$u_{sq}^{-} = \pi j H_{\nu_s}^{(2)}(k\rho) Q_s e^{-j\nu_s(3\frac{\pi}{2} - \varphi + 2\pi q)} \tag{6.136}$$

and

$$Q_s = \frac{2\kappa J_{\nu_s}(ka)}{\frac{\partial}{\partial \nu}\kappa H_\nu^{(2)}(ka)\Big|_{\nu=\nu_s}} = \frac{\kappa H_{\nu_s}^{(1)}(ka)}{\frac{\partial}{\partial \nu}\kappa H_\nu^{(2)}(ka)\Big|_{\nu=\nu_s}}. \tag{6.137}$$

Assuming that $k(\rho - a) > (k\rho)^{1/3}$, that is, the observation point is not too close to the cylinder surface, we may use the Debye approximation (6.103) for the Hankel function $H_{\nu_s}^{(2)}(k\rho)$, which gives

$$u_{sq}^{\pm} = \sqrt{\frac{2\pi}{k\rho}} \frac{Q_s}{[1 - (\nu_s/k\rho)^2]^{1/4}} \mathrm{e}^{-j\Phi_{sq}^{\pm} + j\frac{3}{4}\pi}, \tag{6.138}$$

where

$$\Phi_{sq}^{+} = k\rho\sqrt{1 - \left(\frac{\nu_s}{k\rho}\right)^2} + \nu_s\left[\varphi - \frac{\pi}{2} - \arccos\left(\frac{\nu_s}{k\rho}\right) + 2\pi q\right] \tag{6.139}$$

$$\Phi_{sq}^{-} = k\rho\sqrt{1 - \left(\frac{\nu_s}{k\rho}\right)^2} + \nu_s\left[\frac{3}{2}\pi - \varphi - \arccos\left(\frac{\nu_s}{k\rho}\right) + 2\pi q\right]. \tag{6.140}$$

Since in the leading terms of (6.90) ν_s is close to ka, the amplitude and phase factors in (6.138) can be approximated by the Taylor series in the powers of $\nu_s - ka$, and retaining the leading term in the amplitude and the first two terms in the phase gives

$$u_{sq}^{\pm} = \sqrt{\frac{2\pi}{k}} \frac{Q_s}{[\rho^2 - a^2]^{1/4}} \mathrm{e}^{-j\Phi_{sq}^{\pm} + j\frac{3}{4}\pi} \tag{6.141}$$

with

$$\Phi_{sq}^{+} \approx k\sqrt{\rho^2 - a^2} + \nu_s\left[\varphi - \frac{\pi}{2} - \arccos\left(\frac{a}{\rho}\right) + 2\pi q\right] \tag{6.142}$$

$$\Phi_{sq}^{-} \approx k\sqrt{\rho^2 - a^2} + \nu_s\left[\frac{3}{2}\pi - \varphi - \arccos\left(\frac{a}{\rho}\right) + 2\pi q\right], \tag{6.143}$$

where $\sqrt{\rho^2 - a^2} > 0$ and $0 < \arccos(a/\rho) < \pi$.

The phase constants (6.142) and (6.143) in formula (6.141) have a clear geometric interpretation as the phase change over the shortest paths between the source and the observer around the cylinder. Indeed, $\sqrt{\rho^2 - a^2}$ is the length of the straight pieces joining the observation point P and the points $L_\pm$, at which the tangential to the surface points to the observation point P, and the quantities

$$a\theta_+ = a\left(\varphi - \arccos\frac{a}{\rho} - \frac{\pi}{2}\right), \quad a\theta_- = a\left(\frac{3\pi}{2} - \arccos\frac{a}{\rho} - \varphi\right) \tag{6.144}$$

are the lengths of two arcs between the points $L_\pm$ and the points $A_\pm$, at which the incidence direction is tangential to the cylinder surface (see Figure 6.16). Thus, (6.141) with $q = 0$ describes "rays" touching the cylinder surface at the points $A_\pm$, traveling along the arcs $a\theta_\pm$, and tangentially leaving the surface at the points $L_\pm$ toward the observer. These "rays" are called *creeping waves* (Franz and Deppermann 1952), and the points $A_\pm$ and $L_\pm$ are called the attachment and launching points, respectively. Another name for creeping waves is the surface-diffracted rays, which is used in the framework of the GTD to describe the high-frequency mechanism of the transportation of energy in the shadow of smoothly shaped opaque bodies (Keller 1956, 1962; Levy and Keller 1959).

When traveling along the arcs $a\theta_\pm$, the rays $u_{s0}^{\pm}$ undergo an exponential attenuation because ν_s are complex numbers with negative imaginary parts. This attenuation can be seen as the result of the loss of energy due to the rays shedding

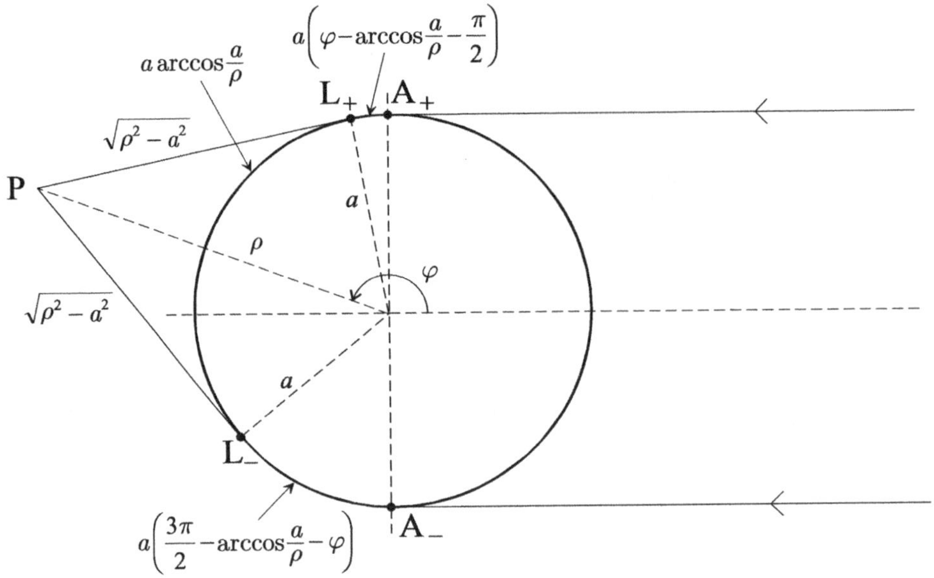

Figure 6.16 Creeping waves: path lengths for the shadow region.

off tangentially to the surface. After being launched from the surface, the creeping waves propagate as cylindrical waves with the apparent source at the launching point.

The meaning of the terms with $q \geq 1$ is also quite apparent: these are the creeping waves circumventing the cylinder q times before being launched toward the observer. The contribution from these rays is small compared with the rays with $q = 0$ and can typically be neglected once ka is larger than unity. Thus, the field in the shadow behind the cylinder consists of two contributions, creeping waves traveling above and below the cylinder (Figure 6.16),

$$u_{\text{tot}} \approx u_+ + u_-, \tag{6.145}$$

and these contributions can be written as

$$u_\pm = u_{\text{inc}}(\text{A}_\pm) T_\pm \frac{\text{e}^{-jk\tau}}{\sqrt{\tau}}, \tag{6.146}$$

where $\tau = \sqrt{\rho^2 - a^2} = |\text{PL}_\pm|$ is the distance between the launching points and the observer, $u_{\text{inc}}(\text{A}_\pm) = u_{\text{inc}}(a, \pm\pi/2) = 1$ is the incident field at the attachment points, and $T_\pm$ is the transformation coefficient for the cylinder which relates the amplitude of the launched cylindrical waves to the incident field at the attachment points. Comparing (6.146) with (6.141), the transformation coefficient can be expressed as

$$T_\pm = \sqrt{\frac{2\pi}{k}} \text{e}^{j\frac{3}{4}\pi} \sum_{s=1}^{+\infty} Q_s \text{e}^{-j\nu_s \theta_\pm}. \tag{6.147}$$

The imaginary part of ν_s grows with increasing s, implying that the rays with $s = 1$ form the dominant contributions at the observation point. Thus, the field in the shadow of an electrically large PEC cylinder is approximately given by the sum of the two creeping waves with $s = 1$ and $q = 0$,

$$u_{\text{tot}} \approx u_{10}^+ + u_{10}^-. \tag{6.148}$$

To obtain a physically clear representation of the field in the region illuminated by the incident wave, for example in the backscattering region, the series (6.90) is inconvenient as it diverges with $-\pi/2 < \varphi < \pi/2$. For the illuminated region we return to the integral representation (6.89) of the total field and perform the transformation

$$\cos[\nu(\pi - \varphi)] = \mathrm{e}^{-j\nu\pi}\cos(\nu\varphi) + j\mathrm{e}^{-j\nu\varphi}\sin(\nu\pi) \tag{6.149}$$

with $0 \le \varphi < \pi$. Substituting (6.149) into (6.89) we get

$$u_{\mathrm{tot}} = \int_{\Gamma_{\mathrm{IV}}} \mathrm{e}^{-j\nu\varphi} A(\nu)\,\mathrm{d}\nu - j\int_{\Gamma_{\mathrm{IV}}} \frac{\mathrm{e}^{-j\nu\pi}\cos\nu\varphi}{\sin\nu\pi} A(\nu)\,\mathrm{d}\nu = u' + u''. \tag{6.150}$$

The first integral coincides with (6.106) and, as we have seen, gives the GO solution consisting of the incident and reflected waves

$$u' = u_{\mathrm{inc}} + u_{\mathrm{ref}}, \tag{6.151}$$

provided that $ka \gg 1$ and the observation point lies in the exterior of the shadow region and not too close to the surface of the cylinder.

The second integral can be evaluated as a sum of residues at the points ν_s,

$$u'' = \pi \sum_s H^{(2)}_{\nu_s}(k\rho) Q_s \frac{\cos(\nu_s\varphi)}{\sin\pi\nu_s} \mathrm{e}^{-j\pi\nu_s/2} \tag{6.152}$$

with Q_s defined in (6.137). When $\mathrm{Im}\,\nu < 0$, the trigonometric factor can be expanded as

$$\frac{\cos(\nu\varphi)}{\sin\nu\pi} \mathrm{e}^{-j\nu\frac{\pi}{2}} = j\left[\mathrm{e}^{-j\nu(\frac{3}{2}\pi+\varphi)} + \mathrm{e}^{-j\nu(\frac{3}{2}\pi-\varphi)}\right] \sum_{q=0}^{+\infty} \mathrm{e}^{-2jq\nu\pi}, \tag{6.153}$$

implying that u'' can be represented in the form similar to (6.134), the only difference being the replacement of φ with $\varphi + 2\pi$ in the terms u^{+}_{sq}. The paths of the corresponding creeping rays are shown in Figures 6.17 and 6.18. Thus, the total field in the lit region is the sum of GO (incident and reflected) and creeping waves. Since $ka \gg 1$, the circumventing creeping waves and the creeping waves with $s \ge 2$ can typically be neglected.

Figure 6.19 shows the backscattering width of a circular PEC cylinder as a function of the electrical size of the cylinder in both polarization cases. The scattering width is normalized to πa to make it a function of ka only (see (6.34)). With

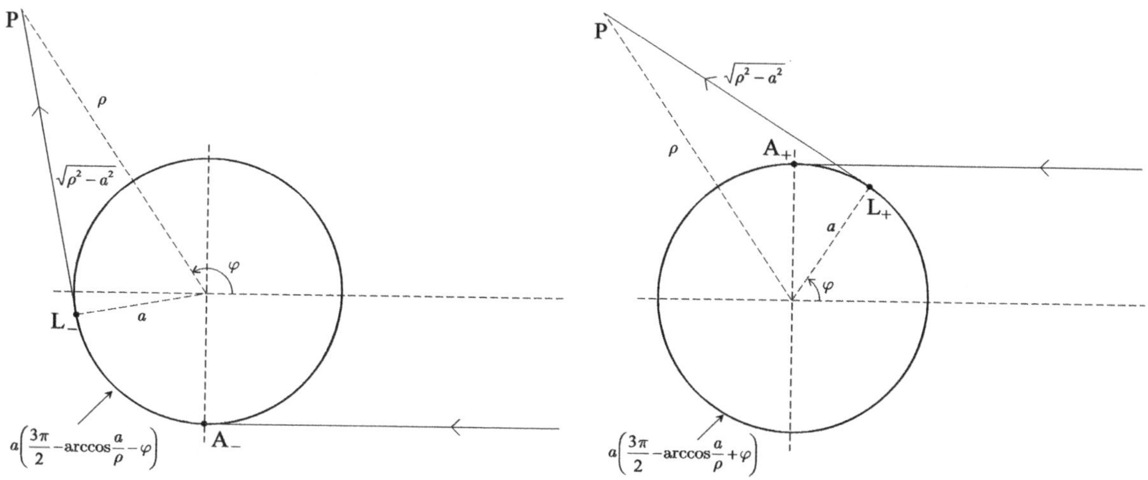

Figure 6.17 Path lengths of the principal creeping waves (without those multiply circumventing the cylinder) for the observer in the illuminated region when $\pi/2 < \varphi < \pi$.

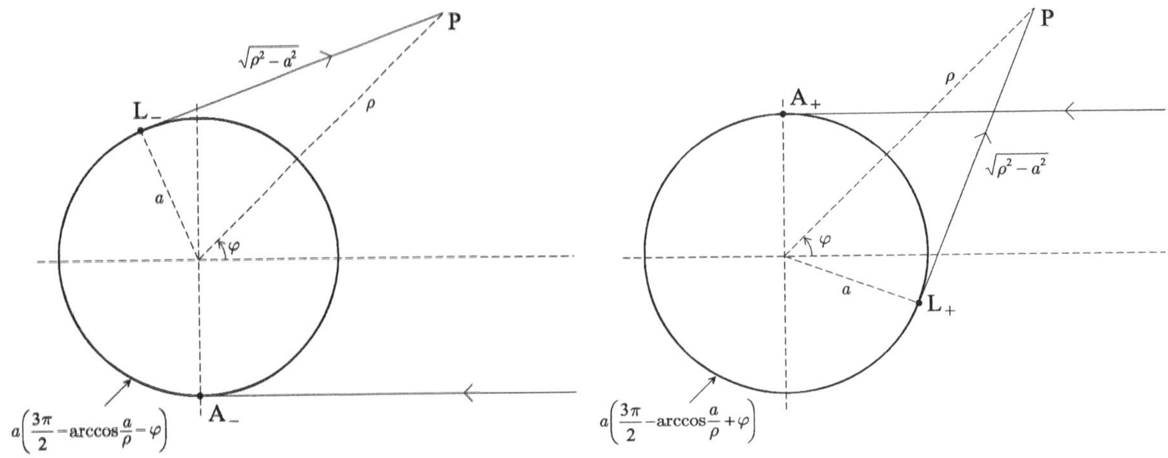

Figure 6.18 The same as Figure 6.17 but with $0 \leq \varphi < \pi/2$.

$ka \geq 10$ the scattering width is already quite close to its optical limit πa. For smaller values of ka, polarization differences are significant. Furthermore, the curve for H polarization shows significant oscillations. This can be physically explained by the contribution from creeping waves since the period of the oscillations corresponds to the path difference $(2 + \pi)a$ between the reflected wave and the creeping waves traveling around the shadowed half of the cylinder. The need to explain these oscillations in the measured backscattering data has led Franz and Deppermann (1952) to the introduction of the concept of the creeping waves. The oscillations are hardly seen in the curve for E polarization, which is due to the much stronger attenuation of the creeping waves on PEC surfaces for this polarization. The behavior of the scattering width when $ka \to 0$ is addressed in the problems section (problem 4).

The high-frequency representation of the scattered field in the lit region as a sum of the creeping waves and the GO reflected wave is appealing physically as it permits a clear physical interpretation of scattering not only from circular

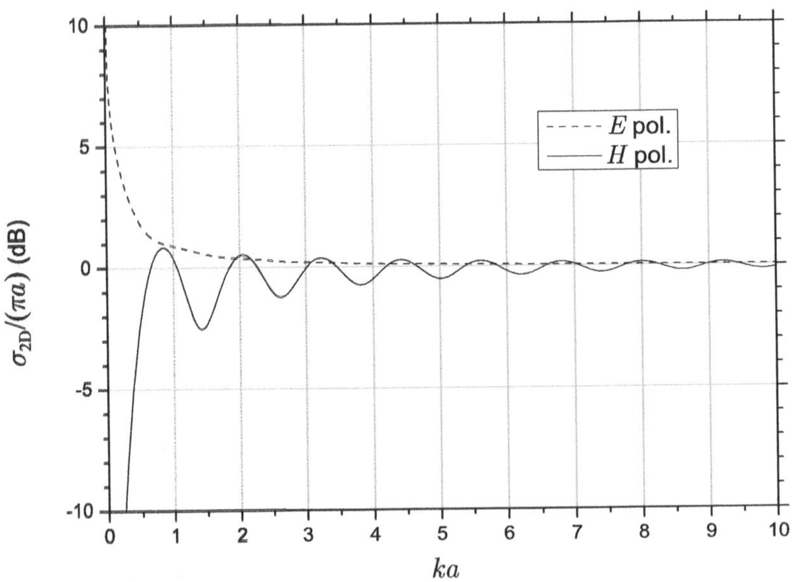

Figure 6.19 Normalized backscattering width of a PEC circular cylinder.

cylinders but also from generally shaped convex bodies. However, the representation is not suitable for accurate numerical calculations since the creeping waves are exponentially small quantities, implying the need to account for higher-order terms in the asymptotic expansion of the integral for the reflected wave. A further problem, common for all high-frequency approximations developed in this section, is their failure within the transition region between the lit and shadowed portions of the space and in the immediate vicinity of the surface of the cylinder. In all these cases, the Debye approximations used to approximate the Hankel functions are no longer applicable. The field in these regions has a more complicated structure, and its accurate asymptotic description requires a more sophisticated analysis.

In section 6.4.4 we derive high-frequency asymptotic representations which are valid on the whole surface of an electrically large circular PEC cylinder, including the transition region between the lit and shadowed portions of the surface, the so-called penumbra region. These representations are approximate, but they can be extended to much more general geometries; this extension will be the subject of section 6.6. However, as long as the scattering from an ideally shaped circular cylinder is of interest, the exact series solution is in most cases sufficient for accurate numerical calculations.

6.4.4 Surface Currents in the Penumbra Region. Fock's Functions

In this section we study the distribution of currents or, which is the same thing, the tangential components of the magnetic field on the surface of a perfectly electrically conducting (PEC) cylinder assuming that the wavelength λ is much shorter than the radius a ($ka \gg 1$). In the optical limit, when $a/\lambda \to \infty$, there is a sharp shadow boundary at $\varphi = \pm\pi/2$, and the currents are zero at the shadowed portion of the cylinder ($\pi/2 < |\varphi| \le \pi$), while taking the ray optical values on the illuminated side of the cylinder ($|\varphi| < \pi/2$). Since in reality the parameter a/λ is always finite, instead of the sharp shadow boundary there is always a transition region within which the currents smoothly vary between their values on the lit and shadowed portions of the surface. This transition region is often referred to as the penumbra or semi-shadow region. GO does not apply in such regions as the shadow boundary is a caustic of the ray continuum. The exact series solutions (6.24) are valid on the whole surface of the cylinder, but they converge too slowly when $a \gg \lambda$. The creeping wave series of the form (6.90), which are obtained by the Watson transformation, diverge on the illuminated portion of the cylinder and are therefore also unsuitable in the penumbra region. By using an asymptotic approach described by Goriainov (1958) it is nevertheless possible to obtain simple and instructive approximate representations for the currents in the penumbra region on the surface of a PEC cylinder, which recover the results by Fock for general convex surfaces (Fock 1965, Chapter 2). It can be shown that the tangential components of the total magnetic field on the surface of the cylinder are equal to the magnetic field components in the incident wave multiplied by a Fock function, either $F(\xi)$ or $G(\xi)$, of the argument $\xi = l/d$, where l is the distance from the boundary of the geometric shadow and

$$d = \left(\frac{\lambda a^2}{\pi}\right)^{1/3} = \left(\frac{2a^2}{k}\right)^{1/3} \tag{6.154}$$

is the width of the penumbra region. In this section, these asymptotic representations are derived in the particular case of a PEC circular cylinder by using the contour integral representations for the tangential components of the magnetic field on the surface of the cylinder.

The only tangential components of the electromagnetic field that do not vanish on the surface of a PEC cylinder are H_φ and H_z. The former is excited by an incident E-polarized wave,

$$E_z^{\text{inc}}(\rho, \varphi) = E_0 \exp(jk\rho\cos\varphi) \tag{6.155}$$

and the latter by an H-polarized wave

$$H_z^{\text{inc}}(\rho, \varphi) = H_0 \exp(jk\rho\cos\varphi). \tag{6.156}$$

Specializing the series solutions (6.24), (6.25), and (6.26) to $\rho = a$ and using the Wronskian formula (E.8)

$$J_n(ka)H_n^{(2)\prime}(ka) - J_n^\prime(ka)H_n^{(2)}(ka) = -\frac{2j}{\pi ka}, \tag{6.157}$$

we obtain

$$H_z(a,\varphi) = -\frac{2jH_0}{\pi ka}\sum_{n=-\infty}^{\infty}\frac{\mathrm{e}^{-jn(\varphi-\pi/2)}}{H_n^{(2)\prime}(ka)} \tag{6.158}$$

$$H_\varphi(a,\varphi) = \frac{2E_0}{\pi kaZ}\sum_{n=-\infty}^{\infty}\frac{\mathrm{e}^{-jn(\varphi-\pi/2)}}{H_n^{(2)}(ka)}, \tag{6.159}$$

where Z is the wave impedance of the medium surrounding the cylinder. In (6.159) we have used the relation (6.1) to express H_φ through E_z. By using the Watson transformation as described in section 6.4.1 and limiting φ to the interval $0 \le \varphi \le \pi$ (which is sufficient because of the problem symmetry), the components can be represented by the contour integrals

$$H_z(a,\varphi) = -\frac{2H_0}{\pi ka}\int_{\Gamma_{\mathrm{IV}}}\frac{\cos[\nu(\pi-\varphi)]\mathrm{e}^{j\nu\pi/2}}{\sin(\nu\pi)H_\nu^{(2)\prime}(ka)}\,\mathrm{d}\nu \tag{6.160}$$

$$H_\varphi(a,\varphi) = -\frac{2jE_0}{\pi kaZ}\int_{\Gamma_{\mathrm{IV}}}\frac{\cos[\nu(\pi-\varphi)]\mathrm{e}^{j\nu\pi/2}}{\sin(\nu\pi)H_\nu^{(2)}(ka)}\,\mathrm{d}\nu \tag{6.161}$$

with the integration path Γ_{IV} shown in Figure 6.14.

Consider the integrals (6.160) and (6.161) in the vicinity of the shadow boundary $\varphi = \pi/2$, using the fact that ka and therefore the parameter m (see (6.129)) are large compared to unity. In the lower half-plane of the complex ν plane the integration contour Γ_{IV} can be freely deformed between the real axis and the points ν_s (Figure 6.14) and since $\mathrm{Im}\,\nu_1 < -m\sqrt{3}/2$ (see (6.131)), we can choose $\delta = m$. Because $m \gg 1$, the parameter δ is also large, and the trigonometric part in the integrands of (6.160) and (6.161) can be approximated over the whole integration contour with the expression

$$\frac{\cos[\nu(\pi-\varphi)]\mathrm{e}^{j\nu\pi/2}}{\sin(\nu\pi)} \approx j\mathrm{e}^{-j\nu(\varphi-\pi/2)}, \tag{6.162}$$

giving

$$H_z(a,\varphi) \approx -\frac{2jH_0}{\pi ka}\int_{\Gamma_{\mathrm{IV}}}\frac{\mathrm{e}^{-j\nu(\varphi-\pi/2)}}{H_\nu^{(2)\prime}(ka)}\,\mathrm{d}\nu \tag{6.163}$$

$$H_\varphi(a,\varphi) \approx \frac{2E_0}{\pi kaZ}\int_{\Gamma_{\mathrm{IV}}}\frac{\mathrm{e}^{-j\nu(\varphi-\pi/2)}}{H_\nu^{(2)}(ka)}\,\mathrm{d}\nu. \tag{6.164}$$

The creeping wave series

$$H_z(a,\varphi) = -\frac{4H_0}{ka}\sum_{s=1}^{+\infty}\left[\frac{\partial}{\partial\nu}H_\nu^{(2)\prime}(ka)\Big|_{\nu=\nu_s^H}\right]^{-1}\mathrm{e}^{-j\nu_s^H(\varphi-\pi/2)} \tag{6.165}$$

$$H_\varphi(a,\varphi) = -\frac{4jE_0}{kaZ}\sum_{s=1}^{+\infty}\left[\frac{\partial}{\partial\nu}H_\nu^{(2)}(ka)\Big|_{\nu=\nu_s^E}\right]^{-1}\mathrm{e}^{-j\nu_s^E(\varphi-\pi/2)} \tag{6.166}$$

are obtained by evaluation of the integrals (6.163) and (6.164) as sums of residues at the points ν_s in the fourth quadrant of the complex ν plane (ν_s are given by (6.131)). However, the series converge only with $\varphi > \pi/2$ because along the line of their zeros in the fourth quadrant, functions $H_\nu^{(2)}(ka)$ and $H_\nu^{(2)\prime}(ka)$ decay as $O(\nu^{-1/2})$ and $O(\nu^{-3/2})$, respectively (see (6.394)). On the shadowed portion of the cylinder ($\varphi > \pi/2$) the residue series converge but at a decreasing rate as the observation point approaches the shadow boundary. Thus, the creeping wave series are not suitable for the analysis of fields in the penumbra.

Now we are going to evaluate the integrals (6.163) and (6.164) asymptotically. Approximating the Hankel function by the Debye formula (6.103), we find the equation for the critical point

$$\arccos\left(\frac{\nu_c}{ka}\right) = \frac{\pi}{2} - \varphi \tag{6.167}$$

and conclude that the critical point $\nu_c = ka \sin\varphi$ approaches the point $\nu = ka$ as $\varphi \to \pi/2$. This means that the main contribution to the integrals gives an asymptotically small vicinity of the point $\nu = ka$. However, in this vicinity the Debye approximation for the Hankel function is not valid, but the Hankel functions can be approximated by the Airy functions (6.126) as in (6.127) and (6.128) (see section E.3). With the help of a variable t defined by

$$\nu = ka + mt, \tag{6.168}$$

where m is given by (6.129), the asymptotic approximations (6.127) and (6.128) can be conveniently expressed in terms of the normalized Airy function

$$w_2(t) = 2\sqrt{\pi}\mathrm{e}^{-j\pi/6}\mathrm{Ai}(t\mathrm{e}^{-2j\pi/3}) \tag{6.169}$$

as

$$H_\nu^{(2)}(ka) \approx \frac{j}{\sqrt{\pi}m} w_2(t) \tag{6.170}$$

$$H_\nu^{(2)\prime}(ka) \approx -\frac{j}{\sqrt{\pi}m^2} w_2'(t). \tag{6.171}$$

Since the main contribution to the integrals (6.163) and (6.164) comes from the small vicinity around the point $\nu = ka$, we can truncate the integration contour Γ_{IV} to an interval $ka - \Delta\nu \le \nu \le ka + \Delta\nu$ with a small real-valued $\Delta\nu$ and replace the Hankel functions with (6.170) and (6.171) on this interval. Allowing a small error, the ends of the integration contour can be let to go to infinity, $\Delta\nu \to +\infty$, and in terms of the new integration variable t the integrals (6.163) and (6.164) read

$$H_z(a,\varphi) \approx \frac{H_0}{\sqrt{\pi}}\mathrm{e}^{-jka(\varphi-\pi/2)} \int_{-\infty}^{+\infty} \frac{\mathrm{e}^{-jmt(\varphi-\pi/2)}}{w_2'(t)}\,\mathrm{d}t \tag{6.172}$$

$$H_\varphi(a,\varphi) \approx \frac{-jE_0}{\sqrt{\pi}mZ}\mathrm{e}^{-jka(\varphi-\pi/2)} \int_{-\infty}^{+\infty} \frac{\mathrm{e}^{-jmt(\varphi-\pi/2)}}{w_2(t)}\,\mathrm{d}t. \tag{6.173}$$

The most rapid convergence of the integrals (6.172) and (6.173) is achieved upon deformation of the integration contours onto a contour Γ which begins at $t = \infty\exp(j4\pi/3)$ and ends at $t = +\infty$ (Figure 6.20). Along this contour, as $|t|$

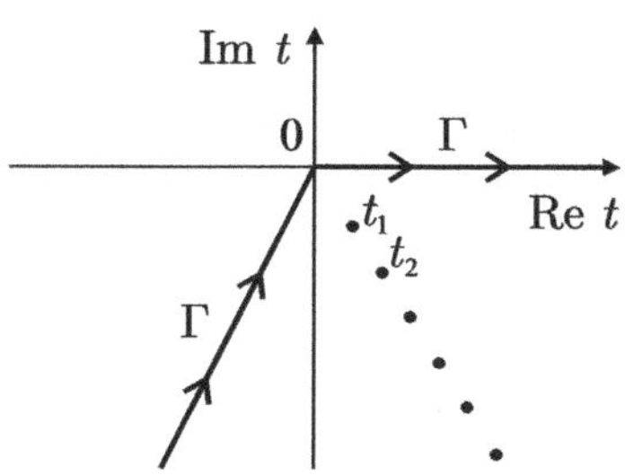

Figure 6.20 Integration contour Γ and the poles t_s of the integrand in the Fock functions.

grows, functions $w_2(t)$ and $w_2'(t)$ exhibit the fastest growth as $\exp(2|t|^{3/2}/3)$, which follows from the relation (6.169) and the large argument asymptotics of the Airy function (see (D.30) and (D.31))

$$w_2(t) \approx t^{-1/4} \exp\left(\frac{2}{3}t^{3/2}\right), \quad w_2'(t) \approx t^{1/4} \exp\left(\frac{2}{3}t^{3/2}\right), \tag{6.174}$$

where $-\pi/3 < \arg t < 5\pi/3$.

Introducing the new parameter

$$\xi = m\left(\varphi - \frac{\pi}{2}\right), \tag{6.175}$$

we obtain

$$H_z(a,\varphi) \approx H_0 g(\xi) \mathrm{e}^{-jka(\varphi-\pi/2)} \tag{6.176}$$

$$H_\varphi(a,\varphi) \approx \frac{-jE_0}{mZ} f(\xi) \mathrm{e}^{-jka(\varphi-\pi/2)} \tag{6.177}$$

with

$$f(\xi) = \frac{1}{\sqrt{\pi}} \int_\Gamma \frac{\mathrm{e}^{-j\xi t}}{w_2(t)} \mathrm{d}t \tag{6.178}$$

$$g(\xi) = \frac{1}{\sqrt{\pi}} \int_\Gamma \frac{\mathrm{e}^{-j\xi t}}{w_2'(t)} \mathrm{d}t, \tag{6.179}$$

which are referred to as *Fock functions*. The integrand functions in (6.178) and (6.179) have pole singularities at

$$t_s^E = -p_s^E \exp(-j\pi/3), \quad t_s^H = -p_s^H \exp(-j\pi/3), \quad (s = 1, 2, \ldots) \tag{6.180}$$

which are zeros of $w_2(t)$ and $w_2'(t)$, respectively. Table 6.1 gives the numerical values of p_s^E and p_s^H with $s = 1, 2, 3$.

In the vicinity of the shadow boundary where $|\varphi - \pi/2| = O(1/m)$, we can use the approximate relation

$$ka\left(\frac{1}{2}\pi - \varphi\right) = ka\cos\varphi - \frac{1}{3}\xi^3 + O(m^{-2}) \tag{6.181}$$

and rewrite (6.176) and (6.177) in terms of the components of the incident fields (6.155) and (6.156) as

$$H_z(a,\varphi) \approx H_z^{\mathrm{inc}}(a,\varphi) G(\xi) \tag{6.182}$$

$$H_\varphi(a,\varphi) \approx -\frac{j}{mZ} E_z^{\mathrm{inc}}(a,\varphi) F(\xi), \tag{6.183}$$

where the new functions

$$F(\xi) = f(\xi)\mathrm{e}^{-j\xi^3/3}, \quad G(\xi) = g(\xi)\mathrm{e}^{-j\xi^3/3} \tag{6.184}$$

are variants of the Fock functions which include the complete dependence of the tangential magnetic fields on ξ. Note that the component $H_\varphi(a,\varphi)$ in (6.183) is asymptotically small (as m^{-1}) and vanishes in the limit of a flat surface ($a \to \infty$), as expected.

The behavior of the functions (6.184) is illustrated in Figures 6.21 and 6.22, and reference numerical values, obtained by numerical integration with the eight-digit precision setting, are collected in Tables 6.2 and 6.3. More detailed tables of the Fock functions (for the alternative time dependence $\exp(-i\omega t)$) can be found in Fock (1965) and Ruck *et al.* (1970).

Equations (6.182) and (6.183) describe the tangential components of the magnetic field on the surface of a PEC cylinder in the neighborhood of the shadow boundary. The components are proportional to the incident field multiplied by functions $F(\xi)$ or $G(\xi)$ which account for the influence of the surface and describe a continuous transition of the surface currents between the illuminated ($\varphi < \pi/2$) and the shadowed ($\varphi > \pi/2$) portions of the cylinder surface. Indeed, when

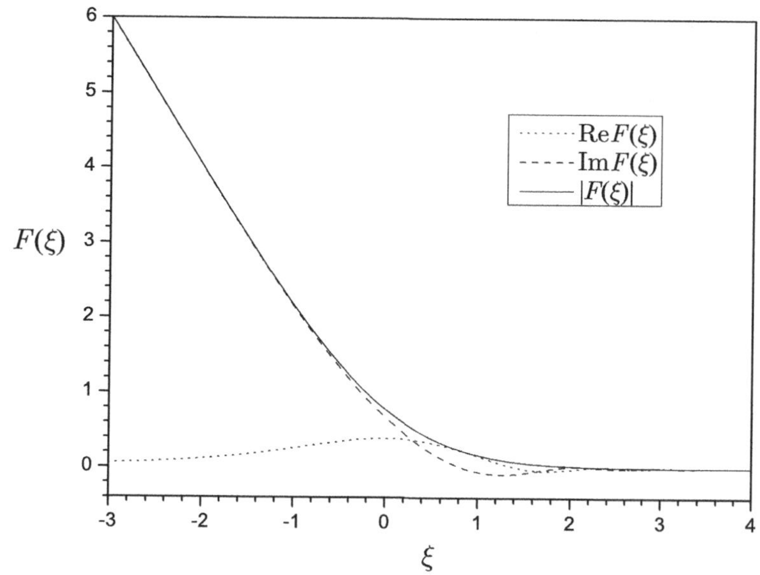

Figure 6.21 The magnitude, the real part, and the imaginary part of the Fock function $F(\xi)$.

$\xi \to -\infty$ (deep lit region), the Fock functions can be evaluated at the saddle point $t = -\xi^2$ (problem 15), which leads to the asymptotic approximations

$$F(\xi) \approx -2j\xi \left[1 + \frac{j}{4\xi^3} + O(\xi^{-6})\right] \tag{6.185}$$

$$G(\xi) \approx 2 \left[1 - \frac{j}{4\xi^3} + O(\xi^{-6})\right] \tag{6.186}$$

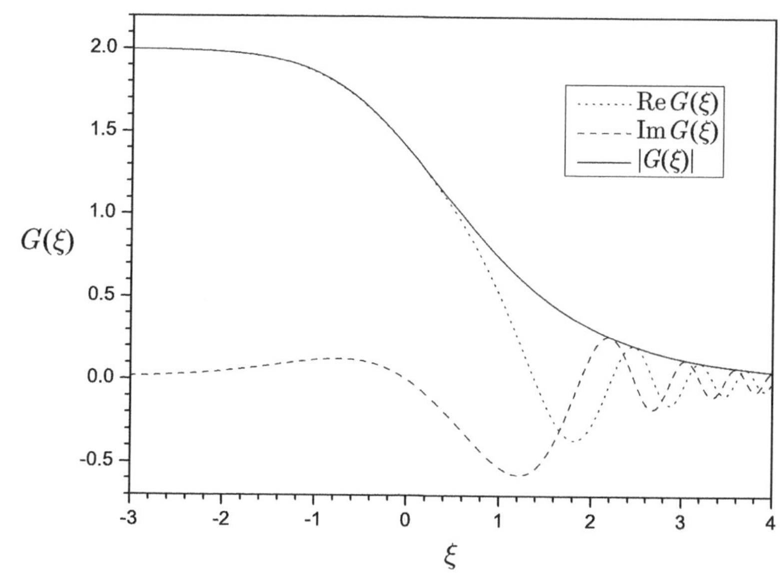

Figure 6.22 The magnitude, the real part, and the imaginary part of the Fock function $G(\xi)$.

Table 6.2 The function $F(\xi) = e^{-j\frac{1}{3}\xi^3} f(\xi)$.

ξ	Re $F(\xi)$	Im $F(\xi)$	$\lvert F(\xi)\rvert$	arg $F(\xi)$ (deg.)
4.	0.00042772	0.0000732791	0.000433952	9.72182
3.	0.00146724	0.00296412	0.00330738	63.6645
2.	−0.0178501	0.0174785	0.0249824	135.603
1.	0.157986	−0.0549614	0.167273	−19.1822
0.	0.387911	0.671881	0.775821	60.
−1.	0.255098	2.14539	2.16051	83.2191
−2.	0.113864	4.02223	4.02384	88.3785
−3.	0.0548182	6.00389	6.00414	89.4769
−4.	0.0311687	8.00097	8.00103	89.7768

and transforms (6.182) and (6.183) into the ray-optics formulas

$$H_z(a,\varphi) \approx 2H_z^{\text{inc}}(a,\varphi) \tag{6.187}$$

$$H_\varphi(a,\varphi) \approx 2H_\varphi^{\text{inc}}(a,\varphi), \tag{6.188}$$

where in the latter equation we have used the relations:

$$\xi \approx -m\cos\varphi, \quad H_\varphi^{\text{inc}}(a,\varphi) = H_y^{\text{inc}}(a,\varphi)\cos\varphi, \quad ZH_y^{\text{inc}}(a,\varphi) = E_z^{\text{inc}}(a,\varphi).$$

In the deep shadow region, when $\xi \to +\infty$, the Fock functions (6.178) and (6.179) are approximated by the residues of the respective integrand functions at the nearest poles $t = t_1^E$ or $t = t_1^H$ (see equation (6.180)), which leads to the approximations

$$F(\xi) \approx \frac{2\sqrt{\pi}}{jw_2'\left(t_1^E\right)} e^{-j\frac{1}{3}\xi^3 - j\xi t_1^E} \tag{6.189}$$

$$G(\xi) \approx \frac{2\sqrt{\pi}}{jt_1^H w_2\left(t_1^H\right)} e^{-j\frac{1}{3}\xi^3 - j\xi t_1^H}, \tag{6.190}$$

where the Airy differential equation is employed to calculate $w_2''(t_1^H) = t_1^H w_2(t_1^H)$. The field components become

$$H_z(a,\varphi) \approx -\frac{2j\sqrt{\pi}H_0}{t_1^H w_2\left(t_1^H\right)} e^{-jv_1^H(\varphi-\pi/2)} \tag{6.191}$$

$$H_\varphi(a,\varphi) \approx -\frac{2\sqrt{\pi}E_0}{mZw_2'\left(t_1^E\right)} e^{-jv_1^E(\varphi-\pi/2)}, \tag{6.192}$$

Table 6.3 The function $G(\xi) = e^{-j\frac{1}{3}\xi^3} g(\xi)$.

ξ	Re $G(\xi)$	Im $G(\xi)$	$\lvert G(\xi)\rvert$	arg $G(\xi)$ (deg.)
4.	−0.0101952	0.0527686	0.0537445	100.935
3.	−0.0585545	0.116091	0.130022	116.766
2.	−0.27095	0.161315	0.315335	149.232
1.	0.515216	−0.528701	0.738223	−45.7401
0.	1.39938	0.	1.39938	0.
−1.	1.85686	0.118593	1.86065	3.65438
−2.	1.98111	0.0497968	1.98173	1.43987
−3.	1.99749	0.017883	1.99757	0.512941
−4.	1.99952	0.00775852	1.99954	0.222317

where the relations (6.131) and (6.181) have been used. These recover the leading terms in the creeping wave series (6.165) and (6.166) if the Hankel functions are approximated by (6.170) and (6.171).

Thus, in the lit and penumbra regions, $F(\xi)$ and $G(\xi)$ are slowly varying functions of ξ. Analysis of the numerical and graphical data in Figures 6.21 and 6.22, and Tables 6.2 and 6.3 suggests that functions $F(\xi)$ and $G(\xi)$ begin to approach their asymptotic approximations (6.185), (6.186), (6.189), and (6.190) already with $|\xi| > 2$, which gives an estimation of the characteristic size of the transition region between the illuminated and shadowed portions of the cylinder. Because of (6.175), the angular size $\delta\varphi$ of the transition region between the shadowed and lit portions can be estimated by

$$\delta\varphi = \frac{4}{m}, \tag{6.193}$$

which goes to zero as $(ka)^{-1/3}$ as the electrical size of the cylinder grows ($ka \to \infty$). The linear width $a\delta\varphi$ of the penumbra region decays as $k^{-1/3}$ with increasing frequency but grows as $a^{2/3}$ with increasing radius.

On the introduction of the arc length $l = a(\varphi - \pi/2)$, which is the distance between the shadow boundary and the observer measured along the surface of the cylinder, considered positive in the direction of the shadow and negative in the direction of the lit portion of the cylinder, we have

$$\xi = m\frac{l}{a} = \frac{l}{d} \tag{6.194}$$

with d given by (6.154). Thus, d defines the characteristic width of the transition region, and when the distance from the shadow boundary is such that $|l| > d$, the surface currents are already close to either the ray optical or the creeping wave approximations.

Summarizing the expressions for the fields in the penumbra region on the surface of the PEC circular cylinder, we have components that vanish because of the boundary conditions

$$E_z(a,\varphi) = E_\varphi(a,\varphi) = 0, \quad H_\rho(a,\varphi) = 0,$$

and components that do not vanish,

$$H_z(a,\varphi) \approx H_z^{\text{inc}}(a,\varphi)G(\xi) = -YE_y^{\text{inc}}(a,\varphi)G(\xi) \tag{6.195}$$

$$H_\varphi(a,\varphi) \approx -j(Y/m)E_z^{\text{inc}}(a,\varphi)F(\xi) = -jm^{-1}H_y^{\text{inc}}(a,\varphi)F(\xi) \tag{6.196}$$

$$E_\rho(a,\varphi) \approx -ZH_z^{\text{inc}}(a,\varphi)G(\xi) = E_y^{\text{inc}}(a,\varphi)G(\xi), \tag{6.197}$$

where $Y = 1/Z$ is the wave admittance of the medium surrounding the cylinder. The expression (6.197) is obtained from (6.195) by using (6.3) and neglecting the derivative $\partial G(\xi)/\partial\varphi$ compared with $\partial H_z^{\text{inc}}(a,\varphi)/\partial\varphi$ since

$$\frac{\partial}{\partial\varphi}H_z^{\text{inc}}(a,\varphi) = O(ka), \quad \frac{\partial}{\partial\varphi}G(\xi) = O(m). \tag{6.198}$$

In writing (6.195)–(6.197), we have used the relations between the electric and magnetic fields in the incident plane wave

$$E_z^{\text{inc}}(a,\varphi) = ZH_y^{\text{inc}}(a,\varphi), \quad H_z^{\text{inc}}(a,\varphi) = -YE_y^{\text{inc}}(a,\varphi) \tag{6.199}$$

to express the solution in terms of the y components of the incident field.

6.5 Coated and Impedance Cylinders under Oblique Illumination

This section is devoted to scattering from circular cylinders with imperfectly conducting surfaces. Moreover, the incidence direction is no longer limited to the plane perpendicular to the axis of the cylinder (Figure 6.23). The oblique illumination of a non-ideally conducting cylinder results in polarization coupling, which means that an incident TM-polarized wave ($H_z^{\text{inc}} = 0$) generates a finite H_z component in the scattered field and, similarly, an incident TE-polarized wave ($E_z^{\text{inc}} = 0$) leads to a finite E_z component in the scattered field. The construction of an exact solution of Maxwell's equations is complicated by the coupling but is technically possible. An exact solution for a homogeneous dielectric cylinder under oblique illumination has been derived by Wait (1955) (see also Wait (1959, Chapter 17)), some 78 years after the solutions

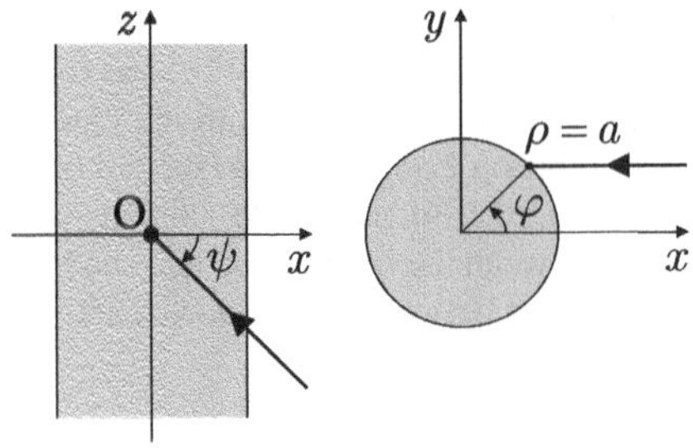

Figure 6.23 A plane wave is incident perpendicular to the y axis at an oblique angle with the axis of an infinite cylinder of radius a.

for normal incidence (Lord Rayleigh 1881). In principle, the exact solutions can be derived by separation of variables for an arbitrary number of concentric cylindrical layers, each filled with a different isotropic material, but because of the polarization coupling the procedure is much more involved compared with the cases of stratified planar media (section 4.3.1) or concentric spheres (section 7.2). In the case of normal incidence on the cylinder, the procedure has been described by Kerker and Matijevich (1961).

In this section we study one of the most important cases, namely, a perfectly conducting (PEC) cylinder coated with a layer of a homogeneous magneto-dielectric material (Figure 6.24), which includes the homogeneous ($a_1 = 0$) and PEC ($a_1 = a$) cylinders as special cases. Furthermore, the exact solution for a cylinder with impedance boundary conditions is presented and its relation to that for the coated cylinder is analyzed. The impedance boundary conditions were introduced in section 2.3 to describe the electromagnetic properties of a flat interface of a high-refractive-index material or a thin material coating on a PEC ground plane. In this section we show that the impedance boundary conditions can be successfully applied to curved material surfaces, provided that the cylinder radius is greater than about $\lambda/4$.

We begin the section with exact series solutions for the coated cylinder and study characteristic features of electromagnetic scattering from such cylinders under oblique illumination (section 6.5.1). Next, in section 6.5.2, an exact series solution for a cylinder with impedance boundary conditions is presented, and the relation to the case of the coated PEC cylinder is established. This permits an estimation of the accuracy of the impedance boundary conditions for cylindrically curved surfaces. The high-frequency description of the fields in the penumbra and deep shadow regions on the surface of an impedance cylinder is similar to the case of the PEC cylinder (section 6.4.4) but is much more complicated because of the polarization coupling. The approach we follow in this section is to derive the high-frequency approximations in the special case of normal incidence, which is much simpler, and then to extend the solutions to the case of oblique incidence by relying on the principle of the local field in the penumbra region or by following the physical argument of the GTD. These topics are addressed in section 6.6.

6.5.1 PEC Cylinder with Magneto-dielectric Coating

Exact Series Solution for the Scattered Field

Consider a cylindrical structure which consists of a perfectly conducting cylinder of radius a_1 coated with a homogeneous layer of thickness $h = a - a_1$ of a material with the relative permittivity $\epsilon_r = \epsilon_1/\epsilon_0$ and the relative permeability

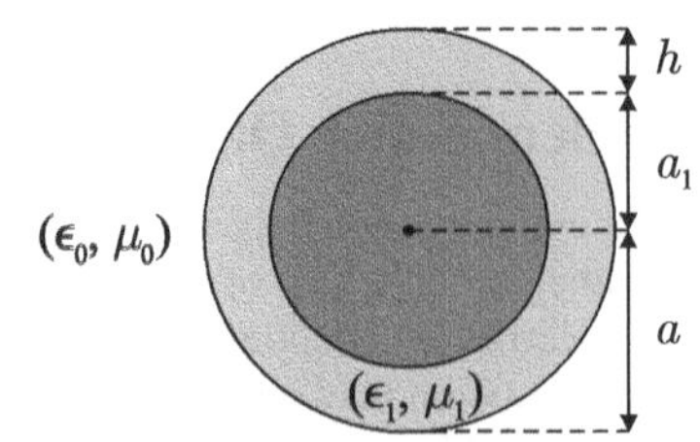

Figure 6.24 PEC cylinder of radius a_1 coated with a layer of material of thickness $h = a - a_1$.

$\mu_r = \mu_1/\mu_0$, where ϵ_0 and μ_0 are the permittivity and permeability in the surrounding medium (Figure 6.24). A plane electromagnetic wave is incident on the cylinder and the direction of propagation lies in the xOz plane, making an angle ψ with the x axis (Figure 6.23) so that $\psi = 0$ corresponds to the illumination perpendicular to the axis of the cylinder and $\psi = \pi/2$ to the incidence in the direction of the positive z axis. The z components in the incident wave can be expressed as

$$E_z^{\text{inc}} = E_0 \cos\psi \exp[-jk_0(z\sin\psi - x\cos\psi)] \tag{6.200}$$
$$H_z^{\text{inc}} = H_0 \cos\psi \exp[-jk_0(z\sin\psi - x\cos\psi)], \tag{6.201}$$

where $k_0 = \omega\sqrt{\epsilon_0\mu_0}$ is the wavenumber in the surrounding medium and the factors E_0 and H_0 depend on the polarization of the incident plane wave. In the case of TM polarization, the magnetic field in the incident wave has just the y component

$$\mathbf{H}_{\text{inc}} = \hat{y}H_{0y}^{\text{TM}} \exp[-jk_0(z\sin\psi - x\cos\psi)], \tag{6.202}$$

implying that

$$E_0 = Z_0 H_{0y}^{\text{TM}}, \quad H_0 = 0. \tag{6.203}$$

Alternatively, in the case of TE polarization,

$$\mathbf{E}_{\text{inc}} = \hat{y}E_{0y}^{\text{TE}} \exp[-jk_0(z\sin\psi - x\cos\psi)] \tag{6.204}$$

and

$$E_0 = 0, \quad H_0 = -Y_0 E_{0y}^{\text{TE}}. \tag{6.205}$$

Because the configuration is translation-invariant in the z direction, the dependence of the fields on the z coordinate must be the same as in the incident wave, which permits expressing the field components in terms of two scalar functions E_z and H_z (section 2.5.5). Inserting $k_z = k_0 \sin\psi$ in (2.313)–(2.318) gives the equations

$$E_\rho = \frac{1}{jk_0\cos^2\psi}\left(\sin\psi\frac{\partial}{\partial\rho}E_z + \frac{Z_0}{\rho}\frac{\partial}{\partial\varphi}H_z\right) \tag{6.206}$$
$$E_\varphi = \frac{1}{jk_0\cos^2\psi}\left(\frac{\sin\psi}{\rho}\frac{\partial}{\partial\varphi}E_z - Z_0\frac{\partial}{\partial\rho}H_z\right) \tag{6.207}$$
$$H_\rho = \frac{1}{jk_0\cos^2\psi}\left(\sin\psi\frac{\partial}{\partial\rho}H_z - \frac{Y_0}{\rho}\frac{\partial}{\partial\varphi}E_z\right) \tag{6.208}$$
$$H_\varphi = \frac{1}{jk_0\cos^2\psi}\left(\frac{\sin\psi}{\rho}\frac{\partial}{\partial\varphi}H_z + Y_0\frac{\partial}{\partial\rho}E_z\right), \tag{6.209}$$

where $Z_0 = \sqrt{\mu_0/\epsilon_0} = 1/Y_0$ is the wave impedance of the surrounding medium, and E_z and H_z are solutions of the Helmholtz equations:

$$\left(\frac{\partial^2}{\partial\rho^2} + \frac{1}{\rho}\frac{\partial}{\partial\rho} + \frac{1}{\rho^2}\frac{\partial^2}{\partial\varphi^2} + k_0^2\cos^2\psi\right)E_z = 0, \tag{6.210}$$
$$\left(\frac{\partial^2}{\partial\rho^2} + \frac{1}{\rho}\frac{\partial}{\partial\rho} + \frac{1}{\rho^2}\frac{\partial^2}{\partial\varphi^2} + k_0^2\cos^2\psi\right)H_z = 0. \tag{6.211}$$

In the material layer, the field components satisfy relations (2.313)–(2.318) with $k_z = k_0 \sin\psi$ and k, Z, and $k_\perp$ replaced with $k_1 = \omega\sqrt{\epsilon_1\mu_1}$, $Z_1 = \sqrt{\mu_1/\epsilon_1}$, and

$$k_\perp^{(1)} = \sqrt{k_1^2 - k_z^2} = k_0\sqrt{\epsilon_r\mu_r - \sin^2\psi}. \tag{6.212}$$

The field components tangential to the material interfaces must be continuous at $\rho = a$,

$$E_z(a-0,\varphi) = E_z(a+0,\varphi), \quad H_z(a-0,\varphi) = H_z(a+0,\varphi) \tag{6.213}$$

$$E_\varphi(a-0,\varphi) = E_\varphi(a+0,\varphi), \quad H_\varphi(a-0,\varphi) = H_\varphi(a+0,\varphi), \tag{6.214}$$

where the symbols ± 0 mean that ρ approaches a from the exterior (interior) of the material boundary. The components E_z and E_φ must vanish at $\rho = a_1$, implying the conditions:

$$E_z(a_1,\varphi) = 0, \quad \frac{\partial H_z}{\partial \rho}(a_1,\varphi) = 0. \tag{6.215}$$

Equations (6.213)–(6.215) are conditions on the components E_z and H_z because E_φ and H_φ are expressed through E_z and H_z.

Every incident plane wave can be represented as a combination of the TM and TE plane waves, so the solution can be constructed separately for TM and TE excitation. In the exterior of the cylinder ($a \le \rho < \infty$) we have

$$E_z = E_z^{\rm inc} + E_z^{\rm sc}, \quad H_z = H_z^{\rm sc} \tag{6.216}$$

and

$$E_z = E_z^{\rm sc}, \quad H_z = H_z^{\rm inc} + H_z^{\rm sc} \tag{6.217}$$

in the TM and TE cases, respectively.

The fields must be 2π periodic functions of φ, and similarly to the case of normal incidence (sections 6.2.1 and 6.3.2) the scattered field can be sought in the form of the Fourier series:

$$E_z^{\rm sc} = E_0 \cos\psi \mathrm{e}^{-jk_0 z \sin\psi} \sum_{n=-\infty}^{+\infty} \mathrm{e}^{-jn\varphi} j^n A_n^{\rm TM} H_n^{(2)}(k_0\rho\cos\psi) \tag{6.218}$$

$$H_z^{\rm sc} = Y_0 E_0 \cos\psi \mathrm{e}^{-jk_0 z \sin\psi} \sum_{n=-\infty}^{\infty} \mathrm{e}^{-jn\varphi} j^n B_n^{\rm TM} H_n^{(2)}(k_0\rho\cos\psi) \tag{6.219}$$

and

$$E_z^{\rm sc} = -Z_0 H_0 \cos\psi \mathrm{e}^{-jk_0 z \sin\psi} \sum_{n=-\infty}^{\infty} \mathrm{e}^{-jn\varphi} j^n A_n^{\rm TE} H_n^{(2)}(k_0\rho\cos\psi) \tag{6.220}$$

$$H_z^{\rm sc} = H_0 \cos\psi \mathrm{e}^{-jk_0 z \sin\psi} \sum_{n=-\infty}^{\infty} \mathrm{e}^{-jn\varphi} j^n B_n^{\rm TE} H_n^{(2)}(k_0\rho\cos\psi) \tag{6.221}$$

in the TM and TE cases, respectively. Representations (6.218)–(6.221) by construction satisfy the Helmholtz equations (6.210) and (6.211), and describe outgoing waves at $\rho \to \infty$. Expansion coefficients $A_n^{\rm TM}$, $B_n^{\rm TM}$, $A_n^{\rm TE}$, and $B_n^{\rm TE}$ are to be determined from the boundary conditions.

To this end, we express the total field in the material layer ($a_1 \le \rho \le a$) as

$$E_z = \mathrm{e}^{-jk_0 z \sin\psi} \sum_{n=-\infty}^{\infty} \mathrm{e}^{-jn\varphi} \left[C_n^{\rm TM} J_n\left(k_\perp^{(1)}\rho\right) + D_n^{\rm TM} H_n^{(2)}\left(k_\perp^{(1)}\rho\right)\right] \tag{6.222}$$

$$H_z = \mathrm{e}^{-jk_0 z \sin\psi} \sum_{n=-\infty}^{\infty} \mathrm{e}^{-jn\varphi} \left[E_n^{\rm TM} J_n\left(k_\perp^{(1)}\rho\right) + F_n^{\rm TM} H_n^{(2)}\left(k_\perp^{(1)}\rho\right)\right] \tag{6.223}$$

in the case of TM polarization and as

$$E_z = \mathrm{e}^{-jk_0 z \sin\psi} \sum_{n=-\infty}^{\infty} \mathrm{e}^{-jn\varphi} \left[C_n^{\rm TE} J_n\left(k_\perp^{(1)}\rho\right) + D_n^{\rm TE} H_n^{(2)}\left(k_\perp^{(1)}\rho\right)\right] \tag{6.224}$$

$$H_z = \mathrm{e}^{-jk_0 z \sin\psi} \sum_{n=-\infty}^{\infty} \mathrm{e}^{-jn\varphi} \left[E_n^{\rm TE} J_n\left(k_\perp^{(1)}\rho\right) + F_n^{\rm TE} H_n^{(2)}\left(k_\perp^{(1)}\rho\right)\right] \tag{6.225}$$

in the case of TE polarization. The representations have the required z dependence of the fields, satisfy the Helmholtz equations, and include both in- and outgoing waves in the material of the coating.

On representing the incident field components $E_z^{\rm inc}$ and $H_z^{\rm inc}$ as Fourier expansions with the help of the formula

$$e^{jk_0\rho\cos\psi\cos\varphi} = \sum_{n=-\infty}^{\infty} e^{-jn\varphi} j^n J_n(k_0\rho\cos\psi),$$

the expansion coefficients A_n, B_n, C_n, D_n, E_n, and F_n for a respective polarization are obtained by insertion of (6.216), (6.218), (6.219), (6.222), and (6.223) (TM case) or (6.217), (6.220), (6.221), (6.224), and (6.225) (TE case) in the boundary conditions (6.213)–(6.215), and since these are to be satisfied for every harmonic in the Fourier expansions, the six conditions (6.213)–(6.215) give six linear algebraic equations for the six coefficients A_n, B_n, C_n, D_n, E_n, and F_n, from which all the expansion coefficients are determined. Omitting for brevity the other coefficients, the coefficients in the expansions of the scattered field read (Ruck *et al.* 1970)[8]

$$A_n^{\rm TM} = -\frac{\delta_n V_n P_n + q_n^2 J_n(x_0) H_n^{(2)}(x_0)}{\delta_n N_n P_n + \left[q_n H_n^{(2)}(x_0)\right]^2} \tag{6.226}$$

$$B_n^{\rm TM} = \frac{-2s_0 q_n}{\pi x_0\left\{\delta_n N_n P_n + \left[q_n H_n^{(2)}(x_0)\right]^2\right\}} \tag{6.227}$$

$$A_n^{\rm TE} = B_n^{\rm TM} \tag{6.228}$$

$$B_n^{\rm TE} = -\frac{\delta_n M_n N_n + q_n^2 J_n(x_0) H_n^{(2)}(x_0)}{\delta_n N_n P_n + \left[q_n H_n^{(2)}(x_0)\right]^2}, \tag{6.229}$$

where

$$M_n = J_n(x_0) - y_n J_n'(x_0), \quad N_n = H_n^{(2)}(x_0) - z_n H_n^{(2)\prime}(x_0) \tag{6.230}$$

$$P_n = H_n^{(2)}(x_0) - y_n H_n^{(2)\prime}(x_0), \quad V_n = J_n(x_0) - z_n J_n'(x_0) \tag{6.231}$$

$$\delta_n = -\frac{s_0^2}{y_n z_n}, \quad q_n = -\frac{n\sin\psi}{k_0 a\cos^2\psi}\left(1 - \frac{\cos^2\psi}{\epsilon_r\mu_r - \sin^2\psi}\right) \tag{6.232}$$

$$s_0 = \frac{1}{\cos\psi}, \quad s_1 = \frac{\epsilon_r}{\sqrt{\epsilon_r\mu_r - \sin^2\psi}}, \quad r_1 = \frac{\mu_r}{\sqrt{\epsilon_r\mu_r - \sin^2\psi}} \tag{6.233}$$

$$x_0 = k_0 a\cos\psi, \quad x_1 = k_0 a\sqrt{\epsilon_r\mu_r - \sin^2\psi}, \quad x_2 = \frac{a_1}{a}x_1 \tag{6.234}$$

$$y_n = \frac{s_0}{r_1}\frac{J_n(x_1)H_n^{(2)\prime}(x_2) - H_n^{(2)}(x_1)J_n'(x_2)}{J_n'(x_1)H_n^{(2)\prime}(x_2) - H_n^{(2)\prime}(x_1)J_n'(x_2)} \tag{6.235}$$

$$z_n = \frac{s_0}{s_1}\frac{J_n(x_1)H_n^{(2)}(x_2) - H_n^{(2)}(x_1)J_n(x_2)}{J_n'(x_1)H_n^{(2)}(x_2) - H_n^{(2)\prime}(x_1)J_n(x_2)} \tag{6.236}$$

and the prime denotes the derivative with respect to the argument of a Bessel function.

[8] In Ruck *et al.* (1970) parameter δ_n is missing from equations (4.2-66) and (4.2-67) and the right-hand side of (4.2-68) has an extra factor i.

The solution for the coated cylinder includes several important special cases. The case of a PEC cylinder under oblique illumination results in the limit $a_1 \to a$, which means that $x_2 \to x_1$, $y_n \to \infty$, and $z_n \to 0$. It can be shown (see the problems section) that these lead to

$$A_n^{\mathrm{TM}} \to -\frac{J_n(x_0)}{H_n^{(2)}(x_0)}, \quad B_n^{\mathrm{TM}} \to 0, \quad A_n^{\mathrm{TE}} \to 0, \quad B_n^{\mathrm{TE}} \to -\frac{J_n'(x_0)}{H_n^{(2)\prime}(x_0)}. \tag{6.237}$$

The special case of a homogeneous cylinder is obtained with $a_1 \to 0$. In this limit, $x_2 \to 0$ and since $|H_n^{(2)}(x_2)|$, $|H_n^{(2)\prime}(x_2)| \to \infty$, and $|J_n(x_2)|$, $|J_n'(x_2)| = O(1)$, expressions for δ_n, y_n, and z_n simplify to

$$\delta_n = -s_1 r_1 \left[\frac{J_n'(x_1)}{J_n(x_1)}\right]^2, \quad y_n = \frac{s_0 J_n(x_1)}{r_1 J_n'(x_1)}, \quad z_n = \frac{s_0 J_n(x_1)}{s_1 J_n'(x_1)}. \tag{6.238}$$

The exact solution of the electromagnetic problem for a plane wave obliquely illuminating a homogeneous cylinder is attributed to Wait (1955).

The solution for the coated cylinder significantly simplifies in the limit of normal illumination. If $\psi \to 0$, then $q_n \to 0$ and

$$A_n^{\mathrm{TM}} \to \frac{V_n}{N_n} = -\frac{J_n(x_0) - z_n J_n'(x_0)}{H_n^{(2)}(x_0) - z_n H_n^{(2)\prime}(x_0)}, \quad B_n^{\mathrm{TM}} \to 0 \tag{6.239}$$

$$A_n^{\mathrm{TE}} \to 0, \quad B_n^{\mathrm{TE}} \to \frac{M_n}{P_n} = -\frac{J_n(x_0) - y_n J_n'(x_0)}{H_n^{(2)}(x_0) - y_n H_n^{(2)\prime}(x_0)} \tag{6.240}$$

with

$$y_n = \sqrt{\frac{\epsilon_r}{\mu_r}}\, \frac{J_n(k_1 a) H_n^{(2)\prime}(k_1 a_1) - H_n^{(2)}(k_1 a) J_n'(k_1 a_1)}{J_n'(k_1 a) H_n^{(2)\prime}(k_1 a_1) - H_n^{(2)\prime}(k_1 a) J_n'(k_1 a_1)} \tag{6.241}$$

$$z_n = \sqrt{\frac{\mu_r}{\epsilon_r}}\, \frac{J_n(k_1 a) H_n^{(2)}(k_1 a_1) - H_n^{(2)}(k_1 a) J_n(k_1 a_1)}{J_n'(k_1 a) H_n^{(2)}(k_1 a_1) - H_n^{(2)\prime}(k_1 a) J_n(k_1 a_1)}. \tag{6.242}$$

The solution for the homogeneous cylinder under normal illumination studied in section 6.3.2 follows from (6.238) with $\psi = 0$ or from (6.239)–(6.242) with $a_1 = 0$ as

$$A_n^{\mathrm{TM}} = -\frac{Z_0 J_n(k_0 a) J_n'(k_1 a) - Z_1 J_n'(k_0 a) J_n(k_1 a)}{Z_0 H_n^{(2)}(k_0 a) J_n'(k_1 a) - Z_1 H_n^{(2)\prime}(k_0 a) J_n(k_1 a)}, \quad B_n^{\mathrm{TM}} = 0 \tag{6.243}$$

$$A_n^{\mathrm{TE}} = 0, \quad B_n^{\mathrm{TE}} = -\frac{Z_1 J_n(k_0 a) J_n'(k_1 a) - Z_0 J_n'(k_0 a) J_n(k_1 a)}{Z_1 H_n^{(2)}(k_0 a) J_n'(k_1 a) - Z_0 H_n^{(2)\prime}(k_0 a) J_n(k_1 a)}. \tag{6.244}$$

Regardless of the value of ψ, the expansion coefficients (6.226)–(6.229) have the symmetry properties

$$A_{-n}^{\mathrm{TM}} = A_n^{\mathrm{TM}}, \quad B_{-n}^{\mathrm{TM}} = -B_n^{\mathrm{TM}}, \quad A_{-n}^{\mathrm{TE}} = -A_n^{\mathrm{TE}}, \quad B_{-n}^{\mathrm{TE}} = B_n^{\mathrm{TE}}, \tag{6.245}$$

which result from the properties of the integer-order Bessel functions $J_{-n}(z) = (-1)^n J_n(z)$ and $H_{-n}^{(2)}(z) = (-1)^n H_n^{(2)}(z)$. Using this symmetry and the fact that $B_0^{\mathrm{TM}} = A_0^{\mathrm{TE}} = 0$, the sums in (6.218)–(6.221) can be rewritten in the form of trigonometric semi-infinite series

$$E_z^{sc} = E_0 \cos\psi e^{-jk_0 z \sin\psi} \sum_{n=0}^{\infty} \nu_n j^n A_n^{TM} H_n^{(2)}(k_0\rho\cos\psi)\cos(n\varphi) \quad (6.246)$$

$$H_z^{sc} = -2jY_0E_0 \cos\psi e^{-jk_0 z \sin\psi} \sum_{n=1}^{\infty} j^n B_n^{TM} H_n^{(2)}(k_0\rho\cos\psi)\sin(n\varphi) \quad (6.247)$$

in the TM case and

$$E_z^{sc} = 2jZ_0H_0 \cos\psi e^{-jk_0 z \sin\psi} \sum_{n=1}^{\infty} j^n A_n^{TE} H_n^{(2)}(k_0\rho\cos\psi)\sin(n\varphi) \quad (6.248)$$

$$H_z^{sc} = H_0 \cos\psi e^{-jk_0 z \sin\psi} \sum_{n=0}^{\infty} \nu_n j^n B_n^{TE} H_n^{(2)}(k_0\rho\cos\psi)\cos(n\varphi) \quad (6.249)$$

in the TE case, where

$$\nu_n = \begin{cases} 1, & n = 0 \\ 2, & n \geq 1 \end{cases}. \quad (6.250)$$

It follows from (6.247) and (6.248) that when $\varphi = 0$ (backscattering) or $\varphi = \pi$ (forward scattering), $H_z^{sc} = 0$ for TM excitation and $E_z^{sc} = 0$ for TE excitation, regardless of the thickness and material parameters of the coating.

Scattering Amplitudes and Scattering Widths of the Coated Cylinder

If the observation point is removed far from the cylinder in the sense that $\rho \gg \lambda$ and $\rho \gg D_\perp^2/\lambda$, where $D_\perp = 2a$ is the diameter of the cylinder, then the scattering from the cylinder is characterized by the scattering amplitudes $F_{E,H}$, the bistatic scattering width σ_{2D}, the total scattering width σ_T^{2D}, the extinction width σ_{ext}^{2D}, and the absorption width σ_{abs}^{2D}. We can directly use the definitions from sections 3.2.2 and 3.4.4, noting that E polarization is TM polarization and H polarization is TE polarization, $E_{0z} = E_0 \cos\psi$, $H_{0z} = H_0 \cos\psi$, and $\sin\beta = \cos\psi$.

The scattering amplitudes defined in (3.156) and (3.157) are obtained from the series solution (6.246)–(6.249) for the scattered field by letting $\rho \to \infty$ and using the large-argument approximations for the Hankel functions (6.31)

$$H_n^{(2)}(k_0\rho\cos\psi) \approx \sqrt{\frac{2}{\pi k_0\rho\cos\psi}} e^{-jk_0\rho\cos\psi + jn\frac{\pi}{2} + j\frac{\pi}{4}}, \quad (6.251)$$

which gives the expressions:

$$F_{EE}(\varphi) = -E_0\cos\psi \sum_{n=0}^{+\infty} \nu_n (-1)^n A_n^{TM} \cos(n\varphi) \quad (6.252)$$

$$F_{HE}(\varphi) = 2jY_0E_0\cos\psi \sum_{n=1}^{+\infty} (-1)^n B_n^{TM} \sin(n\varphi) \quad (6.253)$$

$$F_{EH}(\varphi) = -2jZ_0H_0\cos\psi \sum_{n=1}^{+\infty} (-1)^n A_n^{TE} \sin(n\varphi) \quad (6.254)$$

$$F_{HH}(\varphi) = -H_0\cos\psi \sum_{n=0}^{+\infty} \nu_n (-1)^n B_n^{TE} \cos(n\varphi). \quad (6.255)$$

Here, F_{EE} and F_{HE} mean the scattering amplitudes F_E and F_H in the case of TM excitation; F_{EH} and F_{HH} are the scattering amplitudes F_E and F_H in the case of TE excitation. The elements D_{ee}, D_{eh}, D_{he}, and D_{hh} of the scattering matrix (3.179) follow from (3.180) and (6.252)–(6.255).

The scattering width, defined in (3.20)–(3.23) for various polarization cases, follows from (3.181), (6.200), (6.201), and (6.252)–(6.255) as

$$\sigma_{EE} = \frac{4}{k_0 \cos^2\psi}\left|\sum_{n=0}^{\infty}(-1)^n \nu_n A_n^{\mathrm{TM}} \cos(n\varphi)\right|^2 \tag{6.256}$$

$$\sigma_{EH} = \sigma_{HE} = \frac{16}{k_0 \cos^2\psi}\left|\sum_{n=1}^{\infty}(-1)^n B_n^{\mathrm{TM}} \sin(n\varphi)\right|^2 \tag{6.257}$$

$$\sigma_{HH} = \frac{4}{k_0 \cos^2\psi}\left|\sum_{n=0}^{\infty}(-1)^n \nu_n B_n^{\mathrm{TE}} \cos(n\varphi)\right|^2 . \tag{6.258}$$

The total scattering width is obtained on integration of the bistatic scattering width over all scattering directions as in (3.182), giving

$$\sigma_{\mathrm{T}}^{E} = \frac{2}{\pi k_0 |E_0|^2 \cos^2\psi}\int_0^{2\pi}\left[|F_{EE}(\varphi)|^2 + |Z_0 F_{HE}(\varphi)|^2\right] \mathrm{d}\varphi = \frac{4}{k_0}\left(\sum_{n=0}^{\infty}\nu_n |A_n^{\mathrm{TM}}|^2 + 2\sum_{n=1}^{\infty}|B_n^{\mathrm{TM}}|^2\right) \tag{6.259}$$

in the TM case and

$$\sigma_{\mathrm{T}}^{H} = \frac{2}{\pi k_0 |H_0|^2 \cos^2\psi}\int_0^{2\pi}\left[|Y_0 F_{EH}(\varphi)|^2 + |F_{HH}(\varphi)|^2\right] \mathrm{d}\varphi = \frac{4}{k_0}\left(\sum_{n=0}^{\infty}\nu_n |B_n^{\mathrm{TE}}|^2 + 2\sum_{n=1}^{\infty}|A_n^{\mathrm{TE}}|^2\right) \tag{6.260}$$

in the TE case.

Finally, the extinction widths result from (3.204)

$$\sigma_{\mathrm{ext}}^{E} = \frac{4}{k_0 \cos\psi}\mathrm{Re}\left[\frac{F_{EE}(\pi)}{E_0}\right], \quad \sigma_{\mathrm{ext}}^{H} = \frac{4}{k_0 \cos\psi}\mathrm{Re}\left[\frac{F_{HH}(\pi)}{H_0}\right], \tag{6.261}$$

where the scattering amplitudes are evaluated for the forward-scattering direction $\varphi = \pi$. The difference between the extinction and total scattering widths is the absorption width

$$\sigma_{\mathrm{abs}}^{E} = \sigma_{\mathrm{ext}}^{E} - \sigma_{\mathrm{T}}^{E}, \quad \sigma_{\mathrm{abs}}^{H} = \sigma_{\mathrm{ext}}^{H} - \sigma_{\mathrm{T}}^{H}, \tag{6.262}$$

which is the portion of the incident power lost due to the absorption in the material of the coating.

Figures 6.25–6.27 illustrate the influence of the thickness of the coating on the behavior of the scattered fields. The scattering width is made dimensionless by dividing by πa, which is the GO backscattering cross section per unit length of a PEC cylinder. The radius a of the cylinder is equal to λ, the coating is made of an absorbing material with $\epsilon_r = 15 - j5$ and $\mu_r = 1$, and the illumination direction makes an angle of 45° with the axis of the cylinder. The thickness varies from $h = 0$ (no coating, PEC cylinder) through 0.04λ, 0.07λ, and 0.2λ to $h = a$, in which case the PEC core disappears ($a_1 = 0$) and the configuration becomes a homogeneous cylinder. The co-polar scattered fields have a pronounced maximum in the direction of the forward scattering $\varphi = 180°$ (Figures 6.25 and 6.26), which is a general property of scattering from electrically large smoothly shaped bodies. With $\varphi = 0$ (backscattering) the observation point lies in the plane of incidence and the scattered wave arrives at the observer on reflection from the surface of the cylinder. The thickness of the coating only moderately influences the level of the forward scattering but strongly affects the backscattering. Adjusting the coating thickness ($h = 0.07\lambda$ in this case) may lead to a significant reduction in backscattering. The cross-polarized components are typically smaller than the co-polarized ones, vanishing at $\varphi = 0$ and 180° (Figure 6.27).

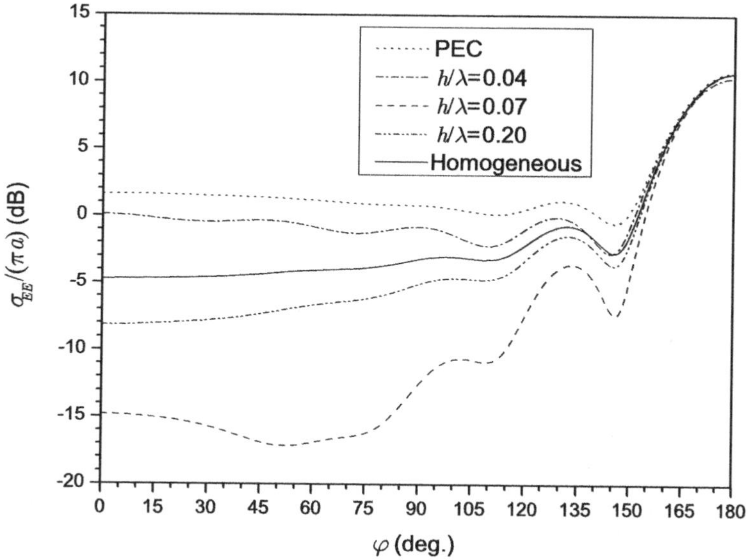

Figure 6.25 Co-polar bistatic scattering width of coated cylinders with $a = \lambda$ and $a_1 = a - h$, normalized to πa. TM polarization, $\psi = \pi/4$. The coating is an absorbing material of thickness h with $\epsilon_r = 15 - j5$ and $\mu_r = 1$.

6.5.2 Impedance Cylinder

In sections 2.3.1 and 4.6 we have seen that the impedance boundary conditions provide a useful approximation that may considerably simplify the solution of a problem of scattering from material bodies, while maintaining an acceptable accuracy of the solution. In this section an exact solution for an impedance cylinder is derived and compared with the exact solutions for a PEC cylinder coated with a magneto-dielectric layer from section 6.5.1.

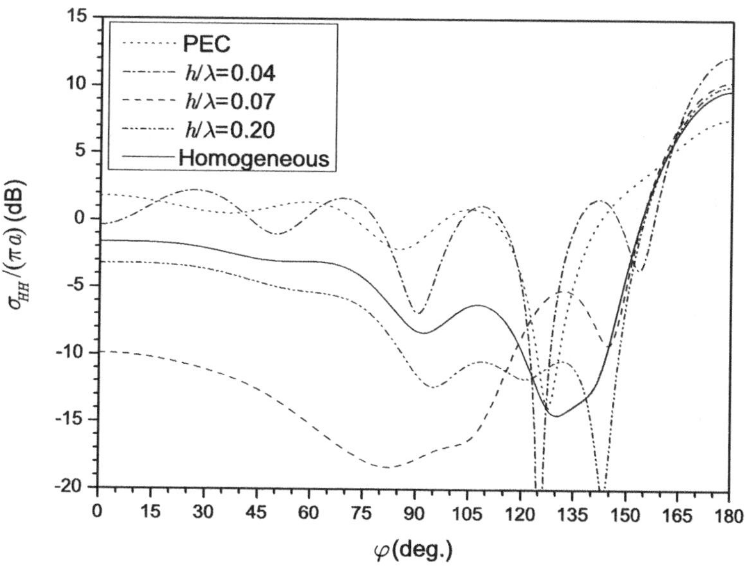

Figure 6.26 The same as Figure 6.25 but for TE polarization.

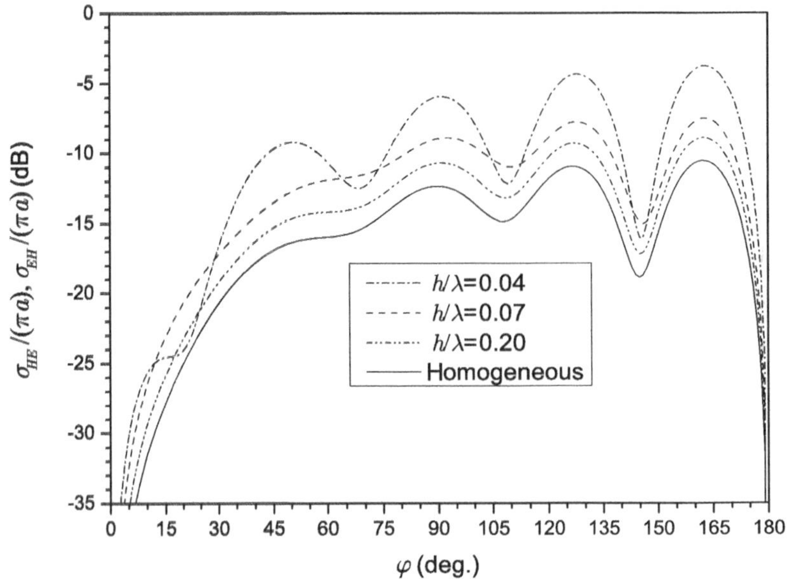

Figure 6.27 The same as Figure 6.25 but for cross-polarization.

Series Solution for the Scattered Field

Consider the geometry shown in Figure 6.23 and assume that on the surface of the cylinder ($\rho = a$, $0 \le \varphi \le 2\pi$) the fields are subject to the impedance boundary conditions (2.74). In the cylindrical coordinates, the boundary conditions are expressed as

$$H_\varphi = \tilde{Z}_s^{-1} E_z, \qquad E_\varphi = -\tilde{Z}_s H_z, \tag{6.263}$$

with $\tilde{Z}_s$ being the equivalent surface impedance. Since both polarizations are related to each other via the duality of Maxwell's equations ($\mathbf{E} \to \mathbf{H}$, $\mathbf{H} \to -\mathbf{E}$, $Z_0 \to 1/Z_0$, and $\tilde{Z}_s \to 1/\tilde{Z}_s$), it is sufficient to derive a solution for one polarization, let us say for the TM polarization ($H_0 = 0$, incident wave is given by (6.202)).

To this end, we express the z components of the TM-polarized field as in (6.216), (6.218), and (6.219), and obtain series representations for the components E_φ and H_φ from (6.207) and (6.209). Inserting these into the boundary conditions (6.263) and using the expansion (see (6.23))

$$E_z^{\text{inc}} = E_0 \cos\psi e^{-jk_0 z \sin\psi} \sum_{n=-\infty}^{\infty} j^n e^{-jn\varphi} J_n(k_0 \rho \cos\psi) \tag{6.264}$$

and the orthogonality of the Fourier harmonics, a system of two algebraic equations with two unknowns A_n^{TM} and B_n^{TM} for each value of n is obtained, from which the expansion coefficients are determined. Omitting the straightforward but cumbersome details of the derivations, we present only the result:

$$A_n^{\text{TM}} = -\frac{\tilde{V}_n \tilde{P}_n + \tilde{q}_n^2 J_n(x_0) H_n^{(2)}(x_0)}{\tilde{N}_n \tilde{P}_n + \left[\tilde{q}_n H_n^{(2)}(x_0)\right]^2} \tag{6.265}$$

$$B_n^{\text{TM}} = \frac{-2 s_0 \tilde{q}_n}{\pi x_0 \left\{ \tilde{N}_n \tilde{P}_n + \left[\tilde{q}_n H_n^{(2)}(x_0)\right]^2 \right\}}, \tag{6.266}$$

where

$$\tilde{q}_n = -\frac{n \sin\psi}{ka\cos^2\psi}, \quad s_0 = \frac{1}{\cos\psi}, \quad x_0 = k_0 a \cos\psi \tag{6.267}$$

$$\tilde{V}_n = J_n(x_0) + \frac{j\eta}{\cos\psi} J_n'(x_0) \tag{6.268}$$

$$\tilde{N}_n = H_n^{(2)}(x_0) + \frac{j\eta}{\cos\psi} H_n^{(2)\prime}(x_0) \tag{6.269}$$

$$\tilde{P}_n = H_n^{(2)}(x_0) + \frac{j}{\eta\cos\psi} H_n^{(2)\prime}(x_0), \tag{6.270}$$

and $\eta = \tilde{Z}_s/Z_0$ is the normalized surface impedance of the cylinder.

The solution for the TE-polarized incident wave (6.204) is expressed by (6.217), (6.220), and (6.221), and the expansion coefficients A_n^{TE} and B_n^{TE} are obtained from B_n^{TM} and A_n^{TM} by substituting $\eta \to 1/\eta$ in (6.265)–(6.270), which leads to the expressions

$$A_n^{TE} = B_n^{TM}, \quad B_n^{TE} = -\frac{\tilde{M}_n\tilde{N}_n + \tilde{q}_n^2 J_n(x_0) H_n^{(2)}(x_0)}{\tilde{N}_n\tilde{P}_n + \left[\tilde{q}_n H_n^{(2)}(x_0)\right]^2}, \tag{6.271}$$

where

$$\tilde{M}_n = J_n(x_0) + \frac{j}{\eta\cos\psi} J_n'(x_0) \tag{6.272}$$

and the other quantities are defined in (6.267), (6.269), and (6.270).

The far-field scattering characteristics of impedance cylinders (far-field coefficients and cross sections) are formally given by the same equations as for the coated cylinder (see (6.252)–(6.262)). An example is presented in Figure 6.28. In this example, $\eta = 1$ and the curves are polarization independent as $\eta = 1/\eta$. The extinction cross section approaches its limit $4a\cos\psi$ as $k_0 a$ grows. At every value of $k_0 a$, $\sigma_{ext}^{2D} = \sigma_T^{2D} + \sigma_{abs}^{2D}$ by definition of the absorption cross section. Since

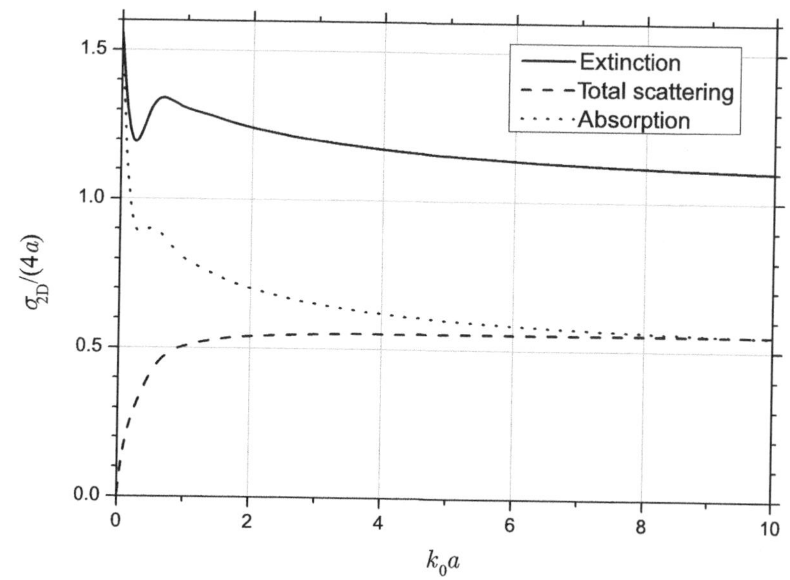

Figure 6.28 Extinction, total scattering, and absorption widths of a cylinder with $\eta = 1$ at $\psi = 0$ as a function of the electrical size of the cylinder. The widths are normalized to the asymptotic limit of σ_{ext}^{2D} for $k_0 a \to \infty$.

the surface impedance is matched to the wave impedance of the surrounding medium, the scattering is relatively weak and the shares of scattering and losses in $\sigma_{\rm ext}^{\rm 2D}$ are comparable. Compare this with the curves in Figure 6.5 for the PEC cylinder. In this case, $\sigma_{\rm abs}^{\rm 2D} = 0$ and $\sigma_{\rm ext}^{\rm 2D} = \sigma_{\rm T}^{\rm 2D}$ as there are no losses in the material of the cylinder. The scattering depends on the polarization and is much stronger for E (TM) polarization than for H (TE) polarization as the electric field vector in the former is parallel to the axis of the cylinder and excites, therefore, stronger currents on the surface of the cylinder.

Relation to the Solution for the Coated Cylinder

This solution is an exact solution of an approximately formulated scattering problem since boundary conditions (6.263) only approximately describe the material properties of the scattering body. The accuracy of the impedance boundary conditions for flat material interfaces has been studied in section 4.6, but in the case of a cylinder a further assumption is made, namely that the boundary conditions are applicable to a curved surface. It is instructive therefore to compare the solution for the impedance cylinder with the solutions for a homogeneous cylinder and a coated PEC cylinder, whose material properties have been modeled exactly in section 6.5.1. Comparing the expressions for the expansion coefficients in the exact and impedance formulations, that is, (6.226)–(6.236) with (6.265)–(6.272), it is seen that the former recover the latter when

$$|\epsilon_r\mu_r| \gg 1, \quad |k_0 a| \gg 1, \quad |k_1 a_1| \gg |k_0 a|. \tag{6.273}$$

Indeed, in this case $|x_{1,2}| \gg 1$, and in the terms with $|n| \leq |k_0 a|$ we can replace the Bessel functions in (6.235) and (6.236) with their large-argument approximations

$$J_n(x_{1,2}) \approx \sqrt{\frac{2}{\pi x_{1,2}}} \cos\left(x_{1,2} - \frac{n\pi}{2} - \frac{\pi}{4}\right) \tag{6.274}$$

$$J_n'(x_{1,2}) \approx -\sqrt{\frac{2}{\pi x_{1,2}}} \sin\left(x_{1,2} - \frac{n\pi}{2} - \frac{\pi}{4}\right) \tag{6.275}$$

$$H_n^{(2)}(x_{1,2}) \approx \sqrt{\frac{2}{\pi x_{1,2}}} \mathrm{e}^{-j\left(x_{1,2} - n\frac{\pi}{2} - \frac{\pi}{4}\right)} \tag{6.276}$$

$$H_n^{(2)\prime}(x_{1,2}) \approx -j\sqrt{\frac{2}{\pi x_{1,2}}} \mathrm{e}^{-j\left(x_{1,2} - n\frac{\pi}{2} - \frac{\pi}{4}\right)}, \tag{6.277}$$

which leads to the relations

$$y_n \approx \frac{-j}{\eta \cos\psi}, \qquad z_n \approx \frac{-j\eta}{\cos\psi} \tag{6.278}$$

with

$$\eta = j\sqrt{\frac{\mu_r}{\epsilon_r}} \tan[k_1(a - a_1)]. \tag{6.279}$$

Thus, $\delta_n \approx 1$, $M_n \approx \tilde{M}_n$, $N_n \approx \tilde{N}_n$, $P_n \approx \tilde{P}_n$, and $V_n \approx \tilde{V}_n$, and since $q_n \approx \tilde{q}_n$ when $|\epsilon_r\mu_r| \gg 1$, the expansion coefficients $A_n^{\rm TM}$, $B_n^{\rm TM}$, $A_n^{\rm TE}$, and $B_n^{\rm TE}$ in (6.226)–(6.229) reduce to those in (6.265), (6.266), and (6.271) with the impedance (6.279).

In expression (6.279) we recognize the normalized equivalent surface impedance of a PEC-backed flat material layer (4.341) with thickness $h = a - a_1$ and material constants ϵ_1 and μ_1. Thus, the solution for an impedance cylinder with $\tilde{Z}_s = Z_0\eta$ approximates the exact solution for a coated PEC cylinder, provided that the conditions (6.273) are satisfied.

In the case of a homogeneous cylinder ($a_1 = 0$), the third condition in (6.273) is not satisfied, and to arrive at the impedance formulation one has to assume a sufficiently strong absorption in the material of the cylinder,

$$|\epsilon_r\mu_r| \gg 1, \quad |k_0 a| \gg 1, \quad |\mathrm{Im}\, k_1|a \gg 1, \tag{6.280}$$

such that $J_n(x_1)/J_n'(x_1) \approx -j$ and $\delta_n \approx 1$ in (6.238), which reduces the exact solution for a homogeneous cylinder to the solution for an impedance cylinder with $\tilde{Z}_s = Z_1 = \sqrt{\mu_1/\epsilon_1}$. Condition $|\mathrm{Im}\, k_1|a \gg 1$ ensures that the absorption

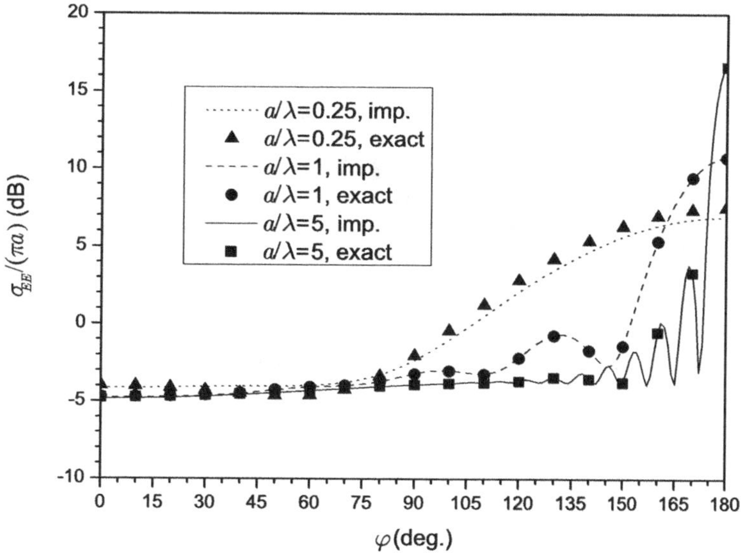

Figure 6.29 Co-polar normalized bistatic scattering width of a homogeneous cylinder: impedance versus exact formulation, TM polarization, incidence angle $\psi = 45°$, material parameters $\epsilon_r = 15 - j5$ and $\mu_r = 1$.

extinguishes the transmission through the cylinder as well as multiple reflections inside the cylinder, effects that cannot be modeled by the impedance boundary conditions.

Despite the limitations on the cylinder radius, impedance boundary conditions provide remarkable accuracy even for a smaller than λ. Figures 6.29–6.31 illustrate this conclusion by comparing the bistatic scattering widths (6.256)–(6.258) calculated by using the exact and the impedance ($\tilde{Z}_s = \sqrt{\mu_1/\epsilon_1}$) formulations for homogeneous absorbing cylinders with various radii. It is seen that the impedance solution gives an excellent approximation for both co- and cross-polarized

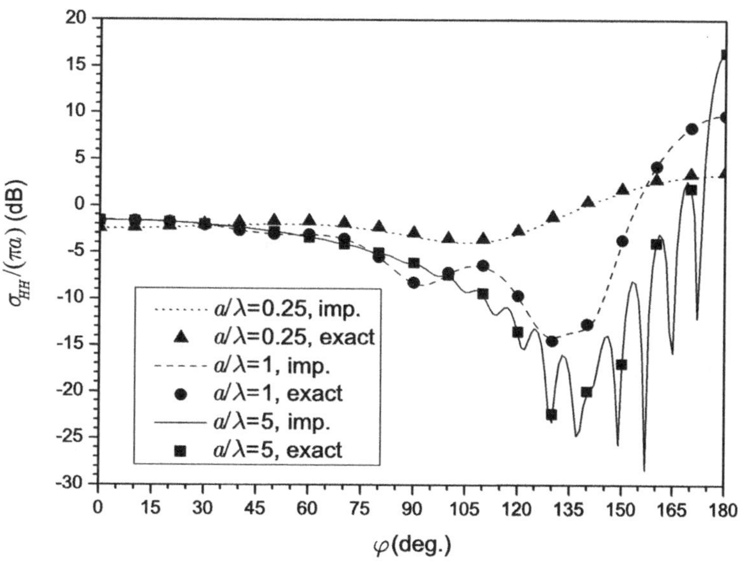

Figure 6.30 The same as Figure 6.29 but for TE polarization.

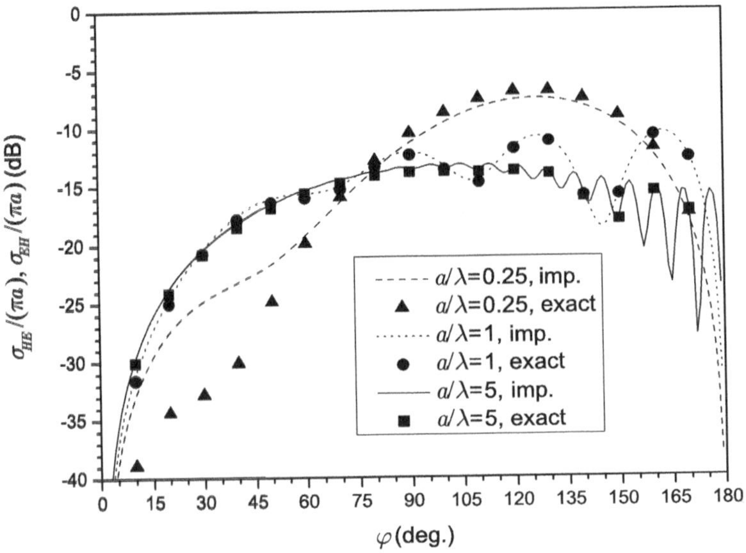

Figure 6.31 The same as Figure 6.29 but for cross-polarization.

components. Even for the smallest cylinder with $a = \lambda/4$, the difference between the exact and approximate solutions appears to be still sufficiently small for practical applications. A further decrease in the cylinder radius may lead, however, to a significant disagreement.

High-frequency Approximations

We conclude this section with approximations of the solution for an electrically large impedance cylinder, with the focus on the Fock-type asymptotic representation of the surface currents in the penumbra region. Compared to the case of the PEC cylinder, application of the Watson transformation to the solution for oblique illumination of an impedance cylinder is much more complicated because of the more complex dependence of the expansion coefficients (6.265) and (6.266) on the index n. However, for the normal incidence ($\psi = 0$) the coefficients greatly simplify and it is possible to derive the penumbra approximations by following the same procedure as in section 6.4.4. Extension to the case of oblique illumination is then achieved by application of the locality principle (section 6.6).

Indeed, if $\psi = 0$, then $\tilde{q}_n = 0$ and equations (6.265) and (6.266) simplify to[9]

$$A_n^{\mathrm{TM}} = -\frac{J_n(ka) + j\eta J_n'(ka)}{H_n^{(2)}(ka) + j\eta H_n^{(2)\prime}(ka)} \tag{6.281}$$

$$B_n^{\mathrm{TM}} = 0. \tag{6.282}$$

Inserting these in the expansions (6.218) and (6.219) and following the approach described in section 6.4.4 for a PEC cylinder, we arrive at the following approximations of the tangential field components:

$$E_z(a,\varphi) \approx E_0 e^{jka\cos\varphi} G(\xi, -jm\eta^{-1}) \tag{6.283}$$

$$H_z(a,\varphi) = 0, \tag{6.284}$$

[9] For the sake of brevity, from this point on we omit the subscript in k_0 and use k for the wave number in the exterior of the impedance cylinder.

Table 6.4 The function $G(\xi, p)$ with $p = -j$.

ξ	Re $G(\xi,p)$	Im $G(\xi,p)$	$\|G(\xi,p)\|$	arg $G(\xi,p)$ (deg.)
4.	0.00437743	−0.00218452	0.00489224	−26.5212
3.	0.0175052	0.00551399	0.0183531	17.484
2.	0.0133818	0.0670507	0.068373	78.7133
1.	0.00853393	−0.233936	0.234092	−87.9108
0.	0.570404	−0.200362	0.604571	−19.3545
−1.	1.03966	−0.0482565	1.04078	−2.65751
−2.	1.33458	−0.00387421	1.33458	−0.166326
−3.	1.49983	0.00170149	1.49983	0.0649995
−4.	1.59991	0.00174173	1.59991	0.0623743

where ξ is given by (6.175) and $G(\xi, p)$ is defined by the formula

$$G(\xi,p) = e^{-j\xi^3/3} \frac{1}{\sqrt{\pi}} \int_\Gamma \frac{e^{-j\xi t}}{w_2'(t) - p w_2(t)} \, dt \tag{6.285}$$

with the integration contour Γ shown in Figure 6.20. Equations (6.283) and (6.284) describe the solution in the case of E (TM) polarization.

Function $G(\xi, p)$ is a generalization of the Fock function $G(\xi)$ (defined in (6.184)) such that $G(\xi) = G(\xi, 0)$ (Fock 1965). It permits the asymptotic approximations (see problem 16)

$$G(\xi,p) \approx \frac{2\xi}{\xi - jp}, \quad \xi \to -\infty \tag{6.286}$$

and

$$G(\xi,p) \approx \frac{-2j\sqrt{\pi}}{w_2(t_1)(t_1 - p^2)} e^{-j\xi t_1 - j\xi^3/3}, \quad \xi \to +\infty, \tag{6.287}$$

where t_1 is the root of the equation $w_2'(t_1) = p w_2(t_1)$ with the greatest negative imaginary part. When $p \to 0$ or ∞, t_1 approaches t_1^H or t_1^E from (6.180), respectively. Table 6.4 presents reference values of $G(\xi, p)$ with $p = -j$ calculated by numerical integration with eight effective digits of precision. Figure 6.32 illustrates the influence of the parameter p on the absolute value of the function. The limiting case $p = 0$ corresponds to the Fock function $G(\xi)$ studied in the section on PEC cylinders. Various approaches to quick numerical evaluation of $G(\xi, p)$ and further numerical results can be found in Pearson (1987) and Osipov *et al.* (1998).

The formulas for the alternative, H (TE) polarization are obtained from (6.283) and (6.284) by using duality, and the result is given by the relations:

$$E_z(a,\varphi) = 0 \tag{6.288}$$

$$H_z(a,\varphi) \approx H_0 e^{jka\cos\varphi} G(\xi, -jm\eta). \tag{6.289}$$

Combining these with (6.283) and (6.284) and using (6.200) and (6.201) give the penumbra approximations for the z components of the electric and magnetic fields on the surface of the impedance cylinder under normal illumination,

$$H_z(a,\varphi) \approx H_z^{\text{inc}}(a,\varphi) G(\xi, -jm\eta) \tag{6.290}$$

$$E_z(a,\varphi) \approx E_z^{\text{inc}}(a,\varphi) G(\xi, -jm\eta^{-1}). \tag{6.291}$$

The remaining tangential components $E_\varphi(a,\varphi)$ and $H_\varphi(a,\varphi)$ immediately follow from the impedance boundary conditions (6.263). This completely defines approximations for the surface currents $\mathbf{K}_e$ and $\mathbf{K}_m$ in the penumbra.

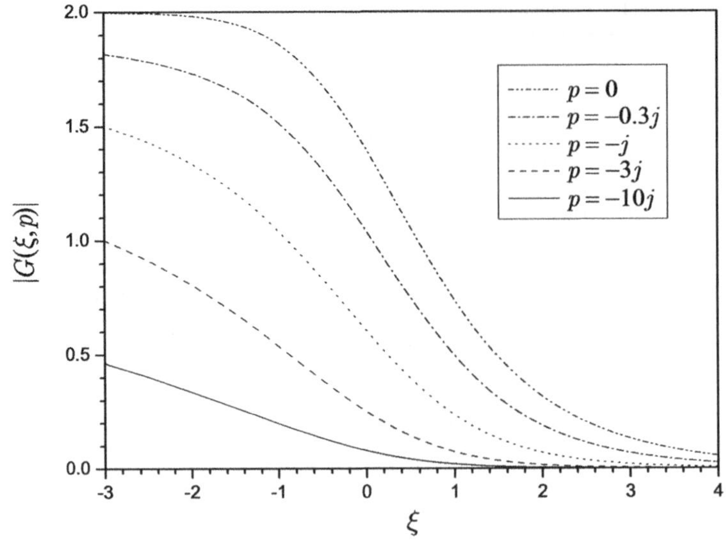

Figure 6.32 Magnitude of the Fock function $G(\xi, p)$ with various values of p.

Consider the vicinity of the shadow boundary $\varphi = \pi/2$. In the lit area, when $\xi \to -\infty$, the use of the asymptotic formula (6.286) in (6.290) and (6.291) leads to the relations

$$H_z(a,\varphi) \approx H_z^{\text{inc}}(a,\varphi)\frac{2\cos\varphi}{\cos\varphi + \eta} = H_z^{\text{inc}}(a,\varphi)\left[1 + \tilde{R}_{\|}(\varphi)\right] \tag{6.292}$$

$$E_z(a,\varphi) \approx E_z^{\text{inc}}(a,\varphi)\frac{2\cos\varphi}{\cos\varphi + 1/\eta} = E_z^{\text{inc}}(a,\varphi)\left[1 + \tilde{R}_{\perp}(\varphi)\right], \tag{6.293}$$

where $\tilde{R}_{\|}$ and $\tilde{R}_{\perp}$ are the reflection coefficients (4.334) and (4.335) of an impedance plane. Since φ is equal to the angle between the normal to the surface and the incidence direction, (6.292) and (6.293) give the fields in the GO approximation. Sometimes these approximations are called *tangent plane approximations* since the fields on a curved surface are described with reflection coefficients for an infinite plane tangent to the curved surface at the observation point.

In the deep shadow ($\xi \to +\infty$), because of (6.287), the penumbra representations (6.290) and (6.291) take the form of creeping waves

$$H_z(a,\varphi) \approx H_z^{\text{inc}}\left(a, \frac{\pi}{2}\right) c_1^{\text{TE}} e^{-j\nu_1^{\text{TE}}(\varphi - \pi/2)} \tag{6.294}$$

$$E_z(a,\varphi) \approx E_z^{\text{inc}}\left(a, \frac{\pi}{2}\right) c_1^{\text{TM}} e^{-j\nu_1^{\text{TM}}(\varphi - \pi/2)}, \tag{6.295}$$

where $\nu_1^{\text{TE}} = ka + mt_1^{\text{TE}}$, $\nu_1^{\text{TM}} = ka + mt_1^{\text{TM}}$,

$$c_1^{\text{TE}} = \frac{-2j\sqrt{\pi}}{w_2\left(t_1^{\text{TE}}\right)\left(t_1^{\text{TE}} + m^2\eta^2\right)} \tag{6.296}$$

$$c_1^{\text{TM}} = \frac{-2j\sqrt{\pi}}{w_2\left(t_1^{\text{TM}}\right)\left(t_1^{\text{TM}} + m^2/\eta^2\right)}, \tag{6.297}$$

and the parameters t_1^{TE} and t_1^{TM} are the zeros of the equations

$$w_2'\left(t_1^{\mathrm{TE}}\right) = -jm\eta w_2\left(t_1^{\mathrm{TE}}\right), \quad w_2'\left(t_1^{\mathrm{TM}}\right) = -j\frac{m}{\eta}w_2\left(t_1^{\mathrm{TM}}\right) \tag{6.298}$$

with the greatest negative imaginary parts, that is, $0 > \operatorname{Im} t_1 > \operatorname{Im} t_2 > \dots$ (Figure 6.20). In (6.294) and (6.295), the Taylor expansion (6.181) has been used to approximate the incident field and only the leading creeping wave with $s = 1$ generated at the shadow boundary $\varphi = \pi/2$ has been retained. The creeping waves (6.294) and (6.295) rapidly decay as the observation point moves away from the shadow boundary, thus approaching the GO value for the field in the shadow, which is zero.

Numerical simulations reveal that the function $G(\xi, p)$ is close to its asymptotic approximations (6.286) and (6.287) already with $|\xi| > 2$, regardless of the value of the parameter p. Thus, the width of the penumbra region on the surface of an impedance cylinder, defined by the parameter ξ, is still equal to d and remains the same as in the case of a PEC cylinder (see (6.194) in section 6.4.4). In the exterior of this region the field on the surface is almost equal to its GO values.

The asymptotic solutions (6.290) and (6.291) for the fields in the penumbra consist of two factors: $\exp(jka\cos\varphi)$ and $G(\xi, p)$. Since $ka \gg 1$, the former is a rapidly varying factor and the latter a slowly varying factor of φ (see (6.198)). Using this property and the relations (6.202), (6.204), (6.206), and (6.208) (with $\psi = 0$) between the field components, (6.290) and (6.291) can be transformed to the expressions

$$E_\rho(a, \varphi) \approx E_y^{\mathrm{inc}}(a, \varphi) G(\xi, -jm\eta) \tag{6.299}$$

$$H_\rho(a, \varphi) \approx H_y^{\mathrm{inc}}(a, \varphi) G(\xi, -jm\eta^{-1}), \tag{6.300}$$

which describe the components normal to the surface of the cylinder in the vicinity of the shadow boundary $\varphi = \pi/2$.

Expressions (6.294) and (6.295) can be used to derive formulas for the field at an observation point with coordinates ρ and φ, offset from the cylinder surface but still in deep shadow (Figure 6.33). To derive these, we simply note that the field in deep shadow has the form of an expansion in terms of $H_{\nu_s}^{(2)}(k\rho)$ (see (6.90)) and the leading term is that with $s = 1$. Thus, the expression for the field at a point distant from the cylinder surface can be obtained by simply multiplying the right-hand sides in (6.294) and (6.295) with the ratio

$$\frac{H_{\nu_1}^{(2)}(k\rho)}{H_{\nu_1}^{(2)}(ka)} \approx \sqrt{\frac{2}{k}}\frac{m}{w_2(t_1)(\rho^2 - a^2)^{1/4}} \mathrm{e}^{-jk\sqrt{\rho^2-a^2}+j\nu_1 \arccos\frac{a}{\rho} - j\frac{\pi}{4}}, \tag{6.301}$$

where we have approximated $H_{\nu_1}^{(2)}(ka)$ with (6.170), $H_{\nu_1}^{(2)}(k\rho)$ with (6.103), and ν_1 with ka in the arccos and square-root functions. Multiplying (6.294) by (6.301) with $t_1 = t_1^{\mathrm{TE}}$ and $\nu_1 = \nu_1^{\mathrm{TE}}$, and (6.295) by (6.301) with $t_1 = t_1^{\mathrm{TM}}$ and $\nu_1 = \nu_1^{\mathrm{TM}}$, we obtain

$$H_z(a, \varphi) \approx H_z^{\mathrm{inc}}\left(a, \frac{\pi}{2}\right) T_1^{\mathrm{TE}} \frac{\mathrm{e}^{-jk\sqrt{\rho^2-a^2}}}{(\rho^2 - a^2)^{1/4}} \tag{6.302}$$

$$E_z(a, \varphi) \approx E_z^{\mathrm{inc}}\left(a, \frac{\pi}{2}\right) T_1^{\mathrm{TM}} \frac{\mathrm{e}^{-jk\sqrt{\rho^2-a^2}}}{(\rho^2 - a^2)^{1/4}}, \tag{6.303}$$

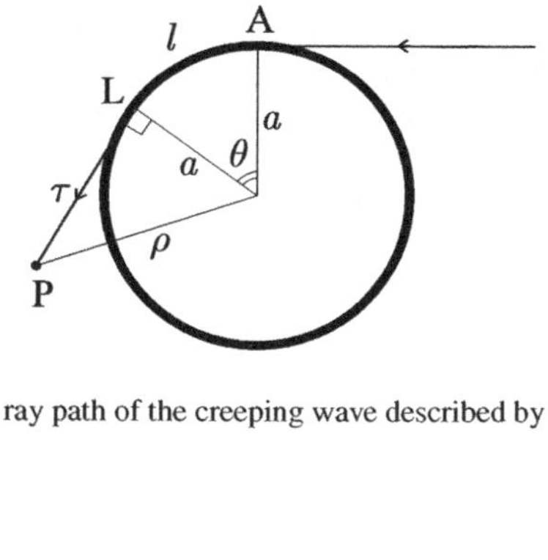

Figure 6.33 The ray path of the creeping wave described by (6.302) and (6.303).

where

$$T_1^{\mathrm{TE}} = \sqrt{\frac{2\pi}{k}} \frac{2me^{-j\theta v_1^{\mathrm{TE}} - j\frac{3}{4}\pi}}{w_2^2\left(t_1^{\mathrm{TE}}\right)\left(t_1^{\mathrm{TE}} + m^2\eta^2\right)} \tag{6.304}$$

$$T_1^{\mathrm{TM}} = \sqrt{\frac{2\pi}{k}} \frac{2me^{-j\theta v_1^{\mathrm{TM}} - j\frac{3}{4}\pi}}{w_2^2\left(t_1^{\mathrm{TM}}\right)\left(t_1^{\mathrm{TM}} + m^2/\eta^2\right)} \tag{6.305}$$

with $\theta = \varphi - \pi/2 - \arccos(a/\rho)$. The incident wave generates a creeping wave at the point A ($\rho = a$, $\varphi = \pi/2$) on the shadow boundary; the creeping wave glides along the surface of the cylinder to a point L ($\rho = a$, $\varphi = \pi/2 + \theta$), at which the wave departs from the surface, propagating along a straight path to the observation point P. The arc length l of the path between A and L is equal to $a\theta$, and the distance τ between L and P is equal to $\sqrt{\rho^2 - a^2}$ (Figure 6.33).

As we shall see in the next section, the penumbra representations (6.299) and (6.300) and the deep shadow representations (6.302) and (6.303) derived for a circular cylinder illuminated at a right angle to its axis, are fundamental to the fields diffracted by smoothly curved convex surfaces. These expressions can be extended to arbitrary incidence directions and generally shaped convex impedance surfaces, which is the subject of section 6.6.

6.6 Extension to Generally Shaped Convex Impedance Bodies

It is remarkable that to the first degree of approximation, the high-frequency penumbra approximations derived in sections 6.4.3, 6.4.4, and 6.5.2 for PEC and impedance circular cylinders in the special case of normal incidence are valid for arbitrary incidence directions and general convex surfaces if a is understood as the radius of curvature of the surface and l as the distance from the shadow boundary, both measured in the plane of incidence (the plane defined by the direction of incidence of the incoming wave and the vector normal to the surface). This is a consequence of the locality of high-frequency fields in the penumbra region on a convex surface, a property described and established around 1945 by Fock (1965, Chapters 2 and 5). The locality means that all bodies with a smoothly varying curvature have nearly the same current distribution in the penumbra region if the curvature in the plane of incidence and the incident wave are the same near the point of observation. The asymptotic expressions for the current in the penumbra region can be derived either indirectly from an exact solution for any suitable convex canonical body, like a cylinder, a sphere or a paraboloid of revolution (Fock 1965, Chapters 3 and 4), or directly from an integral equation for the surface current (Cullen 1958).

Asymptotic solutions in the deep shadow region of the cylinder can be interpreted in terms of the surface-diffracted rays of the GTD, which opens up the possibility of extending the solutions to generally shaped convex bodies by following the physical argumentation due to Keller (Borovikov and Kinber 1994; James 1976; Keller 1956, 1962; Levy and Keller 1959).

In section 6.6.1, following Fock (1965, Chapter 5), we describe the asymptotic structure of electromagnetic fields in the penumbra region near the surface of a smooth convex impedance body which can be an infinite cylinder but also a compact body with electrically large radii of curvature. The locality principle follows from this analysis, implying that the solution for the circular cylinder under normal illumination is a universal high-frequency solution for the currents and fields in the penumbra on generally curved convex surfaces. In particular, by placing the solution for normal incidence in the general context, we extend the solution for the circular impedance cylinder to arbitrary incidence directions ψ (section 6.6.2).[10] In section 6.6.3 the high-frequency solutions are expressed in ray-fixed coordinates, and explicit formulas are given for the fields on the surface of a convex electrically large impedance body (Fock-type solution) and for the fields in the deep shadow, far from the body, in the format of GTD.

[10] In the case of a circular PEC cylinder, the generalization to oblique illumination simply results from the solution for normal illumination on substituting $k \to k\cos\psi$, multiplying with $\exp(-jkz\sin\psi)$ and replacing E_0 with $E_0\cos\psi$ and H_0 with $H_0\cos\psi$ in the expressions for the field components E_z and H_z.

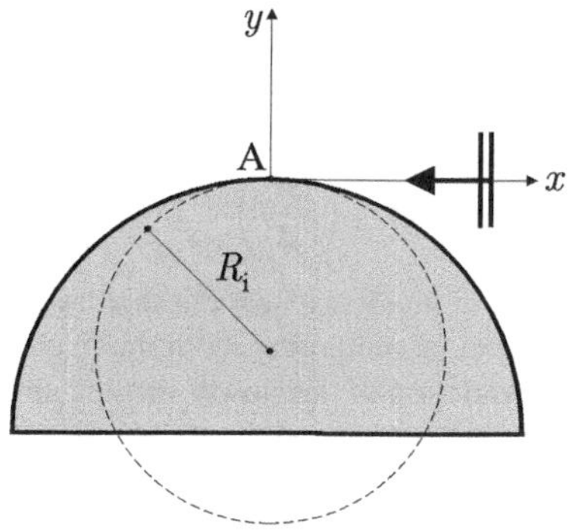

Figure 6.34 A convex impenetrable body illuminated by a plane wave propagating along the x axis. A is a point on the shadow boundary. The y axis is perpendicular to the surface at A. R_i is the radius of curvature of the surface in the incidence plane at A.

6.6.1 Fock's Principle of the Local Field in the Penumbra Region

Consider a plane wave $\exp(jkx)$ incident on a convex smooth impenetrable body (Figure 6.34). The point A, where the incidence direction touches the surface, lies on the shadow boundary. We search for high-frequency solutions of Maxwell's equations in the vicinity of the shadow boundary, where "high-frequency" means that the radii of curvature of the surface in the vicinity of the shadow boundary are much greater than the wavelength of the incident wave. The boundary is described by the impedance boundary conditions (2.74). The y axis is perpendicular to the surface at the shadow boundary. The plane xAy includes the incidence direction and the normal to the surface, and is therefore the incidence plane. The axes x and z are tangent to the surface at the shadow boundary. In the following analysis we will show that the field in the penumbra region can be asymptotically expressed in terms of just two components, which can be any of the pairs E_y and E_z or E_y and H_y or H_y and H_z.

By referring to the solution for the cylinder, we search the solution in the form of the product of a rapidly varying exponent multiplied by a slowly varying amplitude,

$$E_x = e^{jkx}\tilde{E}_x, \quad E_y = e^{jkx}\tilde{E}_y, \ldots \quad H_z = e^{jkx}\tilde{H}_z. \tag{6.306}$$

Inserting these representations in Maxwell's equations gives equations for the amplitudes,

$$-jkZ\tilde{H}_x = \frac{\partial \tilde{E}_z}{\partial y} - \frac{\partial \tilde{E}_y}{\partial z} \tag{6.307}$$

$$-jkZ\tilde{H}_y = \frac{\partial \tilde{E}_x}{\partial z} - jk\tilde{E}_z - \frac{\partial \tilde{E}_z}{\partial x} \tag{6.308}$$

$$-jkZ\tilde{H}_z = jk\tilde{E}_y + \frac{\partial \tilde{E}_y}{\partial x} - \frac{\partial \tilde{E}_x}{\partial y} \tag{6.309}$$

$$jkY\tilde{E}_x = \frac{\partial \tilde{H}_z}{\partial y} - \frac{\partial \tilde{H}_y}{\partial z} \tag{6.310}$$

$$jkY\tilde{E}_y = \frac{\partial \tilde{H}_x}{\partial z} - jk\tilde{H}_z - \frac{\partial \tilde{H}_z}{\partial x} \tag{6.311}$$

$$jkY\tilde{E}_z = jk\tilde{H}_y + \frac{\partial \tilde{H}_y}{\partial x} - \frac{\partial \tilde{H}_x}{\partial y}. \tag{6.312}$$

Here, Z and Y are the wave impedance and admittance of the medium surrounding the obstacle, so that $kZ = \omega\mu$ and $kY = \omega\epsilon$.

Equations (6.307)–(6.312) are exact. In the case that the wavelength is much shorter than the radii of curvature of the scattering surface, the quotient R/λ (R is the curvature radius) is a large parameter. By referring to the solution for circular cylinder, for example (6.129), it is convenient to define the large parameter by

$$m = \left(\frac{kR_{\mathrm{i}}}{2}\right)^{1/3}, \tag{6.313}$$

where R_{i} is the radius of curvature in the plane of incidence near the shadow boundary.

At the next step, by using this large parameter, assumptions about the rate of variation of the amplitudes with respect to the coordinates are made, namely that the variation of the slowly varying amplitude in the y direction (normally to the surface) be more rapid than in the x and z directions (tangentially to the surface), so that

$$\frac{\partial\tilde{\Psi}}{\partial y} \sim \frac{k}{m}\tilde{\Psi}, \quad \frac{\partial\tilde{\Psi}}{\partial x}, \frac{\partial\tilde{\Psi}}{\partial z} \sim \frac{k}{m'}\tilde{\Psi}, \tag{6.314}$$

where $\tilde{\Psi}$ is the slow amplitude of any field component and m and m' are the parameters describing the rate of the variations such that $m' \gg m \gg 1$. To comply with the Helmholtz equation

$$\frac{\partial^2\Psi}{\partial x^2} + \frac{\partial^2\Psi}{\partial y^2} + \frac{\partial^2\Psi}{\partial z^2} + k^2\Psi = 0, \tag{6.315}$$

which is satisfied by every component of the field, the parameters must be related as

$$m' = m^2. \tag{6.316}$$

Indeed, inserting $\Psi = \tilde{\Psi}\exp(jkx)$ in (6.315) we obtain

$$2jk\frac{\partial\tilde{\Psi}}{\partial x} + \frac{\partial^2\tilde{\Psi}}{\partial x^2} + \frac{\partial^2\tilde{\Psi}}{\partial y^2} + \frac{\partial^2\tilde{\Psi}}{\partial z^2} = 0, \tag{6.317}$$

and estimations (6.314) imply (6.316). It can be checked by direct inspection that the asymptotic solutions derived in this chapter for the penumbra region of the circular cylinder comply with (6.314) and (6.316).

By using the estimations (6.314) and (6.316) we can now show that the field components can be expressed through two components, say E_y and E_z. Equation (6.307) defines $\tilde{H}_x$ in terms of $\tilde{E}_y$ and $\tilde{E}_z$ as

$$\tilde{H}_x = \frac{j}{k}Y\left(\frac{\partial\tilde{E}_z}{\partial y} - \frac{\partial\tilde{E}_y}{\partial z}\right). \tag{6.318}$$

In (6.311), the terms with the derivatives are small and can be neglected, giving

$$\tilde{H}_z \approx -Y\tilde{E}_y. \tag{6.319}$$

Similarly, it follows from (6.312) that

$$\tilde{H}_y \approx Y\tilde{E}_z. \tag{6.320}$$

Now that $\tilde{H}_y$ and $\tilde{H}_z$ are expressed through E_y and E_z, we find from (6.310) that

$$\tilde{E}_x \approx \frac{j}{k}\left(\frac{\partial\tilde{E}_y}{\partial y} + \frac{\partial\tilde{E}_z}{\partial z}\right). \tag{6.321}$$

Relations (6.318)–(6.321) do not include derivatives with respect to x, so they are valid not only for the amplitudes but also for the complete components,

$$E_x \approx \frac{j}{k}\left(\frac{\partial E_y}{\partial y} + \frac{\partial E_z}{\partial z}\right) \tag{6.322}$$

$$H_x = \frac{j}{k}Y\left(\frac{\partial E_z}{\partial y} - \frac{\partial E_y}{\partial z}\right) \tag{6.323}$$

$$H_y \approx YE_z \tag{6.324}$$
$$H_z \approx -YE_y. \tag{6.325}$$

These equations show that the fields in the penumbra region can be expressed through E_y and E_z – the two components of the electric field perpendicular to the direction of propagation of the incoming wave. Expressions (6.324) and (6.325), which formally have the form of the relations between the electric and magnetic field components in a plane wave propagating in the direction of the negative x axis, imply that any pair of the components E_y, E_z, H_y, and H_z can serve as a basis pair.

On the surface of the impedance body, the fields satisfy the impedance boundary conditions in the vector form (2.74). Now we show that the basis pair of components satisfy separate boundary conditions. By noting that

$$\hat{n} \times (\hat{n} \times \mathbf{H}) = \hat{x}(n_x H_n - H_x) + \hat{y}(n_y H_n - H_y) + \hat{z}(n_z H_n - H_z), \tag{6.326}$$

where $\hat{n} = n_x\hat{x} + n_y\hat{y} + n_z\hat{z}$ is the outward unit vector normal to the surface of the body and $H_n = \hat{n} \cdot \mathbf{H} = n_x H_x + n_y H_y + n_z H_z$, we rewrite the boundary conditions in the component form as

$$n_y E_z - n_z E_y = \tilde{Z}_s(n_x H_n - H_x) \tag{6.327}$$
$$n_z E_x - n_x E_z = \tilde{Z}_s(n_y H_n - H_y) \tag{6.328}$$
$$n_x E_y - n_y E_x = \tilde{Z}_s(n_z H_n - H_z). \tag{6.329}$$

Using (6.324) and (6.325) in (6.327) allows us to express H_x as

$$ZH_x\left(1 - n_x^2\right) = \left(\frac{1}{\eta} - n_x\right)(n_z E_y - n_y E_z). \tag{6.330}$$

Here $\eta = \tilde{Z}_s/Z$ is the equivalent surface impedance $\tilde{Z}_s$ normalized to the wave impedance Z of the surrounding medium. Multiplying equation (6.328) with n_z and subtracting from it equation (6.329) multiplied with n_y, we exclude H_x from the equations and obtain

$$E_x\left(1 - n_x^2\right) = (n_x - \eta)(n_y E_y + n_z E_z). \tag{6.331}$$

Finally, replacing H_x with (6.323) in (6.330) and E_x with (6.322) in (6.331) we arrive at the boundary conditions for the basis pair of components,

$$\left(1 - n_x^2\right)\frac{j}{k}\left(\frac{\partial E_z}{\partial y} - \frac{\partial E_y}{\partial z}\right) \approx \left(\frac{1}{\eta} - n_x\right)(n_z E_y - n_y E_z) \tag{6.332}$$
$$\left(1 - n_x^2\right)\frac{j}{k}\left(\frac{\partial E_y}{\partial y} + \frac{\partial E_z}{\partial z}\right) \approx (n_x - \eta)(n_y E_y + n_z E_z). \tag{6.333}$$

Since the surface is smooth and we look for the fields in an asymptotically small vicinity of the shadow boundary (see (6.193) for an estimation of the angular width of the penumbra region), the unit normal is close to $\hat{y}$ so that $n_y \approx 1$ and $n_x, n_z = o(1)$. Furthermore, because of (6.314) and (6.316), the tangential derivatives of the fields can be neglected compared with the normal derivatives. These considerations reduce (6.332) and (6.333) to

$$\frac{\partial E_y}{\partial y} - jk\eta E_y = 0 \tag{6.334}$$
$$\frac{\partial E_z}{\partial y} - jk\frac{1}{\eta}E_z = 0. \tag{6.335}$$

These relations have the form of the impedance boundary conditions for the fields that do not depend on z since (6.334) follows from $\tilde{Z}_s H_z = E_x$ and (6.335) from $\tilde{Z}_s H_x = -E_z$ (see (2.306), (2.307), and (2.308) with $k_z = 0$).

Summarizing, we conclude that if the boundary conditions (6.334) and (6.335) are used, the components E_y and E_z can be determined separately: each of them satisfies a separate Helmholtz equation, a separate boundary condition on the surface of the body, and a separate condition at infinity. From these conditions E_y and E_z are determined uniquely.

It is important that only the curvature of the body in the plane of incidence matters as the boundary conditions (6.334) and (6.335) include only the derivative with respect to the coordinate y perpendicular to the surface and the propagation direction, and thus the change in the shape of the surface in the direction perpendicular to the plane of incidence is not present in the leading term of the approximation. This important property of the asymptotic solution has been checked by considering the exact solution for a PEC paraboloid of revolution (Fock 1965, Chapter 4), as this shape allows a general combination of the principal radii of curvature, in contrast to the sphere, which has equal curvature radii, and the cylinder with one of curvature radii being infinite.

The asymptotic solution around the shadow boundary on the surface of every convex body can thus be found by looking at the corresponding solution for a locally conformal body, for example a cylinder or a sphere with the same value of the surface impedance and the curvature radius $R_{\rm i}$ in the penumbra region, illuminated by the same incident wave. For a PEC body we can use the solution for a PEC circular cylinder under normal illumination (section 6.4.4) and for an impedance body the solution for an impedance circular cylinder under normal illumination (section 6.5.2). In both cases, the leading term is equal to the incident wave multiplied by a function of the ratio l/d, where l is the distance from the shadow boundary measured along the surface in the plane of incidence and d is the width of the penumbra region

$$d = \left(\frac{2R_{\rm i}^2}{k}\right)^{1/3} \tag{6.336}$$

obtained from (6.154) on replacing a with $R_{\rm i}$.

A consequence of the locality property is that in the high-frequency limit the fields in the penumbra on the scattering surface are independent of the shape of the rest part of the surface as only the local curvature and the local value of the incident field influence the leading terms in the asymptotic representations.

It should be noted that the Fock solution for penumbra solves the complete problem of high-frequency scattering from a smooth convex body since the currents on the lit portion of the body can always be obtained by application of GO and in the shadow the currents quickly decay and can be typically neglected.[11] Once the surface current is determined, the scattered fields in the whole space are obtained by integrating the current in a Huygens-type representation of solutions of Maxwell's equations (sections 2.6 and 3.4.1).

The condition of applicability of the Fock penumbra solution is apparently that in the vicinity of the point of contact of the incident ray with the scattering surface (attachment point), the surface is homogeneous and convex and both its principal radii of curvature are greater than the wavelength. The solution does not apply if in the vicinity of the attachment point the surface impedance has a jump and/or the scattering surface has an edge or a tip, in which cases there is no smooth transition between the lit and shadowed portions of the surface and asymptotic solutions of other kinds are required, for example Molinet *et al.* (2005).

6.6.2 *Asymptotic Solution for the Field on the Surface of Circular Impedance Cylinders under Oblique Illumination*

By referring to the locality principle, we can immediately extend the asymptotic solutions obtained in sections 6.4.3, 6.4.4, and 6.5.2 for the particular case of normal incidence to the general case of oblique illumination. Consider a circular cylinder illuminated by a plane wave with the incidence direction lying in the plane xOz and making an angle ψ with the x axis, as shown in Figure 6.23. Depending on the polarization case, the incident field is given by (6.202) or (6.204). The components $E_y^{\rm inc}$ and $H_y^{\rm inc}$ and E_ρ and H_ρ are still the components normal to the surface at the shadow boundary and the observation point, respectively. So, according to the locality principle, we can use the solutions (6.197) or (6.299) and (6.300) with adjusted values of the curvature radius in the incidence plane and the distance between the shadow boundary and the observation point.

[11] There are of course exceptions in which the creeping waves have to be considered, e.g. scattering from strongly elongated ogive-shaped scatterers under illumination along the longer axis or electromagnetic coupling between antennas or antenna elements placed close to each other on a cylindrical surface.

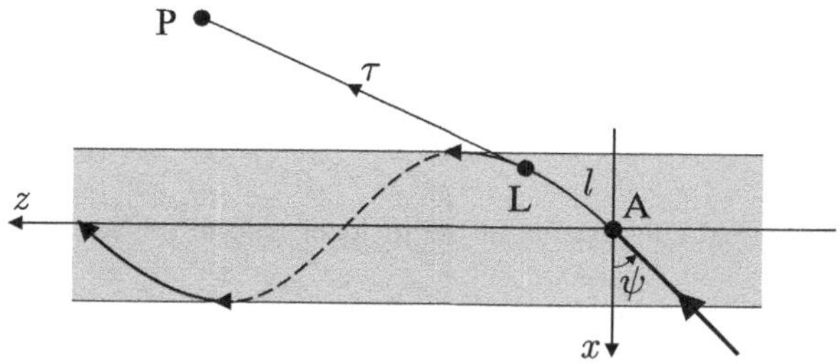

Figure 6.35 The incidence direction of a plane wave lies in the x-z plane and makes an angle ψ with the x axis. At the attachment point A on the shadow boundary on the surface of the cylinder, a creeping wave is excited, which follows a spiral path along the surface of the cylinder until it departs tangentially to the surface at the launching point L toward the observer at the point P.

The radius of curvature of the cylinder in the plane of incidence at the attachment point is related to the radius a of the cylinder by the formula

$$R_{\mathrm{i}} = \frac{a}{\cos^2\psi}, \tag{6.337}$$

and it follows from (6.313) and (6.336) that

$$m = \left(\frac{ka}{2\cos^2\psi}\right)^{1/3} \tag{6.338}$$

$$d = \left(\frac{2a^2}{k\cos^4\psi}\right)^{1/3}, \tag{6.339}$$

where d is the characteristic width of the penumbra region.

Parameter ξ in the Fock functions is defined by l/d with l being the arc length on the cylinder surface between the attachment and observation points. Since the width of the penumbra region is small compared to the radius,

$$l = \frac{a}{\cos\psi}\left(\varphi - \frac{\pi}{2}\right) \tag{6.340}$$

as in a plane right-angled triangle with sides l and $a(\varphi - \pi/2)$ and angle ψ between them (Figure 6.35). Expressions (6.340) and (6.339) result in the formula

$$\xi = \frac{l}{d} = \left(\frac{ka\cos\psi}{2}\right)^{1/3}\left(\varphi - \frac{\pi}{2}\right). \tag{6.341}$$

Equations (6.299) and (6.300) together with the expressions (6.202) or (6.204) for the incident fields E_y^{inc} and H_y^{inc} and with (6.338) and (6.341) completely define asymptotic representations for E_ρ and H_ρ in the penumbra region on the surface of the circular impedance cylinder under oblique illumination. For example, in the shadow region ($\varphi > \pi/2$), approximating the cosine function in the phase of the incident plane wave near the point $\varphi = \pi/2$ by

$$\cos\varphi \approx \frac{\pi}{2} - \varphi - \frac{1}{6}\left(\frac{\pi}{2} - \varphi\right)^3 \tag{6.342}$$

and using the definition (6.285) of $G(\xi, p)$, the penumbra solution can be expressed as

$$E_\rho(\rho,\varphi,z) = E_\rho^{\mathrm{inc}}(\mathrm{A})\mathrm{e}^{-jkz\sin\psi - jk_\perp a(\varphi-\frac{\pi}{2})}\frac{1}{\sqrt{\pi}}\int_\Gamma \frac{\mathrm{e}^{-jm_\perp(\varphi-\frac{\pi}{2})t}\,\mathrm{d}t}{w_2'(t) + jm\eta w_2(t)} \tag{6.343}$$

$$H_\rho(\rho,\varphi,z) = H_\rho^{\mathrm{inc}}(\mathrm{A})\mathrm{e}^{-jkz\sin\psi - jk_\perp a(\varphi-\frac{\pi}{2})}\frac{1}{\sqrt{\pi}}\int_\Gamma \frac{\mathrm{e}^{-jm_\perp(\varphi-\frac{\pi}{2})t}\,\mathrm{d}t}{w_2'(t) + jmw_2(t)/\eta}, \tag{6.344}$$

where $k_\perp = k\cos\psi$,

$$m_\perp = m\cos\psi = \left(\frac{k_\perp a}{2}\right)^{1/3}, \tag{6.345}$$

and $E_\rho^{\rm inc}(\mathrm{A})$ and $H_\rho^{\rm inc}(\mathrm{A})$ are the normal components of the incident field at the attachment point A ($\rho = a$, $\varphi = \pi/2$, $z = 0$).

Deeper in the shadow region, similarly to approximation (6.287) for $G(\xi, p)$, the integrals in (6.343) and (6.344) can be approximated by the residues at the poles of their integrand functions (see Figure 6.20), which gives

$$E_\rho(a,\varphi,z) \approx E_\rho^{\rm inc}(\mathrm{A}) c_1^{\rm TE} \mathrm{e}^{-jm_\perp(\varphi-\frac{\pi}{2})t_1^{\rm TE}} \mathrm{e}^{-jkz\sin\psi - jk_\perp a(\varphi-\frac{\pi}{2})} \tag{6.346}$$

$$H_\rho(a,\varphi,z) \approx H_\rho^{\rm inc}(\mathrm{A}) c_1^{\rm TM} \mathrm{e}^{-jm_\perp(\varphi-\frac{\pi}{2})t_1^{\rm TM}} \mathrm{e}^{-jkz\sin\psi - jk_\perp a(\varphi-\frac{\pi}{2})}, \tag{6.347}$$

where $c_1^{\rm TE}$, $c_1^{\rm TM}$, $t_1^{\rm TE}$, and $t_1^{\rm TM}$ are defined in (6.296), (6.297), and (6.298) with m given by (6.338). In (6.346) and (6.347) we have neglected the contributions from the pole singularities $t_s^{\rm TM}$ and $t_s^{\rm TE}$ with $s \geq 2$ as in the deep shadow the contributions are exponentially small compared with the terms with $s = 1$.

Expressions (6.346) and (6.347) describe creeping waves. Indeed, the phase factors in (6.346) and (6.347)

$$kz\sin\psi + k_\perp a(\varphi - \pi/2) \tag{6.348}$$

suggest that the path of the waves on the surface of the cylinder is a helix (Figure 6.35). It can be shown (problems 2 through 5 in section A.5) that (1) the helix can be represented in the parametric form as

$$x = a\cos\varphi, \quad y = a\sin\varphi, \quad z = a(\varphi - \pi/2)\tan\psi \tag{6.349}$$

with $-\infty < \varphi < \infty$; (2) the arc length between the attachment point A and the observation point P is exactly given by (6.340) (i.e., the expression (6.340), though obtained in a small vicinity of the attachment point, is valid at arbitrary distances from the shadow boundary); (3) the helix is a *geodesic* on the surface of the cylinder implying that the wave propagates over the shortest path from A to P. It follows from (6.340) and (6.349) that the phase (6.348) of the creeping waves is equal to kl. Attenuation of the creeping waves on their way from A to P is described in (6.346) and (6.347) by the factor

$$\mathrm{e}^{-jm_\perp(\varphi-\frac{\pi}{2})t_1} = \mathrm{e}^{-j\xi t_1} \tag{6.350}$$

with ξ defined by (6.341) and the complex-valued parameter t_1 equal to either $t_1^{\rm TE}$ or $t_1^{\rm TM}$.

6.6.3 *Fock- and GTD-type Solutions for Electrically Large Convex Impedance Bodies*

In this section we extend the asymptotic solutions for the circular impedance cylinder to generally shaped convex impedance bodies. We begin by formulating the locality principle in the ray-fixed coordinates. Then, the Fock-type asymptotic solution for the penumbra region is expressed in the new coordinates and studied in the deep lit and shadow regions on the surface of a general scatterer.

Furthermore, the asymptotic solution for the circular impedance cylinder in the deep shadow region, but far from the surface of the cylinder, is cast in the format of GTD and extended to the case of smooth convex electrically large impedance bodies by following Keller (1956) (see also Levy and Keller (1959) and Keller (1962)).

To describe the high-frequency fields scattered by curved surfaces, it is convenient to introduce the ray-fixed coordinate basis $\hat{e}_1$, $\hat{e}_2$, and $\hat{e}_3$, where $\hat{e}_1$ is the unit tangent to the ray, $\hat{e}_2$ is the outward normal $\hat{n}$ to the surface, and $\hat{e}_3 = \hat{e}_2 \times \hat{e}_1$ is the unit vector tangential to the surface and perpendicular to the ray. The plane defined by $\hat{e}_1$ and $\hat{e}_2$ is therefore the incidence plane. The components of the electric and magnetic fields in the directions $\hat{e}_1$, $\hat{e}_2$, and $\hat{e}_3$ will be denoted by E_1, E_2, E_3, H_1, H_2, and H_3.

In terms of the ray-fixed coordinates, Fock's locality principle can be formulated as follows. The components E_2 and E_3 propagate independently from each other and satisfy the impedance boundary conditions (6.334) and (6.335)

$$\frac{\partial E_2}{\partial n} - jk\eta E_2 = 0 \tag{6.351}$$

$$\frac{\partial E_3}{\partial n} - jk\frac{1}{\eta}E_3 = 0 \tag{6.352}$$

on the scattering surface. In the penumbra region, E_2 and E_3 are asymptotically close to the product of the incident field with a universal function which depends on the curvature radius R_i of the surface in the incidence plane, the normalized surface impedance η, the wave number (frequency) k and the distance l between the observation point and the shadow boundary measured along the scattering surface. Because of (6.324) and (6.325), E_3 and H_3 are related to E_2 and H_2 by $E_2 = -ZH_3$ and $E_3 = ZH_2$, where Z is the wave impedance of the surrounding medium, so the principle can be formulated for other combinations of the components, that is, for E_2 and H_2. Compared with the other components of the fields, the components E_1 and H_1 are asymptotically small as $O(m^{-1})$ because of (6.313), (6.314), and (6.316) with (6.322) and (6.323).

The universal distribution of the field components E_2 and H_2 in the penumbra region on the scattering surface can be taken from the solution (6.343) and (6.344) for a circular cylinder. Assuming that $E_1 = 0$, the electric field on the surface can therefore be represented by

$$\mathbf{E}(P) = E_2^{\mathrm{inc}}(\mathrm{A})V(\mathrm{P},\mathrm{A})\hat{e}_2(\mathrm{P}) + E_3^{\mathrm{inc}}(\mathrm{A})W(\mathrm{P},\mathrm{A})\hat{e}_3(\mathrm{P}), \tag{6.353}$$

where

$$V(\mathrm{P},\mathrm{A}) = \mathrm{e}^{-jkl}\frac{1}{\sqrt{\pi}}\int_\Gamma \frac{\mathrm{e}^{-j\xi t}\,\mathrm{d}t}{w_2'(t) + jm\eta w_2(t)} \tag{6.354}$$

with $\xi = l/d$, l being the arc length between the attachment point A and the observer P, and d and m defined by (6.313) and (6.336). By definition, the arc length l is positive in the shadow and negative on the lit side of the shadow boundary. Function $W(\mathrm{P},\mathrm{A})$ is given by (6.354) with η replaced by $1/\eta$. Since functions $V(\mathrm{P},\mathrm{A})$ and $W(\mathrm{P},\mathrm{A})$ relate (or transform) components of the incident field at the attachment point to the components at the observation point, they can be called transformation functions.

Deep in the lit region ($\xi < -1$), (6.353) with (6.354) recover the GO solution for the field

$$\mathbf{E}(\mathrm{P}) \approx \mathbf{E}_{\mathrm{inc}}(\mathrm{P}) + \mathbf{E}_{\mathrm{ref}}(\mathrm{P}). \tag{6.355}$$

Indeed, it follows from the saddle point approximation (6.286) of the integral in (6.354) that

$$V(\mathrm{P},\mathrm{A}) \approx \frac{2\xi}{\xi - m\eta}\mathrm{e}^{-jkl+j\xi^3/3}, \quad W(\mathrm{P},\mathrm{A}) \approx \frac{2\xi}{\xi - m/\eta}\mathrm{e}^{-jkl+j\xi^3/3}, \tag{6.356}$$

which, using the definitions of the parameters ξ, d, and m, can be rewritten as

$$V(\mathrm{P},\mathrm{A}) \approx \frac{2l/R_\mathrm{i}}{(l/R_\mathrm{i}) - \eta}\mathrm{e}^{-jkl\left[1-\frac{1}{6}(l/R_\mathrm{i})^2\right]} \tag{6.357}$$

$$W(\mathrm{P},\mathrm{A}) \approx \frac{2l/R_\mathrm{i}}{(l/R_\mathrm{i}) - 1/\eta}\mathrm{e}^{-jkl\left[1-\frac{1}{6}(l/R_\mathrm{i})^2\right]}. \tag{6.358}$$

Now introducing the incidence angle θ_i as the angle between the normal to the surface and the incidence direction, which is close to $\pi/2$ near the shadow boundary (Figure 6.36), we express the arc length l as

$$l \approx -R_\mathrm{i}\left(\frac{\pi}{2} - \theta_\mathrm{i}\right) \approx -R_\mathrm{i}\cos\theta_\mathrm{i} \tag{6.359}$$

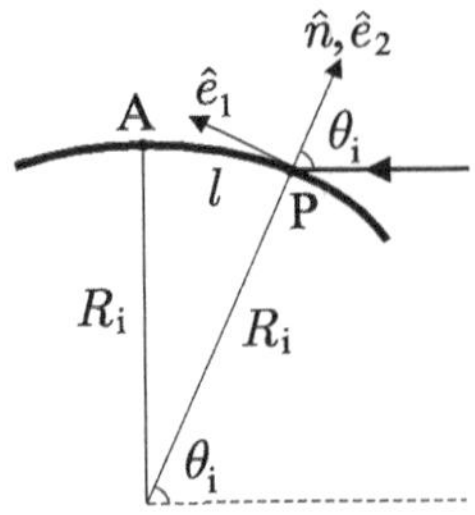

Figure 6.36 Observation point P located on the curved scattering surface in the lit region.

and rewrite (6.357) and (6.358) as

$$V(\mathrm{P,A}) \approx \frac{2\cos\theta_i}{\cos\theta_i + \eta} e^{jkR_i\cos\theta_i} = \left[1 + \tilde{R}_{\parallel}(\theta_i)\right] e^{jkR_i\cos\theta_i} \tag{6.360}$$

$$W(\mathrm{P,A}) \approx \frac{2\cos\theta_i}{\cos\theta_i + 1/\eta} e^{jkR_i\cos\theta_i} = \left[1 + \tilde{R}_{\perp}(\theta_i)\right] e^{jkR_i\cos\theta_i}, \tag{6.361}$$

where $\tilde{R}_{\parallel}(\theta_i)$ and $\tilde{R}_{\perp}(\theta_i)$ are the reflection coefficients (4.334) and (4.335) of an impedance plane tangent to the scattering surface at the point P. Since the incident wave is a plane wave, we have that

$$E_2^{\mathrm{inc}}(\mathrm{A})e^{jkR_i\cos\theta_i} = E_2^{\mathrm{inc}}(\mathrm{P}), \quad E_3^{\mathrm{inc}}(\mathrm{A})e^{jkR_i\cos\theta_i} = E_3^{\mathrm{inc}}(\mathrm{P}), \tag{6.362}$$

and equation (6.353) reduces to

$$\mathbf{E}(\mathrm{P}) \approx E_2^{\mathrm{inc}}(\mathrm{P})\left[1 + \tilde{R}_{\parallel}(\theta_i)\right]\hat{e}_2(\mathrm{P}) + E_3^{\mathrm{inc}}(\mathrm{P})\left[1 + \tilde{R}_{\perp}(\theta_i)\right]\hat{e}_3(\mathrm{P}), \tag{6.363}$$

which is the GO approximation of the electric field on the surface of the scatterer (section 4.2.6).

In the deep shadow ($\xi > 1$), $V(\mathrm{P,A})$, and $W(\mathrm{P,A})$ are evaluated as sums of residues, and retaining the leading terms gives for the transformation functions

$$V(\mathrm{P,A}) \approx c_1^{\mathrm{TE}} e^{-jkl - j\xi t_1^{\mathrm{TE}}}, \quad W(\mathrm{P,A}) \approx c_1^{\mathrm{TM}} e^{-jkl - j\xi t_1^{\mathrm{TM}}} \tag{6.364}$$

with c_1^{TE}, c_1^{TM}, t_1^{TE}, and t_1^{TM} defined in (6.296), (6.297), and (6.298). Equations (6.364) together with (6.353) describe the electric field of the creeping wave at an observation point located on the scattering surface.

As we have seen in the case of the circular cylinder in section 6.4.3, far from the scattering surface the contributions of the creeping waves have the form of rays launched at such points on the scattering surface where the tangential to the path of the creeping wave exactly points to the observer. On launching from the surface, every ray propagates along a straight path connecting the launching point L and the observation point P (Figure 6.35). These rays are called the surface-diffracted rays (Keller 1956). In contrast to conventional GO rays, surface-diffracted rays circumvent opaque bodies by traveling along the shortest path on the surface of the scatterer, that is, along the geodesics.

To describe the contributions of the surface-diffracted rays for generally shaped convex impedance bodies, we define the vectors $\hat{e}_1$, $\hat{e}_2$, and $\hat{e}_3$ such that at a point P offset from the surface they are obtained by translation along the straight line connecting the launching and observation points, that is, $\hat{e}_1(\mathrm{P}) = \hat{e}_1(\mathrm{L})$, $\hat{e}_2(\mathrm{P}) = \hat{e}_2(\mathrm{L})$, and $\hat{e}_3(\mathrm{P}) = \hat{e}_3(\mathrm{L})$. On launching, the field components E_2 and E_3 propagate through the surrounding homogeneous medium in the direction $\hat{e}_1(\mathrm{L})$ independently from each other, so that we can write

$$\mathbf{E}(\mathrm{P}) = E_2^{\mathrm{inc}}(\mathrm{A})V(\mathrm{P,A})\hat{e}_2(\mathrm{L}) + E_3^{\mathrm{inc}}(\mathrm{A})W(\mathrm{P,A})\hat{e}_3(\mathrm{L}). \tag{6.365}$$

To determine the transformation functions $V(\mathrm{P,A})$ and $W(\mathrm{P,A})$, we resort to the solution (6.302)–(6.305) for the circular cylinder. Noticing that (6.302) describes E_2 and (6.303) describes E_3, and replacing $\sqrt{\rho^2-a^2}$ with $\tau = |\mathrm{LP}|$, we obtain

$$V(\mathrm{P,A}) \approx \frac{\mathrm{e}^{-jk\tau}}{\sqrt{\tau}} T_1^{\mathrm{TE}}, \quad W(\mathrm{P,A}) \approx \frac{\mathrm{e}^{-jk\tau}}{\sqrt{\tau}} T_1^{\mathrm{TM}}, \tag{6.366}$$

where

$$T_1^{\mathrm{TE}} = \sqrt{\frac{2\pi}{k}} \frac{2m\mathrm{e}^{-jkl-j\xi t_1^{\mathrm{TE}}-j\frac{3}{4}\pi}}{w_2^2\left(t_1^{\mathrm{TE}}\right)\left(t_1^{\mathrm{TE}}+m^2\eta^2\right)} \tag{6.367}$$

$$T_1^{\mathrm{TM}} = \sqrt{\frac{2\pi}{k}} \frac{2m\mathrm{e}^{-jkl-j\xi t_1^{\mathrm{TM}}-j\frac{3}{4}\pi}}{w_2^2\left(t_1^{\mathrm{TM}}\right)\left(t_1^{\mathrm{TE}}+m^2/\eta^2\right)}. \tag{6.368}$$

By using (6.298), the two latter expressions can be also represented as

$$T_1^{\mathrm{TE}} = \sqrt{\frac{2\pi}{k}} \frac{2m\mathrm{e}^{-jkl-j\xi t_1^{\mathrm{TE}}+j\frac{\pi}{4}}}{\left[w_2'\left(t_1^{\mathrm{TE}}\right)\right]^2 - t_1^{\mathrm{TE}} w_2^2\left(t_1^{\mathrm{TE}}\right)} \tag{6.369}$$

$$T_1^{\mathrm{TM}} = \sqrt{\frac{2\pi}{k}} \frac{2m\mathrm{e}^{-jkl-j\xi t_1^{\mathrm{TM}}+j\frac{\pi}{4}}}{\left[w_2'\left(t_1^{\mathrm{TM}}\right)\right]^2 - t_1^{\mathrm{TM}} w_2^2\left(t_1^{\mathrm{TM}}\right)} \tag{6.370}$$

which have the advantage of including only finite parameters in the limits of $\eta \to 0$ and $\eta \to \infty$.

A physical interpretation of the solution results from representing T_1^{TE} and T_1^{TM} as

$$T_1 = D_1(\mathrm{L})\mathrm{e}^{-jkl-j\xi t_1} D_1(\mathrm{A}) \tag{6.371}$$

with

$$D_1(\mathrm{L})D_1(\mathrm{A}) = \sqrt{\frac{2\pi}{k}} \frac{2m\mathrm{e}^{j\frac{\pi}{4}}}{\left[w_2'(t_1)\right]^2 - t_1 w_2^2(t_1)} \tag{6.372}$$

and t_1 being either t_1^{TE} or t_1^{TM}. The quantities $D_1(\mathrm{A})$ and $D_1(\mathrm{L})$ can be referred to as the attachment and launching coefficients, as $D_1(\mathrm{A})$ describes the excitation of the leading creeping wave with $s = 1$ by the incident plane wave at the attachment point A, and $D_1(\mathrm{L})$ describes the re-radiation of the creeping wave at the launching point L toward the observer. For a circular cylinder with constant surface impedance, the attachment and launching coefficients must coincide because of the reciprocity. Finally, the exponential factor

$$\mathrm{e}^{-jkl-j\xi t_1} \tag{6.373}$$

in (6.371) describes the change of phase and the attenuation of the creeping wave on the way from A to L.

Representation (6.365) with (6.366) and (6.371) describes diffraction by smooth convex bodies in the framework of GTD (Keller 1956; Keller 1962; Levy and Keller 1959). The representation is explicitly built on the high-frequency mechanisms involved in the diffraction and scattering of electromagnetic waves by convex impedance surfaces and is therefore of great physical importance. GTD is a ray-based concept, which interprets diffraction by electrically large smooth convex bodies in terms of surface-diffracted rays that are excited on shadow boundaries, propagating along the shortest paths on the scattering surfaces before being launched to the observer. In the framework of GTD, excitation and launching of the surface-diffracted rays are assumed to be local processes in the sense that the attachment and launching coefficients are postulated to be functions of the local geometric and material properties of the surface at the attachment and launching points only. Attenuation of the surface-diffracted rays is interpreted as the result of energy losses due to the rays shedding off tangentially to the surface all along their propagation path on the surface.

Furthermore, the GTD format is of great heuristic power as it suggests a way of extending the high-frequency solutions to more general configurations. GTD derives its knowledge of the high-frequency diffraction and scattering mechanisms from interpreting exact solutions of diffraction problems for simple geometries in terms of various diffraction rays and corresponding local diffraction coefficients. It is then postulated that these mechanisms apply to more general geometries, for which no exact solution is available.

For example, the extension of the solution (6.365), (6.366), and (6.371) obtained for a circular impedance cylinder to generally shaped convex impedance bodies with varying curvature consists in the use of local values of the curvature radius R_i. The attachment and launching coefficients, which may differ on a generally curved surface, are determined with local values of R_i at the attachment and launching points as

$$D_1^2 = \sqrt{\frac{2\pi}{k}} \frac{2m e^{j\frac{\pi}{4}}}{\left[w_2'(t_1)\right]^2 - t_1 w_2^2(t_1)} \tag{6.374}$$

with $R_i(\mathrm{A})$ for $D_1(\mathrm{A})$ and $R_i(\mathrm{L})$ for $D_1(\mathrm{L})$. The phase and attenuation factor (6.373) becomes

$$\exp\left(-jkl - j\int_{\mathrm{A}}^{\mathrm{L}} \frac{t_1}{d} \mathrm{d}l\right), \tag{6.375}$$

where d and t_1 are calculated with R_i at the integration point, $\mathrm{d}l$ is the infinitesimal arc length, and l is the length of the geodesic between A and P,

$$l = \int_{\mathrm{A}}^{\mathrm{L}} \mathrm{d}l. \tag{6.376}$$

As a result, the field of the surface-diffracted ray is given by (6.365) with $V(\mathrm{P}, \mathrm{A})$ and $W(\mathrm{P}, \mathrm{A})$ having the form

$$\mathrm{e}^{-jk(\tau+l)} \sqrt{\frac{\chi}{\tau}} D_1(\mathrm{A}) D_1(\mathrm{L}) \exp\left(-j\int_{\mathrm{A}}^{\mathrm{L}} \frac{t_1}{d} \mathrm{d}l\right). \tag{6.377}$$

The factor χ accounts for the geometric properties of the scattering surface and is similar to the spread factor for waves with curved wavefronts. It follows from the conservation of the power flow within every strip bounded by two adjacent diffraction rays and is equal to the ratio of the strip width $\delta(\mathrm{A})$ between two adjacent rays at A to the strip width $\delta(\mathrm{L})$ between the same two rays at L in the limit of vanishing strip width,

$$\chi = \frac{\delta(\mathrm{A})}{\delta(\mathrm{L})}. \tag{6.378}$$

For circular cylinders, $\chi = 1$ as the distance between two adjacent rays is the same at every point on the propagation path of the creeping wave.

In obtaining the complete diffracted field, the contribution from all the rays that pass through the observation point must be taken into account, including the diffraction rays generated at the shadow boundary on the opposite side of the scatterer and the rays circumventing the scatterer before being launched to the observer (Figure 6.16). These diffraction rays should be added to the GO rays, which include the rays reflected from the scatterer (Figures 6.17 and 6.18).

GTD is a beautiful concept, enabling an appealing physical interpretation of the structure of high-frequency electromagnetic fields and in some cases simple numerical estimations. However, as we already saw at the end of section 6.4.3 with respect to the particular case of a PEC circular cylinder, GTD is based on the concept of rays and therefore it is inapplicable at shadow boundaries or caustics of the ray fields, which makes GTD unsuitable for practical calculations.

One approach that is commonly used nowadays in practical calculations of fields at finite distances around scattering bodies is the uniform theory of diffraction (UTD) (McNamara *et al.* 1990; Pathak 1979; Pathak *et al.* 1980). UTD describes the diffracted field by a single formula which is valid at the shadow boundaries and caustics while recovering GTD solutions away from these regions. A disadvantage of this method is the absence of a universal description for all types of caustics, which leads to the need to derive a separate uniform solution for every specific caustic type, which is not always possible.

Another approach to calculation of the high-frequency scattered fields everywhere in space, and particularly in the far-field region and around caustics, is the use of the Huygens principle (section 2.6) with an integration surface chosen so that a high-frequency approximation for the tangential field components be applicable on the whole surface of integration (the surface of the scatterer or a mathematical surface enclosing the scatterer). This approach is called physical optics (PO) when the fields on the integration surface are determined in the GO approximation. The PO method, including its application to smooth electrically large convex bodies, is the subject of Chapters 8 and 9.

It is also possible to integrate more accurate asymptotic solutions. For example, for convex bodies, the Fock-type penumbra solution supplemented with the GO solution on the lit portion of the scatterer can be used to describe the field on the whole integration surface and, on integration, everywhere in space (e.g., Osipov *et al.* (1998)). However, this approach, which is sometimes called the method of asymptotic currents (e.g., Molinet *et al.* (2005)), requires a time-consuming two-dimensional integration, so in practice it has so far been limited to simple scatterers.

For non-convex shapes, for example a cavity or concave reflector, neither the Fock-type solutions nor the GTD representations apply, as an observer on the concave portion of the scattering surface may be illuminated not only by the direct incident wave but also by the waves reflected at adjacent portions of the scattering surface, implying that the field is no more local in nature. New types of asymptotic solutions have to be accounted for, most notably the whispering gallery wave which occurs near the concave surface as a result of the superposition of waves multiply reflected at the surface (e.g., see Babič and Buldyrev (1991) and Bouche *et al.* (1997)).

6.7 Problems

1. Is it possible to solve the problem of plane-wave scattering by a PEC cylinder as a scalar problem in terms of E- and H-polarized waves, if the wave is incident obliquely with respect to the cylinder axis?

   Answer: Yes.

   Solution: The oblique incidence of a plane wave means the dependence of the fields on the z coordinate as $\exp(-jk_z z)$ (section 2.5.5), where z is the coordinate along the axis of the cylinder and k_z is the z component of the wave vector in the incident wave. Maxwell's equations reduce to the Helmholtz equations (2.317) and (2.318) for the components E_z and H_z and to the equations (2.313)– (2.316), which express the other components of the field through E_z and H_z. It follows from (2.313)–(2.316) that the boundary conditions $E_z = 0$ and $E_\varphi = 0$ on the surface of the PEC cylinder are satisfied if E_z and H_z are subject to the boundary conditions $E_z = 0$ and $\partial H_z/\partial\rho = 0$. The conditions at infinity ($\rho \to \infty$) are also to be separately satisfied by E_z and H_z. Thus, the scattering problem is reduced to two decoupled scalar problems for E- and H-polarized waves.

2. The same question as above but for an isotropic magneto-dielectric cylinder.

   Answer: No.

3. Study the frequency dependence for the scattering width of a silver cylinder of radius 50 nm in the visible frequency range (normal plane-wave incidence, H polarization).

   Hint: Data on the electromagnetic properties of silver in the visible can be found in section 2.2.2.

4. Show that the backscattering width of an electrically small ($ka \to 0$) circular PEC cylinder under normal illumination behaves as

   $$\frac{\sigma}{\pi a} \approx \frac{\pi}{ka \ln^2 ka}$$

   for E polarization and as

   $$\frac{\sigma}{\pi a} \approx \frac{9}{4}\pi(ka)^3$$

   for H polarization.

   Hint: Use approximations (6.37) and (6.45) for the scattered field.

5. Generalize the solution given in section 6.2 for ideally conducting cylinders to the case of plane waves incident at an oblique angle with respect to the cylinder axis.

Solution: Assume that the incident field is described by (6.200) and (6.201) with ψ being the angle between the incidence direction and the plane perpendicular to the axis of the cylinder. The problem can be split up into two decoupled problems for E_z and H_z (see problem 1).

Function E_z depends on z as $\exp(-jkz\sin\psi)$, is excited by a plane wave with the amplitude $E_0\cos\psi$, satisfies the Helmholtz equation with $k_\perp = k\cos\psi$, and vanishes at $\rho = a$, where a is the radius of the cylinder. Thus, the solution of this problem can be obtained from the solution for E polarization and normal incidence by multiplying (6.24) with $E_0\cos\psi\exp(-jkz\sin\psi)$ and substituting $k \to k_\perp$ in (6.24) and (6.25), which gives for the scattered field:

$$E_z^{\rm sc}(\rho,\varphi) = -E_0\cos\psi e^{-jkz\sin\psi}\sum_{n=-\infty}^{\infty}\frac{J_n(k_\perp a)}{H_n^{(2)}(k_\perp a)}H_n^{(2)}(k_\perp\rho)e^{-jn(\varphi-\pi/2)}. \tag{6.379}$$

Function H_z results from the solution for the normal incidence upon the substitution $k \to k_\perp$ in (6.24) and (6.26) and multiplication of (6.24) by $H_0\cos\psi\exp(-jkz\sin\psi)$, which leads to the scattered field

$$H_z^{\rm sc}(\rho,\varphi) = -H_0\cos\psi e^{-jkz\sin\psi}\sum_{n=-\infty}^{\infty}\frac{J_n'(k_\perp a)}{H_n^{(2)\prime}(k_\perp a)}H_n^{(2)}(k_\perp\rho)e^{-jn(\varphi-\pi/2)} \tag{6.380}$$

with the prime denoting differentiation with respect to the argument of the functions.

6. What is the minimum electrical size of the cross section of a dielectric cylinder, such that resonant phenomena can be observed in its scattering response?

   Answer: The radius can be arbitrarily small if one of the material parameters of the cylinder has the real part different in sign to the corresponding parameter of the surrounding space.

7. Consider a cylinder of an isotropic material with the parameters ϵ_2 and μ_2 illuminated by an H-polarized plane wave (normal incidence). Assume the radius to be electrically small and find the amplitude and spatial distribution of the circulating electric field induced inside the cylinder.

   Solution: Due to the electrically small size and the problem symmetry, circulating electric field lines will be circles, that is, we are interested in the part of the φ component of the electric field inside the cylinder that does not depend on φ. In (6.57) this corresponds to the term A_0. Estimating the value of A_0 for electrically small cylinders we find that $A_0 \approx 1$, with the next expansion term of the order of $O[(k_1a)^2\ln(k_1a)]$. Neglecting these higher-order terms, we find from (6.57)

$$E_\varphi = \frac{1}{j\omega\epsilon_2}\frac{\partial u_{\rm tot}}{\partial\rho} \approx -j\omega\mu_2\frac{\rho}{2}. \tag{6.381}$$

   This can be considered as the quasi-static circulating electric field induced by the time-harmonic magnetic field H_z of unit amplitude.

8. Find the electromagnetic fields created by external current I flowing along an infinitely long ideally conducting cylinder of radius ρ_0 covered by an isotropic material cylinder of radius a. The material parameters of the cylinder are ϵ and μ. The current is uniform along the cylinder axis and the surrounding medium is free space.

   Hint: The boundary condition on the surface of the current line can be written as

$$2\pi\rho_0 H_\varphi|_{\rho=\rho_0} = I,$$

   where H_φ is the φ component of the magnetic field (it does not depend on φ) and I is the current in the exciting line.

   Solution: Using the assumed cylindrical symmetry we write for the electric field:

$$E_z = \begin{cases} AH_0^{(2)}(k\rho) + BH_0^{(1)}(k\rho), & \rho_0 \le \rho \le a \\ CH_0^{(2)}(k_0\rho), & \rho \ge a \end{cases}$$

and for the magnetic field:

$$H_\varphi = -\frac{j}{\omega\mu}\frac{\partial E_z}{\partial\rho} = \begin{cases} \frac{jk}{\omega\mu}\left[AH_1^{(2)}(k\rho) + BH_1^{(1)}(k\rho)\right], & \rho_0 \le \rho \le a \\ \frac{jk_0}{\omega\mu_0}CH_1^{(2)}(k_0\rho), & \rho \ge a \end{cases}.$$

Here (ρ, φ, z) form a usual set of cylindrical coordinates.

On the surface $\rho = a$ we have the continuity boundary conditions for the E_z and H_φ field components:

$$AH_0^{(2)}(ka) + BH_0^{(1)}(ka) = CH_0^{(2)}(k_0a)$$
$$\frac{\mu_0 k}{\mu k_0}\left[AH_1^{(2)}(ka) + BH_1^{(1)}(ka)\right] = CH_1^{(2)}(k_0a).$$

The relation between the wire current I and the magnetic field H_φ at the wire surface $\rho = \rho_0$ gives

$$2\pi j\frac{k\rho_0}{\omega\mu}\left[AH_1^{(2)}(k\rho_0) + BH_1^{(1)}(k\rho_0)\right] = I.$$

From the last three algebraical equations we determine the unknown constants A, B, and C.

9. Assume that in the previous problem the radius of the exciting cylinder is electrically small: $|k|\rho_0 \ll 1$. Find the excited fields in this case.

Hint: Use the following approximation for the Hankel function of small arguments:

$$H_1^{(1,2)}(z) \approx \mp\frac{2j}{\pi z}.$$

Answer:

$$A = -\frac{\omega\mu I}{8}\frac{H_1^{(1)}(ka)H_0^{(2)}(k_0a) - \frac{\mu k_0}{\mu_0 k}H_0^{(1)}(ka)H_1^{(2)}(k_0a)}{J_1(ka)H_0^{(2)}(k_0a) - \frac{\mu k_0}{\mu_0 k}J_0(ka)H_1^{(2)}(k_0a)}$$
$$C = -\frac{j\omega\mu I}{4}\frac{J_0(ka)Y_1(ka) - J_1(ka)Y_0(ka)}{J_1(ka)H_0^{(2)}(k_0a) - \frac{\mu k_0}{\mu_0 k}J_0(ka)H_1^{(2)}(k_0a)}$$
$$B = A + \frac{\omega\mu}{4}I.$$

10. Assume that in the previous problem also the radius of the covering cylinder is electrically small: $k_0a \ll 1$, $|k|a \ll 1$. Find the excited fields and the power radiated by the current line per unit length.

Hint: Use the following approximations of Bessel functions with small arguments:

$$J_0(z) \approx 1, \quad J_1(z) \approx \frac{z}{2}, \quad Y_0(z) \approx \frac{2}{\pi}\ln\frac{\gamma' z}{2}, \quad Y_1(z) \approx -\frac{2}{\pi z}$$

(here $\gamma' = e^\gamma \approx 1.781$ with γ being the Euler constant).

Answer:

$$A = -\frac{\omega I}{8}\left[\mu_0 + \mu + \frac{2j}{\pi}\left(\mu\ln\frac{\gamma' ka}{2} - \mu_0\ln\frac{\gamma' k_0 a}{2}\right)\right]$$
$$B = -\frac{\omega I}{8}\left[\mu_0 - \mu + \frac{2j}{\pi}\left(\mu\ln\frac{\gamma' ka}{2} - \mu_0\ln\frac{\gamma' k_0 a}{2}\right)\right]$$
$$C = -\frac{\omega\mu_0 I}{4}.$$

The radiated power per unit length is equal to $\omega\mu_0|I|^2/4$ and does not depend on the material parameters of the covering cylinder.

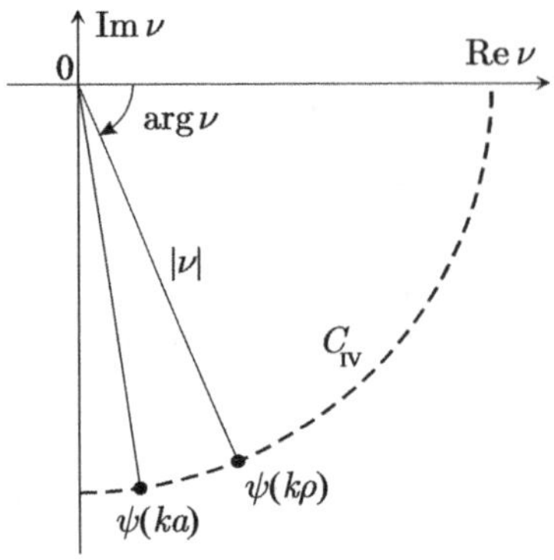

Figure 6.37 Subdivision of the integration path $C_{\rm IV}$.

11. Prove that when $|\varphi| > \pi/2$ and the radius of the quarter-circle $C_{\rm IV}$ (Figure 6.37) goes to infinity,

$$\left|\int_{C_{\rm IV}} \frac{\cos[\nu(\pi-|\varphi|)]}{\sin\nu\pi} A(\nu)\,\mathrm{d}\nu\right| \to 0 \tag{6.382}$$

with $A(\nu)$ defined in (6.73) and (6.75).

Hint: Subdivide $C_{\rm IV}$ into two arcs, C_1 with $\psi(k\rho) \le \arg\nu < 0$ and C_2 with $-\pi/2 \le \arg\nu < \psi(k\rho)$, where $\psi(k\rho)$ is a solution of the equation

$$\psi\tan\psi = \ln\frac{2|\nu|}{ex} \tag{6.383}$$

with $x = k\rho$ and $\nu \in C_{\rm IV}$, and estimate the asymptotic behavior of the integrand by using (6.73) on C_1 and (6.75) on C_2.

Solution: Denote for brevity $\psi = \arg\nu$. On the arc C_1 we use representation (6.73) for $A(\nu)$ and estimate the terms by using (6.82) and (6.84) as

$$|J_\nu(k\rho)| \approx \frac{1}{\sqrt{2\pi|\nu|}}\exp\left[|\nu|\left(\psi\sin\psi - \cos\psi\ln\frac{2|\nu|}{ek\rho}\right)\right] \tag{6.384}$$

$$\left|\frac{\kappa J_\nu(ka)}{\kappa H_\nu^{(2)}(ka)}H_\nu^{(2)}(k\rho)\right| \approx \frac{1}{\sqrt{2\pi|\nu|}}\exp\left[|\nu|\left(\psi\sin\psi - \cos\psi\ln\frac{2|\nu|}{ek\rho} + 2\cos\psi\ln\frac{a}{\rho}\right)\right].$$

With $|\nu| \to \infty$, both terms vanish as $O(|\nu|^{-1/2})$ or faster on C_1 because $\rho \ge a$ and $\psi \ge \psi(k\rho)$. This means that the part of the integral (6.382) taken over the arc C_1 is guaranteed to vanish if $|\varphi| > \pi/2$, thanks to the remaining trigonometric functions in the integrand, which are estimated by

$$\left|\frac{\cos[\nu(\pi-|\varphi|)]}{\sin\nu\pi}e^{j\nu\frac{\pi}{2}}\right| \approx e^{|\nu|\sin\psi(|\varphi|-\frac{\pi}{2})}. \tag{6.385}$$

The contribution from the arc C_2 requires a separate estimation since $\psi < \psi(k\rho)$ on C_2 and $J_\nu(k\rho)$ grows (see (6.383) and (6.384)). Furthermore, it follows from (6.383) that

$$\psi(k\rho) \approx -\frac{\pi}{2} + \frac{\pi}{2\ln[2|\nu|/(ek\rho)]} \tag{6.386}$$

so that $\psi(k\rho) \to -\pi/2$ and the angular width of C_2 vanishes as $O(1/\ln|\nu|)$ in the limit $|\nu| \to \infty$. However, the linear size of the arc grows as $O(|\nu|/\ln|\nu|)$, and it is therefore necessary to ensure that the integrand in (6.382) vanishes uniformly on C_2 for $|\nu| \to \infty$.

It is convenient to use the alternative representation (6.75) for $A(\nu)$ on C_2 because in contrast to $J_\nu(k\rho)$, function $H_\nu^{(1)}(k\rho)$ is estimated by

$$|H_\nu^{(1)}(k\rho)| \approx \sqrt{\frac{2}{\pi|\nu|}} \exp\left[|\nu|\left(-\psi \sin\psi + \cos\psi \ln \frac{2|\nu|}{ek\rho}\right)\right] \tag{6.387}$$

(see (6.83)) and vanishes uniformly on C_2 since $\psi \sin\psi > \cos\psi \ln[2|\nu|/(ek\rho)]$ with $\psi < \psi(k\rho)$. The second term in (6.75) is estimated by

$$\left|\frac{\kappa H_\nu^{(1)}(ka)}{\kappa H_\nu^{(2)}(ka)} H_\nu^{(2)}(k\rho)\right| \approx \frac{2}{\pi|\nu|}\left(\frac{\rho}{a}\right)^{|\nu|\cos\psi} \frac{1}{|\kappa H_\nu^{(2)}(ka)|}. \tag{6.388}$$

The analytical form of the asymptotic approximation for $H_\nu^{(2)}(ka)$ (see (6.84)) depends on whether ψ is greater or smaller than $\psi(ka)$, where $\psi(ka)$ is a solution of (6.383) with $x = ka$ from the fourth quadrant on the complex ν plane. Note that the point $\nu = |\nu|\exp[j\psi(ka)]$ is the intersection of the arc C_{IV} with the line of zeros ν_s of $H_\nu^{(2)}(ka)$ and $\psi(ka) \le \psi(k\rho)$ (Figure 6.37) because of $a \le \rho$. When $\psi \ne \psi(ka)$, the function $H_\nu^{(2)}(ka)$ grows, which permits the estimation

$$\left|\frac{\kappa H_\nu^{(1)}(ka)}{\kappa H_\nu^{(2)}(ka)} H_\nu^{(2)}(k\rho)\right| \le O(\mathrm{e}^{|\nu|\cos\psi \ln\frac{\rho}{a}}) \tag{6.389}$$

which grows as $|\nu| \to \infty$ but at a vanishing rate since $\psi \to -\pi/2$. In the integrand of (6.382) this term is multiplied with (6.385), which leads to an exponential decay of the integrand and, therefore, of the integral, once $|\varphi| > \pi/2$.

12. By using Debye's asymptotic representations of cylindrical functions (Appendix E), show that (a) the creeping wave series (6.90) converges when $|\varphi| - \pi/2 = \epsilon$ with $\epsilon > 0$ and (b) its terms begin to decay once $|\nu_s|$ are such that

$$|\nu_s| > ka\left(\frac{\rho}{a}\right)^{\pi/(2\epsilon)}. \tag{6.390}$$

Hint: Zeros ν_s are defined by (6.77) and approximately located on the line described by the equation

$$\psi \sin\psi = \cos\psi \ln \frac{2|\nu|}{eka}, \tag{6.391}$$

where $|\nu| \gg ka$, $\psi = \arg \nu$, and $-\pi/2 \le \psi < 0$. On this line, $H_\nu^{(1)}(ka)$ and $H_\nu^{(2)}(ka)$ are oscillating with amplitudes that are $O(|\nu|^{-1/2})$, which is inessential for the convergence of the series.

Solution: Denote the sth term of (6.90) by q_s. Estimating the factors in the term, we note that $J_{\nu_s}(ka) = H_{\nu_s}^{(1)}(ka)/2$ and thus

$$\left|\frac{\kappa J_\nu(ka)}{(\partial/\partial\nu)\left[\kappa H_\nu^{(2)}(ka)\right]}\right|_{\nu=\nu_s} = O(1). \tag{6.392}$$

The behavior of q_s with $|\nu_s| \gg ka$ is therefore determined by the trigonometric factor, which is estimated in (6.385), and by the Hankel function $H_{\nu_s}^{(2)}(k\rho)$, which is estimated by the formula

$$\left|H_{\nu_s}^{(2)}(k\rho)\right| \approx \sqrt{\frac{2}{\pi|\nu_s|}} \exp\left[|\nu_s|\left(\psi \sin\psi - \cos\psi \ln \frac{2|\nu_s|}{ek\rho}\right)\right], \tag{6.393}$$

which follows from (6.84) with $\nu = \nu_s$. Because of (6.391), this approximation can be rewritten as

$$\left|H_{\nu_s}^{(2)}(k\rho)\right| \approx \sqrt{\frac{2}{\pi|\nu_s|}} \exp\left[|\nu_s| \cos\psi \ln \frac{\rho}{a}\right] \tag{6.394}$$

and leads to the estimation

$$|q_s| = O\left\{\exp\left[|\nu_s| \sin\psi \left(|\varphi| - \frac{\pi}{2}\right) + |\nu_s| \cos\psi \ln\frac{\rho}{a}\right]\right\}. \tag{6.395}$$

The approximate solution of (6.391) is

$$\psi \approx -\frac{\pi}{2} + \frac{\pi}{2\ln[2|\nu|/(eka)]} \tag{6.396}$$

and therefore

$$\sin\psi \to -1, \quad \cos\psi \approx \frac{\pi}{2\ln[2|\nu|/(eka)]} \to 0, \tag{6.397}$$

implying the decay of (6.395) if $|\varphi| > \frac{\pi}{2}$. The decay starts with ν_s such that

$$|\varphi| - \frac{\pi}{2} \geq \frac{\pi \ln(\rho/a)}{2\ln[2|\nu_s|/(eka)]}, \tag{6.398}$$

which results in the estimation (6.390). Thus, the closer the observer to the surface of the cylinder ($\rho = a$) and the deeper the observer in the shadow, the faster the convergence of the creeping wave series.

13. Coefficients in the creeping wave expansion (6.90) include the function Q_s given by (6.137). Show that if $ka \gg 1$, the factor can be represented by either

$$Q_s \approx \frac{m e^{-j5\pi/6}}{2\pi[\mathrm{Ai}'(t_s e^{-j2\pi/3})]^2} \tag{6.399}$$

when $\kappa = 1$ (E polarization, PEC cylinder), or by

$$Q_s \approx \frac{m e^{j5\pi/6}}{2\pi t_s[\mathrm{Ai}(t_s e^{-j2\pi/3})]^2} \tag{6.400}$$

when $\kappa = \partial/\partial(ka)$ (H polarization, PEC cylinder). Here, $\mathrm{Ai}(z)$ is the Airy function (6.126),

$$t_s = \frac{\nu_s - ka}{m}, \tag{6.401}$$

$m = (ka/2)^{1/3}$, and ν_s is a root of the equation $\kappa H^{(2)}_{\nu_s}(ka) = 0$ from the vicinity of the point $\nu = ka$, in which $|\nu_s - ka| \ll ka$.

Hint: Use the Wronskian relation (E.9),

$$H^{(1)}_\nu(ka)H^{(2)\prime}_\nu(ka) - H^{(1)\prime}_\nu(ka)H^{(2)}_\nu(ka) = \frac{4}{j\pi ka}, \tag{6.402}$$

the asymptotic approximation (E.162) for the Hankel function $H^{(2)}_\nu(ka)$ with $\nu \approx ka$, and the differential equation

$$\mathrm{Ai}''(z) = z\mathrm{Ai}(z) \tag{6.403}$$

satisfied by the Airy function.

Solution: Let us start with the case of $\kappa = 1$. Then, ν_s are zeros of $H^{(2)}_\nu(ka)$, and (6.402) with $\nu = \nu_s$ implies that

$$H^{(1)}_{\nu_s}(ka) = \frac{4}{j\pi ka H^{(2)\prime}_{\nu_s}(ka)}, \tag{6.404}$$

so that

$$Q_s = \frac{4}{j\pi ka}\left(H^{(2)\prime}_{\nu_s}(ka)\left.\frac{\partial}{\partial\nu}H^{(2)}_{\nu}(ka)\right|_{\nu=\nu_s}\right)^{-1}. \tag{6.405}$$

The zeros ν_s are located in the vicinity of the point $\nu = ka$, and if $ka \gg 1$ we can approximate the Hankel functions with (E.162), which gives

$$\left.\frac{\partial}{\partial\nu}H^{(2)}_{\nu}(ka)\right|_{\nu=\nu_s} \approx -H^{(2)\prime}_{\nu_s}(ka) \approx \frac{2^{5/3}}{(ka)^{2/3}}\mathrm{e}^{-j\pi/3}\mathrm{Ai}'\left(t_s\mathrm{e}^{-2j\pi/3}\right) \tag{6.406}$$

and leads to the formula (6.399).

In the alternative case of $\kappa = \partial/\partial(ka)$, ν_s are zeros of $H^{(2)\prime}_{\nu}(ka)$ and it follows from (6.402) that

$$H^{(1)\prime}_{\nu_s}(ka) = \frac{4j}{\pi ka H^{(2)}_{\nu_s}(ka)}. \tag{6.407}$$

For the zeros ν_s that are located close to the point $\nu = ka$, (E.162) applies and leads to

$$\left.\frac{\partial}{\partial\nu}\kappa H^{(2)}_{\nu}(ka)\right|_{\nu=\nu_s} \approx \frac{4}{ka}\mathrm{Ai}''\left(t_s\mathrm{e}^{-2j\pi/3}\right) = \frac{4t_s}{ka}\mathrm{e}^{-2j\pi/3}\mathrm{Ai}\left(t_s\mathrm{e}^{-2j\pi/3}\right), \tag{6.408}$$

which together with (6.407) transforms (6.137) to (6.400).

14. Evaluate the expressions (6.369) and (6.370) for the transformation coefficients T_1^{TE} and T_1^{TM} in the limit of a PEC cylinder and show that they recover the expression (6.147).

Hint: In (6.147) neglect the terms with $s \geq 2$, use (6.399) (TM polarization) or (6.400) (TE polarization) for Q_s, and account for the relations

$$\mathrm{Ai}(t\mathrm{e}^{-j2\pi/3}) = \frac{\mathrm{e}^{j\pi/6}}{2\sqrt{\pi}}w_2(t), \quad \mathrm{Ai}'(t\mathrm{e}^{-j2\pi/3}) = \frac{\mathrm{e}^{j5\pi/6}}{2\sqrt{\pi}}w_2'(t). \tag{6.409}$$

15. Derive approximations (6.185) and (6.186) for the Fock functions $F(\xi)$ and $G(\xi)$ with $\xi \to -\infty$.

Hint: In the integrals (6.178) and (6.179) replace the normalized Airy functions with their large argument approximations (6.174) and apply the steepest descent method to evaluate the contribution from the saddle point $t = -\xi^2$.

16. Derive approximations (6.286) and (6.287) of the Fock function $G(\xi, p)$ for $\xi \to \pm\infty$.

Hint: To obtain (6.286), use the large argument approximations of $w(t)$ and $w'(t)$ and apply the steepest descent method to the integral representation (6.285). The deep shadow representation (6.287) is obtained as the residue of (6.285) at the pole $t = t_1$ on accounting for the relations $w_2''(t) = tw_2(t)$ (differential equation for Airy functions) and $w_2'(t_1) = pw_2(t_1)$.

17. Consider the function

$$F(n, x_1, x_2) = \frac{J_n'(x_1)H_n^{(2)\prime}(x_2) - J_n'(x_2)H_n^{(2)\prime}(x_1)}{J_n(x_1)H_n^{(2)}(x_2) - J_n(x_2)H_n^{(2)}(x_1)} \tag{6.410}$$

with $x_2 \to x_1$ and show that

$$F(n, x_1, x_2) \to 1 - \frac{n^2}{x_1^2}. \tag{6.411}$$

Solution: Both the numerator and the denominator vanish in the limit, so we have indeterminacy of type 0/0. Representing the Bessel functions by the Taylor expansions

$$J_n(x_2) = J_n(x_1) + J_n'(x_1)(x_2 - x_1) + \dots$$
$$J_n'(x_2) = J_n'(x_1) + J_n''(x_1)(x_2 - x_1) + \dots$$
$$H_n^{(2)}(x_2) = H_n^{(2)}(x_1) + H_n^{(2)'}(x_1)(x_2 - x_1) + \dots$$
$$H_n^{(2)'}(x_2) = H_n^{(2)'}(x_1) + H_n^{(2)''}(x_1)(x_2 - x_1) + \dots$$

eliminates the indeterminacy and leads to the expression

$$F(n, x_1, x_1) = \frac{J_n'(x_1)H_n^{(2)''}(x_1) - J_n''(x_1)H_n^{(2)'}(x_1)}{J_n(x_1)H_n^{(2)'}(x_1) - J_n'(x_1)H_n^{(2)}(x_1)}, \tag{6.412}$$

which is further simplified by using the Bessel differential equation (E.6) to exclude the second-order derivatives from the numerator.

18. The solution for scattering from a PEC cylinder is a special case of the general solution for a coated PEC cylinder in the limit of the vanishing thickness of the coating (section 6.5.1). Derive expressions (6.237) from the general formulas (6.226)–(6.236).

Hint: Use (6.411) to show that δ_n goes to a finite limit as $x_2 \to x_1$.

19. Check the solutions for a coated PEC cylinder and for a homogeneous cylinder (section 6.5.1) on their compliance with the duality property of Maxwell's equations. Explain the result for the coated cylinder.

Answer: In the case of the coated cylinder, interchanging $\epsilon \leftrightarrow \mu$ and $Z_0 \leftrightarrow 1/Z_0$ does not transform the expressions for E_z and H_z for one polarization into those for H_z and $-E_z$ for the alternative polarization. The interchange leads to a solution for a cylinder with perfectly *magnetically* conducting core.

20. Find the dispersion relation for surface waves propagating along the circular impedance cylinder in the case of electrically large cylinder radii.

Solution: Surface waves are eigenwaves (bound to surfaces) that can exist in the absence of sources. To find the dispersion relation we can use the exact solution for plane-wave scattering from an impedance cylinder (6.265)–(6.266). Nonzero fields in the absence of sources can exist only if these coefficients are infinite. Setting $n = 0$ (waves propagate along the cylinder axis), we see that the denominators are zero if $\tilde{N}_0 = 0$ or $\tilde{P}_0 = 0$ (expressions (6.269) and (6.270)). Replacing the Hankel functions and their derivatives with the corresponding large-argument approximations (6.276) and (6.277) we find the simple relations

$$1 + \frac{\eta}{\cos\psi} = 0, \qquad 1 + \frac{1}{\eta\cos\psi} = 0.$$

Here

$$\cos\psi = \frac{\sqrt{k^2 - k_z^2}}{k}$$

is an imaginary number because for surface waves the propagation constant $k_z > k$, so that the propagation constant along the cylinder radius is imaginary, ensuring exponentially decaying fields away from the cylinder. Obviously, surface waves can exist only for imaginary η (lossless material of the cylinder). Substituting the value of $\cos\psi$ in terms of the propagation constants into the above relations with η gives dispersion relations between k_z and k. They can be seen to coincide with the dispersion relations for surface waves over planar impedance surfaces (e.g., see Tretyakov (2003)).

References

Abramowitz M and Stegun IA (eds) 1965 *Handbook of Mathematical Functions with Formulas, Graphs, and Mathematical Tables*, Dover Publications.

Babič VM and Buldyrev VS 1991 *Short-Wavelength Diffraction Theory*, Springer-Verlag.

Borovikov VA and Kinber BY 1994 *Geometrical Theory of Diffraction*, The Institution of Electrical Engineers, London.

Bouche D, Molinet F, and Mittra R 1997 *Asymptotic Methods in Electromagnetics*, Springer-Verlag.

Bouman JJ, Senior TBA, and Uslenghi PLE (eds) 1987 *Electromagnetic and Acoustic Scattering by Simple Shapes*, Hemisphere Publishing Corporation.

Bremmer H 1949 *Terrestrial Radio Waves*, Elsevier.

Cullen SA 1958 Surface currents induced by short-wavelength radiation. *Phys. Rev.*, **109**, 1863–1867.

Felsen LB and Marcuvitz N 1973 *Radiation and Scattering of Waves*, Prentice-Hall.

Fock VA 1965 *Electromagnetic Diffraction and Propagation Problems*, Pergamon Press.

Franz W and Deppermann K 1952 Theorie der Beugung am Zylinder unter Berücksichtigung der Kriechwelle. *Ann. Phys. (Berlin)*, **10**, 361–373.

Goriainov AS 1958 An asymptotic solution of the problem of diffraction of a plane electromagnetic wave by a conducting cylinder. *Radio Eng. Electron. (USSR)*, **3**, 23–39.

James GL 1976 *Geometrical Theory of Diffraction for Electromagnetic Waves*, Peter Peregrinus, London.

Johnson PB and Christy RW 1972 Optical constants of noble metals. *Phys. Rev. B*, **6**, 4370–4379.

Keller JB 1956 Diffraction by a convex cylinder. *IRE Trans. Antennas Propag.*, **24**, 312–321.

Keller JB 1962 Geometrical theory of diffraction. *J. Opt. Soc. Am.*, **52**, 116–130.

Kerker M and Matijevich E 1961 Scattering of electromagnetic waves from concentric infinite cylinders. *J. Opt. Soc. Am.*, **51**, 506–508.

King RWP and Wu TT 1959 *The Scattering and Diffraction of Waves*, Harvard University Press, Cambridge, MA.

Landau LD, Pitaevskii LP, and Lifshitz EM 1984 *Electrodynamics of Continuous Media (Course of Theoretical Physics, Volume 8)*, Second edition, Butterworth-Heinemann.

Levy BR and Keller JB 1959 Diffraction by a smooth object. *Comm. Pure Appl. Math.*, **12**, 169–209.

Makarov GI, Novikov VV, and Rybachek ST 1991 *Electromagnetic Waves Propagation over Earth's Surface*, Nauka, Moscow (in Russian).

McNamara DA, Pistorius CWI, and Malherbe JAG 1990 *Introduction to the Uniform Geometrical Theory of Diffraction*, Artech House.

Molinet F, Andronov I, and Bouche D 2005 *Asymptotic and Hybrid Methods in Electromagnetics*, The Institution of Electrical Engineers, London.

Osipov AV, Kobayashi H, and Hongo K 1998 Shadow boundary currents in the problem of high-frequency electromagnetic diffraction by a circular impedance cylinder. *IEICE Trans. Electron.*, **E81-C**, 1655–1666.

Pathak PH 1979 An asymptotic analysis of the scattering of plane waves by a smooth convex surface. *Radio Sci.*, **14**, 419–435.

Pathak PH, Burnside WD, and Marhefka RJ 1980 A uniform GTD analysis of the diffraction of electromagnetic waves by a smooth convex surface. *IEEE Trans. Antennas Propag.*, **28**, 631–642.

Pearson LW 1987 A scheme for automatic computation of Fock-type integrals. *IEEE Trans. Antennas Propag.*, **35**, 1111–1118.

Ruck GT, Barrick DE, Stuart WD, and Krichbaum CK 1970 *Radar Cross Section Handbook*, Plenum Press.

Ryshik IM and Gradstein IS 1957 *Tables of Series, Products, and Integrals*, VEB, Berlin.

Strutt JW (Lord Rayleigh) 1881 On the electromagnetic theory of light. *Philos. Mag. Ser. 5*, **12**, 81–101.

Thomson JJ 1893 *Notes on Recent Researches in Electricity and Magnetism*, Clarendon Press, Oxford.

Tretyakov SA 2003 *Analytical Modeling in Applied Electromagnetics*, Artech House.

van de Hulst HC 1957 *Light Scattering by Small Particles*, John Wiley & Sons.

Wait JR 1955 Scattering of a plane wave from a circular dielectric cylinder at oblique incidence. *Can. J. Phys.*, **33**, 189–195.

Wait JR 1959 *Electromagnetic Radiation from Cylindrical Structures*, Pergamon Press.

Watson GN 1918 The diffraction of electric waves by the Earth. *Proc. R. Soc. Lond. A*, **95**, 83–99.

7

Spheres

7.1 Introduction

Scattering of a plane electromagnetic wave by a spherical object is perhaps the most important canonical problem in electromagnetic scattering theory. The sphere serves as a model of a compact smooth scatterer and is one of only two compact geometries (another one is the disc) for which Maxwell's equations can be solved in an explicit form. An exact solution is available not only for PEC but also for material, homogeneous or radially stratified, magneto-dielectric spheres. The fields are expressed in terms of well-known special functions (Bessel functions and associated Legendre polynomials) and can be precisely calculated. Therefore the solution is widely used for validation of other simulation approaches, numerical (like FDTD, FEM, IEs) and approximate (GO, PO). Spherical scatterers serve as calibration targets in radar applications as well as models of rain drops in optics or of the Earth surface in radio wave propagation studies. Quite recently, electromagnetic properties of small spheres have begun attracting attention as such spheres can be used as an important constitutive part in designing advanced materials.

Wave mechanisms governing electromagnetic scattering from spheres are diverse and numerous, particularly when absorption in the sphere is low, and the mechanisms include, for example, resonant absorption and scattering in electrically small spheres and propagation of creeping waves and multiply reflected GO rays in electrically large spheres. These mechanisms are of general value as they may explain scattering from more general scatterers for which no exact solution exists. The exact solution for the sphere includes all these mechanisms, and this is why most of the knowledge about diffraction and scattering of electromagnetic waves by compact material bodies has been gained from solutions for spheres, which are, therefore, of great theoretical and practical importance.

For a detailed and, at the same time, fascinating historical survey of the quest for the solution of the problem of electromagnetic scattering from a sphere, the reader is referred to Logan (1965). We indicate here only the cornerstones. The mathematical apparatus (separation of variables in the vector wave equation in spherical coordinates) is due to Clebsch (1863). An exact solution for a plane wave incident on a PEC sphere was presented in Thomson (1893, pp. 437–452). The first exact solution for a dielectric sphere was published by Lorenz (1890) but remained unnoticed, and equivalent solutions were published later on by other authors (Debye 1909b; Love 1899; Mie 1908; Walker 1900). The work by Mie (1908), although limited to non-magnetic spheres, has become particularly popular in the scientific community, apparently because of its clarity, practical value (the work explained the diverse coloring of colloids of metals), and its completeness as the paper presented both theory and numerical estimations. The solution for a homogeneous sphere is therefore often referred to as the Mie solution, although from the historical point of view it seems to be wrong.

An exact solution for two concentric spheres is attributed to Aden and Kerker (1951). The paper by Güttler (1952) published shortly afterward is limited to the case of non-magnetic materials but presents a quasi-static approximation to the solution. The solution for an arbitrary number of concentric layers is described in Wait (1963).

The problems of electromagnetic scattering from spheres have been extensively studied in the past in the context of light scattering from small particles, most notably from water drops (rainbow), and radio wave propagation over the Earth surface. In the latter case, a point source (mostly a vertical electric dipole) and an observer located on or near the

Modern Electromagnetic Scattering Theory with Applications, First Edition. Andrey V. Osipov and Sergei A. Tretyakov.

surface are assumed (e.g. see Bremmer (1949), Fock (1965), Makarov *et al.* (1991), and Wait (1962)). Since the radius of the sphere in such problems is typically much larger that the wavelength, the series representing the exact solution converges too slowly and a transformation of the series into a different, better convergent form can be quite useful. This is achieved by expressing the series solution as a contour integral in a complex plane followed by deformation of the integration contour onto a different path, over which the integral converges better or can be evaluated asymptotically. The necessary treatment (Watson's transformation) was developed by Watson (1918) in the context of the problem of radio wave propagation over the Earth surface for a point source and observer located close to the strongly absorbing surface of the Earth. In problems of scattering, however, the incidence wave is plane, and the Watson technique has to be modified (White 1922). Applications of the modified Watson transformation to the analysis of high-frequency scattering from PEC spheres are described in Fedorov (1958) and Senior and Goodrich (1964). In Chapter 6 we applied the Watson transformation to the high-frequency analysis of the scattering of a plane electromagnetic wave from impenetrable (PEC and impedance) cylinders. In the case of a low-absorption sphere, Watson's transformation has to be further modified.

In the high-frequency case, one can expect that the exact solution should recover the GO formulas. This relationship has been demonstrated in Nicholson (1911, 1913) and Bromwich (1920) for PEC spheres and in Van der Pol and Bremmer (1937a,b) for homogeneous dielectric spheres.

The standard literature mostly describes the case of a PEC sphere (Bouman *et al.* 1987) or that of a homogeneous sphere, non-magnetic (Born and Wolf 1959; Fock 1965; Grandy 2000; Kerker 1969; Ruck *et al.* 1970; Sommerfeld 1935; Van de Hulst 1957) and magnetic (Bohren and Huffman 1983; Franz 1957; Kong 1986; Newton 1966; Stratton 1941). Solutions for core-shell spheres can be found in Kerker (1969, Chapter 5). Ruck *et al.* (1970, sections 3.4.1 and 3.4.2) also provides the exact solution for a non-magnetic sphere consisting of an arbitrary number of concentric layers. Fortran codes for non-magnetic homogeneous and coated spheres can be found in Bohren and Huffman (1983).

Solutions for spheres are available in a variety of forms, and special care should be taken when using solutions of different authors. For example, solutions may differ in the choice of the direction of propagation of the incident wave (positive or negative direction of the z axis), time dependence ($\exp(j\omega t)$ or $\exp(-i\omega t)$), units (SI or Gaussian), and definitions and notation for the employed special functions. The solutions can be represented in a component form or in a vector form, Cartesian or spherical coordinate systems.

A source of confusion is the factor -1 in the definition of the associate Legendre function

$$P_n^1(x) = -\frac{1}{2^n n!}\sqrt{1-x^2}\frac{d^{n+1}}{dx^{n+1}}(x^2-1)^n \tag{7.1}$$

which is adopted in mathematical and computational literature (e.g., Abramowitz and Stegun (1965), Hobson (1931), Korn and Korn (1968), Ryshik and Gradstein (1957), and Press *et al.* (1994)) but is omitted in the physical treatments (Born and Wolf 1959; Bouman *et al.* 1987; Ruck *et al.* 1970; Stratton 1941; Van de Hulst 1957). We find it convenient to use the notation

$$P_{n1}(x) = -P_n^1(x) \tag{7.2}$$

from Abramowitz and Stegun (1965) to clearly distinguish between the definitions.

Radial functions in the solutions are used in a variety of equivalent forms, for example in terms of cylindrical Bessel functions of fractional order, $J_{n+1/2}(w)$ and $H_{n+1/2}^{(1),(2)}(w)$, or spherical Bessel functions, $j_n(w)$ and $h_n^{(1),(2)}(w)$, or Riccati–Bessel functions, $\psi_n(w)$ and $\zeta_n^{(1),(2)}(w)$. The latter permit more compact representations but lack a universally accepted notation. For example, Debye (1909b) uses $\eta_\beta(w)$ for $\zeta_\beta^{(1)}(w)$ and $\zeta_\beta(w)$ for $\zeta_\beta^{(2)}(w)$, whereas Fock (1965, Chapter 10) employs $\zeta_\beta(w)$ for $\zeta_\beta^{(1)}(w)$. Our notation for the Riccati–Bessel functions can be confused with Sommerfeld's notation for the spherical Bessel functions, where $j_\beta(w)$ and $h_\beta^{(1,2)}(w)$ are denoted by $\psi_\beta(w)$ and $\zeta_\beta^{1,2}(w)$ (Sommerfeld 1935, Chapters 20 and 23).

This chapter is devoted to the scattering of plane electromagnetic waves by spheres of various size and material composition, large or small, magnetic or non-magnetic, homogeneous or multilayered, lossless or lossy. In section 7.2 we derive the exact solution for a plane wave incident on a spherical configuration consisting of an arbitrary number of concentric spheres with arbitrary values of ϵ and μ in the involved materials. The generality of this solution compensates for a slight increase in the complexity of the solution procedure. The solution for an impedance sphere is also presented.

Section 7.3 addresses the physical interpretation of scattering from spheres, depending on the electrical size and the degree of absorption in the material of the sphere. The case of electrically large spheres is treated by using a modified Watson transformation, in terms of radial components of the electric and magnetic fields, which significantly simplifies the analysis and permits extraction of the whole bunch of high-frequency scattering mechanisms, including creeping waves (interior and exterior) and GO rays, not only in the far zone of the sphere but also close to it.

Then, in section 7.4, exact solutions for the scattered field in the far zone of the sphere and for various scattering cross sections are presented, and general properties of the fields scattered by spherically symmetric scatterers are addressed. An emphasis is placed on electrically small low-absorption spheres since scattering from them can be significantly different from conventional Rayleigh scattering (section 3.6). We apply the general solution for the far scattered field to the particular cases of two concentric spheres and their variants, including PEC, impedance, and core-shell spheres, and describe the most important features of low- and high-frequency scattering from these structures.

Electromagnetic scattering from homogeneous magneto-dielectric spheres is treated in a separate section (section 7.5) because of the particular importance of this configuration for applications. The cases of an electrically small and an electrically large sphere, with and without absorption, are addressed. We study mono- and bistatic, total and absorption cross sections of these scatterers. We show the relation of these solutions to the solution for a sphere with impedance boundary conditions. A complete description of low-frequency scattering from homogeneous magneto-dielectric spheres is given, including resonant scattering and absorption. Application of the exact and high-frequency asymptotic solutions to scattering of light from rain, cloud, and fog droplets is also addressed.

Section 7.6 is devoted to electromagnetic scattering from spheres in metamaterial applications, including scattering from nano-spheres and the spherical invisibility cloak.

7.2 Exact Solution for a Multilayered Sphere

The following problem is considered. A plane wave with

$$\mathbf{E}_{\text{inc}} = E_0 \hat{x} e^{jkz}, \quad Z\mathbf{H}_{\text{inc}} = -E_0 \hat{y} e^{jkz} \tag{7.3}$$

is incident in the direction of the negative z axis on a sphere of radius a located at the origin of a coordinate system. The geometry is shown in Figure 7.1. The material constants of the surrounding space are ϵ and μ, $Z = \sqrt{\mu/\epsilon}$ and $k = \omega\sqrt{\epsilon\mu}$. The fixed incidence direction ($\hat{i} = -\hat{z}$) does not restrict the generality because of the spherical symmetry of the scatterer.

The sphere may consist of an arbitrary number L of concentric spheres of radii a_{l-1} with material constants ϵ_l and μ_l, where $l = 1, 2, 3 \ldots, L$ and $a_0 = a$ (Figure 7.2). The configuration can be also seen as a spherical core of radius a_{L-1} and material parameters ϵ_L and μ_L coated with $L-1$ spherical layers of constant, but in general different, thicknesses $a_{l-1} - a_l$ with material parameters ϵ_l and μ_l (here $l = 1, 2, \ldots L-1$) so that $l = 1$ in the outermost and $l = L-1$ in the innermost layer. The configuration is extremely general as it includes, for example, a PEC and a homogeneous sphere ($L = 1$), a PEC sphere coated with a magneto-dielectric layer and a spherical shell of constant thickness ($L = 2$), radially stratified spheres and even radially inhomogeneous spheres with continuous distributions of material parameters since these latter configurations can be modeled by a sufficiently fine step-like radial stratification.

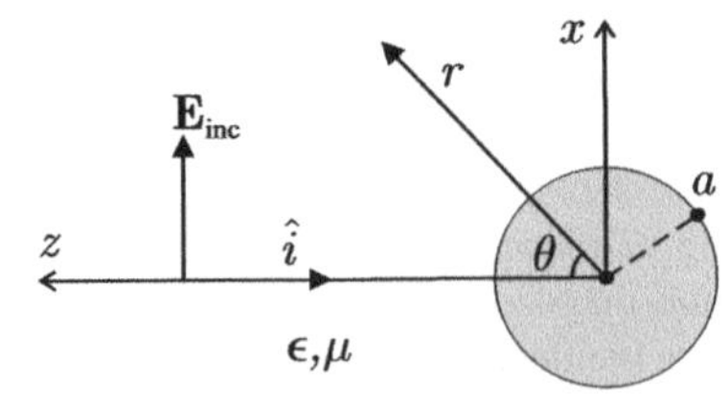

Figure 7.1 Sphere illuminated by a plane wave.

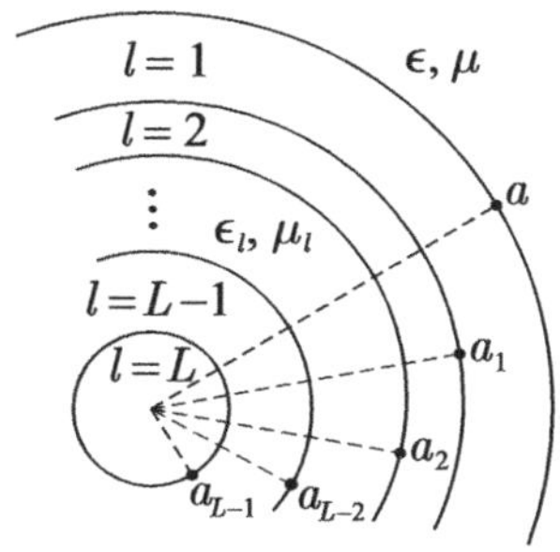

Figure 7.2 Scattering geometry for an L-layer radially stratified sphere.

The problem consists of finding the solution of Maxwell's equations which is compliant with radiation conditions at infinity and ensures the continuity of tangential components of the electric and magnetic fields at all the material interfaces.

It is possible to derive an exact solution of this boundary value problem by the separation of variables. The solution approach uses the separability of the Helmholtz equation in spherical coordinates and the fact that in the spherical coordinate system (r, θ, ϕ) with the origin at the center of the spheres the material interfaces coincide with the coordinate surfaces $r = \text{const}$. So the key issue is to reduce the vector Maxwell equations to the scalar Helmholtz equation. This can be done in different ways,[1] and we follow here a procedure which is based on the use of two scalar potentials u and v (Debye's potentials) (Clebsch 1863; Debye 1909b; Lorenz 1890). This approach transforms the original vector problem to two uncoupled scalar boundary value problems for Helmholtz equations, which are separable. The solution for a sphere consisting of any number of concentric homogeneous regions was derived in Wait (1963) in terms of potential functions that are proportional to the Debye potentials and by using an analogy with the non-uniform transmission line theory. In this section we present an equivalent solution directly for the functions u and v without resorting to the transmission-line theory.

7.2.1 Formulation of the Problem in Terms of Debye's Potentials

The Debye potentials u and v are a particular case of the Hertz vectors $\mathbf{\Pi}_{\mathrm{e,m}}$ defined in section 2.5.2, namely, it is assumed that the Hertz vectors have no components in the θ and ϕ directions so that

$$\mathbf{\Pi}_{\mathrm{e}} = \mathbf{r}u, \quad \mathbf{\Pi}_{\mathrm{m}} = \mathbf{r}v, \tag{7.4}$$

where $\mathbf{r} = r\hat{r}$ is the position vector of the observation point. Inserting these into (2.191) and (2.192) and using the expression (A.56) for the curl in spherical coordinates gives the field components in terms of the potential functions as

$$E_r = \left(\frac{\partial^2}{\partial r^2} + k^2\right)(ru) \tag{7.5}$$

$$E_\theta = \frac{1}{r}\frac{\partial^2}{\partial r \partial \theta}(ru) - \frac{j\omega\mu}{\sin\theta}\frac{\partial v}{\partial \phi} \tag{7.6}$$

$$E_\phi = \frac{1}{r\sin\theta}\frac{\partial^2}{\partial r \partial \phi}(ru) + j\omega\mu\frac{\partial v}{\partial \theta} \tag{7.7}$$

[1] For example, Mie (1908) formulated the problem in terms of radial components E_r and H_r, which can be shown to satisfy the scalar Helmholtz equations. However, finding the other field components from E_r and H_r requires integration of differential equations, which complicates the analysis. Another approach uses the formalism of vector spherical wave functions, e.g. Stratton (1941, section 9.25).

$$H_r = \left(\frac{\partial^2}{\partial r^2} + k^2\right)(rv) \tag{7.8}$$

$$H_\theta = \frac{1}{r}\frac{\partial^2}{\partial r \partial \theta}(rv) + \frac{j\omega\epsilon}{\sin\theta}\frac{\partial u}{\partial \phi} \tag{7.9}$$

$$H_\phi = \frac{1}{r\sin\theta}\frac{\partial^2}{\partial r \partial \phi}(rv) - j\omega\epsilon\frac{\partial u}{\partial \theta}, \tag{7.10}$$

where k, ϵ, and μ are to be taken for the medium in which the fields are calculated. As apparent from the expressions for the field components, the potentials u and v describe the parts of the field that can be excited by a radial electric and a radial magnetic dipole, respectively. The corresponding parts of the solution will be referred to as electric and magnetic.

Equations for the functions u and v are derived from the condition that in the source-free region the Hertz potentials should satisfy the vector Helmholtz equations up to a gradient of an arbitrary function (section 2.5.2). Using equation (A.57) for the vector Laplacian operator in spherical coordinates, we obtain

$$\nabla^2 \mathbf{\Pi}_e + k^2 \mathbf{\Pi}_e = \mathbf{r}(\nabla^2 u + k^2 u) + 2\nabla u \tag{7.11}$$

$$\nabla^2 \mathbf{\Pi}_m + k^2 \mathbf{\Pi}_m = \mathbf{r}(\nabla^2 v + k^2 v) + 2\nabla v. \tag{7.12}$$

It is now apparent that if u and v satisfy the scalar Helmholtz equations

$$\nabla^2 u + k^2 u = 0 \tag{7.13}$$

$$\nabla^2 v + k^2 v = 0, \tag{7.14}$$

then the Hertz potentials (7.4) satisfy the vector Helmholtz equations up to the gradient terms, which in turn ensures that the fields (7.5)–(7.10) are solutions of Maxwell's equations. The analytical form of the scalar Laplacian operators in spherical coordinates in (7.13) and (7.14) is given by the formula (A.54).

Equations (7.5)–(7.10) imply that the tangential field components, which are E_θ, E_ϕ, H_θ, and H_ϕ, are continuous at the material interfaces if u and v are such that

$$[\epsilon(r)u] = 0, \quad \left[\frac{\partial}{\partial r}(ru)\right] = 0 \tag{7.15}$$

and

$$[\mu(r)v] = 0, \quad \left[\frac{\partial}{\partial r}(rv)\right] = 0, \tag{7.16}$$

where the square brackets denote the jump of the respective quantity at the interface and

$$\epsilon(r) = \begin{cases} \epsilon, & a < r < \infty \\ \epsilon_l, & a_l < r < a_{l-1} \\ \epsilon_L, & 0 \le r < a_{L-1} \end{cases}, \quad \mu(r) = \begin{cases} \mu, & a < r < \infty \\ \mu_l, & a_l < r < a_{l-1} \\ \mu_L, & 0 \le r < a_{L-1} \end{cases} \tag{7.17}$$

with $l = 1, 2, \ldots, L-1$.

Thus, we have obtained two decoupled scalar boundary value problems for the functions u and v, each of which satisfies the homogeneous Helmholtz equation (7.13) or (7.14) with

$$k(r) = \begin{cases} k, & a < r < \infty \\ k_l, & a_l < r < a_{l-1} \\ k_L, & 0 \le r < a_{L-1} \end{cases} \tag{7.18}$$

and is subject to continuity conditions (7.15) or (7.16) at all the material interfaces. Once u and v are determined, the fields are obtained from (7.5)–(7.10).

7.2.2 Derivation of the Series Solution

The boundary value problems for the Debye potentials can be solved in an explicit form by separating the variables since the Helmholtz equation is separable in the spherical coordinates and the boundary surfaces coincide with the coordinate surfaces $r = \text{const}$. A partial solution $u_{\alpha\beta}(r, \theta, \phi)$ of the Helmholtz equations (7.13) and (7.14) can be written as

$$u_{\alpha\beta}(r, \theta, \phi) = \Phi_\alpha(\phi) R_\beta(r) \Theta_{\alpha\beta}(\theta), \tag{7.19}$$

where the factors $\Phi_\alpha(\phi)$, $R_\beta(r)$, and $\Theta_{\alpha\beta}(\theta)$ are solutions of the ordinary differential equations

$$\frac{\mathrm{d}^2\Phi_\alpha}{\mathrm{d}\phi^2} = -\alpha^2 \Phi_\alpha(\phi) \tag{7.20}$$

$$\frac{1}{r^2}\frac{\mathrm{d}}{\mathrm{d}r}\left(r^2 \frac{\mathrm{d}}{\mathrm{d}r} R_\beta\right) + k^2 R_\beta(r) = \frac{\beta(\beta+1)}{r^2} R_\beta(r) \tag{7.21}$$

$$\frac{1}{\sin\theta}\frac{\mathrm{d}}{\mathrm{d}\theta}\left(\sin\theta \frac{\mathrm{d}}{\mathrm{d}\theta}\Theta_{\alpha\beta}\right) = \left[\frac{\alpha^2}{\sin^2\theta} - \beta(\beta+1)\right]\Theta_{\alpha\beta}(\theta), \tag{7.22}$$

and the parameters α and β are the separation constants.

The solution of (7.20) is an arbitrary linear combination of functions $\sin(\alpha\phi)$ and $\cos(\alpha\phi)$, and if the solution $\Phi_\alpha(\phi)$ is to be 2π periodic in ϕ, we have to require that $\alpha = m$ with $m = 0, 1, \ldots$

Legendre Functions

Equation (7.22) is the equation for the Legendre functions of the argument $\cos\theta$, the order α and the degree β. Solutions of this equation are in general singular at $\theta = 0$ and π, but if α and β are integers, $\alpha = m$ and $\beta = n$ with $n \geq m$, then there exist bounded solutions, the associated Legendre polynomials,

$$\Theta_{mn}(\theta) = P_n^m(\cos\theta) = (-1)^m P_{nm}(\cos\theta) \tag{7.23}$$

which are related to the ones with $m = 0$ by the formula

$$P_{nm}(\cos\theta) = (1 - x^2)^{\frac{m}{2}} \frac{\mathrm{d}^m}{\mathrm{d}x^m} P_n(x), \tag{7.24}$$

where $x = \cos\theta$ and $P_n(x) = P_{n0}(x)$. Functions $P_n(x)$ with $n = 0, 1, 2, \ldots$ (Legendre polynomials) are polynomials of the nth degree,

$$P_0(x) = 1, \quad P_n(x) = \frac{1}{2^n n!}\frac{\mathrm{d}^n}{\mathrm{d}x^n}(x^2 - 1)^n, \quad n \geq 1. \tag{7.25}$$

It is therefore apparent that $P_{nm}(\cos\theta) = 0$ when $m > n$. The associated Legendre polynomials are orthogonal to each other in the sense that

$$\int_0^\pi P_{nm}(\cos\theta) P_{n'm}(\cos\theta) \sin\theta \, \mathrm{d}\theta = \frac{2\delta_{nn'}}{2n+1}\frac{(n+m)!}{(n-m)!}, \tag{7.26}$$

where $m, n, n' = 0, 1, 2 \ldots$ and $\delta_{nn'} = 1$ when $n = n'$ and $\delta_{nn'} = 0$ when $n \neq n'$. For example, the first three functions $P_{nm}(\cos\theta)$ with $m = 1$ and $n = 1, 2, 3$ are as follows:

$$P_{11}(\cos\theta) = \sin\theta, \quad P_{21}(\cos\theta) = 3\cos\theta\sin\theta, \quad P_{31}(\cos\theta) = \frac{3}{2}\sin\theta(5\cos^2\theta - 1). \tag{7.27}$$

In the analysis of the scattering of plane electromagnetic waves by spheres, only the associated Legendre polynomials of order $m = 1$ and argument $x = \cos\theta$ $(0 \leq \theta \leq \pi)$ are required, and they are related to the Legendre polynomials by the formula

$$P_{\beta 1}(\cos\theta) = -\frac{\partial}{\partial\theta}P_\beta(\cos\theta). \tag{7.28}$$

Their degree β can be arbitrary, however, not necessarily integer, in which case they are no longer polynomials but rather infinite expansions,

$$P_\beta(\cos\theta) = \sum_{s=0}^{\infty}\frac{(-1)^s\Gamma(\beta+s+1)}{(s!)^2\Gamma(\beta-s+1)}\sin^{2s}\left(\frac{\theta}{2}\right) \tag{7.29}$$

$$P_{\beta 1}(\cos\theta) = \frac{1}{2}\sin\theta\sum_{s=0}^{\infty}\frac{(-1)^s\Gamma(\beta+s+2)}{s!(s+1)!\Gamma(\beta-s)}\sin^{2s}\left(\frac{\theta}{2}\right), \tag{7.30}$$

where $\Gamma(w)$ is the Gamma function. With integer β, the infinite series include only a finite number of terms because of the poles of $\Gamma(w)$ at $w = 0, -1, -2, \ldots$ Functions $P_\beta(\cos\theta)$ and $P_{\beta 1}(\cos\theta)$ with non-integer β are called the Legendre function and the associated Legendre function of the first order, respectively.

A useful alternative representation of Legendre functions, applicable to arbitrary β, is the Mehler integral formula

$$P_{\nu-\frac{1}{2}}(\cos\theta) = \frac{\sqrt{2}}{\pi}\int_0^\theta \frac{\cos(\nu\tau)}{\sqrt{\cos\tau - \cos\theta}}d\tau, \tag{7.31}$$

where $\beta = \nu - 1/2$ for brevity. As apparent from (7.31), $P_{\nu-1/2}(\cos\theta)$ is an even entire function of ν on the complex ν plane. If $|\nu| \gg 1$, $|\arg\nu| < \pi$, and θ is such that $|\nu|\theta \gg 1$ and $|\nu|(\pi - \theta) \gg 1$, then

$$P_{\nu-\frac{1}{2}}(\cos\theta) \approx \sqrt{\frac{2}{\pi\nu\sin\theta}}\cos\left(\nu\theta - \frac{\pi}{4}\right) = \sqrt{\frac{2}{\pi\nu\sin\theta}}\sin\left(\nu\theta + \frac{\pi}{4}\right). \tag{7.32}$$

A uniform asymptotic approximation, which includes the limit $\theta \to 0$, is also available (see problem 5 in section C.4). These and many further formulas for Legendre functions can be found in Hobson (1931), Ryshik and Gradstein (1957), Abramowitz and Stegun (1965), Bouman *et al.* (1987), Erdélyi *et al.* (1953), Magnus *et al.* (1966), and Kratzer and Franz (1960).

Riccati–Bessel Functions

Solutions of (7.21) with $\beta = n$ where n is any integer are spherical Bessel functions of the order n, which are related to the ordinary Bessel functions of the order $n + 1/2$ as in (E.13) and (E.14). $R_\beta(r)$ can be an arbitrary linear combination of a pair of linearly independent spherical Bessel functions, for example $j_\beta(kr)$ and $h_\beta^{(2)}(kr)$. Since the field components are expressed through the products ru and rv (see equations (7.5)–(7.10)), it is more convenient to use Riccati–Bessel functions $\psi_\beta(w) = wj_\beta(w)$ and $\zeta_\beta^{(2)}(w) = wh_\beta^{(2)}(w)$ (Debye 1909b; Fock 1965, Chapter 10). The functions are related to Bessel functions of the first and the third kind by (E.15) and (E.16), and satisfy the differential equation

$$\frac{\mathrm{d}^2 z_\beta}{\mathrm{d}w^2} + \left[1 - \frac{\beta(\beta+1)}{w^2}\right] z_\beta(w) = 0, \tag{7.33}$$

where $z_\beta(w)$ is $\psi_\beta(w)$ or $\zeta_\beta^{(2)}(w)$.

All properties of these functions follow from those of spherical Bessel functions when β is integer or, when β is real- or complex-valued, from those of the ordinary Bessel functions (Appendix E). For example, when $\beta = n$ with $n = 0, 1, 2, \ldots$, these functions reduce to elementary functions and can be obtained by using the formulas (Watson 1962, section 3.4)

$$\psi_n(w) = w^{n+1}\left(-\frac{1}{w}\frac{\mathrm{d}}{\mathrm{d}w}\right)^n \frac{\sin w}{w}, \quad \zeta_n^{(2)}(w) = jw^{n+1}\left(-\frac{1}{w}\frac{\mathrm{d}}{\mathrm{d}w}\right)^n \frac{\mathrm{e}^{-jw}}{w}, \tag{7.34}$$

and those of the orders from $n = 0$ to $n = 2$ are

$$\begin{aligned} \psi_0(w) &= \sin w \\ \psi_1(w) &= -\cos w + \frac{1}{w}\sin w \\ \psi_2(w) &= -\frac{3}{w}\cos w - \left(1 - \frac{3}{w^2}\right)\sin w \end{aligned} \tag{7.35}$$

and

$$\begin{aligned} \zeta_0^{(2)}(w) &= j\mathrm{e}^{-jw} \\ \zeta_1^{(2)}(w) &= -\mathrm{e}^{-jw}\left(1 - \frac{j}{w}\right) \\ \zeta_2^{(2)}(w) &= -j\mathrm{e}^{-jw}\left(1 - \frac{3j}{w} - \frac{3}{w^2}\right). \end{aligned} \tag{7.36}$$

When $w \to 0$,

$$\psi_n(w) = \sqrt{\frac{\pi w}{2}} J_{n+1/2}(w) = \frac{w^{n+1}}{(2n+1)!!}\left[1 - \frac{w^2}{2(2n+3)} + \frac{w^4}{8(2n+3)(2n+5)} + O(w^6)\right] \tag{7.37}$$

$$\begin{aligned} \zeta_n^{(2)}(w) &= \sqrt{\frac{\pi w}{2}}[J_{n+1/2}(w) - jY_{n+1/2}(w)] = \frac{w^{n+1}}{(2n+1)!!}[1 + O(w^2)] \\ &+ \frac{j}{w^n}(2n-1)!!\left[1 - \frac{w^2}{2(1-2n)} + \frac{w^4}{8(1-2n)(3-2n)} + O(w^6)\right], \end{aligned} \tag{7.38}$$

where $(2n+1)!! = (2n+1)(2n-1)!! = 1 \cdot 3 \cdot 5 \ldots (2n+1)$. Approximations (7.37) and (7.38) imply that $\psi_n(0) = 0$, whereas $\zeta_n^{(2)}(0)$ is finite with $n = 0$ and infinite with $n \geq 1$. Except for the case $n = 0$, the term $O(w^{n+1})$ on the right-hand side of (7.38) is much smaller than the other terms; nevertheless, it is meaningful to keep it in the approximation since in the case of real-valued w, it describes the leading term in the small-argument expansion of the real part of $\zeta_n^{(2)}(w)$.

When $w \to \infty$, from (E.25) and (E.26) we obtain that

$$\psi_\beta(w) \approx \cos\left[w - \frac{\pi}{2}(\beta+1)\right] \tag{7.39}$$

$$\zeta_\beta^{(2)}(w) \approx \exp\left[-jw + j\frac{\pi}{2}(\beta+1)\right] \tag{7.40}$$

and

$$\psi'_\beta(w) \approx -\sin\left[w - \frac{\pi}{2}(\beta+1)\right] \tag{7.41}$$

$$\zeta_\beta^{(2)\prime}(w) \approx \exp\left(-jw + j\frac{\pi}{2}\beta\right), \tag{7.42}$$

where prime means the derivative with respect to the argument w. Asymptotic estimations (7.39)–(7.42) are valid with arbitrary fixed value of β, integer, real or even complex. Functions $\psi_\beta(w)$ and $\zeta_\beta^{(2)}(w)$ are linearly independent, and their Wronskian is given by the formula

$$\zeta_\beta^{(2)}(w)\psi'_\beta(w) - \zeta_\beta^{(2)\prime}(w)\psi_\beta(w) = j. \tag{7.43}$$

We will also use the function $\zeta_\beta^{(1)}(w)$, which is associated with the Hankel function $H_{\beta+1/2}^{(1)}(w)$ according to (E.16). Since

$$\zeta_\beta^{(1)}(w) = 2\psi_\beta(w) - \zeta_\beta^{(2)}(w), \tag{7.44}$$

its properties follow from those for $\psi_\beta(w)$ and $\zeta_\beta^{(2)}(w)$. In particular, it is singular at $w = 0$ as $O(|w|^{-|\mathrm{Re}\beta|})$, satisfies the relation

$$\zeta_\beta^{(2)}(w)\zeta_\beta^{(1)\prime}(w) - \zeta_\beta^{(2)\prime}(w)\zeta_\beta^{(1)}(w) = 2j, \tag{7.45}$$

and behaves as

$$\zeta_\beta^{(1)}(w) \approx \exp\left[jw - j\frac{\pi}{2}(\beta+1)\right] \tag{7.46}$$

when $|w| \to \infty$. The Riccati–Bessel functions of the third kind $\zeta_\beta^{(1,2)}(w)$ will be called the Riccati–Hankel functions.

In this chapter, when $|w| \gg 1$, we will also use approximations that are more accurate than (7.39), (7.40), and (7.46). They follow from Debye's asymptotic formulas (E.27)–(E.29) and the relations (E.15) and (E.16) between Bessel and Riccati–Bessel functions, and using $\beta = \nu - 1/2$ for convenience they can be written as

$$\zeta_{\nu-1/2}^{(1)}(w) \approx \frac{1}{\sqrt[4]{1-(\nu/w)^2}} \exp[j\Phi(\nu,w) - j\pi/4] \tag{7.47}$$

$$\zeta_{\nu-1/2}^{(1)\prime}(w) \approx j\sqrt[4]{1-(\nu/w)^2} \exp[j\Phi(\nu,w) - j\pi/4] \tag{7.48}$$

$$\zeta_{\nu-1/2}^{(2)}(w) \approx \frac{1}{\sqrt[4]{1-(\nu/w)^2}} \exp[-j\Phi(\nu,w) + j\pi/4] \tag{7.49}$$

$$\zeta_{\nu-1/2}^{(2)\prime}(w) \approx -j\sqrt[4]{1-(\nu/w)^2} \exp[-j\Phi(\nu,w) + j\pi/4] \tag{7.50}$$

$$\psi_{\nu-1/2}(w) \approx \frac{1}{\sqrt[4]{1-(\nu/w)^2}} \cos[\Phi(\nu,w) - \pi/4] \tag{7.51}$$

$$\psi'_{\nu-1/2}(w) \approx -\sqrt[4]{1-(\nu/w)^2} \sin[\Phi(\nu,w) - \pi/4], \tag{7.52}$$

where $\Phi(\nu, w)$ is defined in (E.32). Here the argument w and the parameter ν can be complex-valued but such that $|\nu| < |w|$; precise conditions of applicability on the complex ν plane are described in Appendix E. From symmetry relations (E.17) and (E.18) for the Hankel functions, it follows that

$$\zeta_{-\nu-1/2}^{(1)}(w) = e^{j\nu\pi}\zeta_{\nu-1/2}^{(1)}(w), \quad \zeta_{-\nu-1/2}^{(2)}(w) = e^{-j\nu\pi}\zeta_{\nu-1/2}^{(2)}(w). \tag{7.53}$$

The same relations held for the derivatives with respect to w since (7.53) are valid with arbitrary w.

Finally, we note the useful property

$$\mathrm{Im}\left[\frac{\zeta_\beta^{(2)\prime}(w)}{\zeta_\beta^{(2)}(w)}\right] = \frac{-1}{\left|\zeta_\beta^{(2)}(w)\right|^2}, \tag{7.54}$$

where β and w are real-valued. The property follows from the Wronskian relation (7.45) and the fact that $\zeta_\beta^{(1)}(w)$ and $\zeta_\beta^{(2)}(w)$ are complex conjugates with real-valued β and w (see (E.2)). The Riccati–Bessel function $\psi_\beta(w)$ with real-valued β and w is always real-valued.

Multipole Expansion of the Incident Field

Thus, the solutions of the Helmholtz equation, which are 2π periodic in ϕ and bounded at $\theta = 0$ and $\theta = \pi$, can be searched for as linear combinations of the partial solutions (7.19) with $\alpha = m$ and $\beta = n$, where m and n are integer and $n \geq m$, and with functions $\Phi_m(\phi)$, $\Theta_{mn}(\theta)$, and $R_n(r)$ defined above.[2] The solutions have the form of a series, which is referred to as multipole expansion. The terms of such expansions are called multipole moments. The initial term with $n = 0$, called the monopole moment, is independent of the angles θ and ϕ. The following terms – called dipole ($n = 1$), quadrupole ($n = 2$), octupole ($n = 3$), etc. moments – vary with the angles at a rate that increases with growing n.

The construction of the solution begins with finding the expansions for the portions of the Debye potentials that are responsible for the incident wave. To this end, it is convenient to consider the radial components of the incident electric and magnetic fields (7.3)

$$E_r^{\mathrm{inc}} = E_0 \cos\phi \sin\theta e^{jkr\cos\theta}, \quad ZH_r^{\mathrm{inc}} = -E_0 \sin\phi \sin\theta e^{jkr\cos\theta} \tag{7.55}$$

which can be rewritten as

$$E_r^{\mathrm{inc}} = \frac{j}{kr} E_0 \cos\phi \frac{\partial}{\partial\theta} e^{jkr\cos\theta}, \quad ZH_r^{\mathrm{inc}} = -\frac{j}{kr} E_0 \sin\phi \frac{\partial}{\partial\theta} e^{jkr\cos\theta}. \tag{7.56}$$

The exponential term in (7.56) can be represented by the series of Legendre polynomials (e.g., see Born and Wolf (1959) and Watson (1962))

$$e^{jkr\cos\theta} = \frac{1}{kr}\sum_{n=0}^{\infty} j^n (2n+1)\psi_n(kr) P_n(\cos\theta), \tag{7.57}$$

and using (7.28) we obtain for the fields

$$E_r^{\mathrm{inc}} = -\frac{jE_0}{k^2 r^2} \cos\phi \sum_{n=1}^{\infty} j^n (2n+1)\psi_n(kr) P_{n1}(\cos\theta) \tag{7.58}$$

$$ZH_r^{\mathrm{inc}} = \frac{jE_0}{k^2 r^2} \sin\phi \sum_{n=1}^{\infty} j^n (2n+1)\psi_n(kr) P_{n1}(\cos\theta). \tag{7.59}$$

[2] The angular part of $u_{mn}(r, \theta, \phi)$ given by the product of $\Phi_m(\phi)$ with $\Theta_{mn}(\theta)$ leads to the so-called spherical harmonics $Y_n^m(\theta, \phi)$.

The radial field components are simply related by (7.5) and (7.8) to the potentials u and v, and using the differential equation

$$\frac{d^2\psi_n}{dr^2}+\left[k^2-\frac{n(n+1)}{r^2}\right]\psi_n(kr)=0 \tag{7.60}$$

which follows from (7.33), we find that the Debye potentials

$$u_{\text{inc}} = \frac{E_0}{k^2 r}\cos\phi\sum_{n=1}^{\infty}c_n P_{n1}(\cos\theta)\psi_n(kr) \tag{7.61}$$

$$v_{\text{inc}} = -\frac{E_0}{Zk^2 r}\sin\phi\sum_{n=1}^{\infty}c_n P_{n1}(\cos\theta)\psi_n(kr) \tag{7.62}$$

with

$$c_n = j^{n-1}\frac{2n+1}{n(n+1)} \tag{7.63}$$

represent the incident wave (7.3).

Multipole Expansion of the Total Field

Now we are ready to write down expansions for Debye's potentials representing the total field. These expansions include partial solutions $u_{mn}(r,\theta,\phi)$ with $m=1$ only, which follows from the form of expansions (7.61) and (7.62) for the incident field and from the orthogonality of the trigonometric functions $\sin(m\phi)$ and $\cos(m\phi)$ with different values m on the interval $0\le\phi<2\pi$. Thus, we search for the solution in the following form:

$$u = \frac{E_0}{r}\cos\phi\sum_{n=1}^{\infty}P_{n1}(\cos\theta)R_n^u(r) \tag{7.64}$$

$$v = -\frac{E_0}{r}\sin\phi\sum_{n=1}^{\infty}P_{n1}(\cos\theta)R_n^v(r), \tag{7.65}$$

where $R_n^u(r)$ and $R_n^v(r)$ are suitable linear combinations of Riccati–Bessel functions. Functions $R_n^u(r)$ and $R_n^v(r)$ are chosen so as to make (7.64) and (7.65) satisfy the Helmholtz equations (7.13) and (7.14) in every subregion and the boundary conditions (7.15) and (7.16) at every material interface. Furthermore, it is required that the field components, which are obtained from ru and rv according to (7.5)–(7.10), satisfy the radiation conditions at $r=\infty$ and are bounded everywhere, including the origin $r=0$ of the coordinate system. These conditions must hold for every term in the expansions (7.64) and (7.65) because of the orthogonality property (7.26).

Referring to the analytical properties (7.37)–(7.40) of Riccati–Bessel functions, we write

$$R_n^u(r)=\begin{cases}A_n\zeta_n^{(2)}(kr)+C_n\psi_n(kr), & a\le r<\infty\\ A_n^{(l)}\zeta_n^{(2)}(k_l r)+C_n^{(l)}\psi_n(k_l r), & a_l\le r\le a_{l-1}\\ C_n^{(L)}\psi_n(k_L r), & 0\le r\le a_{L-1}\end{cases} \tag{7.66}$$

and

$$R_n^v(r)=\begin{cases}B_n\zeta_n^{(2)}(kr)+D_n\psi_n(kr), & a\le r<\infty\\ B_n^{(l)}\zeta_n^{(2)}(k_l r)+D_n^{(l)}\psi_n(k_l r), & a_l\le r\le a_{l-1}\ ,\\ D_n^{(L)}\psi_n(k_L r), & 0\le r\le a_{L-1}\end{cases} \tag{7.67}$$

where l enumerates the subregions in the interior of the sphere. In all the subregions, apart from the central one $(0 \leq r \leq a_{L-1})$, $R_n^u(r)$ and $R_n^v(r)$ may include both kinds of the linearly independent solutions of the Riccati–Bessel equation (with the respective wavenumber). In the central subregion, however, the functions of the third kind, $\zeta_n^{(2)}(k_L r)$, are not allowed because of their singularity at $r = 0$. In the exterior of the sphere $(a \leq r < \infty)$, the coefficients in front of $\psi_n(kr)$ must be those from the expansions (7.61) and (7.62) for the incident wave,

$$C_n = \frac{c_n}{k^2}, \quad D_n = \frac{C_n}{Z}, \quad n = 1, 2, 3, \ldots \tag{7.68}$$

to ensure that the total field consists of the incident wave and an outgoing scattered wave which is described by the terms with $\zeta_n^{(2)}(kr)$ (see (7.40)).

Each of the representations (7.66) and (7.67) involve, therefore, $2L$ unknown constants. For their determination we have $2L$ conditions (7.15) and (7.16), respectively, which are to be satisfied for every value of n separately, so that the functions $R_n^u(r)$ and $R_n^v(r)$ should satisfy the conditions

$$\left[\epsilon(r)R_n^u(r)\right] = 0, \quad \left[\frac{\mathrm{d}}{\mathrm{d}r}R_n^u(r)\right] = 0 \tag{7.69}$$

and

$$\left[\mu(r)R_n^v(r)\right] = 0, \quad \left[\frac{\mathrm{d}}{\mathrm{d}r}R_n^v(r)\right] = 0 \tag{7.70}$$

at every material interface. Insertion of (7.66) in (7.69) and (7.67) in (7.70) leads to two systems of $2L$ algebraic equations, from which all the unknown coefficients can be determined. This is the most general procedure, but it can hardly be treated analytically if the number L of the involved media is large. In the scattering problems, the source and the observation point are typically located in the exterior of the scattering body, which permits an elegant recurrent procedure for finding the coefficients B_n and D_n, and, therefore, for the complete specification of the scattered field in the exterior of the sphere.

Consider, for example, the determination of the coefficient A_n. Let us introduce the function

$$\delta_n^u(r) = \frac{1}{\epsilon(r)R_n^u(r)}\frac{\mathrm{d}}{\mathrm{d}r}R_n^u(r). \tag{7.71}$$

This function, measured in units of impedance, is continuous at the material interfaces by virtue of the boundary conditions (7.69). Now, assume that its value $\delta_n^u(a)$ at the surface of the sphere is known. Then, from (7.66) and (7.71) it follows that

$$\delta_n^u(a) = \frac{k}{\epsilon}\frac{A_n\zeta_n^{(2)\prime}(ka) + C_n\psi_n'(ka)}{A_n\zeta_n^{(2)}(ka) + C_n\psi_n(ka)}, \tag{7.72}$$

which depends on the ratio of the coefficients A_n and C_n only. Denoting this ratio by

$$r_n^u = \frac{A_n}{C_n}, \tag{7.73}$$

and solving (7.72) for it, we find that

$$r_n^u = -\frac{k\psi_n'(ka) - \epsilon\delta_n^u(a)\psi_n(ka)}{k\zeta_n^{(2)\prime}(ka) - \epsilon\delta_n^u(a)\zeta_n^{(2)}(ka)}. \tag{7.74}$$

Since C_n is known from (7.68), equations (7.73) and (7.74) specify the expansion coefficient A_n. This reduces the problem of finding the coefficients A_n to finding the value of the function $\delta_n^u(r)$ at $r = a$.

This value can be derived by the following recursive procedure. First of all, note that in the central sub-region $(0 \le r \le a_{L-1})$, the function (7.71) is known in an explicit form

$$\delta_n^u(r) = \frac{k_L}{\epsilon_L} \frac{\psi_n'(k_L r)}{\psi_n(k_L r)}, \tag{7.75}$$

implying the limiting value

$$\delta_n^u(a_{L-1}) = \frac{k_L}{\epsilon_L} \frac{\psi_n'(k_L a_{L-1})}{\psi_n(k_L a_{L-1})} \tag{7.76}$$

at the interface $r = a_{L-1}$. In the adjacent layer $a_{L-1} \le r \le a_{L-2}$, the function $\delta_n^u(r)$ is given by

$$\delta_n^u(r) = \frac{k_{L-1}}{\epsilon_{L-1}} \frac{\zeta_n^{(2)\prime}(k_{L-1} r)\left(A_n^{L-1}/C_n^{L-1}\right) + \psi_n'(k_{L-1} r)}{\zeta_n^{(2)}(k_{L-1} r)\left(A_n^{L-1}/C_n^{L-1}\right) + \psi_n(k_{L-1} r)}, \tag{7.77}$$

and because of its continuity, the limiting value of (7.77) at $r = a_{L-1}$ is equal to the known expression (7.76). Thus, from (7.77) at $r = a_{L-1}$ we can determine the ratio

$$\frac{A_n^{L-1}}{C_n^{L-1}} = -\frac{k_{L-1}\psi_n'(k_{L-1}a_{L-1}) - \epsilon_{L-1}\delta_n^u(a_{L-1})\psi_n(k_{L-1}a_{L-1})}{k_{L-1}\zeta_n^{(2)\prime}(k_{L-1}a_{L-1}) - \epsilon_{L-1}\delta_n^u(a_{L-1})\zeta_n^{(2)}(k_{L-1}a_{L-1})}, \tag{7.78}$$

which, because of (7.77), specifies the function $\delta_n^u(r)$ within the whole layer $L-1$. Inserting (7.78) in (7.77) and setting $r = a_{L-2}$ in (7.77), we express the value of $\delta_n^u(r)$ at the interface $r = a_{L-2}$ through the known value of $\delta_n^u(r)$ at the preceding interface $r = a_{L-1}$. This procedure can be continued further, through all the subsequent layers, until we arrive at the value of $\delta_n^u(r)$ at the surface of the sphere, $r = a$, which allows us to determine the coefficient A_n and, therefore, the function $R_n^u(r)$ and the Debye potential u in the exterior of the sphere from (7.64), (7.66), (7.68), (7.73), and (7.74).

The solution for the function $R_n^v(r)$ and the Debye potential v in the exterior of the sphere is derived in a similar way, by introducing the function

$$\delta_n^v(r) = \frac{1}{\mu(r)R_n^v(r)} \frac{\mathrm{d}}{\mathrm{d}r} R_n^v(r) \tag{7.79}$$

which has units of admittance and, according to the boundary conditions (7.16), is continuous at the material interfaces. The coefficient B_n in (7.67) is given by

$$B_n = r_n^v D_n, \tag{7.80}$$

where

$$r_n^v = -\frac{k\psi_n'(ka) - \mu\delta_n^v(a)\psi_n(ka)}{k\zeta_n^{(2)\prime}(ka) - \mu\delta_n^v(a)\zeta_n^{(2)}(ka)} \tag{7.81}$$

and $\delta_n^v(a)$ is obtained by a recurrent procedure starting from its value at the innermost interface,

$$\delta_n^v(a_{L-1}) = \frac{k_L}{\mu_L} \frac{\psi_n'(k_L a_{L-1})}{\psi_n(k_L a_{L-1})}. \tag{7.82}$$

The quantities r_n^v and r_n^u will be referred to as electric and magnetic multipole coefficients, respectively. It is apparent that they are related to each other through the formal interchange $\epsilon(r) \leftrightarrow \mu(r)$.

Multipole Expansion of the Scattered Field

The Debye potentials consist of the two parts,

$$u = u_{\text{inc}} + u_{\text{sc}}, \quad v = v_{\text{inc}} + v_{\text{sc}}, \tag{7.83}$$

where u_{inc} and v_{inc}, represented by (7.61) and (7.62), describe the incident field (7.3). Functions u_{sc} and v_{sc} are responsible for the scattered field, and they are expressed as

$$u_{\text{sc}} = \frac{E_0}{rk^2} \cos\phi \sum_{n=1}^{\infty} c_n r_n^u P_{n1}(\cos\theta) \zeta_n^{(2)}(kr) \tag{7.84}$$

$$v_{\text{sc}} = -\frac{E_0}{Zrk^2} \sin\phi \sum_{n=1}^{\infty} c_n r_n^v P_{n1}(\cos\theta) \zeta_n^{(2)}(kr), \tag{7.85}$$

where c_n is given by (7.63) and r_n^u and r_n^v by (7.74) and (7.81). The values $\delta_n^u(a)$ and $\delta_n^v(a)$ are determined from $\delta_n^u(a_{L-1})$ and $\delta_n^v(a_{L-1})$, which are given in (7.76) and (7.82), through successive application of the formulas

$$\delta_n^u(a_{l-1}) = \frac{k_l}{\epsilon_l} \frac{k_l T_{21}^{nl} + \epsilon_l T_{22}^{nl} \delta_n^u(a_l)}{k_l T_{11}^{nl} + \epsilon_l T_{12}^{nl} \delta_n^u(a_l)} \tag{7.86}$$

$$\delta_n^v(a_{l-1}) = \frac{k_l}{\mu_l} \frac{k_l T_{21}^{nl} + \mu_l T_{22}^{nl} \delta_n^v(a_l)}{k_l T_{11}^{nl} + \mu_l T_{12}^{nl} \delta_n^v(a_l)} \tag{7.87}$$

so that one begins with $l = L - 1$ and then decrements the value of l by one, until $l = 1$ is achieved. In (7.86) and (7.87),

$$T_{11}^{nl} = \zeta_n^{(2)}(k_l a_{l-1}) \psi_n'(k_l a_l) - \zeta_n^{(2)\prime}(k_l a_l) \psi_n(k_l a_{l-1}) \tag{7.88}$$

$$T_{12}^{nl} = \zeta_n^{(2)}(k_l a_l) \psi_n(k_l a_{l-1}) - \zeta_n^{(2)}(k_l a_{l-1}) \psi_n(k_l a_l) \tag{7.89}$$

$$T_{21}^{nl} = \zeta_n^{(2)\prime}(k_l a_{l-1}) \psi_n'(k_l a_l) - \zeta_n^{(2)\prime}(k_l a_l) \psi_n'(k_l a_{l-1}) \tag{7.90}$$

$$T_{22}^{nl} = \zeta_n^{(2)}(k_l a_l) \psi_n'(k_l a_{l-1}) - \zeta_n^{(2)\prime}(k_l a_{l-1}) \psi_n(k_l a_l) \tag{7.91}$$

and the prime denotes, as usual, the derivative with respect to the argument. Replacing ψ_n with $(\zeta_n^{(1)} + \zeta_n^{(2)})/2$ in equations (7.88)–(7.91) gives the alternative representations:

$$2T_{11}^{nl} = \zeta_n^{(2)}(k_l a_{l-1}) \zeta_n^{(1)\prime}(k_l a_l) - \zeta_n^{(2)\prime}(k_l a_l) \zeta_n^{(1)}(k_l a_{l-1}) \tag{7.92}$$

$$2T_{12}^{nl} = \zeta_n^{(2)}(k_l a_l) \zeta_n^{(1)}(k_l a_{l-1}) - \zeta_n^{(2)}(k_l a_{l-1}) \zeta_n^{(1)}(k_l a_l) \tag{7.93}$$

$$2T_{21}^{nl} = \zeta_n^{(2)\prime}(k_l a_{l-1}) \zeta_n^{(1)\prime}(k_l a_l) - \zeta_n^{(2)\prime}(k_l a_l) \zeta_n^{(1)\prime}(k_l a_{l-1}) \tag{7.94}$$

$$2T_{22}^{nl} = \zeta_n^{(2)}(k_l a_l) \zeta_n^{(1)\prime}(k_l a_{l-1}) - \zeta_n^{(2)\prime}(k_l a_{l-1}) \zeta_n^{(1)}(k_l a_l). \tag{7.95}$$

Note that in the limit of vanishing thickness of the lth layer ($a_{l-1} - a_l \to 0$), because of the Wronskian relations (7.43) and (7.45),

$$T_{11}^{nl}, T_{22}^{nl} \to j, \quad T_{12}^{nl}, T_{21}^{nl} \to 0, \tag{7.96}$$

and $\delta_n^u(a_{l-1}) \to \delta_n^u(a_l)$ and $\delta_n^v(a_{l-1}) \to \delta_n^v(a_l)$, as expected.

A further important property of recursions (7.86) and (7.87) is as follows (aggregation property). Assume that $\epsilon_{L-1} = \epsilon_L$ and $\mu_{L-1} = \mu_L$, that is, the material properties of the core and the layer adjacent to the core are the same. Then

$$\delta_n^u(a_{L-2}) = \frac{k_L}{\epsilon_L}\frac{\psi_n'(k_L a_{L-2})}{\psi_n(k_L a_{L-2})}, \quad \delta_n^v(a_{L-2}) = \frac{k_L}{\mu_L}\frac{\psi_n'(k_L a_{L-2})}{\psi_n(k_L a_{L-2})}, \tag{7.97}$$

which corresponds to a homogeneous sphere of radius a_{L-2} with the material properties of the core. Relations (7.97) are derived by insertion of (7.76) and (7.82) in (7.86) and (7.87) with $l = L - 1$, followed by the use of the Wronskian relation (7.43).

At the surface of the sphere, it is convenient to introduce modal surface impedance z_n and admittance y_n

$$y_n = \frac{jk}{\epsilon}\frac{1}{\delta_n^u(a)} \tag{7.98}$$

$$z_n = \frac{jk}{\mu}\frac{1}{\delta_n^v(a)} \tag{7.99}$$

and express the quantities (7.74) and (7.81) as

$$r_n^u = -\frac{\psi_n(ka) + jy_n\psi_n'(ka)}{\zeta_n^{(2)}(ka) + jy_n\zeta_n^{(2)\prime}(ka)} \tag{7.100}$$

$$r_n^v = -\frac{\psi_n(ka) + jz_n\psi_n'(ka)}{\zeta_n^{(2)}(ka) + jz_n\zeta_n^{(2)\prime}(ka)}. \tag{7.101}$$

In the trivial case of a homogeneous sphere of the radius a with material properties of the surrounding medium (homogeneous space),

$$y_n = z_n = j\frac{\psi_n(ka)}{\psi_n'(ka)} \tag{7.102}$$

and the coefficients r_n^u and r_n^v and, therefore, the potentials u_{sc} and v_{sc} and the field scattered by the sphere vanish.

Once the potentials u_{sc} and v_{sc} are determined, the components of the scattered electromagnetic field in the exterior of the multilyered sphere are derived by insertion of (7.84) and (7.85) in the relations (7.5)–(7.10), which gives the formulas:

$$E_r^{sc} = \frac{E_0}{k^2r^2}\cos\phi\sum_{n=1}^{\infty} n(n+1)c_n r_n^u P_{n1}(\cos\theta)\zeta_n^{(2)}(kr) \tag{7.103}$$

$$E_\theta^{sc} = \frac{E_0}{kr}\cos\phi\sum_{n=1}^{\infty} c_n\left[r_n^u\tau_n(\theta)\zeta_n^{(2)\prime}(kr) + jr_n^v\pi_n(\theta)\zeta_n^{(2)}(kr)\right] \tag{7.104}$$

$$E_\phi^{sc} = -\frac{E_0}{kr}\sin\phi\sum_{n=1}^{\infty} c_n\left[r_n^u\pi_n(\theta)\zeta_n^{(2)\prime}(kr) + jr_n^v\tau_n(\theta)\zeta_n^{(2)}(kr)\right] \tag{7.105}$$

$$ZH_r^{sc} = -\frac{E_0}{k^2r^2}\sin\phi\sum_{n=1}^{\infty} n(n+1)c_n r_n^v P_{n1}(\cos\theta)\zeta_n^{(2)}(kr) \tag{7.106}$$

$$ZH_\theta^{sc} = -\frac{E_0}{kr}\sin\phi\sum_{n=1}^{\infty} c_n\left[jr_n^u\pi_n(\theta)\zeta_n^{(2)}(kr) + r_n^v\tau_n(\theta)\zeta_n^{(2)\prime}(kr)\right] \tag{7.107}$$

$$ZH_\phi^{sc} = -\frac{E_0}{kr}\cos\phi\sum_{n=1}^{\infty} c_n\left[jr_n^u\tau_n(\theta)\zeta_n^{(2)}(kr) + r_n^v\pi_n(\theta)\zeta_n^{(2)\prime}(kr)\right]. \tag{7.108}$$

Here, c_n is given by (7.63), and

$$\pi_n(\theta) = \frac{1}{\sin\theta} P_{n1}(\cos\theta), \quad \tau_n(\theta) = \frac{\mathrm{d}}{\mathrm{d}\theta} P_{n1}(\cos\theta) \tag{7.109}$$

which are bounded on the interval $0 \leq \theta \leq \pi$ with the limiting values

$$\pi_n(0) = \frac{1}{2}n(n+1), \quad \tau_n(0) = \frac{1}{2}n(n+1) \tag{7.110}$$

and

$$\pi_n(\pi) = (-1)^{n-1}\frac{1}{2}n(n+1), \quad \tau_n(\pi) = (-1)^n \frac{1}{2}n(n+1) \tag{7.111}$$

at the endpoints $\theta = 0$ and $\theta = \pi$. The first several functions follow from (7.27) as

$$\pi_1(\theta) = 1, \quad \pi_2(\theta) = 3\cos\theta, \quad \pi_3(\theta) = \frac{3}{2}(5\cos^2\theta - 1) \tag{7.112}$$

and

$$\tau_1(\theta) = \cos\theta, \quad \tau_2(\theta) = 3\cos(2\theta), \quad \tau_3(\theta) = \frac{3}{2}\cos\theta(15\cos^2\theta - 11). \tag{7.113}$$

The solution (7.103)–(7.108) describes the scattered electromagnetic field in the exterior ($a \leq r < \infty$) of a sphere. The total field in the exterior ($a \leq r < \infty$) of the sphere is obtained by adding the electromagnetic field of the incident wave (7.3),

$$\mathbf{E}(\mathbf{r}) = \mathbf{E}_{\text{inc}}(\mathbf{r}) + \mathbf{E}_{\text{sc}}(\mathbf{r}), \quad \mathbf{H}(\mathbf{r}) = \mathbf{H}_{\text{inc}}(\mathbf{r}) + \mathbf{H}_{\text{sc}}(\mathbf{r}). \tag{7.114}$$

To find the fields inside the sphere, one has to determine the remaining coefficients in (7.66) and (7.67), of which only the coefficients A_n and B_n in the exterior of the sphere have been determined so far. From these, further coefficients describing fields inside the sphere can be subsequently obtained by using the continuity relations (7.69) and (7.70), starting from the sphere surface. For example, from the continuity conditions at $r = a$ one determines the coefficients $A_n^{(1)}$, $B_n^{(1)}$, $C_n^{(1)}$, and $D_n^{(1)}$, and thus the Debye potentials (7.64) and (7.65) for the total field in the outermost layer $l = 1$. Then, using the continuity conditions at the next interface $r = a_1$ and the known values of $A_n^{(1)}$, $B_n^{(1)}$, $C_n^{(1)}$, and $D_n^{(1)}$, one determines the coefficients $A_n^{(2)}$, $B_n^{(2)}$, $C_n^{(2)}$, and $D_n^{(2)}$, and the Debye potentials in the layer $l = 2$, and so forth, until the last coefficients $C_n^{(L)}$ and $D_n^{(L)}$ and the Debye potentials in the central region are specified. The fields follow from the Debye potentials on application of (7.5)–(7.10). Of course, if one is interested in the fields inside the sphere, an alternative, direct approach is always possible, in which the linear algebraic equations that result from the continuity conditions at all the material interfaces are solved simultaneously.

7.2.3 Solution for Impedance Boundary Conditions

The impedance boundary conditions (2.74) are widely used in various applications to simulate a material interface. The use of the boundary conditions may significantly simplify the solution of a scattering problem as they eliminate the need to search for fields beyond the interface. As we have seen in sections 2.3.1 and 4.6.1, when the interface is lossy or consists of a thin PEC-backed magneto-dielectric layer, the impedance boundary conditions may provide a good approximation to simulate scattering and reflection from flat material interfaces. In section 6.5.2 we saw that they are applicable to cylindrically curved scattering surfaces and lead to sufficient simulation accuracy for a cylinder radius greater than a quarter of the wavelength. In contrast to the circular cylinder, the spherical surface is an example of a doubly curved surface since both curvature radii are finite. It is therefore important to estimate the approximation quality of the boundary conditions as applied to a spherical surface. In this section we derive the exact solution for a sphere with impedance boundary conditions. In section 7.5 the solution will be compared with the solution for a homogeneous penetrable sphere and the conditions of proximity of the two will be presented.

Consider scattering of a plane wave (7.3) from a spherical surface of radius a (Figure 7.1). On the surface of the sphere the total electromagnetic field is required to satisfy the impedance boundary conditions (2.74) with the equivalent surface impedance $\tilde{Z}_s$. The scattered portion of the field should comply with the radiation conditions as $r \to \infty$.

We begin with specialization of the impedance boundary conditions (2.74) to the spherical surface $r = a$. In spherical coordinates, the unit vector of the normal directed into the exterior of the sphere is given by $\hat{n} = \hat{r}$, the vectors of the total electric and magnetic fields are

$$\mathbf{E} = E_r\hat{r} + E_\theta\hat{\theta} + E_\phi\hat{\phi}, \quad \mathbf{H} = H_r\hat{r} + H_\theta\hat{\theta} + H_\phi\hat{\phi} \tag{7.115}$$

and, thus,

$$\hat{n} \times \mathbf{E} = -E_\phi\hat{\theta} + E_\theta\hat{\phi}, \quad \hat{n} \times (\hat{n} \times \mathbf{H}) = -H_\theta\hat{\theta} - H_\phi\hat{\phi}. \tag{7.116}$$

As expected, the impedance boundary conditions relate the tangential components of the electric and magnetic fields, and projecting (2.74) on the vectors $\hat{\theta}$ and $\hat{\phi}$ gives the boundary conditions

$$E_\phi = \tilde{Z}_s H_\theta, \quad E_\theta = -\tilde{Z}_s H_\phi \tag{7.117}$$

that must be satisfied on the surface $r = a$ for $0 \leq \theta \leq \pi$ and $0 \leq \phi < 2\pi$.

The next step consists of rewriting (7.117) in terms of the Debye potentials. Using (7.6), (7.7), (7.9), and (7.10), we obtain

$$\frac{1}{\sin\theta}\frac{\partial}{\partial\phi}\left[\frac{\partial(ru)}{\partial r} - jk\frac{\tilde{Z}_s}{Z}ru\right] = \tilde{Z}_s\frac{\partial}{\partial\theta}\left[\frac{\partial(rv)}{\partial r} - jk\frac{Z}{\tilde{Z}_s}rv\right] \tag{7.118}$$

$$\frac{\partial}{\partial\theta}\left[\frac{\partial(ru)}{\partial r} - jk\frac{\tilde{Z}_s}{Z}ru\right] = -\frac{\tilde{Z}_s}{\sin\theta}\frac{\partial}{\partial\phi}\left[\frac{\partial(rv)}{\partial r} - jk\frac{Z}{\tilde{Z}_s}rv\right], \tag{7.119}$$

where $r = a$. The boundary conditions (7.118) and (7.119) are apparently satisfied if the Debye potentials are subject to the boundary conditions:

$$\left.\frac{\partial}{\partial r}(ru) - jk\frac{\tilde{Z}_s}{Z}ru\right|_{r=a} = 0 \tag{7.120}$$

$$\left.\frac{\partial}{\partial r}(rv) - jk\frac{Z}{\tilde{Z}_s}rv\right|_{r=a} = 0. \tag{7.121}$$

The boundary conditions (7.120) and (7.121) include only the potentials from the exterior of the sphere, implying that the boundary value problem is now posed on the interval $a \leq r < \infty$ only, and we do not need to search for the potentials in the interior of the sphere. We therefore represent them by series expansions of the form (7.64) and (7.65) with

$$R_n^u(r) = \tilde{A}_n\zeta_n^{(2)}(kr) + C_n\psi_n(kr) \tag{7.122}$$

$$R_n^v(r) = \tilde{B}_n\zeta_n^{(2)}(kr) + D_n\psi_n(kr), \tag{7.123}$$

where coefficients C_n and D_n, which represent the incident field, are the same as in (7.68), but $\tilde{A}_n$ and $\tilde{B}_n$ are, in general, different from A_n and B_n from (7.66) and (7.67).

The boundary conditions (7.120) and (7.121) when applied to the expansions (7.64) and (7.65) should be satisfied for every n separately, leading to the conditions

$$\left.\frac{\partial}{\partial r}R_n^u - jk\frac{\tilde{Z}_s}{Z}R_n^u\right|_{r=a} = 0 \tag{7.124}$$

$$\left.\frac{\partial}{\partial r}R_n^v - jk\frac{Z}{\tilde{Z}_s}R_n^v\right|_{r=a} = 0, \tag{7.125}$$

from which $\tilde{A}_n$ and $\tilde{B}_n$ can be determined. The conditions are similar to those that we used for radially stratified spheres, and replacing

$$\delta_n^u(a) \to j\frac{k\tilde{Z}_s}{\epsilon Z}, \qquad \delta_n^v(a) \to j\frac{kZ}{\mu\tilde{Z}_s} \tag{7.126}$$

in (7.100) and (7.101) we immediately arrive at the solution

$$\tilde{A}_n = r_n^u C_n, \qquad \tilde{B}_n = r_n^v D_n, \tag{7.127}$$

where

$$r_n^u = -\frac{\psi_n(ka) + j(Z/\tilde{Z}_s)\psi_n'(ka)}{\zeta_n^{(2)}(ka) + j(Z/\tilde{Z}_s)\zeta_n^{(2)\prime}(ka)} \tag{7.128}$$

$$r_n^v = -\frac{\psi_n(ka) + j(\tilde{Z}_s/Z)\psi_n'(ka)}{\zeta_n^{(2)}(ka) + j(\tilde{Z}_s/Z)\zeta_n^{(2)\prime}(ka)}. \tag{7.129}$$

The components of the scattered field are obtained by inserting r_n^u and r_n^v from (7.128) and (7.129) in (7.103)–(7.108).

The impedance boundary conditions are an approximation, and their use results in an approximate solution of the scattering problem. The formal relations (7.126) can be seen as the conditions of proximity of the solutions for exact and impedance boundary conditions. The conditions take a particularly simple form in terms of the modal admittances and impedances y_n and z_n. Using (7.98) and (7.99) in (7.126) gives the conditions

$$y_n \approx \frac{Z}{\tilde{Z}_s}, \qquad z_n \approx \frac{\tilde{Z}_s}{Z} \tag{7.130}$$

which should hold for all the modes in the expansions (7.64) and (7.65) that essentially contribute to the fields. As we shall see later, conditions (7.130) are satisfied when the sphere either is strongly absorbing or consists of a PEC kernel coated with a thin magneto-dielectric layer.

7.3 Physics of Scattering from Spheres

The physical interpretation of solutions from section 7.2 depends on the electrical size and material composition of the sphere. Far fields scattered by spheres of various compositions and sizes are studied in detail in sections 7.4 and 7.5. In this section we describe fields at near and intermediate distances from spherical scatterers, and the focus is on the physical interpretation of the solutions. We begin (section 7.3.1) with a classification of scattering regimes based on characteristic numerical illustrations and on the concepts already introduced in previous chapters, including low-frequency scattering (section 3.6), GO and creeping waves (section 6.6.3). Then the exact series solutions are interpreted in terms of spiral waves (section 7.3.2) as subsequent transmissions and reflections of these waves at the exterior and, when present, interior boundaries of spherical scatterers (section 7.3.3). Finally, in section 7.3.4, the case of electrically large spheres is studied, and by using Watson's transformation and the methods of asymptotic evaluation of integrals of rapidly oscillating functions in the complex plane (Appendices C and D) various types of high-frequency asymptotic contributions are extracted from the exact solutions for homogeneous spheres.

7.3.1 *Classification of Scattering*

Scattering from spheres can be classified according to two attributes: (a) electrical size, small, intermediately sized or large, and (b) material composition, either opaque (PEC, strongly absorbing) or penetrable (weakly absorbing). The specific combination of these attributes determines the characteristic features of the scattered field.

An electrically small ($ka \ll 1$) sphere scatters in a significantly different way compared with an electrically large ($ka \gg 1$) sphere. Scattering from intermediately sized spheres with, say, $1 < ka < 10$, may combine in a weaker form the features of the two limiting cases, providing a smooth transition between the two extremes and exhibit resonances. Scattering from electrically large penetrable spheres excels in its rapidly oscillating behavior when a small change in frequency, incidence or observation directions, material or geometric parameters may significantly change the scattered field. This classification scheme corresponds to the scattering regimes for the far fields (section 3.5).

Electrically Small Spheres

A sphere much smaller than the wavelength is seen by the incident wave as a point scatterer, and to the lowest degree of approximation the scattered field can therefore be expected to be similar to the field radiated by a pair of electric and magnetic dipoles and described in terms of polarizability tensors (section 3.6). In some cases, however, more refined approximations are necessary. An interesting case is that of an electrically dense sphere when the sphere is electrically small compared with the wavelength in the background medium but electrically large compared with the wavelength in the interior of the sphere, which may lead to resonances and an untypical frequency dependence, different from Rayleigh's law (3.215).

We refer to all these cases as low-frequency scattering and study them in sections on far-field scattering since the far zone for an electrically small scatterer begins at distances greater than a couple of wavelengths (see (3.103)). The next sections of the chapter provide a general description of low-frequency scattering from general spherical configurations (section 7.4.2), with applications to several specific types of spheres (PEC, core-shell, lossy and low-absorption homogeneous) in sections 7.4.3, 7.4.4, 7.5.2, and 7.5.3, respectively.

Intermediately Sized Spheres

As the electrical size of a sphere increases, scattering starts to show features typical for electrically large spheres. Figures 7.3 through 7.5 illustrate this for a PEC sphere with $a = 2\lambda$ illuminated by a plane wave of unit amplitude coming from the right. Though plotted for a fixed value of frequency ($\lambda = 0.03$ m), the presented field distributions are of universal value since in coordinates scaled to λ, they become frequency independent.

Figures 7.3 and 7.4 show amplitude distributions of the total electric field near the sphere in the E and H planes,[3] respectively. Though the field distributions look similar, there are apparent differences in the details, which is the result of polarization dependence in electromagnetic scattering. In both cases, the region to the right of the sphere is the region where the incident wave and the wave reflected from the sphere superimpose, which results in a typical interference pattern. Since the sphere is already quite large ($ka \approx 12.57$), the region shadowed by the sphere is clearly seen to the left of the sphere. The shadow region gradually disappears as the observation point moves away from the sphere. With the electrical size of the sphere getting greater, the finite extent of the shadow region, the finite field level in the shadow,

[3] The notation relates to the special cases in which the observation point lies in the plane defined by the incidence direction and either by the direction of the electric field vector in the incident wave (the so-called E plane) or by the direction of the magnetic field vector in the incident wave (the so-called H plane).

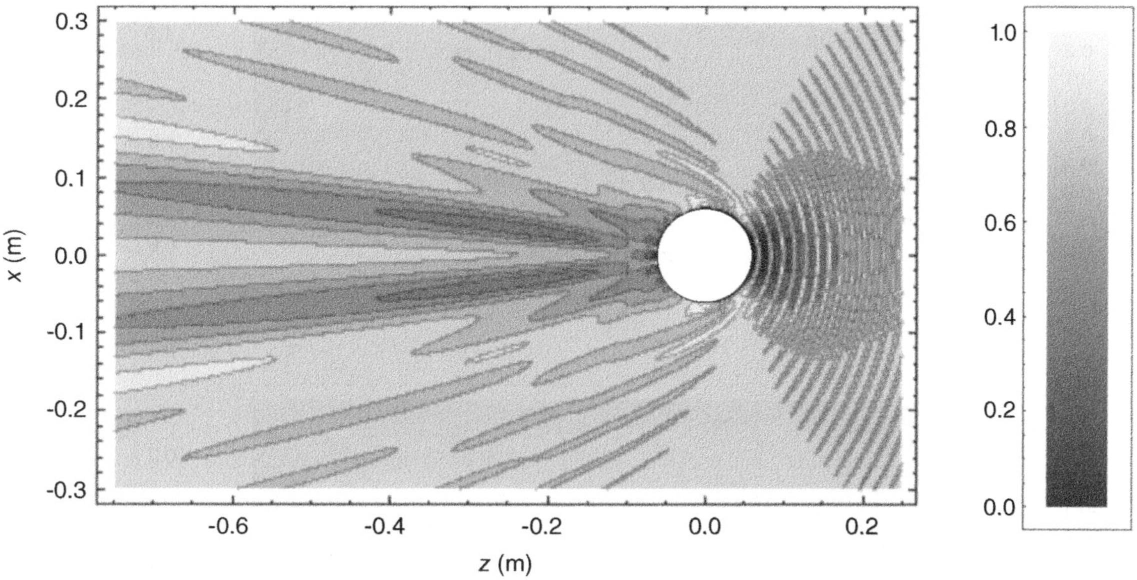

Figure 7.3 The E-plane normalized amplitude distribution of the total electric field with $\lambda = 0.03$ m around a PEC sphere of 2λ radius. The interior of the sphere is colored white. The incident plane wave comes from the right.

and the smooth transition between the lit and shadowed portions of space gradually approach the GO picture: an infinite straight strip of shadow of width $2a$ with vanishing fields inside.

It is instructive to compare the distributions of the total field from Figures 7.3 and 7.4 with those for the scattered field, shown in Figure 7.5. The scattered field exhibits a protruded maximum in the direction of forward scattering, whereas

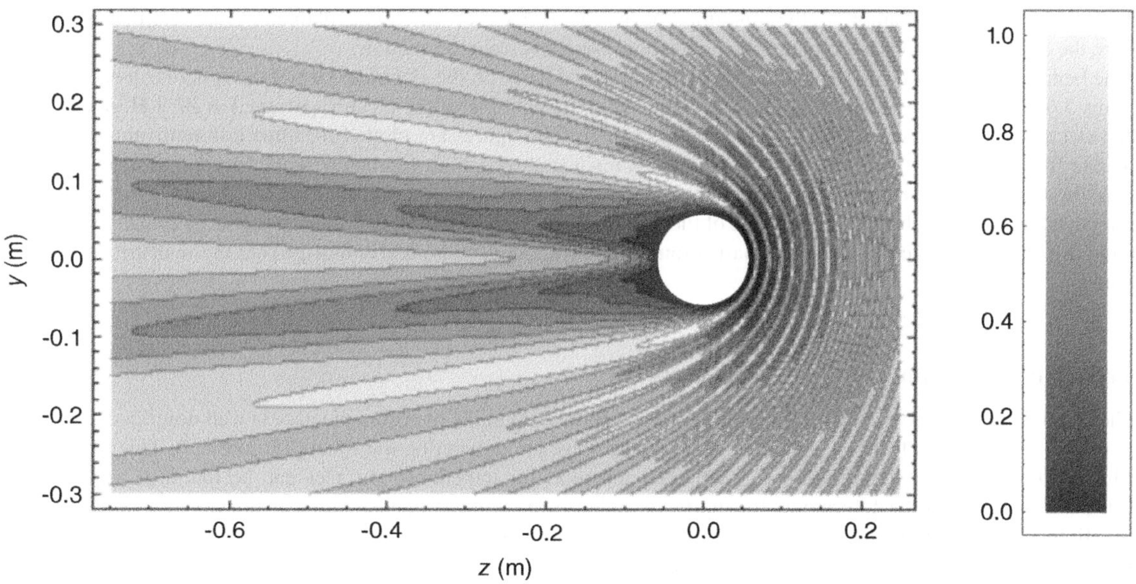

Figure 7.4 The same as Figure 7.3 but in the H plane.

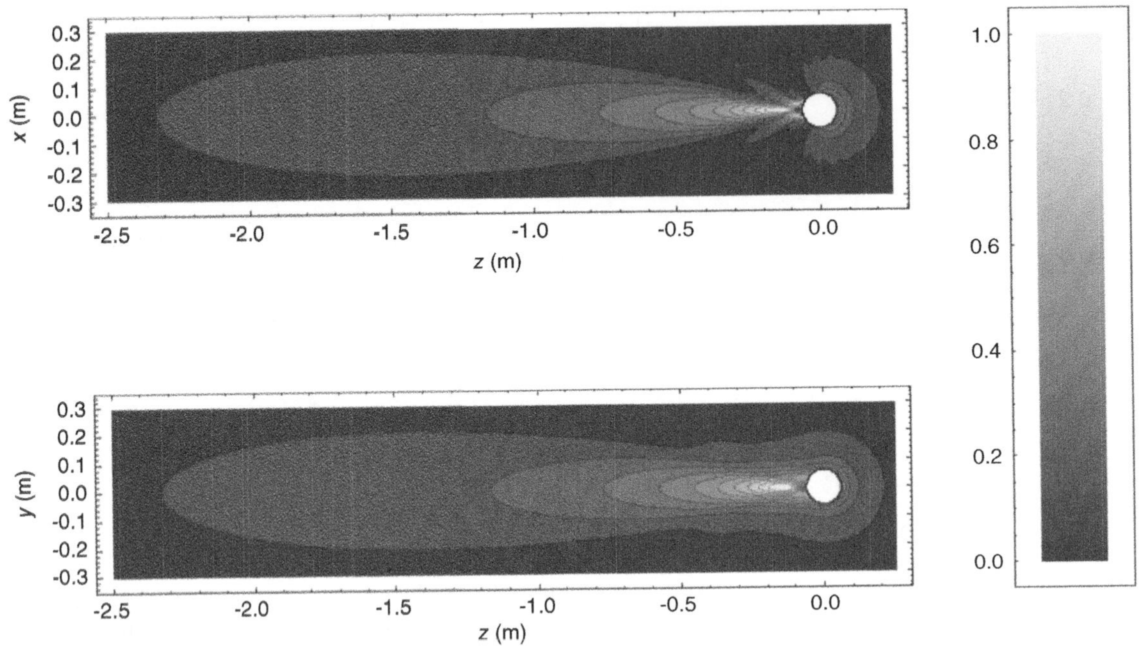

Figure 7.5 The E-plane (top) and H-plane (bottom) amplitude distributions of the scattered electric field with $\lambda = 0.03$ m around a PEC sphere of 2λ radius. The interior of the sphere is colored white.

scattering in other directions is much weaker, which is a characteristic feature of scattering from large smooth objects made of ordinary materials. For example, back- and forward scattering from scatterers with flat faces, like metal plates, are almost equal (section 9.3). Note that in the shadow, near the surface of an electrically large sphere, the amplitude of the scattered field is comparable to the amplitude of the incident wave, which is explained by the fact that behind the sphere the scattered field has to compensate the incident wave to create the shadow.

The behavior of the fields on the scattering surface in dependence on the electrical size of the sphere can be seen from Figures 7.6, 7.7, and 7.8, which show the distribution of the amplitude of the electric current $\mathbf{J} = Z\hat{r} \times \mathbf{H}$ induced by a plane electromagnetic wave on the surface of a PEC sphere. The figures are plotted for three different values of a/λ to illustrate the current distribution in the low-frequency, resonance, and high-frequency scattering regimes. As expected, increasing the electrical size of the sphere leads to decreasing field values on the shadowed side and to formation of a pronounced lit region on the illuminated side of the sphere. Note that the current distribution is not rotationally symmetric around the incidence direction, and the polarization of the incident wave is clearly foot-printed in the distribution of the current.

Electrically Large Spheres

In the case of an electrically large opaque sphere, the scattered field in the lit region of space is well described by the sum of a plane wave reflected from the sphere and a creeping wave circumventing the sphere (Figure 7.9). This corresponds to the GTD representation of scattering; asymptotic formulas for impedance surfaces can be found in section 6.6.3. Furthermore, in section 8.4.1 formulas for the fields reflected by smooth opaque surfaces are derived, and numerical estimations show that in the lit region, far enough from shadow boundaries, the reflection formula provides acceptable accuracy for spheres as small as 3λ in diameter (see Figures 8.23 and 8.24). This estimation is in agreement with the behavior of RCS of the PEC sphere plotted in Figure 3.16 as a function of ka, which approaches its geometric limit πa^2 when $ka > 10$.

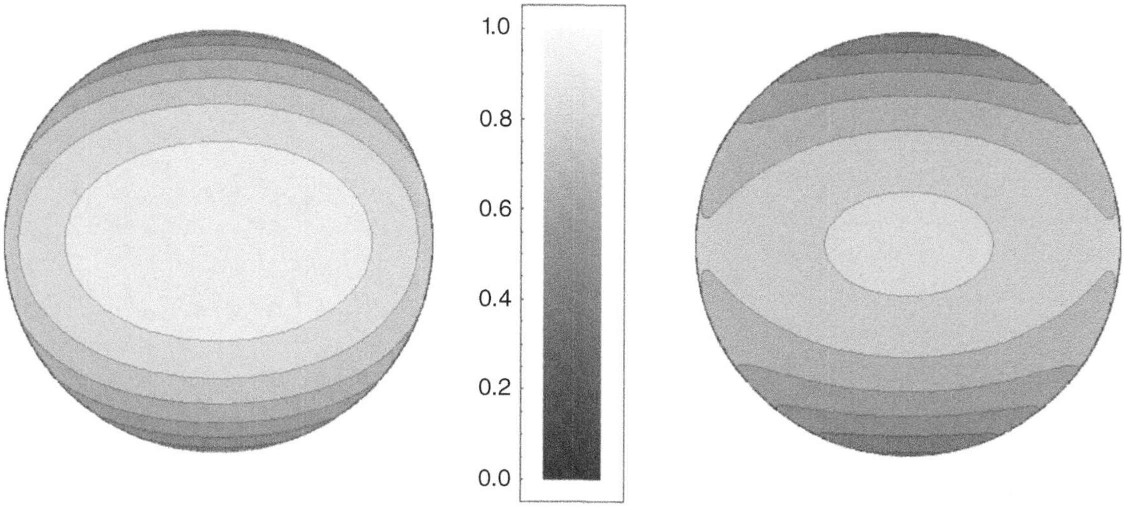

Figure 7.6 Normalized amplitude distribution of the induced total electric current over the lit (left) and shadowed (right) portions of the surface of a PEC sphere with $a = \lambda/10$. The incidence direction is perpendicular to the page. The electric field in the incident wave lies in the plane of the picture and is horizontally oriented. The larger amplitude values are shown lighter.

For an observer in the shadow region behind the sphere the field is entirely due to creeping waves. The waves get excited at the shadow boundary on the surface of the sphere and then propagate along the surface in the shadow; they can circumvent the sphere several times before becoming detached from the surface toward the observation point. These contributions explain the oscillations of RCS of a PEC sphere in Figure 3.16 in the region $1 \leq ka \leq 10$. However, the creeping waves decay exponentially as they travel along the surface of the sphere, and for spheres with $ka > 10$ and for observation points outside of the penumbra region their contribution is negligible. A similar case for a PEC circular cylinder is studied in section 6.4.3.

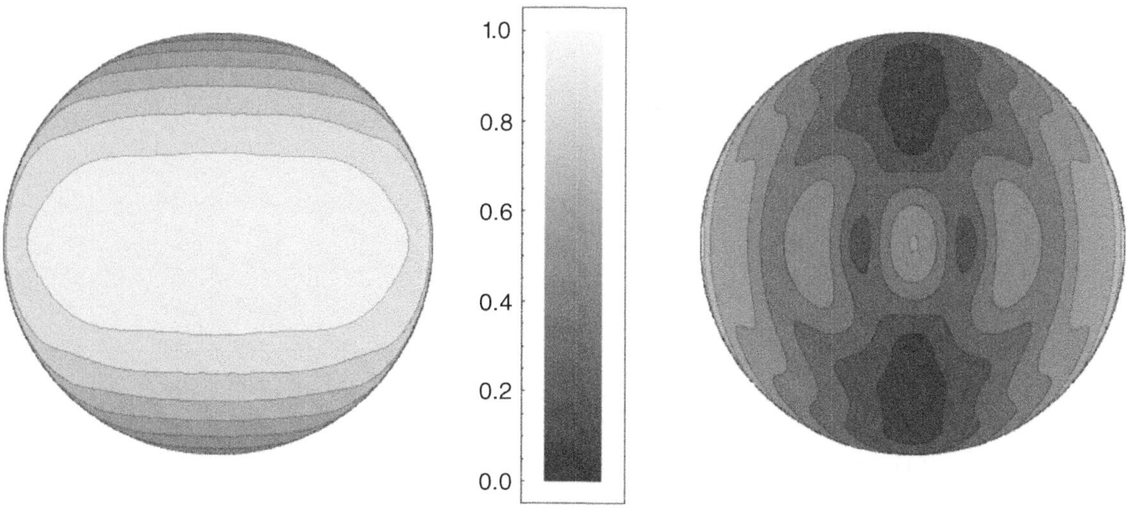

Figure 7.7 The same as Figure 7.6 but with $a = \lambda$.

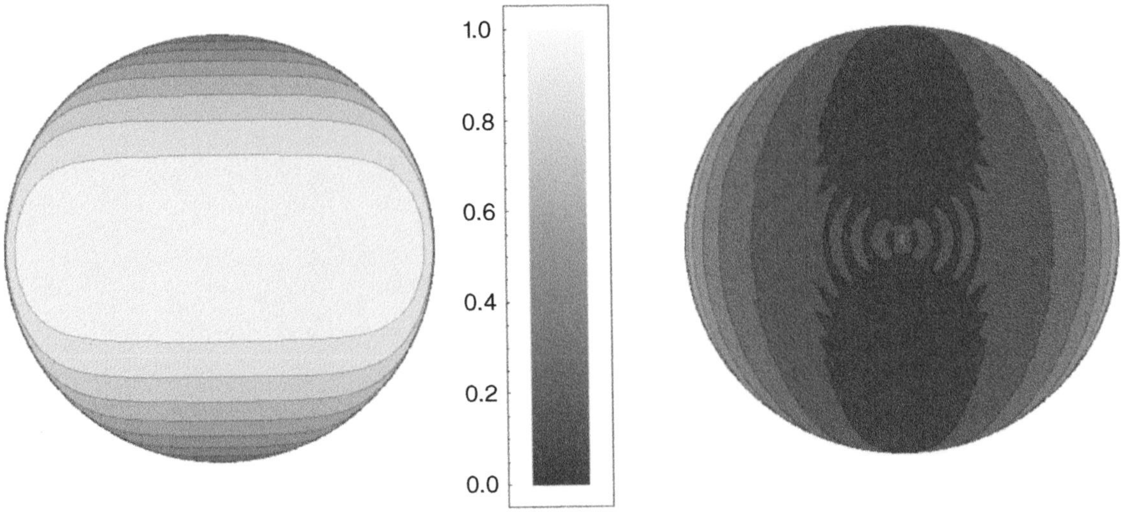

Figure 7.8 The same as Figure 7.6 but with $a = 5\lambda$.

In the penumbra region, the GTD representations are inapplicable and have to be modified to eliminate infinities in the amplitudes of the rays. Furthermore, on the surface of the sphere, close to the shadow boundary, the GTD representations become unpractical as many creeping waves need to be accounted for. Analytical description of fields in these regions requires more refined asymptotic approximations, like Fock-type penumbra formulas or uniform theories of diffraction, for example UTD (see section 6.6.3).

The ray picture becomes even more complicated for penetrable spheres as the GTD representation has to include a multitude of other rays, for example a ray transmitted through the sphere and the rays multiply reflected inside the sphere (Figure 7.10). If absorption in the interior of the sphere is small, then these new rays can undergo a large number of reflections and circumvent the sphere many times. Furthermore, creeping waves of a new kind, propagating in the interior of the sphere on the concave side of the boundary, are excited in penetrable spheres. Much more weakly attenuated than the creeping waves propagating in the exterior of the sphere, they are responsible for the ripple in scattering diagrams and

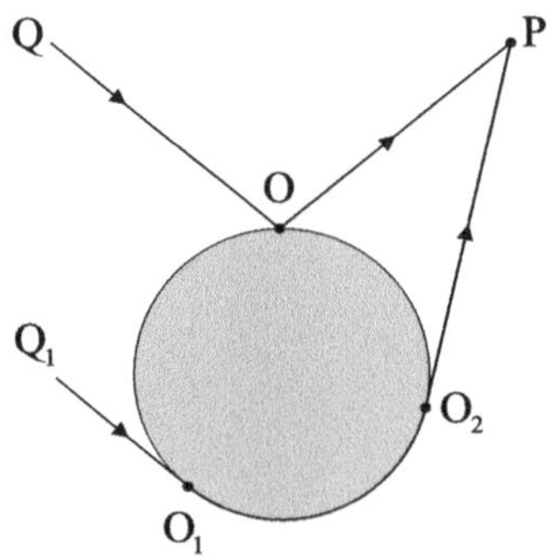

Figure 7.9 Scattering of a plane wave from an electrically large impenetrable sphere according to GTD. The ray-optics contribution is the ray that starts from a point Q, is reflected at a point O and goes to the observation point P. A diffracted ray starts from a point Q_1, touches the surface of the sphere at a point O_1, travels along an arc of a geodesic on the surface up to a detachment point O_2, from which it propagates toward the observer.

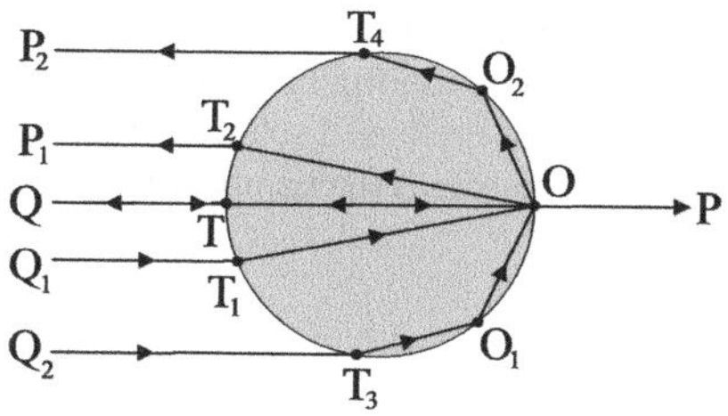

Figure 7.10 Examples of additional GO rays in the case of backscattering from a penetrable electrically large sphere: QTOP – the transmitted ray; QTOTQ and $Q_1T_1OT_2P_1$ – rays singly reflected in the interior of the sphere; $Q_2T_3O_1OO_2T_4P_2$ – a ray multiply reflected in the interior of the sphere.

play a major role in backscattering from low-absorption spheres. In section 7.3.4 we extract all the asymptotic components from the exact solution for a homogeneous sphere.

With the exception of the close vicinity of some scattering directions, for example back- and forward scattering, caustics, and shadow boundaries, the GO rays asymptotically dominate the creeping wave contributions. The GO part of the solution can be obtained either analytically from available asymptotic formulas or numerically by subdividing the front of the incident wave into small pieces and tracing them individually until the ray leaves the sphere toward the observer (Figure 7.11). For sufficiently small pieces of the wavefront the boundary of the sphere appears almost flat, so the laws of reflection and refraction at planar material interfaces (section 4.2.6) can be used to trace the rays.

From the computational point of view, however, the ray description is typically less advantageous compared to the direct summation of the series solution. Modern computers can treat series with a large number of terms, and calculation of the associate Legendre polynomials and Riccati–Bessel functions is not a problem anymore. Furthermore, every ray has to be traced separately, and the resulting ray structure, particularly in penetrable spheres, can be extremely complex and involve multiple caustics. Contrastingly, the series solution is exact and its evaluation is straightforward. When $ka \gg 1$, the required number of terms can be estimated by the formula

$$n_{\max} \geq ka + C(ka)^{1/3}, \tag{7.131}$$

where $C = O(1)$, which follows from the asymptotic behavior of Bessel functions of large order and argument (Appendix E, particularly (E.34)). The value of C depends on the required accuracy, and we have found $C = 3$ to be sufficient in most of our calculations. Further information about numerical computation of the multipole solution for spheres can be found in Ruck *et al.* (1970, section 3.1.3).

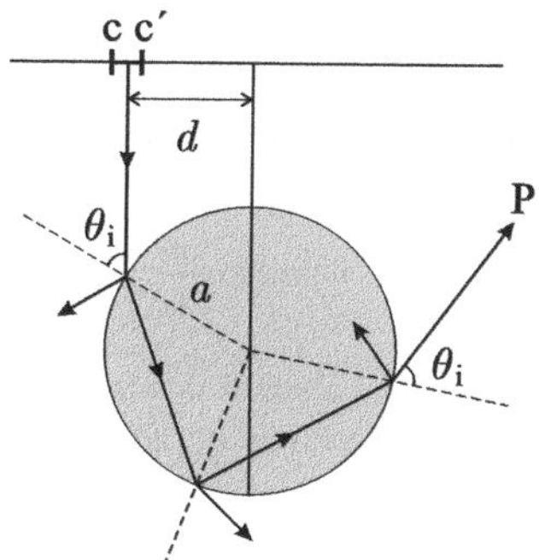

Figure 7.11 Tracing rays in a homogeneous sphere. A small piece cc′ of the wavefront at a distance d from the axis of the sphere is traced through the sphere using the reflection and refraction laws for flat material interfaces until it leaves the sphere toward the observer P. The incidence angle θ_i at the point of injection is defined by $d = a \sin \theta_i$ and is equal to the ejection angle.

The main advantage of the asymptotic solutions is that they are not limited to spheres and can be applied to more general geometries. The ray description explains the scattering mechanisms and therefore provides physical insight into scattering from electrically large bodies. Asymptotic ray-based solutions may have no alternative if simple engineering formulas are necessary or if the sphere is extremely large, as the number of terms to be accounted for in the series solution can become too big for a direct summation.

7.3.2 Spiral Waves

Series solutions (7.61), (7.62), (7.84), and (7.85) consist of terms, or modes, each of which is in general a superposition of elementary waves propagating in opposite directions with respect to the coordinates r and θ. Indeed, the modes can be represented as combinations of the functions

$$\chi_n^{(1)}(kr,\theta) = P_{n1}(\cos\theta)\zeta_n^{(1)}(kr) \tag{7.132}$$

$$\chi_n^{(2)}(kr,\theta) = P_{n1}(\cos\theta)\zeta_n^{(2)}(kr), \tag{7.133}$$

and by looking at the asymptotic formulas (7.40) and (7.46) for the Riccati–Hankel functions of large arguments, one concludes that $\chi_n^{(1)}(kr,\theta)$ and $\chi_n^{(2)}(kr,\theta)$ represent incoming and outgoing waves, respectively. Each of these, in turn, consists of two components, $\chi_{n\pm}^{(1)}(kr,\theta)$ and $\chi_{n\pm}^{(2)}(kr,\theta)$, propagating around the sphere in opposite directions, which follows from the asymptotic formula

$$P_{n1}(\cos\theta) \approx -\left(\frac{n'}{2\pi\sin\theta}\right)^{\frac{1}{2}}\left[\exp\left(jn'\theta + j\frac{\pi}{4}\right) + \exp\left(-jn'\theta - j\frac{\pi}{4}\right)\right], \tag{7.134}$$

where

$$n' = n + \frac{1}{2} \tag{7.135}$$

and is such that $n'\theta \gg 1$ and $n'(\pi - \theta) \gg 1$. Formula (7.134) results from (7.28) and (7.32). So, the series solution is a superposition of four types of elementary waves propagating from or to the center of the sphere, counter- or clockwise around the sphere.

To describe the shape of the wavefronts of these elementary waves we need Debye's approximations for $\zeta_n^{(1)}(kr)$ and $\zeta_n^{(2)}(kr)$, and using (7.47), (7.49), and (7.134) in (7.132) and (7.133) we express the phase factors of the elementary waves as

$$\exp[\pm j\Phi(n', kr) \pm jn'\theta] \tag{7.136}$$

with $\Phi(n', kr)$ given in (E.32). The phase is constant on the lines defined by the equation

$$kr\sqrt{1 - \left(\frac{n'}{kr}\right)^2} - n'\arccos\frac{n'}{kr} \pm n'\theta = \text{const} \tag{7.137}$$

which, depending on the sign in front of $n'\theta$, describes a spiral unwinding clockwise or counterclockwise on the $r - \theta$ plane. With const = 0, (7.137) describes an involute of a circle of radius n'/k with the parametric representation

$$r = \frac{n'}{k\cos\psi}, \quad \theta = \mp(\tan\psi - \psi), \tag{7.138}$$

where ψ is a parameter ($0 \le \psi \le \pi/2$) (Figure 7.12). This is why $\chi_{n\pm}^{(1)}(kr,\theta)$ and $\chi_{n\pm}^{(2)}(kr,\theta)$ are sometimes referred to as spiral waves (e.g., Makarov *et al.* (1991, p. 144)).

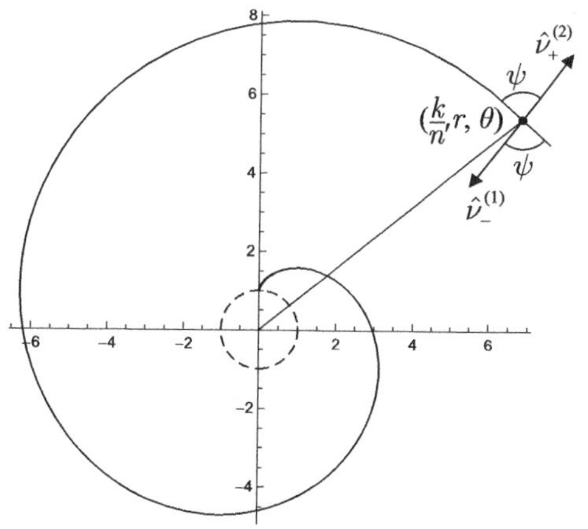

Figure 7.12 Wavefront of spiral waves $\chi^{(1)}_{n-}(kr, \theta)$ and $\chi^{(2)}_{n+}(kr, \theta)$ in coordinates scaled by the factor k/n'. The waves propagate in the opposite directions as $\hat{\nu}^{(1)}_- = -\hat{\nu}^{(2)}_+$ at every point of the wavefront. The propagation directions make the angle ψ with the front. The dashed circle is the caustic of the spiral waves.

The direction of propagation of the wavefronts is given by the normals $\hat{\nu}^{(1)}_\pm$ and $\hat{\nu}^{(2)}_\pm$ to the wavefronts, and evaluating the gradient of the phase functions in (7.136) leads to the formulas

$$\hat{\nu}^{(1)}_\pm = -\sqrt{1 - \left(\frac{n'}{kr}\right)^2}\,\hat{r} \mp \frac{n'}{kr}\hat{\theta} \tag{7.139}$$

$$\hat{\nu}^{(2)}_\pm = \sqrt{1 - \left(\frac{n'}{kr}\right)^2}\,\hat{r} \mp \frac{n'}{kr}\hat{\theta}. \tag{7.140}$$

At $r = \infty$, the vectors $\hat{\nu}^{(1)}_\pm$ and $\hat{\nu}^{(2)}_\pm$ are purely radial, but as r decreases, they get a transversal component to make with the circle of the radius r a finite angle ψ defined by $n' = kr\cos\psi$ (Figure 7.12). At the circle of the radius $r = n'/k$ (dashed in Figure 7.12) they are purely tangent ($\psi = 0$). In the interior of the circle, when $r < n'/k$, the spiral waves have complex-valued phase functions and grow indefinitely as r approaches 0.

At every circle of radius r ($r > r_0$) the directions of propagation $\hat{\nu}^{(1)}_-$ and $\hat{\nu}^{(1)}_+$ of the incoming spiral waves are related to the directions of propagation of the outgoing waves $\hat{\nu}^{(2)}_-$ and $\hat{\nu}^{(2)}_+$ by a spherical analogue of the reflection law:

$$\hat{\nu}^{(1)}_- \cdot \hat{r} = -\hat{\nu}^{(2)}_- \cdot \hat{r}, \quad \hat{\nu}^{(1)}_- \cdot \hat{\theta} = \hat{\nu}^{(2)}_- \cdot \hat{\theta} \tag{7.141}$$

$$\hat{\nu}^{(1)}_+ \cdot \hat{r} = -\hat{\nu}^{(2)}_+ \cdot \hat{r}, \quad \hat{\nu}^{(1)}_+ \cdot \hat{\theta} = \hat{\nu}^{(2)}_+ \cdot \hat{\theta}. \tag{7.142}$$

Thus, the series solution (7.61), (7.62), (7.84), and (7.85) can be interpreted as superposition of spiral waves, and by replacing $2\psi_n(kr)$ with $\zeta^{(1)}_n(kr) + \zeta^{(2)}_n(kr)$ in the incident wave and $2\psi_n(ka)$ with $\zeta^{(1)}_n(ka) + \zeta^{(2)}_n(ka)$ in the scattered wave, it can be rewritten as

$$u = \frac{E_0}{2rk^2}\cos\phi \sum_{n=1}^{\infty} c_n P_{n1}(\cos\theta)\left[\zeta^{(1)}_n(kr) + \tilde{r}^u_n \zeta^{(2)}_n(kr)\right] \tag{7.143}$$

$$v = -\frac{E_0}{2Zrk^2}\sin\phi \sum_{n=1}^{\infty} c_n P_{n1}(\cos\theta)\left[\zeta^{(1)}_n(kr) + \tilde{r}^v_n \zeta^{(2)}_n(kr)\right], \tag{7.144}$$

where c_n is given by (7.63) and

$$\tilde{r}_n^u = 2r_n^u + 1 = -\frac{\zeta_n^{(1)}(ka) + jy_n\zeta_n^{(1)\prime}(ka)}{\zeta_n^{(2)}(ka) + jy_n\zeta_n^{(2)\prime}(ka)} \tag{7.145}$$

$$\tilde{r}_n^v = 2r_n^v + 1 = -\frac{\zeta_n^{(1)}(ka) + jz_n\zeta_n^{(1)\prime}(ka)}{\zeta_n^{(2)}(ka) + jz_n\zeta_n^{(2)\prime}(ka)}. \tag{7.146}$$

Representations (7.143) and (7.144) suggest the following interpretation of the scattering from spheres. The incident field consists of an infinite number of incoming and outgoing modes, as in (7.61) and (7.62). The modes with different values of n do not interact with each other on the spherically shaped material interfaces and can be considered separately. An incoming mode from the incident field, $\chi_n^{(1)}(kr, \theta)$, arrives at the exterior boundary of the sphere, where it gets reflected with the reflection coefficient $\tilde{r}_n^u$ or $\tilde{r}_n^v$, depending on the polarization case, to become an outgoing wave $\chi_n^{(2)}(kr, \theta)$. Superimposing all incoming and reflected modes over n gives the total field in the presence of the sphere.

Coefficients $\tilde{r}_n^u$ and $\tilde{r}_n^v$ can therefore be understood as reflection coefficients that relate the incoming and outgoing spiral waves, $\chi_n^{(1)}(kr, \theta)$ and $\chi_n^{(2)}(kr, \theta)$, on the surface of the sphere. We will refer to them as reflection coefficients for spiral waves or spherical reflection coefficients. When $n > ka$, $\zeta_n^{(1)}(ka) \approx -\zeta_n^{(2)}(ka)$ (see (E.133) and (E.134)), and regardless of the material composition of the sphere, (7.145) and (7.146) simplify to

$$\tilde{r}_n^u, \tilde{r}_n^v \approx 1, \tag{7.147}$$

and the corresponding terms in (7.143) and (7.144) recover those of the expansions (7.61) and (7.62) for the incident field, implying vanishing contributions of the terms with $n > ka$ to the scattered field. A more precise criterion for the truncation of the series for the scattered field follows from the estimation (E.34) of the size of the transition region for Bessel functions and leads to the formula (7.131).

This property is related to the so-called *localization principle* (e.g., Van de Hulst (1957, section 12.31)). The angle which the normal $\hat{v}_{\pm}^{(1)}$ to the wavefront of the incoming spiral waves makes with the surface of the sphere (see (7.139) with $r = a$) is the same as in the ray propagating at the distance $d = n'/k$ from the origin of the sphere (Figure 7.11). When $n' > ka$, the associated ray misses the sphere and therefore generates no scattering.

We conclude this section with an important property of $\tilde{r}_n^u$ and $\tilde{r}_n^v$, which is that

$$|\tilde{r}_n^u| = |\tilde{r}_n^v| = 1 \tag{7.148}$$

for every lossless sphere in a lossless background medium. This follows from the facts that $\zeta_n^{(1)}(w) = [\zeta_n^{(2)}(w)]^*$ when w and n are real-valued (property (E.2) in Appendix E) and the parameters y_n and z_n are purely imaginary (problem 3), so that the numerators in (7.145) and (7.146) are complex conjugates of the respective denominators.

7.3.3 Debye's Expansions for Homogeneous Spheres

Spherical reflection coefficients $\tilde{r}_n^u$ or $\tilde{r}_n^v$ describe the response of the whole configuration behind the spherical boundary at $r = a$. For a multilayered sphere they include the multiple reflections and transmissions between the interior material interfaces that the spiral waves undergo in the interior of the sphere before they finally leave the sphere. This suggests a possibility of representing the spherical reflection coefficients of the whole, possibly multilayered configuration in terms of spherical reflection and transmission coefficients of individual spherical material interfaces, similarly to the Airy's solution (4.284) and (4.285) for a flat homogeneous layer that we studied in section 4.5. For a homogeneous penetrable sphere such representations can be derived in a relatively straightforward manner because of the simple logical structure of reflections and transmissions inside the sphere (Van der Pol and Bremmer 1937b). The idea of the representation was proposed by Debye (1908) in the context of scattering from an electrically large homogeneous penetrable circular cylinder. Here we derive this representation for a more general case of a sphere with a magnetic permeability μ_1 not necessarily equal to that in the surrounding medium. The material in the interior of the sphere may be slightly lossy, so we permit k_1 to be a complex number.

Expansions in Terms of Reflection and Transmission Coefficients of Spiral Waves

Let us start with potential u and write down the nth incoming mode in the exterior of the sphere as

$$ru_n^{\mathrm{inc}} = \chi_n^{(1)}(kr, \theta). \tag{7.149}$$

Representing the reflected and transmitted (or refracted) modes by

$$ru_n^{\mathrm{ref}} = R_{u,n}^{00} \chi_n^{(2)}(kr, \theta) \quad (r \geq a) \tag{7.150}$$

$$ru_n^{\mathrm{tr}} = T_{u,n}^{01} \chi_n^{(1)}(k_1 r, \theta) \quad (0 < r \leq a) \tag{7.151}$$

and imposing the boundary conditions (7.15), we determine the modal reflection and refraction coefficients at a spherical boundary between two homogeneous media as

$$R_{u,n}^{00} = -\frac{\zeta_n^{(1)}(k_1 a)\zeta_n^{(1)\prime}(ka) - (Z_1/Z)\zeta_n^{(1)\prime}(k_1 a)\zeta_n^{(1)}(ka)}{\zeta_n^{(1)}(k_1 a)\zeta_n^{(2)\prime}(ka) - (Z_1/Z)\zeta_n^{(1)\prime}(k_1 a)\zeta_n^{(2)}(ka)} \tag{7.152}$$

$$T_{u,n}^{01} = \frac{\epsilon}{\epsilon_1} \frac{\zeta_n^{(1)}(ka)\zeta_n^{(2)\prime}(ka) - \zeta_n^{(1)\prime}(ka)\zeta_n^{(2)}(ka)}{\zeta_n^{(1)}(k_1 a)\zeta_n^{(2)\prime}(ka) - (Z_1/Z)\zeta_n^{(1)\prime}(k_1 a)\zeta_n^{(2)}(ka)}, \tag{7.153}$$

where $Z = \sqrt{\mu/\epsilon}$, $Z_1 = \sqrt{\mu_1/\epsilon_1}$, and the superscripts in $R_{u,n}^{00}$ and $T_{u,n}^{01}$ indicate the incidence from medium "0" with material parameters ϵ and μ on the interface with medium "1" with material parameters ϵ_1 and μ_1. Formula (7.153) can be simplified by using the Wronskian relation (7.45) to give

$$T_{u,n}^{01} = -2j\frac{\epsilon}{\epsilon_1} \frac{1}{\zeta_n^{(1)}(k_1 a)\zeta_n^{(2)\prime}(ka) - (Z_1/Z)\zeta_n^{(1)\prime}(k_1 a)\zeta_n^{(2)}(ka)}. \tag{7.154}$$

The refracted portion (7.151) travels farther toward the center of the sphere until it arrives there. The point $r = 0$ is a singular point for ru_n^{tr} because of the presence of $\zeta_n^{(1)}(k_1 r)$, and in order for the total field to remain bounded an outgoing wave

$$T_{u,n}^{01} \chi_n^{(2)}(k_1 r, \theta) \tag{7.155}$$

has to be generated, which can be seen as reflection at the center of the sphere. This reflected wave compensates the singularity of ru_n^{tr} at $r = 0$ and propagates back toward the surface of the sphere, at which it is partially transmitted in the exterior of the sphere and partially reflected back in the interior of the sphere, with the contributions that can be respectively represented by

$$T_{u,n}^{10} T_{u,n}^{01} \chi_n^{(2)}(kr, \theta) \tag{7.156}$$

and

$$R_{u,n}^{11} T_{u,n}^{01} \chi_n^{(1)}(k_1 r, \theta). \tag{7.157}$$

Here $R_{u,n}^{11}$ and $T_{u,n}^{10}$ are the modal reflection and refraction coefficients for the material interface at $r = a$ illuminated from medium "1". These can be directly obtained from (7.152) and (7.153) on interchanging $\epsilon \leftrightarrow \epsilon_1$, $\mu \leftrightarrow \mu_1$, and $\zeta_n^{(1)} \leftrightarrow \zeta_n^{(2)}$, which gives

$$R^{11}_{u,n} = -\frac{\zeta_n^{(2)}(ka)\zeta_n^{(2)\prime}(k_1 a) - (Z/Z_1)\zeta_n^{(2)\prime}(ka)\zeta_n^{(2)}(k_1 a)}{\zeta_n^{(2)}(ka)\zeta_n^{(1)\prime}(k_1 a) - (Z/Z_1)\zeta_n^{(2)\prime}(ka)\zeta_n^{(1)}(k_1 a)} \tag{7.158}$$

$$T^{10}_{u,n} = 2j\frac{\epsilon_1}{\epsilon}\frac{1}{\zeta_n^{(2)}(ka)\zeta_n^{(1)\prime}(k_1 a) - (Z/Z_1)\zeta_n^{(2)\prime}(ka)\zeta_n^{(1)}(k_1 a)}. \tag{7.159}$$

The reflected portion (7.157) goes back to the center of the sphere, where the outgoing contribution

$$R^{11}_{u,n}T^{01}_{u,n}\chi_n^{(2)}(k_1 r, \theta) \tag{7.160}$$

is generated, which again undergoes a partial reflection and partial transmission at the surface $r = a$, etc. Numbering with an integer p the reflections in the interior of the sphere and summing up the outgoing contributions in the exterior of the sphere, we arrive at

$$\tilde{r}^{\mu}_n = R^{00}_{u,n} + T^{10}_{u,n}T^{01}_{u,n}\sum_{p=0}^{\infty}(R^{11}_{u,n})^p, \tag{7.161}$$

which represents the spherical reflection coefficient (7.145) as a succession of internal reflections from the surface of the sphere.

Reflection coefficient (7.146) for the v potential can be represented in a form similar to (7.161),

$$\tilde{r}^{v}_n = R^{00}_{v,n} + T^{10}_{v,n}T^{01}_{v,n}\sum_{p=0}^{\infty}(R^{11}_{v,n})^p, \tag{7.162}$$

where coefficients $R^{00}_{v,n}$, $T^{10}_{v,n}$, $T^{01}_{v,n}$, and $R^{11}_{v,n}$ are obtained from (7.152), (7.154), (7.158), and (7.159) by $\epsilon \leftrightarrow \mu$ and $\epsilon_1 \leftrightarrow \mu_1$, which follows from comparing the boundary conditions (7.15) and (7.16). Representations (7.161) and (7.162) can be derived directly from (7.145) and (7.146) by expanding them in a geometric progression (see problem 4).

Expansions in Terms of Planar Reflection and Transmission Coefficients

In the case of an electrically large sphere ($ka \gg 1$, $|k_1|a \gg 1$), the surface of the sphere and the wavefronts of the spiral waves are almost flat on the scale of the wavelength, and it is natural to expect a reduction of the transmission and reflection coefficients of the spiral waves to those for plane waves incident on a flat material interface. Indeed, this can be done by using the Debye asymptotic approximations (7.47)–(7.50) for the Riccati–Hankel functions in the expressions for $R^{00}_{u,n}$, $T^{10}_{u,n}$, $T^{01}_{u,n}$, $R^{11}_{u,n}$, $R^{00}_{v,n}$, $T^{10}_{v,n}$, $T^{01}_{v,n}$, and $R^{11}_{v,n}$ with the following result:

$$\tilde{r}^{\mu}_n = -je^{2j\Phi(n',ka)}\left\{R^{00}_{\parallel} + T^{01}_{\parallel}T^{10}_{\parallel}\sum_{p=0}^{\infty}\left(R^{11}_{\parallel}\right)^p e^{j(p+1)\left[\frac{\pi}{2}-2\Phi(n',k_1 a)\right]}\right\} \tag{7.163}$$

$$\tilde{r}^{v}_n = -je^{2j\Phi(n',ka)}\left\{R^{00}_{\perp} + T^{01}_{\perp}T^{10}_{\perp}\sum_{p=0}^{\infty}\left(R^{11}_{\perp}\right)^p e^{j(p+1)\left[\frac{\pi}{2}-2\Phi(n',k_1 a)\right]}\right\}, \tag{7.164}$$

where

$$R_{\parallel}^{00} = -R_{\parallel}^{11} = \frac{Z\sqrt{1-\left(\frac{n'}{ka}\right)^2} - Z_1\sqrt{1-\left(\frac{n'}{k_1 a}\right)^2}}{Z\sqrt{1-\left(\frac{n'}{ka}\right)^2} + Z_1\sqrt{1-\left(\frac{n'}{k_1 a}\right)^2}} \tag{7.165}$$

$$R_{\perp}^{00} = -R_{\perp}^{11} = \frac{Y\sqrt{1-\left(\frac{n'}{ka}\right)^2} - Y_1\sqrt{1-\left(\frac{n'}{k_1 a}\right)^2}}{Y\sqrt{1-\left(\frac{n'}{ka}\right)^2} + Y_1\sqrt{1-\left(\frac{n'}{k_1 a}\right)^2}} \tag{7.166}$$

$$T_{\parallel}^{01} T_{\parallel}^{10} = \frac{4ZZ_1\sqrt{1-\left(\frac{n'}{ka}\right)^2}\sqrt{1-\left(\frac{n'}{k_1 a}\right)^2}}{\left[Z\sqrt{1-\left(\frac{n'}{ka}\right)^2} + Z_1\sqrt{1-\left(\frac{n'}{k_1 a}\right)^2}\right]^2} \tag{7.167}$$

$$T_{\perp}^{01} T_{\perp}^{10} = \frac{4YY_1\sqrt{1-\left(\frac{n'}{ka}\right)^2}\sqrt{1-\left(\frac{n'}{k_1 a}\right)^2}}{\left[Y\sqrt{1-\left(\frac{n'}{ka}\right)^2} + Y_1\sqrt{1-\left(\frac{n'}{k_1 a}\right)^2}\right]^2}, \tag{7.168}$$

$Y = Z^{-1}$ and $Y_1 = Z_1^{-1}$.

Introducing the angles ψ_0 and ψ_1 between the propagation direction of a spiral wave of the nth order with the circles of the radii ka and $k_1 a$, respectively,

$$\cos\psi_0 = \frac{n'}{ka}, \quad \cos\psi_1 = \frac{n'}{k_1 a}, \tag{7.169}$$

we find that

$$\sqrt{1-\left(\frac{n'}{ka}\right)^2} = \cos\theta_0, \quad \sqrt{1-\left(\frac{n'}{k_1 a}\right)^2} = \cos\theta_1, \tag{7.170}$$

where $\theta_0 = \pi/2 - \psi_0$ and $\theta_1 = \pi/2 - \psi_1$ (Figure 7.13). It follows from (7.169) that θ_0 and θ_1 can be interpreted as the incidence and refraction angles as they are related by $k \sin\theta_0 = k_1 \sin\theta_1$, which coincides with the refraction law for a

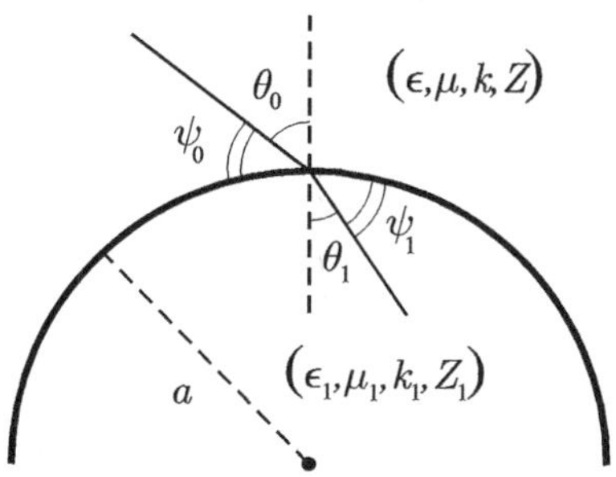

Figure 7.13 Interpretation of the parameters from Debye's expansions for electrically large homogeneous spheres.

plane wave at a planar material interface. It is now apparent that $R^{im}_{\parallel,\perp}$ and $T^{im}_{\parallel,\perp}$ with $i, m = 0, 1$ in (7.165)–(7.168) are reflection and transmission coefficients at a planar interface between medium i and medium m illuminated from the medium i under the incidence angle θ_i (see (4.17), (4.18), and (4.27)–(4.30)).

Expressions (7.163)–(7.168) are limited to spiral waves with $n < ka$ and $n < |k_1|a$. Otherwise, when the order is greater than the argument, Riccati–Hankel functions are large, the modal refraction coefficients (7.154) and (7.159) are small, and (7.161) and (7.162) reduce to $\tilde{r}^u_n \approx R^{00}_{u,n}$ and $\tilde{r}^v_n \approx R^{00}_{v,n}$. When $n > ka$, the spherical reflection coefficients $\tilde{r}^u_n$ and $\tilde{r}^v_n$ are close to unity (see (7.147)).

7.3.4 Waves in Electrically Large Homogeneous Low-absorption Spheres

The series solutions (7.143) and (7.144) together with Debye's representations (7.161) and (7.162) of spherical reflection coefficients permit a physical interpretation of scattering from multilayered spheres in terms of spiral waves multiply reflected in the interior of the sphere. However, when the sphere is electrically large ($ka \gg 1$), many spiral waves of comparable magnitude have to be accounted for (see (7.131)) and summation of the series and physical interpretation of the result become difficult.

A powerful method of analysis of series solutions for electrically large scatterers is the Watson transformation. This transformation is not only a tool for a more efficient numerical calculation but also an approach to extracting asymptotic contributions due to various scattering mechanisms. The contributions can be often represented in a simple form, thus leading to a clear physical interpretation of the solution. The method and its application to impenetrable cylindrical scatterers (PEC circular cylinders) are addressed in section 6.4, and in section 6.4.3 it is shown that the residue contributions describe creeping waves on the convex side of the cylindrical boundary, whereas the stationary point contributions represent the incident and reflected waves in the GO approximation.

In contrast to PEC spheres (e.g., White (1922), Fedorov (1958), Senior and Goodrich (1964), and Beckmann and Franz (1957)), application of the Watson method to high-frequency analysis of scattering of plane electromagnetic waves from penetrable spheres is scarcely addressed in the literature. In his pioneering work, Watson (1918) studied the field of a radial electric dipole placed over a lossy sphere and, neglecting the waves in the interior of the sphere, represented the solution by a sum of residues rapidly converging in the shadow (without interpreting the series members as creeping waves). Van der Pol and Bremmer (1937a, 1938) used the transformation to study creeping waves excited by a dipole on the convex side of a large lossy sphere (the Earth), but to derive the GO waves due to a plane wave excitation of a large lossless sphere they used the original multipole series solution (Van der Pol and Bremmer 1937b). On replacing the spherical reflection coefficients with Debye's expansions and the Riccati–Hankel functions with their integral representations (E.36) and (E.37), asymptotic formulas for internally reflected waves of arbitrary order, including rainbow approximations for scattered intensities, have been derived by applying a multidimensional version of the stationary point method.

Franz (1957, section 19) applied the Watson method to the dyadic Green function of a homogeneous sphere, and associated the creeping and GOs wave contributions with integrals over specific paths in the complex plane of the summation index.

A significant amount of research has been devoted to asymptotic analysis of the far fields scattered by large homogeneous transparent spheres in the context of optical scattering from raindrops under typical assumptions of lossless and non-magnetic media (Grandy 2000; Newton 1966; Van de Hulst 1957). The GO part can be obtained by approximate summation of the multipole series solution (e.g., Bromwich (1920), Van de Hulst (1957, section 12.3), and Newton (1966, section 3.5), which does not allow for extracting the creeping wave contributions.

Nussenzveig (1965) introduced a modified Watson transformation, which is based on Poisson's summation formula (Morse and Feshbach 1953, p. 467) but is equivalent to the conventional Watson transform (Grandy 2000, p. 49), and applied it to an asymptotic analysis of scalar scattering from large transparent spheres ((Nussenzveig 1969a,b); extension to the electromagnetic case is presented in Khare and Nussenzveig (1974), see also Nussenzveig (1979)). Grandy (2000, Chapter 5) followed the method to describe the contributions from the first three Debye terms (direct reflection, direct transmission, single internal reflection) to the far electromagnetic field.

In this section we apply the Watson method to the case in which the scattering body, a homogeneous sphere, is transparent so that the waves propagating through the scatterer play an important role. It is assumed that the material of the sphere possesses small but finite losses, $0 < |\mathrm{Im}\, k_1|a \ll (ka)^{1/3}$, while μ_1 is not necessarily equal to μ and the ratio

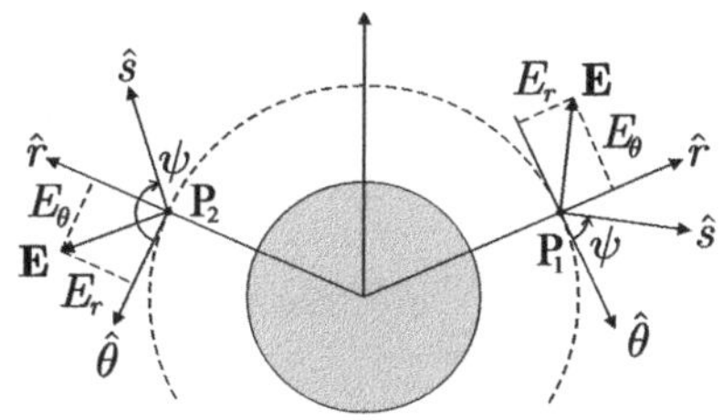

Figure 7.14 Relations between the electric field components in GO rays propagating clockwise. Here $\hat{s}$ is the propagation direction and the electric field **E** in the wave is perpendicular to $\hat{s}$. The arrival angle at the point P_1 lies in the range $0 \le \psi \le \pi/2$ and at the point P_2 in the range $\pi/2 \le \psi \le \pi$.

$|k_1|/k$ (refractive index) can be greater or smaller than unity. We will show that new types of waves, namely creeping waves propagating on the concave side of the spherical interface and GO rays multiply reflected in the interior of the sphere, enter the asymptotic solution.

The basic idea of Watson's transformation for spheres is that the series solution is transformed into a contour integral on the complex plane of the summation index n'. The integration path on the complex plane can be deformed to capture pole singularities of the integrand function, which leads to residue contributions describing the waves creeping on both sides of the spherical boundary. The GO waves, both reflected by and passed through the sphere, are obtained upon using Debye's expansions (7.163) and (7.164) in the integrand function followed by application of the stationary phase method. The result for the GO part of the wave field is in agreement with what can be constructed by simply applying the laws of GO (e.g. Van de Hulst (1957, section 12.2)), and provides a justification of the method of GO.

The analysis is performed for the radial components of the fields, which is a significant simplification since compared with the Debye potentials and other field components, E_r and H_r are given by simpler expressions[4] (see (7.84), (7.85), and (7.103)–(7.108)). It suffices to perform all the constructions for E_r since the corresponding expressions for H_r are obtained by the replacements:

$$\epsilon \to \mu, \qquad \mu \to \epsilon, \qquad E_0 \cos\phi \to -YE_0 \sin\phi. \tag{7.171}$$

Asymptotic solutions for E_r and H_r include all the wave types propagating in electrically large low-absorption spheres and, once they are known, the other field components are obtained by specializing the general relations (7.5)–(7.10) to the specific wave types. For example, for GO rays, the specialization gives relations as in plane waves,

$$E_\theta = \mp \tan\psi E_r, \quad ZH_\phi = \csc\psi E_\theta \tag{7.172}$$
$$H_\theta = \mp \tan\psi H_r, \quad YE_\phi = -\csc\psi H_\theta, \tag{7.173}$$

where the upper (lower) sign corresponds to the waves propagating clockwise (counterclockwise). The arrival angle ψ is assumed to be in the range $0 \le \psi \le \pi/2$ on the right side of the sphere and $\pi/2 \le \psi \le \pi$ on the left side of the sphere (Figure 7.14). In this way we provide a complete high-frequency analysis of the electric and magnetic parts of the solution. The relations between the field components are not limited to homogeneous spheres and hold for every spherically symmetric configuration.

[4] In particular, the spiral mode series for E_r and H_r can be exactly transformed to a path integral enclosing the whole positive real axis on the complex plane ν, without the need for special treatment of the pole singularity at $\nu = 1/2$. Residue at this pole makes application of Watson's transformation more cumbersome (see Franz (1957, section 19) and Newton (1966, section 3.8)).

Watson's Transformation

The series solution for the radial component of the total electric field in the exterior of the sphere ($r \geq a$) follows from (7.58), (7.103), and (7.114) as

$$E_r = \frac{jE_0}{(kr)^2} \cos\phi \frac{\partial}{\partial\theta} \sum_{n=1}^{\infty} e^{jn\pi/2}(2n+1)P_n(\cos\theta)\Pi_n^u(r), \tag{7.174}$$

where

$$\Pi_n^u(r) = \psi_n(kr) + r_n^u \zeta_n^{(2)}(kr) = \frac{1}{2}\left[\zeta_n^{(1)}(kr) + \tilde{r}_n^u \zeta_n^{(2)}(kr)\right] \tag{7.175}$$

with r_n^u and $\tilde{r}_n^u$ given by (7.100) and (7.145); the associated Legendre polynomials $P_{n1}(\cos\theta)$ are expressed through the Legendre polynomials $P_n(\cos\theta)$ by using (7.28). The modal admittances y_n for the homogeneous sphere follow from (7.75) and (7.98) as

$$y_n = j\frac{Z}{Z_1}\frac{\psi_n(k_1 a)}{\psi_n'(k_1 a)}. \tag{7.176}$$

The sum (7.174) can be expressed as an integral in a complex plane over a path enclosing the positive real axis (Figure 7.15),

$$E_r = -\frac{E_0}{(kr)^2} e^{-j\frac{\pi}{4}} \cos\phi \frac{\partial}{\partial\theta} \int_\Gamma e^{j\nu\frac{\pi}{2}} P_{\nu-\frac{1}{2}}[\cos(\pi-\theta)]\Pi_{\nu-\frac{1}{2}}^u(r)\frac{\nu\, d\nu}{\cos(\nu\pi)}. \tag{7.177}$$

Indeed, on the positive real axis, zeros of the denominator are at $\nu = n + 1/2$ with $n = 0, 1, 2, \ldots$ while the other functions in the integrand are holomorphic, and evaluating the integral by the residue theorem recovers (7.174) since according to the definition (7.25) of the Legendre polynomials we have

$$P_n[\cos(\pi-\theta)] = P_n(-\cos\theta) = (-1)^n P_n(\cos\theta) \tag{7.178}$$

with $P_0(\cos\theta) = 1$.

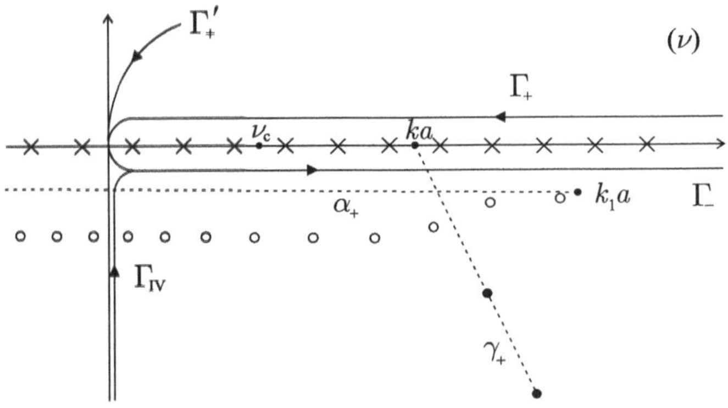

Figure 7.15 Integration on the complex ν plane when $|k_1| > k$. Contours Γ_+ and Γ_- form the contour Γ. Γ'_+ is a contour at which the integrand function is negligible. Γ_{IV} goes along the negative imaginary axis and then along and slightly below the positive real axis to enclose the fourth quadrant. The horizontal branch of poles (circles without filling) follows the anti-Stokes line α_+ that starts at the point $k_1 a$. The vertical branch of poles (fat black circles) follows the anti-Stokes line γ_+ that starts from the point ka. ν_c is the stationary point.

The contour integral (7.177) is the starting point for physical interpretation and asymptotic evaluation of the solution. In the analysis of scattering from impenetrable circular cylinders (section 6.4), similar integrals have been studied and it has been shown that the integral over Γ can be exactly transformed in an integral over a contour Γ_{IV} enclosing the forth quadrant on the complex ν plane (Figures 6.13 and 6.14). This integral, either directly or on extraction of a GO portion, can be evaluated as a sum of contributions from the residues at the poles enclosed by Γ_{IV}. The residue contributions describe creeping waves which are excited by the incident wave at the shadow boundary of the scatterer and propagate along the surface on the shadowed portion of surface of the the scatterer before detaching from the surface and leaving along the shortest path to the observer (Figures 6.16, 6.17, and 6.18). However, for penetrable spheres (and for penetrable cylinders), Watson's transformation turns out to be more complicated since the contours Γ and Γ_{IV} are no longer equivalent and the integrand function has an additional branch of poles, resulting in additional contributions that do not appear in the solutions for impenetrable scatterers. Let us take a closer look at that.

For convenience, rewrite (7.177) as

$$E_r = -\frac{E_0}{(kr)^2} e^{-j\frac{\pi}{4}} \cos\phi \frac{\partial}{\partial\theta} \int\limits_{\Gamma} F_u(\nu, r, \theta)\, d\nu \tag{7.179}$$

with

$$F_u(\nu, r, \theta) = \nu \sec(\nu\pi) e^{j\nu\frac{\pi}{2}} P_{\nu-\frac{1}{2}}[\cos(\pi-\theta)] \Pi^u_{\nu-\frac{1}{2}}(r). \tag{7.180}$$

Deforming Γ_+, the upper portion of Γ, on the positive imaginary axis[5] (Figure 7.15) and changing the integration variable according to $\nu = -\nu$ to integrate over the negative imaginary axis results in the formula

$$\int\limits_{\Gamma} F_u(\nu, r, \theta)\, d\nu = \int\limits_{\Gamma_{IV}} F_u(\nu, r, \theta)\, d\nu - \int\limits_{-j\infty}^{0} [F_u(\nu, r, \theta) + F_u(-\nu, r, \theta)]\, d\nu. \tag{7.181}$$

The integral over the imaginary half-axis vanishes for impenetrable scatterers, and its meaning is addressed in a small section a bit later. Let us first discuss reduction of the integral over Γ_{IV} to a sum of residues at the poles of $F(\nu, r, \theta)$ in the fourth quadrant. With $\pi/2 < \theta < \pi$, the integral is equivalent to the sum of residues since the integral over the path closing the ends of Γ_{IV} at infinity in the fourth quadrant vanishes. The poles of $F(\nu, r, \theta)$ in the fourth quadrant are zeros of the denominator of the spherical reflection coefficient (7.147)

$$\tilde{r}^u_{\nu-1/2} = 2r^u_{\nu-1/2} + 1 = -\frac{\zeta^{(1)}_{\nu-1/2}(ka) + jy_{\nu-1/2}\zeta^{(1)\prime}_{\nu-1/2}(ka)}{\zeta^{(2)}_{\nu-1/2}(ka) + jy_{\nu-1/2}\zeta^{(2)\prime}_{\nu-1/2}(ka)}, \tag{7.182}$$

and on accounting for (7.176), the corresponding equation can be written as

$$\frac{\zeta^{(2)\prime}_{\nu-1/2}(ka)}{\zeta^{(2)}_{\nu-1/2}(ka)} = \frac{Z_1}{Z} \frac{\psi'_{\nu-1/2}(k_1 a)}{\psi_{\nu-1/2}(k_1 a)}. \tag{7.183}$$

Compared with the cases of PEC or impedance scatterers, equation (7.183) implies a more complicated pole structure as in addition to the creeping wave poles, which asymptotically follow an anti-Stokes line γ_+ of the function $\zeta^{(2)}_{\nu-1/2}(ka)$, this branch will be referred to as vertical, there is a further branch of poles in the fourth quadrant, referred to as horizontal, which follows an anti-Stokes line α_+ of the function $\psi_{\nu-1/2}(k_1 a)$ (Figure 7.15). The anti-Stokes lines are the lines with oscillating behavior of the respective Bessel function (see Figures E.11– E.13).

[5] Here and in what follows we omit estimations of the behavior of integrands at $|\nu| \to \infty$. A similar and detailed analysis can be found in section 6.4.1; see also problem 11 in section 6.7.

When $0 < \theta < \pi/2$, the application of Cauchy's residue theorem requires a preliminary transformation of the integrand. For circular cylinders it was the transformation (6.149), and an analogous transformation for spheres reads

$$P_{\nu-\frac{1}{2}}[\cos(\pi-\theta)] = je^{-j\nu\pi}P_{\nu-\frac{1}{2}}(\cos\theta) - 2j\cos(\nu\pi)Q^{(1)}_{\nu-\frac{1}{2}}(\cos\theta), \tag{7.184}$$

where $Q^{(1)}_{\nu-\frac{1}{2}}(\cos\theta)$ is the Legendre function of the second kind with the asymptotics

$$Q^{(1)}_{\nu-\frac{1}{2}}(\cos\theta) \approx \frac{e^{-j\nu\theta+j\pi/4}}{\sqrt{2\pi\nu\sin\theta}} \tag{7.185}$$

for $|\nu| \to \infty$ ($\mathrm{Re}\,\nu > -1/2$). A more detailed description of the function can be found in Franz (1957, section 18).

Substituting (7.184) transforms the integral over Γ_{IV} to the form

$$\int_{\Gamma_{\mathrm{IV}}} F_u(\nu, r, \theta)\,\mathrm{d}\nu = -2j\int_{\Gamma_{\mathrm{IV}}} \nu e^{j\nu\pi/2}Q^{(1)}_{\nu-\frac{1}{2}}(\cos\theta)\Pi^u_{\nu-\frac{1}{2}}(r)\,\mathrm{d}\nu + j\int_{\Gamma_{\mathrm{IV}}} \nu\sec(\nu\pi)e^{-j\nu\frac{\pi}{2}}P_{\nu-\frac{1}{2}}(\cos\theta)\Pi^u_{\nu-\frac{1}{2}}(r)\,\mathrm{d}\nu. \tag{7.186}$$

The first integral on the right-hand side is convergent but cannot be evaluated as a sum of residues in the fourth quadrant. This integral describes the incident and the externally reflected waves (see (7.242)). The second integral on the right-hand side of (7.186) is equal to the sum of residues since its integrand rapidly decays in the fourth quadrant. Expressions for the residue contributions to (7.179) in the cases $0 < \theta < \pi/2$ and $\pi/2 < \theta \le \pi$ are apparently related by the formal substitution

$$P_{\nu-\frac{1}{2}}[\cos(\pi-\theta)] \leftrightarrow je^{-j\nu\pi}P_{\nu-\frac{1}{2}}(\cos\theta). \tag{7.187}$$

Watson's transformation of the series solution for H_r is completely similar as replacements (7.171) do not change the behavior of the integrand functions on the complex ν plane. The poles of the corresponding integrand function $F_\nu(\nu, r, \theta)$ are determined by the equation

$$\frac{\zeta^{(2)\prime}_{\nu-1/2}(ka)}{\zeta^{(2)}_{\nu-1/2}(ka)} = \frac{Y_1}{Y}\frac{\psi'_{\nu-1/2}(k_1a)}{\psi_{\nu-1/2}(k_1a)} \tag{7.188}$$

and also belong to either a vertical or a horizontal branch associated with the points $\nu = ka$ and $\nu = k_1a$, respectively (Figure 7.15).

Equations (7.183) and (7.188) describe the poles of spherical reflection coefficients and, therefore, resonances, or eigenmodes, of homogeneous spheres. The resonances related to (7.183) will be referred to as *electric resonances* and those related to (7.188) as *magnetic resonances*. A solution of either equation with $\nu = n + 1/2$ where $n = 1, 2, 3, \ldots$ in terms of, say, ka would mean infinite value of either r^u_n or r^v_n at that particular value of ka, without any energy supply from the source. Such resonances are unphysical even for lossless spheres since the sphere is an open structure that allows leakage of energy to the surrounding space. These unphysical cases are, however, excluded thanks to the analytical properties of Riccati–Bessel and Hankel functions of real-valued argument and order. Indeed, in the absence of losses, the right-hand sides of (7.183) and (7.188) are real-valued, whereas the left-hand sides have always a finite (negative) imaginary part (see (7.54)), implying that the equations have no solutions.[6] For complex-valued ν the solutions are possible, and so this section continues with a study of the position of the poles on the complex ν plane and physical interpretation of the corresponding contributions to the scattered field.

[6] This property holds also for spherically symmetric multilayer configurations as jy_n and jz_n in the resonance equations are purely real in the absence of losses (see discussion around equation (7.148)).

In what follows we derive mathematical expressions and study the physical meaning of the residue and stationary point contributions to the electromagnetic fields scattered by electrically large low-absorption spheres. A complete description of solutions of equations (7.183) and (7.188) for electrically small ($ka \ll 1$) homogeneous spheres is given in sections 7.4.2, 7.5.2, and 7.5.3. Zeros of (7.183) and (7.188) located close to the interval $ka \ll \nu \ll k_1 a$ are addressed in section 7.5.4 in the context of the ripple pattern in the dependence of various cross sections on ka.

Creeping Waves on the Convex Side of the Spherical Interface

Let us start with contributions from the vertical branch of poles. Simple approximations for zeros of (7.183) and (7.188) can be derived by using Debye's approximations (E.132) and (E.134) for the involved Riccati–Bessel functions (Appendix E). On the anti-Stokes line γ_+,

$$2\psi_{\nu-1/2}(k_1 a) \approx \sqrt{\frac{\pi k_1 a}{2}} S_\nu^{(2)}(k_1 a) \tag{7.189}$$

$$\zeta_{\nu-1/2}^{(2)}(ka) \approx \sqrt{\frac{\pi ka}{2}} \left[S_\nu^{(2)}(ka) - S_\nu^{(1)}(ka)\right], \tag{7.190}$$

and using (E.30), (E.31), (E.73), and (E.74), we obtain an equation for the poles of the vertical branch,

$$\sqrt{1-\left(\frac{\nu}{ka}\right)^2} \cot\left[\Phi(\nu, ka) - \frac{\pi}{4}\right] \approx -j\frac{Z_1}{Z}\sqrt{1-\left(\frac{\nu}{k_1 a}\right)^2}, \tag{7.191}$$

where $\Phi(\nu, ka)$ is given by (E.32). With $ka \gg 1$, the solutions of (7.191), nearest to the point $\nu = ka$, can be approximated as (problem 5)

$$\nu_m \approx ka + \left(\frac{ka}{2}\right)^{1/3} \left[\frac{3\pi}{2}\left(m - \frac{1}{4}\right)\right]^{2/3} e^{-j\pi/3} - \frac{j\epsilon_r}{\sqrt{\epsilon_r \mu_r - 1}}, \tag{7.192}$$

where $m = 1, 2, \ldots$ and $\sqrt{\epsilon_r \mu_r - 1} = j\sqrt{1 - \epsilon_r \mu_r}$ if $\epsilon_r \mu_r < 1$.

Zeros of (7.188) are described by (7.192) with ϵ_r interchanged with μ_r. In both cases, regardless of the material composition of the sphere, it is apparent that $|\text{Im}\, \nu_m| = O[(ka)^{1/3}]$. The residues at the poles of the vertical branch are therefore the usual creeping waves, similar to those on PEC cylinders (see section 6.4.3). The waves are generated at every point, where the incidence direction is tangent to the surface of the sphere (attachment point), follow a circular arc on the surface, and leave the sphere at the points at which the direction to the observer is tangent to the surface (launching points). On their way along the convex boundary, they constantly lose energy due to radiation to the surrounding medium, which results in the decay at the exponential rate defined by $|\text{Im}\, \nu_m|$. Since the sphere is electrically large ($ka \gg 1$), the rate of decay is large too, and as we shall see, the waves decay much faster than other contributions to the solution, which are much less attenuated in low-absorption spheres and which we study next.

Creeping Waves on the Concave Side of the Spherical Interface

Let us next take a look at the horizontal branch of poles and the meaning of associated residue contributions. We consider E_r given by (7.179) with the poles defined by (7.183); contributions of the poles to H_r follow from duality of Maxwell's equations according to (7.171). We will use Debye's approximations for the Riccati–Bessel functions, which combine analytical simplicity with excellent accuracy, including a quite close proximity of the points $\nu = ka$ and $\nu = k_1 a$ (e.g., see Table E.2).

For real-valued or almost real-valued $k_1 a$, the poles of the horizontal branch approximately follow the semi-infinite line $-\infty \leq \nu < k_1 a$ (Figure E.11). The poles that are enclosed by Γ_{IV} are located close to the interval $0 \leq \nu < k_1 a$. On this interval, the Riccati–Bessel function oscillates according to the formula

$$2\psi_{\nu-1/2}(k_1 a) \approx \sqrt{\frac{\pi k_1 a}{2}} \left[S_\nu^{(1)}(k_1 a) + S_\nu^{(2)}(k_1 a)\right] \tag{7.193}$$

and the Riccati–Hankel function is given by the single-term formula[7]

$$\zeta^{(2)}_{\nu-1/2}(ka) \approx \sqrt{\frac{\pi ka}{2}} \begin{cases} S_\nu^{(2)}(ka), & \nu < ka \\ -S_\nu^{(1)}(ka), & \nu > ka \end{cases} \tag{7.194}$$

(see Table E.1), so that equation (7.183) can be approximately written as

$$\pm j\sqrt{1-\left(\frac{\nu}{ka}\right)^2} \approx \frac{Z_1}{Z}\sqrt{1-\left(\frac{\nu}{k_1 a}\right)^2} \tan\left[\Phi(\nu, k_1 a) - \frac{\pi}{4}\right] \tag{7.195}$$

(see relations (E.73) and (E.74)). In the left-hand side of (7.195), the upper sign corresponds to $\nu < ka$ and the lower sign to $\nu > ka$.

Let us now analytically describe the position of zeros of (7.195) on the complex ν plane. The following approximations depend on whether ν is smaller or greater than ka as this makes the left-hand side of (7.195) either imaginary or real-valued. We consider both cases in succession, noting that for spheres with $|k_1| < k$ only the former case is possible.

We start by assuming that $0 < \nu < ka$. Then the left-hand side of (7.195) is purely imaginary and the real part of the tangent function on the right-hand side must vanish, which means that

$$\sin(2\mathrm{Re}\, w) = 0, \tag{7.196}$$

where $w = \Phi(\nu, k_1 a) - \pi/4$, since

$$\tan w = \frac{\sin(2\mathrm{Re}\, w) + j\sinh(2\mathrm{Im}\, w)}{\cos(2\mathrm{Re}\, w) + \cosh(2\mathrm{Im}\, w)}. \tag{7.197}$$

Equation (7.196) implies that $2\mathrm{Re}\, w = \pi l$ with an integer l, which can be rewritten as

$$\mathrm{Re}\, \Phi(\nu, k_1 a) = \frac{\pi}{2}\left(l + \frac{1}{2}\right) \tag{7.198}$$

so that (7.195) becomes

$$\frac{Z}{Z_1}\frac{\sqrt{1-[\nu/(ka)]^2}}{\sqrt{1-[\nu/(k_1 a)]^2}} = \begin{cases} \tanh[\mathrm{Im}\, \Phi(\nu, k_1 a)], & l \text{ even} \\ \coth[\mathrm{Im}\, \Phi(\nu, k_1 a)], & l \text{ odd} \end{cases}. \tag{7.199}$$

Since $\mathrm{Re}\, \Phi(\nu, k_1 a)$ is a monotonically decreasing function limited to the range $1 \geq \mathrm{Re}\, \Phi(\nu, k_1 a) \geq 0$ when $0 \leq \nu \leq k_1 a$ (see Figure E.1), equation (7.198) has a unique solution for every l from the interval $0 \leq l \leq L$, where $L = \mathrm{floor}(2k_1 a/\pi - 1/2)$. On the other hand, equation (7.199) has a solution for either even or odd l, depending on whether the absolute value of its left-hand side is smaller or greater than unity, since $|\tanh w| \leq 1$ and $|\coth w| \geq 1$ when w is real-valued.

[7] Here we assume that ka is real-valued and ν has an infinitesimal negative imaginary part so that when $\nu > ka$, ν lies on the lower side of the branch cut for the multivalued functions in Debye's formulas (see (E.81), (E.83), and Figure E.5).

Let m numerate the values of ν, for which both (7.198) and (7.199) are solvable. Separating real and imaginary parts, $\nu_m = \nu'_m - j\nu''_m$, and assuming that ν''_m is small, we can approximate $\Phi(\nu_m, k_1 a)$ by the first two terms of its Taylor expansion

$$\Phi(\nu_m, k_1 a) \approx \Phi(\nu'_m, k_1 a) + j\nu''_m \arccos \frac{\nu'_m}{k_1 a} \tag{7.200}$$

(see (E.76)) and determine ν'_m from (7.198) with $\nu = \nu'_m$. Then, ν''_m follows from (7.199) as

$$\nu''_m = \frac{1}{\arccos[\nu'_m/(k_1 a)]} \begin{cases} \operatorname{arctanh} \rho_m, & 0 < \rho_m < 1 \\ \operatorname{arccoth} \rho_m, & 1 < \rho_m < \infty \end{cases}, \tag{7.201}$$

where

$$\rho_m = \frac{Z}{Z_1} \frac{\sqrt{1 - [\nu'_m/(ka)]^2}}{\sqrt{1 - [\nu'_m/(k_1 a)]^2}}. \tag{7.202}$$

The distance between the neighboring poles is estimated by

$$|\nu_{m+1} - \nu_m| \approx \frac{\pi}{\arccos(\nu'_m/k_1 a)} \tag{7.203}$$

(see problem 6). So, the poles of the horizontal branch fill the interval $0 \le \nu < k_1 a$ in a non-uniform manner, as the distance between them is the smallest $|\nu_{m+1} - \nu_m| \approx 2$ when $|\nu/k_1 a| \ll 1$, increasing continuously with growing ν'_m.

Furthermore, it follows from (7.201) that even in the lossless case the poles of the horizontal branch always have a finite negative imaginary part, which is $|\mathrm{Im}\, \nu_m| = O(1)$, so that they are located closer to the real axis on the complex ν plane than the poles of the vertical branch. The location of the poles is schematically illustrated in Figure 7.15.

The imaginary part of the poles gets even smaller if $ka < \nu < k_1 a$ (this requires that $k_1 > k$). If $\nu > ka$, then $\sqrt{1 - [\nu/(ka)]^2} = j\sqrt{[\nu/(ka)]^2 - 1}$ and equation (7.195) has only real-valued solutions when the material parameters are real-valued. Such solutions are unphysical, and the necessary correction is obtained on recalling that the logarithmic derivative of $\zeta^{(2)}_{\nu-1/2}(ka)$ in (7.183) has an imaginary part even with real-valued ka and ν (see (7.54)). The corrected equation (7.195) has the form

$$\sqrt{\left(\frac{\nu}{ka}\right)^2 - 1} + j\delta \approx \frac{Z_1}{Z} \sqrt{1 - \left(\frac{\nu}{k_1 a}\right)^2} \tan\left[\Phi(\nu, k_1 a) - \frac{\pi}{4}\right], \tag{7.204}$$

where $\delta = |\zeta^{(2)}_{\nu-1/2}(ka)|^{-2}$ if ka and ν are real-valued. When $\nu > ka$, parameter δ is asymptotically small because

$$\zeta^{(2)}_{\nu-1/2}(ka) \approx \frac{j}{\sqrt[4]{[\nu/(ka)]^2 - 1}} e^{j\Phi(\nu, ka)} \tag{7.205}$$

with

$$\Phi(\nu, ka) = jka\sqrt{\left(\frac{\nu}{ka}\right)^2 - 1} - j\nu \ln\left[\frac{\nu}{ka} + \sqrt{\left(\frac{\nu}{ka}\right)^2 - 1}\right] \tag{7.206}$$

(see (7.194), (E.81), and (E.83)), which grows at a factorial rate as $[\nu/(ka)]^\nu$ with increasing $\nu/(ka)$, implying the decrease of δ at a double rate. Figure 7.16 illustrates the rate of decay of the factor $\exp[-j\Phi(\nu, ka)]$.

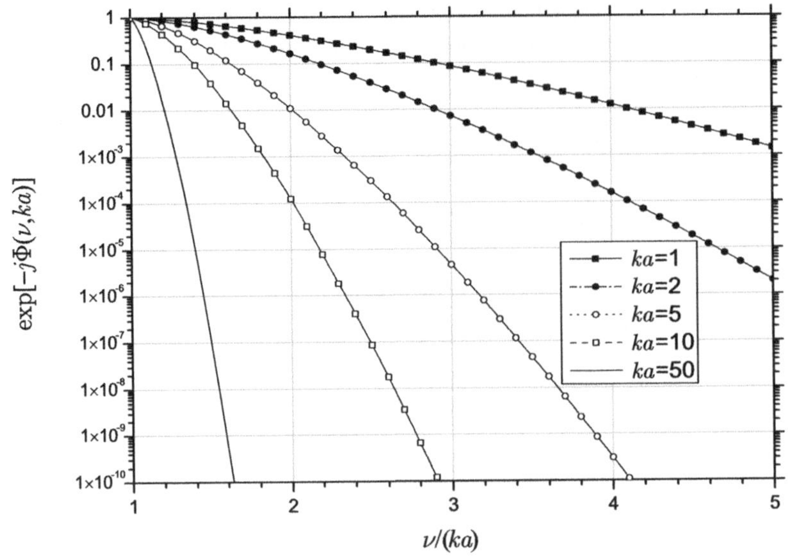

Figure 7.16 Function $\exp[-j\Phi(\nu, ka)]$ of the normalized argument $\nu/(ka)$ for several values of ka.

Using the smallness of δ, we can again write $\nu_m = \nu'_m - j\nu''_m$, where ν'_m is a solution of (7.204) with $\delta = 0$, and determine the small correction ν''_m by applying the perturbation method to (7.204), which gives

$$\nu''_m = \frac{1}{C_m} |\zeta^{(2)}_{\nu'_m-1/2}(ka)|^{-2} \tag{7.207}$$

with

$$C_m = \frac{\arccos[\nu'_m/(k_1 a)]}{\sqrt{1-[\nu'_m/(k_1 a)]^2}} \left\{ \frac{Z_1}{Z}\left[1-\left(\frac{\nu'_m}{k_1 a}\right)^2\right] + \frac{Z}{Z_1}\left[\left(\frac{\nu'_m}{ka}\right)^2 - 1\right]\right\}. \tag{7.208}$$

Since $C_m = O(1)$ and $C_m > 0$, the imaginary parts of zeros with $ka < \nu'_m < k_1 a$ are asymptotically small negative quantities, which suggests the possibility of quasi-resonance effects in the waves associated with the poles.

Let us take a look at the contributions of the poles of the horizontal branch to the field. Deforming the integration contour Γ_{IV} in (7.179) to cross the corresponding poles and denoting the residue sum by E_r^{hb}, we obtain

$$E_r^{\mathrm{hb}} = E_0 \frac{\pi \cos\phi}{(kr)^2} e^{j\frac{\pi}{4}} \sum_m \tau^u_m \frac{\nu_m e^{j\nu_m \pi/2}}{\cos(\nu_m \pi)} \zeta^{(2)}_{\nu_m-\frac{1}{2}}(kr) \frac{\partial}{\partial\theta} P_{\nu_m-\frac{1}{2}}[\cos(\pi-\theta)], \tag{7.209}$$

where the summation includes all the poles of the horizontal branch located in the fourth quadrant and

$$\tau^u_m = 2\,\mathrm{res}\, r^u_{\nu_m-\frac{1}{2}} = \mathrm{res}\, \tilde{r}^u_{\nu_m-\frac{1}{2}}. \tag{7.210}$$

To evaluate the residues (7.210), we rewrite the spherical reflection coefficient (7.182) as

$$\tilde{r}^u_{\nu-\frac{1}{2}} = \frac{j}{y_{\nu-\frac{1}{2}}\zeta^{(2)}_{\nu-\frac{1}{2}}(ka)} \frac{\zeta^{(1)}_{\nu-\frac{1}{2}}(ka) + jy_{\nu-\frac{1}{2}}\zeta^{(1)\prime}_{\nu-\frac{1}{2}}(ka)}{\left[\zeta^{(2)\prime}_{\nu-\frac{1}{2}}(ka)/\zeta^{(2)}_{\nu-\frac{1}{2}}(ka) - j/y_{\nu-\frac{1}{2}}\right]} \tag{7.211}$$

and, noting that $y_{\nu_m-1/2} = j\zeta^{(2)}_{\nu_m-\frac{1}{2}}(ka)/\zeta^{(2)\prime}_{\nu_m-\frac{1}{2}}(ka)$ and using the Wronskian relation (7.45), arrive at

$$\tau^u_m = \frac{-2j}{\left[\zeta^{(2)}_{\nu_m-\frac{1}{2}}(ka)\right]^2} \frac{1}{\frac{\partial}{\partial\nu}\Big|_{\nu=\nu_m}\left[\zeta^{(2)\prime}_{\nu-\frac{1}{2}}(ka)/\zeta^{(2)}_{\nu-\frac{1}{2}}(ka) - j/y_{\nu-\frac{1}{2}}\right]}. \tag{7.212}$$

Inserting Debye's approximations (7.193) and (7.194) in (7.212), taking the derivative of the rapidly varying tangent function in the denominator, and using the formulas

$$\frac{\mathrm{d}}{\mathrm{d}w}\tan w = 1 + \tan^2 w \tag{7.213}$$

and (7.195) to exclude the tangent function, we obtain

$$\tau^u_m \approx \frac{2s_m\sqrt{1-\left(\frac{\nu_m}{ka}\right)^2}\sqrt{1-\left(\frac{\nu_m}{k_1a}\right)^2}\exp[2js_m\Phi(\nu_m, ka)]}{\arccos\left(\frac{\nu_m}{k_1a}\right)\left\{\frac{Z_1}{Z}\left[1-\left(\frac{\nu_m}{k_1a}\right)^2\right] - \frac{Z}{Z_1}\left[1-\left(\frac{\nu_m}{ka}\right)^2\right]\right\}}, \tag{7.214}$$

where $s_m = 1$ for $\nu_m < ka$ and $s_m = -1$ for $\nu_m > ka$.

The Legendre function in (7.209) is approximated by

$$\frac{\partial}{\partial\theta}P_{\nu_m-\frac{1}{2}}[\cos(\pi-\theta)] \approx -\sqrt{\frac{2\nu_m}{\pi\sin\theta}}\cos\left[\nu_m(\pi-\theta)+\frac{\pi}{4}\right], \tag{7.215}$$

which results from the asymptotic formula (7.32) and assumes that θ is not too close to the directions of forward and backscattering so that $\operatorname{Re}\nu_m\sin\theta \gg 1$.

When $\nu_m < kr$ (see (7.194)), the Riccati–Hankel function in (7.209) can be expressed as

$$\zeta^{(2)}_{\nu_m-\frac{1}{2}}(kr) \approx \frac{1}{\sqrt[4]{1-[\nu_m/(kr)]^2}}\mathrm{e}^{-j\Phi(\nu_m,kr)+j\pi/4}, \tag{7.216}$$

and we obtain

$$E^{\mathrm{hb}}_r = 2\frac{E_0\cos\phi}{(kr)^2}\sqrt{\frac{2\pi}{\sin\theta}}\sum_m A^u_m\frac{\mathrm{e}^{-j\nu_m\frac{\pi}{2}-j\Phi(\nu_m,kr)}}{1+\mathrm{e}^{-2j\nu_m\pi}}\cos\left[\nu_m(\pi-\theta)+\frac{\pi}{4}\right], \tag{7.217}$$

where

$$A^u_m = \frac{-j\nu_m^{3/2}\tau^u_m}{\sqrt[4]{1-[\nu_m/(kr)]^2}}. \tag{7.218}$$

Since $\mathrm{Im}\,\nu_m < 0$, we can always use the expansion

$$\frac{1}{1+\mathrm{e}^{-2j\nu_m\pi}} = \sum_{q=0}^{+\infty}(-1)^q \mathrm{e}^{-2j\nu_m\pi q}. \tag{7.219}$$

Let us discuss the physical meaning of the residue contributions by looking at their phases. By definition, $0 < \nu_m < k_1 a$ and $a \leq r$ so that depending on the ratio k_1/k, the quantities $\nu_m/(ka)$ and $\nu_m/(kr)$ can be smaller or greater than unity, which associates the factors $\exp[j\Phi(\nu_m, ka)]$ and $\exp[j\Phi(\nu_m, kr)]$ in (7.214) and (7.217) either with phase or amplitude. If $\nu_m < ka$, then functions $\Phi(\nu_m, kr)$ and $\Phi(\nu_m, ka)$ are almost real-valued (imaginary parts of ν_m are small) and the contributions of the poles can be expressed as

$$E_r^{(m)} = E_{r,m}^{+} + E_{r,m}^{-} \tag{7.220}$$

with

$$E_{r,m}^{\pm} = \frac{E_0\cos\phi}{(kr)^2}\sqrt{\frac{2\pi}{\sin\theta}}\tilde{A}_m^u \sum_{q=0}^{+\infty}(-1)^q \mathrm{e}^{jS_{mq}^{\pm} \pm j\pi/4}, \tag{7.221}$$

where $\tilde{A}_m^u = A_m^u \mathrm{e}^{-2j\Phi(\nu_m, ka)}$ and the phase factors are given by

$$S_{mq}^{+} = \nu_m\left(\frac{\pi}{2} - \theta - 2\pi q\right) + 2\Phi(\nu_m, ka) - \Phi(\nu_m, kr) \tag{7.222}$$

$$S_{mq}^{-} = \nu_m\left(\theta - 3\frac{\pi}{2} - 2\pi q\right) + 2\Phi(\nu_m, ka) - \Phi(\nu_m, kr). \tag{7.223}$$

Using the auxiliary angles ψ_m and $\psi_m^{(2)}$ $(0 \leq \psi_m, \psi_m^{(2)} \leq \pi/2)$ defined by the relations $\nu_m = ka\cos\psi_m = kr\cos\psi_m^{(2)}$ gives

$$\Phi(\nu_m, ka) = ka\sin\psi_m - \nu_m\psi_m \tag{7.224}$$

$$\Phi(\nu_m, kr) = kr\sin\psi_m^{(2)} - \nu_m\psi_m^{(2)} \tag{7.225}$$

so that

$$S_{mq}^{+} = \nu_m\left(\frac{\pi}{2} - \theta - 2\psi_m + \psi_m^{(2)} - 2\pi q\right) + 2ka\sin\psi_m - kr\sin\psi_m^{(2)} \tag{7.226}$$

$$S_{mq}^{-} = \nu_m\left(\theta - 3\frac{\pi}{2} - 2\psi_m + \psi_m^{(2)} - 2\pi q\right) + 2ka\sin\psi_m - kr\sin\psi_m^{(2)}. \tag{7.227}$$

It can be now seen that S_{m0}^{+} is equal to the phase of a ray (see problem 7) that arrives at the surface of the sphere at a point $\mathrm{P}_0^+(a, \theta_0^+)$ with $\theta_0^+ = \theta_m = \pi/2 - \psi_m$ (attachment point), travels along a circular arc to a point $\mathrm{P}_1^+(a, \theta_1^+)$ with $\theta_1^+ = \theta + \psi_m - \psi_m^{(2)}$ (launching point), and leaves the surface under the angle θ_m toward the observation point $\mathrm{P}(r, \theta)$ to arrive at P at an angle $\theta_m^{(2)} = \pi/2 - \psi_m^{(2)}$ measured from the direction PO pointing to the center of the sphere (Figure 7.17). The angles between the direction of propagation of the ray and the normal to the surface of the sphere at the attachment and launching points are therefore equal to θ_m. If $q \geq 1$, the phase S_{mq}^{+} includes a q-fold circumvention of the sphere before the ray departs from the sphere to the observer.

S_{mq}^{-} differs from S_{mq}^{+} by the change $2\pi - \theta$, and defines waves propagating counterclockwise. The corresponding contributions exist for every value of θ $(0 \leq \theta \leq \pi)$. The attachment point P_0^- is located symmetrically to P_0^+ with respect to the symmetry axis QO, and the launching point P_1^- symmetrically to P_1^+ with respect to the direction PO (Figure 7.17). Again, the angles at the attachment and launching points are equal to θ_m.

The interpretation of the terms $E_{r,m}^{+}$ applies as long as $\theta_1^+ > \theta_0^+$ (Figure 7.17) or $\theta > \psi_m^{(2)} - 2\psi_m + \pi/2$. With smaller values of θ, the integral over Γ_{IV} in (7.181) must be transformed to the form (7.186) before applying the residue theorem. Using the substitution (7.187) in (7.209) leads to expressions similar to (7.220) and (7.221), the only difference being

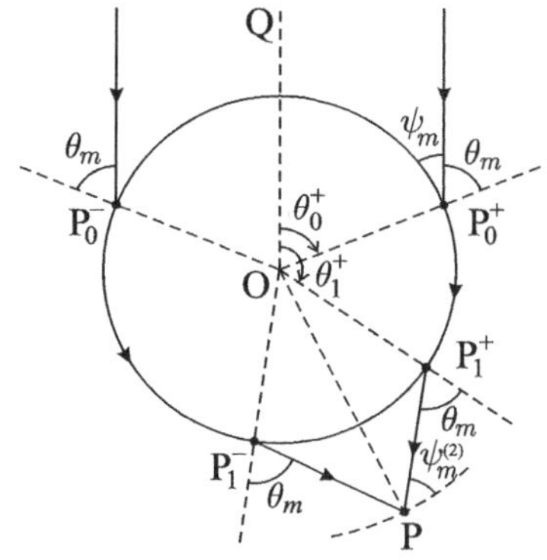

Figure 7.17 Creeping waves on the concave side of the spherical boundary: the ray paths correspond to S_{m0}^+ and S_{m0}^-.

the missing term with $q = 0$ in the expression for $E_{r,m}^+$, which ensures the physically plausible interpretation of the contributions. We return to this case in section 7.5.5 in our analysis of the atmospheric glory. In the rest of the section it will be assumed for brevity that $\theta > \pi/2$ and the representation (7.209) is applicable.

The waves with a given value of m can be thought of as the rays generated by the portions of the front of the incident wave that are located at a distance $a \sin\theta_m = a\cos\psi_m = \nu_m/k$ from the symmetry axis QO. Those with $\nu_m \ll ka$ enter and leave the sphere almost at the right angle to its surface as $\theta_m \approx 0$. In contrast, the rays with $\nu_m \approx ka$ have $\theta_m \approx \pi/2$ and propagate therefore almost tangentially to the sphere at the attachment and launching points, resembling in this respect the usual creeping waves on convex opaque surfaces.

If $ka < \nu_m < k_1 a$ and $\nu_m < kr$, then $\Phi(\nu_m, kr)$ is still real-valued, whereas $\Phi(\nu_m, ka)$ is an almost imaginary quantity given by (7.206), implying that the factor $\exp[-2j\Phi(\nu_m, ka)]$ in (7.214) describes not the phase but rather the amplitude. The residue contributions (7.221) should therefore be rewritten as

$$E_{r,m}^{\pm} = \frac{E_0 \cos\phi}{(kr)^2}\sqrt{\frac{2\pi}{\sin\theta}} A_m^u \sum_{q=0}^{+\infty} (-1)^q e^{j\tilde{S}_{mq}^{\pm} \pm j\pi/4}, \tag{7.228}$$

where

$$\tilde{S}_{mq}^{+} = \nu_m\left(\frac{\pi}{2} - \theta - 2\pi q\right) - \Phi(\nu_m, kr) \tag{7.229}$$

$$\tilde{S}_{mq}^{-} = \nu_m\left(\theta - 3\frac{\pi}{2} - 2\pi q\right) - \Phi(\nu_m, kr). \tag{7.230}$$

Comparing these phase factors with (7.222) and (7.223) suggests that (7.228) describes the rays that are grazing to the sphere at the attachment and launching points ($\psi_m = 0$), propagating between the points along the boundary of the sphere in the interior of the sphere since $ka < \nu_m < k_1 a$ (Figure 7.18). The amplitude A_m^u is strongly damped by the small factor $\exp[-2j\Phi(\nu_m, ka)]$ (see Figure 7.16). This allows the conclusion that contributions of this kind correspond to lateral waves. Similar waves on flat surfaces excited by a line source are described in section 4.4.1 (see the discussion around equation (4.232)). In the case of the sphere the incident field is a plane wave but the boundary is curved, and the evanescent modes are excited due to the curvature of the boundary rather than of the incident wavefront.

Furthermore, since $\nu_m > ka$, $\psi_m = \arccos(\nu_m/ka) = -j|\psi_m|$ and the impact factor of the corresponding ray is formally equal to $a\cosh|\psi_m|$ which is greater than a. A quantum mechanics analogue of this situation is tunneling of a particle into the target (see Grandy (2000, section 7.3)).

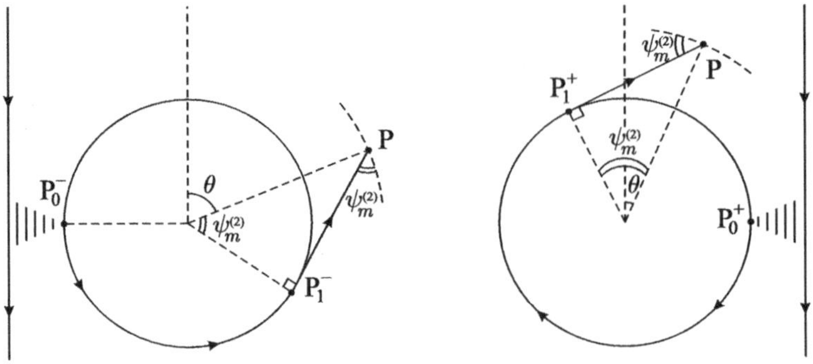

Figure 7.18 Lateral creeping waves on the concave side of the spherical boundary: the ray paths correspond to S^-_{m0} (to the left) and S^+_{m1} (to the right).

Finally, if $\nu_m > kr$, which is possible when $k_1 > k$ and $a \leq r < k_1 a/k$, approximation (7.216) for the Riccati–Hankel function should be replaced with

$$\zeta^{(2)}_{\nu_m - \frac{1}{2}}(kr) \approx \frac{j}{\sqrt[4]{[\nu_m/(kr)]^2 - 1}} e^{j\Phi(\nu_m, kr)}. \tag{7.231}$$

In this case, the factor

$$e^{j\Phi(\nu_m, kr)} = \left[\frac{\nu_m}{kr} + \sqrt{\left(\frac{\nu_m}{kr}\right)^2 - 1}\right]^{\nu_m} \exp\left[-kr\sqrt{\left(\frac{\nu_m}{kr}\right)^2 - 1}\right] \tag{7.232}$$

is an amplitude factor which grows as the observer approaches the sphere, a physical explanation being concentration of the wave in the interior of the sphere.

Thus, the waves associated with the horizontal branch are creeping waves propagating in the interior of the sphere along the concave side of the spherical interface on their way from the attachment to the launching point. The propagation on the concave side explains the much smaller attenuation rate, which is $\nu''_m = O(1)$, compared with $\nu''_m = O[(ka)^{1/3}]$ for the creeping waves on the convex side of the boundary. In this respect they resembles the whispering gallery modes observed near concave boundaries, e.g. (Babič and Buldyrev 1991).

The attenuation of the modes becomes particularly weak in the spheres with $k_1 \gg k$ when $ka < \nu'_m < k_1 a$ (see (7.207)). In this case, ν_m may closely approach a value equal to half an odd number, which will lead to a large value of the factor (7.219) and, therefore, to a sharp increase in the contribution of the corresponding lateral wave to the field (7.217), which may even outweigh the small excitation coefficients of the modes. The presence of such poles of the spherical reflection coefficients on the complex ν plane explains the ripple structure that is observed in various scattering cross sections of low-absorption spheres when the electrical size of the sphere is varied (section 7.5.4).

However, even in lossless spheres, attenuation of the modes is always present, which can be attributed to a constant leakage of energy through the interface into the exterior of the sphere.

Physical interpretation of contributions from the poles of the horizontal branch remains the same for the magnetic part of the solution. The duality transformation (7.171) slightly shifts the poles on the complex ν plane, without changing the whole picture.

The Role of Continuous Spectrum

From mathematical point of view, the imaginary axis and the poles of $F_u(\nu, r, \theta)$ and $F_v(\nu, r, \theta)$ in the forth quadrant on the complex ν plane correspond to continuous and discreet parts of the spectrum of the radial operator (Makarov *et al.*

(1991); see also section 6.4.2). Let us discuss the meaning of the integral over the negative imaginary axis in (7.181). The even part of the integrand function as a function of ν can be estimated by using the evenness of $P_{\nu-1/2}[\cos(\pi-\theta)]$, which follows from the Mehler formula (7.31), and the symmetry properties (7.53) of the Riccati–Hankel functions,

$$F_u(\nu,r,\theta)+F_u(-\nu,r,\theta)=\frac{\nu\sec(\nu\pi)e^{j\nu\frac{\pi}{2}}\left[y_{\nu-\frac{1}{2}}-y_{-\nu-\frac{1}{2}}\right]P_{\nu-\frac{1}{2}}[\cos(\pi-\theta)]\zeta^{(2)}_{\nu-\frac{1}{2}}(kr)}{4\left[\zeta^{(2)}_{\nu-\frac{1}{2}}(ka)+jy_{\nu-\frac{1}{2}}\zeta^{(2)\prime}_{\nu-\frac{1}{2}}(ka)\right]\left[\zeta^{(2)}_{\nu-\frac{1}{2}}(ka)+jy_{-\nu-\frac{1}{2}}\zeta^{(2)\prime}_{\nu-\frac{1}{2}}(ka)\right]}. \tag{7.233}$$

Thus, the even part of $F_u(\nu,r,\theta)$ is zero when the modal admittance $y_{\nu-1/2}$ is an even function of ν. For example, this is the case for every impedance sphere since the admittance $Z/\tilde{Z}_s$ is independent of ν. This includes the PEC case ($\tilde{Z}_s\to 0$) as well. More generally, the admittance of every multilayered sphere is even when the innermost sphere is impedance or PEC, which can be shown by a closer inspection of the recursive relations (7.86)–(7.91). The admittance of spheres with a penetrable core (innermost sphere), for example that of a homogeneous sphere (see (7.176)), is neither even nor odd, implying the finite value of the integral over the negative imaginary axis in (7.181).

The even part of the function $F_v(\nu,r,\theta)$ describing the magnetic part of the fields is estimated by (7.233) with $y_{\nu-\frac{1}{2}}$ replaced with the modal impedance $z_{\nu-\frac{1}{2}}$. The impedance has the same symmetry properties with respect to the change $\nu\to-\nu$, which implies a finite contribution from the continuous spectrum once the innermost sphere is neither PEC nor impedance.

The integral over the imaginary axis does not seem to have any stand-alone physical interpretation as its extraction is always accompanied by capturing poles of the integrand in the fourth quadrant on the complex ν plane. In problems with large absorption in the sphere, it is the superposition of the integral and the residues at the poles of the horizontal branch which are physically meaningful (Makarov and Osipov 1987). The presence of an integral over the imaginary axis has been noticed already in the pioneering work of Watson (1918) for the vertical transmitting and receiving dipoles placed on the surface of an electrically large sphere of finite conductivity (the Earth). Watson estimated this contribution by $O[\exp(-2\mathrm{Im}\,k_1a)]$, that is, as the waves which pass through the sphere; the same estimation can also be found in Bremmer (1949) and Fock (1965). A more accurate estimation $O[\exp(-0.52\mathrm{Im}\,k_1a\theta)]$, where θ is the angle between the transmitter and receiver, both on the surface of the sphere, was obtained by Makarov and Ryzhkov (1977) on noting that the main contribution to the integral comes from the vicinity of the point of intersection of the horizontal branch of poles with the imaginary axis. It is still difficult to interpret this contribution as the attenuation is less than that of the wave propagating over the shortest pass from the source to the receiver. A further analysis has shown that the integral and the contributions from the poles of the horizontal branch almost compensate each other so that their sum is $O\{\exp[-2\mathrm{Im}\,k_1a\sin(\theta/2)]\}$, which corresponds to the wave following the shortest pass through the sphere from the source to the receiver (Makarov and Osipov 1987).

The need to integrate the functions $F_u(\nu,r,\theta)$ and $F_v(\nu,r,\theta)$ over the imaginary axis (or their even parts over the negative imaginary axis) and to account for the presence of multiple pole singularities in the fourth quadrant on the complex ν plane may significantly complicate the evaluation and interpretation of the results of the Watson transformation for transparent spheres. This is particularly true for electrically large low-absorption spheres because of the great many comparable contributions from the poles of the horizontal branch. The pole structure is even more complicated for multilayered spheres, with the residue terms describing various resonance modes in the interior of the spheres.

Representation According to the Number of Reflections

A more preferential way of asymptotic evaluation and physical interpretation of the solution for electrically large low-absorption spheres consists of application of the stationary phase method (Appendix D) to the path integral (7.179). In combination with the Debye expansions of the spherical reflection coefficients, this approach leads to physically clear and mathematically simple estimations of the various contributions forming the field scattered by such spheres. In what follows, a complete GO solution for transparent spheres is derived.

We begin by noting that in the first quadrant of the complex ν plane the integrand of (7.179) has no singularities and is asymptotically small, which can be proved by using the Debye approximations for the Riccati–Bessel functions and the asymptotic formula (7.215) for the Legendre function. By deforming Γ_+, the upper portion of Γ, into any suitable contour Γ'_+ in the first quadrant (Figure 7.15), we obtain an integral which can be neglected.

The remaining portion of Γ, the integration path Γ_-, by construction, goes below the positive real axis, where the poles of $\sec(\nu\pi)$ lie, but above the poles of the spherical reflection coefficients (Figure 7.15). The latter poles consist of the vertical and horizontal branches, both of which, even for lossless spheres, have some negative imaginary parts, asymptotically large as $O[(ka)^{1/3}]$ in the vertical branch and of the order $O(1)$ in the horizontal branch. So, excluding the $O(1)$ vicinity of $\nu = 0$, the function $\sec(\nu\pi)$ can be represented by the geometric progression

$$\sec(\nu\pi) = 2\mathrm{e}^{-j\nu\pi} \sum_{q=0}^{+\infty} (-1)^q \mathrm{e}^{-2j\nu\pi q}, \tag{7.234}$$

which together with (7.215) transforms (7.179) to

$$E_r \approx \sum_{q=0}^{+\infty} \left(E_q^+ + E_q^- \right), \tag{7.235}$$

where

$$E_q^{\pm} = \frac{\sqrt{2}E_0 \mathrm{e}^{-j\frac{\pi}{4}} \cos\phi}{(kr)^2 \sqrt{\pi \sin\theta}} \mathrm{e}^{\pm j\frac{\pi}{4}} (-1)^q \int_0^{+\infty} \nu^{\frac{3}{2}} \Pi^u_{\nu-1/2}(r) \mathrm{e}^{j\nu(\pm\pi\mp\theta-\frac{\pi}{2}-2\pi q)} \,\mathrm{d}\nu. \tag{7.236}$$

Integration in (7.236) can be performed along the positive real axis since on expansion (7.234) the integrand functions have no pole singularities there and Γ_- can be deformed on the positive real axis. Furthermore, the integration range in (7.236) is effectively limited to a number ν_∞ slightly greater than kr since the functions $\psi_{\nu-1/2}(kr)$ and $\psi_{\nu-1/2}(ka)$ in $\Pi^u_{\nu-1/2}(r)$ rapidly decay when $\nu > kr$.

In a vicinity of the origin, approximations (7.215) and (7.234) are inaccurate. As we shall see, main contributions to the integral come from critical points ν_c, which are positive real numbers located on the interval

$$0 < \nu_c < \nu_\infty. \tag{7.237}$$

The case of a critical point approaching the origin requires special treatment. Physically, this corresponds to the case when all reflected rays, either that reflected from the surface of the sphere or those multiply reflected in the interior of the sphere, leave the sphere in the backscattering direction. This results in an increased backscattering, which, in the case of electrically large low-absorption spheres, contributes to the optical phenomenon of glory (section 7.5.5).

At the next step, the function $\Pi^u_{\nu-1/2}(r)$ in the integrand of (7.236) is expressed in the form that includes $\tilde{r}^u_{\nu-1/2}$ (see (7.175)). Expanding $\tilde{r}^u_{\nu-1/2}$ according to (7.161) gives

$$E_r \approx \sum_{p=-1}^{+\infty} \sum_{q=0}^{+\infty} \left(E_{pq}^+ + E_{pq}^- \right), \tag{7.238}$$

where

$$E_{pq}^{\pm} = \frac{E_0 \mathrm{e}^{-j\frac{\pi}{4}} \cos\phi}{(kr)^2 \sqrt{2\pi \sin\theta}} \mathrm{e}^{\pm j\frac{\pi}{4}} (-1)^q I_{pq}^{\pm} \tag{7.239}$$

with

$$I^{\pm}_{-1,q} = \int_0^{\nu_\infty} \nu^{\frac{3}{2}} \left[\zeta^{(1)}_{\nu-\frac{1}{2}}(kr) + R^{00}_{u,\nu-\frac{1}{2}} \zeta^{(2)}_{\nu-\frac{1}{2}}(kr) \right] \mathrm{e}^{j\nu(\pm\pi\mp\theta-\frac{\pi}{2}-2\pi q)} \, \mathrm{d}\nu \tag{7.240}$$

and

$$I^{\pm}_{pq} = \int_0^{\nu_\infty} \nu^{\frac{3}{2}} T^{01}_{u,\nu-\frac{1}{2}} T^{10}_{u,\nu-\frac{1}{2}} \left[R^{11}_{u,\nu-\frac{1}{2}} \right]^p \zeta^{(2)}_{\nu-\frac{1}{2}}(kr) \mathrm{e}^{j\nu(\pm\pi\mp\theta-\frac{\pi}{2}-2\pi q)} \, \mathrm{d}\nu \tag{7.241}$$

for $p \geq 0$.

As apparent from (7.239)–(7.241), the terms E^{+}_{pq} describe waves propagating clockwise with respect to θ, circumventing the sphere q times; the waves E^{-}_{pq} propagate counterclockwise. In what follows the integrals (7.240) and (7.241) will be evaluated asymptotically with the one-dimensional method of the stationary phase (section D.2). We shall see that every term $E^{\pm}_{pq}$ permits a clear physical interpretation.

Functions $\zeta^{(1)}_{\nu-1/2}(w)$, $\zeta^{(2)}_{\nu-1/2}(w)$, and their derivatives with respect to w at $w = ka$, $k_1 a$, and kr will be approximated by the Debye formulas (7.47)–(7.50). Approximations (7.49) and (7.50) for $\zeta^{(2)}_{\nu-1/2}(w)$ and $\zeta^{(2)'}_{\nu-1/2}(w)$ are valid for $0 \leq \nu < +\infty$; functions $\zeta^{(1)}_{\nu-1/2}(w)$ and $\zeta^{(1)'}_{\nu-1/2}(w)$ are described by (7.47) and (7.48) when $0 \leq \nu < w$ and by $\zeta^{(1)}_{\nu-1/2}(w) \approx -\zeta^{(2)}_{\nu-1/2}(w)$ when $w < \nu < +\infty$. The approximations are invalid in an $O(w^{1/3})$ vicinity of the point $\nu = w$ on the complex ν plane (Appendix E), which excludes the penumbra region from the analysis that is to follow.

Incident Wave and the Wave Reflected from the Surface of the Sphere

We start by showing that $E^{+}_{-1,0}$ describes the radial components of the incident wave and the wave reflected from the surface of the sphere. The term $E^{-}_{-1,0}$ does not contribute to the results of asymptotic evaluations because θ is limited to the range $0 \leq \theta \leq \pi$. Asymptotic evaluations for these components of the field are similar to those for PEC cylinders (section 6.4.3), so we just summarize the results of evaluations.

By noting that $R^{00}_{u,\nu-1/2} \approx 1$ when $ka < \nu$, we rewrite (7.239) and (7.240) as

$$E^{+}_{-1,0} = e_0 + e_1, \tag{7.242}$$

where

$$e_0 = \frac{E_0 \cos\phi}{(kr)^2 \sqrt{2\pi \sin\theta}} \left[\int_0^{\nu_\infty} \nu^{\frac{3}{2}} \zeta^{(1)}_{\nu-\frac{1}{2}}(kr) \mathrm{e}^{-j\nu(\theta-\frac{\pi}{2})} \, \mathrm{d}\nu + \int_{ka}^{\nu_\infty} \nu^{\frac{3}{2}} \zeta^{(2)}_{\nu-\frac{1}{2}}(kr) \mathrm{e}^{-j\nu(\theta-\frac{\pi}{2})} \, \mathrm{d}\nu \right] \tag{7.243}$$

$$e_1 = \frac{E_0 \cos\phi}{(kr)^2 \sqrt{2\pi \sin\theta}} \int_0^{ka} \nu^{\frac{3}{2}} R^{00}_{u,\nu-1/2} \zeta^{(2)}_{\nu-\frac{1}{2}}(kr) \mathrm{e}^{-j\nu(\theta-\frac{\pi}{2})} \, \mathrm{d}\nu. \tag{7.244}$$

The term e_0 describes the radial component of the incident wave (7.55),

$$e_0 \approx E^{\mathrm{inc}}_r = E_0 \cos\phi \sin\theta \mathrm{e}^{jkr\cos\theta}, \tag{7.245}$$

which can be checked by evaluating the integrals in (7.243) according to (D.21) at the stationary point $\nu_c = kr \sin\theta$. When $0 \leq \theta < \pi/2$, it is the integral with $\zeta^{(1)}_{\nu-1/2}(kr)$ that has the stationary point with with the contribution (7.245). The second integral in (7.243) has the stationary point $\nu_c = kr \sin\theta$ with the contribution (7.245) in the complimentary case when $\pi/2 < \theta < \theta_s$, where $\theta_s = \pi - \arcsin(a/r)$. When $\theta \to \theta_s$, the stationary point approaches the end of the integration range

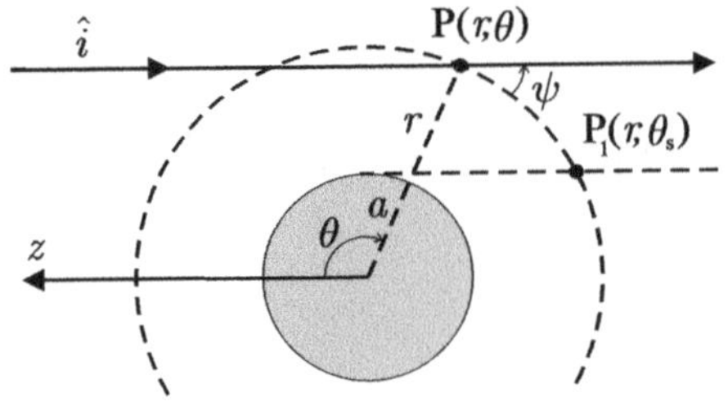

Figure 7.19 Contribution from the incident wave. Point P_1 is located at the shadow boundary.

and the observation point the boundary of the shadow behind the sphere. When $\theta > \theta_s$, the integrals in (7.243) have no stationary points and their asymptotic contribution is equal to zero, as expected, since the observation point is shadowed by the sphere (Figure 7.19).

The other components of the incident field are obtained from (7.245) by using (7.171), (7.172), and (7.173) with $\psi = \theta - \pi/2$, which recovers the incident plane wave (7.3).

The wave reflected from the surface of the sphere is described by the term e_1. On the integration range in (7.244), the function $\zeta^{(2)}_{\nu-1/2}(kr)$ can be approximated by (7.49) and the spherical reflection coefficient, with the exception of the $O[(ka)^{1/3}]$ vicinity of $\nu = ka$, by

$$R^{00}_{u,\nu-1/2} \approx -jR^{00}_{\parallel}(\nu)e^{2j\Phi(\nu,ka)}, \tag{7.246}$$

where $R^{00}_{\parallel}(\nu)$ is given by (7.165) and $\Phi(\nu, w)$ by (E.32). The result is an integral of the form

$$e_1 = \int_0^{ka} A(\nu,k)e^{jS(\nu,k)}\, d\nu \tag{7.247}$$

with

$$S(\nu,k) = \nu\left(\frac{\pi}{2} - \theta\right) + 2\Phi(\nu,ka) - \Phi(\nu,kr) \tag{7.248}$$

and

$$A(\nu,k) = \frac{E_0\cos\phi e^{-j\pi/4}}{(kr)^2\sqrt{2\pi\sin\theta}}\,\frac{\nu^{3/2}R^{00}_{\parallel}(\nu)}{\sqrt[4]{1-\nu^2/(kr)^2}}. \tag{7.249}$$

The equation of the stationary point

$$\frac{\partial S}{\partial \nu}(\nu_c,k) = \frac{\pi}{2} - \theta - 2\arccos\left(\frac{\nu_c}{ka}\right) + \arccos\left(\frac{\nu_c}{kr}\right) = 0 \tag{7.250}$$

has the solution

$$\nu_c = kr\cos\psi = ka\sin\theta_c, \tag{7.251}$$

where the angles ψ and θ_c are associated with a triangle OPC formed by the center of the sphere, the observation point (r,θ,ϕ) and a point (a,θ_c,ϕ) on the surface of the sphere (reflection point), at which the angle between the incidence direction and the direction to the observation point is bisected by the normal to the surface (Figure 7.20). θ_c is therefore

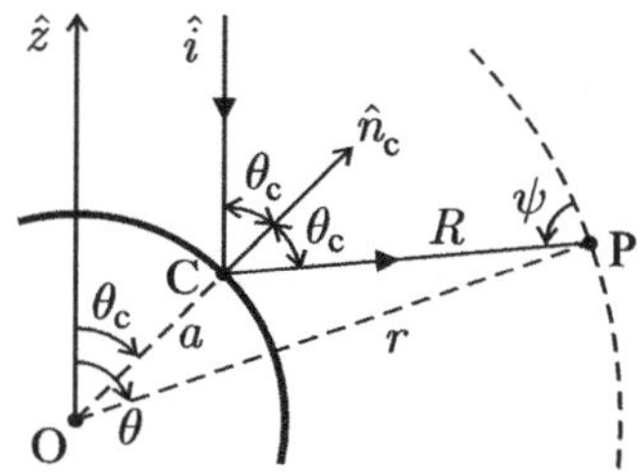

Figure 7.20 Reflection from a sphere: O is the center of the sphere, C is the point of reflection, and P is the observation point.

the incidence angle at the point of reflection and ψ is the angle at which the ray arrives at the observation point (arrival angle).

The remaining quantities in the formula (D.21) for the stationary point contribution are

$$S(v_c, k) = 2ka\cos\theta_c - kr\sin\psi \tag{7.252}$$

$$\frac{\partial^2 S}{\partial v^2}(v_c, k) = \frac{2}{ka\cos\theta_c} - \frac{1}{kr\sin\psi} \tag{7.253}$$

$$\operatorname{sgn}\left[\frac{\partial^2 S}{\partial v^2}(v_c, k)\right] = 1, \tag{7.254}$$

leading to the asymptotic formula

$$e_1 \approx \frac{E_0\cos\phi(\cos\psi)^{\frac{3}{2}}R_{\parallel}^{00}(\theta_c)\mathrm{e}^{j2ka\cos\theta_c - jkr\sin\psi}}{\sqrt{r\sin\theta\sin\psi}\sqrt{2/(a\cos\theta_c) - 1/(r\sin\psi)}} \tag{7.255}$$

which describes a wave reflected with the reflection coefficient $R_{\parallel}^{00}(\theta_c)$ in the specular direction. Its phase is equal to the phase of the incident wave $\exp(jkz)$ at the point of reflection $z = a\cos\theta_c$ plus the phase change in the reflected plane wave over the distance $R = r\sin\psi - a\cos\theta_c$ from the reflection point to the observation point (Figure 7.20). The other factors in (7.255) are geometric in nature and describe the influence of the curvature of the spherical surface on the amplitude of the reflected wave. The physical meaning of (7.255) is addressed more closely in connection with the reflection formula in sections 8.4.1 and 8.4.2 (see also problem 9 in section 8.6).

The components of the electric field in the reflected wave follow from (7.255), (7.171), (7.172), and (7.173) as

$$E_r^{\mathrm{ref}} = \frac{a}{2r}\sqrt{\frac{\sin(2\theta_c)}{\sin\theta}}\frac{E_0R_{\parallel}^{00}(\theta_c)\cos\phi\cos\psi}{\sqrt{\sin\psi - (a/2r)\cos\theta_c}}\mathrm{e}^{j2ka\cos\theta_c - jkr\sin\psi} \tag{7.256}$$

$$E_\theta^{\mathrm{ref}} = -\tan\psi E_r^{\mathrm{ref}} \tag{7.257}$$

$$E_\phi^{\mathrm{ref}} = \sec\psi Z H_r^{\mathrm{ref}}, \tag{7.258}$$

where

$$H_r^{\mathrm{ref}} = -\frac{a}{2r}\sqrt{\frac{\sin(2\theta_c)}{\sin\theta}}\frac{YE_0R_{\perp}^{00}(\theta_c)\sin\phi\cos\psi}{\sqrt{\sin\psi - (a/2r)\cos\theta_c}}\mathrm{e}^{j2ka\cos\theta_c - jkr\sin\psi}. \tag{7.259}$$

The position of the reflection point can be determined from the position of the observation point and the size of the sphere (problem 8). In general, $0 \le \theta_c \le \theta$, the special cases being that $\theta_c = \theta$ when $r = a$, $\theta_c = \theta/2$ when $r \to \infty$, and $\theta_c = 0$ when $\theta = 0$. The latter property ensures the boundedness of (7.256) and (7.259) in the backscattering limit.

If $r \to \infty$ and $\theta \to \pi$, then $\theta_c \to \pi/2$, implying that $\nu_c \to ka$ and the derived asymptotic formulas for the reflected field fail since the asymptotic approximation (7.215) for the Legendre function and Debye's approximations for $\zeta^{(1,2)}_{\nu-1/2}(ka)$ in $R^{00}_{u,\nu-1/2}$ do not apply at $\nu \approx \nu_c$. Physically, this means a situation close to the forward scattering, in which the reflected field is a contribution not from a single reflection point but rather from a ring of reflection points distributed along the shadow boundary on the surface of the sphere. Fields and induced currents in vicinities of shadow boundaries on opaque scattering surfaces are addressed in sections 6.4.4 and 6.6.1.

Waves Passed through the Sphere

We will now show that the terms $E^{\pm}_{pq}$ with $p \geq 0$ and $q \geq 0$ in (7.235) describe the waves that enter the interior of the sphere where they undergo p reflections before leaving the sphere toward the observer. The contributions are given by (7.239) and (7.241). The upper integration range in (7.241) can be truncated to $\nu = ka$ since $|T^{01}_{u,\nu-1/2}T^{10}_{u,\nu-1/2}|$ is negligibly small when $\nu > ka$. Approximating $\zeta^{(2)}_{\nu-1/2}(kr)$ by (7.49) and the spherical reflection and transmission coefficients according to (7.163) by

$$T^{01}_{u,\nu-\frac{1}{2}}T^{10}_{u,\nu-\frac{1}{2}}\left(R^{11}_{u,\nu-\frac{1}{2}}\right)^p \approx T^{01}_{\|}T^{10}_{\|}\left(R^{11}_{\|}\right)^p e^{2j\Phi(\nu,ka)-2j(p+1)\Phi(\nu,k_1a)+jp\pi/2} \tag{7.260}$$

leads to the representation

$$E^{\pm}_{pq} = \int_0^{ka} A^{\pm}_{pq}(\nu,k)e^{jS^{\pm}_{pq}(\nu,k)}\,d\nu \tag{7.261}$$

with

$$S^{\pm}_{pq}(\nu,k) = \nu\left(\pm\pi \mp \theta - \frac{\pi}{2} - 2\pi q\right) + 2\Phi(\nu,ka) - 2(p+1)\Phi(\nu,k_1a) - \Phi(\nu,kr) \tag{7.262}$$

and

$$A^{\pm}_{pq}(\nu,k) = \frac{E_0\cos\phi(-1)^q}{(kr)^2\sqrt{2\pi\sin\theta}}\frac{\nu^{3/2}T^{01}_{\|}T^{10}_{\|}}{\sqrt[4]{1-\nu^2/(kr)^2}}\left(R^{11}_{\|}\right)^p e^{jp\pi/2\pm j\pi/4}. \tag{7.263}$$

When $ka \gg 1$, the integral can be asymptotically evaluated by the stationary phase method (Appendix D.2). The equation for the stationary point ν_c

$$\frac{\partial S^{\pm}_{pq}}{\partial\nu}(\nu_c,k) = -\frac{\pi}{2} - 2\pi q \pm (\pi-\theta) - 2\arccos\left(\frac{\nu_c}{ka}\right) + 2(p+1)\arccos\left(\frac{\nu_c}{k_1a}\right) + \arccos\left(\frac{\nu_c}{kr}\right) = 0 \tag{7.264}$$

can be rewritten as

$$-\frac{\pi}{2} - 2\pi q \pm (\pi-\theta) - 2\psi_c + 2(p+1)\psi_1 + \psi = 0, \tag{7.265}$$

where the parameters ψ_c, ψ_1, and ψ are defined by

$$\nu_c = ka\cos\psi_c = k_1a\cos\psi_1 = kr\cos\psi \tag{7.266}$$

with $0 \leq \psi_c, \psi_1, \psi \leq \pi/2$ (the latter conditions correspond to the choice of the branch of the arccos function in Debye's approximations).

The physical meaning of the parameters is best seen from the expression for the phase function at the stationary point,

$$S_{pq}^{\pm}(\nu_c, k) = 2ka\sqrt{1-\left(\frac{\nu_c}{ka}\right)^2} - 2(p+1)k_1a\sqrt{1-\left(\frac{\nu_c}{k_1a}\right)^2} - kr\sqrt{1-\left(\frac{\nu_c}{kr}\right)^2} \tag{7.267}$$

which follows from (7.262) and (7.264). Introducing for convenience two auxiliary parameters θ_c and θ_1 ($0 \le \theta_c, \theta_1 \le \pi/2$)

$$\psi_c = \frac{\pi}{2} - \theta_c, \quad \psi_1 = \frac{\pi}{2} - \theta_1, \tag{7.268}$$

we obtain

$$S_{pq}^{\pm}(\nu_c, k) = 2ka\cos\theta_c - 2(p+1)k_1a\cos\theta_1 - kr\sin\psi. \tag{7.269}$$

Assuming that θ_c is the incidence angle at the point P_0, at which an incident ray crosses the surface of the sphere and enters the interior of the sphere (injection point), we recognize the phase of the incident wave at P_0 in the quantity $ka\cos\theta_c$ (Figure 7.21). In terms of θ_c and θ_1, relations (7.266) take the form of the Snell law, $k\sin\theta_c = k_1\sin\theta_1$, implying that θ_1 is the refraction angle at P_0 (Figure 7.21). The quantity $2k_1a\cos\theta_1$ in (7.269) is the phase variation over the distance from P_0 to P_1, where P_1 is the intersection point of the refracted ray with the surface of the sphere. This is also the phase variation over the length of the base of every other triangle equal to P_0OP_1, for example P_1OP_2, where O is the center of the sphere (Figure 7.21). Every subsequent reflection in the interior of the sphere therefore changes the phase by $2k_1a\cos\theta_1$. Finally, $kr\sin\psi - ka\cos\theta_c$ is the phase variation over the distance between a point at which the ray leaves the sphere (ejection point), for example P_1 or P_2, and the observation point P, provided that the ray makes the angle θ_c with the normal to the surface of the sphere at the ejection point (Figure 7.21).

Thus, θ_c is the angle at which a ray gets injected in and ejected from the sphere, θ_1 is the refraction angle at which the ray enters and illuminates the interior side of the spherical boundary at every subsequent reflection, and ψ is the angle at which the ray arrives at the observer (Figure 7.21). $E_{pq}^{\pm}$ is the radial component of the electric field in a ray that undergoes p reflections in the interior of the sphere. The case $p = 0$ corresponds to a directly transmitted ray. The parameter q describes the number of times the ray circumvents the sphere before being ejected toward the observer. There are rays circumventing the sphere clockwise and counterclockwise, and they are generated by the opposite parts of the incident wavefront with respect to the z axis (above and below the z axis on Figure 7.21, respectively). The rays remain in the plane that includes the z axis and the observation point, which can be explained by the fact that at every reflection and transmission the directions of the reflected and transmitted rays remain in the plane of incidence.

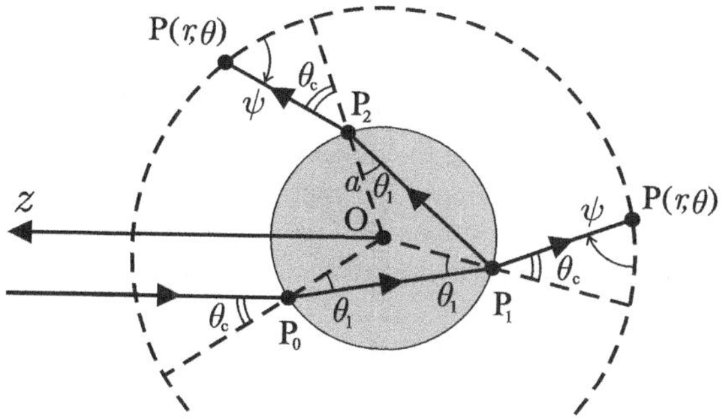

Figure 7.21 Interpretation of parameters θ_c, θ_1, and ψ from the stationary phase equation. Shown is the cross section of the sphere in the plane that includes both the z axis (direction of incidence) and the observation point.

To apply the formula (D.21) for the stationary point contribution, we need the second-order derivative of the phase function at the stationary point. This follows from (7.264) as

$$\frac{\partial^2 S^{\pm}_{pq}}{\partial \nu^2}(\nu_c, k) = \frac{2}{ka\sqrt{1-\nu_c^2/(ka)^2}} - \frac{2(p+1)}{k_1 a\sqrt{1-\nu_c^2/(k_1 a)^2}} - \frac{1}{kr\sqrt{1-\nu_c^2/(kr)^2}} = \frac{2}{ka\cos\theta_c} - \frac{2(p+1)}{k_1 a\cos\theta_1} - \frac{1}{kr\sin\psi}. \tag{7.270}$$

In contrast to (7.254), the sign of this expression can be arbitrary. Introducing the factor

$$\sigma^{\pm}_{pq} = (-1)^q \exp\left\{ j\frac{\pi}{4}\operatorname{sgn}\left[\frac{\partial^2 S^{\pm}_{pq}}{\partial \nu^2}(\nu_c, k)\right] + jp\frac{\pi}{2} \pm j\frac{\pi}{4} \right\}, \tag{7.271}$$

the stationary point contribution is expressed as

$$E^{\pm}_{pq} = \frac{a}{2r}\sqrt{\frac{\sin(2\theta_c)}{\sin\theta\sin\psi}}\, \frac{E_0\cos\phi\cos\psi\,\sigma^{\pm}_{pq} T^{01}_{\parallel}(\theta_c) T^{10}_{\parallel}(\theta_1)\left[R^{11}_{\parallel}(\theta_1)\right]^p}{\sqrt{|1 - a\cos\theta_c/(2r\sin\psi) - (p+1)k\cos\theta_c/(k_1\cos\theta_1)|}} \cdot e^{2jka\cos\theta_c - 2j(p+1)k_1 a\cos\theta_1 - jkr\sin\psi}. \tag{7.272}$$

The radial component of the magnetic field in the wave is obtained from (7.272) by using the duality relations (7.171),

$$H^{\pm}_{pq} = -\frac{a}{2r}\sqrt{\frac{\sin(2\theta_c)}{\sin\theta\sin\psi}}\, \frac{YE_0\sin\phi\cos\psi\,\sigma^{\pm}_{pq} T^{01}_{\perp}(\theta_c) T^{10}_{\perp}(\theta_1)\left[R^{11}_{\perp}(\theta_1)\right]^p}{\sqrt{|1 - a\cos\theta_c/(2r\sin\psi) - (p+1)k\cos\theta_c/(k_1\cos\theta_1)|}} \cdot e^{2jka\cos\theta_c - 2j(p+1)k_1 a\cos\theta_1 - jkr\sin\psi}, \tag{7.273}$$

and the remaining components of the electric field follow from (7.173) as

$$\hat{r}\cdot\mathbf{E}^{\pm}_{pq} = E^{\pm}_{pq}, \quad \hat{\theta}\cdot\mathbf{E}^{\pm}_{pq} = \mp\tan\psi E^{\pm}_{pq}, \quad \hat{\phi}\cdot\mathbf{E}^{\pm}_{pq} = \pm\sec\psi Z H^{\pm}_{pq}. \tag{7.274}$$

The angle θ that defines the position of the point, at which the ray ejected from the sphere intersects the circle of the radius r, is related to the incidence angle θ_c at the injection point. Taking the derivative of the stationary point equation (7.264) with respect to θ_c results in the relation

$$\frac{\partial^2 S^{\pm}_{pq}}{\partial \nu^2}(\nu_c, k) = \pm\frac{1}{ka\cos\theta_c}\frac{\partial\theta}{\partial\theta_c}, \tag{7.275}$$

an implication of which is that the asymptotic expressions (7.272) and (7.273) get infinite when $\partial\theta/(\partial\theta_c) = 0$. Let $\tilde{\theta}_c$ denote the angle θ_c at which the derivative vanishes. Then the vanishing partial derivative means that rays with injection angles from a small but finite vicinity of $\tilde{\theta}_c$ intersect at the same observation point, which results in greater values of the scattered field. In such cases a more accurate evaluation of the integral (7.261) is necessary (see sections 7.5.5 and D.2.3).

Furthermore, when the parameter θ_c crosses the value of $\tilde{\theta}_c$, the second derivative of $S^{\pm}_{pq}$ changes its sign, implying a $\pi/2$ jump in the phase of the factor (7.271). This is a characteristic property of GO ray solutions that are multiplied by the factor j on every contact with a caustic.

Further singularities in (7.272) and (7.273) are at $\theta = 0$ and $\theta = \pi$, which are the directions of back- and forward scattering, respectively. The exact solution for the scattered field is everywhere finite, including back- and forward scattering, and the infinities are due to the employed asymptotic approximation (7.215) for the Legendre function which is limited to the case where $|\nu\sin\theta| \gg 1$. This anomalous back- and forward scattering from electrically large transparent spheres leads to the optical phenomena of glory and corona, which we discuss in section 7.5.5.

For a given position (r, θ) of the observation point, radius a of the sphere, and the ratio k_1/k (refractive index when the background medium is vacuum) there are (infinitely) many solutions of the stationary point equation (7.264). It is convenient to introduce a new variable

$$\tau = \frac{\nu}{ka} \tag{7.276}$$

and rewrite the equation as

$$F_{pq}\left(\tau, \frac{k}{k_1}, \frac{a}{r}\right) = \pm\frac{\pi - \theta}{\pi}, \tag{7.277}$$

where

$$F_{pq}(\tau, \alpha, \beta) = \frac{2}{\pi}\arccos\tau + 2q - \frac{2}{\pi}(p+1)\arccos(\alpha\tau) + \frac{1}{2} - \frac{1}{\pi}\arccos(\beta\tau) \tag{7.278}$$

and $0 \le \arccos\tau,\ \arccos(\alpha\tau),\ \arccos(\beta\tau) \le \pi/2$. One searches for solutions of (7.277) with $0 \le \theta \le \pi, p = 0, 1, 2, \ldots$ and $q = 0, 1, 2, \ldots$ on the interval $0 \le \tau \le 1$. A solution of (7.277) exists when p and q are such that $-1 \le F_{pq}(\tau, k/k_1, a/r) \le 1$, at least on a part of the interval $0 \le \tau \le 1$. There can be one or two solutions for given values of θ, p, q, k/k_1, and a/r. Examples can be found in section 7.5.5.

7.4 Scattered Field in the Far Zone

In this section we use the solutions and results from sections 7.2 and 7.3 to study far fields scattered by spherically symmetric scatterers. The section describes multipole expansions of the far-field coefficient and of various scattering cross sections, the polarization structure of the scattered field, and a simple approximation for far-field coefficients of electrically large ($ka \gg 1$) opaque spheres (section 7.4.1). A general theory of scattering from electrically small ($ka \ll 1$) spheres is presented in section 7.4.2, including the spheres that are large compared with the wavelength in the interior of the sphere.

Furthermore, the general solution for the far scattered field is specialized to several important spherical configurations: the PEC sphere (section 7.4.3), the core-shell sphere (section 7.4.4), and the impedance sphere (section 7.4.5). We describe the characteristic features of scattering from these geometries, including backscattering (RCS), bistatic and forward-scattering cross sections, extinction, absorption, and total scattering cross sections. For every geometry, with the exception of the impedance sphere,[8] low-frequency approximations for the polarizabilities and scattering cross sections are given that include dynamic and radiation corrections up to the terms $O[(ka)^6]$. The particularly important special case of a homogeneous sphere is studied in section 7.5.

7.4.1 Far-field Coefficient, Scattering Cross Sections, and Polarization Structure. Approximations for Electrically Large Spheres

Multipole expansions (7.103)–(7.108) for scattered fields have the same mathematical form for every spherically symmetric scatterer, the only difference being in the coefficients r_n^u and r_n^v, which suggests the possibility of a general description of electromagnetic scattering from a broad variety of spherical configurations. Of course, this is a manifestation of the spherical symmetry of the scattering configuration. In this section we stay on this level of abstraction and describe the scattering properties of the spheres without specifying the coefficients r_n^u and r_n^v.

We begin by writing down multipole expansions for the far-field coefficient and the various scattering cross sections. Then, by using spherical coordinates and plane-of-scattering (PoS) bases, we study general polarization properties of

[8] Impedance approximation is inapplicable to electrically small scatterers (see section 2.3.1).

the scattered field and show the absence of cross-polarization in the PoS bases. Finally, high-frequency approximations ($ka \gg 1$) for the far fields scattered in the specular and forward directions are presented by using the results from sections 3.4.3 and 7.3.4.

Far-field Coefficients and Scattering Cross Sections

Far from the sphere, where $kr \to \infty$ (the physical condition is, of course, that $r \gg a^2/\lambda$), we can simplify the exact expressions (7.103)–(7.108) for the scattered fields by replacing $\zeta_n^{(2)}(kr)$ and $\zeta_n^{(2)\prime}(kr)$ with the asymptotic representations (7.40) and (7.42), which gives for the far-field coefficient (3.86) of the sphere

$$\mathbf{F} = E_0[\hat{\theta} S_1(\theta)\cos\phi - \hat{\phi} S_2(\theta)\sin\phi], \tag{7.279}$$

where

$$S_1(\theta) = j\sum_{n=1}^{\infty}(-1)^{n-1}\frac{2n+1}{n(n+1)}\left[r_n^{\mu}\tau_n(\theta) - r_n^{\nu}\pi_n(\theta)\right] \tag{7.280}$$

$$S_2(\theta) = j\sum_{n=1}^{\infty}(-1)^{n-1}\frac{2n+1}{n(n+1)}\left[r_n^{\mu}\pi_n(\theta) - r_n^{\nu}\tau_n(\theta)\right]. \tag{7.281}$$

The radial component F_r is not present in (7.279) because it decays as $O[(kr)^{-2}]$ and, therefore, faster than the θ and ϕ components of the scattered field. Equation (7.279) implies that the scattered field in the far zone has the amplitude, phase, and polarization behavior of a spherical wave, as expected.

The scattered field in (7.279) is determined by the two functions $S_1(\theta)$ and $S_2(\theta)$, which are sometimes referred to as far-field functions or scattering coefficients and have the meaning of the complex far-field amplitudes of the scattered electric field in the planes parallel and perpendicular to the electric field vector in the incident wave. In the particular case that the observation point lies in the plane defined by the incidence direction and the direction of the electric field vector in the incident wave (the so-called E plane), $\sin\phi = 0$ and the scattered field depends entirely on $S_1(\theta)$. On the other hand, it is the function $S_2(\theta)$ that defines the scattered field if $\cos\phi = 0$ and the observation point lies in the plane defined by the incidence direction and the direction of the magnetic field vector in the incident wave (the so-called H plane). As we shall see later, $S_1(\theta)$ and $S_2(\theta)$ are the only entries in the scattering matrix of every sphere in the PoS bases.

Functions $S_1(\theta)$ and $S_2(\theta)$ depend not only on θ but also on all the geometric and material parameters of the spherical scatterer. As shown in section 7.2.2, interchanging ϵ and μ in every involved medium transforms r_n^{μ} to r_n^{ν} and vice versa, implying that

$$S_1(\theta) \to -S_2(\theta), \quad S_2(\theta) \to -S_1(\theta). \tag{7.282}$$

Denoting by σ_E and σ_H the bistatic scattering cross sections in the E and H planes, we obtain from (3.110) and (7.279)

$$\sigma_E(\theta) = \frac{\lambda^2}{\pi}|S_1(\theta)|^2 = \frac{4\pi}{k^2}|S_1(\theta)|^2 \tag{7.283}$$

$$\sigma_H(\theta) = \frac{\lambda^2}{\pi}|S_2(\theta)|^2 = \frac{4\pi}{k^2}|S_2(\theta)|^2. \tag{7.284}$$

It is often convenient to normalize (7.283) and (7.284) to πa^2, the area of the geometric cross section of a sphere, which leads to the dimensionless quantities

$$\frac{\sigma_{E,H}(\theta)}{\pi a^2} = \frac{4}{(ka)^2}|S_{1,2}(\theta)|^2 \tag{7.285}$$

that in the case of a PEC sphere approach unity at $\theta = 0$ as $ka \to \infty$.

Because of the properties (7.110) and (7.111) of the coefficients $\pi_n(\theta)$ and $\tau_n(\theta)$ with $\theta = 0$ and $\theta = \pi$, it follows from (7.280) and (7.281) that in the limiting cases of back- and forward scattering[9]

$$S_1(0) = S_2(0) \tag{7.286}$$
$$S_1(\pi) = -S_2(\pi), \tag{7.287}$$

implying that

$$\sigma_{\text{back}} = \sigma_E(0) = \sigma_H(0), \tag{7.288}$$
$$\sigma_{\text{fwd}} = \sigma_E(\pi) = \sigma_H(\pi). \tag{7.289}$$

A further implication of (7.286) and (7.287) is that

$$\lim_{\theta \to 0} \mathbf{F} = S_1(0)\mathbf{E}_0 \tag{7.290}$$
$$\lim_{\theta \to \pi} \mathbf{F} = -S_1(\pi)\mathbf{E}_0 \tag{7.291}$$

with $\mathbf{E}_0 = E_0\hat{x}$, the incident electric field at the origin of the coordinate system.

As physically apparent and also as seen from (7.288) and (7.289), back- and forward-scattering cross sections of spherically symmetric scatterers are independent of the polarization of the incident wave. The corresponding series representations follow from (7.280) and (7.283) (or (7.281) and (7.284)) with (7.110) and (7.111) as

$$\sigma_{\text{back}} = \frac{\pi}{k^2}\left|\sum_{n=1}^{\infty}(-1)^{n-1}(2n+1)\left(r_n^u - r_n^v\right)\right|^2 \tag{7.292}$$
$$\sigma_{\text{fwd}} = \frac{\pi}{k^2}\left|\sum_{n=1}^{\infty}(2n+1)\left(r_n^u + r_n^v\right)\right|^2 . \tag{7.293}$$

The extinction cross section σ_{ext} is obtained by using (7.291) in the forward-scattering theorem (3.147),

$$\sigma_{\text{ext}} = \frac{4\pi}{k^2}\operatorname{Im} S_1(\pi), \tag{7.294}$$

and the corresponding series expansion follows from (7.280) with (7.110) and (7.111) as

$$\sigma_{\text{ext}} = -\frac{2\pi}{k^2}\sum_{n=1}^{\infty}(2n+1)\operatorname{Re}\left(r_n^u + r_n^v\right). \tag{7.295}$$

The optical ($ka \to \infty$) limit of σ_{ext} is equal to $2\pi a^2$ (see (3.12)).

The total scattering cross section σ_{T} of a sphere follows from the definition (3.67) with the components of the scattered field given by (7.279) as

$$\sigma_{\text{T}} = \frac{\pi}{k^2}\int_0^{\pi}\left[|S_1(\theta)|^2 + |S_2(\theta)|^2\right]\sin\theta\,\mathrm{d}\theta. \tag{7.296}$$

[9] Relations (7.286) and (7.287) differ in sign from similar relations (10.192) and (10.193) in Bouman *et al.* (1987) because of the definition of $S_2(\theta)$ with an alternative sign in (10.189).

This can be equivalently written as the series expansion

$$\sigma_{\mathrm{T}} = \frac{2\pi}{k^2} \sum_{n=1}^{\infty} (2n+1) \left(|r_n^{\mu}|^2 + |r_n^{\nu}|^2 \right). \tag{7.297}$$

To derive equation (7.297), $S_1(\theta)$ and $S_2(\theta)$ in (7.296) are represented by the series (7.280) and (7.281), which expresses σ_{T} as the double series

$$\sigma_{\mathrm{T}} = \frac{\pi}{k^2} \sum_{n=1}^{\infty} \sum_{n'=1}^{\infty} (-1)^{n+n'-2} \frac{(2n+1)(2n'+1)}{nn'(n+1)(n'+1)} \left\{ \left[r_n^{\mu} (r_{n'}^{\mu})^* + r_n^{\nu} (r_{n'}^{\nu})^* \right] \int_0^{\pi} \left[\tau_n(\theta)\tau_{n'}(\theta) + \pi_n(\theta)\pi_{n'}(\theta) \right] \sin\theta \, \mathrm{d}\theta \right.$$
$$\left. - \left[r_n^{\mu} (r_{n'}^{\nu})^* + (r_{n'}^{\mu})^* r_n^{\nu} \right] \int_0^{\pi} \left[\tau_n(\theta)\pi_{n'}(\theta) + \pi_n(\theta)\tau_{n'}(\theta) \right] \sin\theta \, \mathrm{d}\theta \right\}. \tag{7.298}$$

The integrals in (7.298) are explicitly evaluated as

$$\int_0^{\pi} \left[\tau_n(\theta)\tau_{n'}(\theta) + \pi_n(\theta)\pi_{n'}(\theta) \right] \sin\theta \, \mathrm{d}\theta = \int_0^{\pi} \left[\frac{\mathrm{d}P_{n1}}{\mathrm{d}\theta} \frac{\mathrm{d}P_{n'1}}{\mathrm{d}\theta} + \frac{P_{n1}P_{n'1}}{\sin^2\theta} \right] \sin\theta \, \mathrm{d}\theta = 2\delta_{nn'} n^2 \frac{(n+1)^2}{2n+1} \tag{7.299}$$

(see problem 11) and

$$\int_0^{\pi} \left[\tau_n(\theta)\pi_{n'}(\theta) + \pi_n(\theta)\tau_{n'}(\theta) \right] \sin\theta \, \mathrm{d}\theta = \int_0^{\pi} \left[P_{n1} \frac{\mathrm{d}P_{n'1}}{\mathrm{d}\theta} + P_{n'1} \frac{\mathrm{d}P_{n1}}{\mathrm{d}\theta} \right] \mathrm{d}\theta = P_{n1}(-1)P_{n'1}(-1) - P_{n1}(1)P_{n'1}(1) = 0 \tag{7.300}$$

since $P_{n1}(\pm 1) = 0$ for all $n \geq 1$.

In accordance with the general definition (3.11),

$$\sigma_{\mathrm{abs}} = -\frac{2\pi}{k^2} \sum_{n=1}^{\infty} (2n+1) \left(\mathrm{Re}\, r_n^{\mu} + |r_n^{\mu}|^2 + \mathrm{Re}\, r_n^{\nu} + |r_n^{\nu}|^2 \right). \tag{7.301}$$

The absorption cross section σ_{abs} vanishes when there is no absorption in the interior of the sphere, for example the sphere is PEC or lossless.

In terms of the spherical reflection coefficients $\tilde{r}_n^{\mu}$ and $\tilde{r}_n^{\nu}$ (see (7.145) and (7.146)), the cross sections (7.297) and (7.301) can be rewritten as (problem 12)

$$\sigma_{\mathrm{T}} = \frac{\pi}{2k^2} \sum_{n=1}^{\infty} (2n+1) \left(|1 - \tilde{r}_n^{\mu}|^2 + |1 - \tilde{r}_n^{\nu}|^2 \right) \tag{7.302}$$

$$\sigma_{\mathrm{abs}} = \frac{\pi}{2k^2} \sum_{n=1}^{\infty} (2n+1) \left(2 - |\tilde{r}_n^{\mu}|^2 - |\tilde{r}_n^{\nu}|^2 \right). \tag{7.303}$$

Representation (7.303) makes it obvious that because of (7.148), $\sigma_{\mathrm{abs}} = 0$ for every lossless sphere in a lossless background medium. On the other hand, $\sigma_{\mathrm{abs}} \approx \sigma_{\mathrm{T}}$ if the sphere is such that $|\tilde{r}_n^{\mu}| \to 0$ and $|\tilde{r}_n^{\nu}| \to 0$. This is the case for large impedance-matched (section 7.4.5) and high-absorption spheres (section 7.5.4), or at resonances of electrically small spheres (section 7.4.2).

Normalizing the scattering cross sections to the area of the geometric cross section of the sphere gives the so-called efficiency factors:

$$Q_{\text{ext}} = \frac{\sigma_{\text{ext}}}{\pi a^2}, \quad Q_{\text{T}} = \frac{\sigma_{\text{T}}}{\pi a^2}, \quad Q_{\text{abs}} = \frac{\sigma_{\text{abs}}}{\pi a^2}. \tag{7.304}$$

Equations of this section are applicable to spheres of arbitrary size and material composition (PEC, impedance, material with arbitrary radial stratification). Because of the spherical symmetry of the scatterer, the cross sections σ_{back}, σ_{ext}, σ_{T}, and σ_{abs} (and the corresponding efficiency factors) are independent of the direction of incidence and polarization of the incoming wave.

Polarization Structure of the Scattered Field

For every isotropic spherically symmetric scatterer, the cross sections σ_{back}, σ_{ext}, σ_{T}, and σ_{abs} are independent of the polarization and incidence direction of the incident wave. The polarization dependence takes place, however, in the functions of the bistatic angle, including such quantities as the bistatic scattering cross section and the scattering matrix.

In the limiting cases of back- and forward scattering, the scattered field (7.279) is polarized in the same direction as the incident field (7.3), which follows from the formulas (7.290) and (7.291). For general values of the bistatic angle θ, however, the polarization of the scattered wave differs from that of the incident wave. Let us write down the scattering matrix resulting from the solution (7.279) for the scattered field. We begin with the scattering matrix in spherical coordinates as described in (3.61). Choosing the z axis to be the zenith axis of the coordinate system, and assuming that $\theta_{\text{inc}} = 0$ and $\phi_{\text{inc}} = 0$, we note that $\hat{x} = \hat{\theta}_{\text{inc}}$ and referring to the definition of the incident wave (7.3), write $E^{\text{inc}}_{\theta_{\text{inc}}}(0) = E_0$ and $E^{\text{inc}}_{\phi_{\text{inc}}}(0) = 0$. The components of the scattered field can be written as

$$E^{\text{sc}}_{\theta} = \frac{\mathrm{e}^{-jkr}}{kr} S_1(\theta) \cos\phi E^{\text{inc}}_{\theta_{\text{inc}}}(0) \tag{7.305}$$

$$E^{\text{sc}}_{\phi} = -\frac{\mathrm{e}^{-jkr}}{kr} S_2(\theta) \sin\phi E^{\text{inc}}_{\theta_{\text{inc}}}(0) \tag{7.306}$$

and comparing with the definition of the scattering matrix in (3.61), we obtain that

$$A_{\theta\theta_{\text{inc}}} = \frac{\sqrt{4\pi}}{k} S_1(\theta) \cos\phi, \quad A_{\phi\theta_{\text{inc}}} = -\frac{\sqrt{4\pi}}{k} S_2(\theta) \sin\phi. \tag{7.307}$$

To determine the other two elements of the scattering matrix, a solution for an orthogonally polarized incident wave, $\mathbf{E}_{\text{inc}} = E_0 \hat{y} \exp(jkz)$, is necessary. The solution follows from that for an x-polarized plane wave on noting that the angle ϕ is in fact counted from the direction of polarization of the incident wave. Thus, substituting $\phi \to \phi - \pi/2$ in (7.279) we obtain the solution for the incident wave polarized in the $\hat{y} = \hat{\phi}_{\text{inc}}$ direction,

$$E^{\text{sc}}_{\theta} = \frac{\mathrm{e}^{-jkr}}{kr} S_1(\theta) \sin\phi E^{\text{inc}}_{\phi_{\text{inc}}}(0) \tag{7.308}$$

$$E^{\text{sc}}_{\phi} = \frac{\mathrm{e}^{-jkr}}{kr} S_2(\theta) \cos\phi E^{\text{inc}}_{\phi_{\text{inc}}}(0), \tag{7.309}$$

from which it follows that

$$A_{\theta\phi_{\text{inc}}} = \frac{\sqrt{4\pi}}{k} S_1(\theta) \sin\phi, \quad A_{\phi\phi_{\text{inc}}} = \frac{\sqrt{4\pi}}{k} S_2(\theta) \cos\phi \tag{7.310}$$

and, thus,

$$\overline{\overline{A}} = \frac{\sqrt{4\pi}}{k} \begin{bmatrix} S_1(\theta)\cos\phi & S_1(\theta)\sin\phi \\ -S_2(\theta)\sin\phi & S_2(\theta)\cos\phi \end{bmatrix}. \tag{7.311}$$

Equation (7.311) implies that the scattering matrix of a sphere is fully populated, and becomes diagonal in the special cases of $\phi = 0$, π, and 2π only. For a general scattering angle ϕ, it includes both diagonal and off-diagonal elements, implying the presence of cross-polarization components in the scattered field if the bases of the incidence and scattered waves are defined in spherical coordinates. In section 3.3 we have introduced the definition of co- and cross-polarizations with respect to a reference plane called the PoS and have shown the construction of the PoS bases for the incident and scattered waves and the relation between scattering matrices in spherical and PoS coordinates. Now we will show that the scattering matrix of a sphere, which can be homogeneous (PEC or material) or radially stratified, is always diagonal in the PoS bases. To this end we use the formula (3.81), which relates the scattering matrices in different coordinate systems.

The scattering plane is defined by the vectors

$$\hat{i} = -\hat{z}, \quad \hat{s} = \hat{r} = \hat{x}\sin\theta\cos\phi + \hat{y}\sin\theta\sin\phi - \hat{z}\cos\theta \tag{7.312}$$

and using the definitions (3.70) and (3.71) of the PoS bases, we find that

$$\hat{\eta}_{\text{inc}} = -\hat{x}\sin\phi + \hat{y}\cos\phi = \hat{\phi}, \quad \hat{\xi}_{\text{inc}} = \hat{x}\cos\phi + \hat{y}\sin\phi \tag{7.313}$$

and

$$\hat{\eta} = \hat{\phi}, \quad \hat{\xi} = \hat{\phi}\times\hat{r} = \hat{\theta}. \tag{7.314}$$

Using (7.314) in (3.80) and (7.313) in (3.79) with $\hat{\theta}_{\text{inc}} = \hat{x}$ and $\hat{\phi}_{\text{inc}} = \hat{y}$, we obtain

$$\overline{\overline{V}} = \begin{bmatrix} \cos\phi & -\sin\phi \\ \sin\phi & \cos\phi \end{bmatrix} \tag{7.315}$$

and

$$\overline{\overline{U}} = \begin{bmatrix} 1 & 0 \\ 0 & 1 \end{bmatrix}. \tag{7.316}$$

Inserting (7.311), (7.315), and (7.316) in (3.81) and multiplying the matrices, we arrive at the result

$$\overline{\overline{A}}_{\text{PoS}} = \frac{\sqrt{4\pi}}{k} \begin{bmatrix} S_1(\theta) & 0 \\ 0 & S_2(\theta) \end{bmatrix}. \tag{7.317}$$

Thus, the scattering matrix of spheres is diagonal in the PoS coordinates for the arbitrary scattering direction, regardless of the wavelength, sphere size, and interior structure (homogeneous or radially stratified).

The absence of cross-polarized components in the PoS coordinates can be seen from the geometric decomposition of the electric field vectors into the parallel and perpendicular components (Van de Hulst 1957), which is illustrated in Figure 7.22. The vector of the electric field in the incident wave can be decomposed as

$$E_{\parallel}^{\text{inc}} = \mathbf{E}_{\text{inc}}\cdot\hat{\xi}_{\text{inc}} = E_0\cos\phi, \quad E_{\perp}^{\text{inc}} = \mathbf{E}_{\text{inc}}\cdot\hat{\eta}_{\text{inc}} = -E_0\sin\phi, \tag{7.318}$$

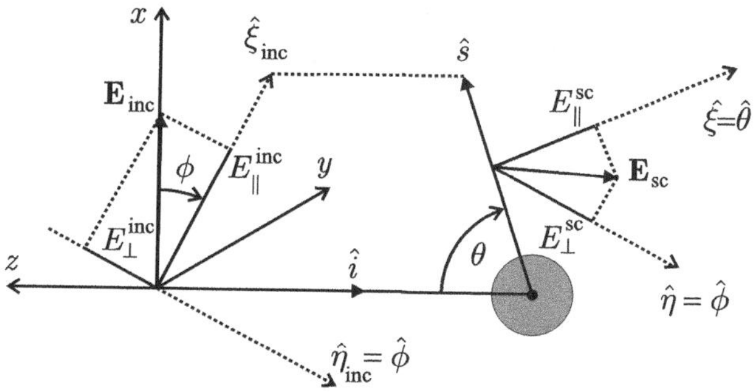

Figure 7.22 Decomposition of electric vectors in the parallel and perpendicular components in the PoS coordinates.

and since

$$E_{\parallel}^{\text{sc}} = E_{\theta}^{\text{sc}}, \quad E_{\perp}^{\text{sc}} = E_{\phi}^{\text{sc}}, \tag{7.319}$$

it follows from (7.279) that

$$E_{\parallel}^{\text{sc}} = S_1(\theta)\frac{\mathrm{e}^{-jkr}}{kr}E_{\parallel}^{\text{inc}}, \quad E_{\perp}^{\text{sc}} = S_2(\theta)\frac{\mathrm{e}^{-jkr}}{kr}E_{\perp}^{\text{inc}}, \tag{7.320}$$

implying that the parallel (perpendicular) component in the incident wave defines the parallel (perpendicular) component in the scattered wave.

Scattering Amplitudes and Cross Sections of Electrically Large Spheres

As shown in section 7.3.4, fields scattered by electrically large ($ka \gg 1$) spheres involve multiple GO and creeping ray contributions. The GO rays typically play the main role, so it is of interest to describe their contributions to the far scattered fields.

According to (7.279), the far-field coefficients $S_{1,2}(\theta)$ are related to the θ and ϕ components of the scattered electric field in the far zone. The asymptotic approach developed in section 7.3.4 yields the radial components of the electric and magnetic fields at an observation point located at an arbitrary distance from the sphere. Denoting the components of the electric and magnetic fields in a GO ray by $E_r^{\pm}$, $E_{\theta}^{\pm}$, $E_{\phi}^{\pm}$, and $H_r^{\pm}$, where $\pm$ indicates the direction of propagation of the ray around the sphere center (clockwise and counterclockwise), we obtain from (7.172) and (7.173) that

$$E_{\theta}^{\pm} = \mp \tan\psi E_r^{\pm}, \quad E_{\phi}^{\pm} = \pm \sec\psi Z H_r^{\pm}, \tag{7.321}$$

where ψ is the arrival angle (Figure 7.14), which depends on r and tends to $\pi/2$ as $r \to \infty$. The contribution of the ray to the scattering coefficients is then obtained from (3.86) and (7.279) by letting the observer go to infinity,

$$S_1^{\text{ray}}(\theta) = \mp\frac{k}{E_0\cos\phi}\lim_{r\to\infty}\left(r\tan\psi\,\mathrm{e}^{jkr}E_r^{\pm}\right) \tag{7.322}$$

$$S_2^{\text{ray}}(\theta) = \mp\frac{kZ}{E_0\sin\phi}\lim_{r\to\infty}\left(r\sec\psi\,\mathrm{e}^{jkr}H_r^{\pm}\right). \tag{7.323}$$

If the sphere is impenetrable, for example PEC, impedance, or lossy (so that the waves directly transmitted through the interior of the sphere can be neglected), then over a broader range of the scattering angle θ the scattered field is mainly formed by the specular reflection from the surface of the sphere. Asymptotic expressions for E_r^+ and H_r^+ in the reflected ray, which propagates clockwise when $0 < \theta < \pi$, are given in (7.256) and (7.259), and inserting these in (7.322) and (7.323), going to the limit $r \to \infty$, and using the relation $\theta_c \to \theta/2$ gives the formulas

$$S_1^{\text{ref}}(\theta) = -\frac{ka}{2} R_{\parallel}\left(\frac{\theta}{2}\right) e^{2jka\cos(\theta/2)} \tag{7.324}$$

$$S_2^{\text{ref}}(\theta) = \frac{ka}{2} R_{\perp}\left(\frac{\theta}{2}\right) e^{2jka\cos(\theta/2)}, \tag{7.325}$$

where $R_{\parallel}$ and $R_{\perp}$ are the reflection coefficients for a flat surface with the material properties of the sphere. For example, for an impedance sphere with the equivalent surface impedance $\tilde{Z}_s$, the reflection coefficients are given by (4.334) and (4.335).

Formulas (7.324) and (7.325) describe the specular component in the far field scattered by opaque spheres. The component provides a good approximation to the far scattered field,

$$S_1(\theta) \approx S_1^{\text{ref}}(\theta), \quad S_2(\theta) \approx S_2^{\text{ref}}(\theta) \tag{7.326}$$

for arbitrary values of θ, except for an asymptotically small vicinity of the forward-scattering direction, where $\pi - \theta = O[(ka)^{-1/3}]$, as not a single reflection point but rather the whole equator ($\theta = \pi/2, 0 \le \phi < 2\pi$) of the sphere contributes to the field reflected in the forward direction. The bistatic scattering cross sections are then approximated by the formulas:

$$\frac{\sigma_E(\theta)}{\pi a^2} \approx \left|R_{\parallel}\left(\frac{\theta}{2}\right)\right|^2 \tag{7.327}$$

$$\frac{\sigma_H(\theta)}{\pi a^2} \approx \left|R_{\perp}\left(\frac{\theta}{2}\right)\right|^2. \tag{7.328}$$

Approximations (7.324), (7.325), (7.326), (7.327), and (7.328) work particularly well in the backscattering direction ($\theta = 0$). For example, because of the general property $R_{\parallel}(0) = -R_{\perp}(0)$ (see (4.168)), equations (7.324) and (7.325) comply with the property (7.286) of exact scattering coefficients. The radar cross section of the sphere is given by

$$\sigma_{\text{back}} \approx \pi a^2 |R_{\parallel,\perp}(0)|^2. \tag{7.329}$$

For the forward-scattering direction ($\theta = \pi$), approximations (7.324), (7.325), (7.326), (7.327), and (7.328) are invalid. Instead, the following simple approximation can be used:

$$S_1(\pi) = -S_2(\pi) \approx -\frac{j}{2}(ka)^2, \tag{7.330}$$

which results from the definition (7.279) of scattering coefficients and the asymptotic formula (3.123) for the far-field coefficient $\mathbf{F}_{\text{fwd}}$. Because of (7.289) and (7.294), equation (7.330) implies that in the high-frequency limit ($ka \gg 1$)

$$\sigma_{\text{fwd}} \approx \pi a^2 (ka)^2 \tag{7.331}$$

$$\sigma_{\text{ext}} \approx 2\pi a^2, \tag{7.332}$$

which are special cases of the more general formulas (3.124) and (3.12).

In the case of very low absorption, as in water in the visible, the contributions of the waves passed through the sphere must be accounted for; this case is addressed in section 7.5.5.

7.4.2 Electrically Small Spheres: Dipole, Quasi-static, and Resonance Approximations

In this section we study electromagnetic scattering from spheres that are small compared with the wavelength, that is $ka \ll 1$. The material composition of the scatterer can be arbitrary (but spherically symmetric), which may include magnetic or non-magnetic, lossless or lossy, PEC or dielectric, homogeneous or multilayered configurations.

When $ka \ll 1$, the far-field functions $S_{1,2}(\theta)$ can often be described with sufficient accuracy by the terms with $n = 1$ in the multipole expansions (7.280) and (7.281). In this approximation the dependence of the scattered field on θ is similar to that of an infinitesimal dipole, and that is why this case is referred to as dipole approximation.

The dependence on frequency can be categorized in two cases. Once the frequency is so low or the size of the sphere so small that not only $ka \ll 1$ but also $|k_l a_{l-1}| \ll 1$ with $l = 1, \ldots, L$ (see Figure 7.2), we arrive at the quasi-static limit, also known as Rayleigh approximation.[10] It is still different from the static case as the scattered field must be an outgoing wave at infinity.

If the sphere involves materials with high values of relative permittivity and/or permeability, then the smallness of ka does not necessarily mean smallness of the sphere on the scale of the wavelength in the involved materials, which may imply the presence of interior resonances and, therefore, a complicated dependence of the scattering parameters on the frequency.

The resonance behavior is particularly pronounced when the losses in the involved materials are weak. The dipole approximation may be insufficiently accurate in this case. For example, a correct description of the absorption cross section σ_{abs} requires retaining the terms with $n = 1, 2$, that is, dipole and quadrupole, in (7.280) and (7.281).

In general, multipole coefficients of every order n may have resonances, the scattering is then dominated by a single multipole term of the nth order. We provide simple closed-form approximations for this case (resonance approximation) and show that even for spheres with low absorption, σ_{abs} has strong maxima under the resonance conditions.

The approximations for the scattering coefficients and scattering cross sections of electrically small spheres are of great theoretical and practical value because of their analytical simplicity and the fact that electrically small spheres are often encountered in applications, for example as building blocks of metamaterials.

Far-field Coefficients in Dipole Approximations

If ka is small, then the far-field functions (7.280) and (7.281) can be approximated by the lowest-order terms with $n = 1$ (dipole approximation) as

$$S_1(\theta) \approx \frac{3}{2} j \left(r_1^u \cos\theta - r_1^v \right) \tag{7.333}$$

$$S_2(\theta) \approx \frac{3}{2} j \left(r_1^u - r_1^v \cos\theta \right) \tag{7.334}$$

since

$$r_n^u, r_n^v = O[(ka)^{2n+1}]. \tag{7.335}$$

We first prove the estimation (7.335) and then study the scattered fields in the dipole approximation.

To prove estimations (7.335), express r_n^u and r_n^v, which are given by (7.100) and (7.101), as

$$r_n^u = -\frac{\psi_n(ka)}{\zeta_n^{(2)}(ka)} \frac{1 + jy_n \psi_n'(ka)/\psi_n(ka)}{1 + jy_n \zeta_n^{(2)\prime}(ka)/\zeta_n^{(2)}(ka)} \tag{7.336}$$

$$r_n^v = -\frac{\psi_n(ka)}{\zeta_n^{(2)}(ka)} \frac{1 + jz_n \psi_n'(ka)/\psi_n(ka)}{1 + jz_n \zeta_n^{(2)\prime}(ka)/\zeta_n^{(2)}(ka)} \tag{7.337}$$

[10] The Rayleigh approximation is the dipole approximation for very small spheres.

and note the approximations

$$\frac{\psi_n'(ka)}{\psi_n(ka)} \approx \frac{n+1}{ka} \tag{7.338}$$

$$\frac{\zeta_n^{(2)\prime}(ka)}{\zeta_n^{(2)}(ka)} \approx -\frac{n}{ka} \tag{7.339}$$

$$\frac{\psi_n(ka)}{\zeta_n^{(2)}(ka)} \approx -\frac{j(ka)^{2n+1}}{(2n+1)[(2n-1)!!]^2}, \tag{7.340}$$

which follow from the small-argument approximations (7.37) and (7.38) of Riccati–Bessel functions. Inserting (7.338), (7.339), and (7.340) in (7.336) and (7.337) gives

$$r_n^u \approx \frac{j(ka)^{2n+1}}{(2n+1)[(2n-1)!!]^2} \frac{1+\tilde{y}_n(n+1)}{1-\tilde{y}_n n} \tag{7.341}$$

$$r_n^v \approx \frac{j(ka)^{2n+1}}{(2n+1)[(2n-1)!!]^2} \frac{1+\tilde{z}_n(n+1)}{1-\tilde{z}_n n}, \tag{7.342}$$

where

$$\tilde{y}_n = \frac{jy_n}{ka}, \quad \tilde{z}_n = \frac{jz_n}{ka}. \tag{7.343}$$

Parameters $\tilde{y}_n$ and $\tilde{z}_n$, as well as the modal admittances and impedances y_n and z_n defined in (7.98) and (7.99), are configuration dependent and functions of ka, but it is evident from (7.341) and (7.342) that regardless of their precise dependence on ka, the estimations (7.335) hold as long as neither $\tilde{y}_n$ nor $\tilde{z}_n$ tend to n^{-1}. The latter is a resonance case which requires special treatment (see section 7.5.2). On the other hand, if $\tilde{y}_n = -(n+1)^{-1}$ or $\tilde{z}_n = -(n+1)^{-1}$, then r_n^u or, respectively, r_n^v is $o[(ka)^{2n+1}]$.

Coefficients r_1^u and r_1^v are directly related to the polarizabilities of the sphere. To see this, let us apply the general formulas from section 3.6 to the case of a sphere. Because of the rotational symmetry of the sphere with respect to any of the Cartesian axes, the polarizabiliy tensors are diagonal (see (3.238)),

$$\overline{\overline{P}} = P_{11}\overline{\overline{I}}, \quad \overline{\overline{M}} = M_{11}\overline{\overline{I}} \tag{7.344}$$

and, thus, fully characterized by the two scalar quantities P_{11} and M_{11}, which are related to the electric and magnetic dipole moments that are induced on the sphere by any incident field. The θ and ϕ components of the scattered field (3.61) are obtained on multiplication of the incident wave (7.3) with the elements of the scattering matrix in (3.300) and (3.302) ($\theta_{\mathrm{inc}} = 0$, $\phi_{\mathrm{inc}} = 0$), which gives

$$\mathbf{E}_{\mathrm{sc}} = k^2 E_0 \frac{e^{-jkr}}{4\pi r}[\hat{\theta}(P_{11}\cos\theta + M_{11})\cos\phi - \hat{\phi}(P_{11} + M_{11}\cos\theta)\sin\phi] \tag{7.345}$$

and therefore by comparison with (7.279)

$$S_1(\theta) = \frac{k^3}{4\pi}(P_{11}\cos\theta + M_{11}) \tag{7.346}$$

$$S_2(\theta) = \frac{k^3}{4\pi}(P_{11} + M_{11}\cos\theta). \tag{7.347}$$

Now, comparing (7.346) and (7.347) with (7.333) and (7.334) we arrive at

$$P_{11} = \frac{6\pi j}{k^3} r_1^u, \quad M_{11} = -\frac{6\pi j}{k^3} r_1^v, \tag{7.348}$$

which express the elements of the polarizability tensors through the expansion coefficients with $n = 1$ from the series solution for scattering of a plane wave by a sphere. Using (7.341) and (7.342) in (7.348) gives

$$P_{11} \approx -2\pi a^3 \frac{1+2\tilde{y}_1}{1-\tilde{y}_1} \tag{7.349}$$

$$M_{11} \approx 2\pi a^3 \frac{1+2\tilde{z}_1}{1-\tilde{z}_1}. \tag{7.350}$$

The parameters $\tilde{y}_1$ and $\tilde{z}_1$ in (7.349) and (7.350) can be found by using recursive procedures for the admittance y_1 and impedance z_1 (section 7.2), and the result includes various Riccati–Bessel functions of the arguments $k_l a_l$ and $k_l a_{l-1}$ with $l = 1, 2, 3, \ldots, L$. If these parameters are not small, then $\tilde{y}_1$ and $\tilde{z}_1$ may strongly depend on ka, thus leading to frequency-dependent polarizabilities P_{11} and M_{11}. In the quasi-static limit, $\tilde{y}_1$ and $\tilde{z}_1$ tend to some limiting values, which are independent of ka, implying that P_{11} and M_{11} are proportional to a^3 and therefore to the volume of the sphere, $V = 4\pi a^3/3$, which is a general feature of the electric and magnetic polarizabilities of voluminous bodies (section 3.6.3).

Once P_{11} and M_{11} are determined, the scattering functions $S_{1,2}(\theta)$ follow from (7.346) and (7.347), bistatic scattering cross sections $\sigma_{E,H}$ from (7.283) and (7.284), and the other cross sections are expressed as

$$\sigma_{\text{back}} = \frac{k^4}{4\pi}|M_{11} + P_{11}|^2 \tag{7.351}$$

$$\sigma_{\text{fwd}} = \frac{k^4}{4\pi}|M_{11} - P_{11}|^2 \tag{7.352}$$

$$\sigma_{\text{ext}} = k\,\text{Im}(M_{11} - P_{11}) \tag{7.353}$$

$$\sigma_{\text{T}} = \frac{k^4}{6\pi}\left(|P_{11}|^2 + |M_{11}|^2\right). \tag{7.354}$$

The absorption cross section can be determined from the general formula (7.301).

Due to the spherical symmetry of the scatterer, the cross sections are independent of the polarization and the direction of illumination. Equations (7.353) and (7.354) are in agreement with the low-frequency approximations (3.304), (3.305), (3.306), and (3.307) since $M_{33} = M_{11}$ and $P_{33} = P_{11}$ for bodies of revolution.

Since $P_{11}, M_{11} = O(V)$, it follows from (7.351)–(7.354) that σ_{back} and σ_{T} are quantities of the order k^4V^2, whereas σ_{ext} of the order kV, implying that $\sigma_{\text{ext}} \gg \sigma_{\text{T}}$ and $\sigma_{\text{ext}} \approx \sigma_{\text{abs}}$ for electrically small lossy spheres.

Low-frequency Approximations for Low-absorption Spheres

For spheres involving materials with low losses, σ_{ext} is small and comparable with σ_{T}. In order to correctly describe σ_{abs}, which is the difference of the two and vanishes for lossless spheres, an approximation of σ_{ext} more accurate than (7.353) is necessary. This is achieved by accounting for the terms with $n = 2$ in (7.295),

$$\sigma_{\text{ext}} = -\frac{2\pi}{k^2}\left[3\text{Re}(r_1^u + r_1^v) + 5\text{Re}(r_2^u + r_2^v)\right], \tag{7.355}$$

and expanding the coefficients in the power series

$$r_1^u = \alpha_1^u(ka)^3 + \beta_1^u(ka)^4 + \gamma_1^u(ka)^5 + \delta_1^u(ka)^6 \tag{7.356}$$

$$r_1^v = \alpha_1^v(ka)^3 + \beta_1^v(ka)^4 + \gamma_1^v(ka)^5 + \delta_1^v(ka)^6 \tag{7.357}$$

$$r_2^u = \gamma_2^u(ka)^5 + \delta_2^u(ka)^6 \tag{7.358}$$

$$r_2^u = \gamma_2^v(ka)^5 + \delta_2^v(ka)^6, \tag{7.359}$$

which ensures the required accuracy of the approximation of σ_{ext}. The terms of the order $(ka)^3$ in the expansions (7.356) and (7.357) are sometimes referred to as static terms, the terms of the order higher than $(ka)^3$ as dynamic corrections, and the terms proportional to $(ka)^6$ as radiation corrections since they describe radiation losses in scattering from lossless spheres. Expressions (7.358) and (7.359) describe the contributions of quadrupole moments to the scattered fields.

The expansions (7.356)–(7.359) are derived from (7.336) and (7.337) with $n = 1, 2$ by using the following small-argument approximations of the ratios of the involved Riccati–Bessel functions:

$$\frac{\psi_1'(ka)}{\psi_1(ka)} = \frac{2}{ka} - \frac{ka}{5} - \frac{(ka)^3}{175} + O[(ka)^5] \tag{7.360}$$

$$\frac{\zeta_1^{(2)\prime}(ka)}{\zeta_1^{(2)}(ka)} = -\frac{1}{ka} + ka - j(ka)^2 - (ka)^3 + O[(ka)^4] \tag{7.361}$$

$$\frac{\psi_1(ka)}{\zeta_1^{(2)}(ka)} = -\frac{j}{3}(ka)^3 + \frac{j}{5}(ka)^5 + \frac{1}{9}(ka)^6 + O[(ka)^7] \tag{7.362}$$

$$\frac{\psi_2'(ka)}{\psi_2(ka)} = \frac{3}{ka} - \frac{ka}{7} + O[(ka)^3] \tag{7.363}$$

$$\frac{\zeta_2^{(2)\prime}(ka)}{\zeta_2^{(2)}(ka)} = -\frac{2}{ka} + \frac{ka}{3} + O[(ka)^3] \tag{7.364}$$

$$\frac{\psi_2(ka)}{\zeta_2^{(2)}(ka)} = -\frac{j}{45}(ka)^5 + \frac{j}{189}(ka)^7 + O[(ka)^{10}]. \tag{7.365}$$

Inserting (7.360)–(7.362) in (7.336) with $n = 1$ gives

$$r_1^u \approx (ka)^3 \left[\frac{j}{3} - \frac{j}{5}(ka)^2 - \frac{1}{9}(ka)^3\right] \frac{1 + \tilde{y}_1[2 - (ka)^2/5]}{1 - \tilde{y}_1[1 - (ka)^2 + j(ka)^3]}, \tag{7.366}$$

where $\tilde{y}_1$ is defined in (7.343). If $|1 - \tilde{y}_1| \gg (ka)^2$, then the ratio in the right-hand side can be expanded in a geometric progression to give (7.356) with the coefficients:

$$\alpha_1^u = \frac{j}{3}\frac{1 + 2\tilde{y}_1}{1 - \tilde{y}_1}, \quad \beta_1^u = 0, \quad \gamma_1^u = -\frac{j}{5}\frac{1 + 3\tilde{y}_1 + \tilde{y}_1^2}{(1 - \tilde{y}_1)^2}, \quad \delta_1^u = (\alpha_1^u)^2. \tag{7.367}$$

The coefficients in (7.357) can be obtained by simply replacing $\tilde{y}_1$ with $\tilde{z}_1$,

$$\alpha_1^v = \frac{j}{3}\frac{1 + 2\tilde{z}_1}{1 - \tilde{z}_1}, \quad \beta_1^v = 0, \quad \gamma_1^v = -\frac{j}{5}\frac{1 + 3\tilde{z}_1 + \tilde{z}_1^2}{(1 - \tilde{z}_1)^2}, \quad \delta_1^v = (\alpha_1^v)^2. \tag{7.368}$$

Furthermore, using (7.363)–(7.365) in (7.336) and (7.337) with $n = 2$ results in the formulas

$$\gamma_2^u = \frac{j}{45}\frac{1 + 3\tilde{y}_2}{1 - 2\tilde{y}_2}, \quad \delta_2^u = 0, \quad \gamma_2^v = \frac{j}{45}\frac{1 + 3\tilde{z}_2}{1 - 2\tilde{z}_2}, \quad \delta_2^v = 0 \tag{7.369}$$

for the expansions of r_2^u and r_2^v in (7.358) and (7.359). Parameters $\tilde{z}_1$, $\tilde{y}_2$, and $\tilde{z}_2$ are defined in (7.343).

The obtained expansions are very general as they apply to spheres of arbitrary material constitution. Note two features of these expansions. The first feature is the vanishing coefficients β_1^u, β_1^v, δ_2^u, and δ_2^v, which can be seen as a consequence of the form of the expansions (7.360)–(7.365), in which the term that is by a factor ka smaller than the leading term is always zero. The second feature is the simple relation of δ_1^u and δ_1^v to α_1^u and α_1^v, respectively.

The importance of this latter property can be seen from the expansions:

$$\sigma_{\text{ext}} = -2\pi k a^3 \left\{3\text{Re}(\alpha_1^u + \alpha_1^v) + \left[3\text{Re}(\gamma_1^u + \gamma_1^v) + 5\text{Re}(\gamma_2^u + \gamma_2^v)\right](ka)^2 + 3\text{Re}\left[(\alpha_1^u)^2 + (\alpha_1^v)^2\right](ka)^3\right\} + O(k^5 a^7) \tag{7.370}$$

$$\sigma_{\text{T}} = 6\pi k^4 a^6 \left(|\alpha_1^u|^2 + |\alpha_1^v|^2\right) + O(k^5 a^7). \tag{7.371}$$

For lossless spheres (see problem 14),

$$\mathrm{Re}(\alpha_1^u) = \mathrm{Re}(\alpha_1^v) = \mathrm{Re}(\gamma_1^u) = \mathrm{Re}(\gamma_1^v) = \mathrm{Re}(\gamma_2^u) = \mathrm{Re}(\gamma_2^v) = 0 \tag{7.372}$$

and

$$\mathrm{Re}\left[(\alpha_1^u)^2\right] = -|\alpha_1^u|^2, \quad \mathrm{Re}\left[(\alpha_1^v)^2\right] = -|\alpha_1^v|^2, \tag{7.373}$$

which ensures that $\sigma_{\text{ext}} = \sigma_{\text{T}}$ and $\sigma_{\text{abs}} = 0$ in this limit.

Insertion of (7.356) and (7.357) in (7.348) gives corresponding expansions for the dipole polarizabilities,

$$P_{11} = 6\pi j a^3 \left[\alpha_1^u + \gamma_1^u (ka)^2 + (\alpha_1^u)^2 (ka)^3\right] \tag{7.374}$$

$$M_{11} = -6\pi j a^3 \left[\alpha_1^v + \gamma_1^v (ka)^2 + (\alpha_1^v)^2 (ka)^3\right]. \tag{7.375}$$

It can be seen from (7.370) that knowledge of P_{11} and M_{11} is insufficient to specify σ_{ext} to the order of k^3a^5 and higher because of the presence of the coefficients γ_2^u and γ_2^v, which are due to the quadrupole terms r_2^u and r_2^v. For most practical cases, however, the coefficients α_1^u and α_1^v (and therefore P_{11} and M_{11}) are most relevant as they determine the leading terms in σ_{ext} and σ_{T}.

In the analysis of scattering by dipole particles, inverse polarizabilities are often of interest. The corresponding approximations follow from (7.374) and (7.375) as

$$\begin{aligned}\frac{1}{P_{11}} &= \frac{\alpha_1^u - \gamma_1^u (ka)^2}{6\pi j a^3 (\alpha_1^u)^2} + \frac{jk^3}{6\pi} \\ \frac{1}{M_{11}} &= -\frac{\alpha_1^v - \gamma_1^v (ka)^2}{6\pi j a^3 (\alpha_1^v)^2} - \frac{jk^3}{6\pi}.\end{aligned} \tag{7.376}$$

The terms of the order k^3 describe the radiation losses and are in agreement with the formula (3.308) for rotationally symmetric lossless scatterers.

The multipole coefficients r_n^u and r_n^v transform into each other on interchanging ϵ and μ (section 7.2.2), implying the same symmetry of the expansion coefficients in (7.356)–(7.359) and the relations $P_{11} \to -M_{11}$ and $M_{11} \to -P_{11}$ between the polarizabilities.

The expansions (7.356)–(7.359) describe spheres that are not necessarily small compared with the wavelength in the material of the spheres, implying that the coefficients α_1^u, α_v^u, γ_1^u, γ_1^v, γ_2^u, and γ_2^v through the parameters $\tilde{y}_{1,2}$ and $\tilde{z}_{1,2}$ can be complicated functions of ka. In the quasi-static limit, when $|k_l a_{l-1}| \ll 1$ with $l = 1, 2, \ldots, L$, the expansion coefficients can be again expanded in power series in the powers of ka, and on collecting terms with the same power of ka, one obtains expansions similar to (7.356)–(7.359) with

$$\beta_1^u = \beta_1^v = \delta_2^u = \delta_2^v = 0, \quad \delta_1^u = (\alpha_1^u)^2, \quad \delta_1^v = (\alpha_1^v)^2. \tag{7.377}$$

Later in this chapter we give examples of such expansions for several spherical configurations.

The approximations developed in this section fail when $|n\tilde{y}_n - 1| = O[(ka)^2]$ and/or $|n\tilde{z}_n - 1| = O[(ka)^2]$. In these cases the behavior of r_n^u and r_n^v is determined by pole singularities which correspond to electric and magnetic multipole ($n = 1$ dipole, $n = 2$ quadrupole, etc.) resonances in the scattered field. The scattering coefficients and cross sections are dominated by the nth term in their multipole expansions.

Resonance Approximations of Multipole Coefficients

When all material parameters are real-valued or almost real-valued, the multipole coefficients $r_n^{u,v}$ and, therefore, the scattering cross sections may have sharp peaks, which are referred to as resonances. The peaks are always bounded, even in lossless configurations, since material spheres as all open resonators always scatter power, as soon as they can be coupled to external fields. The approximations developed in the previous sections are inapplicable close to the resonance values of the material parameters. Provided that all but one material parameter of the configuration are real-valued, a simple approximation of the resonance case can be derived by considering the multipole coefficients as functions on a complex plane of the selected material parameter. The resonances are then simply poles on the complex plane, which results in useful representations of the multipole coefficients in the resonance case.

Assume that p is a material parameter, for example a permittivity or permeability of the sphere or of a spherical layer involved in the spherical configuration if the sphere is multilayered. Consider the parameter as a complex variable $p = p' - jp''$ ($p'' \geq 0$ for physically passive media). The other parameters must be real-valued in order for the approach to be valid.

Now assume that p is close to a pole $p_\infty = p'_\infty - jp''_\infty$ ($p''_\infty < 0$) of a multipole coefficient r_n, where r_n can be either r_n^u or r_n^v. Even in a lossless sphere, because of the scattering losses, p cannot be equal to p_∞ but can be quite close to it when $p = p'_\infty$ since $|p''_\infty| \ll 1$.

In the close proximity of the pole, where $|p - p_\infty| = O(p''_\infty)$, scattering cross sections are dominated by the resonance multipole term, so we can write

$$\sigma_{\text{back}} \approx \frac{\pi}{k^2}(2n+1)^2|r_n|^2 \tag{7.378}$$

$$\sigma_{\text{fwd}} \approx \frac{\pi}{k^2}(2n+1)^2|r_n|^2 \tag{7.379}$$

$$\sigma_{\text{ext}} \approx -\frac{2\pi}{k^2}(2n+1)\text{Re}\, r_n \tag{7.380}$$

$$\sigma_{\text{T}} \approx \frac{2\pi}{k^2}(2n+1)|r_n|^2 \tag{7.381}$$

$$\sigma_{\text{abs}} \approx -\frac{2\pi}{k^2}(2n+1)\left(\text{Re}\, r_n + |r_n|^2\right). \tag{7.382}$$

Note the relations

$$\sigma_{\text{back}} \approx \sigma_{\text{fwd}} \approx \left(n + \frac{1}{2}\right)\sigma_{\text{T}}, \tag{7.383}$$

valid in the resonance region.

Close to the resonance,

$$r_n = \frac{A}{p - p_\infty}, \tag{7.384}$$

where A, the residue of r_n at the pole, is independent of p. The parameter A can be determined by requiring that $\sigma_{\text{abs}} = 0$ when $p'' = 0$. Indeed, since

$$|r_n|^2 = \frac{(\text{Re}\, A)^2 + (\text{Im}\, A)^2}{(p' - p'_\infty)^2 + (p'' - p''_\infty)^2}$$

$$\text{Re}\, r_n = \frac{\text{Re}\, A(p' - p'_\infty) - \text{Im}\, A(p'' - p''_\infty)}{(p' - p'_\infty)^2 + (p'' - p''_\infty)^2},$$

the absorption cross section is proportional to

$$\text{Re}\, r_n + |r_n|^2 = \frac{(\text{Re}\, A)^2 + \text{Re}\, A(p' - p'_\infty) - \text{Im}\, A(p'' - p''_\infty) + (\text{Im}\, A)^2}{(p' - p'_\infty)^2 + (p'' - p''_\infty)^2}. \tag{7.385}$$

With $p'' = 0$ and arbitrary p', this value is zero when $\mathrm{Re}\,A = 0$ and $\mathrm{Im}\,A = -p''_\infty$, implying the approximations:

$$r_n \approx \frac{-jp''_\infty}{p - p_\infty} \tag{7.386}$$

$$\sigma_{\text{back}} \approx \sigma_{\text{fwd}} \approx \frac{\pi}{k^2}\frac{(2n+1)^2(p''_\infty)^2}{(p'-p'_\infty)^2 + (p''-p''_\infty)^2} \tag{7.387}$$

$$\sigma_{\text{ext}} \approx \frac{2\pi}{k^2}\frac{(2n+1)p''_\infty(p''_\infty - p'')}{(p'-p'_\infty)^2 + (p''-p''_\infty)^2} \tag{7.388}$$

$$\sigma_{\text{T}} \approx \frac{2\pi}{k^2}\frac{(2n+1)(p''_\infty)^2}{(p'-p'_\infty)^2 + (p''-p''_\infty)^2} \tag{7.389}$$

$$\sigma_{\text{abs}} \approx -\frac{2\pi}{k^2}\frac{(2n+1)p''_\infty p''}{(p'-p'_\infty)^2 + (p''-p''_\infty)^2}. \tag{7.390}$$

So, in the $O(p''_\infty)$ vicinity of the pole, the resonant multipole coefficient and the cross sections are completely determined by the value of p_∞.

In the limit of a lossless configuration, when $p'' = 0$,

$$r_n \approx \frac{-jp''_\infty}{p' - p'_\infty - jp''_\infty} \tag{7.391}$$

$$\sigma_{\text{back}} \approx \sigma_{\text{fwd}} \approx \frac{\pi}{k^2}\frac{(2n+1)^2(p''_\infty)^2}{(p'-p'_\infty)^2 + (p''_\infty)^2} \tag{7.392}$$

$$\sigma_{\text{ext}} = \sigma_{\text{T}} \approx \frac{2\pi}{k^2}\frac{(2n+1)(p''_\infty)^2}{(p'-p'_\infty)^2 + (p''_\infty)^2} \tag{7.393}$$

$$\sigma_{\text{abs}} = 0. \tag{7.394}$$

Considered as functions of p', cross sections (7.392) and (7.393) have a peak at $p = p'_\infty$ with the characteristic width proportional to $|p''_\infty|$. For example, when $|p' - p'_\infty| = |p''_\infty|$, the values of the cross sections equal one half of their maximum values. As we shall see in section 7.5.3, the imaginary part of the pole can be extremely small, particularly for multipoles of higher orders, which results in extremely sharp peaks in scattering characteristics. At $p = p'_\infty$,

$$r_n \approx -1 \tag{7.395}$$

$$\sigma_{\text{back}} \approx \sigma_{\text{fwd}} \approx \frac{\pi}{k^2}(2n+1)^2 \tag{7.396}$$

$$\sigma_{\text{ext}} = \sigma_{\text{T}} \approx \frac{2\pi}{k^2}(2n+1). \tag{7.397}$$

Let us take a closer look at the determination of p_∞ for the multipole coefficients $r_n^{u,v}$ with arbitrary n. According to (7.341) and (7.342), the poles occur when

$$n\tilde{y}_n \to 1, \quad n\tilde{z}_n \to 1. \tag{7.398}$$

The normalized admittances and impedances of multilayered spheres are real-valued when all material parameters are real-valued, and equations (7.398) imply real-valued solutions for p_∞, which is unphysical since scattering losses take place even for lossless spheres.

To determine the small but fundamentally important imaginary part of p_∞, note that equations (7.398) are approximations resulted from the simple approximation (7.339). By using a more accurate approximation

$$\frac{\zeta_n^{(2)\prime}(ka)}{\zeta_n^{(2)}(ka)} = -\frac{n}{ka}[1 + \delta_n(ka)], \tag{7.399}$$

where

$$\delta_n(ka) = \frac{(ka)^2}{n(1-2n)} + \frac{(ka)^4}{n(3-2n)(1-2n)^2} + \cdots + \frac{j(ka)^{2n+1}}{n[(2n-1)!!]^2}, \tag{7.400}$$

which follows from (7.38), one obtains more accurate conditions on the poles,

$$n\tilde{y}_n[1 + \delta_n(ka)] = 1, \quad n\tilde{z}_n[1 + \delta_n(ka)] = 1. \tag{7.401}$$

It is the imaginary term in (7.400) that leads to solutions of the resonance equations which correctly accounts for the scattering losses.

Strictly speaking, in order to determine p''_∞, solutions of (7.401) up to the terms of $O[(ka)^{2n+1}]$ are required, which can be quite laborious when n is larger than unity. There is, however, a simpler way of determining p''_∞, for arbitrary n, which is based on reduction of (7.341) or (7.342) to the form (7.386). If, for example, $r_n = r_n^u$ (electric resonances), then expanding $\tilde{y}_n(p)$ in (7.341) in the Taylor series

$$\tilde{y}_n(p) \approx \tilde{y}_n(p_0) + \frac{\partial \tilde{y}_n}{\partial p}(p_0)(p - p_0), \tag{7.402}$$

where p_0 denotes a solution of (7.398) for which $\tilde{y}_n(p_0) = n^{-1}$, transforms (7.341) to the formula

$$r_n^u \approx \frac{-j(ka)^{2n+1}}{n^2[(2n-1)!!]^2 \frac{\partial \tilde{y}_n}{\partial p}(p_0)(p - p_0)}. \tag{7.403}$$

Comparing (7.403) with (7.386) we obtain

$$p'_\infty \approx p_0, \quad p''_\infty \approx \frac{(ka)^{2n+1}}{n^2[(2n-1)!!]^2 \frac{\partial \tilde{y}_n}{\partial p}(p_0)}. \tag{7.404}$$

In the case of a magnetic resonance ($r_n = r_n^v$), $\tilde{y}_n$ is replaced with $\tilde{z}_n$.

Resonant Absorption in Low-absorption Spheres

When applied to the absorption cross section, the resonance representations lead to general conclusions about the resonance peaks of σ_{abs}. It follows from (7.390) that σ_{abs} vanishes as $p'' \to 0$ and $+\infty$, which implies a maximum of the cross section in the range $0 < p'' < +\infty$. The partial derivatives

$$\frac{\partial \sigma_{\text{abs}}}{\partial p'} \approx \frac{4\pi(2n+1)p''_\infty p''(p' - p'_\infty)}{k^2\left[(p' - p'_\infty)^2 + (p'' - p''_\infty)^2\right]^2} \tag{7.405}$$

$$\frac{\partial \sigma_{\text{abs}}}{\partial p''} \approx -\frac{2\pi(2n+1)p''_\infty\left[(p' - p'_\infty)^2 - (p'')^2 + (p''_\infty)^2\right]}{k^2\left[(p' - p'_\infty)^2 + (p'' - p''_\infty)^2\right]^2} \tag{7.406}$$

vanish when $p' = p'_\infty$ and $p'' = \pm p''_\infty$, and since $p''_\infty < 0$, the physical maximum of σ_{abs} is located at

$$p = p'_\infty + jp''_\infty, \tag{7.407}$$

that is, at the value that is a complex conjugate of p_∞. At that point

$$\sigma_{\text{abs}} = \frac{\pi}{2k^2}(2n+1), \tag{7.408}$$

and the absorption efficiency (see (7.304)) is large

$$Q_{\text{abs}} = \frac{(2n+1)}{2(ka)^2} \tag{7.409}$$

in spite of the low losses in the material of the sphere. The other quantities from (7.386)–(7.390) are equal to

$$r_n \approx -\frac{1}{2} \tag{7.410}$$

$$\sigma_{\text{back}} \approx \sigma_{\text{fwd}} \approx \frac{\pi}{4k^2}(2n+1)^2 \tag{7.411}$$

$$\sigma_{\text{ext}} \approx \frac{\pi}{k^2}(2n+1) \tag{7.412}$$

$$\sigma_{\text{T}} \approx \frac{\pi}{2k^2}(2n+1) \tag{7.413}$$

so that

$$\sigma_{\text{abs}} = \sigma_{\text{T}} = \frac{1}{2}\sigma_{\text{ext}}. \tag{7.414}$$

In section 7.5.3 the resonance formulas are applied to the case of a small homogeneous sphere.

7.4.3 PEC Spheres

A PEC sphere is an important particular case because of the simplicity of the solution and its practical relevance. A metal sphere serves, for example, as a reference target for calibration of radars and antenna measurement facilities.

The exact solution for the field scattered by a PEC sphere results from the solution (7.128) and (7.129) for the impedance sphere with $\tilde{Z}_s = 0$,

$$r_n^u = -\frac{\psi_n'(ka)}{\zeta_n^{(2)\prime}(ka)}, \quad r_n^v = -\frac{\psi_n(ka)}{\zeta_n^{(2)}(ka)}. \tag{7.415}$$

These formulas can be also obtained from the general formulas (7.100) and (7.101) for r_n^u and r_n^v, in which $y_n \to \infty$ and $z_n \to 0$. The limit of the PEC sphere can be derived from solutions for any radially stratified sphere by letting the conductivity of the material in the uppermost layer go to infinity, that is, $\text{Im}\,\epsilon_1 \to -\infty$, while keeping the permeability μ_1 finite.

When inserted in (7.103)–(7.108), the formulas (7.415) lead to the well-known solution for a PEC sphere (e.g., Ruck *et al.* (1970), Bouman *et al.* (1987), and Van de Hulst (1957)). The far scattered field is given by (7.280) and (7.281) with (7.415).

Electrically Small Spheres

The Rayleigh scattering from PEC spheres has been already addressed in section 3.6 by using a low-frequency solution for an ellipsoid. In this section the formulas for polarizabilities and scattering cross sections of the PEC sphere are rigorously derived from the exact solution of the scattering problem in the limit $ka \to 0$.

When ka is small, (7.415) with the small-argument approximations (7.37) and (7.38) of Riccati–Bessel functions gives

$$r_1^u \approx -\frac{2j}{3}(ka)^3, \quad r_1^v \approx \frac{j}{3}(ka)^3, \tag{7.416}$$

and it follows from (7.348) that

$$P_{11} = 4\pi a^3, \quad M_{11} = 2\pi a^3, \tag{7.417}$$

which is the well-known result that we used in section 3.6.

More accurate approximations for the polarizabilities

$$P_{11} = 4\pi a^3 \left[1 + \frac{3}{10}(ka)^2 - j\frac{2}{3}(ka)^3\right] \tag{7.418}$$

$$M_{11} = 2\pi a^3 \left[1 - \frac{3}{5}(ka)^2 + \frac{j}{3}(ka)^3\right] \tag{7.419}$$

follow from (7.374) and (7.375) with the coefficients

$$\alpha_1^u = -\frac{2j}{3}, \quad \gamma_1^u = \gamma_1^v = -\frac{j}{5}, \quad \alpha_1^v = \frac{j}{3}, \tag{7.420}$$

which result from (7.367) and (7.368) in the limits $y_1 \to \infty$ and $z_1 \to 0$. Imaginary parts of P_{11} and M_{11} are due to the scattering losses and are by a factor of $(ka)^3$ smaller than the real parts. This result is in agreement with the general formulas (3.308) (problem 15).

Using (7.417) in (7.346) and (7.347) gives the low-frequency approximation for the far-field functions:

$$S_1(\theta) \approx (ka)^3 \left(\frac{1}{2} + \cos\theta\right) \tag{7.421}$$

$$S_2(\theta) \approx (ka)^3 \left(1 + \frac{1}{2}\cos\theta\right). \tag{7.422}$$

It follows from (7.421) and (7.422) that the field scattered by electrically small PEC spheres is fully polarized at $\theta = 120°$ as no field polarized in the scattering plane is scattered.

Using (7.421) and (7.422) in (7.283) and (7.284) leads to the simple approximations for the bistatic scattering cross sections of small PEC spheres as

$$\sigma_E(\theta) \approx \pi a^2 (ka)^4 (1 + 2\cos\theta)^2 \tag{7.423}$$

$$\sigma_H(\theta) \approx \pi a^2 (ka)^4 (2 + \cos\theta)^2, \tag{7.424}$$

implying the values

$$\sigma_E(0) = \sigma_H(0) \approx 9\pi a^2 (ka)^4 \tag{7.425}$$

and

$$\sigma_E(\pi) = \sigma_H(\pi) \approx \pi a^2 (ka)^4 \tag{7.426}$$

in the back- and forward-scattering limits.

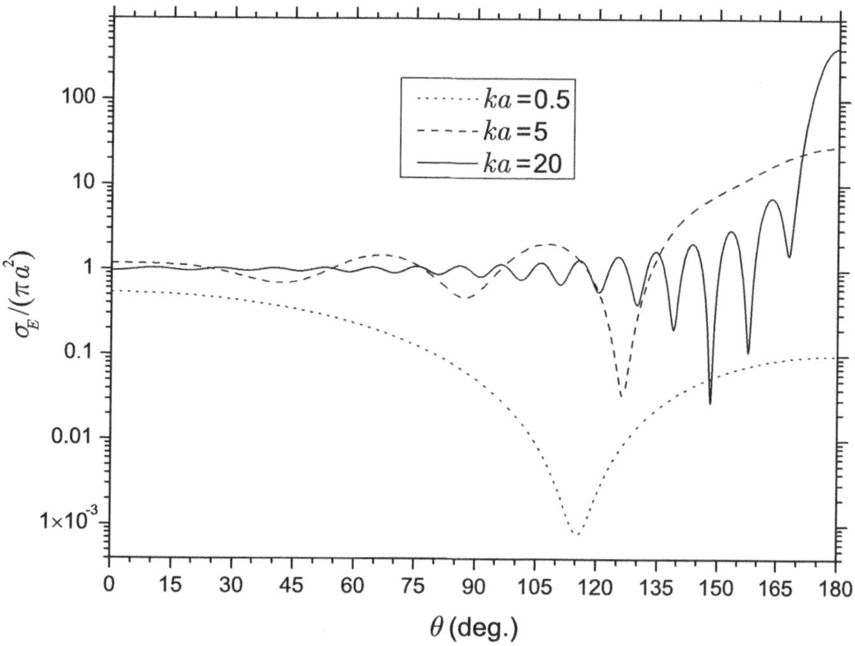

Figure 7.23 Normalized E-plane bistatic cross section of variously sized PEC spheres as a function of the bistatic angle.

Since PEC scatterers do not absorb energy, the extinction cross section σ_{ext} of every PEC sphere is equal to σ_{T}. The cross sections of a small PEC sphere

$$\sigma_{\text{ext}} = \sigma_{\text{T}} \approx \frac{10}{3}\pi a^2 (ka)^4 \tag{7.427}$$

can be obtained from (7.371) with α_1^u and α_1^v given in (7.420) or by inserting (7.417) in (7.354). Because of the scatterer's symmetry, the formula (7.427) is valid for arbitrary direction of incidence and polarization of the incidence wave and, on normalization to λ^2, coincides with (3.326). In Figure 3.24 the low-frequency approximation for the total cross section is plotted as a function of frequency and compared with the exact dependence. Further features of scattering from electrically small PEC spheres are illustrated in Figures 7.23, 7.24, and 7.25 in comparison with larger PEC spheres.

Electrically Large Spheres

In the alternative case of an electrically large ($ka \gg 1$) PEC sphere, approximations for the scattering coefficients and bistatic cross sections result from equations (7.324), (7.325), (7.326), (7.327), and (7.328) with $R_{\parallel} \to 1$ and $R_{\perp} \to -1$ as

$$S_1(\theta), S_2(\theta) \approx -\frac{ka}{2} e^{2jka\cos(\theta/2)} \tag{7.428}$$

$$\sigma_E(\theta), \sigma_H(\theta) \approx \pi a^2. \tag{7.429}$$

The asymptotic approximation (7.429) does not depend on the bistatic angle θ, which is a manifestation of a more general asymptotic result, the so-called monostatic-bistatic equivalence theorem (section 8.4.3) valid for arbitrarily curved convex PEC surfaces.

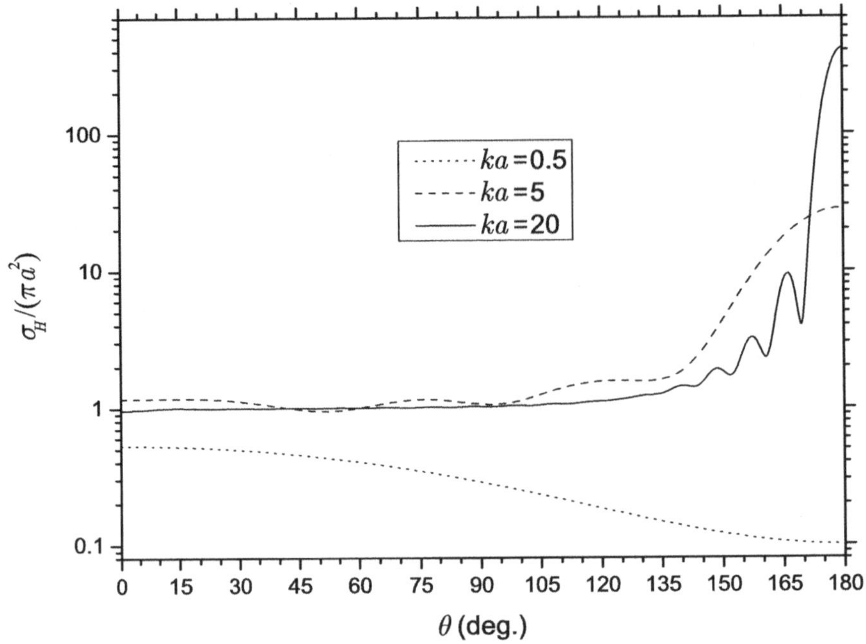

Figure 7.24 The same as Figure 7.23 but in the H plane.

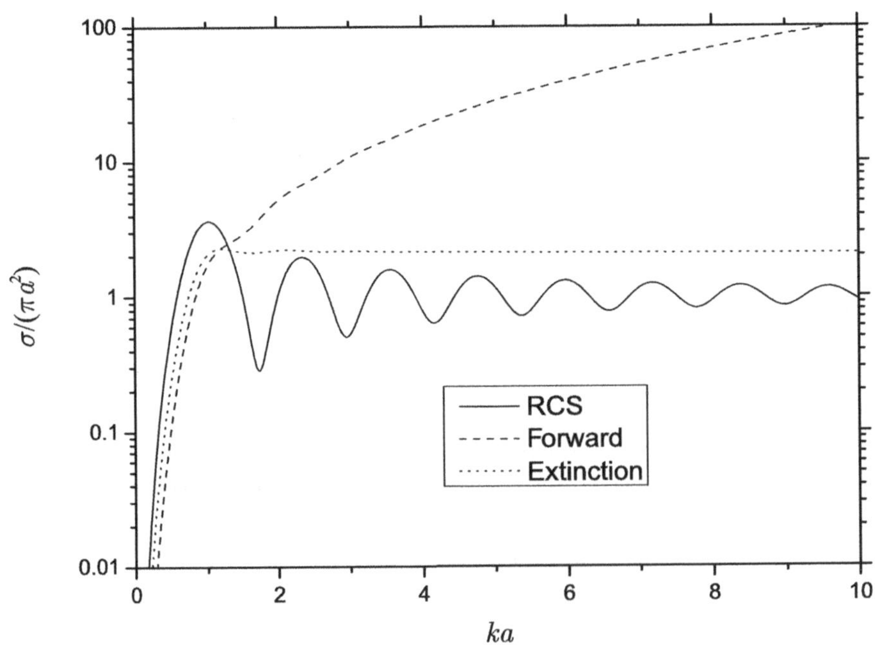

Figure 7.25 Various scattering cross sections of a PEC sphere as functions of ka.

The approximation (7.428) is inapplicable in the asymptotically small vicinity of the forward-scattering direction, when $\pi - \theta = O[(ka)^{-1/3}]$, where the approximations used in the derivation of (7.324) and (7.325) are no longer valid. The scattered field in this region is formed not by a single reflection point but rather by an asymptotically narrow ring-shaped vicinity of the shadow boundary on the sphere surface around the equator of the sphere. The contributions from the various portions of this ring are comparable to each other and their overlapping makes the scattered field quickly oscillating in the region close to the forward scattering. An asymptotic description of the scattered field in this region requires special treatment (Senior and Goodrich 1964), except for the case of forward scattering ($\theta = \pi$) in which the simple formulas (7.330), (7.331), and (7.332) are applicable when ka is sufficiently large.

Numerical Illustrations

Figures 7.23 and 7.24 present $\sigma_E(\theta)$ and $\sigma_H(\theta)$ in dependence on the bistatic angle θ for an electrically small ($ka = 0.5$), resonance ($ka = 5$), and large ($ka = 20$) PEC spheres. The cross sections are normalized to πa^2, the cross-section area of the sphere, which makes them functions of ka only. The deep minimum close to $\theta = 120°$ in the curve for $\sigma_E(\theta)$ with $ka = 0.5$ corresponds to the general property of scattering from electrically small PEC spheres discussed around equations (7.421) and (7.422). It has been seen that already with $ka = 5$, the cross sections at $\theta = 0$ are close to πa^2, which is the GO limit (7.429) of $\sigma_E(\theta)$ and $\sigma_H(\theta)$. The values at the forward-scattering direction ($\theta = \pi$) quickly grow as ka increases, in accordance with the estimation (7.331). It is also seen that $\sigma_E(\theta)$ and $\sigma_H(\theta)$ of the spheres with $ka = 5, 20$ are indeed almost constant over an extended range of values of θ; the oscillations in the curves are concentrated in a vicinity of the forward-scattering direction, which gets narrower with growing ka. Finally, we observe that the variations in $\sigma_E(\theta)$ (Figure 7.23) are more pronounced than in $\sigma_H(\theta)$ (Figure 7.24).

Figure 7.25 summarizes the information about the frequency dependence of the various scattering cross sections of PEC spheres. The plotted quantities are the backscattering cross section (7.292) (RCS), the forward-scattering cross section (7.293), and the extinction cross section (7.294); the latter coincides with the total scattering cross section for PEC scatterers. As seen from the figure, the curves asymptotically follow their low-frequency ($ka \ll 1$)

$$\frac{\sigma_{\text{back}}}{\pi a^2} \approx 9(ka)^4, \quad \frac{\sigma_{\text{fwd}}}{\pi a^2} \approx (ka)^4, \quad \frac{\sigma_{\text{ext}}}{\pi a^2} \approx \frac{10}{3}(ka)^4 \tag{7.430}$$

and high-frequency ($ka \gg 1$)

$$\frac{\sigma_{\text{back}}}{\pi a^2} \approx 1, \quad \frac{\sigma_{\text{fwd}}}{\pi a^2} \approx (ka)^2, \quad \frac{\sigma_{\text{ext}}}{\pi a^2} \approx 2 \tag{7.431}$$

behaviors. Note that σ_{fwd} is nine times smaller than σ_{back} for electrically small spheres but greater by the factor of $(ka)^2$ for electrically large spheres. The dominance of σ_{fwd} over σ_{back} is clearly seen from Figure 7.5. Further curves illustrating the back- and total scattering cross sections of PEC spheres as functions of the electrical size of the sphere can be also found in Figures 3.16 and 3.24.

7.4.4 *Core-shell Spheres*

Magneto-dielectric Core inside a Magneto-dielectric Shell

The core-shell sphere consists of an inner sphere of radius a_1 with material constants ϵ_2 and μ_2 placed inside a concentric sphere of radius a with material constants ϵ_1 and μ_1 (Figure 7.26). The configuration is quite general as it includes as special cases the PEC sphere ($\text{Im}\,\epsilon_1 = \infty$), the coated PEC sphere ($\text{Im}\,\epsilon_2 = \infty$), and a hollow spherical shell ($\epsilon_2 = \epsilon$, $\mu_2 = \mu$). The homogeneous sphere is obtained when either $\epsilon_2 \to \epsilon_1$ and $\mu_2 \to \mu_1$ or $a_1 \to 0$. When $\epsilon_1 = \epsilon$, $\mu_1 = \mu$, the configuration becomes a homogeneous sphere of radius a_1.

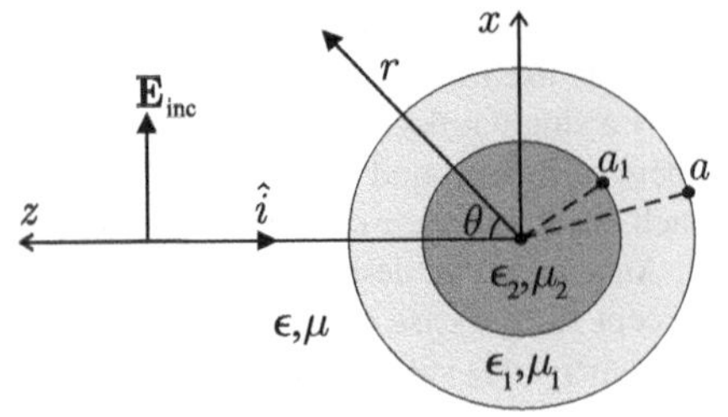

Figure 7.26 Core-shell sphere illuminated by a plane wave.

Expansion coefficients r_n^u and r_n^v in the series solutions for the core-shell sphere are obtained by specializing the general formulas (7.86)–(7.101) to the case $L = 2$. With $l = 1$ we obtain from (7.86) and (7.87)

$$\delta_n^u(a) = \frac{k_1}{\epsilon_1} \frac{k_1 T_{21}^{n1} + \epsilon_1 T_{22}^{n1} \delta_n^u(a_1)}{k_1 T_{11}^{n1} + \epsilon_1 T_{12}^{n1} \delta_n^u(a_1)} \tag{7.432}$$

$$\delta_n^v(a) = \frac{k_1}{\mu_1} \frac{k_1 T_{21}^{n1} + \mu_1 T_{22}^{n1} \delta_n^v(a_1)}{k_1 T_{11}^{n1} + \mu_1 T_{12}^{n1} \delta_n^v(a_1)} \tag{7.433}$$

and from (7.88)–(7.91)

$$T_{11}^{n1} = \psi_n'(k_1 a_1)\zeta_n^{(2)}(k_1 a) - \psi_n(k_1 a)\zeta_n^{(2)\prime}(k_1 a_1) \tag{7.434}$$

$$T_{12}^{n1} = \psi_n(k_1 a)\zeta_n^{(2)}(k_1 a_1) - \psi_n(k_1 a_1)\zeta_n^{(2)}(k_1 a) \tag{7.435}$$

$$T_{21}^{n1} = \psi_n'(k_1 a_1)\zeta_n^{(2)\prime}(k_1 a) - \psi_n'(k_1 a)\zeta_n^{(2)\prime}(k_1 a_1) \tag{7.436}$$

$$T_{22}^{n1} = \psi_n'(k_1 a)\zeta_n^{(2)}(k_1 a_1) - \psi_n(k_1 a_1)\zeta_n^{(2)\prime}(k_1 a). \tag{7.437}$$

From (7.76) and (7.82) with $L = 2$ we get

$$\delta_n^u(a_1) = \frac{k_2}{\epsilon_2} \frac{\psi_n'(k_2 a_1)}{\psi_n(k_2 a_1)} \tag{7.438}$$

$$\delta_n^v(a_1) = \frac{k_2}{\mu_2} \frac{\psi_n'(k_2 a_1)}{\psi_n(k_2 a_1)}. \tag{7.439}$$

Inserting (7.438) and (7.439) in (7.432) and (7.433) specifies $\delta_n^u(a)$ and $\delta_n^v(a)$. The modal admittances y_n and impedances z_n are then determined from (7.98) and (7.99) and the coefficients r_n^u and r_n^v from (7.100) and (7.101), which completely determines the exact series solutions for the scattered fields (7.103)–(7.108) and scattering amplitudes (7.280) and (7.281).

Figure 7.27 shows various cross sections of a hollow glass shell in air, $\epsilon = \epsilon_2 = \epsilon_0$, $\mu = \mu_1 = \mu_2 = \mu_0$. The shell is made from an electrically dense material with small losses, $\epsilon_1 = (6.7 - j0.1)\epsilon_0$ (standard glass at 10 GHz). With ka greater than unity, the scattering from the shell shows resonant behavior and $\sigma_{ext} \approx \sigma_T$. However, as the electrical size of the sphere decreases, σ_{abs} gets greater than σ_T so that $\sigma_{ext} \approx \sigma_{abs}$ when $ka < 0.2$. These are typical features of cross sections of particles with small losses.

The core-shell sphere is a useful building block in the design of metamaterials because of the larger number of the degrees of freedom compared with the homogeneous sphere. This application is described in more detail in section 7.6.1. Here we just present the quasi-static approximation ($ka \ll 1$, $|k_1 a| \ll 1$, and $|k_2 a_1| \ll 1$). The case of non-magnetic core-shell spheres was addressed by Güttler (1952).

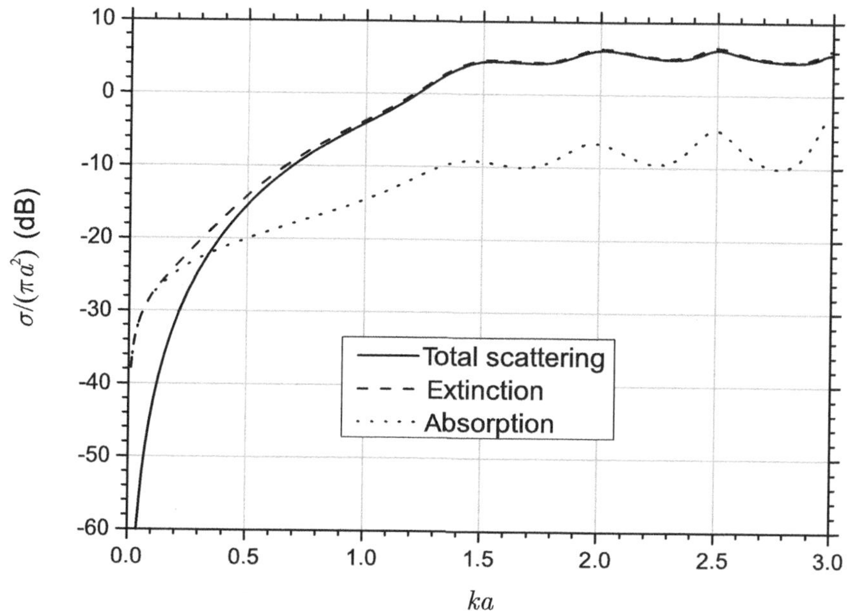

Figure 7.27 Normalized cross sections of a hollow glass shell: $a_1 = 0.813a$, $\epsilon_1 = (6.7 - j0.1)\epsilon_0$.

Expanding r_n^u and r_n^v with $n = 1, 2$ in the power series of the form (7.356)–(7.359) gives

$$\alpha_1^u = -\frac{2j}{3}\frac{(\epsilon_{1r}-1)(2\epsilon_{1r}+\epsilon_{2r}) - a_{10}^3(1+2\epsilon_{1r})(\epsilon_{1r}-\epsilon_{2r})}{(\epsilon_{1r}+2)(2\epsilon_{1r}+\epsilon_{2r}) - 2a_{10}^3(\epsilon_{1r}-1)(\epsilon_{1r}-\epsilon_{2r})} \tag{7.440}$$

$$\gamma_1^u = -\frac{j}{5}\frac{C_u}{\left[(\epsilon_{1r}+2)(2\epsilon_{1r}+\epsilon_{2r}) - 2a_{10}^3(\epsilon_{1r}-1)(\epsilon_{1r}-\epsilon_{2r})\right]^2} \tag{7.441}$$

$$\begin{aligned} C_u = {} & (2\epsilon_{1r}+\epsilon_{2r})^2\left[4 - 6\epsilon_{1r} + \epsilon_{1r}^2(1+\mu_{1r})\right] + 2a_{10}^3(\epsilon_{1r}-\epsilon_{2r})(2\epsilon_{1r}+\epsilon_{2r})\left[4 + 3\epsilon_{1r} - \epsilon_{1r}^2(2+5\mu_{1r})\right] \\ & + 9a_{10}^5\epsilon_{1r}^2\left[2\mu_{1r}\epsilon_{1r}(2\epsilon_{1r}-3\epsilon_{2r}) + \epsilon_{2r}^2(\mu_{1r}+\mu_{2r})\right] + 4a_{10}^6(\epsilon_{1r}-\epsilon_{2r})^2\left[1 + 3\epsilon_{1r} + \epsilon_{1r}^2(1-5\mu_{1r})\right] \end{aligned} \tag{7.442}$$

$$\gamma_2^u = -\frac{j}{15}\frac{(\epsilon_{1r}-1)(3\epsilon_{1r}+2\epsilon_{2r}) - a_{10}^5(2+3\epsilon_{1r})(\epsilon_{1r}-\epsilon_{2r})}{(2\epsilon_{1r}+3)(3\epsilon_{1r}+2\epsilon_{2r}) - 6a_{10}^5(\epsilon_{1r}-1)(\epsilon_{1r}-\epsilon_{2r})} \tag{7.443}$$

$$\beta_1^u = 0, \quad \delta_1^u = (\alpha_1^u)^2, \quad \delta_2^u = 0. \tag{7.444}$$

Here, $\epsilon_{1r} = \epsilon_1/\epsilon$, $\epsilon_{2r} = \epsilon_2/\epsilon$, $\mu_{1r} = \mu_1/\mu$, and $\mu_{2r} = \mu_2/\mu$ are the permittivities and permeabilities of the shell and the core relative to those of the background medium, and $a_{10} = a_1/a$. The coefficients α_1^v, β_1^v, γ_1^v, δ_1^v, and δ_2^v are obtained from (7.440)–(7.444) by the substitutions $\epsilon_{1r} \leftrightarrow \mu_{1r}$ and $\epsilon_{2r} \leftrightarrow \mu_{2r}$. The electric and magnetic polarizabilities of two small concentric spheres are given by (7.374) and (7.375), and the leading terms can be represented as

$$P_{11} = -4\pi a^3\frac{\epsilon(\beta\epsilon_1+\alpha\epsilon_2) - \epsilon_1(2\alpha\epsilon_1+\gamma\epsilon_2)}{2\epsilon(\beta\epsilon_1+\alpha\epsilon_2) + \epsilon_1(2\alpha\epsilon_1+\gamma\epsilon_2)} \tag{7.445}$$

$$M_{11} = 4\pi a^3\frac{\mu(\beta\mu_1+\alpha\mu_2) - \mu_1(2\alpha\mu_1+\gamma\mu_2)}{2\mu(\beta\mu_1+\alpha\mu_2) + \mu_1(2\alpha\mu_1+\gamma\mu_2)} \tag{7.446}$$

with

$$\alpha = 1 - a_{10}^3, \quad \beta = 2 + a_{10}^3, \quad \gamma = 1 + 2a_{10}^3. \tag{7.447}$$

The materials inside the sphere cannot be highly conductive as it is assumed that $|k_1 a|$ and $|k_2 a_1|$ are also small. The cases of a PEC sphere or a PEC-coated sphere require, therefore, special treatment.

Coated PEC Spheres

This case corresponds to the limit $\text{Im}\,\epsilon_2 \to -\infty$ with a fixed value of μ_2. In this limit, $\psi_n'(k_2a_1)/\psi_n(k_2a_1) \to j$ (see (7.39) and (7.41)), and $\delta_n^u(a_1) \to 0$ and $\delta_n^v(a_1) \to \infty$, which reduces (7.432) and (7.433) to

$$\delta_n^u(a) = \frac{k_1}{\epsilon_1}\frac{T_{21}^{n1}}{T_{11}^{n1}}, \quad \delta_n^v(a) = \frac{k_1}{\mu_1}\frac{T_{22}^{n1}}{T_{12}^{n1}}. \tag{7.448}$$

The modal admittances and impedances (7.98) and (7.99) can be therefore expressed as

$$y_n = j\sqrt{\frac{\epsilon_r}{\mu_r}}\frac{T_{11}^{n1}}{T_{21}^{n1}}, \quad z_n = j\sqrt{\frac{\mu_r}{\epsilon_r}}\frac{T_{12}^{n1}}{T_{22}^{n1}}, \tag{7.449}$$

where $\epsilon_r = \epsilon_1/\epsilon$ and $\mu_r = \mu_1/\mu$.

If the PEC core is electrically large on the scale of the wavelength in the coating, that is, $|k_1 a_1| \gg 1$, then asymptotic approximations (7.39)–(7.42) for Riccati–Bessel functions in (7.434)–(7.437) lead to

$$T_{11}^{n1} \approx T_{22}^{n1} \approx j\cos[k_1(a - a_1)], \quad T_{12}^{n1} \approx -T_{21}^{n1} \approx j\sin[k_1(a - a_1)], \tag{7.450}$$

and (7.449) dramatically simplifies to

$$y_n^{-1} \approx z_n \approx j\sqrt{\frac{\mu_r}{\epsilon_r}}\tan[k_1(a - a_1)], \tag{7.451}$$

in which we recognize the expression (4.341) for the equivalent surface impedance $\tilde{Z}_s$ of a magneto-dielectric layer of thickness $d = a - a_1$ normalized to the wave impedance $Z = \sqrt{\mu/\epsilon}$ of the surrounding medium. Equations (7.451) are valid for the modes with $n < |k_1 a_1|$, implying that when

$$|k_1 a_1| \gg 1, \tag{7.452}$$

the solution for an impedance sphere of the radius a with $\tilde{Z}_s/Z$ given by the right-hand side of (7.451) may be a good approximation of the exact solution for the coated PEC sphere. A comparison of the exact backscattering cross section σ with the cross section in the impedance approximation σ_{imp} for a PEC sphere with a thin ($kd = 0.1$) Teflon coating is given in Figure 7.28, and it is seen that once $ka > 1$, the impedance formulation achieves the level of accuracy for the planar interface of the same material composition. This is in agreement with what is shown in Figures 6.29–6.31 for a coated PEC cylinder and in Figures 7.40–7.43 for a lossy homogeneous sphere: impedance formulation applies to surfaces with curvature radii greater than $\lambda/4$.

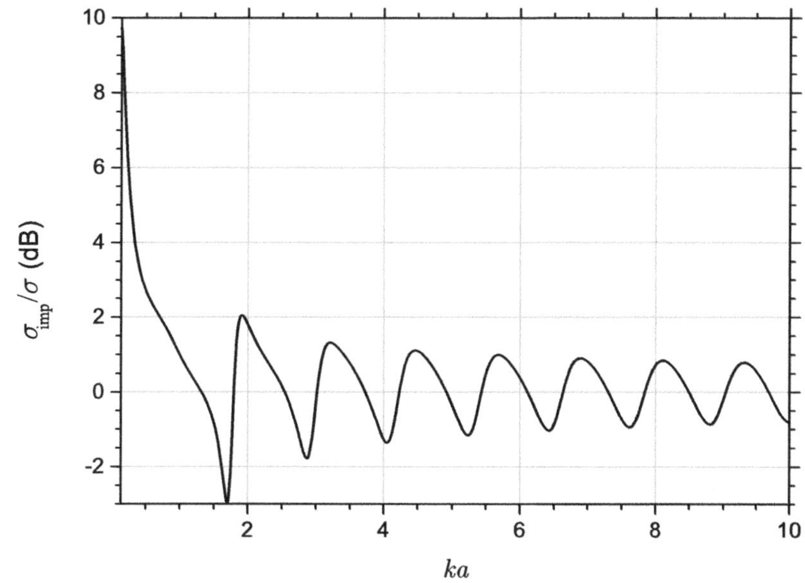

Figure 7.28 Accuracy of impedance formulation for backscattering from a PEC sphere coated with a thin layer of Teflon ($\epsilon_r = 2.08 - j0.0008$, $\mu_r = 1$ at 10 GHz) as a function of ka. Oscillations in the accuracy when $ka > 1$ are explained by generally lower accuracy of the impedance boundary conditions for thin metal backed coatings (see Figures 4.34–4.36).

Expansion of r_1^u and r_1^v for the electrically small coated PEC sphere in the Taylor series in powers of ka leads to low-frequency approximations of the form (7.356)–(7.359) with the coefficients:

$$\alpha_1^u = -\frac{2j}{3}\frac{\epsilon_r - 1 + a_{10}^3(2\epsilon_r + 1)}{2 + \epsilon_r + 2a_{10}^3(\epsilon_r - 1)}, \quad \beta_1^u = 0 \tag{7.453}$$

$$\gamma_1^u = -\frac{j}{5}\frac{c_u}{\left[2 + \epsilon_r + 2a_{10}^3(\epsilon_r - 1)\right]^2}, \quad \delta_1^u = (\alpha_1^u)^2 \tag{7.454}$$

$$c_u = 4 + \epsilon_r(\epsilon_r + \epsilon_r\mu_r - 6) - 2a_{10}^3[4 - \epsilon_r(2\epsilon_r + 5\epsilon_r\mu_r - 3)] + 9a_{10}^5\epsilon_r^2\mu_r + 4a_{10}^6[1 + \epsilon_r(\epsilon_r - 5\epsilon_r\mu_r + 3)] \tag{7.455}$$

$$\alpha_1^v = \frac{j}{3}\frac{2 - 2\mu_r + a_{10}^3(2\mu_r + 1)}{2 + \mu_r - a_{10}^3(\mu_r - 1)}, \quad \beta_1^v = 0 \tag{7.456}$$

$$\gamma_1^v = -\frac{j}{5}\frac{c_v}{\left[2 + \mu_r - a_{10}^3(\mu_r - 1)\right]^2}, \quad \delta_1^v = (\alpha_1^v)^2 \tag{7.457}$$

$$c_v = 4 + \mu_r(\mu_r + \epsilon_r\mu_r - 6) + a_{10}^3[4 + \mu_r(3 - 2\mu_r - 5\epsilon_r\mu_r)] + 9a_{10}^5\epsilon_r\mu_r^2 + a_{10}^6[1 + \mu_r(\mu_r - 5\epsilon_r\mu_r + 3)] \tag{7.458}$$

$$\gamma_2^u = -\frac{j}{30}\frac{2\epsilon_r - 2 + a_{10}^5(2 + 3\epsilon_r)}{2\epsilon_r + 3 + 3a_{10}^5(\epsilon_r - 1)}, \quad \delta_2^u = 0 \tag{7.459}$$

$$\gamma_2^v = -\frac{j}{45}\frac{3\mu_r - 3 - a_{10}^5(2 + 3\mu_r)}{2\mu_r + 3 - 2a_{10}^5(\mu_r - 1)}, \quad \delta_2^v = 0. \tag{7.460}$$

The electric and magnetic dipole polarizabilities are given by (7.374) and (7.375), and retaining just the leading terms gives

$$P_{11} = 4\pi a^3\frac{\epsilon_r\gamma - \alpha}{\epsilon_r\gamma + 2\alpha}, \quad M_{11} = 2\pi a^3\frac{\beta - 2\mu_r\alpha}{\beta + \mu_r\alpha}, \tag{7.461}$$

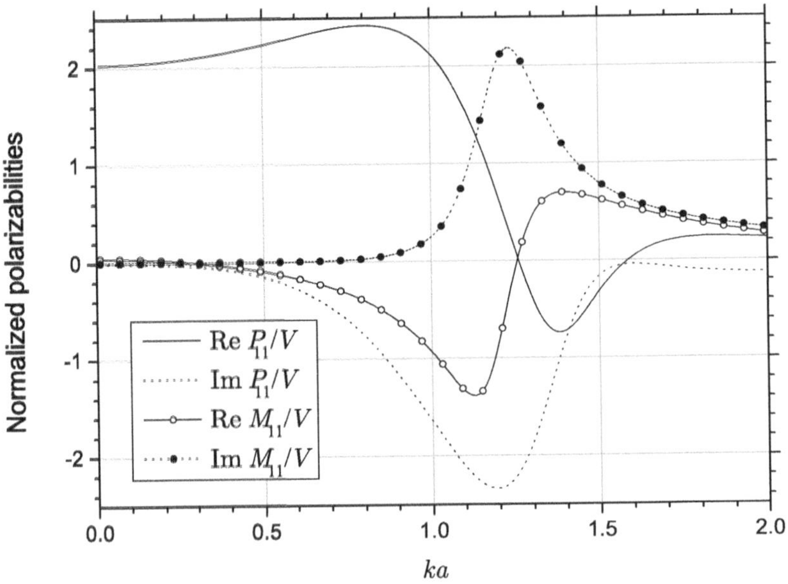

Figure 7.29 Normalized electric and magnetic polarizabilities of a PEC sphere coated with a layer of glass ($\epsilon_r = 6.7 - j0.1$, $\mu_r = 1$) with an adjusted ratio of the radii ($a_1 = 0.305a$). Normalization to the volume of the sphere ($V = 4\pi a^3/3$).

where α, β, and γ are defined in (7.447). Note that even for non-magnetic coatings the PEC core leads to a finite magnetic polarizability as $M_{11} = 2\pi a_1^3$ when $\mu_r = 1$.

When ka is not too small and the quasi-static approximation does not apply, then the polarizabilities P_{11} and M_{11} should be described by more accurate formulas (7.348) with r_1^u and r_1^v given by (7.349) and (7.350), which may have pole singularities (resonances) at particular values of the configuration parameters. For example, by adjusting a_{10}, the ratio of the core radius a_1 to the shell radius a, it is possible to make the lowest resonance frequency in the electric and magnetic moments coincide. Around the resonance frequency the electric and magnetic moments are large but inversely phased, implying a resonance in σ_{T} and σ_{fwd} but a low value of σ_{back} (see (7.351), (7.352), and (7.354)). An example is shown in Figures 7.29 and 7.30 for a PEC core of the radius $a_1 = 0.305a$ and a shell with $\epsilon_r = 6.7 - j0.1$, $\mu_r = 1$ (standard glass at 10 GHz, see Table 2.1). The electric and magnetic moments are at resonance around $ka = 1.2$ (Figure 7.29), which leads to a maximum in total and forward-scattering cross sections and cancellation of backscattering around that value of ka (Figure 7.30). This property makes the tuned sphere similar to the Huygens source (see section 2.6).

The cross sections in Figure 7.30 are plotted by using the exact multipole solution and the dipole approximation. The latter means the use of approximations (7.351) and (7.354) with P_{11} and M_{11} given by (7.348) with exact r_1^u and r_1^v, and it is seen that the approximation works well for $ka < 1.5$.

7.4.5 *Impedance Spheres*

The impedance sphere is a simple but often sufficiently accurate model of spheres of various material compositions, provided that the boundary can be adequately modeled by the boundary condition (2.74) with an equivalent surface impedance $\tilde{Z}_s$. Examples include lossy homogeneous spheres and PEC spheres coated with thin magneto-dielectric layers (sections 7.4.4 and 7.5.4). The case of a PEC sphere is included in the limit $\tilde{Z}_s \to 0$. Furthermore, the impedance sphere serves as a canonical configuration for estimating the accuracy of the impedance formulation for arbitrarily shaped scatterers.

As an exact solution of an approximate boundary value problem, the solution for the impedance sphere is physically meaningful only within the applicability area of the impedance formulation. Since the formulation applies to the spheres

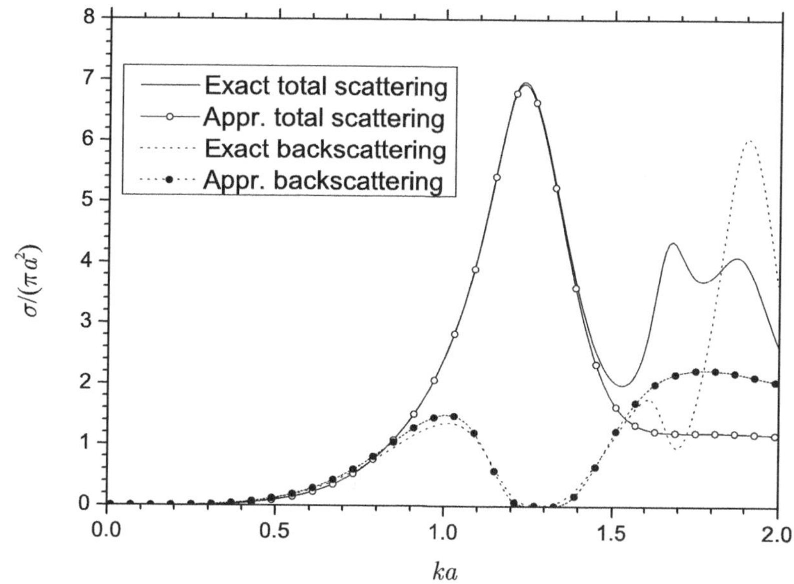

Figure 7.30 Backscattering and total scattering cross sections of the coated PEC sphere with $\epsilon_r = 6.7 - j0.1$, $\mu_r = 1$, and $a_1 = 0.305a$ as functions of the electrical size of the sphere.

larger than half a wavelength in diameter, no results on the low-frequency scattering from the impedance sphere are presented in this section.

Accuracy of the Impedance Model

The conditions of applicability of impedance boundary conditions for general material boundaries are listed in (2.76)–(2.81). In addition to conditions (2.76) and (2.77), which define the area of applicability of the boundary conditions for planar boundaries, conditions (2.78) and (2.80) ensure that the curvature of the boundary does not introduce unacceptable errors in the impedance formulation. When applied to spheres, these mean that the radius of the sphere must be sufficiently large compared with the wavelength and the sphere must be almost impenetrable for incident waves, for example include a large PEC core coated with a thin magneto-dielectric layer or be sufficiently lossy to absorb most of the incident power penetrated in its interior.

In sections 7.4.4 and 7.5.4, analytical conditions are derived under which the exact solutions for coated and homogeneous spheres approach the solution for an impedance sphere with the equivalent surface impedance of a corresponding planar configuration. Furthermore, numerical comparisons of the exact solutions for a coated PEC sphere and a lossy homogeneous sphere with their impedance approximations in Figures 7.28, 7.40, 7.41, 7.42, and 7.43 suggest that the radius of the sphere should be at least a quarter of the wavelength large and either the skin depth of the material of the sphere should be smaller than the radius (lossy homogeneous spheres) or the thickness and material properties of the coating should satisfy condition (7.452) (coated PEC spheres). Similar conditions apply to impedance models for homogeneous and coated PEC cylinders, conditions (6.273) and (6.280).

Electromagnetic Scattering as a Function of Surface Impedance

Except for the $O[(ka)^{-1/3}]$ vicinity of the forward-scattering direction, bistatic far-field scattering from impedance spheres is well described by the GO approximations (7.324)–(7.329) with reflection coefficients (4.334) and (4.335).

At the direction of forward scattering, scattering coefficients and the bistatic scattering cross section are described by equations (7.330) and (7.331), which are material independent in the main terms with respect to ka.

The exact solution for scattering of a plane wave from an impedance sphere is obtained in section 7.2.3. Let us study the solution as a function of η where $\eta = \tilde{Z}_s/Z$ is the equivalent surface impedance normalized to the wave impedance of the surrounding medium. The focus will be on the symmetry of scattering with respect to the interchange $\eta \to 1/\eta$ and on some extremal features of scattering from impedance-matched spheres ($\eta = 1$).

The multipole coefficients (7.128) and (7.129) can be rewritten in the form

$$r_n^u = -\frac{\psi_n(ka) + j\eta^{-1}\psi_n'(ka)}{\zeta_n^{(2)}(ka) + j\eta^{-1}\zeta_n^{(2)\prime}(ka)} \tag{7.462}$$

$$r_n^v = -\frac{\psi_n(ka) + j\eta\psi_n'(ka)}{\zeta_n^{(2)}(ka) + j\eta\zeta_n^{(2)\prime}(ka)}, \tag{7.463}$$

and it is apparent that interchanging η with $1/\eta$ transforms r_n^u to r_n^v and vice versa. It follows therefore from (7.280) and (7.281) that

$$S_1(\theta, \eta^{-1}) = -S_2(\theta, \eta), \quad S_2(\theta, \eta^{-1}) = -S_1(\theta, \eta) \tag{7.464}$$

and from (7.283) and (7.284) that

$$\sigma_E(\theta, \eta^{-1}) = \sigma_H(\theta, \eta), \quad \sigma_H(\theta, \eta^{-1}) = \sigma_E(\theta, \eta), \tag{7.465}$$

implying polarization independence of scattering when $\eta = 1$. Furthermore it follows from (7.292), (7.293), (7.295), (7.297), and (7.301) that σ_{back}, σ_{fwd}, σ_{ext}, σ_{T}, and σ_{abs} are invariant under the transformation $\eta \to 1/\eta$,

$$\sigma_{\text{back}}(\eta^{-1}) = \sigma_{\text{back}}(\eta) \tag{7.466}$$

$$\sigma_{\text{fwd}}(\eta^{-1}) = \sigma_{\text{fwd}}(\eta) \tag{7.467}$$

$$\sigma_{\text{ext}}(\eta^{-1}) = \sigma_{\text{ext}}(\eta) \tag{7.468}$$

$$\sigma_{\text{T}}(\eta^{-1}) = \sigma_{\text{T}}(\eta) \tag{7.469}$$

$$\sigma_{\text{abs}}(\eta^{-1}) = \sigma_{\text{abs}}(\eta). \tag{7.470}$$

An important consequence of this invariance is that the cross sections have an extremum at $\eta = 1$ (Osipov 2015). Indeed, the invariance means that the cross sections can be represented as functions of the argument $t = \eta^{-1} + \eta$ and, therefore, their derivative with respect to η must vanish at $\eta = 1$, which is seen from the formula

$$\frac{\partial}{\partial \eta}\sigma(t) = \left(1 - \frac{1}{\eta^2}\right)\frac{\partial}{\partial t}\sigma(t). \tag{7.471}$$

Since η is a complex variable, the extremum can be a minimum (σ_{back}, σ_{T}), a maximum (σ_{abs}), or a saddle point (σ_{fwd}, σ_{ext}). To see that, we start with the case of backscattering. If $\eta = 1$, then it follows from (7.462) and (7.463) that regardless of the value of ka, $r_n^u = r_n^v = r_n$ and the general expressions (7.280) and (7.281) simplify to the relation

$$S_1(\theta) = -S_2(\theta) = j\sum_{n=1}^{\infty}(-1)^{n-1}\frac{2n+1}{n(n+1)}r_n[\tau_n(\theta) - \pi_n(\theta)], \tag{7.472}$$

implying that $\sigma_{\text{back}} = 0$ since $\tau_n(0) = \pi_n(0)$ (see (7.110)). Since σ_{back} by definition is non-negative, the point $\eta = 1$ is a global minimum of σ_{back} on the complex η plane. The vanishing backscattering from impedance-matched spheres is a special case of a more general property (Weston's theorem) described in section 3.2.3.

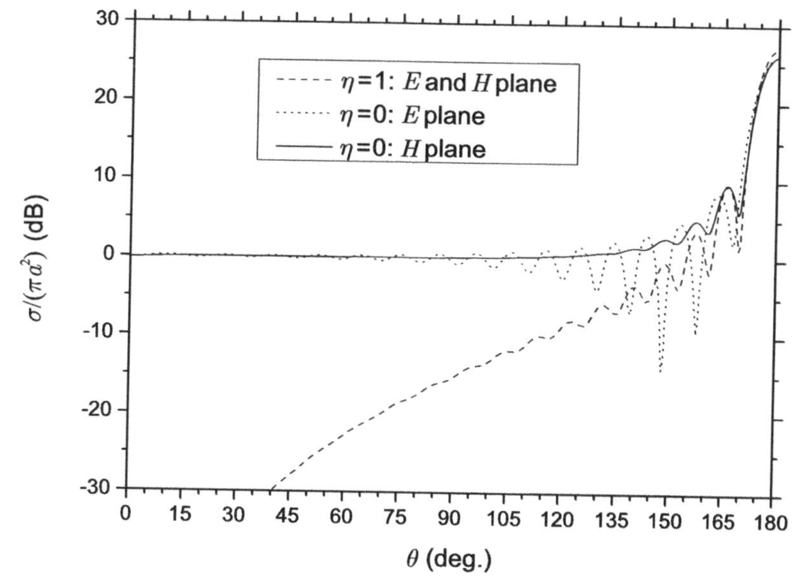

Figure 7.31 Bistatic scattering cross section of an impedance-matched sphere ($\eta = 1$) compared with the E- and H-plane cross sections of a PEC sphere of the same size ($ka = 20$).

In Figure 7.31, the bistatic scattering cross section of a sphere with $\eta = 1$ and $ka = 20$ is plotted as a function of the bistatic angle θ and compared with the cross section of a PEC sphere. The curve for $\eta = 1$ clearly shows the polarization independence of bistatic scattering and vanishing backscattering from impedance-matched spheres. Note that at some values of θ in the range $135° < \theta \leq 180°$ the impedance-matched sphere can scatter stronger than the PEC sphere.

The extinction cross section σ_{ext} of electrically large impedance spheres is close to its optical limit $2\pi a^2$ (see (7.332)) and therefore almost independent of the surface impedance. This is illustrated in Figure 7.32 for a sphere with $ka = 20$

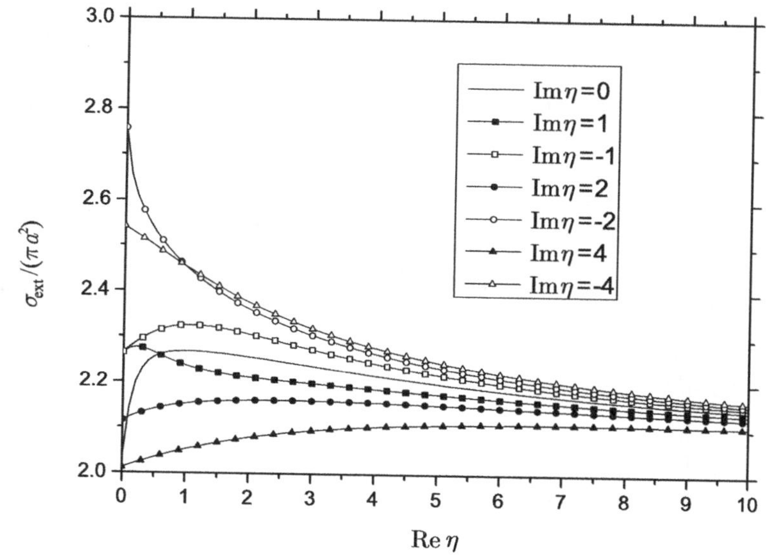

Figure 7.32 Normalized extinction cross section of an impedance sphere with $ka = 20$ as a function of Re η for several values of Im η.

and a variety of impedance values. A closer analysis on the complex η plane reveals that σ_{ext} has a saddle point at $\eta = 1$, that is, the function σ_{ext} may grow or decay as a function of $|\eta - 1|$ depending on $\arg(\eta - 1)$.

Similarly, the forward-scattering cross section σ_{fwd} asymptotically follows the formula $\pi a^2 (ka)^2$ for $ka \gg 1$ (see (7.331)) and only weakly depends on η. On the complex η plane, the point $\eta = 1$ is a saddle point of σ_{fwd}.

In contrast to that, σ_{abs} and σ_{T} have a maximum and a minimum at $\eta = 1$, respectively. A surface impedance with $\text{Re}\,\eta > 0$ implies a flow of the incident power in the interior of the sphere (see (2.133) and (2.134)), where a part of the power can be absorbed, which means that $\sigma_{\text{abs}} > 0$. If $\text{Re}\,\eta = 0$, then $\sigma_{\text{abs}} = 0$ (see (7.148) and (7.303)). When $\eta = 1$, the spherical reflection coefficients $\tilde{r}_n^u$ and $\tilde{r}_n^v$, defined in (7.145) and (7.146), are such that

$$\tilde{r}_n^u = \tilde{r}_n^v = -\frac{\zeta_n^{(1)}(ka) + j\zeta_n^{(1)\prime}(ka)}{\zeta_n^{(2)}(ka) + j\zeta_n^{(2)\prime}(ka)}, \tag{7.473}$$

vanishing in the limit $ka \to \infty$ for $n < ka$ because of the asymptotic behavior of the Riccati–Hankel functions (see (7.40) and (7.46)). As apparent from (7.303), σ_{abs} has a maximum when $\tilde{r}_n^u = \tilde{r}_n^v = 0$, and its maximum value can be estimated by

$$\sigma_{\text{abs}} \approx \frac{\pi}{k^2} \sum_{n=1}^{ka} (2n+1) \approx \pi a^2. \tag{7.474}$$

The dependence of σ_{abs} of a sphere with $ka = 20$ on $\text{Re}\,\eta$ for various values of $\text{Im}\,\eta$ is illustrated in Figure 7.33.

The total scattering cross section σ_{T} is the difference between σ_{ext}, which is almost independent of η, and σ_{abs}, which is at maximum at $\eta = 1$, and has therefore a minimum at $\eta = 1$, where $\sigma_{\text{T}} \approx \sigma_{\text{abs}} \approx \pi a^2$ (see Figure 7.34). The physical explanation for the minimum in σ_{T} and the maximum in σ_{abs} is apparently that the impedance-matched boundary of the sphere minimizes reflection from the boundary while allowing the best possible transmission of the incident power into the sphere, where the energy is dissipated due to the losses in the interior of the sphere assumed by the impedance formulation.

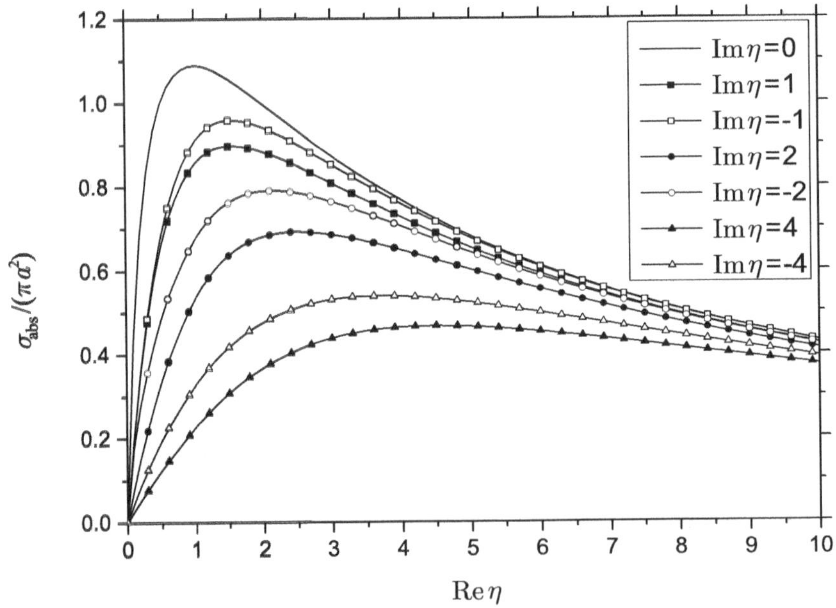

Figure 7.33 The same as Figure 7.32 but for the normalized absorption cross section.

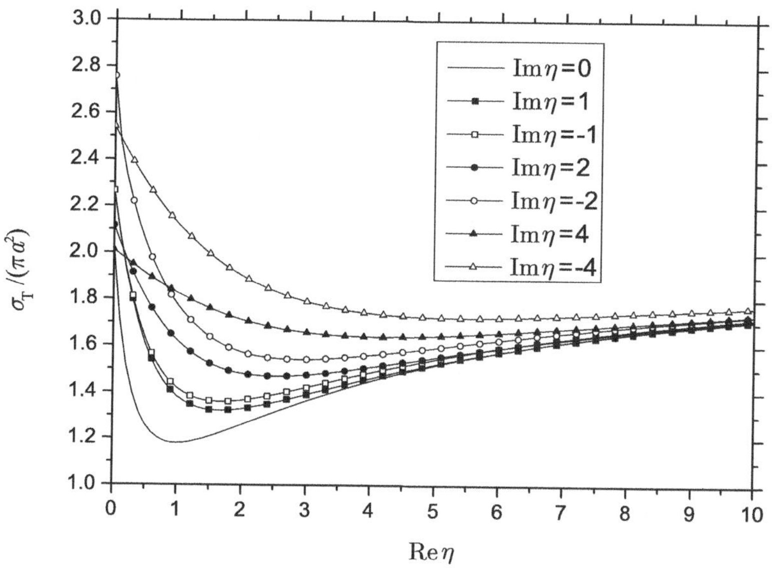

Figure 7.34 The same as Figure 7.32 but for the normalized total scattering cross section.

7.5 Far-field Scattering from Homogeneous Spheres

In this section we examine the far fields scattered by spheres with the constitutive parameters ϵ_1 and μ_1 which are finite and constant throughout the sphere volume. The value of μ_1 can differ from μ, the permeability of the surrounding space. The material inside the sphere may be lossy, in which case ϵ_1 or μ_1 or both may have an imaginary part. Most of the equations that follow can be expressed in terms of relative permittivity $\epsilon_r = \epsilon_1/\epsilon$ and permeability $\mu_r = \mu_1/\mu$ of the sphere.

The homogeneous sphere is one of the most important models in scattering theory. The relatively simple analytical solution permits detailed studies of the features of the scattered fields, with further simplifications in the cases of electrically small ($ka \ll 1$) and electrically large spheres ($ka \gg 1$). So, the model has been widely used over the whole electromagnetic spectrum, from optical to radio wave frequencies, for example to model scattering of light from metal particles in colloidal suspensions (Mie 1908), from raindrops in the atmosphere (Van der Pol and Bremmer 1937a,b), and propagation of radio waves over the Earth surface (Fock 1965; Watson 1918). Furthermore, the homogeneous sphere is a building block in the design of metamaterials (Vendik and Gashinova 2004).

The model of a homogeneous and penetrable sphere includes perfectly penetrable ($\text{Im}\,\epsilon_1 = 0, \text{Im}\,\mu_1 = 0$) and perfectly opaque, for example PEC, spheres as the limiting cases and permits a continuous and detailed analysis of the role of absorption in scattering from penetrable scatterers.

A particularly exciting but at the same time complicated case is that in which the absorption in the material of the sphere is small and the sphere is electrically large ($ka \gg 1$). The simulation and physical interpretation of the scattering from such spheres becomes a challenge. The number of terms to be accounted for in the series solutions (7.280) and (7.281) is estimated by the formula (7.131) and is, therefore, large for electrically large spheres. If the losses are small, the scattering functions and the various scattering cross sections become rapidly oscillating functions of the scattering angle θ and the ratio λ/a, which makes a denser sampling necessary to catch all the variations of the functions. Furthermore, if the losses are small so that multiple reflections in the interior of the sphere cannot be neglected, then a multitude of various ray contributions (see Figure 7.10) forms the scattered field, which complicates a physical interpretation of the scattering.

This is the case in the scattering of light by rain droplets, which features such phenomena as rainbows, glories, and coronae. Water has extraordinary low absorption in the visible range of the electromagnetic spectrum, and the wavelength of light is much shorter than the diameter of a typical raindrop. For example, for the wavelength 0.55 μm (green light) and droplet radius 1 mm, $ka = 11424$ and the number of terms needed in the summation of the series (7.280) and (7.281) is of the order of 11500. The step in θ, which is of the order of $(ka)^{-1}$, is about 0.005°, implying that about 36000 sample points in θ are necessary to adequately represent the bistatic scattering cross section of a water drop.

We start this section by specializing the general series solution to the case of a homogeneous sphere and taking a look at the limiting cases of the homogeneous sphere, including the peculiar scattering properties of spheres made of a material with the wave impedance equal to that of the surrounding medium (section 7.5.1). Then, in section 7.5.2 we discuss scattering from spheres which are small compared to the wavelength in the surrounding medium, $ka \ll 1$, but not necessary small on the scale of the wavelength in the interior of the sphere. In addition to the classic Rayleigh approximation, we study further important special cases, including dipole approximations for spheres with strong absorption (section 7.5.2) and improved approximations for low-absorption spheres (section 7.5.3). The latter include dynamic and radiation corrections in the dipole polarizabilities, quadrupole corrections in extinction cross sections, and resonance approximation for spherical reflection coefficients of arbitrary multipole order. A complete set of approximations for the various cross sections of the electrically small homogeneous sphere is presented on the complex plane of ϵ_r.

In section 7.5.4 we examine scattering from electrically large lossy spheres, study the backscattering, extinction, absorption, and total scattering cross sections of these, and determine the conditions under which the scattering can be adequately described by the impedance boundary conditions. Finally, in section 7.5.5 scattering of light from spherical water droplets is addressed as an example of scattering from electrically large low-absorption spheres. We present the bistatic scattering cross sections and focus on the features that are responsible for such optical atmospheric phenomena as rainbows, glories, and coronae.

7.5.1 Exact Solution and Limiting Cases

Let us specialize the general solution from section 7.2.2 to the case of a homogeneous sphere. The material parameters ϵ_1 and μ_1 are assumed constant throughout the interior of the sphere and, in general, are different from ϵ and μ in the surrounding medium. Since the material structure has no stratification and the innermost and outermost material boundaries coincide, no recursion is needed to determine the parameters $\delta_n^u(a)$ and $\delta_n^v(a)$. The latter are equal to their initial values

$$\delta_n^u(a) = \frac{k_1}{\epsilon_1} \frac{\psi_n'(k_1 a)}{\psi_n(k_1 a)} \tag{7.475}$$

$$\delta_n^v(a) = \frac{k_1}{\mu_1} \frac{\psi_n'(k_1 a)}{\psi_n(k_1 a)}, \tag{7.476}$$

which correspond to the admittances and impedances

$$y_n = j\frac{Z}{Z_1} \frac{\psi_n(k_1 a)}{\psi_n'(k_1 a)} \tag{7.477}$$

$$z_n = j\frac{Z_1}{Z} \frac{\psi_n(k_1 a)}{\psi_n'(k_1 a)}. \tag{7.478}$$

Insertion of these in (7.100) and (7.101) completely specifies the coefficients r_n^u and r_n^v, and, therefore, the multipole expansion for the field scattered by a homogeneous sphere.

The PEC sphere is a limiting case of a homogeneous sphere with the conductivity of the material going to infinity, that is, $|\text{Im}\, \epsilon_1| \to \infty$ ($\text{Im}\, \epsilon_1 < 0$), while the permeability μ_1 remains finite. Going to the PEC limit and using (7.39) and (7.41) we have

$$\frac{\psi_n'(k_1 a)}{\psi_n(k_1 a)} \to j, \tag{7.479}$$

which transforms (7.475) and (7.476) to

$$\delta_n^u(a) \to 0, \quad \delta_n^v(a) \to \infty \tag{7.480}$$

and (7.477) and (7.478) to

$$y_n \to \infty, \quad z_n \to 0. \tag{7.481}$$

Electromagnetic scattering from a homogeneous sphere posses remarkable features if the wave impedance in the interior of the sphere is equal to that in the surrounding medium, that is, $Z_1 = Z$ and $\epsilon_1/\epsilon = \mu_1/\mu$ (Kerker *et al.* 1983). In this case it follows from (7.100), (7.101), (7.280), (7.281), (7.477), and (7.478) that $y_n = z_n$ and $r_n^u = r_n^v = r_n$, and the coefficients $S_{1,2}(\theta)$ are given by the formula (7.472) since $S_1(\theta) = -S_2(\theta)$, $\sigma_E(\theta) = \sigma_H(\theta)$ for such spheres.

Furthermore, when $\theta = 0$ (backscattering), it follows from (7.110) that $\tau_n(0) = \pi_n(0)$ and the scattering amplitudes $S_{1,2}(0)$ given by (7.472) vanish and so does the backscattering cross section. This is a particular case of the general property of the vanishing axial backscattering from rotationally symmetric scatterers with $Z = Z_1$ (Weston 1963). Note that this property holds for scatterers of any electrical size.

All general properties and relations described in sections 7.4.1 and 7.4.2 hold for the homogeneous sphere as well, and specializing these results to the homogeneous sphere we will now take a closer look at the characteristic features of scattering from homogeneous spheres with the electrical size varying from small to large and with absorption from negligibly weak to strong.

7.5.2 Electrically Small Lossy Spheres

Dipole Scattering

Consider electromagnetic scattering from electrically small spheres ($ka \ll 1$) with arbitrary, including large, values of ϵ_1 and μ_1 so that the parameter $|k_1 a| = |\epsilon_r \mu_r|^{1/2} ka$ is not necessarily small. Furthermore, sufficiently high losses in the material of the sphere are assumed so that absorption dominates scattering in the sense that $\sigma_{\text{abs}} \gg \sigma_{\text{T}}$. Under these conditions the scattered field is determined by the leading dipole terms ($n = 1$) in the multipole expansions (7.280) and (7.281). The case of low-absorption spheres, in which σ_{abs} is comparable with or even smaller than σ_{T} and there can be resonances, requires accounting for the multipole terms ($n \geq 2$) and, therefore, special treatment (see below in this section).

As we saw in section 7.4.2, in the dipole approximation, the cross sections are given by

$$\sigma_{\text{back}} \approx \frac{9\pi}{k^2} |r_1^u - r_1^v|^2 \tag{7.482}$$

$$\sigma_{\text{fwd}} \approx \frac{9\pi}{k^2} |r_1^u + r_1^v|^2 \tag{7.483}$$

$$\sigma_{\text{ext}} \approx -\frac{6\pi}{k^2} \left(\text{Re}\, r_1^u + \text{Re}\, r_1^v\right) \tag{7.484}$$

$$\sigma_{\text{T}} \approx \frac{6\pi}{k^2} \left(|r_1^u|^2 + |r_1^v|^2\right) \tag{7.485}$$

$$\sigma_{\text{abs}} \approx -\frac{6\pi}{k^2} \left(\text{Re}\, r_1^u + |r_1^u|^2 + \text{Re}\, r_1^v + |r_1^v|^2\right) \tag{7.486}$$

and the scattering functions $S_{1,2}(\theta)$ by (7.333) and (7.334). The coefficients r_1^u and r_1^v are also directly related to the dipole polarizabilities P_{11} and M_{11} by (7.348).

The polarizabilities P_{11} and M_{11} are related by (7.349) and (7.350) to the modal admittance y_1 and impedance z_1. For a homogeneous sphere, y_1 and z_1 are given by (7.477) and (7.478) with $n = 1$, and by using the explicit formula (7.35) for the function $\psi_1(w)$ they can be expressed as

$$y_1 = j\frac{ka}{2}\epsilon_r Q_1(k_1 a) \tag{7.487}$$

$$z_1 = j\frac{ka}{2}\mu_r Q_1(k_1 a), \tag{7.488}$$

where

$$Q_1(w) = \frac{2}{w}\frac{\psi_1(w)}{\psi_1'(w)} = 2\frac{1 - w\cot w}{w^2 - 1 + w\cot w}. \tag{7.489}$$

Function $Q_1(w)$ behaves as

$$Q_1(w) = 1 + \frac{w^2}{10} + \frac{9w^4}{700} + O(w^6) \tag{7.490}$$

when $w \to 0$, and as

$$Q_1(w) = -\frac{2}{w}\cot w + O\left(\frac{1}{w^2}\right) \tag{7.491}$$

when $w \to \infty$ and $w \neq m\pi$ with $m = 0, \pm 1, \pm 2, \ldots$. It is an even meromorphic function of a complex variable w with an infinite number of interchanging poles and zeros on the real axis (Figure 7.35) approaching the poles and zeros of $\cot w$ when $|w|$ is large and real.

When $|\mathrm{Im}\, w| > 1$, then cotangent functions in (7.489) simplify as $\cot w \approx -j\mathrm{sgn}(\mathrm{Im}\, w)$, and

$$Q_1(w) \approx \frac{2(1 \pm jw)}{w^2 - 1 \mp jw}, \tag{7.492}$$

where the upper and lower signs correspond to the cases $\mathrm{Im}\, w > 1$ and $\mathrm{Im}\, w < -1$, respectively.

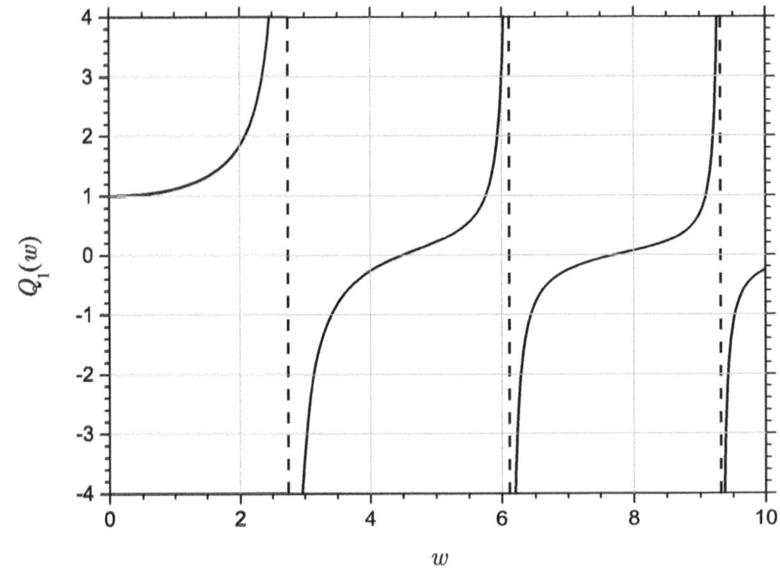

Figure 7.35 Function $Q_1(w)$.

Inserting (7.487) and (7.488) in (7.349) and (7.350) leads to the polarizabilities:

$$P_{11} \approx 4\pi a^3 \frac{\epsilon_r Q_1(k_1 a) - 1}{\epsilon_r Q_1(k_1 a) + 2} \tag{7.493}$$

$$M_{11} \approx -4\pi a^3 \frac{\mu_r Q_1(k_1 a) - 1}{\mu_r Q_1(k_1 a) + 2}. \tag{7.494}$$

Using these in (7.348) gives

$$r_1^\mu \approx \frac{2}{3} j(ka)^3 \frac{1 - \epsilon_r Q_1(k_1 a)}{2 + \epsilon_r Q_1(k_1 a)} \tag{7.495}$$

$$r_1^\nu \approx \frac{2}{3} j(ka)^3 \frac{1 - \mu_r Q_1(k_1 a)}{2 + \mu_r Q_1(k_1 a)} \tag{7.496}$$

for the coefficients with $n = 1$ in the series solutions (7.280) and (7.281).

Rayleigh Approximation

In the simplest case, when the sphere is small compared with the wavelength in the interior of the sphere, $|k_1|a < 1$ and, according to (7.490), $Q_1(k_1 a) \approx 1$ so that (7.493) and (7.494) result in the well-known quasi-static formulas for polarizabilities (e.g., Stratton (1941)):

$$P_{11} = 4\pi a^3 \frac{\epsilon_r - 1}{\epsilon_r + 2}, \quad M_{11} = 4\pi a^3 \frac{1 - \mu_r}{\mu_r + 2}. \tag{7.497}$$

This leads to scattering cross sections that follow the Rayleigh law (3.215). For example, from (7.351),

$$\sigma_{\text{back}} = 4\pi k^4 a^6 \left| \frac{\epsilon_r - 1}{\epsilon_r + 2} + \frac{1 - \mu_r}{\mu_r + 2} \right|^2, \tag{7.498}$$

which is proportional to the fourth power of the frequency, the squared volume of the sphere, and a squared deviation of the material parameters of the sphere from those in the background medium.

The formulas (7.497) do not account for dynamic effects and are therefore of limited accuracy. For example, the quasi-static approximation for M_{11} vanishes when $\mu_r = 1$. However, according to the more accurate formula (7.494), M_{11} is not zero, having even resonances when $Q_1(k_1 a) = -2$. The latter can be physically explained by the resonances in the polarization current circumventing the sphere. Dielectric non-magnetic spheres have an infinite number of magnetic resonances, see (7.489), (7.494), and Figure 7.35, with the lowest one at $k_1 a = \pi$.

The insufficient accuracy of (7.497) leads to a paradox noted by Kerker *et al.* (1983). If the sphere is such that

$$\epsilon_r = \frac{4 - \mu_r}{2\mu_r + 1} \quad \text{or} \quad \mu_r = \frac{4 - \epsilon_r}{2\epsilon_r + 1}, \tag{7.499}$$

then it follows from (7.497) that $M_{11} = P_{11}$, and according to (7.353) and (7.354), $\sigma_{\text{ext}} = 0$, whereas $\sigma_{\text{T}} > 0$. The paradox is solved by accounting for the higher-order terms in the expansions of the dipole polarizabilities in powers of ka (Alù and Engheta 2010; see also problem 20).

Spheres with High Losses

Despite the condition $ka \ll 1$, it is possible that $|k_1 a| > 1$ when the refractive index in the sphere is high, that is, $|\epsilon_r \mu_r| > (ka)^{-2}$. In this case, the sphere is larger than the wavelength in its interior, the quasi-static approximations (7.497) do not apply and more accurate formulas (7.493) and (7.494), or (7.495) and (7.496), should be used. For example, when the material of the sphere approaches the PEC limit ($\text{Im}\,\epsilon_1 \to -\infty$ with any finite μ_1), (7.493) and (7.494) with (7.491) recover the expressions (7.417) for the polarizabilities of small PEC spheres. The quasi-static approximations

(7.497) lead to the correct limit for P_{11}, but to recover the correct limiting value of M_{11}, μ_r has to go to a peculiar limit $\mu_r \to 0$.

Metals at microwave and radio frequencies have very large values of $|\epsilon_r|$ (Table 2.2), for example $\epsilon_r \approx -j1.13 \times 10^9$ for silver at 1 GHz. So, it appears at the first glance that scattering from such spheres should be almost the same as from PEC spheres. On the other hand, we can expect significant deviations from the PEC model when the diameter of a lossy sphere is smaller than the skin depth in the material of the sphere as such sphere cannot be regarded as opaque. More specifically, it can be shown that once ka is sufficiently small, highly lossy spheres significantly deviate from PEC spheres in the values of σ_{ext} and σ_{abs} while having almost the same values of σ_{T}.

Let us start with the case of an opaque sphere, in which the absorption in the material of the sphere is significant so that the skin depth is smaller than the radius of the sphere, implying that $\text{Im}\, k_1 a < -1$ ($\text{Im}\, k_1$ is always negative in passive media) and $|k_1 a| > 1$. Assuming further that $|\mu_r| < |k_1 a| < |\epsilon_r|$ (as $|\epsilon_r| \gg 1$ and $|\mu_r| \approx 1$ in most metals) and approximating $Q_1(k_1 a)$ by (7.492) transform (7.495) and (7.496) to

$$r_1^u \approx -\frac{2j}{3}(ka)^3 - (ka)^4\sqrt{\frac{\mu_r}{\epsilon_r}} \tag{7.500}$$

$$r_1^v \approx \frac{j}{3}(ka)^3 - (ka)^2\sqrt{\frac{\mu_r}{\epsilon_r}}, \tag{7.501}$$

implying that

$$|r_1^u| \approx \frac{2}{3}(ka)^3, \quad \text{Re}\, r_1^u \approx -(ka)^4 \text{Re}\sqrt{\frac{\mu_r}{\epsilon_r}} \tag{7.502}$$

$$|r_1^v| \approx \frac{1}{3}(ka)^3, \quad \text{Re}\, r_1^v \approx -(ka)^2 \text{Re}\sqrt{\frac{\mu_r}{\epsilon_r}}, \tag{7.503}$$

with the result that the electric part dominates scattering ($|r_1^u| > |r_1^v|$), but the magnetic part dominates absorption ($|\text{Re}\, r_1^u| \ll |\text{Re}\, r_1^v|$).

It follows from (7.484) and (7.485) that

$$\frac{\sigma_{\text{ext}}}{\pi a^2} \approx 6\text{Re}\sqrt{\frac{\mu_r}{\epsilon_r}}, \quad \frac{\sigma_{\text{T}}}{\pi a^2} \approx \frac{10}{3}(ka)^4, \tag{7.504}$$

and comparing (7.504) with the respective formulas (7.427) for PEC spheres, we note that the values of σ_{T} are the same, but the values of σ_{ext} can be dramatically different. For metal spheres with $ka < |\mu_r/\epsilon_r|^{1/8}$, $\sigma_{\text{ext}} \gg \sigma_{\text{T}}$ and, therefore,

$$\sigma_{\text{abs}} \gg \sigma_{\text{T}}. \tag{7.505}$$

In the quasi-static region ($|k_1 a| < 1$) it follows from (7.353), (7.354), and (7.497) with $\epsilon_r \approx -j\epsilon_r''$ and $\mu_r = 1$ that

$$\frac{\sigma_{\text{ext}}}{\pi a^2} \approx \frac{12ka}{\epsilon_r''}, \quad \frac{\sigma_{\text{T}}}{\pi a^2} \approx \frac{8}{3}(ka)^4, \tag{7.506}$$

and

$$\frac{\sigma_{\text{ext}}}{\sigma_{\text{T}}} \approx \frac{9}{2\epsilon_r''(ka)^3}, \tag{7.507}$$

which grows indefinitely with decreasing ka.

The dependence of $\sigma_{\text{ext}}/\sigma_{\text{T}}$ on ka is illustrated in Figure 7.36 for silver and lead spheres at several radio- and microwave frequencies. The value $\sigma_{\text{ext}}/\sigma_{\text{T}} = 1$ corresponds to the PEC case. The common feature of the curves is that the smaller ka

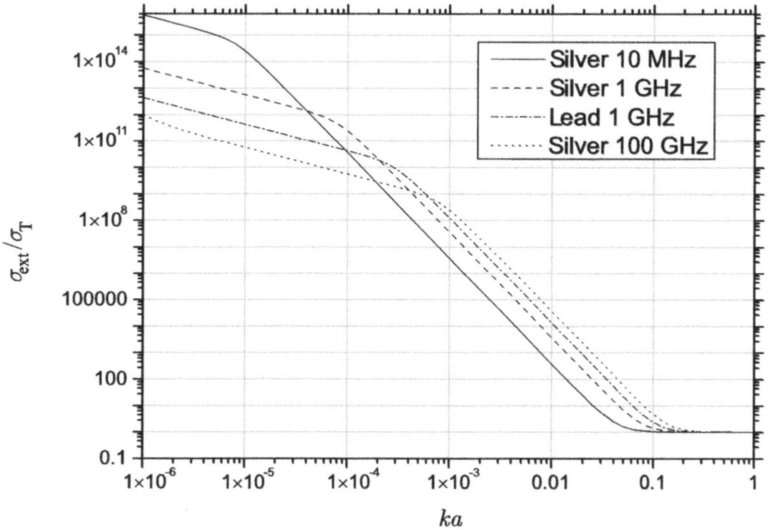

Figure 7.36 Ratio of extinction and total scattering cross sections of metal spheres as a function of the electrical size of the spheres. It is assumed that $\mu_r = 1$ and $\epsilon_r = 1 - j\sigma/(\omega\epsilon_0)$. The values of conductivity σ are taken from Table 2.2. The curves are calculated from exact series solutions (7.295) and (7.297).

the stronger the dominance of absorption over scattering and therefore the deviation from the PEC case. For example, for silver spheres at 1 GHz, the ratio $\sigma_{\text{ext}}/\sigma_{\text{T}}$ is significantly greater than unity already when $ka < 0.07$. When $ka < 3 \times 10^{-5}$, the ratio follows the quasi-static formula (7.507).

In the case of low absorption ($k_1 a$ almost real-valued), the presence of $Q_1(k_1 a)$ in formulas (7.493) and (7.494) suggests the resonance behavior of electric and magnetic polarizabilities when $|k_1 a| > 1$ (see Figure 7.35). This case requires special treatment and is addressed next. A further example can be found in section 7.6 (Figure 7.81).

7.5.3 *Electrically Small Low-absorption Spheres*

As the absorption in the sphere gets smaller, the dominance of absorption over scattering weakens and the scattering losses begin to dominate so that $\sigma_{\text{abs}} = 0$ and $\sigma_{\text{ext}} = \sigma_{\text{T}}$ in the limit of a lossless sphere. In this section we study the case in which σ_{abs} is comparable with or much smaller than σ_{T} and resonance effects, including resonant absorption, may take place.

The dipole approximations (7.482)–(7.486) and (7.493)–(7.496) are insufficient for low-loss spheres. As we have seen in section 7.4.2, the case of low absorption requires determining σ_{ext} with the accuracy of the order of $\sigma_{\text{T}} = O(k^4 a^6)$, which implies accounting for quadrupole contributions as in (7.355) and retaining the terms up to $O[(ka)^6]$ in the quasi-static expansions of r_1^u, r_1^v, r_2^u, and r_2^v, as in (7.356)–(7.359). The scattered field and the cross sections are then determined from (7.280), (7.281), (7.292), (7.295), and (7.297) with $n = 1, 2$.

Small-sphere Approximations

The coefficients $\alpha_1^{u,v}$, $\beta_1^{u,v}$, $\gamma_{1,2}^{u,v}$, and $\delta_{1,2}^{u,v}$ in the expansions (7.356)–(7.359) of $r_{1,2}^{u,v}$ are given by (7.367)–(7.369). For electrically small spheres, when $|k_1 a| < 1$, the coefficients can be simplified by approximating the normalized admittances and impedances, which are defined in (7.343), (7.477), and (7.478), by

$$\tilde{y}_1 \approx -\frac{\epsilon_r}{2}\left[1 + \frac{\epsilon_r \mu_r}{10}(ka)^2\right], \quad \tilde{y}_2 \approx -\frac{\epsilon_r}{3} \tag{7.508}$$

and

$$\tilde{z}_1 \approx -\frac{\mu_r}{2}\left[1 + \frac{\epsilon_r \mu_r}{10}(ka)^2\right], \quad \tilde{z}_2 \approx -\frac{\mu_r}{3}. \tag{7.509}$$

Using (7.508) and (7.509) in (7.367)–(7.369), expanding $\alpha_1^{u,v}$, $\gamma_{1,2}^{u,v}$, and $\delta_1^{u,v}$ in geometric progressions in the powers of ka, inserting these geometric progressions in (7.356)–(7.359), and collecting the terms with equal powers of ka lead to expansions of the form (7.356)–(7.359) with the coefficients:

$$\alpha_1^u = \frac{2j}{3}\frac{1-\epsilon_r}{2+\epsilon_r}, \quad \alpha_1^v = \frac{2j}{3}\frac{1-\mu_r}{2+\mu_r}, \quad \beta_1^u = \beta_1^v = 0 \tag{7.510}$$

$$\gamma_1^u = \frac{\epsilon_r^2(1+\mu_r) - 6\epsilon_r + 4}{5j(2+\epsilon_r)^2}, \quad \gamma_1^v = \frac{\mu_r^2(1+\epsilon_r) - 6\mu_r + 4}{5j(2+\mu_r)^2} \tag{7.511}$$

$$\delta_1^u = (\alpha_1^u)^2, \quad \delta_1^v = (\alpha_1^v)^2 \tag{7.512}$$

$$\gamma_2^u = \frac{j}{15}\frac{1-\epsilon_r}{3+2\epsilon_r}, \quad \gamma_2^v = \frac{j}{15}\frac{1-\mu_r}{3+2\mu_r}, \quad \delta_2^u = \delta_2^v = 0. \tag{7.513}$$

Expansions (7.370) and (7.371) for σ_{ext} and σ_{T} with the coefficients (7.510)–(7.513) result in the approximations that remain valid even in the limit of vanishing absorption, in which $\sigma_{\text{ext}} = \sigma_{\text{T}}$ and $\sigma_{\text{abs}} = 0$.

More terms in the higher-order expansions of σ_{ext} and σ_{T} for non-magnetic dielectric spheres ($\mu_r = 1$) can be found in Penndorf (1962).

Resonances and Resonant Absorption

As apparent from (7.510)–(7.513), coefficients α_1^u, γ_1^u, and δ_1^u are singular when $\epsilon_r \to -2$, α_1^v, γ_1^v, and δ_1^v when $\mu_r \to -2$, γ_2^u when $\epsilon_r \to -3/2$, and γ_2^v when $\mu_r \to -3/2$. This is a manifestation of the pole singularities in the multipole coefficients r_1^u, r_1^v, r_2^u, and r_2^v (see (7.341) and (7.342)) with $n = 1, 2$. The pole singularities mean infinite scattered fields and can therefore be interpreted as resonances. The resonance values of ϵ_r and μ_r cannot be purely real as the absorption and scattering losses are always present.

Let us take a closer look at the resonances of electrically small homogeneous spheres, that is, on the poles of $r_n^{u,v}$ with arbitrary n. Equations for the poles (7.401) are formulated in terms of the normalized admittances and impedances $\tilde{y}_n$ and $\tilde{z}_n$ (see (7.343)). For homogeneous spheres they follow from (7.477) and (7.478) as

$$\tilde{y}_n = -\frac{\epsilon_r \psi_n(k_1 a)}{k_1 a\, \psi_n'(k_1 a)}, \quad \tilde{z}_n = -\frac{\mu_r \psi_n(k_1 a)}{k_1 a\, \psi_n'(k_1 a)}. \tag{7.514}$$

Representing them by

$$\tilde{y}_n = -\frac{\epsilon_r}{n+1} Q_n(k_1 a), \quad \tilde{z}_n = -\frac{\mu_r}{n+1} Q_n(k_1 a) \tag{7.515}$$

with

$$Q_n(w) = (n+1)\frac{\psi_n(w)}{w\psi_n'(w)}, \tag{7.516}$$

where $w = k_1 a$ and

$$Q_n(w) = 1 + \frac{w^2}{(n+1)(2n+3)} + O(w^4), \tag{7.517}$$

the resonance equations (7.401) can be expressed as

$$\epsilon_r Q_n(k_1 a) = -\frac{n+1}{n[1+\delta_n(ka)]} \tag{7.518}$$

$$\mu_r Q_n(k_1 a) = -\frac{n+1}{n[1+\delta_n(ka)]}. \tag{7.519}$$

Equations (7.518) and (7.519) describe resonances of electric and magnetic multipoles, respectively. The parameter $\delta_n(ka)$, defined in (7.400), is a small parameter with the real and imaginary parts of the orders $O[(ka)^2]$ and $O[(ka)^{2n+1}]$, respectively; the imaginary part of $\delta_n(ka)$ must be accounted for in order to determine the imaginary parts of the resonance values of the material parameters.

Function $Q_1(w)$ has already been studied at the beginning of this section (see (7.489), (7.490), (7.491), (7.492), and Figure 7.35). Functions $Q_n(w)$ with $n = 2, 3, \ldots$ have similar properties: they are even functions of w, equal to untity at $w = 0$, with an infinite number of alternating poles and zeros on the real axis $-\infty < w < +\infty$.

Let us first describe general properties of the solutions of the resonance equations. When ϵ_r and μ_r are such that a resonance equation is satisfied, a corresponding multipole term in the multipole expansions of the scattered field is at resonance. The solutions for ϵ_r and μ_r, which will be denoted by $\epsilon_{r\infty}$ and $\mu_{r\infty}$, have positive imaginary parts, which excludes infinite scattered fields as $\mathrm{Im}\,\epsilon_r$ and $\mathrm{Im}\,\mu_r$ are negative in every physically passive material. For a given value of n and a given value of μ_r, either equation has an infinite number of solutions in ϵ_r, and vice versa when n and ϵ_r are given, there is infinite number of μ_r solutions of (7.518) and (7.519). The solutions of every equation fall in two categories: (a) small-sphere solutions, for which $|k_1 a| < 1$, and (b) high-epsilon (high-mu) solutions when $|k_1 a| > 1$.

As an example we describe resonances of a sphere free of magnetic losses ($\mu_r' > 0$, $\mu_r'' = 0$) on the complex plane of ϵ_r. In this case the resonance approximations described in section 7.4.2, most notably the simplified approximations (7.403) and (7.404), with $p = \epsilon_r$ and $p_\infty = \epsilon_{r\infty}$ can be used. The complex permittivity is written as $\epsilon_r = \epsilon_r' - j\epsilon_r''$ with $\epsilon_r'' \geq 0$.

We begin with the electric resonances, that is, with the solutions of (7.518). The real part of $\epsilon_{r\infty}$ can be approximately determined from (7.518) when $\delta_n(ka)$ is neglected, which gives

$$w_\infty^2 Q_n(w_\infty) = -\frac{n+1}{n}\mu_r(ka)^2, \tag{7.520}$$

where

$$w_\infty = ka\sqrt{\mu_r \epsilon_{r\infty}'}, \quad \epsilon_{r\infty}' = \frac{1}{\mu_r}\left(\frac{w_\infty}{ka}\right)^2. \tag{7.521}$$

The imaginary part of $\epsilon_{r\infty}$ follows from the second equation in (7.404) as

$$\epsilon_{r\infty}'' = \frac{(ka)^{2n+1}}{n^2[(2n-1)!!]^2 \frac{\partial \tilde{y}_n}{\partial \epsilon_r}(\epsilon_{r\infty}')} \tag{7.522}$$

with

$$\frac{\partial \tilde{y}_n}{\partial \epsilon_r}(\epsilon_{r\infty}') = \frac{1}{n\epsilon_{r\infty}'} - \frac{w_\infty}{2(n+1)} Q_n'(w_\infty), \tag{7.523}$$

where $\tilde{y}_n$ is given in (7.515) and the prime in $Q_n'(w)$ denotes differentiation with respect to w.

The small-sphere solutions are those complying with the condition $|w_\infty| \ll 1$. In this case, $Q_n(w_\infty) \approx 1$, the second term on the right-hand side of (7.523) can be neglected, and it follows from (7.520) and (7.522) that

$$\epsilon_{r\infty}' = -\frac{n+1}{n} \tag{7.524}$$

$$\epsilon_{r\infty}'' = -\frac{(n+1)(ka)^{2n+1}}{n^2[(2n-1)!!]^2}. \tag{7.525}$$

These equations describe the location of resonances of the electric multipoles with $n = 1, 2, \ldots$ on the complex ϵ_r plane. The resonances, all with small imaginary parts ($\epsilon''_{r\infty} < 0$), lie close to the interval $-2 < \epsilon'_r < -1$, accumulating at the point $\epsilon_r = -1$ as the multipole order n grows. This type of resonances is practically relevant as negative permittivity values are possible, for example in noble metals at optical frequencies (see Table 2.3). The use of these resonances is in the core of plasmonics (Schuller *et al.* 2010; Zayats *et al.* 2005).

Equation (7.524) neglects the terms of the order $O[(ka)^2]$. To find the correction terms, the small parameter $\delta_n(ka)$ should be accounted for in equation (7.518). Note that the derived approximations (7.524) and (7.525) are independent of μ_r; this dependence is, however, present in the correction terms.

Equation (7.520) has solutions also when $|w_\infty| \gg 1$. These are solutions with large values of ϵ_r (see (7.521)), which is why we call them high-epsilon resonances. It is easy to see that the solutions of (7.520) must be close to the points $w_{n,0}$ where $Q_n(w_{n,0}) = 0$; these points coincide with zeros of the function $\psi_n(w)$, which follows from (7.516) and the fact that $\psi'_n(w)$ is bounded on the whole complex w plane. For a fixed order n, $\psi_n(w)$ has an infinite number of zeros, all real-valued (e.g., see Figure 7.35), and the smallest positive zeros are $w_{10} = 4.49341$ and $w_{20} = 5.76346$.

Thus, the real part of the resonance values of ϵ_r must be close to

$$\epsilon'_{r\infty} = \frac{1}{\mu_r}\left(\frac{w_{n0}}{ka}\right)^2. \tag{7.526}$$

The imaginary part is given by (7.522), in which the derivative (7.523) is dominated by the second term on the right-hand side, so that we have

$$\epsilon''_{r\infty} = -\frac{2(n+1)(ka)^{2n+1}}{n^2[(2n-1)!!]^2 w_{n0} Q'_n(w_{n0})}. \tag{7.527}$$

It follows immediately from the definition (7.516) of $Q_n(w)$ that

$$w_{n0} Q'_n(w_{n0}) = n + 1, \tag{7.528}$$

implying the simple formula

$$\epsilon''_{r\infty} = -\frac{2(ka)^{2n+1}}{n^2[(2n-1)!!]^2}. \tag{7.529}$$

The magnetic resonances are described by (7.519), and neglecting $\delta_n(ka)$ gives a simplified equation for ϵ'_∞,

$$Q_n(w_\infty) = -\frac{n+1}{n\mu_r}. \tag{7.530}$$

The real and imaginary parts of ϵ_∞ are related to the solutions w_∞ of (7.530) by

$$\epsilon'_{r\infty} = \frac{1}{\mu_r}\left(\frac{w_\infty}{ka}\right)^2 \tag{7.531}$$

$$\epsilon''_{r\infty} = \frac{(ka)^{2n+1}}{n^2[(2n-1)!!]^2 \frac{\partial \tilde{z}_n}{\partial \epsilon_r}(\epsilon'_{r\infty})}, \tag{7.532}$$

where

$$\frac{\partial \tilde{z}_n}{\partial \epsilon_r}(\epsilon'_{r\infty}) = -\frac{ka\mu_r^{3/2} Q'_n(w_\infty)}{2(n+1)\sqrt{\epsilon'_{r\infty}}} \tag{7.533}$$

(see (7.515)).

It follows from the definition (7.516) of $Q_n(w)$ and the Riccati–Bessel differential equation (7.33) that

$$Q'_n(w) = \frac{n+1-Q_n(w)}{w} - Q_n^2(w)\left(\frac{n}{w} - \frac{w}{n+1}\right), \tag{7.534}$$

and using (7.530) in (7.534) gives

$$\frac{\partial \tilde{z}_n}{\partial \epsilon_r}(\epsilon'_{r\infty}) = -\frac{(ka)^2}{2n^2}\left[1 + \frac{n}{w_\infty^2}(\mu_r - 1)(n\mu_r + n + 1)\right], \tag{7.535}$$

which results in the formula

$$\epsilon''_{r\infty} = -\frac{2(ka)^{2n-1}}{[(2n-1)!!]^2\left[1 + \frac{n}{w_\infty^2}(\mu_r - 1)(n\mu_r + n + 1)\right]}. \tag{7.536}$$

In contrast to (7.520), equation (7.530) for magnetic resonances has no small-sphere solutions on the complex ϵ_r plane since $Q_n(w) \approx 1$ when $|w| \ll 1$ (see (7.517)) and ϵ_r disappears from the equation.[11] The high-epsilon solutions still exist and, in general, they have to be numerically determined from the transcendental equation (7.530). However, when $\mu_r = 1$, the solutions of (7.530) coincide with zeros of $\psi_{n-1}(w)$ (problem 21). The function $\psi_{n-1}(w)$ with a fixed value of n has an infinite number of zeros, which we denote by $w_{n-1,0}$. For example, if $n = 1$, then $\psi_0(w) = \sin w$ (see (7.35)) and $w_{00} = m\pi$ where $m = 0, \pm1, \pm2 \ldots$. Zeros of $\psi_{n-1}(w)$ with $n \geq 2$ cannot be expressed in a simple form, but their numerical values are well known (e.g., Abramowitz and Stegun (1965, Table 10.6)) and the smallest positive zeros are $w_{10} = 4.49341$ ($n = 2$) and $w_{20} = 5.76346$ ($n = 3$). Thus, the high-epsilon magnetic resonances of a non-magnetic ($\mu_r = 1$) sphere are approximated by the formulas:

$$\epsilon'_{r\infty} = \left(\frac{w_{n-1,0}}{ka}\right)^2 \tag{7.537}$$

$$\epsilon''_{r\infty} = -\frac{2(ka)^{2n-1}}{[(2n-1)!!]^2}. \tag{7.538}$$

For magnetic spheres ($\mu_r \neq 1$, $\mu''_r = 0$) the resonance values of ϵ_r are given by (7.531) and (7.536), with w_∞ being a solution of (7.530).

Approximate formulas for the location of the resonances on the complex plane of ϵ_r permit useful practical estimations of the parameters of small resonant homogeneous spheres. More accurate approximations that include higher-order terms with respect to the parameter ka can be derived from equations (7.518) and (7.519). The latter equations can be also solved numerically by using analytical approximations of this section as initial values in root-finding procedures.

Table 7.1 describes the position of several poles $\epsilon_{r\infty}$, electric and magnetic (e/m), dipole and quadrupole ($n = 1, 2$), small-sphere and high-epsilon, on the complex plane ϵ_r for a non-magnetic ($\mu_r = 1$) sphere with $2a = \lambda/10$. The values, numerically obtained from the exact expressions (7.100) and (7.101) for the multipole coefficients $r_{1,2}^{\mu,\nu}$, are quite close to the approximate values though ka is only moderately small ($ka \approx 0.3$). The close proximity of electric dipole and magnetic quadrupole high-epsilon resonances follows from the comparison of (7.537) and (7.526) with $\mu_r = 1$.

As shown in section 7.4.2, the poles of the multipole coefficients are associated with the phenomenon of resonance absorption, namely, provided that μ_r is real-valued, the absorption cross section σ_{abs} and the absorption efficiency $Q_{\text{abs}} = \sigma_{\text{abs}}/(\pi a^2)$ have sharp peaks at the values of ϵ_r that are complex conjugates of $\epsilon_{r\infty}$; the amplitudes of the peaks depend on n and ka but are independent of ϵ_r and μ_r (see (7.407), (7.408), and (7.409)).

Table 7.2 illustrates the location of the points of the resonant absorption on the complex ϵ_r plane and gives the corresponding values of Q_{abs} for a non-magnetic sphere with $ka = \pi/10$. The values of ϵ'_r, ϵ''_r, and the exact values of

[11] The case in which $\mu_r \approx -(n+1)/n$ is an exception. This case can be usually ignored since isotropic materials with $\mu_r < 0$ are unavailable in nature.

Table 7.1 Resonances $\epsilon_{r\infty} = \epsilon'_{r\infty} - j\epsilon''_{r\infty}$ of a non-magnetic ($\mu_r = 1$) sphere with $2a = \lambda/10$, exact vs. approximate.

Type	Order	Exact		Approximate	
(e/m)	n	$\epsilon'_{r\infty}$	$\epsilon''_{r\infty}$	$\epsilon'_{r\infty}$	$\epsilon''_{r\infty}$
e	1	−2.24354	−0.0710863	−2.	−0.0620126
e	2	−1.53687	−0.000258904	−1.5	−0.000255016
m	1	98.1571	−0.587729	100.	−0.628319
e	1	202.374	−0.0684024	204.575	−0.0620126
m	2	203.885	−0.00672008	204.575	−0.00689028
e	2	335.545	−0.000170392	336.563	−0.000170011

max Q_{abs} are obtained by maximizing the exact multipole expansion for σ_{abs} (three to four terms suffice for the sphere of the chosen size). The location of the maxima can be compared with the location of the associated poles $\epsilon_{r\infty}$ (see Table 7.1) and the values of max Q_{abs} with the approximate values from the simple formula (7.409). It is seen that the resonance approximations (7.407) and (7.409) are quite accurate. Decreasing the value of ka increases the accuracy of the resonance approximation and leads to even sharper maxima of Q_{abs}.

The behavior of the normalized cross sections (efficiencies) Q_{abs}, Q_{T}, and Q_{ext} (see (7.304)) in the vicinity of the quasi-static electric dipole resonance is illustrated in Figures 7.37 and 7.38. Relations (7.414) between the values of σ_{abs}, σ_{ext}, and σ_{T} are clearly observed in Figure 7.37: σ_{abs} and σ_{T} are almost equal to each other and to one half of σ_{ext} when $\epsilon'_r = \epsilon'_{r\infty}$ and $\epsilon''_r = -\epsilon''_{r\infty}$. Figure 7.38 shows that only σ_{abs} has a maximum, whereas σ_{ext} and σ_{T} continue increasing as ϵ''_r gets smaller. When $\epsilon''_r = 0$ (no absorption), $\sigma_{\text{abs}} = 0$ and $\sigma_{\text{ext}} = \sigma_{\text{T}}$.

Summary of Low-frequency Approximations

Figure 7.39 summarizes the several examples of low-frequency ($ka \ll 1$) scattering from homogeneous spheres. The figure is a computer-generated image of the region defined by $-20 < \text{Re}\,\epsilon_r < 60$ and $-50 < \text{Im}\,\epsilon_r < 0$ on the complex ϵ_r plane with the three lines $|k_1 a| = 1$, $\text{Im}\,k_1 a = -1$, and $\sigma_{\text{abs}} = \sigma_{\text{T}}$ that separate sub-regions with different low-frequency approximations. At the line $\sigma_{\text{abs}} = \sigma_{\text{T}}$ the dipole approximation (D), valid below the line, must be complemented with quadrupole (Q) terms in order to correctly describe σ_{abs} in the region between the line and the real axis $\text{Im}\,\epsilon_r = 0$. This is also the line at which the points of resonance absorption may lie. However, there can be further lines at which $\sigma_{\text{abs}} = \sigma_{\text{T}}$; in contrast to the solid line, they are very small closed contours that are not seen on the scale of the plot.

In the interior of the semi-circle, at which $|k_1 a| = 1$ (dotted line), but below the solid line, the quasi-static dipole (Rayleigh) approximation is valid; otherwise, between the solid line and the real axis one needs dynamic and radiation corrections in the dipole moments and quadrupole terms. The line $\text{Im}\,k_1 a = -1$ defines the size of the sphere equal to the skin depth in the material of the sphere, thus separating the cases of opaque and transparent spheres.

Table 7.2 Maxima of Q_{abs} and their location $\epsilon_r = \epsilon'_r - j\epsilon''_r$ for a sphere with $2a = \lambda/10$ and $\mu_r = 1$.

				max Q_{abs}	
Type	Order	ϵ'_r	ϵ''_r	Exact	Approximate
e	1	−2.24353	0.0711228	15.2022	15.1982
e	2	−1.53687	0.000258949	25.3325	25.3303
m	1	98.1571	0.58777	15.1987	15.1982
e	1	202.374	0.0685882	15.2189	15.1982
m	2	203.885	0.00672658	25.3425	25.3303
e	2	335.545	0.000170392	25.3303	25.3303

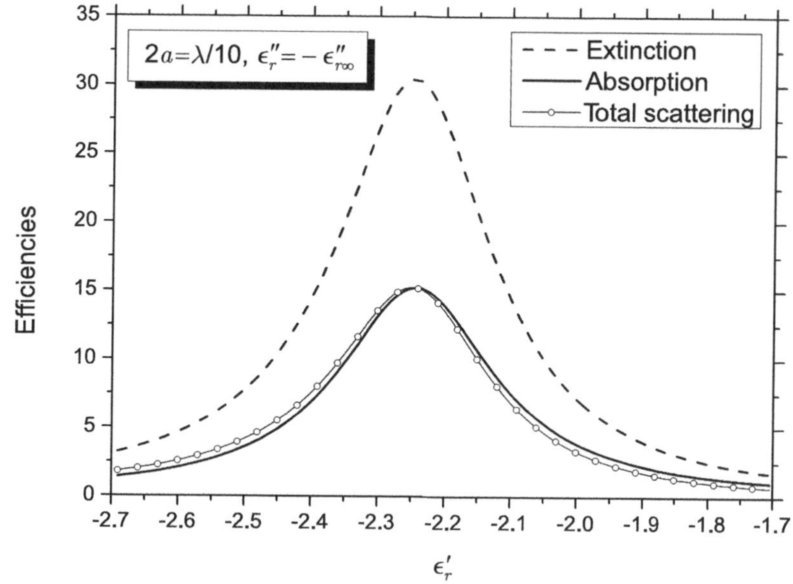

Figure 7.37 Normalized cross sections Q_{ext}, Q_{abs}, and Q_{T} (efficiencies) of an electrically small non-magnetic spherical particle as functions of ϵ_r' in the vicinity of an electric dipole resonance.

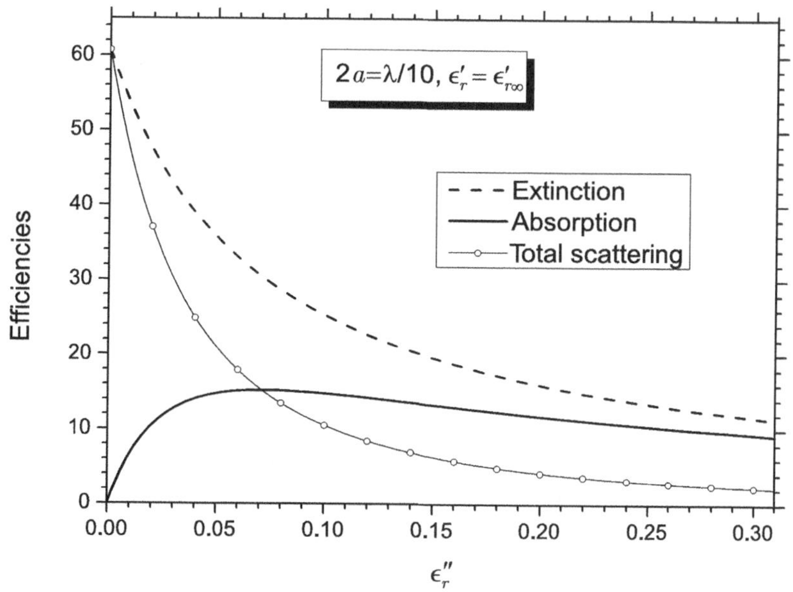

Figure 7.38 The same as Figure 7.37 but as functions of ϵ_r''.

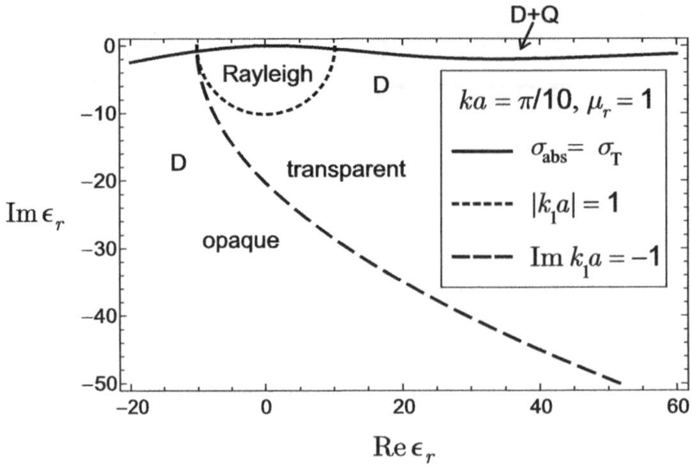

Figure 7.39 Zones on the complex ϵ_r plane that correspond to different types of low-frequency scattering from small homogeneous spheres. Only the lower half-plane $\mathrm{Im}\,\epsilon_r \leq 0$ that corresponds to physically passive materials is shown.

7.5.4 *Electrically Large Lossy Spheres: Relation to the Impedance Sphere and the Role of Absorption*

Scattering from electrically large lossy spheres is similar to scattering from PEC and impedance spheres (sections 7.4.3 and 7.4.5) once the losses over distances comparable to the sphere diameter make the waves propagating in the interior of the sphere negligible. In this section we vary the losses in the material of the sphere and study the transmission between the opaque and transparent sphere cases.

After a brief description of asymptotic approximations for the scattering coefficients of lossy homogeneous spheres, we establish the relation between the exact solution and the solution for a sphere with impedance boundary condition and examine the accuracy of the impedance approximation. The section concludes with an analysis of the role of absorption in the sphere by looking at the various types of scattering cross section depending on the electrical size of the sphere. It is shown that weak absorption results in an irregular behavior (ripple) of the scattered field due to the resonances of the creeping waves propagating on the interior side of the spherical boundary. These resonances, associated with the horizontal branch of poles of the spherical reflection coefficients, are theoretically described in section 7.3.4.

GO Approximation for the Scattered Far Field

The multitude of waves propagating in electrically large ($ka \gg 1$) homogeneous spheres is described in section 7.3.4. If the sphere is lossy so that

$$|\mathrm{Im}\, k_1 a| \gg 1, \tag{7.539}$$

then the waves transmitted through the sphere can be neglected and the sphere can be considered as opaque. The scattered far field is then formed mainly by the specular reflection component. The scattering coefficients of a lossy homogeneous sphere are given by (7.324) and (7.325) with the reflection coefficients

$$R_{\parallel}(\chi) = \frac{\cos\chi - \epsilon_r^{-1}\sqrt{\epsilon_r\mu_r - \sin^2\chi}}{\cos\chi + \epsilon_r^{-1}\sqrt{\epsilon_r\mu_r - \sin^2\chi}} \tag{7.540}$$

$$R_{\perp}(\chi) = \frac{\cos\chi - \mu_r^{-1}\sqrt{\epsilon_r\mu_r - \sin^2\chi}}{\cos\chi + \mu_r^{-1}\sqrt{\epsilon_r\mu_r - \sin^2\chi}}, \tag{7.541}$$

where $\chi = \theta/2$.

The specular reflection approximation is invalid in a $(ka)^{-1/3}$ vicinity of the forward-scattering direction as it does not account for creeping waves excited at the shadow boundary on the surface of the sphere and circumventing the sphere one or more times before being launched toward the observation point. The creeping wave contributions are, however, asymptotically small (see section 6.6.3); they lead to weak oscillations in scattering cross sections that quickly decay with growing electrical size of the sphere. For example, for PEC cylinders the oscillations in RCS are almost insignificant once $ka > 10$ where a is the radius of the cylinder (see Figure 6.19).

Relation to Impedance Boundary Conditions

The condition (7.539) means that the curvature radius of the surface is much greater than the depth of penetration of the incident wave into the sphere. All the conditions (2.78)–(2.81) of applicability of the impedance boundary condition (7.117) are therefore satisfied, and we can expect a good approximation quality of the boundary conditions in this class of scattering problems.

Let us consider the exact solution for homogeneous spheres (section 7.5.1) to see its relation to the solution for the impedance sphere (section 7.4.5). When the interior of the sphere is electrically large ($|k_1 a| \gg 1$), the modal admittances and impedances (7.477) and (7.478) with $n < |k_1 a|$ can be simplified by using the asymptotic formulas (7.39) and (7.41),

$$y_n \approx j\frac{Z}{Z_1}\tan\left(k_1 a - \frac{\pi}{2}n\right) \tag{7.542}$$

$$z_n = j\frac{Z_1}{Z}\tan\left(k_1 a - \frac{\pi}{2}n\right) \tag{7.543}$$

(see Appendix E for more precise conditions of the applicability of the large-argument approximation for Riccati–Bessel functions of the first kind). If, furthermore, the absorption in the sphere is significant, so that $|\mathrm{Im}\, k_1 a| > 1$, then the tangent functions in (7.542) and (7.543) are approximately equal to $-j$, and we obtain

$$\frac{1}{y_n} \approx z_n \approx \frac{Z_1}{Z}. \tag{7.544}$$

Comparing (7.544) with (7.130) suggests that, under made assumptions, the solution for homogeneous spheres approaches the solution for an impedance sphere with the equivalent surface impedance

$$\tilde{Z}_s = Z_1 = \sqrt{\frac{\mu_1}{\epsilon_1}}. \tag{7.545}$$

Summarizing, the boundary condition (7.117) with surface impedance (7.545) leads to a good approximation if (1) the interior of the sphere is electrically large and (2) the absorption losses over distances of the order of the sphere radius are strong. The former condition alone is insufficient as it leads to approximations (7.542) and (7.543), which depend on the multipole order n. Although requiring that the refractive index $\sqrt{\epsilon_r \mu_r}$ be large transform reflections coefficients (7.540) and (7.541) to (4.334) and (4.335) with $\tilde{Z}_s$ from (7.545), one still needs the condition on the absorption to fully eliminate waves passing through the sphere.

Figures 7.40 through 7.43 illustrate the accuracy of the impedance approximation in describing E- and H-plane bistatic scattering cross sections of variously sized spheres made from a lossy material ($\epsilon_r = 15 - j5$, $\mu_r = 1$). The curves are plotted by using multipole expansions for the exact and impedance formulations. In Figures 7.40 and 7.41, the error of the impedance formulation is hardly seen for spheres with $a \geq \lambda/2$ but becomes noticeable when $a = \lambda/4$. The error can be better estimated from Figures 7.42 and 7.43. With $a \geq \lambda/2$ the error is less than 2 dB in the E plane and 0.5 dB in the H plane.

The Role of Absorption. Ripple

Absorption in the material of the sphere significantly influences the scattering behavior. Strong absorption excludes a great variety of waves that may propagate in the interior of the sphere, thus making the sphere opaque, whereas low

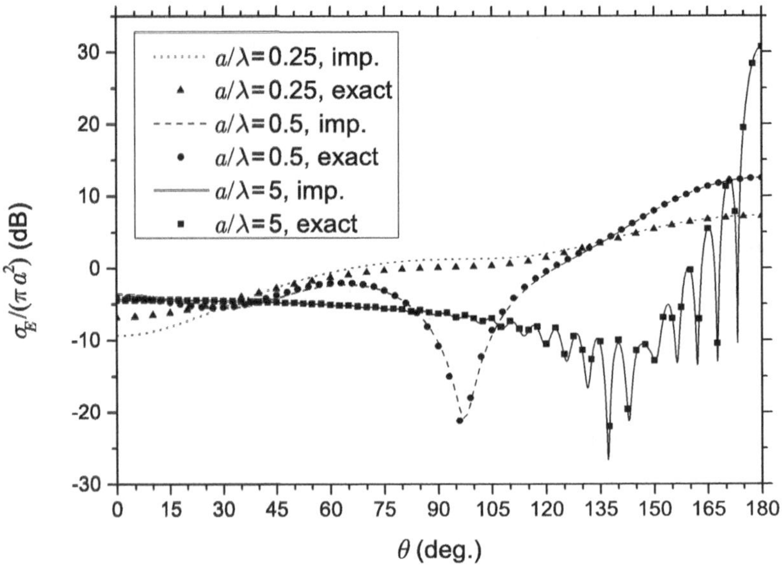

Figure 7.40 E-plane bistatic scattering cross section of several homogeneous lossy spheres: impedance vs. exact formulation.

absorption results in resonance effects in scattering. Let us compare the scattering behavior of opaque and transparent spheres.

Figure 7.44 shows the dependence of backscattering (RCS) on the electrical size of the sphere for several materials. As ka grows, the RCS of a PEC sphere quickly approaches its GO limit πa^2, slightly oscillating with decreasing amplitude around the limit. Materials RAM 1 and RAM 2 are some radar absorbing materials with high losses, non-magnetic

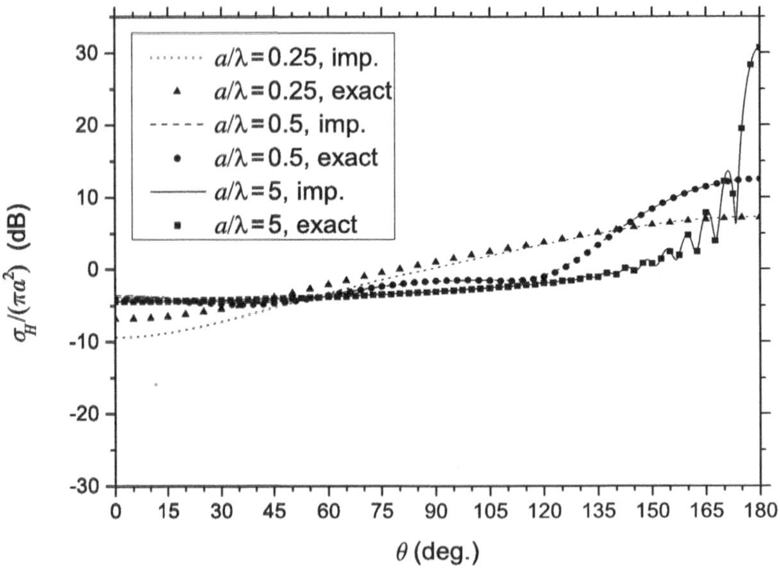

Figure 7.41 The same as Figure 7.40 but in the H plane.

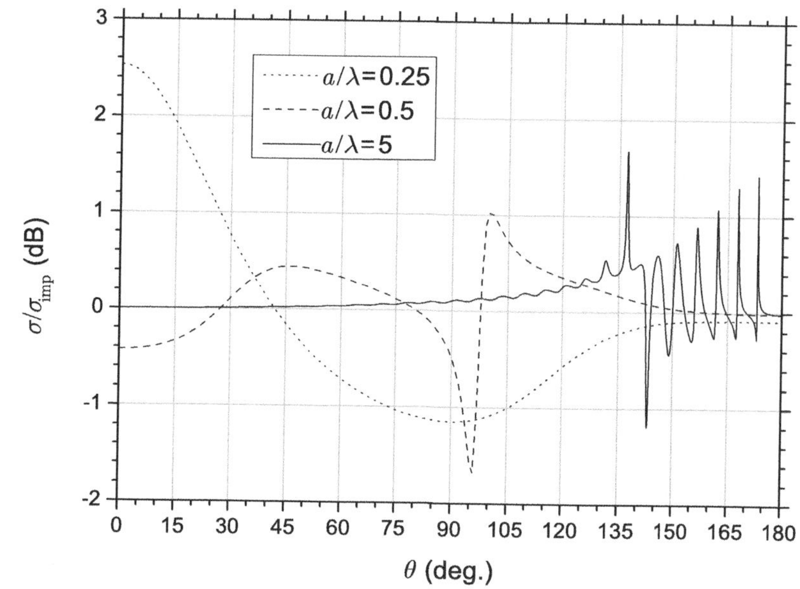

Figure 7.42 Accuracy of impedance approximation for several homogeneous lossy spheres: *E* plane.

(RAM 1) and magnetic (RAM 2) (see Table 2.1). Backscattering from the lossy non-magnetic sphere is lower than in the PEC case since a part of the incident power is absorbed by the sphere and, therefore, less power is reflected back. Magnetic material leads to even lower RCS values since the wave impedance of the material is closer to free space impedance, which further decreases reflection from the surface of the sphere so that more power gets transmitted and absorbed in the interior of the sphere. Plexiglas is an example of a low-loss material; compared with PEC and RAM-spheres,

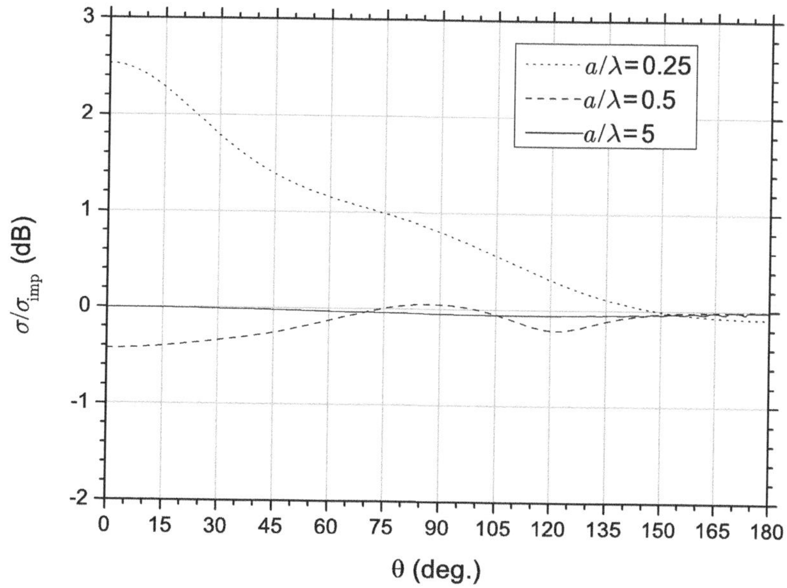

Figure 7.43 The same as Figure 7.42 but in the *H* plane.

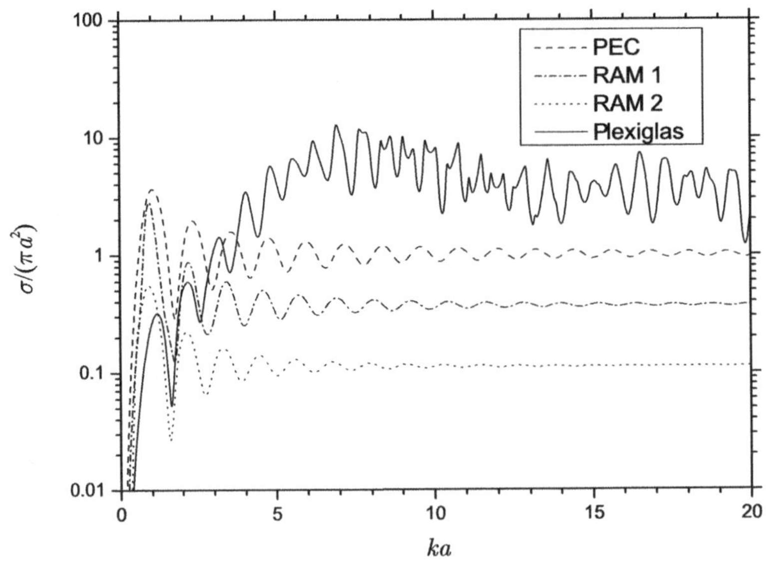

Figure 7.44 Backscattering from homogeneous spheres as a function of electrical size for several materials: RAM 1, $|\epsilon_r| = 16$, $\arg \epsilon_r = -20°$, $\mu_r = 1$; RAM 2, $|\epsilon_r| = 16$, $\arg \epsilon_r = -20°$, $|\mu_r| = 4$, $\arg \mu_r = -20°$; Plexiglas, $\epsilon_r = 2.59 - j0.02$, $\mu_r = 1$.

Plexiglas spheres with $ka > 4$ scatter stronger and in a much less regular manner, with sharp peaks of moderate to small amplitude.

This ripple structure is typical for low-loss spheres and seen not only in backscattering but also in extinction, absorption, and total scattering cross sections. Figures 7.45 and 7.46 show these cross sections for high- (RAM) and low- (Plexiglas) absorption materials. It is seen that for the RAM sphere with $ka > 5$, the cross sections are smooth functions of ka with $\sigma_{\text{abs}} \approx \sigma_{\text{T}} \approx \sigma_{\text{ext}}/2$.

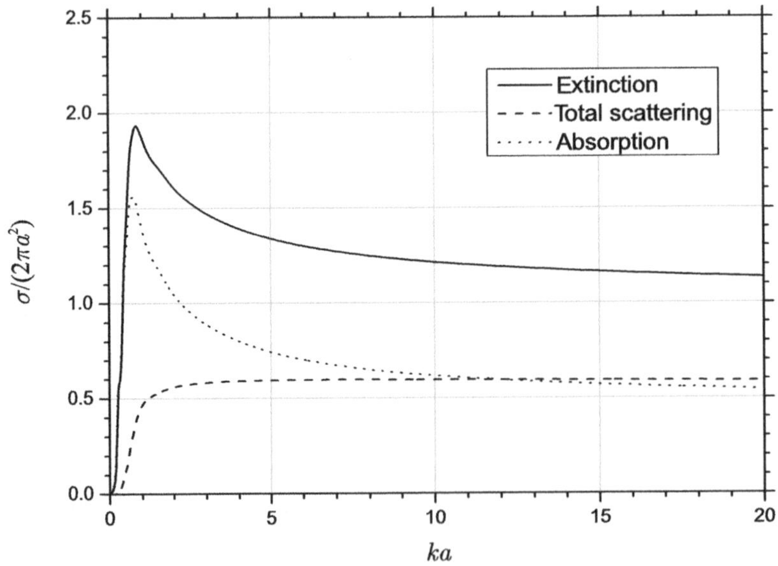

Figure 7.45 Normalized extinction, total scattering, and absorption cross sections for an ideal RAM with $|\epsilon_r| = 16$, $\arg \epsilon_r = -15°$, and $\epsilon_r = \mu_r$.

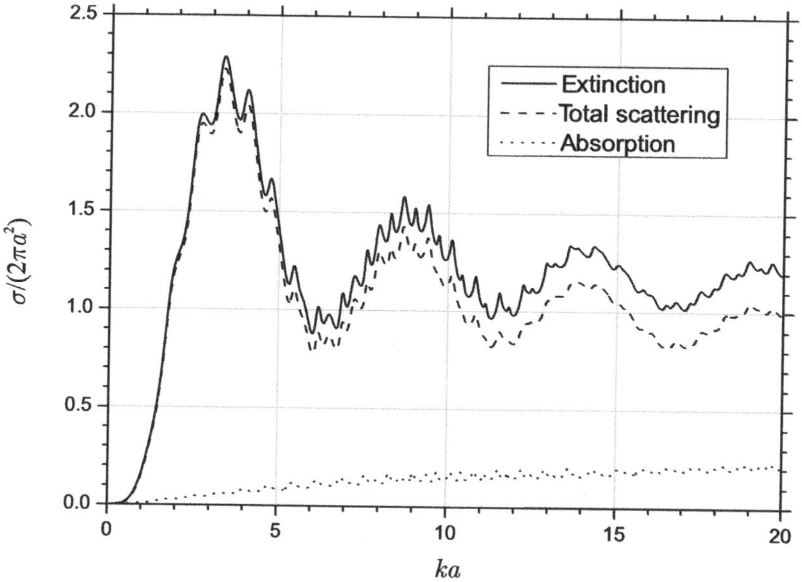

Figure 7.46 The same as Figure 7.45 but for Plexiglas ($\epsilon_r = 2.59 - j0.02$, $\mu_r = 1$).

For the weakly absorbing Plexiglas sphere, however, over the range of ka plotted in Figure 7.46, losses of energy of the incident wave are mainly due to scattering, whereas the absorption losses are much smaller, $\sigma_{\text{abs}} \ll \sigma_{\text{ext}} \approx \sigma_{\text{T}}$. Figure 7.46 reveals the pronounced ripple structure in every cross section. Of course, with growing ka, attenuation losses over the distance comparable with the size of the sphere grow as well, and once $|\text{Im}\, k_1|a \geq 1$ the sphere starts approaching the opaque limit. For the Plexiglas sphere this condition reaches $ka \geq 161$.

That the ripple is determined by the waves passed through the sphere is also seen in the limit of PEC sphere, which does not absorb ($\sigma_{\text{abs}} = 0$) but fully reflects the incident radiation ($\sigma_{\text{ext}} = \sigma_{\text{T}}$); again, no ripple is observed in Figure 7.25.

As $ka \to \infty$, the extinction cross section approaches its theoretical limit $2\pi a^2$ (see (3.12)) but rather slowly. For example, at $ka = 10$, 100, and 1000, the quantity $\sigma_{\text{ext}}/(2\pi a^2)$ takes values 1.0312, 1.0041, and 1.0007 (PEC), 1.213, 1.0459, and 1.0099 (RAM), and 1.2808, 1.0412, and 1.0099 (Plexiglas), respectively.

Let us now explain the origin of the ripple. The relation of this phenomenon to peaks in multipole coefficients r_n^u and r_n^v was suggested in Mevel (1958), Metz and Dettmar (1963), and Chýlek (1976). Grandy (2000, Chapter 7) provides further insight into the ripple by considering the poles of the spherical reflection coefficients on complex planes of ka and ν, and using analogies from quantum mechanics.

A first attempt of physical interpretation of the ripple dates back to Van de Hulst (1957, section 17.5), who suggested an explanation by means of surface waves on the convex side of the spherical boundary reinforced by waves traveling through the sphere. However, as we have seen in section 7.3.4, the ripple should be rather related to the creeping waves propagating on the concave side of the spherical boundary. The waves are associated with the poles ν_m of the horizontal branch of poles of the functions $r_{\nu-1/2}^{u,v}$ on the complex ν plane, which may have anomalously small imaginary parts when $ka \ll \nu_m \ll k_1 a$ (see Figure 7.16). When the poles approach half odd numbers, $\nu_m \approx n + 1/2$ with n being an arbitrary integer (exact equality is prohibited by conservation of energy), the multipole reflection coefficients r_n^u or r_n^v are almost at resonance.

Figure 7.47 shows the normalized extinction, total scattering, and absorption cross sections of a Plexiglas sphere over the range $6 \leq ka \leq 13$. Every maximum in the curves is associated with an extremum of an electric r_n^u or magnetic r_n^v multipole coefficient. This can be nicely seen from Figure 7.48, which shows the extinction cross section of the Plexiglas sphere plotted by using the exact solution and a solution with the mode e10 turned off ($r_{10}^u = 0$). Turning on the mode results in a sharp maximum of σ_{ext} around $ka = 8.7$ and in greater values of σ_{ext} in the range $10 \leq ka \leq 13$ (see Figure 7.47).

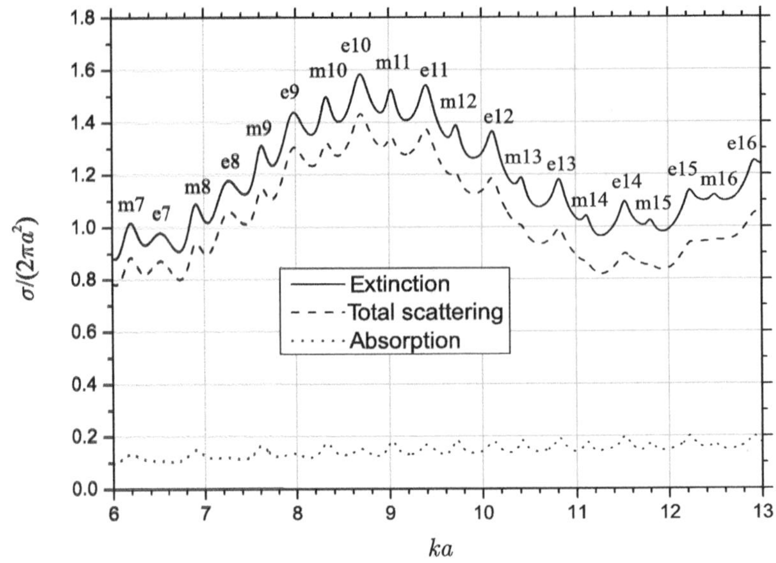

Figure 7.47 Cross sections of a Plexiglas sphere. The peaks are attributed to extrema in electric (e) and magnetic (m) multipole moments, r_n^u and r_n^v.

Such behavior of σ_{ext} immediately follows from the dependence of r_{10}^u on ka (Figure 7.49). When $ka < 7$, the reflection coefficient is almost zero, but its amplitude grows with ka approaching the multipole order, culminating in a sharp minimum of $\text{Re}\, r_{10}^u$ at $ka = 8.69$ (and in a maximum of $|r_{10}^u|$ since $\text{Im}\, r_{10}^u = 0$ at that value of ka). This results in the peak e10 that is observed in all cross sections on Figure 7.47. With larger values of ka, the coefficient r_{10}^u shows further maxima and minima, which are, however, not so sharp and lead to smoother background corrections of σ_{ext} on the interval $10 < ka < 13$ (Figure 7.48).

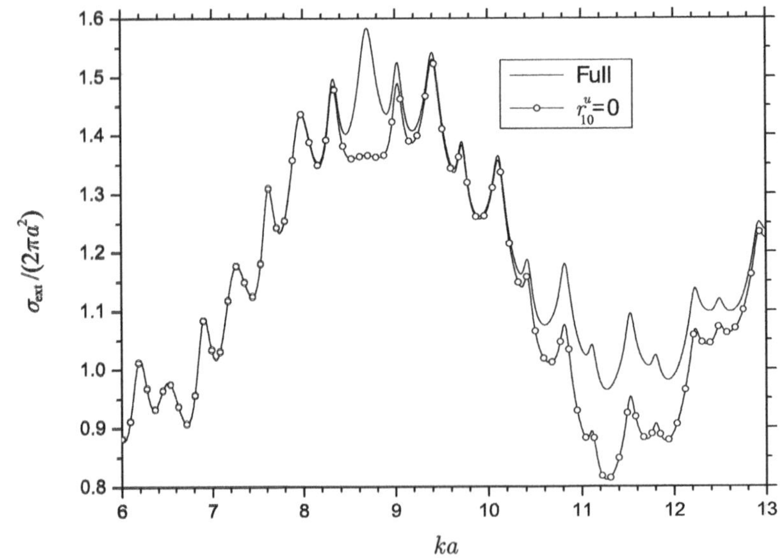

Figure 7.48 Extinction cross section of a Plexiglas sphere with and without contribution from the electric multipole of the 10th order.

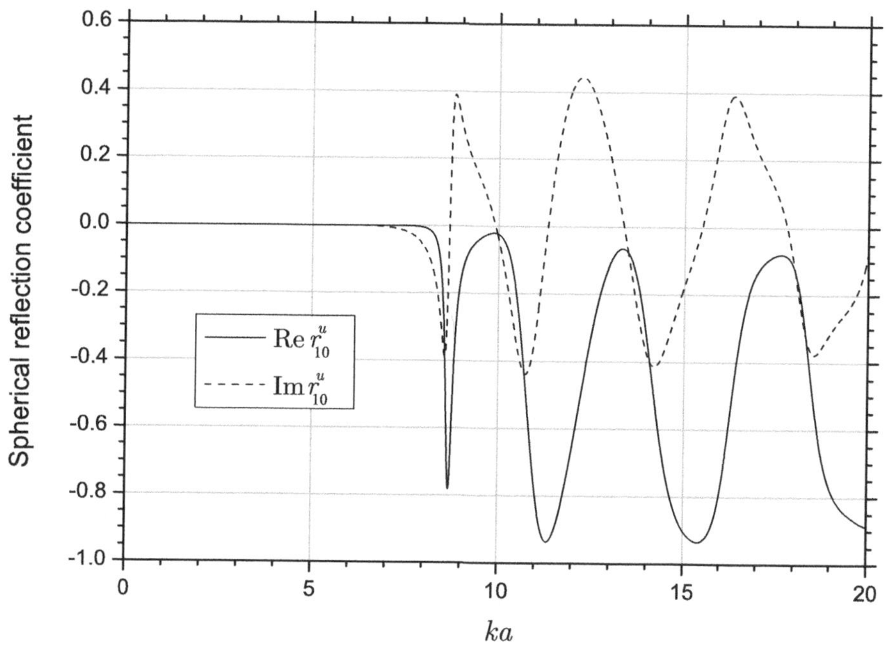

Figure 7.49 Real and imaginary parts of r^u_{10} for a Plexiglas sphere as a function of *ka*.

7.5.5 Electrically Large Low-absorption Spheres: Light Scattering from Water Droplets

Electromagnetic scattering from electrically large spheres with low absorption can be well understood in terms of GO. The field scattered in any direction is typically the result of overlapping of many rays due to various scattering mechanisms. Since absorption is low, in addition to the external reflection there can be further significant contributions, most notably from the waves passed through the sphere, directly or on a single reflection. Waves multiply reflected inside the sphere contribute much more weakly to the scattered field because of generally small reflection from interfaces of transparent materials (e.g., see Figure 4.7). Because of the absence of the total shadow behind penetrable spheres, the ray optical contributions dominate the creeping waves, though those propagating on the concave side of the spherical boundary result in clearly noticeable ripple in scattering diagrams.

In this section the formulas for GO and creeping wave contributions to the fields scattered by low-absorption electrically large spheres derived in section 7.3.4 are applied to the analysis of light scattering from rain, cloud, and fog droplets as this is an instructive example with interesting applications in atmospheric optics. In the visible part of the electromagnetic spectrum, the wavelength of light ranges from 0.4 μm (violet light) to 0.7 μm (red light) and water is a weakly dispersive medium with extremely low absorption (Table 2.3). The size distribution of water droplets encountered in the atmosphere is rather broad, with diameters varying from 100 μm to approximately 9 mm in rain droplets[12] and from 0.1 μm to 100 μm in clouds and fog. The presence of drops smaller than the wavelength of light makes clouds and fog opaque and white due to diffuse scattering. However, it is the presence of electrically large droplets that gives rise to such spectacular phenomena as rainbows, glories, and coronae.

The high-frequency formulas (7.324) and (7.325) describe the contribution from the externally reflected waves only and, therefore, are insufficient, and we start the section by deriving contributions of multiple reflected rays to the scattering

[12]Raindrops smaller than 100 μm are insufficiently heavy to overcome the air resistance and to fall as rain; raindrops larger than 9 mm are unstable and break up. The droplets with diameter smaller than 2 mm are approximately spherical; larger droplets get deformed, becoming increasingly flattened on the bottom due to the airflow.

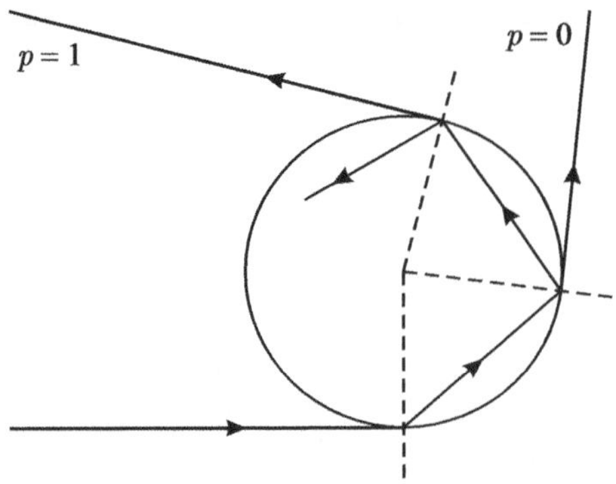

Figure 7.50 Directly transmitted ($p = 0$) and interiorly reflected rays in a sphere with $k_1/k = 4/3$: counterclockwise propagation.

coefficients. We study the behavior of rays passing through the sphere, including their caustics, and take a look at typical bistatic scattering diagrams from electrically large water spheres. At the caustics the GO formulas are inapplicable, and we derive caustic approximations in terms of the Airy function by applying a modified version of the stationary phase method to the Watson-type representation of the scattered field. Near the caustics the scattered field is large, which, combined with dispersion, leads to the phenomenon of rainbow. By using the exact solution, the ray optical and caustic approximations we describe the fundamental properties of atmospheric rainbows. Finally, anomalously strong back- and forward scattering by electrically large penetrable spheres and the related atmospheric phenomena of glory and corona are addressed. In particular, it is shown the glory is mainly formed by contributions from several creeping waves propagating clock- and counterclockwise in the interior of water droplets.

GO of Homogeneous Spheres: Bistatic Scattering

We use the GO solution from section 7.3.4. The asymptotic solutions for the radial components of the electric and magnetic fields of a ray should be specialized to the case of the far field ($r \to \infty$), and using relations (7.322) and (7.323) we can arrive at the asymptotic approximations for the contribution of the ray to the scattering coefficients $S_{1,2}(\theta)$. The contribution of the externally reflected ray (specular component) has already been determined and given by (7.326). Let us study the contributions from the rays passed though the sphere (Figure 7.50). The radial components $E^{\pm}_{pq}$ and $H^{\pm}_{pq}$ are given by (7.272) and (7.273), where p is the number of internal reflections ($p = 0, 1, 2, \ldots$), q is the number of times the ray circumvents the sphere before being ejected ($q = 0, 1, 2, \ldots$), and the plus/minus sign denotes the direction of propagation, clock- or counterclockwise, respectively.

Inserting (7.272) in (7.322) and (7.273) in (7.323) and evaluating the limit $r \to \infty$ in which $\chi \to \pi/2$, we arrive at the formulas:

$$S_1^{\text{ray}}(\theta) = \mp \frac{ka\,\sigma^{\pm}_{pq}\sqrt{\sin(2\theta_c)}\,T_{\parallel}^{01}(\theta_c)T_{\parallel}^{10}(\theta_1)\left[R_{\parallel}^{11}(\theta_1)\right]^p}{2\sqrt{\sin\theta}|1-(p+1)k\cos\theta_c/(k_1\cos\theta_1)|^{1/2}} e^{2ja[k\cos\theta_c-(p+1)k_1\cos\theta_1]} \tag{7.546}$$

$$S_2^{\text{ray}}(\theta) = \pm \frac{ka\,\sigma^{\pm}_{pq}\sqrt{\sin(2\theta_c)}\,T_{\perp}^{01}(\theta_c)T_{\perp}^{10}(\theta_1)\left[R_{\perp}^{11}(\theta_1)\right]^p}{2\sqrt{\sin\theta}|1-(p+1)k\cos\theta_c/(k_1\cos\theta_1)|^{1/2}} e^{2ja[k\cos\theta_c-(p+1)k_1\cos\theta_1]}. \tag{7.547}$$

Here $\sigma^{\pm}_{pq}$ is the limit of (7.271),

$$\sigma^{\pm}_{pq} = (-1)^q \exp\left\{jp\frac{\pi}{2} \pm j\frac{\pi}{4} + j\frac{\pi}{4}\operatorname{sgn}\left[1-(p+1)\frac{k\cos\theta_c}{k_1\cos\theta_1}\right]\right\}. \tag{7.548}$$

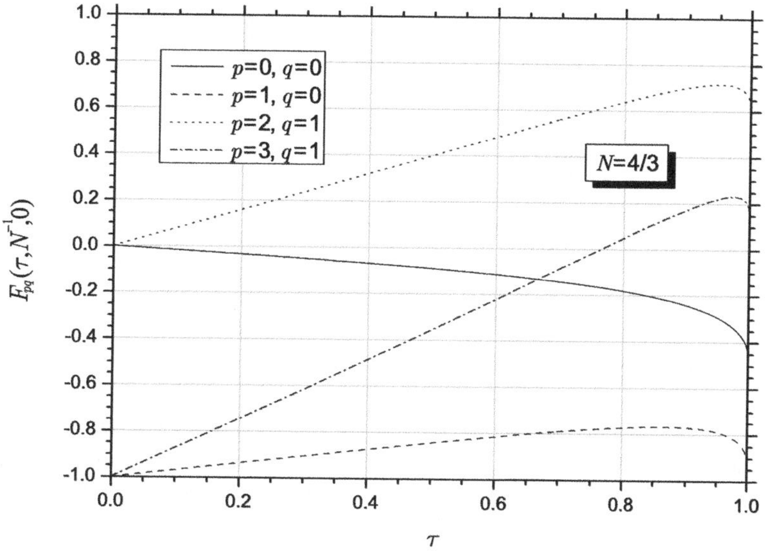

Figure 7.51 The left-hand side of equation (7.549) for water in the visible.

To determine the parameters of the formulas, the incidence angle θ_c, and the refraction angle θ_1 that correspond to a given bistatic angle θ, one has to solve the stationary point equation (7.277), which in the limit $r \to \infty$ has the form

$$\frac{2}{\pi} \arccos \tau + 2q - \frac{2}{\pi}(p+1) \arccos \left(\frac{k}{k_1} \tau \right) = \pm \frac{\pi - \theta}{\pi}, \tag{7.549}$$

where the left-hand side is equal to $F_{pq}(\tau, k/k_1, 0)$ defined in (7.278). One searches for a solution on the interval $0 \le \tau \le 1$ with given values of the integer parameters p and q. Equation (7.549) with positive (negative) right-hand side describes rays propagating clockwise (counterclockwise). The incidence and refraction angles are related to each other and to the solution of (7.549) by the formula

$$\tau = \sin \theta_c = \frac{k_1}{k} \sin \theta_1. \tag{7.550}$$

As an example, let us study solutions of (7.549) up to triple reflection for non-magnetic spheres with refractive index $N = k_1/k = 4/3$, which is quite close to that of water in the visible. Figure 7.51 shows the left-hand side of (7.549) as a function of τ for (a) $p = 0$, $q = 0$ (direct transmission), (b) $p = 1$, $q = 0$ (single interior reflection), (c) $p = 2$, $q = 1$ (double interior reflection), and (d) $p = 3$, $q = 1$ (triple interior reflection).

In (a), the left-hand side of (7.549), $F_{00}(\tau, k/k_1, 0)$, is a monotonously decreasing function of τ with the values in the range $-0.46 \le F_{00}(\tau, k/k_1, 0) \le 0$, and equation (7.549) implies that the ray of the type E_{00}^- and H_{00}^- arrives at the observation point if $97.2° \le \theta \le 180°$. This is the ray directly transmitted through the sphere. With $\theta = \pi$, the solution of (7.549) is given by $\tau = 0$, which corresponds to the axial ray ($\theta_c = \theta_1 = 0$). If $\theta = 97.2°$, then $\tau = 1$, which is the ray tangential to the sphere at the injection and ejection points (Figure 7.50). The parameter σ_{00}^- in (7.546) and (7.547) is equal to unity since

$$1 > \frac{k}{k_1} \ge \frac{k \cos \theta_c}{k_1 \cos \theta_1} \ge 0 \tag{7.551}$$

when $0 \le \theta_c \le \pi/2$. There is no contribution from the directly transmitted ray when $\theta < 97.2°$.

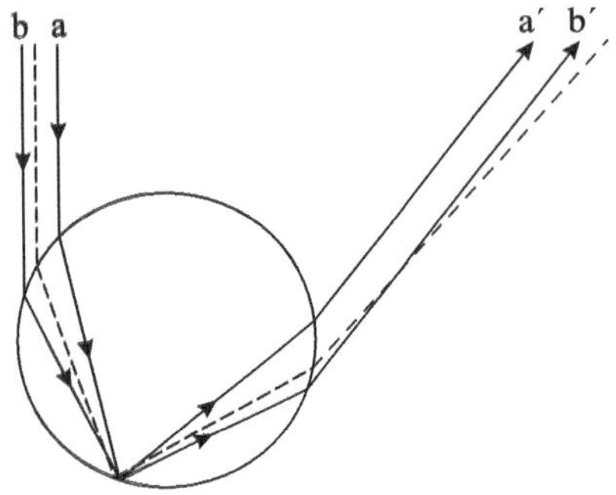

Figure 7.52 Rainbow ray (dashed) and two rays with different injection angles but equal ejection angles for a water drop with $N = 4/3$.

In (b), (7.549) permits a solution of the type E_{10}^- and H_{10}^- if $0 \leq \theta \leq \tilde{\theta}$ where $\tilde{\theta} \approx 42.0°$ (the dashed curve in Figure 7.51). The curve has a maximum at $\tau = \tilde{\tau} \approx 0.86$ which corresponds to $\theta_c = \tilde{\theta}_c \approx 59.4°$; a ray incident upon this angle at the sphere reaches the observer located at $\theta = \tilde{\theta}$. When $14.4° \leq \theta \leq \tilde{\theta}$, there are two solutions, $\tau_1 \leq \tilde{\tau} \leq \tau_2$, of (7.549), which means that two rays of this type arrive at the observation point (Figure 7.52). With $\theta_c = \tilde{\theta}_c$, the two solutions coincide, $\tau_1 = \tau_2 = \tilde{\tau}$. When $\theta = 14.4°$, one of these rays (bb′ in Figure 7.52) illuminates the sphere at the grazing angle ($\tau_2 = 1$, $\theta_c = \pi/2$) and disappears when θ decreases further, so that only one ray arrives at observation points with $0 \leq \theta < 14.4°$.

The maximum of $F_{10}(\tau, k/k_1, 0)$ at $\tau = \tilde{\tau}$ corresponds to a caustic of the rays singly reflected in the interior of the sphere as no ray of this type can arrive at observation points with $\theta > \tilde{\theta}$, and the angle $\tilde{\theta}$ is the maximum possible bistatic angle for rays of this type.

The presence of similar caustics in the rays of the types E_{21}^+ and H_{21}^+ (double reflection in the interior of the sphere, clockwise propagation) and E_{31}^+ and H_{31}^+ (triple reflection in the interior of the sphere, clockwise propagation) is apparent from the other curves in Figure 7.51. The value of $\tilde{\tau}$, at which $F_{pq}(\tau, k/k_1, 0)$ has an extremum, can be determined by differentiation of (7.549) with respect to τ, which leads to the simple formula

$$\tilde{\tau} = \sqrt{\frac{(p+1)^2 - N^2}{(p+1)^2 - 1}}. \qquad (7.552)$$

The formula defines the value of $\tilde{\tau}$ for caustics of the rays with p interior reflections.[13] The parameters $\tilde{\theta}_c$ and $\tilde{\theta}_1$ are obtained from (7.550) and $\tilde{\theta}$ from (7.549) as

$$\tilde{\theta} = \pi \mp 2\pi q \mp 2 \arccos \tilde{\tau} \pm 2(p+1) \arccos(\tilde{\tau}/N). \qquad (7.553)$$

The parameters $\tilde{\tau}$, $\tilde{\theta}_c$, $\tilde{\theta}_1$, and $\tilde{\theta}$ describe the extremal ray, or caustic ray, which bounds the rays with a given value of p.

Parameters $\sigma_{pq}^{\pm}$ in (7.546) and (7.547) are piecewise constant functions of θ_c which change their values by a factor j at the values of θ_c at which the second derivative of the phase function $S_{pq}^{\pm}(\nu, k)$ with respect to ν vanishes (see (7.262), (7.270), and (7.271)). This value of θ_c is equal to $\tilde{\theta}_c$ since by definition $\nu = ka\tau$ and $\tilde{\tau}$ is the point at which the derivative of $F_{pq}(\tau, k/k_1, 0)$ with respect to τ vanishes, while $F_{pq}(\tau, k/k_1, 0)$ itself is proportional to the first derivative of the phase function.

At a given bistatic angle θ, there is always a contribution due to exterior reflection (see (7.324) and (7.325)) and there can be further contributions from interiorly reflected rays of various types. With the help of Figure 7.51, one can easily determine whether a ray of a given type arrives at the observation point. For example, when $\theta = \pi/2$, there are two solutions of (7.549), $\tau \approx 0.39$ ($p = 3$, counterclockwise propagation) and $\tau \approx 0.62$ ($p = 2$, clockwise propagation). On

[13]For water at visible wavelengths, $1 < N < 2$ and $0 < \tilde{\tau} < 1$ for all $p = 1, 2, \ldots$. For materials with $2 < N < 3$, equation (7.552) with $p = 1$ gives a purely imaginary $\tilde{\tau}$, which means that (7.549) has no solution when $p = 1$. For materials with $3 < N < 4$ there are no caustics with $p = 1, 2$, and so on.

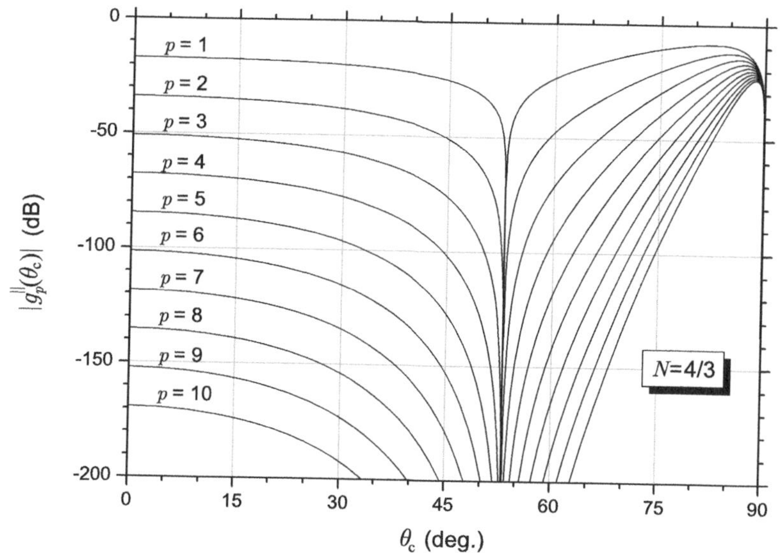

Figure 7.53 Amplitude factors of internally reflected rays as functions of the incidence angle θ_c at injection point for water in the visible ($N = 4/3$). Parallel polarization.

the contrary, the rays with $p = 0$ and $p = 1$ do not reach the observer at that bistatic angle. There can be further rays with $p \geq 4$ that may contribute.

Every interior reflection reduces the amplitude of the respective ray, and it is instructive to compare the amplitudes of the rays with various values of p. The absolute values of the GO factors

$$g_p^{\|,\perp}(\theta_c) = T_{\|,\perp}^{01}(\theta_c) T_{\|,\perp}^{10}(\theta_1) \left[R_{\|,\perp}^{11}(\theta_1) \right]^p \tag{7.554}$$

for water in the visible are plotted in Figures 7.53 and 7.54. The factors depend only on the incidence angle θ_c at the injection point since the refraction angle θ_1 is related to θ_c by the Snell law (7.550). The minimum in the curves for parallel polarization at $\theta_c = 53.1°$ corresponds to Brewster's angle of $R_{\|}^{00}(\theta_c)$ at the air–water interface, which is at the same time a zero of $R_{\|}^{11}(\theta_1)$. Every reflection makes the amplitude factors significantly smaller, except for the range $80° < \theta_c < 90°$ where all the curves are at maximum with the values decaying significantly slower with growing p than on the rest of the curves (Table 7.3).

Figures 7.55 and 7.56 show the amplitudes of the ray contributions (7.546) and (7.547) to scattering coefficients, normalized to the common factor $ka/(2\sqrt{\sin\theta})$, as functions of θ_c. An interesting feature is that as long as $0 \leq \theta_c < 75°$, the externally reflected ray is weaker than the directly transmitted and singly reflected rays, which is apparently due to the weakness of external reflection compared with the transmission and interior reflections in a medium with the refractive index only slightly different from unity. Another noticeable feature in the figures is the singularity of the amplitudes of the rays with $p = 1$ and $p = 2$ at $\theta_c \approx 59°$ and $72°$, respectively. The singularities occur at the caustics of the respective rays, implying the failure of the ray optical approximation (7.546) and (7.547) at the caustics since the exact solution for the scattered field is finite everywhere.

Let us take a look at exact scattering diagrams which illustrate features typical for bistatic scattering from electrically large low-absorption spheres. The curves are plotted for spherical water droplets of radius $a = 10$ μm (Figures 7.57 and 7.58) and $a = 200$ μm (Figures 7.59 and 7.60) at a frequency in the middle of the optical range, $\lambda = 0.55$ μm, at which $\epsilon_r = 1.776 - j5.22 \times 10^{-9}$ (see Table 2.3). The chosen size of the droplets corresponds to typical values for cloud or fog droplets and for raindrops.

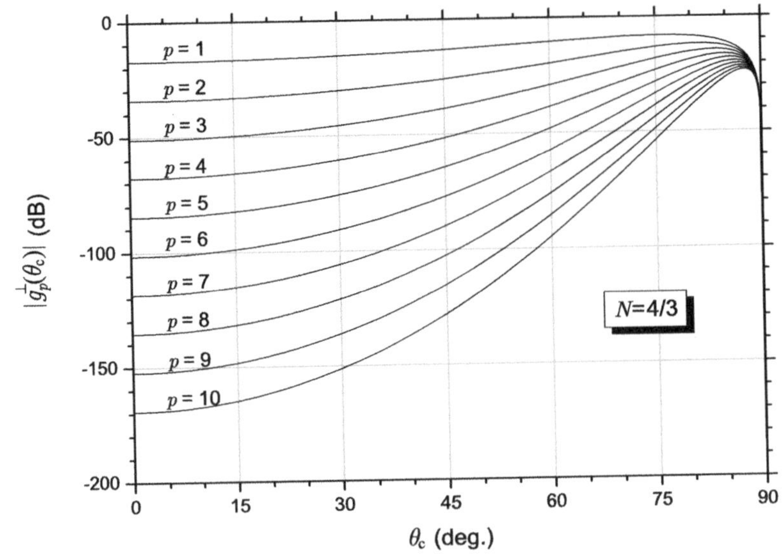

Figure 7.54 The same as Figure 7.53 but for perpendicular polarization.

In both cases the parameter *ka* is large, 114 and 2280, respectively, and the cross sections are quickly varying functions of θ (the variation scale $O[(ka)^{-1}]$) and achieve very large values (of the order $O[(ka)^2]$) in the forward-scattering direction ($\theta = \pi$). In Figures 7.59 and 7.60, a step in θ as small as 0.01° or less is necessary to catch all the oscillations, and the dynamic range is by four orders of magnitude larger than in Figures 7.57 and 7.58.

For every size of droplet, the bistatic scattering cross section is plotted in two mutually orthogonal planes $\phi = 0$ (E plane) and $\phi = \pi/2$ (H plane). The apparent differences between $\sigma_E(\theta)$ and $\sigma_H(\theta)$ can be explained by the polarization difference in the reflection and transmission coefficients at the air–water interface (see Figures 4.4, 4.5, 4.6, and 4.7). The deviations between $\sigma_E(\theta)$ and $\sigma_H(\theta)$ are most apparent around $\theta \approx 42°$, where σ_H is considerably greater than σ_E, and around $\theta \approx 105°$, where σ_H oscillates much stronger than σ_E. Both features are related to the Brewster effect. $R_{\parallel}^{00}(\theta/2)$ vanishes at $\theta \approx 106.2°$ (see Figure 4.5), and according to (7.324) and (7.325) the specular component, which is the reason for strong oscillations in this range of values of θ, is much smaller in $S_1(\theta)$ than in $S_2(\theta)$. The generally higher level of σ_H around $\theta \approx 42°$ is due to the caustic of the rays singly reflected in the interior of the sphere; the corresponding angle

Table 7.3 Maxima of the GO factors ($\max|g_p^{\parallel,\perp}| = |g_p^{\parallel}(\theta_{\max}^{\parallel})| = |g_p^{\perp}(\theta_{\max}^{\perp})|$) for water in the visible $\mu_r = 1$ and $\epsilon_r = 1.778$ ($N = 4/3$).

p	$\max\|g_p^{\parallel,\perp}\|$	$\theta_{\max}^{\parallel}$ (deg.)	$\theta_{\max}^{\perp}$ (deg.)
1	0.385	82.3	75.8
2	0.250	85.1	81.2
3	0.186	86.4	83.5
4	0.148	87.1	84.9
5	0.123	87.6	85.7
6	0.105	87.9	86.4
7	0.092	88.2	86.8
8	0.082	88.4	87.2
9	0.074	88.6	87.5
10	0.067	88.7	87.7

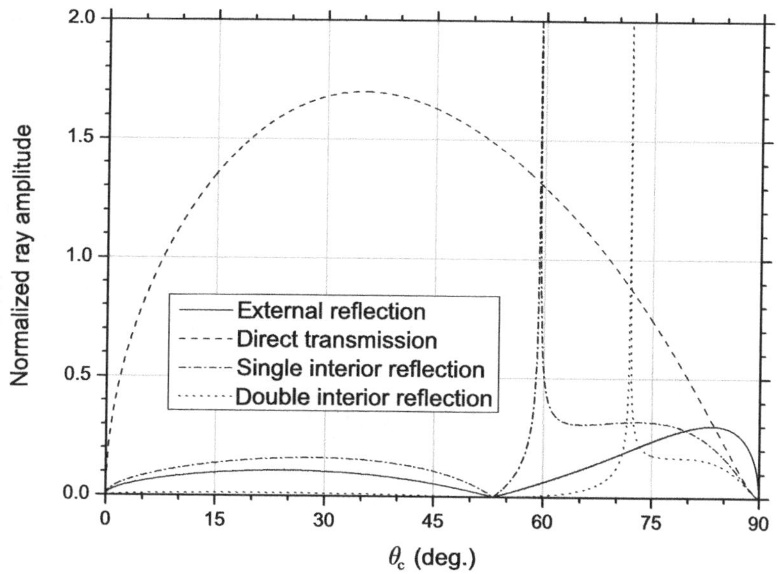

Figure 7.55 Normalized amplitudes of several rays in the E-plane as functions of the incidence angle at the injection point for spheres with $N = 4/3$.

of interior reflection $\theta_1 \approx 40.2°$ is close to the Brewster angle of $R_{\parallel}^{11}(\theta_1)$ at 37.5° (see Figure 4.7), which leads to lower values of σ_E compared with σ_H.

Finally, comparing Figures 7.58 and 7.60, we note that the maxima at about 42° and 52° in Figure 7.60 are more pronounced than in Figure 7.58. These are the maxima that correspond to the caustics of singly and doubly internally reflected rays (main and secondary rainbows), a discussion of which now follows.

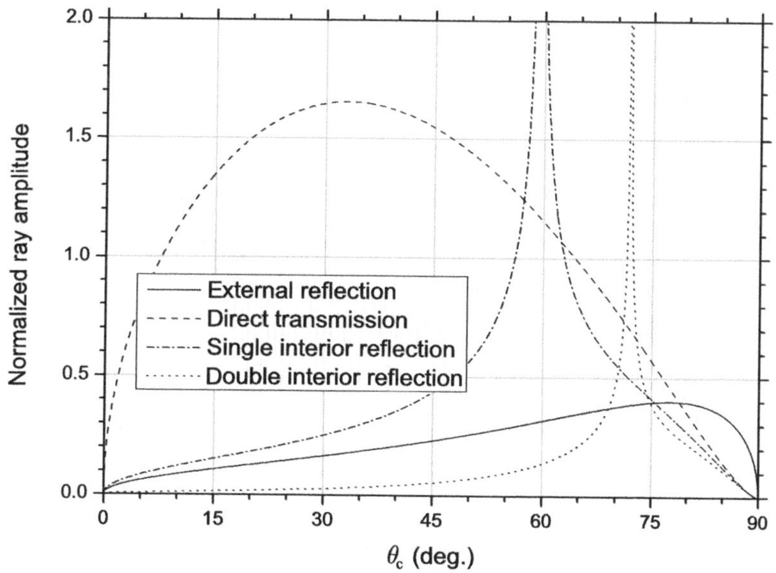

Figure 7.56 The same as Figure 7.55 but in the H plane.

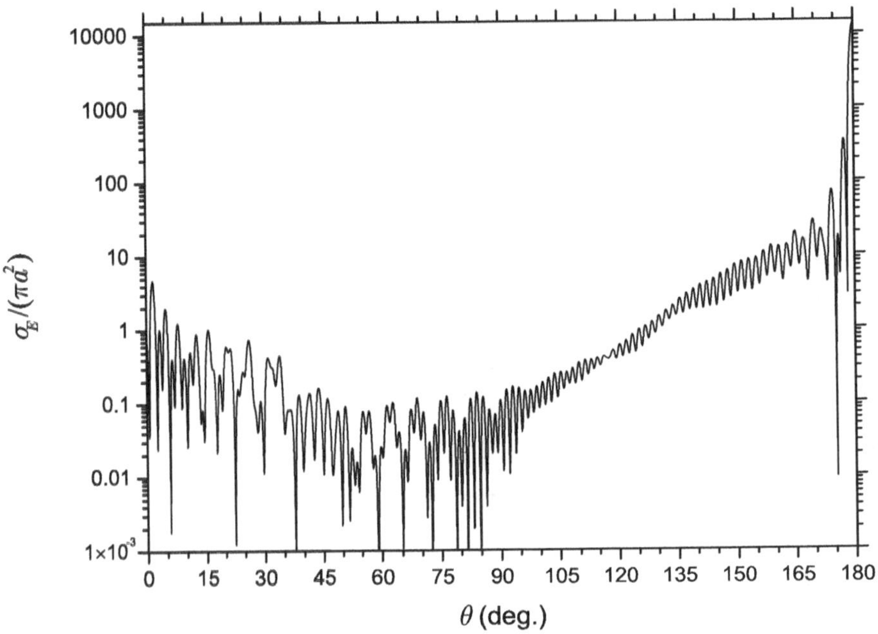

Figure 7.57 Normalized E-plane bistatic cross section of a spherical water droplet with $a = 10$ μm as a function of the bistatic angle.

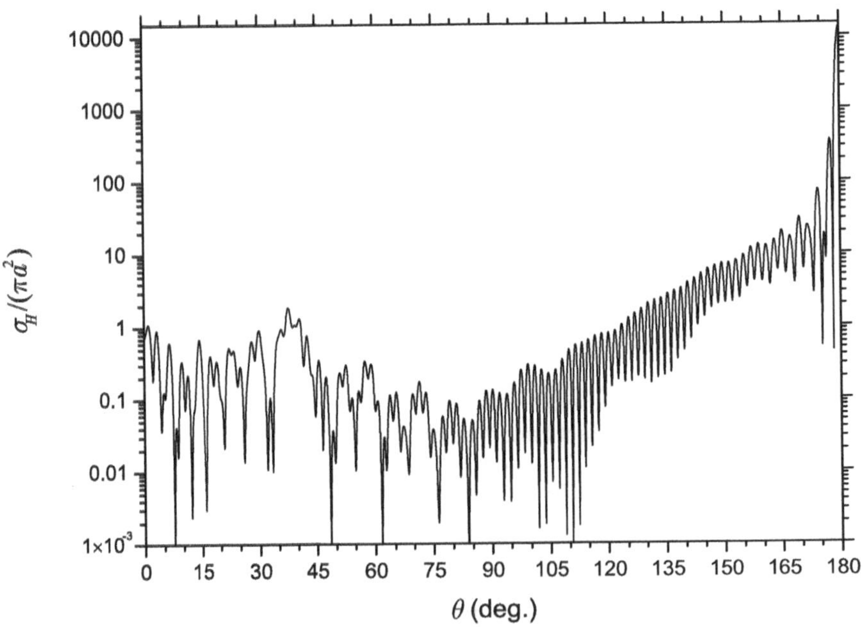

Figure 7.58 The same as Figure 7.57 but in the H plane.

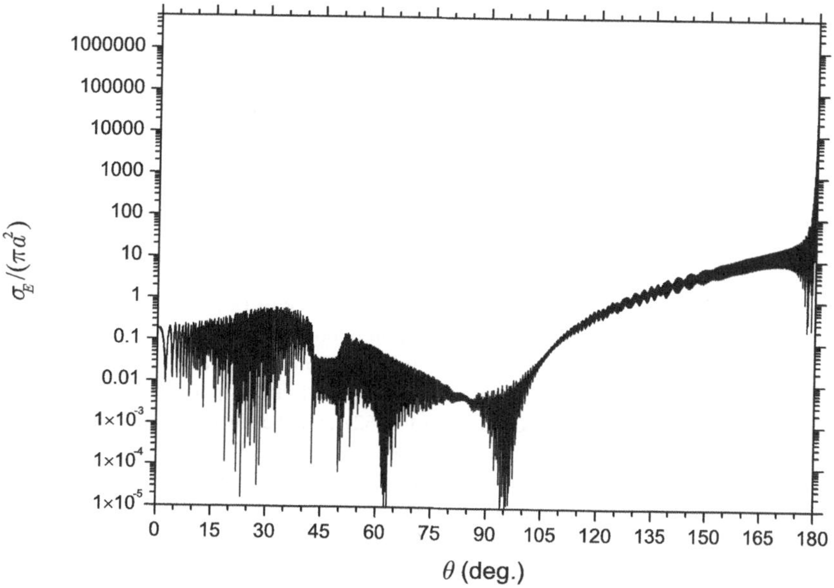

Figure 7.59 The same as Figure 7.57 but for a greater sphere radius ($a = 200$ μm).

Caustics of Interiorly Reflected Rays: Caustic Approximations

As the electrical size of a penetrable sphere grows, the caustic regions become more pronounced and start dominating the background in the scattering diagram. The strong scattering near the caustic directions leads to noticeable effects, for example to well-observable atmospheric rainbows. To describe the caustic regions, we need an approximation which, in contrast to the ray approximations, is applicable there.

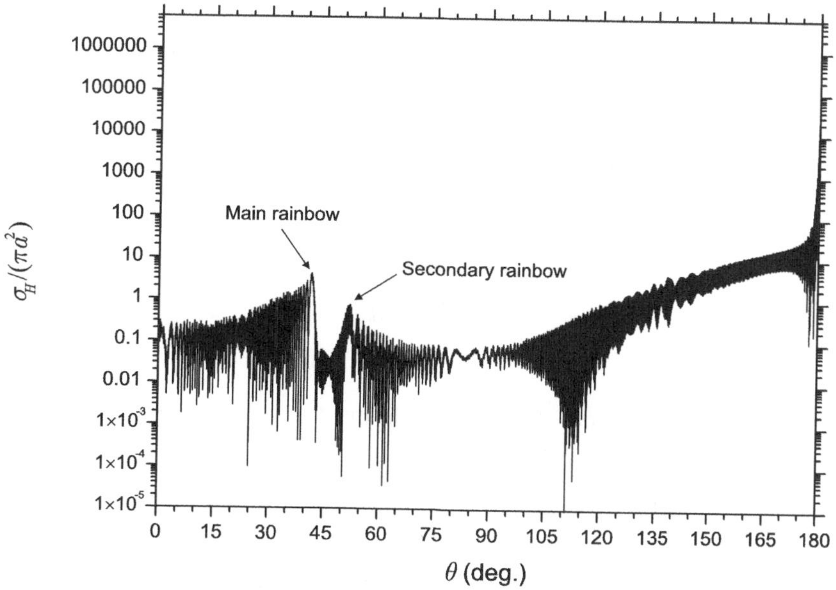

Figure 7.60 The same as Figure 7.59 but in the H plane.

Here we derive the caustic approximations which are valid around the caustics of the internally reflected rays with arbitrary order p. In contrast to those available in the literature (Khare and Nussenzveig 1974), the formulas can be applied to magnetic materials, as μ_r is not necessarily unity, and explicitly include the reflection and transmission coefficients of the respective caustic rays. The formulas are derived from the contour integral representations (7.239) and (7.241) of the exact solution for the fields by applying the stationary phase method modified to the case of two coalescing critical points (section D.2.3). The scattering coefficients $S_{1,2}(\theta)$ are obtained in the far-field limit by letting r go to infinity. The presented caustic approximations include only the leading terms of the asymptotic expansions to make the result physically appealing. The caustic formulas match with the ray-optics approximations away from the caustic. The resulting formulas are particularly useful in the study of rainbows, permitting us to explain all the features of the atmospheric rain- and fog bows.

The failure of ray approximations at the caustics is due the coalescence of two stationary points, denoted ν_{c1} and ν_{c2}. At certain values of the bistatic angle $\theta = \tilde{\theta}$, the stationary points coalesce at a point $\tilde{\nu}$, which is a maximum in the curves in Figure 7.51. In such cases, the stationary point approximations are invalid and are replaced with asymptotic approximations that include the Airy function $\mathrm{Ai}(q)$ (section D.2.3). Let us apply the formula (D.32) to asymptotic evaluation of the integral (7.261) with the phase function (7.262) and the amplitude function (7.263). The resulting formula for $E^{\pm}_{pq}$, referred to as the caustic approximation, will be inserted in (7.322) to describe $S_1(\theta)$ in the caustic region. Formulas for $H^{\pm}_{pq}$ and $S_2(\theta)$ are obtained from $E^{\pm}_{pq}$ and $S_1(\theta)$ by the substitutions (7.171). Since the scattering coefficients $S_{1,2}(\theta)$ are obtained from the scattered fields in the limit of $r \to \infty$, the terms vanishing in the limit are omitted in the following constructions for brevity.

To write down the caustic approximation, one needs the first three derivatives of the phase function (7.262) with respect to ν at the point $\tilde{\nu}$,

$$\frac{\partial S^{\pm}_{pq}}{\partial \nu}(\tilde{\nu}, k) = \pm(\pi - \theta) - 2\pi q - 2\arccos\left(\frac{\tilde{\nu}}{ka}\right) + 2(p+1)\arccos\left(\frac{\tilde{\nu}}{k_1 a}\right) \tag{7.555}$$

$$\frac{\partial^2 S^{\pm}_{pq}}{\partial \nu^2}(\tilde{\nu}, k) = \frac{2}{ka\sqrt{1 - \tilde{\nu}^2/(ka)^2}} - \frac{2(p+1)}{k_1 a\sqrt{1 - \tilde{\nu}^2/(k_1 a)^2}} = 0 \tag{7.556}$$

$$\frac{\partial^3 S^{\pm}_{pq}}{\partial \nu^3}(\tilde{\nu}, k) = \frac{2\tilde{\nu}}{(ka)^3[1 - \tilde{\nu}^2/(ka)^2]^{3/2}} - \frac{2(p+1)\tilde{\nu}}{(k_1 a)^3\left[1 - \tilde{\nu}^2/(k_1 a)^2\right]^{3/2}}. \tag{7.557}$$

By definition, $\tilde{\nu}$ is the point at which the second derivative of the phase function vanishes. The solution of (7.556) is therefore given by $\tilde{\nu} = ka\tilde{\tau}$ with $\tilde{\tau}$ defined by (7.552). Using (7.556), equation (7.555) can be rewritten as

$$\frac{\partial S^{\pm}_{pq}}{\partial \nu}(\tilde{\nu}, k) = \mp\theta \pm \tilde{\theta}, \tag{7.558}$$

where $\tilde{\theta}$, the bistatic angle of the caustic, is given by (7.553). The expression (7.557) for the third derivative simplifies to

$$\frac{\partial^3 S^{\pm}_{pq}}{\partial \nu^3}(\tilde{\nu}, k) = \frac{\omega_p(N)}{(ka)^2}, \tag{7.559}$$

where

$$\omega_p(N) = \frac{2p(p+2)\tilde{\tau}}{(p+1)^2(1-\tilde{\tau}^2)^{3/2}} = \frac{2p^2(p+2)^2}{(p+1)^2(N^2-1)^{3/2}}\sqrt{(p+1)^2 - N^2}. \tag{7.560}$$

For media with $1 < N < 2$, which includes water in the visible, $\omega_p(N) > 0$ for all $p = 1, 2, \ldots$ and it follows from (7.559) that

$$\operatorname{sgn}\left[\frac{\partial^3 S^{\pm}_{pq}}{\partial \nu^3}(\tilde{\nu}, k)\right] = 1. \tag{7.561}$$

Now we are ready to apply the uniform formula (D.32), which gives the following approximation for the radial component of the scattererd electric field near the caustic:

$$E^{\pm}_{pq} \approx 2\pi A^{\pm}_{pq}(\tilde{\nu}, k)\tilde{m}\,\mathrm{Ai}[\mp\tilde{m}(\theta - \tilde{\theta})]\mathrm{e}^{jS^{\pm}_{pq}(\tilde{\nu},k)}, \tag{7.562}$$

where

$$\tilde{m} = \left[\frac{2(ka)^2}{\omega_p(N)}\right]^{1/3}. \tag{7.563}$$

The value of the phase function at $\tilde{\nu}$ follows from (7.262) as

$$S^{\pm}_{pq}(\tilde{\nu}, k) = -kr \mp \tilde{\nu}(\theta - \tilde{\theta}) + 2ka\cos\tilde{\theta}_{\mathrm{c}} - 2(p+1)k_1 a\cos\tilde{\theta}_1, \tag{7.564}$$

where $\Phi(\tilde{\nu}, kr)$ is replaced with the two terms $kr - \tilde{\nu}\pi/2$ that do not vanish in the limit $r \to \infty$. The angles $\tilde{\theta}_{\mathrm{c}}$ and $\tilde{\theta}_1$ are defined by

$$\tilde{\nu} = ka\sin\tilde{\theta}_{\mathrm{c}} = k_1 a\sin\tilde{\theta}_1 \tag{7.565}$$

and have the meaning of the incidence and refraction angles at the injection point for the caustic ray.

Inserting (7.562) in (7.322) and using (7.263) give the caustic approximation of $S_1(\theta)$,

$$S_1(\theta) \approx \mp(-1)^q \mathrm{e}^{jp\frac{\pi}{2} \pm j\frac{\pi}{4}} \sqrt{2\pi ka\frac{\sin\tilde{\theta}_{\mathrm{c}}}{\sin\theta}}\, T^{01}_{\parallel}(\tilde{\theta}_{\mathrm{c}})T^{10}_{\parallel}(\tilde{\theta}_1)\left[R^{11}_{\parallel}(\tilde{\theta}_1)\right]^p \tilde{m}\,\mathrm{Ai}[\mp\tilde{m}(\theta - \tilde{\theta})]\mathrm{e}^{\mp j\tilde{\nu}(\theta-\tilde{\theta})+2jka\cos\tilde{\theta}_{\mathrm{c}}-2j(p+1)k_1 a\cos\tilde{\theta}_1}. \tag{7.566}$$

The corresponding approximation for $S_2(\theta)$ follows from (7.566) on the substitutions (7.171),

$$S_2(\theta) \approx \pm(-1)^q \mathrm{e}^{jp\frac{\pi}{2} \pm j\frac{\pi}{4}} \sqrt{2\pi ka\frac{\sin\tilde{\theta}_{\mathrm{c}}}{\sin\theta}}\, T^{01}_{\perp}(\tilde{\theta}_{\mathrm{c}})T^{10}_{\perp}(\tilde{\theta}_1)\left[R^{11}_{\perp}(\tilde{\theta}_1)\right]^p \tilde{m}\,\mathrm{Ai}[\mp\tilde{m}(\theta - \tilde{\theta})]\mathrm{e}^{\mp j\tilde{\nu}(\theta-\tilde{\theta})+2jka\cos\tilde{\theta}_{\mathrm{c}}-2j(p+1)k_1 a\cos\tilde{\theta}_1}. \tag{7.567}$$

In contrast to ray approximations (7.546) and (7.547), the caustic approximations (7.566) and (7.567) predict finite values when $\theta = \tilde{\theta}$ (see (D.29)). Furthermore, the behavior of the Airy function of a large real-valued argument suggests a rapid decay of the approximate scattering coefficients when the argument is positive and oscillations when the argument is negative (see Figure D.3). In fact when $\tilde{m}|\theta - \tilde{\theta}| \geq 1$, the caustic fields are either small (the argument is positive) or fit well with ray formulas (7.546) and (7.547) (see section D.2.3). From the physical viewpoint, the decay beyond the caustic means that the rays cannot reach there, whereas oscillations on the other side are the result of constructive and destructive interference of the two rays (Figure 7.52).

On the illuminated side of the caustic, the absolute values of the approximations (7.566) and (7.567) have a number of maxima at the angles

$$\theta = \tilde{\theta} \pm \frac{w_s}{\tilde{m}}, \tag{7.568}$$

where w_s ($s = 1, 2, \ldots$ and $-\infty < \ldots < w_2 < w_1 < 0$) are the points at which $|\mathrm{Ai}(w)|$ has a maximum. The values of w_s and $\mathrm{Ai}(w_s)$ for $s = 1, 2, 3, 4$ can be found in section D.2.3. The second term in the right-hand side of (7.568) is the wave correction to the position of the caustic as it depends on ka, vanishing in the optical limit $ka \to \infty$. The GO value $\tilde{\theta}$ for the bistatic angle of the caustic is independent of a but may depend on k if the material of the sphere is dispersive (see (7.553)).

Table 7.4 The normalized characteristic width and the degree of polarization for the first 10 caustics for $\mu_r = 1$ and $\epsilon_r = 1.778$ ($N = 4/3$).

p	1	2	3	4	5	6	7	8	9	10
$(ka)^{\frac{2}{3}}\delta\theta$ (deg.)	97.3	174	244	312	379	446	512	578	643	709
$\rho_p(\epsilon_r, \mu_r)$	0.20	0.33	0.35	0.36	0.36	0.37	0.37	0.37	0.37	0.37

At the caustic, the scattering coefficients $S_{1,2}(\tilde{\theta})$ are $O[(ka)^{7/6}]$ and the cross sections $\sigma_{E,H}/(\pi a^2) = O[(ka)^{1/3}]$. The characteristic scale of variation of the caustic approximations with θ can be estimated as

$$\delta\theta = \frac{1}{\tilde{m}} = \left[\frac{\omega_p(N)}{2(ka)^2}\right]^{1/3}, \tag{7.569}$$

so that $\delta\theta = O[(ka)^{-2/3}]$ and is therefore asymptotically small. The characteristic scale strongly depends on the number of internal reflections p. Table 7.4 shows the parameters $\omega_p(N)$ and the normalized widths $(ka)^{2/3}\delta\theta = [\omega_p(N)/2]^{2/3}$ for $p = 1, 2, \ldots, 10$ and $N = 4/3$. The width of the caustics therefore grows with increasing p; at the same time the amplitude at the caustic direction decreases as $|R^{11}_{\perp,\parallel}|^p$ since $|R^{11}_{\perp,\parallel}| < 1$, implying that the higher-order maxima are much less pronounced. Outside the $\delta\theta$ vicinity of the caustic rays, scattering coefficients and cross sections reduce to the background level, $S_{1,2}(\theta) = O(ka)$ and $\sigma_{E,H}/(\pi a^2) = O(1)$ (apart from the back- and forward-scattering directions $\theta = 0$ and π, which are caustics of a different kind; this case is addressed at the end of this section).

As seen from (7.283), (7.284), (7.566), and (7.567), scattering in the E and H planes is different due to the polarization difference in reflection and transmission coefficients. For non-magnetic materials, reflection of parallel polarization is generally weaker because of the Brewster effect (section 4.2.2). The caustic fields are therefore often strongly polarized so that the caustic effects are more pronounced in the H plane than in the E plane. The degree of polarization at a caustic can be defined by

$$\rho_p(\epsilon_r, \mu_r) = \left|\frac{S_1(\tilde{\theta})}{S_2(\tilde{\theta})}\right| \approx \left|\frac{T^{01}_{\parallel}(\tilde{\theta}_c)T^{10}_{\parallel}(\tilde{\theta}_1)}{T^{01}_{\perp}(\tilde{\theta}_c)T^{10}_{\perp}(\tilde{\theta}_1)}\right|\left|\frac{R^{11}_{\perp}(\tilde{\theta}_1)}{R^{11}_{\parallel}(\tilde{\theta}_1)}\right|^p. \tag{7.570}$$

The more $\rho_p(\epsilon_r, \mu_r)$ deviates from unity, the greater is the polarization difference. For scattering intensities and bistatic cross sections the number should be squared. In the high-frequency approximation, (7.570) is independent of a but may depend on k for a dispersive material.

Table 7.4 shows the values of $\rho_p(\epsilon_r, \mu_r)$ at several caustics, up to the 10th internal reflection, for a sphere with $\mu_r = 1$ and $\epsilon_r = 1.778$, which corresponds to water in the visible range (Table 2.3). The degree of polarization has a maximum for $p = 1$, which means that the power scattered in the direction of the caustic in the H plane is 25 times that in the E plane.

Let us take a closer look at the caustic fields for water at $\lambda = 0.55$ μm, which is a wavelength in the middle of the visible range. The relative permittivity of water depends on the wavelength, so the value of ϵ_r is taken from Table 2.3 at a corresponding wavelength. The size of the sphere ($a = 1$ mm) corresponds to a middle-sized raindrop and leads to the electrical size $ka = 11424$. Figures 7.61 and 7.62 show the bistatic scattering cross sections in the E and H planes for bistatic angles around the caustic with $p = 1$. The cross section in the H plane is significantly greater than in the E plane, as expected. Both $\sigma_E(\theta)$ and $\sigma_H(\theta)$ have a maximum at $\theta \approx 41.9°$, quickly decay for larger values of θ, and oscillate for smaller values of θ. The three maxima plotted in Figures 7.61 and 7.62 are correctly predicted by (7.568) with $s = 1, 2, 3$.

The curves show a great many ripple peaks. The amplitudes of the peaks are comparable in the E and H planes, and the ripple is much less pronounced in σ_H than in σ_E because of the larger cross-section values. The curves plotted by using the caustic approximations do not show the ripple but describe well the mean level of the cross sections. The position of ripple peaks depends on the size of the sphere, so in applications in which variously sized raindrops are present, for example atmospheric rainbows, the ripple must be much less prominent because of the averaging effect.

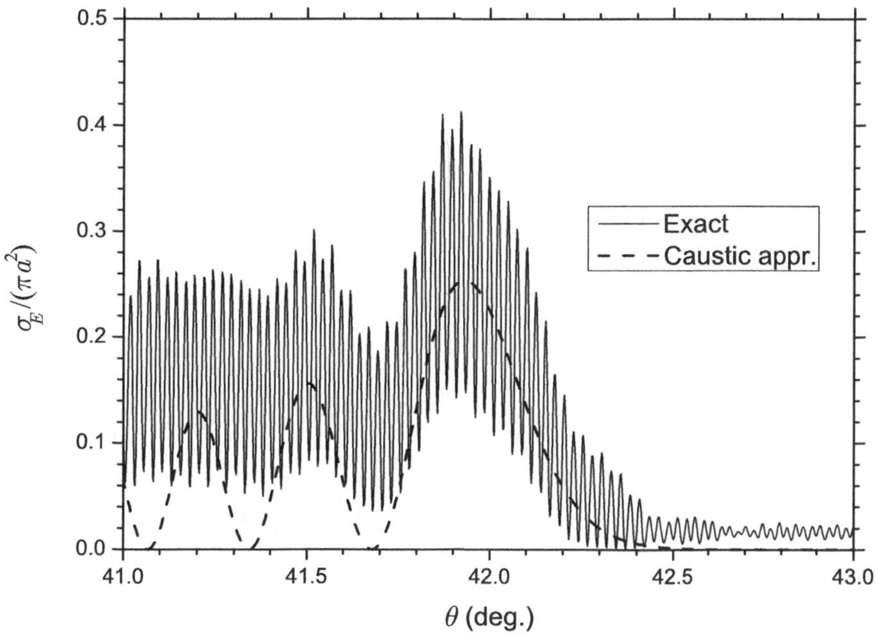

Figure 7.61 Normalized E-plane bistatic cross section of a spherical water droplet as a function of the bistatic angle around the caustic due to the single internal reflection: exact solution vs. caustic approximation.

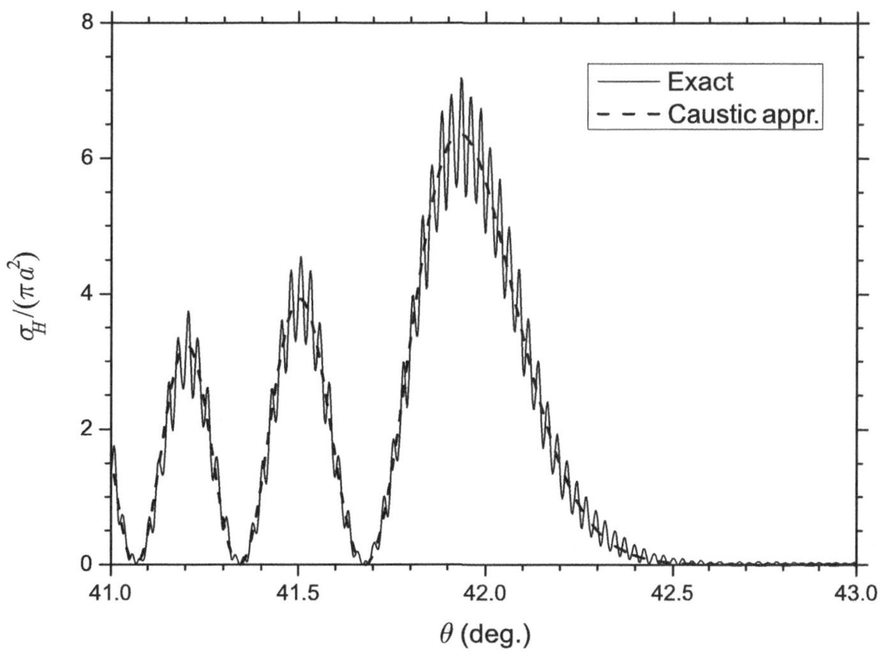

Figure 7.62 The same as Figure 7.61 but in the H plane.

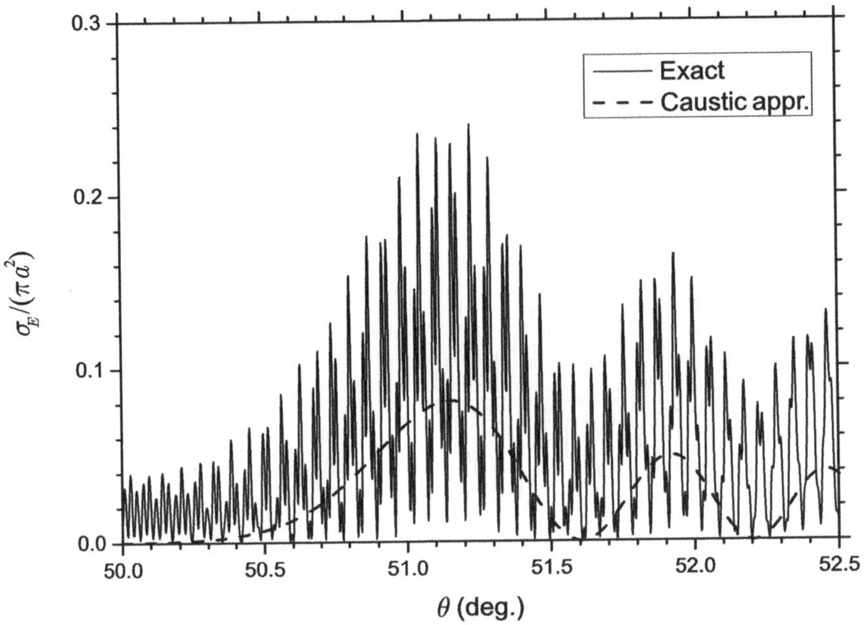

Figure 7.63 The same as Figure 7.61 but for the doubly internally reflected rays.

The fields close to caustic for $p = 2$ are plotted in Figures 7.63 and 7.64. Compared with $p = 1$ the level is lower because of one additional interior reflection. The H plane scattering is stronger than that in E plane. The ripple is comparable with the contribution of the caustic rays, nevertheless it is clearly seen that the asymptotic formulas provide a good approximation for the middle level of the caustic fields also for $p = 2$. The cross sections have a maximum at $\theta \approx 51.2°$ and in contrast to the case $p = 1$ show oscillations for *larger* values of θ and a rapid decay for *smaller* values of θ.

Atmospheric Rainbows

The previous analysis has shown that bistatic scattering cross sections of electrically large low-absorption spheres have prominent maxima in the direction of the caustics of the internally reflected rays. If the material of the sphere is dispersive, then the position of the maxima in the scattering diagrams depends on the frequency, and when the sphere is illuminated by a spectrum of waves with different wavelengths, the maxima are split according to the wavelength. Water is dispersive in the visible, natural light has a continuous frequency spectrum, and raindrops are much larger than the wavelength of light, all these leading to rainbows, a spectacular phenomenon in atmospheric optics. The appearance of rainbows as rings or arcs is the result of the spherical symmetry of the raindrops[14] (see Figure 7.65). The power scattered at a larger bistatic angle comes to the observer from a circular arc of larger radius and vice versa.

Typically a single (primary) rainbow is observed, but under particularly good conditions a secondary rainbow above the primary rainbow and weaker rainbows attached to the primary and secondary rainbows (supernumeraries) are visible. In a primary rainbow, the arc is red on the outer part and violet on the inner side. In a secondary rainbow, the order of colors is reversed, with red on the inner side of the arc. In addition to the common primary and secondary rainbows, there exist further (higher-order) rainbows with arcs of their own size. They are hardly visible to the naked eye but can be photographed and extracted from the images. Photographs of third-, forth-, and fifth-order bows have been published in

[14] Raindrops larger than 2 mm in diameter deviate from the spherical form, which may complicate the picture of the observed rainbows.

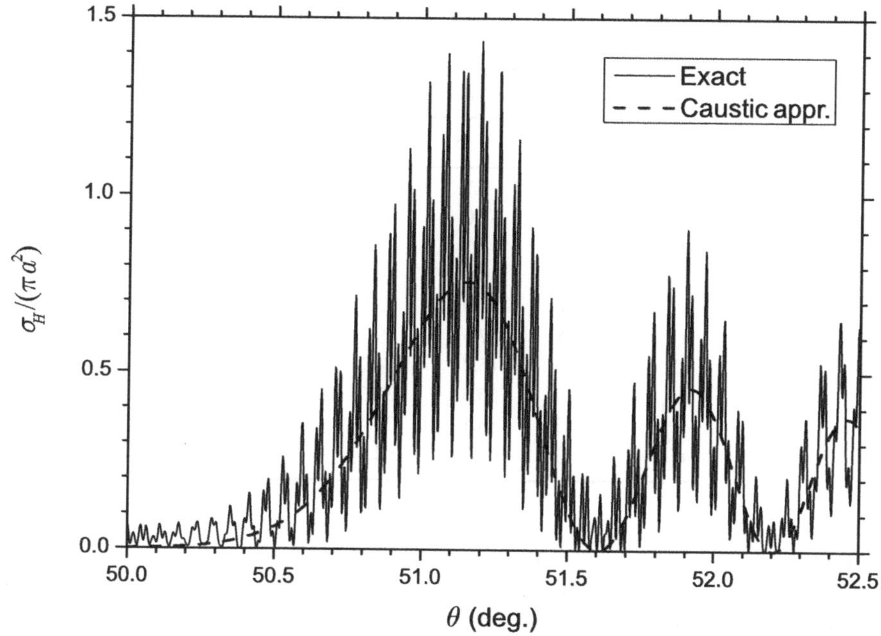

Figure 7.64 The same as Figure 7.63 but in the H plane.

the literature (Edens 2015; Großmann *et al.* 2011; Theusner 2011). In a laboratory setting it is possible to observe many more rainbows, for example Walker (1976) describes the observation of 13 rainbows in water.

The bistatic angles of the caustics are given by equations (7.553) and (7.568), and the bistatic cross section near those directions by the caustic approximations (7.566) and (7.567). We will now show that these equations explain the principal features of atmospheric rainbows.

Visible light approximately covers wavelengths from 0.4 μm (violet light) to 0.7 μm (red light), which corresponds to a frequency range of 430–750 THz. Inside the visible spectrum, the intensity of sunlight at sea level relatively weakly depends on wavelength and can be roughly considered as frequency independent. Sunlight can be thought of as a continuum of randomly polarized plane waves so that every polarization is equally probable. To describe atmospheric rainbows, it makes sense to work with a polarization-averaged bistatic scattering cross section,

$$\bar{\sigma}(\theta) = \frac{1}{2}[\sigma_E(\theta) + \sigma_H(\theta)]. \tag{7.571}$$

This expression follows from (3.360), (7.283), (7.284), and (7.311) with $\phi = 0$.

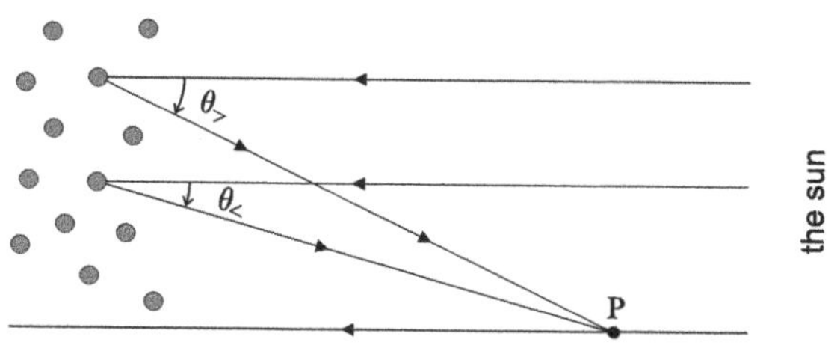

Figure 7.65 For a cloud of spherical droplets a fixed value of the bistatic angle θ corresponds to a circular arc centered at the axis joining the sun and the observer P. A larger bistatic angle corresponds to a larger radius of the arc. Here $\theta_< < \theta_>$.

Table 7.5 First 10 rainbow angles $\tilde{\theta}$ (deg.) for water droplets in the visible.

Type	Wavelength (μm)			Notes
$\tau^{\pm}_{pq}$	0.4	0.55	0.7	
$\tau^{+}_{10,3}$	12.0	5.7	3.7	
$\tau^{+}_{6,2}$	36.0	32.0	30.8	
$\tau^{-}_{1,0}$	41.3	42.1	42.4	Primary
$\tau^{+}_{2,1}$	52.4	50.8	50.3	Secondary
$\tau^{-}_{5,1}$	48.6	52.0	53.0	
$\tau^{-}_{9,2}$	71.8	77.5	79.3	
$\tau^{+}_{7,2}$	120.3	115.7	114.2	
$\tau^{-}_{4,1}$	133.7	136.5	137.3	Quaternary
$\tau^{+}_{3,1}$	140.3	138.1	137.4	Tertiary
$\tau^{-}_{8,2}$	155.7	160.8	162.5	

We begin with the position of maxima. As seen from (7.553) and (7.568), to the leading degree of approximation, the bistatic angle of the caustics is equal to $\tilde{\theta}$, which is independent of the droplet radius a, and it is this feature that permits the formation of rainbows despite the presence of variously sized droplets in every rainfall.[15] Table 7.5 shows the values of $\tilde{\theta}$ for the first 10 rainbows. The quantity $\tau^{\pm}_{pq}$ denotes a solution of (7.553) for the specific choice of the sign and the values of p and q, and every row in the table lists the corresponding values of $\tilde{\theta}$ at the middle and the ends of the visible frequency range. The frequency-dependent values of ϵ_r (with neglected imaginary parts) are taken from Table 2.3.

The parameter p is equal to the number of reflections in the interior of the water droplet, directly corresponding with the strength of a rainbow as more and more light is lost with each internal reflection (see (7.566) and (7.567)). The upper six rows in the table correspond to scattering in the back hemisphere (observer between the sun and water drops), whereas the lower four rows assume the water drops between the sun and observer. The latter scattering configuration complicates observation of relatively strong tertiary and quaternary rainbows, which are located about 40° and 45° from the sun, respectively; their observation requires that the sun is covered by a cloud or by a terrain feature, like a building or a peak of a mountain. A further complication is overlapping of rainbows, for example the secondary rainbow overlaps with the rainbow of the fifth order, which make them appear less pronounced and more blurred compared with the primary rainbow. Later we will take a closer look at overlapping rainbows.

Rainbows are strongly polarized tangentially to the arcs. For example, the light of the primary rainbow arc is 96% polarized, and the light of second arc is 90% polarized (see Table 7.4). This is explained by the small value of $R^{11}_{\parallel}(\theta_1)$ compared with $R^{11}_{\perp}(\theta_1)$ in (7.566) and (7.567) since θ_1, the incidence angle at every internal reflection, is close to the Brewster angle of the water–air interface, particularly when $p = 1$.

The data in Table 7.5 can be used to estimate the angular widths of the respective rainbows as the variation of $\tilde{\theta}$ when λ changes from 0.4 μm to 0.7 μm. Estimated in that way, the widths are equal to 1.1° ($p = 1$), 2.1° ($p = 2$), 2.9° ($p = 3$), ..., 8.3° ($p = 10$). Slightly different values can be found in the literature (e.g., Grandy (2000, Table 6.1)), which is due to a slightly different definition of the visible spectrum.

However, as follows from (7.566), (7.567), and (7.568), the positions of maxima and the widths of the rainbows depend on the radius of the sphere and on the wave number (diffraction corrections). The dependence on a is particularly important as the size of water drops in a rainfall, a fog, or a cloud may broadly vary. In what follows we study the implications of the diffraction corrections for the appearance of rainbows.

[15] The size distribution approximately follows the exponential law $N_a = N_0 \exp(-2\Lambda a)$, where $N_a da$ is the number of drops per cubic meter with the radius between a and $a + da$, and if a is measured in μm, then $N_0 = 8\ \mathrm{m}^{-3}\,\mu\mathrm{m}^{-1}$, $\Lambda = 0.0041R^{-0.21}\ \mu\mathrm{m}^{-1}$, and R is the rate of rainfall in millimeters per hour or, which is the same, in liters per square meter (Marshall and Palmer 1948).

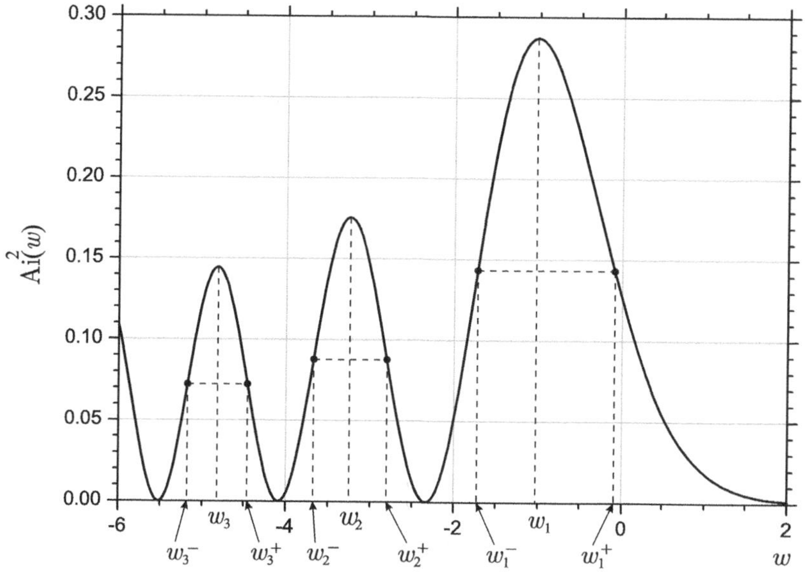

Figure 7.66 Magnitude of the Airy function around the first three maxima.

According to the caustic approximations (7.566) and (7.567), the intensity of light in a vicinity of a caustic is described by $\mathrm{Ai}(w)^2$ (Figure 7.66). In fact, there is an infinite number of maxima located on $-\infty < w < 0$, which we numerate by an integer parameter s, so that $s = 1$ corresponds to the main maximum and $s \geq 2$ to the maxima in the side lobes of $\mathrm{Ai}(w)^2$. The maxima are caused by interference between two rays of light following slightly different paths within the water drops. The values of w and $\mathrm{Ai}(w)^2$ at the first three maxima can be found in Table 7.6.

According to (7.568), the position of the main maximum ($s = 1$) is slightly shifted from the GO value $\tilde{\theta}$ by a quantity approximately equal to $\delta\theta$ since $|w_1| \approx 1$. The shift, given by (7.569), is asymptotically small but can become significant for smaller water drops and higher rainbow orders. For example, for $\lambda = 0.55$ μm and $p = 1$, $\delta\theta$ is equal to 0.19° ($a = 1$ mm), 0.56° ($a = 200$ μm), and 4.1° ($a = 10$ μm). With $p = 2$, these values are almost doubled, amounting to 0.34°, 1°, and 7.3°, respectively.

Similarly to the main maximum, the position of the maxima described by $s = 2, 3, \ldots$ depends on the wavelength, which results in the appearance of faint rainbows contiguous to the main maximum. These rainbows are called supernumerary rainbows, or supernumeraries. Natural supernumerary rainbows are sometimes observed on the inner side of the primary rainbow and very rarely also outside the secondary rainbow.

As seen from (7.566) and (7.567), the intensity of every rainbow, including supernumeraries, depends on the frequency and the droplet size as $O(k^{1/3}a^{7/3})$. Note that GO predicts just the position of the main bows but fails to predict the intensities of light near caustics and even the existence of supernumeraries. The latter was historically a first indication of the wave nature of light (Young 1804).

According to the caustic approximation, the position and the widths of the lobes depend on a, which is a broadly varying parameter. For example, cloud, mist, or fog droplets are smaller than 100 μm in diameter, but raindrops can be several millimeters large. Let us therefore discuss the appearance of rainbows in dependence on the droplet size.

Table 7.6 The first three maxima of $\mathrm{Ai}(w)^2$ and the points at the half maximum level.

	$s = 3$			$s = 2$			$s = 1$		
w	−5.17	−4.82	−4.46	−3.67	−3.25	−2.80	−1.72	−1.02	−0.09
$\mathrm{Ai}(w)^2$	0.072	0.144	0.072	0.088	0.176	0.088	0.143	0.286	0.143

We define the width of a lobe at the half maximum intensity level. In Figure 7.66, these points are denoted by $w_1^\pm$ (main lobe), $w_2^\pm$ (first supernumerary), and $w_3^\pm$ (second supernumerary), and their values can be found in Table 7.6. We will now plot the bistatic angles of the points w_s and $w_s^\pm$ with $s = 1, 2, 3$ for water spheres with a varying between 10 μm and 2000 μm at two wavelengths at the ends of the visible spectrum, $\lambda = 0.4$ μm (blue) and 0.7 μm (red).

Permittivity of water is a monotonous function of λ in the visible range (see Table 2.3), so the whole visible spectrum maps to the gap between the maxima for the red and blue light. Such regions with a continuous variation of colors between the blue and red ends are referred to as bands. Every band is associated with a maximum of the magnitude of the Airy function, implying the existence of main and supernumerary bands ($s = 1, 2, \ldots$) in rainbows of any order ($p = 1, 2, \ldots$). However, as we shall see now, the bands have varying angular widths and positions depending on a, which may result in overlapping of bands with different values of p and s. The implication is the variable appearance of rainbows with emerging or disappearing supernumeraries and mixed, impure, sometimes pale (eventually white) colors. These properties are illustrated in Figures 7.67, 7.68, and 7.69 calculated by using (7.568) and the data from Table 7.6. Again the small imaginary part in the permittivity of water at $\lambda = 0.4$ μm and $\lambda = 0.7$ μm is neglected (see Table 2.3).

Figure 7.67 shows the band structure (main bow and two supernumeraries) of the primary rainbow for $100\ \mu\text{m} \leq a \leq 2000\ \mu\text{m}$ (larger raindrops may significantly deviate from spherical form). With $a > 1000$ μm, the GO picture is approached as the position of maxima is almost independent of a, supernumeraries are fused to the main bow, and a unified bow of an almost constant angular width filled with an almost continuous variation of colors from blue to red is formed. For smaller a, say $a < 1000$ μm, the dependence on a is significant as the bands widen and supernumeraries emerge from the main band as a decreases. With $a < 500$ μm the band associated with the second supernumerary leaves the main band and with $a < 250$ μm is separated from the first supernumerary. With $a < 200$ μm the first supernumerary is fully separated from the main band so that the primary rainbow consists of separated main and supernumerary bands.

A further feature seen in Figure 7.67 is the variable width of the bands depending on a. For example, the widths of the supernumerary bands decrease with decreasing a so that their red and blue ends start coalescing when $a < 250$ μm for $s = 2$ and $a < 200$ μm for $s = 1$, implying pale colors of the supernumeraries. The width of the main band, however, increases with decreasing a from 1.3° at $a = 2000$ μm to 2.2° at $a = 100$ μm. On the other hand, the red and blue ends of the band approach each other and partially overlap when $a < 200$ μm. Thus, as a decreases, the main band becomes wider but at the same time paler and more blurred.

This trend continues and becomes more pronounced for water drops with $a < 100$ μm. Figure 7.68 shows the angular width of the main band ($s = 1$) of the primary bow ($p = 1$) as a function of a for $a < 100$ μm. As a decreases, the distance between the red and blue ends of the band gets smaller so that at $a \approx 15$ μm they overlap. The overlapping of colors creates a white bow, which is typical for fogbows. On the other hand, the width of the band increases from 2.3° at $a = 100$ μm to 8° at $a = 10$ μm. The widths of the lobes at the red and blue ends of the band grow at different rates as a decreases, so that when $a < 35$ μm, the red end lobe completely covers the blue end lobe, implying a white bow with reddish sides. When $a > 35$ μm, the lower side of the bow becomes bluish.

Intersection of the red and blue ends of a band, as in Figure 7.68 at $a \approx 15$ μm, is in fact a typical feature. For example, in the primary rainbow, the intersection takes place at $a \approx 15$ μm in the main band, at $a \approx 80$ μm in the first supernumerary band, and at $a \approx 140$ μm in the second supernumerary. It is interesting to note that such intersections imply inversion of the order of the colors in the bands. However, this effect must be hardly observable since around the intersection points the red and blue ends of the band are close to each other and the droplets must be small, so that the colors appear pale and blurred.

Features similar to those shown in Figure 7.67 for the primary rainbow are encountered in the secondary rainbow as well (Figure 7.69). With $a > 600$ μm the first supernumerary band overlaps with the main band and one single band is formed. With $a < 200$ μm the first supenumerary band is separated from the main bow and starts to implode so that the red and blue ends overlap at $a \approx 100$ μm. The main band widens with decreasing a, but the blue and red maxima approach each other. With a between 100 μm and 200 μm the main and supernumerary bands are separated, but both must appear pale.

Water droplets with $a = 200$ μm are expected to lead to well-separated main and supernumerary bands in the primary and secondary rainbows (see Figure 7.67 and 7.69). Figure 7.70 shows the polarization-averaged bistatic scattering cross section, obtained from the series solution at $\lambda = 0.45$ μm and $\lambda = 0.65$ μm, in the region around the primary ($38° < \theta < 44°$) and secondary ($49° < \theta < 57°$) rainbows. It is seen that in the primary and secondary rainbows the colors in the main and first supernumerary bands do not overlap. The scattering cross section in the region between the

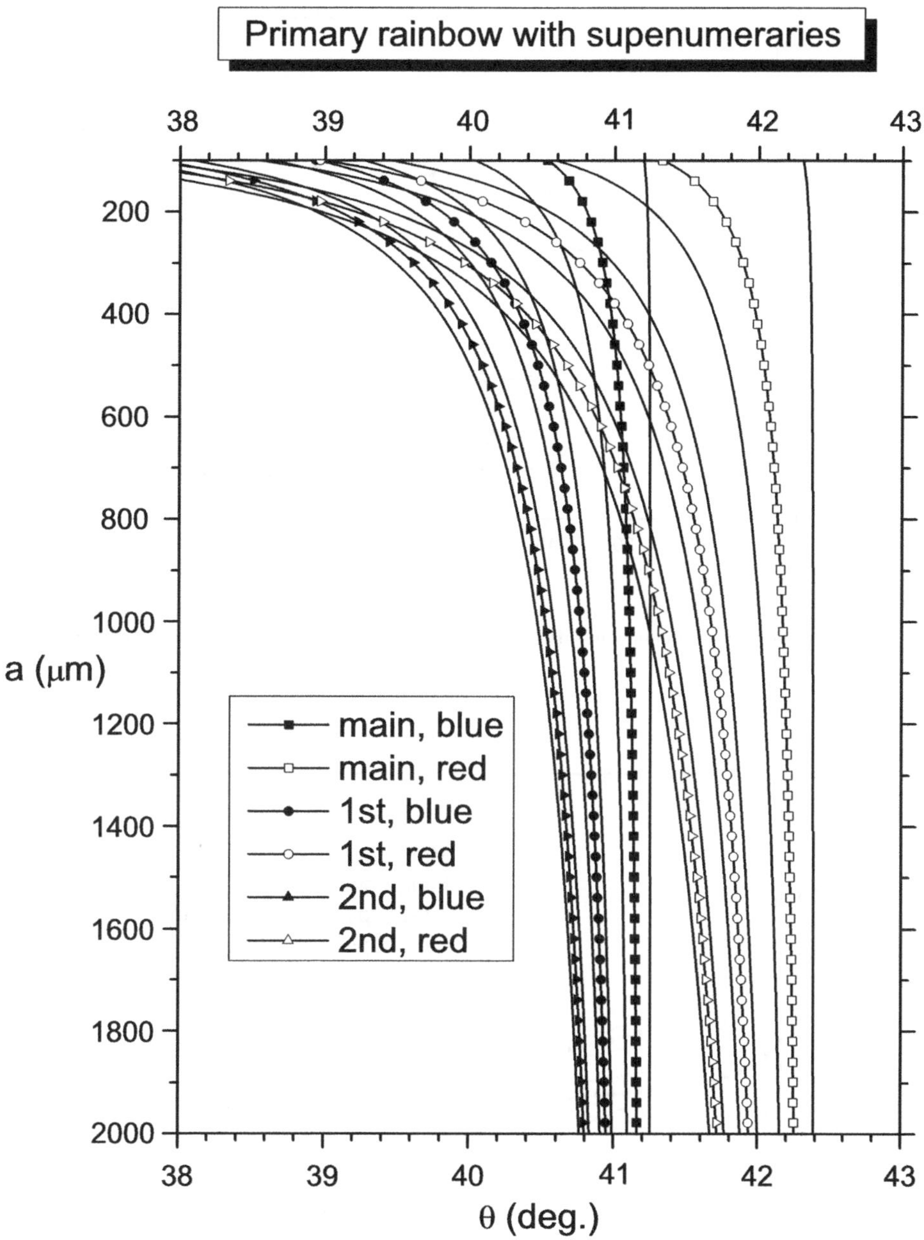

Figure 7.67 Positions of the lobes (main bow and two supernumeraries) in the primary rainbow at two wavelengths from opposite ends of the visible spectrum (blue 0.4 μm, red 0.7 μm). Solid lines correspond to half maximum intensity levels.

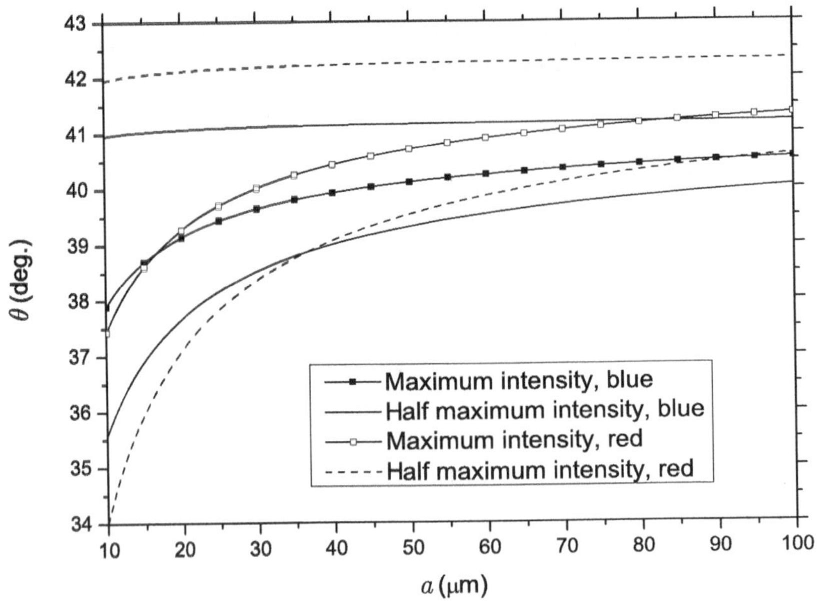

Figure 7.68 Appearance of the primary rainbow in smaller water droplets at two wavelengths from the opposite ends of the visible spectrum (blue 0.4 μm, red 0.7 μm).

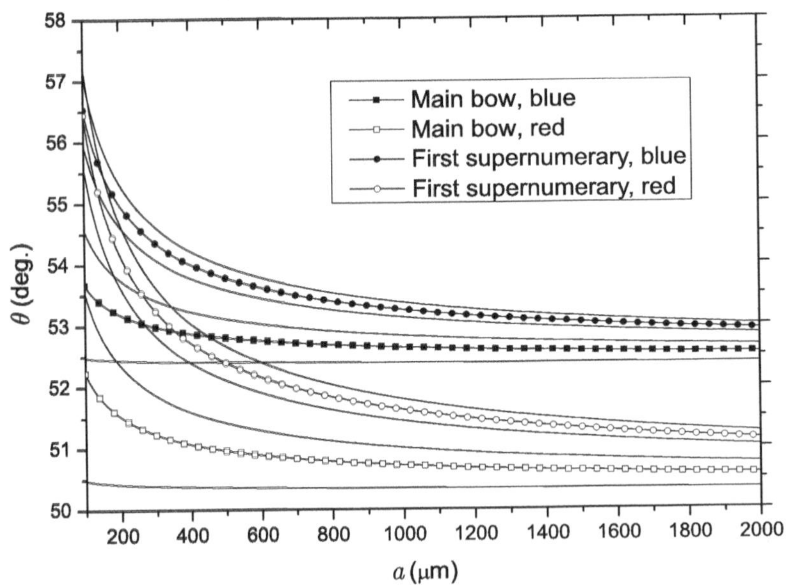

Figure 7.69 Positions of the lobes (main order and first supernumerary) in the secondary rainbow at two wavelengths from the opposite ends of the visible spectrum (blue 0.4 μm, red 0.7 μm). Solid lines correspond to the half maximum intensity levels.

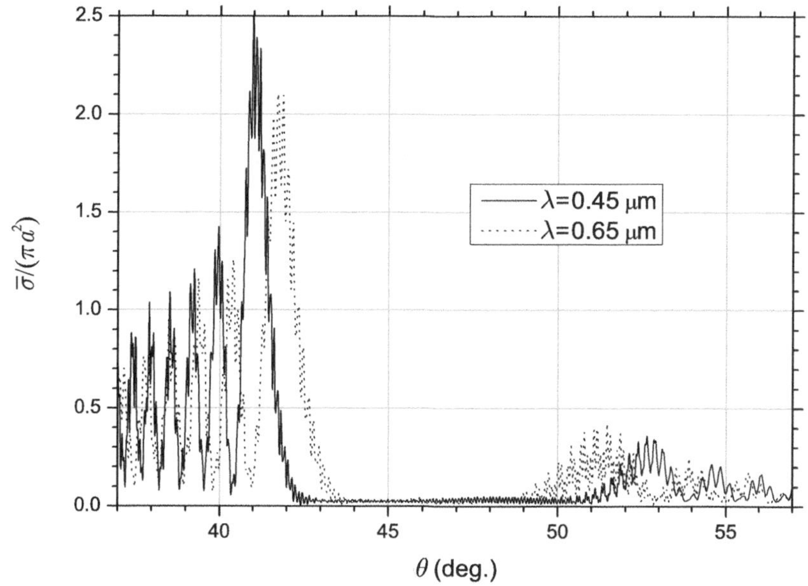

Figure 7.70 Polarization-averaged bistatic scattering cross section of a spherical water droplet with $a = 200$ μm around primary and secondary rainbows at two wavelengths from the opposite ends of the visible spectrum.

primary and secondary rainbows ($44° < \theta < 49°$) is quite low, which is a feature observed in nature (Alexander's dark band). The singly and doubly internally reflected rays cannot penetrate into this region, and it is the weaker externally reflected and triply internally reflected rays that illuminate the region.

Back- and Forward Scattering: Glory and Corona

Asymptotic formulas for ray contributions as well as approximations near caustics include the factor $(\sin\theta)^{-1/2}$ (see (7.546), (7.547), (7.566), and (7.567)), implying a singularity at back- ($\theta = 0$) and forward- ($\theta = \pi$) scattering directions. Mathematically, the singularity is a consequence of using the asymptotic approximation (7.32) for the Legendre function, which is invalid when $\theta \to 0, \pi$. The exact solutions for the scattering coefficients (7.280) and (7.281) are not singular at those directions as $\tau_n(\theta)$ and $\pi_n(\theta)$ are bounded at $\theta = 0, \pi$ (see (7.110) and (7.111)). Nevertheless, stronger scattering in the back- and forward directions is physically meaningful and follows from the rotational symmetry of spheres with respect to the z axis (the direction of incidence through the center of the sphere). Indeed, every ray circumventing the sphere, for example clockwise, in a given incidence plane has a counterpart propagating counterclockwise in the same plane, both being in phase in the back- and forward-scattering directions, and it is not just these two rays but rather a continuum of similar rays propagating in the incidence planes rotated around the z axis that constructively interfere in the back- and forward-scattering directions. Higher levels of back- and forward scattering compared with the other scattering directions are clearly seen in Figure 7.57. Furthermore, for electrically large spheres, forward scattering is much stronger than backscattering, and this difference grows as the electrical size increases: compare Figure 7.57 ($ka \approx 114$) and Figure 7.59 ($ka \approx 2285$).

The anomalous back- and forward scattering by spheres manifests itself in two clearly observable atmospheric phenomena: glory and corona. Both appear as concentric colored rings around either the shadow cast by an observer (usually from an aircraft) on a cloud (glory) or around the cloud-covered moon or sun (corona). As we shall see, these features are present in the scattering diagrams of individual water droplets.[16] In this section we take a closer look at back- and

[16]As wave interference phenomena, rings forming corona and glory are similar to side lobes of large aperture antennas.

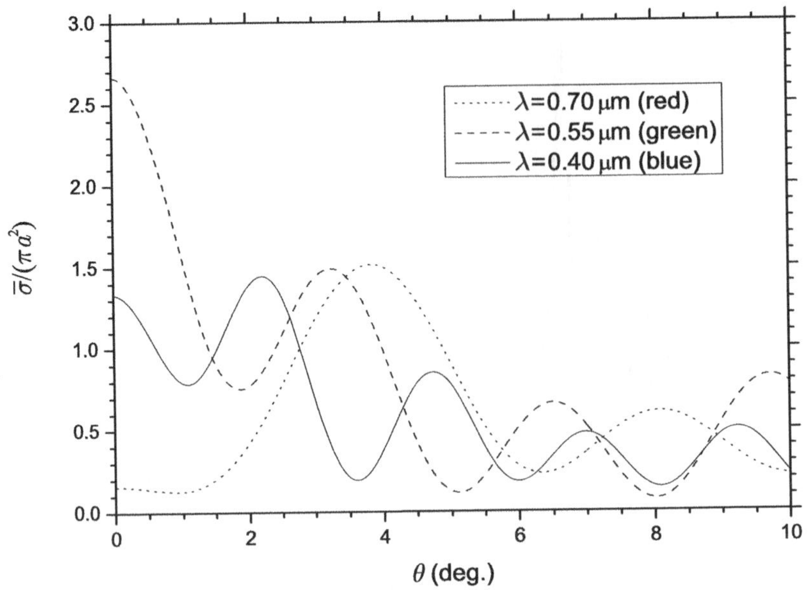

Figure 7.71 Polarization-averaged normalized bistatic scattering cross section of a spherical water droplet near the backscattering direction for three different wavelengths.

forward scattering from water droplets, paying particular attention to backscattering and the associated phenomenon of glory, for which we provide a simple theory and physical explanation. The case of forward scattering will be only briefly addressed.

Glory is associated with backscattering of light from water droplets in clouds or a fog with the size of droplets ranging from several micrometers to about 100 micrometers ($10 \leq ka < 1000$). Glory is observed as one or more concentric, successively dimmer rings, each of which is red on the outside and bluish towards the center. The angular size of the inner and brightest ring is much smaller than that of a rainbow, about 4° to 20° in diameter.

Figure 7.71 shows the bistatic scattering cross section of a water drop with $a = 5$ μm near backscattering direction at three wavelengths from the red, green, and blue regions of the visible spectrum. The curves are obtained by averaging the cross section over the randomly distributed polarization in the incident light waves (see (7.571) and (7.705)). Due to dispersion in water, the position of minima/maxima in the curves depends on the wavelength, which leads to the separation of colors and eventually to colored rings around the backscattering direction in a cloud of droplets. In Figure 7.71, all curves have a maximum at $\theta = 0$, whereas the position of the side lobes is frequency dependent, for example the maximum of the first side lobe is located at $\theta \approx 2.2°$, 3.2°, and 3.8° for the blue, green, and red light, respectively. There are further maxima at greater bistatic angles θ. Note that the cross section at the maxima is greater than the geometric cross-section area of the droplet (anomalous scattering).

An explanation of the phenomenon is not obvious as GO contributions are too weak to explain the glory by an individual contribution. Contributions from the externally reflected ray (Figure 7.20) and the singly internally reflected ray aa′ (Figure 7.52) to the backscattering, the former being about four times smaller than the latter, are roughly by an order of magnitude smaller than the observed backscattering values. Because of the factor $\sqrt{\sin(2\theta_c)}$ in the respective ray optical approximations (7.546) and (7.547), their contributions are bounded at $\theta = 0$ since θ_c, the incidence angle of the respective ray at the reflection (injection) point, is also zero as both rays propagate along the z axis in the backscattering case. Axial rays with higher number of internal reflections ($p \geq 2$) are even smaller and cannot significantly contribute (see Figures 7.53 and 7.54).

Non-axial rays, for which $\theta_c \neq 0$ when $\theta = 0$, may bring greater contributions to backscattering because of the uncompensated singularity at $\theta = 0$. The first candidate is the singly internally reflected ray bb′ (Figure 7.52), but as we

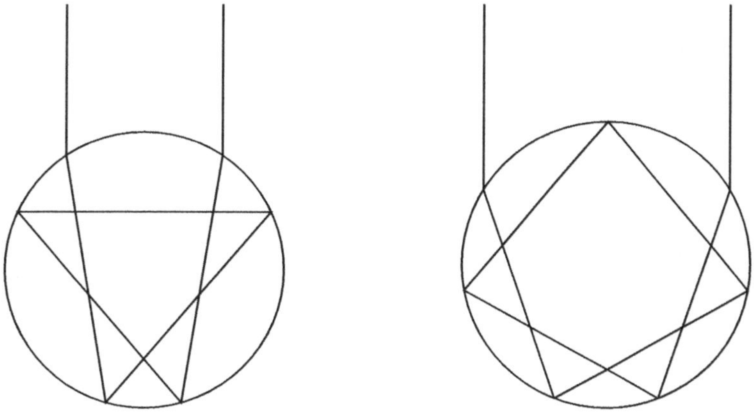

Figure 7.72 Computer-generated paths of the non-axial rays contributing to backscattering on four- and fivefold internal reflections in a water droplet with $N = 4/3$.

have seen in Figure 7.51, the ray cannot leave the water droplet in the backscattering direction, the smallest achievable bistatic angle being around 14.4°.

A closer look at the ray equation (7.549) reveals that for water in the visible ($N \approx 4/3$), backscattering of non-axial rays is first possible for the rays with $p \geq 4$. The trajectories of the rays backscattered on four- and fivefold internal reflections are shown in Figure 7.72. However, apart from the direction $\theta = 0$, their amplitudes are negligibly small since they loose too much power at reflections at the water-air interface (see Figures 7.53 and 7.54).

The most obvious diffraction mechanisms are also insufficient to explain the phenomenon of glory. As seen from Table 7.5, the caustic of the tenfold internally reflected wave is located close to the backscattering direction and the incidence angle at the injection point $\theta_c \approx 85.4°$ falls within the range with minimum reflection losses (Figures 7.53 and 7.54). However, reflection losses are significant (see Table 7.3), the backscattering direction is located on the dark side of the rainbow, and the caustic is blurred as the electrical size of water droplets ($a < 50$ μm, $ka < 800$) is insufficiently large to form a pronounced caustic (see Table 7.4). Numerical estimations show that the contribution from this mechanism is at least 15 times smaller than that from the single internal reflection.

Classical creeping waves propagating on the convex side of the spherical boundary may contribute to backscattering on circumventing the scatterer as we have already seen in the case of PEC circular cylinders (section 6.4.3). However, characteristic exponents ν_m, which define damping of the waves between attachment and launching points (see (7.192)), have large imaginary parts of the order $O[(ka)^{1/3}]$ so that the waves are strongly attenuated on a 180° trip over the shadowed portion of the sphere. For example, for a water droplet with $a = 2.5$ μm at $\lambda = 0.7$ μm, the damping exponent $\exp(\mathrm{Im}\, \nu_1 \pi)$ for the lowest order magnetic creeping wave is equal to 1.3×10^{-8}; for greater orders m, electric modes, shorter wavelength, and larger spheres the factor is even smaller. This is in agreement with our observations for PEC cylinders; the contribution of the creeping waves for cylinders with $ka > 10$ are hardly noticeable (Figure 6.19).

Van de Hulst (1957) conjectured that the non-axial singly internally reflected wave (bb′ in Figure 7.52) can be backscattered, provided that it makes the missing 14.4° arc as a creeping wave. Numerical estimations of the damping exponent based on (7.192) suggest, however, attenuation by an order of magnitude over the arc length of 14.4°. Additionally, the wave is still damped by its reflection at the water–air interface, which makes this contribution smaller than that from the axial singly internally reflected ray as soon as θ slightly deviates from $\theta = 0$.

Despite multiple attempts, no convincing explanation of the physics of the phenomenon of glory has been given in the literature so far (see literature notes at the end of this section). In what follows we provide a new explanation by showing that the glory is mainly formed by contributions of the creeping waves propagating in the interior of the sphere, that is, on the *concave* side of the spherical boundary. These waves, which are similar to whispering-gallery modes of spherical resonators, mathematically correspond to contributions from the poles of the horizontal branch (section 7.3.4). Let us denote such poles by ν_m where m is an integer numbering the poles. The poles approximately fill the interval

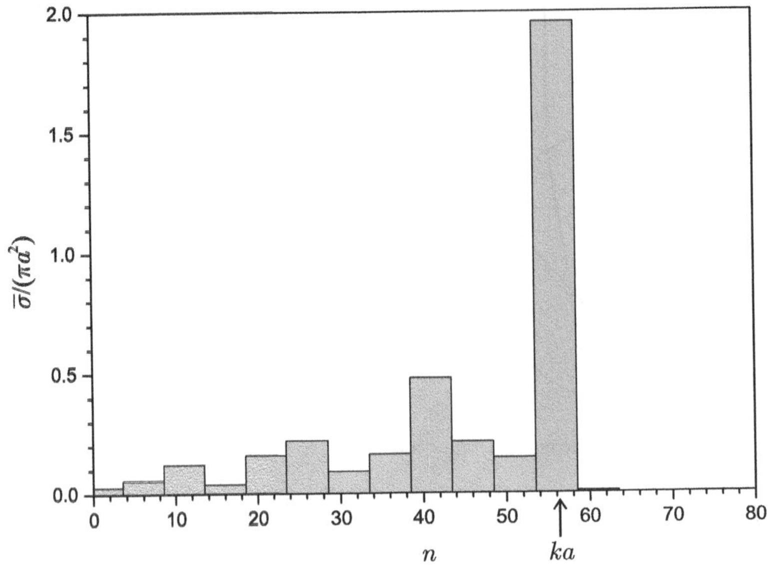

Figure 7.73 Partial sums of the multipole expansion for polarization-averaged normalized bistatic scattering cross section of a spherical water droplet near the backscattering direction. The height of every bar is equal to the contribution of the five terms with $1 \leq n \leq 5$, $6 \leq n \leq 10$, etc., to the expansion.

$0 < \nu_m < k_1 a$ (see Figure 7.15), and those located close to the point $\nu = ka$ (we assume that $k_1 > k$) have small imaginary parts, about 10 times smaller than for creeping waves on the convex side of the boundary, and can be efficiently excited by the incident plane wave. As we shall see, these are the poles whose contribution dominate backscattering from water spheres in the visible.

Let us start by noting that the main contribution to backscattering must be related to waves or rays incident on the sphere at an almost grazing angle (Bryant and Cox 1966). This is clearly seen from partial sums of the exact multipole expansions (7.280) and (7.281). Figure 7.73 shows the partial sums of five subsequent terms of the series for a water droplet of $a = 5$ μm at $\lambda = 0.55$ μm and $\theta = 3.2°$ (which is the first maximum, see Figure 7.71). It is clearly seen that the main contribution comes from the terms with $n \approx ka$, implying that backscattering is formed by some contributions excited by the incident plane wave at almost grazing incidence in a vicinity of the equator, since $n \approx ka \sin\theta_c$ (see (7.169)).

In the case of backscattering, bistatic angles θ are small, and contributions of the residues to the radial component of the electric field at an observation point (r, θ, ϕ) can be obtained from (7.209) on the substitution (7.187). According to (7.220), the contribution from every pole ν_m consists of the two parts propagating clock- and counterclockwise,

$$E^{\pm}_{r,m} = \sqrt{\frac{2\pi}{\sin\theta}} \frac{-jE_0 \cos\phi A^u_m}{(kr)^2(1 + e^{-2j\nu_m\pi})} e^{-j3\frac{\pi}{2}\nu_m - j\Phi(\nu_m, kr) \mp j\left(\nu_m\theta + \frac{\pi}{4}\right)}. \tag{7.572}$$

Here ν_m are zeros of (7.183) (electric modes). The corresponding contributions to H_r are obtained from (7.572) by the substitutions (7.171) and taking ν_m from (7.188) (magnetic modes).

Since the waves in (7.572) can be considered as locally plane at sufficiently large distance r, the scattering coefficients $S_{1,2}(\theta)$ can be derived from (7.322) and (7.323). The arrival angle ψ is equal to $\psi^{(2)}_m$ (see Figures 7.17 and 7.18). As $r \to \infty$, the arrival angle approaches $\pi/2$ as

$$\psi^{(2)}_m \approx \frac{\pi}{2} - \frac{a}{r}\delta_m \tag{7.573}$$

with

$$\delta_m = \begin{cases} \csc\theta_m, & \nu_m < ka \\ 1, & \nu_m > ka \end{cases}, \tag{7.574}$$

where θ_m is the incidence angle at the injection point for the mth mode, defined by $\theta_m = \arcsin(\nu_m/ka)$. It therefore follows from (7.322) and (7.572) that

$$S^{\pm}_{1,m}(\theta) = \frac{1}{ka}\sqrt{\frac{2\pi}{\sin\theta}}\frac{B^u_m}{1+\mathrm{e}^{-2j\nu_m\pi}}\mathrm{e}^{-j\nu_m(\pi\pm\theta)\pm j\frac{\pi}{4}}, \tag{7.575}$$

where

$$B^u_m = -j\delta_m\nu_m^{3/2}\tau^u_m \tag{7.576}$$

and τ^u_m is given by (7.214).

The total contribution from the mth mode to the scattering coefficient consists of two contributions circumventing the sphere in opposite directions:

$$S_{1,m}(\theta) = S^+_{1,m}(\theta) + S^-_{1,m}(\theta) \approx \frac{2}{ka}\sqrt{\frac{2\pi}{\sin\theta}}\frac{B^u_m\mathrm{e}^{-j\nu_m\pi}}{1+\mathrm{e}^{-2j\nu_m\pi}}\cos\left(\nu_m\theta - \frac{\pi}{4}\right). \tag{7.577}$$

A similar expression for the contribution of the mth mode to $S_2(\theta)$ follows from the duality relations (7.282) on interchanging ϵ and μ in (7.577) and using solutions of (7.188) as ν_m.

The contributions $S^{\pm}_{1,m}$ and $S^{\pm}_{2,m}$ have a simple physical interpretation. Those with $\nu_m < ka$ describe the waves that enter the sphere at the incidence angle θ_m, propagate in the interior of the sphere as creeping waves on the concave side of the boundary and get ejected at the angle θ_m toward the observer (Figure 7.17). Waves with $\nu_m > ka$ are lateral in the sense that they are excited by the parts of the wavefront of the incident plane wave that miss the sphere (tunneling); these waves get excited at the equator of the sphere ($\theta = \pi/2$) and propagate over an arc of 180° along the concave side of the spherical boundary before being ejected to the observer (Figure 7.18). Dominating contributions must come from the poles with ν_m close to ka since they combine smaller attenuation with larger excitation coefficients.

At the backscattering direction, contributions from the waves running around the sphere in the opposite directions are equal and interfere constructively, which leads to a maximum in the bistatic scattering cross section at $\theta = 0$. The values given by (7.575) and (7.577) are infinite at $\theta = 0$, which is due to the use of asymptotic approximation (7.215) for the Legendre function. However, they become applicable as soon as $\nu_m\theta > 1$, suggesting the scattered intensity oscillating as $\cos^2(ka\theta - \pi/4)$, with maxima at

$$\theta \approx \frac{\pi}{ka}\left(l + \frac{1}{4}\right), \quad l = 0, 1, 2, \ldots \tag{7.578}$$

The formula gives a good estimation for the position of maxima, and its accuracy improves with growing bistatic angle.

Numerical simulations over various droplet sizes and the whole visible spectrum strongly support the assumption that glory is mainly formed by just several interior creeping waves with $\nu_m \approx ka$ with the axial singly interiorly reflected ray as the main correction. Typical results are presented in Figures 7.74 and 7.75, which compare exact and approximate bistatic scattering cross section of a water droplet of the size typical for water droplets in clouds and at the wavelength in the middle of the visible spectrum. Just two modes closest to $\nu = ka$, one with $\nu_m < ka$ and another with $\nu_m > ka$, are used. By construction, approximation (7.577) is not applicable in a close proximity of $\theta = 0$, but it seems to work fine already for the first side lobe. Accounting for the axial single internal reflection and for the external reflection importantly corrects the curves around the second and third side lobes. Contributions from other GO rays with $p = 4$ and $p = 5$ as well from the caustic with $p = 10$ have been found to be negligibly small.

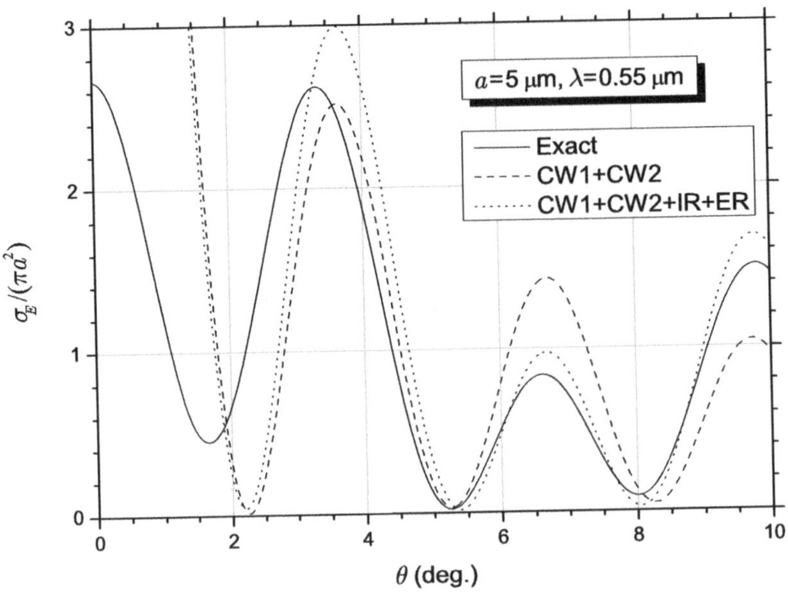

Figure 7.74 Normalized bistatic scattering cross section of a spherical water droplet near the backscattering direction in the E plane: exact vs. internal creeping wave approximation. CW1 and CW2, internal creeping waves; IR, internal reflection; ER, external reflection.

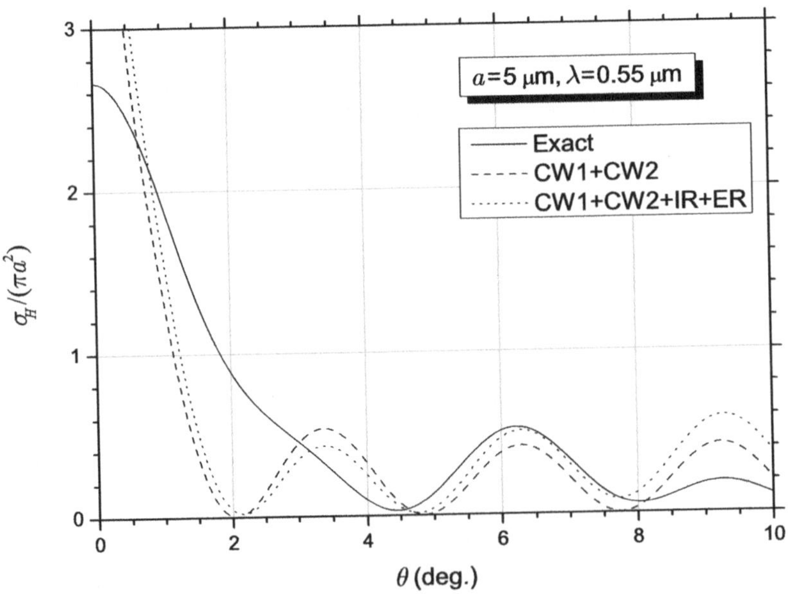

Figure 7.75 The same as Figure 7.74 but in the H plane.

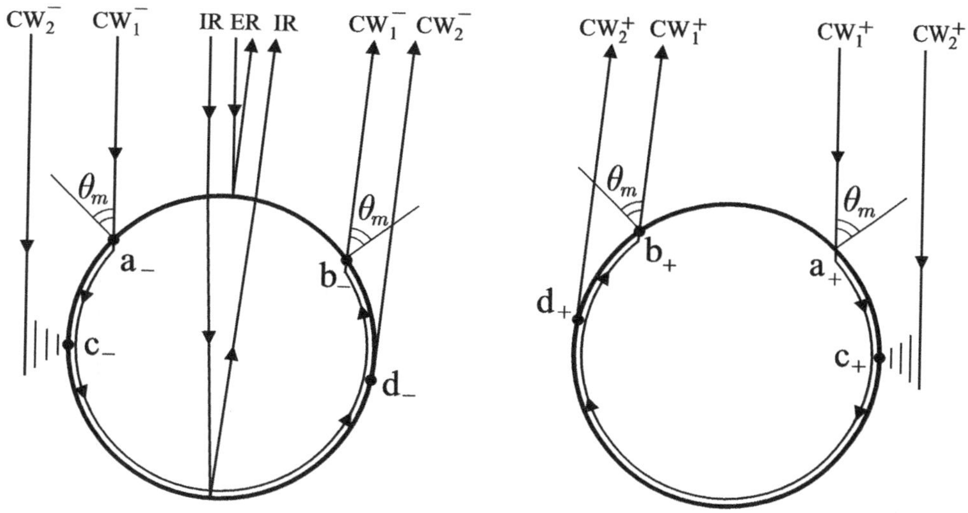

Figure 7.76 Waves forming the glory, propagating counterclockwise (to the left) and clockwise (to the right). $CW_1^{\pm}$, interior creeping waves with an injection angle θ_m closest to grazing ($\theta_m < \pi/2$); $CW_2^{\pm}$, lateral interior creeping waves; IR, internally reflected ray; ER, externally reflected ray.

Figure 7.76 illustrates the waves predominantly forming the glory. Main contributions come from interior creeping waves with eigenvalues ν_m closest to ka: Re $\nu_m < ka$ for $CW_1^{\pm}$ and Re $\nu_m > ka$ for $CW_2^{\pm}$. Every creeping wave consists of two components propagating in the opposite directions, indicated by ± signs, which are separately shown to the left and to the right in Figure 7.76. Points $a_{\pm}$ and $c_{\pm}$ are the injection (excitation) points of the waves, and $b_{\pm}$ and $d_{\pm}$ are the points of ejection of the respective waves, with ejection angles equal to θ_m and $\pi/2$, respectively. Contributions from GO rays are small corrections, and the internally reflected ray dominates the externally reflected ray.

Now let us take a look at the forward scattering from electrically large spheres and the associated phenomenon of atmospheric corona. A corona consists of a central bright and almost white disc (aureole) surrounded by small number of concentric faint colored rings. Coronae are observed around the sun or the moon when small, as in a cloud, water droplets, or ice crystals are in the air. The corona can be as large as 15° in diameter and therefore much larger then the disc of the sun or the moon.[17] The angular size of the aureole and the rings may change as clouds move in front of the observer. In what follows, we discuss general properties of forward scattering and explain the features of coronae.

An interesting property is that in contrast to backscattering, the level of the forward-scattered field is asymptotically independent of the material constitution of the sphere. Figure 7.77 compares bistatic scattering cross sections of a water sphere and a PEC sphere of the same, typical for cloud droplets, radius within the range $170° \leq \theta \leq 180°$. The curves, obtained from the exact multipole solutions, indicate the weak polarization and material dependence around the main and first side lobes in the scattering diagram. In the forward direction, in accordance with the formula (3.124), the cross sections grow as $\pi a^2 (ka)^2$ and approach each other in the sense of relative error (though the absolute value of their difference grows with increasing ka). This can be explained by looking at the derivation of the optical limit (3.124) for the far-field coefficient given in section 3.4.3 for impenetrable bodies. For a penetrable body, like a water droplet in the visible, equation (3.126) should be modified by including the transmitted field on the surface Ω_1, which is the geometric cross section of the shadow region behind the sphere (Figure 3.13). Because of refraction in the sphere, the direction of propagation of the directly transmitted wave is, in general, different from the incidence direction (Figure 7.21), implying a vanishing asymptotic contribution from the transmitted field to the forward scattering (similarly to the contribution from $\mathbf{E}_{\text{ref}}$ on the complementary surface Ω_2).

[17]The angular diameter of the sun and the moon on the sky is about 0.5°.

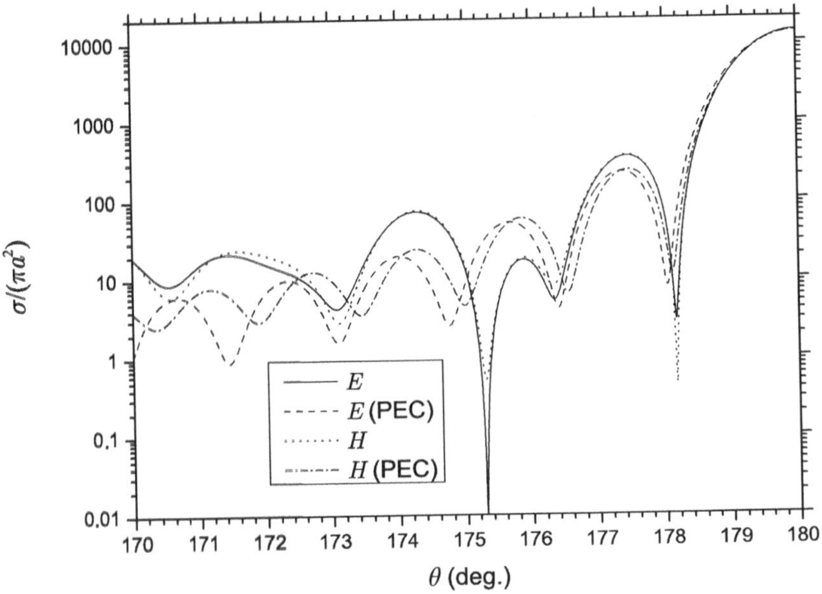

Figure 7.77 Comparison of the normalized bistatic scattering cross sections of PEC and water spheres with $a = 10$ μm near the forward-scattering direction at $\lambda = 0.55$ μm.

Forward-scattering contributions specific to penetrable spheres can come from the axial ray propagating along the z axis and from non-axial rays if they are scattered in the forward direction. In the case of water in the visible, it is the rays with $p = 3$ and $p \geq 5$ that can be scattered in the forward direction (Figure 7.78). Since these rays intersect Ω_1 at a point (axial ray) or across rings of infinitesimal width (non-axial rays), their contributions are infinitesimally small compared with the main contribution given by integration of the incident field over the surface Ω_2 complimentary to Ω_1.

Applied to corona formation, this property means that to a good approximation the pattern near forward-scattering direction will be almost the same for spheres of different material constitution. This explains why coronae are also observed in a cold air filled with frozen water droplets.

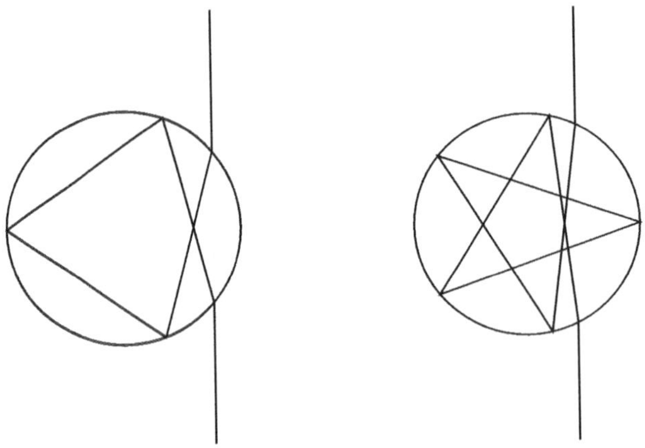

Figure 7.78 Computer-generated paths of non-axial rays contributing to forward scattering from a water droplet with $N = 4/3$ on three- and fivefold internal reflections. Non-axial rays with $p = 1, 2, 4$ cannot be scattered in the forward direction for the given value of refractive index.

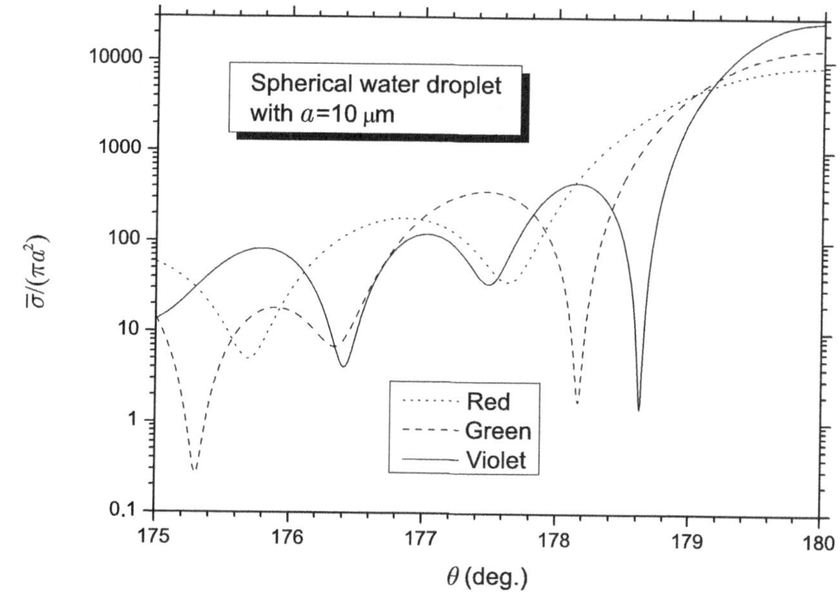

Figure 7.79 Polarization-averaged normalized bistatic scattering cross section of a spherical water droplet near the forward-scattering direction for three different wavelengths.

Thus, the appearance of a corona depends primarily on the electrical size ka of the droplets. Smaller droplets produce wider coronae and vice versa. The strong dependence of the position of the side lobes in scattering pattern on ka makes the appearance of coronae more sensitive to the varying size of water droplets under natural conditions than that of rainbows.

For a fixed droplet size, the position of the side lobes in scattering diagram depends on the wavelength, which leads to colored rings around a central disc if the incident field, as the natural light, includes waves with different wavelengths. The resulting corona is the sum of all the coronae contributions from each spectral color. The central disc corresponds to the main lobe, and since its position is independent of the wavelength, the disc is white and fringed with red because the main lobe in red is wider than in the rest of colors. The colored rings correspond to the side lobes, implying a bluish inside and reddish outside of every ring with the rest of the colors grading between the two.

These features are indeed observed in atmospheric coronae. A simulation result in Figure 7.79, obtained with the exact multipole solution, exhibits all those features. Polarization-averaged bistatic cross sections plotted for three different wavelengths (red, green, and blue light) overlap in the range $179.1° < \theta \leq 180°$ (white aureole); the slightly different levels of the main lobes are due to different values of the wavelengths. At larger scattering angles, scattering is dominated by the red light when $178.1° < \theta < 179.1°$ (red fringe of the aureole), by the blue light when $177.8° < \theta < 178.1°$ (the blue inside of the first ring), by the green light when $177° < \theta < 177.8°$, and by the red light when $176.1° < \theta < 177°$, before the second blue ring appears between 176.1° and 175.3°, and so on. The dominating color is always mixed with weaker colors, which leads to the blurred appearance of the corona rings. The intensity of the corona is only a small fraction (about 1% in the case considered) of the aureole intensity.

Literature Notes

Good introductory reading about optical atmospheric phenomena, well illustrated and with clear explanations of underlying physics, is provided by the internet resources at http://www.atoptics.co.uk/ and https://en.wikipedia.org/wiki/Rainbow. For historic surveys of the development of the theory of rainbow see Boyer (1987) and Nussenzveig (1977). More detailed overviews of the theories can be found in Nussenzveig (1979), Adam (2002), and Grandy (2000).

The first quantitative theory of rainbows was developed by Airy (1838) by using a scalar PO-based approach. Van der Pol and Bremmer (1937b) presented an electromagnetic theory of rainbows based on Debye's representation of the multipole series solution. Asymptotic approximations for rays with an arbitrary number of reflections, including the caustic approximation, were derived for the intensity of the scattered field by replacing Bessel functions in the mulitpole expansion with their integral representations followed by application of the multidimensional stationary point approximation. An electromagnetic theory of rainbows by Khare and Nussenzveig (1974) was formulated in terms of the far-field coefficients by using a modified Watson transformation and higher-order asymptotic formulas for the caustic region.

A physical explanation of glory has been the subject of debates and research for more than half of a century. Van de Hulst (1957, sections 13.3 and 17.42) speculated that the colored rings of the glory may be caused by interference between two rays, both singly reflected in the interior of the water droplet but entering the droplet at diametrically opposite points, provided that the one following the longer path makes a part of its travel as a wave creeping on the exterior side of the spherical boundary.

A theory of glory due to Khare and Nussenzveig (1977) and Nussenzveig (1979) states that within the range $100 < ka < 1000$, which is most relevant to observation of the glory, several mechanisms may importantly contribute to backscattering, including the van de Hulst term, the 10th order rainbow term ($p = 10$) and the GO axial ray ($p = 1$). It is further suggested that a part of the light energy scattered back may originate from wave tunneling, which occurs when light rays that missed a droplet can still transfer energy into it. The results imply in fact that formation of a glory cannot be associated with a specific term in the Debye representation of the scattered field.

The theory of glory presented in this section bears similarities to both concepts, being nevertheless substantially different. Similarly to the van de Hulst concept, creeping waves are involved, but those traveling in the *interior* of the sphere. Similarly to the Nussenzveig concept, tunneling (lateral excitation) plays a role but only for a pair of major contributions ($CW_2^{\pm}$), whereas another pair ($CW_1^{\pm}$) is excited by the light rays directly meeting the sphere. Furthermore, in the interior of the sphere the waves $CW_{1,2}^{\pm}$ are creeping waves and not GO rays as assumed by Nussenzveig.

7.6 Metamaterial Effects in Scattering from Spheres

7.6.1 Small Spheres

Electrically small spheres made of various materials play an important role in the design of artificial materials with unusual electromagnetic properties: metamaterials. There are two aspects of this application. First, electrically small spherical particles can exhibit strong resonant response to time-harmonic electromagnetic fields, and, thus, can be used as "artificial molecules" of metamaterial composites. Second, scattering from coated spheres made of materials with unusual properties, including metamaterials, show interesting and important applications properties.

Let us first consider small spheres as individual inclusions in composites designed to exhibit the desired effective electromagnetic properties. Artificial dielectrics formed by randomly or regularly arranged electrically small metal spheres have been known and used in microwave antennas for a long time, for example see Collin (1960, Chapter 12) and references therein. This allows realization of lightweight artificial dielectrics with permittivity about 1.5–2.5 in the microwave frequency range. Use of electrically small but resonant spheres opens a way to design artificial materials with more exotic properties, including double-negative composites. The use of spherical particles is advantageous if isotropy of the effective response is required.

Considering a single uniform isotropic dielectric sphere (permittivity ϵ_1) of radius a positioned in an isotropic dielectric host (permittivity ϵ), we note that its electric polarizability P_{11} (7.497) has a resonance at the frequency where $\mathrm{Re}\,(\epsilon_1 + 2\epsilon) = 0$. The resonance behavior of the various cross sections, σ_{ext}, σ_{abs}, σ_{T}, and σ_{back}, is addressed in section 7.5.2 and illustrated in Figures 7.37 and 7.38. As seen from Table 2.3 for silver and already mentioned in section 3.6, this phenomenon takes place, for example, for noble metals in the visible, and it is called the plasmon resonance.

The left-hand part of Figure 7.80 shows the electric polarizability of a small silver sphere at optical frequencies. The values of the permittivity of silver are according to the experimental data (Johnson and Chrysty 1972); see Table 2.3. The plasmon resonance of a single sphere can be utilized to create artificial materials with resonant permittivity. The effective permittivity can be estimated using, for example, the Maxwell Garnett mixing rule (e.g., Sihvola (1999)). The right-hand part of Figure 7.80 shows an example result for a composite of silver spheres with the polarizability shown on the left in Figure 7.80. Obviously, such composites can be used as materials with unusually high permittivity values (below the plasmon resonance) or as materials with permittivity close to zero (above the resonance).

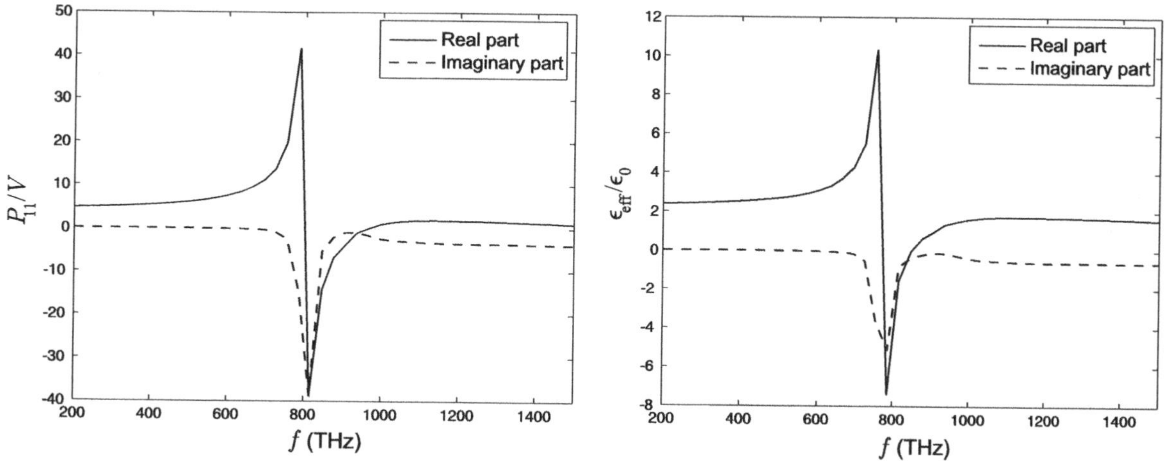

Figure 7.80 Left: Polarizability of a small silver sphere (7.497) in the visible frequency range. The permittivity of the background medium $\epsilon = 1.5\epsilon_0$. The polarizability is normalized to the sphere volume $V = 4\pi a^3/3$. Right: The Maxwell–Garnett estimate for the effective permittivity of a random mixture of small silver spheres in a dielectric host medium. The sphere radius is 10 nm, the average distance between spheres is 30 nm, and the permittivity of the background medium $\epsilon = 1.5\epsilon_0$.

In the microwave frequency range, the availability of low-loss dielectric materials with high values of the permittivity allows realization of isotropic resonant scatterers whose size is small compared with the wavelength in the surrounding space. Typical dependence of the polarizabilities of a small high-permittivity sphere is illustrated in Figure 7.81. We see that the lowest resonance is of the magnetic type. Far from the resonance of this mode the magnetic response is negligible. At a higher frequency, the same sphere resonates as an electric dipole, with a considerably weaker response. Outside of

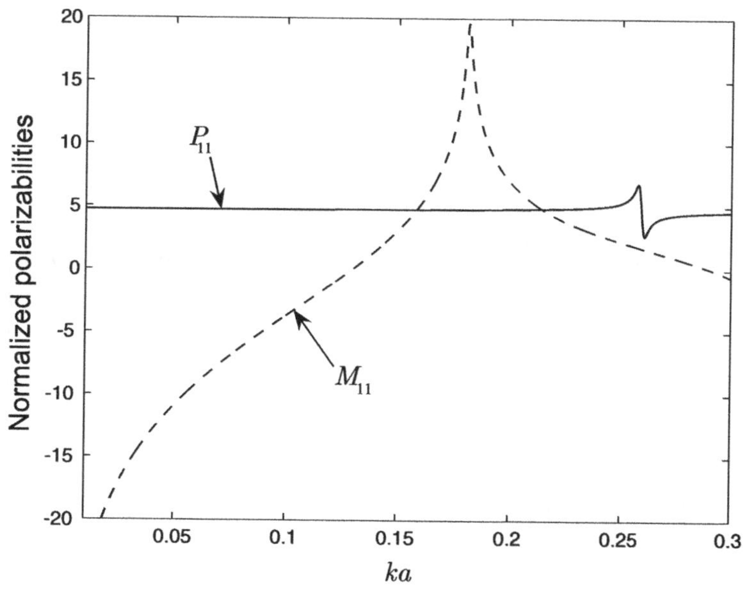

Figure 7.81 Electric (solid line) and magnetic (dashed line) polarizabilities of a small dielectric sphere (7.493) and (7.494). The permittivity of the sphere material is $\epsilon_1 = 300(1 - j0.01)\epsilon_0$. The background medium $\epsilon = \epsilon_0$. The polarizability is normalized to the sphere volume $V = 4\pi a^3/3$. Logarithmic scale is used: $10\log_{10}(|P_{11}|/V)$ and $10\log_{10}(|M_{11}|/V)$ are plotted.

the resonant band, the electric polarizability is approximately a constant, corresponding to the quasi-static polarizability of the sphere. The analysis of this plot suggests that a two-phase composite formed by spheres of different radii or/and permittivities can be engineered to realize artificial materials with the desired effective permittivity and permeability. For example, choosing the radii of the two sets of spheres properly, the resonance frequency of the magnetic mode in one set of spheres can be designed to be close to the resonance frequency of the electric mode of the second set. This arrangement was proposed in Vendik and Gashinova (2004) as a route for the realization of double-negative materials.

To estimate the effective parameters of such two-phase mixtures, one can use the generalized Clausius–Mossotti relation (Sihvola 1999). Let us assume that the spheres used to create desired magnetic response have the polarizabilities M_{11}^{m} and P_{11}^{m}, while the spheres in the other set are characterized by M_{11}^{e} and P_{11}^{e}. Then we can estimate the effective parameters of the composite from (Jylhä *et al.* 2006)

$$\frac{\epsilon_{\mathrm{eff}} - \epsilon}{\epsilon_{\mathrm{eff}} + 2\epsilon} = n_{\mathrm{e}} \frac{P_{11}^{\mathrm{e}}}{3} + n_{\mathrm{m}} \frac{P_{11}^{\mathrm{m}}}{3} \tag{7.579}$$

$$\frac{\mu_{\mathrm{eff}} - \mu}{\mu_{\mathrm{eff}} + 2\mu} = -n_{\mathrm{m}} \frac{M_{11}^{\mathrm{m}}}{3}. \tag{7.580}$$

Here, $n_{\mathrm{e,m}}$ are the numbers of the two types of inclusions in the unit volume of the composite. The contribution of the magnetic polarization of spheres far from the magnetic resonance (missing M_{11}^{e} in (7.580)) can be neglected because the magnetic effect arises only as a resonant phenomena.

Another approach to the design of media with the desired values of permittivity and permeability is to use coated spheres, or core-shell spheres, as they provide the necessary freedom to achieve the desired values of the polarizabilities and therefore ϵ_{eff} and μ_{eff} at a given frequency (Morits and Simovski 2012; Paniagua-Dominguez *et al.* 2011). The polarizabilities can be tuned by adjusting the radii of the shell and the core. Figure 7.82 shows an example for a core-shell sphere with the silver core of the radius $a_1 = 27.3$ nm and the silicon shell of the radius $a = 130$ nm. The sphere has the magnetic and electric resonances at 310 THz. Note that in a range of frequencies close to the resonance, $\mathrm{Re}P_{11} < 0$ and

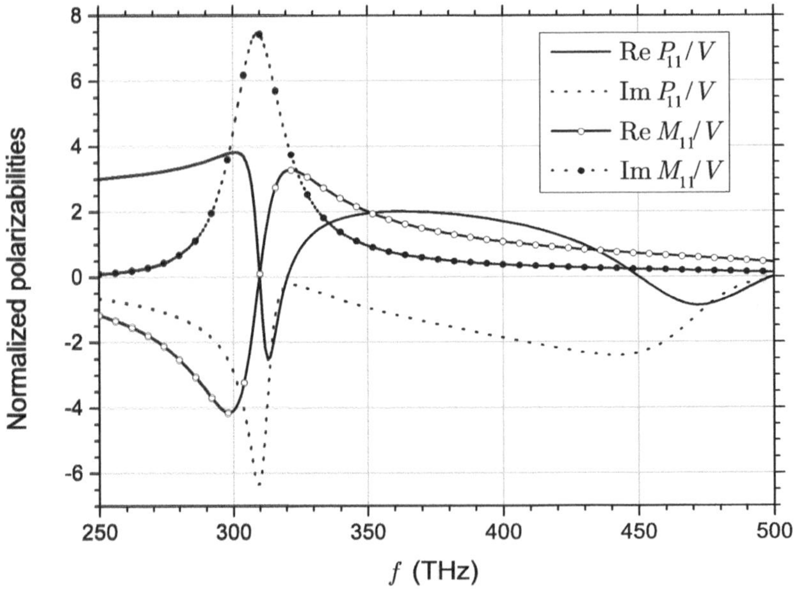

Figure 7.82 Polarizabilities of a core-shell sphere in near infrared. A silicon shell and a silver core with $a = 130$ nm and $a_1 = 27.3$ nm. Permittivity of silver according to Table 2.3, permittivity of silicon (intrinsic silicon) from (Green and Keevers 1995).

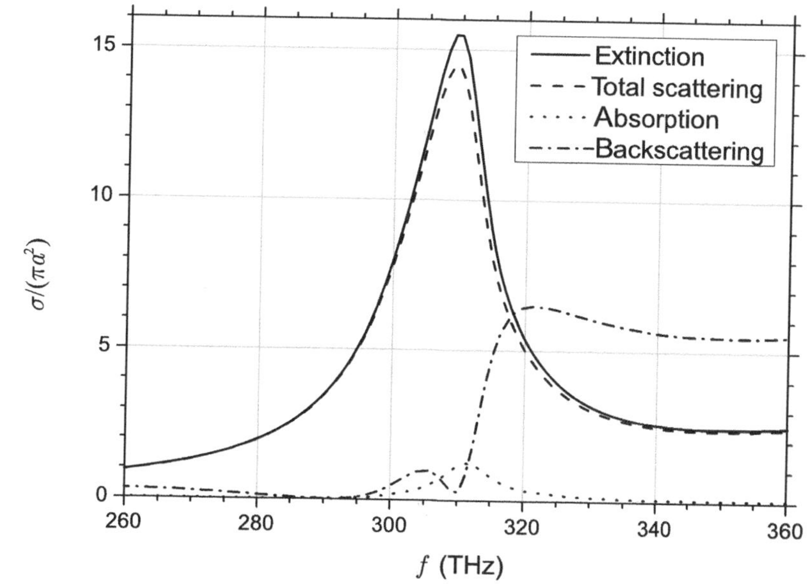

Figure 7.83 Cross sections of a balanced core-shell sphere.

Re$M_{11} > 0$, implying the possibility of obtaining a material with Re$\epsilon_{\text{eff}} < 0$ and Re$\mu_{\text{eff}} < 0$ (double-negative material), which is the kind of material required for superlensing (see section 4.5.3).

Another possible application of arrays of core-shell spheres is related to the fact that, as illustrated in Figure 7.82, the resonance frequencies in the electric and magnetic polarizabilities can be tuned to coincide. At the resonance, P_{11} and M_{11} achieve extreme values but with opposite signs, a consequence of which is a maximum in the extinction cross section and a minimum in the backscattering cross section (see (7.351) and (7.353)). Figure 7.83 presents cross sections of the core-shell sphere with the polarizabilities from Figure 7.82. At the resonance frequency, $P_{11} \approx -M_{11}$ and $\sigma_{\text{back}} \approx 0$, whereas σ_{ext} has a maximum. Since the wavelength at the resonance is about 3.7 times greater than the diameter of the sphere, it seems to be possible to build a metamaterial array from the spheres to obtain, for example, a planar structure that is non-reflective and strongly absorptive around the resonance frequency.

Scattering from coated spheres and multilayered spheres can be dramatically reduced if the lowest-mode polarizabilites are minimized by proper design. The issue of "invisible particles" was discussed in the literature for a long time (Bohren and Huffman (1983); Kerker (1975)) and gained more attention recently (e.g., Alù and Engheta (2005a)) when artificial materials with exotic values of material parameters were realized, first for microwave frequencies. For spherical particles, the main idea is to compensate the dipole moment induced in the core by the out-of-phase moment induced in the shell. For example, equating the numerator of P_{11} in (7.461) to zero (see section 7.4.4), we can find the ratio of the two radii at which the induced electric dipole moment of a small coated PEC sphere vanishes (Alù and Engheta 2005a),

$$\left(\frac{a_1}{a}\right)^3 = \frac{\epsilon - \epsilon_1}{\epsilon + 2\epsilon_1}. \tag{7.581}$$

This formula gives physical values for the ratio of the radii if $0 < \epsilon_1 < \epsilon$.[18] If the background material is free space ($\epsilon = \epsilon_0$), this requirement implies frequency dispersion of the material properties (see section 2.2.2) and limitations on the frequency bandwidth of the low-scattering regime.

[18]Note that the required shell thickness becomes very small when $\epsilon_1 \to 0$.

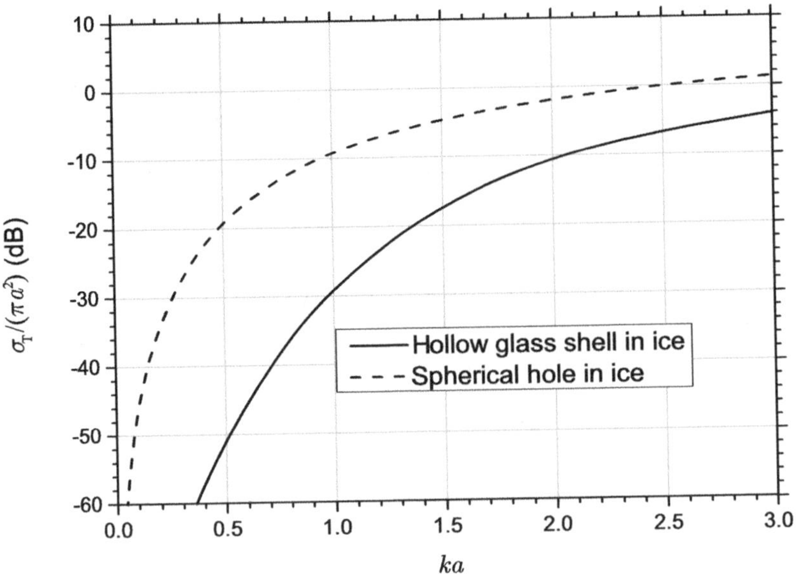

Figure 7.84 Normalized total scattering cross section of a hollow glass shell with $a_1 = 0.813a$ in ice vs. the cross section of a spherical hole of the radius a in ice.

More generally, from (7.445) we find that the electric dipole moment of a small coated sphere is zero if

$$\left(\frac{a_1}{a}\right)^3 = \frac{(\epsilon_1 - \epsilon)(\epsilon_2 + 2\epsilon_1)}{(\epsilon_1 - \epsilon_2)(\epsilon + 2\epsilon_1)} \tag{7.582}$$

and $M_{11} = 0$ when either the materials are non-magnetic ($\mu_2 = \mu_1 = \mu$) or

$$\left(\frac{a_1}{a}\right)^3 = \frac{(\mu_1 - \mu)(\mu_2 + 2\mu_1)}{(\mu_1 - \mu_2)(\mu + 2\mu_1)}. \tag{7.583}$$

For example, a non-magnetic material with $0 < \epsilon_2 < \epsilon < \epsilon_1$ permits a physical solution ($0 < a_1 < a$) of (7.582). Figure 7.84 gives an example: a hollow glass shell ($\epsilon_2 = \epsilon_0$, $\epsilon_1/\epsilon_0 = 6.7 - j0.1$) is surrounded by ice $\epsilon/\epsilon_0 = 3.17 - j0.002$; the material parameters from Table 2.1. The quasi-static magnetic polarizability is zero since the materials are non-magnetic while the quasi-static electric polarizability vanishes due to the choice of the ratio a_1/a according to (7.582). In Figure 7.84, the total scattering cross section σ_T of the shell is plotted as a function of the electrical size ka and compared with the σ_T of a hole of the same radius in the ice. It is seen that P_{11} of the shell vanishes much faster than that of the hole as ka goes to zero.

If the real parts of the material parameters of the core and the shell spheres are allowed to take values smaller than unity or negative (plasmonic materials in the visible or metamaterials in the microwave range), it is also possible to realize the regime of very strong scattering from electrically small scatterers. The design freedom for plasmonic resonances in the case of coated spheres is much broader than for single plasmonic spheres (Alù and Engheta 2005b). The resonance conditions for small core-shell particles can be found by equating the denominators of (7.445) and (7.446) to zero.

7.6.2 *Invisibility Cloak*

A spherical configuration that is studied in this section is a spectacular example of the possibilities that metamaterials offer to the scattering theory and its applications. By using a layer of artificial material, generally anisotropic and inhomogeneous, it is possible, at least in theory, to create regions in space that do not scatter and may hide other scatterers, thus leading to something that is similar to an invisibility cloak (Pendry *et al.* 2006). This theoretical discovery and the appealing applications that it promises have stimulated a lot of research work. However, the required material properties of the masking layer cannot be achieved with natural materials, and the use of metamaterials is in general necessary.

The basic principle is based on the observation that the change of the analytical form of Maxwell's equations on a coordinate transformation is equivalent to a change in the permittivity and permeability tensors. By transforming coordinates one can construct a whole family of configurations which are electromagnetically equivalent to the infinite space filled with a homogeneous and isotropic material. The configurations in the family differ in material composition and in geometry, but in every configuration the shape of the cloak and the material composition of the cloak match each other in the sense that Maxwell's equation in the configuration can be reduced to Maxwell's equations in the homogeneous space by a coordinate transformation. The idea has become a foundation for a new theoretical discipline called transformation optics.

In this section we describe the construction of cloaks in arbitrary orthogonal coordinates and then specify the result to the spherically shaped cloak.

Consider Maxwell's equations in orthogonal coordinates (ξ, η, ζ), assuming that in these coordinates the permittivity and permeability tensors of the medium are diagonal

$$\overline{\overline{\epsilon}} \cdot \mathbf{E} = \epsilon_\xi E_\xi + \epsilon_\eta E_\eta + \epsilon_\zeta E_\zeta, \quad \overline{\overline{\mu}} \cdot \mathbf{H} = \mu_\xi H_\xi + \mu_\eta H_\eta + \mu_\zeta H_\zeta. \tag{7.584}$$

Using (7.584) and the expression (A.35) for the curl in orthogonal coordinates, Maxwell's equations (2.7) and (2.8) are expressed as

$$j\omega\epsilon_\xi \frac{h_\eta h_\zeta}{h_\xi} \tilde{E}_\xi = \frac{\partial}{\partial \eta}\tilde{H}_\zeta - \frac{\partial}{\partial \zeta}\tilde{H}_\eta \tag{7.585}$$

$$j\omega\epsilon_\eta \frac{h_\zeta h_\xi}{h_\eta} \tilde{E}_\eta = \frac{\partial}{\partial \zeta}\tilde{H}_\xi - \frac{\partial}{\partial \xi}\tilde{H}_\zeta \tag{7.586}$$

$$j\omega\epsilon_\zeta \frac{h_\xi h_\eta}{h_\zeta} \tilde{E}_\zeta = \frac{\partial}{\partial \xi}\tilde{H}_\eta - \frac{\partial}{\partial \eta}\tilde{H}_\xi \tag{7.587}$$

$$-j\omega\mu_\xi \frac{h_\eta h_\zeta}{h_\xi} \tilde{H}_\xi = \frac{\partial}{\partial \eta}\tilde{E}_\zeta - \frac{\partial}{\partial \zeta}\tilde{E}_\eta \tag{7.588}$$

$$-j\omega\mu_\eta \frac{h_\zeta h_\xi}{h_\eta} \tilde{H}_\eta = \frac{\partial}{\partial \zeta}\tilde{E}_\xi - \frac{\partial}{\partial \xi}\tilde{E}_\zeta \tag{7.589}$$

$$-j\omega\mu_\zeta \frac{h_\xi h_\eta}{h_\zeta} \tilde{H}_\zeta = \frac{\partial}{\partial \xi}\tilde{E}_\eta - \frac{\partial}{\partial \eta}\tilde{E}_\xi, \tag{7.590}$$

where $\tilde{\mathbf{E}} = (h_\xi E_\xi, h_\eta E_\eta, h_\zeta E_\zeta)$ and $\tilde{\mathbf{H}} = (h_\xi H_\xi, h_\eta H_\eta, h_\zeta H_\zeta)$, with h_ξ, h_η, and h_ζ being the scale factors of the coordinates (see Appendix A.3).

It follows from (7.585)–(7.590) that the analytical form of Maxwell's equations effectively depends on the combinations

$$\epsilon_\xi \frac{h_\eta h_\zeta}{h_\xi}, \quad \epsilon_\eta \frac{h_\zeta h_\xi}{h_\eta}, \quad \epsilon_\zeta \frac{h_\xi h_\eta}{h_\zeta}, \quad \mu_\xi \frac{h_\eta h_\zeta}{h_\xi}, \quad \mu_\eta \frac{h_\zeta h_\xi}{h_\eta}, \quad \mu_\zeta \frac{h_\xi h_\eta}{h_\zeta}, \tag{7.591}$$

implying that every coordinate transformation in Maxwell's equations can be represented as a change of the permittivity and permeability tensors, and vice versa. This fact suggests the existence of families of electromagnetically equivalent configurations, in the sense that they are described by Maxwell's equations of the same analytical form despite different geometric shape and material composition. Of particular interest are configurations that are equivalent to an infinite space filled with a homogeneous material, as these configurations should not scatter fields.

In a homogeneous isotropic medium ($\bar{\bar{\epsilon}} = \epsilon \bar{\bar{I}}, \bar{\bar{\mu}} = \mu \bar{\bar{I}}$) and in spherical coordinates $\xi \to r, \eta \to \theta, \zeta \to \phi$ with $h_\xi = 1$, $h_\eta = r$, and $h_\zeta = r \sin\theta$, (7.585)–(7.590) take the form

$$j\omega\epsilon r^2 \sin\theta \tilde{E}_r = \frac{\partial}{\partial\theta}\tilde{H}_\phi - \frac{\partial}{\partial\phi}\tilde{H}_\theta \tag{7.592}$$

$$j\omega\epsilon \sin\theta \tilde{E}_\theta = \frac{\partial}{\partial\phi}\tilde{H}_r - \frac{\partial}{\partial r}\tilde{H}_\phi \tag{7.593}$$

$$j\omega\epsilon \csc\theta \tilde{E}_\phi = \frac{\partial}{\partial r}\tilde{H}_\theta - \frac{\partial}{\partial\theta}\tilde{H}_r \tag{7.594}$$

$$-j\omega\mu r^2 \sin\theta \tilde{H}_r = \frac{\partial}{\partial\theta}\tilde{E}_\phi - \frac{\partial}{\partial\phi}\tilde{E}_\theta \tag{7.595}$$

$$-j\omega\mu \sin\theta \tilde{H}_\theta = \frac{\partial}{\partial\phi}\tilde{E}_r - \frac{\partial}{\partial r}\tilde{E}_\phi \tag{7.596}$$

$$-j\omega\mu \csc\theta \tilde{H}_\phi = \frac{\partial}{\partial r}\tilde{E}_\theta - \frac{\partial}{\partial\theta}\tilde{E}_r \tag{7.597}$$

with $\tilde{E}_r = E_r$, $\tilde{E}_\theta = rE_\theta$, $\tilde{E}_\phi = r\sin\theta E_\phi$, $\tilde{H}_r = H_r$, $\tilde{H}_\theta = rH_\theta$, and $\tilde{H}_\phi = r\sin\theta H_\phi$. In the infinite space ($0 \le r < +\infty$, $0 \le \theta \le \pi$, $0 \le \phi < 2\pi$), the solution of (7.592)–(7.597) is simply the incident field as no scattering takes place in the homogeneous space.

A spherical cloak is an example of a configuration electromagnetically equivalent to a homogeneous space. Consider a spherical shell (Figure 7.85) defined by $R_1 \le r \le R_2$, $0 \le \theta \le \pi$, and $0 \le \phi < 2\pi$ and filled with some material with tensors

$$\bar{\bar{\epsilon}} = \begin{pmatrix} \epsilon_r & 0 & 0 \\ 0 & \epsilon_\theta & 0 \\ 0 & 0 & \epsilon_\phi \end{pmatrix}, \quad \bar{\bar{\mu}} = \begin{pmatrix} \mu_r & 0 & 0 \\ 0 & \mu_\theta & 0 \\ 0 & 0 & \mu_\phi \end{pmatrix}. \tag{7.598}$$

The surrounding medium ($R_2 < r < +\infty$) is assumed to be homogeneous and isotropic with permittivity ϵ and permeability μ.

Our goal is to choose the elements of tensors in (7.598) so as to make the shell electromagnetically equivalent to the homogeneous sphere $0 \le r \le R_2$ with ϵ and μ. To this end, we note that

$$r = R_1 + r' \frac{R_2 - R_1}{R_2} \tag{7.599}$$

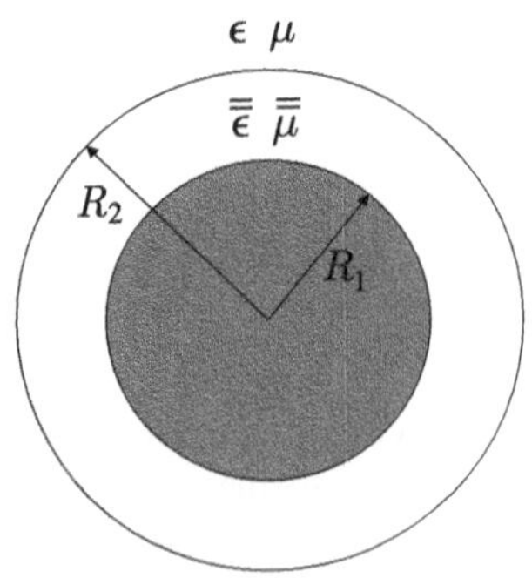

Figure 7.85 Spherical shell with tensor permittivity and permeability in homogeneous infinite space.

transforms the sphere $0 \le r' \le R_2$ into the spherical shell $R_1 \le r \le R_2$, and introducing new coordinates

$$r' = \frac{R_2(r - R_1)}{R_2 - R_1}, \quad \theta' = \theta, \quad \phi' = \phi \tag{7.600}$$

we write Maxwell's equations as

$$j\omega\epsilon_r \frac{h'_\theta h'_\phi}{h'_r}\tilde{E}'_r = \frac{\partial}{\partial\theta'}\tilde{H}'_\phi - \frac{\partial}{\partial\phi'}\tilde{H}'_\theta \tag{7.601}$$

$$j\omega\epsilon_\theta \frac{h'_\phi h'_r}{h'_\theta}\tilde{E}'_\theta = \frac{\partial}{\partial\phi'}\tilde{H}'_r - \frac{\partial}{\partial r'}\tilde{H}'_\phi \tag{7.602}$$

$$j\omega\epsilon_\phi \frac{h'_r h'_\theta}{h'_\phi}\tilde{E}'_\phi = \frac{\partial}{\partial r'}\tilde{H}'_\theta - \frac{\partial}{\partial\theta'}\tilde{H}'_r \tag{7.603}$$

$$-j\omega\mu_r \frac{h'_\theta h'_\phi}{h'_r}\tilde{H}'_r = \frac{\partial}{\partial\theta'}\tilde{E}'_\phi - \frac{\partial}{\partial\phi'}\tilde{E}'_\theta \tag{7.604}$$

$$-j\omega\mu_\theta \frac{h'_\phi h'_r}{h'_\theta}\tilde{H}'_\theta = \frac{\partial}{\partial\phi'}\tilde{E}'_r - \frac{\partial}{\partial r'}\tilde{E}'_\phi \tag{7.605}$$

$$-j\omega\mu_\phi \frac{h'_r h'_\theta}{h'_\phi}\tilde{H}'_\phi = \frac{\partial}{\partial r'}\tilde{E}'_\theta - \frac{\partial}{\partial\theta'}\tilde{E}'_r, \tag{7.606}$$

where the prime denotes the dependence on the new coordinates, for example $\tilde{E}'_r = \tilde{E}_r(r', \theta', \phi')$, $\tilde{E}'_\theta = \tilde{E}_\theta(r', \theta', \phi')$, and so on, and

$$h'_r = h_r(r', \theta', \phi') = \sqrt{\left(\frac{\partial x}{\partial r'}\right)^2 + \left(\frac{\partial y}{\partial r'}\right)^2 + \left(\frac{\partial z}{\partial r'}\right)^2} = \frac{R_2 - R_1}{R_2} \tag{7.607}$$

$$h'_\theta = h_\theta(r', \theta', \phi') = \sqrt{\left(\frac{\partial x}{\partial\theta'}\right)^2 + \left(\frac{\partial y}{\partial\theta'}\right)^2 + \left(\frac{\partial z}{\partial\theta'}\right)^2} = r \tag{7.608}$$

$$h'_\phi = h_\phi(r', \theta', \phi') = \sqrt{\left(\frac{\partial x}{\partial\phi'}\right)^2 + \left(\frac{\partial y}{\partial\phi'}\right)^2 + \left(\frac{\partial z}{\partial\phi'}\right)^2} = r\sin\theta' \tag{7.609}$$

with r given by (7.599). Now by comparing (7.592)–(7.597) with (7.601)–(7.606) and requiring that

$$\epsilon_r \frac{h'_\theta h'_\phi}{h'_r} = \epsilon r'^2 \sin\theta', \quad \epsilon_\theta \frac{h'_\phi h'_r}{h'_\theta} = \epsilon\sin\theta', \quad \epsilon_\phi \frac{h'_r h'_\theta}{h'_\phi} = \epsilon\csc\theta'$$

and

$$\mu_r \frac{h'_\theta h'_\phi}{h'_r} = \mu r'^2 \sin\theta', \quad \mu_\theta \frac{h'_\phi h'_r}{h'_\theta} = \mu\sin\theta', \quad \mu_\phi \frac{h'_r h'_\theta}{h'_\phi} = \mu\csc\theta',$$

the system (7.601)–(7.606) is reduced to

$$j\omega\epsilon r'^2 \sin\theta' \tilde{E}'_r = \frac{\partial}{\partial\theta'}\tilde{H}'_\phi - \frac{\partial}{\partial\phi'}\tilde{H}'_\theta \tag{7.610}$$

$$j\omega\epsilon \sin\theta' \tilde{E}'_\theta = \frac{\partial}{\partial\phi'}\tilde{H}'_r - \frac{\partial}{\partial r'}\tilde{H}'_\phi \tag{7.611}$$

$$j\omega\epsilon \csc\theta' \tilde{E}'_\phi = \frac{\partial}{\partial r'}\tilde{H}'_\theta - \frac{\partial}{\partial \theta'}\tilde{H}'_r \tag{7.612}$$

$$-j\omega\mu r'^2 \sin\theta' \tilde{H}'_r = \frac{\partial}{\partial \theta'}\tilde{E}'_\phi - \frac{\partial}{\partial \phi'}\tilde{E}'_\theta \tag{7.613}$$

$$-j\omega\mu \sin\theta' \tilde{H}'_\theta = \frac{\partial}{\partial \phi'}\tilde{E}'_r - \frac{\partial}{\partial r'}\tilde{E}'_\phi \tag{7.614}$$

$$-j\omega\mu \csc\theta' \tilde{H}'_\phi = \frac{\partial}{\partial r'}\tilde{E}'_\theta - \frac{\partial}{\partial \theta'}\tilde{E}'_r, \tag{7.615}$$

where $0 \le r' \le R_2$, $0 \le \theta' \le \pi$, $0 \le \phi' < 2\pi$, which formally coincides with (7.592)–(7.597). Thus, the spherical shell occupying the region $R_1 \le r \le R_2$, $0 \le \theta \le \pi$, $0 \le \phi < 2\pi$ and filled with the material

$$\frac{\epsilon_r}{\epsilon} = \frac{\mu_r}{\mu} = \frac{R_2 - R_1}{R_2}\left(\frac{r'}{r}\right)^2, \quad \frac{\epsilon_\theta}{\epsilon} = \frac{\epsilon_\phi}{\epsilon} = \frac{\mu_\theta}{\mu} = \frac{\mu_\phi}{\mu} = \frac{R_2}{R_2 - R_1} \tag{7.616}$$

is electromagnetically equivalent to the homogeneous sphere of radius R_2 with the material parameters ϵ and μ.

Assume that the spherical shell with material parameters (7.616) is immersed in the infinite homogeneous space with permittivity ϵ and permeability μ and an incident field with $\mathbf{E}_{\text{inc}}$ and $\mathbf{H}_{\text{inc}}$ propagates toward the shell. Because the shell is electromagnetically equivalent to the homogeneous sphere of radius R_2 filled with the same material as the surrounding medium, there will be no field scattered by the shell so that in the exterior of the shell ($R_2 \le r < +\infty$) the incident field coincides with the total field $\mathbf{E}$ and $\mathbf{H}$,

$$E_r(r,\theta,\phi) = E_r^{\text{inc}}(r,\theta,\phi) \tag{7.617}$$

$$E_\theta(r,\theta,\phi) = E_\theta^{\text{inc}}(r,\theta,\phi) \tag{7.618}$$

$$E_\phi(r,\theta,\phi) = E_\theta^{\text{inc}}(r,\theta,\phi) \tag{7.619}$$

$$H_r(r,\theta,\phi) = H_r^{\text{inc}}(r,\theta,\phi) \tag{7.620}$$

$$H_\theta(r,\theta,\phi) = H_\theta^{\text{inc}}(r,\theta,\phi) \tag{7.621}$$

$$H_\phi(r,\theta,\phi) = H_\theta^{\text{inc}}(r,\theta,\phi). \tag{7.622}$$

Let us write down the explicit expressions for the fields in the interior of the cloak. The coordinate change

$$r' = \begin{cases} \frac{R_2(r-R_1)}{R_2-R_1}, & R_1 \le r \le R_2 \\ r, & R_2 < r < +\infty \end{cases}, \quad \theta' = \theta, \quad \phi' = \phi \tag{7.623}$$

transforms Maxwell's equations to the form (7.592)–(7.597) in terms of the coordinates (r', θ', ϕ'). Since these latter equations describe the fields in the infinite homogeneous space, their solution is simply the incident field in the transformed coordinates so that for $0 \le r' \le R_2$ we have

$$h'_r E_r(r',\theta',\phi') = E_r^{\text{inc}}(r',\theta',\phi') \tag{7.624}$$

$$h'_\theta E_\theta(r',\theta',\phi') = r' E_\theta^{\text{inc}}(r',\theta',\phi') \tag{7.625}$$

$$h'_\phi E_\phi(r',\theta',\phi') = r' \sin\theta' E_\theta^{\text{inc}}(r',\theta',\phi') \tag{7.626}$$

$$h'_r H_r(r',\theta',\phi') = H_r^{\text{inc}}(r',\theta',\phi') \tag{7.627}$$

$$h'_\theta H_\theta(r',\theta',\phi') = r' H_\theta^{\text{inc}}(r',\theta',\phi') \tag{7.628}$$

$$h'_\phi H_\phi(r',\theta',\phi') = r' \sin\theta' H_\theta^{\text{inc}}(r',\theta',\phi') \tag{7.629}$$

or, using (7.607)–(7.609),

$$E_r(r',\theta',\phi') = \frac{R_2}{R_2 - R_1} E_r^{\text{inc}}(r',\theta',\phi') \tag{7.630}$$

$$E_\theta(r',\theta',\phi') = \frac{r'}{r} E_\theta^{\text{inc}}(r',\theta',\phi') \tag{7.631}$$

$$E_\phi(r',\theta',\phi') = \frac{r'}{r} E_\theta^{\text{inc}}(r',\theta',\phi') \tag{7.632}$$

$$H_r(r',\theta',\phi') = \frac{R_2}{R_2 - R_1} H_r^{\text{inc}}(r',\theta',\phi') \tag{7.633}$$

$$H_\theta(r',\theta',\phi') = \frac{r'}{r} H_\theta^{\text{inc}}(r',\theta',\phi') \tag{7.634}$$

$$H_\phi(r',\theta',\phi') = \frac{r'}{r} H_\theta^{\text{inc}}(r',\theta',\phi') \tag{7.635}$$

with r given by (7.599). In the original coordinate system (r, θ, ϕ), (7.630)–(7.635) have the form

$$E_r(r,\theta,\phi) = \frac{R_2}{R_2 - R_1} E_r^{\text{inc}}(r,\theta,\phi) \tag{7.636}$$

$$E_\theta(r,\theta,\phi) = \frac{R_2(r - R_1)}{r(R_2 - R_1)} E_\theta^{\text{inc}}(r,\theta,\phi) \tag{7.637}$$

$$E_\phi(r,\theta,\phi) = \frac{R_2(r - R_1)}{r(R_2 - R_1)} E_\theta^{\text{inc}}(r,\theta,\phi) \tag{7.638}$$

$$H_r(r,\theta,\phi) = \frac{R_2}{R_2 - R_1} H_r^{\text{inc}}(r,\theta,\phi) \tag{7.639}$$

$$H_\theta(r,\theta,\phi) = \frac{R_2(r - R_1)}{r(R_2 - R_1)} H_\theta^{\text{inc}}(r,\theta,\phi) \tag{7.640}$$

$$H_\phi(r,\theta,\phi) = \frac{R_2(r - R_1)}{r(R_2 - R_1)} H_\theta^{\text{inc}}(r,\theta,\phi) \tag{7.641}$$

and describe the fields in the interior of the shell ($R_1 \le r \le R_2$).

It follows from (7.637), (7.638), (7.640), and (7.641) that the tangential field components E_θ, E_ϕ, H_θ, and H_ϕ are continuous on the exterior boundary of the shell ($r = R_2$) and vanish on the interior boundary ($r = R_1$). The latter implies that the incident field does not penetrate into the spherical volume of radius R_1 enclosed by the shell. Objects placed in this volume are fully shielded by the cloak and since the cloak does not scatter, are electromagnetically undetectable from the exterior of the cloak. Thus, a spherical shell $R_1 \le r \le R_2$ with material parameters (7.616) may be considered as an invisibility cloak that hides objects placed in its interior. Physically this means that the shell guides any incident field around its interior in such a way that beyond the cloak the incident field is fully restored and propagates undistorted farther away from the cloak as if no shell were present. Figure 7.86 illustrates this by showing the simulation results for the power flow in the interior and exterior of a spherical cloak illuminated by a plane wave.

The required permittivity and permeability of the cloak material are tensors and functions of the position (see (7.598) and (7.616)). In contrast to the θ and ϕ components of the tensors, the components ϵ_r and μ_r are smaller than ϵ and μ, vanishing on the interior boundary of the cloak ($r = R_1$), which means a greater wave speed in the cloak than in the surrounding medium. There is no natural material that can be used to build the cloak, but it does seem possible to build the cloak by using metamaterials, which are artificial materials with engineered electromagnetic parameters. An experimental realization of a metamaterial two-dimensional version of the cloak has been reported in Schurig *et al.* (2006). A problem inherit to metamaterials is their resonant behavior so that any metamaterial realization will unavoidably be limited to a specific frequency band, though the theoretical solution is valid at all frequencies.

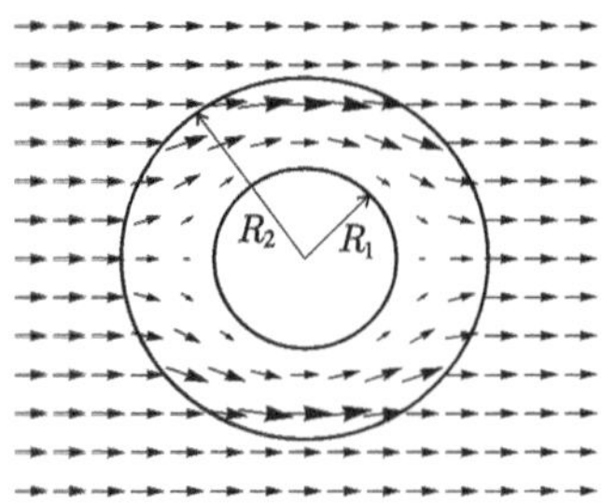

Figure 7.86 Vector plot of the Pointing vector for a plane wave incident from the left on a spherical invisibility cloak.

The shape of the cloak is not limited to the spherical shell; arbitrarily shaped cloaks can be designed by finding a coordinate transformation that transforms the given shape into the homogeneous sphere and choosing the material parameters according to (7.591).

Cloaks with relaxed requirements on $\overline{\overline{\epsilon}}$ and $\overline{\overline{\mu}}$ can be designed by limiting the shielding quality of the cloak to the accuracy of GO (Leonhardt 2006). Another cloaking technique, which uses the so-called carpet cloak, may help to hide objects placed on a flat conductive surface (Li and Pendry 2008). Cloaks of these kinds can be made from non-magnetic isotropic gradient index dielectric materials, and since the frequency dispersion of the dielectrics can be made very small, they are more broadband in nature.

7.7 Problems

1. Prove the expressions (7.5)–(7.10) for the field components.

2. Prove the expressions (7.11) and (7.12) for the vector Helmholtz operators.

3. Prove that regardless of the material composition of a sphere, modal admittances y_n and impedances z_n ($n = 1, 2, \ldots$) are purely imaginary quantities when the sphere and the background medium are lossless.

   Hint: Use the definitions (7.98) and (7.99) of y_n and z_n and study the recurrence procedure for the determination of the parameters $\delta_n^{u,v}(a)$ from $\delta_n^{u,v}(a_{L-1})$ (section 7.2.2).

   Solution: It follows from (7.76) and (7.82) that if ϵ_L and μ_L are real-valued, then $\delta_n^{u,v}(a_{L-1})$ are also real-valued, as are functions $\psi_n(k_L a_{L-1})$ and $\psi_n'(k_L a_{L-1})$. The values of $\delta_n^{u,v}(a_{L-1})$ initialize the recursions (7.86) and (7.87) for determining $\delta_n^{u,v}(a_{l-1})$ from $\delta_n^{u,v}(a_l)$, starting with $l = L-1$ and ending with $l = 1$. The relations in (7.86) and (7.87) include the factors T_{ik}^{nl} with $i, k = 1, 2$, and it is seen from (7.92)–(7.95) and from the property (E.2) of the Hankel functions that $(T_{ik}^{nl})^* = -T_{ik}^{nl}$, implying that $\operatorname{Re} T_{ik}^{nl} = 0$. A consequence of this is that if $\delta_n^{u,v}(a_l)$, ϵ_l, and μ_l are real-valued, then $\operatorname{Im} \delta_n^{u,v}(a_{l-1}) = 0$. This means that if all spherical layers are lossless, then $\delta_n^{u,v}(a)$ are real-valued, and we have from (7.98) and (7.99) that $\operatorname{Re} y_n = \operatorname{Re} z_n = 0$, provided that the background medium is lossless too.

4. Derive Debye's representations (7.161) and (7.162) for a homogeneous penetrable sphere by direct expansion of the reflection coefficients (7.145) and (7.146).

   Hint: Since $\tilde{r}_n^u$ and $\tilde{r}_n^v$ are related by duality ($\epsilon \leftrightarrow \mu$), it is sufficient to find the transformation for one of them. The expansion is based on the principle that the ratios $|\zeta_n^{(2)}(k_1 a)/\zeta_n^{(1)}(k_1 a)|$ and $|\zeta_n^{(2)\prime}(k_1 a)/\zeta_n^{(1)\prime}(k_1 a)|$ are "small", which is obvious when $|\operatorname{Im} k_1| a > 1$ and the waves in the interior of the sphere get significantly attenuated between two subsequent reflections.

   Solution: Consider, for example, $\tilde{r}_n^u$ and express (7.145) as

$$\tilde{r}_n^u = -\frac{\zeta_n^{(1)}(ka)\psi_n'(k_1 a) - (Z/Z_1)\zeta_n^{(1)\prime}(ka)\psi_n(k_1 a)}{\zeta_n^{(2)}(ka)\psi_n'(k_1 a) - (Z/Z_1)\zeta_n^{(2)\prime}(ka)\psi_n(k_1 a)}. \tag{7.642}$$

In the denominator, $\psi_n(k_1a)$ is replaced through $\zeta_n^{(1)}(k_1a)$ and $\zeta_n^{(2)}(k_1a)$, and the dominant terms with $\zeta_n^{(1)}(k_1a)$ and $\zeta_n^{(1)\prime}(k_1a)$ are factored out to yield

$$\tilde{r}_n^u = -\frac{\zeta_n^{(1)}(ka)\psi_n'(k_1a) - (Z/Z_1)\zeta_n^{(1)\prime}(ka)\psi_n(k_1a)}{\zeta_n^{(2)}(ka)\zeta_n^{(1)\prime}(k_1a) - (Z/Z_1)\zeta_n^{(2)\prime}(ka)\zeta_n^{(1)}(k_1a)} \frac{2}{1 - R_{u,n}^{11}}, \tag{7.643}$$

where $R_{u,n}^{11}$ is given by (7.158). Replacing $\psi_n(k_1a)$ in the numerator of (7.643) through $\zeta_n^{(1)}(k_1a)$ and $\zeta_n^{(2)}(k_1a)$, using the definition (7.152) of $R_{u,n}^{00}$, and expanding the outermost right-hand factor in (7.643) in a geometric progression in powers of $R_{u,n}^{11}$, one arrives at the expression

$$\tilde{r}_n^u = \left[R_{u,n}^{00} - \frac{\zeta_n^{(1)}(ka)\zeta_n^{(2)\prime}(k_1a) - (Z/Z_1)\zeta_n^{(1)\prime}(ka)\zeta_n^{(2)}(k_1a)}{\zeta_n^{(2)}(ka)\zeta_n^{(1)\prime}(k_1a) - (Z/Z_1)\zeta_n^{(2)\prime}(ka)\zeta_n^{(1)}(k_1a)} \right] \sum_{p=0}^{\infty} \left(R_{u,n}^{11} \right)^p. \tag{7.644}$$

The next step consists of rewriting (7.644) as

$$\tilde{r}_n^u = R_{u,n}^{00} + \left[R_{u,n}^{00} R_{u,n}^{11} - \frac{\zeta_n^{(1)}(ka)\zeta_n^{(2)\prime}(k_1a) - (Z/Z_1)\zeta_n^{(1)\prime}(ka)\zeta_n^{(2)}(k_1a)}{\zeta_n^{(2)}(ka)\zeta_n^{(1)\prime}(k_1a) - (Z/Z_1)\zeta_n^{(2)\prime}(ka)\zeta_n^{(1)}(k_1a)} \right] \sum_{p=0}^{\infty} \left(R_{u,n}^{11} \right)^p. \tag{7.645}$$

By using (7.152), (7.158), and the Wronskian relation (7.45) the expression in the square brackets in (7.645) can be dramatically simplified and reduced to the product of the transmission coefficients (7.154) and (7.159),

$$\begin{aligned}[\ldots] &= \frac{4}{\zeta_n^{(1)}(k_1a)\zeta_n^{(2)\prime}(ka) - (Z_1/Z)\zeta_n^{(1)\prime}(k_1a)\zeta_n^{(2)}(ka)} \cdot \frac{1}{\zeta_n^{(2)}(ka)\zeta_n^{(1)\prime}(k_1a) - (Z/Z_1)\zeta_n^{(2)\prime}(ka)\zeta_n^{(1)}(k_1a)} \\ &= T_{u,n}^{01} T_{u,n}^{10},\end{aligned} \tag{7.646}$$

which results in the formula (7.161).

5. Derive formula (7.192) for the poles of the vertical branch in a vicinity of the point $\nu = ka$ when $ka \gg 1$.
Solution: When $\nu \approx ka$, the factor $\sqrt{1 - (\nu/ka)^2}$ in front of the cotangent-function is small, implying that the zeros ν_m of (7.191) should approach the zeros $\nu_m^{(0)}$ of $\sin[\Phi(\nu, ka) - \pi/4]$. Close to $\nu = ka$, the function $\Phi(\nu, ka)$ can be approximated by (E.106), and when $\nu - ka = |\nu - ka| \exp(-j\pi/3)$,

$$\Phi(\nu, ka) \approx -\frac{2\sqrt{2}}{3\sqrt{ka}} |\nu - ka|^{3/2} \tag{7.647}$$

so that

$$\nu_m^{(0)} = ka + e^{-j\frac{\pi}{3}} \left(\frac{ka}{2} \right)^{1/3} \left[\frac{3\pi}{2} (m - \frac{1}{4}) \right]^{2/3} \tag{7.648}$$

with $m = 1, 2, \ldots$

To improve the solution, we now represent it as $\nu_m = \nu_m^{(0)} + \delta_m$ where δ_m is a correction. If δ_m is small, then

$$\tan \left[\Phi(\nu, ka) - \frac{\pi}{4} \right] \approx -\delta_m \arccos \left(\frac{\nu_m^{(0)}}{ka} \right), \tag{7.649}$$

and it follows from (7.191) that

$$\delta_m \approx \frac{-jZ\sqrt{1 - (\nu_m^{(0)}/ka)^2}}{Z_1 \arccos(\nu_m^{(0)}/ka)\sqrt{1 - (\nu_m^{(0)}/k_1a)^2}}. \tag{7.650}$$

Simplifying in (7.650) according to

$$\frac{\sqrt{1-(\nu_m^{(0)}/ka)^2}}{\arccos(\nu_m^{(0)}/ka)} \approx 1, \quad \sqrt{1-(\nu_m^{(0)}/k_1 a)^2} \approx \frac{1}{N}\sqrt{N^2-1}, \tag{7.651}$$

where $N = k_1/k$, results in the formula

$$\delta_m \approx \frac{-jNZ}{Z_1\sqrt{N^2-1}} = \frac{-j\epsilon_r}{\sqrt{\epsilon_r\mu_r - 1}}. \tag{7.652}$$

6. Show that the distance between two neighboring poles of the horizontal branch (section 7.3.4) can be estimated by (7.203).
Solution: Use the Taylor approximation (7.200) and write

$$\Phi(\nu_{m+1}, k_1 a) \approx \Phi(\nu_m, k_1 a) + (\nu_{m+1} - \nu_m)\arccos\frac{\nu_m}{k_1 a}. \tag{7.653}$$

The estimation (7.203) follows from (7.198).

7. A wave of the form $\exp(jkz)$ arrives at a point $P_{in}(a, \theta_{in})$ on the surface of a sphere of radius a (azimuthal coordinate ϕ is the same for all the points and is omitted). From P_{in} it propagates as a wave of the form $\exp(-j\nu\theta)$ along the surface of the sphere up to a point $P_{out}(a, \theta_{out})$, at which it gets launched toward an observation point $P(r, \theta)$ as a wave of the form $\exp(-jk|P_{out}P|)$, where $|P_{out}P|$ is the distance between the launching and observation points. Calculate the phase of the wave at P in terms of the attachment angle θ_0, the launching angle θ_1, and the arrival angle θ_2 (Figure 7.87). Specialize the answer to the case of a creeping wave with $\nu = ka\sin\theta_m$ and $\theta_0 = \theta_1 = \theta_m$.
Solution: The phase of the incident wave at P_{in} is equal to $ka\cos\theta_{in}$; the phase difference between P_{in} and P_{out} is equal to $\nu(\theta_{out} - \theta_{in})$ and between P_{out} and P to $k|P_{out}P|$. Thus, at P the phase of the wave is equal to

$$ka\cos\theta_{in} + \nu(\theta_{in} - \theta_{out}) - k|P_{out}P|. \tag{7.654}$$

At the attachment point, $\theta_{in} = \theta_0$ since the incident wave propagates parallel to the z axis. From triangle POP_{out}, as the sum of angles in every triangle is equal to π, $\theta_{out} = \theta - \theta_1 + \theta_2$. From the right triangle POO′, $|P_{out}P| = r\cos\theta_2 - a\cos\theta_1$. Inserting these in (7.654) gives

$$ka\cos\theta_0 + \nu(\theta_0 - \theta + \theta_1 - \theta_2) - kr\cos\theta_2 + ka\cos\theta_1. \tag{7.655}$$

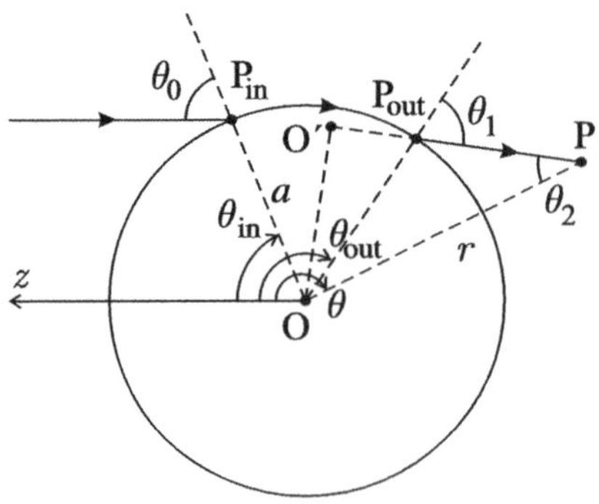

Figure 7.87 A creeping wave on the surface of a sphere.

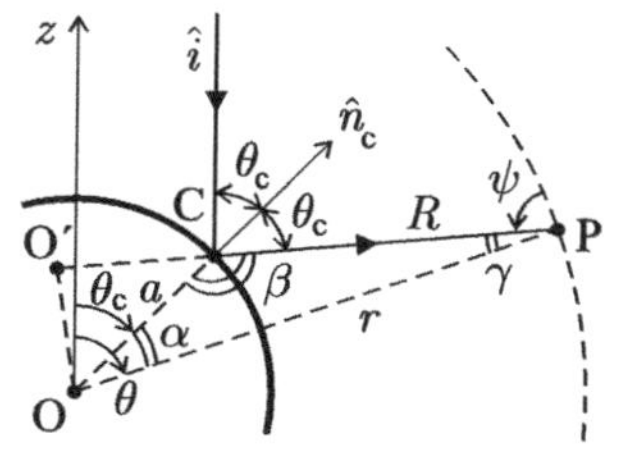

Figure 7.88 Reflection from a sphere: O is the center of the sphere, C is the point of reflection, P is the observation point.

A creeping wave with $\nu = ka\sin\theta_m$ and $\theta_0 = \theta_1 = \theta_m$ is therefore described by the exponential function

$$\exp[2jka\cos\theta_m + jka\sin\theta_m(2\theta_m - \theta - \theta_2) - jkr\cos\theta_2]. \tag{7.656}$$

8. Consider reflection of a plane wave from an impenetrable sphere (Figure 7.88). Assuming that the radius a of the sphere and the coordinates (r, θ, ϕ) of the observer are known, determine the coordinates (a, θ_c, ϕ_c) of the reflection point and the distance between the reflection point and the observer.

Solution: Assuming the incidence along the z axis and referring to the rotational symmetry of the geometry, we find that the reflection point lies in the plane defined by the incidence and scattering directions, i.e. $\phi_c = \phi$.

To determine the incidence angle θ_c at the reflection point, consider the triangle OPC formed by the center of the sphere and the reflection and observation points. The parameters ψ and θ_c define internal angles of the triangle as

$$\alpha = \theta - \theta_c, \quad \beta = \pi - \theta_c, \quad \gamma = \frac{\pi}{2} - \psi \tag{7.657}$$

and since in every plane triangle $\alpha + \beta + \gamma = \pi$, the following relation is satisfied:

$$\psi + 2\theta_c = \frac{\pi}{2} + \theta. \tag{7.658}$$

The sine rule $r/\sin\beta = a/\sin\gamma$ can be written as

$$r\cos\psi = a\sin\theta_c \tag{7.659}$$

and using (7.658), we arrive at

$$\sin(2\theta_c - \theta) = \frac{a}{r}\sin\theta_c, \tag{7.660}$$

from which θ_c can be determined as a zero on the interval $\theta/2 \le \theta_c \le \theta$.

Once θ_c is known, the angles α, β, γ, and ψ are determined from (7.657) and (7.658). The length R of the side CP is then easily calculated. We note a useful representation

$$R = r\sin\psi - a\cos\theta_c, \tag{7.661}$$

which is immediately seen from the right triangle OO′P.

9. Prove the relation (7.275) between the injection and ejection angles.

Solution: Rewriting the stationary point equation (7.264) as

$$\pm\theta = \pm\pi - \frac{\pi}{2} - 2\pi q + \arccos\left(\frac{\nu_c}{kr}\right) - 2\arccos\left(\frac{\nu_c}{ka}\right) + 2(p+1)\arccos\left(\frac{\nu_c}{k_1 a}\right) \tag{7.662}$$

and taking the derivative of the identity with respect to θ_c give

$$\pm\frac{\partial\theta}{\partial\theta_c} = \frac{\partial v_c}{\partial\theta_c}\frac{\partial}{\partial v_c}\left[\arccos\left(\frac{v_c}{kr}\right) - 2\arccos\left(\frac{v_c}{ka}\right) + 2(p+1)\arccos\left(\frac{v_c}{k_1 a}\right)\right], \tag{7.663}$$

which transforms to the formula

$$\pm\frac{\partial\theta}{\partial\theta_c} = ka\cos\theta_c\frac{\partial^2 S_{pq}^{\pm}}{\partial v^2}(v_c, k) \tag{7.664}$$

on using the relations $\partial v_c/(\partial\theta_c) = ka\cos\theta_c$ and (7.270).

10. Prove relations (7.290) and (7.291).

Solution: When $\theta \to 0$, because of (7.286) the formula (7.279) reduces to

$$\mathbf{E}_{sc} \to E_0 S_1(0)\frac{e^{-jkr}}{kr}(\hat{\theta}\cos\phi - \hat{\phi}\sin\phi) \tag{7.665}$$

and using (A.51) and (A.52), we obtain

$$\hat{\theta}\cos\phi - \hat{\phi}\sin\phi \to \hat{x}, \tag{7.666}$$

which transforms (7.665) to (7.290).

Similarly, in the limit $\theta \to \pi$, by using (7.287) and the relation

$$\hat{\theta}\cos\phi + \hat{\phi}\sin\phi \to -\hat{x}, \tag{7.667}$$

one shows that (7.291) results from (7.279).

11. Derive the formula (7.299).

Hint: Multiply the differential equation for $P_{n1}(\cos\theta)$ with $P_{n'1}(\cos\theta)\sin\theta$ and integrate from $\theta = 0$ to $\theta = \pi$.

Solution: The differential equation (7.22) with $\alpha = 1$ and $\beta = n$ reads

$$\frac{1}{\sin\theta}\frac{d}{d\theta}\left(\sin\theta\frac{d}{d\theta}P_{n1}\right) = \left[\frac{1}{\sin^2\theta} - n(n+1)\right]P_{n1}(\cos\theta). \tag{7.668}$$

Multiplying with $P_{n'1}(\cos\theta)\sin\theta$ and integrating from $\theta = 0$ to $\theta = \pi$ give the relations

$$\int_0^\pi \left[\frac{1}{\sin^2\theta} - n(n+1)\right]P_{n1}(\cos\theta)P_{n'1}(\cos\theta)\sin\theta\,d\theta = \int_0^\pi P_{n'1}(\cos\theta)\frac{d}{d\theta}\left(\sin\theta\frac{d}{d\theta}P_{n1}\right)d\theta$$
$$= -\int_0^\pi \frac{dP_{n1}}{d\theta}\frac{dP_{n'1}}{d\theta}\sin\theta\,d\theta, \tag{7.669}$$

where in (7.669) we have integrated by parts and used the properties of the associated Legendre polynomials at $\theta = 0, \pi$. Thus,

$$\int_0^\pi \left[\frac{P_{n1}P_{n'1}}{\sin^2\theta} + \frac{dP_{n1}}{d\theta}\frac{dP_{n'1}}{d\theta}\right]\sin\theta\,d\theta = n(n+1)\int_0^\pi P_{n1}(\cos\theta)P_{n'1}(\cos\theta)\sin\theta\,d\theta = 2\delta_{nn'}n^2\frac{(n+1)^2}{2n+1} \tag{7.670}$$

because of the orthogonality relation (7.26) with $m = 1$.

12. Derive formulas (7.302) and (7.303).
Hint: Use the relation

$$\mathrm{Re}\, Z + |Z|^2 = \frac{1}{4}(|2Z + 1|^2 - 1), \tag{7.671}$$

which is valid for every complex number Z.

13. Derive the asymptotic formulas (7.324) and (7.325).
Hint: Specialize the formula (8.206) to the particular case of a sphere and the incident wave (7.3).
Solution: The Gaussian curvature of a sphere

$$|K_c| = \frac{1}{a^2} \tag{7.672}$$

is the same at every point on the spherical surface. Because the surface of a sphere is convex, the reflected ray intersects no caustics on the way from the reflection point to the observer and, therefore,

$$\sigma_c^\infty = -j. \tag{7.673}$$

In spherical coordinates with the polar axis along the incidence direction, the directions of incidence and scattering are

$$\hat{\imath} = -\hat{z}, \quad \hat{s} = \hat{x}\sin\theta\cos\phi + \hat{y}\sin\theta\sin\phi + \hat{z}\cos\theta. \tag{7.674}$$

The point of reflection is located at

$$\mathbf{r}_c = \hat{x}\sin\frac{\theta}{2}\cos\phi + \hat{y}\sin\frac{\theta}{2}\sin\phi + \hat{z}\cos\frac{\theta}{2} \tag{7.675}$$

and the unit normal to the surface at the reflection point is

$$\hat{n}_c = \frac{1}{a}\mathbf{r}_c. \tag{7.676}$$

The bases of the incident and reflected rays are

$$\hat{\alpha}_{\mathrm{inc}} = \hat{\alpha}_{\mathrm{ref}} = \hat{x}\sin\phi - \hat{y}\cos\phi = -\hat{\phi} \tag{7.677}$$

$$\hat{\beta}_{\mathrm{inc}} = -\hat{x}\cos\phi - \hat{y}\sin\phi \tag{7.678}$$

$$\hat{\beta}_{\mathrm{ref}} = \hat{x}\cos\theta\cos\phi + \hat{y}\cos\theta\sin\phi - \hat{z}\sin\theta = \hat{\theta} \tag{7.679}$$

and, therefore,

$$E_{\mathrm{inc}}^{\perp}(\mathbf{r}_c) = \hat{\alpha}_{\mathrm{inc}} \cdot \mathbf{E}_{\mathrm{inc}}(\mathbf{r}_c) = E_0 \sin\phi\, \mathrm{e}^{jka\cos(\theta/2)} \tag{7.680}$$

$$E_{\mathrm{inc}}^{\|}(\mathbf{r}_c) = \hat{\beta}_{\mathrm{inc}} \cdot \mathbf{E}_{\mathrm{inc}}(\mathbf{r}_c) = -E_0 \cos\phi\, \mathrm{e}^{jka\cos(\theta/2)}, \tag{7.681}$$

where we have used that $z_c = a\cos(\theta/2)$.
Evaluating the scalar products

$$\hat{\theta}\cdot\hat{\alpha}_{\mathrm{ref}} = 0, \quad \hat{\theta}\cdot\hat{\beta}_{\mathrm{ref}} = 1, \quad \hat{\phi}\cdot\hat{\alpha}_{\mathrm{ref}} = -1, \quad \hat{\phi}\cdot\hat{\beta}_{\mathrm{ref}} = 0 \tag{7.682}$$

and using (7.680) and (7.681) transform (8.206) to

$$\begin{bmatrix} E_\theta^{\rm sc}(\mathbf{r}) \\ E_\phi^{\rm sc}(\mathbf{r}) \end{bmatrix} = E_0 \frac{a}{2r} \mathrm{e}^{-jkr+2jka\cos(\theta/2)} \begin{pmatrix} 0 & 1 \\ -1 & 0 \end{pmatrix} \begin{bmatrix} R_\perp \sin\phi \\ -R_\parallel \cos\phi \end{bmatrix} \tag{7.683}$$

since $\hat{s} \cdot \mathbf{r}_c = a\cos(\theta/2)$. It follows from (7.683) that the scattered field can be expressed as

$$\mathbf{E}_{\rm sc} = -E_0 \frac{\mathrm{e}^{-jkr}}{kr} \mathrm{e}^{2jka\cos(\theta/2)} \frac{ka}{2} (\hat{\theta} R_\parallel \cos\phi + \hat{\phi} R_\perp \sin\phi), \tag{7.684}$$

where the argument of reflection coefficients $R_\parallel$ and $R_\perp$ is equal to $\theta/2$, which is the incidence angle at the point of reflection. Comparing (7.684) with (7.279) gives the asymptotic formulas (7.324) and (7.325).

14. Prove the properties (7.372) and (7.373) of the coefficients of the low-frequency expansions (7.356)–(7.359) for lossless spheres.
Hint: Use the fact that the modal admittances and impedances, y_n and z_n, of any lossless sphere are purely imaginary (problem 3).
Solution: In the limit of a lossless sphere it follows from (7.343) that $\tilde{y}_{1,2}$ and $\tilde{z}_{1,2}$ are real-valued and, therefore, from (7.367)–(7.369) that the coefficients α_1^u, α_1^v, γ_1^u, γ_1^v, γ_2^u, and γ_2^v are purely imaginary, which leads to the property (7.372). Since α_1^u and α_1^v are purely imaginary quantities, their squares are negative real numbers, which proves the property (7.373).

15. Show that the dipole polarizabilities P_{11} and M_{11} of a PEC sphere comply with the general formula (3.308).
Hint: Use the formulas (7.418) and (7.419).

16. By specializing formulas (7.367), (7.368), (7.369), (7.370), and (7.371) to the case of a PEC sphere prove that $\sigma_{\rm abs} = 0$.
Hint:

$$\gamma_2^u = -\frac{j}{30}, \quad \gamma_2^v = \frac{j}{45}. \tag{7.685}$$

17. Show that $\mathrm{Im}\,(1/P_{11})$ and $\mathrm{Im}\,(1/M_{11})$ of a lossless ($\epsilon_r'' = 0$, $\mu_r'' = 0$) homogeneous sphere comply with the general formulas (3.308).
Hint: Note that

$$\mathrm{Im}\frac{1}{P_{11}} = -\frac{\mathrm{Im}\,P_{11}}{|P_{11}|^2}, \quad \mathrm{Im}\frac{1}{M_{11}} = -\frac{\mathrm{Im}\,M_{11}}{|M_{11}|^2} \tag{7.686}$$

and use expansions (7.374) and (7.375) for P_{11} and M_{11} with the coefficients (7.510)–(7.512).
Solution: Consider, for example, P_{11}; the case of M_{11} is studied in a similar way. It follows from (7.374) that

$$|P_{11}|^2 = (6\pi a^3)^2 |\alpha_1^u|^2 \tag{7.687}$$

and

$$\mathrm{Im}\,P_{11} = -6\pi k^3 a^6 |\alpha_1^u|^2 \tag{7.688}$$

since $\mathrm{Re}\,\alpha_1^u = \mathrm{Re}\,\beta_1^u = \mathrm{Re}\,\gamma_1^u = 0$ when ϵ_r and μ_r are purely real. Thus,

$$\mathrm{Im}\frac{1}{P_{11}} = \frac{k^3}{6\pi}. \tag{7.689}$$

18. Determine M_{11} of a homogeneous non-magnetic ($\mu_r = 1$) dielectric sphere. Determine $\mathrm{Im}\, M_{11}$ for a lossless non-magnetic sphere ($\mathrm{Im}\epsilon_r = 0$).

Solution: Letting $\mu_r \to 1$ in the expansion coefficients (7.510)–(7.512) transforms the low-frequency expansion (7.375) of the magnetic polarizability of a dielectric sphere to the formula

$$M_{11} = -\frac{2\pi}{15}(\epsilon_r - 1)k^2a^5. \tag{7.690}$$

This polarizability is real-valued for lossless media and to obtain the imaginary part of M_{11}, higher-order terms in the expansion are necessary. A direct derivation of the imaginary part of M_{11} would require expanding r_1^{ν} in a series up to the order $(ka)^{10}$. This can be avoided by using the general relation (3.308), and the result reads

$$\mathrm{Im}\, M_{11} = \frac{6\pi}{2025}(\epsilon_r - 1)^2k^7a^{10}. \tag{7.691}$$

19. Show that electrically small homogeneous spheres with $\epsilon_r = 1 - 3j$ exhibit the maximum absorption in the class of materials with $\mathrm{Re}\,\epsilon_r \geq 1$ and $\mu_r = 1$.

Solution: For such spheres, $\sigma_{\mathrm{abs}} \approx \sigma_{\mathrm{ext}}$ and $|M_{11}| \ll |P_{11}|$, and it follows from (7.353) and (7.497) that

$$\sigma_{\mathrm{abs}} \approx \frac{12\pi k a^3 \epsilon_r''}{(\epsilon_r' + 1)^2 + (\epsilon_r'')^2}. \tag{7.692}$$

It is easy to see that when $\epsilon_r' = 1$ and $\epsilon_r'' = 3$, the expression is at its maximum $2\pi k a^3$.

20. Assume that the material parameters of a homogeneous magneto-dielectric sphere are subject to the Kerker conditions (7.499). Calculate the extinction cross section of the sphere.

Hint: Note that (7.499) can be satisfied only with real-valued values of ϵ_r and μ_r.

Solution: Separating the real and imaginary parts according to $\epsilon_r = \epsilon_r' - j\epsilon_r''$ and $\mu_r = \mu_r' - j\mu_r''$, the imaginary part of (7.499) is given by the formula

$$-\epsilon_r'' = \frac{9\mu_r''}{(1 + 2\mu_r')^2 + 4(\mu_r'')^2}, \tag{7.693}$$

which implies that $\epsilon_r'' = 0$ and $\mu_r'' = 0$ since ϵ_r'' and μ_r'' cannot be negative in a passive medium. The extinction cross section is therefore equal to the total scattering cross section, and using (7.354) with (7.497) gives

$$\sigma_{\mathrm{ext}} = \sigma_{\mathrm{T}} = \frac{16}{3}\pi k^4 a^6 \left(\frac{\mu_r - 1}{\mu_r + 2}\right)^2. \tag{7.694}$$

The formula (7.694) can be also obtained by inserting the higher-order approximations (7.374) and (7.375) for P_{11} and M_{11} in the formula (7.353) on taking into account that ϵ_r and μ_r are real-valued. Yet another way is to use (7.370) with (7.510)–(7.513).

21. Show that solutions of the equation

$$Q_n(w) = -\frac{n+1}{n} \tag{7.695}$$

with integer n and $Q_n(w)$ defined in (7.516) coincide with w_{n-1}, the zeros of the Riccati–Bessel functions $\psi_{n-1}(w)$.

Hint: Use the recurrence relation (e.g., Abramowitz and Stegun (1965, 10.1.21))

$$\frac{n+1}{w}j_n(w) + \frac{\mathrm{d}j_n(w)}{\mathrm{d}w} = j_{n-1}(w) \tag{7.696}$$

and the relation $\psi_n(w) = wj_n(w)$ between the Riccati–Bessel and spherical Bessel functions.

22. Study the ranges of material permittivities for which the polarizability of a small coated dielectric sphere can be designed to be zero (see (7.582)). Discuss the conditions under which the cover can be made thin as compared with the sphere radius.

Answer: $0 < a_1/a < 1$ if the parameters $\epsilon_{1r} = \epsilon_1/\epsilon$ and $\epsilon_{2r} = \epsilon_2/\epsilon$ are purely real and belong to any of the following five regions on the plane $(\epsilon_{1r}, \epsilon_{2r})$:

$$\epsilon_{1r} < -\frac{1}{2}, \quad 1 < \epsilon_{2r} < -2\epsilon_{1r} \tag{7.697}$$

$$-\frac{1}{2} < \epsilon_{1r} < 0, \quad -2\epsilon_{1r} < \epsilon_{2r} < 1 \tag{7.698}$$

$$0 < \epsilon_{1r} < 1, \quad \epsilon_{2r} < -2\epsilon_{1r} \tag{7.699}$$

$$0 < \epsilon_{1r} < 1, \quad \epsilon_{2r} > 1 \tag{7.700}$$

$$\epsilon_{1r} > 1, \quad -2\epsilon_{1r} < \epsilon_{2r} < 1. \tag{7.701}$$

If the background medium is denser than vacuum ($\epsilon > \epsilon_0$), then conditions (7.700) and (7.701) can be satisfied with conventional natural materials: (7.700) with $\epsilon_0 < \epsilon_1 < \epsilon < \epsilon_2$ and (7.701) with $\epsilon_0 < \epsilon_2 < \epsilon < \epsilon_1$ (see Figure 7.84).

23. Small PEC spheres coated with a concentric dielectric layer have a quasi-static electric dipole resonance when $\epsilon_r \to -2\alpha/\gamma$ (see (7.461)). Derive the resonance approximation for P_{11}, assuming a non-magnetic sphere ($\mu_r = 1$).

Hint: Consider P_{11} as a function of a complex variable ϵ_r, determine the imaginary part of the pole $\epsilon_{r\infty}$, and use the resonance formula (7.386).

Answer: Using quasi-static approximations for the Riccati–Bessel functions in the electric dipole coefficient r_1^u gives

$$\epsilon_{r\infty} = 2\frac{a_{10}^3 - 1}{1 + 2a_{10}^3}\left[1 + \frac{3(2 + 10a_{10}^3 + 3a_{10}^5)}{5(1 + 2a_{10}^3)^2}(ka)^2 - j(ka)^3\right], \tag{7.702}$$

and it follows from (7.348) and (7.386) with

$$\epsilon''_{r\infty} = 2\frac{a_{10}^3 - 1}{1 + 2a_{10}^3}(ka)^3 \tag{7.703}$$

that

$$P_{11} \approx \frac{12\pi a^3(a_{10}^3 - 1)}{(1 + 2a_{10}^3)(\epsilon_r - \epsilon_{r\infty})}. \tag{7.704}$$

24. Consider the plasmonic resonance of an empty spherical shell made of silver. Assume that the sphere radius is small compared with the free-space wavelength. Using the Drude model of silver permittivity (2.23), study the frequency dependence of the shell polarizability. Compare the results with a solid silver sphere of the same diameter.

25. Study the caustics of the internally reflected rays for a homogeneous electrically large sphere with real-valued refractive index $n = k_1/k$. Show that there is no caustic if $p < n - 1$ and just one otherwise.

Solution: The caustics correspond to zeros of the function $1 - (p + 1)\cos\theta_c/(n\cos\theta_1)$. The function is a monotonous function of θ_c, equal to unity at $\theta_c = \pi/2$ and to $1 - (p + 1)k/k_1$ at $\theta_c = 0$ (see (7.551)). So, if $p < n - 1$, then the function is always positive and has therefore no zeros on the range $0 \leq \theta_c \leq \pi/2$. If $p > n - 1$, then there is one zero of the function.

26. Assume that a sphere is illuminated by a natural light which is a superposition of non-coherent, linearly polarized waves with randomly distributed orientations of polarization. Prove the relation

$$\bar{\sigma}_\psi = \frac{1}{2}[\sigma_E(\theta) + \sigma_H(\theta)], \tag{7.705}$$

where $\bar{\sigma}_\psi$ denotes the average of the bistatic scattering cross section over the polarization angle ψ (see (3.359)).
Hint: Use the results (3.66), (3.360), (7.283), (7.284), and (7.311).

27. A linearly polarized plane wave is incident on the spherical invisibility cloak described in section 7.6.2. Derive the explicit expressions for the energy flow

$$\mathbf{w} = \frac{1}{2}\mathrm{Re}(\mathbf{E} \times \mathbf{H}^*)$$

in the whole space and show that (a) the energy flow into the interior of the cloak ($0 \leq r < R_1$) vanishes, (b) the radial component of the energy flow is continuous across the exterior boundary of the shell ($r = R_2$), (c) the distribution of the energy flow is the same in E and H planes (the E plane is the plane that includes the incidence direction and the electric field vector in the incident wave, and the H plane includes the incidence direction and the magnetic field vector in the incident wave).
Solution: Let the cloak in the form of a spherical shell ($R_1 \leq r \leq R_2$) be illuminated by the plane wave

$$\mathbf{E}_{\mathrm{inc}}(\mathbf{r}) = E_0 \hat{z} e^{-jkx}, \quad \mathbf{H}_{\mathrm{inc}}(\mathbf{r}) = -YE_0 \hat{y} e^{-jkx}, \tag{7.706}$$

where E_0 is a positive amplitude. The energy flow in the plane wave

$$\mathbf{w} = \frac{\hat{x}}{2} YE_0^2 \tag{7.707}$$

was obtained in section 2.5.1, and in the exterior of the cloak ($r > R_2$) it coincides with the energy flow in the total field (section 7.6.2).

To find the energy flow within the shell ($R_1 \leq r \leq R_2$), the solution (7.636)–(7.641) can be used. The field components in the spherical coordinates are as follows:

$$\begin{aligned}
E_r^{\mathrm{inc}} &= E_0 \cos\theta e^{-jkr\sin\theta\cos\phi}, \quad E_\theta^{\mathrm{inc}} = -E_0 \sin\theta e^{-jkr\sin\theta\cos\phi}, \quad E_\phi^{\mathrm{inc}} = 0 \\
H_r^{\mathrm{inc}} &= -YE_0 \sin\theta \sin\phi e^{-jkr\sin\theta\cos\phi}, \quad H_\theta^{\mathrm{inc}} = -YE_0 \cos\theta \sin\phi e^{-jkr\sin\theta\cos\phi} \\
H_\phi^{\mathrm{inc}} &= -YE_0 \cos\phi e^{-jkr\sin\theta\cos\phi}.
\end{aligned}$$

Representing the fields as

$$\mathbf{E} = \hat{r}E_r + \hat{\theta}E_\theta + \hat{\phi}E_\phi, \quad \mathbf{H} = \hat{r}H_r + \hat{\theta}H_\theta + \hat{\phi}H_\phi \tag{7.708}$$

and using the relations

$$\hat{r} \times \hat{\theta} = \hat{\phi}, \quad \hat{\theta} \times \hat{\phi} = \hat{r}, \quad \hat{r} \times \hat{\phi} = -\hat{\theta}, \tag{7.709}$$

one obtains

$$w_r = \frac{1}{2}\left(E_\theta H_\phi^* - E_\phi H_\theta^*\right) = \frac{1}{2} YE_0^2 \sin\theta \cos\phi \left(\frac{r'}{r}\right)^2 \tag{7.710}$$

$$w_\theta = \frac{1}{2}\left(E_\phi H_r^* - E_r H_\phi^*\right) = \frac{1}{2} YE_0^2 \cos\theta \cos\phi \frac{r' R_2}{r(R_2 - R_1)} \tag{7.711}$$

$$w_\phi = \frac{1}{2}\left(E_r H_\theta^* - E_\theta H_r^*\right) = -\frac{1}{2} YE_0^2 \sin\phi \frac{r' R_2}{r(R_2 - R_1)} \tag{7.712}$$

with

$$\frac{r'}{r} = \frac{R_2(r - R_1)}{r(R_2 - R_1)}. \tag{7.713}$$

Thus, the energy flow in the shell is given by the expression

$$\mathbf{w} = \frac{1}{2} Y E_0^2 \left(\frac{R_2}{R_2 - R_1} \right)^2 \left(1 - \frac{R_1}{r} \right) \left[\hat{r} \sin\theta \cos\phi \left(1 - \frac{R_1}{r} \right) + \hat{\theta} \cos\theta \cos\phi - \hat{\phi} \sin\phi \right]. \tag{7.714}$$

It is seen from this formula that $\mathbf{w} = 0$ when $r = R_1$ (no energy flow in the concealed volume) and that $w_r = w_r^{\text{inc}}$ when $r = R_2$ (the energy flow through the exterior boundary of the cloak is continuous).

The H plane corresponds to $\theta = \pi/2$ and $0 \le \phi < 2\pi$, which leads to the expression

$$\mathbf{w} = \frac{1}{2} Y E_0^2 \left(\frac{R_2}{R_2 - R_1} \right)^2 \left(1 - \frac{R_1}{r} \right) \left[\hat{x} \left(1 - \frac{R_1}{r} \cos^2\phi \right) - \hat{y} \frac{R_1}{r} \cos\phi \sin\phi \right]. \tag{7.715}$$

The energy flow in the E plane is obtained with $0 \le \theta \le \pi$ and $\phi = 0$ or π,

$$\mathbf{w} = \frac{1}{2} Y E_0^2 \left(\frac{R_2}{R_2 - R_1} \right)^2 \left(1 - \frac{R_1}{r} \right) \left[\hat{x} \left(1 - \frac{R_1}{r} \sin^2\theta \right) \mp \hat{z} \frac{R_1}{r} \cos\theta \sin\theta \right], \tag{7.716}$$

where the upper sign corresponds to $\phi = 0$ and the lower sign to $\phi = \pi$. It follows from these expressions that there is no difference in the energy flows in the E and H planes.

References

Abramowitz M and Stegun IA (eds) 1965 *Handbook of Mathematical Functions with Formulas, Graphs, and Mathematical Tables*, Dover Publications.

Adam JA 2002 The mathematical physics of rainbows and glories. *Phys. Rep.*, **356**, 229–365.

Aden AL and Kerker M 1951 Scattering of electromagnetic waves from two concentric spheres. *J. Appl. Phys.*, **22**, 1242–1246.

Airy GB 1838 On the intensity of light in the neighbourhood of a caustic. *Trans. Cambridge Phil. Soc.*, **6**, 379–402.

Alù A and Engheta N 2005a Achieving transparency with plasmonic and metamaterial coatings. *Phys. Rev. E*, **72**, 016623.

Alù A and Engheta N 2005b Polarizabilities and effective parameters for collections of spherical nanoparticles formed by pairs of concentric double-negative, single-negative, and/or double-positive metamaterial layers. *J. Appl. Phys.*, **97**, 094310.

Alù A and Engheta N 2010 How does zero forward-scattering in magnetodielectric nanoparticles comply with the optical theorem? *J. Nanophotonics*, **4**, 041590.

Babič VM and Buldyrev VS 1991 *Short-Wavelength Diffraction Theory*, Springer-Verlag.

Beckmann P and Franz W 1957 Berechnung der Streuquerschnitte von Kugel und Zylinder unter Anwendung einer modifizierten Watson-Transformation. *Z. Naturforsch. A*, **12**, 533–537.

Bohren CF and Huffman DR 1983 *Absorption and Scattering of Light by Small Particles*, Wiley-Interscience.

Born M and Wolf E (eds) 1959 *Principles of Optics: Electromagnetic Theory of Propagation, Interference and Diffraction of Light*, Pergamon Press.

Borovikov VA and Kinber BY 1994 *Geometrical Theory of Diffraction*, The Institution of Electrical Engineers, London.

Bouche D, Molinet F, and Mittra R 1997 *Asymptotic Methods in Electromagnetics*, Springer-Verlag.

Bouman JJ, Senior TBA, and Uslenghi PLE (eds) 1987 *Electromagnetic and Acoustic Scattering by Simple Shapes*, Hemisphere Publishing Corporation.

Boyer CB 1987 *The Rainbow, from Myth to Mathematics*, Princeton University Press.

Bremmer H 1949 *Terrestrial Radio Waves*, Elsevier.

Bromwich TJI'A 1920 The scattering of plane electric waves by spheres. *Phil. Trans. R. Soc. Lond. A*, **220**, 175–206.

Bryant HC and Cox AJ 1966 Mie theory and the glory. *J. Opt. Soc. Am.*, **56**, 1529–1532.

Chýlek P 1976 Partial-wave resonances and the ripple structure in the Mie normalized extinction cross section. *J. Opt. Soc. Am.*, **66**, 285–287.

Clebsch A 1863 Über die Reflexion an einer Kugelfläche. *J. Reine Angew. Math.*, **61**, 195–262.
Collin RE 1960 *Field Theory of Guided Waves*, McGraw-Hill.
Debye P 1908 Das elektromagnetische Feld um einen Zylinder und die Theorie des Regenbogens. *Physik. Z.*, **9**, 775–778.
Debye P 1909 Der Lichtdruck auf Kugeln von beliebigem Material. *Ann. Phys. (Berlin)*, **335**, 57–136.
Edens H 2015 Photographic observation of a natural fifth-order rainbow. *Appl. Opt.*, **54**, B26–B34.
Erdélyi A, Magnus W, Oberhettinger F, and Tricomi FG 1953 *Higher Transcendental Functions*, vol. 1, McGraw-Hill.
Fedorov AA 1958 Asymptotic solution of the problem of diffraction of plane electromagnetic waves on ideally conducting spheres. *Radio Eng. Electron. Phys.*, **3**, 57–72.
Fock VA 1965 *Electromagnetic Diffraction and Propagation Problems*, Pergamon Press.
Franz W 1957 *Theorie der Beugung elektromagnetischer Wellen*, Springer-Verlag.
Grandy WT 2000 *Scattering of Waves from Large Spheres*, Cambridge University Press.
Green MA and Keevers MJ 1995 Optical properties of intrinsic silicon at 300 K. *Progr. Photovoltaics: Res. Appl.*, **3**, 189–192.
Großmann M, Schmidt E, and Haußmann A 2011 Photographic evidence for the third-order rainbow. *Appl. Opt.*, **50**, F134–F141.
Güttler A 1952 Die Miesche Theorie der Beugung durch dielektrische Kugeln mit absorbierendem Kern und ihre Bedeutung für Probleme der interstellaren Materie und des atmosphärischen Aerosols. *Ann. Phys. (Berlin)*, **446**, 65–98.
Hobson EW 1931 *The Theory of Spherical and Ellipsoidal Harmonics*, Cambridge University Press.
James GL 1976 *Geometrical Theory of Diffraction for Electromagnetic Waves*, Peter Peregrinus, London.
Johnson PB and Christy RW 1972 Optical constants of noble metals. *Phys. Rev. B*, **6**, 4370–4379.
Jylhä L, Kolmakov I, Maslovski S, and Tretyakov S 2006 Modeling of isotropic backward-wave materials composed of resonant spheres. *J. Appl. Phys.*, **99**, 043102.
Keller JB 1962 Geometrical theory of diffraction. *J. Opt. Soc. Am.*, **52**, 116–130.
Kerker M 1969 *The Scattering of Light and Other Electromagnetic Radiation*, Academic Press.
Kerker M 1975 Invisible bodies. *J. Opt. Soc. Am.*, **65**, 376–379. Erratum: *ibid.*, 1085.
Kerker M, Wang DS, and Giles CL 1983 Electromagnetic scattering by magnetic spheres. *J. Opt. Soc. Am.*, **73**, 765–767.
Khare V and Nussenzveig HM 1974 Theory of the rainbow. *Phys. Rev. Lett.*, **33**, 976–980.
Khare V and Nussenzveig HM 1977 Theory of the glory. *Phys. Rev. Lett.*, **38**, 1279–1282.
Kong JA 1986 *Electromagnetic Wave Theory*, John Wiley & Sons.
Korn GA and Korn TM 1968 *Mathematical Handbook for Scientists and Engineers: Definitions, Theorems, and Formulas for Reference and Review*, 2nd revised edn, McGraw-Hill.
Kratzer A and Franz W 1960 *Transzendente Funktionen*, Akademische Verlagsgesellschaft Geest & Portig K.-G., Leipzig (in German).
Leonhardt U 2006 Optical conformal mapping. *Science*, **312**, 1777–1780.
Li J and Pendry JB 2008 Hiding under the carpet: a new strategy for cloaking. *Phys. Rev. Lett.*, **101**, 203901.
Logan NA 1965 Survey of some early studies of the scattering of plane waves by a sphere. *Proc. IEEE*, **53**, 773–785.
Lorenz LV 1890 Sur la lumière réfléchie et réfractée par une sphère transparante. *Vidensk. Selsk. Skr.*, **6**, 1–62 (French translation: *Œuvres Scientifiques de L. Lorenz*, vol 1, Librairie Lehmann & Stage, Copenhagen, 1898, pp. 405–502).
Love AEH 1899 The scattering of electric waves by a dielectric sphere. *Proc. London Math. Soc.*, **30**, 308–321.
Magnus W, Oberhettinger F, and Soni RP 1966 *Formulas and Theorems for the Special Functions of Mathematical Physics*. Springer-Verlag.
Makarov GI and Ryzhkov AV 1977 Problem of electromagnetic wave propagation over a conducting sphere and a study of spectral characteristics of the field. *Problems of Diffraction and Wave Propagation, Leningrad State University, USSR*, **15**, 3–28 (in Russian).
Makarov GI and Osipov AV 1987 Spectral representations of the Green function in the problem of electromagnetic wave diffraction by an imperfectly conducting sphere. *Problems of Diffraction and Wave Propagation, Leningrad State University, USSR*, **21**, 3–18 (in Russian).
Makarov GI, Novikov VV, and Rybachek ST 1991 *Electromagnetic Waves Propagation over Earth's Surface*, Nauka, Moscow (in Russian).
Marshall JS and Palmer WM 1948 The distribution of raindrops with size. *J. Meteorol.*, **5**, 165–166.
McNamara DA, Pistorius CWI, and Malherbe JAG 1990 *Introduction to the Uniform Geometrical Theory of Diffraction*, Artech House.
Metz HJ and Dettmar HK 1963 Zur Berechnung der Mieschen Streukoeffizienten für reelle Brechungsindizes. *Kolloid-Z. u. Z. Polymere*, **192**, 107–114.
Mevel J 1958 Étude de la structure détaillée des courbes de diffusion des ondes électromagnétiques par les sphères diélectriques. *J. Phys. Radium*, **19**, 630–636.
Mie G 1908 Beiträge zur Optik trüber Medien, speziell kolloidaler Metallösungen. *Ann. Phys. (Berlin)*, **330**, 377–445.
Molinet F, Andronov I, and Bouche D 2005 *Asymptotic and Hybrid Methods in Electromagnetics*, The Institution of Electrical Engineers, London.
Morits D and Simovski C 2012 Isotropic negative refractive index at near infrared. *J. Optics*, **14**, 125102 (7pp).
Morse PM and Feshbach H 1953 *Methods of Theoretical Physics*, McGraw-Hill.

Newton RG 1966 *Scattering Theory of Waves and Particles*, McGraw-Hill.
Nicholson JW 1911 The scattering of light by a large conducting sphere. *Proc. London Math. Soc.*, **9**, 67–80.
Nicholson JW 1913 The scattering of light by a large conducting sphere (2nd paper). *Proc. London Math. Soc.*, **11**, 277–284.
Nussenzveig HM 1965 High-frequency scattering by an impenetrable sphere. *Ann. Phys. (N.Y.)*, **34**, 23–95.
Nussenzveig HM 1969a High-frequency scattering by a transparent sphere. I. Direct reflection and transmission. *J. Math. Phys.*, **10**, 82–124.
Nussenzveig HM 1969b High-frequency scattering by a transparent sphere. II. Theory of the rainbow and the glory. *J. Math. Phys.*, **10**, 125–176.
Nussenzveig HM 1977 The theory of the rainbow. *SciAm*, **236**, 116–127.
Nussenzveig HM 1979 Complex angular momentum theory of the rainbow and the glory. *J. Opt. Soc. Am.*, **69**, 1068–1079.
Osipov AV 2015 Scattering cross sections of impedance-matched bodies. *IEEE Trans. Antennas Propag.*, **63**, 3122–3126.
Paniagua-Dominguez R, Lopez-Tejeira F, Marques R, and Sanchez-Gil JA 2011 Metallo-dielectric core-shell nanospheres as building blocks for optical three-dimensional isotropic negative-index metamaterials. *New J. Phys.*, **13**, 123017 (15pp).
Pendry JB, Schurig D, and Smith DR 2006 Controlling electromagnetic fields. *Science*, **312**, 1780–1782.
Penndorf RB 1962 Scattering and extinction coefficients for small absorbing and nonabsorbing aerosols. *J. Opt. Soc. Am.*, **52**, 896–902.
Press WH, Teukolsky SA, Vetterling WT, and Flannery BP 1994 *Numerical Recipies in FORTRAN: the Art of Scientific Computing*, Cambridge University Press.
Ruck GT, Barrick DE, Stuart WD, and Krichbaum CK 1970 *Radar Cross Section Handbook*, Plenum Press.
Ryshik IM and Gradstein IS 1957 *Tables of Series, Products, and Integrals*, VEB, Berlin.
Schurig D, Mock JJ, Justice BJ, Cummer SA, Pendry JB, Starr AF, and Smith DR 2006 Metamaterial electromagnetic cloak at microwave frequencies. *Science*, **314**, 977–980.
Schuller JA, Barnard ES, Cai W, Jun YC, White JS, and Brongersma ML 2010 Plasmonics for extreme light concentration and manipulation. *Nat. Mater.*, **9**, 193–204.
Senior TBA and Goodrich RF 1964 Scattering by a sphere. *Proc. IEE*, **111**, 907–916.
Sihvola A 1999 *Electromagnetic Mixing Formulas and Applications*, The Institution of Electrical Engineers, London.
Sommerfeld A 1935 Elektromagnetische Schwingungen, in *Die Differential- und Integralgleichungen der Mechanik und Physik*, Riemann B, Weber HM, Frank P, and von Mises R (eds), chapters 19–23, Vieweg.
Stratton JA 1941 *Electromagnetic Theory*, McGraw-Hill.
Theusner M 2011 Photographic observation of a natural fourth-order rainbow. *Appl. Opt.*, **50**, F129–F133.
Thomson JJ 1893 *Notes on Recent Researches in Electricity and Magnetism*, Clarendon Press, Oxford.
van de Hulst HC 1957 *Light Scattering by Small Particles*, John Wiley & Sons.
van der Pol B and Bremmer H 1937a The diffraction of electromagnetic waves from an electrical point source round a finitely conducting sphere, with applications to radiotelegraphy and the theory of the rainbow. Part I. *Philos. Mag. Ser. 7*, **24**, 141–176.
van der Pol B and Bremmer H 1937b The diffraction of electromagnetic waves from an electrical point source round a finitely conducting sphere, with applications to radiotelegraphy and the theory of the rainbow. Part II. *Philos. Mag. Ser. 7*, **24**, 825–864.
van der Pol B and Bremmer H 1938 The propagation of radio waves over a finitely conducting spherical earth. *Philos. Mag. Ser. 7*, **25**, 817–834.
Vendik IB and Gashinova M 2004 Artificial double negative (DNG) media composed by two different dielectric sphere lattices embedded in a dielectric matrix. *Proceedings of the 34th European Microwave Conference*, Amsterdam, the Netherlands, 1209–1212.
Wait JR 1962 *Electromagnetic Waves in Stratified Media*, Pergamon Press.
Wait JR 1963 Electromagnetic scattering from a radially inhomogeneous sphere. *Appl. sci. Res. B*, **10**, 441–450.
Walker GW 1900 The scattering of electromagnetic waves by a sphere. *Quart. J. Pure Appl. Math.*, **31**, 36–49.
Walker JD 1976 Multiple rainbows from single drops of water and other liquids. *Am. J. Phys.*, **44**, 421–433.
Watson GN 1918 The diffraction of electric waves by the Earth. *Proc. R. Soc. Lond. A*, **95**, 83–99.
Watson GN 1962 *A Treatise on the Theory of Bessel Functions*, Cambridge.
Weston VH 1963 Theory of absorbers in scattering. *IEEE Trans. Antennas Propag.*, **11**, 578–584.
White FP 1922 The diffraction of plane electromagnetic waves by a perfectly reflecting sphere. *Proc. R. Soc. Lond. A*, **100**, 505–525.
Young T 1804 The Bakerian lecture: experiments and calculations relative to physical optics. *Phil. Trans. R. Soc. Lond.*, **94**, 1–16.
Zayats AV, Smolyaninov II, and Maradudin AA 2005 Nano-optics of surface plasmon polaritons. *Phys. Rep.*, **408**, 131–314.

8

Method of Physical Optics

8.1 Introduction

In the preceding chapters of this book we dealt with canonical diffraction problems that allow exact solutions. However, it is obvious that only a limited number of relatively simple cases can be treated in such rigorous way. Two approaches are feasible for studying complex-shaped or electrically large objects: numerical techniques and analytical approximate solutions. Both approaches are based on approximations, and they have complementary and overlapping areas of applicability.

The existing methods for solving scattering problems can be classified into three broad categories: analytical methods, which provide exact explicit solutions (as introduced in the preceding chapters), approximate analytical models, which lead to approximate analytical solutions for usually simplified structures (this is the subject of this and the following chapter), and numerical methods, in which differential or integral equations are transformed into matrix equations by projective approximations and solved iteratively or by matrix inversion.

While both numerical and approximate analytical models are approximations to the exact solution, the analytical approaches have the unique advantage of providing physical insight and have the potential for handling electrically very large structures. Numerical techniques, on the other hand, have the unique potential for handling very complicated geometries and materials, serving as design tools for practical engineers. The main fundamental limitation of all numerical techniques is their inability to handle electrically very large objects and fundamental differences between electromagnetic properties of discrete and continuous objects, leading to numerical artifacts.

8.1.1 On Numerical Techniques for Studying Scattering from Arbitrary-shaped Bodies

Modern electromagnetic simulators based on the discretization approach to solving Maxwell's equations are modeling tools that solve electromagnetic field problems using numerical methods, extract engineering parameters from the field solution, and visualize fields and parameters. These tools allow engineering design by means of numerical electromagnetic analysis combined with optimization, and enable computer-aided design (CAD) of radio, microwave, and optical components, devices, and systems. For the users of these tools it is of prime importance to understand their position among other methods of solving electromagnetic problems and clearly understand advantages, disadvantages, and limitations of various solution methods. Quite often numerical solvers are used as complementary tools to analytical methods. For these reasons we present here a brief comparative overview of the main existing numerical techniques and outline their main limitations.

The most commonly used numerical techniques are the method of moments (MoM), the finite-element method (FEM), and the finite-difference time domain method (FDTD), although more approaches or variations of the above methods (e.g., finite-difference frequency domain technique) exist. MoM and FEM are most commonly implemented in the frequency domain, where complex amplitudes of the fields are solved, while FDTD is most commonly used as a time-domain technique.

Modern Electromagnetic Scattering Theory with Applications, First Edition. Andrey V. Osipov and Sergei A. Tretyakov.

In MoM, usually only conducting surfaces are discretized. The solution is thought of in terms of a basis-function expansion, and it is possible to choose various functions as the basis. There are also various possibilities to choose the test functions, which lead to the collocation method, Galerkin's method, and other variations. As a result of discretization, one arrives at a set of linear equations with a dense matrix. In FEM, the whole simulation domain is discretized by using polyhedrons. The solution is found as an extremum of a functional, and leads to a set of linear equations with a sparse matrix. This is a suitable method for arbitrary shaped objects, and dielectric materials can be treated easily. However, simulations with large simulation domains take long time. FDTD uses the orthogonal geometry of cells and discretized Maxwell's equations, which leads to artifacts of numerical dispersion. Iterative calculation of field components in time domain is performed, which allows simple treatment of arbitrary material parameter distributions (but there are problems with curved shapes). This method has comparatively low requirements on the amount of computer memory, since there is no need to solve linear equations with huge numbers of unknowns, but long running time is often needed.

Successful use of any numerical tool requires physical understanding of the scattering phenomena under investigation, and to this end the approximate analytical solutions are necessary. Moreover, approximate analytical techniques remain the only available tool for solving electrically very large problems.

8.1.2 PO as one of the Approximate Analytical Techniques

Physical optics (PO) is an approximate analytical technique that allows us to solve scattering problems for electrically large objects. The big advantage of the PO method compared to the discretization-based numerical approaches is that it eliminates the need to solve systems of algebraic equations as the solution of the scattering problem is immediately available in the form of an integral over the scattering surface, to which a variety of efficient integration techniques, numerical and asymptotic, are applicable. This advantage becomes crucial when a solution is desirable not for a single frequency and a single illumination direction but rather for a frequency range (pulse excitation) or even for a range of frequencies and incidence angles (radar imagining, synthetic aperture radars, e.g. Knott *et al.* (1993) and Skolnik (2008)). The PO method is a high-frequency technique in the sense that the accuracy of the PO solutions improves with the increase in frequency so that for scatterers, apertures, lenses, and reflectors greater than several wavelength in diameter PO provides solutions with an accuracy that is sufficient for most practical applications. In contrast to geometrical optics (GO), PO solutions correctly (at least qualitatively) describe a variety of diffraction effects, including the smooth transition between the lit and shadowed portions of space at shadow boundaries, the edge-diffracted waves, and the fields at caustics. All these features make the PO method an indispensable tool in optical and microwave engineering.

A scalar version of the method is known as the Fresnel–Kirchhoff diffraction theory (Baker and Copson 1953; Born and Wolf 1959; Kirchhoff 1891; Rubinowicz 1957; Helmholtz 1897; Walker 1904). The history goes back to the very origins of the wave theory of light (Fresnel 1818; Huygens 1690; Young 1845)[1] that represents the light wave through a single scalar wave function, the value of which is determined by the interference of the spherical waves from the secondary sources distributed over the front of the original wave. A more rigorous mathematical foundation of the scalar diffraction theory rests on the fact that the wave function can be regarded as a solution of the Helmholtz equation and, thus, by the use of Green's theorem, can be represented by integration of the field and its normal derivative over a closed surface surrounding the point of observation (Helmholtz 1860). The Fresnel theory results from the Helmholtz formula if specific, following from GO considerations, values for the field and its normal derivative on the integration surface are assumed, and this is the procedure described by Kirchhoff (1891).

Full vector electromagnetic versions of the Fresnel–Kirchhoff method have received much less attention in the literature, apparently due to the more complicated form of the integral relations for the solutions of Maxwell's equations than for the solutions of the scalar Helmholtz equation. These integral representations are formulated in terms of the boundary values of the electric and magnetic fields and their derivatives on a closed surface (Franz 1948; Kottler 1923; Larmor 1903; Macdonald 1902; Stratton and Chu 1939; Stratton 1941); replacing the fields on the integration surface with their GO values leads to an electromagnetic version of the Fresnel–Kirchhoff method (Macdonald 1913). The integration surface can be a physical (the surface of the scattering body) or mathematical (a wavefront, an aperture in a screen) surface, and the latter approach is predominantly used in optics to account for diffraction in systems transmitting light (Debye

[1] A collection of English translations of the most important works by Huygens, Young, and Fresnel related to the wave theory of light can be found in Crew (1900).

1909; Kline and Kay 1965; Wolf 1959), whereas the former approach is often used in electromagnetics to calculate radar cross sections of large scatterers or radiation patterns of large antennas (Jenn 2005; Knott *et al.* 1993; Ruck *et al.* 1970; Ufimtsev 1999, 2014; Youssef 1989). The PO solutions employed in electromagnetics have been typically limited to PEC scatterers with the source and receiver in the far field of the scatterer, often at the same location (monostatic radar configuration). Extensions of the PO method to calculation of monostatic RCS of non-metallic scatterers are described in Ruck *et al.* (1970) and Klement *et al.* (1988).

In this chapter we develop a fully electromagnetic version of the PO method by using the Franz integral representations (2.346) and (2.347), and the GO solutions for planar interfaces from section 4.2.6. Among various possible surface-integral representations of solutions of Maxwell's equations, the Franz formulation has the advantage of being equally applicable to continuous and discontinuous field distributions, closed and open surfaces, whereas the alternative form, the Stratton–Chu representation, has to be modified by adding line integrals along shadow boundaries and edges in the scattering surface (Baker and Copson 1953; Kottler 1923; Stratton 1941) and is, therefore, less convenient (Tai 1972). A further disadvantage of the Stratton–Chu formulation is the presence of the normal components of the boundary fields under the integration sign, which complicates the GO formulation in the case of imperfectly reflecting surfaces and is the reason for the limitation of the Stratton–Chu-formula-based versions of the PO method to the far zone of the scatterer where the integrand can be simplified to include only the tangential components of the fields (Ruck *et al.* 1970). Contrastingly, the Franz representation involves only the tangential field components, regardless of the location of the observation point, which makes it possible to derive PO solutions applicable at an arbitrary distance (greater than several wavelengths) from a non-metallic scattering surface. The use of different integral representations will lead to different versions of PO, and since approximate values of the surface fields are used, the PO formulations will give in general different results which, however, may deviate only in correction terms (like edge corrections or shadow boundary corrections) but have to coincide in the leading terms.

8.1.3 *Structure of the Chapter*

In this chapter we present PO solutions for electromagnetic fields scattered from three-dimensional objects, PEC or non-metallic, of variable curvature, under the general assumption that the object is illuminated by a spherical wave and the observer is located at a finite distance from the scatterer. These solutions are evaluated asymptotically by using the method of the stationary point (section D.3), which makes the physical interpretation of the involved components apparent. In section 8.2, the PO principles are formulated (section 8.2.1) and the PO solution for a bounded integration surface (a compact body, an aperture in an infinite screen) is derived (section 8.2.2). The case of an infinite cylindrical body is addressed in section 8.2.3. In section 8.3 we apply the PO method to the problem of transmission through an arbitrarily shaped aperture in an impenetrable screen and in section 8.4 to the problem of scattering from an arbitrarily shaped body described by tensor reflection coefficient. As a result, a contour integral representation over the rim of the aperture for the field transmitted through the aperture, the GTD-type representations of the edge-diffracted rays (Keller 1962), and the classical reflection formulas for curved non-metallic surfaces (Fock 1965) are derived. The chapter concludes with a discussion of advantages and disadvantages of the PO method (section 8.5).

8.2 Principles and General Solution

The method of PO is a recipe of constructing approximate solutions of scattering problems in the presence of scatterers that are large compared to the wavelength. The method is based on exact integral representations of scattered fields in which the fields on an integration surface are approximated by using the laws of GO. The integration surface can be physical (the surface of a scatterer, of a lens, etc.) or mathematical (a wavefront, an opening in a screen), and the PO solution has the form of a superposition of secondary sources distributed over this surface.

8.2.1 *Principles of PO*

Let us first discuss the fundamental physical assumption behind the PO method: the GO approximation of the fields or currents on the scatterer surface. When a plane wave illuminates a flat object (with an infinite planar surface), the reflected field is also a plane wave. The propagation direction of the reflected wave is given by the GO rule: the reflection

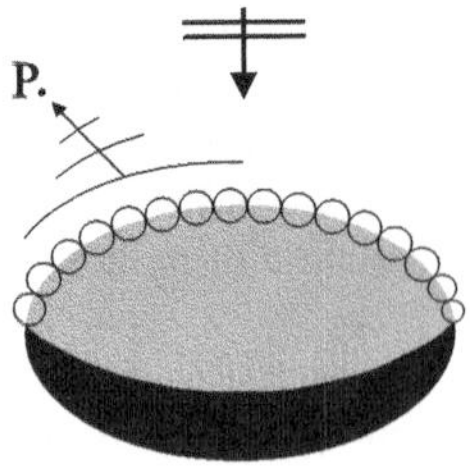

Figure 8.1 Scattering from a material body according to PO. The scattered field at the observation point P is obtained by integration over all secondary sources on the illuminated portion of the scatterer (gray) which are described in the GO approximation. On the shadowed portion of the scatterer (black) the secondary sources are assumed to be zero.

angle is equal to the incidence angle. For this case of an infinite flat object, the reflected field values can be easily found (see Chapter 4). The PO approximation treats the neighborhood of every point on the surface of electrically large arbitrary-shaped bodies as locally flat and tangential to the surface at this point, and uses the solution for the corresponding infinite and flat surface to define the fields or currents on the surface of the scatterer.[2] This leads to the assumption of the local character of the interaction between the incident field and the scatterer: the current at a given point is determined by the incident field at the same point. This is approximately correct if the body is large compared to the wavelength and smooth so that the curvature radii are also large compared to the wavelength. Naturally, near sharp edges this current approximation is not valid, but the results are generally acceptable even for bodies with sharp edges.

The PO solution for a scattering problem starts from choosing a suitable integration surface over which the electromagnetic field can be reasonably approximated in the framework of GO, for example by a combination of incident, reflected, and transmitted plane waves. Then, one writes the GO-based approximations for the tangential components of the electric and magnetic fields on the integration surface or, which is the same, for the equivalent or physical surface currents. Finally, these approximated currents are inserted into the original integral representation of the scattered field and integrated over the surface, which gives the scattered field in the PO approximation. For example, if a scatterer is impenetrable so that the direct transmission through it can be neglected and the integration surface is chosen to coincide with the physical scattering surface (Figure 8.1), the GO approximation for the equivalent currents can be obtained by using the solutions described in section 4.2.6. Knowing the reflection matrix $\overline{\overline{R}}$ for all incidence angles, we then assume that at every point of the surface of the actual scatterer the electric field is approximately

$$\mathbf{E}^{\text{go}} = \begin{cases} \mathbf{E}^{\text{inc}} + \mathbf{E}^{\text{ref}} = (\overline{\overline{I}} + \overline{\overline{R}}) \cdot \mathbf{E}^{\text{inc}}, & \text{scatterer's lit portion} \\ 0, & \text{scatterer's shadowed portion} \end{cases}. \tag{8.1}$$

Here, $\mathbf{E}^{\text{inc}}$ and $\mathbf{E}^{\text{ref}}$ are the electric fields in the incident and reflected waves, and at every point of the scatterer the values of the reflection matrix $\overline{\overline{R}}$, which are needed to specify the reflected field, are taken for the incidence angle defined at that particular point. The local values of the magnetic field $\mathbf{H}^{\text{go}}$ are obtained from $\mathbf{E}^{\text{inc}}$ and $\mathbf{E}^{\text{ref}}$ by using the relations between the electric and magnetic fields in plane waves (section 2.5.1).

Once the fields on the scattering surface are determined, one evaluates their tangential components

$$\mathbf{K}_{\text{e}}^{\text{go}} = \hat{n} \times \mathbf{H}^{\text{go}}, \quad \mathbf{K}_{\text{m}}^{\text{go}} = -\hat{n} \times \mathbf{E}^{\text{go}}, \tag{8.2}$$

where $\hat{n}$ is the outward unit normal vector at the same point of the scattering surface where the fields are evaluated. In case of a PEC body, the tangential components of the magnetic field in the incident and reflected waves at the surface are

[2] That is why the method is sometimes called the tangent plane approximation.

equal to each other in the GO approximation, the tangential electric field is zero (which is not an approximation), and the surface currents become

$$\mathbf{K}_{\mathrm{e}}^{\mathrm{go}} = 2\hat{n} \times \mathbf{H}^{\mathrm{inc}}, \quad \mathbf{K}_{\mathrm{m}}^{\mathrm{go}} = 0 \tag{8.3}$$

(see (4.97)). Now that we have approximate values of the fields on the surface of the scatterer, the scattered fields elsewhere can be found using the exact integral representation for the fields scattered by or radiated from a closed surface (section 2.6). The concept of PO leads to simple approximate solutions for a great variety of problems of scattering and radiation of electromagnetic waves.

The same idea can be used to solve transmission problems for apertures in impenetrable screens. In this case, the GO field in the aperture is simply equal to the incident field, while the field just behind the screen is zero in this approximation. Generally speaking, the integration surface in a PO solution does not have to coincide with the scattering surface or with an aperture in a screen. It can be any surface permissible in an exact integral representation of the scattered/ transmitted field, for example a phase front of the scattered/ transmitted wave is sometimes the best choice.

8.2.2 Derivation of PO Solutions

Derivation of a PO solution consists of three steps: (a) choice of a suitable integration surface Ω in an integral representation of the field (section 2.6), (b) simplification of the exact integral representation through approximate evaluation of the differential operators in the integral representation, and (c) approximation of the fields on the integration surface using the GO laws. These approximations and simplifications are justified in the high-frequency case since it is assumed that (a) the source and the observer are located at least several wavelengths away from the integration surface and (b) the surface is electrically large.

The surface integral representations (see section 2.6) permit a great degree of flexibility in choosing the integration surface, which can be an arbitrary surface enclosing a scattering body or the observation point (Figure 8.2), and the surface is typically chosen to be such that the boundary field can be well approximated by an incident wave (as over a hole in an impenetrable screen) or by a superposition of the incident, reflected, and transmitted waves (as on the surface of a material scatterer). Since the boundary fields are approximated, the choice of the integration surface may influence the accuracy of the PO approximation.

In problems of electromagnetic scattering by compact scatterers the most convenient choice is typically the surface of the scatterer (Ω_1 in Figure 8.2); in problems of transmission through or radiation from an opening in a impenetrable screen a reasonable choice is a surface placed directly behind the screen which coincides either with the aperture or with the front of the transmitted wave (see problems 3 and 14) (Figure 8.5).

Having chosen the integration surface, let us consider the integral representation (2.346) for the electric field and simplify it, assuming that the observation point $\mathbf{r}$ is located at least several wavelengths λ away from the integration surface Ω,

$$d \gg \lambda \tag{8.4}$$

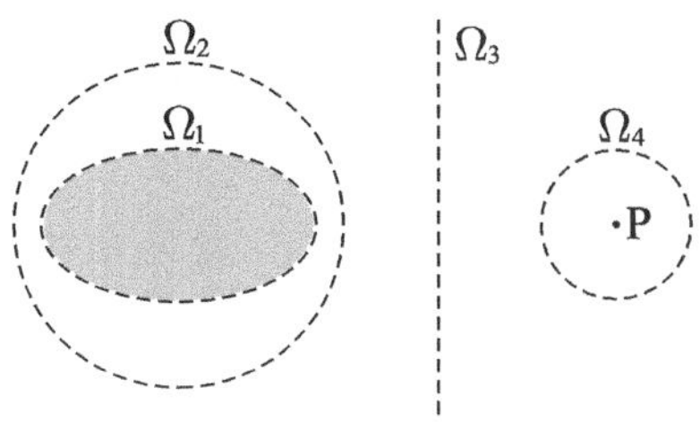

Figure 8.2 Permissible integration surfaces: Ω_1, the physical surface of a scatterer; Ω_2, a surface enclosing the scatterer; Ω_3, an infinite plane separating the scatterer and the observation point P; Ω_4, a surface enclosing the observer.

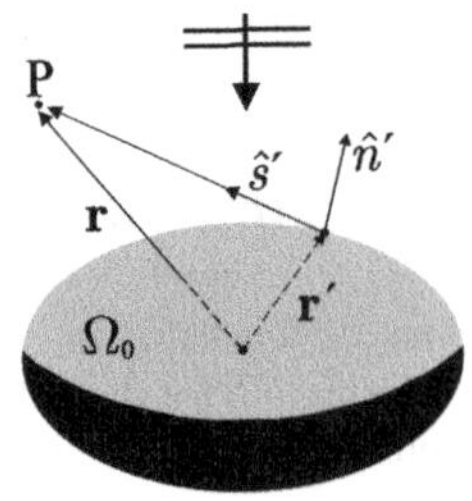

Figure 8.3 Notation used in the derivation of PO solution: P, the observation point; $\mathbf{r}$, the position vector of P; Ω_0, the PO integration surface (e.g., the illuminated portion of an impenetrable scatterer); $\mathbf{r}'$, an integration point on Ω_0; $\hat{n}'$, the unit normal to the surface at $\mathbf{r}'$; $\hat{s}'$, the unit vector pointing from $\mathbf{r}'$ at P.

where $d = \min |\mathbf{r} - \mathbf{r}'|$ with $\mathbf{r}' \in \Omega$ is the distance between the observer and the surface Ω. The coordinates and the parameters are shown in Figure 8.3. The relation of d to the size D of Ω and to the far-field parameter D^2/λ can be arbitrary so that the observation point may reside in the Fresnel zone ($d < D^2/\lambda$) or in the far (Fraunhofer) zone ($d > D^2/\lambda$) of the surface Ω.

The PO solution is now obtained from (3.97) by replacing the exact distributions of the tangential field components on the surface Ω through their GO approximations, $\mathbf{K}_{\mathrm{e}}^{\mathrm{go}}$ and $\mathbf{K}_{\mathrm{m}}^{\mathrm{go}}$, which gives the formula

$$\mathbf{E}_{\Omega}^{\mathrm{po}}(\mathbf{r}) = \frac{jk}{4\pi} \int_{\Omega_0} \frac{\mathrm{e}^{-jk|\mathbf{r}-\mathbf{r}'|}}{|\mathbf{r}-\mathbf{r}'|} \mathbf{c}(\mathbf{r}')\, \mathrm{d}\Omega', \tag{8.5}$$

where

$$\mathbf{c}(\mathbf{r}') = Z\hat{s}' \times \left(\hat{s}' \times \mathbf{K}_{\mathrm{e}}^{\mathrm{go}}\right) + \left(\hat{s}' \times \mathbf{K}_{\mathrm{m}}^{\mathrm{go}}\right). \tag{8.6}$$

The shadowed portion of Ω does not contribute to the scattered field because $\mathbf{K}_{\mathrm{e}}^{\mathrm{go}}$ and $\mathbf{K}_{\mathrm{m}}^{\mathrm{go}}$ vanish there, and the integration in (8.5) is performed over a surface Ω_0 which is the illuminated portion of the original integration surface Ω.

To calculate fields given by the PO integral (8.5), we will often use the stationary phase method, described in Appendix D. This method requires extraction of a rapidly varying factor from the integrand function which oscillates in space on the scale of the wavelength. In the situations when multiple reflections between portions of the integration surface Ω do not exist or can be neglected, these rapidly oscillating factor results from the type of the incident wave. For a spherical incident wave from a point source at $\mathbf{r} = \mathbf{r}_0$ we will do this extraction by writing

$$\mathbf{c}(\mathbf{r}') = \mathbf{C}(\mathbf{r}', \mathbf{r}_0) \frac{\mathrm{e}^{-jk|\mathbf{r}'-\mathbf{r}_0|}}{k|\mathbf{r}'-\mathbf{r}_0|} \tag{8.7}$$

and

$$\mathbf{E}_{\Omega}^{\mathrm{po}}(\mathbf{r}) = \frac{j}{4\pi} \int_{\Omega_0} \frac{\mathrm{e}^{-jk(|\mathbf{r}-\mathbf{r}'|+|\mathbf{r}'-\mathbf{r}_0|)}}{|\mathbf{r}-\mathbf{r}'||\mathbf{r}'-\mathbf{r}_0|} \mathbf{C}(\mathbf{r}', \mathbf{r}_0)\, \mathrm{d}\Omega'. \tag{8.8}$$

For an incident plane wave propagating in the direction $\hat{i}$

$$\mathbf{E}_{\mathrm{inc}}(\mathbf{r}) = \mathbf{E}_0\, \mathrm{e}^{-jk\hat{i}\cdot\mathbf{r}} \tag{8.9}$$

we write

$$\mathbf{c}(\mathbf{r}') = \mathbf{V}(\mathbf{r}')\, \mathrm{e}^{-jk\hat{i}\cdot\mathbf{r}'} \tag{8.10}$$

and

$$\mathbf{E}_{\Omega}^{\mathrm{po}}(\mathbf{r}) = \frac{jk}{4\pi} \int_{\Omega_0} \frac{\mathrm{e}^{-jk|\mathbf{r}-\mathbf{r}'|-jk\hat{\imath}\cdot\mathbf{r}'}}{|\mathbf{r}-\mathbf{r}'|} \mathbf{V}(\mathbf{r}')\,\mathrm{d}\Omega'. \tag{8.11}$$

The exponential factors in (8.8) and (8.11) are the rapidly varying factors.

If the observation point is located in the far-field region of the surface Ω_0, then

$$\frac{\mathrm{e}^{-jk|\mathbf{r}-\mathbf{r}'|}}{|\mathbf{r}-\mathbf{r}'|} \approx \frac{\mathrm{e}^{-jkr}}{r}\mathrm{e}^{jk\hat{s}\cdot\mathbf{r}'} \tag{8.12}$$

with $r = |\mathbf{r}|$ and $\hat{s} = \mathbf{r}/r$ on the whole integration surface, and (8.5) becomes

$$\mathbf{E}_{\Omega}^{\mathrm{po}}(\mathbf{r}) = jk\frac{\mathrm{e}^{-jkr}}{4\pi r} \int_{\Omega_0} \mathrm{e}^{jk\hat{s}\cdot\mathbf{r}'} \mathbf{c}(\mathbf{r}')\,\mathrm{d}\Omega'. \tag{8.13}$$

Note that $\hat{s}' = \hat{s}$ in this limit. Furthermore, if Ω is the physical surface of a scattering body and the excitation is a plane wave (8.9), then (8.13) leads to the useful formula for the scattered far field

$$\mathbf{E}_{\mathrm{sc}}^{\mathrm{po}}(\mathbf{r}) = \frac{\mathrm{e}^{-jkr}}{kr}\mathbf{F}_{\mathrm{po}}(\hat{s},\hat{\imath}) \tag{8.14}$$

with the far-field coefficient

$$\mathbf{F}_{\mathrm{po}}(\hat{s},\hat{\imath}) = \frac{jk^2}{4\pi} \int_{\Omega_0} \mathrm{e}^{-jk\mathbf{q}\cdot\mathbf{r}} \mathbf{V}(\mathbf{r})\,\mathrm{d}\Omega, \tag{8.15}$$

where we have omitted the prime symbol from the position vector of the integration point under the integration sign as this vector cannot be confused anymore with the position vector of the observer. Furthermore,

$$\mathbf{q} = \hat{\imath} - \hat{s} \tag{8.16}$$

and $\mathbf{V}$, which is defined by (8.6) and (8.10), can be expressed as

$$\mathbf{V} = \hat{s} \times (\hat{s} \times \mathbf{v}_{\mathrm{e}} + \mathbf{v}_{\mathrm{m}}) \tag{8.17}$$

with

$$Z\mathbf{K}_{\mathrm{e}}^{\mathrm{go}}(\mathbf{r}) = \mathbf{v}_{\mathrm{e}}\,\mathrm{e}^{-jk\hat{\imath}\cdot\mathbf{r}}, \quad \mathbf{K}_{\mathrm{m}}^{\mathrm{go}}(\mathbf{r}) = \mathbf{v}_{\mathrm{m}}\,\mathrm{e}^{-jk\hat{\imath}\cdot\mathbf{r}}. \tag{8.18}$$

If the surface is impenetrable and can be described by the reflection coefficients $R_{\|}$ and $R_{\perp}$, then using the GO solution for the field on such surface (section 4.2.6) gives

$$\mathbf{v}_{\mathrm{e}} = (R_{\perp} - 1)(\hat{\imath}\cdot\hat{n})(\mathbf{E}_0\cdot\hat{\alpha}_{\mathrm{inc}})\hat{\alpha}_{\mathrm{inc}} - (R_{\|}+1)(\mathbf{E}_0\cdot\hat{\beta}_{\mathrm{inc}})(\hat{n}\times\hat{\alpha}_{\mathrm{inc}}) \tag{8.19}$$

$$\mathbf{v}_{\mathrm{m}} = (1 - R_{\|})(\hat{\imath}\cdot\hat{n})(\mathbf{E}_0\cdot\hat{\beta}_{\mathrm{inc}})\hat{\alpha}_{\mathrm{inc}} - (R_{\perp}+1)(\mathbf{E}_0\cdot\hat{\alpha}_{\mathrm{inc}})(\hat{n}\times\hat{\alpha}_{\mathrm{inc}}), \tag{8.20}$$

where

$$\hat{\alpha}_{\mathrm{inc}} = \frac{\hat{\imath}\times\hat{n}}{|\hat{\imath}\times\hat{n}|}, \quad \hat{\beta}_{\mathrm{inc}} = \hat{\imath}\times\hat{\alpha}_{\mathrm{inc}}$$

are the unit vectors, with respect to which the polarization of the incident wave is described (section 4.2.6).

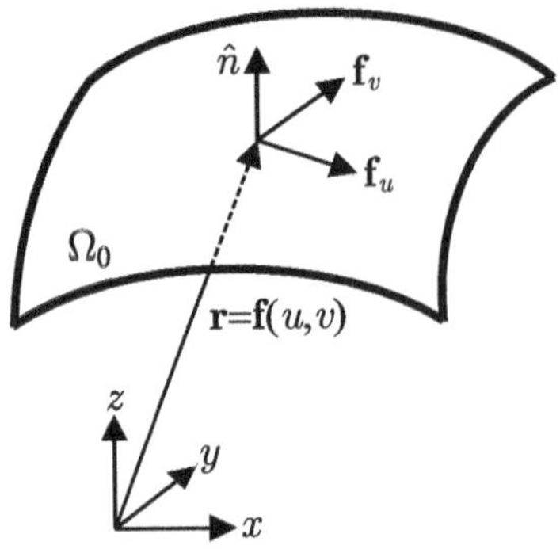

Figure 8.4 Parametrization of the PO integration surface Ω_0: the position vector $\mathbf{f}(u, v)$, the tangent vectors $\mathbf{f}_u$ and $\mathbf{f}_v$, and the unit normal $\hat{n}$.

In the case of a PEC surface, because of (8.3), $\mathbf{v}_{\mathrm{m}} = 0$ and $\mathbf{v}_{\mathrm{e}} = 2Z\hat{n} \times \mathbf{H}_0$, where $\mathbf{H}_0 = Y\hat{\imath} \times \mathbf{E}_0$ is the amplitude of the magnetic field in the incident plane wave. Thus,

$$\mathbf{V} = 2\hat{s} \times \hat{s} \times \hat{n} \times \hat{\imath} \times \mathbf{E}_0. \tag{8.21}$$

This formula follows also from (8.17), (8.19), and (8.20) with $R_{\parallel} = 1$ and $R_{\perp} = -1$.

Numerical or asymptotic evaluation of the PO solutions (8.8) and (8.11) is typically performed by reducing the surface integrals to the usual Riemann integrals. This is done by parametrization of the integration surface as

$$\mathbf{r} = \mathbf{f}(u, v), \tag{8.22}$$

where u and v are the new parameters or coordinates which are defined on a two-dimensional region Ω_{uv} (Figure 8.4). The relation (8.22) maps the three-dimensional integration surface Ω_0 to the region Ω_{uv} on the (u, v) plane. The infinitesimal element $\mathrm{d}\Omega'$ of the surface is expressed through u and v as

$$\mathrm{d}\Omega' = |\mathbf{f}_u(u, v) \times \mathbf{f}_v(u, v)| \, \mathrm{d}u \, \mathrm{d}v, \tag{8.23}$$

where

$$\mathbf{f}_u(u, v) = \frac{\partial \mathbf{f}(u, v)}{\partial u}, \quad \mathbf{f}_v(u, v) = \frac{\partial \mathbf{f}(u, v)}{\partial v}, \tag{8.24}$$

which reduces (8.8) and (8.11) to the canonical form

$$\mathbf{E}_{\Omega}^{\mathrm{po}}(\mathbf{r}) = \int_{\Omega_{uv}} \mathbf{A}(u, v) \, \mathrm{e}^{jkS(u,v)} \, \mathrm{d}u \, \mathrm{d}v \tag{8.25}$$

used in evaluations with the method of stationary phase (Appendix D). The functions here are given by either

$$S(u, v) = -|\mathbf{r} - \mathbf{f}(u, v)| - |\mathbf{f}(u, v) - \mathbf{r}_0| \tag{8.26}$$

$$\mathbf{A}(u, v) = \frac{j}{4\pi} |\mathbf{f}_u(u, v) \times \mathbf{f}_v(u, v)| \frac{\mathbf{C}[\mathbf{f}(u, v), \mathbf{r}_0]}{|\mathbf{r} - \mathbf{f}(u, v)||\mathbf{f}(u, v) - \mathbf{r}_0|} \tag{8.27}$$

or

$$S(u,v) = -|\mathbf{r} - \mathbf{f}(u,v)| - \hat{\imath} \cdot \mathbf{f}(u,v) \tag{8.28}$$

$$\mathbf{A}(u,v) = \frac{jk}{4\pi} |\mathbf{f}_u(u,v) \times \mathbf{f}_v(u,v)| \frac{\mathbf{V}[\mathbf{f}(u,v)]}{|\mathbf{r} - \mathbf{f}(u,v)|} \tag{8.29}$$

for the point source and the plane wave excitation, respectively. The far-field coefficient (8.15) is expressed by (8.25) with

$$S(u,v) = -\mathbf{q} \cdot \mathbf{f}(u,v) \tag{8.30}$$

$$\mathbf{A}(u,v) = \frac{jk^2}{4\pi} |\mathbf{f}_u(u,v) \times \mathbf{f}_v(u,v)| \mathbf{V}[\mathbf{f}(u,v)]. \tag{8.31}$$

8.2.3 PO for Cylindrical Bodies

PO can be applied to scattering configurations with translation symmetry, like a cylinder, a wedge or a slit in a screen. If the z axis is the axis of the translation symmetry, then the field dependence on the z coordinate can be extracted and the scattering problem can be formulated in two dimensions for the components E_z and H_z of the electric and magnetic fields (section 2.5.5). This approach simplifies the PO formulas. The derivation of the PO solutions for the scattered field is similar to the three-dimensional case (section 8.2.2), so here we only briefly describe the constructions.

The $\exp(-jk_z z)$ dependence of the fields is factored out as in (2.321) and (3.154), and the scattering problem is formulated in terms of the complex amplitudes $U_{E,H}$ for the total fields, $U^{\mathrm{inc}}_{E,H}$ for the incidence field and $U^{\mathrm{sc}}_{E,H}$ for the scattered field. The formulas look the same for E_z and H_z, so for the sake of brevity we use U_{sc} and F instead of $U^{\mathrm{sc}}_{E,H}$ and $F_{E,H}$.

In section 2.6.2 an exact integral representation for the field scattered by a group of cylindrical scatterers or by a single cylindrical scatterer has been formulated in terms of the values of the functions U and $\partial U/\partial n$ on a contour enclosing the scatterer(s) (Figure 3.15). By approximating these functions by their GO values on the integration contour in the integral representation (2.367), we obtain the PO approximation for the fields scattered by translation-invariant bodies,

$$U^{\mathrm{po}}_{\mathrm{sc}}(\boldsymbol{\rho}) = \int_C \left[U_{\mathrm{go}}(\boldsymbol{\rho}') \frac{\partial}{\partial n'} G_2(\boldsymbol{\rho},\boldsymbol{\rho}';k_\perp) - G_2(\boldsymbol{\rho},\boldsymbol{\rho}';k_\perp) \frac{\partial}{\partial n'} U_{\mathrm{go}}(\boldsymbol{\rho}') \right] \mathrm{d}l', \tag{8.32}$$

which is applicable at any distances from the scatterer(s), including near- and far-field zones. Here, $k_\perp = \sqrt{k^2 - k_z^2}$, $\boldsymbol{\rho} = \hat{x}\rho\cos\varphi + \hat{y}\rho\sin\varphi$ is the two-dimensional position vector of the observation point, and ρ and φ are the polar coordinates. C is the integration contour, $\boldsymbol{\rho}'$ is the integration point, $\hat{n}'$ is the unit normal pointing to the exterior of the cylinder at the point $\boldsymbol{\rho}'$, and $\mathrm{d}l'$ is the infinitesimal arc length. The two-dimensional Green function $G_2(\boldsymbol{\rho},\boldsymbol{\rho}';k_\perp)$ is given by (3.160) and its normal derivative by (3.161).

If the observer is located at a distance greater than a couple of wavelengths from the integration contour, then the Green function simplifies according to the large-argument approximation (E.25) of the Hankel function, and the formula (8.32) transforms to

$$U^{\mathrm{po}}_{\mathrm{sc}}(\boldsymbol{\rho}) \approx \frac{\mathrm{e}^{3j\pi/4}}{\sqrt{8\pi k_\perp}} \int_C \frac{\mathrm{e}^{-jk_\perp|\boldsymbol{\rho}-\boldsymbol{\rho}'|}}{|\boldsymbol{\rho}-\boldsymbol{\rho}'|^{1/2}} \left[\frac{\partial}{\partial n'} U_{\mathrm{go}}(\boldsymbol{\rho}') - jk_\perp(\hat{n}' \cdot \hat{s}'_\perp) U_{\mathrm{go}}(\boldsymbol{\rho}') \right] \mathrm{d}l', \tag{8.33}$$

where $\hat{s}'_\perp = (\boldsymbol{\rho} - \boldsymbol{\rho}')/|\boldsymbol{\rho} - \boldsymbol{\rho}'|$. This PO formula corresponds to the integral representation (3.163).

For bodies with a compact cross section, for example a circular cylinder, the far zone can be defined as in (3.155), and the PO approximation for the scattered field reads

$$U_{\mathrm{sc}}^{\mathrm{po}}(\boldsymbol{\rho}) \approx \frac{\mathrm{e}^{-jk_\perp|\boldsymbol{\rho}-\boldsymbol{\rho}_{\mathrm{c}}|+j\frac{3}{4}\pi}}{\sqrt{8\pi k_\perp|\boldsymbol{\rho}-\boldsymbol{\rho}_{\mathrm{c}}|}} \int_C \mathrm{e}^{jk_\perp\hat{s}_{\mathrm{c}}\cdot(\boldsymbol{\rho}'-\boldsymbol{\rho}_{\mathrm{c}})} \left[\frac{\partial}{\partial n'}U_{\mathrm{go}}(\boldsymbol{\rho}') - jk_\perp(\hat{n}'\cdot\hat{s}_{\mathrm{c}})U_{\mathrm{go}}(\boldsymbol{\rho}')\right] \mathrm{d}l', \tag{8.34}$$

where $\hat{s}_{\mathrm{c}} = (\boldsymbol{\rho}-\boldsymbol{\rho}_{\mathrm{c}})/|\boldsymbol{\rho}-\boldsymbol{\rho}_{\mathrm{c}}|$ and $\boldsymbol{\rho}_{\mathrm{c}}$ is an arbitrary point in the interior of the integration contour, for example the middle point of the geometric cross section of the cylinder. If the observation point goes to infinity ($\rho \to \infty$), then (8.34) simplifies further

$$U_{\mathrm{sc}}^{\mathrm{po}}(\boldsymbol{\rho}) \approx \frac{\mathrm{e}^{-jk_\perp\rho+j\frac{3}{4}\pi}}{\sqrt{8\pi k_\perp\rho}} \int_C \mathrm{e}^{jk_\perp\hat{s}_\perp\cdot\boldsymbol{\rho}'} \left[\frac{\partial}{\partial n'}U_{\mathrm{go}}(\boldsymbol{\rho}') - jk_\perp(\hat{n}'\cdot\hat{s}_\perp)U_{\mathrm{go}}(\boldsymbol{\rho}')\right] \mathrm{d}l', \tag{8.35}$$

where $\hat{s}_\perp = \boldsymbol{\rho}/\rho$ is the direction from the origin of the coordinate system to the observation point.

Scattering amplitudes F_E and F_H are defined in (3.156) and (3.157), implying that

$$U_{\mathrm{sc}}^{\mathrm{po}}(\boldsymbol{\rho}) = -F_{\mathrm{po}}(\varphi)\sqrt{\frac{2}{\pi k_\perp\rho}}\,\mathrm{e}^{-jk_\perp\rho+j\frac{\pi}{4}}, \tag{8.36}$$

and comparing with (8.35) we obtain the PO approximation for the scattering amplitude

$$F_{\mathrm{po}}(\varphi) = \frac{1}{4j}\int_C \mathrm{e}^{jk_\perp\hat{s}_\perp\cdot\boldsymbol{\rho}'} \left[\frac{\partial}{\partial n'}U_{\mathrm{go}}(\boldsymbol{\rho}') - jk_\perp(\hat{n}'\cdot\hat{s}_\perp)U_{\mathrm{go}}(\boldsymbol{\rho}')\right] \mathrm{d}l'. \tag{8.37}$$

When the incident field is a plane wave, this formula can be used to calculate the scattering matrix (3.180) and the scattering widths (3.181), (3.182), and (3.183) of cylindrical bodies in PO approximation.

The GO approximation means that we use $U_{\mathrm{go}} = 0$ on the shadowed portion of the integration contour; on the illuminated portion, at every point of the contour the field is obtained as a sum of incident and reflected plane waves,

$$U_{\mathrm{go}} = U_{\mathrm{inc}} + U_{\mathrm{ref}}. \tag{8.38}$$

If the integration contour coincides with the boundary of the scatterer, then the reflected field at each point is determined by using the reflection coefficients of an infinite and flat boundary tangential to the boundary at that point. Since the z direction is in general not perpendicular to the plane of incidence defined by the incidence direction $\hat{i}$ of the incoming wave and the normal $\hat{n}$ to the scattering surface at the point of interest, we have to use the matrix reflection coefficient $\overline{\overline{R}}_z(\hat{i},\hat{n})$ as described in (4.359),

$$\begin{bmatrix} U_E^{\mathrm{ref}} \\ ZU_H^{\mathrm{ref}} \end{bmatrix} = \overline{\overline{R}}_z(\hat{i},\hat{n}) \cdot \begin{bmatrix} U_E^{\mathrm{inc}} \\ ZU_H^{\mathrm{inc}} \end{bmatrix}. \tag{8.39}$$

For a plane wave with $\hat{i} = -\hat{x}\cos\varphi_0\sin\beta - \hat{y}\sin\varphi_0\sin\beta + \hat{z}\cos\beta$,

$$U_E^{\mathrm{inc}} = E_{0z}\,\mathrm{e}^{-jk\rho\sin\beta\cos(\varphi-\varphi_0)}, \quad U_H^{\mathrm{inc}} = H_{0z}\,\mathrm{e}^{-jk\rho\sin\beta\cos(\varphi-\varphi_0)}, \tag{8.40}$$

$k_z = k\cos\beta$, and $k_\perp = k\sin\beta$. Here, β is the angle between the axis of the cylinder and the direction of propagation of the incoming wave (Figure 3.4).

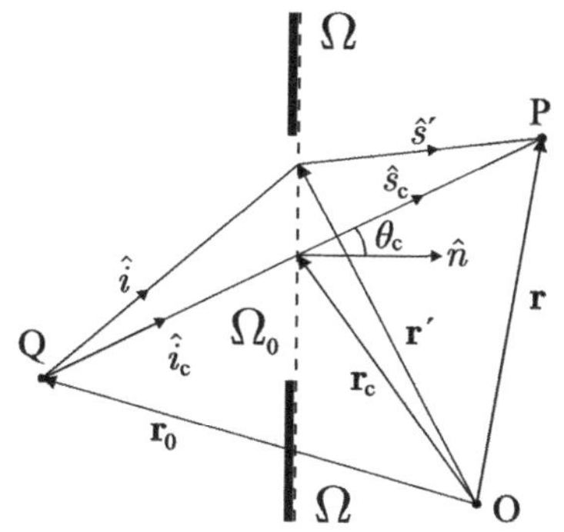

Figure 8.5 The point source Q at $\mathbf{r}_0$ and the observation point P at $\mathbf{r}$ are separated by a screen with an aperture Ω_0. Vectors $\mathbf{r}'$ and $\mathbf{r}_c$ define the position of an integration point and a critical point at the aperture. O is the origin of the coordinate system.

8.3 Transmission through Apertures

Consider the following problem. A point source and an observation point are separated by an impenetrable screen with an opening (or aperture), the size of which is large compared to the wavelength (Figure 8.5). In this section we use the PO solution to study the field transmitted through the aperture. It is assumed for simplicity that the rim of the aperture is a flat curve, that is, a curve which lies entirely in a plane. The screen itself must not be necessarily flat.

Application of the method of stationary phase reduces the solution to a sum of a curve integral along the rim of the aperture and, if the source is visible from the observation point, the incident wave. The curve integral describes a contribution from the rim of the aperture. Its asymptotic evaluation leads to edge-diffracted waves, and the solution of the transmission problem is thus expressed in the form of GTD, that is, as the sum of the incident and edge-diffracted rays. The PO-based expression for the edge-diffracted component includes a spreading factor which accounts for the curvature of the edge and a diffraction coefficient which corresponds to a straight edge and turns out to be different from the exact expression that is known from the exact solution for an infinite wedge (Chapter 5).

8.3.1 PO Solution

We choose the surface Ω to be adjacent to the side of the screen which is opposite to the source, so that the currents $\mathbf{K}_{\mathrm{e}}^{\mathrm{go}}$ and $\mathbf{K}_{\mathrm{m}}^{\mathrm{go}}$ in the PO integral (8.5) are finite on the aperture Ω_0 and zero on the rest portion of Ω where the field is assumed to be fully shadowed by the screen.

The surface Ω_0 is a mathematical surface at which no reflection takes place, and thus the GO approximation for the field in the aperture is simply equal to the incident field. If the point source is located many wavelengths away from the aperture, then the incident field is close to a spherical wave, and we find that

$$\mathbf{E}^{\mathrm{inc}}(\mathbf{r}) = \mathbf{E}_0^{\mathrm{inc}} \frac{\mathrm{e}^{-jk|\mathbf{r}-\mathbf{r}_0|}}{k|\mathbf{r}-\mathbf{r}_0|}, \quad \mathbf{H}^{\mathrm{inc}}(\mathbf{r}) = \mathbf{H}_0^{\mathrm{inc}} \frac{\mathrm{e}^{-jk|\mathbf{r}-\mathbf{r}_0|}}{k|\mathbf{r}-\mathbf{r}_0|}, \tag{8.41}$$

where $\mathbf{E}_0^{\mathrm{inc}}$ and $\mathbf{H}_0^{\mathrm{inc}}$ are the amplitudes of the incident electric and magnetic fields in the incident wave. This gives the PO solution of the transmission problem,

$$\mathbf{E}_{\Omega}^{\mathrm{po}}(\mathbf{r}) = \frac{j}{4\pi} \int_{\Omega_0} \frac{\mathrm{e}^{-jk|\mathbf{r}-\mathbf{r}'|-jk|\mathbf{r}'-\mathbf{r}_0|}}{|\mathbf{r}-\mathbf{r}'||\mathbf{r}'-\mathbf{r}_0|} \mathbf{C}(\mathbf{r}', \mathbf{r}_0)\, \mathrm{d}\Omega', \tag{8.42}$$

where

$$\mathbf{C}(\mathbf{r}', \mathbf{r}_0) = Z\hat{s}' \times \hat{s}' \times \hat{n}' \times \mathbf{H}_0^{\mathrm{inc}} - \hat{s}' \times \hat{n}' \times \mathbf{E}_0^{\mathrm{inc}} \tag{8.43}$$

and the prime sign in $\hat{s}'$ and $\hat{n}'$ (and later in $\hat{\imath}'$) denotes functions of the integration variable $\mathbf{r}'$. For a flat aperture, the unit normal is position independent so that $\hat{n}' = \hat{n}$.

To reduce the surface integral (8.42) to the usual, Riemann-type integral of the form (D.49), the surface of the aperture is represented parametrically, that is,

$$\mathbf{r}' = \mathbf{f}(u, v) \tag{8.44}$$

so that $\mathbf{r}' \in \Omega_0$ when $u, v \in \Omega_{uv}$ (Figure D.5). We do not specify the explicit analytical form of the function $\mathbf{f}(u, v)$ as this choice does not influence the result; we only assume that it is twice differentiable and the coordinates u and v are orthogonal. The infinitesimal surface element $\mathrm{d}\Omega'$ is then replaced with $|\mathbf{f}_u \times \mathbf{f}_v|\,\mathrm{d}u\,\mathrm{d}v$, where for brevity

$$\mathbf{f}_u = \frac{\partial}{\partial u}\mathbf{f}(u, v) = \hat{e}_u|\mathbf{f}_u|, \quad \mathbf{f}_v = \frac{\partial}{\partial v}\mathbf{f}(u, v) = \hat{e}_v|\mathbf{f}_v|, \tag{8.45}$$

which are vectors tangential to the surface Ω_0. Their vector product $\mathbf{f}_u \times \mathbf{f}_v$ is, therefore, a vector normal to the surface.

In the coordinates u and v, the PO solution (8.42) takes the form

$$\mathbf{E}_{\Omega}^{\mathrm{po}}(\mathbf{r}) = \int_{\Omega_{uv}} \mathbf{A}(u, v)\,\mathrm{e}^{jkS(u,v)}\,\mathrm{d}u\,\mathrm{d}v, \tag{8.46}$$

where

$$\mathbf{A}(u, v) = \frac{j|\mathbf{f}_u \times \mathbf{f}_v|\mathbf{C}[\mathbf{f}(u, v), \mathbf{r}_0]}{4\pi|\mathbf{r} - \mathbf{f}(u, v)||\mathbf{f}(u, v) - \mathbf{r}_0|} \tag{8.47}$$

$$S(u, v) = -\chi(u, v) \tag{8.48}$$

and

$$\chi(u, v) = |\mathbf{r} - \mathbf{f}(u, v)| + |\mathbf{f}(u, v) - \mathbf{r}_0|. \tag{8.49}$$

The representation (8.46) has the same form as the canonical integral (D.49), and the results of its asymptotic evaluation by the method of stationary phase are immediately available from Appendix D.

8.3.2 GO Rays and Fresnel Zones

We begin with the contribution from the critical point of the first kind, the position of which $\mathbf{r}_c = \mathbf{f}(u_c, v_c)$ is determined from the stationary phase conditions (D.50). We will show that this contribution represents the GO portion of the transmitted field. The relation (8.48) and the geometric interpretation of $\chi(u, v)$ as the distance from the source though the integration point to the observer suggest a solution at a point $\mathbf{r}' = \mathbf{r}_c$ which lies at the intersection of the aperture Ω_0 with the direct ray from the source to the observer along which the distance $\chi(u, v)$ is at its minimum. At the critical point the unit vectors

$$\hat{\imath}' = \frac{\mathbf{f}(u, v) - \mathbf{r}_0}{|\mathbf{f}(u, v) - \mathbf{r}_0|} \tag{8.50}$$

$$\hat{s}' = \frac{\mathbf{r} - \mathbf{f}(u, v)}{|\mathbf{r} - \mathbf{f}(u, v)|}, \tag{8.51}$$

which describe the directions of incidence and scattering, coincide, so that $\hat{\imath}_c = \hat{s}_c$ where the subscript denotes the value at the critical point. If the source is not visible from the observation point through the aperture, then the stationary phase equations have no solution on Ω_0 and the critical point disappears.

To find the contribution from the critical point, we need the values of the second derivatives of the phase function at the critical point

$$\alpha = -\frac{\partial^2 \chi}{\partial u^2}(u_c, v_c) \tag{8.52}$$

$$\beta = -\frac{\partial^2 \chi}{\partial v^2}(u_c, v_c) \tag{8.53}$$

$$\gamma = -\frac{\partial^2 \chi}{\partial u \partial v}(u_c, v_c). \tag{8.54}$$

Since

$$\frac{\partial \chi}{\partial u} = \mathbf{q} \cdot \mathbf{f}_u, \quad \frac{\partial \chi}{\partial v} = \mathbf{q} \cdot \mathbf{f}_v \tag{8.55}$$

with $\mathbf{q} = \hat{i}' - \hat{s}'$, we have

$$\frac{\partial^2 \chi}{\partial u^2} = \mathbf{q} \cdot \mathbf{f}_{uu} + \mathbf{f}_u \cdot \frac{\partial \mathbf{q}}{\partial u} \tag{8.56}$$

$$\frac{\partial^2 \chi}{\partial v^2} = \mathbf{q} \cdot \mathbf{f}_{vv} + \mathbf{f}_v \cdot \frac{\partial \mathbf{q}}{\partial v} \tag{8.57}$$

$$\frac{\partial^2 \chi}{\partial u \partial v} = \mathbf{q} \cdot \mathbf{f}_{uv} + \mathbf{f}_v \cdot \frac{\partial \mathbf{q}}{\partial u}, \tag{8.58}$$

where

$$\mathbf{f}_{uu} = \frac{\partial^2}{\partial u^2}\mathbf{f}(u, v), \quad \mathbf{f}_{vv} = \frac{\partial^2}{\partial v^2}\mathbf{f}(u, v), \quad \mathbf{f}_{uv} = \frac{\partial^2}{\partial u \partial v}\mathbf{f}(u, v).$$

At the critical point, $\mathbf{q} = 0$ and the first term on the right-hand side of every equation in (8.56)–(8.58) vanishes. Evaluating the partial derivatives of $\mathbf{q}$ in the other terms leads to the expressions

$$\alpha = \left(\frac{1}{R_0} + \frac{1}{R}\right)[(\hat{i}_c \cdot \mathbf{f}_u)^2 - |\mathbf{f}_u|^2] \tag{8.59}$$

$$\beta = \left(\frac{1}{R_0} + \frac{1}{R}\right)[(\hat{i}_c \cdot \mathbf{f}_v)^2 - |\mathbf{f}_v|^2] \tag{8.60}$$

$$\gamma = \left(\frac{1}{R_0} + \frac{1}{R}\right)[(\hat{i}_c \cdot \mathbf{f}_u)(\hat{i}_c \cdot \mathbf{f}_v) - (\mathbf{f}_u \cdot \mathbf{f}_v)], \tag{8.61}$$

where

$$R = |\mathbf{r} - \mathbf{r}_c|, \quad R_0 = |\mathbf{r}_0 - \mathbf{r}_c| \tag{8.62}$$

and we have used relations that follow from the stationary phase equations

$$\hat{i}_c \cdot \mathbf{f}_u = \hat{s}_c \cdot \mathbf{f}_u \tag{8.63}$$

$$\hat{i}_c \cdot \mathbf{f}_v = \hat{s}_c \cdot \mathbf{f}_v. \tag{8.64}$$

Using the orthogonality of the vectors $\mathbf{f}_u$ and $\mathbf{f}_v$ in (8.59)–(8.61) and the definitions of the unit vectors $\hat{e}_u$ and $\hat{e}_v$ in (8.45), we obtain the formula

$$\alpha\beta - \gamma^2 = |\mathbf{f}_u|^2|\mathbf{f}_v|^2 \left(\frac{1}{R_0} + \frac{1}{R}\right)^2 \left[1 - (\hat{i}_c \cdot \hat{e}_u)^2 - (\hat{i}_c \cdot \hat{e}_v)^2\right], \tag{8.65}$$

and since $\hat{i}_c$ is a unit vector and the vectors $\hat{e}_u$, $\hat{e}_v$, and $\hat{n}$ are mutually orthogonal, the expression in the square brackets is equal to

$$1 - (\hat{i}_c \cdot \hat{e}_u)^2 - (\hat{i}_c \cdot \hat{e}_v)^2 = (\hat{i}_c \cdot \hat{n})^2, \tag{8.66}$$

and (8.65) can be rewritten as

$$\alpha\beta - \gamma^2 = |\mathbf{f}_u|^2|\mathbf{f}_v|^2 \left(\frac{1}{R_0} + \frac{1}{R}\right)^2 \cos^2\theta_c. \tag{8.67}$$

Here θ_c is the angle between the incidence direction and the unit normal to the aperture (Figure 8.5).

From (8.59) and (8.67) it follows that $\alpha < 0$ and $\alpha\beta - \gamma^2 > 0$, and thus $\sigma_c = -j$, where σ_c is defined in (D.61). Application of the formula (D.60) to the integral (8.46) gives the critical point contribution

$$\mathbf{E}_c^{po}(\mathbf{r}) = \frac{\mathbf{C}(\mathbf{r}_c, \mathbf{r}_0)}{2k\cos\theta_c(R + R_0)} e^{-jk(R+R_0)}. \tag{8.68}$$

The amplitude factor

$$\mathbf{C}(\mathbf{r}_c, \mathbf{r}_0) = Z\hat{i}_c \times \hat{i}_c \times \hat{n} \times \mathbf{H}_0^{inc} - \hat{i}_c \times \hat{n} \times \mathbf{E}_0^{inc} \tag{8.69}$$

can be greatly simplified. Note that $\mathbf{C}(\mathbf{r}_c, \mathbf{r}_0)$ has no component in the direction $\hat{i}_c$ and thus can be expressed as

$$\mathbf{C}(\mathbf{r}_c, \mathbf{r}_0) = C_\perp \hat{\alpha}_{inc} + C_\parallel \hat{\beta}_{inc}, \tag{8.70}$$

where the unit vectors $\hat{\alpha}_{inc}$ and $\hat{\beta}_{inc}$ are defined in (4.71) (with $\hat{i}$ replaced with $\hat{i}_c$) and together with $\hat{i}_c$ constitute an orthogonal vector basis. The components $C_\perp$ and $C_\parallel$ are determined through multiplication of (8.69) with $\hat{\alpha}_{inc}$ and $\hat{\beta}_{inc}$,

$$C_\perp = \hat{\alpha}_{inc} \cdot \mathbf{C}(\mathbf{r}_c, \mathbf{r}_0), \quad C_\parallel = \hat{\beta}_{inc} \cdot \mathbf{C}(\mathbf{r}_c, \mathbf{r}_0). \tag{8.71}$$

Using the relation between the electric and magnetic field amplitudes in a spherical wave as in (2.165),

$$Z\mathbf{H}_0^{inc} = \hat{i}_c \times \mathbf{E}_0^{inc}, \tag{8.72}$$

after some vector algebra we obtain

$$C_\perp = 2(\hat{i}_c \cdot \hat{n})\left(\mathbf{E}_0^{inc} \cdot \hat{\alpha}_{inc}\right), \quad C_\parallel = 2(\hat{i}_c \cdot \hat{n})\left(\mathbf{E}_0^{inc} \cdot \hat{\beta}_{inc}\right) \tag{8.73}$$

or

$$\mathbf{C}(\mathbf{r}_c, \mathbf{r}_0) = 2\cos\theta_c \mathbf{E}_0^{inc}. \tag{8.74}$$

Insertion of (8.74) into (8.68) gives

$$\mathbf{E}_{\mathrm{c}}^{\mathrm{po}}(\mathbf{r}) = \mathbf{E}_0^{\mathrm{inc}} \frac{\mathrm{e}^{-jk(R+R_0)}}{k(R+R_0)} = \mathbf{E}_0^{\mathrm{inc}} \frac{\mathrm{e}^{-jk|\mathbf{r}-\mathbf{r}_0|}}{k|\mathbf{r}-\mathbf{r}_0|} \tag{8.75}$$

which, because of (8.41), is simply the incident spherical wave at the observation point. Thus, we conclude that the contribution to the PO solution (8.42) from the critical point of the first kind is equal to $\mathbf{E}^{\mathrm{inc}}(\mathbf{r})$ if $\mathbf{r}_{\mathrm{c}} \in \Omega_0$ and zero otherwise. Since the critical point is located at the intersection of the aperture with a straight line connecting the source and the observer, the result has a simple interpretation as the GO portion of the transmitted field

$$\mathbf{E}_{\mathrm{c}}^{\mathrm{po}}(\mathbf{r}) = \mathbf{E}^{\mathrm{go}}(\mathbf{r}) = \begin{cases} \mathbf{E}^{\mathrm{inc}}(\mathbf{r}), & \text{ray from P to Q goes through aperture} \\ 0, & \text{otherwise} \end{cases}. \tag{8.76}$$

The contribution from the critical point is independent of the aperture size and shape since the aperture was assumed to be sufficiently large. As the aperture gets smaller and smaller, it starts eventually blocking the free transmission through the aperture. Let us formulate a condition of applicability of equation (8.75). As shown in section D.2, the main contribution to integrals with rapidly oscillating kernels comes from a small vicinity of the critical point of the respective phase function. In the case of the integral (8.42), the phase function is proportional (with the minus sign) to $\chi(\mathbf{r}, \mathbf{r}', \mathbf{r}_0)$, which is the length of the path from the source through an integration point on the aperture and to the observer. At the critical point $\mathbf{r} = \mathbf{r}_{\mathrm{c}}$, which lies at the direct line between the source and the observer, the length of the path is at minimum $\chi(\mathbf{r}, \mathbf{r}_{\mathrm{c}}, \mathbf{r}_0) = \chi_{\mathrm{c}} = |\mathbf{r} - \mathbf{r}_0|$. Any deviation of the integration point from this line increases the length of the path and, therefore, the deviation of the phase from its stationary value. Once the phase difference becomes comparable to π, the exponential factor in the integrand begins to oscillate and effectively cancels out the contributions from the more distant vicinity of the critical point. Thus, the integration domain, which essentially contributes to the integral, can be determined from the condition

$$k[\chi(\mathbf{r}, \mathbf{r}', \mathbf{r}_0) - \chi_{\mathrm{c}}] \leq \pi. \tag{8.77}$$

This equation describes the region on the aperture around the critical point $\mathbf{r}_{\mathrm{c}}$, which is often referred to as the *field-forming region*.

Since the integration surface is flat, it is easy to determine the shape of the region. With $\mathbf{r}'$ not restricted to the integration surface, equation (8.77) describes the interior of a spheroid with the foci at the points $\mathbf{r}' = \mathbf{r}_0$ and $\mathbf{r}' = \mathbf{r}$ because by the definition of spheroids, the sum of the distances from any point on the surface of the spheroid to the focal points is constant. In the case of equation (8.77) this sum is equal to π/k. The intersection of the spheroid with the aperture plane is an ellipse (Figure 8.6) which is centered at the critical point and described by the equation

$$\frac{1}{2}\alpha(u-u_{\mathrm{c}})^2 + \frac{1}{2}\beta(v-v_{\mathrm{c}})^2 + \gamma(u-u_{\mathrm{c}})(v-v_{\mathrm{c}}) = -\frac{\pi}{k}. \tag{8.78}$$

If u and v are chosen to be the Cartesian coordinates in the aperture plane with the origin at the critical point and with the direction $\hat{v}$ perpendicular to $\hat{i}_{\mathrm{c}}$ (which is always possible), then $|\mathbf{f}_u| = 1$, $|\mathbf{f}_v| = 1$, $\mathbf{f}_u \cdot \hat{i}_{\mathrm{c}} = \sin\theta_{\mathrm{c}}$, $\mathbf{f}_v \cdot \hat{i}_{\mathrm{c}} = 0$, and the equation (8.78) takes the form

$$\cos^2\theta_{\mathrm{c}} u^2 + v^2 = F_1^2 \tag{8.79}$$

with

$$F_1 = \left(\frac{\lambda R R_0}{R+R_0}\right)^{1/2}. \tag{8.80}$$

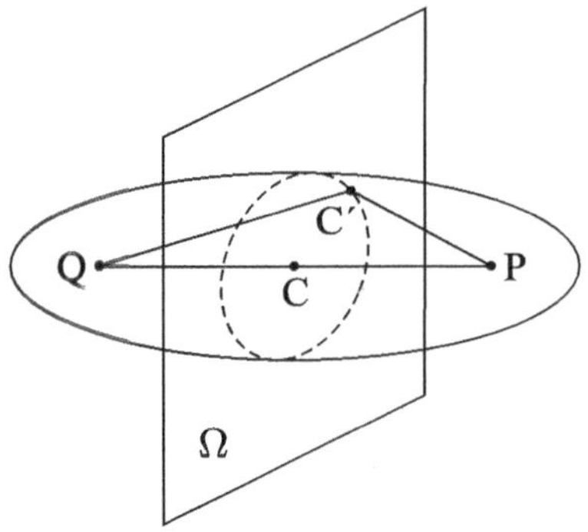

Figure 8.6 Definition of the field-forming region and Fresnel zones: spheroid with the foci at the source Q and the observer P is the locus of all points C′ in the space whose distances to Q and P add to the same constant (Fresnel's spheroid). The intersection of the spheroid with the aperture plane Ω is an ellipse (Fresnel zone). The shortest distance from the source to the observer goes through the center C of the ellipse.

The parameter F_1 has the dimension of length and defines the semi-minor (F_1) and semi-major ($F_1/\cos\theta_c$) axes of the ellipse on the (u, v) coordinate plane (Figure 8.7). If the aperture plane is perpendicular to the incidence direction, then the ellipse becomes a circle with the radius F_1.

This, in general, elliptic region on the aperture plane is called the first Fresnel zone, and F_1 is the key parameter that characterizes the transmission of electromagnetic waves through apertures. Apertures smaller than the first Fresnel zone significantly distort the transmission. To describe transmission through apertures, it is sometimes useful to introduce higher-order Fresnel zones by considering contours on the aperture plane at which the phase difference in (8.77) is equal to $m\pi$ with $m = 2, 3, \ldots$ These contours are ellipses with the semi-minor axes $F_m = \sqrt{m}F_1$, which define a set of ring-shaped regions concentric with the first Fresnel zone (Figure 8.7). An aperture should include at least several Fresnel zones to ensure an almost undisturbed transmission.

When the integration area is subdivided into the Fresnel zones, the PO integral becomes a sum of contributions from the zones and because of the phase difference, every next contribution is about 180° out of phase with the preceding one, which results in oscillating behavior of the sum as the number of zones varies. One consequence of this behavior is that an aperture with the size of the first Fresnel zone transmits more than a larger aperture covering the first and the second zones. This phenomenon is illustrated in Figure 8.8. At a distance 4λ behind the hole, there

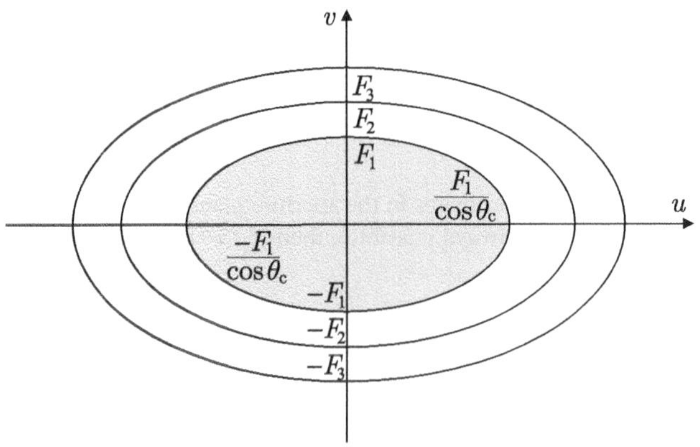

Figure 8.7 The shape of Fresnel zones in the aperture plane. The first Fresnel zone is shown in gray. The incidence direction $\hat{i}_c$ is perpendicular to the v axis and makes the angle θ_c with the normal to the aperture plane.

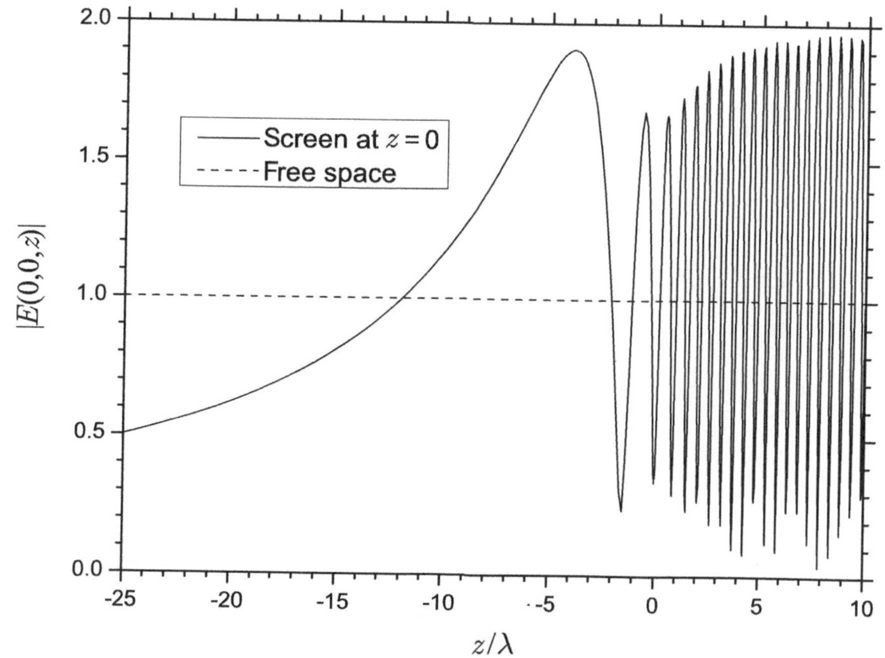

Figure 8.8 Transmission through a circular hole of 2λ radius in a infinite PEC screen at $z = 0$. The magnitude of the total electric field is sampled on the z axis, which goes through the center of the hole. The incident wave propagates toward the negative direction of the z axis.

is a maximum with the field level approximately twice that in the absence of the screen. At this distance, $F_1 = a$ as it follows from (8.80) with $R_0 \to \infty$ (plane wave illumination). As the observer moves farther beyond the field maximum point ($z \to -\infty$), the field level begins to decay monotonically since the aperture is smaller than the first Fresnel zone at these distances. The field oscillations in front of the opening ($z > 0$) result from interference of the incident wave with the wave reflected from the screen, and the amplitude of the reflected wave increases with growing z as the influence of the hole gets weaker.

Equation (8.80) suggests that the size of the Fresnel zones depends on the wavelength and on the distances from the aperture to the source and to the observer. The shorter the wavelength, or the closer the source or the observer to the aperture, the smaller the zones. The dependence of the spatial distribution of the transmitted field on the distance from the aperture is illustrated in Figures 8.9 and 8.10. The contour plots are obtained by numerical integration of the PO solution (8.42) for a plane wave incident perpendicularly to the square-shaped aperture with the side length $D = 10\lambda$. The magnitude of the transmitted field is plotted over an image plane located at distances 20λ and 200λ from the screen in Figures 8.9 and 8.10, respectively. The image plane at the 20λ distance resides in the near field and that at 200λ in the far field of the aperture ($D^2/\lambda = 100\lambda$), which explains the significant differences in the field distributions. The aperture and its quadratic shape are clearly visible in Figure 8.9. Contrastingly, the far-field distribution in Figure 8.10 looks completely different in that the main maximum appears as a protruded circular spot surrounded by several side lobes.

In terms of the Fresnel zones, one can say that the degree of deviation of the transmitted field from the GO expectations is determined by the parameter

$$P = \frac{2F_1}{D} \tag{8.81}$$

which is the ratio of the characteristic diameters of the first Fresnel zone to that of the opening. When $P \ll 1$, the influence of diffraction is hardly noticeable and the PO integral can be asymptotically evaluated with the stationary point

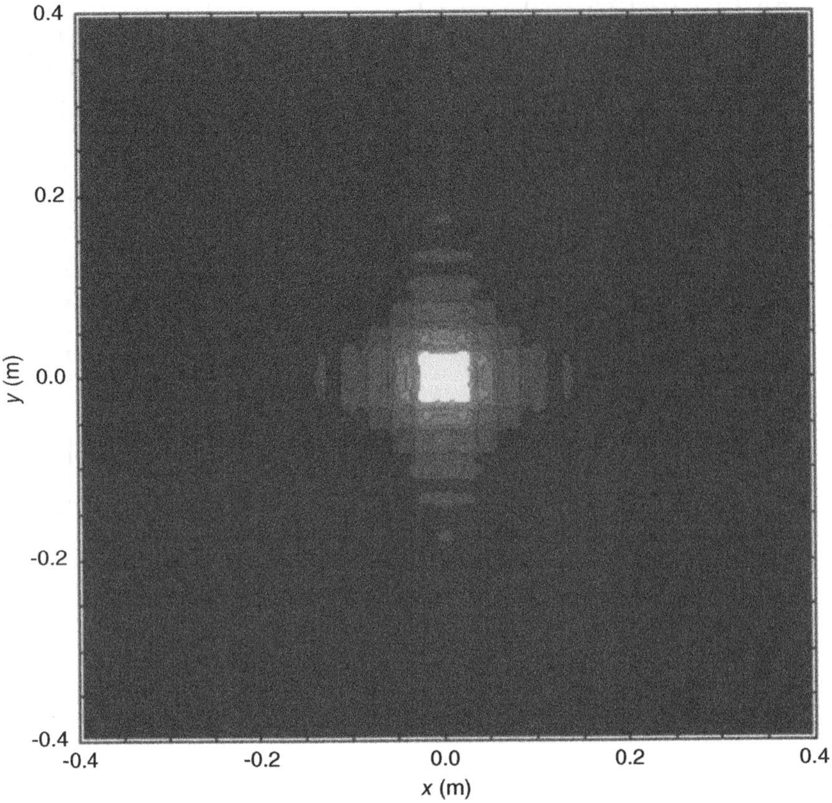

Figure 8.9 Field distribution behind a square $10\lambda \times 10\lambda$ aperture over an $80\lambda \times 80\lambda$ image plane at a distance 20λ. The incident field is a plane wave with $\lambda = 0.01$ m, polarized along a side of the aperture and coming perpendicularly to the aperture plane.

method, which leads to the GO representation for the transmitted field. Otherwise, when $P > 1$, diffraction effects play an important role and the GO solution is insufficient. This criterion is not limited to transmission through apertures and is applicable to the general problem of scattering of a wave with the wavelength λ from a compact scatterer of the characteristic size D. As expected, $P \to 0$ when $\lambda \to 0$, which is the known condition for the applicability of ray optics.

Notice that the parameter P and, therefore, the applicability of GO to scattering problems, depends not only on the ratio between λ and D but also on the distances from the observer to the scatterer and from the source to the scatterer (R and R_0 in the case of an aperture). When both the source and the observer go away from the aperture (plane wave illumination, observer in the far zone of the aperture), even the smallest, first Fresnel zone eventually becomes larger than the aperture, implying that the estimation (8.75) needs to be corrected by accounting for the contribution from the rim of the aperture.

8.3.3 Contribution from the Rim of the Aperture: Edge-diffracted Rays

The PO solution (8.42) for the transmitted field has the form of the canonical integral (D.49) and its asymptotic evaluation for large values of k, which involves the integration by parts, necessarily leads to the appearance of an integral over the boundary of the integration region. This contour integral, which is given by (D.80) or (D.82) with the amplitude and phase functions defined in (8.47), (8.48), and (8.49), describes a contribution from the rim of the aperture. Representation of the field transmitted through an aperture as a sum of the incident wave (if the source is visible from the observation

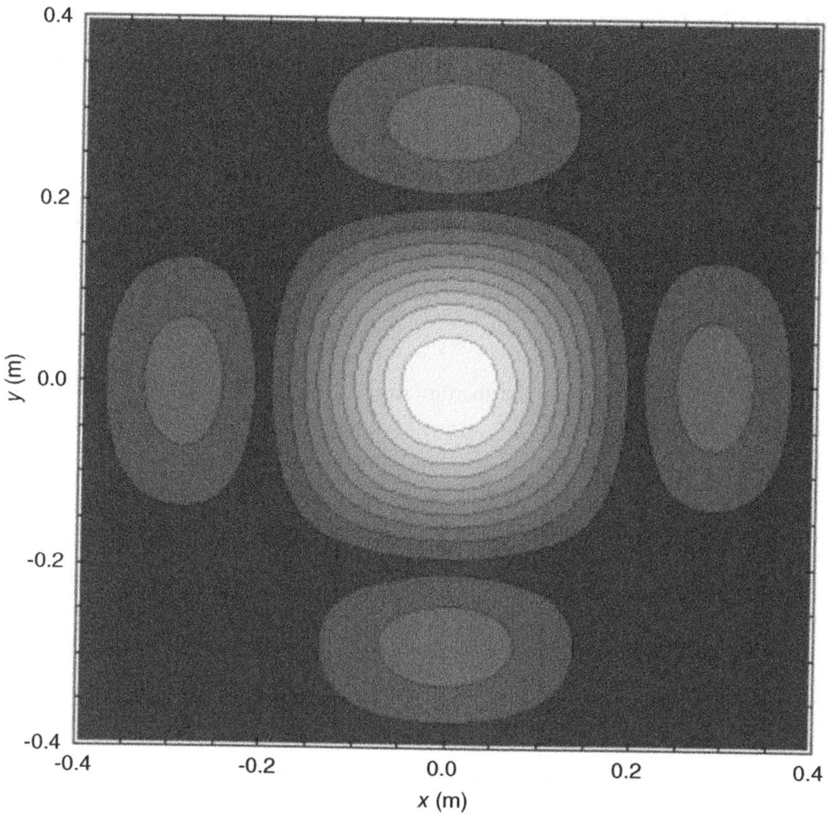

Figure 8.10 The same as Figure 8.9 but with the image plane at a distance 200λ behind the aperture.

point) and a wave that appears to emanate from the boundary of the aperture (the corresponding contribution has been labeled as the boundary diffraction wave) can be considered as a mathematical formulation of Young's interpretation of the diffraction phenomenon (Young 1845). Furthermore, the conversion of the PO surface integral into an integral over the rim of the integration area may significantly simplify numerical evaluation of the PO solutions, particularly when the size of the integration area is large compared to the wavelength. The search for the mathematically exact transformation has been the topic of many publications, mostly in the framework of the scalar, Kirchhoff's diffraction theory (Baker and Copson 1953; Born and Wolf 1959; Maggi 1888; Rubinowicz 1924, 1957, 1965). Electromagnetic extensions have been described for a number of specific configurations, including an arbitrarily shaped aperture in a flat impenetrable screen and a PEC polygonal flat plate, illuminated by a plane or a spherical wave (Asvestas 1985a,b, 1986; Johansen and Breinbjerg 1995; Karczewsky 1963; Kottler 1923; Laport and Meixner 1958; Meneghini *et al.* 1982; Rubinowicz 1965).

In this section, the PO integral over an aperture in an impenetrable screen is reduced to a contour integral over the rim of the aperture using the integration by parts, as described in sections D.3.3 and D.3.4. This asymptotic approach is justified since the PO integral is by definition a high-frequency approximation. The integral over the rim of the aperture is then evaluated by the method of stationary phase to show that the contribution from the critical point of the second kind has the form of an edge-diffracted wave scattered in accordance with the GTD diffraction laws (Borovikov and Kinber 1994; James 1976; Keller 1962; Kouyoumjian and Pathak 1974; McNamara *et al.* 1990). We estimate the size of the vicinity of the critical point which is essential for the formation of the edge-diffracted wave. Finally, we extract the PO diffraction coefficient for the rim of an aperture and compare this result with the diffraction coefficient of the PEC half-plane.

Transmitted Field as a Line Integral

Since the stationary point responsible for the direct ray is of elliptic type (see (8.67)), the Kontorovich formula (D.101) can be applied to the PO integral to express the PO integral by a line integral over the boundary of the integration domain,

$$\mathbf{E}^{\mathrm{po}}_{\Omega}(\mathbf{r}) = \mathbf{E}^{\mathrm{po}}_{\mathrm{c}}(\mathbf{r}) + \mathbf{E}^{\mathrm{po}}_{\delta\Omega}(\mathbf{r}), \tag{8.82}$$

where $\mathbf{E}^{\mathrm{po}}_{\mathrm{c}}(\mathbf{r})$ is given by (8.76) and $\mathbf{E}^{\mathrm{po}}_{\delta\Omega}(\mathbf{r})$ is given by any of the line integrals, (D.80) or (D.82), with the amplitude and phase functions defined in (8.47), (8.48), and (8.49).

Edge-diffracted Rays

Let us take a closer look at the contribution from the aperture rim to the transmitted field. We will suppose that the rim is a curve described in the parametric form by

$$\mathbf{r} = \tilde{\mathbf{f}}(t) = \mathbf{f}[U(t), V(t)] \tag{8.83}$$

with functions $U(t)$ and $V(t)$ from the parametrization (D.81). Consequently, the contribution from the rim of the aperture is described by the formulas (D.82), (D.83), and (D.84) where the phase function is expressed as

$$\tilde{S}(t) = -\tilde{\chi}(t) \tag{8.84}$$

with

$$\tilde{\chi}(t) = |\mathbf{r} - \tilde{\mathbf{f}}(t)| + |\tilde{\mathbf{f}}(t) - \mathbf{r}_0| \tag{8.85}$$

and the amplitude function $A(u, v)$ as in (8.47). Since k is a large parameter, the integral over the rim is asymptotically equal to a sum of contributions from the stationary points of the phase function (8.84), which are the critical points of the second kind for the PO integral (8.46).

In the special case that $\tilde{S}(t) = \mathrm{const}$ within $0 \leq t \leq 2\pi$, there are no stationary points and the result is given by the formula (D.85), which is of the same asymptotic order of magnitude as the incident field. This special case takes place if the aperture has the shape of the intersection of the Fresnel spheroid (Figure 8.6) with the aperture plane, for example when the aperture is a circle and the line connecting the source and the observer goes through the center of the circle, perpendicularly to the aperture plane. Evaluation of the rim contribution in this case typically requires numerical integration in (D.85), which is straightforward because the integrand is a smoothly varying function.

In general, $\tilde{S}(t)$ is not a constant and only several points $\mathbf{r}_{em}$ $(m = 1, 2, \ldots, M)$ on the integration contour are important for the rim contribution. From the stationary phase condition (D.86) one finds solutions t_m and, therefore, the position of the critical points on the rim as $\mathbf{r}_{em} = \tilde{\mathbf{f}}(t_m)$. The boundary of the aperture is a closed curve and thus the phase function takes the same value at the ends of the integration range, $\tilde{S}(0) = \tilde{S}(2\pi)$, implying that the stationary phase equation has at least two solutions. Consider one of these points, denoting, for brevity, its position vector and the corresponding parameter value by $\mathbf{r}_e$ and t_e, respectively.

The derivative of the phase function is given by

$$\frac{\mathrm{d}\tilde{S}}{\mathrm{d}t} = -\frac{\mathrm{d}\tilde{\chi}}{\mathrm{d}t} = -(\hat{\imath} - \hat{s}) \cdot \frac{\mathrm{d}\tilde{\mathbf{f}}}{\mathrm{d}t}, \tag{8.86}$$

where

$$\hat{\imath} = \frac{\tilde{\mathbf{f}}(t) - \mathbf{r}_0}{|\tilde{\mathbf{f}}(t) - \mathbf{r}_0|} \tag{8.87}$$

and

$$\hat{s} = \frac{\mathbf{r} - \tilde{\mathbf{f}}(t)}{|\mathbf{r} - \tilde{\mathbf{f}}(t)|} \tag{8.88}$$

are the unit vectors directed from the source to a point $\tilde{\mathbf{f}}(t)$ at the rim and from the point at the rim to the observer, respectively. The derivative of $\tilde{\mathbf{f}}(t)$ is a vector proportional to the tangent $\hat{\tau}$ to the rim,

$$\frac{\mathrm{d}\tilde{\mathbf{f}}}{\mathrm{d}t} = \left|\frac{\mathrm{d}\tilde{\mathbf{f}}}{\mathrm{d}t}\right| \hat{\tau}, \tag{8.89}$$

which allows us to express the stationary phase condition (D.86) as

$$\hat{\imath}_{\mathrm{e}} \cdot \hat{\tau}_{\mathrm{e}} = \hat{s}_{\mathrm{e}} \cdot \hat{\tau}_{\mathrm{e}}. \tag{8.90}$$

The subscript in $\hat{\tau}_{\mathrm{e}}$, $\hat{\imath}_{\mathrm{e}}$ and $\hat{s}_{\mathrm{e}}$ (and any other functions encountered below) means that the functions are taken at the stationary point $t = t_{\mathrm{e}}$. Equation (8.90) shows that at the stationary point the vectors $\hat{\imath}$ and $\hat{s}$ make the same angle with the vector $\hat{\tau}$, the geometric interpretation of which is that the directions of incidence and scattering belong to a conical surface rotationally symmetric about the direction tangential to the rim. In the framework of the GTD, such a conical surface is known as the Keller cone, which embraces all possible directions of the rays diffracted at an edge in a scattering surface (Figure 5.32a).

The contribution of the stationary point to the field at the observation point $\mathbf{r}$ is given by the formula

$$\mathbf{E}_{\mathrm{e}}^{\mathrm{po}}(\mathbf{r}) = \tilde{A}_{\mathrm{e}} \sqrt{\frac{2\pi}{k^3 \left|\tilde{\chi}_{\mathrm{e}}''\right|}} \exp\left[-jk\tilde{\chi}_{\mathrm{e}} - j\frac{\pi}{4}\operatorname{sgn}\tilde{\chi}_{\mathrm{e}}'' - j\frac{\pi}{2}\right], \tag{8.91}$$

which results from (D.87) on using the relation (8.84). We now calculate in succession all the quantities in (8.91). Denoting $R_{\mathrm{e}} = |\mathbf{r} - \mathbf{r}_{\mathrm{e}}|$ and $R_{0\mathrm{e}} = |\mathbf{r}_0 - \mathbf{r}_{\mathrm{e}}|$ gives

$$\tilde{\chi}_{\mathrm{e}} = R_{\mathrm{e}} + R_{0\mathrm{e}}. \tag{8.92}$$

To calculate $\tilde{\chi}_{\mathrm{e}}''$, it is convenient to introduce the vector $\mathbf{q} = \hat{\imath} - \hat{s}$ with $\hat{\imath}$ and $\hat{s}$ defined in (8.87) and (8.88). Then,

$$\tilde{\chi}'(t) = \mathbf{q} \cdot \frac{\mathrm{d}\tilde{\mathbf{f}}}{\mathrm{d}t} \tag{8.93}$$

$$\tilde{\chi}''(t) = \frac{\mathrm{d}\mathbf{q}}{\mathrm{d}t} \cdot \frac{\mathrm{d}\tilde{\mathbf{f}}}{\mathrm{d}t} + \mathbf{q} \cdot \frac{\mathrm{d}^2\tilde{\mathbf{f}}}{\mathrm{d}t^2} \tag{8.94}$$

and, therefore,

$$\tilde{\chi}_{\mathrm{e}}'' = \mathbf{q}_{\mathrm{e}}' \cdot \tilde{\mathbf{f}}_{\mathrm{e}}' + \mathbf{q}_{\mathrm{e}} \cdot \tilde{\mathbf{f}}_{\mathrm{e}}''. \tag{8.95}$$

Expressing the first derivative of $\mathbf{q}$ with respect to t at $t = t_{\mathrm{e}}$ as

$$\mathbf{q}_{\mathrm{e}}' = \left.\frac{\mathrm{d}\mathbf{q}}{\mathrm{d}t}\right|_{t=t_{\mathrm{e}}} = \tilde{\mathbf{f}}_{\mathrm{e}}' \left(\frac{1}{R_{\mathrm{e}}} + \frac{1}{R_{0\mathrm{e}}}\right) - \left(\frac{\hat{s}_{\mathrm{e}}}{R_{\mathrm{e}}} + \frac{\hat{\imath}_{\mathrm{e}}}{R_{0\mathrm{e}}}\right) (\hat{\imath}_{\mathrm{e}} \cdot \tilde{\mathbf{f}}_{\mathrm{e}}') \tag{8.96}$$

and multiplying this expression with $\tilde{\mathbf{f}}_{\mathrm{e}}'$ lead, on using the stationary phase condition (8.90), to the relation

$$\mathbf{q}_{\mathrm{e}}' \cdot \tilde{\mathbf{f}}_{\mathrm{e}}' = \left(\frac{1}{R_{\mathrm{e}}} + \frac{1}{R_{0\mathrm{e}}}\right) \left[|\tilde{\mathbf{f}}_{\mathrm{e}}'|^2 - (\hat{\imath}_{\mathrm{e}} \cdot \tilde{\mathbf{f}}_{\mathrm{e}}')^2\right] \tag{8.97}$$

or, in terms of the unit tangent vector $\hat{\tau}$,

$$\mathbf{q}'_e \cdot \tilde{\mathbf{f}}'_e = T_e \left(\frac{1}{R_e} + \frac{1}{R_{0e}} \right) \left[1 - (\hat{\imath}_e \cdot \hat{\tau}_e)^2 \right], \tag{8.98}$$

where $T_e = |\tilde{\mathbf{f}}'_e|^2$ for brevity. At an arbitrary point, the parameter T is related to the infinitesimal arc length along the rim, $dl = \sqrt{T}\, dt = |\tilde{\mathbf{f}}'|\, dt$. In terms of dl, the relation (8.89) becomes

$$\hat{\tau} = \frac{d\tilde{\mathbf{f}}}{dl}. \tag{8.99}$$

To calculate the second term in the right-hand side of (8.95), we use the derivatives with respect to the arc length,

$$\tilde{\mathbf{f}}'' = T \frac{d^2\tilde{\mathbf{f}}}{dl^2} + (\sqrt{T})' \frac{d\tilde{\mathbf{f}}}{dl}, \tag{8.100}$$

and multiplying (8.100) with $\mathbf{q}$ at $t = t_e$ obtain

$$\mathbf{q}_e \cdot \tilde{\mathbf{f}}''_e = T_e \mathbf{q}_e \cdot \left. \frac{d^2\tilde{\mathbf{f}}}{dl^2} \right|_{t=t_e} \tag{8.101}$$

since $\mathbf{q}_e \cdot \tau_e = 0$ at the stationary phase point. The second derivative of the position vector with respect to the arc length relates to the curvature of the curve according to the formula

$$\frac{d^2\tilde{\mathbf{f}}}{dl^2} = \frac{1}{a} \hat{N}, \tag{8.102}$$

where a is the curvature radius in the plane of the curve and $\hat{N}$ is the curvature vector, which is the unit vector normal to the curve, in the plane of the curve, and directed to the center of curvature (see (A.62)).

Inserting (8.98) and (8.101) into (8.95) gives

$$\tilde{\chi}''_e = T_e \left(\frac{1}{R_e} + \frac{1}{R_{0e}} \right) \left[1 - (\hat{\imath}_e \cdot \hat{\tau}_e)^2 \right] + T_e \frac{\mathbf{q}_e \cdot \hat{N}_e}{a_e}, \tag{8.103}$$

and introducing the angle β between the incidence direction and the tangent to the rim

$$\cos\beta = \hat{\imath}_e \cdot \hat{\tau}_e \tag{8.104}$$

and a parameter ρ_e defined by

$$\frac{1}{\rho_e} = \frac{1}{R_{0e}} + \frac{\mathbf{q}_e \cdot \hat{N}_e}{a_e \sin^2\beta} \tag{8.105}$$

we arrive at

$$\tilde{\chi}''_e = T_e \sin^2\beta \left(\frac{1}{R_e} + \frac{1}{\rho_e} \right). \tag{8.106}$$

It follows from (8.106) that

$$\operatorname{sgn} \tilde{\chi}_e'' = \operatorname{sgn} \left(\frac{1}{R_e} + \frac{1}{\rho_e} \right) \tag{8.107}$$

which can be equal to 1 or −1 because the product $\mathbf{q}_e \cdot \hat{N}_e$ and thereby the parameter ρ_e can be positive or negative.

Substituting (D.84), (8.47), (8.48), (8.92), and (8.106) in (8.91) results in the formula

$$\mathbf{E}_e^{po}(\mathbf{r}) = \gamma_e \mathbf{C}(\mathbf{r}_e, \mathbf{r}_0) \left| \frac{\rho_e}{R_e + \rho_e} \right|^{1/2} \frac{\exp\left[-jk(R_e + R_{0e}) - j\frac{\pi}{4} \operatorname{sgn} \tilde{\chi}_e'' \right]}{2kR_{0e}\sqrt{2\pi k R_e} \sin\beta} \tag{8.108}$$

where

$$\gamma_e = -\frac{1}{\sqrt{T_e}} \sqrt{U_e'^2 + V_e'^2} \left[|\mathbf{f}_u \times \mathbf{f}_v| \frac{\hat{w} \cdot \nabla_{uv}\chi(u,v)}{|\nabla_{uv}\chi(u,v)|^2} \right]_{u=U_e, v=V_e} \tag{8.109}$$

with $U_e = U(t_e)$ and $V_e = V(t_e)$. Vector $\hat{w}$ is the unit normal vector on the $u - v$ plane (Figure D.5).

The factor γ_e can be represented in a simpler form if we account for the fact that only an asymptotically small vicinity of the critical point contributes to the value of the integral. In this vicinity, we can define the parameters u and v as the coordinates tangential and orthogonal to the rim, respectively, and approximate the function (8.44) that describes the surface of the aperture by a linear function of u and v,

$$\mathbf{f}(u,v) \approx \mathbf{r}_e + (u - U_e)\hat{\tau}_e + (v - V_e)\hat{\nu}_e, \tag{8.110}$$

where $\hat{\nu}_e$ is a unit vector in the aperture plane, perpendicular to the rim at the critical point and directed into the interior of the aperture[3] (Figure 8.12). Thus,

$$\mathbf{f}_u \approx \hat{\tau}_e, \quad \mathbf{f}_v \approx \hat{\nu}_e \tag{8.111}$$

and since the vectors $\hat{\tau}_e$ and $\hat{\nu}_e$ are unit and orthogonal,

$$|\mathbf{f}_u \times \mathbf{f}_v| \approx 1. \tag{8.112}$$

The shape of the rim is approximately described by

$$\tilde{\mathbf{f}}(t) \approx \mathbf{r}_e + (U(t) - U_e)\hat{\tau}_e + (V(t) - V_e)\hat{\nu}_e, \tag{8.113}$$

and since

$$T = \tilde{\mathbf{f}}' \cdot \tilde{\mathbf{f}}' = U'^2(t) + V'^2(t), \tag{8.114}$$

we find that

$$T_e = U_e'^2 + V_e'^2. \tag{8.115}$$

[3] Vector $\hat{\nu}_e$ is in general different from $\hat{N}_e$, which is also perpendicular to the rim at the critical point but lies in the plane that includes the curve in the vicinity of the critical point. Both vectors coincide when the rim is a flat curve and integration in the PO integral is performed in the plane of the rim.

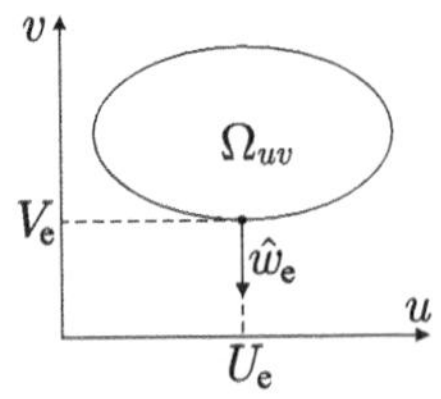

Figure 8.11 Projection of the aperture on the u–v plane. At the critical point $u = U_e$, $v = V_e$, and the vector $\hat{w}$ points in the direction of the negative v axis.

To calculate the remaining factor in (8.109) we express it as

$$\frac{\hat{w} \cdot \nabla_{uv} \chi(u, v)}{|\nabla_{uv} \chi(u, v)|^2} = \frac{w_u \frac{\partial \chi}{\partial u} + w_v \frac{\partial \chi}{\partial v}}{\left(\frac{\partial \chi}{\partial u}\right)^2 + \left(\frac{\partial \chi}{\partial v}\right)^2} \tag{8.116}$$

and use (8.55) and (8.111) to evaluate the derivatives

$$\frac{\partial \chi}{\partial u} \approx \mathbf{q} \cdot \hat{t}_e, \quad \frac{\partial \chi}{\partial v} \approx \mathbf{q} \cdot \hat{v}_e. \tag{8.117}$$

At the stationary point, the tangential derivative vanishes because of (8.90), and since $w_v = -1$ (Figure 8.11), we obtain

$$\left.\frac{\hat{w} \cdot \nabla_{uv} \chi(u, v)}{|\nabla_{uv} \chi(u, v)|^2}\right|_{u=U_e, v=V_e} \approx \frac{-1}{\mathbf{q}_e \cdot \hat{v}_e}. \tag{8.118}$$

Inserting (8.112), (8.115), and (8.118) in (8.109) gives

$$\gamma_e \approx \frac{1}{\mathbf{q}_e \cdot \hat{v}_e}. \tag{8.119}$$

By using (8.41) and (8.119) we can express the contribution of the critical point on the rim of the aperture as

$$\mathbf{E}_e^{\text{po}}(\mathbf{r}) = \mathbf{c}(\mathbf{r}_e, \mathbf{r}_0) \left| \frac{\rho_e}{R_e + \rho_e} \right|^{1/2} \frac{\exp\left[-jkR_e - j\frac{\pi}{4} \operatorname{sgn} \tilde{\chi}_e''\right]}{2(\mathbf{q}_e \cdot \hat{v}_e)\sqrt{2\pi k R_e} \sin\beta}, \tag{8.120}$$

where

$$\mathbf{c}(\mathbf{r}_e, \mathbf{r}_0) = Z\hat{s}_e \times \hat{s}_e \times \hat{n} \times \mathbf{H}^{\text{inc}}(\mathbf{r}_e) - \hat{s}_e \times \hat{n} \times \mathbf{E}^{\text{inc}}(\mathbf{r}_e). \tag{8.121}$$

The expression (8.120) describes a ray which emanates from a point source at $\mathbf{r}_0$, arrives at the point of scattering $\mathbf{r}_e$ at the rim of the aperture, and gets scattered toward the observation point at $\mathbf{r}$ (Figure 8.12). The direction of scattering $\hat{s}_e$ has the same projection on the direction $\hat{t}_e$ tangential to the rim at the point of scattering as the direction of the incidence wave $\hat{i}_e$ but otherwise can be arbitrary. This property constitutes the law of edge diffraction, the basic principle of the GTD (Keller 1962).

As seen from (8.120), the amplitude of the ray is described by the factor

$$\left| \frac{\rho_e}{R_e(R_e + \rho_e)} \right|^{1/2} \tag{8.122}$$

which may vary from $O(R_e^{-1/2})$ as in a cylindrical wave when $|\rho_e| \gg R_e$ to $O(R_e^{-1})$ as in a spherical wave when $|\rho_e| \ll R_e$. The factor (8.122) is known in GTD/UTD as the spread (or spreading) factor and can be obtained by employing the

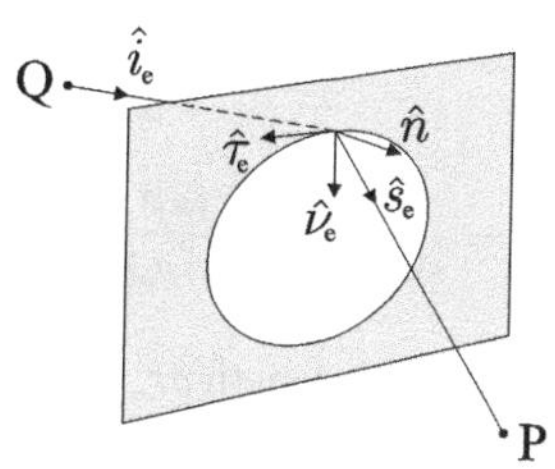

Figure 8.12 A ray propagating from a point source at Q in the direction $\hat{i}_e$ is diffracted at the rim of the aperture in the direction $\hat{s}_e$ toward the observation point P. The unit vectors $\hat{n}$ (normal to the aperture), $\hat{\nu}_e$ (normal to the rim, inward the aperture, in the aperture plane), and $\hat{\tau}_e$ (tangent to the rim, in the aperture plane) are mutually orthogonal.

energy principle according to which the amplitude of a ray varies inversely to the square root of the cross-sectional area of a tube of rays (Keller 1962). The parameter ρ_e defined in (8.105) is the distance between the scattering point on the rim and the caustic of diffracted rays, which is illustrated in Figure 8.13. If, for example, the incident field is a plane wave coming at a right angle to the aperture ($R_{0e} = \infty$, $\hat{i}_e = \hat{n}$, $\beta = \pi/2$), then $\rho_e = -a_e/(\hat{s}_e \cdot \hat{N}_e)$ and the spread factor (8.122) and, therefore, the scattered wave (8.120) become unbounded at $R_e = a_e/(\hat{s}_e \cdot \hat{N}_e)$, where the intersecting scattered rays form the caustic in this case. In the limit of a straight rim ($a_e = \infty$), $\rho_e = R_{0e}$ suggesting the presence of a point source "in the past" at $R_e = -R_{0e}$. Finally, for plane wave incidence ($R_{0e} = \infty$) and a straight rim ($a_e = \infty$), $\rho_e = \infty$ and the scattered wave reduces to a purely cylindrical wave.

In addition to the singularity of the amplitude of the edge-diffracted rays at caustics, there is also a $\pi/2$ phase discontinuity when the ray crosses a caustic. This is best seen from the formula (8.91) as $\tilde{\chi}_e'' = 0$ at caustics, which causes the infinity in the amplitude and the jump in the phase factor. The physical fields are of course bounded and smooth everywhere in space, implying that the asymptotic formulas (8.91) and (8.120) for the edge-diffracted rays are invalid in an immediate vicinity of caustics. However, sufficiently far from the caustics the asymptotic formulas are applicable; for example, the $\pi/2$ phase jump in the phase of an edge-diffracted ray crossing a caustic is a phenomenon that can be observed.

In the process of asymptotic evaluation of the integral over the rim it was assumed that the amplitude factor $\tilde{A}(t)$ is a slowly varying function within the domain essentially contributing to the integral. Let us formulate the condition of applicability of the approximation (8.120). The essential domain of integration can be described by the condition

$$\frac{k}{2}\,|\tilde{\chi}_e''|\,(t - t_e)^2 \le \pi, \tag{8.123}$$

and using (8.106) leads to the estimation

$$|t - t_e| \le \left(\frac{\lambda R_e |\rho_e|}{T_e \sin^2\beta |R_e + \rho_e|}\right)^{1/2}. \tag{8.124}$$

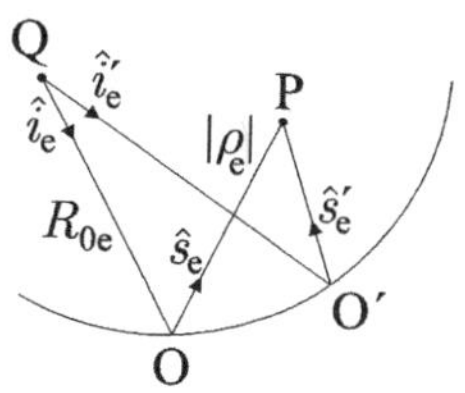

Figure 8.13 Two neighboring rays from a point source Q propagate in the directions $\hat{i}_e$ and $\hat{i}'_e$, and are scattered at points O and O′ on the curved rim of an aperture in the directions $\hat{s}_e$ and $\hat{s}'_e$, respectively. The scattered rays intersect at a point P which belongs to a caustic. The distance between O and P is equal to $|\rho_e|$.

The arc length $\delta l_{\rm es}$ of the portion of the rim around the critical point essentially contributing to scattering is obtained by multiplication of (8.124) with $2\sqrt{T_{\rm e}}$,

$$\delta l_{\rm es} = \frac{2}{\sin\beta}\left(\frac{\lambda R_{\rm e}|\rho_{\rm e}|}{|R_{\rm e}+\rho_{\rm e}|}\right)^{1/2}. \tag{8.125}$$

Due to the presence of the spread factor in (8.125), the length $\delta l_{\rm es}$ becomes infinite for observation points at caustics, which violates the assumption of locality made in the process of the asymptotic evaluation of the PO integral, and this explains why the simple ray approximations (8.91) and (8.120) do not apply at the caustics. The asymptotic formula (8.120) is valid (see approximation (8.110)) if $\delta l_{\rm es}$ is much smaller than the curvature radius at the scattering point,

$$\delta l_{\rm es} \ll a_{\rm e}. \tag{8.126}$$

For example, when $R_{\rm e}, R_{0\rm e} \to \infty$,

$$\frac{\delta l_{\rm es}}{a_{\rm e}} = \frac{2}{\sqrt{|\mathbf{q}_{\rm e}\cdot\hat{N}_{\rm e}|}}\sqrt{\frac{\lambda}{a_{\rm e}}} \tag{8.127}$$

and the applicability condition (8.126) simply suggests that the wavelength should be much shorter than the curvature radius at the scattering point. If, on the contrary, $a_{\rm e} = \infty$ but $R_{\rm e}$ and $R_{0\rm e}$ are finite (e.g., a point source illuminates a polygonal aperture), then (8.125) gives

$$\delta l_{\rm es} = \frac{2}{\sin\beta}F_1, \tag{8.128}$$

where F_1 is the radius of the first Fresnel zone defined in (8.80), implying that the scattering segment of the polygon should be greater than the first Fresnel zone in order (8.120) to apply. The Fresnel zone grows as the source and the observer go away from the aperture, and if, for example, $R_{0\rm e} = \infty$ (plane wave illumination of a polygonal aperture), the size of the essential domain is estimated by the formula

$$\delta l_{\rm es} = \frac{2}{\sin\beta}\sqrt{\lambda R_{\rm e}} \tag{8.129}$$

which grows indefinitely as $R_{\rm e} \to \infty$, implying that the asymptotic formulas of the type (8.120) are not applicable in the far zone of polygonal apertures (the far-field scattering by polygonal structures is studied in section 9.3.3).

The expression (8.120) has the form of a wave diffracted at the rim. The PO solution (8.42) and (8.43) has been derived under the general assumptions that the screen is impenetrable and the aperture is flat, without any specification of the geometric and material properties of the configuration at the rim. It is therefore instructive to compare (8.120) with the wave diffracted at the edge of a half-plane. To this end we let $a_{\rm e}$ and $R_{0\rm e}$ go to infinity (straight rim, plane wave illumination) and rewrite the formula (8.120) in the edge-fixed coordinates (5.449) and (5.450) to obtain the PO diffraction coefficient of the rim of the aperture and compare it with the exact diffraction coefficient of a PEC half-plane ((5.88) and (5.89) with $\nu = 1/2$).

By definition (5.459), the diffraction coefficient relates the $\hat{\beta}$ and $\hat{\varphi}$ components of the scattered electric field with the $\hat{\beta}_{\rm inc}$ and $\hat{\varphi}_{\rm inc}$ components of the electric field in the incident wave (Figure 5.33). Thus, multiplying (8.120) with the basis vectors $\hat{\beta}$ and $\hat{\varphi}$ and using the relation $\operatorname{sgn}\tilde{\chi}''_{\rm e} = \operatorname{sgn}(R_{\rm e}^{-1}) = 1$ we find

$$\begin{bmatrix}\hat{\beta}\cdot\mathbf{E}^{\rm po}_{\rm e}(\mathbf{r})\\ \hat{\varphi}\cdot\mathbf{E}^{\rm po}_{\rm e}(\mathbf{r})\end{bmatrix} = \frac{\exp(-jkR_{\rm e}-j\pi/4)}{2(\mathbf{q}_{\rm e}\cdot\hat{v}_{\rm e})\sqrt{2\pi kR_{\rm e}}\sin\beta}\begin{bmatrix}\hat{\beta}\cdot\mathbf{c}(\mathbf{r}_{\rm e},\mathbf{r}_0)\\ \hat{\varphi}\cdot\mathbf{c}(\mathbf{r}_{\rm e},\mathbf{r}_0)\end{bmatrix}. \tag{8.130}$$

In the formula (8.121) for the vector $\mathbf{c}(\mathbf{r}_{\rm e},\mathbf{r}_0)$, the magnetic field can be expressed through the electric field by using the plane wave relation (2.156),

$$\mathbf{c}(\mathbf{r}_{\rm e},\mathbf{r}_0) = \hat{s}_{\rm e}\times\hat{s}_{\rm e}\times\hat{n}\times\hat{i}_{\rm e}\times\mathbf{E}^{\rm inc}(\mathbf{r}_{\rm e}) - \hat{s}_{\rm e}\times\hat{n}\times\mathbf{E}^{\rm inc}(\mathbf{r}_{\rm e}), \tag{8.131}$$

and representing the incident electric field by

$$\mathbf{E}^{\text{inc}} = \hat{\beta}_{\text{inc}}(\hat{\beta}_{\text{inc}} \cdot \mathbf{E}^{\text{inc}}) + \hat{\varphi}_{\text{inc}}(\hat{\varphi}_{\text{inc}} \cdot \mathbf{E}^{\text{inc}}) \tag{8.132}$$

we can rewrite (8.130) as

$$\begin{bmatrix} \hat{\beta} \cdot \mathbf{E}_{\text{e}}^{\text{po}}(\mathbf{r}) \\ \hat{\varphi} \cdot \mathbf{E}_{\text{e}}^{\text{po}}(\mathbf{r}) \end{bmatrix} = \frac{\exp(-jkR_{\text{e}} - j\pi/4)}{\sqrt{2\pi k R_{\text{e}}}\sin\beta} \begin{bmatrix} D^{\text{po}}_{\beta\beta_{\text{inc}}} & D^{\text{po}}_{\beta\varphi_{\text{inc}}} \\ D^{\text{po}}_{\varphi\beta_{\text{inc}}} & D^{\text{po}}_{\varphi\varphi_{\text{inc}}} \end{bmatrix} \begin{bmatrix} \hat{\beta}_{\text{inc}} \cdot \mathbf{E}^{\text{inc}}(\mathbf{r}_{\text{e}}) \\ \hat{\varphi}_{\text{inc}} \cdot \mathbf{E}^{\text{inc}}(\mathbf{r}_{\text{e}}) \end{bmatrix}, \tag{8.133}$$

where

$$D^{\text{po}}_{\beta\beta_{\text{inc}}} = \frac{1}{2(\mathbf{q}_{\text{e}} \cdot \hat{\nu}_{\text{e}})}\hat{\beta} \cdot (\hat{s}_{\text{e}} \times \hat{s}_{\text{e}} \times \hat{n} \times \hat{i}_{\text{e}} \times \hat{\beta}_{\text{inc}} - \hat{s}_{\text{e}} \times \hat{n} \times \hat{\beta}_{\text{inc}}) \tag{8.134}$$

$$D^{\text{po}}_{\beta\varphi_{\text{inc}}} = \frac{1}{2(\mathbf{q}_{\text{e}} \cdot \hat{\nu}_{\text{e}})}\hat{\beta} \cdot (\hat{s}_{\text{e}} \times \hat{s}_{\text{e}} \times \hat{n} \times \hat{i}_{\text{e}} \times \hat{\varphi}_{\text{inc}} - \hat{s}_{\text{e}} \times \hat{n} \times \hat{\varphi}_{\text{inc}}) \tag{8.135}$$

$$D^{\text{po}}_{\varphi\beta_{\text{inc}}} = \frac{1}{2(\mathbf{q}_{\text{e}} \cdot \hat{\nu}_{\text{e}})}\hat{\varphi} \cdot (\hat{s}_{\text{e}} \times \hat{s}_{\text{e}} \times \hat{n} \times \hat{i}_{\text{e}} \times \hat{\beta}_{\text{inc}} - \hat{s}_{\text{e}} \times \hat{n} \times \hat{\beta}_{\text{inc}}) \tag{8.136}$$

$$D^{\text{po}}_{\varphi\varphi_{\text{inc}}} = \frac{1}{2(\mathbf{q}_{\text{e}} \cdot \hat{\nu}_{\text{e}})}\hat{\varphi} \cdot (\hat{s}_{\text{e}} \times \hat{s}_{\text{e}} \times \hat{n} \times \hat{i}_{\text{e}} \times \hat{\varphi}_{\text{inc}} - \hat{s}_{\text{e}} \times \hat{n} \times \hat{\varphi}_{\text{inc}}). \tag{8.137}$$

Now we express the elements of the diffraction coefficient tensor in (8.134)–(8.137) in terms of the angles β, φ, and φ_0 that are used to describe the diffraction coefficient of the edge of a wedge in (5.88) and (5.89). Let us define the direction of the vector $\hat{\tau}_{\text{e}}$ tangent to the rim at the diffraction point by the relation $\hat{\tau}_{\text{e}} = \hat{\nu}_{\text{e}} \times \hat{n}$ (Figure 8.12). The diffraction angle β is then given by $\cos\beta = \hat{i}_{\text{e}} \cdot \hat{\tau}_{\text{e}}$ and if the angles φ and φ_0 are measured from the direction $\hat{\nu}_{\text{e}}$ in the $\hat{\nu}_{\text{e}}$–$\hat{n}$ plane (Figure 8.14), then

$$\hat{i}_{\text{e}} = \hat{\tau}_{\text{e}}\cos\beta - \hat{\nu}_{\text{e}}\sin\beta\cos\varphi_0 - \hat{n}\sin\beta\sin\varphi_0 \tag{8.138}$$

$$\hat{s}_{\text{e}} = \hat{\tau}_{\text{e}}\cos\beta + \hat{\nu}_{\text{e}}\sin\beta\cos\varphi + \hat{n}\sin\beta\sin\varphi \tag{8.139}$$

and therefore

$$\mathbf{q}_{\text{e}} \cdot \hat{\nu}_{\text{e}} = -\sin\beta(\cos\varphi + \cos\varphi_0). \tag{8.140}$$

Inserting vectors (8.138) and (8.139) in the formulas (5.449) and (5.450) for the bases of the edge-fixed coordinates and using these in (8.134)–(8.137) lead to the simple expression for the PO diffraction coefficient of the rim,

$$\overline{\overline{D}}^{\text{po}} = -\frac{1}{2}\begin{bmatrix} \tan\dfrac{\varphi - \varphi_0}{2} & \cos\beta \\ -\cos\beta & \tan\dfrac{\varphi - \varphi_0}{2} \end{bmatrix}. \tag{8.141}$$

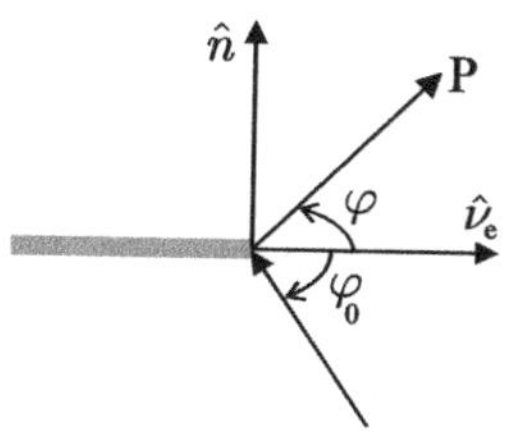

Figure 8.14 Definition of the coordinates φ and φ_0. The screen is shown as a half-plane. The incident wave comes from the half-space in front of the aperture where $-\pi < \varphi_0 < 0$. The unit normal $\hat{n}$ is directed into the half-space $0 < \varphi < \pi$ where the observation point P is located.

Compare (8.141) with the exact diffraction coefficient of a PEC half-plane,

$$\overline{\overline{D}} = -\frac{1}{2}\begin{bmatrix} \sec\frac{\varphi-\varphi_0}{2} + \sec\frac{\varphi+\varphi_0}{2} & 0 \\ 0 & \sec\frac{\varphi-\varphi_0}{2} - \sec\frac{\varphi+\varphi_0}{2} \end{bmatrix}, \tag{8.142}$$

which is obtained by specializing equations (5.88) and (5.89) to the case $\Phi = \pi$. Expressions (8.141) and (8.142) are clearly different. However, at the shadow boundary of the direct wave when $\varphi = \varphi_0 + \pi$, their diagonal elements have the same pole singularity with the unit residue, so that the difference $\overline{\overline{D}} - \overline{\overline{D}}^{\mathrm{po}}$ is a bounded function of φ in the half-plane $0 < \varphi < \pi$. On the other hand, the PO diffraction coefficient does not include the pole singularity associated with the shadow boundary of the reflected wave at $\varphi = -\pi - \varphi_0$, which is present in the exact half-plane diffraction coefficient. This disagreement is a consequence of the choice of the integration surface in the PO solution (Figure 8.5), which assumes that the observation point is located *behind* the aperture. A PO solution valid in front of the screen can be obtained by integrating over a surface that consists of the aperture and the *illuminated* side of the screen. As we shall see in section 8.4, PO solutions correctly describe the reflected fields if the integration surface includes the reflecting portions of the scattering surface.

The contribution of the stationary point of the second kind to the PO solution – function $\mathbf{E}_{\mathrm{e}}^{\mathrm{po}}$ in (8.120) – gives a qualitatively accurate description of the wave scattered at the rim of an aperture in the half-space behind the screen because its phase structure in the half-space and its singularity at the shadow boundary of the direct wave (where $\hat{i}_{\mathrm{e}} = \hat{s}_{\mathrm{e}}$) are correct. The amplitude of the wave (8.120) is however inaccurate as the PO diffraction coefficient is "unaware" of the properties of the screen near the edge. When such information is available, one can always improve the accuracy of the PO solution for the transmitted field by replacing the PO diffraction coefficient in (8.120) with an exact diffraction coefficient of a locally similar canonical wedge-shaped configuration. The result is a quite general GTD-type expression for the edge-diffracted rays, which is applicable to curved edges and to the plane or spherical wave excitation,

$$\begin{bmatrix} E_{\beta}^{\mathrm{d}}(\mathbf{r}) \\ E_{\varphi}^{\mathrm{d}}(\mathbf{r}) \end{bmatrix} = \left|\frac{\rho_{\mathrm{e}}}{R_{\mathrm{e}} + \rho_{\mathrm{e}}}\right|^{1/2} \overline{\overline{D}} \cdot \begin{bmatrix} E_{\beta_{\mathrm{inc}}}^{\mathrm{inc}}(\mathbf{r}_{\mathrm{e}}) \\ E_{\varphi_{\mathrm{inc}}}^{\mathrm{inc}}(\mathbf{r}_{\mathrm{e}}) \end{bmatrix} \frac{\exp\left(-jkR_{\mathrm{e}} - j\frac{\pi}{4}\operatorname{sgn}\tilde{\chi}_{\mathrm{e}}''\right)}{\sqrt{2\pi kR_{\mathrm{e}}}\sin\beta}. \tag{8.143}$$

This formula is not limited to the transmission problem studied here but can be used independently to describe the edge diffraction in general scattering problems.

The PO (and GTD) edge-diffraction coefficients are singular at the shadow boundary of the direct wave, which is a well-known feature of ray-optical solutions. It is important to understand that the singularity is associated with the asymptotic evaluation of the line integral which describes the contribution of the rim of the aperture. The PO integral itself is free of this singularity and remains finite for all scattering directions. The mathematical reason for the singularity in (8.120) and (8.143) at $\hat{i}_{\mathrm{e}} = \hat{s}_{\mathrm{e}}$ is simply the coalescence of the two critical points, one of the first kind at $\mathbf{r} = \mathbf{r}_{\mathrm{c}}$ and another of the second kind at $\mathbf{r} = \mathbf{r}_{\mathrm{e}}$, which violates the assumption of isolated critical points employed in our analysis of the stationary phase method in section D.2. A uniform treatment of such cases in terms of Fresnel integrals is described in, for example, Borovikov (1994) and Karatygin and Rosov (1972). Another approach is simply to calculate the PO line integral (D.85) numerically, which can be done at a relatively low computational cost even for large apertures.

The asymptotic formula (8.120) for the PO integral has further singularities at the caustics of the edge-diffracted rays. For example, if the incidence direction is perpendicular to the aperture plane and the aperture is circular, then the spread factor in (8.120) is unbounded at the symmetry axis of the aperture. In this case the phase function in the PO line integral is constant over the rim and has no separate stationary points (see (D.85)), so that all points of the rim contribute constructively to the field at the symmetry axis. The PO solution in the integral form remains, however, bounded and describes correctly the field at the caustic. The validity of PO solutions at shadow boundaries, caustics, focal points, etc., constitutes an important theoretical and practical advantage of the PO method over other high-frequency methods in the electromagnetic scattering.

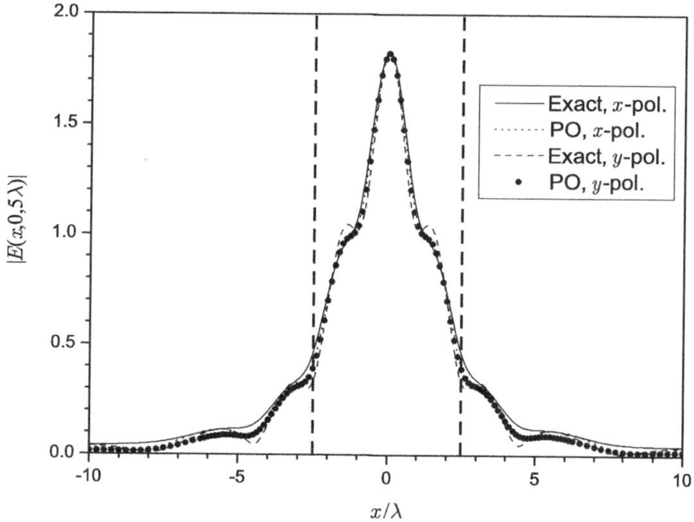

Figure 8.15 Magnitude of the electric field transmitted through a circular hole, 5λ in diameter, in a PEC screen at $z = 0$. The field is calculated behind the aperture on the line $-10 \leq x/\lambda \leq 10$, $y = 0$, $z = 5\lambda$. A plane wave of a unit amplitude is incident normally to the screen with the electric field vector along $\hat{x}$ or $\hat{y}$ directions. The vertical dashed lines indicate the size of the aperture.

We conclude this section with a numerical illustration of the accuracy of the PO method. Figures 8.15 and 8.16 compare the exact solution (solid and dashed lines) with the PO results (dotted line and dots) obtained by numerical integration of the PO solution. A plane wave polarized in $\hat{x}$ or $\hat{y}$ directions is incident along the z axis perpendicularly to a circular aperture, 5λ in diameter, in a PEC screen at $z = 0$, and the transmitted field is sampled behind the aperture over the x axis at $z = 5\lambda$ (Figure 8.15) and at $z = 50\lambda$ (Figure 8.16). The exact solution of this problem can be obtained by using Babinet's principle (section 9.3.4) from the exact solution of a complimentary problem, which is scattering of a plane

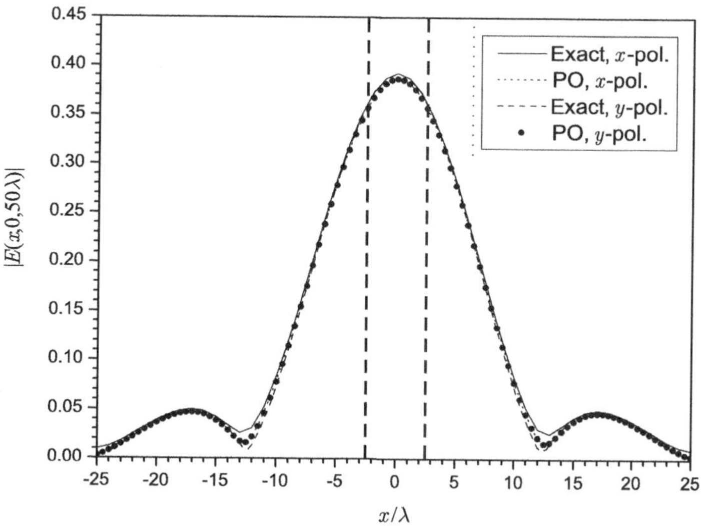

Figure 8.16 The same as Figure 8.15 but over a line $-25 \leq x/\lambda \leq 25$, $y = 0$, and $z = 50\lambda$.

wave at a circular PEC disc (Meixner and Andrejewski 1950). Despite the moderate size of the aperture compared to the wavelength, the accuracy of the PO solution is excellent both in the Fresnel (Figure 8.15) and the far (Figure 8.16) zone of the aperture. The PO solution predicts no polarization difference (the dotted line and the dots coincide within the numerical accuracy); the corresponding error is, nevertheless, small as the solid and dashed curves, which represent the exact results for alternative polarization cases, are only slightly different. An increase in the size of the aperture will lead to a further improvement in the accuracy. The PO solution becomes insufficient for apertures of one wavelength or smaller in size as the basic PO assumption about the smoothness of the geometry on the scale of λ is no longer valid.

8.4 Scattering from Curved Surfaces

In this section the PO solution (8.5) is applied to the problem of reflection and scattering from impenetrable scatterers with the boundary described by a tensor reflection coefficient (Figure 8.3). The incident field is a spherical wave from a point source at $\mathbf{r}_0$. The observation point $\mathbf{r}$ is located at a distance greater than several wavelengths from the scattering surface Ω but otherwise arbitrarily, including the Fresnel and Fraunhofer (far-field) zones. The shape of the scattering surface Ω and the positions of the source and observer are assumed to be such that no multiple reflections between different portions of the surface take place.

The PO solution is evaluated by the method of stationary phase, and the contribution from the critical points of the first kind is the focus of section 8.4.1. It will be shown that the asymptotic evaluation of the PO integral recovers the reflection laws of GO, including the Fermat shortest propagation path principle, the coplanarity of the normal vector to the scattering surface at the point of scattering and the incidence and scattering directions, the identity of the incidence and reflection angles (reflection law), and the Fresnel formulas for the amplitude of the reflected wave. We show that the contribution from the critical point of the first kind leads to the reflection formula that can be obtained for a curved surface (PEC or impedance) by combination of the Fresnel reflection laws with ray optics (Fock 1965). A PO-based derivation of the reflection formula for a PEC surface is described in Kontorovich and Muravév (1952).

The case of a scattering surface with an edge is studied asymptotically in section 8.4.3, and the contribution from a critical point of the second kind is interpreted as the edge-diffracted field in the GTD approximation. We present the PO tensor diffraction coefficient of the edge in a surface described by tensor reflection coefficients, which has the correct pole singularities at the shadow boundaries of the incident and reflected waves and can be useful in PTD-based RCS simulations and in UTD-based simulations of wireless propagation in an urban environment.

8.4.1 Fock's Reflection Formula

Let an incident spherical wave (8.41) illuminate an impenetrable scattering body Ω. The PO solution of the reflection and scattering problem is given by the formula (8.5) in which the surface currents are determined in the GO approximation. In this approximation, the currents are zero on the shadowed portion of the surface and the integration in (8.5) is, therefore, limited to Ω_0, the illuminated portion of the scattering surface. The integration surface Ω_0 is represented in the parametric form as in (8.44), with the coordinates u and v being not necessarily orthogonal. This reduces the PO integral to the form (8.46)–(8.49) with

$$\mathbf{C}(\mathbf{r}', \mathbf{r}_0) = Z\hat{s}' \times \hat{s}' \times \hat{n}' \times \mathbf{H}_0^{\mathrm{go}} - \hat{s}' \times \hat{n}' \times \mathbf{E}_0^{\mathrm{go}}, \tag{8.144}$$

where $\mathbf{H}_0^{\mathrm{go}}$ and $\mathbf{E}_0^{\mathrm{go}}$ are the amplitudes of the GO fields at a point $\mathbf{r}'$ on the scattering surface, which are related to the surface currents according to the formulas:

$$\mathbf{K}_{\mathrm{e}}^{\mathrm{go}}(\mathbf{r}') = \hat{n}' \times \mathbf{H}_0^{\mathrm{go}} \frac{\mathrm{e}^{-jk|\mathbf{r}'-\mathbf{r}_0|}}{k|\mathbf{r}'-\mathbf{r}_0|}, \quad \mathbf{K}_{\mathrm{m}}^{\mathrm{go}}(\mathbf{r}') = -\hat{n}' \times \mathbf{E}_0^{\mathrm{go}} \frac{\mathrm{e}^{-jk|\mathbf{r}'-\mathbf{r}_0|}}{k|\mathbf{r}'-\mathbf{r}_0|}. \tag{8.145}$$

The amplitudes $\mathbf{H}_0^{\mathrm{go}}$ and $\mathbf{E}_0^{\mathrm{go}}$ may depend on $\mathbf{r}'$ if reflecting properties of the scattering surface vary over the surface, but in order for the stationary phase method to apply, their variation on the scale of the wavelength must be slow (no jumps in material properties of the scattering surface near the critical point are permitted). Since the scattering surface is in general

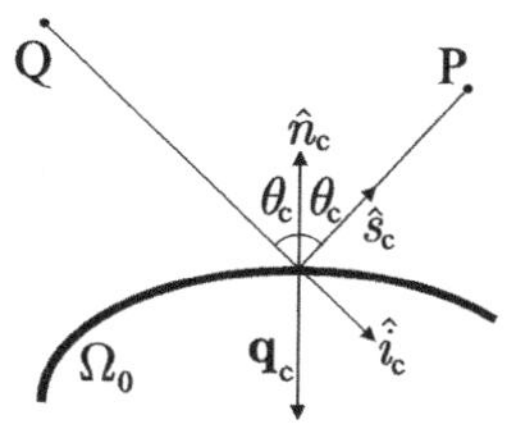

Figure 8.17 Vicinity of a critical point on the scattering surface Ω_0. The incidence and scattering directions $\hat{i}_c$ and $\hat{s}_c$ at the critical point are such that their difference $\mathbf{q}_c = \hat{i}_c - \hat{s}_c$ is orthogonal to the surface and the incidence and scattering angles are equal to θ_c.

curved, the normal vector $\hat{n}'$ is a function of the integration variables u and v and it is required that its variation be slow compared to that of the exponent factor in the integrand of (8.46).

With that said, the PO integral (8.46) can be evaluated asymptotically for $k \to \infty$ by using the stationary phase method described in section D.2. We begin with the contributions from the critical points of the first kind, which are determined by the conditions of stationarity of the phase function (8.48),

$$\frac{\partial \chi}{\partial u} = 0, \quad \frac{\partial \chi}{\partial v} = 0, \tag{8.146}$$

leading to the equations

$$\mathbf{q} \cdot \mathbf{f}_u = 0, \quad \mathbf{q} \cdot \mathbf{f}_v = 0 \tag{8.147}$$

with $\mathbf{q} = \hat{i}' - \hat{s}'$. The unit vectors $\hat{i}'$ and $\hat{s}'$, given by (8.50) and (8.51), define the direction from the source $\mathbf{r}_0$ to the integration point $\mathbf{r}'$ and from $\mathbf{r}'$ to the observation point $\mathbf{r}$, respectively. Let $u = u_c$ and $v = v_c$ satisfy (8.147) and define an isolated critical point $\mathbf{r}_c = \mathbf{f}(u_c, v_c)$ on Ω_0,[4] and let us denote the values of the vectors $\hat{n}'$, $\mathbf{q}$, $\hat{i}'$, and $\hat{s}'$ at the critical point by $\hat{n}_c$, $\mathbf{q}_c$, $\hat{i}_c$, and $\hat{s}_c$, respectively. Since vectors $\mathbf{f}_u(u, v)$ and $\mathbf{f}_v(u, v)$ are tangential to the surface, equations (8.147) imply that (a) vector $\mathbf{q}_c$ is orthogonal to the surface at the point $\mathbf{r}_c$, (b) vectors $\hat{n}_c$, $\hat{i}_c$, and $\hat{s}_c$ lie in the same plane, and (c) if θ_c $(0 \le \theta_c \le \pi/2)$ is the angle between $\hat{n}_c$ and $\hat{s}_c$ (Figure 8.17), then

$$\hat{n}_c \cdot \hat{s}_c = -\hat{n}_c \cdot \hat{i}_c = \cos\theta_c \tag{8.148}$$

and

$$\mathbf{q}_c = -2\cos\theta_c \hat{n}_c. \tag{8.149}$$

The geometric relations that take place at the critical point $\mathbf{r}_c$ can be illustrated by the following elegant construction (Fedoryuk 1989). Consider two points P and Q located outside of a convex body Ω and an ellipse with the foci at P and Q. Rotating the ellipse around the axis PQ defines a spheroid, the size of which we will now vary until the spheroid touches the surface of Ω (Figure 8.18). Since both bodies are convex, there is only one such point O. The sum of distances to the foci is the same for all points of a given spheroid, and since the spheroid E is the smallest one that still intersects Ω, the point O is such point of Ω for which the sum of distances to Q and P is at minimum. The point O apparently coincides with the critical point because $\mathbf{r}_c$ is determined from the conditions of minimum of the function $\chi = |\mathbf{r} - \mathbf{r}'| + |\mathbf{r}' - \mathbf{r}_0|$, which is the total length of the path from Q to P through a point on Ω.

Since the surfaces of E and Ω are tangent to each other at the point O, the normal $\hat{n}$ to the surface of Ω at O is at the same time orthogonal to the surface of the spheroid and, as such, points to the QP axis and lies in the plane defined by the directions QO and OP. This is a geometric explanation of why the vectors $\hat{n}_c$, $\hat{i}_c$, and $\hat{s}_c$ lie in the same plane, the plane of incidence. Finally, the identity of the angle of incidence to the angle of reflection corresponds to a basic property

[4] For compact scatterers, equations (8.147) have always a second,"parasitic" solution which corresponds to a point on the shadowed portion of the scatterer. This critical point apparently brings no contribution to the scattered field in the PO approximation.

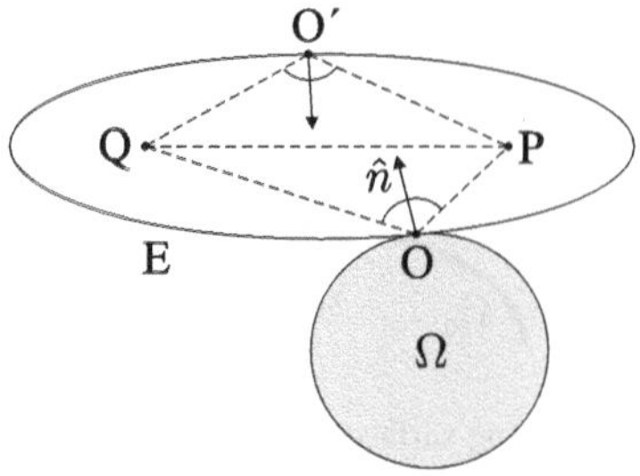

Figure 8.18 Spheroid E with foci at P and Q touches the surface of a body Ω at a point O. At any point of the spheroid's boundary (a) the sum of the distances to the foci is the same, QO+OP=QO′+O′P, and (b) the normal to the boundary bisects the angle between the directions to the foci.

of ellipses and spheroids, according to which every normal to the boundary of an ellipse or a spheroid bisects the angle between the directions to the foci (e.g., Korn and Korn (1968, section 2.5)).

Thus, the asymptotic analysis of the PO solution recovers the reflection laws of GO. Next we evaluate the contribution from the critical point $\mathbf{r}_c$ by using the asymptotic formula (D.60). The evaluation of coefficients α, β, and γ is similar to that described in section 8.3 since equations (8.52)–(8.58) are still applicable. In contrast, equations (8.59)–(8.61) have to be modified by including the terms proportional to $\mathbf{q}_c$ since in contrast to the transmission problem, $\mathbf{q}_c$ does not vanish in problems of reflection and scattering (see (8.149)). The modified equations read

$$\alpha = p[(\hat{i}_c \cdot \mathbf{f}_u)^2 - |\mathbf{f}_u|^2] - (\mathbf{q}_c \cdot \mathbf{f}_{uu}) \tag{8.150}$$
$$\beta = p[(\hat{i}_c \cdot \mathbf{f}_v)^2 - |\mathbf{f}_v|^2] - (\mathbf{q}_c \cdot \mathbf{f}_{vv}) \tag{8.151}$$
$$\gamma = p[(\hat{i}_c \cdot \mathbf{f}_u)(\hat{i}_c \cdot \mathbf{f}_v) - (\mathbf{f}_u \cdot \mathbf{f}_v)] - (\mathbf{q}_c \cdot \mathbf{f}_{uv}), \tag{8.152}$$

in which all the functions of coordinates are taken at the critical point. Here,

$$p = \frac{1}{R} + \frac{1}{R_0} \tag{8.153}$$

with

$$R = |\mathbf{r} - \mathbf{r}_c|, \quad R_0 = |\mathbf{r}_c - \mathbf{r}_0|. \tag{8.154}$$

It is now convenient to express the coefficients in (8.150)–(8.152) in terms of the fundamental forms which are used in differential geometry to define the curvature of a surface in space (section A.4). Using (8.149) and the definitions of the fundamental forms (A.67) and (A.71), equations (8.150)–(8.152) can be expressed as

$$\alpha = p\left[(\hat{i}_c \cdot \mathbf{f}_u)^2 - E\right] \pm 2L\cos\theta_c \tag{8.155}$$
$$\beta = p[(\hat{i}_c \cdot \mathbf{f}_v)^2 - G] \pm 2N\cos\theta_c \tag{8.156}$$
$$\gamma = p[(\hat{i}_c \cdot \mathbf{f}_u)(\hat{i}_c \cdot \mathbf{f}_v) - F] \pm 2M\cos\theta_c, \tag{8.157}$$

where the upper sign in expressions (8.155)–(8.157) is to be taken when the unit normal defined in (A.70) is directed into the outside of the scattering body and, thus, $\hat{n} = \hat{n}_c$. Otherwise, $\hat{n} = -\hat{n}_c$ and the lower sign should be used. This sign convention is assumed in all the expressions that follow in this section.

To apply the formula (D.60), the quantity $\alpha\beta - \gamma^2$ has to be calculated. Using the expressions (8.155)–(8.157), combined with definitions (A.72)–(A.74) for the curvature radii, one can arrive at the following expression

$$\frac{\alpha\beta - \gamma^2}{EG - F^2} = p^2\cos^2\theta_c \mp 2p\cos\theta_c\left(2H\cos^2\theta_c + \frac{1}{R_i}\sin^2\theta_c\right) + 4K\cos^2\theta_c, \tag{8.158}$$

where R_i is the curvature radius of the surface in the plane of incidence, which is the plane defined by the vectors $\hat{\imath}_c$ and $\hat{n}_c$. The values of the surface parameters E, G, F, H, K, and R_i are taken at the critical point. The radius R_i can be calculated from (A.72) by using the formula

$$\frac{dv}{du} = -\frac{\mathbf{f}_u \cdot (\hat{\imath}_c \times \hat{n}_c)}{\mathbf{f}_v \cdot (\hat{\imath}_c \times \hat{n}_c)} = \frac{\hat{\imath}_c \cdot (\mathbf{f}_v E - \mathbf{f}_u F)}{\hat{\imath}_c \cdot (\mathbf{f}_u G - \mathbf{f}_v F)}. \tag{8.159}$$

Thus, the contribution from the critical point of the first kind to the PO integral describing the electric field reflected/scattered from a curved surface reads

$$\mathbf{E}_c(\mathbf{r}) = j\sigma_c \frac{\mathbf{C}(\mathbf{r}_c, \mathbf{r}_0)}{2k \cos\theta_c R_0} \left(\frac{D(0)}{D(R)}\right)^{\frac{1}{2}} e^{-jk(R+R_0)}, \tag{8.160}$$

where the parameter σ_c is defined in (D.61),

$$D(R) = R^2 \left| p^2 \cos^2\theta_c \mp 2p \cos\theta_c \left(2H \cos^2\theta_c + R_i^{-1} \sin^2\theta_c\right) + 4K \cos^2\theta_c \right|, \tag{8.161}$$

and $D(0) = \cos^2\theta_c$ is the limiting value of (8.161) at $R = 0$.

Expression (8.160) can be transformed into the formula derived for the field reflected from a curved surface in the GO approximation by Fock (1965). To this end, we will now show that

$$\frac{\mathbf{C}(\mathbf{r}_c, \mathbf{r}_0)}{kR_0} e^{-jkR_0} = 2\cos\theta_c \mathbf{E}_{ref}(\mathbf{r}_c), \tag{8.162}$$

where $\mathbf{E}_{ref}(\mathbf{r}_c)$ is the surface value of the reflected field obtained from the Fresnel formula (4.79). We begin by noting that

$$\frac{\mathbf{C}(\mathbf{r}_c, \mathbf{r}_0)}{kR_0} e^{-jkR_0} = Z\hat{s}_c \times \hat{s}_c \times \mathbf{K}_e^{go}(\mathbf{r}_c) + \hat{s}_c \times \mathbf{K}_m^{go}(\mathbf{r}_c), \tag{8.163}$$

where the equivalent surface currents in the tangent plane approximation are these given in (4.93) and (4.96) for a flat surface illuminated by a plane wave,

$$Z\mathbf{K}_e^{go}(\mathbf{r}_c) = (R_\perp - 1)(\hat{\imath}_c \cdot \hat{n}_c) E_{inc}^{\perp} \hat{\alpha}_{inc} - (1 + R_\parallel) E_{inc}^{\parallel} \hat{\imath}_c \tag{8.164}$$

$$\mathbf{K}_m^{go}(\mathbf{r}_c) = (1 - R_\parallel)(\hat{\imath}_c \cdot \hat{n}_c) E_{inc}^{\parallel} \hat{\alpha}_{inc} - (1 + R_\perp) E_{inc}^{\perp} \hat{\imath}_c. \tag{8.165}$$

The mutually orthogonal vectors $\hat{\alpha}_{inc}$ and $\hat{\imath}_c$, defined in (4.71) and (4.81), are tangential to the surface at the critical point, first of which is perpendicular to the plane of incidence while the second one lies in the plane of incidence.

It is apparent from (8.163) that the vector $\mathbf{C}(\mathbf{r}_c, \mathbf{r}_0)$ has no component in the direction of $\hat{s}_c$, so it can be represented as the linear combination

$$\frac{\mathbf{C}(\mathbf{r}_c, \mathbf{r}_0)}{kR_0} e^{-jkR_0} = A\hat{\alpha}_{ref} + B\hat{\beta}_{ref} \tag{8.166}$$

with $\hat{\alpha}_{ref} = \hat{\alpha}_{inc}$ and $\hat{\beta}_{ref} = \hat{s}_c \times \hat{\alpha}_{inc}$, which are mutually orthogonal unit vectors defined in (4.75) and constituting the basis of the reflected/scattered wave. To determine coefficients A and B in (8.166), we insert (8.164) and (8.165) in (8.163) and multiply (8.163) with $\hat{\alpha}_{ref}$ and $\hat{\beta}_{ref}$, respectively. Evaluation of the vector products is based on relations (8.148), the property $\hat{s}_c \cdot \hat{\alpha}_{inc} = 0$, and the formulas for the scalar and vector triple products (A.11) and (A.12). When multiplying with $\hat{\alpha}_{ref}$, the vector identities

$$\hat{\alpha}_{inc} \cdot (\hat{s}_c \times \hat{\alpha}_{inc}) = 0 \tag{8.167}$$

$$\hat{\alpha}_{inc} \cdot (\hat{s}_c \times \hat{n}_c \times \hat{\alpha}_{inc}) = -\cos\theta_c \tag{8.168}$$

$$\hat{\alpha}_{inc} \cdot (\hat{s}_c \times \hat{s}_c \times \hat{\alpha}_{inc}) = -1 \tag{8.169}$$

$$\hat{\alpha}_{inc} \cdot (\hat{s}_c \times \hat{s}_c \times \hat{n}_c \times \hat{\alpha}_{inc}) = 0 \tag{8.170}$$

are used, which yields

$$A = 2\cos\theta_c R_\perp E^\perp_{inc}. \tag{8.171}$$

To calculate the product of (8.163) with $\hat{\beta}_{ref}$, we note that

$$\hat{\beta}_{ref} \cdot \mathbf{C} = \mathbf{C} \cdot (\hat{s}_c \times \hat{\alpha}_{inc}) = -\hat{\alpha}_{inc} \cdot (\hat{s}_c \times \mathbf{C}) \tag{8.172}$$

and using (8.169), (8.170), and the relations

$$\hat{\alpha}_{inc} \cdot (\hat{s}_c \times \hat{s}_c \times \hat{s}_c \times \hat{\alpha}_{inc}) = 0 \tag{8.173}$$

$$\hat{\alpha}_{inc} \cdot (\hat{s}_c \times \hat{s}_c \times \hat{s}_c \times \hat{n}_c \times \hat{\alpha}_{inc}) = \cos\theta_c \tag{8.174}$$

arrive at the formula

$$B = 2\cos\theta_c R_\parallel E^\parallel_{inc}. \tag{8.175}$$

Inserting (8.171) and (8.175) in (8.166) and using (4.79) lead to the relation (8.162).

The contribution (8.160) from the stationary point of the first kind can therefore be expressed as

$$\mathbf{E}_c(\mathbf{r}) = j\sigma_c \mathbf{E}_{ref}(\mathbf{r}_c) \left(\frac{D(0)}{D(R)}\right)^{\frac{1}{2}} e^{-jkR}. \tag{8.176}$$

This expression, similar to the Fock reflection formula (Fock 1965),[5] has a simple physical interpretation as the field reflected from a curved surface in the GO approximation. Indeed, in a vicinity of the reflection point where $R \ll |R_{1,2}|$ and the surface appears almost flat, equation (8.161) gives $D(R) \approx R^2 p^2 \cos^2\theta_c$. The value of σ_c is determined by (D.61), and from (8.155) and (8.158) it follows that

$$\alpha\beta - \gamma^2 \approx (EG - F^2)p^2\cos^2\theta_c > 0, \quad \alpha \approx -pE\left[1 - \left(\hat{i}_c, \frac{\mathbf{f}_u}{|\mathbf{f}_u|}\right)^2\right] < 0 \tag{8.177}$$

and, therefore, $\sigma_c = -j$. Thus, the reflection formula (8.176) simplifies to

$$\mathbf{E}_c(\mathbf{r}) = \mathbf{E}_{ref}(\mathbf{r}_c)\frac{R_0}{R + R_0} e^{-jkR}, \tag{8.178}$$

[5] The Fock reflection formula does not contain the factor $j\sigma_c$, which describes the jump in the phase of a ray at a caustic. The reason is apparently that Fock considered the reflected field between the reflection point and the nearest caustic where $j\sigma_c = 1$.

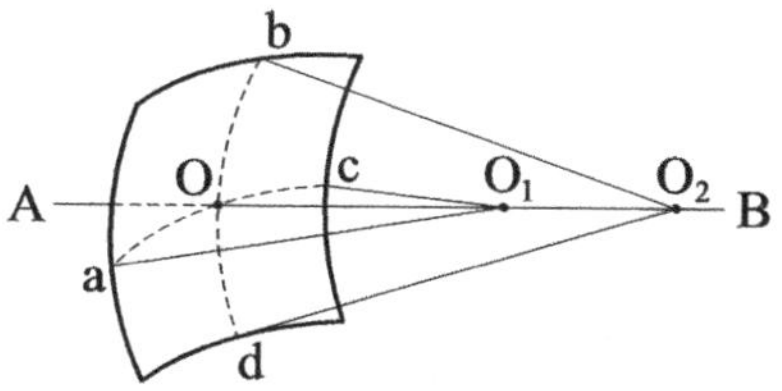

Figure 8.19 An infinitesimal piece of a curved wavefront. The arcs ac and bd are elements of the two circles centered at the points O_1 and O_2, respectively. The circles define the two principal curvature radii OO_1 and OO_2 of the wavefront element. In the framework of GO, the amplitude of the ray AB is infinite at O_1 and O_2. These points belong to the caustics, which are the loci of all the curvature centers of a given wavefront.

which is the GO approximation of the field of a point source reflected from an infinite flat surface. For a plane incident wave ($R_0 \to \infty$), the formula (8.178) simplifies further

$$\mathbf{E}_c(\mathbf{r}) = \mathbf{E}_{ref}(\mathbf{r}_c)e^{-jkR} \tag{8.179}$$

and describes the reflected field as a plane wave propagating in the specular direction.

Thus, close to the reflection point ($R \ll R_0$, $R \ll |R_{1,2}|$), the reflected field looks like a plane wave reflected from a flat surface tangential to the scattering surface at the point of reflection. As the distance R increases, the amplitude factor $D(0)/D(R)$ begins to deviate from unity and the reflected field from the plane wave. Hence, the quantity $D(0)/D(R)$ accounts for the spread or contraction of the reflected beam due to the curvature of the scattering surface and is often referred to as the spread factor.

Reflection at a curved surface results in a curved wavefront of the reflected wave. Every infinitesimal piece of the wavefront propagates along the direction of a local normal to the wavefront and since the front is curved, different pieces of the wavefront may propagate in different directions, which leads to local contractions or broadenings of the wavefront and, therefore, to the increase or decay of the field amplitude along the rays. Figure 8.19 illustrates the relation between the curvature of the front and the amplitude of the reflected wave (Landau and Lifshitz 1980). Consider an infinitesimal piece of the wavefront abcd that propagates along the direction AB. Since the selected wavefront element is small, it can be approximated by a quadratic surface with, generally speaking, two different principal curvature radii $\rho_1 = OO_1$ and $\rho_2 = OO_2$. The area of the surface element is proportional to the product $\rho_1\rho_2$ and it is this factor that defines the variation of the wavefront element area as the wavefront propagates. The intensity, which is the density of the energy flow, is inversely proportional to the area of the selected wavefront element and, hence, the amplitude varies as $(\rho_1\rho_2)^{-1/2}$.

These simple intensity considerations can be used to write down an explicit formula for the field of a wave with the curved wavefront. The field at a point O′ shifted from O by a distance s is apparently related to its value at O by the formula

$$\mathbf{E}(O') = \mathbf{E}(O)\sqrt{\frac{\rho_1\rho_2}{(\rho_1 + s)(\rho_2 + s)}}e^{-jks}, \tag{8.180}$$

which is the well-known expression for the GO field (e.g., Born and Wolf (1959, p.115) and Kline and Kay (1965, p.189)). This expression is in agreement with the reflection formula (8.176) if we choose the point O to be the reflection point $\mathbf{r}_c$, in which case $R = s$. The spread factors in (8.176) and (8.180) agree as the factor $D(R)$ given by (8.161) is indeed a second-degree polynomial in R. The curvature radii ρ_1 and ρ_2 of the wavefront of the reflected wave can be extracted from the formula (8.161) as functions of the curvatures R_1 and R_2 of the reflecting surface at the reflection point and of the curvature R_0 of the incident wave. Expressions for the curvatures of the reflected wave in the case of a more general incident wave are derived in Deschamps (1972).

As it is apparent from the previous construction, the amplitude of a curved wavefront element abcd is infinite when $\rho_1 = 0$ or $\rho_2 = 0$, that is, at the centers of curvature of the surface element. Applying this result to the whole wavefront

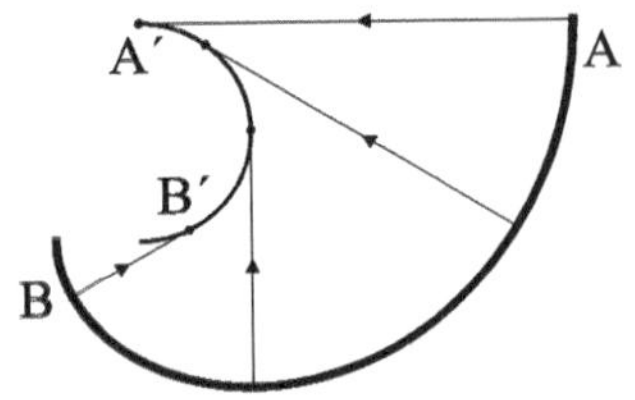

Figure 8.20 Cross section AB of a wavefront and the corresponding caustic A′B′. Every ray leaves the wavefront at the right angle and touches the caustic as a tangent at the distance equal to the curvature radius of the wavefront at the starting point of the ray. The cross sections of the wavefront and the caustic relate to each other as the evolute (wavefront) and the involute (caustic).

of the reflected wave, which consists of many surface elements, we conclude that the reflected wave may become infinite at two surfaces which are the loci of curvature centers of all the surface elements. These surfaces are called caustics. For wavefronts with a rotational symmetry, a caustic surface can degenerate into a line or even a point (a focus).

A geometric construction of a caustic is often simplified by building a cross section of the corresponding wavefront with a plane perpendicular to the wavefront, which is not always possible. Such a cross section is then a plane curve that can be considered as the evolute and the corresponding cross section of the caustic as the involute (Figure 8.20). In differential geometry, the evolute of a curve is defined as the locus of all its centers of curvature or, equivalently, as the envelope of the normals to a curve. The original curve is an involute of its evolute. An important property of the evolute/involute relationship is that the rays (i.e., the normals to a wavefront) are always tangent to the caustic surfaces.

A curved wavefront does not necessarily lead to caustic singularities, for example this is the case when the front is convex and propagates in the direction opposite to the location of the curvature centers (from B to A in Figure 8.19). Analytical determination of the location of caustics for a reflected wave given by the formula (8.176) requires finding zeros of $D(R)$. If R_c is a zero of $D(R)$ for the reflection point $\mathbf{r}_c$ and the reflection direction $\hat{s}_c$, then $\mathbf{r} = \mathbf{r}_c + \hat{s}_c R_c$ is a point on a caustic.

Now let us discuss the role of the parameter σ_c in the reflection formula (8.176). The parameter is a piecewise constant given by (D.61), which has a discontinuity when the combination $\alpha\beta - \gamma^2$ changes its sign, that is, when $\alpha\beta - \gamma^2 = 0$. This is exactly the condition of vanishing $D(R)$, as is apparent from the general formula (D.60), and thus the parameter σ_c describes a $\pi/2$ jump in the phase of the reflected ray when it touches a caustic surface.

The infinity in the amplitude and the discontinuity in the phase predicted by the reflection formula (8.176) is a consequence of its asymptotic nature. The physical fields must be bounded and continuous functions in space. This suggests that at caustics and in asymptotically small vicinities of them the reflection formula is inapplicable and the original PO integral cannot be evaluated by using the simple formula (D.60) for the contribution of an isolated critical point of the first kind. To describe the field near a caustic, the PO integral should be evaluated with a more sophisticated asymptotic approximation (Borovikov 1994) or simply by numerical integration. A more accurate analysis shows that at caustics the field amplitudes achieve high but always bounded values and the phases undergo the $\pi/2$ change in a rapid but continuous manner within the caustic layer (see problems 3 and 15). Examples of the field distributions near caustics are presented in Figures 8.21, 8.34, 8.35, and D.3.

8.4.2 *Application to a Spherical Segment*

It is instructive to apply the general formulas developed in this section to the important particular case of a spherical segment. It is necessary to differentiate between the two cases in which the reflection point lies either at the convex or concave side of the segment (Figure 8.22). The surface can be represented in the parametric form

$$\mathbf{f}(u, v) = a(\hat{x}\sin u\cos v + \hat{y}\sin u\sin v + \hat{z}\cos u), \tag{8.181}$$

where a denotes the radius of the segment. Parameters u and v are the spherical angles defined over suitable intervals. For segments rotationally symmetric with respect to the z axis, $0 \le v < 2\pi$. The u variable describes the angular width of the

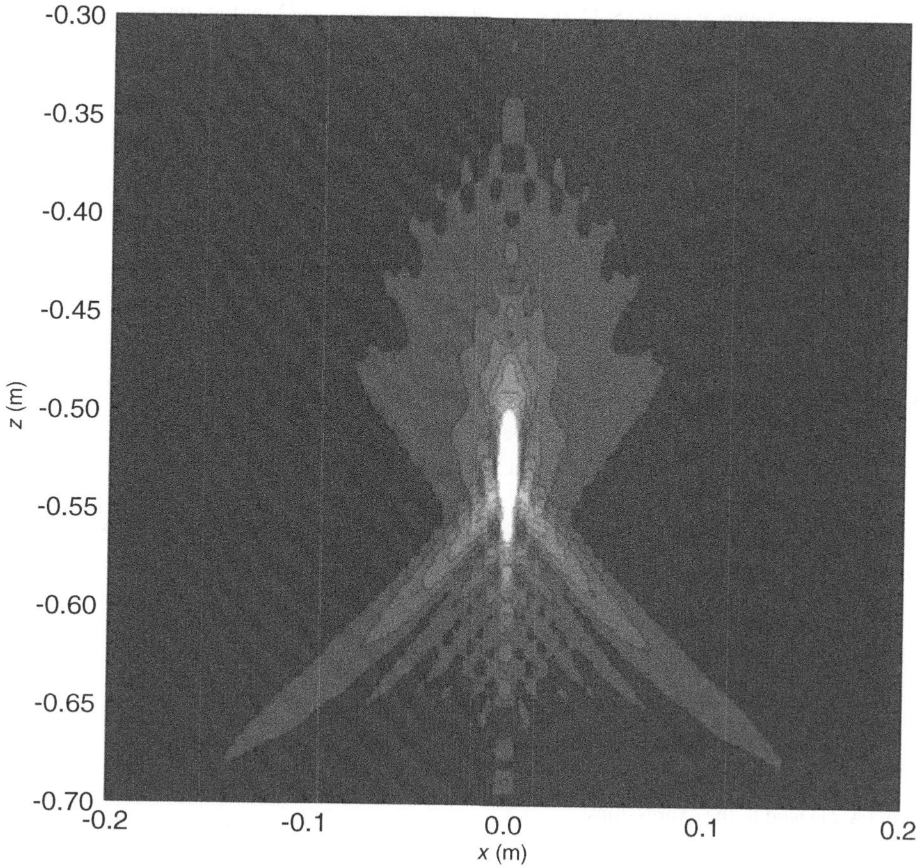

Figure 8.21 Magnitude distribution of the electric field in the focal region of a PEC spherical reflector under axial illumination. The geometry is shown in Figure 8.40 and the parameters are as follows: $a = 1$ m, $\theta_0 = 150°$, $\lambda = 0.01$ m. The illuminated side of the reflector is represented as a set of 9720 triangular facets (panels) with the side length smaller than 0.036 m. The plot shows the field distribution in the plane $y = 0$ within a square region ($-0.2 \leq x$ (m) ≤ 0.2 and $-0.7 \leq z$ (m) ≤ -0.3) around the cusp of the caustics at $x = 0$ and $z = -0.5$ m. To make the fine structure of the field distribution visible, the points with the field magnitude exceeding the half maximum level are colored white.

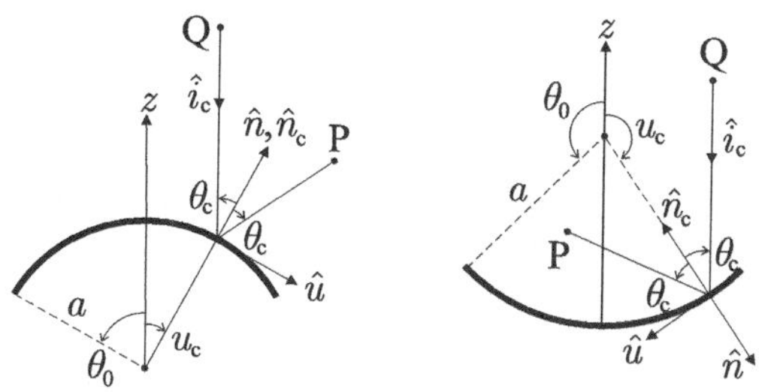

Figure 8.22 Reflection from the convex (to the left) and concave (to the right) side of a spherical segment.

segments and for the shapes shown in Figure 8.22, $0 \leq u \leq \theta_0$ with $0 < \theta_0 \leq \pi/2$ for the convex segment and $\theta_0 \leq u \leq \pi$ with $\pi/2 < \theta_0 < \pi$ for the concave segment.

The partial derivatives of $\mathbf{f}(u, v)$ with respect to u and v are

$$\mathbf{f}_u(u, v) = a(\hat{x}\cos u \cos v + \hat{y}\cos u \sin v - \hat{z}\sin u) \tag{8.182}$$

$$\mathbf{f}_v(u, v) = a(-\hat{x}\sin u \sin v + \hat{y}\sin u \cos v), \tag{8.183}$$

and from (A.67) we find the first fundamental forms of the surface

$$E = a^2, \quad G = a^2 \sin^2 u, \quad F = 0. \tag{8.184}$$

Using (8.182)–(8.184) in (A.70) permits specification of the unit vector $\hat{n}$ normal to the surface

$$\hat{n} = \hat{x}\sin u \cos v + \hat{y}\sin u \sin v + \hat{z}\cos u \tag{8.185}$$

which is directed out of the curvature center in the convex case and to the curvature center in the concave case (Figure 8.22).

Determination of the second fundamental forms requires knowledge of the second-order partial derivatives of $\mathbf{f}(u, v)$, and since

$$\mathbf{f}_{uu}(u, v) = -a(\hat{x}\sin u \cos v + \hat{y}\sin u \sin v + \hat{z}\cos u) \tag{8.186}$$

$$\mathbf{f}_{uv}(u, v) = \mathbf{f}_{vu}(u, v) = a\cos u(-\hat{x}\sin v + \hat{y}\cos v) \tag{8.187}$$

$$\mathbf{f}_{vv}(u, v) = -a\sin u(\hat{x}\cos v + \hat{y}\sin v), \tag{8.188}$$

the definitions in (A.71) lead to the expressions:

$$L = -a, \quad M = 0, \quad N = -a\sin^2 u. \tag{8.189}$$

Once the fundamental forms are determined, the principal curvature radii result from (A.73) and (A.74) as

$$R_1 = R_2 = -a \tag{8.190}$$

and the Gaussian and mean curvatures from (A.75) as

$$K = \frac{1}{a^2}, \quad H = -\frac{1}{a}. \tag{8.191}$$

The curvature radii in (8.190) are negative because the vector $\hat{n}$ defined in (8.185) points in the direction opposite to the direction to the curvature center. The curvature radius R_i for a spherical surface results from (A.72), (8.184), and (8.189) as

$$R_i = \frac{a^2(du)^2 + a^2\sin^2 u(dv)^2}{-a(du)^2 - a\sin^2 u(dv)^2} = -a \tag{8.192}$$

and is independent of the orientation of the plane of incidence, as expected.

Now we are ready to calculate the quantities α and $\alpha\beta - \gamma^2$ to completely specify the reflection formula. We begin with the convex segment. The vector $\hat{n}$ is directed out of the curvature center of the reflecting surface in this case. We choose therefore the upper sign in (8.155) and (8.158) to get

$$\alpha = -pa^2\left[1 - (\hat{i}_c \cdot \hat{e}_u)^2\right] - 2a\cos\theta_c \tag{8.193}$$
$$\alpha\beta - \gamma^2 = a^2\cos^2\theta_c \sin^2 u_c(pa + 2\cos\theta_c)(pa + 2\sec\theta_c), \tag{8.194}$$

where $\hat{e}_u = \mathbf{f}_u/a$ is a unit vector and all functions of u and v are evaluated at the reflection point, at which $u = u_c$ and $v = v_c$. According to its definition (8.153), the parameter p is always positive and, thus, for arbitrary values of R and R_0, equations (8.193) and (8.194) lead to the estimations $\alpha < 0$ and $\alpha\beta - \gamma^2 > 0$ which, because of (D.61), imply that $\sigma_c = -j$ everywhere on the reflected ray. The function $D(R)$ follows from (8.161), (8.191), and (8.192) as

$$D(R) = \left(\frac{R}{a}\right)^2 \cos^2\theta_c |pa + 2\cos\theta_c||pa + 2\sec\theta_c|, \tag{8.195}$$

which gives for the spread factor

$$\sqrt{\frac{D(0)}{D(R)}} = \frac{a}{R}|pa + 2\cos\theta_c|^{-1/2}|pa + 2\sec\theta_c|^{-1/2}. \tag{8.196}$$

The spread factor (8.196) is equal to unity at the reflection point ($R = 0$) and, as R grows, monotonically decays, approaching the $O(R^{-1})$ dependence at $R \to \infty$.

The case of reflection from the concave side of the spherical segment is more complicated because the reflected ray may touch caustics. The vector $\hat{n}$, which is defined in (8.185), is directed to the curvature center of of the reflecting surface, and choosing therefore the lower sign in (8.155) and (8.158) we obtain

$$\alpha = -pa^2\left[1 - (\hat{i}_c \cdot \hat{e}_u)^2\right] + 2a\cos\theta_c \tag{8.197}$$
$$\alpha\beta - \gamma^2 = a^2\cos^2\theta_c \sin^2 u_c(pa - 2\cos\theta_c)(pa - 2\sec\theta_c). \tag{8.198}$$

As p varies from 0 to $+\infty$, the quantities in (8.197) and (8.198) change their sign. The quantity $\alpha\beta - \gamma^2$ is negative on the interval $2\cos\theta_c < pa < 2\sec\theta_c$ and positive otherwise. The parameter α is negative in a vicinity of the reflection point ($p \to \infty$) but becomes positive for sufficiently small values of p, namely when $p < p_0$ with

$$p_0 = \frac{1}{a}\frac{2\cos\theta_c}{1 - (\hat{i}_c \cdot \hat{e}_u)^2}. \tag{8.199}$$

Since $\cos^2\theta_c \leq 1 - (\hat{i}_c \cdot \hat{e}_u)^2 \leq 1$, (8.199) implies that $2\cos\theta_c \leq p_0 a \leq 2\sec\theta_c$, and thus, depending on the value of p, we have the four cases: (1) $\alpha > 0$ and $\alpha\beta - \gamma^2 > 0$ when $0 \leq pa < 2\cos\theta_c$; (2) $\alpha > 0$ and $\alpha\beta - \gamma^2 < 0$ when $2\cos\theta_c < pa < p_0 a$; (3) $\alpha < 0$ and $\alpha\beta - \gamma^2 < 0$ when $p_0 a < pa < 2\sec\theta_c$; (4) $\alpha < 0$ and $\alpha\beta - \gamma^2 > 0$ when $2\sec\theta_c < pa \leq +\infty$. Using (D.61) and summarizing, we find that

$$\sigma_c = \begin{cases} j & \text{for } 0 \leq pa < 2\cos\theta_c \\ 1 & \text{for } 2\cos\theta_c < pa < 2\sec\theta_c \\ -j & \text{for } 2\sec\theta_c < pa \leq +\infty \end{cases} \tag{8.200}$$

and, thus, in the case of a concave surface the parameter σ_c is a piecewise constant with stepwise discontinuities at particular values of p.

These values of p corresponds exactly to the points of contact of the reflected ray with the caustics. Indeed, the spread factor in the concave case is given by formula

$$\sqrt{\frac{D(0)}{D(R)}} = \frac{a}{R}|pa - 2\cos\theta_c|^{-1/2}|pa - 2\sec\theta_c|^{-1/2} \tag{8.201}$$

and in contrast to (8.196), this expression is infinite at $pa = 2\sec\theta_c$ and $pa = 2\cos\theta_c$, which coincide with the position of the discontinuities in (8.200). Using the definition of the parameter p in (8.153), we can express the distance R between the reflection point and the points of contact with the caustics as

$$R = \begin{cases} \dfrac{aR_0}{2R_0\sec\theta_c - a} \\ \dfrac{aR_0}{2R_0\cos\theta_c - a} \end{cases}. \tag{8.202}$$

At these distances, the amplitude of the reflected field in the GO approximation becomes infinite and the parameter σ_c, which describes the jumps in the phase of the reflected wave on touching a caustic, is multiplied by the factor j if the observer moves away from the reflection point and by $-j$ if the observer approaches the reflection point.

Equations (8.202) define, in general, two caustic surfaces which, depending on the position of the source, may have common points, degenerate in a line or disappear at infinity; in the latter case the distance R in (8.202) is negative.

The reflection formula (8.176) combines the laws of reflection and the laws of ray optics as it consists of the GO approximation for the reflected field on the surface, the beam-broadening factor, and the phase factor. Despite its asymptotic nature, the reflection formula can provide good numerical accuracy when its applicability conditions (smooth surface on the scale of the wavelength within the whole field-forming region) are satisfied. Figure 8.23 compares the exact and approximate solutions for a PEC sphere. The spread factor in the GO formula (8.196) specialized to the case of plane wave excitation ($R_0 \to \infty$) and backscattering ($\theta_c = 0$) accurately describes the decay of the reflected field with growing distance from the reflection point. The amplitudes of the GO and exact fields are quite close to each other, even for smaller spheres. At the surface of the sphere ($R = 0$), the GO value for the tangential component of the electric field is exact. To present the accuracy of the reflection formula better, the relative error, which accounts for the errors in the

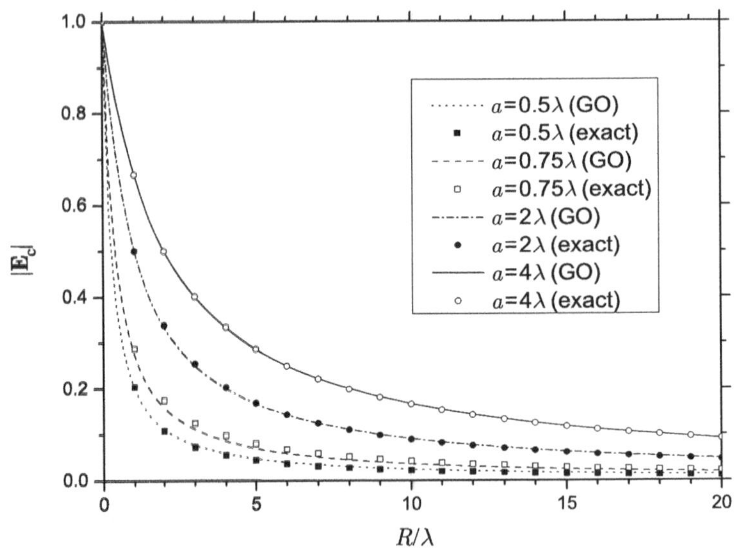

Figure 8.23 Amplitude of the field reflected from a PEC sphere in the backscattering direction as a function of the distance from the reflection point for several values of the sphere radius.

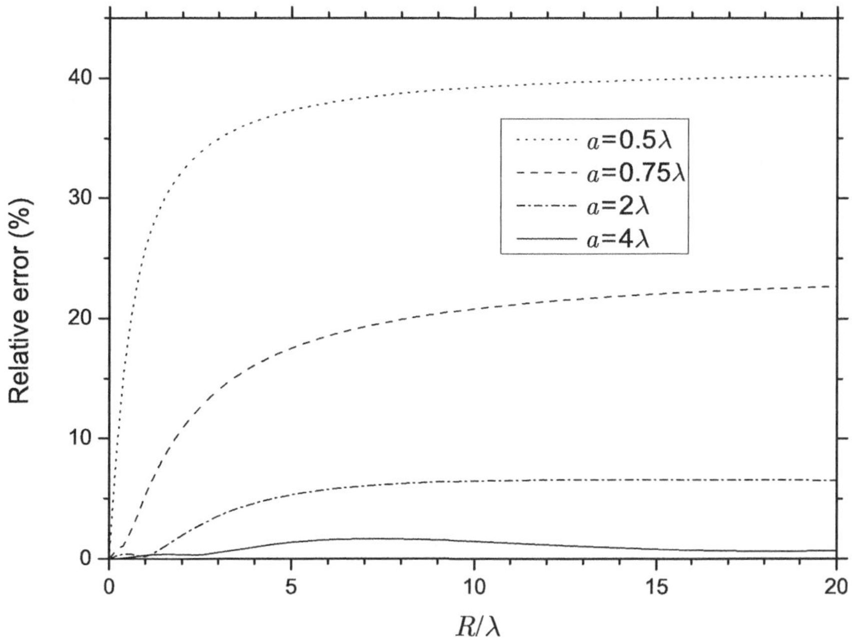

Figure 8.24 Relative accuracy of the reflection formula for the curves in Figure 8.23.

phase and in the amplitude, is plotted in Figure 8.24. Note that the amplitude is modeled by the reflection formula more accuratly than the phase. The figure suggests a good approximation accuracy already for spheres greater than 1.5λ in diameter.

8.4.3 Reflection Formula in the Far-field Region

Outside of caustics the reflection formula (8.176) applies again, and we now consider the important case in which the incident field is a plane wave ($R_0 \to \infty$) and the observation point is located far from the reflection point ($R \to \infty$). In the formula (8.161) for the spread factor the terms proportional to p and p^2 become small, and neglecting them we obtain

$$\mathbf{E}_{\mathrm{c}}(\mathbf{r}) = \frac{j\sigma_{\mathrm{c}}^{\infty}}{2R|K|^{1/2}}\mathbf{E}_{\mathrm{ref}}(\mathbf{r}_{\mathrm{c}})e^{-jkR}, \tag{8.203}$$

where $\sigma_{\mathrm{c}}^{\infty}$ denotes the limiting value of the parameter σ_{c} for $R, R_0 \to \infty$. This quantity can take any of the three possibles values from the formula (D.61) since if $R, R_0 \to \infty$, then $\alpha \to \pm 2L\cos\theta_{\mathrm{c}}$ and $\alpha\beta - \gamma^2 \to 4\cos^2\theta_{\mathrm{c}}(LN - M^2) = 4\cos^2\theta_{\mathrm{c}}(EG - F^2)K$, so that

$$\operatorname{sgn}\alpha = \pm\operatorname{sgn}L, \quad \operatorname{sgn}(\alpha\beta - \gamma^2) = \operatorname{sgn}K$$

and $\sigma_{\mathrm{c}}^{\infty}$ can be $-j$ or 1 or j, depending on the signs of the surface parameters K and L. The value of $\sigma_{\mathrm{c}}^{\infty}$ can be determined by simply counting the number of times the ray connecting the reflection and observation points has touched the caustics. On every contact with the caustic, $\sigma_{\mathrm{c}}^{\infty}$ is multiplied by j. If the ray does not touch any caustic (as it is the case for an arbitrary convex surface) then $\sigma_{\mathrm{c}}^{\infty} = -j$, which is the value of σ_{c} at the reflection point (see (8.177)).

Formula (8.203) is obtained by application of the stationary phase method to the PO solution of the scattering problem and describes the contribution of a critical point to the scattered field. Consequently (8.203) describes the scattered field in the GO approximation.

In the case of a convex scatterer, there apparently exists only one critical point contributing to the scattering, and (8.203) represents the contribution of the whole scattering surface to the scattered field. If the scatterer is non-convex, then there can be further contributing critical points, in which case the scattered field is given by a coherent sum of these contributions. Formula (8.203) will fail in the limit of forward scattering ($\hat{s}_c = \hat{i}_c$) because in this limit the critical point approaches the shadow boundary, at which the PO approximation is incorrect, and moreover contributions from other points on the shadow boundary or even from the whole shadow boundary become equally important and should be accounted for.

The formula (8.203) shows that in the GO approximation the scattered field is determined by the value of the GO reflected field $\mathbf{E}_{\text{ref}}(\mathbf{r}_c)$ at the point of reflection and this is the mechanism through which reflecting properties of the scattering surface affect the scattered field. The geometric properties of the scattering surface influence the phase, which is retarded by $\pi/2$ or π if the reflected ray has touched caustics, and the amplitude of the scattered field through the presence of the Gaussian curvature K at the point of reflection $\mathbf{r}_c$. If at the reflection point the scattering surface is flat or cylindrical, then the Gaussian curvature vanishes and the formula (8.203) predicts infinite field values. The reason for that is the failure of the basic assumption of the stationary point method about the asymptotic smallness of the field-forming region around the reflection point.

The field-forming region is a region on the scattering surface centered at the critical point and bounded by a curve which is a locus of points for which the value of the phase of the exponent in the PO representation of the scattered field differs from the phase value at the reflection point by π. The definition of the field-forming region (8.77) given in section 8.3.2 for the case of transmission through a flat aperture applies to the case of reflection from a curved surface as well, the only difference being that the shape of the region may be arbitrary, not necessarily elliptic, as the intersection of the Fresnel spheroid with a curved surface can be arbitrarily shaped.

Figure 8.25 illustrates the difference between the field forming regions on a flat ($R_{1,2} = \infty$) and a curved ($R_{1,2} = a$) surface in the limit $R, R_0 \to \infty$. A monostatic configuration ($R = R_0$) is assumed. In the case of a flat reflecting surface, the field-forming region is a disc with the diameter $\sqrt{2\lambda R}$ equal to the diameter of the first Fresnel zone (see formula (8.80)) which increases indefinitely as $R \to \infty$. This implies that at sufficiently large distances from the reflection point, the size of the field-forming region becomes comparable or even larger than the size of the scatterer, which makes the stationary point method inapplicable. On the contrary, on a curved surface with bounded principal radii of curvature at the point of reflection, the diameter of the field-forming region is bounded for $R, R_0 \to \infty$ and asymptotically small when $\lambda \to 0$. For example, if the reflecting surface in a vicinity of the reflecting point is spherical ($R_1 = R_2 = a$), then the diameter of the region tends to the bound $\sqrt{2\lambda a}$ (see problem 16).

In section 9.3 we describe an alternative method for calculating contributions from flat pieces of a scattering surface to the scattered field when $R, R_0 \to \infty$.

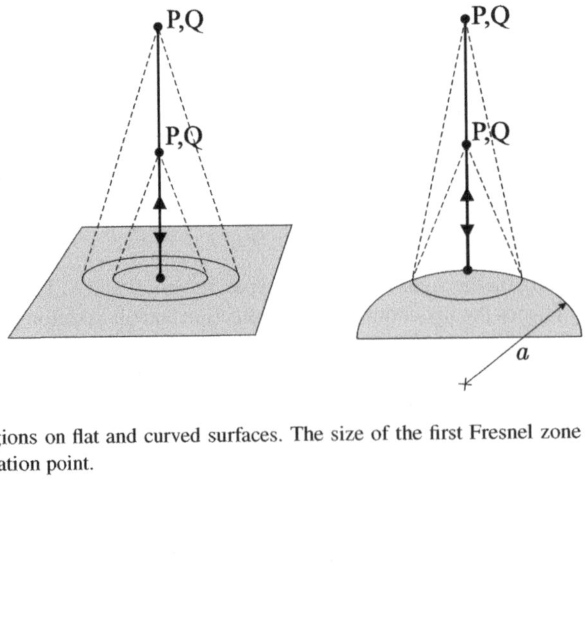

Figure 8.25 Field-forming regions on flat and curved surfaces. The size of the first Fresnel zone on the flat surface increases with increasing distance to the observation point.

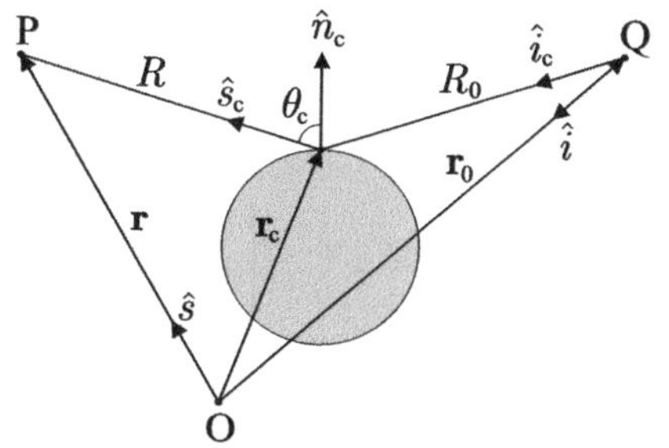

Figure 8.26 Coordinate systems and related parameters used in the derivation of scattering matrix from the reflection formula.

Scattering Matrix of Electrically Large Impenetrable Convex Scatterers

Let us derive the scattering matrix and the bistatic scattering cross section from the formula (8.203). We define the scattering matrix in an arbitrarily oriented spherical coordinate system with an origin at an arbitrary point O, which can be different from the coordinate system used in the reflection formula, which is centered at the reflection point $\mathbf{r}_c$ and oriented according to the unit vector $\hat{n}_c$ (Figure 8.26). By definition (3.61), the scattering matrix relates the transversal components of the scattered and incident waves in the bases $(\hat{s}, \hat{\theta}, \hat{\phi})$ and $(\hat{i}, \hat{\theta}_{inc}, \hat{\phi}_{inc})$. On the other hand, the reflected wave $\mathbf{E}_{ref}(\mathbf{r}_c)$ and the incident plane wave $\mathbf{E}_{inc}(\mathbf{r}_c)$ are expressed in the bases $(\hat{s}_c, \hat{\alpha}_{ref}, \hat{\beta}_{ref})$ and $(\hat{i}_c, \hat{\alpha}_{inc}, \hat{\beta}_{inc})$ as

$$\mathbf{E}_{inc}(\mathbf{r}_c) = \hat{\alpha}_{inc} E^{\perp}_{inc}(\mathbf{r}_c) + \hat{\beta}_{inc} E^{\parallel}_{inc}(\mathbf{r}_c) \tag{8.204}$$

$$\mathbf{E}_{ref}(\mathbf{r}_c) = \hat{\alpha}_{ref} R_{\perp} E^{\perp}_{inc}(\mathbf{r}_c) + \hat{\beta}_{ref} R_{\parallel} E^{\parallel}_{inc}(\mathbf{r}_c), \tag{8.205}$$

where $R_\perp$ and $R_\parallel$ are the values of the reflection coefficients for the perpendicular and parallel polarization at the reflection point $\mathbf{r}_c$ (section 4.2.6).

Using (8.205) and multiplying (8.203) with the unit vectors $\hat{\theta}$ and $\hat{\phi}$, we obtain

$$\begin{bmatrix} E^{sc}_{\theta}(\mathbf{r}) \\ E^{sc}_{\phi}(\mathbf{r}) \end{bmatrix} = \frac{j\sigma^{\infty}_{c} e^{-jkR}}{2R|K|^{1/2}} \begin{pmatrix} \hat{\theta}\cdot\hat{\alpha}_{ref} & \hat{\theta}\cdot\hat{\beta}_{ref} \\ \hat{\phi}\cdot\hat{\alpha}_{ref} & \hat{\phi}\cdot\hat{\beta}_{ref} \end{pmatrix} \begin{pmatrix} R_\perp & 0 \\ 0 & R_\parallel \end{pmatrix} \begin{bmatrix} E^{\perp}_{inc}(\mathbf{r}_c) \\ E^{\parallel}_{inc}(\mathbf{r}_c) \end{bmatrix}, \tag{8.206}$$

where $E^{sc}_{\theta}(\mathbf{r}) = \hat{\theta}\cdot\mathbf{E}_c(\mathbf{r})$ and $E^{sc}_{\phi}(\mathbf{r}) = \hat{\phi}\cdot\mathbf{E}_c(\mathbf{r})$. The radial component $\hat{s}\cdot\mathbf{E}_c(\mathbf{r})$ of the scattered field vanishes because of the transversality of the reflected field (see (8.205)) and the fact that $\hat{s}_c \to \hat{s}$ when the observation point goes to infinity $(r, R \to \infty)$.

The next step is to find the relation between the components of the incident wave (8.204) in the bases $(\hat{i}_c, \hat{\alpha}_{inc}, \hat{\beta}_{inc})$ and $(\hat{i}, \hat{\theta}_{inc}, \hat{\phi}_{inc})$. With the source going to infinity $(R_0, r_0 \to \infty)$, $\hat{i}_c \to \hat{i}$ implying that the unit vectors $\hat{\alpha}_{inc}$ and $\hat{\beta}_{inc}$ can differ from $\hat{\theta}_{inc}$ and $\hat{\phi}_{inc}$ by a rotation around the incidence direction only. The incident wave can, therefore, be represented as

$$\mathbf{E}_{inc}(\mathbf{r}_c) = \hat{\theta}_{inc} E^{inc}_{\theta_{inc}}(\mathbf{r}_c) + \hat{\phi}_{inc} E^{inc}_{\phi_{inc}}(\mathbf{r}_c) \tag{8.207}$$

and multiplying (8.207) with the mutually orthogonal unit vectors $\hat{\alpha}_{inc}$ and $\hat{\beta}_{inc}$ and using the representation (8.204), we arrive at the relation

$$\begin{bmatrix} E^{\perp}_{inc}(\mathbf{r}_c) \\ E^{\parallel}_{inc}(\mathbf{r}_c) \end{bmatrix} = \begin{pmatrix} \hat{\theta}_{inc}\cdot\hat{\alpha}_{inc} & \hat{\phi}_{inc}\cdot\hat{\alpha}_{inc} \\ \hat{\theta}_{inc}\cdot\hat{\beta}_{inc} & \hat{\phi}_{inc}\cdot\hat{\beta}_{inc} \end{pmatrix} \begin{bmatrix} E^{inc}_{\theta_{inc}}(\mathbf{r}_c) \\ E^{inc}_{\phi_{inc}}(\mathbf{r}_c) \end{bmatrix}. \tag{8.208}$$

Inserting (8.208) in (8.206) and using the formulas

$$\frac{e^{-jkR}}{R} \approx \frac{e^{-jk(r-\hat{s}\cdot\mathbf{r}_c)}}{r}, \qquad r \to \infty \tag{8.209}$$

$$\begin{bmatrix} E^{\text{inc}}_{\theta_{\text{inc}}}(\mathbf{r}_c) \\ E^{\text{inc}}_{\phi_{\text{inc}}}(\mathbf{r}_c) \end{bmatrix} = e^{-jk\hat{\imath}\cdot\mathbf{r}_c} \begin{bmatrix} E^{\text{inc}}_{\theta_{\text{inc}}}(O) \\ E^{\text{inc}}_{\phi_{\text{inc}}}(O) \end{bmatrix} \tag{8.210}$$

transform (8.206) into a relation of the form (3.61) with the scattering matrix

$$\overline{\overline{A}} = j\sigma_c^\infty \sqrt{\frac{\pi}{|K|}}\, e^{jk(\hat{s}-\hat{\imath})\cdot\mathbf{r}_c} \cdot \begin{pmatrix} \hat{\theta}\cdot\hat{\alpha}_{\text{ref}} & \hat{\theta}\cdot\hat{\beta}_{\text{ref}} \\ \hat{\phi}\cdot\hat{\alpha}_{\text{ref}} & \hat{\phi}\cdot\hat{\beta}_{\text{ref}} \end{pmatrix} \begin{pmatrix} R_\perp & 0 \\ 0 & R_\parallel \end{pmatrix} \begin{pmatrix} \hat{\theta}_{\text{inc}}\cdot\hat{\alpha}_{\text{inc}} & \hat{\phi}_{\text{inc}}\cdot\hat{\alpha}_{\text{inc}} \\ \hat{\theta}_{\text{inc}}\cdot\hat{\beta}_{\text{inc}} & \hat{\phi}_{\text{inc}}\cdot\hat{\beta}_{\text{inc}} \end{pmatrix}. \tag{8.211}$$

Here, the reflection coefficients $R_\perp$ and $R_\parallel$ are taken at the reflection point $\mathbf{r}_c$ and are functions of the incidence angle θ_c, which is the angle between the unit normal $\hat{n}_c$ and the reflection/scattering direction $\hat{s} = \hat{\imath} - 2\hat{n}_c(\hat{n}_c \cdot \hat{\imath})$ (Figure 8.26). The unit vectors $\hat{\theta}_{\text{inc}}$, $\hat{\phi}_{\text{inc}}$, $\hat{\theta}$, and $\hat{\phi}$ are calculated from the vectors $\hat{\imath}$ and $\hat{s}$ by using (3.51), (3.54), (3.55), (3.56), (A.51), and (A.52). The vectors $\hat{\alpha}_{\text{inc}}$, $\hat{\beta}_{\text{inc}}$, $\hat{\alpha}_{\text{ref}}$, and $\hat{\beta}_{\text{ref}}$ are determined by the vectors $\hat{\imath}$ and $\hat{n}_c$ according to the formulas (4.71), (4.74), and (4.75).

Monostatic-bistatic Equivalence Theorem

The expression (8.211) for the scattering matrix considerably simplifies if we choose the z axis of the global spherical coordinate system to be parallel to the unit normal $\hat{n}_c$ (Figure 8.27). In this case, $\hat{\alpha}_{\text{inc}} = \hat{\phi}_{\text{inc}}$, $\hat{\beta}_{\text{inc}} = \hat{\theta}_{\text{inc}}$, $\hat{\alpha}_{\text{ref}} = -\hat{\phi}$, and $\hat{\beta}_{\text{ref}} = \hat{\theta}$, the matrices in (8.211) are easily evaluated, and we obtain the simple expression

$$\overline{\overline{A}} = j\sigma_c^\infty \sqrt{\frac{\pi}{|K|}}\, e^{jk(\hat{s}-\hat{\imath})\cdot\mathbf{r}_c} \begin{bmatrix} R_\parallel(\theta_c) & 0 \\ 0 & -R_\perp(\theta_c) \end{bmatrix} \tag{8.212}$$

in which, for convenience, the dependence of the reflection coefficients on the incidence angle θ_c is explicitly indicated.

The corresponding bistatic scattering cross sections result from (3.66) as

$$\sigma_{\theta\theta_{\text{inc}}} = \pi|R_1 R_2||R_\parallel(\theta_c)|^2 \tag{8.213}$$

$$\sigma_{\phi\phi_{\text{inc}}} = \pi|R_1 R_2||R_\perp(\theta_c)|^2, \tag{8.214}$$

where the definition (A.75) of the Gaussian curvature is used to emphasize the relation of the GO scattering cross sections to the principal curvature radii $R_{1,2}$ at the point of reflection. The cross-polar cross sections $\sigma_{\theta\phi_{\text{inc}}}$ and $\sigma_{\phi\theta_{\text{inc}}}$ vanish in the GO approximation with the chosen orientation of the global coordinate system. The co-polar cross sections in (8.213) and (8.214) depend on the reflection coefficients at the point of reflection and are, in general, different.

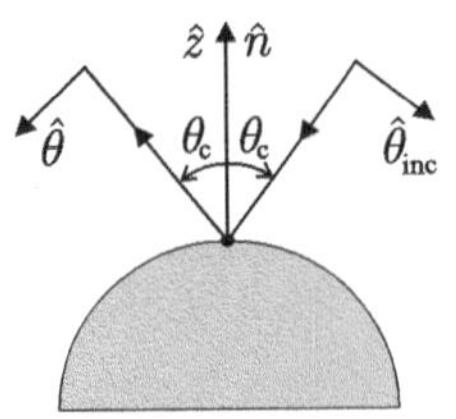

Figure 8.27 Geometry assumed in the formulation of the monostatic–bistatic equivalence theorem.

When the angle $2\theta_c$ between and incidence and scattering directions (separation angle) (Figure 8.27) vanishes, the bistatic cross sections reduce to the radar cross section

$$\sigma = \pi |R_1 R_2| |R_\parallel(0)|^2 \tag{8.215}$$

which is polarization-independent because $R_\parallel(0) = -R_\perp(0)$ for normal incidence on an arbitrary isotropic material surface (see (4.168)).

Bistatic and monostatic cross sections generally take different numerical values, but in the special case of a perfectly conducting surface, when $R_\parallel(\theta_c) \to 1$ and $R_\perp(\theta_c) \to -1$, the GO expressions (8.213) and (8.214) become independent of the bistatic angle and coincide numerically with the radar cross section (8.215),

$$\sigma_{\theta\theta_{\rm inc}} = \sigma_{\phi\phi_{\rm inc}} = \sigma = \pi |R_1 R_2|. \tag{8.216}$$

This property of scattering cross sections of perfectly conducting bodies in the limit of vanishing wavelength is known as the monostatic-bistatic equivalence theorem, according to which the bistatic cross section is equal to the monostatic cross section at the bisector of the bistatic angle between the direction to the transmitter and receiver (Crispin and Siegel 1968). The theorem can be useful to estimate, for example, the monostatic cross section of a target from quasi-monostatic measurements, in which two different but closely located antennas are employed in order to separate the strong transmitted signal from the weak received signal and increase the accuracy of RCS measurements.

RCS of Impedance-matched Targets

An interesting consequence of (8.215) is that if the reflection coefficient of the scattering surface vanishes at normal incidence, then the monostatic cross section of the scatterer decays with growing frequency, eventually vanishing in the optical limit ($\lambda \to 0$) (Osipov 2015). It follows from (4.170) that $R_{\parallel,\perp}(0)$ vanish if the input impedance of the interface is equal to the free space impedance (impedance-matched boundary). So, RCS of every impedance-matched convex target vanishes in the optical limit. In this case, asymptotically smaller contributions from surface singularities (e.g., edges or vertices) may become important (Osipov 2014). In the case of a non-convex body, this property may disappear when multiple reflections between distant portions of the scattering surface take place. If an impedance-matched body is symmetric with respect to the 90° rotation around an axis and is illuminated along this axis, then RCS is identically zero at all frequencies (Weston's theorem, section 3.2.3).

8.4.4 Diffraction by an Edge in a Non-metallic Surface

When the illuminated portion of the scattering surface has an edge, the asymptotic evaluation of the PO solution for the field scattered/reflected by the surface leads to the appearance of contributions from critical points of the second kind located at the edge (Figure 8.28). These contributions describe the rays diffracted at the edge of the surface. The importance of these contributions results from the fact that for a broad variety of positions of the source and receiver they are the only contributions that arrive at the receiver since the contributions reflected from the sides of the surface are bounced away from the observer (e.g., nose-on illumination of a wedge with a monostatic radar).

In section 8.3.3 we have studied the edge diffraction for an aperture in an impenetrable screen. We have chosen the aperture to be the integration surface in the PO solution and have asymptotically evaluated the edge-diffraction contribution from a generally curved edge for spherical or plane wave excitation. We have seen, however, that this approach does not allow the description of the fields near the discontinuities of the GO rays reflected from the screen as the PO solution by the choice of the integration surface does not include information about the material (reflecting) properties of the screen. This is not important for transmission problems but is important for the problems of scattering from compact objects. In this section we show that integration over the surface of the scatterer in the PO solution leads to a representation of the edge-diffracted field which correctly recovers the required singularities at the transmission regions across all the

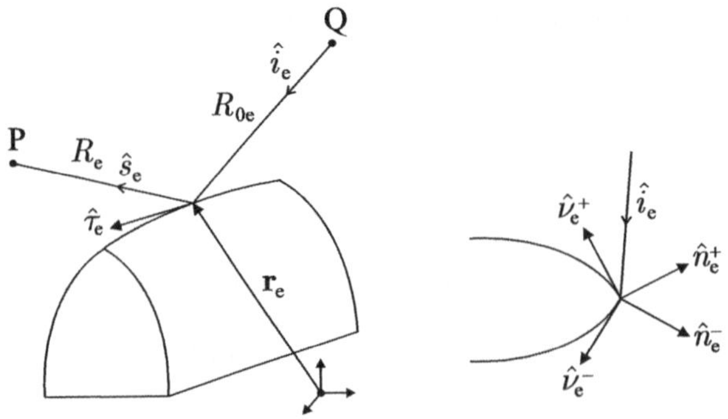

Figure 8.28 The curved edge and geometric parameters describing diffraction by the edge.

jumps in the ray-optical fields, including the shadow boundaries for the direct and reflected rays. Similarly to the case of an aperture, the asymptotic evaluation of the PO solution gives the correct phase structure of the edge-diffracted rays but includes an error in the amplitude (diffraction coefficient), which can be corrected by replacing the PO diffraction coefficient with an exact one if this latter is available from a solution of a corresponding canonical problem for a locally conformal wedge. With this difference in the choice of the integration surface, the analysis of this section is similar to that of section 8.3.3. The main result of this section is the PO diffraction coefficient of a wedge with impenetrable faces with given reflection coefficients which can be different on the opposite sides of the wedge. The presented result includes as a special case the case of a wedge with faces described by the standard impedance boundary conditions (Senior and Volakis 1995). The PO diffraction coefficient provides a useful approximation for configurations for which no exact solution is available. It is also a necessary building block used in the framework of PTD to account for edge corrections.

PO Solution for a Curved Edge in a Non-metallic Surface

We begin with specializing the PO solution (8.5) to the curved wedge shown in Figure 8.28. The edge is an intersection of two smooth surfaces, the faces of the wedge, and denoting them by $\Omega_0^\pm$, we express the PO solution as

$$\mathbf{E}^{\mathrm{po}}_{\pm}(\mathbf{r}) = \mathbf{E}^{\mathrm{po}}_{+}(\mathbf{r}) + \mathbf{E}^{\mathrm{po}}_{-}(\mathbf{r}), \tag{8.217}$$

where

$$\mathbf{E}^{\mathrm{po}}_{\pm}(\mathbf{r}) = \frac{jk}{4\pi}\int\limits_{\Omega_0^\pm} \frac{\mathrm{e}^{-jk|\mathbf{r}-\mathbf{r}'|}}{|\mathbf{r}-\mathbf{r}'|}\mathbf{c}(\mathbf{r}',\mathbf{r}_0)\,\mathrm{d}\Omega' \tag{8.218}$$

$$\mathbf{c}(\mathbf{r}',\mathbf{r}_0) = Z\hat{s}'\times\hat{s}'\times\mathbf{K}^{\mathrm{go}}_{\mathrm{e}}(\mathbf{r}') + \hat{s}'\times\mathbf{K}^{\mathrm{go}}_{\mathrm{m}}(\mathbf{r}') \tag{8.219}$$

and

$$\mathbf{K}^{\mathrm{go}}_{\mathrm{e}}(\mathbf{r}') = \hat{n}(\mathbf{r}')\times\mathbf{H}^{\mathrm{go}}(\mathbf{r}'), \quad \mathbf{K}^{\mathrm{go}}_{\mathrm{m}}(\mathbf{r}') = -\hat{n}(\mathbf{r}')\times\mathbf{E}^{\mathrm{go}}(\mathbf{r}'), \tag{8.220}$$

the latter being the equivalent surface currents in the GO approximation ((4.93) and (4.96)).

The next step is to represent the surfaces $\Omega_0^\pm$ in the parametric form by using suitable functions $\mathbf{f}^\pm(u,v)$ which are such that $\mathbf{r} = \mathbf{f}^+(u,v)$ with $u,v\in\Omega^+_{uv}$ when $\mathbf{r}\in\Omega_0^+$ and $\mathbf{r} = \mathbf{f}^-(u,v)$ with $u,v\in\Omega^-_{uv}$ when $\mathbf{r}\in\Omega_0^-$. This reduces the integrals (8.218) to the basic form (D.49) studied in Appendix D on the stationary phase method. Contributions from critical points of the first kind, which may exist or may not exist on the smooth surfaces $\Omega_0^\pm$, describe the rays reflected from the faces of the wedge (sections 8.4.1 and 8.4.3). In this section we concentrate on the contribution from the critical point(s) of the second kind located on the edge of the wedge and responsible for the edge-diffracted rays.

Let $\mathbf{r}_e$ be one such point, and let us consider the contribution from it. If there are further critical points of the second kind, then their contributions should be added together. At the point $\mathbf{r}_e$, we introduce the unit vector $\hat{\tau}_e$ of the tangent to the edge, which is defined in (8.99), the unit vectors $\hat{n}_e^{\pm}$ of the exterior normals to the faces $\Omega_0^{\pm}$, and the unit vectors $\hat{\nu}_e^{\pm}$ which are perpendicular to $\hat{\tau}_e$ and $\hat{n}_e^{\pm}$ and directed into the interior of the respective faces (Figure 8.28). We will suppose that $\hat{n}_e^{\pm} \times \hat{\nu}_e^{\pm} = \pm\hat{\tau}_e$ on $\Omega_0^{\pm}$.

The evaluation of the contribution from the critical point $\mathbf{r}_e$ to the integral (8.218) is similar to that described in section 8.3.3, the only differences being that there are two integration surfaces and every surface acts as a reflecting boundary. Assuming that the edge is a curve described by the function $\tilde{\mathbf{f}}(t)$ as in (8.83) and using (8.120), we write

$$\mathbf{E}_e^{po}(\mathbf{r}) = \left|\frac{\rho_e}{R_e + \rho_e}\right|^{1/2} \frac{\exp\left[-jkR_e - j\frac{\pi}{4}\operatorname{sgn}\tilde{\chi}_e''\right]}{2\sqrt{2\pi kR_e}\sin\beta} \cdot \left[\frac{\mathbf{c}_e^+\Theta_+}{\left(\mathbf{q}_e\cdot\hat{\nu}_e^+\right)} + \frac{\mathbf{c}_e^-\Theta_-}{\left(\mathbf{q}_e\cdot\hat{\nu}_e^-\right)}\right], \tag{8.221}$$

where $\mathbf{q}_e = \hat{\imath}_e - \hat{s}_e$ and R_e, β, ρ_e, and $\operatorname{sgn}\tilde{\chi}_e''$ are defined in (8.92), (8.104), (8.105), and (8.107), respectively. The quantities $\mathbf{c}_e^{\pm}$ are given by

$$\mathbf{c}_e^{\pm} = Z\hat{s}_e \times \hat{s}_e \times \mathbf{K}_e^{\pm}(\mathbf{r}_e) + \hat{s}_e \times \mathbf{K}_m^{\pm}(\mathbf{r}_e) \tag{8.222}$$

in which $\mathbf{K}_e^{\pm}(\mathbf{r}_e)$ and $\mathbf{K}_m^{\pm}(\mathbf{r}_e)$ are the limiting values of $\mathbf{K}_e^{go}(\mathbf{r}')$ and $\mathbf{K}_m^{go}(\mathbf{r}')$ when $\mathbf{r}' \to \mathbf{r}_e$ with $\mathbf{r}' \in \Omega_0^{\pm}$. The parameters $\Theta_{\pm}$ are equal to 1 or 0, depending on whether the respective face is illuminated by the incident wave or not, and since the scattering surface is assumed to be impenetrable, they can be expressed through the unit step function (B.12) as

$$\Theta_{\pm} = \Theta\left(-\hat{\imath}_e \cdot \hat{n}_e^{\pm}\right). \tag{8.223}$$

The formula (8.221) represents a ray diffracted at the point $\mathbf{r}_e$. The spread factor $|\rho_e/(R_e+\rho_e)|^{1/2}$ and the phase factor $\exp(-j\pi\operatorname{sgn}\tilde{\chi}_e''/4)$ account for the curvature of the edge, and their role has been studied in section 8.3.3. Next we will derive from (8.221) the expression for the diffraction coefficient.

The diffraction coefficient relates the β_{inc} and φ_{inc} components of the incident wave at the diffraction point with the β and φ components of the edge-diffracted wave at the observation point. The bases of the incident and diffracted waves, $(\hat{\imath}_e, \hat{\beta}_{inc}, \hat{\varphi}_{inc})$ and $(\hat{s}_e, \hat{\beta}, \hat{\varphi})$, are uniquely defined by the incidence and scattering directions $\hat{\imath}_e$ and $\hat{s}_e$, and by the tangent $\hat{\tau}_e$ at the point of diffraction $\mathbf{r}_e$ according to (5.449) and (5.450).

The β and φ components of the edge-diffracted wave are determined by multiplying (8.221) with $\hat{\beta}$ and $\hat{\varphi}$, and combining the components in a vector we obtain

$$\begin{bmatrix}\mathbf{E}_e^{po}(\mathbf{r})\cdot\hat{\beta}\\ \mathbf{E}_e^{po}(\mathbf{r})\cdot\hat{\varphi}\end{bmatrix} = \left|\frac{\rho_e}{R_e+\rho_e}\right|^{1/2} \frac{\exp\left[-jkR_e - j\frac{\pi}{4}\operatorname{sgn}\tilde{\chi}_e''\right]}{2\sqrt{2\pi kR_e}\sin\beta} \cdot \left\{\frac{\Theta_+}{\left(\mathbf{q}_e\cdot\hat{\nu}_e^+\right)}\begin{bmatrix}\mathbf{c}_e^+\cdot\hat{\beta}\\ \mathbf{c}_e^+\cdot\hat{\varphi}\end{bmatrix} + \frac{\Theta_-}{\left(\mathbf{q}_e\cdot\hat{\nu}_e^-\right)}\begin{bmatrix}\mathbf{c}_e^-\cdot\hat{\beta}\\ \mathbf{c}_e^-\cdot\hat{\varphi}\end{bmatrix}\right\}. \tag{8.224}$$

Coefficients $\mathbf{c}_e^{\pm}$ in (8.224) can be written as linear combinations of the β_{inc} and φ_{inc} components of the incident wave $\mathbf{E}_e^{inc} = \mathbf{E}^{inc}(\mathbf{r}_e)$ because according to the relations (4.93) and (4.96) we can express the equivalent currents in (8.222) as

$$Z\mathbf{K}_e^{\pm}(\mathbf{r}_e) = \left(R_\perp^{\pm} - 1\right)\left(\hat{\imath}_e\cdot\hat{n}_e^{\pm}\right)\left(\mathbf{E}_e^{inc}\cdot\hat{\alpha}_{inc}^{\pm}\right)\hat{\alpha}_{inc}^{\pm} - \left(1 + R_\parallel^{\pm}\right)\left(\mathbf{E}_e^{inc}\cdot\hat{\beta}_{inc}^{\pm}\right)\hat{t}^{\pm} \tag{8.225}$$

$$\mathbf{K}_m^{\pm}(\mathbf{r}_e) = \left(1 - R_\parallel^{\pm}\right)\left(\hat{\imath}_e\cdot\hat{n}_e^{\pm}\right)\left(\mathbf{E}_e^{inc}\cdot\hat{\beta}_{inc}^{\pm}\right)\hat{\alpha}_{inc}^{\pm} - \left(1 + R_\perp^{\pm}\right)\left(\mathbf{E}_e^{inc}\cdot\hat{\alpha}_{inc}^{\pm}\right)\hat{t}^{\pm} \tag{8.226}$$

and replace $\mathbf{E}_{\mathrm{e}}^{\mathrm{inc}}$ in (8.225) and (8.226) with the linear combination (5.451). The result has the desired form of a linear relation between the components of the incident and edge-diffracted waves in the ray-fixed coordinates,

$$\begin{bmatrix} \mathbf{E}_{\mathrm{e}}^{\mathrm{po}}(\mathbf{r}) \cdot \hat{\beta} \\ \mathbf{E}_{\mathrm{e}}^{\mathrm{po}}(\mathbf{r}) \cdot \hat{\varphi} \end{bmatrix} = \left| \frac{\rho_{\mathrm{e}}}{R_{\mathrm{e}} + \rho_{\mathrm{e}}} \right|^{1/2} \frac{\exp\left[-jkR_{\mathrm{e}} - j\frac{\pi}{4}\operatorname{sgn}\tilde{\chi}_{\mathrm{e}}''\right]}{\sqrt{2\pi k R_{\mathrm{e}}}\sin\beta} \overline{\overline{D}}^{\mathrm{po}} \cdot \begin{bmatrix} \mathbf{E}_{\mathrm{e}}^{\mathrm{inc}} \cdot \hat{\beta}_{\mathrm{inc}} \\ \mathbf{E}_{\mathrm{e}}^{\mathrm{inc}} \cdot \hat{\varphi}_{\mathrm{inc}} \end{bmatrix}, \tag{8.227}$$

where

$$\overline{\overline{D}}^{\mathrm{po}} = \Theta_+ \begin{bmatrix} D_{11}^+ & D_{12}^+ \\ D_{21}^+ & D_{22}^+ \end{bmatrix} + \Theta_- \begin{bmatrix} D_{11}^- & D_{12}^- \\ D_{21}^- & D_{22}^- \end{bmatrix} \tag{8.228}$$

is the diffraction coefficient. Expressions for the diffraction coefficient are given in the next subsection ((8.231)–(8.234)).

By construction, the PO diffraction coefficient (8.228) is a sum of two components due to the GO equivalent currents induced by the incident wave on the faces Ω_0^+ and Ω_0^-. Letting the curvature of the edge a_{e} and the distance $R_{0\mathrm{e}}$ between the diffraction point and the source go to infinity makes the spread and the phase factors equal to unity, and the formula (8.227) reduces to the form of (5.459), which describes the field diffracted at the edge of an infinite wedge that is locally conformal with the intersection of the two surfaces $\Omega_0^\pm$ in the immediate vicinity of the diffraction point $\mathbf{r}_{\mathrm{e}}$. The faces of the wedge are flat surfaces that include the point $\mathbf{r}_{\mathrm{e}}$, the vector $\hat{\tau}_{\mathrm{e}}$, and the vectors $\hat{\nu}_{\mathrm{e}}^+$ or $\hat{\nu}_{\mathrm{e}}^-$, respectively, and their material properties are described by the reflection coefficients $R_\perp^\pm$ and $R_\parallel^\pm$.

PO Diffraction Coefficient of an Impedance Wedge in Ray-fixed Coordinates

It is convenient to express the elements of the PO diffraction coefficient in the coordinate system used in the chapter on diffraction by wedges (Figure 5.24), in which $\hat{\tau}_{\mathrm{e}} = \hat{z}$, the polar angle φ is measured from the plane bisecting the exterior of the wedge, and the position of the wedge faces is given by the conditions $\varphi = \pm\Phi$. The parameter Φ is equal to half the angle between the vectors $\hat{\nu}_{\mathrm{e}}^+$ and $\hat{\nu}_{\mathrm{e}}^-$ measured in the exterior of the wedge. In this coordinate system the directions of the incidence and scattering can be written as

$$\hat{i}_{\mathrm{e}} = -\hat{x}\cos\varphi_0\sin\beta - \hat{y}\sin\varphi_0\sin\beta + \hat{z}\cos\beta \tag{8.229}$$

$$\hat{s}_{\mathrm{e}} = \hat{x}\cos\varphi\sin\beta + \hat{y}\sin\varphi\sin\beta + \hat{z}\cos\beta, \tag{8.230}$$

and using these formulas we express the vectors $\hat{\phi}$, $\hat{\beta}$, $\hat{\phi}_{\mathrm{inc}}$, $\hat{\beta}_{\mathrm{inc}}$ of the ray-fixed coordinate system and the vectors $\alpha_{\mathrm{inc}}^\pm$ and $\beta_{\mathrm{inc}}^\pm$ associated with the planes of incidence in terms of the coordinates β, φ_0, and φ, insert these vectors into (8.224)–(8.226), calculate all the vector products, and (omitting the lengthy calculations) arrive at the following explicit expressions for the elements of the PO diffraction coefficient (8.228):

$$D_{11}^\pm = \frac{A^\pm + B^\pm R_\parallel^\pm + E^\pm R_\perp^\pm}{2[1 - \sin^2\beta\sin^2(\Phi \mp \varphi_0)]} \tag{8.231}$$

$$D_{12}^\pm = -\cos\beta\,\frac{F^\pm + G^\pm R_\parallel^\pm + H^\pm R_\perp^\pm}{2[1 - \sin^2\beta\sin^2(\Phi \mp \varphi_0)]} \tag{8.232}$$

$$D_{21}^\pm = \cos\beta\,\frac{F^\pm + H^\pm R_\parallel^\pm + G^\pm R_\perp^\pm}{2[1 - \sin^2\beta\sin^2(\Phi \mp \varphi_0)]} \tag{8.233}$$

$$D_{22}^\pm = \frac{A^\pm + E^\pm R_\parallel^\pm + B^\pm R_\perp^\pm}{2[1 - \sin^2\beta\sin^2(\Phi \mp \varphi_0)]} \tag{8.234}$$

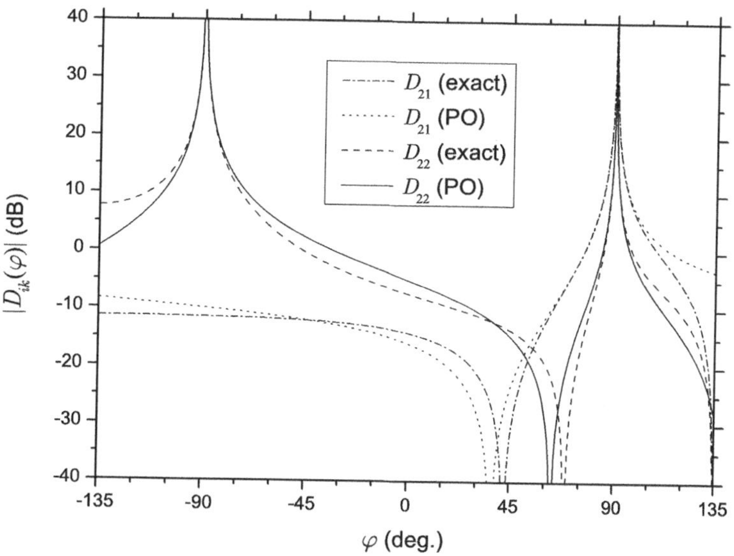

Figure 8.29 Diffraction coefficient: PO vs. exact. A right-angled wedge ($\Phi = 3\pi/4$) with the normalized face impedances $\eta_+ = 1$ and $\eta_- = 0$ is illuminated from the direction $\beta = 45°$ and $\varphi_0 = 0°$.

with

$$A^{\pm} = \pm\tan\left(\frac{\varphi - \varphi_0}{2}\right)[1 - \sin^2\beta\sin^2(\Phi \mp \varphi_0)] \tag{8.235}$$

$$B^{\pm} = \cos^2\beta\sin(\Phi \mp \varphi_0)\cos\left(\frac{\varphi - \varphi_0}{2}\right)\sec\left(\Phi \mp \frac{\varphi + \varphi_0}{2}\right) \tag{8.236}$$

$$E^{\pm} = \cos(\Phi \mp \varphi_0)\left[\pm\sin\left(\frac{\varphi - \varphi_0}{2}\right)\sec\left(\Phi \mp \frac{\varphi + \varphi_0}{2}\right) - \sin^2\beta\sin(\Phi \mp \varphi_0)\right] \tag{8.237}$$

$$F^{\pm} = \mp[1 - \sin^2\beta\sin^2(\Phi \mp \varphi_0)] \tag{8.238}$$

$$G^{\pm} = \mp\cos(\Phi \mp \varphi_0)\cos\left(\frac{\varphi - \varphi_0}{2}\right)\sec\left(\Phi \mp \frac{\varphi + \varphi_0}{2}\right) \tag{8.239}$$

$$H^{\pm} = \sin(\Phi \mp \varphi_0)\left[\sin\left(\frac{\varphi - \varphi_0}{2}\right)\sec\left(\Phi \mp \frac{\varphi + \varphi_0}{2}\right) \mp \sin^2\beta\sin(\Phi \mp \varphi_0)\right]. \tag{8.240}$$

As apparent from (8.231)–(8.240), the PO diffraction coefficient does not comply with the symmetry properties (5.418)–(5.421) required by reciprocity. On the other hand, it has the correct pole singularities (5.416) and (5.417) at the GO shadow boundaries $\varphi = \varphi_0 \pm \pi$ (incident wave) and $\varphi = -\varphi_0 \pm (2\Phi - \pi)$ (reflected waves).

Figures 8.29 and 8.30 illustrate the accuracy of the PO approximation (8.231)–(8.240) for the tensor diffraction coefficient of an impedance wedge with $\Phi = 3\pi/4$, $\eta_+ = 1$, and $\eta_- = 0$, assuming the incidence from the direction $\beta = 45°$ and $\varphi_0 = 0°$. Reflection coefficients are given by the formulas (section 4.6)

$$R_{\perp}^{\pm} = \frac{\eta_{\pm}\cos\theta_{\pm} - 1}{\eta_{\pm}\cos\theta_{\pm} + 1} \tag{8.241}$$

$$R_{\parallel}^{\pm} = \frac{\cos\theta_{\pm} - \eta_{\pm}}{\cos\theta_{\pm} + \eta_{\pm}}, \tag{8.242}$$

where $\theta_{\pm} = \pi - \arccos(\hat{i}_e \cdot \hat{n}_e^{\pm})$ is the angle of incidence on the respective face of the wedge.

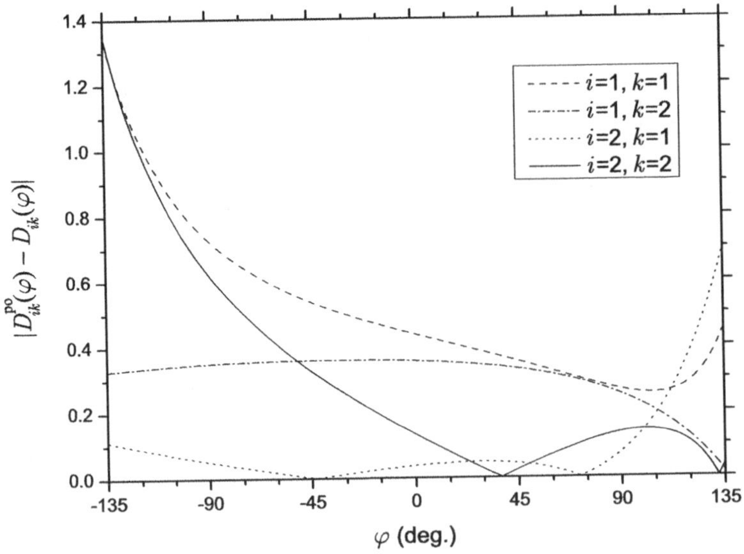

Figure 8.30 Absolute accuracy of the PO approximation for the elements of the tensor diffraction coefficient of a right-angled wedge ($\Phi = 3\pi/4$) with $\eta_+ = 1$ and $\eta_- = 0$ illuminated from the direction $\beta = 45°$ and $\varphi_0 = 0°$.

Figure 8.29 compares the exact and PO-based results for the elements D_{21} and D_{22} of the diffraction coefficient tensor. It is seen that the PO curves for both co- and cross-polar components have the correct pole singularities at the directions of the shadow boundaries of the waves reflected from the faces of the wedge ($\varphi = \pm 90°$). The PO curves may have a significant error near the wedge faces $\varphi = \pm 135°$, which is explained by the fact that the PO solutions only approximately (in GO approximation) satisfy the boundary conditions on the scattering surface. The accuracy of the PO diffraction coefficient is seen better if we subtract from it the exact diffraction coefficient. These curves are plotted in Figure 8.30. The pole singularities are canceled and the error is a bounded function of the scattering angle φ. The error is at maximum near the wedge faces and can be significant there. However, slightly off the faces, the PO approximation can be considered acceptable for applications (e.g., wireless propagation studies).

In the PEC limit, $R_{\perp}^{\pm} \to -1$ and $R_{\parallel}^{\pm} \to 1$, the PO diffraction coefficient significantly simplifies and reduces to the expression

$$\overline{\overline{D}}{}^{\mathrm{po}}_{\mathrm{pec}} = \Theta_+ \begin{bmatrix} \dfrac{\sin(\Phi - \varphi_0)}{\cos(\Phi - \varphi) + \cos(\Phi - \varphi_0)} & \cos\beta \\ 0 & -\dfrac{\sin(\Phi - \varphi)}{\cos(\Phi - \varphi) + \cos(\Phi - \varphi_0)} \end{bmatrix} + \Theta_- \begin{bmatrix} \dfrac{\sin(\Phi + \varphi_0)}{\cos(\Phi + \varphi) + \cos(\Phi + \varphi_0)} & -\cos\beta \\ 0 & -\dfrac{\sin(\Phi + \varphi)}{\cos(\Phi + \varphi) + \cos(\Phi + \varphi_0)} \end{bmatrix}, \tag{8.243}$$

which has a cross-polar component ($\pm\cos\beta$). The component vanishes in the particular cases of normal incidence ($\beta = \pi/2$) or when both faces of the wedge are illuminated but otherwise is present in the PO diffraction coefficient. The exact diffraction coefficient of a PEC wedge (equations (5.88) and (5.89)) is diagonal as no polarization coupling occurs at PEC wedges. The coupling in (8.243), which was noted in Ufimtsev (1975) and Breinbjerg *et al.* (1987), is the result of the approximate nature of the PO method.

The formula (8.243) can be further specialized to the case of half-plane ($\Phi \to \pi$). The result will differ from the PO diffraction coefficient (8.141) derived for the edge of an aperture in an impenetrable screen. The reason is the use of different integration surfaces in the PO solutions. Integration over the physical boundaries of the scatterer performed in this section permits the correct description of all the GO singularities in the diffraction coefficient, including those for reflected waves.

8.5 Advantages and Limitations of Physical Optics

The PO method gives an approximate solution of scattering problems. Why do we use it and what are the limitations of the method? This section is a short summary of advantages and limitations of the PO method. An extensive comparison of PO with GTD can be found in Borovikov and Kinber (1994) (see also Asvestas (1980)).

In contrast to the Fresnel–Kirchhoff diffraction theory, the PO method that we describe in this book uses exact integral representations for the electromagnetic fields, so that inserting the exact equivalent currents in the integral representations would give the exact solution for the fields. Hence the accuracy of the PO method is completely determined by the accuracy of the basic assumption, the tangent plane approximation for the equivalent currents on the integration surface. The tangent plane approximation fails if the fields on the integration surface significantly deviate from zero on the shadowed portion of the obstacle and from the simple exponential plane-wave dependence on the illuminated portion. This is the case, for example, in the following situations:

- Penetrable scatterer such that the waves transmitted through the scatterer cannot be neglected.
- Nose-on illumination of a strongly elongated smooth body when the penumbra region associated with the shadow boundary covers a considerable portion of the scattering surface.
- Multiple reflections between various portions of the scattering surface as in a corner reflector.
- Caustics of the incident wavefront on the scattering surface. The surface currents are no longer given by a simple plane-wave exponent but rather by more complex functions, for example the Airy function. This kind of problem can often be alleviated by simply choosing a different integration surface that is free from caustics.
- Discontinuities in the geometric and/or material properties of the scattering surface, for example an edge, a conical or corner point, or an abrupt change in the reflection properties of the scattering surface. Diffraction of the incident wave by the discontinuities brings additional correction contributions to the surface currents, including the shadowed portion of the scattering surface.
- Forward scattering from a smooth body, in which case the main contribution to the scattered field comes from the vicinity of the shadow boundary on the scattering surface. The shadow boundary is the line of an abrupt change in the PO currents, producing unphysical edge-diffraction waves.
- The scatterer as a whole or some of the elements of the scattering configuration are small or comparable to the wavelength; strong, non-GO interaction between the elements (typical for antenna arrays). Such configurations are beyond the applicability area of PO, which is a high-frequency, GO-related method. Other approaches, like discretization-based methods, are necessary.

The PO assumption of the vanishing currents on the shadowed portion of an impenetrable scatterer may lead to seemingly paradoxical results, as illustrated in Figure 8.31. However, if the basic assumptions of the PO method are satisfied (electrically large and smooth scatterers), the contribution from the shadowed portion of the scatterer is much smaller than the contributions from the illuminated portions, and the fields scattered by the two scatterers shown in Figure 8.31 will be close to each other.

On the other hand, the PO method possesses many very significant advantages.

- When the applicability requirements are satisfied, PO solutions typically provide acceptable numerical accuracy (e.g., Figures 8.15 and 8.16), most importantly in those regions of space where the scattered fields are particularly strong, for example near a caustic of a reflector or around the main and first side lobes of a scattering or radiation diagram.
- As we have seen in sections 8.3.2, 8.3.3, 8.4.1, and 8.4.4, asymptotic evaluation of PO solutions with the stationary phase method recovers the correct ray structure, including the reflected rays and the edge-diffracted rays. PO correctly reproduces the amplitude and the phase of the reflected rays. The edge-diffracted rays in the PO approximation have correct phases but inaccurate amplitudes, and the PO diffraction coefficients differ from the exact diffraction

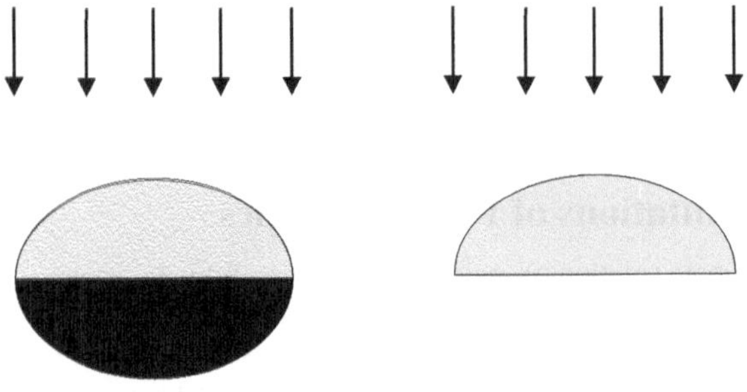

Figure 8.31 Two scatterers illuminated by the same incident wave. Cutting off the shadowed portion of the scatterer does not change the PO solution for the scattered field.

coefficients that may be derived from solutions of canonical diffraction problems. However, the pole singularities in the PO diffraction coefficients are correct. The same is true for the tip-diffracted rays which are obtained by asymptotic evaluation of PO solutions at the critical points of the third kind and at conical points (section 9.2). (PO diffraction coefficients for a vertex seem to be more accurate than for an edge.) PO confirms the postulates used in the framework of GTD and UTD and provides useful approximations for the diffraction coefficients.

- PO solutions are well suited to numerical simulations. No three-dimensional integration or matrix inversion is required. In contrast to the ray-based approaches (GO/GTD/UTD), PO solutions in an integral form are applicable to caustics, giving finite and correct values for the fields.
- The PO algorithm is simple, universal, and extremely versatile. It is used in solving problems of scattering, transmission, reflection, diffraction, and radiation of waves. Typical applications include RCS of diverse vehicles (aircraft, ships, cars, etc.), directivity patterns of large reflector and lens antennas, diffraction losses in wireless propagation, and aberrations of optical components (apertures, lenses).
- Thanks to its clear physical foundations, the PO method is flexible and extensible. Extensions can be achieved by correcting the equivalent surface currents to account for particular scattering mechanisms. For example, Ufimtsev's PTD introduces the so-called *fringe currents* to correctly describe edge diffraction (Ufimtsev 2014).
- In some cases, most notably when extremely large scatterers with the characteristic size much greater than 100λ are involved, no alternative approach can be available as discretization-based methods may fail because of the problem size. This concerns many microwave, infrared, and optical applications. For example, radar imaging typically requires simulation of scattering from a very large scene over a range of frequencies and aspect angles, a task which can hardly be handled with discretization-based methods as every new frequency or aspect angle would require a separate solution of an extremely big (say, more than 10^7 degrees of freedom) system of linear algebraic equations. Furthermore, PO solutions can be very useful in inverse and optimization problems, allowing a reliable initialization of the solution procedure.

8.6 Problems

1. A plane wave is incident on a PEC half-plane at the right angle to the edge. Derive the expressions for the PO surface currents $\mathbf{K}_{\mathrm{e}}^{\mathrm{go}}$ and $\mathbf{K}_{\mathrm{m}}^{\mathrm{go}}$ and compare these with the exact surface currents.

   Hint: Use the results of problems 3 and 4 in section 5.7.

   Answer: Because of the PEC boundary conditions, $\mathbf{K}_{\mathrm{m}}^{\mathrm{go}} = 0$. The expression for the equivalent electric current depends on polarization,

$$\mathbf{K}_{\mathrm{e}}^{\mathrm{go}} = 2\Theta(\varphi_0)\mathrm{e}^{-jk\rho\cos\varphi_0}\begin{cases}\hat{x}H_{0z}, & H \text{ polarization}\\ \hat{z}Z^{-1}E_{0z}\sin\varphi_0, & E \text{ polarization}\end{cases}, \tag{8.244}$$

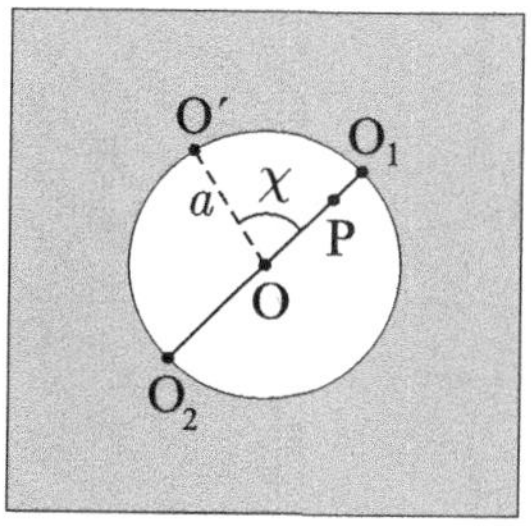

Figure 8.32 A circular aperture in an impenetrable screen.

where the unit step function $\Theta(\varphi_0)$ with φ_0 defined as in Figure 5.4 selects the illuminated face of the half-plane. Thus, it follows from (5.522), (5.523), and (8.244) that when $k\rho \gg 1$,

$$\mathbf{K}_e - \mathbf{K}_e^{go} \approx \hat{x} H_{0z} \csc \frac{\varphi_0}{2} \frac{e^{-jk\rho + j3\pi/4}}{\sqrt{2\pi k\rho}} \tag{8.245}$$

for the H-polarized field and

$$\mathbf{K}_e - \mathbf{K}_e^{go} \approx \hat{z} Z^{-1} E_{0z} \csc \frac{\varphi_0}{2} \cot \frac{\varphi_0}{2} \frac{e^{-jk\rho - j3\pi/4}}{\sqrt{\pi}(2k\rho)^{3/2}} \tag{8.246}$$

for the E-polarized field.

2. Assume that an incident plane wave propagates perpendicularly to the aperture in an impenetrable plane screen ($\hat{i} = \hat{n}$, Figure 8.12) and the aperture is a circle of radius a. Determine the position of the diffraction points on the rim and prove that these are the nearest and the farthest points of the rim with respect to the observation point P.

Hint: Construct a plane that includes the symmetry axis of the configuration and the point P.

Solution: Since the incidence direction $\hat{i}$ is perpendicular to the rim at every point, the stationary phase condition (8.90) implies that the scattering direction is perpendicular to the rim at the point of diffraction. Now introduce a cylindrical coordinate system with the z axis perpendicular to the aperture plane and with the origin O at the center of the aperture. The plane that includes the z axis and the observation point P intersects the rim at two points O_1 and O_2 (Figure 8.32). At these points the plane is perpendicular to the rim since the aperture is a circle, implying that the directions from O_1 and O_2 to P are perpendicular to the rim and thus the points O_1 and O_2 are the points of diffraction.

To see that the points O_1 and O_2 are the nearest and the farthest points on the circle with respect to the point P, consider the distance from P to a test point O′ on the rim. If χ is the angle between the planes crossing the z axis and the points P and O′ respectively, then the distance |PO′| is given by the relation

$$|\mathrm{PO}'| = \sqrt{(\rho_P - a\cos\chi)^2 + (a\sin\chi)^2 + z_P^2}, \tag{8.247}$$

where ρ_P and z_P are the cylindrical coordinates of the observation point P, and to find the maximum and minimum distances we differentiate (8.247) with respect to χ,

$$\frac{\partial|\mathrm{PO}'|}{\partial\chi} = \frac{a\rho_P \sin\chi}{\sqrt{(\rho_P - a\cos\chi)^2 + (a\sin\chi)^2 + z_P^2}}. \tag{8.248}$$

This is zero when $\chi = 0$ or π, implying that the diffraction points O_1 and O_2 are the nearest and the farthest points of the rim with respect to the observation point P.

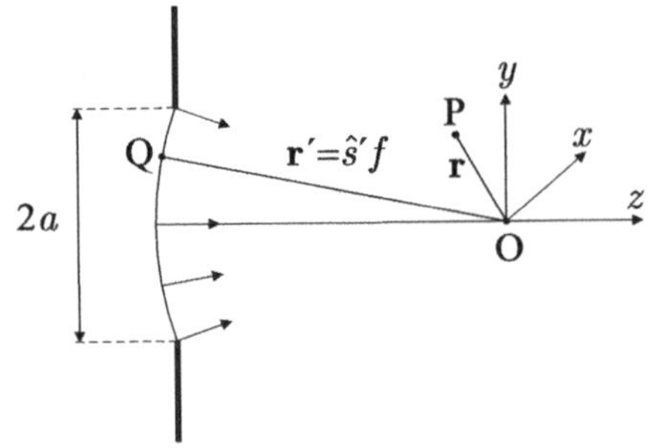

Figure 8.33 Transmission of a convergent spherical wave with the focus at the point O through a circular aperture in an impenetrable screen. P is an observation point near the focus. Q is a point on the front of the incoming wave.

3. Consider transmission of a converging spherical wave through a circular aperture in an impenetrable screen (Figure 8.33). Derive a PO solution for the field distribution near the focus under the assumptions that $f \gg a \gg \lambda$, where f is the distance between the screen and the focus and a is the radius of the aperture. Show that the amplitude of the field is a function of the two variables $u = kz(a/f)^2$ and $v = k\rho(a/f)$ only, where z and ρ are cylindrical coordinates measured with respect to the optical axis of the configuration which goes through the center of the aperture and the focus (Born and Wolf 1959).

   Hint: Use the PO formula (8.5) in which the integration surface Ω_0 is the portion of the front of the incoming wave covering the aperture. Reduce the PO integral to the form

$$E(u,v) = 2E(0,0)\mathrm{e}^{-ju(f/a)^2} \int_0^1 \mathrm{e}^{jut^2/2} J_0(vt)t\,\mathrm{d}t, \tag{8.249}$$

   where E is an arbitrary component of the electric field tangential to the front of the spherical wave.

   Solution: If Ω_0 is the portion of the spherical wavefront, then the electric and magnetic fields of the incident wave and therefore the currents $\mathbf{K}^{\mathrm{go}}_{\mathrm{e,m}}$ are constant over Ω_0 and an arbitrary field component tangential to Ω_0 can be expressed as

$$E(\mathbf{r}) \approx \frac{kA}{f}\mathrm{e}^{jkf} \int_{\Omega_0} \frac{\mathrm{e}^{-jk|\mathbf{r}-\mathbf{r}'|}}{|\mathbf{r}-\mathbf{r}'|}\,\mathrm{d}\Omega', \tag{8.250}$$

   where A is a coordinate-independent coefficient. When $a \ll f$, we can use the approximations

$$|\mathbf{r}-\mathbf{r}'| \approx f - \mathbf{r}\cdot\hat{s}' \tag{8.251}$$

   in the phase and $|\mathbf{r}-\mathbf{r}'| \approx f$ in the amplitude of the integrand function in (8.250) to obtain

$$E(\mathbf{r}) \approx \frac{kA}{f^2} \int_{\Omega_0} \mathrm{e}^{jk\mathbf{r}\cdot\hat{s}'}\,\mathrm{d}\Omega'. \tag{8.252}$$

   The next step is the introduction of cylindrical coordinates with the polar axis along the optical axis of the configuration so that $\mathbf{r} = (\rho\cos\phi, \rho\sin\phi, z)$ and $\mathbf{r}' = (\rho'\cos\phi', \rho'\sin\phi', z')$ with $z' = -\sqrt{f^2-\rho'^2}$ since $\mathbf{r}'$ belongs to the wavefront. The phase factor in (8.252) can then be represented as

$$\begin{aligned}\mathbf{r}\cdot\hat{s}' &= \frac{1}{f}\left[\rho\rho'\cos(\phi-\phi') - z\sqrt{f^2-\rho'^2}\right] \\ &\approx \frac{\rho\rho'}{f}\cos(\phi-\phi') - z\left(1-\frac{\rho'^2}{2f^2}\right),\end{aligned} \tag{8.253}$$

and substituting $\rho' = at$ yields

$$E(\mathbf{r}) \approx kA\frac{a^2}{f^2}e^{-jkz}\int_0^1\int_0^{2\pi} \exp\left[jk\rho\frac{a}{f}t\cos(\phi-\phi') + jkz\frac{a^2t^2}{2f^2}\right] t\,d\phi'\,dt, \tag{8.254}$$

which can be conveniently rewritten as

$$E(u,v) \approx kA\frac{a^2}{f^2}e^{-ju(f/a)^2}\int_0^1\int_0^{2\pi} e^{jvt\cos(\phi-\phi')+jut^2/2}\,t\,d\phi'\,dt \tag{8.255}$$

if we introduce the following dimensionless variables:

$$u = kz\left(\frac{a}{f}\right)^2, \quad v = k\rho\frac{a}{f}. \tag{8.256}$$

In these coordinates the focus is located at $u = 0$ and $v = 0$; the position of the geometric shadow boundary is given by the equation $v = |u|$.

The integral with respect to ϕ' can be reduced to a definition of the Bessel function,

$$\int_0^{2\pi} e^{jvt\cos(\phi-\phi')}\,d\phi' = 2\pi J_0(vt), \tag{8.257}$$

which leads to the formula

$$E(u,v) \approx 2\pi kA\left(\frac{a}{f}\right)^2 e^{-ju(f/a)^2}\int_0^1 e^{jut^2/2}J_0(vt)t\,dt. \tag{8.258}$$

The value at the focus is then given by

$$E(0,0) = \pi kA\left(\frac{a}{f}\right)^2, \tag{8.259}$$

and using this result in (8.258) we obtain the formula (8.249).

The field given by (8.249) has the following apparent symmetry properties: $E(u,-v) = E(u,v)$ and $E(-u,v) = E^*(u,v)$, where the asterisk denotes the complex conjugate. The integral in (8.249) can be expressed through Lommel's functions of two variables (Born and Wolf 1959; Watson 1962), but in practical calculations this approach does not seem to be necessary as the integral can be easily calculated numerically. Simulation results presented in Figures 8.34 and 8.35 show the behavior of the field near the focus in a meridional plane ($x = 0$) and the focal plane ($z = 0$), respectively. Due to diffraction, the focal spot has a finite size (Airy disc) (Airy 1838). It fits approximately into the interval $-\pi \le v \le \pi$ in the focal plane and $-2\pi \le u \le 2\pi$ in the meridional plane. In terms of the original, unscaled coordinates, the focal spot has the size of about $\lambda f/a$ in the focal plane and $2\lambda(f/a)^2$ in the meridional plane. As expected, the spot becomes a point in the GO limit ($\lambda \to 0$) but spreads to infinity when $a \to 0$.

4. A scatterer is illuminated by a plane wave and is located 50 m away from an observer. Estimate the role of diffraction in the following two cases: (a) $\lambda = 500$ nm (green light), characteristic size of the scatterer 0.01 m; (b) $\lambda = 0.05$ m (microwave frequency, C band), the characteristic size of the scatterer 10 m.

Hint: Estimate parameter P from (8.81).

Answer: In (a), $P = 1$ and the diffraction effects may considerably influence the scattered field at the point of observation. In (b), $P \approx 0.32 < 1$, and the effect of diffraction is small.

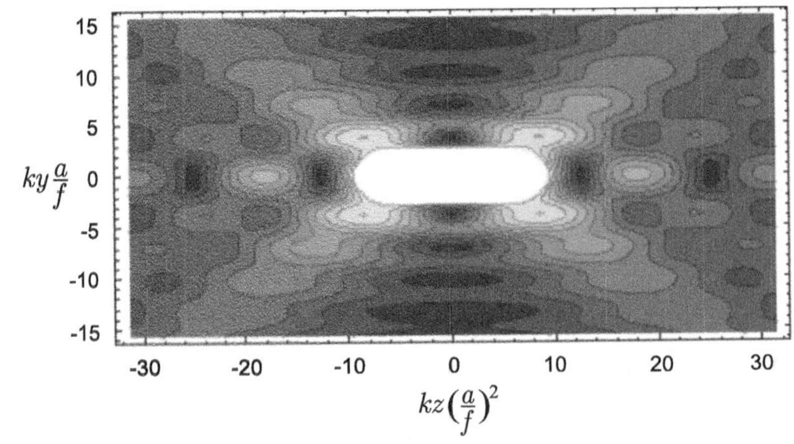

Figure 8.34 The field magnitude distribution in a meridional plane near the focus of a converging spherical wave transmitted through a circular aperture. The distribution is plotted in the scaled coordinates. To make the finer details more visible, the image is "overexposed" in the sense that values higher than 32% of the maximum value at the focus are cut off and the corresponding area is colored white.

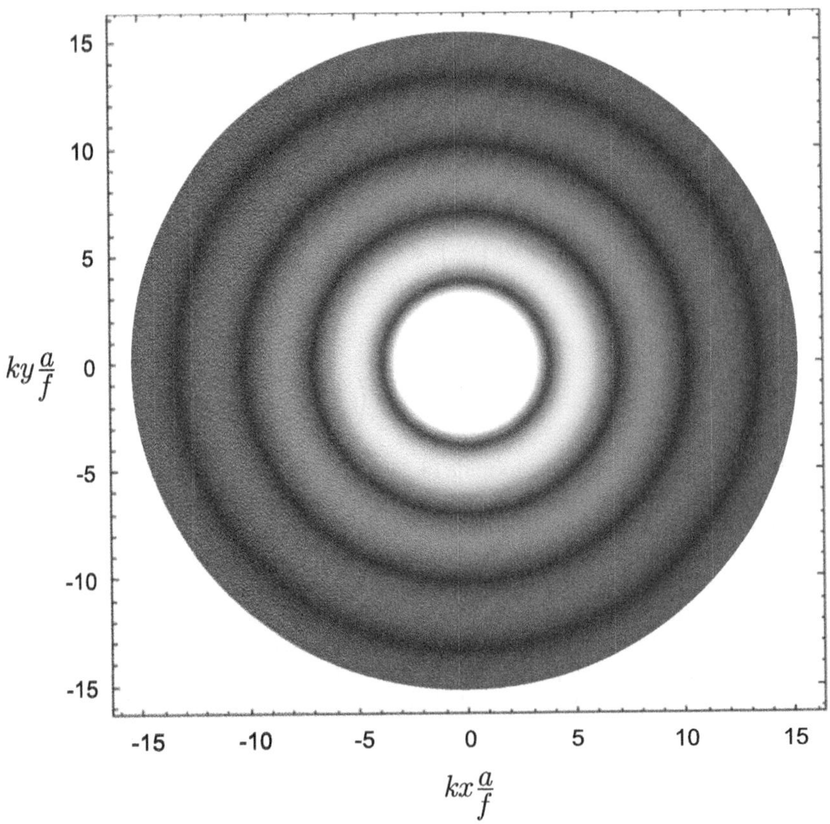

Figure 8.35 The same as Figure 8.34 but in the focal plane $z = 0$. The series of concentric rings around the bright spot (Airy pattern) is the result of diffraction. The field magnitude values are clipped at the 14% level.

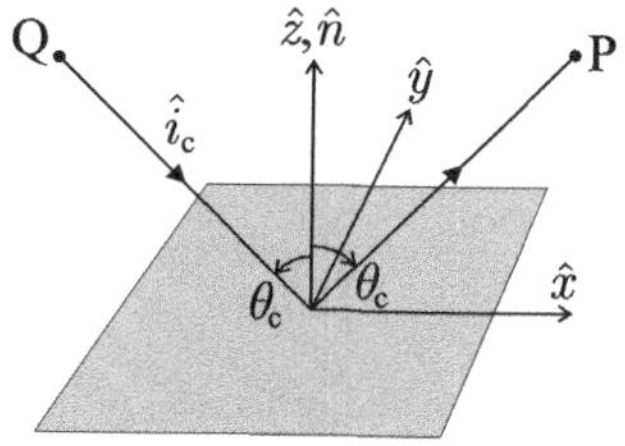

Figure 8.36 A spherical wave from a point source at Q is reflected from a flat plate toward a receiver at P.

5. Describe reflection of a spherical wave from a right-angled plate (Figure 8.36) by using the reflection formula (8.176). Does the reflection formula apply when $R, R_0 \to \infty$?

Answer: The result is given by (8.178) for arbitrary values of R and R_0. However, the formula is not applicable when $R, R_0 \to \infty$. It applies as long as the field-forming region is located entirely inside the plate surface.

Solution: Assuming that the plate lies in the $x - y$ plane, we can represent the reflecting surface by the parametric equation

$$\mathbf{f}(u, v) = \hat{x}u + \hat{y}v, \tag{8.260}$$

where $u_- \leq u \leq u_+$, $v_- \leq u \leq v_+$, and $u_\pm$ and $v_\pm$ describe the position of the rim of the plate. From (A.67) and (A.70) we find that

$$E = 1, \quad F = 0, \quad G = 1, \quad \hat{n} = \hat{z}. \tag{8.261}$$

All second-order derivatives of the linear function (8.260) vanish, which implies that

$$L, M, N, K, H = 0, \quad R_1 = R_2 = R_i = \infty. \tag{8.262}$$

The parameters (8.155)–(8.157) can therefore be expressed as

$$\alpha = -p[1 - (\hat{i}_c \cdot \hat{x})^2] \tag{8.263}$$

$$\beta = -p[1 - (\hat{i}_c \cdot \hat{y})^2] \tag{8.264}$$

$$\gamma = p(\hat{i}_c \cdot \hat{x})(\hat{i}_c \cdot \hat{y}) \tag{8.265}$$

which leads to the formula

$$\alpha\beta - \gamma^2 = p^2(\hat{i}_c \cdot \hat{z})^2 \tag{8.266}$$

in which we have used the relation

$$(\hat{i}_c \cdot \hat{x})^2 + (\hat{i}_c \cdot \hat{y})^2 + (\hat{i}_c \cdot \hat{z})^2 = 1. \tag{8.267}$$

From (8.263) and (8.266) it follows that $\alpha < 0$ and $\alpha\beta - \gamma^2 > 0$ and, therefore, $\sigma_c = -j$ for arbitrary values of R and R_0. The spread factor results from (8.161) and (8.262) as

$$\sqrt{\frac{D(0)}{D(R)}} = \frac{R_0}{R + R_0}, \tag{8.268}$$

and using these results in (8.176) we arrive at the formula (8.178).

6. Derive the spread factor for the field of a point source reflected from a segment of a circular cylindrical surface.

Answer: If a is the radius, then

$$\sqrt{\frac{D(0)}{D(R)}} = \frac{\sqrt{a}}{R\sqrt{p}}|pa \pm p_1 a|^{-1/2}, \tag{8.269}$$

where the upper (lower) sign is to be taken when the reflection point resides on the convex (concave) side of the segment and

$$p_1 = \frac{2}{a}\sec\theta_c\left(\cos^2\theta_c + \frac{a}{R_i}\sin^2\theta_c\right). \tag{8.270}$$

Solution: Assume that the cylinder axis is the y axis. The surface of a cylindrical segment can then be described by the parametric equation

$$\mathbf{f}(u,v) = a\hat{x}\cos u + v\hat{y} + a\hat{z}\sin u \tag{8.271}$$

with $-\infty < v < +\infty$ and $u_- \leq u \leq u_+$, where $u_\pm$ define the angular width of the segment. It follows from (A.67), (A.71), and (A.70) that

$$E = a^2, \quad F = 0, \quad G = 1, \quad L = a, \quad M = 0, \quad N = 0 \tag{8.272}$$

and

$$\hat{n} = -\hat{x}\cos u - \hat{z}\sin u. \tag{8.273}$$

Using (8.272) in (A.73)–(A.75) gives

$$K = 0, \quad H = \frac{1}{2a} \tag{8.274}$$

and

$$R_1 = a, \quad R_2 = \infty. \tag{8.275}$$

The curvature radii are positive since the unit normal (8.273) is directed to the center of curvature of the reflecting surface. Insertion of (8.274) in (8.161) leads to the formulas (8.269) and (8.270) for the spread factor.

The curvature radius in the incidence plane follows from (A.72) and (8.159) as

$$R_i = \frac{1}{a}\left[a^2 + \left(\frac{dv}{du}\right)^2\right] \tag{8.276}$$

with

$$\frac{dv}{du} = a^2\frac{\hat{\imath}_c\cdot\mathbf{f}_v}{\hat{\imath}_c\cdot\mathbf{f}_u}. \tag{8.277}$$

If the direction of incidence, that is, the direction form the source to the reflection point, is written in the form

$$\hat{\imath}_c = -\hat{x}\sin\theta_0\cos\phi_0 - \hat{y}\sin\theta_0\sin\phi_0 - \hat{z}\cos\theta_0, \tag{8.278}$$

then (8.276) becomes

$$\frac{R_i}{a} = 1 + \left(\frac{\sin\theta_0 \sin\phi_0}{\sin u_c \sin\theta_0 \cos\phi_0 - \cos u_c \cos\theta_0} \right)^2, \tag{8.279}$$

where u_c is the coordinate of the reflection point. Depending on the incidence direction and the position of the reflection point, the radius R_i varies between the principal curvature radii R_1 and R_2, that is, $a \leq R_i < \infty$.

It is seen from (8.270) that the value of $p_1 a$, which defines the position of the caustic in the concave case, lies between $2\cos\theta_c$ ($R_i = \infty$) and $2\sec\theta_c$ ($R_i = a$). Using (8.279) and the relation $\hat{i}_c \cdot \hat{n} = \pm\cos\theta_c$ in (8.270), the parameter p_1 can be expressed as

$$p_1 = \frac{1}{2a\cos\theta_c}[3 + \cos(2\theta_0) + 2\cos(2\phi_0)\sin^2\theta_0]. \tag{8.280}$$

7. Determine the position of the reflection point on a spherical surface of radius a for a plane wave illumination (Figure 8.37). The position of the observer can be arbitrary ($a \leq r < \infty$).

Hint: Specialize equations (8.147) to the case of a spherical surface.

Solution: We use spherical coordinates in which the position of the observer is given by $\mathbf{r} = r(\hat{x}\sin\theta\cos\phi + \hat{y}\sin\theta\sin\phi + \hat{z}\cos\theta)$. Without loss of generality, we can assume that the incident wave propagates in the direction of the negative z axis and since this is a plane wave, $\hat{i}_c = -\hat{z}$ for all points on the surface. The spherical surface can be described by the parametric equation (8.181). The position of the reflection point is given by $\mathbf{r}_c = a(\hat{x}\sin u_c \cos v_c + \hat{y}\sin u_c \sin v_c + \hat{z}\cos u_c)$. Because of the rotational symmetry of the scatterer with respect to the incidence direction, the angle u_c, regardless of the value of v_c, is equal to θ_c, which is the angle between the normal $\hat{n}_c$ to the surface at the reflection point and the incidence/scattering directions.

The equations (8.147) for the coordinates u_c and v_c of the reflection point can be expressed as

$$\hat{s}_c \cdot \mathbf{f}_u(u_c, v_c) = -\hat{z} \cdot \mathbf{f}_u(u_c, v_c) \tag{8.281}$$

$$\hat{s}_c \cdot \mathbf{f}_v(u_c, v_c) = -\hat{z} \cdot \mathbf{f}_v(u_c, v_c), \tag{8.282}$$

where $\mathbf{f}_u(u, v)$ and $\mathbf{f}_v(u, v)$ are given by (8.182) and (8.183) and

$$\hat{s}_c = \frac{1}{R}(\mathbf{r} - \mathbf{r}_c) \tag{8.283}$$

with $R = |\mathbf{r} - \mathbf{r}_c| = \sqrt{r^2 - 2ra\cos(\theta - u_c) + a^2}$. It is convenient to begin with equation (8.282) which, upon elementary manipulations, simplifies to

$$\sin(\phi - v_c) = 0. \tag{8.284}$$

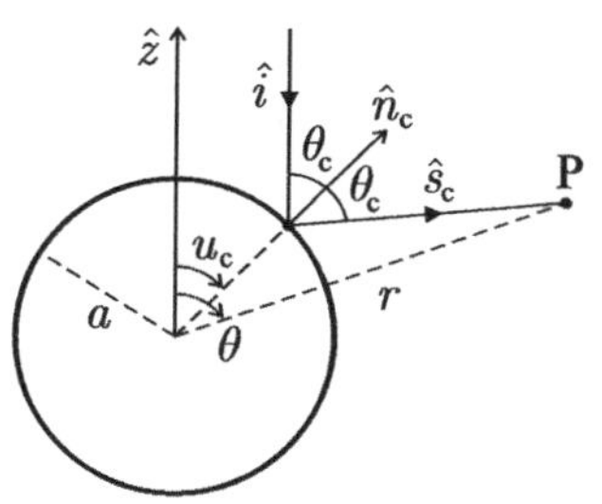

Figure 8.37 Geometry assumed in finding the reflection point on a spherical surface.

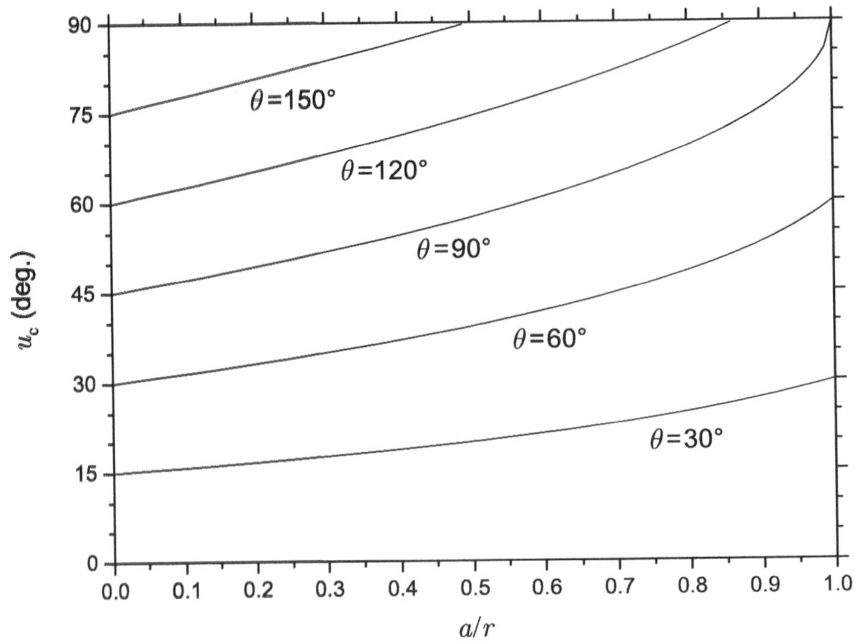

Figure 8.38 Coordinate u_c of the reflection point on a spherical surface illuminated by a plane wave as a function of the ratio a/r for several values of θ.

The solution $v_c = \phi$ describes the reflection point that lies in the same azimuthal plane as the observation point. Other possible solutions $v_c = \phi \pm \pi$ are unphysical and should be dropped.

The remaining parameter u_c can be determined from the remaining equation (8.281) in which v_c is replaced with ϕ. This leads to

$$\sin(\theta - u_c) = \sin u_c \sqrt{1 - 2\frac{a}{r}\cos(\theta - u_c) + \left(\frac{a}{r}\right)^2}. \tag{8.285}$$

The equation has a simple solution in several limiting cases: $u_c = \theta$ when $r = a$ (observation point lies on the scattering surface); $u_c = \theta/2$ when $r \to \infty$ (observer at infinity, the direction from the center of the sphere to the reflection point bisects the angle between the incidence and scattering directions); $u_c = 0$ when $\theta = 0$ (monostatic configuration).

In general, equation (8.285) has no explicit solution (substitution $\tan(u_c/2)$ transforms it to a polynomial equation of the sixth degree).[6] It is, however, amenable to numerical solution by various root-finding iterative methods by using the fact that the solution always lies on the interval $\theta/2 \le u_c \le \theta$. Figure 8.38 illustrates the dependence of u_c on the ratio a/r for several values of θ. The values of u_c greater than 90° are not shown because in this case the reflection point is located on the shadowed portion of the sphere and lies outside the integration range in the PO solution.

8. Specialize the reflection formula (8.176) to the case of an impenetrable spherical surface with reflection coefficients $R_\perp$ and $R_\parallel$ illuminated by a plane linearly polarized wave.
Hint: Use the vector solution of the reflection problem from section 4.2.6 to describe the reflected electric field at the point of reflection.

[6] For a simpler equation, equivalent to a polynomial equation of the forth degree, see problem 8 in section 7.7.

Solution: Assuming that the incident field is a plane wave with $\mathbf{E}_{\text{inc}} = E_0\hat{x}\exp(jkz)$ and the point of reflection $\mathbf{r}_c$ is located on the surface of the sphere at $r = a$, $\theta = \theta_c$ and $\phi = \phi_c$, we have $\hat{\imath} = -\hat{z}$ for the incidence direction and

$$\hat{n}_c = \sin\theta_c \cos\phi_c\hat{x} + \sin\theta_c \sin\phi_c\hat{y} + \cos\theta_c\hat{z} \tag{8.286}$$

for the unit normal at the point of reflection. In the basis of the incident wave,

$$\hat{\alpha}_{\text{inc}} = \frac{\hat{\imath}\times\hat{n}_c}{|\hat{\imath}\times\hat{n}_c|} = \sin\phi_c\hat{x} - \cos\phi_c\hat{y} \tag{8.287}$$

$$\hat{\beta}_{\text{inc}} = \hat{\imath}\times\hat{\alpha}_{\text{inc}} = -\cos\phi_c\hat{x} - \sin\phi_c\hat{y} \tag{8.288}$$

$$E^{\perp}_{\text{inc}} = \hat{\alpha}_{\text{inc}}\cdot\mathbf{E}_{\text{inc}} = E_0\sin\phi_c e^{jka\cos\theta_c} \tag{8.289}$$

$$E^{\parallel}_{\text{inc}} = \hat{\beta}_{\text{inc}}\cdot\mathbf{E}_{\text{inc}} = -E_0\cos\phi_c e^{jka\cos\theta_c}. \tag{8.290}$$

The direction of reflection is given by

$$\hat{s}_c = \hat{\imath} - 2\hat{n}_c(\hat{\imath}\cdot\hat{n}_c) = \sin(2\theta_c)\cos\phi_c\hat{x} + \sin(2\theta_c)\sin\phi_c\hat{y} + \cos(2\theta_c)\hat{z} \tag{8.291}$$

and the polarization directions in the basis of the reflected wave are

$$\hat{\alpha}_{\text{ref}} = \hat{\alpha}_{\text{inc}} \tag{8.292}$$

$$\hat{\beta}_{\text{ref}} = \hat{s}_c\times\hat{\alpha}_{\text{ref}} = \cos(2\theta_c)\cos\phi_c\hat{x} + \cos(2\theta_c)\sin\phi_c\hat{y} - \sin(2\theta_c)\hat{z}. \tag{8.293}$$

The basis vectors associated with the reflection problem are shown in Figure 8.39.

The general formula (4.79) for the reflected wave at the reflection point leads to the expression

$$\begin{aligned}\mathbf{E}_{\text{ref}}(\mathbf{r}_c) = E_0 e^{jka\cos\theta_c}\{&\hat{x}[R_\perp\sin^2\phi_c - R_\parallel\cos(2\theta_c)\cos^2\phi_c] \\ &- \hat{y}\sin\phi_c\cos\phi_c[R_\perp + R_\parallel\cos(2\theta_c)] + \hat{z}R_\parallel\sin(2\theta_c)\cos\phi_c\}.\end{aligned} \tag{8.294}$$

The incident wave is plane, which means that $R_0 \to \infty$ and $p = R^{-1}$, where R is the distance between the point of reflection and observation point and the parameter p is defined in (8.153). The spread factor (8.196) becomes therefore

$$\sqrt{\frac{D(0)}{D(R)}} = \frac{a}{\sqrt{(a + 2R\cos\theta_c)(a + 2R\sec\theta_c)}}. \tag{8.295}$$

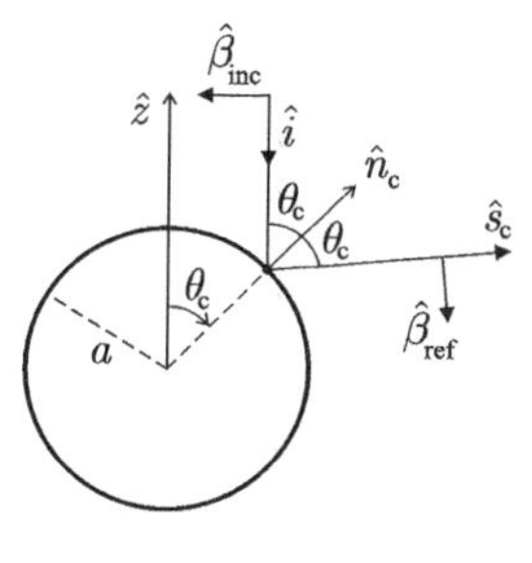

Figure 8.39 Reflection of a plane wave from a spherical surface: the bases of the incident and reflected waves. Vectors $\hat{\alpha}_{\text{inc}}$ and $\hat{\alpha}_{\text{ref}}$ point toward the reader.

The sphere is a convex body, thus $\sigma_c = -j$ (section 8.4.2), which together with (8.294) and (8.295) completely specifies the reflection formula for the sphere illuminated by a plane wave. The position of the reflection point can be determined by (8.284) and (8.285) from the previous problem.

9. In problem 8 an approximation (reflection formula) for the field of a plane wave reflected from a spherical surface with reflection coefficients $R_\parallel$ and $R_\perp$ is presented based on a stationary point approximation of the PO solution. Show that this result coincides with the asymptotic formulas (7.256), (7.257), and (7.258) obtained by stationary point evaluation of the exact solution for the sphere.

Solution: First, multiplying the reflected field (8.294) in the reflection formula with $\hat{r} = \hat{x}\sin\theta\cos\phi + \hat{y}\sin\theta\sin\phi + \hat{z}\cos\theta$, which is the unit radial vector at the observation point, and using the fact that the reflection and observation points lie in the same azimuthal plane, $\phi_c = \phi$, give

$$\hat{r}\cdot\mathbf{E}_{\mathrm{ref}}(\mathbf{r}_c) = E_r^{\mathrm{ref}}(\mathbf{r}_c) = E_0\cos\phi\sin(2\theta_c - \theta)R_\parallel e^{jka\cos\theta_c}, \tag{8.296}$$

and it follows from the reflection formula that

$$E_r^{\mathrm{ref}}(\mathbf{r}) = \frac{aE_r^{\mathrm{ref}}(\mathbf{r}_c)e^{-jkR}}{\sqrt{(a+2R\cos\theta_c)(a+2R\sec\theta_c)}}, \tag{8.297}$$

where R is the distance between the reflection and observation points and θ_c is the incidence angle at the reflection point.

Now, using (7.659) and (7.661) and noting that

$$1 + 2\frac{R}{a}\cos\theta_c = \frac{\sin\theta}{\cos\psi} \tag{8.298}$$

$$1 + 2\frac{R}{a}\sec\theta_c = r\sin\psi\left(\frac{2}{a\cos\theta_c} - \frac{1}{r\sin\psi}\right) \tag{8.299}$$

and that because of (7.658), $\sin(2\theta_c - \theta) = \cos\psi$, the PO-based formula (8.297) is transformed to the formula (7.255).

Multiplying (8.294) with $\hat{\theta} = \hat{x}\cos\theta\cos\phi + \hat{y}\cos\theta\sin\phi - \hat{z}\sin\theta$ leads to

$$\hat{\theta}\cdot\mathbf{E}_{\mathrm{ref}}(\mathbf{r}_c) = -\tan\psi E_{\mathrm{ref}}(\mathbf{r}_c), \tag{8.300}$$

which recovers the relation (7.257). In a similar way, multiplication of (8.294) with $\hat{\phi} = -\hat{x}\sin\phi + \hat{y}\cos\phi$ results in the relation (7.258) on accounting for (7.259).

10. Consider reflection of a spherical wave from a concave segment of radius a and determine the number of times the reflected ray may touch the caustics depending on the distance R_0 between the source and the reflection point.

Hint: Use formula (8.202) to determine conditions under which the caustics disappear at infinity.

Answer: The reflected ray may touch two caustics when $2R_0 > a\sec\theta_c$, a single caustic when $a\cos\theta_c < 2R_0 < a\sec\theta_c$, and no caustic when $0 \le 2R_0 < a\sec\theta_c$.

11. Describe the shape of the caustics that occur when a spherical segment defined in spherical coordinates by $r = a, 0 \le \phi < 2\pi$, and $\theta_0 \le \theta \le \pi$ with $\theta_0 > \pi/2$ is illuminated by a plane wave incident in the direction $\hat{i} = -\hat{z}$ (Figure 8.40).

Hint: From zeros of function $D(R)$ determine the distances from the reflection point to the points of contact of the reflected ray with the caustics.

Solution: There are two caustics: a piece $a/(2\cos\theta_0) \le z \le -a/2$ of the z axis and a surface of revolution with a cusp at $z = -a/2$ (Figure 8.40).

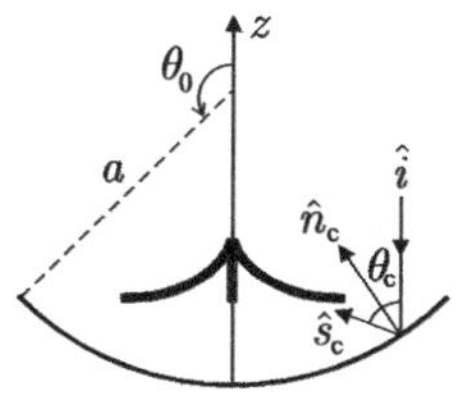

Figure 8.40 Caustics of a spherical segment under axial illumination.

The spread factor is infinite when $pa = 2\sec\theta_c$ and $pa = 2\cos\theta_c$. Since the source is at infinity ($R_0 \to \infty$), $p = R^{-1}$ and the distances between the reflection point

$$\mathbf{r}_c = \hat{x}a\sin\theta\cos\phi + \hat{y}a\sin\theta\sin\phi + \hat{z}a\cos\theta$$

and the caustics are given by $R = (a/2)\cos\theta_c$ and $R = (a/2)\sec\theta_c$. The position vectors $\mathbf{r}_{1,2}$ of the touch points can be therefore expressed as

$$\mathbf{r}_1 = \mathbf{r}_c + \hat{s}_c\frac{a}{2}\cos\theta_c, \quad \mathbf{r}_2 = \mathbf{r}_c + \hat{s}_c\frac{a}{2}\sec\theta_c. \tag{8.301}$$

The angle of reflection θ_c and the spherical coordinate θ of the reflection point are related by the formula $\theta_c = \pi - \theta$. The direction of reflection $\hat{s}_c$ is determined by the reflection law

$$\hat{s}_c = \hat{i} - 2\hat{n}_c(\hat{i}\cdot\hat{n}_c)$$

with

$$\hat{n}_c = -\hat{x}\sin\theta\cos\phi - \hat{y}\sin\theta\sin\phi - \hat{z}\cos\theta.$$

Using these results in (8.301) gives

$$\mathbf{r}_1 = a\sin^3\theta(\hat{x}\cos\phi + \hat{y}\sin\phi) + a\hat{z}\cos\theta\left(\frac{1}{2} + \sin^2\theta\right) \tag{8.302}$$

$$\mathbf{r}_2 = \frac{a}{2\cos\theta}\hat{z}. \tag{8.303}$$

Equation (8.302) with $\theta_0 \le \theta \le \pi$ and $0 \le \phi \le 2\pi$ describes a surface of revolution, and equation (8.303) with $\theta_0 \le \theta \le \pi$ a segment of the z axis. The reflected ray first touches the caustic surface at a point $\mathbf{r}_1$ and then crosses the z axis at a point $\mathbf{r}_2$. Because of the rotational symmetry of the reflector, every reflected ray crosses the z axis if the incident wave comes along the z axis. The degeneration of one of the caustic surfaces in a focal line is a common feature of axially symmetric fields. Figure 8.21 presents the results of numerical simulations of the field distribution inside a spherical reflector.

12. Paraboloid of revolution or circular paraboloid is the surface generated by a parabola revolving around its axis. Derive the spread factor in the reflection formula (8.176). Consider both concave and convex cases (Figure 8.41).

Hint: Let ρ_0 be the curvature radius at the vertex of the paraboloid. Then, the parametric representation of the surface is as follows:

$$\mathbf{f}(u,v) = \sqrt{\rho_0}u\cos v\hat{x} + \sqrt{\rho_0}u\sin v\hat{y} + \frac{1}{2}u^2\hat{z}, \tag{8.304}$$

where $0 \le v < 2\pi$ and $0 \le u \le u_+$, with u_+ defining the height of the paraboloid.

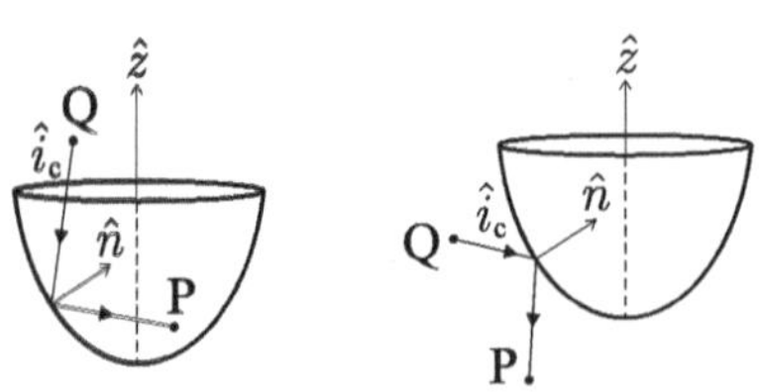

Figure 8.41 Spherical wave from a point source at Q is reflected from the surface of a circular paraboloid toward a receiver at P. Left: the reflection point on the concave side of the surface; right: the reflection point on the convex side of the surface.

Solution: The first fundamental forms:

$$E = \rho_0 + u^2, \quad F = 0, \quad G = \rho_0 u^2. \tag{8.305}$$

The unit normal to the surface:

$$\hat{n} = \frac{-u\cos v\hat{x} - u\sin v\hat{y} + \sqrt{\rho_0}\hat{z}}{\sqrt{\rho_0 + u^2}}. \tag{8.306}$$

The second fundamental forms:

$$L = \frac{\sqrt{\rho_0}}{\sqrt{\rho_0 + u^2}}, \quad M = 0, \quad N = \frac{\sqrt{\rho_0}u^2}{\sqrt{\rho_0 + u^2}}. \tag{8.307}$$

The Gaussian and the mean curvatures:

$$K = \frac{1}{(\rho_0 + u^2)^2}, \quad H = \frac{2\rho_0 + u^2}{2\sqrt{\rho_0}(\rho_0 + u^2)^{3/2}}. \tag{8.308}$$

The principal curvature radii:

$$R_1 = \sqrt{\rho_0}\sqrt{\rho_0 + u^2}, \quad R_2 = \frac{1}{\sqrt{\rho_0}}(\rho_0 + u^2)^{3/2}. \tag{8.309}$$

Note that in contrast to the cases of a plate, a circular cylinder, and a spherical segment, the radii are not constants but rather functions of u. At the vertex of paraboloid ($u = 0$), $R_1 = R_2 = \rho_0$. As a point on the surface moves further away from the origin, the radii grow, becoming infinite for $u \to \infty$.

The curvature radius in the plane of incidence:

$$R_i = \frac{\sqrt{\rho_0}\left(i_u^2 + i_v^2\right)(\rho_0 + u^2)^{3/2}}{\rho_0 i_u^2 + i_v^2(\rho_0 + u^2)}, \tag{8.310}$$

where

$$i_u = E^{-1}\hat{\imath}_c \cdot \mathbf{f}_u(u_c, v_c), \quad i_v = G^{-1}\hat{\imath}_c \cdot \mathbf{f}_v(u_c, v_c) \tag{8.311}$$

and u_c and v_c are the coordinates of the reflection point.

Equations (8.308) (with $u = u_c$) and (8.310) completely specify the spread factor. Because of the direction of the vector $\hat{n}$, equation (8.161) is to be taken with the upper sign in the concave case and with the lower sign in the convex case.

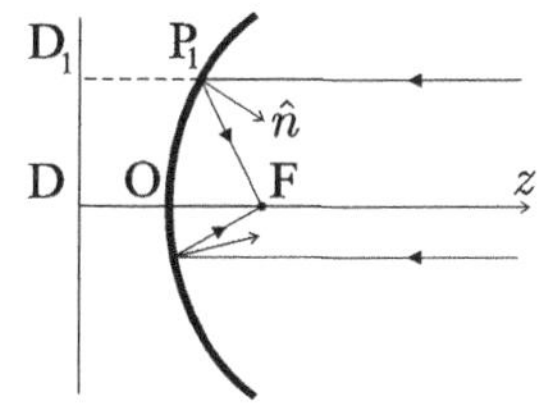

Figure 8.42 Plane wave illuminating the interior of a circular paraboloid.

13. Let a plane wave be incident on the concave side of a circular paraboloid reflector along the symmetry axis of the reflector (Figure 8.42). Prove that the spread factor is infinite at the focus of the paraboloid. Explain the physics of the singularity.

Hint: Use the rotational symmetry of the problem and the following geometric properties of every parabola in space (e.g.. Korn and Korn (1968)).

(a) The parabola is the locus of points that are equidistant from a given straight line (the directrix) and a given point (the focus) and lie in the plane defined by the line and the focus (Figure 8.42).

(b) If F is the focus, DD_1 is the directrix, P_1 is an arbitrary point on the parabola, and the symmetry axis of the parabola is the coordinate axis z with the origin at the parabola vertex O, then $|P_1F| = |P_1D_1| = |OD| + z$. By construction of the parabola, the distance |OD| between the vertex and the directrix is equal to the distance |OF| between the vertex and the focus.

(c) The distance |OF| is equal to half the curvature radius ρ_0 of the parabola at the vertex.

(d) At any point on the parabola, a vector $\hat{n}$ normal to the curve halves the angle between the direction from this point to the focus and the direction parallel to the symmetry axis of the parabola.

Solution: For a plane wave, $R_0 \to \infty$ and, thus, $p \to 1/R$ and $\hat{i}_c = -\hat{z}$. The incidence angle θ_c between the normal $\hat{n}$ to the surface and the direction of the incoming/reflected wave is determined from the relation $\cos\theta_c = -\hat{i}_c \cdot \hat{n}$, and using (8.306) we obtain

$$\cos\theta_c = \frac{\sqrt{\rho_0}}{\sqrt{\rho_0 + u^2}}. \tag{8.312}$$

The quantity $i_v = 0$, which is defined in (8.311), vanishes because the incidence direction $\hat{i}_c$ is perpendicular to the vector $\mathbf{f}_v$, and the expression (8.310) for the curvature radius R_i simplifies to

$$R_i = \frac{1}{\sqrt{\rho_0}}(\rho_0 + u^2)^{3/2}. \tag{8.313}$$

Inserting (8.308), (8.312), and (8.313) in (8.161) gives

$$D(R) = \frac{\rho_0 R^2}{\rho_0 + u^2}\left|\frac{1}{R} - \frac{2}{\rho_0 + u^2}\right|^2 \tag{8.314}$$

and leads to the spread factor

$$\sqrt{\frac{D(0)}{D(R)}} = \frac{\rho_0 + u^2}{2}\left|R - \frac{\rho_0 + u^2}{2}\right|^{-1} \tag{8.315}$$

which becomes infinite when

$$R = \frac{\rho_0 + u^2}{2} = \frac{\rho_0}{2} + z, \tag{8.316}$$

where z is the z coordinate of the reflection point (see the parametric equation (8.304)). The distance (8.316) is the distance between the reflection point and the focus of the parabola because of the parabola properties (b) and (c). Furthermore, property (d) ensures that the direction of the reflected ray coincides with the direction to the focus, which implies that the singularity of the spread factor is located at the focus of the parabola.

Thus, the geometric properties of parabolic surfaces ensure that in the case of axial illumination of a circular parabolic reflector, every incident ray, irrespective of the position of the reflection point, is reflected toward a common point, the focus of the paraboloid, at which all the reflected rays meet each other with the same value of phase. The constructive interference of the reflected rays leads to infinite field values at the focus. Note that the singularity of the spread factor is that of a point source located at the focus, which means that a parabolic reflector transforms a spherical wave emitted by a point source at the focus into an outgoing plane wave.

14. Show that the Airy function $\mathrm{Ai}(\xi\sqrt[3]{2k^2/\rho})$ (section D.2.3) describes the fields near a caustic surface, where ρ is the curvature of the caustic and ξ is a coordinate measured transversely to the caustic so that $\xi > 0$ on the shadowed side of the caustic (Figure 8.43).

Hint: Select an arc on the caustic surface and consider the PO integral over the corresponding arc of the wavefront. Account for the exponential factor and neglect the rest, slowly varying factors.

Solution: We follow here an elegant solution of a similar problem described in Landau and Lifshitz (1980). Let ab be a cross section of a caustic surface and a′b′ the corresponding cross section of the wavefront. We are interested in the dependence of the field on the transversal coordinate ξ in a vicinity of the point O at which the ray OQ touches the caustic. The coordinate ξ denotes the distance from O along the normal to the caustic and is assumed positive in the direction toward the center of curvature of the caustic.

The field can be represented by a PO integral over the wavefront with an exponential factor which is a function of the distance between an arbitrary integration point O′ and the observation point P. Let us approximate this distance assuming that the angle θ defining the position of the integration point O′ with respect to O and the coordinate ξ are small parameters. The wavefront is the involute of the caustic (section 8.4) and, thus, $|\mathrm{OQ}| = |\mathrm{O'Q'}| + \rho\theta$, where O′ is the point on the caustic where a ray from point Q′ is tangent to the caustic (remember that every ray is perpendicular to the wavefront since this is a surface with a constant phase value). Next we note that $|\mathrm{OQ'}| \approx |\mathrm{O'Q'}| + \rho\sin\theta$ (see formula (A.15)) and denoting $D = |\mathrm{OQ}|$, we obtain

$$|\mathrm{OQ'}| \approx D - \rho\theta + \rho\sin\theta \approx D - \frac{1}{6}\rho\theta^3. \tag{8.317}$$

Again using (A.15) and the smallness of θ and ξ, we find that

$$|\mathrm{PQ'}| \approx |\mathrm{OQ'}| - \xi\sin\theta \approx |\mathrm{OQ'}| - \xi\theta, \tag{8.318}$$

and combining this with (8.317) gives

$$|\mathrm{PQ'}| \approx D - \xi\theta - \frac{1}{6}\rho\theta^3. \tag{8.319}$$

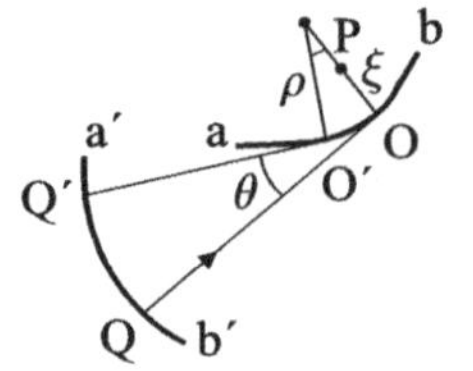

Figure 8.43 A caustic and a corresponding wavefront.

Thus, the field can be approximated by an integral over the arc a′b′ of the wavefront,

$$E \approx C \int_{a'}^{b'} e^{-jk\left(D-\xi\theta-\frac{1}{6}\rho\theta^3\right)} \, d\theta, \tag{8.320}$$

where E is an arbitrary field component and C is a coefficient composed of all the slowly varying factors in the PO integral. Since the main contribution to the integral in (8.320) comes from the vicinity of the point Q and the integral is convergent, the integration region can be extended to infinity, which allows us to express (8.320) in terms of the Airy function (6.126) as

$$E \approx 2\pi C e^{-jkD} \mathrm{Ai}\left(\xi \sqrt[3]{\frac{2k^2}{\rho}}\right). \tag{8.321}$$

15. By using the solution of the previous problem and the properties of the Airy function, describe the field behavior near the caustic for an observation point P in the deep shadow ($\xi \gg 1$), on the caustic ($\xi = 0$), and in the lit region ($-\xi \gg 1$). Estimate the width $\delta\xi$ of the transition zone between the lit and shadowed portions of space. How does the field on the caustic depend on the wave number k?

Hint: The properties of the Airy function can be found in (D.29), (D.30), and (D.31) (see also Figure D.3).

Solution: The field near a caustic is proportional to $\mathrm{Ai}(\xi \sqrt[3]{2k^2/\rho})$ and therefore its behavior is that of the Airy function with the appropriately scaled argument. Behind the caustic, when $\xi > 0$, the Airy function quickly decays as the observation point moves away from the caustic, closely following the asymptotic formula (D.30) already when $\xi \sqrt[3]{2k^2/\rho} \geq 1$. On the caustic, when $\xi = 0$, the Airy function takes the finite value (D.29). As the argument ξ becomes negative, the function grows, achieves its maximum value 0.536 at $\xi \sqrt[3]{2k^2/\rho} = -1.02$, and begins to oscillate beyond that point with a decreasing amplitude and an oscillation period in accordance with the formula (D.31). Thus, the width of the transition region can be estimated as $\delta\xi \approx \rho^{1/3} k^{-2/3}$.

The dependence of the field magnitude at the caustic on the wavenumber can be determined from (8.321) by requiring that far from the caustic in the illuminated region the field should approach the free-space wave $E \approx E_0 e^{-jkD}$, implying that

$$C \sim k^{1/6} \tag{8.322}$$

in order to compensate the factor $k^{-1/6}$ which results from application of (D.31) to $\mathrm{Ai}(\xi \sqrt[3]{2k^2/\rho})$. Thus, the fields behave as $O(k^{1/6})$ in front of the caustics and as $\exp(-\alpha k)$ with $\alpha > 0$ beyond the caustics. The increase (8.322) in the field amplitude with the growing frequency explains the different intensities of light at the caustics depending on the wavelength (Figures 7.70 and 7.79).

16. Describe the shape of the field-forming region in the case of a locally spherical reflecting surface (Figure 8.25) with $a \gg \lambda$ (a the radius) in the monostatic case with $R, R_0 \to \infty$. Is there a difference between the convex and concave cases (Figure 8.22)?

Answer: The region is a spherical segment with the solid angle $\sqrt{2\lambda/a}$ measured from the center of curvature of the segment. The diameter of the base of the segment equals $\sqrt{2a\lambda}$ and the height of the segment is $\lambda/4$.

Solution: Without loss of generality, we can describe the surface by equation (8.181), assuming that the z axis of the spherical coordinate system points from the curvature center of the surface in the direction of the source/receiver. From the definition of the field-forming region (8.77) it follows that in coordinates u and v the boundary of the field-forming region is determined by the equation

$$\left|\frac{1}{2}\alpha(u-u_c)^2 + \gamma(u-u_c)(v-v_c) + \frac{1}{2}\beta(v-v_c)^2\right| = \frac{\pi}{k}. \tag{8.323}$$

Specializing (8.155)–(8.157) to the spherical surface and to $R, R_0 \to \infty$, we obtain $\alpha \to \alpha_\infty = \mp 2a\cos\theta_c$, $\beta \to \beta_\infty = \mp 2a\cos\theta_c \sin^2 u_c$, and $\gamma \to \gamma_\infty = 0$ (the upper/lower sign corresponds to the convex/concave case, respectively) and since at the point of reflection $u_c = 0$ and $\theta_c = 0$, one finds that $\alpha_\infty = \mp 2a$, $\beta_\infty = \gamma_\infty = 0$. Equation (8.323) simplifies to

$$\frac{1}{2}|\alpha_\infty|u^2 = \frac{\pi}{k} \tag{8.324}$$

and gives the solution $u = \sqrt{\lambda/(2a)}$; the value of v can be arbitrary. Inserting these into (8.181) leads to an equation of the curve that encloses the field-forming region on the reflecting surface. Using the smallness of λ/a, we conclude that the field-forming region is a spherical segment with $0 \le \theta \le \sqrt{\lambda/(2a)}$ and $0 \le \phi < 2\pi$ for the convex segment and with $\pi - \sqrt{\lambda/(2a)} \le \theta \le \pi$ for the concave segment.

17. The PO approximation (8.227) and (8.243) for the field diffracted at the edge of a PEC wedge has singularities at shadow boundaries $\phi = \pi \pm \phi_0$. Prove that the difference between the GTD approximation in (5.460)–(5.463) with (5.88) and (5.89) and the PO approximation for the edge-diffracted field is bounded at the shadow boundaries.

References

Airy GB 1838 On the intensity of light in the neighborhood of a caustic. *Trans. Cambridge Phil. Soc.*, **6**, 379–402.

Asvestas JS 1980 The physical optics method in electromagnetic scattering. *J. Math. Phys.*, **21**, 290–299.

Asvestas JS 1985a Line integrals and physical optics. Part I. The transformation of the solid-angle surface integral to a line integral. *J. Opt. Soc. Am. A*, **2**, 891–895.

Asvestas JS 1985b Line integrals and physical optics. Part II. The conversion of the Kirchhoff surface integral to a line integral. *J. Opt. Soc. Am. A*, **2**, 896–902.

Asvestas JS 1986 The physical optics fields of an aperture on a perfectly conducting screen in terms of line integrals. *IEEE Trans. Antennas Propag.*, **34**, 1155–1159.

Baker BB and Copson ET 1953 *The Mathematical Theory of Huygens' Principle*, Clarendon Press.

Born M and Wolf E (eds) 1959 *Principles of Optics: Electromagnetic Theory of Propagation, Interference and Diffraction of Light*, Pergamon Press.

Borovikov VA 1994 *Uniform Stationary Phase Method*, The Institution of Electrical Engineers, London.

Borovikov VA and Kinber BY 1994 *Geometrical Theory of Diffraction*, The Institution of Electrical Engineers, London.

Breinbjerg O, Rahmat-Samii Y, and Appel-Hansen J 1987 *A theoretical examination of the physical theory of diffraction and related equivalent currents*, Report 339, Electromagnetics Institute, Technical University of Denmark, Lyngby.

Crew H (ed) 1900 *The Wave Theory of Light. Memoirs by Huygens, Young and Fresnel*, American Book Company.

Crispin JW and Siegel KM (eds) 1968 *Methods of Radar Cross-Section Analysis*, Academic Press.

Debye P 1909 Das Verhalten von Lichtwellen in der Nähe eines Brennpunktes oder einer Brennlinie. *Ann. Phys. (Berlin)*, **335**, 755–776.

Deschamps GA 1972 Ray techniques in electromagnetics. *Proc. IEEE*, **60**, 1022–1035.

Fedoryuk MV 1989 Asymptotic Methods in Analysis. In *Analysis I*, volume 13 of the series Encyclopaedia of Mathematical Sciences (ed. Gamkrelidze RV), pp. 83–191, Springer.

Fock VA 1965 *Electromagnetic Diffraction and Propagation Problems*, Pergamon Press.

Franz W 1948 Zur Formulierung des Huygenschen Prinzips. *Z. Naturforsch. A*, **3**, 500–506.

Fresnel A 1818 Mémoire sur la diffraction de la lumière. In *Œuvres complètes d'Augustin Fresnel, Tome premier: Théorie de la Lumière* (eds De Sénarmont H, Verdet É, and Fresnel L), pp. 247–382, Imprimerie impériale, Paris.

Huygens C 1690 *Traite de la Lumiere*, Leyden.

James GL 1976 *Geometrical Theory of Diffraction for Electromagnetic Waves*, Peter Peregrinus, London.

Jenn DC 2005 *Radar and Laser Cross Section Engineering*, American Institute of Aeronautics and Astronautics, Reston.

Johansen PM and Breinbjerg O 1995 An exact line integral representation of the physical optics scattered field: the case of a perfectly conducting polyhedral structure illuminated by electric Hertzian dipoles. *IEEE Trans. Antennas Propag.*, **43**, 689–696.

Karatygin VA and Rosov VA 1972 Method of stationary phase for double integral with arbitrary located stationary point. *J. Comput. Math. Phys.*, **12**, 1391–1405.

Karczewski B 1963 Boundary wave in electromagnetic theory of diffraction. *J. Opt. Soc. Am.*, **53**, 878–879.

Keller JB 1962 Geometrical theory of diffraction. *J. Opt. Soc. Am.*, **52**, 116–130.

Kirchhoff G 1891 *Mathematische Optik (Vorlesungen über mathematische Physik, 2. Band)*, Teubner Verlag, Leipzig.

Klement D, Preissner J, and Stein V 1988 Special problems in applying the Physical Optics method for backscatter computations of complicated objects. *IEEE Trans. Antennas Propag.*, **36**, 228–237.

Kline M and Kay IW 1965 *Electromagnetic Theory and Geometrical Optics*, John Wiley & Sons.

Knott EF, Shaeffer JF, and Tuley MT 1993 *Radar Cross Section*, Artech House.

Kontorovich MI and Muravév YK 1952 Derivation of the laws of reflection of geometrical optics on the basis of an asymptotic treatment of the diffraction problem. *Zhurn. Techn. Fiz.*, **22**, no. 3, 394–407 (in Russian).

Korn GA and Korn TM 1968 *Mathematical Handbook for Scientists and Engineers: Definitions, Theorems, and Formulas for Reference and Review*, 2nd revised edn, McGraw-Hill.

Kottler F 1923 Elektromagnetiche Theorie der Beugung an schwarzen Schirmen. *Ann. Phys. (Berlin)*, **376**, 457–508.

Kouyoumjian RG and Pathak PH 1974 A uniform geometrical theory of diffraction for an edge in a perfectly conducting surface. *Proc. IEEE*, **62**, 1448–1461.

Landau LD and Lifshitz EM 1980 *The Classical Theory of Fields (Course of Theoretical Physics, Volume 2)*, 4th rev. English edn. Butterworth-Heinemann.

Laport O and Meixner J 1958 Kirchhoff–Youngsche Theorie der Beugung elektromagnetischer Wellen. *Z. Phys.*, **153**, 129–148.

Larmor J 1904 On the mathematical expression of the principle of Huygens. *Proc. London Math. Soc.*, **1**, 1–13.

Macdonald HM 1902 *Electric Waves*, Cambridge University Press.

Macdonald HM 1913 The effect produced by an obstacle on a train of electric waves. *Phil. Trans. R. Soc. Lond. A*, **212**, 299–337.

Maggi GA 1888 Sulla propagazione libera e perturbata delle onde luminose in un mezzo isotropo. *Ann. Mat. Pura Appl.*, **16**, 21–48.

McNamara DA, Pistorius CWI, and Malherbe JAG 1990 *Introduction to the Uniform Geometrical Theory of Diffraction*, Artech House.

Meixner J and Andrejewski W 1950 Strenge Theorie der Beugung ebener elektromagnetischer Wellen an der vollkommen leitenden Kreisscheibe und an der kreisförmigen Öffnung im vollkommen leitenden ebenen Schirm. *Ann. Phys. (Berlin)*, **442**, 157–168.

Meneghini R, Shu P, and Bay J 1982 Several Maggi–Rubinowicz representations of the electric field. *IEEE Trans. Antennas Propag.*, **30**, 516–520.

Osipov AV 2014 Physical theory of diffraction for scatterers with low reflection surface. *Radio Sci.*, **49**, 1052–1064.

Osipov AV 2015 Scattering cross sections of impedance-matched bodies. *IEEE Trans. Antennas Propag.*, **63**, 3122–3126.

Rubinowicz A 1924 Zur Kirchhoffschen Beugungstheorie. *Ann. Phys. (Berlin)*, **378**, 339–364.

Rubinowicz A 1957 *Die Beugungswelle in der Kirchhoffschen Theorie der Beugung*, Panstwowe Wydawnictwo Naukowe, Warsaw.

Rubinowicz A 1965 The Miyamoto–Wolf diffraction wave. *Progress in Optics*, **4**, 199–240.

Ruck GT, Barrick DE, Stuart WD, and Krichbaum CK 1970 *Radar Cross Section Handbook*, Plenum Press.

Senior TBA and Volakis JL 1995 *Approximate Boundary Conditions in Electromagnetics*, The Institution of Electrical Engineers, London.

Skolnik MI (ed.) 2008 *Radar Handbook*, 3rd edition. McGraw-Hill Professional.

Stratton JA and Chu LJ 1939 Diffraction of electromagnetic waves. *Phys. Rev.*, **56**, 99–107.

Stratton JA 1941 *Electromagnetic Theory*, McGraw-Hill.

Tai C-T 1972 Kirchhoff Theory: Scalar, Vector, or Dyadic? *IEEE Trans. Antennas Propag.*, **20**, 114–115.

Ufimtsev PY and Senior TBA 1975 Comments on "Comparison of three high-frequency diffraction techniques". *Proc. IEEE*, **63**, 1734–1737.

Ufimtsev PY 1999 Backscatter. In *Wiley Encyclopedia of Electrical and Electronics Engineering, vol. 2* (ed. Webster JG), pp. 175–187, John Wiley & Sons.

Ufimtsev PY 2014 *Fundamentals of the Physical Theory of Diffraction*, Wiley.

von Helmholtz H 1860 Theorie der Luftschwingungen in Röhren mit offenen Enden. *J. Reine Angew. Math.*, **57**, 1–72.

von Helmholtz H 1897 *Vorlesungen über elektromagnetische Theorie des Lichtes*, Voss Verlag, Hamburg und Leipzig.

Walker J 1904 *The Analytical Theory of Light*, Cambridge.

Watson GN 1962 *A Treatise on the Theory of Bessel Functions*, Cambridge.

Wolf E 1959 Electromagnetic diffraction in optical systems I. An integral representation of the image field. *Proc. R. Soc. Lond. A*, **253**, 349–357.

Young Th 1845 *A Course of Lectures on Natural Philosophy and Mechanical Arts*, 2nd edition, Taylor and Walton, London, lecture XXXIX.

Youssef NN 1989 Radar cross-section of complex targets. *Proc. IEEE*, **77**, 722–734.

9

Physical Optics Solutions of Canonical Problems

9.1 Introduction

This chapter shows how to apply the physical optics (PO) method to study electromagnetic scattering from electrically large bodies of simple shapes. PO solutions are given by a two-dimensional integral which is explicit but difficult to calculate if the body is electrically large. One approach is to use the stationary phase method, which leads to an asymptotic approximation of the PO solution. In Chapter 8 this approach was used to describe reflection and scattering from generally curved convex bodies, transmission through apertures in opaque screens, and diffraction by curved edges. We start this chapter by applying the method to infinite cones with variously shaped cross sections and derive approximations for the tip-diffracted wave (section 9.2). It should be pointed out that the PO solution itself is an approximation and its asymptotic evaluation leads to a further approximation. However, comparison with scarcely available numerical and experimental data for scattering from vertices suggests that in many cases the stationary point approximation of the PO solution provides a good approximation.

The method of stationary phase is not always applicable, for example because a stationary point may not exist or be not isolated or be close to a singularity of the slowly varying integrand. For simply shaped finite bodies, by using simplicity of the scattering surface, the PO integral can be sometimes reduced to one-dimensional integrals or even evaluated in a closed form, particularly when the incident wave is a plane wave and the observer is located in the far-field region of the body. This is another approach to evaluation of PO solutions, and in sections 9.3 and 9.4 we study two relevant cases, namely, scattering by flat plates and by bodies of revolution.

In section 9.3 it is shown that the PO solution for scattering from opaque plates can always be reduced to a line integral, and if the plate has the form of a polygon the PO integral simplifies even further to a sum of contributions from the tips of the polygon. The tip contributions are in agreement with the results of section 9.2 obtained with the stationary phase method. The explicit solutions for polygonal plates are particularly important since they serve as building blocks in practical calculations of PO solutions for generally shaped bodies (e.g., Youssef (1989)). Furthermore, by invoking the Babinet principle the solutions for PEC plates can be used to describe transmission of fields through complimentary PEC screens. Note that the stationary point method is inapplicable to plates illuminated by plane waves and for an observer in the far-field region as in this case the phase function of the PO integral is linear and has therefore no stationary points.

Section 9.4 is devoted to PO solutions for bodies of revolution, and we show that the transformation to a one-dimensional integral is possible for the incidence along the symmetry axis of the body. This kind of solution is particularly important when the scattering direction is also along the symmetry axis (back- or forward scattering) as this formation leads to caustics in the scattered field, so that the corresponding PO integral cannot be calculated with the conventional stationary phase method. In the case of axial backscattering from bodies of revolution, the one-dimensional PO solution can be evaluated asymptotically by integrating by parts, and this approach permits a clear physical interpretation of the solution.

Modern Electromagnetic Scattering Theory with Applications, First Edition. Andrey V. Osipov and Sergei A. Tretyakov.

We will see that the axial backscattering from smoothly shaped pointed scatterers, like an ogive, is particularly weak, with the main contribution coming from the tip; a broad variety of tip diffraction coefficients is described in section 9.2.

The bodies studied in this section are assumed impenetrable with the material properties described by reflection coefficients or surface impedances.

9.2 Vertices

A vertex is a point on a surface at which every tangent to the surface is discontinuous. In this section we use the PO approximation to study electromagnetic scattering from vertices of various kinds, including a corner point on an edge of a thin plate, the apex of a pyramid, and the conical point (as the tip of a cone) (Figure 9.1). Mathematically, the analysis consists of evaluation of contributions to the PO integral from critical points of the third kind or from conical points. As we shall see, all these contributions are of the order $O(k^{-2})$ and describe a spherical wave, which in a spherical coordinate system with the origin at the vertex can be expressed as

$$\begin{bmatrix} \mathbf{E}_{sc}(\mathbf{r}) \cdot \hat{\theta} \\ \mathbf{E}_{sc}(\mathbf{r}) \cdot \hat{\phi} \end{bmatrix} = \frac{e^{-jkr}}{kr} \overline{\overline{D}} \cdot \begin{bmatrix} \mathbf{E}_{inc}(0) \cdot \hat{\theta}_{inc} \\ \mathbf{E}_{inc}(0) \cdot \hat{\phi}_{inc} \end{bmatrix}, \tag{9.1}$$

where r is the distance from the vertex to the observation point. $\overline{\overline{D}}$ is the matrix diffraction coefficient

$$\overline{\overline{D}} = \begin{bmatrix} D(\hat{\theta}, \hat{\theta}_{inc}) & D(\hat{\theta}, \hat{\phi}_{inc}) \\ D(\hat{\phi}, \hat{\theta}_{inc}) & D(\hat{\phi}, \hat{\phi}_{inc}) \end{bmatrix}, \tag{9.2}$$

which depends on the kind of vertex.

The expression (9.1) has the form of a spherical wave decaying as $O(r^{-1})$ as the distance r increases. So, for this particular scattering geometry the diffraction coefficient is directly related to the scattering matrix (3.61)

$$\overline{\overline{A}} = \frac{\sqrt{4\pi}}{k} \overline{\overline{D}} \tag{9.3}$$

and therefore to the scattering cross section (3.66)

$$\sigma(\hat{e}, \hat{e}_{inc}) = \frac{4\pi}{k^2} |D(\hat{e}, \hat{e}_{inc})|^2, \tag{9.4}$$

where $\hat{e} = \hat{\theta}$ or $\hat{\phi}$ and $\hat{e}_{inc} = \hat{\theta}_{inc}$ or $\hat{\phi}_{inc}$.

For PEC and impedance bodies the vertex diffraction coefficient is independent of k, which results from dimensionality considerations (the diffraction coefficient is dimensionless and there is no parameter with the dimensionality of length

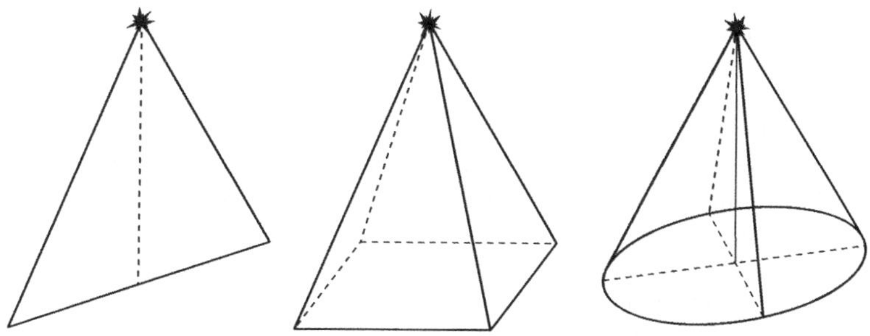

(a) (b) (c)

Figure 9.1 Canonical scatterers with a vertex: (a) a flat angular sector, (b) a pyramid, and (c) an elliptic cone.

to build a dimensionless product of it with k, since we consider an isolated vertex at an infinitely large body). It follows from (9.1) that compared to the incident field, the asymptotic order of the field scattered by the vertex is $O(k^{-1})$, which is smaller than the contributions from the critical points associated with the specular reflection from the surface $O(1)$ or with the edge diffraction $O(k^{-1/2})$. In general, therefore, the tip scattering is weak, but under special circumstances when no points of specular reflection appear on the scattering surface and the scatterer has no edges or the edges lie in the shadow, it can dominate the scattered signal.

Diffraction coefficients of a vertex can be in principle derived from an exact solution of the problem of scattering of a plane electromagnetic wave, arbitrarily polarized and incident from an arbitrary direction, from a correspondingly shaped canonical scatterer. Exact expressions for dyadic Green's functions, obtained by separation of variables in spherical and conical (or spheroconal) coordinates, are available for PEC cones of circular (Bailin and Silver 1956; Bouman *et al.* 1987; Felsen 1957) and elliptic (Blume 1971; Kraus and Levine 1961; Vafiadis and Sahalos 1983) cross sections, including a plane angular sector (Hansen 1990; Satterwhite 1974) (the latter can be regarded as a limiting case of an elliptic cone). By letting the source and the observation point go to infinity in these expressions, one can obtain infinite series expansions for the diffraction coefficients which, however, in the case of an elliptic cone involve less studied special functions, periodic and non-periodic Lamé functions, and do not converge in the usual sense, so that sequence transformation techniques have to be applied to determine the diffraction coefficient numerically (Blume *et al.* 1993; Blume and Uschkerat 1995; Blume and Krebs 1998; Bouman *et al.* 1987; Klinkenbusch 2007). A further limitation of the separation of variables is that it is only applicable to cones with elliptic cross section and PEC surface. An overview of research on scattering from the elliptic cone can be found in Blume (1996).

To overcome these difficulties, analytical-numerical approaches, which combine analytical considerations with the numerical solution of an integral equation, have been applied to the determination of the diffraction coefficients of a PEC quarter plane (Hansen 1991), PEC cones of arbitrary cross section (Babich *et al.* 1996; Smyshlyaev 1993), and a circular impedance cone (Bernard *et al.* 2008). The disadvantages of this approach are that the diffraction coefficient is available only in a numerical form and the numerical costs are still significant as one needs to solve either a two-dimensional integral equation (Hansen 1991) or a one-dimensional integration equation over a range of values of one (Babich *et al.* 1996; Smyshlyaev 1993) or two spectral parameters (Bernard *et al.* 2008).

It should be noted that the determination of diffraction coefficients of vertices significantly simplifies in the particular case of incidence along the cone axis, and most of the numerical data available in the literature are limited to this case (Bouman *et al.* 1987; Goriainov 1961; Ruck *et al.* 1970).

The use of PO solutions for canonical scatterers with a vertex of a required type is free of the limitations inherent to exact formulations. Asymptotic evaluation of the contribution to the PO integral from the vicinity of the vertex immediately leads to a closed-form expression for the diffraction coefficient, which is possible for cones with virtually every shape of cross section, including angular sectors (Albani *et al.* 2011; Taket and Burge 1991), elliptic cones, and cones with polygonal cross sections (pyramids). The surface of the canonical scatterer can be PEC but also non-metallic, described by reflection coefficients or surface impedance. The only limitations are that in the vicinity of the vertex the scatterer should be convex to exclude multiple reflections and impenetrable to exclude transmission through the cone. The latter limitations can be overcome for a thin transparent plate as shown in Albani *et al.* (2011).

An apparent deficiency of the PO diffraction coefficients is their approximate nature. Special emphasis should therefore be placed on estimation of the accuracy of the PO diffraction coefficients. Experimental verification of the theoretical results is complicated by the fact that the vertex scattering is masked by much stronger GO (primary and reflected) and edge-diffracted contributions, for example from the lateral surface and the rim of the base of a truncated cone. Comparisons with several available exact solutions for conical geometries reveal that despite the physically apparent deviation of the PO currents from the real ones near a vertex (the latter may even become infinite at the vertex), the approximation quality of PO solutions is typically good. It is particularly good for the vertex of a circular PEC cone illuminated along the symmetry axis.

Of course, there are cases in which the PO approximation appears to fail, for example for a plane angular sector illuminated in the plane of the sector the PO currents cannot be defined as GO fails to discriminate between lit and shadowed portions of the scatterer. However, the exact scattered field is small in this case, so that even in this situation the error of PO solutions turns out to be not so significant.

We have checked the PO solution for the tip diffracted field against exact and numerical solutions for several configurations, including the PEC circular cone (Goriainov 1961), the PEC quarter plane and elliptic cone (Blume and Krebs

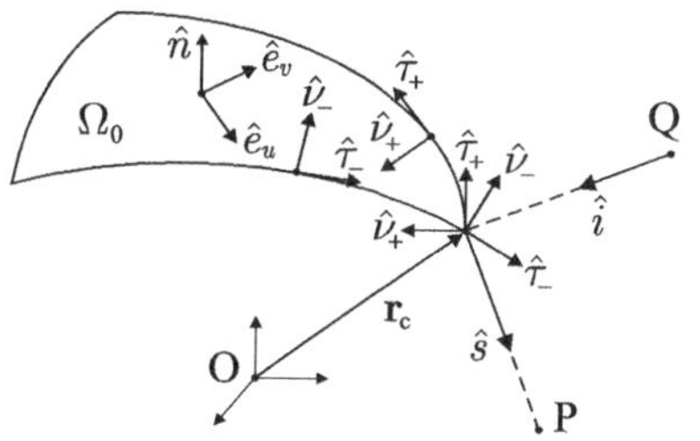

Figure 9.2 Scattering from a vertex of a thin plate.

1998), and the impedance circular cone (Bernard *et al.* 2008). In all these cases, provided that the scattered field was entirely due to the tip diffraction, the agreement was good. This is of course only a partial confirmation of the validity of the PO solutions for the scattering from vertices. Further checks are desirable and possible.

The section begins with the derivation of a PO-based solution for a vertex on an edge of a thin plate. The PO solution for an angular segment of a thin impenetrable (PEC or non-metallic) plate with material properties described by reflection coefficients is asymptotically evaluated by applying the results of section D.2 to obtain the diffraction coefficient of the vertex. A pyramid can be considered as a combination of angular segments meeting at an apex, which makes the extension of the results obtained for a single segment to the case of a pyramid straightforward. The section concludes with the analysis of scattering from the tip of an impenetrable elliptic cone. In this case the vertex is a conical point. We derive the PO diffraction coefficient of the cone, estimate the accuracy of the PO solution and study the limiting cases of a circular cone and an angular sector. In the case of a fully illuminated PEC elliptic cone, closed-form expressions for the matrix diffraction coefficient and the bistatic scattering cross section are presented.

9.2.1 Vertex on an Edge of a Thin Plate

Consider a thin impenetrable, in general curved plate whose edge has a vertex at a point $\mathbf{r}_c$ and where the curvature of the edge is discontinuous (Figure 9.2). To describe the geometry, we introduce two kinds of unit vectors tangent to the surface. Vectors $\hat{\tau}_\pm$ are tangent to the edges forming the vertex, while $\hat{\nu}_\pm$ are directed into the interior of the scattering surface and are perpendicular to $\hat{\tau}_\pm$, respectively. For curved edges, these vectors are functions of position.

Let a point source Q at $\mathbf{r}_0$ illuminate the side of the plate denoted by Ω_0. This illuminated side of the plate is described by the unit normal vector $\hat{n}$ and by reflection coefficients $R_\perp$ and $R_\parallel$. In the PO approximation, the field scattered by the plate is given by the surface integral (8.5) as

$$\mathbf{E}^{\rm po}(\mathbf{r}) = \frac{jk}{4\pi}\int\limits_{\Omega_0}\frac{\mathrm{e}^{-jk|\mathbf{r}-\mathbf{r}'|}}{|\mathbf{r}-\mathbf{r}'|}\mathbf{c}(\mathbf{r}',\mathbf{r}_0)\,\mathrm{d}\Omega' \tag{9.5}$$

with

$$\mathbf{c}(\mathbf{r}',\mathbf{r}_0) = Z\hat{s}'\times\hat{s}'\times\mathbf{K}_{\rm e}^{\rm go}(\mathbf{r}') + \hat{s}'\times\mathbf{K}_{\rm m}^{\rm go}(\mathbf{r}'), \tag{9.6}$$

where $\mathbf{K}_{\rm e}^{\rm go}$ and $\mathbf{K}_{\rm m}^{\rm go}$ are the equivalent surface currents in the GO approximation ((4.93) and (4.96)). Since the incident wave is a spherical wave, we can factor out the corresponding amplitudes and phase factors, and write

$$\mathbf{c}(\mathbf{r},\mathbf{r}_0) = \mathbf{C}(\mathbf{r},\mathbf{r}_0)\frac{\mathrm{e}^{-jk|\mathbf{r}-\mathbf{r}_0|}}{k|\mathbf{r}-\mathbf{r}_0|}, \tag{9.7}$$

which helps to extract in the explicit form the rapidly oscillating exponent in the integrand of (9.5).

We are interested in the contribution of the vertex to the asymptotics of the PO integral (9.5) as $k \to \infty$. The vertex is a critical point of the third kind, and the corresponding equation has been derived in section D.2 (see (D.89)). We now specialize the formula (D.89) to the geometry in Figure 9.2. Let $\mathbf{r} = \mathbf{f}(u, v)$ with $u, v \in \Omega_{uv}$ be a parametric representation of the scattering surface with Ω_{uv} being the image of the scattering surface on the u–v plane. Let the edge of the plate be described in the parametric form $\mathbf{r} = \tilde{\mathbf{f}}(t) = \mathbf{f}(U(t), V(t))$, where $u = U(t)$ and $v = V(t)$ define the image of the edge on the u–v plane. The position of the vertex is then given by the relations $\mathbf{r}_c = \mathbf{f}(u_c, v_c) = \tilde{\mathbf{f}}(t_c)$.

According to (D.89), the contribution of the vertex is expressed as

$$\mathbf{E}_c^{po}(\mathbf{r}) = \frac{\tilde{\mathbf{A}}(t)}{k^2\tilde{S}'(t)} e^{jk\tilde{S}(t)}\Big|_{t_c-0}^{t_c+0}, \tag{9.8}$$

where

$$\tilde{\mathbf{A}}(t) = \sqrt{U'^2(t) + V'^2(t)}\, \mathbf{A}(U(t), V(t)) \left(\frac{\hat{w} \cdot \nabla_{uv} S}{|\nabla_{uv} S|^2} \right)_{u=U(t), v=V(t)} \tag{9.9}$$

$$\mathbf{A}(u, v) = \frac{j|\mathbf{f}_u \times \mathbf{f}_v| \mathbf{C}(\mathbf{f}(u, v), \mathbf{r}_0)}{4\pi |\mathbf{f}(u, v) - \mathbf{r}||\mathbf{f}(u, v) - \mathbf{r}_0|} \tag{9.10}$$

$$\tilde{S}(t) = S(U(t), V(t)) \tag{9.11}$$

$$S(u, v) = -|\mathbf{f}(u, v) - \mathbf{r}| - |\mathbf{f}(u, v) - \mathbf{r}_0| \tag{9.12}$$

and $\mathbf{f}_u$ and $\mathbf{f}_v$ are the partial derivatives of $\mathbf{f}(u, v)$ with respect to u and v. Using (9.7) and (9.9) in (9.8) and factoring out functions continuous at $t = t_c$ (these are functions that do not include derivatives with respect to t or the unit normal $\hat{w}$), we arrive at

$$\mathbf{E}_c^{po}(\mathbf{r}) = \frac{e^{-jk|\mathbf{r}-\mathbf{r}_c|}}{k|\mathbf{r} - \mathbf{r}_c|} \frac{|\mathbf{f}_u \times \mathbf{f}_v| \mathbf{c}(\mathbf{r}_c, \mathbf{r}_0)}{|\nabla_{uv} S(u_c, v_c)|^2} \left[\sqrt{U'^2(t) + V'^2(t)} \frac{\hat{w} \cdot \nabla_{uv} S}{\tilde{S}'(t)} \right]_{t_c-0}^{t_c+0}. \tag{9.13}$$

In the asymptotic integration, only an asymptotically small vicinity of the vertex is important, so we can use the approximation

$$\mathbf{f}(u, v) \approx \mathbf{r}_c + (u - u_c)\hat{e}_u + (v - v_c)\hat{e}_v, \tag{9.14}$$

where $\hat{e}_u$ and $\hat{e}_v$ are unit vectors tangential to the surface and orthogonal to each other and to the vector $\hat{n}$. Thus, in the vicinity of the vertex,

$$\mathbf{f}_u \approx \hat{e}_u, \quad \mathbf{f}_v \approx \hat{e}_v, \tag{9.15}$$

implying that $|\mathbf{f}_u \times \mathbf{f}_u| \approx 1$ and

$$\frac{\partial S}{\partial u} = -\mathbf{q} \cdot \mathbf{f}_u \approx -\mathbf{q} \cdot \hat{e}_u \tag{9.16}$$

$$\frac{\partial S}{\partial v} = -\mathbf{q} \cdot \mathbf{f}_v \approx -\mathbf{q} \cdot \hat{e}_v \tag{9.17}$$

$$|\nabla_{uv} S(u, v)|^2 = \left(\frac{\partial S}{\partial u}\right)^2 + \left(\frac{\partial S}{\partial v}\right)^2 \approx (\mathbf{q} \cdot \hat{e}_u)^2 + (\mathbf{q} \cdot \hat{e}_v)^2, \tag{9.18}$$

where $\mathbf{q} = \hat{i} - \hat{s}$.

Since $\hat{e}_u$, $\hat{e}_v$, and $\hat{n}$ constitute an orthonormal basis,

$$(\mathbf{q} \cdot \hat{e}_u)^2 + (\mathbf{q} \cdot \hat{e}_v)^2 + (\mathbf{q} \cdot \hat{n})^2 = \mathbf{q} \cdot \mathbf{q}, \tag{9.19}$$

formula (9.18) can be expressed as

$$|\nabla_{uv} S(u, v)|^2 \approx \mathbf{q} \cdot \mathbf{q} - (\mathbf{q} \cdot \hat{n})^2. \tag{9.20}$$

Furthermore,

$$\hat{w} \cdot \nabla_{uv} S = w_u \frac{\partial S}{\partial u} + w_v \frac{\partial S}{\partial v} \approx -\mathbf{q} \cdot (w_u \hat{e}_u + w_v \hat{e}_v), \tag{9.21}$$

where $\hat{w}$ is the unit vector on the u–v plane, normal to the boundary of Ω_{uv}. Because of (9.14), the components of $\hat{w}$ can be expressed through the components of the unit vector $\hat{\nu}$ orthogonal to the edge and directed in the interior of the plate as $w_u = -\hat{\nu} \cdot \hat{e}_u$ and $w_v = -\hat{\nu} \cdot \hat{e}_v$, which leads to the formula

$$\hat{w} \cdot \nabla_{uv} S = \mathbf{q} \cdot \hat{\nu}. \tag{9.22}$$

Finally,

$$\tilde{S}'(t) = -\mathbf{q} \cdot \tilde{f}' = -\sqrt{U'^2(t) + V'^2(t)}\, \mathbf{q} \cdot \hat{\tau}, \tag{9.23}$$

where $\hat{\tau}$ is the unit tangent to the edge pointing to the direction of increasing values of t. In (9.22) and (9.23), the vectors $\hat{\nu}$ and $\hat{\tau}$ are equal to $\hat{\nu}_-$ and $\hat{\tau}_-$ or to $\hat{\nu}_+$ and $\hat{\tau}_+$ (Figure 9.2), depending on whether $t < t_c$ or $t > t_c$.

Using (9.20), (9.22), and (9.23) in (9.13), we obtain an asymptotic estimation for the contribution of the vertex to the field scattered by the plate,

$$\mathbf{E}_c^{po}(\mathbf{r}) = \frac{j}{4\pi k} \frac{e^{-jk|\mathbf{r}-\mathbf{r}_c|}}{|\mathbf{r}-\mathbf{r}_c|} \frac{\mathbf{c}(\mathbf{r}_c, \mathbf{r}_0)}{\mathbf{q}\cdot\mathbf{q} - (\mathbf{q}\cdot\hat{n})^2} \left(\frac{\mathbf{q}\cdot\hat{\nu}_-}{\mathbf{q}\cdot\hat{\tau}_-} - \frac{\mathbf{q}\cdot\hat{\nu}_+}{\mathbf{q}\cdot\hat{\tau}_+} \right), \tag{9.24}$$

in which the quantities $\mathbf{q}$, $\tau_\pm$, and $\nu_\pm$ are taken at the vertex.

Note the following properties of expression (9.24). First, compared to the incident field, the scattered field is of the order $O(k^{-1})$ and therefore asymptotically smaller than the GO reflected field and the edge diffracted field, which behave as $O(1)$ and $O(k^{-1/2})$, respectively. Furthermore, (9.24) vanishes when $\hat{\tau}_- = \hat{\tau}_+$ and $\hat{\nu}_- = \hat{\nu}_+$, which means simply that the edge is continuous and the vertex disappears. Finally, (9.24) is singular when $\hat{i} = \hat{s}$ and therefore $\mathbf{q} = 0$ (forward scattering) or $\mathbf{q}$ is perpendicular to the surface (GO reflection, in which case $\mathbf{q} = 2\hat{n}(\hat{n} \cdot \hat{i})$) or $\mathbf{q} \cdot \tau_\pm = 0$ (scattering direction $\hat{s}$ lies on a Keller's cone associated with either of the two edges meeting at the vertex). These singularities imply that the assumption made in the process of asymptotic evaluation, namely that there are no other critical points in an asymptotic vicinity of the vertex, are invalid and the simple formula (D.89) does not apply. In this case, more sophisticated, so-called uniform asymptotic approximations are required (see the references at the end of section D.2).

Despite the asymptotic smallness of the vertex contribution, the formula (9.24) is meaningful since there are situations in which the vertex contribution is the only contribution to the field that comes from the scatterer to the observation point. Expression (9.24) can be written in the GTD form as in (9.1) and (9.2). In the spherical coordinate system with the origin at the vertex ($\mathbf{r}_c = 0$), the directions of incidence and scattering are defined through the spherical angles θ_{inc}, ϕ_{inc}, θ, and ϕ as

$$\hat{i} = -\hat{x}\sin\theta_{inc}\cos\phi_{inc} - \hat{y}\sin\theta_{inc}\sin\phi_{inc} - \hat{z}\cos\theta_{inc} \tag{9.25}$$

$$\hat{s} = \hat{x}\sin\theta\cos\phi + \hat{y}\sin\theta\sin\phi + \hat{z}\cos\theta \tag{9.26}$$

and the incident and scattered fields can be expressed as

$$\mathbf{E}_{inc}(\mathbf{r}) = (\mathbf{E}_{inc} \cdot \hat{\theta}_{inc})\hat{\theta}_{inc} + (\mathbf{E}_{inc} \cdot \hat{\phi}_{inc})\hat{\phi}_{inc} \tag{9.27}$$

$$\mathbf{E}_{sc}(\mathbf{r}) = (\mathbf{E}_{sc} \cdot \hat{\theta})\hat{\theta} + (\mathbf{E}_{sc} \cdot \hat{\phi})\hat{\phi} \tag{9.28}$$

with

$$\hat{\theta}_{\text{inc}} = \hat{x}\cos\theta_{\text{inc}}\cos\phi_{\text{inc}} + \hat{y}\cos\theta_{\text{inc}}\sin\phi_{\text{inc}} - \hat{z}\sin\theta_{\text{inc}} \tag{9.29}$$

$$\hat{\phi}_{\text{inc}} = -\hat{x}\sin\phi_{\text{inc}} + \hat{y}\cos\phi_{\text{inc}} \tag{9.30}$$

$$\hat{\theta} = \hat{x}\cos\theta\cos\phi + \hat{y}\cos\theta\sin\phi - \hat{z}\sin\theta \tag{9.31}$$

$$\hat{\phi} = -\hat{x}\sin\phi + \hat{y}\cos\phi. \tag{9.32}$$

The diffraction coefficient can be determined from (9.24) on representing the incident field as in (9.27), followed by multiplying (9.24) with $\hat{\theta}$ and $\hat{\phi}$, which gives the expression

$$D(\hat{e},\hat{e}_{\text{inc}}) = \frac{j}{4\pi}\frac{1}{\mathbf{q}\cdot\mathbf{q}-(\mathbf{q}\cdot\hat{n})^2}\left(\frac{\mathbf{q}\cdot\hat{v}_-}{\mathbf{q}\cdot\hat{\tau}_-} - \frac{\mathbf{q}\cdot\hat{v}_+}{\mathbf{q}\cdot\hat{\tau}_+}\right)\sum_{m=1}^{4}(\hat{e}\cdot\mathbf{a}_m)(\mathbf{b}_m\cdot\hat{e}_{\text{inc}}), \tag{9.33}$$

where $\hat{e} = \hat{\theta}$ or $\hat{\phi}$ and $\hat{e}_{\text{inc}} = \hat{\theta}_{\text{inc}}$ or $\hat{\phi}_{\text{inc}}$. Vectors $\mathbf{a}_m$ and $\mathbf{b}_m$ with $m = 1, 2, 3, 4$ result from (4.93) and (4.96) as

$$\mathbf{a}_1 = \hat{s}\times\hat{s}\times\hat{\alpha}_{\text{inc}} \tag{9.34}$$

$$\mathbf{a}_2 = \hat{s}\times\hat{s}\times\hat{\imath} \tag{9.35}$$

$$\mathbf{a}_3 = \hat{s}\times\hat{\alpha}_{\text{inc}} \tag{9.36}$$

$$\mathbf{a}_4 = \hat{s}\times\hat{\imath} \tag{9.37}$$

$$\mathbf{b}_1 = -\hat{\alpha}_{\text{inc}}(1-R_\perp)\hat{\imath}\cdot\hat{n} \tag{9.38}$$

$$\mathbf{b}_2 = -\hat{\beta}_{\text{inc}}(1+R_\parallel) \tag{9.39}$$

$$\mathbf{b}_3 = \hat{\beta}_{\text{inc}}(1-R_\parallel)\hat{\imath}\cdot\hat{n} \tag{9.40}$$

$$\mathbf{b}_4 = -\hat{\alpha}_{\text{inc}}(1+R_\perp), \tag{9.41}$$

where the unit vectors $\hat{\alpha}_{\text{inc}}$, $\hat{\beta}_{\text{inc}}$, and $\hat{\imath}$ are defined in (4.71) and (4.81).

The obtained solution for the tip-diffracted field involves several approximations. In addition to the use of the PO approximation for the equivalent surface currents, the PO integral has been estimated asymptotically, under the assumptions that the source and the observation point are located at least several wavelengths away from the vertex and the corresponding critical point in the PO integral is isolated (i.e., has no further critical points in its asymptotically small vicinity). A consequence of these assumptions is the divergence of (9.33) at the shadow boundary of the reflected wave and at the Keller cones of the edges. The original PO integral (9.5) is always bounded and its direct numerical evaluation would lead to finite values of the scattered field everywhere in the exterior of the plate, in which case the scattered field is calculated as a whole, without splitting in the GO, edge-, and vertex-diffracted components.

Figure 9.3 illustrates scattering from the vertex of an impedance quarter plane in the PO approximation. The quarter plane is described by boundary conditions (2.74), and η is the normalized surface impedance. The quarter plane lies in the x–z plane and is described in spherical coordinates by the conditions $3\pi/4 \le \theta \le \pi$ with $0 \le r < \infty$ when $\phi = 0$ and $\phi = \pi$. The PO diffraction coefficient vanishes when the incidence/scattering direction lies in the plane of the quarter plane ($\phi = \phi_0 = 0, \pi$). Figure 9.3 shows the normalized RCS, σ/λ^2, in the plane perpendicular to the quarter plane ($\phi = \phi_0 = \pi/2$). In the PO approximation, there are no cross-polar components, whereas the co-polar components are in general finite and different, coinciding in the particular cases when $\eta = 0$ (PEC) and $\eta = 1$. The cross section is infinite at the direction of specular reflection $\theta = \theta_{\text{inc}} = \pi/2$, except for the case $\eta = 1$ in which case the GO reflected field vanishes and RCS remains bounded.

The exact value of RCS may deviate from these approximate PO values. For example, exact solutions for a thin angular PEC plate suggest a finite value (about 10 times smaller than that for a circular cone) of $\sigma_{\phi\phi_{\text{inc}}}/\lambda^2$ for the incidence along the vertex bisector ($\theta_{\text{inc}} = \theta = 0$) (Blume 1996, Figure 8). However, despite the approximate nature of the diffraction coefficient (9.33), the latter is expected to be useful in practical applications because of its simplicity and

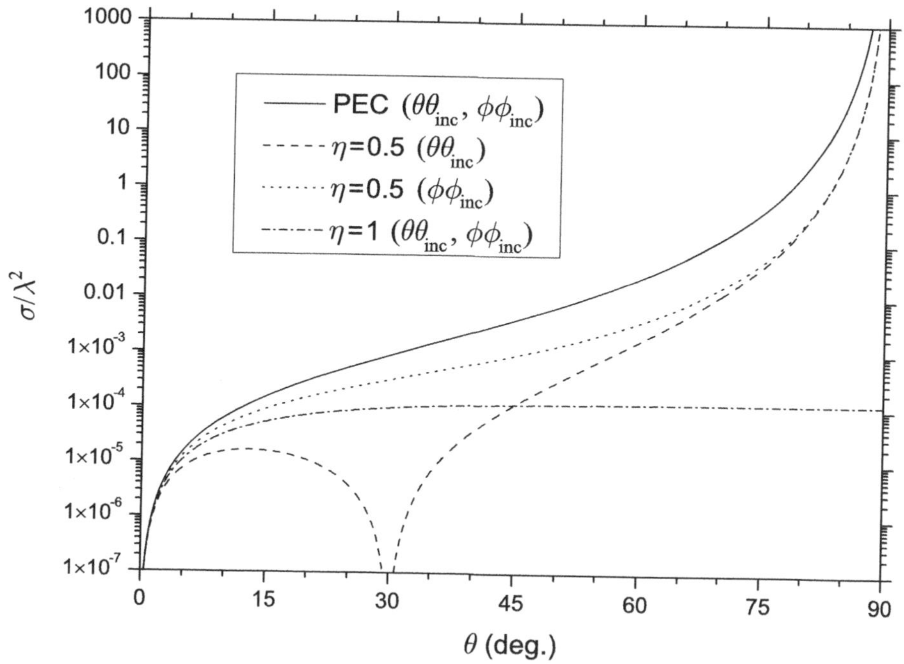

Figure 9.3 Normalized RCS of a quarter plane as a function of the incidence/scattering aspect in the plane perpendicular to the quarter plane for several values of the surface impedance.

flexibility. The derived PO solution can be applied to obtuse and acute sectors cut from thin impenetrable plates, PEC or imperfectly reflecting, flat or curved as the curvature of the plate does not influence the leading term of the approximation. Furthermore, the solution can be used as a building block to construct PO-based solutions for pyramids with an arbitrary number of faces. This subject is considered next.

9.2.2 *Apex of a Pyramid*

A pyramid consists of three or more faces meeting at the apex (Figure 9.1b). In the PO approximation, there is no electromagnetic interaction between the faces, and the PO solution for the field scattered by a pyramid is a superposition of PO integrals of the form (9.5) over all illuminated faces. Subsequently, the electric field scattered by the apex is given in the PO approximation by the formula (9.1) with the diffraction coefficient being a sum of the PO diffraction coefficients (9.33) derived in the previous section for the vertex of an angular sector. Let us write down the PO diffraction coefficient of a pyramid with an arbitrary number of faces. For the sake of simplicity, the pyramid is assumed to be a convex body to avoid multiple reflections between neighboring faces of the pyramid.

In a spherical coordinate system with the origin at the apex, a pyramid with L edges can be described by the angles θ_l ($\pi/2 < \theta_l < \pi$) and ϕ_l ($0 \le \phi_l < 2\pi$) where $l = 1, 2, \ldots, L$, implying that the unit vectors tangent to the edges and directed from the apex are given by

$$\hat{e}_l = \hat{x}\sin\theta_l\cos\phi_l + \hat{y}\sin\theta_l\sin\phi_l + \hat{z}\cos\theta_l. \tag{9.42}$$

To apply the formula (9.33) we need to define, at every face of the pyramid, the unit vector $\hat{n}_l$ normal to the face and the unit vectors $\hat{t}_l^{\pm}$ and $\hat{v}_l^{\pm}$ tangent and normal to either of the two edges of the face (Figure 9.4). These vectors are

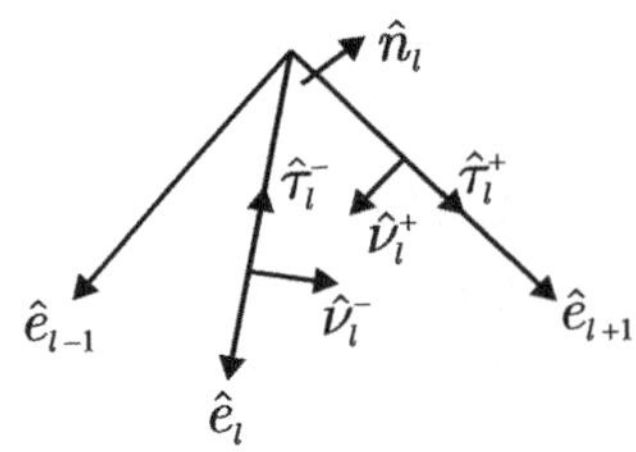

Figure 9.4 Defining the unit vectors on the lth face of a pyramid.

expressed through $\hat{e}_l$ as

$$\hat{n}_l = \frac{\hat{e}_l \times \hat{e}_{l+1}}{|\hat{e}_l \times \hat{e}_{l+1}|} \tag{9.43}$$

$$\hat{\tau}_l^- = -\hat{e}_l, \quad \hat{\tau}_l^+ = \hat{e}_{l+1} \tag{9.44}$$

$$\hat{\nu}_l^- = \hat{n}_l \times \hat{e}_l, \quad \hat{\nu}_l^+ = \hat{e}_{l+1} \times \hat{n}_l, \tag{9.45}$$

where $l = 1, 2, \ldots, L$ and $\hat{e}_{L+1} = \hat{e}_1$.

The PO diffraction coefficient of the apex of the L-sided pyramid is therefore given by the expressions

$$\overline{\overline{D}} = \sum_{l=1}^{L} p_l \overline{\overline{D}}_l \tag{9.46}$$

with

$$\overline{\overline{D}}_l = \begin{bmatrix} D_l(\hat{\theta}, \hat{\theta}_{\text{inc}}) & D_l(\hat{\theta}, \hat{\phi}_{\text{inc}}) \\ D_l(\hat{\phi}, \hat{\theta}_{\text{inc}}) & D_l(\hat{\phi}, \hat{\phi}_{\text{inc}}) \end{bmatrix}, \tag{9.47}$$

where

$$D_l(\hat{e}, \hat{e}_{\text{inc}}) = \frac{j}{4\pi} \frac{1}{\mathbf{q} \cdot \mathbf{q} - (\mathbf{q} \cdot \hat{n}_l)^2} \left(\frac{\mathbf{q} \cdot \hat{\nu}_l^-}{\mathbf{q} \cdot \hat{\tau}_l^-} - \frac{\mathbf{q} \cdot \hat{\nu}_l^+}{\mathbf{q} \cdot \hat{\tau}_l^+} \right) \sum_{m=1}^{4} (\hat{e} \cdot \mathbf{a}_{lm})(\mathbf{b}_{lm} \cdot \hat{e}_{\text{inc}}) \tag{9.48}$$

and $\mathbf{a}_{lm} = \mathbf{a}_m(\hat{n}_l)$, $\mathbf{b}_{lm} = \mathbf{b}_m(\hat{n}_l)$ with $\mathbf{a}_m$ and $\mathbf{b}_m$ defined in (9.34)–(9.41). Parameter p_l is equal to 1 when the lth face is illuminated by the incident wave; otherwise, $p_l = 0$. For a convex body, the parameter can be expressed as

$$p_l = \Theta(-\hat{n}_l \cdot \hat{i}), \tag{9.49}$$

where $\Theta(x)$ represents the Heaviside theta function, equal to 0 for $x < 0$ and 1 for $x > 0$. For $L = 2$ the pyramid reduces to an angular sector cut from a thin plate. Thus, (9.46) consists of a single term (as only one face of the sector can be illuminated at a time), and (9.46)–(9.48) recover the diffraction coefficient of a vertex of a thin plate given in (9.2) and (9.33).

The PO formulas (9.1) and (9.46)–(9.48) are asymptotic ($k \to \infty$) and therefore applicable to pyramids with curved faces as well, in which case the vectors $\hat{e}_l$ are the local tangents to the edges at the apex and $p_l = 1$ when the lth face is illuminated in a vicinity of the apex.

Figure 9.5 shows the influence of the surface impedance on backscattering from a vertex of a cube at the incidence along the diagonal going through the interior of the cube. This layout simulates RCS of box-shaped non-metallic geometries, for example a corner of a building. In this case, $L = 3$, $\theta_1 = \theta_2 = \theta_3 = \pi - \arcsin(2/\sqrt{5}) \approx 116.565°$, $\phi_1 = 0$, $\phi_2 = 2\pi/3$,

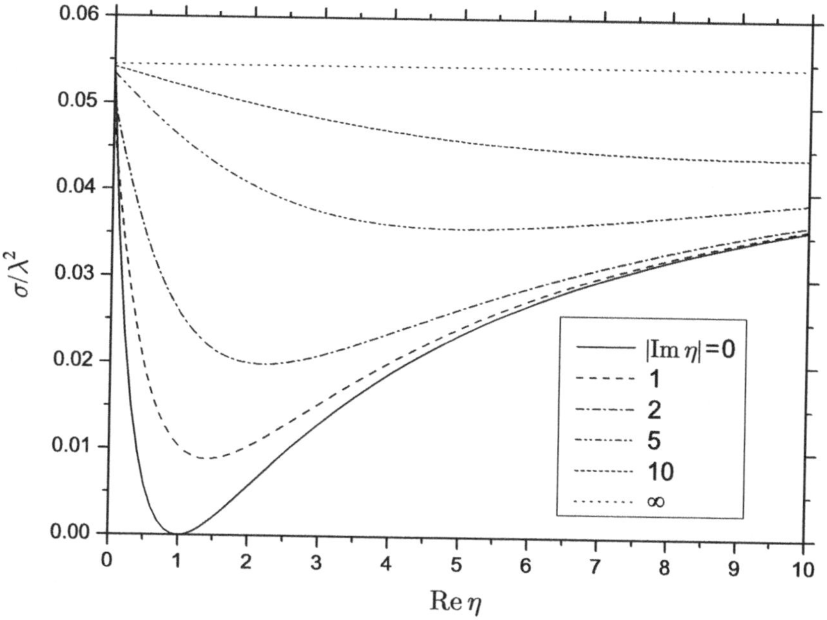

Figure 9.5 Normalized RCS of an apex of an impedance cube as a function of Re η for several values of $|\mathrm{Im}\,\eta|$ at the incidence along the diagonal going through the body of the cube. η is the normalized surface impedance.

$\phi_2 = 4\pi/3$, and $\theta = \theta_{\mathrm{inc}} = 0$. The normalized surface impedance η is assumed to be the same on all faces of the pyramid. The RCS is calculated from the diffraction coefficient according to (9.4). Simulations have shown that RCS values are independent of rotation of the pyramid around the incidence/scattering direction, are the same for co-polar components, $\sigma(\hat{\theta}, \hat{\theta}_{\mathrm{inc}}) = \sigma(\hat{\phi}, \hat{\phi}_{\mathrm{inc}}) = \sigma$, and vanish for cross-polar components, $\sigma(\hat{\theta}, \hat{\phi}_{\mathrm{inc}}) = \sigma(\hat{\phi}, \hat{\theta}_{\mathrm{inc}}) = 0$. Though RCS does not depend on the sign of Im η, Figure 9.5 suggests a strong dependence of RCS on $|\mathrm{Im}\,\eta|$ and Re η. The curves with Im $\eta = 0$ and Im $\eta = \infty$ mark two extremes, with all other curves for intermediate values of Im η uniformly filling the gap between them. Note that with $\eta = 1$, RCS is exactly zero (section 3.2.3).

If all edges of a pyramid make the same angle with the z axis, that is, $\theta_l = \theta_1$ with $l = 2, 3 \ldots, L$, it is natural to expect that as the number of faces of the pyramid increases, the scattered field approaches the field scattered by the tip of a circular cone. This is indeed so and is illustrated in Figure 9.6. RCS of every pyramid is plotted as a function of θ in a plane perpendicular to a face of the respective pyramid. All pyramids are PEC, and in the PO approximation, $\sigma(\hat{\theta}, \hat{\theta}_{\mathrm{inc}}) = \sigma(\hat{\phi}, \hat{\phi}_{\mathrm{inc}}) = \sigma$ and $\sigma(\hat{\theta}, \hat{\phi}_{\mathrm{inc}}) = \sigma(\hat{\phi}, \hat{\theta}_{\mathrm{inc}}) = 0$. The case $L = 2$ corresponds to an infinitely thin PEC plate with an apex angle 20°. It is seen that RCS of a pyramid with $L = 6$ is already quite close to that of a circular cone.

The dependence of RCS on the angle ϕ, or on rotation of the pyramid around the direction of incidence and scattering, is illustrated in Figure 9.7. As expected, the dependence is strong when $L = 2$ (angular sector) but gets weaker as the number of faces increases, approaching a constant in the limit $L = \infty$ (circular cone).

9.2.3 *Tip of an Elliptic Cone*

A cone with an elliptic cross section is a useful and flexible geometry which includes the circular cone and the angular sector of a thin plate as limiting cases (Figure 9.1). The PO vertex diffraction coefficient can be derived by asymptotically evaluating the contribution of the apex of the cone to the scattered field. Again we assume that the cone is impenetrable and the material properties of the surface are described by reflection coefficients $R_{\parallel}$ and $R_{\perp}$.

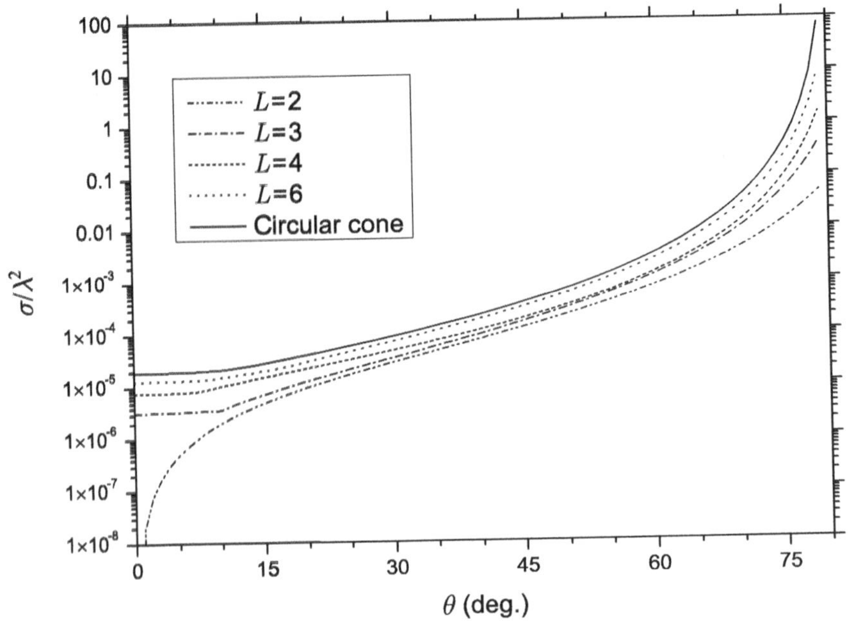

Figure 9.6 Normalized RCS of PEC pyramids with $\theta_1 = 170°$ and different number L of faces as a function of the zenith angle θ in a plane perpendicular to a face of the respective pyramid.

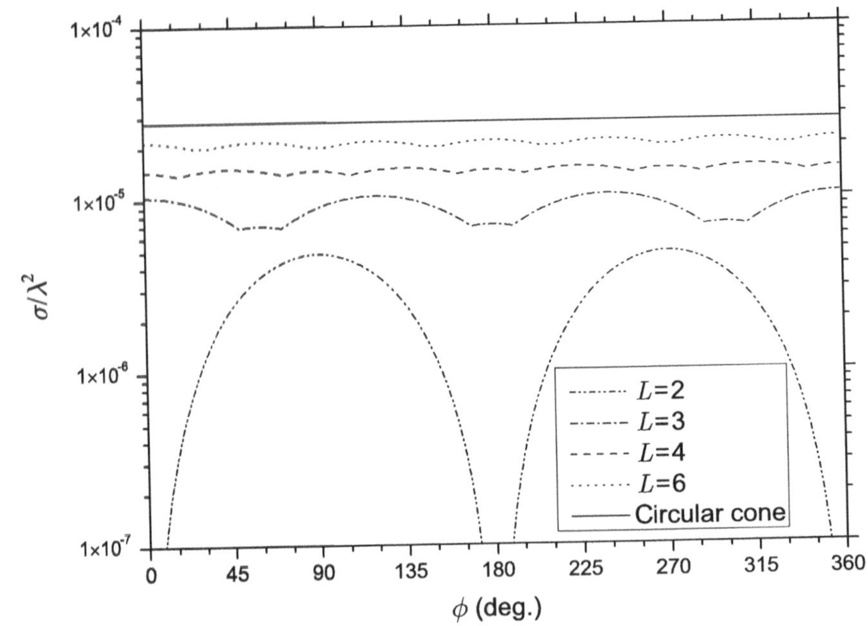

Figure 9.7 The same as Figure 9.6 but as a function of azimuth angle at $\theta = 15°$.

If the z axis is the symmetry axis of the cone, which goes through the middle of the cross section, and the vertex of the cone is at the origin of the coordinate system, then the conical surface is described by the equation

$$\frac{x^2}{a^2} + \frac{y^2}{b^2} = \frac{z^2}{c^2} \tag{9.50}$$

with $a, b, c > 0$. At every value of z the cross section of the cone is an ellipse with semi-axes $a|z|/c$ and $b|z|/c$. A convenient description of the geometry uses two opening angles that characterize the angular width of the cone in the xOz and yOz planes,

$$\frac{a}{c} = -\tan\theta_x, \quad \frac{b}{c} = -\tan\theta_y, \tag{9.51}$$

where θ_x and θ_y are zenith angles measured between the positive z axis and the tangents to the cone at the tip in the principal planes $\phi = 0$ and $\phi = \pi/2$, respectively. Alternatively, one can use half-cone angles $\alpha_x = \pi - \theta_x$ and $\alpha_y = \pi - \theta_y$ which are measured from the negative z axis in the interior of the cone so that $2\alpha_x$ and $2\alpha_y$ are equal to the angular widths of the cone in the two principal planes. Our analysis neglects multiple scattering and is, therefore, limited to the conical scatterers with $\pi/2 \le \theta_x, \theta_y \le \pi$ or $0 \le \alpha_x, \alpha_y \le \pi/2$. The limiting cases of the elliptic cone include a plane ($\theta_x = \theta_y = \pi/2$), a wedge (either θ_x or θ_y equals to $\pi/2$), a circular cone ($\theta_x = \theta_y$), an angular plane sector (either α_x or α_y vanishes), and a semi-infinite line ($\alpha_x = \alpha_y = 0$).

The portion of the conical surface which resides in the half-space $z < 0$ can be represented in the parametric form

$$\mathbf{r} = \mathbf{f}(u, v) = u(\hat{x}a\cos v + \hat{y}b\sin v - \hat{z}c) \tag{9.52}$$

with $u \ge 0$ and $0 \le v < 2\pi$. The unit normal directed into the exterior of the cone is given by

$$\hat{n} = \frac{\mathbf{f}_u \times \mathbf{f}_v}{|\mathbf{f}_u \times \mathbf{f}_v|} = \frac{1}{V(v)}(\hat{x}bc\cos v + \hat{y}ac\sin v + \hat{z}ab) \tag{9.53}$$

with

$$V(v) = \sqrt{(ab)^2 + (bc)^2\cos^2 v + (ac)^2\sin^2 v} \tag{9.54}$$

and is independent of u.

To derive the diffraction coefficient, it is sufficient to consider a plane wave as the incident field. Then, assuming that the source is located sufficiently far from the vertex so that in a vicinity of the vertex the front of the incident wave can be regarded as plane, we can describe the range of integration in the PO solution in relatively simple terms. Namely, the condition

$$\hat{\imath} \cdot \hat{n} \le 0, \tag{9.55}$$

where $\hat{\imath}$ is the direction of incidence, describes the illuminated portion of the cone, with equality at the shadow boundary. Using (9.25) and (9.53), this condition can be written as

$$F(v; \alpha, \beta, \gamma) \ge 0, \tag{9.56}$$

where

$$F(v; \alpha, \beta, \gamma) = \alpha\cos v + \beta\sin v + \gamma \tag{9.57}$$

and

$$\alpha = bc\cos\phi_{\text{inc}}\sin\theta_{\text{inc}}, \quad \beta = ac\sin\phi_{\text{inc}}\sin\theta_{\text{inc}}, \quad \gamma = ab\cos\theta_{\text{inc}}. \tag{9.58}$$

In the case of the incidence in the negative direction of the z axis (axial or nose-on incidence), the whole surface of the cone is illuminated, implying that $F(v;\alpha,\beta,\gamma) > 0$ for every value of v. Under oblique incidence, however, only a portion $v_- \leq v \leq v_+$ of the surface is illuminated in general, where $v_\pm$ are two zeros of $F(v;\alpha,\beta,\gamma)$ chosen so that $F(v;\alpha,\beta,\gamma) > 0$ between them. It can be shown (see problem 1) that when

$$\gamma^2 > \alpha^2 + \beta^2, \tag{9.59}$$

the whole surface is illuminated and therefore $v_- = 0$ and $v_+ = 2\pi$. Otherwise, $F(v;\alpha,\beta,\gamma)$ has two real-valued zeros $v_<$ and $v_>$ $(-\pi < v_< < v_> < \pi)$ such that $v_- = v_<$ and $v_+ = v_>$ if $F(v;\alpha,\beta,\gamma) > 0$ on the interval $v_< < v < v_>$ and $v_- = v_>$ and $v_+ = v_< + 2\pi$ if $F(v;\alpha,\beta,\gamma) < 0$ on the interval $v_< < v < v_>$.

Using these findings, the PO expression (9.5) for the scattered field can be rewritten as a double integral

$$\mathbf{E}^{\mathrm{po}}(\mathbf{r}) = \frac{jk}{4\pi}\int_{v_-}^{v_+}\int_{0}^{+\infty} \frac{\mathrm{e}^{-jk|\mathbf{r}-\mathbf{f}(u,v)|}}{|\mathbf{r}-\mathbf{f}(u,v)|}\mathbf{c}(\mathbf{f}(u,v),\mathbf{r}_0)uV(v)\,\mathrm{d}u\,\mathrm{d}v, \tag{9.60}$$

where the relation $|\mathbf{f}_u \times \mathbf{f}_v| = uV(v)$ has been used. Since $v_\pm$ is independent of u, this is an integral of the type (D.102) studied in section D.2. We can use the asymptotic result (D.104) to obtain the contribution from the vertex,

$$\mathbf{E}_{\mathrm{c}}^{\mathrm{po}}(\mathbf{r}) = \frac{\mathrm{e}^{-jkr}}{4\pi jkr}\int_{v_-}^{v_+} \frac{V(v)\mathbf{c}(v)}{(\mathbf{q}\cdot\mathbf{f}_u(0,v))^2}\,\mathrm{d}v, \tag{9.61}$$

where $\mathbf{c}(v)$ denotes the limiting value of $\mathbf{c}(\mathbf{f}(u,v),\mathbf{r}_0)$ for $u \to 0$. By using the vectors defined in (9.34)–(9.41), this limiting value can be expressed as

$$\mathbf{c}(v) = \sum_{m=1}^{4} \mathbf{a}_m(\mathbf{b}_m \cdot \mathbf{E}_{\mathrm{inc}}(0)), \tag{9.62}$$

where $\mathbf{E}_{\mathrm{inc}}(0)$ is the value of the incident wave at the vertex. The vectors $\mathbf{a}_m$ and $\mathbf{b}_m$ with $m = 1,2,3,4$ depend on $\hat{n}$ and are therefore functions of v. Expressing $\mathbf{E}_{\mathrm{inc}}(0)$ according to (9.27) and multiplying (9.61) with the unit vectors $\hat{\theta}$ and $\hat{\phi}$ lead to the PO approximation for the elements of matrix diffraction coefficient (9.2),

$$D(\hat{e},\hat{e}_{\mathrm{inc}}) = \frac{1}{4\pi j}\sum_{m=1}^{4}\int_{v_-}^{v_+} \frac{V(v)(\hat{e}\cdot\mathbf{a}_m)(\mathbf{b}_m\cdot\hat{e}_{\mathrm{inc}})}{(q_x a\cos v + q_y b\sin v - q_z c)^2}\,\mathrm{d}v. \tag{9.63}$$

The components of the vector $\mathbf{q} = \hat{\imath} - \hat{s}$ result from (9.25) and (9.26) as

$$q_x = -\cos\phi\sin\theta - \cos\phi_{\mathrm{inc}}\sin\theta_{\mathrm{inc}} \tag{9.64}$$

$$q_y = -\sin\phi\sin\theta - \sin\phi_{\mathrm{inc}}\sin\theta_{\mathrm{inc}} \tag{9.65}$$

$$q_z = -\cos\theta - \cos\theta_{\mathrm{inc}}. \tag{9.66}$$

Once the matrix diffraction coefficient is known, corresponding expressions for the scattering matrix and the scattering cross section of the tip immediately follow from relations (9.3) and (9.4).

Formula (9.63) is very general as it expresses the elements of the matrix diffraction coefficient of the tip of an arbitrary elliptic cone whose material properties are described by a pair of reflection coefficients $R_\perp$ and $R_\parallel$. The limitations are that the field at the observation point is completely due to the tip diffraction so that no rays reflected at the lateral surface of the cone arrive at the observer. At the shadow boundary between the lit and shadowed portions of space, the integral in (9.63) diverges since at some value of v the vector $\mathbf{q}$ is perpendicular to the surface and the denominator, which is the squared scalar product of $\mathbf{q}$ with the tangent vector $\mathbf{f}_u(0, v)$, vanishes. The integral becomes infinite also for the forward-scattering direction when $\hat{s} = \hat{i}$ and therefore $\mathbf{q} = 0$.

PEC Elliptic Cone

Let us now study the PO diffraction coefficient in the most important particular cases. If the cone is PEC and fully illuminated, then the integral in (9.63) can be evaluated in a closed form (problem 2), which leads to the following formulas:

$$D(\hat{\theta}, \hat{\theta}_{\text{inc}}) = D(\hat{\phi}, \hat{\phi}_{\text{inc}}), \quad D(\hat{\theta}, \hat{\phi}_{\text{inc}}) = -D(\hat{\phi}, \hat{\theta}_{\text{inc}}), \tag{9.67}$$

where

$$D(\hat{\theta}, \hat{\theta}_{\text{inc}}) = \frac{j \tan \alpha_x \tan \alpha_y}{Q^{3/2}} \cdot [\cos(\phi - \phi_{\text{inc}})(1 + \cos\theta \cos\theta_{\text{inc}}) + \sin\theta \sin\theta_{\text{inc}}] \tag{9.68}$$

$$D(\hat{\theta}, \hat{\phi}_{\text{inc}}) = \frac{j \tan \alpha_x \tan \alpha_y}{Q^{3/2}} \sin(\phi - \phi_{\text{inc}})(\cos\theta + \cos\theta_{\text{inc}}) \tag{9.69}$$

and

$$Q = (\cos\theta + \cos\theta_{\text{inc}})^2 - \tan^2\alpha_x(\cos\phi \sin\theta + \cos\phi_{\text{inc}} \sin\theta_{\text{inc}})^2 - \tan^2\alpha_y(\sin\phi \sin\theta + \sin\phi_{\text{inc}} \sin\theta_{\text{inc}})^2. \tag{9.70}$$

Expressions (9.67)–(9.70) comply with the reciprocity (section 3.3) as interchanging the scattering and incidence directions, $\theta \leftrightarrow \theta_{\text{inc}}$ and $\phi \leftrightarrow \phi_{\text{inc}}$, results in the transposed diffraction coefficient matrix, that is,

$$\overline{\overline{D}} \rightarrow \overline{\overline{D}}^T. \tag{9.71}$$

If, however, the surface of the cone is partially illuminated by an incident wave or is imperfectly reflecting, in which case (9.67)–(9.70) do not apply anymore and the elements of the matrix are given by the general formula (9.63), then the PO diffraction coefficient satisfies the reciprocity relation (9.71) only approximately. The degree of compliance of matrix diffraction coefficients or scattering matrices with the reciprocity requirement may serve as a measure of accuracy of PO solutions. It should be noted that in the case of backscattering, (9.63) gives a matrix with equal off-diagonal elements, which is what the reciprocity requires; this property of the PO diffraction coefficient, which has been checked by us numerically, holds for imperfectly reflecting and arbitrarily illuminated cones. For backscattering from PEC cones (including the limiting case of an angular sector), $D(\hat{\theta}, \hat{\theta}_{\text{inc}}) = D(\hat{\phi}, \hat{\phi}_{\text{inc}})$ and $D(\hat{\theta}, \hat{\phi}_{\text{inc}}) = D(\hat{\phi}, \hat{\theta}_{\text{inc}}) = 0$.

In the limit of nose-on incidence ($\theta_{\text{inc}} = 0$, $\phi_{\text{inc}} = 0$) on a PEC elliptic cone, (9.67)–(9.70) recover the results by Blume and Kahl (1987) for the PO bistatic cross sections in the principal planes of the cone,

$$\sigma(\hat{\theta}, \hat{\theta}_{\text{inc}}) = \sigma(\hat{\phi}, \hat{\phi}_{\text{inc}}) = \sigma_0 \frac{2[1 + \cos(2\alpha_x)]^3}{(1 + \cos\theta)[\cos\theta + \cos(2\alpha_x)]^3} \tag{9.72}$$

$$\sigma(\hat{\theta}, \hat{\phi}_{\text{inc}}) = \sigma(\hat{\phi}, \hat{\theta}_{\text{inc}}) = 0 \tag{9.73}$$

when $\phi = 0$ and

$$\sigma(\hat{\theta}, \hat{\theta}_{\text{inc}}) = \sigma(\hat{\phi}, \hat{\phi}_{\text{inc}}) = 0 \tag{9.74}$$

$$\sigma(\hat{\theta}, \hat{\phi}_{\text{inc}}) = \sigma(\hat{\phi}, \hat{\theta}_{\text{inc}}) = \sigma_0 \frac{2[1 + \cos(2\alpha_y)]^3}{(1 + \cos\theta)[\cos\theta + \cos(2\alpha_y)]^3} \tag{9.75}$$

when $\phi = \pi/2$. Here

$$\sigma_0 = \frac{\lambda^2}{16\pi} \tan^2 \alpha_x \tan^2 \alpha_y \tag{9.76}$$

stands for the nose-on RCS of an elliptic PEC cone in PO approximation, which is an extension of the widely quoted nose-on backscattering cross section of a circular PEC cone ($\alpha_x = \alpha_y = \alpha$)

$$\sigma_0 = \frac{\lambda^2}{16\pi} \tan^4 \alpha \tag{9.77}$$

(e.g., Bouman *et al.* (1987) and Ruck *et al.* (1970)).

Figure 9.8 illustrates the polarization behavior of the bistatic cross section of a PEC circular cone under nose-on illumination. The observation point moves in the plane $\theta = 90°$ around the axis of the cone. In this case, equations (9.67) apply and predict both co- and cross-polar components (dashed and dotted lines, respectively). However, transformation of the matrix diffraction coefficient $\overline{\overline{D}}$ to the PoS reference frame (section 3.3) results in a diagonal diffraction coefficient matrix $\overline{\overline{D}}_{\text{PoS}}$

$$D_{\text{PoS}}(\hat{\theta}, \hat{\theta}_{\text{inc}}) = D_{\text{PoS}}(\hat{\phi}, \hat{\phi}_{\text{inc}}) = D_{\text{PoS}}, \quad D_{\text{PoS}}(\hat{\theta}, \hat{\phi}_{\text{inc}}) = D_{\text{PoS}}(\hat{\phi}, \hat{\theta}_{\text{inc}}) = 0, \tag{9.78}$$

provided that the cone is PEC and fully illuminated. On this transformation, the diffraction coefficient, the scattering matrix, and the bistatic scattering cross section of the circular cone become independent of the azimuth (solid line), which is what one may intuitively expect for a rotationally symmetric scatterer.

Figure 9.9 compares the PoS bistatic scattering cross sections

$$\sigma_{\text{PoS}} = \frac{\lambda^2}{\pi} |D_{\text{PoS}}|^2$$

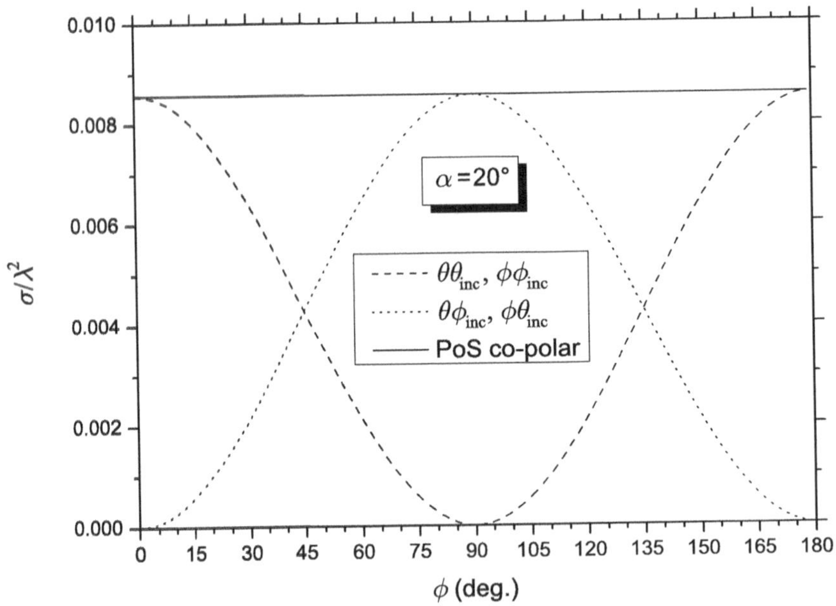

Figure 9.8 Normalized bistatic scattering cross section of a circular PEC cone at axial illumination as a function of the azimuth ϕ in the plane perpendicular to the cone axis ($\theta = 90°$). The solid line shows the co-polar component of the scattering cross section in the PoS reference frame.

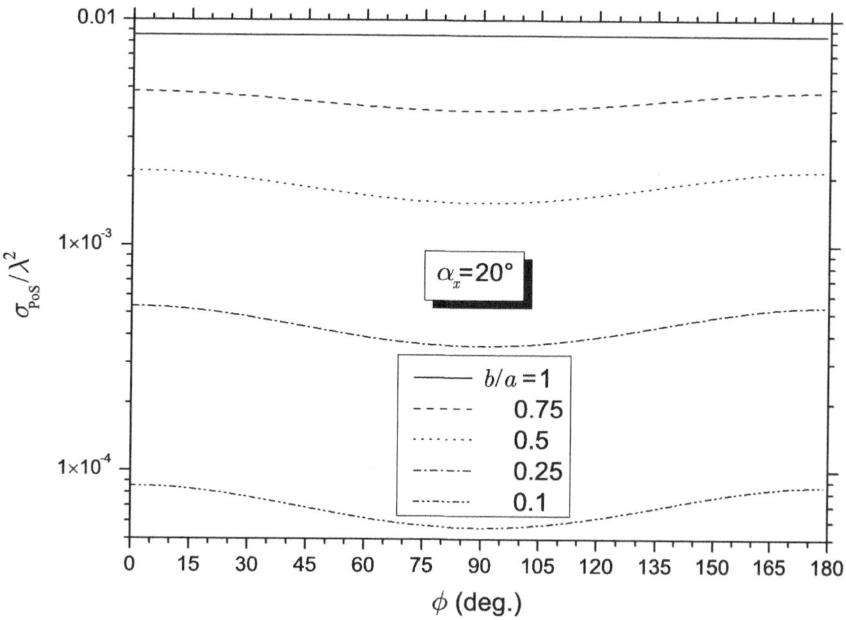

Figure 9.9 Normalized bistatic scattering cross section in the PoS bases for various elliptic PEC cones at axial illumination as a function of the azimuth ϕ in the plane perpendicular to the cone axis ($\theta = 90°$).

of PEC cones with varying degree of ellipticity b/a. As the ratio of the semi-axes of the cones deviates stronger from unity, the area of the geometric cross section of the cone decreases and so does the level of σ_{PoS}; at the same time the cone gets more asymmetric, so the dependence on ϕ in the curves becomes more pronounced. Thus, the use of the PoS reference frame helps to reveal the degree of ellipticity of the cones.

In the limit of a circular cone ($\alpha_x = \alpha_y = \alpha$), (9.67)–(9.70) reduce to

$$D(\hat{\theta}, \hat{\theta}_{\mathrm{inc}}) = \frac{j \tan^2 \alpha}{\Pi^{3/2}} [\cos(\phi - \phi_{\mathrm{inc}})(1 + \cos\theta \cos\theta_{\mathrm{inc}}) + \sin\theta \sin\theta_{\mathrm{inc}}] \tag{9.79}$$

$$D(\hat{\theta}, \hat{\phi}_{\mathrm{inc}}) = \frac{j \tan^2 \alpha}{\Pi^{3/2}} \sin(\phi - \phi_{\mathrm{inc}})(\cos\theta + \cos\theta_{\mathrm{inc}}) \tag{9.80}$$

with

$$\Pi = (\cos\theta + \cos\theta_{\mathrm{inc}})^2 - \tan^2 \alpha [\sin^2\theta + \sin^2\theta_{\mathrm{inc}} + 2\cos(\phi - \phi_{\mathrm{inc}}) \sin\theta \sin\theta_{\mathrm{inc}}], \tag{9.81}$$

thus recovering formulas published in Trott *et al.* (1990). A corresponding expression for $\sigma(\hat{\phi}, \hat{\phi}_{\mathrm{inc}})$ can be found in Bouman *et al.* (1987, eq. 18.220) (with the factor $\cos(\phi - \phi_{\mathrm{inc}})$ in the numerator apparently misplaced).

For axial incidence ($\theta_{\mathrm{inc}} \to 0$) on a circular PEC cone, (9.79)–(9.81) further simplify to give

$$\overline{\overline{D}} = jL(\theta, \alpha) \begin{bmatrix} \cos(\phi - \phi_{\mathrm{inc}}) & \sin(\phi - \phi_{\mathrm{inc}}) \\ -\sin(\phi - \phi_{\mathrm{inc}}) & \cos(\phi - \phi_{\mathrm{inc}}) \end{bmatrix} \tag{9.82}$$

with

$$L(\theta, \alpha) = \frac{\tan^2 \alpha (1 + \cos \theta)}{[(1 + \cos \theta)^2 - \tan^2 \alpha \sin^2 \theta]^{3/2}}. \tag{9.83}$$

Using simple trigonometric identities, the latter can be expressed as

$$L(\theta, \alpha) = \frac{\sin^2 \alpha \cos \alpha}{4 \cos\left(\frac{\theta}{2}\right) \left[\cos\left(\alpha + \frac{\theta}{2}\right) \cos\left(\alpha - \frac{\theta}{2}\right)\right]^{3/2}}. \tag{9.84}$$

The dependence of the PO solution (9.82) for the field scattered at the tip of the cone on the angles θ and α is described by a single function $L(\theta, \alpha)$. However, an exact solution to this problem involves two different functions $L_1(\theta, \alpha)$ and $L_2(\theta, \alpha)$ (Goriainov 1961). For example, if the incident wave is polarized in the x direction, then

$$E_\theta^{\mathrm{inc}}(\mathbf{r}) = j \frac{e^{-jkr}}{kr} \cos \phi L_1(\theta, \alpha) E_x^{\mathrm{inc}}(0) \tag{9.85}$$

$$E_\phi^{\mathrm{inc}}(\mathbf{r}) = -j \frac{e^{-jkr}}{kr} \sin \phi L_2(\theta, \alpha) E_x^{\mathrm{inc}}(0) \tag{9.86}$$

and the dependence of the scattered field on θ is determined by $L_1(\theta, \alpha)$ in the E plane ($\phi = 0$) and by $L_2(\theta, \alpha)$ in the H plane ($\phi = \pi/2$)[1]. The PO solution (9.82), however, is proportional to the same function $L(\theta, \alpha)$ in both planes, which is due to the approximate nature of the PO solution. However, it turns out that the deviation of $L(\theta, \alpha)$ from $L_{1,2}(\theta, \alpha)$ is rather small. Numerical estimations of the relative errors $|L_{1,2}(\theta, \alpha)/L(\theta, \alpha) - 1|$ for cones with $0 < \alpha < \pi/2$ over $0 \le \theta < \pi - 2\alpha$ have shown that the errors do not exceed 10% (Goriainov 1961), implying a smaller than 1 dB error in the scattering cross sections.

The quality of the PO approximation is illustrated in Figure 9.10, which shows the bistatic scattering cross section of a circular PEC cone under axial illumination for several values of the half-cone angle α. The exact values in the E and H planes are indicated by the black and white circles, while the PO values are shown by the dotted ($\alpha = 60°$), dot-dashed ($\alpha = 30°$), dashed ($\alpha = 10°$), and solid ($\alpha = 1°$) lines. The cross section becomes infinite at the shadow boundary of reflected waves when $\theta = \pi - 2\alpha$. The difference between the exact and PO values is so small that the results are almost indistinguishable on the scale of the plot. It is interesting to note that when $\theta = 0$, $L_1(0, \alpha) = L_2(0, \alpha) \neq L(0, \alpha)$ and there is still an error in the PO solution, which disproves the conjecture, put forward in Siegel *et al.* (1955), that PO may give the exact result for the axial backscattering from a semi-infinite circular PEC cone.

Axial Backscattering from Circular Imperfectly Reflecting Cones

In the case that the cone is circular ($\alpha_x = \alpha_y = \alpha$) and the incidence and scattering directions are along the symmetry axis ($\hat{z} = \hat{s} = -\hat{\imath}$), the integral in (9.61) can be explicitly calculated since the dependence of the integrand on the azimuthal coordinate v significantly simplifies as the incidence angle, which is the angle between the incidence direction and the normal to the surface, is constant over the whole surface of the cone, implying the constant values of the reflections coefficients. The result, the backscattered field for the axial illumination of the cone, can be expressed as

$$\mathbf{E}_{\mathrm{c}}^{\mathrm{po}}(\mathbf{r}) = \mathbf{E}_{\mathrm{inc}}(0) \frac{e^{-jkr}}{kr} \frac{j}{8} \tan^2 \alpha \left[R_{\parallel}\left(\frac{\pi}{2} - \alpha\right) - R_{\perp}\left(\frac{\pi}{2} - \alpha\right)\right]. \tag{9.87}$$

The derivation of this equation from (9.61) is left to the problems section. The circular cone is a particular case of bodies of revolution, and scattering from such bodies is studied in a more general framework in section 9.4.

[1] The E plane is the plane that includes the vector of electric field in the incident wave and the direction of incidence. The H plane is defined by the vector of magnetic field in the incident wave and by the incidence direction.

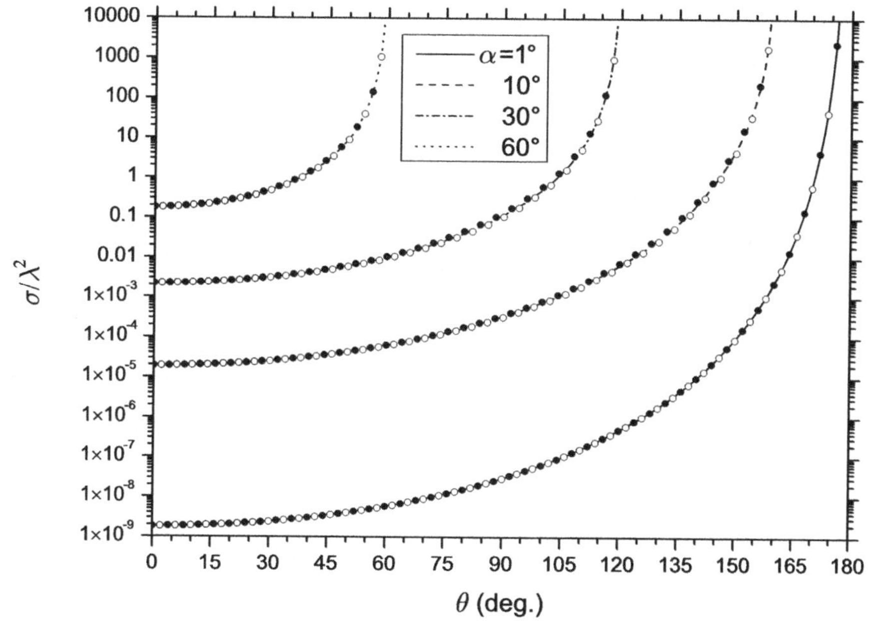

Figure 9.10 Normalized bistatic scattering cross section of a circular PEC cone at axial illumination as a function of the bistatic scattering angle for several values of the half-cone angle α. The black and white circles indicate the exact values of the cross section in the E and H planes, respectively. The lines are the polarization-independent PO results.

Figure 9.12 shows how the normalized nose-on RCS of a circular cone depends on the opening angle $\theta_1 = \pi - \alpha$ of the cone for several values of the normalized surface impedance η. The range of impedances is limited to $0 \leq \eta < 1$ as the values of the RCS of cones with impedances η and η^{-1} coincide in the PO approximation. Again, the PO (solid line) and exact (squares) results for a PEC cone are seen to agree closely. Further curves are obtained with a finite value of η. As η approaches unity, the value of the RCS monotonically decreases, and with $\eta = 1$, the RCS vanishes. This conclusion follows from (9.87) and the fact that $R_\perp = R_\parallel$ when $\eta = 1$, and is in agreement with Weston's theorem (section 3.2.3). When $\alpha = 90°$, the cone reduces to an infinite plane, and the RCS becomes infinite in this limit, provided that $\eta \neq 1$.

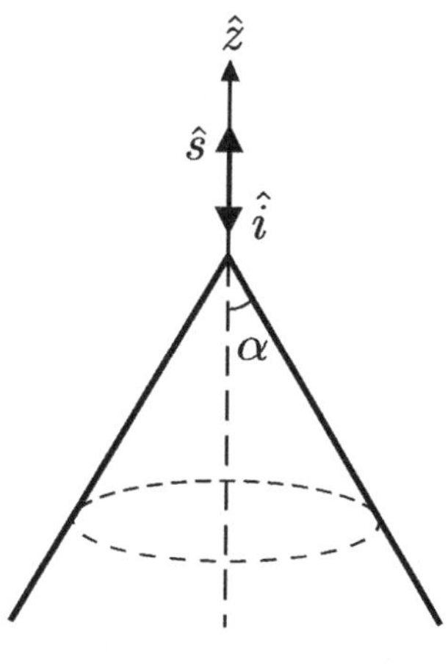

Figure 9.11 Axial backscattering from a circular cone.

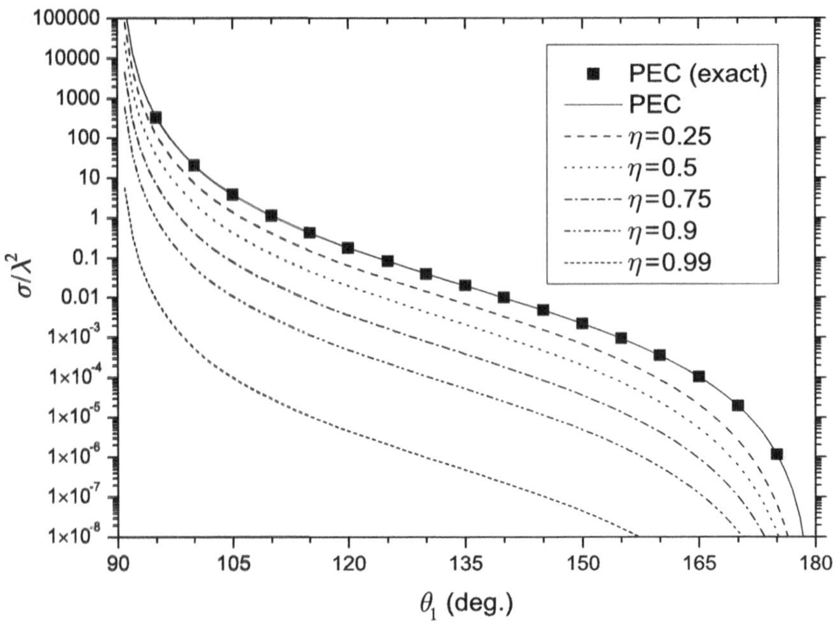

Figure 9.12 Normalized RCS of a circular cone as a function of the cone angular width for several values of the surface impedance.

The Limit of a Plane Angular Sector

It can be shown (see the problems section) that in the limit of the plane angular sector, when either α_x or α_y goes to zero, (9.63) reduces to (9.33) This limit cannot be derived from explicit formulas (9.67)–(9.70) because the condition of the full illumination of the surface does not hold for the thin sector as for any incidence direction other than axial, one of its sides is always shadowed.

Figure 9.13 numerically illustrates this limit by showing the axial RCS ($\theta_{\text{inc}} = \theta = 0$) in the PO approximation for a sequence of PEC elliptic cones as they approach the angular plate limit. The PO solution leads to equal co-polar components $\sigma_{\theta\theta_{\text{inc}}} = \sigma_{\phi\phi_{\text{inc}}} = \sigma$ and vanishing cross-polar components $\sigma_{\theta\phi_{\text{inc}}} = \sigma_{\phi\theta_{\text{inc}}} = 0$. As $\alpha_x \to 0$, the RCS decreases and disappears in the limit of a thin sector $\alpha_x = 0$. This dependence can be compared with similar curves for the exact solution presented in Blume (1996, Figure 9). It turns out that the exact RCS is subject to different polarization relations, $\sigma_{\theta\theta_{\text{inc}}} \neq \sigma_{\phi\phi_{\text{inc}}}$ and $\sigma_{\theta\phi_{\text{inc}}} = \sigma_{\phi\theta_{\text{inc}}} \neq 0$. Furthermore, in the angular sector limit ($\alpha_x \to 0$), three of the four cross-section components $\sigma_{\theta\theta_{\text{inc}}}$, $\sigma_{\theta\phi_{\text{inc}}}$, and $\sigma_{\phi\theta_{\text{inc}}}$ vanish, but the component $\sigma_{\phi\phi_{\text{inc}}}$, which corresponds to the electric field vectors in the incident and scattered waves lying in the plane of the sector, remains finite. However, this limiting value of $\sigma_{\phi\phi_{\text{inc}}}$ is approximately by an order of magnitude smaller than for the circular cone (Blume 1996, Figure 8), implying that the error of the PO solution is not so significant even in this case.

9.3 Electrically Large Plates

In this section the PO solutions for the fields scattered by impenetrable flat plates illuminated by an incident plane wave are derived. The PO solution for the scattered field has the form of an integral over the illuminated side of the plate, and we show that the integral can be reduced to a line integral over the rim of the plate. For polygonal plates the PO solution can be further reduced to a sum over all vertices of the polygon.

In contrast to the cases of the curved edge of an impedance wedge and the tip of the impedance cone, where the stationary point method is used, the evaluations of this section are exact, and since the scatterer (plate) is finite in size the derived PO solutions do not have singularities as functions of incidence and scattering angles.

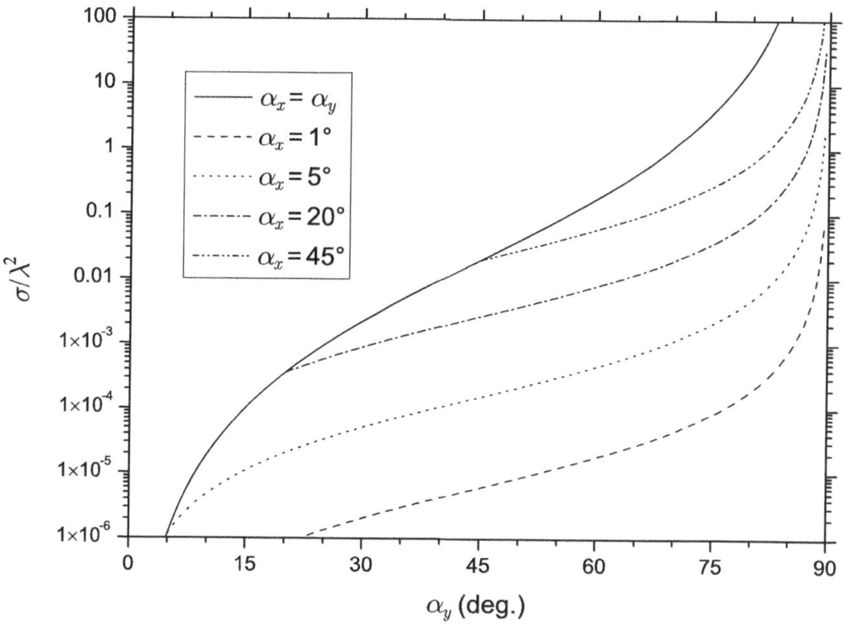

Figure 9.13 Normalized nose-on RCS of PEC elliptic cones as a function of α_y for various values of α_x.

We start by presenting the PO solution as an integral over a two-dimensional area, and the formula is specialized to the limiting cases of backscattering ($\hat{s} = -\hat{\imath}$), forward scattering ($\hat{s} = \hat{\imath}$), and specular scattering ($\hat{s} = \hat{\imath} - 2\hat{n}(\hat{\imath} \cdot \hat{n})$). For the incident field in form of a plane wave the PO integral for the scattered field significantly simplifies and requires evaluation of a canonical scalar integral $I_{\mathrm{po}}(\mathbf{q}, k)$ over the illuminated side of the plate. In the case of forward and specular scattering the integral is evaluated exactly. For backscattering and general orientation of the incidence and scattering directions the integral has to be evaluated numerically. We show that the integral can be exactly transformed into a contour integral over the rim of the plate for a smooth rim and even reduced to a finite sum of contributions from vertices for polygons. Next the Gordon formula is derived by noting that the formula is a result of two-dimensional integration by parts in the PO integral $I_{\mathrm{po}}(\mathbf{q}, k)$. Next, we study the case of a plate with a smoothly shaped rim, including the special case of the circular disc. Finally, the important case of scattering from polygonal plates is studied. We study the phenomenology of forward-scattering patterns for variously shaped polygonal plates, noting that the forward-scattering and backscattering cross sections coincide for PEC plates. Furthermore, the Babinet principle permits extension of the forward-scattering solutions to the solutions for the field transmitted through a complimentary PEC screen.

The reduction of the PO integral to calculate the Kirchhoff integral of the scalar theory of diffraction for a flat PEC plate was described by Gordon (1975). The general electromagnetic case of a planar structure described by reflection and transmission coefficients was studied in Vesnik and Ufimtsev (1992) by referring to an extension of the PO to flat semitransparent plates (Ufimtsev 1968). The line integral representation of the PO solution for an aperture in a PEC screen is discussed by Asvestas (1986). Reduction of the PO integral for the field of a dipole source scattered by a plane PEC plate is described in Johansen and Breinbjerg (1995). Further generalizations of the line integral representations of PO solutions for structures with zero curvature are considered in Pelosi *et al.* (2000) and Gordon (2003).

9.3.1 *Arbitrarily Shaped Plates*

Let Ω be an impenetrable plate with arbitrarily shaped rim $\partial\Omega$. Since this is a flat surface, the unit normal $\hat{n}$ is constant. The PO solution for the scattered field is given by (8.5) where the integration surface can be an arbitrary closed surface enclosing the plate. Let us choose the physical surface of the plate as the integration surface. For a single source, only one side of the plate is illuminated, which is denoted by Ω_0. The contribution from the currents on the shadowed side of the impenetrable plate are neglected in the framework of the PO approximation. Let $R_\perp$ and $R_\parallel$ be reflection coefficients of the illuminated side. The PO surface currents $\mathbf{K}_\mathrm{m}^\mathrm{go}$ and $\mathbf{K}_\mathrm{e}^\mathrm{go}$ can be calculated by using (4.93) and (4.96).

The PO solution for the plate permits significant analytical simplifications when the observation point is located far from the scatterer and the incident field is a plane wave

$$\mathbf{E}_\mathrm{inc} = \mathbf{E}_0\, \mathrm{e}^{-jk\hat{i}\cdot\mathbf{r}} \tag{9.88}$$

with $\mathbf{E}_0$ being the amplitude of the wave at the origin of a global coordinate system, which is such that $\mathbf{E}_0 \cdot \hat{i} = 0$. Indeed, in this case the scattering direction $\hat{s}$ can be considered almost constant over the whole plate, and the kernel of the PO integral reduces to a simple exponential function. The coordinate dependence of the surface currents is the same as in the incident plane wave,

$$Z\mathbf{K}_\mathrm{e}^\mathrm{go}(\mathbf{r}) = \mathbf{v}_\mathrm{e}\, \mathrm{e}^{-jk\hat{i}\cdot\mathbf{r}}, \quad \mathbf{K}_\mathrm{m}^\mathrm{go}(\mathbf{r}) = \mathbf{v}_\mathrm{m}\, \mathrm{e}^{-jk\hat{i}\cdot\mathbf{r}},$$

with constant vector coefficients

$$\mathbf{v}_\mathrm{e} = (R_\perp - 1)(\hat{i}\cdot\hat{n})(\mathbf{E}_0\cdot\hat{\alpha}_\mathrm{inc})\hat{\alpha}_\mathrm{inc} - (R_\parallel + 1)(\mathbf{E}_0\cdot\hat{\beta}_\mathrm{inc})(\hat{n}\times\hat{\alpha}_\mathrm{inc}) \tag{9.89}$$

$$\mathbf{v}_\mathrm{m} = (1 - R_\parallel)(\hat{i}\cdot\hat{n})(\mathbf{E}_0\cdot\hat{\beta}_\mathrm{inc})\hat{\alpha}_\mathrm{inc} - (R_\perp + 1)(\mathbf{E}_0\cdot\hat{\alpha}_\mathrm{inc})(\hat{n}\times\hat{\alpha}_\mathrm{inc}), \tag{9.90}$$

where

$$\hat{\alpha}_\mathrm{inc} = \frac{\hat{i}\times\hat{n}}{|\hat{i}\times\hat{n}|}, \quad \hat{\beta}_\mathrm{inc} = \hat{i}\times\hat{\alpha}_\mathrm{inc}$$

are the unit vectors, with respect to which the polarization of the incident wave is defined (section 4.2.6).

Thus, the solution can be written as

$$\mathbf{E}_\mathrm{sc}^\mathrm{po}(\mathbf{r}) = \frac{jk}{4\pi}\mathbf{V}\int_{\Omega_0} \frac{\mathrm{e}^{-jk|\mathbf{r}-\mathbf{r}'|-jk\hat{i}\cdot\mathbf{r}'}}{|\mathbf{r}-\mathbf{r}'|}\,\mathrm{d}\Omega' \tag{9.91}$$

with

$$\mathbf{V} = \hat{s}\times(\hat{s}\times\mathbf{v}_\mathrm{e} + \mathbf{v}_\mathrm{m}). \tag{9.92}$$

If, furthermore, the observation point is located in the far-field region of the plate, then

$$\frac{\mathrm{e}^{-jk|\mathbf{r}-\mathbf{r}'|-jk\hat{i}\cdot\mathbf{r}'}}{|\mathbf{r}-\mathbf{r}'|} \approx \frac{\mathrm{e}^{-jkr-jk(\hat{i}-\hat{s})\cdot\mathbf{r}'}}{r} \tag{9.93}$$

with $r = |\mathbf{r}|$, and (9.91) takes the form

$$\mathbf{E}_\mathrm{sc}^\mathrm{po} = \frac{\mathrm{e}^{-jkr}}{r}\frac{jk}{4\pi}I_\mathrm{po}(\mathbf{q}, k)\mathbf{V}, \tag{9.94}$$

where $\mathbf{q} = \hat{i} - \hat{s}$ and

$$I_\mathrm{po}(\mathbf{q}, k) = \int_{\Omega_0} \mathrm{e}^{-jk\mathbf{q}\cdot\mathbf{r}'}\,\mathrm{d}\Omega'. \tag{9.95}$$

The latter quantity will be referred to as the shape factor. Note that in the antenna theory this quantity has the meaning of the array factor of an electrically large aperture antenna.

The PO approximation for the far-field coefficient (3.86) is therefore given by

$$\mathbf{F}_{\text{po}} = \frac{jk^2}{4\pi} I_{\text{po}}(\mathbf{q}, k)\mathbf{V}. \tag{9.96}$$

Equations (9.91) and (9.94) are the PO solutions for the electric field scattered by an impenetrable plate illuminated by a plane wave for an observation point in the near- and far-field regions, respectively. An interesting feature is the separation of the PO solution into two factors. The factor **V** includes the complete dependence on the material properties of the plate, whereas the factor $I_{\text{po}}(\mathbf{q}, k)$, the shape factor, describes the dependence on the shape and electrical size of the plate. Calculation of **V** is simple and requires only multiplication of several vectors. Evaluation of $I_{\text{po}}(\mathbf{q}, k)$ can be more challenging, particularly for electrically very large plates. Fortunately, the shape factor can be reduced to a line integral or even evaluated in a closed form, and this will be our main focus in the following analysis.

Special Cases of Back-, Forward, and Specular Scattering

Let us start the analysis with a study of the PO solution in several important special cases. If $\mathbf{q} = 0$ (forward scattering), then (9.95) simplifies to

$$I_{\text{po}}(\mathbf{q}, k) = \int_{\Omega_0} \mathrm{d}\Omega' = |\Omega_0|, \tag{9.97}$$

where $|\Omega_0|$ denotes the area of the illuminated side of the plate. Equation (9.92) with $\hat{s} = \hat{\imath}$ after some vector algebra (problem 5) reduces to

$$\mathbf{V} = 2(\hat{\imath} \cdot \hat{n})\mathbf{E}_0, \tag{9.98}$$

and the PO approximation (9.94) for the forward-scattered field reads

$$\mathbf{E}_{\text{sc}}^{\text{po}} = \frac{\mathrm{e}^{-jkr}}{r} \frac{jk}{2\pi} (\hat{\imath} \cdot \hat{n})|\Omega_0|\mathbf{E}_0. \tag{9.99}$$

The scattered field in (9.99) is polarized as the incident wave, implying no cross-polarization components in the PO approximation. Furthermore, the result is independent of the material properties of the plate and depends only on the area $S_0 = -(\hat{\imath} \cdot \hat{n})|\Omega_0|$ of the geometric cross section of the shadow tube behind the plate.

The corresponding far-field coefficient

$$\mathbf{F}_{\text{fws}}^{\text{po}} = \frac{k^2 S_0}{2\pi j} \mathbf{E}_0 \tag{9.100}$$

coincides with the optical ($\lambda \to 0$) limit (3.123) of the coefficient, implying that in the PO approximation the extinction cross section is equal to

$$\sigma_{\text{ext}} = 2S_0 \tag{9.101}$$

(see the optical theorem (3.12)).

In the case of scattering in the specular direction, when $\hat{s} = \hat{r} = \hat{\imath} - 2\hat{n}(\hat{\imath} \cdot \hat{n})$, we have $\mathbf{q} = 2\hat{n}(\hat{\imath} \cdot \hat{n})$, so that

$$\mathrm{e}^{-jk\mathbf{q}\cdot\mathbf{r}'} = \mathrm{e}^{-jk\mathbf{q}\cdot\mathbf{r}_c},$$

where $\mathbf{r}_c$ is an arbitrary point in the plane of the plate, and

$$I_{po}(\mathbf{q}, k) = |\Omega_0| \, e^{-jk\mathbf{q}\cdot\mathbf{r}_c}. \tag{9.102}$$

Furthermore,

$$\mathbf{V} = -2(\hat{\imath}\cdot\hat{n})[R_{\perp}(\mathbf{E}_0\cdot\hat{\alpha}_{inc})\hat{\alpha}_{ref} + R_{\parallel}(\mathbf{E}_0\cdot\hat{\beta}_{inc})\hat{\beta}_{ref}], \tag{9.103}$$

where $\hat{\alpha}_{ref} = \hat{\alpha}_{inc}$ and $\hat{\beta}_{ref} = \hat{r}\times\hat{\alpha}_{inc}$ form the polarization basis of the reflected wave (problem 6). Choosing $\mathbf{r}_c$ to be the point of GO reflection, we can write

$$\mathbf{E}_{ref} = e^{-jk\hat{q}\cdot\mathbf{r}_c}[R_{\perp}(\mathbf{E}_0\cdot\hat{\alpha}_{inc})\hat{\alpha}_{ref} + R_{\parallel}(\mathbf{E}_0\cdot\hat{\beta}_{inc})\hat{\beta}_{ref}], \tag{9.104}$$

where $\mathbf{E}_{ref}$ is the reflected plane wave at the point of reflection, and (9.94) gives

$$\mathbf{E}^{po}_{sc} = \frac{e^{-jkr}}{r}\frac{k|\Omega_0|}{2\pi j}(\hat{\imath}\cdot\hat{n})\mathbf{E}_{ref}. \tag{9.105}$$

In the limit of normal incidence ($\hat{\imath} = -\hat{n}$), (9.105) can be expressed as

$$\mathbf{E}^{po}_{sc} = \frac{e^{-jkr}}{r}\frac{k|\Omega_0|}{2\pi j}R_{\parallel}(0)\mathbf{E}_0\, e^{-2jk(\hat{\imath}\cdot\mathbf{r}_c)}, \tag{9.106}$$

where $R_{\parallel}(0) = -R_{\perp}(0)$ is the reflection coefficient at normal incidence, implying vanishing scattered field in the PO approximation for an impedance surface with $\tilde{Z}_s = Z$ (normalized impedance $\eta = 1$, low-reflection surface).
In the case of backscattering ($\hat{s} = -\hat{\imath}$), $\mathbf{q} = 2\hat{\imath}$,

$$\mathbf{V} = -2(\hat{\imath}\cdot\hat{n})[R_{\perp}(\mathbf{E}_0\cdot\hat{\alpha}_{inc})\hat{\alpha}_{inc} - R_{\parallel}(\mathbf{E}_0\cdot\hat{\beta}_{inc})\hat{\beta}_{inc}] \tag{9.107}$$

(problem 7), and the backscattered field in the PO approximation is expressed as

$$\mathbf{E}^{po}_{sc} = \frac{e^{-jkr}}{r}\frac{jk}{2\pi}(\hat{\imath}\cdot\hat{n})I_{po}(2\hat{\imath}, k)[-R_{\perp}(\mathbf{E}_0\cdot\hat{\alpha}_{inc})\hat{\alpha}_{inc} + R_{\parallel}(\mathbf{E}_0\cdot\hat{\beta}_{inc})\hat{\beta}_{inc}]. \tag{9.108}$$

For a PEC surface, when $R_{\perp} = -1$ and $R_{\parallel} = 1$, (9.108) gives

$$\mathbf{E}^{po}_{sc} = \frac{e^{-jkr}}{r}\frac{jk}{2\pi}(\hat{\imath}\cdot\hat{n})I_{po}(2\hat{\imath}, k)\mathbf{E}_0 \tag{9.109}$$

since

$$\mathbf{E}_0 = (\mathbf{E}_0\cdot\hat{\alpha}_{inc})\hat{\alpha}_{inc} + (\mathbf{E}_0\cdot\hat{\beta}_{inc})\hat{\beta}_{inc}.$$

Similarly to the case of forward scattering, the PO approximation (9.108) does not predict any cross-polarized component in the backscattered field, in the sense that an incident wave polarized either along $\hat{\alpha}_{inc}$ or $\hat{\beta}_{inc}$ produces a scattered field of the same linear polarization. In contrast to the cases of forward and specular scattering, the PO solution for the backscattering requires evaluation of the surface integral (9.95). The only exception is the case of normal incidence ($\hat{\imath} = -\hat{n}$), in which the expression (9.108) recovers (9.106) and the corresponding scattering cross section is given by the formula

$$\sigma_{po} = \frac{1}{\pi}k^2|\Omega_0|^2|R_{\parallel}(0)|^2. \tag{9.110}$$

Reduction of Shape Factor to Line Integral

The numerical evaluation of the integral (9.95) requires two-dimensional integration over the illuminated surface of the plate, which can be quite laborious. However, a closer analysis shows that the area integral $I_{\mathrm{po}}(\mathbf{q}, k)$ can be reduced to a contour integral over the rim of the plate or even to a closed form if the plate has a polygonal shape (the sides are pieces of straight lines). These particularly simple expressions for the fields scattered by the plates are used in practical calculations of scattering from arbitrarily shaped targets on approximating the shape of the target by a sum of flat facets.

The reduction can be performed in different ways by using Green's theorem (Gordon 1975) or Green's second identity (Vesnik and Ufimtsev 1992). Here, we will show that the two-dimensional integration by parts, which is used in Appendix D for asymptotic evaluation of two-dimensional integrals, can be used as well. We derive a three-dimensional vector version of the line-integral representation which will be transformed in the following sections to the form convenient for calculations of scattering from smoothly shaped and polygonal plates.

We start by transforming the surface integral (9.95) to a two-dimensional Riemann integral over the area of the plate. Introducing the Cartesian coordinates (u, v) in the plane of the plate

$$\mathbf{r} = \mathbf{r}_{\mathrm{c}} + \hat{u}u + \hat{v}v, \tag{9.111}$$

where $\mathbf{r}_{\mathrm{c}}$ is an arbitrary point in the plane of the plate and $\hat{u}$ and $\hat{v}$ are three-dimensional unit vectors orthogonal to each other and to the vector $\hat{n}$, gives

$$I_{\mathrm{po}}(\mathbf{q}, k) = \mathrm{e}^{-jk\mathbf{q}\cdot\mathbf{r}_{\mathrm{c}}} \int\limits_{\Omega_{uv}} \mathrm{e}^{-jk(q_u u + q_v v)}\, \mathrm{d}u\, \mathrm{d}v \tag{9.112}$$

with $q_u = \mathbf{q} \cdot \hat{u}$ and $q_v = \mathbf{q} \cdot \hat{v}$. The area Ω_{uv} is defined on the u–v plane (see Figure D.5).

Now we integrate by parts in (9.112) according to (D.75) with

$$A(u, v) = 1, \quad S(u, v) = -q_u u - q_v v. \tag{9.113}$$

Since $A(u, v)$ is a constant and $S(u, v)$ is a linear function of u and v with

$$\nabla_{uv} S = -q_u \hat{u} - q_v \hat{v}, \quad |\nabla_{uv} S|^2 = q_u^2 + q_v^2, \quad \nabla_{uv}^2 S = 0, \tag{9.114}$$

it follows from (D.79) that

$$A_1(u, v) = -\frac{\nabla_{uv}^2 S}{q_u^2 + q_v^2} = 0, \tag{9.115}$$

the area integral I_1 in (D.75) vanishes, and we arrive at the formula

$$I_{\mathrm{po}}(\mathbf{q}, k) = \mathrm{e}^{-jk\mathbf{q}\cdot\mathbf{r}_{\mathrm{c}}} \frac{j}{k(q_u^2 + q_v^2)} \oint\limits_{\partial\Omega_{uv}} \mathrm{e}^{-jk(q_u u + q_v v)} (q_u w_u + q_v w_v)\, \mathrm{d}l_{uv}, \tag{9.116}$$

where w_u and w_v are the components of the two-dimensional normal $\hat{w}$ directed into the exterior of Ω_{uv} and $\mathrm{d}l_{uv}$ is the infinitesimal arc length of $\partial\Omega_{uv}$ (Figure D.5).

Equation (9.116) involves only line integration and can be rewritten in a vector form as a line integral in space taken over the rim $\partial\Omega_0$ of the plate

$$I_{\mathrm{po}}(\mathbf{q}, k) = \frac{-j}{kq_\perp^2} \oint\limits_{\partial\Omega_0} \mathrm{e}^{-jk\mathbf{q}\cdot\mathbf{r}'} (\hat{\nu} \cdot \mathbf{q})\, \mathrm{d}l', \tag{9.117}$$

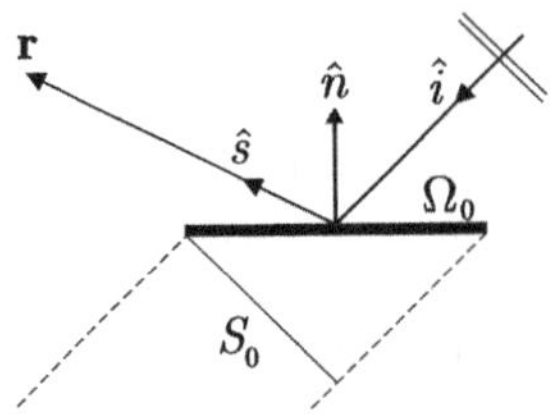

Figure 9.14 An impenetrable plate illuminated by a plane wave. Ω_0 is the illuminated side of the plate and S_0 is the area of the geometric cross section of the shadow tube behind the plate.

where $\mathrm{d}l'$ is the infinitesimal arc length at the integration point $\mathbf{r}'$ and $q_\perp$ is the amplitude of the projection of the vector $\mathbf{q}$ on the plane of the plate

$$q_\perp^2 = q_u^2 + q_v^2 = |\mathbf{q}|^2 - (\hat{n} \cdot \mathbf{q})^2. \tag{9.118}$$

By definition, the unit vector $\hat{\nu}$ lies in the plane of the plate, is normal to the rim and directed into the interior of the plate, so that $\hat{\nu} \cdot \mathbf{q} = -w_u q_u - w_v q_v$. Since the rim of the plate is a flat curve, $\hat{\nu} = \hat{N}$, where $\hat{N}$ is the principal normal vector of the curve $\partial\Omega_0$ (see (A.62)).

The line integral (9.117) is much easier to evaluate than the original two-dimensional integral (9.95). A practical way of doing that is to describe the boundary $\partial\Omega_{uv}$ in the parametric form

$$u = U(t), \quad v = V(t) \tag{9.119}$$

with $0 \le t \le 2\pi$, $U(0) = U(2\pi)$ and $V(0) = V(2\pi)$. Then,

$$\mathrm{d}l = \sqrt{\dot{U}^2(t) + \dot{V}^2(t)}\,\mathrm{d}t, \quad \hat{\nu} = -\frac{\dot{V}(t)\hat{u} - \dot{U}(t)\hat{v}}{\sqrt{\dot{U}^2(t) + \dot{V}^2(t)}}, \tag{9.120}$$

where the dot means the derivative with respect to t, which results in the formula

$$I_{\text{po}}(\mathbf{q}, k) = \mathrm{e}^{-jk\mathbf{q}\cdot\mathbf{r}_\text{c}} \frac{j}{k(q_u^2 + q_v^2)} \int_0^{2\pi} \mathrm{e}^{-jk[q_u U(t) + q_v V(t)]} [\dot{V}(t) q_u - \dot{U}(t) q_v]\,\mathrm{d}t. \tag{9.121}$$

The formula assumes a counterclockwise movement of the point $(U(t), V(t))$ on the u–v plane as the parameter t increases.

9.3.2 Circular Disc

Let us apply the general solution (9.121) to the particular case of a circular disc. We will see that the PO solution for the circular disc looks particularly simple. Once we have the PO solution for the scattered field, we will compare it with the exact solution to estimate the accuracy of the PO solution.

Evaluation of Shape Factor

Assume that the disc has the radius a, lies in the plane $z = 0$, and its center is at the origin of the global coordinate system, that is, $\mathbf{r}_\text{c} = 0$. The disc is illuminated from the upper half-space (Figure 9.15). Then, the normal to the illuminated side

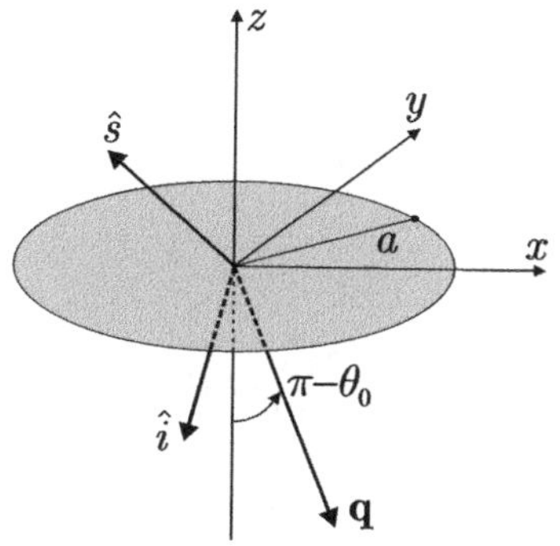

Figure 9.15 A circular disc is illuminated by a plane wave propagating in the direction $\hat{i}$, the scattering direction is $\hat{s}$, and θ_0 is the zenith angle of the vector $\mathbf{q} = \hat{i} - \hat{s}$.

is given by $\hat{n} = \hat{z}$. We can furthermore choose $\hat{u} = \hat{x}$ and $\hat{v} = \hat{y}$. The rim of the disc is parametrized by the functions

$$U(t) = a\cos t, \quad V(t) = a\sin t \tag{9.122}$$

and (9.121) takes the form

$$I_{\text{po}}(\mathbf{q}, k) = \frac{ja}{k(q_x^2 + q_y^2)} \int_0^{2\pi} \mathrm{e}^{-jka(q_x \cos t + q_y \sin t)} (q_x \cos t + q_y \sin t)\,\mathrm{d}t. \tag{9.123}$$

If we represent the components of the vector $\mathbf{q}$ in a spherical coordinate system with the zenith angle θ_0 (Figure 9.15) as

$$q_x = q\sin\theta_0\cos\phi_0, \quad q_y = q\sin\theta_0\sin\phi_0, \quad q_z = q\cos\theta_0, \tag{9.124}$$

where $q = \sqrt{q_x^2 + q_y^2 + q_z^2}$, then (9.123) becomes

$$I_{\text{po}}(\mathbf{q}, k) = \frac{ja}{kq\sin\theta_0} \int_0^{2\pi} \mathrm{e}^{-jkaq\sin\theta_0\cos(t-\phi_0)} \cos(t-\phi_0)\,\mathrm{d}t, \tag{9.125}$$

and changing the integration variable as $\phi = t - \phi_0$ gives

$$I_{\text{po}}(\mathbf{q}, k) = \frac{ja}{kq\sin\theta_0} \int_{-\phi_0}^{2\pi-\phi_0} \mathrm{e}^{-jkaq\sin\theta_0\cos\phi} \cos\phi\,\mathrm{d}\phi. \tag{9.126}$$

The integrand in (9.126) is a 2π periodic function and the integration is performed over the full period of the integrand. So, the value of the integral remains the same when integration is performed over any interval of the length 2π, which we choose to be from $-\pi$ to π, and since the integrand is an even function of ϕ, we can write

$$I_{\text{po}}(\mathbf{q}, k) = \frac{2ja}{kq\sin\theta_0} \int_0^{\pi} \mathrm{e}^{-jkaq\sin\theta_0\cos\phi} \cos\phi\,\mathrm{d}\phi, \tag{9.127}$$

which is evaluated by using the integral representation of the Bessel function (see Abramowitz and Stegun (1965, section 9.1.21))

$$J_1(z) = \frac{j}{\pi} \int_0^{\pi} \mathrm{e}^{-jz\cos\phi} \cos\phi \, \mathrm{d}\phi \tag{9.128}$$

as

$$I_{\mathrm{po}}(\mathbf{q}, k) = \frac{2\pi a}{kq \sin\theta_0} J_1(kaq \sin\theta_0). \tag{9.129}$$

In antenna theory, this formula corresponds to the far-zone radiation pattern of a circular aperture antenna with a uniform field distribution in the aperture (see Balanis (1997, eqs. 12.54–12.55)).

In the limits of forward scattering ($q \to 0$) and specular scattering ($\theta_0 \to \pi$), the argument of the Bessel function in (9.129) goes to zero, and since $J_1(z) \to z/2$ when $|z| \to 0$, we obtain

$$I_{\mathrm{po}}(\mathbf{q}, k) \to \pi a^2, \tag{9.130}$$

which is the area of the illuminated side of the disc. This result complies with the general properties (9.97) and (9.102).

Accuracy of PO Solution

Now we compare the PO results for the field scattered by a circular PEC disc with the exact solution (Meixner and Andrejewski 1950). Since the dependence on the material properties is separated from the shape factor, we expect that the analysis will be of general meaning, applicable to arbitrarily shaped plates and even to arbitrarily shaped scatterers as every shape can be approximated by a set of flat facets.

Figures from 9.16 to 9.20 present typical results for the extinction cross section (which coincides with the total scattering cross section for PEC scatterers), backscattering cross section (RCS), and bistatic scattering cross section of

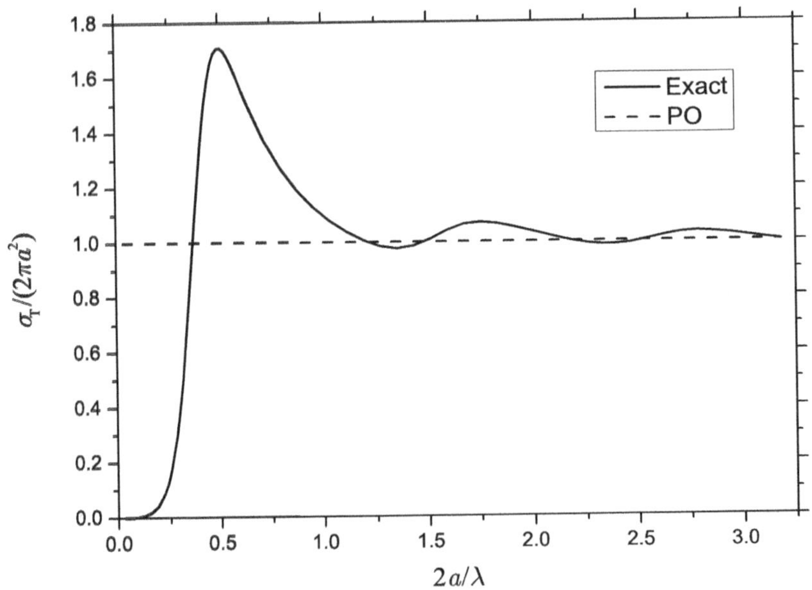

Figure 9.16 Extinction (total scattering) cross section of a PEC disc at the axial illumination as a function of the electrical size of the disc. The cross section is normalized to its optical limit $2\pi a^2$.

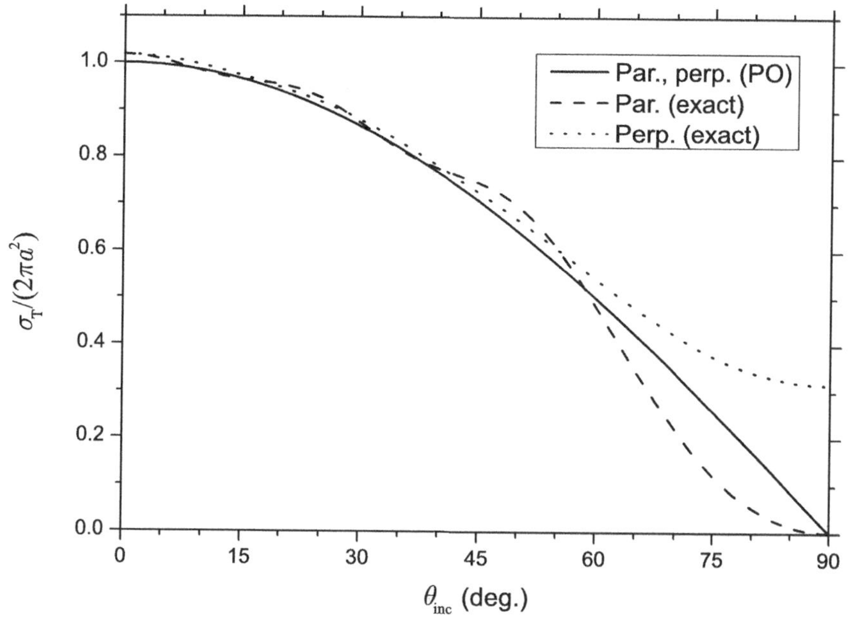

Figure 9.17 Normalized extinction (total scattering) cross section of a PEC disc with 3λ diameter as a function of the incidence angle for the parallel and perpendicular polarization of the incoming plane wave.

the disc. The results suggest that the simple PO solutions for the plates with electrical size greater than one wavelength are useful and in many cases sufficiently accurate.

Figure 9.16 shows the extinction cross section of the disc at the axial illumination in dependence on the electrical size $2a/\lambda$ of the disc. The cross section is normalized to its optical limit $2\pi a^2$, and since the PO cross section is equal to that in the optical limit (see (9.101)), the PO result is simply unity for all values of $2a/\lambda$. The exact solution deviates from the PO curve in the low-frequency range of the diagram, but once $2a/\lambda > 1$ the difference between the two curves reduces, decreasing with the increasing electrical size of the disc.

Generally, the extinction cross section depends on the angle of incidence and polarization of the incident wave. However, the PO approximation for the cross section is independent of the polarization and depends on the incidence angle as $\cos\theta_{inc}$ where $\cos\theta_{inc} = -\hat{i}\cdot\hat{n}$ (see (9.101)), which is shown by the solid line in Figure 9.17. The curves are plotted for the disc with $2a/\lambda = 3$ and normalized to the PO extinction cross section at axial illumination ($\theta_{inc} = 0$). For most of the range of the incidence angle the PO approximation is good, apart from the grazing incidence $\theta_{inc} = \pi/2$ for perpendicular polarization (the electric field vector in the incident wave is perpendicular to the plane of incidence defined by $\hat{n}$ and $\hat{i}$). The exact solution for this polarization leads to a finite value of the extinction cross section. At the grazing incidence the main scattering mechanism is the multiple edge diffraction between the opposite sides of the rim of the disc, which is not accounted for by the PO approximation.

The backscattering cross section (RCS) is another important characteristic of the far-field scattering. In Figures 9.18 and 9.19 the cross section is plotted as a function of the electrical size of the disc and of the incidence angle. In Figure 9.18 the cross section is normalized to its PO value, which is given by $\sigma_{po} = \pi k^2 a^4$ according to the general formula (9.110). Similarly to the case of extinction cross section, the agreement between the PO and exact solutions improves with the growing electrical size of the disc so that the error of the PO approximation is less than 1 dB with $2a/\lambda > 1$ and 0.1 dB with $2a/\lambda > 5$.

Figure 9.19 shows the dependence of RCS on the illumination angle θ_{inc} which is the zenith angle in the spherical coordinate system with the zenith axis perpendicular to the surface of the disc. The disc diameter is equal to 3 cm,

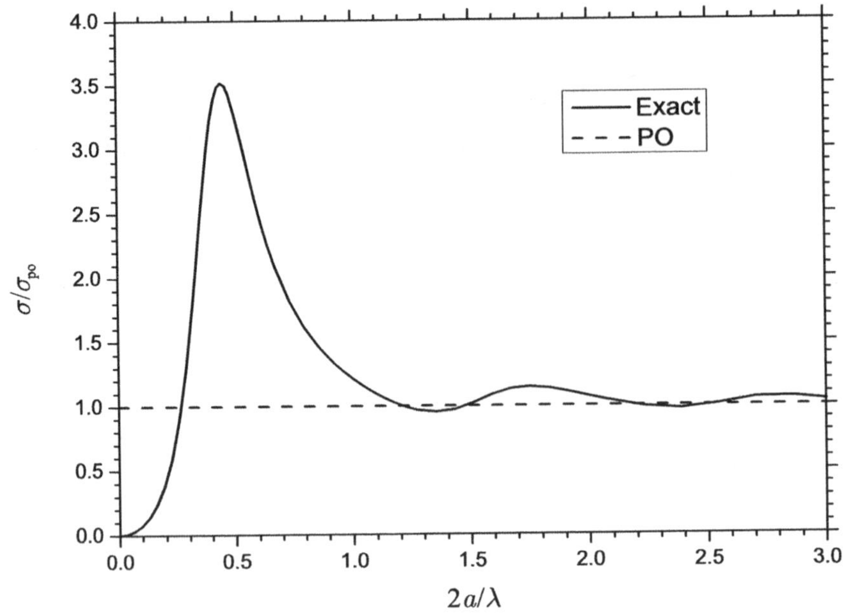

Figure 9.18 Backscattering cross section of a circular PEC disc at axial illumination, normalized to its PO value as a function of the electrical size of the disc.

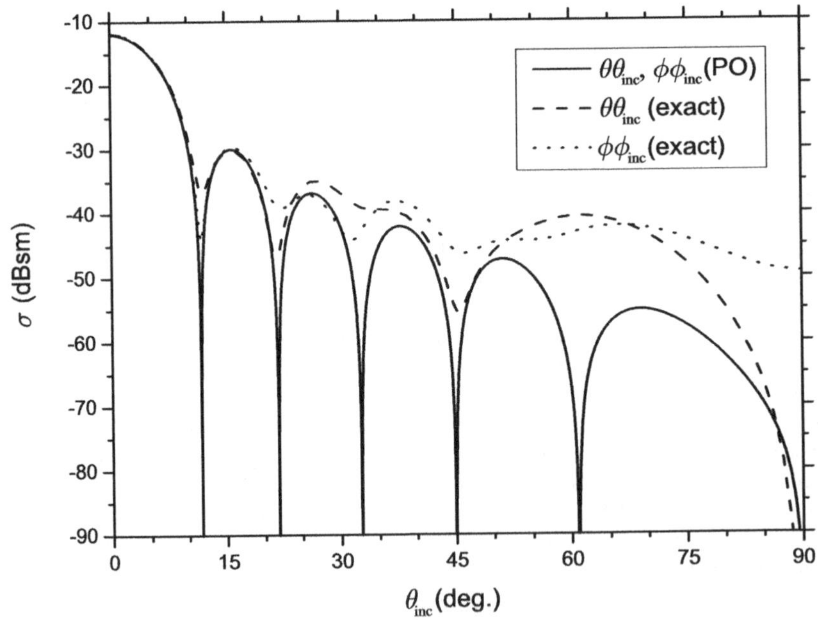

Figure 9.19 Monostatic cross section of a PEC disc with $2a/\lambda = 3$ for $\lambda = 0.01$ m as a function of the illumination angle.

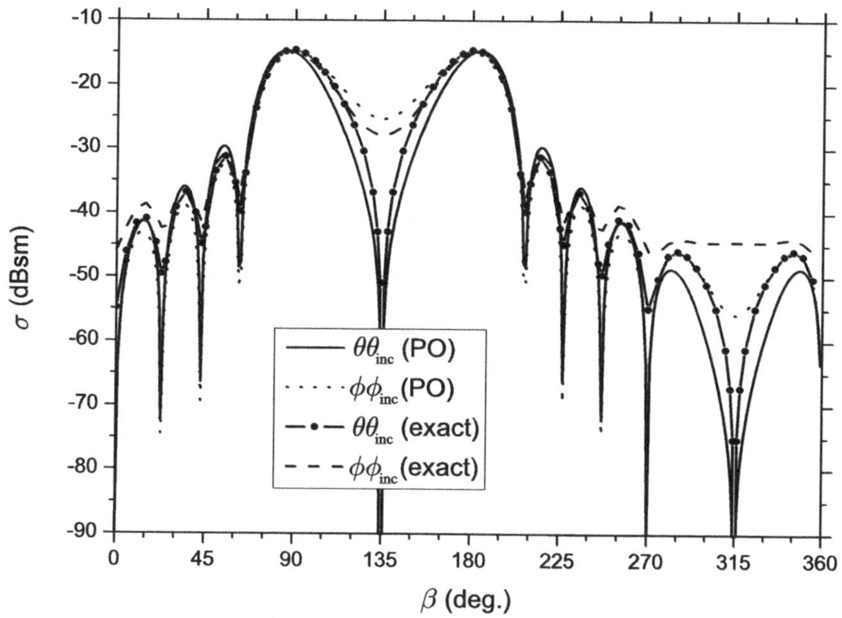

Figure 9.20 Bistatic scattering cross section of a PEC disc of electrical size $2a/\lambda = 3$ with $\lambda = 1$ cm illuminated at an angle of 45° with respect to the normal to the disc surface as a function of the separation angle between the incidence and scattering directions. The scattering direction is in the plane of incidence. The maxima at $\beta = 90°$ and $\beta = 180°$ correspond to the directions of specular and forward scattering, respectively.

$\lambda = 1$ cm, and the RCS values are given in dBsm (i.e., dB value of σ in m^2; see (3.13)). The exact solution is polarization-dependent, and the figure shows two exact curves for two basic polarization cases with the electric field vectors in the meridional ($\theta\theta_{inc}$) and azimuthal ($\phi\phi_{inc}$) planes. The PO curve is the same for both polarizations and shows deep minima as a function of θ_{inc}, which are typical features of PO solutions for plates. Nevertheless, a good agreement with the exact solution is observed around the main lobe and several nearest side lobes of the scattering diagram.

The bistatic scattering cross section of the same disc is presented in Figure 9.20. The incidence direction makes a 45° angle with the normal to the disc surface, and the scattering direction lies in the incidence plane defined by the vectors $\hat{i}$ and $\hat{n}$. The curves are plotted as functions of the separation angle β, which is the angle between the incidence and scattering directions. In this case, the PO approximation gives different curves for the two basic polarizations, which are denoted by $\theta\theta_{inc}$ and $\phi\phi_{inc}$ (electric field vectors of the incident and scattered waves parallel and perpendicular to the incidence plane, respectively). The agreement between the exact and PO values is again good, particularly near the maxima of the scattering diagram in the directions of specular and forward scattering. However, in the region of side lobes there is only a general agreement between exact and PO solutions, similarly to the results presented in Figure 9.19. The reason is that in the side lobe region the edge-diffracted waves, which are only partially accounted for in the framework of PO, play an important role. Techniques for improvement of PO results in the side lobe region are developed in the framework of PTD (Ufimtsev 2014).

9.3.3 Polygonal Plates

If the plate is a polygon, then the shape factor can be further simplified and calculated in a closed form as a sum of contributions from the corner points of the polygon (Figure 9.21). The case of a polygonal plate is important for practice because of the use of polygons (mostly triangles) to approximately represent generally shaped targets in applications of the PO method (Youssef 1989). In what follows we derive two representations of the shape factor.

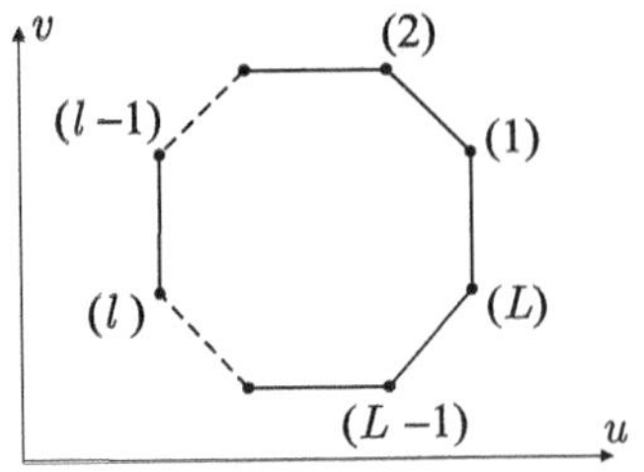

Figure 9.21 Polygon with L vertices on the u–v plane.

Calculation of Shape Factor

Let us specialize the line integral representation (9.121) of the PO solution for arbitrarily shaped plates to the case of a polygon with L vertices (Figure 9.21). On the u–v plane the vertices have coordinates (U_l, V_l) with $l = 1, 2, \ldots, L$.

The vertices of the L-gon are connected by straight segments, and the lth segment is described in the parametric form by

$$U(t) = U_l + \dot{U}_l(t - t_l), \quad V(t) = V_l + \dot{V}_l(t - t_l), \tag{9.131}$$

where $t_l \leq t < t_{l+1}$, $U(t_l) = U_l$, $V(t_l) = V_l$, and $l = 1, 2, \ldots, L$. The constant parameters $\dot{U}_l$ and $\dot{V}_l$, which are the derivatives of $U(t)$ and $V(t)$ with respect to t, describe the slope of the segment. At the vertices the parameter t takes the values t_l such that

$$0 = t_1 < t_2 < \ldots < t_L < t_{L+1} = 2\pi \tag{9.132}$$

with $U_{L+1} = U_1$ and $V_{L+1} = V_1$.

The line integral in (9.121) is the sum of contributions from the segments

$$I_{\text{po}}(\mathbf{q}, k) = \frac{j\,\mathrm{e}^{-jk\mathbf{q}\cdot\mathbf{r}_c}}{k(q_u^2 + q_v^2)} \sum_{l=1}^{L} I_l, \tag{9.133}$$

where

$$I_l = (q_u\dot{V}_l - q_v\dot{U}_l)\,\mathrm{e}^{-jk(q_uU_l+q_vV_l)} \int_{t_l}^{t_{l+1}} \mathrm{e}^{-jk(q_u\dot{U}_l+q_v\dot{V}_l)(t-t_l)}\,\mathrm{d}t, \tag{9.134}$$

which can be easily integrated to give

$$I_l = \frac{j}{k} C_l\,\mathrm{e}^{-jk(q_uU_l+q_vV_l)} \left[\mathrm{e}^{-jk(q_u\dot{U}_l+q_v\dot{V}_l)(t_{l+1}-t_l)} - 1\right] \tag{9.135}$$

with

$$C_l = \frac{q_u\dot{V}_l - q_v\dot{U}_l}{q_u\dot{U}_l + q_v\dot{V}_l}. \tag{9.136}$$

It follows from (9.131) that

$$(t_{l+1} - t_l)\dot{U}_l = U_{l+1} - U_l, \quad (t_{l+1} - t_l)\dot{V}_l = V_{l+1} - V_l. \tag{9.137}$$

Relation (9.135) can be rewritten as

$$I_l = \frac{j}{k} C_l \left[e^{-jk(q_u U_{l+1} + q_v V_{l+1})} - e^{-jk(q_u U_l + q_v V_l)} \right] \tag{9.138}$$

and (9.133) becomes

$$I_{\text{po}}(\mathbf{q}, k) = \frac{e^{-jk\mathbf{q}\cdot\mathbf{r}_c}}{k^2(q_u^2 + q_v^2)} \sum_{l=1}^{L} C_l \left[e^{-jk(q_u U_l + q_v V_l)} - e^{-jk(q_u U_{l+1} + q_v V_{l+1})} \right]. \tag{9.139}$$

The shape factor is given by (9.139) as a sum of L terms, that is, in a closed form. In what follows we express (9.139) in a fully vector form and obtain two representations for the shape factor, referred to as a sum over segments and a sum over vertices, one of which is particularly useful in calculations and another has an appealing physical interpretation.

To obtain the first representation we start by rewriting (9.139) as

$$I_{\text{po}}(\mathbf{q}, k) = \frac{2j\, e^{-jk\mathbf{q}\cdot\mathbf{r}_c}}{k^2(q_u^2 + q_v^2)} \sum_{l=1}^{L} C_l \exp\left\{ -j\frac{k}{2}[q_u(U_{l+1} + U_l) + q_v(V_{l+1} + V_l)] \right\} \sin\left\{ \frac{k}{2}[q_u(U_{l+1} - U_l) + q_v(V_{l+1} - V_l)] \right\}. \tag{9.140}$$

Next we express the arguments of the trigonometric functions in (9.140) through the position vectors of the vertices,

$$\mathbf{r}_l = \mathbf{r}_c + U_l\hat{u} + V_l\hat{v}, \quad l = 1, 2, \ldots, L, \tag{9.141}$$

where $\mathbf{r}_c$ is an arbitrary point in the plane of the plate. Vectors $\hat{u}$ and $\hat{v}$ are orthogonal unit vectors in the plane of the plate, so that $\hat{u} \times \hat{v} = \hat{n}$, where $\hat{n}$ is the unit outward normal to the illuminated side of the plate.

Since $\mathbf{r}_{l+1} - \mathbf{r}_l$ and $\mathbf{r}_c - \mathbf{r}_l$ lie in the plane of the plate,

$$q_u(U_{l+1} - U_l) + q_v(V_{l+1} - V_l) = \mathbf{q} \cdot (\mathbf{r}_{l+1} - \mathbf{r}_l) \tag{9.142}$$

and

$$q_u(U_{l+1} + U_l) + q_v(V_{l+1} + V_l) = \mathbf{q} \cdot (\mathbf{r}_{l+1} + \mathbf{r}_l) - 2\mathbf{q} \cdot \mathbf{r}_c. \tag{9.143}$$

Furthermore, from (9.137) and (9.142)

$$(t_{l+1} - t_l)(q_u\dot{U}_l + q_v\dot{V}_l) = \mathbf{q} \cdot (\mathbf{r}_{l+1} - \mathbf{r}_l), \tag{9.144}$$

and since $\hat{n} \times \mathbf{q} = -q_v\hat{u} + q_u\hat{v}$,

$$(t_{l+1} - t_l)(q_u\dot{V}_l - q_v\dot{U}_l) = \mathbf{q} \cdot (\mathbf{r}_{l+1} - \mathbf{r}_l) \times \hat{n}, \tag{9.145}$$

implying that

$$C_l = \frac{\mathbf{q} \cdot (\mathbf{r}_{l+1} - \mathbf{r}_l) \times \hat{n}}{\mathbf{q} \cdot (\mathbf{r}_{l+1} - \mathbf{r}_l)}. \tag{9.146}$$

Using (9.142), (9.143), and (9.146) in (9.140) gives the representation

$$I_{\text{po}}(\mathbf{q}, k) = \frac{j}{kq_\perp^2} \sum_{l=1}^{L} [\mathbf{q} \cdot (\mathbf{r}_{l+1} - \mathbf{r}_l) \times \hat{n}]\, e^{-jk\mathbf{q}\cdot(\mathbf{r}_{l+1}+\mathbf{r}_l)/2} \operatorname{sinc}\left[\frac{k}{2}\mathbf{q} \cdot (\mathbf{r}_{l+1} - \mathbf{r}_l) \right], \tag{9.147}$$

where $q_\perp$ is the projection of $\mathbf{q}$ on the plane of the plate with the absolute value given by (9.118).

When $q_\perp \to 0$, the shape factor is finite as is apparent from its definition (9.95) (see also (9.102)), implying that the representation (9.147) requires special treatment because of the factor $q_\perp^{-2}$ in front of the sum. The treatment consists of rewriting the trigonometric factors in terms of $q_\perp$. Introducing a unit vector $\hat{e}$ by the formula

$$\mathbf{q} = \hat{n}(\hat{n} \cdot \mathbf{q}) + q_\perp \hat{e} \tag{9.148}$$

and noting that

$$\begin{aligned}
\mathbf{q} \cdot (\mathbf{r}_{l+1} - \mathbf{r}_l) &= q_\perp \hat{e} \cdot (\mathbf{r}_{l+1} - \mathbf{r}_l) \\
\mathbf{q} \cdot (2\mathbf{r}_c - \mathbf{r}_{l+1} - \mathbf{r}_l) &= q_\perp \hat{e} \cdot (2\mathbf{r}_c - \mathbf{r}_{l+1} - \mathbf{r}_l) \\
\mathbf{q} \cdot (\mathbf{r}_{l+1} - \mathbf{r}_l) \times \hat{n} &= q_\perp (\mathbf{r}_{l+1} - \mathbf{r}_l) \cdot \hat{n} \times \hat{e},
\end{aligned}$$

we express (9.147) as

$$I_{\text{po}}(\mathbf{q}, k) = \frac{j}{kq_\perp} \, \mathrm{e}^{-jk\mathbf{q}\cdot\mathbf{r}_c} \sum_{l=1}^{L} [(\mathbf{r}_{l+1} - \mathbf{r}_l) \cdot \hat{n} \times \hat{e}] \mathrm{e}^{j\frac{k}{2} q_\perp \hat{e}\cdot(2\mathbf{r}_c - \mathbf{r}_{l+1} - \mathbf{r}_l)} \operatorname{sinc}\left[\frac{k}{2} q_\perp \hat{e} \cdot (\mathbf{r}_{l+1} - \mathbf{r}_l)\right] \tag{9.149}$$

and using the identity

$$\sum_{l=1}^{L} [(\mathbf{r}_{l+1} - \mathbf{r}_l) \cdot \hat{n} \times \hat{e}] = \hat{n} \times \hat{e} \cdot \sum_{l=1}^{L} (\mathbf{r}_{l+1} - \mathbf{r}_l) = 0, \tag{9.150}$$

rewrite (9.149) in the form

$$I_{\text{po}}(\mathbf{q}, k) = \frac{j}{kq_\perp} \, \mathrm{e}^{-jk\mathbf{q}\cdot\mathbf{r}_c} \sum_{l=1}^{L} [(\mathbf{r}_{l+1} - \mathbf{r}_l) \cdot \hat{n} \times \hat{e}] \left\{ \mathrm{e}^{j\frac{k}{2} q_\perp \hat{e}\cdot(2\mathbf{r}_c - \mathbf{r}_{l+1} - \mathbf{r}_l)} \operatorname{sinc}\left[\frac{k}{2} q_\perp \hat{e} \cdot (\mathbf{r}_{l+1} - \mathbf{r}_l)\right] - 1 \right\}, \tag{9.151}$$

which is explicitly finite in the limit $q_\perp \to 0$ since

$$\mathrm{e}^{j\frac{k}{2} q_\perp \hat{e}\cdot(2\mathbf{r}_c - \mathbf{r}_{l+1} - \mathbf{r}_l)} \operatorname{sinc}\left[\frac{k}{2} q_\perp \hat{e} \cdot (\mathbf{r}_{l+1} - \mathbf{r}_l)\right] - 1 = j\frac{kq_\perp}{2} \hat{e} \cdot (2\mathbf{r}_c - \mathbf{r}_{l+1} - \mathbf{r}_l) + O(q_\perp^2). \tag{9.152}$$

Expressions (9.147) and (9.151) supplemented with (9.152) are very suitable for numerical calculations. The physical interpretation of the expressions is not so straightforward, however. We can only comment that the sinc functions in (9.147) and (9.151) have the same meaning as the sinc array factors of linear antenna arrays with uniform amplitude distribution (see Balanis (1997)). Next we derive an alternative representation for the shape factor of polygons, which permits an appealing physical interpretation of the PO solution for the scattered field.

Shape Factor as a Sum over Vertices

Let us return to the expression (9.139) and rewrite it by collecting the terms with the same exponential factors as

$$\begin{aligned}
\sum_{l=1}^{L} C_l \left[\mathrm{e}^{-jk(q_u U_l + q_v V_l)} - \mathrm{e}^{-jk(q_u U_{l+1} + q_v V_{l+1})}\right] &= \sum_{l=1}^{L} C_l \, \mathrm{e}^{-jk(q_u U_l + q_v V_l)} - \sum_{l=2}^{L+1} C_{l-1} \, \mathrm{e}^{-jk(q_u U_l + q_v V_l)} \\
&= \sum_{l=2}^{L} (C_l - C_{l-1}) \, \mathrm{e}^{-jk(q_u U_l + q_v V_l)} + C_1 \, \mathrm{e}^{-jk(q_u U_1 + q_v V_1)} - C_L \, \mathrm{e}^{-jk(q_u U_{L+1} + q_v V_{L+1})} \\
&= \sum_{l=1}^{L} (C_l - C_{l-1}) \, \mathrm{e}^{-jk(q_u U_l + q_v V_l)},
\end{aligned}$$

where $C_0 = C_L$. Inserting this in (9.139) and noting that

$$q_u U_l + q_v V_l = \mathbf{q} \cdot (\mathbf{r}_l - \mathbf{r}_c) \tag{9.153}$$

we obtain

$$I_{po}(\mathbf{q}, k) = \frac{1}{k^2 q_\perp^2} \sum_{l=1}^{L} (C_l - C_{l-1})\, e^{-jk\mathbf{q}\cdot\mathbf{r}_l}. \tag{9.154}$$

Expressions for C_l in vector notation are given by (9.146), which can be conveniently rewritten as

$$C_l = -\frac{\mathbf{q}\cdot\hat{\nu}_l}{\mathbf{q}\cdot\hat{\tau}_l} \tag{9.155}$$

in terms of the unit vectors

$$\hat{\tau}_l = \frac{\mathbf{r}_{l+1} - \mathbf{r}_l}{|\mathbf{r}_{l+1} - \mathbf{r}_l|}, \quad \hat{\nu}_l = \hat{n} \times \hat{\tau}_l, \tag{9.156}$$

where $\hat{\tau}_l$ is the unit tangent to the lth side of the polygon and $\hat{\nu}_l$ is the unit normal to the side directed in the interior of the polygon. Here, $\mathbf{r}_0 = \mathbf{r}_L$ and $\mathbf{r}_{L+1} = \mathbf{r}_1$.

Combining these in the PO expression for the scattered field leads to the formula

$$\mathbf{E}_{sc}^{po} = \frac{e^{-jkr}}{r} \frac{j\mathbf{V}}{4\pi k q_\perp^2} \sum_{l=1}^{L} \left(\frac{\mathbf{q}\cdot\hat{\nu}_{l-1}}{\mathbf{q}\cdot\hat{\tau}_{l-1}} - \frac{\mathbf{q}\cdot\hat{\nu}_l}{\mathbf{q}\cdot\hat{\tau}_l} \right) e^{-jk\mathbf{q}\cdot\mathbf{r}_l}, \tag{9.157}$$

which can be interpreted as a sum of L spherical waves

$$\frac{e^{-jkr - jk\mathbf{q}\cdot\mathbf{r}_l}}{r} \approx \frac{e^{-jk|\mathbf{r}-\mathbf{r}_l| - jk\hat{i}\cdot\mathbf{r}_l}}{|\mathbf{r}-\mathbf{r}_l|} \tag{9.158}$$

from the tips of the L-gon, illuminated by the plane wave (9.88) and scattering with the far-field coefficients

$$\mathbf{F}_l = \frac{j\mathbf{V}}{4\pi q_\perp^2} \left(\frac{\mathbf{q}\cdot\hat{\nu}_{l-1}}{\mathbf{q}\cdot\hat{\tau}_{l-1}} - \frac{\mathbf{q}\cdot\hat{\nu}_l}{\mathbf{q}\cdot\hat{\tau}_l} \right) e^{-jk\hat{i}\cdot\mathbf{r}_l}. \tag{9.159}$$

The far-field coefficients depend only on the local geometry of the polygon at the vertex, so (9.157) is a sum of independent contributions from every vertex of the polygon. The dependence on the material properties of the polygon is described by the factor **V** and is the same for all vertices. The contribution of every individual vertex in (9.157) is in agreement with the asymptotic contribution (9.24) of a single vertex on an edge of a thin infinite sector derived by asymptotic evaluation of the PO solution for the sector.

9.3.4 *Far-field Patterns of Polygonal Plates and Apertures*

The PO solutions for plates derived in the previous section are easy to evaluate and we use them in this section to study the patterns of the scattered field around the forward direction (Figure 9.22). The patterns show interesting relation to the shape of the plate, and in this section we study these patterns for variously shaped plates, including triangle and right-angle polygons, a star, and a circular disc. The latter is considered as a limiting case of a polygon with an infinite number of vertices.

The plates are assumed to be PEC, which makes the results relevant to other physical situations. For example, the amplitude of the scattered field in the PO approximation is invariant with respect to interchange $s_n \to -s_n$ where $s_n = \hat{s}\cdot\hat{n}$ (problem 12), and the back- and forward-scattering patterns for the absolute values of the fields coincide. For flat PEC

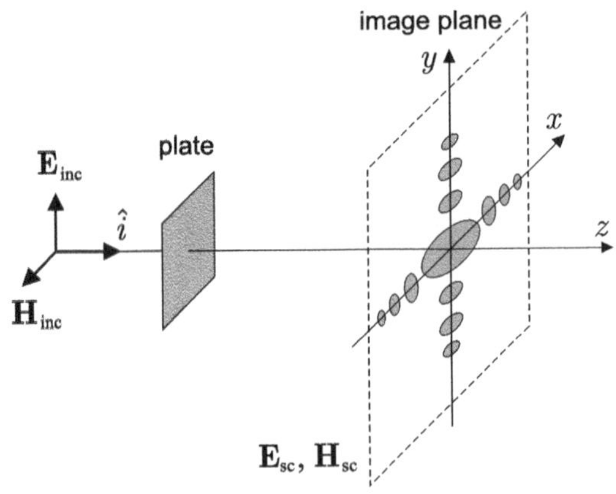

Figure 9.22 Imaging the field scattered in the forward direction. A plate is illuminated by a plane wave propagating along the z axis, and the distribution of the scattered field is plotted on the image plane.

objects, the PO approximation gives equal values for the back- and forward-scattered fields, which is in contrast to smooth objects, like spheres, for which the forward scattering is much greater then the backscattering (Figure 7.5).

Furthermore, the field scattered by a PEC plate to the half-space behind the plate is related by the Babinet principle to the field transmitted through an aperture in an infinite PEC flat screen if the shape of the aperture is the same as the shape of the plate (complementary screen; Figure 9.23). The mathematical formulation of the principle is that if $\mathbf{E}_{sc}$ and $\mathbf{H}_{sc}$ are the scattered fields in the half-space behind the plate due to the incident fields $\mathbf{E}_{inc}$ and $\mathbf{H}_{inc}$ coming from the half-space in front of the plate, then the fields

$$\tilde{\mathbf{E}}_d = -Z\mathbf{H}_{sc}, \quad \tilde{\mathbf{H}}_d = Y\mathbf{E}_{sc} \tag{9.160}$$

are solutions of the problem of transmission through the complementary screen for the incident field

$$\tilde{\mathbf{E}}_{inc} = Z\mathbf{H}_{inc}, \quad \tilde{\mathbf{H}}_{inc} = -Y\mathbf{E}_{inc}. \tag{9.161}$$

The fields $\tilde{\mathbf{E}}_d$ and $\tilde{\mathbf{H}}_d$ have the meaning of the total fields transmitted through the aperture.

If the incident field is a plane wave, then $Z\mathbf{H}_{inc} = \hat{i} \times \mathbf{E}_{inc}$, so that the incident field for the complimentary problem is a plane wave with the polarization plane rotated through a right angle clockwise looking in the direction of propagation.

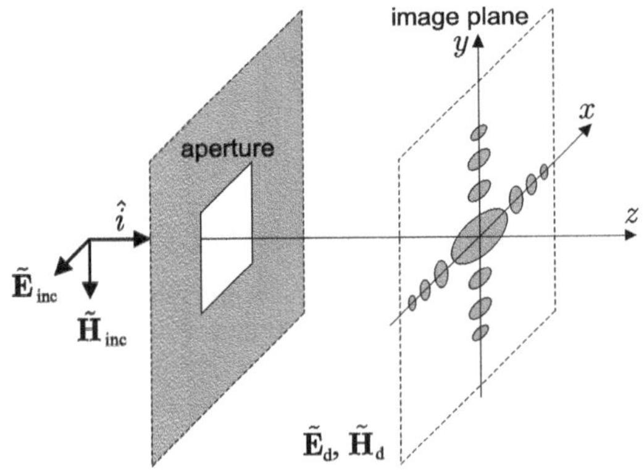

Figure 9.23 Complimentary configuration: an aperture in an infinite PEC flat screen, shaped as the plate in Figure 9.22, is illuminated by a plane wave with a polarization plane rotated around the propagation direction through 90°, and the diffracted field pattern is plotted on the image plane.

Furthermore, if the observation point is located in the far-field region of the plate, then the electric and magnetic fields in the scattered field are related as in a spherical wave by $Z\mathbf{H}_{sc} = \hat{s} \times \mathbf{E}_{sc}$. This implies that

$$\tilde{\mathbf{E}}_{inc} = \hat{i} \times \mathbf{E}_{inc}, \quad \tilde{\mathbf{E}}_{d} = -\hat{s} \times \mathbf{E}_{sc} \tag{9.162}$$

so that the knowledge of the electric field is sufficient to determine the electric field in the complementary problem.

Derivation of the electromagnetic form of Babinet's principle can be found, for example, in Born and Wolf (1959, section 11.3) and Jones (1964, section 9.3). Extension to transparent sheets described by the transition boundary conditions is addressed in Senior and Volakis (1995, section 6.3). No version of Babinet's principle applicable to impenetrable plates, for example described by impedance boundary conditions, is yet available, for fundamental reasons as suggested by Senior (1977).

Figures 9.24–9.27 show the scattering patterns of various PEC polygons. The patterns are two-dimensional distributions of $|\mathbf{E}_{sc}|$ over the area -300 m $\leq x, y \leq 300$ m on an image plane located at a distance 1000 m behind the polygons. The illumination is a plane wave with the incidence direction $\hat{i} = \hat{z}$ perpendicular to the surface of the plates ($\hat{i} = -\hat{n}$) and with the wavelength equal to 0.03 m. All polygons are inscribed in a circle with the diameter $D = 30\lambda$, so the image plane is located well beyond the far-field boundary, which begins at a distance $2D^2/\lambda = 54$ m from the scatterers.

The general features of the presented scattering patterns can be summarized as follows. The pattern of every L-gon consists of $2L$ radial traces with interference patterns along them. The traces lie in the planes perpendicular to the edges of the polygon, which are the directions at which $\hat{s}$ is perpendicular to the respective edge and the respective sinc function in the solution (9.147) is at maximum. Thus, every edge is responsible for two radial traces on the image plane, which

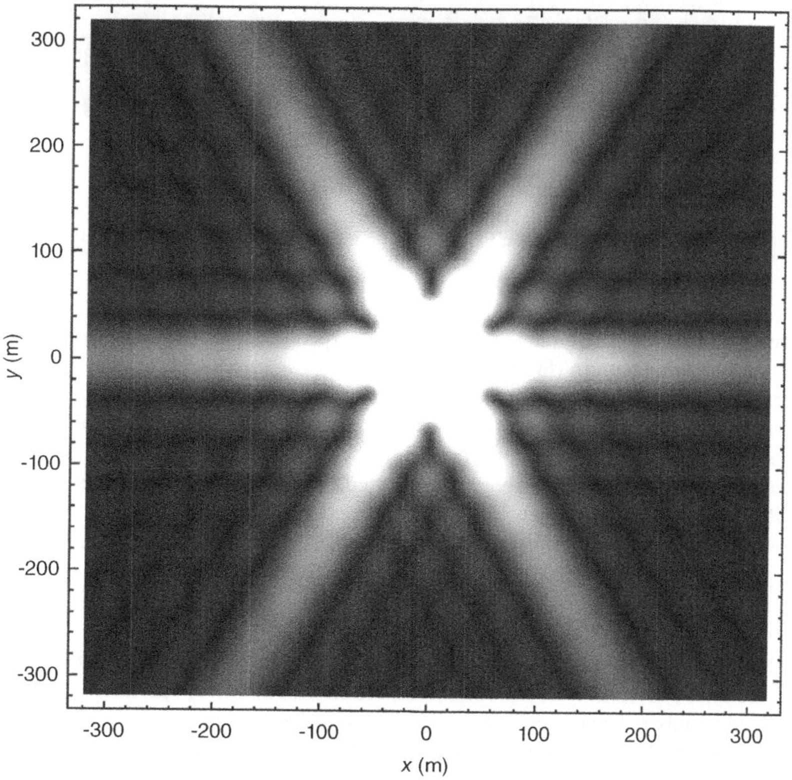

Figure 9.24 Amplitude distribution of the field scattered by an equilateral triangle in the forward direction. The triangle is inscribed in a circle with diameter 30λ. A plane wave with $\lambda = 0.03$ m is incident perpendicularly to the plate. The image screen is located 1000 m behind the plate. Darker regions correspond to lower field levels. In a region around the maximum the plotted values are cut to make faint details visible.

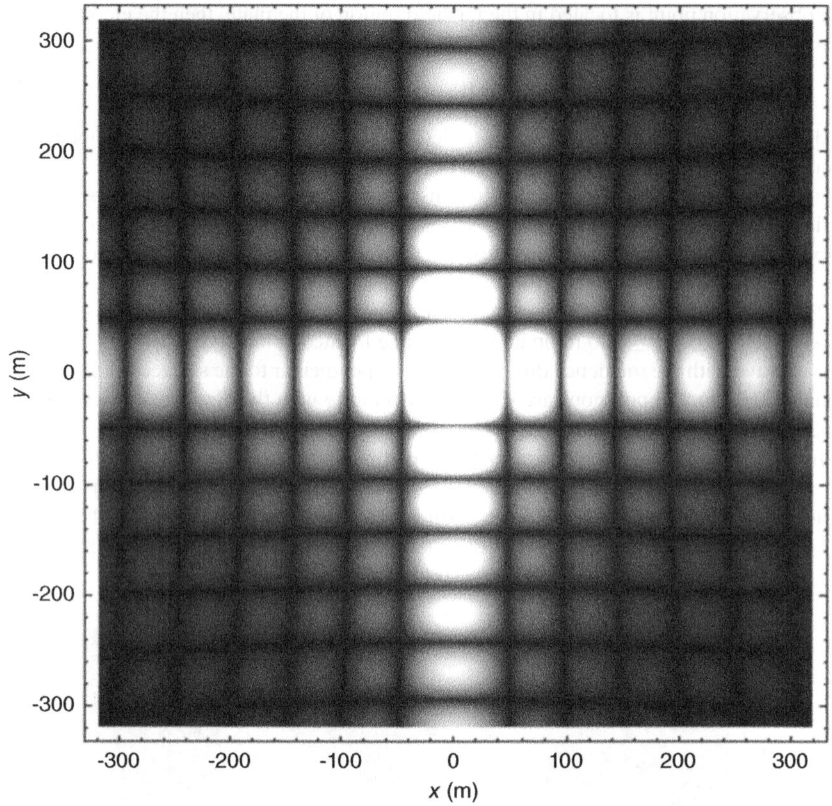

Figure 9.25 The same as Figure 9.24 but for a square PEC plate. The length of the diagonal of the square is equal to 30λ, so the side length is 0.636 m.

are located at the intersection of the image plane and the plane perpendicular to the edge. For example, a triangle leads to six radial traces in the scattering pattern (Figure 9.24).

If a polygon has parallel sides, however, then the traces of the parallel sides are overlapped and the total number of traces observed in the scattering pattern becomes smaller. For example, a general quadrilateral leads to eight radial traces in the scattering pattern but a parallelogram only to four (Figures 9.25 and 9.26).

In general, the more non-parallel sides, the more complicated the pattern. This feature is illustrated in Figure 9.27, which presents the PO forward-scattering pattern of a star-shaped scatterer with 11 arms. The Cartesian coordinates (x_m, y_m) of its vertices are described by the equations: $x_m = \rho_m \cos\varphi_m$, $y_m = \rho_m \sin\varphi_m$,

$$\rho_m = \frac{d}{2}\frac{1-(-1)^m\delta}{1+\delta}, \quad \varphi_m = \frac{\pi}{2M}(2m-1), \tag{9.163}$$

where $m = 1, 2, \ldots, 2M$, $M = 11$ is the number of the arms, and d is the diameter of the enclosing circle. The points with $m = 1, 3, \ldots, 2M-1$ are the outer vertices of the star. Measured field patterns for apertures of this and other shapes can be found in Borovikov and Kinber (1994, section 6.5)). An exception is a polygon inscribed in a circle with the number of vertices going to infinity. This polygon approaches a circle, the radial traces of the sides coalesce, and the scattering pattern becomes rotationally symmetric in the limit (Figure 9.28). Besides the radial traces, there are numerous weaker maxima and minima distributed over the whole image plane off the dominant radial traces, which are seen in Figures 9.24 to 9.26.

The patterns are functions of the electrical size of the polygons and are therefore not limited to the studied case of $\lambda = 0.03$ m. They look the same for other wavelengths if all involved linear dimensions are scaled together with the

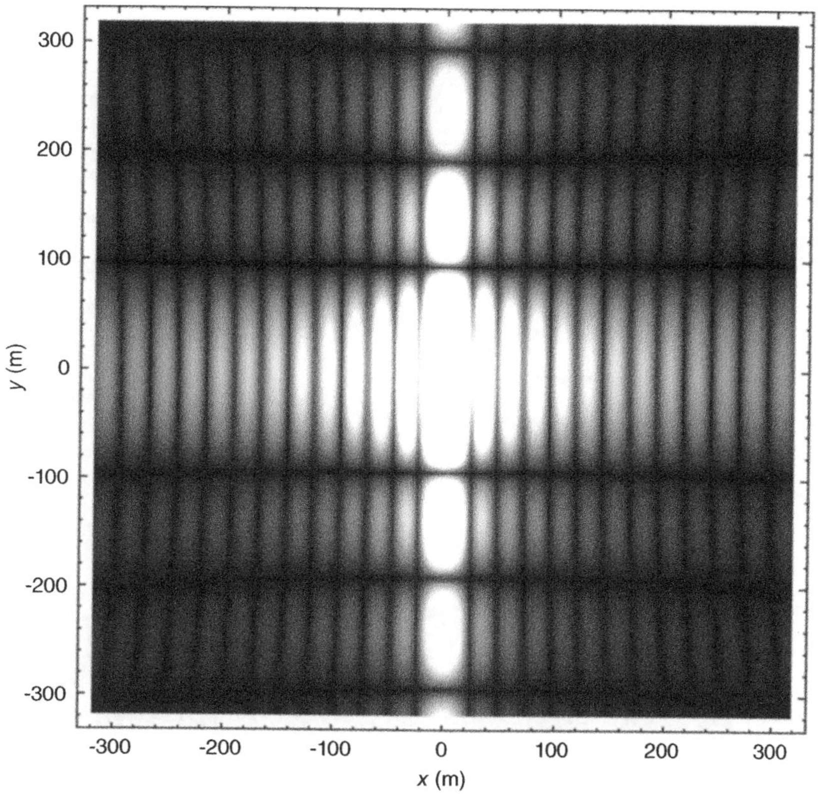

Figure 9.26 The same as Figure 9.25 but for a rectangular plate with the sides twice as large in the Ox direction and twice as small in the Oy direction. The positions of the minima on the image screen, denser along the x axis and sparser along the y direction, inversely correspond to the changes in the side lengths (see problem 11).

wavelength. Furthermore, they apply to the backscattering patterns and to the patterns of the fields transmitted through apertures in the complementary screens. For example, Figure 9.25 may well present the pattern of the field at an optical frequency transmitted through a square aperture with the 30λ diagonal and plotted over correspondingly reduced area on the image screen, which is located correspondingly closer to the aperture. For example, the red light with $\lambda = 0.6\ \mu$m would imply the scaling factor equal to 2×10^{-5}, and to get the same pattern as in Figure 9.25, the image plane should be located at a distance 20 mm behind the aperture and the size of the imaging area would be 12 mm by 12 mm.

The scattering patterns are more complicated when the incidence direction is not perpendicular to the surface of the polygon and therefore to all its edges. In this case, the sinc functions in (9.147) are at maximum when the scattering direction lies on a conical surface around the edge with the angle equal to that between the incidence direction and the edge (similar to Keller's cone in the diffraction by the edge of the wedge).

9.4 Bodies of Revolution

We saw that for simply shaped scattering bodies it is often possible to reduce the PO solution for the scattered field to a line integral or even evaluate it in a closed form. Such solutions are useful since they explicitly show the dependence of the scattered field on the problem parameters. Furthermore, they apply when the stationary phase method fails, for example when no separate critical points exist on the scattering surface. This is a typical situation with bodies of revolution when they are illuminated along their symmetry axis.

In this section we study the PO solution for the far field scattered by bodies of revolution which are assumed to be opaque with the surface described by reflection coefficients. We show that integrating with respect to the azimuth reduces

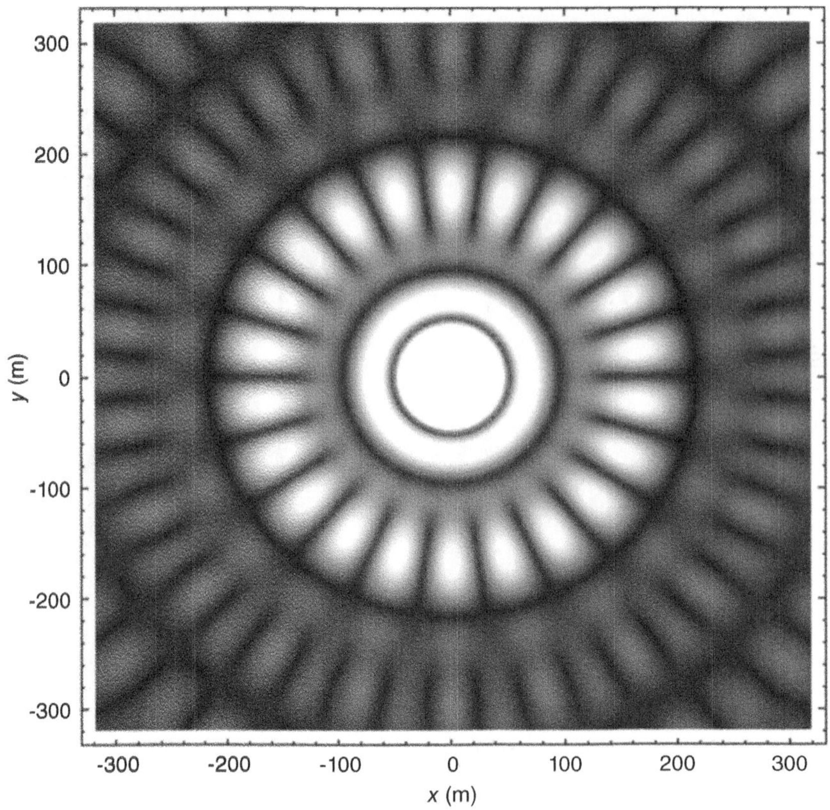

Figure 9.27 The same as Figure 9.24 but for an 11-pointed star with $\delta = 0.235$ inscribed in a circle of diameter 30λ. The meaning of δ is explained in (9.163).

the PO solution to a line integral along the shadow boundary or a circumferential edge on the surface of the body when the body is illuminated by a plane wave incident along its symmetry axis. The restriction to the axial incidence is not necessary for PEC bodies of revolution with circumferential edges, for which the reduction to a line integral is possible for general incidence and scattering directions. Finally, when both incidence and scattering directions are along the symmetry axis (axial backscattering), the PO solution for the backscattered field simplifies even further, and integration by parts allows a physical insight in the solution. The section concludes with examples of simple bodies of practical interest, namely, a sphere, an ogive, a truncated circular cone, and a truncated paraboloid of revolution.

Since the shadowed portion of the body does not contribute to the PO solution, it is only essential that the illuminated portion of the body be rotationally symmetric. For imperfectly reflecting bodies the symmetry axis must coincide with the direction of incidence in order for the results of this section to apply.

Many results on high-frequency electromagnetic scattering from PEC bodies of revolution can be found in Crispin and Siegel (1968, Chapter 4) and Ufimtsev (2003, Chapter 2). The focus of this section is therefore on imperfectly reflecting bodies.

9.4.1 PO Solution for Bodies of Revolution

We begin by specializing the PO solution (8.14) and (8.15) for the far scattered field to the case of a body of revolution. The body is assumed to be impenetrable and described by the reflection coefficients $R_{\parallel}$ and $R_{\perp}$. It is furthermore assumed

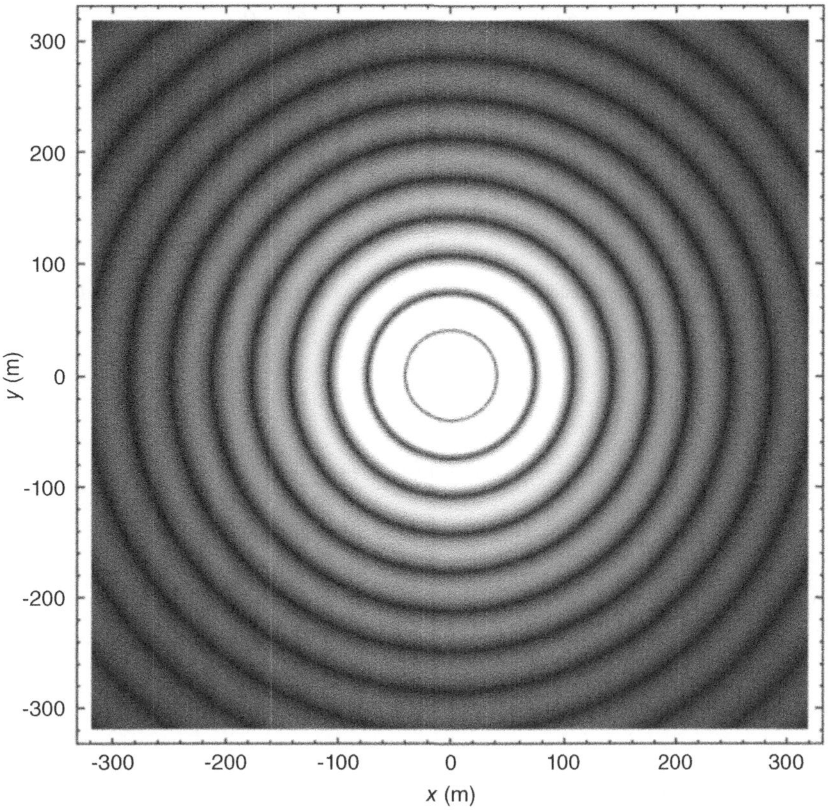

Figure 9.28 Forward-scattering pattern of a PEC disc with diameter 30λ.

that the excitation is a plane wave

$$\mathbf{E}_{\text{inc}}(\mathbf{r}) = \mathbf{E}_0\, e^{-jk\hat{i}\cdot\mathbf{r}} \tag{9.164}$$

and the illuminated portion of the body can be convex or concave. In the latter case, however, it must be assumed that no multiple reflections take place and the fields on the scattering surface can be approximated by the incident and reflected plane waves in the tangent plane approximation.

We assume that the z axis is the symmetry axis and parametrize the illuminated portion of the surface by the equation

$$\mathbf{r} = \rho(z)\cos\phi\hat{x} + \rho(z)\sin\phi\hat{y} + z\hat{z}, \quad z_{\text{min}} \le z \le z_{\text{max}}, \tag{9.165}$$

where ϕ is the azimuth coordinate ($0 \le \phi < 2\pi$) and $\rho(z)$ describes the radius of the surface measured from the symmetry axis so that $\rho \ge 0$ with $\rho = 0$ at $z = z_{\text{min}}$ for concave surfaces and at $z = z_{\text{max}}$ for convex surfaces (see Figure 9.29).

The unit normal vector is determined by

$$\hat{n} = \pm\frac{\mathbf{r}_\phi \times \mathbf{r}_z}{|\mathbf{r}_\phi \times \mathbf{r}_z|} \tag{9.166}$$

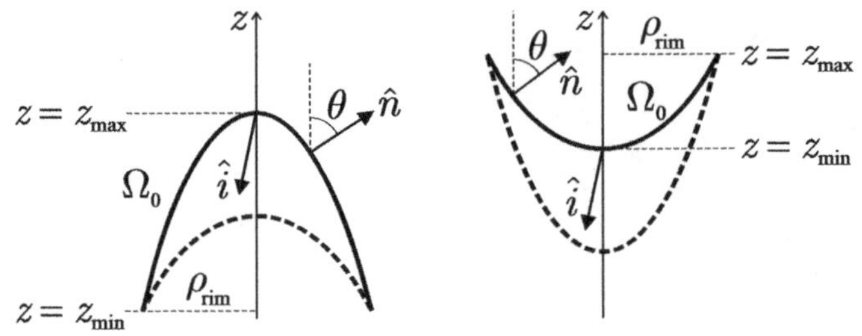

Figure 9.29 Body of revolution with an illuminated convex (left) and concave (right) side. The shadowed portion of the surface of the body (shown by dashing) can be arbitrary.

with

$$\mathbf{r}_\phi = \frac{\partial \mathbf{r}}{\partial \phi} = -\rho(z)\sin\phi\hat{x} + \rho(z)\cos\phi\hat{y} \tag{9.167}$$

$$\mathbf{r}_z = \frac{\partial \mathbf{r}}{\partial z} = \rho'(z)\cos\phi\hat{x} + \rho'(z)\sin\phi\hat{y} + \hat{z}, \tag{9.168}$$

which gives

$$\mathbf{r}_\phi \times \mathbf{r}_z = \rho(z)\cos\phi\hat{x} + \rho(z)\sin\phi\hat{y} - \rho(z)\rho'(z)\hat{z} \tag{9.169}$$

$$|\mathbf{r}_\phi \times \mathbf{r}_z| = \rho(z)\sqrt{1+\rho'(z)^2}, \tag{9.170}$$

where the prime denotes the derivative with respect to z, and it follows from (9.166) that

$$\hat{n} = \pm\frac{\cos\phi\hat{x} + \sin\phi\hat{y} - \rho'(z)\hat{z}}{\sqrt{1+\rho'(z)^2}}. \tag{9.171}$$

In order to have a vector pointing to the exterior of the body and since $\rho'(z) \le 0$ on the convex side and $\rho'(z) \ge 0$ on the concave side, we choose the upper sign in (9.171) if the convex side is illuminated and the lower sign if the concave side is illuminated. The angle θ between the normal to the surface and the z axis can be calculated from the relation

$$\rho'(z) = \mp\cot\theta \tag{9.172}$$

with the upper (lower) sign for the convex (concave) surface.

The PO solution (8.14) and (8.15) for the far field has the form

$$\mathbf{F}_{\mathrm{po}}(\hat{s},\hat{i}) = \frac{jk^2}{4\pi}\int_{\Omega_0} \mathrm{e}^{-jk\mathbf{q}\cdot\mathbf{r}}(\hat{s}\times\hat{s}\times\mathbf{v}_{\mathrm{e}} + \hat{s}\times\mathbf{v}_{\mathrm{m}})\,\mathrm{d}\Omega, \tag{9.173}$$

where Ω_0 denotes the illuminated side of the body and $\mathbf{v}_{\mathrm{e}}$ and $\mathbf{v}_{\mathrm{m}}$ are defined by (8.19) and (8.20). The vector product is distributive over addition or integration, and (9.173) can be written as

$$\mathbf{F}_{\mathrm{po}}(\hat{s},\hat{i}) = \frac{jk^2}{4\pi}\hat{s}\times(\hat{s}\times\mathbf{V}_{\mathrm{e}} + \mathbf{V}_{\mathrm{m}}) \tag{9.174}$$

with

$$\mathbf{V}_{\mathrm{e,m}} = \int_{\Omega_0} \mathrm{e}^{-jk\mathbf{q}\cdot\mathbf{r}} \mathbf{v}_{\mathrm{e,m}}(\mathbf{r})\, \mathrm{d}\Omega. \tag{9.175}$$

Using (8.19) and (8.20), we obtain

$$\mathbf{V}_{\mathrm{e}} = \mathbf{V}_{\mathrm{e}}^{(1)} + \mathbf{V}_{\mathrm{e}}^{(2)}, \quad \mathbf{V}_{\mathrm{m}} = \mathbf{V}_{\mathrm{m}}^{(1)} + \mathbf{V}_{\mathrm{m}}^{(2)} \tag{9.176}$$

with

$$\mathbf{V}_{\mathrm{e}}^{(1)} = \int_{\Omega_0} \mathrm{e}^{-jk\mathbf{q}\cdot\mathbf{r}} (R_\perp - 1)(\hat{\imath}\cdot\hat{n})(\mathbf{E}_0\cdot\hat{\alpha}_{\mathrm{inc}})\hat{\alpha}_{\mathrm{inc}}\, \mathrm{d}\Omega \tag{9.177}$$

$$\mathbf{V}_{\mathrm{e}}^{(2)} = -\int_{\Omega_0} \mathrm{e}^{-jk\mathbf{q}\cdot\mathbf{r}} (R_\parallel + 1)(\mathbf{E}_0\cdot\hat{\beta}_{\mathrm{inc}})(\hat{n}\times\hat{\alpha}_{\mathrm{inc}})\, \mathrm{d}\Omega \tag{9.178}$$

$$\mathbf{V}_{\mathrm{m}}^{(1)} = \int_{\Omega_0} \mathrm{e}^{-jk\mathbf{q}\cdot\mathbf{r}} (1 - R_\parallel)(\hat{\imath}\cdot\hat{n})(\mathbf{E}_0\cdot\hat{\beta}_{\mathrm{inc}})\hat{\alpha}_{\mathrm{inc}}\, \mathrm{d}\Omega \tag{9.179}$$

$$\mathbf{V}_{\mathrm{m}}^{(2)} = -\int_{\Omega_0} \mathrm{e}^{-jk\mathbf{q}\cdot\mathbf{r}} (R_\perp + 1)(\mathbf{E}_0\cdot\hat{\alpha}_{\mathrm{inc}})(\hat{n}\times\hat{\alpha}_{\mathrm{inc}})\, \mathrm{d}\Omega. \tag{9.180}$$

Integrals in (9.177)–(9.180) are surface integrals and parametrization (9.165) transforms them into usual two-dimensional integrals over the axial coordinate z and the azimuthal coordinate ϕ. In what follows we study the cases in which integration with respect to ϕ can be performed analytically.

9.4.2 *Imperfectly Reflecting Bodies under Axial Illumination*

An imperfectly reflecting body is an impenetrable body with a surface different from PEC or PMC. At every point of the scattering surface the reflecting properties of the body are described by reflection coefficients $R_\parallel$ and $R_\perp$ of a locally conformal tangent plane, which are functions of the local material properties and the incidence angle (the angle between the incidence direction and the normal to the surface) and therefore position dependent. This dependence complicates exact analytical integration in (9.177)–(9.180). However, in the case of illumination along the symmetry axis of bodies of revolution ($\hat{\imath} = -\hat{z}$), the incidence angle is independent of the azimuth coordinate ϕ (although it may still depend on the axial coordinate z), which greatly simplifies the calculation and permits an explicit integration with respect to ϕ.

In this section we study this case. The calculations will be performed for convex bodies, that is, with the upper sign in (9.171) for the vector $\hat{n}$; an extension to the concave case can be achieved by simple substitutions, provided that no multiple reflection takes place. The amplitude of the incident plane wave (9.164), which is incident along the z axis, is written as

$$\mathbf{E}_0 = \hat{x}E_{0x} + \hat{y}E_{0y}, \tag{9.181}$$

which corresponds to an elliptically polarized wave (including linear or circular polarizations as special cases).

Using the parametrization (9.165) and specializing to the case $\hat{\imath} = -\hat{z}$ give the formulas

$$\mathrm{d}\Omega = \rho(z)\sqrt{1+\rho'(z)^2}\, \mathrm{d}\phi\, \mathrm{d}z \tag{9.182}$$

$$\hat{\imath}\cdot\hat{n} = \frac{\rho'(z)}{\sqrt{1+\rho'(z)^2}} \tag{9.183}$$

$$\hat{\alpha}_{\mathrm{inc}} = \hat{x}\sin\phi - \hat{y}\cos\phi \tag{9.184}$$

$$\hat{\beta}_{\rm inc} = -\hat{x}\cos\phi - \hat{y}\sin\phi \tag{9.185}$$

$$\hat{n}\times\hat{\alpha}_{\rm inc} = -\frac{\rho'(z)\cos\phi\hat{x} + \rho'(z)\sin\phi\hat{y} + \hat{z}}{\sqrt{1+\rho'(z)^2}} \tag{9.186}$$

$$e^{-jk\mathbf{q}\cdot\mathbf{r}} = e^{-jk\rho(z)(q_x\cos\phi + q_y\sin\phi) - jkq_z z}, \tag{9.187}$$

where $\mathbf{q} = q_x\hat{x} + q_y\hat{y} + q_z\hat{z}$. Inserting these in (9.177)–(9.180) we obtain the following expressions:

$$\mathbf{V}_{\rm e}^{(1)} = \int_{z_{\rm min}}^{z_{\rm max}} e^{-jkq_z z}(R_\perp - 1)\rho(z)\rho'(z)\{\hat{x}[E_{0x}I_{02} - E_{0y}I_{11}] - \hat{y}[E_{0x}I_{11} - E_{0y}I_{20}]\}\,dz \tag{9.188}$$

$$\mathbf{V}_{\rm e}^{(2)} = -\int_{z_{\rm min}}^{z_{\rm max}} e^{-jkq_z z}(R_\parallel + 1)\rho(z)\{\hat{x}\rho'(z)[E_{0x}I_{20} + E_{0y}I_{11}] + \hat{y}\rho'(z)[E_{0x}I_{11} + E_{0y}I_{02}] + \hat{z}[E_{0x}I_{10} + E_{0y}I_{01}]\}\,dz \tag{9.189}$$

$$\mathbf{V}_{\rm m}^{(1)} = \int_{z_{\rm min}}^{z_{\rm max}} e^{-jkq_z z}(R_\parallel - 1)\rho(z)\rho'(z)\{\hat{x}[E_{0x}I_{11} + E_{0y}I_{02}] - \hat{y}[E_{0x}I_{20} + E_{0y}I_{11}]\}\,dz \tag{9.190}$$

$$\mathbf{V}_{\rm m}^{(2)} = \int_{z_{\rm min}}^{z_{\rm max}} e^{-jkq_z z}(R_\perp + 1)\rho(z)\{\hat{x}\rho'(z)[E_{0x}I_{11} - E_{0y}I_{20}] + \hat{y}\rho'(z)[E_{0x}I_{02} - E_{0y}I_{11}] + \hat{z}[E_{0x}I_{01} - E_{0y}I_{10}]\}\,dz, \tag{9.191}$$

where

$$I_{mn} = \int_0^{2\pi} e^{-jk\rho(z)(q_x\cos\phi + q_y\sin\phi)}\cos^m\phi\sin^n\phi\,d\phi \tag{9.192}$$

with $m, n = 0, 1, 2$.

The integrals in (9.192) can be reduced to the integral representation for the Bessel function (Abramowitz and Stegun (1965, section 9.1.21))

$$J_m(p) = \frac{j^m}{\pi}\int_0^{\pi} e^{-jp\cos t}\cos(mt)\,dt \tag{9.193}$$

with $m = 0, 1, 2$. To this end, it is convenient to express $\mathbf{q}$ in spherical coordinates as

$$q_x = |\mathbf{q}|\sin\theta_q\cos\phi_q, \quad q_y = |\mathbf{q}|\sin\theta_q\sin\phi_q, \quad q_z = |\mathbf{q}|\cos\theta_q \tag{9.194}$$

and rewrite (9.192) as

$$I_{mn} = \int_0^{2\pi} e^{-jp\cos(\phi-\phi_q)}\cos^m\phi\sin^n\phi\,d\phi \tag{9.195}$$

with

$$p = k\rho(z)|\mathbf{q}|\sin\theta_q. \tag{9.196}$$

It can be shown that

$$I_{02} = \pi J_0(p) + \pi \cos(2\phi_q)J_2(p) \tag{9.197}$$

$$I_{11} = -\pi \sin(2\phi_q)J_2(p) \tag{9.198}$$

$$I_{20} = \pi J_0(p) - \pi \cos(2\phi_q)J_2(p) \tag{9.199}$$

$$I_{01} = -2\pi j \sin\phi_q J_1(p) \tag{9.200}$$

$$I_{10} = -2\pi j \cos\phi_q J_1(p). \tag{9.201}$$

Consider, for example, the integral I_{10}. Using a new integration variable $\phi' = \phi - \phi_q$ and noting that the integrand is a 2π periodic function we can write

$$I_{10} = \int_{-\phi_q}^{2\pi-\phi_q} \mathrm{e}^{-jp\cos\phi'} \cos(\phi' + \phi_q)\, \mathrm{d}\phi' = \int_{-\pi}^{\pi} \mathrm{e}^{-jp\cos\phi'} \cos(\phi' + \phi_q)\, \mathrm{d}\phi'. \tag{9.202}$$

Representation $\cos(\phi' + \phi_q) = \cos\phi' \cos\phi_q - \sin\phi' \sin\phi_q$ expands the integrand in even and odd functions of ϕ', and since the integration range in (9.202) is symmetric, the contribution of the odd part vanishes and we obtain

$$I_{10} = 2\cos\phi_q \int_0^{\pi} \mathrm{e}^{-jp\cos\phi'} \cos\phi'\, \mathrm{d}\phi' = -2\pi j \cos\phi_q J_1(p). \tag{9.203}$$

The other equations (9.197)–(9.200) are derived in a similar manner (see problem 16).

Thus, equations (9.188)–(9.191) together with (9.197)–(9.201) and (9.174) describe the far field scattered by convex imperfectly reflecting bodies of revolution under axial illumination in terms of one-dimensional integrals with respect to the axial coordinate z. Extension to the case of a concave scattering surface is obtained by simply multiplying the right-hand sides of (9.188)–(9.191) by -1. This follows from the fact that for a concave surface an alternative (lower) sign in the formula (9.171) for the unit normal $\hat{n}$ should be taken, which implies substitutions

$$\hat{\alpha}_{\text{inc}} \to -\hat{\alpha}_{\text{inc}}, \quad \hat{\beta}_{\text{inc}} \to -\hat{\beta}_{\text{inc}}, \quad \hat{\imath} \cdot \hat{n} \to -\hat{\imath} \cdot \hat{n}, \quad \hat{n} \times \hat{\alpha}_{\text{inc}} \to \hat{n} \times \hat{\alpha}_{\text{inc}}, \tag{9.204}$$

alternating the sign in the right-hand sides of (9.177)–(9.180). Care should be taken to make sure that multiple reflections, which may take place at concave surfaces, are not present in order for the PO assumption to apply.

Expressions (9.188)–(9.191) simplify if the incident wave is linearly polarized. Then choosing the polarization direction to be parallel, say, to the x axis makes $E_{0y} = 0$ in the equations. Another limit in which (9.188)–(9.191) significantly simplify is the case of backscattering $\hat{s} = \hat{z}$. In this important special case the scattered field can be expressed in a particularly simple form which permits asymptotic or even exact evaluation, providing physical insight into the solution. This case is studied in section 9.4.4, but prior to that we will consider the special case of PEC scatterers, where the far scattered field can be expressed in a simple and elegant form, not limited to the case of axial illumination.

9.4.3 PEC Bodies under Oblique Illumination

For PEC bodies, reflection coefficients are independent of the incidence direction ($R_\perp = -1$ and $R_\parallel = 1$) and integration with respect to ϕ in (9.175) can be performed for incidence directions not necessarily parallel to the symmetry axis (the only restriction is that the shape of the illuminated portion of the body is independent of $\hat{i}$). With (8.21), the far-field coefficient (8.15) becomes

$$\mathbf{F}_{\mathrm{po}}(\hat{s}, \hat{i}) = \frac{jk^2}{2\pi} \int_{\Omega_0} \mathrm{e}^{-jk\mathbf{q}\cdot\mathbf{r}} \hat{s} \times \hat{s} \times \hat{n} \times \hat{i} \times \mathbf{E}_0 \, \mathrm{d}\Omega \tag{9.205}$$

and since only $\hat{n}$ depends on the position, (9.205) can be expressed as

$$\mathbf{F}_{\mathrm{po}}(\hat{s}, \hat{i}) = \frac{jk^2}{2\pi} \hat{s} \times \hat{s} \times \mathbf{I}_n \times \hat{i} \times \mathbf{E}_0, \tag{9.206}$$

where

$$\mathbf{I}_n = \int_{\Omega_0} \mathrm{e}^{-jk\mathbf{q}\cdot\mathbf{r}} \hat{n} \, \mathrm{d}\Omega. \tag{9.207}$$

On parametrization of the integral and integration with respect to ϕ,

$$\mathbf{I}_n = \int_{z_{\min}}^{z_{\max}} \mathrm{e}^{-jkq_z z} \rho(z)[\hat{x} I_{10} + \hat{y} I_{01} - \hat{z}\rho'(z) I_{00}] \, \mathrm{d}z \tag{9.208}$$

with I_{01} and I_{10} given by (9.200) and (9.201). The integral I_{00} is calculated similarly,

$$I_{00} = 2\pi J_0(p), \tag{9.209}$$

where p is given by (9.196). The sign of the right-hand side of (9.208) alternates if the illuminated surface is concave, which follows from the expression (9.171) for the unit normal.

The solution is applicable to arbitrary incidence directions as long as the illuminated portion of the body is independent of the incidence direction, which requires a circumferential edge as the shadow boundary. The formulas (9.206) and (9.208) agree with (9.174) and (9.188)–(9.191) under the assumptions that $\hat{i} = -\hat{z}$ and the scattering body is PEC.

9.4.4 Axial Backscattering

The PO solution for the far scattered field is particularly simple in the case of axial backscattering, that is, when the incidence and scattering directions are parallel to the symmetry axis of the body of revolution. In this case, $\hat{z} = \hat{s} = -\hat{i}$, $\mathbf{q} = -2\hat{z}$, $\sin\theta_q = 0$, $p = 0$,

$$I_{10} = I_{01} = I_{11} = 0, \quad I_{20} = I_{02} = \pi, \tag{9.210}$$

and equations (9.188)–(9.191) reduce to

$$\mathbf{V}_{\mathrm{e}}^{(1)} = \pi \mathbf{E}_0 \int_{z_{\min}}^{z_{\max}} \mathrm{e}^{2jkz}(R_\perp - 1)\rho(z)\rho'(z) \, \mathrm{d}z \tag{9.211}$$

$$\mathbf{V}_{\mathrm{e}}^{(2)} = -\pi \mathbf{E}_0 \int_{z_{\min}}^{z_{\max}} \mathrm{e}^{2jkz}(R_\parallel + 1)\rho(z)\rho'(z) \, \mathrm{d}z \tag{9.212}$$

$$\mathbf{V}_{\mathrm{m}}^{(1)} = \pi(\hat{x}E_{0y} - \hat{y}E_{0x}) \int_{z_{\min}}^{z_{\max}} \mathrm{e}^{2jkz}(R_{\parallel} - 1)\rho(z)\rho'(z)\,\mathrm{d}z \tag{9.213}$$

$$\mathbf{V}_{\mathrm{m}}^{(2)} = -\pi(\hat{x}E_{0y} - \hat{y}E_{0x}) \int_{z_{\min}}^{z_{\max}} \mathrm{e}^{2jkz}(R_{\perp} + 1)\rho(z)\rho'(z)\,\mathrm{d}z, \tag{9.214}$$

implying that

$$\mathbf{V}_{\mathrm{e}} = \pi\mathbf{E}_0 \int_{z_{\min}}^{z_{\max}} \mathrm{e}^{2jkz}(R_{\perp} - R_{\parallel} - 2)\rho(z)\rho'(z)\,\mathrm{d}z \tag{9.215}$$

$$\mathbf{V}_{\mathrm{m}} = \pi(\hat{x}E_{0y} - \hat{y}E_{0x}) \int_{z_{\min}}^{z_{\max}} \mathrm{e}^{2jkz}(R_{\parallel} - R_{\perp} - 2)\rho(z)\rho'(z)\,\mathrm{d}z. \tag{9.216}$$

By using

$$\hat{s} \times \mathbf{E}_0 = -\hat{x}E_{0y} + \hat{y}E_{0x}, \tag{9.217}$$

we obtain

$$\hat{s} \times \mathbf{V}_{\mathrm{e}} + \mathbf{V}_{\mathrm{m}} = 2\pi(\hat{x}E_{0y} - \hat{y}E_{0x}) \int_{z_{\min}}^{z_{\max}} \mathrm{e}^{2jkz}(R_{\parallel} - R_{\perp})\rho(z)\rho'(z)\,\mathrm{d}z, \tag{9.218}$$

and multiplying again with $\hat{s}$ gives for the far-field coefficient (9.174)

$$\mathbf{F}_{\mathrm{po}}^{\mathrm{bs}} = \frac{jk^2}{2}\mathbf{E}_0 \int_{z_{\min}}^{z_{\max}} \mathrm{e}^{2jkz}(R_{\parallel} - R_{\perp})\rho(z)\rho'(z)\,\mathrm{d}z, \tag{9.219}$$

where $\mathbf{F}_{\mathrm{po}}^{\mathrm{bs}} = \mathbf{F}_{\mathrm{po}}(\hat{z}, -\hat{z})$. For concave surfaces, the right-hand side of (9.219) should be multiplied by -1.

For impedance bodies with

$$R_{\parallel} = \frac{\cos\theta - \eta}{\cos\theta + \eta}, \quad R_{\perp} = \frac{\eta\cos\theta - 1}{\eta\cos\theta + 1}, \tag{9.220}$$

where θ is the incidence angle and η is the normalized surface impedance (see (4.334) and (4.335)), substituting η with $1/\eta$ is equivalent to multiplication of the right-hand side of (9.219) by -1. This implies, in particular, that the axial backscattering cross section of impedance bodies of revolution are invariant under the inversion of the relative surface impedance (see problem 11 in section 3.7).

For generally shaped bodies and for general dependencies of reflection coefficients on the incidence angle the integral in (9.219) can be evaluated numerically. Furthermore, for sufficiently large bodies an asymptotic evaluation of (9.219) is possible, which results in simple but general formulas for the backscattered field, revealing the physical mechanisms that contribute to the axial backscattering from bodies of revolution.

We start the analysis by noting that the axial backscattering vanishes if $R_{\parallel} = R_{\perp}$ on the whole illuminated portion of the scattering surface, which is the case for impedance bodies with the surface impedance equal to the wave impedance of surrounding medium, the result compliant with the Weston theorem (section 3.2.3).

Furthermore it can be seen that (9.219) agrees with (9.106) in the limit of a surface approaching the circular disc. Indeed, with $|z_{\rm max} - z_{\rm min}| \to 0$, on the whole scattering surface $e^{2jkz} \to 1$, $\hat{n} \to -\hat{\imath}$, and

$$\mathbf{F}^{\rm bs}_{\rm po} \to \frac{jk^2}{2}\mathbf{E}_0[R_{\|}(0) - R_{\perp}(0)] \int\limits_{z_{\rm min}}^{z_{\rm max}} \rho(z)\rho'(z)\,{\rm d}z, \tag{9.221}$$

where $R_{\|}(0)$ and $R_{\perp}(0)$ are the reflection coefficients at normal incidence such that $R_{\|}(0) = -R_{\perp}(0)$. Since $\rho(z_{\rm max}) = 0$ and $\rho(z_{\rm min}) = \rho_{\rm max}$,

$$\int\limits_{z_{\rm min}}^{z_{\rm max}} \rho(z)\rho'(z)\,{\rm d}z = \int\limits_{\rho_{\rm max}}^{0} \rho\,{\rm d}\rho = -\frac{1}{2}\rho^2_{\rm max} \tag{9.222}$$

and we obtain for the axial backscattering coefficient

$$\mathbf{F}^{\rm bs}_{\rm po} = \frac{k^2\rho^2_{\rm max}}{2j}\mathbf{E}_0 R_{\|}(0). \tag{9.223}$$

This result agrees with (9.106) for $|\Omega_0| = \pi\rho^2_{\rm max}$, which is the area of the circular disc of radius $\rho_{\rm max}$.

Let us evaluate the integral in (9.219) asymptotically assuming that k is a large parameter. The integrand function consists of a slowly varying factor

$$A(z) = (R_{\|} - R_{\perp})\rho(z)\rho'(z) \tag{9.224}$$

and a rapidly varying exponent, so that we can apply the method described in Appendix D. The phase of the exponent function is a linear function of z and has, therefore, no critical points. Integrating by parts twice as in (D.4) and (D.9) we obtain the asymptotic approximation

$$\mathbf{F}^{\rm bs}_{\rm po} = \mathbf{E}_0 \left\{ \frac{k}{4}\left[A(z_{\rm max})\,e^{2jkz_{\rm max}} - A(z_{\rm min})\,e^{2jkz_{\rm min}}\right] + \frac{j}{8}\left[A'(z_{\rm max})\,e^{2jkz_{\rm max}} - A'(z_{\rm min})\,e^{2jkz_{\rm min}}\right] + o(1)\right\}. \tag{9.225}$$

The approximation includes the leading term of the order $O(k)$ and a correction term of the order $O(1)$. The error of the approximation (9.225) is therefore of the order $o(1)$, which is asymptotically vanishing as $k \to \infty$. The term with the derivative of the slowly varying function is the correction term of the approximation, and its calculation requires evaluation of the derivative

$$A'(z) = (R_{\|} - R_{\perp})'\rho(z)\rho'(z) + (R_{\|} - R_{\perp})[\rho'(z)^2 + \rho(z)\rho''(z)], \tag{9.226}$$

where

$$(R_{\|} - R_{\perp})' = \frac{\rm d}{{\rm d}z}\{R_{\|}[\theta(z)] - R_{\perp}[\theta(z)]\}. \tag{9.227}$$

As we shall see later, there are various cases in which the leading term vanishes and the correction term becomes the main term of the asymptotic approximation.

In the convex case shown on the left in Figure 9.29, the value $z = z_{\rm max}$ corresponds to the apex of the scatterer, and without loss of generality the coordinate system can be always defined so that $z_{\rm max} = 0$. The points with $z = z_{\rm min}$ lie on the shadow boundary, which can be either a circumferential edge or a rim on the surface or, in the case of a smooth surface, a line, at which the incidence direction is tangential to the surface.

Thus, the expression (9.225) can be split up into two contributions

$$\mathbf{F}^{\rm bs}_{\rm po} = \mathbf{F}^{\rm bs}_{\rm po}\Big|_{\rm apex} + \mathbf{F}^{\rm bs}_{\rm po}\Big|_{\rm sb}, \tag{9.228}$$

one from the apex

$$\mathbf{F}^{\rm bs}_{\rm po}\Big|_{\rm apex} = \mathbf{E}_0\left[\frac{k}{4}A(0) + \frac{j}{8}A'(0)\right] \tag{9.229}$$

and another from the shadow boundary

$$\mathbf{F}^{\rm bs}_{\rm po}\Big|_{\rm sb} = -\mathbf{E}_0\,{\rm e}^{2jkz_{\rm min}}\left[\frac{k}{4}A(z_{\rm min}) + \frac{j}{8}A'(z_{\rm min})\right]. \tag{9.230}$$

So, the axial backscattering from electrically large bodies of revolution is determined by scattering from just two regions, the apex and the shadow boundary, provided that the geometric and material properties of the scattering surface in between vary continuously and slowly on the scale of the wavelength.

In the concave case, shown on the right in Figure 9.29, the integral (9.219) should be multiplied by -1 and the physical meaning of contributions from the points $z = z_{\rm max}$ and $z = z_{\rm min}$ alternates compared with the convex case: the specular reflection comes from $z = z_{\rm min}$ and the rim or shadow boundary contribution from $z = z_{\rm max}$. Thus, the expressions for the concave case are obtained from those for the convex case by simply replacing $z_{\rm min}$ with $z_{\rm max}$. If we set $z_{\rm min} = 0$, then the formula for the specular contribution in the concave case coincides with (9.229). The contribution from the rim or shadow boundary results from (9.230) on replacing $z_{\rm min}$ with $z_{\rm max}$,

$$\mathbf{F}^{\rm bs}_{\rm po}\Big|_{\rm sb} = -\mathbf{E}_0\,{\rm e}^{2jkz_{\rm max}}\left[\frac{k}{4}A(z_{\rm max}) + \frac{j}{8}A'(z_{\rm max})\right]. \tag{9.231}$$

In what follows we study these contributions in more detail, assuming the convex case. The specifics of the concave case are discussed at the end of the session.

Specular Scattering

Specular scattering corresponds to the contribution of the apex when the surface in the vicinity of the apex is smooth (Figure 9.30, right). At the apex, $\rho(0) = 0$ but $\rho'(0)$ is infinite, and if $\rho(z)\rho'(z)$ goes to a finite limit at $z = 0$, then the contribution from the apex is given by the leading term in (9.229).

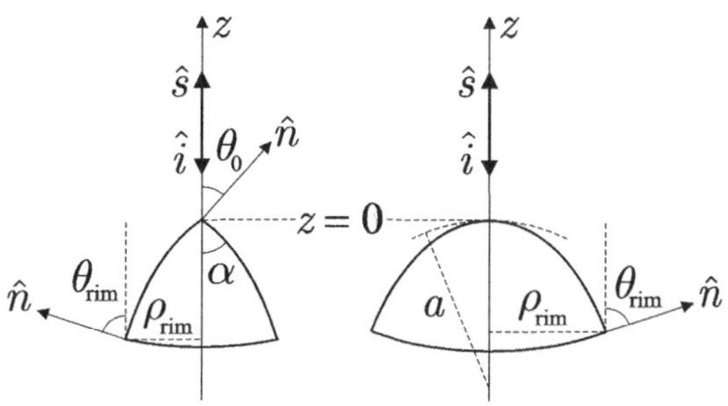

Figure 9.30 Convex body of revolution with a conical point at the apex (left) and smooth at the apex (right).

To calculate this term, we note that the smooth surface near the apex of every body of revolution can be approximated by a sphere of a suitable curvature radius a (Figure 9.30) so that close to the apex

$$\rho(z) = \sqrt{a^2 - (z+a)^2}, \quad \rho'(z) = -\frac{a+z}{\rho(z)} \tag{9.232}$$

and $\rho(z)\rho'(z) \to -a$ for $z \to 0$. Furthermore, the incidence direction is perpendicular to the surface at the apex and reflection coefficients are to be taken at the normal incidence, at which $R_{\|}(0) = -R_{\perp}(0)$, so that

$$A(0) = 2aR_{\perp}(0) \tag{9.233}$$

and the leading term in the apex contribution to (9.229) is given by

$$\mathbf{F}^{\text{bs}}_{\text{po}}\Big|_{\text{sr}} = \frac{ka}{2} R_{\perp}(0)\mathbf{E}_0. \tag{9.234}$$

This contribution is apparently the contribution from the specular reflection and a special case of the reflection formula (8.203) as $\mathbf{E}_{\text{ref}}(\mathbf{r}_c) = R_{\perp}(0)\mathbf{E}_0$ at normal incidence, the Gaussian curvature at the reflection point $\mathbf{r}_c$ is $K = a^{-2}$ and $\sigma_c^{\infty} = -j$ for convex surfaces.

Calculation of the correction term in (9.229) is simplified by the fact that reflection coefficients of every isotropic surface are even functions of the incidence angle θ, implying that $(R_{\|} - R_{\perp})' = 0$ at the reflection point. It follows from (9.226) that

$$A'(0) = -R_{\perp}(0) \left.\frac{\mathrm{d}^2(\rho^2)}{\mathrm{d}^2 z}\right|_{z=0}. \tag{9.235}$$

Contribution from the Tip

If the apex is a conical point, then $\rho(0) = 0$ but $\rho'(0)$ is finite, implying that

$$A(0) = 0, \quad A'(0) = [R_{\|}(\theta_0) - R_{\perp}(\theta_0)][\rho'(0)]^2, \tag{9.236}$$

where θ_0 is the incidence angle as the observation point approaches the tip (Figure 9.30).

In the vicinity of the apex the scattering surface can be locally approximated by a circular cone

$$\rho(z) = -z \tan\alpha, \tag{9.237}$$

where α is the half-width of the cone, implying that

$$\rho'(0) = -\tan\alpha, \quad \theta_0 = \frac{\pi}{2} - \alpha. \tag{9.238}$$

It follows then from (9.229) that the contribution from the tip is given by the formula

$$\mathbf{F}^{\text{bs}}_{\text{po}}\Big|_{\text{tip}} = \frac{j}{8} \tan^2\alpha \left[R_{\|}\left(\frac{\pi}{2} - \alpha\right) - R_{\perp}\left(\frac{\pi}{2} - \alpha\right)\right] \mathbf{E}_0, \tag{9.239}$$

which agrees with the formula (9.87) for the contribution to the backscattering from the tip of an infinite circular cone. The contribution is asymptotically small compared with the contribution (9.234) of the specular reflection.

Contribution from the Rim

The contribution from the circumferential edge of the body, which separates the illuminated and shadowed portions of the body (Figure 9.30), follows from (9.230) with

$$A(z_{\min}) = -\rho_{\text{rim}} \cot\theta_{\text{rim}} [R_{\|}(\theta_{\text{rim}}) - R_{\perp}(\theta_{\text{rim}})] \tag{9.240}$$

$$A'(z_{\min}) = -\rho_{\text{rim}} \cot\theta_{\text{rim}} \left.\frac{\mathrm{d}}{\mathrm{d}z}(R_{\|} - R_{\perp})\right|_{z=z_{\min}} + [R_{\|}(\theta_{\text{rim}}) - R_{\perp}(\theta_{\text{rim}})] \left[\rho'(z_{\min})^2 + \rho_{\text{rim}}\rho''(z_{\min})\right], \tag{9.241}$$

where $\rho(z_{\min}) = \rho_{\text{rim}}$ and $\theta_{\text{rim}} = \theta(z_{\min})$ is the incidence angle as the observation point approaches the rim on the illuminated side of the body, which can be found from the relation $\rho'(z_{\min}) = -\cot\theta_{\text{rim}}$.

The derivative of the reflection coefficients in (9.241) vanishes if the vicinity of the rim is locally a piece of a conical surface. Otherwise, the derivative has to be evaluated, and for impedance surfaces with reflection coefficients (9.220) it can be obtained from the general formula

$$\frac{\mathrm{d}}{\mathrm{d}z}\{R_{\|}[\theta(z)] - R_{\perp}[\theta(z)]\} = \frac{-\rho''(z)}{2\rho'(z)[1+\rho'(z)^2]} \left\{R_{\|}^2[\theta(z)] - R_{\perp}^2[\theta(z)]\right\} \tag{9.242}$$

(see problem 18).

Summarizing and retaining for brevity just the leading term, we obtain

$$\left.\mathbf{F}_{\text{po}}^{\text{bs}}\right|_{\text{rim}} = \mathbf{E}_0 \frac{k}{4} \rho_{\text{rim}} \cot\theta_{\text{rim}} [R_{\|}(\theta_{\text{rim}}) - R_{\perp}(\theta_{\text{rim}})]\, \mathrm{e}^{2jkz_{\min}}. \tag{9.243}$$

The axial backscattering from the rim is therefore of the same order of magnitude as the specular reflection (9.234). This is because the z axis is the focal line and all the points of the rim contribute. Once the incidence or scattering direction deviates from the focal line, the rim contribution comes from just two critical points on the rim and gets asymptotically smaller than the specular reflection. The rim contribution in the PO approximation is independent of the shape and material properties of the scattering body on the shadowed portion of the scatterer. This dependence is, however, present in accurate solutions of scattering problems. To account for that, it is necessary to describe more accurately the equivalent surface currents close to the rim by using the currents induced by the same incident wave at a locally conformal edge (Ufimtsev 2014).

Contribution from the Shadow Boundary

If the body is smooth and convex, then the points with $z = z_{\min}$ correspond to the shadow boundary. At the shadow boundary the incidence direction is tangent to the surface, implying that $\rho'(z_{\min}) = 0$. With z approaching the shadow boundary, the main term in (9.230) vanishes, and the correction term has to be accounted for to describe the contribution from the shadow boundary. With $\rho'(z_{\min}) = 0$ and $\theta_{\text{rim}} = \pi/2$ it follows from (9.226) that

$$A'(z_{\min}) = \left[R_{\|}\left(\frac{\pi}{2}\right) - R_{\perp}\left(\frac{\pi}{2}\right)\right] \rho(z_{\min})\rho''(z_{\min}) \tag{9.244}$$

since $\rho''(z_{\min})$ is finite in general.

For PEC bodies with $R_{\|}(\pi/2) = 1$ and $R_{\perp}(\pi/2) = -1$, (9.244) leads to the shadow boundary contribution

$$\left.\mathbf{F}_{\text{po}}^{\text{bs}}\right|_{\text{sb}} = \mathbf{E}_0 \frac{1}{4j} \rho(z_{\min})\rho''(z_{\min})\, \mathrm{e}^{2jkz_{\min}}. \tag{9.245}$$

For imperfectly reflecting bodies, however, because of the general property $R_{\|}(\pi/2) = R_{\perp}(\pi/2) = -1$ of reflection coefficients at the grazing incidence (see (4.183)), $A'(z_{\min})$ vanishes, implying an asymptotically vanishing shadow boundary contribution to the axial backscattering. Since the PO solution itself is an approximation, it does not make sense to calculate the higher-order correction terms. The situation is different for the forward scattering, in which case the shadow boundary plays an important role (e.g., Osipov *et al.* (1998)).

Contribution from the Shadowed Portion of the Body

In the PO approximation the shadowed portion of the body does not contribute to the scattered field, so that the same PO solution applies to a great variety of bodies with identical illuminated portions but variously shaped shadowed portions. In the PO approximation they all scatter the same. This approach may seem to be a crude drawback of the PO approximation but often leads to good approximations for the scattered fields as this method is based on the physically sound assumption that the currents on the shadowed portions of electrically large bodies can be neglected. These currents (and therefore the shape of the shadowed portion of the scatterer) are accounted for in the framework of PTD through the use of more accurate approximations for the currents in asymptotically small vicinities of edges and shadow boundaries on the scattering surface (e.g., Ufimtsev (2014)).

Concave Case

In the case of concave scattering surfaces (Figure 9.29) the specular reflection at the apex $z_{\min} = 0$ is described by the formula (9.229) and the contribution from the rim at $z = z_{\max}$ by the formula (9.231), where $A(z)$ and $A'(z)$ are given by (9.224) and (9.226). Compared with the results for the same surface illuminated from its convex side, evaluation of ρ' will lead to an alternative sign in the leading terms, proportional to $A(0)$ and $A(z_{\max})$, but to the same result for the correction terms, proportional to $A'(0)$ and $A'(z_{\max})$. For example, the leading term in the specular scattering from a concave surface is given by

$$\mathbf{F}^{\mathrm{bs}}_{\mathrm{po}}\Big|_{\mathrm{sr}} = -\frac{ka}{2} R_{\perp}(0)\mathbf{E}_0, \tag{9.246}$$

which agrees with the general formula (8.203) for the specular scattering since $\sigma_{\mathrm{c}}^{\infty} = j$ for concave surfaces. From the physical point of view, the sign difference between (9.246) and (9.234) can be explained by the fact that the scattered wave touches the caustics of the concave reflector.

The leading term in the contribution from the rim of a concave surface takes the form

$$\mathbf{F}^{\mathrm{bs}}_{\mathrm{po}}\Big|_{\mathrm{rim}} = -\mathbf{E}_0 \frac{k}{4} \rho_{\mathrm{rim}} \cot\theta_{\mathrm{rim}} [R_{\parallel}(\theta_{\mathrm{rim}}) - R_{\perp}(\theta_{\mathrm{rim}})]\, \mathrm{e}^{2jkz_{\max}}. \tag{9.247}$$

9.4.5 *Examples*

In this section the general results of the previous sections for the axial backscattering from bodies of revolution are specialized to simply shaped surfaces: truncated paraboloid, sphere, truncated cone, and ogive. The curves are plotted for the axial backscattering cross section, which is related to the far-field coefficients by the formula (Chapter 3)

$$\sigma = \frac{4\pi}{k^2} \frac{|\mathbf{F}^{\mathrm{bs}}_{\mathrm{po}}|^2}{|\mathbf{E}_0|^2}. \tag{9.248}$$

The results for PEC and impedance scatterers are obtained by numerical integration in (9.219) and compared with each other. The maximum geometric cross section of all the bodies is the same, a circle of radius a, so that the numerical results are comparable. It will be shown in particular that increasing the electrical size of the body does not necessarily increase the backscattering, while the shape of the target strongly influences the level and behavior of the backscattered field.

Paraboloid of Revolution, Full and Truncated

The convex paraboloid of revolution with the curvature radius a at the apex is described by the parametric equation (9.165) with

$$\rho(z) = \sqrt{-2az}, \tag{9.249}$$

where $z_{\min} \leq z \leq 0$. With $z_{\min} = -\infty$, (9.249) describes an infinite paraboloid. Otherwise, it is a truncated paraboloid and its geometric cross section at the level $z = z_{\min}$ is a circle with radius $\rho_{\text{rim}} = \sqrt{2a|z_{\min}|}$. Axial backscattering from an infinite paraboloid is asymptotically dominated by the specular reflection and described by (9.234). Only the leading term contributes, whereas the correction term (9.235) vanishes as $(\rho^2)'' = 0$ for every paraboloid.

In the case of a truncated paraboloid, the contribution from the rim has to be added to the axial backscattering. This contribution for the convex paraboloid is described by (9.243) with $\rho_{\text{rim}} \cot\theta_{\text{rim}} = a$ because $\cot\theta_{\text{rim}} = -\rho'(z_{\min})$ and $\rho'(z) = -a/\rho(z)$. The total far-field coefficient for the axial backscattering from the truncated paraboloid is therefore given by

$$\mathbf{F}_{\text{prb}} = -\mathbf{E}_0 \frac{ka}{4}\left\{R_{\|}(0) - R_{\perp}(0) - \mathrm{e}^{2jkz_{\min}}[R_{\|}(\theta_{\text{rim}}) - R_{\perp}(\theta_{\text{rim}})]\right\}. \tag{9.250}$$

For a PEC paraboloid, this expression becomes

$$\mathbf{F}^{\text{PEC}}_{\text{prb}} = -\mathbf{E}_0 \frac{ka}{2}\left(1 - \mathrm{e}^{2jkz_{\min}}\right), \tag{9.251}$$

implying vanishing axial backscattering when $kz_{\min} = -\pi m$ with $m = 1, 2, \ldots$. It should be pointed out that the field vanishes only in the PO approximation; the exact backscattering is small but not zero, and to account for that, an improved PO solution should be used that describes more accurately the contribution from the rim (e.g., Ufimtsev (2014)).

For non-PEC paraboloids the axial scattering does not vanish but is still a sum of two terms of the same order of magnitude, so that RCS has maxima and deep minima as a function of $kz_{\min}$. Figure 9.31 shows the axial backscattering cross section σ normalized to πa^2 for PEC and impedance paraboloids. The paraboloids are truncated at $z_{\min} = -a/2$ which makes the ratio $\sigma/(\pi a^2)$ a function of ka and the base of the paraboloid a circle of the radius a. In contrast to PEC paraboloid (solid line), the minima in the cross section of the impedance paraboloid (dotted line) are finite and lie at the level of −38 dB for this particular value of the normalized surface impedance.

Sphere

In contrast to the paraboloid, the field distribution on the surface of a sphere has a shadow boundary. As we have seen from (9.244) and (9.245), the contribution from the shadow boundary to backscattering is asymptotically smaller than the contribution from the specular reflection.

The spherical surface of the radius a is described by the equation

$$\rho(z) = \sqrt{a^2 - (a+z)^2} \tag{9.252}$$

with $-2a \leq z \leq 0$, implying that

$$\rho'(z) = -\frac{a+z}{\rho(z)}, \quad \rho''(z) = -\frac{a^2}{\rho^3(z)}. \tag{9.253}$$

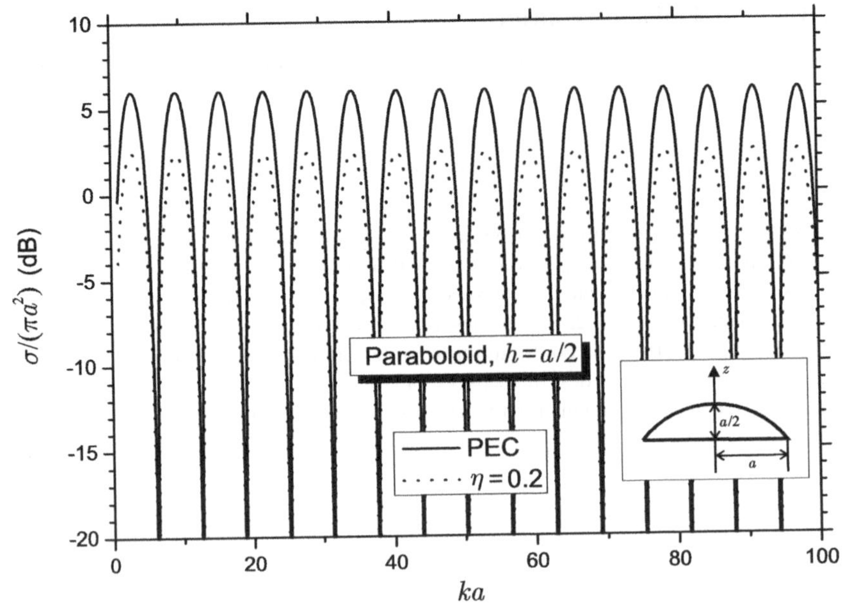

Figure 9.31 Normalized axial backscattering from a truncated paraboloid with the curvature radius a at the apex. The height h of the paraboloid is equal to $a/2$, so that the base of the paraboloid is a circle of radius $\rho_{\text{rim}} = a$.

For an imperfectly reflecting sphere it follows from (9.233) and (9.235) that

$$\mathbf{F}_{\text{sp}} = \mathbf{E}_0 \frac{ka}{2} R_{\perp}(0) \left(1 + \frac{j}{2ka}\right) + o(1), \tag{9.254}$$

where the contribution from the shadow boundary at $z = -a$ is of the order $o(1)$ and goes to zero as ka increases. For a PEC sphere, the shadow boundary contribution (9.245) is of the order $O(1)$ and the far-field coefficient is asymptotically described by the approximation

$$\mathbf{F}_{\text{sp}}^{\text{PEC}} = -\mathbf{E}_0 \frac{ka}{2} \left[1 + \frac{j}{2ka}\left(1 - \mathrm{e}^{-2jka}\right)\right]. \tag{9.255}$$

The axial backscattering cross sections of a PEC sphere and an impedance sphere normalized to πa^2 are plotted in Figure 9.32 as functions of ka. The curves rapidly approach the limits, unity in the PEC case and $|R_{\perp}(0)|^2$ in the impedance case. Oscillations are more pronounced for the PEC sphere because of the stronger shadow boundary contribution. Exact backscattering cross sections of a PEC sphere can be found in Figures 3.16 and 7.25.

Truncated Cone

The characteristic feature of axial backscattering from truncated cones is the asymptotic smallness of the contribution from the apex compared with the contribution from the rim of the base of the truncated cone. A truncated cone can be described by the equations

$$\rho(z) = -z \tan\alpha, \quad \rho'(z) = -\tan\alpha, \quad \rho''(z) = 0, \tag{9.256}$$

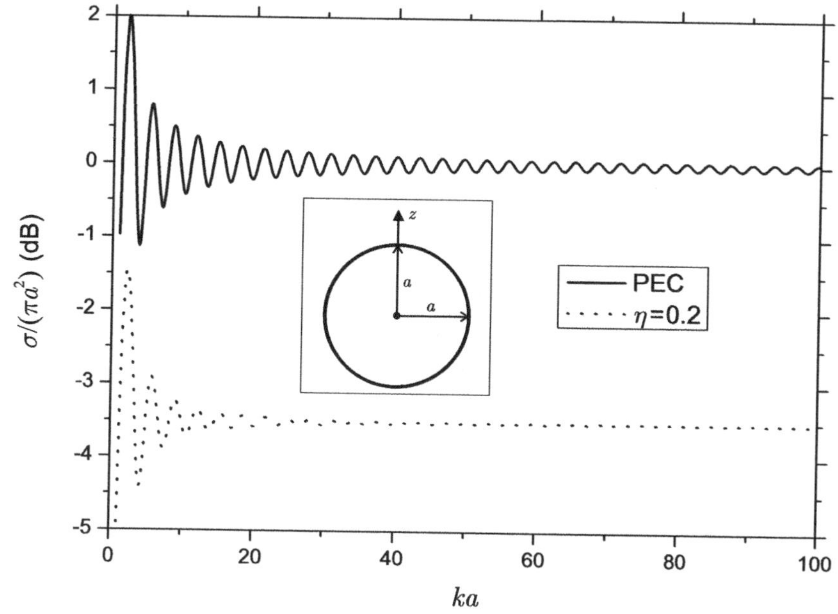

Figure 9.32 Backscattering from a sphere (PEC and impedance).

where $z_{min} \le z \le 0$. The angle α is the half-angular width of the cone, $|z_{min}|$ is the height of the cone, and $\rho_{rim} = \rho(z_{min})$ is the radius of the circular base of the cone.

The far-field coefficient is the sum of the contributions from the tip (9.239) and from the rim (9.230). In the rim contribution, in addition to the main term (9.243), we retain the correction term which is of the same order of magnitude as the tip contribution. For the cone the quantity $A'(z_{min})$ given by (9.241) can be expressed in the simple form

$$A'(z_{min}) = \left[R_{\parallel}\left(\frac{\pi}{2} - \alpha\right) - R_{\perp}\left(\frac{\pi}{2} - \alpha\right)\right] \tan^2 \alpha, \tag{9.257}$$

because the reflection coefficients are constant over the whole illuminated surface of the cone and $\rho'' = 0$. Combining the tip and rim contributions leads to the far-field coefficient

$$\mathbf{F}_{cone} = \frac{1}{4}\mathbf{E}_0 \tan \alpha \left[R_{\parallel}\left(\frac{\pi}{2} - \alpha\right) - R_{\perp}\left(\frac{\pi}{2} - \alpha\right)\right] \cdot \left[\frac{j}{2} \tan \alpha + e^{2jkz_{min}} \left(k\rho_{rim} - \frac{j}{2} \tan \alpha\right)\right]. \tag{9.258}$$

Figure 9.33 shows the normalized axial backscattering cross section of a 90° cone as a function of ka where $a = \rho_{rim} = |z_{min}|$. The axial backscattering from an impedance cone differs from that for a PEC cone by the factor $|R_{\parallel}(\pi/2 - \alpha) - R_{\perp}(\pi/2 - \alpha)|^2$, and this relation is independent of ka as long as the reflection coefficients are frequency independent. It is interesting to note that the curve for the 90° PEC cone coincides with that of a PEC sphere of the radius a, shown in Figure 9.32. This identity can be mathematically proved (problem 20).

Ogive

As we have seen in this section, the contributions to, backscattering from a conical point at the apex and from a shadow boundary on a smooth surface are particularly low, so it is logical to expect that such bodies, conical at the nose and otherwise smooth, may feature particularly low levels of backscattering. In this subsection we consider an example. A body of revolution of the required shape can be constructed by joining two arbitrary points on the z axis with an arc of a

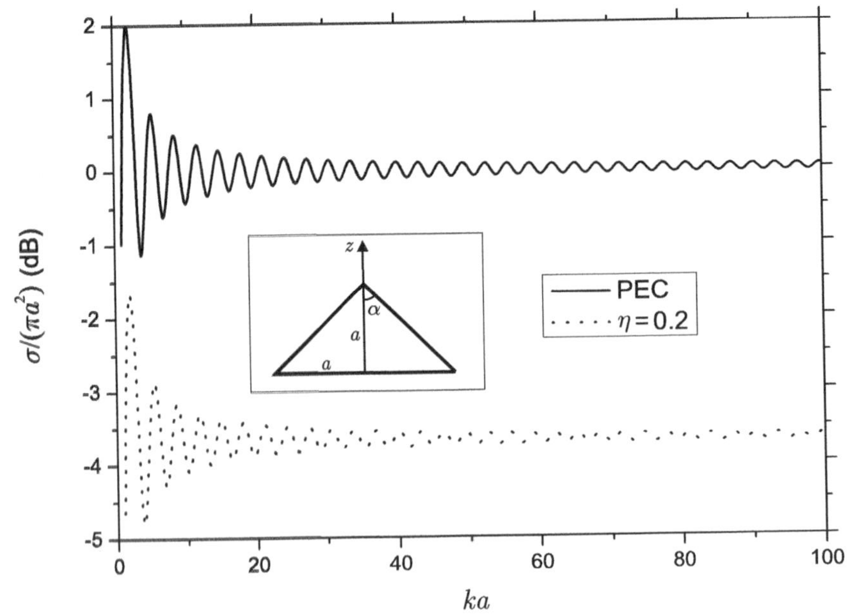

Figure 9.33 Axial backscattering from a truncated cone with the base radius $\rho_{\mathrm{rim}} = a$ and height $h = a$ (the angular width 2α of the cone is 90°).

circle, whose diameter is greater than the distance between the points, and forming the surface by rotating the arc around the z axis. This pointed, curved surface is called ogive and is used, for example, to form the nose of a rocket.

Let us study the axial backscattering from an ogive shown in the inset in Figure 9.34. We will define the shape by two parameters: the angular width 2α of the cone at the nose and the length $2|z_0|$ of the ogive along the z axis. The origin of the z axis is chosen at the illuminated apex, so that the geometric shadow is located at $z = z_0$. The parametric equation of the shape reads

$$\rho(z) = z_0 \cot\alpha + \sqrt{z_0^2 \csc^2\alpha - (z - z_0)^2}, \tag{9.259}$$

where $2z_0 \leq z \leq 0$. In the middle of the shape, at $z = z_0$, the radius is at maximum,

$$\rho_{\max} = \rho(z_0) = |z_0| \tan\frac{\alpha}{2} \tag{9.260}$$

and

$$\rho'(z_0) = 0, \quad \rho''(z_0) = \frac{\sin\alpha}{z_0}. \tag{9.261}$$

With $\alpha = \pi/2$, the shape becomes spherical.

The axial backscattering is asymptotically a sum of contributions from the nose at $z = 0$ and from the shadow boundary at $z = z_0$. The result depends on whether the surface is PEC or imperfectly conducting. For an imperfectly conducting surface, the contribution from the shadow boundary to the far-field coefficient is $o(1)$, that is, it goes to zero as the

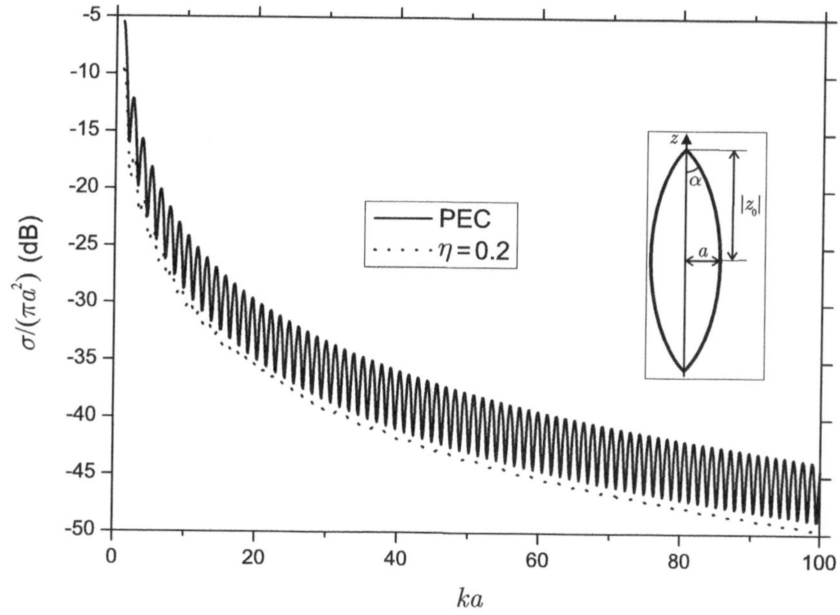

Figure 9.34 Axial backscattering from an ogive with the half-angle of 45° at the tip; the shadow boundary is a circle of radius a.

electrical size of the shape goes to infinity. Thus, in this case, the backscattering level is asymptotically the same as that of a single conical point,

$$\mathbf{F}_{\text{ogive}} = \mathbf{E}_0 \frac{j}{8} \tan^2 \alpha \left[R_\| \left(\frac{\pi}{2} - \alpha \right) - R_\perp \left(\frac{\pi}{2} - \alpha \right) \right] + o(1). \tag{9.262}$$

For a PEC ogive the contribution from the shadow boundary is of the same order of magnitude as that from the tip, so that

$$\mathbf{F}_{\text{ogive}}^{\text{PEC}} = \mathbf{E}_0 \frac{j}{4} \left[\tan^2 \alpha + 2 \sin^2 \left(\frac{\alpha}{2} \right) e^{2jkz_0} \right], \tag{9.263}$$

where the first term in the brackets corresponds to the contribution from the conical point and the second term to that from the shadow boundary. Note that the latter is always smaller than the former as $\tan \alpha > \sqrt{2} \sin(\alpha/2)$ for $0 < \alpha < \pi/2$.

Figure 9.34 shows the axial backscattering cross section of the ogive with $\rho_{\text{max}} = a$, PEC vs. impedance. Compared with the curves plotted before, except for the case of a truncated paraboloid, backscattering from the ogive is much weaker and rapidly decays with the increase in the electrical size of the geometric cross section. Backscattering from the truncated paraboloid may be even lower but only in a narrow vicinity of the values of ka, at which the involved contributions (from the rim and from the specular point) destructively interfere; on average the ogive scatters much less than the paraboloid. The oscillations in the curve for the PEC ogive (solid line) are due to the interference of the contributions from the tip and from the shadow boundary. In the case of an impedance ogive, the contribution from the shadow boundary is asymptotically small, and backscattering is dominated by the tip contribution.

9.5 Problems

1. Find the zeros $v_\pm$ of function $F(v; \alpha, \beta, \gamma)$ from (9.57) that determine the position of the shadow boundary on the surface of an elliptic cone illuminated by a plane wave.

Solution: Equation $F(v;\alpha,\beta,\gamma)=0$ can be formulated as a quadratic equation with respect to $\exp(jv)$. The equation has two solutions

$$e^{jv}=\frac{1}{\alpha-j\beta}(-\gamma\pm j\sqrt{\alpha^2+\beta^2-\gamma^2}), \tag{9.264}$$

where α, β, and γ are defined in (9.58). When $\alpha^2+\beta^2>\gamma^2$, the absolute value of the right-hand side in (9.264) equals unity, and in the interval $-\pi<v<\pi$ there are two real-valued solutions $v=\chi_\pm$ with

$$\chi_\pm=\arg\left[\frac{1}{\alpha-j\beta}(-\gamma\pm j\sqrt{\alpha^2+\beta^2-\gamma^2})\right]. \tag{9.265}$$

Denoting $v_<=\min\chi_\pm$ and $v_>=\max\chi_\pm$, we obtain two zeros $v=v_<$ and $v=v_>$ of $F(v;\alpha,\beta,\gamma)$ on the interval $-\pi<v<\pi$. If $F(v;\alpha,\beta,\gamma)>0$ with $v_<<v<v_>$, then $v_-=v_>$ and $v_+=v_>$ are the lower and upper ends of the range of integration in the PO integral (9.61). Otherwise, if $F(v;\alpha,\beta,\gamma)<0$ between $v_<$ and $v_>$, then we use 2π periodicity of the integrand in (9.61) and take $v_-=v_>$ and $v_+=v_<+2\pi$ as the ends of the integration range.

2. Derive the closed-form expressions (9.67)–(9.70) for the PO diffraction coefficient of a fully illuminated elliptic PEC cone.

   Solution: Specializing (9.63) to this case gives that $v_-=0$ and $v_+=2\pi$, $R_\perp=-1$ and $R_\parallel=1$, coefficients $\mathbf{b}_3$ and $\mathbf{b}_4$ vanish, whereas $\mathbf{b}_1=-2\hat{\alpha}_{\text{inc}}(\hat{i}\cdot\hat{n})$ and $\mathbf{b}_2=-2\hat{\beta}_{\text{inc}}$. On evaluation of vector products (9.34) and (9.35), the numerator in (9.63) can be represented as a linear combination of $\cos v$ and $\sin v$, which permits explicit integration in (9.63) by using the relations:

$$\int_0^{2\pi}\frac{dv}{(p+q_1\cos v+q_2\sin v)^2}=\frac{2\pi p\,\mathrm{sgn}(p)}{(p^2-q_1^2-q_2^2)^{3/2}} \tag{9.266}$$

$$\int_0^{2\pi}\frac{\cos v\,dv}{(p+q_1\cos v+q_2\sin v)^2}=-\frac{2\pi q_1\mathrm{sgn}(p)}{(p^2-q_1^2-q_2^2)^{3/2}} \tag{9.267}$$

$$\int_0^{2\pi}\frac{\sin v\,dv}{(p+q_1\cos v+q_2\sin v)^2}=-\frac{2\pi q_2\mathrm{sgn}(p)}{(p^2-q_1^2-q_2^2)^{3/2}}. \tag{9.268}$$

   The parameters in (9.266)–(9.268) are as follows:

$$p=-c(\cos\theta+\cos\theta_{\text{inc}}) \tag{9.269}$$
$$q_1=a(\cos\phi\sin\theta+\cos\phi_{\text{inc}}\sin\theta_{\text{inc}}) \tag{9.270}$$
$$q_2=b(\sin\phi\sin\theta+\sin\phi_{\text{inc}}\sin\theta_{\text{inc}}), \tag{9.271}$$

   implying that $\mathrm{sgn}(p)=-1$ and $p^2-q_1^2-q_2^2>0$.

3. Show that the general expression (9.63) for the diffraction coefficient of an elliptic cone reduces to (9.33) when the elliptic cone degenerates to a flat angular sector, e.g. when $b\to 0$.

   Solution: Consider first (9.63). In this limit, $\mathbf{f}(u,v)\to u(\hat{x}a\cos v-\hat{z}c)$ and $V(v)\to ac|\sin v|$. Assuming (without loss of generality) that the incident wave illuminates the side of the cone at which $0<v<\pi$, we find that $v_-\to 0$ and

$v_+ \to \pi$. Furthermore, $\hat{n} \to \hat{y}$, which implies that the vectors $\mathbf{a}_m$ and $\mathbf{b}_m$ become independent of the integration variable v, and (9.63) reduces to

$$D(\hat{e}, \hat{e}_{\text{inc}}) \to \frac{ac}{4\pi j} \sum_{m=1}^{4} (\hat{e} \cdot \mathbf{a}_m)(\mathbf{b}_m \cdot \hat{e}_{\text{inc}}) \int_0^{\pi} \frac{\sin v \, dv}{(q_x a \cos v - q_z c)^2}. \tag{9.272}$$

Integration in (9.272) can be performed in closed form, yielding

$$D(\hat{e}, \hat{e}_{\text{inc}}) = \frac{j}{2\pi} \frac{ac}{(q_x a)^2 - (q_z c)^2} \sum_{m=1}^{4} (\hat{e} \cdot \mathbf{a}_m)(\mathbf{b}_m \cdot \hat{e}_{\text{inc}}). \tag{9.273}$$

This is equal to what equation (9.33) gives for an angular sector with the edges

$$\hat{e}_1 = \frac{a\hat{x} - c\hat{z}}{\sqrt{a^2 + c^2}}, \quad \hat{e}_2 = \frac{-a\hat{x} - c\hat{z}}{\sqrt{a^2 + c^2}}. \tag{9.274}$$

Indeed, using these relations in (9.43)–(9.45) (with $L = 2$) leads to the unit vectors

$$\hat{n}_1 = \hat{y}, \quad \hat{\tau}_1^- = -\hat{e}_1, \quad \hat{\tau}_1^+ = \hat{e}_2, \quad \hat{v}_1^- = \frac{-c\hat{x} - a\hat{z}}{\sqrt{a^2 + c^2}}, \quad \hat{v}_1^+ = \frac{c\hat{x} - a\hat{z}}{\sqrt{a^2 + c^2}}, \tag{9.275}$$

which implies that

$$\frac{1}{|\mathbf{q}|^2 - (\hat{n}_1 \cdot \mathbf{q})^2} \left[\frac{\mathbf{q} \cdot \hat{v}_1^-}{\mathbf{q} \cdot \hat{\tau}_1^-} - \frac{\mathbf{q} \cdot \hat{v}_1^+}{\mathbf{q} \cdot \hat{\tau}_1^+} \right] = \frac{2ac}{(q_x a)^2 - (q_z c)^2}. \tag{9.276}$$

Inserting (9.276) in (9.33) recovers (9.273).

4. The field backscattered from an infinite imperfectly reflecting circular cone under axial illumination (Figure 9.11) is given by the formula (9.87). Derive this formula from the general formula (9.61) for elliptic cones.

Solution: For axial backscattering, $\hat{z} = \hat{s} = -\hat{i}$, implying that $\mathbf{E}_{\text{inc}}(0) = E_{0x}\hat{x} + E_{0y}\hat{y}$, $\mathbf{q} = -2\hat{z}$, $\mathbf{q} \cdot \mathbf{f}_u(0, v) = 2c$, and since the whole surface of the cone is illuminated, $v_+ = 2\pi$ and $v_- = 0$ in the integral (9.87). Furthermore, the cone is circular, which means that $b = a$, $\alpha_x = \alpha_y = \alpha$, and

$$V(v) = \frac{ac}{\cos \alpha}, \quad \hat{n} = \cos \alpha (\hat{x} \cos v + \hat{y} \sin v + \tan \alpha \hat{z}), \tag{9.277}$$

which transforms (9.87) to the form

$$\mathbf{E}_c^{\text{po}}(\mathbf{r}) = \frac{e^{-jkr}}{kr} \frac{\tan \alpha}{16\pi j \cos \alpha} \sum_{m=1}^{4} \mathbf{I}_m \tag{9.278}$$

with

$$\mathbf{I}_m = \int_0^{2\pi} \mathbf{a}_m (\mathbf{b}_m \cdot \mathbf{E}_{\text{inc}}(0)) \, dv, \tag{9.279}$$

where $\mathbf{a}_m$ and $\mathbf{b}_m$ are defined by (9.34)–(9.37) and (9.38)–(9.41), respectively.

The polarization basis of the incident wave is defined by the unit vectors

$$\hat{\alpha}_{\text{inc}} = \frac{\hat{\imath} \times \hat{n}}{|\hat{\imath} \times \hat{n}|} = \hat{x} \sin v - \hat{y} \cos v, \quad \hat{\beta}_{\text{inc}} = \hat{\imath} \times \hat{\alpha}_{\text{inc}} = -\hat{x} \cos v - \hat{y} \sin v, \tag{9.280}$$

implying that $\hat{\imath} = \hat{n} \times \alpha_{\text{inc}} = \hat{x} \sin \alpha \cos v + \hat{y} \sin \alpha \sin v - \hat{z} \cos \alpha$ and

$$\mathbf{a}_1 = -\hat{x} \sin v + \hat{y} \cos v, \quad \mathbf{a}_2 = -\sin \alpha (\hat{x} \cos v + \hat{y} \sin v) \tag{9.281}$$

$$\mathbf{a}_3 = \hat{x} \cos v + \hat{y} \sin v, \quad \mathbf{a}_4 = -\sin \alpha (\hat{x} \sin v - \hat{y} \cos v). \tag{9.282}$$

Since $\hat{\imath} \cdot \hat{n} = -\sin \alpha$ and the reflection coefficients $R_{\parallel}$ and $R_{\perp}$ are independent of the coordinate v, $\mathbf{I}_m$ are the integrals of simple trigonometric polynomials which are easily evaluated. For example,

$$\mathbf{I}_1 = (1 - R_{\perp}) \sin \alpha \int_0^{2\pi} (E_{0x} \sin v - E_{0y} \cos v)(-\hat{x} \sin v + \hat{y} \cos v) \, dv \tag{9.283}$$

with $E_{0x} = \mathbf{E}_{\text{inc}}(0) \cdot \hat{x}$ and $E_{0y} = \mathbf{E}_{\text{inc}}(0) \cdot \hat{y}$, which evaluates to

$$\mathbf{I}_1 = -\pi \sin \alpha (1 - R_{\perp}) \mathbf{E}_{\text{inc}}(0) \tag{9.284}$$

because of the identities

$$\int_0^{2\pi} \cos^2 v \, dv = \int_0^{2\pi} \sin^2 v \, dv = \pi, \quad \int_0^{2\pi} \cos v \sin v \, dv = 0. \tag{9.285}$$

In similar way,

$$\mathbf{I}_2 = -\pi \sin \alpha (1 + R_{\parallel}) \mathbf{E}_{\text{inc}}(0) \tag{9.286}$$

$$\mathbf{I}_3 = \pi \sin \alpha (1 - R_{\parallel}) \mathbf{E}_{\text{inc}}(0) \tag{9.287}$$

$$\mathbf{I}_4 = \pi \sin \alpha (1 + R_{\perp}) \mathbf{E}_{\text{inc}}(0). \tag{9.288}$$

Inserting these in (9.278) gives the formula (9.87).

5. Derive equation (9.98) for the material factor $\mathbf{V}$ of a plate in the limit of forward scattering.
Solution: With $\hat{s} = \hat{\imath}$, the vector products involved in (9.89), (9.90), and (9.92) can be evaluated in the simple form

$$\hat{s} \times (\hat{n} \times \hat{\alpha}_{\text{inc}}) = -\hat{\alpha}_{\text{inc}} (\hat{n} \cdot \hat{\imath}), \quad \hat{s} \times \hat{\alpha}_{\text{inc}} = \hat{\beta}_{\text{inc}}, \quad \hat{s} \times \hat{\beta}_{\text{inc}} = -\hat{\alpha}_{\text{inc}}. \tag{9.289}$$

These relations lead to the formula

$$\mathbf{V} = 2(\hat{n} \cdot \hat{\imath})[\hat{\alpha}_{\text{inc}}(\mathbf{E}_0 \cdot \hat{\alpha}_{\text{inc}}) + \hat{\beta}_{\text{inc}}(\mathbf{E}_0 \cdot \hat{\beta}_{\text{inc}})], \tag{9.290}$$

which is equivalent to (9.98) since $\mathbf{E}_0 \cdot \hat{\imath} = 0$ and the unit vectors $\hat{\imath}$, $\hat{\alpha}_{\text{inc}}$, and $\hat{\beta}_{\text{inc}}$ form an orthonormal basis.

6. Derive expression (9.103) for the coefficient $\mathbf{V}$ in the case of specular scattering ($\hat{s} = \hat{r}$) from an impenetrable plate.
Solution: Using the definitions $\hat{r} = \hat{\imath} - 2\hat{n}(\hat{\imath} \cdot \hat{n})$, $\hat{\alpha}_{\text{ref}} = \hat{\alpha}_{\text{inc}}$, and $\hat{\beta}_{\text{ref}} = \hat{r} \times \hat{\alpha}_{\text{inc}}$ one obtains the relations

$$\hat{s} \times (\hat{n} \times \hat{\alpha}_{\text{inc}}) = \hat{\alpha}_{\text{ref}} (\hat{n} \cdot \hat{\imath}), \quad \hat{s} \times \hat{\alpha}_{\text{inc}} = \hat{\beta}_{\text{ref}}, \quad \hat{s} \times \hat{\beta}_{\text{ref}} = -\hat{\alpha}_{\text{ref}}, \tag{9.291}$$

which transforms (9.92) in (9.103).

7. Derive expression (9.107) for the vector coefficient **V** in the case of backscattering from an impenetrable plate.
Hint: Since $\hat{s} = -\hat{i}$, relations (9.289) can be used on alternating the sign on their right-hand sides.

8. Evaluate the shape factor $I_{po}(\mathbf{q}, k)$ for an elliptic plate.
Hint: Specialize the line integral (9.121) to the case of an ellipse and find a way of evaluating this integral analytically.
Solution: Let a and b be the semi-major and semi-minor axes of an ellipse in the x–y plane, so that the parametric representation of the ellipse is given by

$$x = a\cos t, \quad y = b\sin t \tag{9.292}$$

with $0 \le t \le 2\pi$. Specializing the line-integral representation (9.121) for the shape factor to the case of an ellipse gives

$$I_{po}(\mathbf{q}, k) = \frac{j}{k(q_x^2 + q_y^2)} \int_0^{2\pi} e^{-jk(aq_x \cos t + bq_y \sin t)} (bq_x \cos t + aq_y \sin t)\, dt. \tag{9.293}$$

The linear combination of trigonometric functions in the phase can be expressed as

$$aq_x \cos t + bq_y \sin t = \rho_0 \cos(t - \theta_0) \tag{9.294}$$

in terms of the new parameters ρ_0 and θ_0 defined by

$$aq_x = \rho_0 \cos\theta_0, \quad bq_y = \rho_0 \sin\theta_0. \tag{9.295}$$

The new parameters can be found as

$$\rho_0 = \sqrt{a^2 q_x^2 + b^2 q_y^2}, \quad \theta_0 = \arctan\left(\frac{bq_y}{aq_x}\right). \tag{9.296}$$

Similarly,

$$bq_x \cos t + aq_y \sin t = \rho_1 \cos(t - \theta_1) \tag{9.297}$$

if ρ_1 and θ_1 are defined by

$$bq_x = \rho_1 \cos\theta_1, \quad aq_y = \rho_1 \sin\theta_1 \tag{9.298}$$

with

$$\rho_1 = \sqrt{b^2 q_x^2 + a^2 q_y^2}, \quad \theta_1 = \arctan\left(\frac{aq_y}{bq_x}\right). \tag{9.299}$$

These representations transform (9.293) to the form

$$I_{po}(\mathbf{q}, k) = \frac{j\rho_1}{k(q_x^2 + q_y^2)} \int_0^{2\pi} e^{-jk\rho_0 \cos(t-\theta_0)} \cos(t - \theta_1)\, dt, \tag{9.300}$$

which can be reduced to an integral representation of the Bessel function (9.128).

To this end, introduce a new integration variable $t' = t - \theta_0$ to obtain

$$\int_0^{2\pi} e^{-jk\rho_0\cos(t-\theta_0)}\cos(t-\theta_1)\,dt = \int_{-\theta_0}^{2\pi-\theta_0} e^{-jk\rho_0\cos t'}\cos(t'-\delta\theta)\,dt' \tag{9.301}$$

with $\delta\theta = \theta_1 - \theta_0$, and since the integration is performed over the period of the integrand, the integral can be rewritten as

$$\int_{-\pi}^{\pi} e^{-jk\rho_0\cos t}\cos(t-\delta\theta)\,dt. \tag{9.302}$$

Now, writing $\cos(t-\delta\theta) = \cos t\cos\delta\theta + \sin t\sin\delta\theta$ and noting that

$$\int_{-\pi}^{\pi} e^{-jk\rho_0\cos t}\cos t\,dt = -2\pi j J_1(k\rho_0), \quad \int_{-\pi}^{\pi} e^{-jk\rho_0\cos t}\sin t\,dt = 0 \tag{9.303}$$

transform (9.300) to the expression

$$I_{po}(\mathbf{q},k) = 2\pi\rho_1\frac{\cos(\theta_1-\theta_0)}{k(q_x^2+q_y^2)}J_1(k\rho_0), \tag{9.304}$$

where the parameters ρ_0, θ_0, ρ_1, and θ_1 are determined from (9.296) and (9.299).

9. Prove that relations (9.151) and (9.152) comply with the formula (9.102).
Hint: In the limit $q_\perp = 0$, (9.151) and (9.152) lead to the relation

$$I_{po}(\mathbf{q},k) \to -\frac{1}{2}e^{-jk\mathbf{q}\cdot\mathbf{r}_c}\sum_{l=1}^{L}[(\mathbf{r}_{l+1}-\mathbf{r}_l)\cdot\hat{n}\times\hat{e}][(2\mathbf{r}_c-\mathbf{r}_{l+1}-\mathbf{r}_l)\cdot\hat{e}], \tag{9.305}$$

and in order to comply with (9.102) the relation

$$|\Omega_L| = -\frac{1}{2}\sum_{l=1}^{L}[(\mathbf{r}_{l+1}-\mathbf{r}_l)\cdot\hat{n}\times\hat{e}][(2\mathbf{r}_c-\mathbf{r}_{l+1}-\mathbf{r}_l)\cdot\hat{e}] \tag{9.306}$$

must hold for any vector $\hat{e}$ lying in the plane of the polygon. Here $|\Omega_L|$ denotes the area of the polygon with L vertices and $\mathbf{r}_c$ is a point in the plane of the polygon which can be freely chosen. Thus, the problem is to prove that the area of any L-gon can be expressed as (9.306) where $\mathbf{r}_l$ are the position vectors of the vertices of the polygon ($\mathbf{r}_{L+1} = \mathbf{r}_1$).
Solution: Since any polygon can be represented as a combination of triangles, it is sufficient to prove (9.306) for $L = 3$. Writing down the sum and combining the terms with $l = 2$ and $l = 3$ give

$$\begin{aligned}&\sum_{l=1}^{3}[(\mathbf{r}_{l+1}-\mathbf{r}_l)\cdot\hat{n}\times\hat{e}][(2\mathbf{r}_c-\mathbf{r}_{l+1}-\mathbf{r}_l)\cdot\hat{e}]\\ &= [(\mathbf{r}_2-\mathbf{r}_1)\cdot\hat{n}\times\hat{e}][(2\mathbf{r}_c-\mathbf{r}_2-\mathbf{r}_1)\cdot\hat{e}] + [(\mathbf{r}_3-\mathbf{r}_1)\cdot\hat{n}\times\hat{e}][(\mathbf{r}_1-\mathbf{r}_2)\cdot\hat{e}],\end{aligned} \tag{9.307}$$

and choosing $\mathbf{r}_c = (\mathbf{r}_2+\mathbf{r}_3)/2$, which is a point on the segment connecting the vertices with $l = 2$ and $l = 3$, transforms the right-hand side of (9.307) into the form

$$[(\mathbf{r}_2-\mathbf{r}_1)\cdot\hat{n}\times\hat{e}][(\mathbf{r}_3-\mathbf{r}_1)\cdot\hat{e}] - [(\mathbf{r}_3-\mathbf{r}_1)\cdot\hat{n}\times\hat{e}][(\mathbf{r}_2-\mathbf{r}_1)\cdot\hat{e}]. \tag{9.308}$$

Since $\hat{e}$ is a unit vector in the plane of the polygon, it is orthogonal to the unit normal $\hat{n}$ and $\hat{e} \times (\hat{n} \times \hat{e}) = \hat{n}$. The vectors connecting the vertices can be represented as linear combinations of the unit orthogonal vectors $\hat{e}$ and $\hat{n} \times \hat{e}$,

$$\mathbf{r}_2 - \mathbf{r}_1 = \hat{e}(\mathbf{r}_2 - \mathbf{r}_1) \cdot \hat{e} + (\hat{n} \times \hat{e})[(\mathbf{r}_2 - \mathbf{r}_1) \cdot \hat{n} \times \hat{e}]$$
$$\mathbf{r}_3 - \mathbf{r}_1 = \hat{e}(\mathbf{r}_3 - \mathbf{r}_1) \cdot \hat{e} + (\hat{n} \times \hat{e})[(\mathbf{r}_3 - \mathbf{r}_1) \cdot \hat{n} \times \hat{e}],$$

and it is then seen that

$$\hat{n} \cdot (\mathbf{r}_3 - \mathbf{r}_1) \times (\mathbf{r}_2 - \mathbf{r}_1) \tag{9.309}$$

is equal to (9.308), implying that the right-hand side of (9.306) is equal to

$$\frac{1}{2}\hat{n} \cdot (\mathbf{r}_2 - \mathbf{r}_1) \times (\mathbf{r}_3 - \mathbf{r}_1), \tag{9.310}$$

which reduces to a known formula for the area of a triangle with vertices at $\mathbf{r}_1$, $\mathbf{r}_2$, and $\mathbf{r}_3$,

$$\frac{1}{2}|\mathbf{r}_2 - \mathbf{r}_1||\mathbf{r}_3 - \mathbf{r}_1| \sin A, \tag{9.311}$$

where A is the angle between the sides of the triangle at the vertex $l = 1$.

10. Derive the shape factor of a right-angle plate.

Hint: Specialize the general expression for the shape factor of plates to the case of a plate with four vertices and two pairs of parallel sides.

Solution: If a and b are the side lengths, $\mathbf{r}_l$ are the position vectors of the vertices ($l = 1, 2, 3, 4$), and $\mathbf{r}_0 = (\mathbf{r}_1 + \mathbf{r}_2 + \mathbf{r}_3 + \mathbf{r}_4)/4$ is the middle point of the plate, then

$$I_{\text{po}}(\mathbf{q}, k) = ab\, \mathrm{e}^{-jkQ_0} \operatorname{sinc}\left(\frac{k}{2}Q_1\right) \operatorname{sinc}\left(\frac{k}{2}Q_2\right) \tag{9.312}$$

with $Q_0 = \mathbf{q} \cdot \mathbf{r}_0$, $Q_1 = \mathbf{q} \cdot (\mathbf{r}_2 - \mathbf{r}_1)$, and $Q_2 = \mathbf{q} \cdot (\mathbf{r}_3 - \mathbf{r}_2)$.

11. Figures 9.25 and 9.26 show the forward-scattering patterns of a square and rectangular plate, respectively. Explain the position of minima in the patterns.

Hint: Use the formula for the shape factor of a right-angled plate.

Solution: The minima correspond to zeros of the shape factor, which means that

$$\operatorname{sinc}\left(\frac{k}{2}Q_1\right) \operatorname{sinc}\left(\frac{k}{2}Q_2\right) = 0, \tag{9.313}$$

where $Q_1 = \hat{s} \cdot (\mathbf{r}_2 - \mathbf{r}_1)$, $Q_2 = \hat{s} \cdot (\mathbf{r}_3 - \mathbf{r}_2)$ ($\hat{i}$ is perpendicular to the plate), and $\mathbf{r}_l$ with $l = 1, 2, 3$ are any three adjacent corner points. Thus, a minimum is a solution of either equation

$$\frac{k}{2}Q_1 = m_1\pi, \quad \frac{k}{2}Q_2 = m_2\pi, \tag{9.314}$$

where $m_1, m_2 = 0, \pm 1, \pm 2, \ldots$ (the case $m_1 = 0$ and $m_2 = 0$ is excluded as it corresponds to the main maximum of the scattering pattern).

Let the sides of the right-angled plate be such that

$$\mathbf{r}_2 - \mathbf{r}_1 = l_1\hat{y}, \quad \mathbf{r}_3 - \mathbf{r}_2 = -l_2\hat{x}, \tag{9.315}$$

where $l_1 = |\mathbf{r}_2 - \mathbf{r}_1|$ and $l_2 = |\mathbf{r}_3 - \mathbf{r}_2|$. On the image screen, $\hat{s} = (x\hat{x} + y\hat{y} + z_a\hat{z})/r$, where $r = \sqrt{x^2 + y^2 + z_a^2} \approx |z_a|$. Thus,

$$Q_1 \approx \frac{y}{|z_a|} l_1, \quad Q_2 \approx \frac{x}{|z_a|} l_2, \tag{9.316}$$

and it follows from (9.314) that the shape factor vanishes at the two periodic families of lines:

$$\text{any } x, \quad y = 2\pi \frac{|z_a|}{kl_1} m_1, \quad m_1 = 0, \pm 1, \pm 2, \ldots \tag{9.317}$$

and

$$x = 2\pi \frac{|z_a|}{kl_2} m_2, \quad \text{any } y, \quad m_2 = 0, \pm 1, \pm 2, \ldots \tag{9.318}$$

The period of the lines parallel to the x (y) axis is determined by the length of the sides perpendicular to the x (y) axis, which fully corresponds to the shape and position of the black regions on Figures 9.25 and 9.26.

12. Prove that in the PO approximation the fields scattered by an arbitrarily shaped PEC plate in the directions given by the unit vectors $\hat{s}$ and $\hat{s}_1 = \hat{s} - 2(\hat{n} \cdot \hat{s})\hat{n}$ have equal amplitudes. Here $\hat{n}$ is the unit normal to the plate.

Hint: Start with the PO solution for the scattered field (9.94) and show that the substitution $\hat{s} \leftrightarrow \hat{s}_1$ does not change (a) the absolute value of the shape factor and (b) the amplitude $|\mathbf{V}|$ of the material factor if the plate is PEC.

Solution: The shape factor (9.95) can be expressed as

$$I_{\text{po}}(\mathbf{q}, k) = e^{-jk\mathbf{q}\cdot\mathbf{r}_c} \int_{\Omega_0} e^{-jk\mathbf{q}\cdot(\mathbf{r}'-\mathbf{r}_c)} \, d\Omega', \tag{9.319}$$

where $\mathbf{r}_c$ is an arbitrary point in the plane of the plate, e.g. a vertex or the middle point of the plate, and since the vector $\mathbf{r}' - \mathbf{r}_c$ lies in the plane of the plate, the integral in (9.319) depends only on the projection of the vector $\mathbf{q}$ on the plane of the plate. Vectors $\mathbf{q} = \hat{i} - \hat{s}$ and $\mathbf{q}_1 = \hat{i} - \hat{s}_1$ have equal projection on the plane of the plate, implying that $|I_{\text{po}}(\mathbf{q}, k)|$ is invariant under the interchange $\hat{s} \leftrightarrow \hat{s}_1$.

Next consider the material factor $\mathbf{V}$ defined in (9.92) as $\mathbf{V} = \hat{s} \times \hat{s} \times \mathbf{v}_e + \hat{s} \times \mathbf{v}_m$ with $\mathbf{v}_e$ and $\mathbf{v}_m$ from (9.89) and (9.90). The latter vectors are proportional to the equivalent surface currents and therefore tangential to the plate, $\mathbf{v}_e = (v_x^e, v_y^e, 0)$ and $\mathbf{v}_m = (v_x^m, v_y^m, 0)$, where x and y are Cartesian coordinates in the plane of the plate. Calculating vector products with the unit vector $\hat{s} = (s_x, s_y, s_z)$, where z is the coordinate perpendicular to the plate, we obtain

$$\hat{s} \times \mathbf{v}_m = s_z\left(-\hat{x}v_y^m + \hat{y}v_x^m \right) + \hat{z}\left(s_x v_y^m - s_y v_x^m \right) \tag{9.320}$$

$$\hat{s} \times \hat{s} \times \mathbf{v}_e = \hat{x}\left[s_y\left(s_x v_y^e - s_y v_x^e \right) - s_z^2 v_x^e \right] - \hat{y}\left[s_x\left(s_x v_y^e - s_y v_x^e \right) + s_z^2 v_y^e \right] + \hat{z}\left[s_x v_x^e + s_y v_y^e \right] s_z. \tag{9.321}$$

Since $\hat{s}_1 = (s_x, s_y, -s_z)$, the products $\hat{s}_1 \times \mathbf{v}_m$ and $\hat{s}_1 \times \hat{s}_1 \times \mathbf{v}_e$ are given by (9.320) and (9.321) with $s_z \to -s_z$. Since $\mathbf{V}$ is a complex-valued vector, its amplitude is defined by $|\mathbf{V}| = (\mathbf{V} \cdot \mathbf{V}^*)^{1/2}$, and it follows from (9.320) and (9.321) that in general $|\mathbf{V}|$ is not an even function of s_z. For PEC plates, however, $\mathbf{v}_m = 0$, $\mathbf{V} = \hat{s} \times \hat{s} \times \mathbf{v}_e$, and

$$|\mathbf{V}|^2 = \left| s_y\left(s_x v_y^e - s_y v_x^e \right) - s_z^2 v_x^e \right|^2 + \left| s_x\left(s_x v_y^e - s_y v_x^e \right) + s_z^2 v_y^e \right|^2 + s_z^2 \left| s_x v_x^e + s_y v_y^e \right|^2 \tag{9.322}$$

is an even function of s_z, implying the invariance of $|\mathbf{V}|$ with respect to the interchange $\hat{s} \leftrightarrow \hat{s}_1$.

13. Calculate the Gaussian curvature (A.75) of a body of revolution described in the parametric form (9.165).
Answer:

$$K = -\frac{\rho''(z)}{\rho(z)[1+\rho'(z)^2]^2}. \tag{9.323}$$

Solution: Assuming that $\mathbf{f}(u,v) = \mathbf{r}(\phi, z)$, we obtain for the first fundamental forms (A.67) with the help of (9.167) and (9.168)

$$E = \rho^2(z), \quad F = 0, \quad G = 1 + \rho'(z)^2. \tag{9.324}$$

The second fundamental forms (A.71) include the unit vector $\hat{n}$, which can be exterior or interior to the body of revolution; however, since the expression (A.73) is quadratic with respect to $\hat{n}$, the result is independent of whether the chosen normal is interior or exterior. We use for $\hat{n}$ the expression (9.171) with the upper sign and find that

$$L = -\frac{\rho(z)}{\sqrt{1+\rho'(z)^2}}, \quad M = 0, \quad N = \frac{\rho''(z)}{\sqrt{1+\rho'(z)^2}}. \tag{9.325}$$

Inserting (9.324) and (9.325) in (A.73) gives (9.323).

For example, the Gaussian curvature of a spherical surface of radius a is independent of the position on the surface and equal to $K = a^{-2}$. For a parabolic surface with $\rho(z) = \sqrt{2az}$, where $0 \leq z$ and a is the curvature radius at the vertex $z = 0$, the formula (9.323) gives

$$K(z) = \frac{1}{(a+2z)^2}, \tag{9.326}$$

which is independent of the azimuth ϕ, as expected for every body of revolution.

14. Prove that the PO solution (9.188)–(9.191) for the scattered field of bodies of revolution illuminated along the symmetry axis is compliant with the asymptotic property (3.123) of the far-field coefficient for the forward scattering.
Hint: Forward scattering means that the incidence and scattering directions coincide. Use this definition to simplify the PO solution.

Solution: The body is illuminated along the z axis and in the forward-scattering case $\hat{z} = \hat{i} = \hat{s}$, implying that $\mathbf{q} = 0$, $p = 0$, and

$$I_{20} = I_{02} = \pi, \quad I_{10} = I_{01} = I_{11} = 0. \tag{9.327}$$

These relations simplify the integrals to

$$\mathbf{V}_e^{(1)} = \pi \mathbf{E}_0 \int_{z_{min}}^{z_{max}} (R_\perp - 1)\rho(z)\rho'(z)\,dz \tag{9.328}$$

$$\mathbf{V}_e^{(2)} = -\pi \mathbf{E}_0 \int_{z_{min}}^{z_{max}} (R_\parallel + 1)\rho(z)\rho'(z)\,dz \tag{9.329}$$

$$\mathbf{V}_{\mathrm{m}}^{(1)} = \pi(\hat{x}E_{0y} - \hat{y}E_{0x}) \int_{z_{\min}}^{z_{\max}} (R_{\|} - 1)\rho(z)\rho'(z)\,\mathrm{d}z \tag{9.330}$$

$$\mathbf{V}_{\mathrm{m}}^{(2)} = \pi(-\hat{x}E_{0y} + \hat{y}E_{0x}) \int_{z_{\min}}^{z_{\max}} (R_{\perp} + 1)\rho(z)\rho'(z)\,\mathrm{d}z \tag{9.331}$$

and leads to

$$\mathbf{V}_{\mathrm{e}} = \pi\mathbf{E}_0 \int_{z_{\min}}^{z_{\max}} (R_{\perp} - R_{\|} - 2)\rho(z)\rho'(z)\,\mathrm{d}z \tag{9.332}$$

$$\mathbf{V}_{\mathrm{m}} = \pi(\hat{x}E_{0y} - \hat{y}E_{0x}) \int_{z_{\min}}^{z_{\max}} (R_{\|} - R_{\perp} - 2)\rho(z)\rho'(z)\,\mathrm{d}z. \tag{9.333}$$

Since $\hat{s} = -\hat{z}$ and

$$\hat{s} \times \mathbf{E}_0 = -\hat{x}E_{0y} + \hat{y}E_{0x}, \tag{9.334}$$

we obtain

$$\hat{s} \times \mathbf{V}_e + \mathbf{V}_m = -4\pi(\hat{x}E_{0y} - \hat{y}E_{0x}) \int_{z_{\min}}^{z_{\max}} \rho(z)\rho'(z)\,\mathrm{d}z \tag{9.335}$$

in which the reflection coefficients are canceled out. The integral is easily evaluated as

$$\int_{z_{\min}}^{z_{\max}} \rho(z)\rho'(z)\,\mathrm{d}z = \int_{\rho_{\mathrm{rim}}}^{0} \rho\,\mathrm{d}\rho = -\frac{1}{2}\rho_{\mathrm{rim}}^2 \tag{9.336}$$

since $\rho(z_{\max}) = 0$ and $\rho(z_{\min}) = \rho_{\mathrm{rim}}$ for any convex body of revolution (Figure 9.29, left). Inserting in (9.174) gives

$$\mathbf{F}_{\mathrm{po}}(-\hat{z}, -\hat{z}) = -\frac{j}{2}(k\rho_{\mathrm{rim}})^2\mathbf{E}_0, \tag{9.337}$$

which recovers the optical limit (3.123) for the far-field coefficient with $S_0 = \pi\rho_{\mathrm{rim}}^2$.

15. Does (9.337) hold in a concave case?

Answer: Yes. In the concave case (Figure 9.29, right), the integrals (9.328)–(9.331) should be multiplied with -1 (which follows from (9.171)). On the other hand, in contrast to the convex case, $\rho(z_{\max}) = \rho_{\mathrm{rim}}$ and $\rho(z_{\min}) = 0$, and the value of the integral (9.336) alternates its sign as well, implying that the expression (9.337) for the far-field coefficient remains unchanged.

16. Prove the relations

$$\frac{1}{\pi}\int_0^{2\pi} \mathrm{e}^{-jp\cos(\phi-\phi_q)} \cos^2\phi\,\mathrm{d}\phi = J_0(p) - \cos(2\phi_q)J_2(p) \tag{9.338}$$

$$\frac{1}{\pi}\int_{0}^{2\pi} e^{-jp\cos(\phi-\phi_q)}\cos\phi\sin\phi\, d\phi = -\sin(2\phi_q)J_2(p) \tag{9.339}$$

$$\frac{1}{\pi}\int_{0}^{2\pi} e^{-jp\cos(\phi-\phi_q)}\sin^2\phi\, d\phi = J_0(p) + \cos(2\phi_q)J_2(p), \tag{9.340}$$

where $J_0(p)$ and $J_2(p)$ are the Bessel functions.
Hint: Use the identities

$$\cos^2\phi = \frac{1+\cos(2\phi)}{2}, \quad \sin\phi\cos\phi = \frac{1}{2}\sin(2\phi), \quad \sin^2\phi = \frac{1-\cos(2\phi)}{2}$$

and the guidelines around equation (9.202).

17. Expressions (9.174) and (9.188)–(9.191) describe the far-field coefficient of imperfectly reflecting bodies of revolution illuminated along their symmetry axis, and the expression (9.206) with (9.208) the far-field coefficient of PEC bodies of revolution under oblique incidence. Prove the agreement between these two solutions in the limit of axial illumination and PEC scattering bodies.
Hint: Show that

$$2\mathbf{I}_n \times \hat{i} \times \mathbf{E}_0|_{\hat{i}=-\hat{z}} = V_e^{(1)} + V_e^{(2)}|_{R_\parallel=-R_\perp=1}\,. \tag{9.341}$$

Solution: With $\hat{i} = -\hat{z}$ and $\mathbf{E}_0 = \hat{x}E_{0x} + \hat{y}E_{0y}$, we get from (9.206) and (9.208) that

$$2\mathbf{I}_n \times \hat{i} \times \mathbf{E}_0 = -2\int_{z_{min}}^{z_{max}} e^{-jkq_z z}\rho(z)[\rho'(z)I_{00}\mathbf{E}_0 + \hat{z}(I_{10}E_{0x} + I_{01}E_{0y})]\, dz. \tag{9.342}$$

On the other hand, equations (9.188) and (9.189) with $R_\parallel = 1$ and $R_\perp = -1$ give

$$V_e^{(1)} + V_e^{(2)} = -2\int_{z_{min}}^{z_{max}} e^{-jkq_z z}\rho(z)[\rho'(z)(I_{20} + I_{02})\mathbf{E}_0 + \hat{z}(I_{10}E_{0x} + I_{01}E_{0y})]\, dz, \tag{9.343}$$

which coincides with (9.342) since

$$I_{20} + I_{02} = 2\pi J_0(p) = I_{00} \tag{9.344}$$

because of (9.197), (9.199), and (9.209).

18. Derive the formula (9.242).

Hint: The reflection coefficients of an impedance surface, which will be denoted by $\widetilde{R}_{\parallel}(\theta)$ and $\widetilde{R}_{\perp}(\theta)$, are given by (9.220). They depend on $\cos\theta$ only. For bodies of revolution it follows from (9.171) that

$$\cos\theta = \mp\frac{\rho'(z)}{\sqrt{1+\rho'(z)^2}}, \tag{9.345}$$

where the sign on the right-hand side ensures that $\cos\theta \geq 0$.

Solution: Since $\widetilde{R}_{\perp,\parallel}$ are functions of $\cos\theta$, we can write

$$\frac{\mathrm{d}}{\mathrm{d}z}\{\widetilde{R}_{\parallel}[\theta(z)] - \widetilde{R}_{\perp}[\theta(z)]\} = \frac{\mathrm{d}\cos\theta}{\mathrm{d}z}\frac{\mathrm{d}}{\mathrm{d}\cos\theta}(\widetilde{R}_{\parallel} - \widetilde{R}_{\perp}) \tag{9.346}$$

and using (9.220) obtain

$$\frac{\mathrm{d}}{\mathrm{d}\cos\theta}(\widetilde{R}_{\parallel} - \widetilde{R}_{\perp}) = \frac{2\eta(1-\eta^2)\sin^2\theta}{(\cos\theta+\eta)^2(\eta\cos\theta+1)^2}. \tag{9.347}$$

To calculate $(\cos\theta)'$, we use (9.345) and write

$$\frac{\mathrm{d}\cos\theta}{\mathrm{d}z} = \frac{\mathrm{d}\cos\theta}{\mathrm{d}\rho'}\rho''(z) = \mp\frac{\rho''(z)}{[1+\rho'(z)^2]^{3/2}}. \tag{9.348}$$

It follows from (9.345) that

$$\sin^2\theta = \frac{1}{1+\rho'(z)^2}, \tag{9.349}$$

and combining the results we express (9.346) as

$$\frac{\mathrm{d}}{\mathrm{d}z}\{\widetilde{R}_{\parallel}[\theta(z)] - \widetilde{R}_{\perp}[\theta(z)]\} = \frac{2\eta(1-\eta^2)\rho''(z)\cos\theta\sin^4\theta}{\rho'(z)(\cos\theta+\eta)^2(\eta\cos\theta+1)^2}. \tag{9.350}$$

Finally, noting that

$$\widetilde{R}_{\parallel}(\theta) - \widetilde{R}_{\perp}(\theta) = \frac{2(1-\eta^2)\cos\theta}{(\cos\theta+\eta)(\eta\cos\theta+1)} \tag{9.351}$$

$$\widetilde{R}_{\parallel}(\theta) + \widetilde{R}_{\perp}(\theta) = -\frac{2\eta\sin^2\theta}{(\cos\theta+\eta)(\eta\cos\theta+1)}, \tag{9.352}$$

we obtain the relation

$$\frac{\mathrm{d}}{\mathrm{d}z}\{\widetilde{R}_{\parallel}[\theta(z)] - \widetilde{R}_{\perp}[\theta(z)]\} = -\frac{\rho''(z)\sin^2\theta}{2\rho'(z)}\left\{\widetilde{R}^2_{\parallel}[\theta(z)] - \widetilde{R}^2_{\perp}[\theta(z)]\right\}, \tag{9.353}$$

which is equivalent to (9.242).

19. Specify the terms of the order $O(1)$ in the far-field coefficient (9.250) for the axial backscattering from a truncated convex impedance paraboloid.

Hint: Use the formula (9.242) to calculate the derivative of the reflection coefficients.

Solution: Correction terms are determined by the derivative of the integrand function as given in (9.226). Specializing to an impedance surface,

$$A'(z) = -\frac{1}{2}\frac{\rho(z)\rho''(z)}{1+\rho'(z)^2}\left(\widetilde{R}_\parallel^2 - \widetilde{R}_\perp^2\right) + \frac{1}{2}(\widetilde{R}_\parallel - \widetilde{R}_\perp)[\rho(z)^2]''. \tag{9.354}$$

In the case of paraboloids, $\rho(z)^2$ is a linear function of z (see (9.249)), implying that $[\rho(z)^2]'' = 0$ and $\rho(z)\rho''(z) = -\rho'(z)^2$. Thus,

$$A'(z) = \frac{1}{2}\frac{\rho'(z)^2}{1+\rho'(z)^2}(\widetilde{R}_\parallel^2 - \widetilde{R}_\perp^2) \tag{9.355}$$

or using the relation (9.172),

$$A'(z) = \frac{1}{2}\cos^2\theta\left[\widetilde{R}_\parallel^2(\theta) - \widetilde{R}_\perp^2(\theta)\right]. \tag{9.356}$$

It follows from (9.356) that because of $\widetilde{R}_\parallel(0) = \widetilde{R}_\perp(0)$,

$$A'(0) = 0 \tag{9.357}$$

and

$$A'(z_{\text{min}}) = \frac{1}{2}\cos^2\theta_{\text{rim}}[\widetilde{R}_\parallel^2(\theta_{\text{rim}}) - \widetilde{R}_\perp^2(\theta_{\text{rim}})]. \tag{9.358}$$

Thus, the contribution from the specular reflection is still given by the formula (9.234), but the expression (9.243) for the rim contribution is supplemented with the correction term to give

$$\mathbf{F}_{\text{po}}^{\text{bs}}\Big|_{\text{rim}} = \mathbf{E}_0\frac{k}{4}\rho_{\text{rim}}\cot\theta_{\text{rim}}[R_\parallel(\theta_{\text{rim}}) - R_\perp(\theta_{\text{rim}})]\,\mathrm{e}^{2jkz_{\text{min}}}\left\{1 + \frac{\sin(2\theta_{\text{rim}})}{8jk\rho_{\text{rim}}}[\widetilde{R}_\parallel(\theta_{\text{rim}}) + \widetilde{R}_\perp(\theta_{\text{rim}})]\right\} + o(1). \tag{9.359}$$

Note that the term $O(1)$ vanishes for PEC boundary conditions as

$$\widetilde{R}_\parallel(\theta_{\text{rim}}) + \widetilde{R}_\perp(\theta_{\text{rim}}) = 0.$$

20. Prove the identity of the PO solutions for the axial backscattering cross section of a PEC sphere and a 90° PEC cone with the height equal to the radius of the sphere.

Hint: Specialize the PO solution (9.219) for the far-field coefficient to a PEC cone with $\alpha = 45°$ and $z_{\text{min}} = -a$, and compare with the formula that results from (9.219) for a PEC sphere of the radius a.

Solution: For a PEC sphere equation (9.219) reads

$$\mathbf{F}_{\text{sphere}}^{\text{PEC}} = -jk^2\mathbf{E}_0\int_{-a}^{0}\mathrm{e}^{2jkz}(a+z)\,\mathrm{d}z, \tag{9.360}$$

which is exactly evaluated as (9.255).

Equation (9.219) for a PEC cone with $\alpha = 45°$ and height a gives

$$\mathbf{F}_{\text{cone}}^{\text{PEC}} = jk^2\mathbf{E}_0\int_{-a}^{0}\mathrm{e}^{2jkz}z\,\mathrm{d}z, \tag{9.361}$$

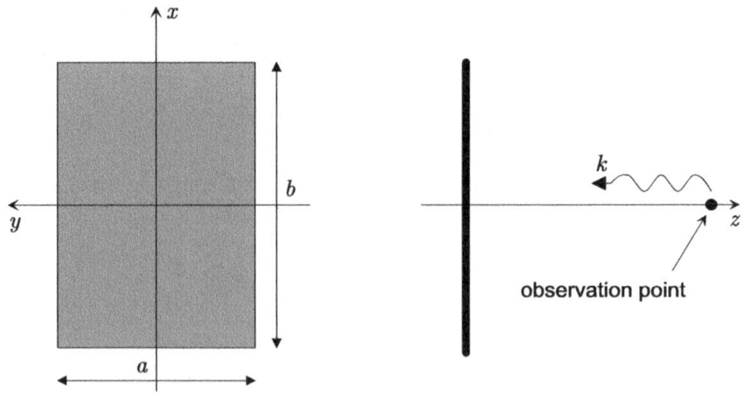

Figure 9.35 A rectangular PEC plate is illuminated by a plane wave generated by a distant antenna, normal incidence.

which can be exactly evaluated as

$$\mathbf{F}^{\mathrm{PEC}}_{\mathrm{cone}} = \mathbf{E}_0 \frac{ka}{2} \mathrm{e}^{-2jka} \left[1 - \frac{j}{2ka}(1 - \mathrm{e}^{2jka})\right]. \tag{9.362}$$

The cross section (9.248) is determined by the magnitude of the far-field coefficient and since

$$|\mathbf{F}^{\mathrm{PEC}}_{\mathrm{sphere}}| = \frac{ka}{2} |\mathbf{E}_0| \left|1 + \frac{j}{2ka}(1 - \mathrm{e}^{-2jka})\right| \tag{9.363}$$

and

$$|\mathbf{F}^{\mathrm{PEC}}_{\mathrm{cone}}| = \frac{ka}{2} |\mathbf{E}_0| \left|1 - \frac{j}{2ka}(1 - \mathrm{e}^{2jka})\right|, \tag{9.364}$$

the cross sections of the sphere and the cone coincide as absolute values of complex conjugated quantities.

21. In this and the following problems find the scattered fields by direct integration of induced surface currents.

Using the PO approximation, study far-zone backscattering from a square PEC plate illuminated by a normally incident plane wave (Figure 9.35).

Solution: An ideally conducting metal plate is illuminated by a plane wave

$$\mathbf{E}_{\mathrm{inc}} = \hat{y} E_0 \, \mathrm{e}^{jkz} \tag{9.365}$$

(Figure 9.35), and we are interested in the backscattered field at point z. The distance from the plate to the observation point is large compared to the plate size: $z \gg a, b$.

The induced current density in the PO approximation follows from (8.3) as

$$\mathbf{K}^{\mathrm{go}}_{\mathrm{e}} = 2\hat{n} \times \mathbf{H}_{\mathrm{inc}} = \frac{2}{Z} \hat{y} E_0. \tag{9.366}$$

The electric field generated by the induced current is given by (8.5)

$$\mathbf{E}_{\mathrm{sc}} = -\frac{jkZ}{4\pi} \int_{\Omega_0} \frac{\mathrm{e}^{-jk|\mathbf{r}-\mathbf{r}'|}}{|\mathbf{r}-\mathbf{r}'|} \mathbf{K}^{\mathrm{go}}_{\mathrm{e}} \, \mathrm{d}\Omega. \tag{9.367}$$

In this case in formula (8.5) $\hat{s}' \times (\hat{s}' \times \mathbf{K}_e^{go}) = -\mathbf{K}_e^{go}$ since far from the plate $\hat{s}' \approx \hat{z}$ and is perpendicular to $\mathbf{K}_e^{go}$. Thus, in the far zone, scattered fields are transverse and the current component along $\hat{s}'$ does not contribute to the fields propagating along z. In (9.367) Ω_0 is the illuminated surface of the plate. For the normal incidence $\mathbf{K}_e^{go}$ is constant over the plate surface, and the distance from a point at the plate with the coordinates x', y' to the observation point is

$$|\mathbf{r} - \mathbf{r}'| = \sqrt{z^2 + x'^2 + y'^2} \approx z + \frac{1}{2z}(x'^2 + y'^2). \tag{9.368}$$

Thus, the scattered field (9.367) can be written as

$$\mathbf{E}_{sc} = -j\hat{y}E_0 \frac{k}{2\pi} \frac{e^{-jkz}}{z} \int_{\Omega_0} e^{-j\frac{k}{2z}(x'^2+y'^2)} \, d\Omega. \tag{9.369}$$

In the denominator we have approximated $r \approx z$. Furthermore, if

$$\frac{k}{2z}\frac{a^2}{4} \ll 1, \quad \frac{k}{2z}\frac{b^2}{4} \ll 1, \tag{9.370}$$

the integrand can be approximated by unity, and the result is

$$\mathbf{E}_{sc} = -j\hat{y}E_0 \frac{kS}{2\pi} \frac{e^{-jkz}}{z}, \tag{9.371}$$

where $S = ab$ is the area of the plate. Finally, the scattering cross section reads

$$\sigma_0 = 4\pi z^2 \frac{|E_{sc}|^2}{|E_0|^2} = \frac{k^2 S^2}{\pi} = 4\pi \frac{S^2}{\lambda^2}. \tag{9.372}$$

This is a familiar expression from the antenna theory,

$$4\pi \frac{S^2}{\lambda^2} = D_0 S, \tag{9.373}$$

where D_0 is the directivity of an aperture of area S with a uniform field distribution (Balanis 1997, eq. (12-37)).

22. Solve the previous problem for the case of oblique incidence (Figure 9.36).

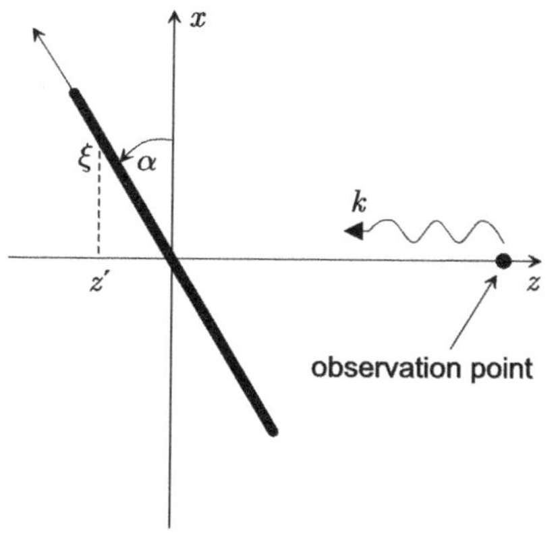

Figure 9.36 Oblique incidence of a plane wave on a rectangular PEC plate.

Solution: The induced current density is now

$$\mathbf{K}_{\mathrm{e}}^{\mathrm{go}} = \hat{y}\cos\alpha \frac{2}{Z} E_0\, \mathrm{e}^{jkz'}. \tag{9.374}$$

Under the same assumptions (9.370),

$$|\mathbf{r}-\mathbf{r}'| = \sqrt{(z-z')^2 + x'^2 + y'^2} \approx z - z' \tag{9.375}$$

and

$$E_{\mathrm{sc}} = \frac{-jk\cos\alpha}{2\pi z} E_0 \int_{\Omega_0} e^{jkz'}\, \mathrm{e}^{-jk(z-z')}\, \mathrm{d}\Omega. \tag{9.376}$$

Since $z' = \xi \sin\alpha$,

$$E_{\mathrm{sc}} = \frac{-jk\cos\alpha}{2\pi z} \mathrm{e}^{-jkz} b E_0 \int_{-a/2}^{a/2} e^{2jk\xi \sin\alpha}\, \mathrm{d}\xi. \tag{9.377}$$

The results is

$$E_{\mathrm{sc}} = \frac{-jkS\cos\alpha}{2\pi} \frac{\mathrm{e}^{-jkz}}{z} \frac{\sin X}{X} E_0, \tag{9.378}$$

where $X = ka\sin\alpha$. The factor $\frac{\sin X}{X}$ is again familiar from the antenna theory: it is the same as the pattern of an aperture antenna with a uniform amplitude at the aperture (Balanis 1997, section 12.5). The scattering cross section reads

$$\sigma = \sigma_0 \cos^2\alpha \frac{\sin^2 X}{X^2}, \tag{9.379}$$

where σ_0 is given by (9.372).

23. Study diffraction by an ideally conducting cylinder (Figure 9.37, incident electric field along the cylinder axis). Assume that the radius is large compared to the wavelength ($ka \gg 1$) and make use of the PO approximation.

Solution: The incident electric field is a plane wave

$$E_{\mathrm{inc}} = \mathrm{e}^{jkx}. \tag{9.380}$$

The induced current density is, in the PO approximation,

$$\mathbf{K}_{\mathrm{e}}^{\mathrm{go}} = 2\hat{n} \times \mathbf{H}_{\mathrm{inc}} = \frac{2}{Z}\hat{n} \times \hat{y}\, e^{jkx'} = \frac{2}{Z}\hat{z}\cos\phi'\, e^{jkx'}. \tag{9.381}$$

The scattered electric field is

$$\mathbf{E}_{\mathrm{sc}} = -\frac{kZ}{4} \int_{-\pi/2}^{\pi/2} \mathbf{K}_{\mathrm{e}}^{\mathrm{go}} H_0^{(2)}(k\rho)\, a\, \mathrm{d}\phi', \tag{9.382}$$

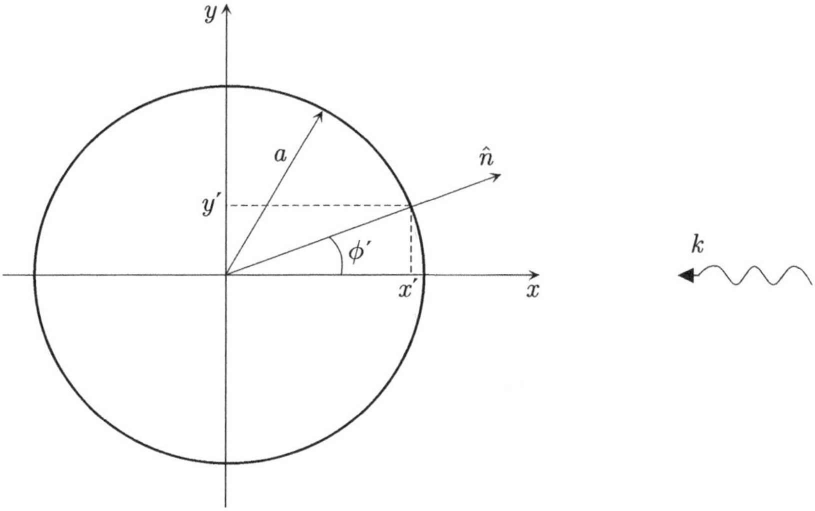

Figure 9.37 Large PEC cylinder and related coordinate system.

where $\rho = \sqrt{(x - x')^2 + (y - y')^2}$ is the distance between the observation point and the point (x', y'), and the integral is taken over the illuminated side of the cylinder. In the far zone, substituting the asymptotic expression of the Hankel function and the expression for the induced current surface density, we get

$$E_{sc} = -\frac{ka}{2}\sqrt{\frac{2j}{\pi k}} \int_{-\pi/2}^{\pi/2} \frac{e^{-jk(\rho - x')}}{\sqrt{\rho}} \cos\phi' \, d\phi'. \tag{9.383}$$

Let us calculate the field at the x axis ($y = 0$) using the stationary phase method. The phase function is

$$\Phi = \rho - x' = \sqrt{(x - a\cos\phi')^2 + a^2\sin^2\phi'} - a\cos\phi'. \tag{9.384}$$

The derivative is

$$\frac{d\Phi}{d\phi'} = \frac{2xa\sin\phi'}{2R} + a\sin\phi'. \tag{9.385}$$

Equating this to zero, we see that the stationary point is at $\phi' = 0$ (as expected, at the specular reflection point). The second derivative at $\phi' = 0$ reads

$$\frac{d^2\Phi}{d\phi'^2} = \frac{a(2x - a)}{x - a}. \tag{9.386}$$

The scattered field becomes (using (D.19))

$$E_{sc} = -\sqrt{\frac{a}{2x - a}}\, e^{-jk(x - 2a)}. \tag{9.387}$$

The contributions from the end points (at $\phi' = \pm\pi/2$) are zero because the surface current density tends to zero at these points.

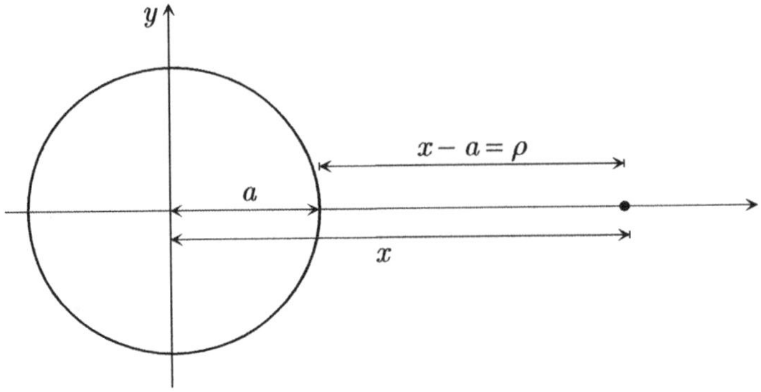

Figure 9.38 Distances in the formula for the scattered field.

The simple physical meaning of this result can be understood with the help of Figure 9.38. The "reflection coefficient" is

$$\Gamma = -\sqrt{\frac{a}{2x-a}} = -\sqrt{\frac{a}{2\rho + a}}. \tag{9.388}$$

For large $a \gg \rho$ and a fixed distance ρ, $\Gamma \to -1$ (as from a conducting plane). For large $\rho \gg a$,

$$\Gamma \approx -\sqrt{\frac{a}{2\rho}} \ll 1. \tag{9.389}$$

Comparing with the exact solution (section 6.4.3) we see that the main contribution (the reflected field) has been calculated correctly, but of course the creeping waves are totally missing in the PO model.

24. Study diffraction by a PEC wedge (Figure 9.39) in the PO approximation. Assume the incidence of a plane electromagnetic wave polarized along the edge. Derive the diffraction coefficient in the PO approximation and compare with the exact diffraction coefficient.

Solution: Let us consider the E-polarized plane-wave excitation (Figure 9.39). The incident electric field (z component) is

$$E_{\text{inc}} = e^{jkx\cos\phi_0 + jky\sin\phi_0}. \tag{9.390}$$

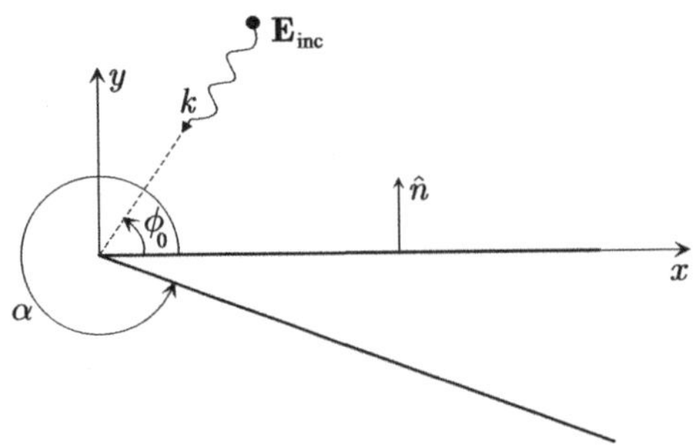

Figure 9.39 Geometry of the problem. The wedge occupies the region $\alpha < \phi < 2\pi$. The incidence angle is ϕ_0.

Suppose that $\phi_0 < \alpha - \pi$, that is, only one of the faces is illuminated. In the PO approximation the induced current density is approximated by

$$\mathbf{K}_{\mathrm{e}}^{\mathrm{go}} = 2\hat{n} \times \mathbf{H}_{\mathrm{inc}} = 2\hat{y} \times \hat{x} H_x^{\mathrm{inc}} = -2\hat{z} H_x^{\mathrm{inc}}. \tag{9.391}$$

At the wedge surface

$$H_x^{\mathrm{inc}}|_{y=0} = -\frac{1}{Z} \sin\phi_0 E_{\mathrm{inc}}|_{y=0}. \tag{9.392}$$

Thus,

$$\mathbf{K}_{\mathrm{e}}^{\mathrm{go}} = \frac{2}{Z} \sin\phi_0 \, \mathrm{e}^{jkx\cos\phi_0} \hat{z}. \tag{9.393}$$

Because there is no dependence on z and the induced current is directed along z, the scattered field is

$$\mathbf{E}_{\mathrm{sc}} = -\frac{kZ}{4} \int_0^\infty \mathbf{K}_{\mathrm{e}}^{\mathrm{go}}(x') H_0^{(2)}[k\sqrt{(x-x')^2 + y^2}] \, \mathrm{d}x'. \tag{9.394}$$

Let us denote

$$R = \sqrt{(x-x')^2 + y^2} \tag{9.395}$$

and calculate the field in the far zone, assuming that $kR \gg 1$. To do that, we replace the Hankel function in (9.394) by its asymptotic expression for large arguments and substitute $\mathbf{K}_{\mathrm{e}}^{\mathrm{go}}$ (9.393). The integral (9.394) becomes

$$E_{\mathrm{sc}} \approx -\frac{\sqrt{jk}\sin\phi_0}{\sqrt{2\pi}} \int_0^\infty \frac{\mathrm{e}^{-jk(R - x'\cos\phi_0)}}{\sqrt{R}} \, \mathrm{d}x'. \tag{9.396}$$

This integral can be estimated using the stationary phase method. The phase of the exponent is

$$\Phi = R - x'\cos\phi_0. \tag{9.397}$$

To find the stationary point we calculate its derivative and equate that to zero:

$$\frac{\mathrm{d}\Phi}{\mathrm{d}x'} = \frac{x - x'}{R} - \cos\phi_0 = \cos\psi - \cos\phi_0 = 0. \tag{9.398}$$

The solution of this equation is illustrated in Figure 9.40. The stationary point, which corresponds to the mirror reflection of the incident wave, does not exist for all observation positions. The coordinate of the stationary point is

$$x'_0 = x + y\cot\phi_0. \tag{9.399}$$

To use the stationary phase formula (D.19) we need the second derivative of the phase function

$$\Phi''|_{x'=x'_0} = \frac{1}{R_0} - \frac{(x - x'_0)^2}{R_0^3} = \frac{y^2}{R_0^3} = \frac{\sin^2\phi_0}{R_0}, \tag{9.400}$$

where $R_0 = \sqrt{y^2 + (x - x'_0)^2}$.

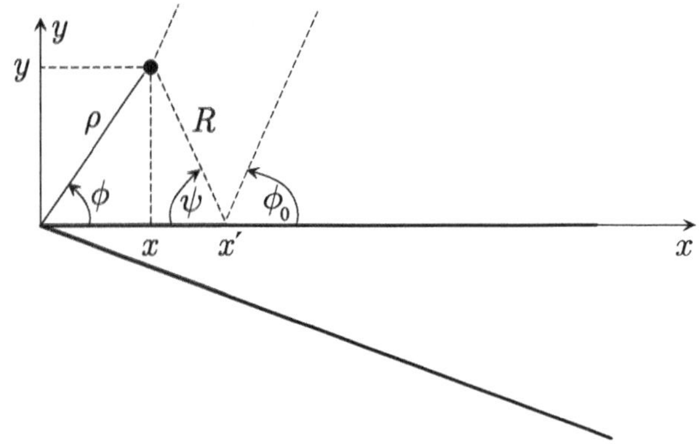

Figure 9.40 The stationary point at $\psi = \phi_0$ corresponds to the specular reflection point. The observation point has the coordinates (ρ, ϕ).

Using formula (D.19), we get for the stationary-point contribution

$$E_{sc} = f_0 \, e^{-jk\Phi_0} \sqrt{\frac{2\pi}{jk|\Phi''|}} = -\frac{\sqrt{jk}\sin\phi_0}{\sqrt{2\pi}} \frac{1}{\sqrt{R_0}} e^{-jk(R_0 - x_0' \cos\phi_0)} \frac{\sqrt{R_0}}{\sin\phi_0} \sqrt{\frac{2\pi}{jk}}$$
$$= -e^{-jk(R_0 - x_0' \cos\phi_0)}.$$

This is the reflected plane wave because

$$k(R_0 - x_0 \cos\phi_0) = -k\rho\cos(\phi + \phi_0). \tag{9.401}$$

Finally, we will estimate the contribution from the end point (the edge). Here we use formula (D.7)

$$\int_0^\infty f(x)\, e^{-jk\Phi(x)}\, dx \approx \frac{f(0)}{jk\Phi'(0)} e^{-jk\Phi(0)}. \tag{9.402}$$

The derivative

$$\Phi'(0) = \cos\psi - \cos\phi_0|_{x'=0} = -(\cos\phi + \cos\phi_0), \tag{9.403}$$

because $\psi = \pi - \phi$ at $x' = 0$. Thus, we have

$$E_{edge} = \sqrt{\frac{-j}{2\pi k\rho}}\, e^{-jk\rho} \frac{\sin\phi_0}{\cos\phi + \cos\phi_0}. \tag{9.404}$$

This can be compared with with the exact solution (Chapter 5). Let us recall that the corresponding asymptotic approximation (5.135) of the exact solution for a half-plane gives

$$E_{edge} = \sqrt{\frac{-j}{2\pi k\rho}}\, e^{-jk\rho} \frac{2\sin\frac{\phi}{2}\sin\frac{\phi_0}{2}}{\cos\phi + \cos\phi_0}. \tag{9.405}$$

Interestingly, at the shadow boundary $\phi = \pi \pm \phi_0$ the results are the same. For other observation angles the error can be significant.

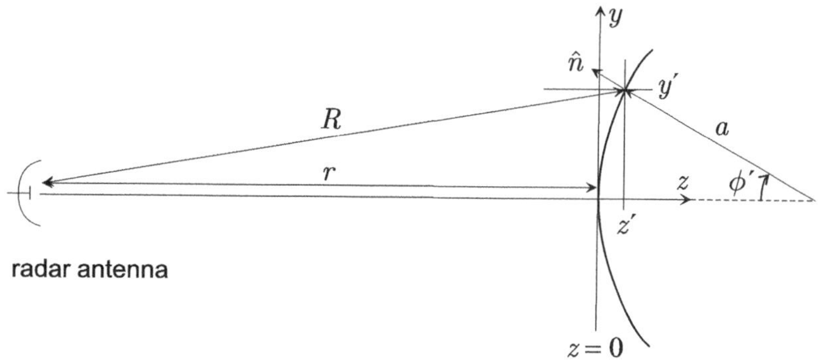

Figure 9.41 A PEC sphere illuminated by a plane wave generated by a distant antenna ($r \gg a$).

25. Study backscattering from a large (in wavelengths) ideally conducting sphere in the PO approximation. The geometry is illustrated in Figure 9.41. The sphere radius is a, and the distance from the source to the sphere surface is r. The receiver is at the same place as the transmitter.

Solution: The incident fields are

$$\mathbf{E}_{\mathrm{inc}} = \mathbf{E}_0 e^{-jkz}, \quad \mathbf{H}_{\mathrm{inc}} = \frac{1}{Z}\hat{z} \times \mathbf{E}_{\mathrm{inc}}. \tag{9.406}$$

Thus, the induced current density is approximately

$$\mathbf{K}_{\mathrm{e}}^{\mathrm{go}} = \frac{2}{Z}\hat{n} \times (\hat{z} \times \mathbf{E}_{\mathrm{inc}}). \tag{9.407}$$

Expanding

$$\hat{n} \times (\hat{z} \times \mathbf{E}_{\mathrm{inc}}) = \hat{z}(\hat{n} \cdot \mathbf{E}_{\mathrm{inc}}) - \mathbf{E}_{\mathrm{inc}}(\hat{n} \cdot \hat{z}) \tag{9.408}$$

we note that the first term on the right-hand side is of no importance for us since it corresponds to longitudinal (z-directed) current component. This current does not radiate along the z axis. With this in view, the current density becomes

$$\mathbf{K}_{\mathrm{e}}^{\mathrm{go}} = -\frac{2}{Z}(\hat{n} \cdot \hat{z})\mathbf{E}_{\mathrm{inc}}. \tag{9.409}$$

The scattered field (8.5) reads

$$\mathbf{E}_{\mathrm{sc}} = \frac{jk}{2\pi r}\mathbf{E}_0 \int_{\Omega_0} (\hat{n} \cdot \hat{z})\, e^{-jk(z'+R)}\, \mathrm{d}\Omega', \tag{9.410}$$

where $R = |\mathbf{r} - \mathbf{r}'|$.

Consider the phase function

$$\Phi = z' + R = z' + \sqrt{(r+z')^2 + y'^2 + x'^2} \approx r + 2z' = r + 2(a - a\cos\phi'). \tag{9.411}$$

Because of the symmetry, we can integrate in the plane $x' = 0$ and then take the square of the result. Here the z' coordinate of a point on the sphere (see Figure 9.41) is $z' = a - a\cos\phi'$, and its derivative with respect to ϕ' is $\mathrm{d}z'/\mathrm{d}\phi' = a\sin\phi'$. Obviously, there is a stationary point at $\phi' = 0$, where $z' = 0$. Physically this corresponds to the

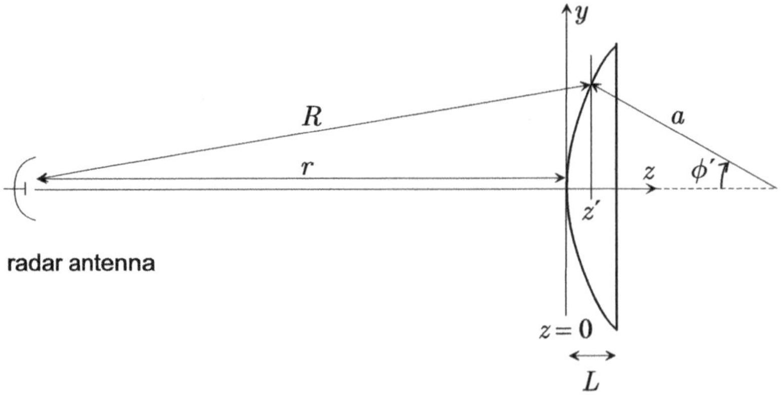

Figure 9.42 A spherical segment (depth L) illuminated by a plane wave generated by a distant antenna ($r \gg a$).

specular reflection point. To apply the stationary phase method we need to know the second derivative of the phase function at the stationary point: it equals to $2a$.

There is no first-order contribution from the end points because the current density at the shadow boundary tends to zero. In this situation we can approximately write $\mathrm{d}\Omega' = a^2 \mathrm{d}\phi' \mathrm{d}\psi'$, where ψ' is measured from the source direction in the orthogonal plane. Applying (D.19) for both linear integrations we get

$$\mathbf{E}_{\mathrm{sc}} = -\frac{a}{2}\frac{e^{-jkr}}{r}\mathbf{E}_0, \tag{9.412}$$

which is the spherical wave scattered by the object. The radar cross section of the sphere reads

$$\sigma = 4\pi r^2 \frac{|E_{\mathrm{sc}}|^2}{|E_0|^2} = \pi a^2. \tag{9.413}$$

As we can see, for electrically large ($a \gg \lambda$) PEC spheres this is simply the geometric cross section area of the sphere.

One might think that the radar cross section of large bodies simply increases with the body size. This is, however, wrong. Consider now a spherical segment (Figure 9.42). If $L \ll a$, the size of this body is much smaller than that of the complete sphere. However, the main contribution to the radar cross section which comes from the stationary point is still the same as in the case of the complete sphere (9.413). Indeed, the value of the integral is mainly determined by a small area around the stationary point, and it depends on the surface *curvature* at this point but not on the integration area. The difference is that now there are also additional contributions from the body edges. The result for the segment is of the form

$$\sigma = \pi a^2 |1 + \mathrm{const}\, e^{-j2kL}|^2, \tag{9.414}$$

and it can be smaller or larger than that of the complete sphere of the same radius.

26. Find the backscattered field from a PEC cone of length l excited by a plane wave traveling along the z axis

$$\mathbf{E}_{\mathrm{inc}} = \hat{y} E_0 e^{-jkz} \tag{9.415}$$

(see Figure 9.43).

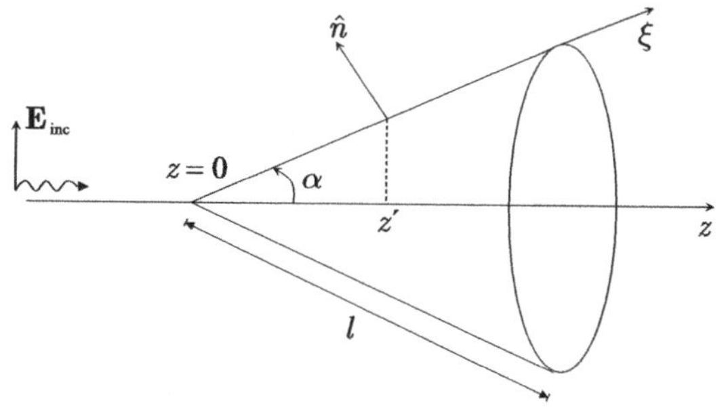

Figure 9.43 A PEC cone of finite length l is illuminated by a plane wave generated by a distant antenna.

Solution: Calculating the induced current as

$$\mathbf{K}_{\mathrm{e}}^{\mathrm{go}} = 2\hat{n} \times \mathbf{H}_{\mathrm{inc}} \tag{9.416}$$

we need to know only the component of the surface current density vector that is orthogonal to the axis z because the longitudinal component does not radiate in the backward direction (along z). This component is directed along y (the incident electric field direction), and its amplitude equals

$$K_{\mathrm{e}}^{\mathrm{go}} = \frac{2}{Z} e^{-jkz'} E_0 \sin \alpha. \tag{9.417}$$

Denoting the distance from the cone tip to the observation point by r, and assuming that the cone dimensions are very small compared to r, the distance from an arbitrary point on the illuminated side of the cone to the observation point is $|\mathbf{r} - \mathbf{r}'| \approx r + z'$. The scattered electric field can then be written as

$$E_{\mathrm{sc}} = -\frac{jk \sin \alpha}{2\pi} \frac{e^{-jkr}}{r} E_0 \int_{\Omega_0} e^{-2jkz'} \, \mathrm{d}\Omega', \tag{9.418}$$

where the integral is taken over the illuminated surface. Here, $z' = \xi \cos \alpha$. Introducing the integration angle ϕ around the cone, we have

$$\mathrm{d}\Omega' = \xi \sin \alpha \, \mathrm{d}\phi \, \mathrm{d}\xi. \tag{9.419}$$

The integral becomes

$$E_{\mathrm{sc}} = -\frac{jk \sin^2 \alpha}{2\pi} \frac{e^{-jkr}}{r} E_0 \int_0^{2\pi} d\phi \int_0^{l} \xi e^{-2jk\xi \cos \alpha} \, \mathrm{d}\xi. \tag{9.420}$$

Integrating by parts, we finally get

$$E_{\mathrm{sc}} = j \tan^2 \alpha \frac{e^{-jkr}}{4kr} [1 - e^{-2jkl \cos \alpha}(1 + 2jkl \cos \alpha)] E_0. \tag{9.421}$$

In this formula, the first term in the brackets can be interpreted as a “contribution from the tip”, and the second one (that depends on l), as a “contribution from the rim of the cone bottom”.

The backscattering cross section is

$$\sigma = 4\pi r^2 \frac{|E_{\mathrm{sc}}|^2}{|E_0|^2} = \sigma_0 |1 - e^{-2jkl \cos \alpha}(1 + 2jkl \cos \alpha)|^2, \tag{9.422}$$

where

$$\sigma_0 = \frac{\pi \tan^4 \alpha}{4k^2} = \frac{\lambda^2}{16\pi} \tan^4 \alpha \tag{9.423}$$

is the nose-on backscattering cross section of the infinite PEC cone (9.77).

References

Abramowitz M and Stegun IA (eds) 1965 *Handbook of Mathematical Functions with Formulas, Graphs, and Mathematical Tables*, Dover Publications.

Albani M, Carluccio G, and Pathak PH 2011 Uniform ray description for the PO scattering by vertices in curved surface with curvilinear edges and relatively general boundary conditions. *IEEE Trans. Antennas Propag.*, **59**, 1587–1596.

Asvestas JS 1986 The physical optics fields of an aperture on a perfectly conducting screen in terms of line integrals. *IEEE Trans. Antennas Propag.*, **34**, 1155–1159.

Babich VM, Smyshlyaev VP, Dement'ev DB, and Samokish BA 1996 Numerical calculation of the diffraction coefficients for an arbitrary shaped perfectly conducting cone. *IEEE Trans. Antennas Propag.*, **44**, 740–747.

Bailin LL and Silver S 1956 Exterior electromagnetic boundary value problems for spheres and cones. *IRE Trans. Antennas Propag.*, **4**, 5–16; errata **5**, 313.

Balanis CA 1997 *Antenna Theory: Analysis and Design*, 2nd edition, J. Wiley & Sons.

Bernard JML, Lyalinov MA, and Ning Yan Zhu 2008 Analytical-numerical calculation of diffraction coefficients for a circular impedance cone. *IEEE Trans. Antennas Propag.*, **56**, 1616–1623.

Blume S 1971 Theorie der Beugung elektromagnetischer Wellen am Kegel elliptischen Querschnitts. *Ann. Phys. (Berlin)*, **481**, 302–308.

Blume S 1996 Spherical-multipole analysis of electromagnetic and acoustical scattering by a semi-infinite elliptic cone. *IEEE Antennas Propag. Mag.*, **38**, no. 2, 33–44.

Blume S and Kahl G 1987 The physical optics radar cross section of an elliptic cone. *IEEE Trans. Antennas Propag.*, **35**, 457–460.

Blume S and Krebs V 1998 Numerical evaluation of dyadic diffraction coefficients and bistatic radar cross sections for a perfectly conducting semi-infinite elliptic cone. *IEEE Trans. Antennas Propag.*, **46**, 414–424.

Blume S and Uschkerat U 1995 The radar cross section of the semi-infinite elliptic cone: numerical evaluation. *Wave Motion*, **22**, 311–324.

Blume S, Klinkenbusch L, and Uschkerat U 1993 The radar cross section of the semi-infinite elliptic cone. *Wave Motion*, **17**, 365–389.

Born M and Wolf E (eds) 1959 *Principles of Optics: Electromagnetic Theory of Propagation, Interference and Diffraction of Light*, Pergamon Press.

Borovikov VA and Kinber BY 1994 *Geometrical Theory of Diffraction*, The Institution of Electrical Engineers, London.

Bouman JJ, Senior TBA, and Uslenghi PLE (eds) 1987 *Electromagnetic and Acoustic Scattering by Simple Shapes*, Hemisphere Publishing Corporation.

Crispin JW and Siegel KM (eds) 1968 *Methods of Radar Cross-Section Analysis*, Academic Press.

Felsen LB 1957 Plane-wave scattering by small-angle cones. *IRE Trans. Antennas Propag.*, **5**, 121–129.

Gordon WB 1975 Far-field approximations to the Kirchhoff–Helmholtz representations of scattered fields. *IEEE Trans. Antennas Propag.*, **13**, 590–592.

Gordon WB 2003 Calculating scatter from surfaces with zero curvature. *IEEE Trans. Antennas Propag.*, **51**, 2506–2508.

Goriainov AS 1961 Diffraction of a plane electromagnetic wave propagating along the axis of a cone. *Radio Eng. Electron. Phys. (USSR)*, **6**, 65–81 (English transl. of *Radiotekhnika i Elektronika*, **6**, no. 1, 47–57).

Hansen TB 1990 Diffraction by a plane angular sector, a new derivation. *IEEE Trans. Antennas Propag.*, **38**, 1892–1894.

Hansen TB 1991 Corner diffraction coefficients for the quarter plane. *IEEE Trans. Antennas Propag.*, **39**, 976–984.

Johansen PM and Breinbjerg O 1995 An exact line integral representation of the physical optics scattered field: the case of a perfectly conducting polyhedral structure illuminated by electric Hertzian dipoles. *IEEE Trans. Antennas Propag.*, **43**, 689–696.

Jones DS 1964 *The Theory of Electromagnetism*, Pergamon Press.

Klinkenbusch L 2007 Electromagnetic scattering by semi-infinite circular and elliptic cones. *Radio Sci.*, **42**, RS6S10, doi:10.1029/2007RS003649.

Kraus L and Levine LM 1961 Diffraction by an elliptic cone. *Comm. Pure Appl. Math.*, **14**, 49–68.

Lyalinov MA and Zhu NY 2013 *Scattering of Waves by Wedges and Cones with Impedance Boundary Conditions*, SciTech Publishing.

Meixner J and Andrejewski W 1950 Strenge Theorie der Beugung ebener elektromagnetischer Wellen an der vollkommen leitenden Kreisscheibe und an der kreisförmigen Öffnung im vollkommen leitenden ebenen Schirm. *Ann. Phys. (Berlin)*, **442**, 157–168.

Osipov AV, Kobayashi H, and Hongo K 1998 Shadow boundary currents in the problem of high-frequency electromagnetic diffraction by a circular impedance cylinder. *IEICE Trans. Electron.*, **E81-C**, 1655–1666.

Pelosi G, Toso G, and Martini E 2000 PO near-field expression of a penetrable planar structure in terms of a line integral. *IEEE Trans. Antennas Propag.*, **48**, 1274–1276.

Ruck GT, Barrick DE, Stuart WD, and Krichbaum CK 1970 *Radar Cross Section Handbook*, Plenum Press.

Satterwhite R 1974 Diffraction by a quarter plane, the exact solution, and some numerical results. *IEEE Trans. Antennas Propag.*, **22**, 500–503.

Senior TBA 1977 Some extensions of Babinet's principle in electromagnetic theory. *IEEE Trans. Antennas Propag.*, **25**, 417–420.

Senior TBA and Volakis JL 1995 *Approximate Boundary Conditions in Electromagnetics*, The Institution of Electrical Engineers, London.

Siegel KM, Crispin JW, and Schensted CE 1955 Electromagnetic and acoustical scattering from a semi-infinite cone. *J. Appl. Phys.*, **26**, 309–313.

Smyshlyaev VP 1993 The high-frequency diffraction of electromagnetic waves by cones of arbitrary cross-section. *SIAM J. Appl. Math.*, **53**, 670–688.

Taket ND and Burge RE 1991 A physical optics version of the geometrical theory of diffraction. *IEEE Trans. Antennas Propag.*, **39**, 719–731.

Trott KD, Pathak PH, and Molinet FA 1990 A UTD type analysis of the plane wave scattering by a fully illuminated perfectly conducting cone. *IEEE Trans. Antennas Propag.*, **38**, 1150–1160.

Ufimtsev PY 1968 Diffraction of electromagnetic waves at black bodies and semi-transparent plates. *Radiophys. Quantum Electron.*, **11**, 527–538.

Ufimtsev PY 2003 *Theory of Edge Diffraction in Electromagnetics*, Tech Science Press.

Ufimtsev PY 2014 *Fundamentals of the Physical Theory of Diffraction*, Wiley.

Vafiadis E and Sahalos JN 1983 Diffraction by a perfectly conducting elliptic cone. *Arch. Elektrotechn.*, **66**, 117–124.

Vesnik MV and Ufimtsev PY 1992 An asymptotic feature of corner waves scattered by polygonal plates. *Electromagnetics*, **12**, 265–272.

Weston VH 1963 Theory of absorbers in scattering. *IEEE Trans. Antennas Propag.*, **11**, 578–584.

Youssef NN 1989 Radar cross section of complex targets. *Proc. IEEE*, **77**, 722–734.

A

Definitions and Useful Relations of Vector Analysis and Differential Geometry

This appendix is a collection of definitions and useful relations that are used throughout the book. Formulas from vector algebra and vector analysis can be found in sections A.1 and A.2, respectively. For a more extensive collection of vector identities and a systematic exposition of the vector analysis the reader is referred to the many good books on the subject, for example Korn and Korn (1968), Marsden and Tromba (2003), and Schey (2005).

Orthogonal curvilinear coordinates, particularly the spherical coordinates, are used in this book on many occasions. Section A.3 is a collection of expressions for vectors and vector differential operators in such coordinates. A more detailed presentation of the subject can be found, for example, in Stratton (1941, sections 1.14–1.18), Morse and Feshbach (1953, sections 1.3 and 1.4), Bouman *et al.* (1987, Appendix C), and Korn and Korn (1968, Chapter 6).

Section A.4 gives the most essential definitions and formulas from differential geometry that are related to curves and surfaces in three-dimensional space. These formulas and definitions are extensively used in Chapter 8. For further information about the subject see, for example Korn and Korn (1968, Chapter 17) and Weatherburn (1955).

A.1 Vector Algebra

In this book, vectors are typeset in bold face, for example **A**. In the Cartesian coordinate system, a vector can be represented as

$$\mathbf{A} = A_x\hat{x} + A_y\hat{y} + A_z\hat{z} \tag{A.1}$$

with real- or complex-valued components A_x, A_y, and A_z.

The scalar product (or dot product) of two vectors **A** and **B** is a scalar defined by the formula

$$\mathbf{A} \cdot \mathbf{B} = A_xB_x + A_yB_y + A_zB_z. \tag{A.2}$$

The magnitude of a vector **A** is defined by the Euclidean norm

$$|\mathbf{A}| = (\mathbf{A} \cdot \mathbf{A}^*)^{\frac{1}{2}} = \sqrt{|A_x|^2 + |A_y|^2 + |A_z|^2}, \tag{A.3}$$

Modern Electromagnetic Scattering Theory with Applications, First Edition. Andrey V. Osipov and Sergei A. Tretyakov.

where $\mathbf{A}^* = A_x^*\hat{x} + A_y^*\hat{y} + A_z^*\hat{z}$ and the asterisk denotes the complex conjugate. For a vector with real-valued components, $|\mathbf{A}|$ gives the Pythagorean length. Real-valued vectors of unit length are indicated by a hat and typeset in italic, that is,

$$\mathbf{A} = \hat{A}, \quad \text{if} \quad |\mathbf{A}| = 1. \tag{A.4}$$

A vector

$$\frac{\mathbf{A}}{|\mathbf{A}|} = \frac{A_x}{|\mathbf{A}|}\hat{x} + \frac{A_y}{|\mathbf{A}|}\hat{y} + \frac{A_z}{|\mathbf{A}|}\hat{z} \tag{A.5}$$

is apparently a unit vector.

The scalar product of two real-valued vectors can be also written as

$$\mathbf{A} \cdot \mathbf{B} = |\mathbf{A}||\mathbf{B}| \cos\theta, \tag{A.6}$$

where θ is the angle between the vectors.

The vector product (or cross product) is a vector defined by

$$\mathbf{A} \times \mathbf{B} = (A_y B_z - A_z B_y)\hat{x} + (A_z B_x - A_x B_z)\hat{y} + (A_x B_y - A_y B_x)\hat{z}. \tag{A.7}$$

For real-valued vectors,

$$\mathbf{A} \times \mathbf{B} = \hat{e}|\mathbf{A}||\mathbf{B}| \sin\theta, \tag{A.8}$$

where θ is the angle between the vectors. The unit vector $\hat{e}$ is perpendicular to both **A** and **B** and its direction is such that the vectors **A**, **B**, and $\mathbf{A} \times \mathbf{B}$ build the right-handed system of vectors. In contrast to the scalar product, the vector product is anticommutative

$$\mathbf{A} \times \mathbf{B} = -\mathbf{B} \times \mathbf{A}. \tag{A.9}$$

Similarly to the scalar product, the vector product is distributive over addition, which means that

$$\mathbf{A} \times (\mathbf{B} + \mathbf{C}) = \mathbf{A} \times \mathbf{B} + \mathbf{A} \times \mathbf{C}. \tag{A.10}$$

The scalar triple product is a combination of the scalar and vector products, and is invariant under the cyclic interchange of its components

$$\mathbf{A} \cdot (\mathbf{B} \times \mathbf{C}) = \mathbf{B} \cdot (\mathbf{C} \times \mathbf{A}) = \mathbf{C} \cdot (\mathbf{A} \times \mathbf{B}). \tag{A.11}$$

A vector triple product can be expressed as

$$\mathbf{A} \times (\mathbf{B} \times \mathbf{C}) = \mathbf{B}(\mathbf{A} \cdot \mathbf{C}) - \mathbf{C}(\mathbf{A} \cdot \mathbf{B}). \tag{A.12}$$

Note that the order of multiplication is important and

$$\mathbf{A} \times (\mathbf{B} \times \mathbf{C}) \neq (\mathbf{A} \times \mathbf{B}) \times \mathbf{C} \tag{A.13}$$

in general.

In a triangle with the vertices at **A**, **B**, and **C** the side lengths are related by the formula

$$|\mathbf{B} - \mathbf{A}| = \sqrt{|\mathbf{C} - \mathbf{A}|^2 + |\mathbf{B} - \mathbf{C}|^2 + 2(\mathbf{C} - \mathbf{A}) \cdot (\mathbf{B} - \mathbf{C})}. \tag{A.14}$$

If $|\mathbf{C}-\mathbf{A}| \gg |\mathbf{B}-\mathbf{C}|$, then this relation can be linearized

$$|\mathbf{B}-\mathbf{A}| \approx |\mathbf{C}-\mathbf{A}| + \frac{1}{|\mathbf{C}-\mathbf{A}|}(\mathbf{C}-\mathbf{A})\cdot(\mathbf{B}-\mathbf{C})\,. \tag{A.15}$$

A.2 Vector Analysis

A vector can be a function of the position vector $\mathbf{r} = \hat{x}x + \hat{y}y + \hat{z}z$ in space, in which case we have a vector function or a vector field. A vector field can be obtained, for example, by applying the gradient operator ∇ to any scalar differentiable function $f(\mathbf{r})$ and in the Cartesian coordinate system the result can be written as

$$\nabla f = \hat{x}\frac{\partial f}{\partial x} + \hat{y}\frac{\partial f}{\partial y} + \hat{z}\frac{\partial f}{\partial z}. \tag{A.16}$$

By using the gradient operator, one can also introduce two kinds of differentiation of a vector function $\mathbf{A}(\mathbf{r})$, the divergence and the curl, denoted by $\nabla\cdot\mathbf{A}(\mathbf{r})$ and $\nabla\times\mathbf{A}(\mathbf{r})$, respectively. These operations can be defined in arbitrary coordinate systems, and in Cartesian coordinates they are given by the relations:

$$\nabla\cdot\mathbf{A} = \frac{\partial A_x}{\partial x} + \frac{\partial A_y}{\partial y} + \frac{\partial A_z}{\partial z} \tag{A.17}$$

$$\nabla\times\mathbf{A} = \hat{x}\left(\frac{\partial A_y}{\partial z} - \frac{\partial A_z}{\partial y}\right) + \hat{y}\left(\frac{\partial A_z}{\partial x} - \frac{\partial A_x}{\partial z}\right) + \hat{z}\left(\frac{\partial A_x}{\partial y} - \frac{\partial A_y}{\partial x}\right). \tag{A.18}$$

Application of the divergence operator gives a scalar, whereas the curl operator produces a vector.

Note the following useful formulas:

$$\nabla(fg) = f\nabla g + g\nabla f \tag{A.19}$$

$$\nabla\cdot(f\mathbf{A}) = \mathbf{A}\cdot\nabla f + f\nabla\cdot\mathbf{A} \tag{A.20}$$

$$\nabla\times(f\mathbf{A}) = \nabla f\times\mathbf{A} + f\nabla\times\mathbf{A} \tag{A.21}$$

$$\nabla\cdot(\mathbf{A}\times\mathbf{B}) = \mathbf{B}\cdot(\nabla\times\mathbf{A}) - \mathbf{A}\cdot\nabla\times\mathbf{B}. \tag{A.22}$$

Application of the divergence operator to the gradient of a scalar function gives the Laplace operator (or Laplacian):

$$\nabla^2 f = \nabla\cdot\nabla f = \frac{\partial^2 f}{\partial x^2} + \frac{\partial^2 f}{\partial y^2} + \frac{\partial^2 f}{\partial z^2}. \tag{A.23}$$

Extension to vector fields is called the vector Laplacian and defined by the relation

$$\nabla^2\mathbf{A} = \nabla\nabla\cdot\mathbf{A} - \nabla\times\nabla\times\mathbf{A}, \tag{A.24}$$

which is valid in arbitrary coordinate systems. In Cartesian coordinates the vector Laplacian is equivalent to the application of the scalar Laplace operator to each component of the vector,

$$\nabla^2\mathbf{A} = \hat{x}\nabla^2 A_x + \hat{y}\nabla^2 A_y + \hat{z}\nabla^2 A_z. \tag{A.25}$$

Application of the gradient to a curl of a vector and of the curl to a gradient of a scalar function give zeros,

$$\nabla\cdot\nabla\times\mathbf{A} = 0 \tag{A.26}$$

$$\nabla\times\nabla f = 0. \tag{A.27}$$

An important integral relation is the divergence theorem[1] (e.g., Morse and Feshbach (1953, section 1.4)). Suppose that V is a volume bounded by a closed piecewise smooth surface Ω. If $\mathbf{A}$ is a continuously differentiable vector field defined on V, then

$$\int_V \nabla \cdot \mathbf{A}\,\mathrm{d}V = \int_\Omega \mathbf{A} \cdot \hat{n}\,\mathrm{d}\Omega, \tag{A.28}$$

where $\hat{n}$ is the unit vector normal to Ω and directed into the exterior of V. The left-hand side is a volume integral of the divergence of the vector field $\mathbf{A}$; the right-hand side is the surface integral of the component of $\mathbf{A}$ normal to Ω, where $\mathrm{d}\Omega$ denotes an infinitesimal element of the bounding surface. The theorem therefore relates the flow of a vector field into or from a volume to the behavior of the field inside the volume, most notably to the presence of sources or sinks of the field in V.

The formula (A.28) produces a variety of useful relations through choosing specific forms of $\mathbf{A}$. For example, with $\mathbf{A} = f\nabla g$ where f and g are arbitrary scalar functions, (A.28) becomes

$$\int_V (f\nabla^2 g + \nabla f \cdot \nabla g)\,\mathrm{d}V = \int_\Omega f\frac{\partial g}{\partial n}\,\mathrm{d}\Omega, \tag{A.29}$$

where $\partial g/(\partial n) = \hat{n} \cdot \nabla g$. Another useful relation is obtained on the substitution $\mathbf{A} = f\mathbf{c}$ where f is an arbitrary scalar function and $\mathbf{c}$ is an arbitrary constant vector. Then, $\nabla \cdot \mathbf{A} = \nabla f \cdot \mathbf{c}$ and (A.28) leads to the formula

$$\mathbf{c} \cdot \int_V \nabla f\,\mathrm{d}V = \mathbf{c} \cdot \int_\Omega f\hat{n}\,\mathrm{d}\Omega, \tag{A.30}$$

and since $\mathbf{c}$ is an arbitrary vector, we obtain

$$\int_V \nabla f\,\mathrm{d}V = \int_\Omega f\hat{n}\,\mathrm{d}\Omega, \tag{A.31}$$

a formula that reduces a volume integral of a gradient to a surface integral.

In the two-dimensional case, V and Ω simply change their meaning to a bounded area on a plane and a contour enclosing the area, respectively. The ∇ operator is replaced with its two-dimensional analogue and the surface integral on the right-hand side of (A.28) becomes a line integral over the boundary of the area. The formula (A.28) is then referred to as Green's theorem.

A.3 Vectors and Vector Differential Operators in Orthogonal Curvilinear Coordinates

A.3.1 General Orthogonal Curvilinear Coordinates

Let the Cartesian coordinates of a point in space be described by the functions $x(\xi, \eta, \zeta)$, $y(\xi, \eta, \zeta)$, and $z(\xi, \eta, \zeta)$, the unit vectors $\hat{\xi}$, $\hat{\eta}$, and $\hat{\zeta}$ be tangent to the coordinate lines, building a right-handed basis such that $\hat{\xi} \times \hat{\eta} = \hat{\zeta}$. The vectors may change their orientation in space, but at every point they remain orthogonal to each other.

[1] The theorem is also called Gauss's theorem, Ostrogradsky's theorem, or the Gauss–Ostrogradsky theorem.

Now, if f is a differentiable scalar function of the coordinates and $\mathbf{A}$ is a differentiable vector field with components $A_\xi = \hat{\xi} \cdot \mathbf{A}$, $A_\eta = \hat{\eta} \cdot \mathbf{A}$, and $A_\zeta = \hat{\zeta} \cdot \mathbf{A}$, then

$$\nabla f = \hat{\xi} \frac{1}{h_\xi} \frac{\partial f}{\partial \xi} + \hat{\eta} \frac{1}{h_\eta} \frac{\partial f}{\partial \eta} + \hat{\zeta} \frac{1}{h_\zeta} \frac{\partial f}{\partial \zeta} \tag{A.32}$$

$$\nabla^2 f = \frac{1}{h_\xi h_\eta h_\zeta} \left[\frac{\partial}{\partial \xi} \left(\frac{h_\eta h_\zeta}{h_\xi} \frac{\partial f}{\partial \xi} \right) + \frac{\partial}{\partial \eta} \left(\frac{h_\xi h_\zeta}{h_\eta} \frac{\partial f}{\partial \eta} \right) + \frac{\partial}{\partial \zeta} \left(\frac{h_\xi h_\eta}{h_\zeta} \frac{\partial f}{\partial \zeta} \right) \right] \tag{A.33}$$

and

$$\nabla \cdot \mathbf{A} = \frac{1}{h_\xi h_\eta h_\zeta} \left[\frac{\partial}{\partial \xi} (h_\eta h_\zeta A_\xi) + \frac{\partial}{\partial \eta} (h_\xi h_\zeta A_\eta) + \frac{\partial}{\partial \zeta} (h_\xi h_\eta A_\zeta) \right] \tag{A.34}$$

$$\nabla \times \mathbf{A} = \frac{\hat{\xi}}{h_\eta h_\zeta} \left[\frac{\partial}{\partial \eta} (h_\zeta A_\zeta) - \frac{\partial}{\partial \zeta} (h_\eta A_\eta) \right] + \frac{\hat{\eta}}{h_\zeta h_\xi} \left[\frac{\partial}{\partial \zeta} (h_\xi A_\xi) - \frac{\partial}{\partial \xi} (h_\zeta A_\zeta) \right] + \frac{\hat{\zeta}}{h_\xi h_\eta} \left[\frac{\partial}{\partial \xi} (h_\eta A_\eta) - \frac{\partial}{\partial \eta} (h_\xi A_\xi) \right]. \tag{A.35}$$

The parameters h_ξ, h_η, and h_ζ are the so-called scale factors or metric coefficients of the coordinate system, defined by the equations:

$$h_\xi^2 = \left(\frac{\partial x}{\partial \xi} \right)^2 + \left(\frac{\partial y}{\partial \xi} \right)^2 + \left(\frac{\partial z}{\partial \xi} \right)^2 \tag{A.36}$$

$$h_\eta^2 = \left(\frac{\partial x}{\partial \eta} \right)^2 + \left(\frac{\partial y}{\partial \eta} \right)^2 + \left(\frac{\partial z}{\partial \eta} \right)^2 \tag{A.37}$$

$$h_\zeta^2 = \left(\frac{\partial x}{\partial \zeta} \right)^2 + \left(\frac{\partial y}{\partial \zeta} \right)^2 + \left(\frac{\partial z}{\partial \zeta} \right)^2. \tag{A.38}$$

For example, in Cartesian coordinates the scale factors are unit, $h_x = h_y = h_z = 1$. In circular cylindrical coordinates (ρ, φ, z) with $x = \rho \cos \varphi$, $y = \rho \sin \varphi$, and $z = z$,

$$h_\rho = 1, \quad h_\varphi = \rho, \quad h_z = 1. \tag{A.39}$$

A.3.2 Spherical Coordinates

In the spherical coordinate system the position of a point is described by the three parameters: the distance r $(0 \leq r < \infty)$ from the origin of the coordinate system, the polar (also zenith, inclination) angle θ $(0 \leq \theta \leq \pi)$, and the azimuth angle ϕ $(0 \leq \phi < 2\pi)$. The angle θ is measured from the positive z axis so that the points with $\theta = 0$ lie on the positive and with $\theta = \pi$ on the negative z axis. The angle ϕ is measured from the positive x axis in the x–y plane so that the points with $\phi = \pi/2$ lie on the positive y axis.

If x, y, and z are the Cartesian coordinates of the point, then

$$x = r \sin\theta \cos\phi, \quad y = r \sin\theta \sin\phi, \quad z = r\cos\theta \tag{A.40}$$

and

$$r = \sqrt{x^2 + y^2 + z^2}, \quad \theta = \arccos(z/r), \quad \phi = \arctan(y/x), \tag{A.41}$$

where the value of the arc tangent of y/x takes into account which quadrant the point (x, y) is in.

At every point in space one has three mutually orthogonal unit vectors $\hat{r}$, $\hat{\theta}$, and $\hat{\phi}$ which build an orthogonal vector basis. In contrast to the Cartesian vector basis $\hat{x}$, $\hat{y}$, and $\hat{z}$, the directions of the vectors $\hat{r}$, $\hat{\theta}$, and $\hat{\phi}$ depend on the point position. The Cartesian

$$A_x = \hat{x} \cdot \mathbf{A}, \quad A_y = \hat{y} \cdot \mathbf{A}, \quad A_z = \hat{z} \cdot \mathbf{A} \tag{A.42}$$

and spherical

$$A_r = \hat{r} \cdot \mathbf{A}, \quad A_\theta = \hat{\theta} \cdot \mathbf{A}, \quad A_\phi = \hat{\phi} \cdot \mathbf{A} \tag{A.43}$$

coordinates of a vector **A** are related by the formulas

$$A_r = A_x \sin\theta\cos\phi + A_y \sin\theta\sin\phi + A_z\cos\theta \tag{A.44}$$

$$A_\theta = A_x \cos\theta\cos\phi + A_y \cos\theta\sin\phi - A_z\sin\theta \tag{A.45}$$

$$A_\phi = -A_x \sin\phi + A_y \cos\phi \tag{A.46}$$

and

$$A_x = A_r \sin\theta\cos\phi + A_\theta\cos\theta\cos\phi - A_\phi\sin\phi \tag{A.47}$$

$$A_y = A_r \sin\theta\sin\phi + A_\theta\cos\theta\sin\phi + A_\phi\cos\phi \tag{A.48}$$

$$A_z = A_r\cos\theta - A_\theta\sin\theta. \tag{A.49}$$

For example, the unit vectors $\hat{r}$, $\hat{\theta}$, and $\hat{\phi}$ at a line specified by the angles θ and ϕ can be expressed in Cartesian coordinates as

$$\hat{r} = \hat{x}\sin\theta\cos\phi + \hat{y}\sin\theta\sin\phi + \hat{z}\cos\theta \tag{A.50}$$

$$\hat{\theta} = \hat{x}\cos\theta\cos\phi + \hat{y}\cos\theta\sin\phi - \hat{z}\sin\theta \tag{A.51}$$

$$\hat{\phi} = -\hat{x}\sin\phi + \hat{y}\cos\phi. \tag{A.52}$$

If $f(r, \theta, \phi)$ is a scalar differentiable function of coordinates, then

$$\nabla f = \frac{\partial f}{\partial r}\hat{r} + \frac{1}{r}\frac{\partial f}{\partial \theta}\hat{\theta} + \frac{1}{r\sin\theta}\frac{\partial f}{\partial \phi}\hat{\phi} \tag{A.53}$$

and

$$\nabla^2 f = \frac{1}{r^2}\frac{\partial}{\partial r}\left(r^2\frac{\partial f}{\partial r}\right) + \frac{1}{r^2\sin\theta}\frac{\partial}{\partial\theta}\left(\sin\theta\frac{\partial f}{\partial\theta}\right) + \frac{1}{r^2\sin^2\theta}\frac{\partial^2 f}{\partial\phi^2}. \tag{A.54}$$

For a differentiable vector-valued function $\mathbf{A}(r, \theta, \phi)$, the definitions of the divergence and curl operators are as follows:

$$\nabla\cdot\mathbf{A} = \frac{1}{r^2}\frac{\partial}{\partial r}(r^2 A_r) + \frac{1}{r\sin\theta}\frac{\partial}{\partial\theta}(\sin\theta A_\theta) + \frac{1}{r\sin\theta}\frac{\partial A_\phi}{\partial\phi} \tag{A.55}$$

$$\begin{aligned}\nabla\times\mathbf{A} = {} & \frac{1}{r\sin\theta}\left[\frac{\partial}{\partial\theta}\left(\sin\theta A_\phi\right) - \frac{\partial A_\theta}{\partial\phi}\right]\hat{r} + \frac{1}{r}\left[\frac{1}{\sin\theta}\frac{\partial A_r}{\partial\phi} - \frac{\partial}{\partial r}\left(rA_\phi\right)\right]\hat{\theta} \\ & + \frac{1}{r}\left[\frac{\partial}{\partial r}\left(rA_\theta\right) - \frac{\partial A_r}{\partial\theta}\right]\hat{\phi}.\end{aligned} \tag{A.56}$$

The vector Laplacian operator in spherical coordinates follows from the general definition (A.24) on application of the formulas (A.53), (A.55), and (A.56),

$$\begin{aligned}\nabla^2\mathbf{A} = {} & \left(\nabla^2 A_r - \frac{2}{r^2}A_r - \frac{2}{r^2}\cot\theta A_\theta - \frac{2}{r^2}\frac{\partial A_\theta}{\partial\theta} - \frac{2}{r^2\sin\theta}\frac{\partial A_\phi}{\partial\phi}\right)\hat{r} \\ & + \left(\nabla^2 A_\theta + \frac{2}{r^2}\frac{\partial A_r}{\partial\theta} - \frac{1}{r^2\sin^2\theta}A_\theta - \frac{2\cot\theta}{r^2\sin\theta}\frac{\partial A_\phi}{\partial\phi}\right)\hat{\theta} + \left(\nabla^2 A_\phi + \frac{2}{r^2\sin\theta}\frac{\partial A_r}{\partial\phi} - \frac{1}{r^2\sin^2\theta}A_\phi + \frac{2\cot\theta}{r^2\sin\theta}\frac{\partial A_\theta}{\partial\phi}\right)\hat{\phi}.\end{aligned} \tag{A.57}$$

A.4 Curves and Surfaces in Space

A.4.1 Curves

In a three-dimensional space, every curve C (close or open) can be described by the parametric equation

$$\mathbf{r} = \mathbf{r}(t), \tag{A.58}$$

where t is a parameter from an interval $t_1 \leq t \leq t_2$ (t_1 and t_2 can be equal to $-\infty$ and $+\infty$, respectively). The vector-valued function

$$\mathbf{r}(t) = x(t)\hat{x} + y(t)\hat{y} + z(t)\hat{z} \tag{A.59}$$

must ensure a one-to-one correspondence between the points on C and the values of t from the interval.

The unit vector $\hat{\tau}$ tangent to the curve is defined by

$$\hat{\tau} = \frac{\mathrm{d}\mathbf{r}}{\mathrm{d}l} = \frac{1}{l'(t)}\frac{\mathrm{d}\mathbf{r}}{\mathrm{d}t}, \tag{A.60}$$

where the prime sign means the derivative with respect to the argument and

$$\mathrm{d}l = \sqrt{\left(\frac{\mathrm{d}x}{\mathrm{d}t}\right)^2 + \left(\frac{\mathrm{d}y}{\mathrm{d}t}\right)^2 + \left(\frac{\mathrm{d}z}{\mathrm{d}t}\right)^2}\,\mathrm{d}t \tag{A.61}$$

is the infinitesimal arc length. The curvature radius a_N of C is determined from

$$\frac{\mathrm{d}^2\mathbf{r}}{\mathrm{d}l^2} = \frac{1}{l'(t)}\frac{\mathrm{d}\hat{\tau}}{\mathrm{d}t} = \frac{1}{a_N}\hat{N}, \tag{A.62}$$

where $\hat{N}$ is the unit vector normal to the curve and called the principal normal vector. By definition, $\hat{N}$ points at the center of curvature. The unit vector defined by $\hat{\tau} \times \hat{N}$ is orthogonal to $\hat{\tau}$ and $\hat{N}$, and is called the binormal vector. This vector is constant if the curve is contained in a plane.

A.4.2 Surfaces

Let Ω be a closed or open surface in space. Then, it can be described in the parametric form

$$\mathbf{r} = \mathbf{f}(u, v), \tag{A.63}$$

where $\mathbf{r}$ is a point on the surface, $\mathbf{f}$ is a vector-function, and u and v are parameters (or coordinates) (see Figure 8.4). The parameters are typically defined on the intervals (bounded or unbounded)

$$u_{\min} \leq u \leq u_{\max}, \quad v_{\min} \leq v \leq v_{\max}$$

so that every point $\mathbf{r}$ on Ω corresponds to a unique pair of coordinates (u, v) from the intervals.

The derivatives

$$\mathbf{f}_u(u, v) = \frac{\partial \mathbf{f}(u, v)}{\partial u}, \quad \mathbf{f}_v(u, v) = \frac{\partial \mathbf{f}(u, v)}{\partial v} \tag{A.64}$$

are vectors tangential to Ω. The unit vector $\hat{n}$ normal to the surface Ω is therefore given by

$$\hat{n} = \pm \frac{\mathbf{f}_u(u, v) \times \mathbf{f}_v(u, v)}{|\mathbf{f}_u(u, v) \times \mathbf{f}_v(u, v)|}, \tag{A.65}$$

where the sign can be arbitrary, provided that the direction of $\hat{n}$ does not have to comply with an additional requirement, for example that $\hat{n}$ is an exterior or interior normal when Ω is a closed surface. The infinitesimal element $\mathrm{d}\Omega$ of the surface is expressed as

$$\mathrm{d}\Omega = |\mathbf{f}_u(u, v) \times \mathbf{f}_v(u, v)| \, \mathrm{d}u \, \mathrm{d}v. \tag{A.66}$$

If u and v are orthogonal coordinates, then $\mathbf{f}_u(u, v) \cdot \mathbf{f}_v(u, v) = 0$ and $|\mathbf{f}_u(u, v) \times \mathbf{f}_v(u, v)| = |\mathbf{f}_u(u, v)||\mathbf{f}_v(u, v)|$.

The so-called first fundamental forms are built on the first derivatives of the vector-function $\mathbf{f}(u, v)$,

$$E = \mathbf{f}_u \cdot \mathbf{f}_u, \quad F = \mathbf{f}_u \cdot \mathbf{f}_v, \quad G = \mathbf{f}_v \cdot \mathbf{f}_v, \tag{A.67}$$

and they have the properties $E > 0$, $G > 0$, $EG - F^2 > 0$, and

$$|\mathbf{f}_u \times \mathbf{f}_v| = \sqrt{EG - F^2}. \tag{A.68}$$

If u and v are orthogonal coordinates, then $F = 0$. In terms of the forms,

$$\mathrm{d}\Omega = \sqrt{EG - F^2} \, \mathrm{d}u \, \mathrm{d}v \tag{A.69}$$

$$\hat{n} = \frac{\mathbf{f}_u \times \mathbf{f}_v}{\sqrt{EG - F^2}}, \tag{A.70}$$

where the direction of $\hat{n}$ defined by (A.70) depends on the definition of coordinates u and v.

The second fundamental forms

$$L = \hat{n} \cdot \mathbf{f}_{uu}, \quad M = \hat{n} \cdot \mathbf{f}_{uv}, \quad N = \hat{n} \cdot \mathbf{f}_{vv} \tag{A.71}$$

are expressed in terms of the second derivatives of $\mathbf{f}(u, v)$ and the unit normal $\hat{n}$. At each point of a smooth surface, one can define a normal plane, that is, a plane which contains the normal vector $\hat{n}$. Intersection of the plane with the surface is a plane curve which has a curvature radius $\tilde{R}$, different for different normal planes. For a curve that goes through the two infinitesimally close points (u, v) and $(u + \mathrm{d}u, v + \mathrm{d}v)$ on the surface, the curvature radius can be determined from the formula

$$\tilde{R} = \frac{E(\mathrm{d}u)^2 + 2F\mathrm{d}u\mathrm{d}v + G(\mathrm{d}v)^2}{L(\mathrm{d}u)^2 + 2M\mathrm{d}u\mathrm{d}v + N(\mathrm{d}v)^2}. \tag{A.72}$$

The principal curvature radii, denoted R_1 and R_2, are the maximum and minimum values of $\tilde{R}$, and they are related to the fundamental forms by the equations

$$\frac{1}{R_1 R_2} = \frac{LN - M^2}{EG - F^2}, \tag{A.73}$$

$$\frac{1}{R_1} + \frac{1}{R_2} = \frac{EN - 2FM + GL}{EG - F^2}. \tag{A.74}$$

The quantities

$$K = \frac{1}{R_1 R_2}, \quad H = \frac{1}{2}\left(\frac{1}{R_1} + \frac{1}{R_2}\right) \tag{A.75}$$

are referred to as the Gaussian curvature and the mean curvature, respectively. The curvature radii R_1, R_2, and $\tilde{R}$ may be positive or negative, depending on whether the normal $\hat{n}$ points toward or outward from the respective center of curvature.

A.5 Problems

1. Let f be a scalar function of $\mathbf{r}$ and $\mathbf{A}$ a constant vector. Show that

$$\nabla^2(f\mathbf{A}) = \mathbf{A}\nabla^2 f. \tag{A.76}$$

Hint: Apply the vector Laplacian (A.24) to Cartesian coordinates of the operand.

2. A helix is a curve in three-dimensional space with the property that at any point its tangent makes a constant angle with a fixed line called the axis. If the axis is chosen to be the z axis, than the helix can be described by the parametric equations

$$x = a\cos t, \quad y = a\sin t, \quad z = bt. \tag{A.77}$$

Show that with $a > 0$ and $b = a\cot\beta$, equation (A.77) defines a curve on the surface of a circular cylinder of radius a which makes the angle β with the z axis.
Solution: Equation (A.77) defines a curve on a circular cylinder of radius a since $x^2 + y^2 = a^2$ for every value of t. The tangent to the curve is obtained from (A.60) as

$$\hat{\tau} = \frac{1}{\sqrt{a^2+b^2}}\left(-a\sin t\hat{x} + a\cos t\hat{y} + b\hat{z}\right). \tag{A.78}$$

The angle β with the z axis follows from the relations

$$\cos\beta = \hat{\tau}\cdot\hat{z} = \frac{b}{\sqrt{a^2+b^2}}, \tag{A.79}$$

implying that β is independent of t and $\sin\beta = a/\sqrt{a^2+b^2}$, from which $b = a\cot\beta$.

3. Evaluate the arc length of the helix between the points with coordinates $t_1 = 0$ and $t_2 = \varphi$.
Hint: The arc length is given by the linear integral

$$l = \int_C \mathrm{d}l = \int_{t_1}^{t_2} \sqrt{\left(\frac{\mathrm{d}x}{\mathrm{d}t}\right)^2 + \left(\frac{\mathrm{d}y}{\mathrm{d}t}\right)^2 + \left(\frac{\mathrm{d}z}{\mathrm{d}t}\right)^2}\,\mathrm{d}t. \tag{A.80}$$

Answer: $l = a(t_2 - t_1)/\sin\beta$.

4. Calculate the radius of curvature of the helix.
Solution: Evaluating

$$\frac{\mathrm{d}\hat{\tau}}{\mathrm{d}l} = \frac{-a}{a^2+b^2}\left(\cos t\hat{x} + \sin t\hat{y}\right) \tag{A.81}$$

and comparing with (A.62) and (A.79), we find that

$$a_N = \frac{a^2+b^2}{a} = \frac{a}{\sin^2\beta}. \tag{A.82}$$

5. Show that the helix is the geodesic line on the cylindrical surface.

Hint: A geodesic is a curve with the shortest arc length between two points on a surface. Its main normal coincides with the normal to the surface at every point.

Solution: It follows from (A.62) and (A.81) that $\hat{N} = -\hat{x}\cos t - \hat{y}\sin t$. The vector on the right-hand side is the unit normal to the cylindrical surface, implying that the helix is the geodesic.

6. Calculate the unit tangent $\hat{\tau}$, the unit normal $\hat{N}$, and the curvature radius a_N of the elliptic curve $x(t) = a\cos t$, $y(t) = b\sin t$, $z = 0$, where $0 \leq t < 2\pi$ and $a, b > 0$. Check that $\hat{N}$ is directed to the center of the curvature. Take a look at the limit $a = b$ (circle).

Answer:

$$\hat{\tau} = \frac{-\hat{x}a\sin t + \hat{y}b\cos t}{\sqrt{a^2\sin^2 t + b^2\cos^2 t}} \tag{A.83}$$

$$\hat{N} = -\frac{\hat{x}b\cos t + \hat{y}a\sin t}{\sqrt{a^2\sin^2 t + b^2\cos^2 t}} \tag{A.84}$$

$$a_N = \frac{1}{ab}(a^2\sin^2 t + b^2\cos^2 t)^{3/2}. \tag{A.85}$$

7. Repeat problem 6 for the astroid defined by the equations $x(t) = a\cos^3 t$, $y(t) = a\sin^3 t$, $z = 0$, where $0 \leq t < 2\pi$ and $a > 0$.

Answer:

$$\hat{\tau} = \mp\hat{x}\cos t \pm \hat{y}\sin t \tag{A.86}$$

$$\hat{N} = \pm\hat{x}\sin t \pm \hat{y}\cos t \tag{A.87}$$

$$a_N = \frac{3a}{2}\left|\sin(2t)\right|, \tag{A.88}$$

where the upper sign is taken for $0 < t < \pi/2$ and $\pi < t < 3\pi/2$, and the lower sign for $\pi/2 < t < \pi$ and $3\pi/2 < t < 2\pi$.

8. Derive the formula (8.159) for the ratio dv/du on a curved surface in the incidence plane.

Hint: Consider the equation of the incidence plane for two infinitesimally close points on the scattering surface.

Solution: If $\mathbf{N}$ is a vector normal to the plane and $\mathbf{r}_1$ and $\mathbf{r}_2$ are two points on the plane, then

$$(\mathbf{r}_2 - \mathbf{r}_1) \cdot \mathbf{N} = 0. \tag{A.89}$$

The incidence plane is defined by the incidence direction $\hat{i}$ and the normal $\hat{n}$ to the scattering surface, and thus $\mathbf{N} = \hat{i} \times \hat{n}$. Now we choose the points $\mathbf{r}_1$ and $\mathbf{r}_2$ to be two infinitesimally close points on the scattering surface, that is, $\mathbf{r}_1 = \mathbf{f}(u, v)$ and $\mathbf{r}_2 = \mathbf{f}(u + du, v + dv)$, and inserting these in (A.89) we obtain the relation

$$\mathbf{f}_u \cdot (\hat{i} \times \hat{n})\, du + \mathbf{f}_v \cdot (\hat{i} \times \hat{n})\, dv = 0. \tag{A.90}$$

In (A.90) we use the formula (A.11) for the scalar triple products

$$\mathbf{f}_u \cdot (\hat{i} \times \hat{n}) = \hat{i} \cdot (\hat{n} \times \mathbf{f}_u) \tag{A.91}$$

$$\mathbf{f}_v \cdot (\hat{i} \times \hat{n}) = \hat{i} \cdot (\hat{n} \times \mathbf{f}_v), \tag{A.92}$$

substitute $\hat{n}$ with the vector product $\mathbf{f}_u \times \mathbf{f}_v$ (normalization factor $\sqrt{EG - F^2}$ can be omitted as (A.90) is a homogeneous relation), and using the formula (A.12) for the vector triple products arrive at the relations

$$\hat{n} \times \mathbf{f}_u = -\mathbf{f}_u(\mathbf{f}_u \cdot \mathbf{f}_v) + \mathbf{f}_v(\mathbf{f}_u \cdot \mathbf{f}_u) \tag{A.93}$$

$$\hat{n} \times \mathbf{f}_v = \mathbf{f}_v(\mathbf{f}_u \cdot \mathbf{f}_v) - \mathbf{f}_u(\mathbf{f}_v \cdot \mathbf{f}_v). \tag{A.94}$$

Combining equations (A.90)–(A.94) and using the definitions (A.67) for the first fundamental forms lead to the formula (8.159).

References

Bouman JJ, Senior TBA, and Uslenghi PLE (eds) 1987 *Electromagnetic and Acoustic Scattering by Simple Shapes*, Hemisphere Publishing Corporation.

Korn GA and Korn TM 1968 *Mathematical Handbook for Scientists and Engineers: Definitions, Theorems, and Formulas for Reference and Review*, 2nd rev. edn, McGraw-Hill.

Marsden JE and Tromba A 2003 *Vector Calculus*, 5th edn. W. H. Freeman.

Morse PM and Feshbach H 1953 *Methods of Theoretical Physics*, McGraw-Hill.

Schey HM 2005 *Div, Grad, Curl, and All That: An Informal Text on Vector Calculus*, 4th edn., W. W. Norton.

Stratton JA 1941 *Electromagnetic Theory*, McGraw-Hill.

Weatherburn CE 1955 *Differential Geometry of Three Dimensions*, Cambridge University Press.

B

Fresnel Integral and Related Functions

B.1 Fresnel Integral

The Fresnel integral has been introduced in the framework of the approximate Fresnel–Kirchhoff theory of diffraction (Fresnel 1818). The form of the Fresnel integral we use in this book slightly differs from the definition in Born and Wolf (1959) and is given by

$$F(X) = \frac{e^{j\pi/4}}{\sqrt{\pi}} \int_{-\infty}^{X} e^{-js^2} \, ds. \tag{B.1}$$

Since

$$\int_{-\infty}^{0} e^{-js^2} \, ds = \frac{1}{2}\sqrt{\pi} e^{-j\pi/4}, \tag{B.2}$$

(B.1) can be rewritten in the form

$$F(X) = \frac{1}{2} + \frac{e^{j\pi/4}}{\sqrt{\pi}} \int_{0}^{X} e^{-js^2} \, ds \tag{B.3}$$

and expressed as

$$F(X) = \frac{1}{2} + \frac{e^{j\pi/4}}{\sqrt{2}} \left[C\left(X\sqrt{\frac{2}{\pi}} \right) - jS\left(X\sqrt{\frac{2}{\pi}} \right) \right] \tag{B.4}$$

through the cosine and sine Fresnel integrals

$$C(z) = \int_{0}^{z} \cos\left(\frac{\pi}{2} t^2 \right) dt, \quad S(z) = \int_{0}^{z} \sin\left(\frac{\pi}{2} t^2 \right) dt, \tag{B.5}$$

which are standard special functions (Abramowitz and Stegun 1965).

Modern Electromagnetic Scattering Theory with Applications, First Edition. Andrey V. Osipov and Sergei A. Tretyakov.

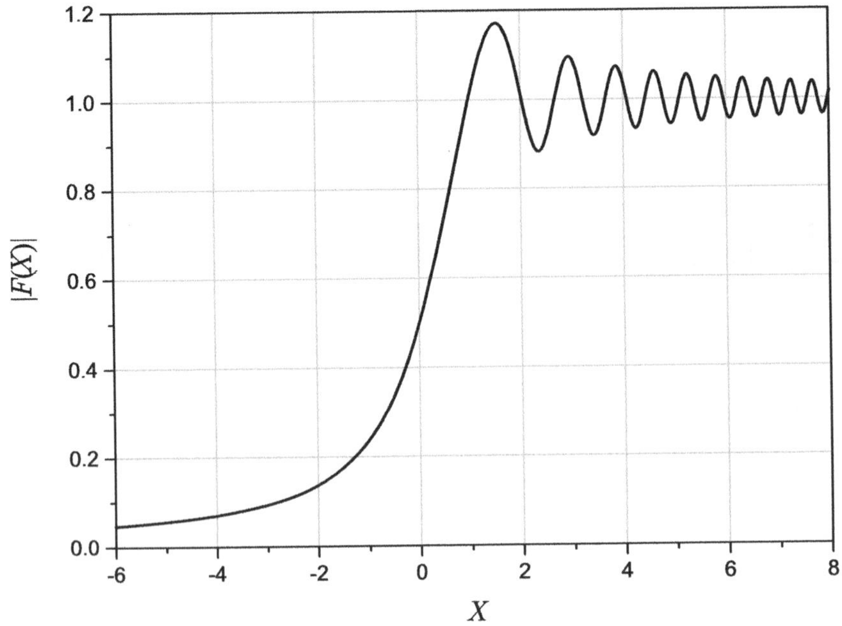

Figure B.1 Amplitude of the Fresnel integral as a function of X.

The Fresnel integral of the form (B.1) has the advantage of being directly present in the exact solution of diffraction problem for a perfectly conducting half-plane (see (5.75)) and in uniform asymptotic representations of fields diffracted by edges (sections 5.5.3 and 5.5.4). Born and Wolf (1959) and Borovikov and Kinber (1994) use similar definitions of the Fresnel integral, the only difference being the replacement of j with $-i$ because of the alternative time-dependence convention; Born and Wolf omit also the factor in front of the integral in (B.1).

We begin by assuming that the argument X is a real-valued variable, which is the case in most applications. Then, by appropriately changing the integration variable in (B.3), the Fresnel integral can be transformed to a superposition of cylindrical waves

$$F(X) = \frac{1}{2} + \operatorname{sgn}(X)\,\frac{e^{j\pi/4}}{2\sqrt{\pi}} \int\limits_0^{X^2} \frac{e^{-jt}}{\sqrt{t}}\,dt, \tag{B.6}$$

which clearly emphasizes its relation to the edge diffraction.

The Fresnel integral has the property

$$F(X) + F(-X) = 1, \tag{B.7}$$

which is easily derived from (B.1). $F(X)$ vanishes when $X \to -\infty$ and approaches unity when $X \to +\infty$, which follows from (B.1) and (B.2). The function can therefore be seen as a smooth version of the unit step function (Figure B.1).

The Fresnel integral (B.1) is a complex-valued function even with the real-valued argument, and a convenient way of representing its behavior is a parametric plot with Re $F(X)$ and Im $F(X)$ serving as x and y coordinates, which leads to a curve (Figure B.2) similar to the classic Cornu spiral (Born and Wolf 1959).

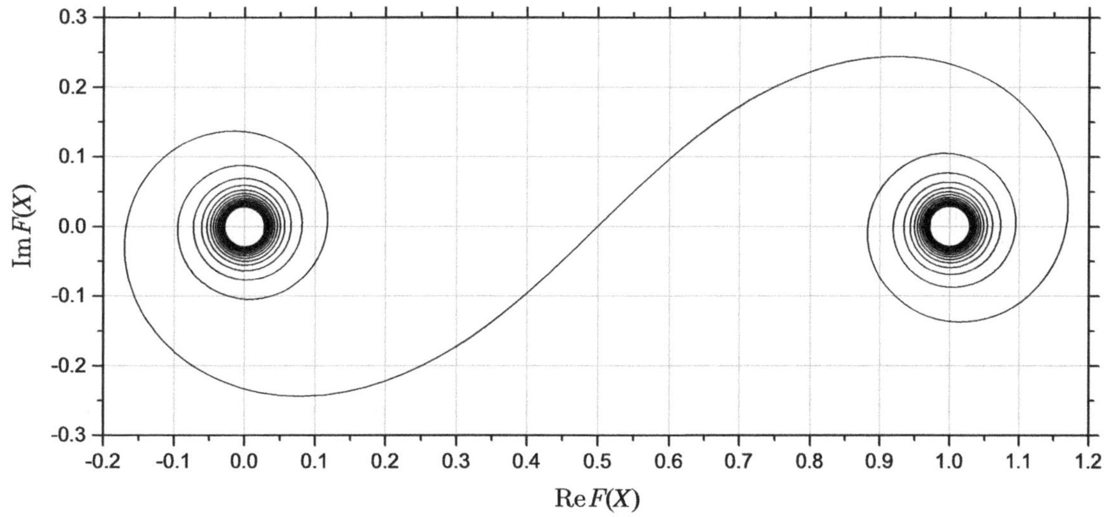

Figure B.2 Parametric representation of $F(X)$ with $-10 < X < 10$.

The asymptotic representation of $F(X)$ for large values of $|X|$ is obtained by integration by parts in (B.1) either directly when $X \to -\infty$ or on rewriting the integral (B.1) as

$$F(X) = 1 - \frac{e^{j\pi/4}}{\sqrt{\pi}} \int_X^\infty e^{-js^2} ds \tag{B.8}$$

when $X \to +\infty$, which leads to the expansions

$$F(X) = \frac{e^{j\frac{3}{4}\pi - jX^2}}{2\pi X} \sum_{n=0}^{\infty} \Gamma\left(n + \frac{1}{2}\right)\left(\frac{j}{X^2}\right)^n \quad (X \to -\infty) \tag{B.9}$$

$$F(X) = 1 + \frac{e^{j\frac{3}{4}\pi - jX^2}}{2\pi X} \sum_{n=0}^{\infty} \Gamma\left(n + \frac{1}{2}\right)\left(\frac{j}{X^2}\right)^n \quad (X \to +\infty) \tag{B.10}$$

or in a unified form

$$F(X) = \Theta(X) + \frac{e^{j\frac{3}{4}\pi - jX^2}}{2\pi X} \sum_{n=0}^{\infty} \Gamma\left(n + \frac{1}{2}\right)\left(\frac{j}{X^2}\right)^n = \Theta(X) + \frac{e^{j\frac{3}{4}\pi - jX^2}}{2\sqrt{\pi}X}\left[1 + \frac{j}{2X^2} + O\left(\frac{1}{X^4}\right)\right], \tag{B.11}$$

where $\Gamma(n)$ is the Gamma function and $\Theta(X)$ is the unit step function defined by

$$\Theta(x) = \begin{cases} 1 & \text{if } x \geq 0 \\ 0 & \text{if } x < 0 \end{cases}. \tag{B.12}$$

It is seen from (B.11) that with $X < 0$ the Fresnel integral is similar to a quickly oscillating exponent with a smoothly varying amplitude, while for $X > 0$ this exponent is superimposed with unity. This property explains the dependence of $|F(X)|$ on X presented in Figure B.1. The curve shows the typical behavior of the fields diffracted at the edge of an impenetrable screen or a wedge, where $X = 0$ corresponds to the position of the shadow boundary, the region $X < 0$ to the geometric shadow and the region $X > 0$ to the lit portion of space with the incident wave overlapped with the edge-diffracted wave.

Function $F(X)$ is well defined by (B.1) for a complex-valued argument. $F(X)$ is an entire function of X, that is, it has no poles and branch points on the whole complex plane. The corresponding power series

$$F(X) = \frac{1}{2} + \frac{e^{j\pi/4}}{\sqrt{\pi}} \sum_{n=0}^{\infty} \frac{(-j)^n X^{2n+1}}{n!\,(2n+1)} = \frac{1}{2} + \frac{e^{j\pi/4}}{\sqrt{\pi}} \left[X - j\frac{X^3}{3} + O\left(X^5\right) \right] \tag{B.13}$$

can be obtained from (B.3) on expanding the integrand $\exp(-js^2)$ in a Taylor series in powers of s^2, followed by integration. The identity (B.7) holds for every complex-valued argument. The asymptotic expansions (B.9) and (B.10) are valid for $\pi/4 < \arg X < 5\pi/4$ and $-3\pi/4 < \arg X < \pi/4$, respectively.

B.2 Relation to the Error Function

One further reason for the important role of the Fresnel integral in the high-frequency diffraction and scattering theory is its connection to the integral

$$w(z) = \frac{1}{\pi j} \int_{-\infty}^{+\infty} \frac{e^{-t^2}}{t - z}\, dt, \tag{B.14}$$

which is the canonical integral of the steepest descent method when the integrand has a pole singularity near a saddle point (section C.3.3). With $\operatorname{Im} z > 0$, equation (B.14) defines the error function for complex arguments (Abramowitz and Stegun 1965)

$$w(z) = e^{-z^2} \operatorname{erfc}(-jz). \tag{B.15}$$

The imaginary part of z can be also negative, and when necessary we will use the notation

$$w(z) = \begin{cases} w_+(z), & \operatorname{Im} z > 0 \\ w_-(z), & \operatorname{Im} z < 0 \end{cases} \tag{B.16}$$

to distinguish between the cases.

When $|z| \gg 1$, the function can be easily expanded in a series in inverse powers of z by expanding the integrand in the geometric progression

$$\frac{1}{t - z} = -\frac{1}{z} \sum_{m=0}^{\infty} \left(\frac{t}{z}\right)^m \tag{B.17}$$

and integrating according to the formula (e.g., Ryshik and Gradstein (1957))

$$\int_{-\infty}^{\infty} e^{-t^2} t^m \, dt = \begin{cases} 0, & m \text{ odd} \\ \Gamma(\frac{m+1}{2}), & m \text{ even} \end{cases}, \tag{B.18}$$

which gives the expansion

$$w(z) = \frac{j}{\pi z} \sum_{n=0}^{\infty} \Gamma\left(n + \frac{1}{2}\right) \frac{1}{z^{2n}}. \tag{B.19}$$

Now let us derive the relation between $F(X)$ and $w(z)$. Consider (B.14) as a contour integral in the complex t plane and assume for a while that $\operatorname{Im} z > 0$, so that the pole singularity at $t = z$ is located in the upper half-plane of the complex plane. When $z \to 0$,

$$w_+(z) \to 1 + \frac{1}{\pi j}\text{v.p.}\int_{-\infty}^{+\infty} e^{-t^2}\frac{dt}{t} = 1, \tag{B.20}$$

where the principal value integral vanishes because of the oddness of the integrand function. Evaluating the derivative of (B.14) with respect to z and integrating by parts give the formula

$$w'_+(z) = \frac{2j}{\pi}\int_{-\infty}^{+\infty} e^{-t^2}\frac{t\,dt}{t-z}, \tag{B.21}$$

and adding and subtracting z in the numerator of the integrand lead to the relation

$$w'_+(z) = \frac{2j}{\sqrt{\pi}} - 2zw_+(z). \tag{B.22}$$

This relation can be considered as a differential equation for $w_+(z)$, and the solution of (B.22) with the initial condition (B.20) is given by

$$w_+(z) = e^{-z^2}\left(1 + \frac{2j}{\sqrt{\pi}}\int_0^z e^{\xi^2}\,d\xi\right). \tag{B.23}$$

A corresponding expression for $w_-(z)$ should satisfy the same differential equation (B.22) but with a different initial condition, $w_-(0) = -1$, since the pole approaches the integration path from beneath, and the result reads

$$w_-(z) = e^{-z^2}\left(-1 + \frac{2j}{\sqrt{\pi}}\int_0^z e^{\xi^2}\,d\xi\right). \tag{B.24}$$

Substituting either $\xi = e^{-j\pi/4}s$ or $\xi = -e^{-j\pi/4}s$ transforms the integrals in (B.23) and (B.24) to a form similar to the definition (B.3) of the Fresnel integral, leading to the expressions

$$w_+(z) = 2e^{-z^2}F(ze^{j\pi/4}) = 2e^{-z^2}[1 - F(-ze^{j\pi/4})] \tag{B.25}$$
$$w_-(z) = 2e^{-z^2}[F(ze^{j\pi/4}) - 1] = -2e^{-z^2}F(-ze^{j\pi/4}). \tag{B.26}$$

The two representations on the right-hand sides of (B.25) and (B.26) correspond to the choice of the sign in the above-mentioned substitution; their identity is apparent from (B.7). Expressions (B.25) and (B.26) apply for $\operatorname{Im} z > 0$ and $\operatorname{Im} z < 0$, respectively.

B.3 Transition Functions of Uniform Theories of Diffraction

Another Fresnel-type integral, which is often encountered in diffraction theory, is the Clemmow transition function (Clemmow 1966)

$$F_C(X) = e^{jX^2} \int_X^{+\infty} e^{-js^2}\, ds. \tag{B.27}$$

Comparing with (B.8) gives the relation

$$F_C(X) = \sqrt{\pi}\,[1 - F(X)]\, e^{jX^2 - j\pi/4}, \tag{B.28}$$

valid on the whole complex plane X.

UTD is based on the transition function (Kouyoumjian and Pathak 1974)

$$F_{KP}(z) = 2j\sqrt{z}e^{jz} \int_{\sqrt{z}}^{+\infty} e^{-js^2}\, ds, \tag{B.29}$$

where one takes the principal branch of the square root, that is, Re $\sqrt{z} \geq 0$. Because of the square root function, $F_{KP}(z)$ is multi-valued and has a branch cut at the negative part of the real axis in the complex z plane, so that

$$F_{KP}(|z|e^{j\pi}) - F_{KP}(|z|e^{-j\pi}) = -2\sqrt{\pi|z|}e^{-j|z| - j\pi/4}. \tag{B.30}$$

Comparing the definitions (B.1) and (B.29), one may find that

$$F_{KP}(X^2) = 2\sqrt{\pi}X\,[1 - F(X)]\, e^{jX^2 + j\pi/4} \tag{B.31}$$

$$F(X) = 1 - F_{KP}(X^2)\frac{\exp(-jX^2 - j\pi/4)}{2\sqrt{\pi}X}, \tag{B.32}$$

and using (B.32) in (B.25) and (B.26) leads to the result

$$w(z) = \frac{j}{\sqrt{\pi}z} F_{KP}(jz^2), \tag{B.33}$$

which is valid for arbitrary values of Im z.

The small and large argument approximations of $F_{KP}(X^2)$ follow from the corresponding approximations for $F(X)$ (see (B.9), (B.10), and (B.11)), namely, for $|X| \to \infty$,

$$F_{KP}(X^2) = \frac{1}{\sqrt{\pi}} \sum_{n=0}^{\infty} \Gamma\left(n + \frac{1}{2}\right)\left(\frac{j}{X^2}\right)^n = 1 + \frac{j}{2X^2} + O\left(\frac{1}{X^4}\right) \tag{B.34}$$

when $-3\pi/4 < \arg X < \pi/4$ and

$$F_{KP}(X^2) = 2\sqrt{\pi}Xe^{jX^2 + j\pi/4} + \frac{1}{\sqrt{\pi}} \sum_{n=0}^{\infty} \Gamma\left(n + \frac{1}{2}\right)\left(\frac{j}{X^2}\right)^n \tag{B.35}$$

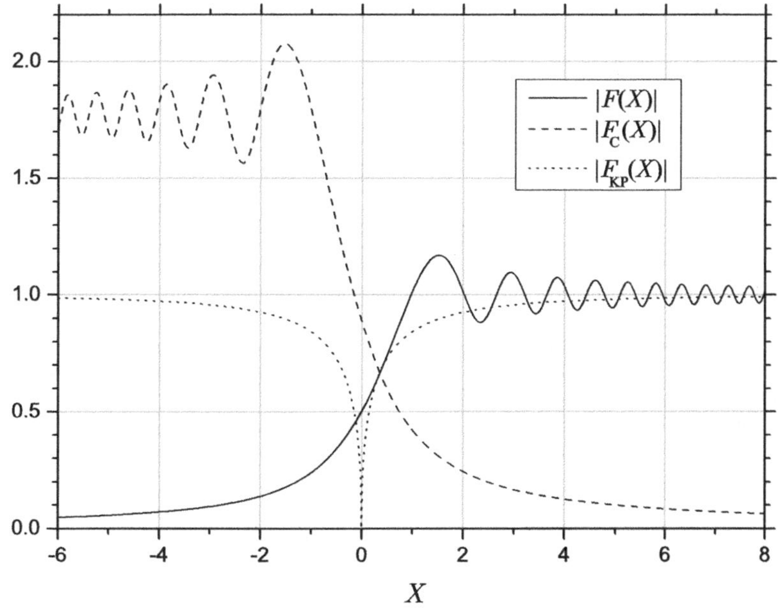

Figure B.3 Amplitudes of various Fresnel-type functions of real-valued arguments. $F_{\rm KP}(X)$ for $X < 0$ is calculated with $X = |X|\exp(-j\pi)$.

when $\pi/4 < \arg X < 5\pi/4$. For $|X| \to 0$,

$$F_{\rm KP}(X^2) = \sqrt{\pi}Xe^{j\pi/4} - 2jX^2 + O(X^3). \tag{B.36}$$

Figure B.3 illustrates the behavior of the amplitudes of the functions F, $F_{\rm C}$, and $F_{\rm KP}$ of a real-valued argument.

B.4 Problems

1. Show that expansions (B.9) and (B.10) are valid for $\pi/4 < \arg X < 5\pi/4$ and $-3\pi/4 < \arg X < \pi/4$, respectively.
   Hint: Combine (B.19) with (B.25) and (B.26).
2. Prove the property (B.30).
   Hint: Deform the integration path in (B.29) to integrate from $\sqrt{z}$ to 0 and then from 0 to $+\infty$.
3. Prove the relation

$$F_{\rm KP}(|z|e^{-j\pi}) = F^*_{\rm KP}(|z|), \tag{B.37}$$

   where the star means the complex conjugation.
4. It follows from (B.34) that $|F_{\rm KP}(X^2)|^2 \approx 1 + 1/(4X^4) > 1$. On the other hand, according to Figure B.3, $|F_{\rm KP}| < 1$ for all arguments. Explain the paradox.
   Hint: One has to account for the next term $O(X^{-4})$ in the large-argument approximation (B.34) to obtain a correct estimation for $|F_{\rm KP}(X^2)|^2$.
5. Show that expansions (B.9), (B.10), (B.19), (B.34), and (B.35) are divergent. Explain why.

Hint: Use Stirling's formula

$$\Gamma(z) \to \sqrt{2\pi}\, e^{-z} z^{z-1/2} \;(|z| \to \infty \text{ in} |\arg z| < \pi). \tag{B.38}$$

References

Abramowitz M and Stegun IA (eds) 1965 *Handbook of Mathematical Functions with Formulas, Graphs, and Mathematical Tables*, Dover Publications.

Born M and Wolf E (eds) 1959 *Principles of Optics: Electromagnetic Theory of Propagation, Interference and Diffraction of Light*, Pergamon Press.

Borovikov VA and Kinber BY 1994 *Geometrical Theory of Diffraction*, The Institution of Electrical Engineers, London.

Clemmow PC 1966 *The Plane Wave Spectrum Representation of Electromagnetic Fields*, Pergamon Press, New York.

Fresnel A 1818 Note sur la théorie de la diffraction. *Œuvres completes*, 171–181.

Kouyoumjian RG and Pathak PH 1974 A uniform geometrical theory of diffraction for an edge in a perfectly conducting surface. *Proc. IEEE*, **62**, 1448–1461.

Ryshik IM and Gradstein IS 1957 *Tables of Series, Products, and Integrals*, VEB, Berlin.

C

Principles of Complex Integration

C.1 Introduction

Solutions of electromagnetic scattering problems are often expressed as integrals over an infinite path, for example over a real axis but also over a curved contour in a complex domain. Examples are the Fourier expansions for the fields excited by line and point sources in free space or in the presence of a material slab (sections 4.4.1 and 4.4.2) or solutions for wedge-shaped regions in terms of Sommerfeld's integrals (section 5.4.1). Even when a solution is given by an infinite series, as for circular cylinders and spheres, it may be beneficial to express the solution as an integral in a complex domain by applying Watson's transformation (sections 6.4.1 and 7.3.4). The material presented in this appendix is essential for understanding the mathematical technique of complex integration widely used in the main body of the book.

This appendix describes methods of evaluation of integrals of the form

$$I(k) = \int_C A(\alpha) \mathrm{e}^{jkS(\alpha)}\, \mathrm{d}\alpha, \tag{C.1}$$

where C is an integration contour of infinite extent, $S(\alpha)$ and $A(\alpha)$ are known complex functions, and k is a real-valued parameter. The integration variable α, even when C is the real axis, is considered as a complex variable and the integral as an integral in the complex α plane. Representation (C.1) can be seen as an expansion in a complex spectrum of plane waves.

Direct numerical evaluation of $I(k)$ can be difficult, particularly when k is large, resulting in rapid oscillations of the integrand over the integration path. Furthermore, simple analytical approximations of (C.1) are often desirable for a physical interpretation of the solution, which may also be complicated as the integral may include a continuum of elementary contributions of comparable amplitude. A powerful method of evaluation uses the fact that on the complex plane the integration contour can be modified, deformed or shifted from the original integration path onto a new path, over which the integration can be performed exactly, approximately or numerically.

Working with complex integrals requires some knowledge of the theory of functions of a complex variable. In this appendix we give the necessary definitions and describe the concepts required for integration in the complex plane, most notably the concept of holomorphic functions, types of singularities on the complex plane, Cauchy's integral theorems, and the calculus of residues (section C.2). This material can be found in many textbooks on complex analysis, for example Brown and Churchill (2013), Carrier *et al.* (1966), and Morse and Feshbach (1953).

In section C.3, the steepest descent method, which is the method for asymptotic evaluation of $I(k)$ for large k, is briefly presented. The method, alternatively known as the saddle-point method, is described in many mathematical and physical texts (e.g., Bleistein and Handelsman (1986), Bouche *et al.* (1997), Born and Wolf (1959), Carrier *et al.* (1966), Copson (1965), Erdélyi (1956), Fedoryuk (1989), Felsen and Marcuvitz (1973), Morse and Feshbach (1953), Senior and Volakis (1995), Collin (1960), and Wong (1989)), so here we just give a short outline of the method and present the formulas

Modern Electromagnetic Scattering Theory with Applications, First Edition. Andrey V. Osipov and Sergei A. Tretyakov.

most relevant to solutions of scattering problems. In contrast to the conventional exposition of the method, which uses the Laplace form of the integral, that is, without j in the exponent, we formulate the results for the superposition of plane waves (C.1) which is the typical form of solutions of scattering problems.

C.2 Deforming the Integration Contour

C.2.1 Basic Facts about Functions of a Complex Variable

This section summarizes the basic definitions and integral theorems from the theory of functions of a complex variable (e.g., Brown and Churchill (2013), Carrier *et al.* (1966), and Morse and Feshbach (1953)).

A complex integral is written as

$$\int_C f(z)\,\mathrm{d}z, \tag{C.2}$$

where $z = \mathrm{Re}\, z + j\mathrm{Im}\, z$ is a complex variable and C is a curve on the plane $(\mathrm{Re}\, z, \mathrm{Im}\, z)$. The curve is called the integration contour or path, and integration is performed in a specific direction. Integration in the opposite direction alternates the sign of the result. The contour can be closed or open, finite or infinite, but necessarily smooth in the sense that it is composed of arcs which join on continuously, each arc having a continuous tangent. Integration over a closed contour is performed in positive direction if with respect to some point inside the domain the contour is circumvented in a counterclockwise direction. The negative direction is just the clockwise one.

Integration in the complex plane can be reduced by parametrization to the conventional, Riemann integral over a real-valued variable,

$$\int_C f(z)\,\mathrm{d}z = \int_{s_1}^{s_2} f[z(s)]\frac{\mathrm{d}z}{\mathrm{d}s}\,\mathrm{d}s, \tag{C.3}$$

where the complex-valued function $z(s)$ describes the integration contour C on the plane $(\mathrm{Re}z, \mathrm{Im}z)$ when $s_1 \le s \le s_2$. Depending on the chosen parametrization, s_1 and s_2 can be finite or infinite. If $f(z)$ is infinite at a point $z = a$ on C and $a = z(s_0)$ with $s_1 < s_0 < s_2$, then the integral is to be understood in the *Cauchy principal value* sense

$$\text{v.p.} \int_C f(z)\,\mathrm{d}z = \lim_{\epsilon\to 0}\left[\int_{s_1}^{s_0-\epsilon} f[z(s)]\frac{\mathrm{d}z}{\mathrm{d}s}\,\mathrm{d}s + \int_{s_0+\epsilon}^{s_2} f[z(s)]\frac{\mathrm{d}z}{\mathrm{d}s}\,\mathrm{d}s\right]. \tag{C.4}$$

A *holomorphic* (or complex analytic) function is a complex-valued function that is equal to its own Taylor series in a neighborhood of every point in its domain. The region $\mathcal{R}$ where the function is holomorphic will be referred to as analyticity region. If $f(z)$ is holomorphic within and on a closed contour C, then

$$\int_C f(z)\,\mathrm{d}z = 0. \tag{C.5}$$

This statement is known as *Cauchy's theorem*. A consequence of this theorem is the fact that a complex integral

$$\int_{z_1}^{z_2} f(z)\,\mathrm{d}z \tag{C.6}$$

over an open contour with the ends at $z = z_1$ and $z = z_2$ located entirely in the analyticity region of $f(z)$ depends only on z_1 and z_2. One may deform the contour within the analyticity region without changing the value of the integral, provided that the ends of the contour remain fixed (Figure C.1). Such contours will be referred to as equivalent. The ends of the contour can be located at infinity, for example $z_1 = -\infty$ and $z_2 = +\infty$.

The absolute value of a holomorphic function has no maximum within the region of analyticity (maximum modulus principle). The maximum value of $|f(z)|$ is always reached on the boundary of $\mathcal{R}$ and never in the interior. If $f(z) \neq 0$ anywhere in $\mathcal{R}$, then the minimum value of $|f(z)|$ occurs on the boundary of $\mathcal{R}$ and never in the interior. Since functions

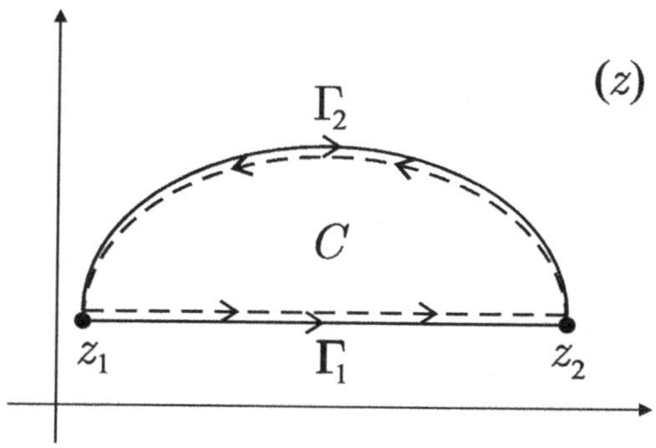

Figure C.1 Deformation of an integration path in the analyticity region of a function $f(z)$. Complex integrals of $f(z)$ over paths Γ_1 and Γ_2 joining the points z_1 and z_2 are equal to each other since integration of $f(z)$ over a closed contour C (dashed) is equal to zero.

$\exp[f(z)]$ and $\exp[jf(z)]$ with a holomorphic $f(z)$ are holomorphic and free of zeros in $\mathcal{R}$, application of the maximum modulus principle suggests that $\mathrm{Re} f$ and $\mathrm{Im} f$ have neither a maximum nor a minimum within $\mathcal{R}$ too.

Functions which are holomorphic in every finite region of the complex z plane are called entire functions. Examples are polynomials, $\sin z$, $\exp z$, but also Bessel functions as functions of their order. Entire functions must grow at infinity, otherwise they are constants. A rigorous formulation is known as *Liouville's theorem*: if a function $f(z)$ is entire and bounded in the complex plane, then $f(z)$ is constant throughout the plane. Our analysis of the Sommerfeld–Maliuzhintes method in section 5.4.1 uses an extended version of this theorem: if $f(z)$ is an entire function of z and if, as z approaches infinity, $|f| \leq F|z|^c$ where F and c are fixed and positive, then f is a polynomial of degree at most $[c]$, where $[c]$ is the integer part of c.

A function is said to have a pole singularity at a point $z = a$ if it can be represented by a series of the form (*Laurent series*),

$$f(z) = \sum_{n=-N}^{+\infty} c_n (z-a)^n, \tag{C.7}$$

with $1 \leq N < \infty$. If $N = \infty$, that is, the series of negative powers in the Laurent expansion does not terminate, then the point $z = a$ is called an essential singularity.

The following integral theorems assume pole singularities. The coefficient c_{-1} is referred to as the residue of $f(z)$ at $z = a$. *The Cauchy integral formula* states that if $f(z)$ is holomorphic inside and on a closed contour C, and if a is a point within C, on C or outside C, then

$$\int_C \frac{f(z)}{z-a}\,\mathrm{d}z = 2\pi j f(a) \begin{cases} 1, & a \text{ within } C \\ \frac{1}{2}, & a \text{ on } C \\ 0, & a \text{ outside } C \end{cases}. \tag{C.8}$$

When a lies on C, the integral is understood in the sense of the Cauchy principal value. The formula (C.8) may be used to evaluate integrals, and the fundamental theorem in the calculus of residues is formulated as follows. If $f(z)$ is holomorphic within a region bounded by C, except for a finite number of poles, then the value of the contour integral around C is equal to $2\pi j$ times the sum of residues at the poles enclosed by the contour,

$$\int_C f(z)\,\mathrm{d}z = 2\pi j \sum_m \mathrm{res} f(\alpha_m), \tag{C.9}$$

where m numerates the enclosed poles α_m. Equations (C.8) and (C.9) assume integration in the positive direction (otherwise the poles are not enclosed in C). If all singularities of a function in a given region of the complex plane are poles, then the function is said to be *meromorphic* in that region. For example, $\tan z$ is a meromorphic function on the whole complex plane.

Another type of singularities of functions of a complex variable are branch points. These are associated with multivalued functions, that is, with functions that are not uniquely specified for a given z. Examples are the square root $\sqrt{z}$, complex logarithm function $\ln z$, and inverse trigonometric functions, for example $\arcsin z$. At a given value of z such functions may take two ($\sqrt{z}$), several ($\sqrt[n]{z}$, n integer) or an infinite number ($\ln z$, $\arcsin z$) of values. Restricting the range of values of z can make $f(z)$ single-valued, in which case one speaks about a branch of $f(z)$. At a branch point a multi-valued function cannot be represented by a Taylor series. The value at the branch point is common to all the branches of the function. Branch points always occur in pairs, for example the branch points of $\sqrt{z}$ are $z = 0$ and $z = \infty$.

A convenient way to graphically represent multi-valued functions is to use the concept of Riemann surfaces. A Riemann surface can be imagined as a stack of multiple complex planes (sheets) arranged vertically and connected through the cuts along lines joining pairs of branch points (branch cuts). The shape of the cuts is not important and is typically determined by convenience reasons, for example by the condition that $\mathrm{Re} f(z) = 0$ or $\mathrm{Im} f(z) = 0$ on the branch cut. Circumventing a branch point leads to another sheet of the Riemann surface, that is, to another branch of the multi-valued function. The values of a multi-valued function obtained by its analytic continuation around a branch point from the same starting point depend on the direction of circumvention (clockwise or counterclockwise) and on the number of times the branch point is circumvented. An example can be found in Appendix E, where the behavior of the functions $\sqrt{1-z^2}$ and $\arccos z$ is analyzed; their branch cut structure is illustrated in Figure E.5.

The integral formulas (C.5), (C.8), and (C.9) do not apply to multi-valued functions. Deforming a contour in the presence of a branch point results in a contour circumventing the branch point and involving integration in the opposite directions over the opposite sides of a branch cut (Figure C.2).

C.2.2 Integrals over Infinite Contours

If an endpoint z_1 of an integration contour is located at infinity ($|z_1| = \infty$, $\arg z_1 = \phi_1$), then the endpoint can be moved, without changing the value of the integral, to another location at infinity with $\arg z_1 = \tilde{\phi}_1$, provided that $f(z)$ decays faster then $|z|^{-1}$ over the arc between ϕ_1 and $\tilde{\phi}_1$, which follows from the fact that

$$\lim_{|z|\to\infty} |z| \int_{\phi_1}^{\tilde{\phi}_1} f\left(|z|e^{j\phi}\right) \mathrm{d}\phi = 0. \tag{C.10}$$

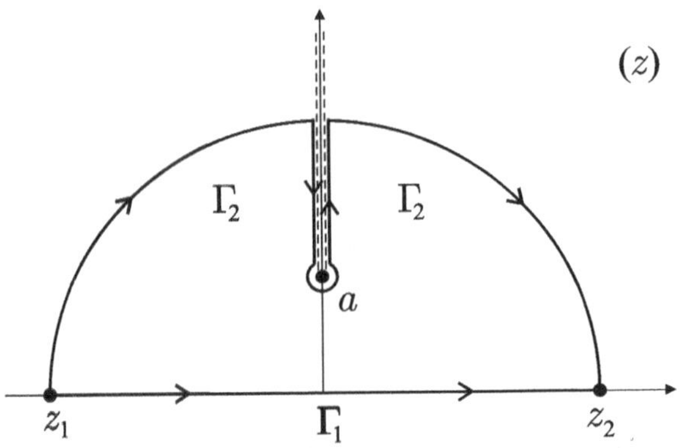

Figure C.2 Deformation of an integration contour Γ_1 into the upper half-plane in the presence of a branch point at $z = a$. Contours Γ_1 and Γ_2 (both connecting the points z_1 and z_2) are equivalent. Contour Γ_2 circumvents the branch point along the opposite sides of the branch cut (dashed).

This fact opens up a possibility of evaluating integrals with endpoints at infinity when the integrand is $o(z^{-1})$ over the whole arc joining the ends of the integration contour at infinity. Adding an integral over this arc does not change the value of the original integral but closes the integration contour so that, for example, the residue theorem (C.9) can be applied if the integrand is meromorphic in the enclosed domain.

An important class of integrals that can be evaluated in this manner are Fourier integrals

$$\int_{-\infty}^{+\infty} e^{jkz} f(z) dz, \tag{C.11}$$

where k is a real-valued parameter. This corresponds to (C.1) with a linear function S. For such integrals the requirement on the decay rate of $f(z)$ can be further relaxed, namely, it is sufficient that $|f(z)| \to 0$ as $|z| \to \infty$ over the arc $0 \le \arg z \le \pi$ in the upper half-plane when $k > 0$ or over the arc $-\pi \le \arg z \le 0$ in the lower half-plane when $k < 0$ (*Jordan's lemma*). Under such conditions the integration contour in (C.11) can be closed with an infinite arc in the respective half-plane. For example, if $f(z)$ is holomorphic and vanishes at infinity in the upper half-plane, then (C.11) vanishes for every $k > 0$. If $f(z)$ is meromorphic and vanishes at infinity in the upper half-plane, then (C.11) is equal to the sum of residues at the poles located in the upper half-plane for every $k > 0$.

In the case of more general functions, for example with branch points, the deformation of integration contour can be used to obtain useful estimations of Fourier integrals as $O[\exp(-|k \mathrm{Im}\, a|)]$, where a is the singularity of $f(z)$ nearest to the real axis.

C.3 Steepest Descent Method

The steepest descent, or saddle-point, method is a technique for obtaining asymptotic approximations for integrals of the form (C.1) when k is a large parameter and $S(\alpha)$ is neither a constant nor a linear function of α. As the parameter k in (C.1) grows, the factor $\exp[jkS(\alpha)]$ makes the integrand a rapidly varying function of α. Depending on the path of integration, the exponent may be purely oscillating (when $\mathrm{Im}\, S(\alpha) = \text{const}$ along the path) or exponentially growing or decaying (when $\mathrm{Re}\, S(\alpha) = \text{const}$ along the path). So, one may look for a point on C or in the vicinity of C, at which $|\exp[jkS(\alpha)]|$ arrives at a "maximum" and then make the contour C go through that point along the path at which $|\exp[jkS(\alpha)]|$ exponentially decays. That would imply a significant simplification as along such a path the main contribution to the integral would come from an asymptotically small vicinity of the maximum, where $A(\alpha)$ and $S(\alpha)$ can be approximated by one or two terms in their Taylor expansions.

One may expect the maximum at a point at which

$$S'(\alpha_c) = 0, \tag{C.12}$$

where the prime denotes the derivative with respect to the argument of the function. However, if $S(\alpha)$ is holomorphic in the vicinity of α_c, the point α_c cannot be the true maximum of the exponent factor as the function $\exp[jkS(\alpha)]$ is also holomorphic and as such does not have any maxima in its analyticity region (see the maximum modulus principle). This means that α_c is a saddle point and implies the existence of two contours, of the steepest ascent and of the steepest descent.[1] Thus, the idea of the approach consists of finding a saddle point α_c and, when necessary, in deforming the original integration contour onto the steepest descent path (SDP), followed by approximate evaluation of the integral.

The first application of the steepest descent method is attributed to Debye (1909), who used an idea by Riemann (1863) to derive asymptotic approximations for Bessel functions. The use of the steepest descent method in order to derive an asymptotic solution for scattering by a PEC wedge is described in Sommerfeld (1927). Extension of the solution to the case of an observer near the shadow boundary requires a generalization of the steepest descent method to the case of a pole of the integrand function close to the saddle point (Pauli 1938). Pauli's approach was further developed by Ott

[1] Two further contours, along which the exponent is purely oscillating, are used for asymptotic estimations in the framework of the stationary phase method (see Appendix D), where the saddle point α_c is called the critical point or the point of stationary phase.

(1943) and Clemmow (1950), and the result was formulated as an expansion in terms of generalized Fresnel integrals. An improved representation, which includes only the standard Fresnel integral and only in the leading term of the asymptotic expansion, was proposed by Van der Waerden (1952). The approach by Pauli, Ott, and Clemmow is sometimes referred to as multiplicative, whereas that by Van der Waerden as additive. A comparison of two approaches is given by Boersma and Rahmat-Samii (1980).

An important feature of the steepest descent method is that the accuracy of the approximation improves with increasing k. The method provides simple approximations for integrals of a quite general form and thus permits physical interpretation of solutions of many important problems in scattering, diffraction, propagation, and radiation of electromagnetic waves. The key ingredients of the steepest descent method are as follows: (a) localization of a suitable saddle point, (b) deformation of the integration contour onto the corresponding SDP, and (c) evaluation of the canonical integral for the saddle point contribution. It should be pointed out that not all functions have a saddle point, for example when $S(\alpha)$ is a constant or a linear function of α. On the other hand, there can be multiple saddle points, for example when $S(\alpha)$ is a polynomial of α, or even an infinite number of saddle points, for example when $S(\alpha)$ is a periodic function of α. Under such circumstances, choosing a suitable saddle point may become a difficult problem; an example is the construction of asymptotic approximations of Debye's type for Bessel functions (see Appendix E). A special case occurs when in addition to (C.12), $S''(\alpha)$ and possibly several higher-order derivatives of $S(\alpha)$ vanish at α_c (higher-order saddle point). Cases in which $S(\alpha)$ and/or $A(\alpha)$ are not holomorphic, for example have branch points, in close proximity to the saddle point require a special treatment.

In what follows we assume that the saddle point is of the first order ($S''(\alpha_c) \neq 0$) and isolated, that is, no other saddle points are located in the asymptotically small region forming the main contribution to the integral. The amplitude function $A(\alpha)$ is assumed to be either holomorphic or with a simple pole in the vicinity of the saddle point. We begin with an analytical description of SDP and deformation of the integration contour in (C.1) onto it (section C.3.1). Then, in section C.3.2, the contribution from the saddle point is described. The case of a simple pole singularity in $A(\alpha)$ approaching the saddle point is addressed in the framework of the additive approach in section C.3.3. The appendix concludes with a list of further special cases which are not treated in the book but may be encountered in practice (section C.3.4).

C.3.1 Steepest Descent Path

Consider the SDP for the integral in (C.1). Along the path, by definition, Re $S(\alpha)$ remains constant and Im $S(\alpha)$ continuously grows, so the path can be defined by the equation

$$\operatorname{Re} S(\alpha) = \operatorname{Re} S(\alpha_c) \tag{C.13}$$

with an additional condition $\operatorname{Im} S(\alpha) \geq \operatorname{Im} S(\alpha_c)$. In parametric form this can be written as

$$S(\alpha) - S(\alpha_c) = j\tau^2 \tag{C.14}$$

with a real-valued parameter τ ($-\infty < \tau < +\infty$). The value $\tau = 0$ corresponds to $\alpha = \alpha_c$. Equation (C.14) defines the SDP over the whole complex α plane.

SDP can terminate either at infinity $|\alpha| = \infty$ or at points at which $S(\alpha)$ looses its analyticity, for example poles and branch points. Indeed, at every point in the analyticity region of $S(\alpha)$ there is always a direction which complies with (C.13). This is the direction of the gradient of the function $v(x, y) = \operatorname{Im} S(\alpha)$ with respect to the variables $x = \operatorname{Re} \alpha$ and $y = \operatorname{Im} \alpha$. Along this direction, $v(x, y)$ shows the fastest growth, whereas $u(x, y) = \operatorname{Re} S(\alpha)$ remains constant. So, starting from α_c, SDP follows the direction of the gradient until it arrives at infinity or at a location where the gradient is infinite or non-uniquely defined, that is, at a singularity of $S(\alpha)$.

The next step is the deformation of C onto SDP. In the process of deformation, singular points of $S(\alpha)$ or $A(\alpha)$, for example poles and branch points, may be captured, which results in additional contributions to $I(k)$. So, we write

$$I(k) = I_{\mathrm{SDP}}(k) + I_{\mathrm{p}}(k) + I_{\mathrm{bc}}(k), \tag{C.15}$$

where $I_{\mathrm{SDP}}(k)$ is the integral over SDP, $I_{\mathrm{p}}(k)$ is the contribution from the captured poles, and $I_{\mathrm{bc}}(k)$ describes contributions due to branch points.

If we denote the captured poles by $\alpha = \alpha_n$ with an integer n numbering the crossed poles, then

$$I_{\mathrm{p}}(k) = \pm 2\pi j \sum_n \operatorname{res} A(\alpha_n) \mathrm{e}^{jkS(\alpha_n)}, \tag{C.16}$$

where the choice of sign in front of the sum depends on the direction of the integration contour (positive or negative with respect to the captured poles). The branch point contributions $I_{\mathrm{bc}}(k)$ are given by integrals over contours enclosing the branch points (see Figure C.2). We will now evaluate the integral over the SDP for $k \to \infty$,

$$I_{\mathrm{SDP}}(k) = \int_{\mathrm{SDP}} A(\alpha) \mathrm{e}^{jkS(\alpha)} \,\mathrm{d}\alpha. \tag{C.17}$$

C.3.2 Saddle Point Contribution

Assume that $S(\alpha)$ and $A(\alpha)$ are holomorphic within at least an asymptotically small vicinity of the saddle point. Then, the integral in (C.17) is mainly determined by the contribution from this saddle point. Changing the integration variable according to (C.14) with τ defined so as to preserve the direction of integration over SDP when τ varies from $-\infty$ to ∞ gives

$$I_{\mathrm{SDP}}(k) = \mathrm{e}^{jkS_{\mathrm{c}}} \int_{-\infty}^{+\infty} \tilde{A}(\tau) \mathrm{e}^{-k\tau^2} \,\mathrm{d}\tau, \tag{C.18}$$

where $S_{\mathrm{c}} = S(\alpha_{\mathrm{c}})$ and

$$\tilde{A}(\tau) = A[\alpha(\tau)] \frac{\mathrm{d}\alpha}{\mathrm{d}\tau}. \tag{C.19}$$

Function $\tilde{A}(\tau)$ can be expanded in the Taylor series

$$\tilde{A}(\tau) = \sum_{m=0}^{\infty} \frac{\tau^m}{m!} \tilde{A}_m, \tag{C.20}$$

where $m! = 1 \cdot 2 \cdot \ldots \cdot m$ with $0! = 1$ and

$$\tilde{A}_m = \left. \frac{\mathrm{d}^m \tilde{A}}{\mathrm{d}\tau^m} \right|_{\tau=0}. \tag{C.21}$$

Inserting the series in (C.18) and integrating according to (B.18) give

$$I_{\mathrm{SDP}}(k) = \frac{\mathrm{e}^{jkS_{\mathrm{c}}}}{\sqrt{k}} \sum_{n=0}^{\infty} \Gamma\left(n + \frac{1}{2}\right) \frac{\tilde{A}_{2n}}{(2n)!\, k^n} \tag{C.22}$$

or retaining just the leading term,

$$I_{\mathrm{SDP}}(k) = \mathrm{e}^{jkS_{\mathrm{c}}} \sqrt{\frac{\pi}{k}} \alpha'(0) A_{\mathrm{c}} + O\left(\frac{1}{k^{3/2}}\right), \tag{C.23}$$

where we have used the identity $\Gamma(1/2) = \sqrt{\pi}$ and denoted $A_{\mathrm{c}} = A(\alpha_{\mathrm{c}})$.

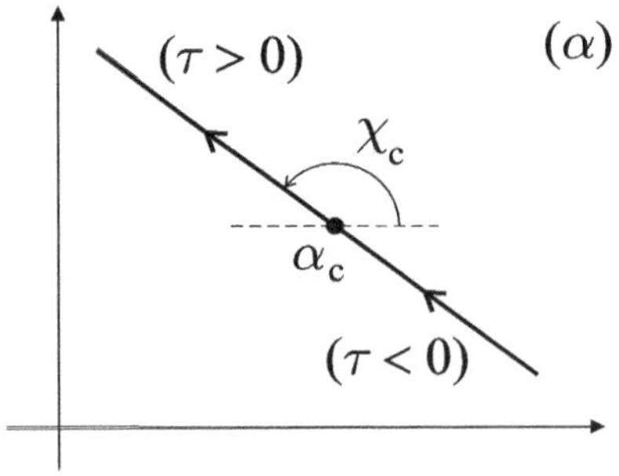

Figure C.3 Steepest descent path on the complex α plane crossing the saddle point α_c. The angle χ_c defines orientation of SDP at α_c.

The series (C.22) is an *asymptotic series*, which means that it is divergent, but when truncated the series may be made as close as desired to the exact value of the integral by making k large enough. Mathematically, this corresponds to the case in which a function $f(z)$ with an essential singularity at $z = \infty$ is expanded in a series in inverse powers of z. Other examples of asymptotic series are large-argument expansions of the Fresnel integral and the error function in Appendix B as well as Debye's expansions for Bessel functions in Appendix E. The degree of proximity of asymptotic expansions to the exact value is described by the order symbols O and o.

The formula (C.23) agrees with the formula that typically appears in the literature,

$$I_{\mathrm{SDP}}(k) = \mathrm{e}^{jkS_c + j\chi_c} \sqrt{\frac{2\pi}{k|S_c''|}} A_c + O\left(\frac{1}{k^{3/2}}\right), \tag{C.24}$$

where S_c'' is the second derivative of $S(\alpha)$ at $\alpha = \alpha_c$ and χ_c is the angle between the positive direction of the real axis on the complex α plane and the direction in which the integration contour leaves the saddle point, measured counterclockwise (Figure C.3). To see that, one can approximate the phase function as

$$S(\alpha) \approx S_c + \frac{1}{2} S_c''(\alpha - \alpha_c)^2 \tag{C.25}$$

and compare it with (C.14) to find that

$$\frac{1}{2} S_c''(\alpha - \alpha_c)^2 \approx j\tau^2 \tag{C.26}$$

and

$$\alpha = \alpha_c + \tau \sqrt{\frac{2}{|S_c''|}} \mathrm{e}^{j\chi_c}. \tag{C.27}$$

Thus,

$$\alpha'(0) = \sqrt{\frac{2}{|S_c''|}} \mathrm{e}^{j\chi_c}, \tag{C.28}$$

and (C.23) recovers (C.24). It is worth pointing out that the approach that is used in derivation of (C.22) is most accurate since the parametrization (C.14) exactly describes the phase function. Any error in approximation of the phase is magnified by the large parameter k in the phase factor, implying a generally lesser accuracy of the approach that relies on the approximation (C.25). Only the leading terms of the two approximations are equivalent.

Let us now discuss the applicability criteria of formulas (C.23) and (C.24). The most apparent condition is that $S''(\alpha_c)$ should not vanish. Furthermore, function $A(\alpha)$ should be slowly varying within the region essential for the integration, which using Taylor expansions for $A(\alpha)$ and $S(\alpha)$ can be formulated as

$$\left|\frac{A'(\alpha_c)}{A(\alpha_c)}\right| \ll \left|\frac{kS''(\alpha_c)}{2}\right|^{\frac{1}{2}}. \tag{C.29}$$

This condition is not satisfied if near the saddle point $A(\alpha)$ has a pole singularity, a branch point or oscillates at a rate comparable or faster than the exponent $\exp[jkS(\alpha)]$. In these cases (C.23) and (C.24) are inapplicable and have to be modified.

Note that the dependence of the exponent in the integral (C.1) on k should not be necessarily explicit; extensions of the formulas (C.23), (C.24), (C.28), and (C.29) to the exponent of the form $\exp[jS(\alpha, k)]$ with an implicit dependence on k are obtained by replacing $kS(\alpha)$ with $S(\alpha, k)$ and $kS''(\alpha)$ with $\partial^2 S(\alpha, k)/\partial\alpha^2$.

C.3.3 Pole Singularity near the Saddle Point

When a pole of $A(\alpha)$ approaches the saddle point, the amplitude cannot be expanded in a Taylor series as in (C.20) and the approach described in section C.3.2 is not applicable. The pole singularity needs special treatment. Assuming that the pole is located at $\tau = \tau_p$ and denoting the residue at the pole by r_p, we can extract the singularity by writing

$$\tilde{A}(\tau) = B(\tau) + \frac{r_p}{\tau - \tau_p}, \tag{C.30}$$

where $B(\tau)$ is a function holomorphic in the vicinity of the saddle point. Thus, from (C.18) and (C.30),

$$I_{SDP}(k) = I_1(k) + I_2(k), \tag{C.31}$$

where

$$I_1(k) = e^{jkS_c} \int_{-\infty}^{+\infty} B(\tau)e^{-k\tau^2}\, d\tau \tag{C.32}$$

$$I_2(k) = e^{jkS_c} r_p \int_{-\infty}^{+\infty} \frac{e^{-k\tau^2}}{\tau - \tau_p}\, d\tau. \tag{C.33}$$

The integrand in the first integral can be expanded in the Taylor series and integrated as in (C.20) and (C.22) to obtain

$$I_1(k) = \frac{e^{jkS_c}}{\sqrt{k}} \sum_{n=0}^{\infty} \Gamma\left(n + \frac{1}{2}\right) \frac{B_{2n}}{(2n)!\, k^n} \tag{C.34}$$

$$= e^{jkS_c}\sqrt{\frac{\pi}{k}}\left[\alpha'(0)A_c + \frac{r_p}{\tau_p}\right] + O\left(\frac{1}{k^{3/2}}\right). \tag{C.35}$$

Changing the integration variable in (C.33) as $\tau = \tilde{\tau}/\sqrt{k}$ reduces $I_2(k)$ to the error function for complex arguments (B.14),

$$I_2(k) = r_p \pi j e^{jkS_c} w(\tau_p\sqrt{k}), \tag{C.36}$$

and we arrive at the approximation

$$I_{\mathrm{SDP}}(k) = \mathrm{e}^{jkS_\mathrm{c}} \sqrt{\frac{\pi}{k}} \left[\alpha'(0)A_\mathrm{c} + \frac{r_\mathrm{p}}{\tau_\mathrm{p}} + jr_\mathrm{p}\sqrt{\pi k}\, w(\tau_\mathrm{p}\sqrt{k}) \right] + O\left(\frac{1}{k^{3/2}}\right). \tag{C.37}$$

Let us check its compliance with limiting cases. The limit of $\mathrm{Im}\,\tau_\mathrm{p} \to 0$ corresponds to a pole crossing the integration contour. If $I^{\pm}_{\mathrm{SDP}}(k)$ are the limiting values of $I_{\mathrm{SDP}}(k)$ in (C.18) with τ_p approaching the integration contour from the upper and lower half-planes on the complex τ plane, then the difference between them is due to the jump in the corresponding limits of $w(\tau_\mathrm{p}\sqrt{k})$,

$$I^{+}_{\mathrm{SDP}}(k) - I^{-}_{\mathrm{SDP}}(k) = jr_\mathrm{p}\pi \mathrm{e}^{jkS_\mathrm{c}} \left[w_{+}(\tau_\mathrm{p}\sqrt{k}) - w_{-}(\tau_\mathrm{p}\sqrt{k}) \right] \tag{C.38}$$

and since $w_{\pm}(\tau_\mathrm{p}\sqrt{k}) = \pm \mathrm{e}^{-k\tau_\mathrm{p}^2}$,

$$I^{+}_{\mathrm{SDP}}(k) - I^{-}_{\mathrm{SDP}}(k) = 2jr_\mathrm{p}\pi \mathrm{e}^{jkS_\mathrm{p}}, \tag{C.39}$$

where $S_\mathrm{p} = S(\alpha_\mathrm{p})$ and we have used the relation $S_\mathrm{p} = S_\mathrm{c} + j\tau_\mathrm{p}^2$, which follows from (C.14). Thus, the asymptotic approximation (C.37) exactly recovers the behavior of the original integral (C.17) when a pole crosses the integration contour. In the opposite limit of a pole moving away from the saddle point ($|\tau_\mathrm{p}\sqrt{k}| \to \infty$), it follows from (B.19) that

$$w(\tau_\mathrm{p}\sqrt{k}) \to \frac{j}{\tau_\mathrm{p}\sqrt{\pi k}}, \tag{C.40}$$

and (C.37) agrees with (C.23).

In applications, instead of the error function $w(z)$, the UTD transition function $F_{\mathrm{KP}}(X^2)$ or the Fresnel integral $F(X)$ is used. The corresponding asymptotic approximations follow from (B.25), (B.26), and (B.33) as

$$I_{\mathrm{SDP}}(k) = \mathrm{e}^{jkS_\mathrm{c}} \sqrt{\frac{\pi}{k}} \left\{ \alpha'(0)A_\mathrm{c} + \frac{r_\mathrm{p}}{\tau_\mathrm{p}} \left[1 - F_{\mathrm{KP}}\left(jk\tau_\mathrm{p}^2 \right) \right] \right\} + O\left(\frac{1}{k^{3/2}}\right) \tag{C.41}$$

and

$$I_{\mathrm{SDP}}(k) = 2\pi jqr_\mathrm{p}\mathrm{e}^{jkS_\mathrm{p}} F\left(q\tau_\mathrm{p}\sqrt{k}\,\mathrm{e}^{j\pi/4} \right) + \mathrm{e}^{jkS_\mathrm{c}} \sqrt{\frac{\pi}{k}} \left[\alpha'(0)A_\mathrm{c} + \frac{r_\mathrm{p}}{\tau_\mathrm{p}} \right] + O\left(\frac{1}{k^{3/2}}\right), \tag{C.42}$$

where $q = \mathrm{sgn}\,(\mathrm{Im}\,\tau_\mathrm{p})$.

C.3.4 Further Cases

There are plenty of other cases that may be encountered in applications. These include, for example,

- a higher-order pole singularity of $A(\alpha)$ near an isolated first-order saddle point;
- a branch point near an isolated first-order saddle point;
- an isolated saddle point of a higher order (higher-order derivatives of $S(\alpha)$ vanish at $\alpha = \alpha_\mathrm{c}$, e.g. $S'(\alpha_\mathrm{c}) = 0$ and $S''(\alpha_\mathrm{c}) = 0$);

- coalescing first-order saddle points;[2]
- a saddle point near an endpoint (a semi-infinite integration contour C in (C.1)).

The interested reader is referred to Felsen and Marcuvitz (1973).

C.4 Problems

1. Evaluate the integral

$$I(k,a) = \int_{-\infty}^{+\infty} \frac{e^{jkz}}{z^2 + a^2} \tag{C.43}$$

with a real-valued k and a complex-valued a ($\mathrm{Im}\, a > 0$).

Answer:

$$I(k,a) = \frac{\pi j}{a} e^{-j|k|a}. \tag{C.44}$$

2. Derive an explicit parametric equation for SDP of the function

$$S(\xi,k) = -\xi|x - x_0| - \gamma(\xi)|y - y_0|, \tag{C.45}$$

where $\gamma(\xi) = \sqrt{k^2 - \xi^2}$ and $\mathrm{Im}\gamma(\xi) \leq 0$.

Hint: Solve (C.14) for ξ.

Answer: Let $|x - x_0| = L \sin\theta_0$ and $|y - y_0| = L\cos\theta_0$, where L is the distance between the points (x_0, y_0) and (x, y) on the x-y plane and θ_0 ($0 < \theta_0 < \pi/2$) is the corresponding angle in the right-angled triangle with the sides $|x - x_0|$, $|y - y_0|$, and L. Then, SDP is given by the parametric equation

$$\xi(\tau) = \sin\theta_0 \left(k - j\frac{\tau^2}{L}\right) + \frac{\tau}{L}\cos\theta_0 \sqrt{\tau^2 + 2jkL}, \tag{C.46}$$

where the branch of the square root function on the complex τ plane is selected by the condition $\mathrm{Re}\sqrt{\tau^2 + 2jkL} \geq 0$.

3. Derive the large-argument approximation for the two-dimensional Green function

$$G_2(x, x_0, y, y_0) = \frac{1}{4j} H_0^{(2)}(kL) \tag{C.47}$$

with $L = \sqrt{(x - x_0)^2 + (y - y_0)^2}$ from the Fourier representation

$$G_2(x, x_0, y, y_0) = \frac{1}{4\pi j} \int_{-\infty}^{+\infty} e^{jS(\xi,k)} \frac{d\xi}{\gamma(\xi)}, \tag{C.48}$$

where $S(\xi, k)$ and $\gamma(\xi)$ are given in (C.45). Use the saddle point method.

---

[2] A similar case of two merging stationary points is studied in section D.2.3 in the framework of the stationary phase method.

Answer:

$$G_2(x, x_0, y, y_0) = \frac{\exp(-jkL - j\pi/4)}{\sqrt{8\pi kL}}. \tag{C.49}$$

4. Show that the SDP associated with Sommerfeld integrals

$$\int A(\alpha) e^{jkr\cos\alpha}\, d\alpha \tag{C.50}$$

can be described as

$$\operatorname{Re}\alpha = m\pi - (-1)^m \operatorname{gd}(\operatorname{Im}\alpha), \tag{C.51}$$

where m is an arbitrary integer number and $\operatorname{gd}(x)$ is the *Gudermannian function*.
Hint: Start by finding the saddle points and the directions of the steepest descent at them. Then write down the SDP equation (C.13) and reduce it to identity (C.52).
The Gudermannian function relates trigonometric and hyperbolic functions. In particular, it satisfies the identity

$$\cos \operatorname{gd}(x) = 1/\cosh x \tag{C.52}$$

and can be defined by the equation

$$\operatorname{gd}(x) = -\frac{\pi}{2} + 2\arctan(e^x) \tag{C.53}$$

so that $\operatorname{gd}(-\infty) = -\pi/2$ and $\operatorname{gd}(\infty) = \pi/2$.

5. Formula (7.32), which describes the asymptotic behavior of the Legendre function $P_\beta(\cos\theta)$ for large $|\beta|$, is inapplicable if $\theta = 0$. Derive the uniform approximation (Szegö 1934)

$$P_\beta(\cos\theta) \approx \left(\frac{\theta}{\sin\theta}\right)^{1/2} J_0\left[\left(\beta + \frac{1}{2}\right)\theta\right], \tag{C.54}$$

where J_0 is the Bessel function of the first kind.
Hint: Consider Mehler's formula (7.31) and note that the main contribution comes from the vicinity of the point $\tau = \theta$. Use the representation

$$J_0(z) = \frac{2}{\pi}\int_0^1 \frac{\cos(zt)}{\sqrt{1-t^2}}\, dt. \tag{C.55}$$

Solution: With $|\beta| \gg 1$ the integrand in (7.31) rapidly oscillates while having an integrable singularity at $\tau = \theta$. Since the integrand can be represented as a sum of two exponents

$$\exp\left[\pm j\beta\tau - \frac{1}{2}\ln(\cos\tau - \cos\theta)\right],$$

its saddle points are located at $\tau \approx \theta \mp j/(2\beta)$, implying that the main contribution to the integral (7.31) comes from an asymptotically small vicinity of the point $\tau = \theta$. In this vicinity, the function $\cos\tau - \cos\theta$ can be approximated by the first term of the Taylor series in powers of τ^2 at point θ^2,

$$\cos\tau - \cos\theta \approx \frac{\sin\theta}{2\theta}(\theta^2 - \tau^2), \tag{C.56}$$

which reduces (7.31) to the expression

$$P_\beta(\cos\theta) \approx \frac{2}{\pi}\left(\frac{\theta}{\sin\theta}\right)^{1/2} \int_0^\theta \frac{\cos\left[\left(\beta+\frac{1}{2}\right)\tau\right]}{\sqrt{\theta^2-\tau^2}}\mathrm{d}\tau. \tag{C.57}$$

Changing the integration variable as $\tau = \theta t$ transforms the integral in (C.57) to the representation (C.55) and Mehler's formula to the approximation (C.54). The latter gives the correct value $P_\beta(1) = 1$ at $\theta = 0$ while recovering the asymptotic formula (7.32) for $|\beta|\theta \gg 1$ due to the large argument approximation (E.26) for the Bessel function.

References

Bleistein N and Handelsman RA 1969 *Asymptotic Expansions of Integrals*, Dover.

Boersma J and Rahmat-Samii Y 1980 Comparison of two leading uniform theories of edge diffraction with the exact uniform asymptotic solution. *Radio Sci.*, **15**, 1179–1194.

Born M and Wolf E (eds) 1959 *Principles of Optics: Electromagnetic Theory of Propagation, Interference and Diffraction of Light*, Appendix III, Pergamon Press.

Bouche D, Molinet F, and Mittra R 1997 *Asymptotic Methods in Electromagnetics*, Appendix 4, Springer-Verlag.

Brown JW and Churchill RV 2013 *Complex Variables and Applications*, 9th edn, McGraw-Hill Education.

Carrier GF, Krook M, and Pearson CE 1966 *Functions of a Complex Variable: Theory and Technique*, McGraw-Hill.

Clemmow PC 1950 Some extensions to the method of integration by steepest descents. *Quart. J. Mech. Appl. Math.*, **3**, 241–256.

Collin RE 1960 *Field Theory of Guided Waves*, sec. 11.8, McGraw-Hill.

Copson ET 1965 *Asymptotic Expansions*, Cambridge University Press.

Debye P 1909 Näherungsformeln für die Zylinderfunktionen für große Werte des Arguments und unbeschränkt veränderliche Werte des Index. *Math. Ann.*, **67**, 535–558.

Fedoryuk MV 1989 *Asymptotic Methods in Analysis*, Springer (in volume 13 of the series Encyclopaedia of Mathematical Sciences, Analysis I, Gamkrelidze RV (ed.), 83–191).

Felsen LB and Marcuvitz N 1973 *Radiation and Scattering of Waves*, Chapter 4, Prentice-Hall.

Erdélyi A 1956 *Asymptotic Expansions*, Dover.

Jones DS 1964 *The Theory of Electromagnetism*, Chapter 8, Pergamon Press.

Morse PM and Feshbach H 1953 *Methods of Theoretical Physics*, Chapter 4, McGraw-Hill.

Ott H 1943 Die Sattelpunktmethode in der Umgebung eines Pols mit Anwendung auf die Wellenoptik und Akustik. *Ann. Phys. (Berlin)*, **435**, 393–403.

Pauli W 1938 On asymptotic series for functions in the theory of diffraction of light. *Phys. Rev.*, **54**, 924–931.

Riemann B 1863 Sullo svolgimento del quoziente di due serie ipergeometriche in frazione continua infinita. In *Gesammelte mathematische Werke und wissenschaftlicher Nachlalss*, Weber H (ed.), Leipzig, 1876, 400–406.

Ryshik IM and Gradstein IS 1957 *Tables of Series, Products, and Integrals*, VEB, Berlin.

Senior TBA and Volakis JL 1995 *Approximate Boundary Conditions in Electromagnetics*, Appendix C, The Institution of Electrical Engineers, London.

Sommerfeld A 1927 Elektromagnetische Schwingungen, in *Die Differential- und Integralgleichungen der Mechanik und Physik. 2. Physikalischer Teil*, Frank P and von Mises R (eds), Chapter 20, Vieweg & Sohn, Braunschweig.

Szegö G 1934 Über einige asymptotische Entwicklungen der Legendreschen Funktionen. *Proc. London Math. Soc.*, **36**, 427–450.

Van der Waerden BL 1952 On the method of saddle points. *Appl. sci. Res. B*, **2**, 33–45.

Wong R 1989 *Asymptotic Approximations of Integrals*, Academic Press.

D

The Stationary Phase Method

D.1 Introduction

PO solutions have the form of a surface integral of a rapidly oscillating function. Such integrals do not have the form required by the method of steepest descent (Appendix C) and are difficult to compute by direct numerical integration. Nevertheless, it is still possible to derive their asymptotic approximation by using the method of stationary phase (e.g., Borovikov (1994), Chako (1965), Cooke (1982), Fedoryuk (1987), Focke (1954), James (1976), Jones and Kline (1958), Kline and Kay (1965), Olver (1974), Van Kampen (1949), and Wong (1989)). The origin of the method, as applied to one-dimensional integrals, is often attributed to William Thomson (Lord Kelvin) (Kelvin 1887), and the basic idea is that the main contribution to such integrals comes from the portions of the domain of integration in which the phase variation slows down and the phase remains nearly constant or stationary.

This appendix is limited to one- and two-dimensional integrals, because of their practical importance, and to the leading terms in asymptotic expansions of the integrals in inverse powers of k. The method can be extended to multiple dimensions (e.g., Fedoryuk (1987) and Chako (1965)) and permits derivation of higher-order terms in the asymptotic expansions. The use of the higher-order terms is, however, hardly justifiable for PO solutions which by construction are approximate high-frequency solutions, meaningful only in the principal orders of k^{-1}.

The mathematical results presented in this appendix are applied in this book to the analysis of multiple reflections in electrically large low-absorption spheres (section 7.3.4), transmission through an aperture in an impenetrable screen (section 8.3), scattering from a smoothly curved surface (section 8.4) and from an edge (sections 8.3 and 8.4), a vertex, and a conical point (section 9.2) in the scattering surface. The results are simple approximations that permit clear physical interpretation and suggest the ways of extension of the asymptotic solutions to more complex scattering problems.

D.2 One-dimensional Integrals

We begin by taking a look at the one-dimensional case, which permits, on the one hand, a simple explanation of the basic principle of the method and, on the other hand, results in useful asymptotic formulas for the contribution from the boundary of the integration domain in the PO solutions as well as for GO contributions to exact solutions of scattering problems when they are given by one-dimensional integrals in spectral domain (as for a sphere on Watson's transformation, section 7.3.4).

Consider the following integral

$$I(k) = \int_a^b A(t)\, e^{jkS(t)}\, dt, \tag{D.1}$$

Modern Electromagnetic Scattering Theory with Applications, First Edition. Andrey V. Osipov and Sergei A. Tretyakov.

where the parameter k is large and positive, the integration range can be finite or infinite, the phase function $S(t)$ is real-valued, and the amplitude function $A(t)$ is complex-valued and integrable. Both functions are assumed to be continuous (otherwise we could subdivide the integration domain to have the discontinuities at the endpoints) and independent[1] of k, so that the factor $\exp[jkS(t)]$ defines the dependence of the integrand on k. We are interested in the approximation of $I(k)$ for $k \to +\infty$[2]. Since k and $S(t)$ are real-valued, the integrand in (D.1) is a rapidly oscillating function without a dominating maximum on the integration range, which prevents application of the standard Laplace method.

With large k the integral (D.1) is expected to vanish,

$$I(k) = o(1) \quad (k \to \infty), \tag{D.2}$$

since its integrand, as a function of t, rapidly oscillates, alternating the sign, and the contributions to the integral from neighboring portions of the integration domain effectively compensate each other. This mechanism of destructive interference does not take place at the endpoints (if they are finite) of the integration range because of the missing symmetry and at the points at which

$$S'(t_c) = 0 \tag{D.3}$$

because of the slower variation of the exponential factor there. These points are referred to as *critical* points. The critical points t_c, at which the phase function is stationary, are called *stationary* points. It is clear that the asymptotic behavior of $I(k)$ with $k \to \infty$ may essentially depend on whether the integration domain is free of stationary points or not, and we consider these cases in succession.

D.2.1 No Stationary Points on the Integration Interval

When the phase function has no critical points in the integration domain, an asymptotic approximation of (D.1) is easily derived by integrating by parts. Integrating in (D.1) once gives

$$I(k) = \frac{1}{jk}\left[\frac{A(t)}{S'(t)}e^{jkS(t)}\Big|_a^b + I_1(k)\right], \tag{D.4}$$

where

$$I_1(k) = \int_a^b A_1(t)\, e^{jkS(t)}\, dt \tag{D.5}$$

with

$$A_1(t) = -\frac{d}{dt}\left[\frac{A(t)}{S'(t)}\right]. \tag{D.6}$$

The integral $I_1(k)$ has a form similar to $I(k)$. If function $A_1(t)$ is integrable, then according to (D.2) $I_1(k) = o(1)$ and the first term in the square brackets on the right-hand side of (D.4) dominates, so that we obtain the approximation

$$I(k) = \frac{1}{jk}\frac{A(t)}{S'(t)}e^{jkS(t)}\Big|_a^b + o\left(\frac{1}{k}\right). \tag{D.7}$$

So, the main contribution to the integral comes from the endpoints of the integration interval, provided that the functions $A(t)$ and $S(t)$ are sufficiently smooth (can be differentiated as required for convergence of $I_1(k)$).

---

[1] Both may depend on k but only in such a way that $A(t, k)$ remains a slowly varying function of t as compared to the exponential factor as $k \to +\infty$. Furthermore, the large parameter can appear implicitly in the phase function; in this case $kS(t)$ should be replaced with $S(t, k)$, $kS'(t)$ with $\partial S(t, k)/(\partial t)$, $kS''(t)$ with $\partial^2 S(t, k)/(\partial t^2)$, etc.

[2] In practical calculations, k is, of course, always finite. We say that k is "large" when $A(t)$ varies slowly as compared with the exponential term $\exp[jkS(t)]$, and the main term in an approximation dominates the correction term. Physical conditions of applicability of such approximations are formulated in terms of Fresnel zones.

The leading term in (D.7) can vanish when, for example, the amplitude function $A(t)$ is zero at the endpoints, or functions $A(t)$ and $S(t)$ are periodic in t with a period equal to the length of the integration interval. In this case,

$$I(k) = \frac{1}{jk} I_1(k) \tag{D.8}$$

and the integration by parts is applied again to obtain the approximation

$$I(k) = \frac{1}{(jk)^2} \frac{A_1(t)}{S'(t)} e^{jkS(t)} \Big|_a^b + o\left(\frac{1}{k^2}\right). \tag{D.9}$$

Apparently, one can continue integrating by parts as long as the integrand remains finite at the endpoints and integrable over the integration domain.

D.2.2 *Isolated Stationary Points*

The asymptotic behavior of $I(k)$ significantly changes if the phase function has a stationary point t_c on the integration interval. The function $A_1(t)$ is then infinite at $t = t_c$ (see (D.3) and (D.6)) and the integral I_1 diverges, making integration by parts inapplicable. Simple qualitative considerations suggest that in this case the main contribution to the integral comes not from the endpoints but rather from the vicinity of the stationary point. We shall now see that in this case $I(k)$ decays as $k^{-1/2}$ with growing k, which can be written as $I(k) = O(k^{-1/2})$.

Consider, for example, an integral of the form (D.1) with $b = -a = T \geq 0$, $A(t) = 1$, and $S(t) = t^2$, which means a critical point $t_c = 0$ on the integration interval. Figure D.1 shows the real part of the integrand

$$\text{Re}\,[e^{jkS(t)}] = \cos(kt^2) \tag{D.10}$$

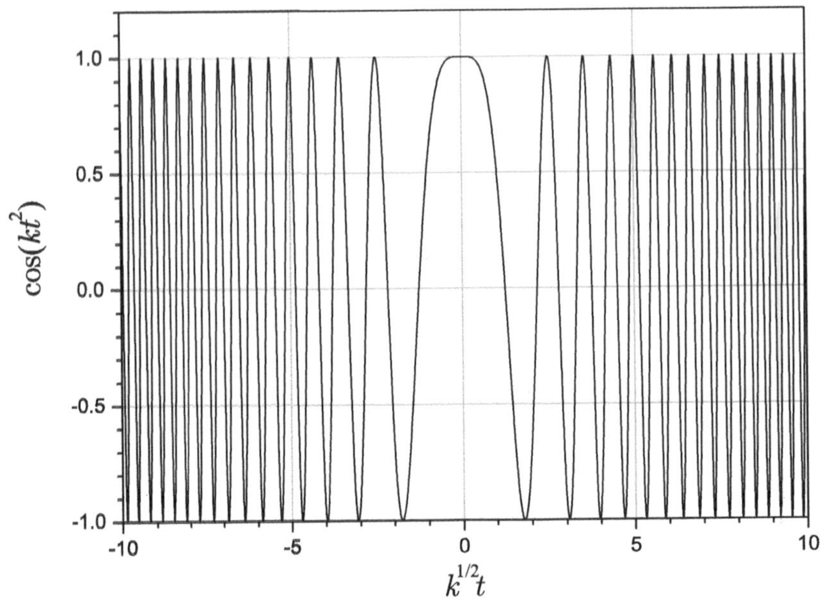

Figure D.1 An example of a rapidly oscillating integrand function with a stationary point at $t = 0$.

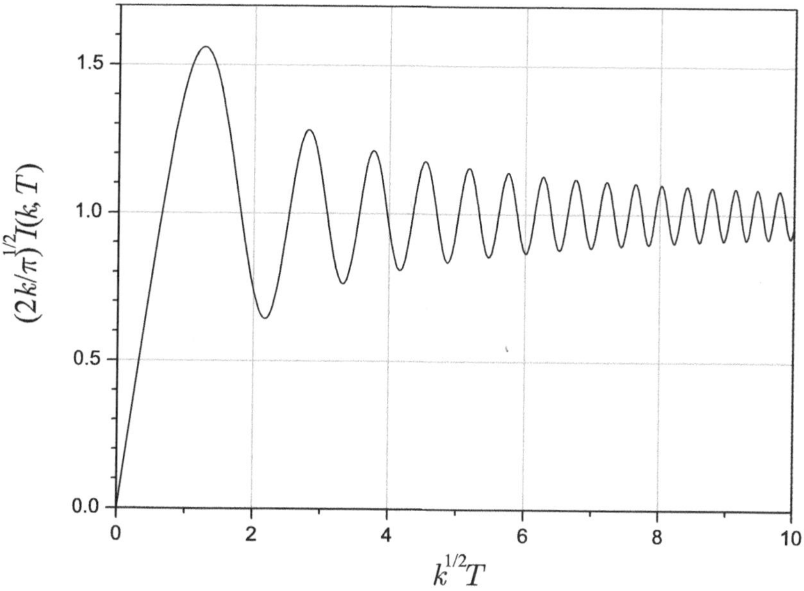

Figure D.2 Integral of the function shown in Figure D.1 over the interval $-T \le t \le T$ normalized to its value at $T = \infty$ as a function of $k^{1/2}T$.

as a function of $t\sqrt{k}$, and it is seen that the oscillation rate is slow near the critical point but rapidly increases as the absolute value of t grows. The oscillation period is proportional to $k^{-1/2}$, so that the greater the parameter k the faster the oscillations. Figure D.2 illustrates the effect of integration on the rapidly oscillating function (D.10) by showing the dependence of the integral

$$I(k, T) = \int_{-T}^{T} \cos(kt^2)\, dt \tag{D.11}$$

on the size T of the integration interval. The function is normalized to $I(k, \infty) = \pi^{1/2}(2k)^{-1/2}$, which is the value of the integral taken over the whole real axis, and with $Tk^{-1/2} \ge 1$ the curve is already close to unity, implying that the main contribution to the integral (D.11) comes from the vicinity of the critical point where $|t - t_c| \le k^{-1/2}$, which is asymptotically small since k is the large parameter.

The observation that small vicinities of critical points determine the value of the integrals (D.1) for large values of k suggests a way of evaluating of such contributions. Consider the contribution from a vicinity of a stationary point t_c

$$I_c(k) = \int_{t_c-\delta}^{t_c+\delta} A(t)\, e^{jkS(t)}\, dt, \tag{D.12}$$

where δ is positive and sufficiently large on the scale of $k^{-1/2}$. For sufficiently large values of k, the size of the integration interval is small and the phase function can be replaced with the Taylor series

$$S(t) \approx S(t_c) + \frac{1}{2} S''(t_c)(t - t_c)^2, \tag{D.13}$$

where the linear term is not present because of the definition (D.3) of the stationary point. The amplitude function $A(t)$ can be considered as slowly varying and replaced with its value at $t = t_c$. By denoting for brevity $A_c = A(t_c)$, $S_c = S(t_c)$, and $S''_c = S''(t_c)$ we obtain

$$I_c(k) \approx A_c e^{jkS_c} \int_{t_c-\delta}^{t_c+\delta} e^{j\frac{k}{2}S''_c(t-t_c)^2}\, dt. \tag{D.14}$$

Since δ has been chosen sufficiently large to comprise the essential vicinity of the stationary point, the parameter δ can be let go to $+\infty$ and (D.14) becomes

$$I_c(k) \approx A_c e^{jkS_c} \int_{-\infty}^{\infty} e^{j\frac{k}{2}S''_c(t-t_c)^2}\, dt, \tag{D.15}$$

which is reduced by the substitution

$$t = t_c + \xi\sqrt{\frac{2}{k\,|S''_c|}} \tag{D.16}$$

to the Fresnel-type integral

$$I_c(k) \approx A_c e^{jkS_c}\sqrt{\frac{2}{k\,|S''_c|}} \int_{-\infty}^{\infty} e^{j\xi^2 \operatorname{sgn} S''_c}\, d\xi. \tag{D.17}$$

The integral is easily evaluated on noting its relation to the Fresnel integrals (B.5),

$$\int_{-\infty}^{\infty} e^{j\xi^2 \operatorname{sgn} S''_c}\, d\xi = \int_{-\infty}^{\infty} \cos\left(\xi^2\right) d\xi + j\operatorname{sgn} S''_c \int_{-\infty}^{\infty} \sin(\xi^2)\, d\xi = \sqrt{\pi} e^{j\frac{\pi}{4}\operatorname{sgn} S''_c}, \tag{D.18}$$

thus arriving at the formula

$$I_c(k) \approx A_c\sqrt{\frac{2\pi}{k\,|S''_c|}} \exp\left[jkS_c + j\frac{\pi}{4}\operatorname{sgn} S''_c\right] \tag{D.19}$$

for the contribution of a stationary point to the integral (D.1). The contribution is of the order $k^{-1/2}$ and thus asymptotically dominant over the endpoint contributions.

In the case that the dependence on the large parameter is more complex and the parameter k cannot be explicitly factored out in the phase function, but the required behavior of the integrand, which is a rapidly oscillating phase factor and a slowly varying amplitude, is still present, instead of (D.1) we have

$$I(k) = \int_a^b A(t,k) e^{jS(t,k)}\, dt, \tag{D.20}$$

and the contribution from a stationary point t_c, which can now be a function of k, is given by

$$I_c(k) \approx A(t_c, k)\sqrt{\frac{2\pi}{|S''(t_c, k)|}} \exp\left[jS(t_c, k) + j\frac{\pi}{4}\operatorname{sgn} S''(t_c, k)\right], \tag{D.21}$$

where $S''(t, k) = \partial^2 S(t, k)/(\partial t^2)$.

Finally, depending on the phase function and the integration interval, there can be multiple stationary points, in which case the result is given by a sum of individual contributions, provided that the stationary points do not reside in an asymptotically small neighborhood of each other or the endpoints of the integration interval. If a stationary point happens to be at one of the endpoints, then its contribution to I_c must be halved. In general, however, approximation (D.19) does not apply any more, and such cases require special treatment (e.g., problem D.2), as do the cases in which the amplitude function has an integrable singularity on the integration interval (e.g., see Borovikov (1994) and Olver (1974)).

D.2.3 Two Coalescing Stationary Points

Perhaps one of the most important special cases is that of two coalescing stationary points. Such integrals describe wave fields near caustics formed by two intersecting rays as in the rainbow (see section 7.3.4). It is convenient to assume that $S(t)$ depends on a parameter ϵ so that for $\epsilon > 0$ the phase function has two stationary points t_{c1} and t_{c2} which merge together at a point t_0 in the limit $\epsilon \to 0$. At that point, $S''(t_0) = 0$ and the formula (D.19) becomes inapplicable as it predicts an infinite value for an apparently bounded integral.

A correct asymptotic approximation is expressed in terms of the function

$$\operatorname{Ai}(w) = \frac{1}{2\pi}\int_{-\infty}^{+\infty} \exp\left(jw\tau + j\frac{1}{3}\tau^3\right) d\tau = \frac{1}{\pi}\int_{0}^{+\infty} \cos\left(w\tau + \frac{1}{3}\tau^3\right) d\tau \tag{D.22}$$

introduced in the seminal paper (Airy 1838). Let us derive this approximation. If ϵ is sufficiently small and t_{c1} and t_{c2} are close to the point t_0, then we can limit the integration to a small vicinity of the point t_0

$$I(k) \approx \int_{t_0-\delta}^{t_0+\delta} A(t)e^{jkS(t)}\, dt \tag{D.23}$$

in which δ is such that the integration range includes both stationary points to ensure an asymptotically small truncation error. Next, the phase function in (D.23) is approximated by the Taylor expansion

$$S(t) \approx S(t_0) + S'(t_0)(t - t_0) + \frac{1}{2}S''(t_0)(t - t_0)^2 + \frac{1}{6}S'''(t_0)(t - t_0)^3, \tag{D.24}$$

where we have to account for the cubical term since $S''(t_0) \to 0$ as $\epsilon \to 0$. On the whole integration range in (D.23), the quadratic term in (D.24) can be neglected, and replacing the slowly varying amplitude $A(t)$ with its value at $t = t_0$ gives

$$I(k) \approx A(t_0)e^{jkS(t_0)} \int_{t_0-\delta}^{t_0+\delta} e^{jk[S'(t_0)(t-t_0)+\frac{1}{6}S'''(t_0)(t-t_0)^3]}\, dt. \tag{D.25}$$

Since the main contribution to the integral comes from a vicinity of t_0, the integration range can be extended to $t = \pm\infty$, and the change of the integration variable

$$t = t_0 + \tau \left[\frac{2}{k|S'''(t_0)|} \right]^{1/3} \tag{D.26}$$

transforms the integral in (D.25) to the Airy function (D.22) with

$$w = \text{sgn}[S'''(t_0)]S'(t_0) \left[\frac{2k^2}{|S'''(t_0)|} \right]^{1/3}. \tag{D.27}$$

The result reads

$$I(k) \approx 2\pi A(t_0) e^{jkS(t_0)} \left[\frac{2}{k|S'''(t_0)|} \right]^{1/3} \text{Ai}(w). \tag{D.28}$$

The Airy function is finite at $w = 0$,

$$\text{Ai}\,(0) = \frac{1}{3^{2/3}\Gamma\left(\frac{2}{3}\right)} \approx 0.355, \tag{D.29}$$

exponentially decays as

$$\text{Ai}\,(w) \approx \frac{1}{2\sqrt{\pi}\, w^{1/4}} \exp\left(-\frac{2}{3} w^{3/2}\right) \tag{D.30}$$

when $w \to +\infty$ and oscillates as

$$\text{Ai}\,(w) \approx \frac{1}{\sqrt{\pi}\, |w|^{1/4}} \cos\left(\frac{2}{3}|w|^{3/2} - \frac{\pi}{4}\right) \tag{D.31}$$

with $w \to -\infty$. At the first four extrema at $w \approx -1.01879$, -3.2482, -4.8201, and -6.16331, $\text{Ai}(w) \approx 0.535657$, -0.419015, 0.380406, and -0.357908, respectively, and the absolute value of the function is therefore greater than Ai(0). In Figure D.3 the function is plotted for $-15 \le w \le 5$. Further properties, relations, and numerical values can be found in, for example, Abramowitz and Stegun (1965, section 10.4).

The formula (D.28) predicts a finite value for $I(k)$ in the limit of two coalescing stationary points ($w \to 0$) because of the property (D.29). This limiting value is of the order $k^{-1/3}$ and asymptotically greater than contributions from isolated stationary points (see (D.19)), which are $O(k^{-1/2})$. Because of (D.31) and (D.30), formula (D.28) describes a function which exponentially decays for $w > 0$ and oscillates for $w < 0$, which suggests that the case of two coalescing stationary points may be relevant to asymptotic approximations of wave fields near caustics, on one side of which the fields are superpositions of two plane waves, whereas on the another side they rapidly vanish. Indeed, this is the universal dependence of wave fields on a coordinate perpendicular to every caustic surface formed by two intersecting rays (see problems 14 and 15 in Chapter 8). Specific examples are given in section 7.3.4, where we study the phenomenon of rainbows in the fields scattering from electrically large low-absorption spheres.

If the large parameter k is unexplicitly present in the phase function and $I(k)$ has the form (D.20), then the asymptotic approximation (D.28) is replaced by

$$I(k) \approx 2\pi m A(t_0, k) \text{Ai}(w) e^{jS(t_0, k)}, \tag{D.32}$$

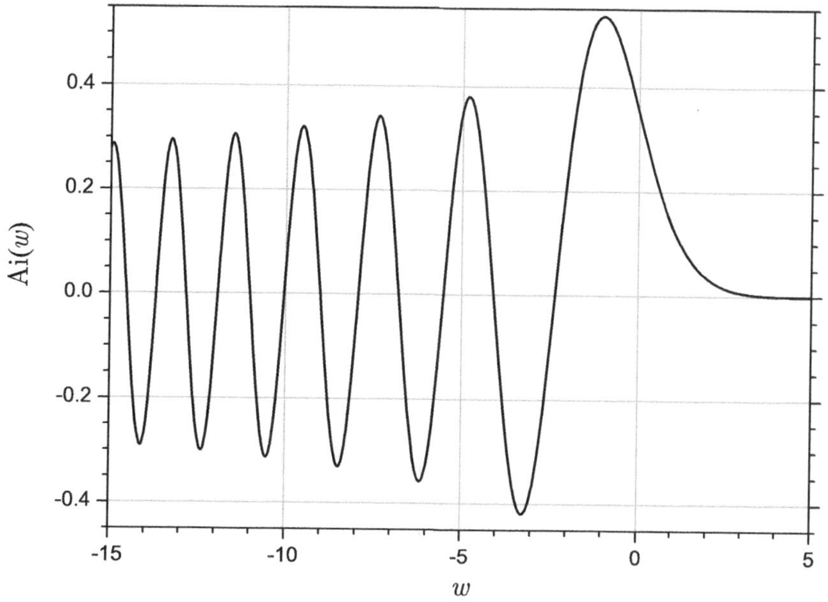

Figure D.3 Airy function of a real-valued argument.

where

$$w = m \,\mathrm{sgn}[S'''(t_0, k)] S'(t_0, k), \quad m = \left[\frac{2}{|S'''(t_0, k)|}\right]^{1/3}, \tag{D.33}$$

$S'(t, k) = \partial S(t, k)/(\partial t)$, and $S'''(t, k) = \partial^3 S(t, k)/(\partial t^3)$.

Once the distance between the stationary points gets larger, it is reasonable to expect that (D.28) and (D.32) should somehow recover the sum of individual contributions from isolated stationary points, given by (D.19) or (D.21). Let us take a closer look at this connection, for example between (D.32) and (D.21).

With $w \to -\infty$, equation (D.32) simplifies to

$$I(k) \approx 2\sqrt{\pi} \frac{m^{3/4}}{|S_0'|^{1/4}} A_0 e^{jS_0} \cos\left(\frac{2}{3}|mS_0'|^{3/2} - \frac{\pi}{4}\right), \tag{D.34}$$

where $A_0 = A(t_0, k)$ and $S_0 = S(t_0, k)$ for brevity. Note that negative values of w are only possible when

$$\mathrm{sgn}\, S_0''' = -\mathrm{sgn}\, S_0'. \tag{D.35}$$

On the other hand, approximating $I(k)$ by a sum of contributions from two isolated critical points leads to

$$I_1 + I_2 \approx A_1 \sqrt{\frac{2\pi}{|S_{c1}''|}} e^{jS_{c1} + j\frac{\pi}{4}\mathrm{sgn}\, S_{c1}''} + A_2 \sqrt{\frac{2\pi}{|S_{c2}''|}} e^{jS_{c2} + j\frac{\pi}{4}\mathrm{sgn}\, S_{c2}''} \tag{D.36}$$

with $A_1 = A(t_{c1}, k), A_2 = A(t_{c2}, k), S_{c1} = S(t_{c1}, k)$, and $S_{c2} = S(t_{c2}, k)$. If the critical points t_{c1} and t_{c2} are located sufficiently close to the point t_0, then A_1 and A_2 can be approximated by A_0 and the phase functions by the Taylor expansions

$$S(t_c) = S_0 + (t_c - t_0)\left[S_0' + \frac{1}{6}(t_c - t_0)^2 S_0'''\right] \tag{D.37}$$

$$S'(t_c) = S_0' + \frac{1}{2}(t_c - t_0)^2 S_0''' \tag{D.38}$$

$$S''(t_c) = (t_c - t_0)S_0''' \tag{D.39}$$

in which t_c is either t_{c1} or t_{c2} and $S_0'' = 0$ by definition of the point t_0. Since t_c is a stationary point, it follows from (D.38) that

$$(t_c - t_0)^2 \approx -\frac{2S_0'}{S_0'''}, \quad t_c - t_0 \approx \pm\left|\frac{2S_0'}{S_0'''}\right|^{1/2}, \tag{D.40}$$

depending on whether t_c is less than or greater than t_0. Then, from (D.35), (D.37), (D.39), and (D.40),

$$S(t_c) \approx S_0 \mp \operatorname{sgn}(S_0''')\frac{2}{3}\left|\frac{2S_0'^3}{S_0'''}\right|^{1/2} \tag{D.41}$$

$$S_c'' \approx \pm\operatorname{sgn}(S_0''')|2S_0'S_0'''|^{1/2} \tag{D.42}$$

and

$$I_1 \approx \sqrt{\pi}\left|\frac{2}{S_0'S_0'''}\right|^{1/4} A_0 e^{jS_0} \exp\left[j\operatorname{sgn}(S_0''')\left(\frac{2}{3}\left|\frac{2S_0'^3}{S_0'''}\right|^{1/2} - \frac{\pi}{4}\right)\right] \tag{D.43}$$

$$I_2 \approx \sqrt{\pi}\left|\frac{2}{S_0'S_0'''}\right|^{1/4} A_0 e^{jS_0} \exp\left[-j\operatorname{sgn}(S_0''')\left(\frac{2}{3}\left|\frac{2S_0'^3}{S_0'''}\right|^{1/2} - \frac{\pi}{4}\right)\right], \tag{D.44}$$

where, without loss of generality, we have assumed that $t_{c1} < t_0 < t_{c2}$. The sum of the contributions is equal to (D.34) since $|S_0'''| = 2/m^3$.

So, we see that approximations (D.19) and (D.21) on the one side and (D.28) and (D.32) on the other side are consistent in the sense that they overlap in the vicinity of the caustic. It should be noted, however, that (D.19), (D.21), (D.28), and (D.32) are local asymptotic approximations as (D.19) and (D.21) fail at the caustic, whereas (D.28) and (D.32) become inaccurate at larger distances from the caustic. Although it is possible to construct a more general formula which is uniformly valid for an arbitrary distance between the stationary points, this formula is more complicated from the analytical viewpoint, and the interested reader is referred to the literature, for example Borovikov (1994). Moreover, approximations (D.19) and (D.28) often overlap with a good accuracy, which may eliminate the need for a uniform formula. This fact is illustrated by the following numerical example.

Consider an integral $I(k)$ with

$$A(t) = 1, \quad S(t) = 1 - \epsilon^2 t + \frac{1}{3}t^3(1 - \epsilon^2) + \frac{1}{5}t^5. \tag{D.45}$$

Since

$$S'(t) = (t^2 - \epsilon^2)(1 + t^2), \tag{D.46}$$

the phase function has two stationary points at $t = \pm\epsilon$ coalescing at a point $t_0 = 0$ in the limit $\epsilon \to 0$. The formula (D.19) applied to the two isolated critical points gives

$$I(k) \approx 2e^{jk}\sqrt{\frac{\pi}{k\epsilon(1+\epsilon^2)}}\cos\left[\frac{\pi}{4} - \frac{2}{3}k\epsilon^3\left(1 + \frac{1}{5}\epsilon^2\right)\right], \tag{D.47}$$

which is singular at $\epsilon = 0$, as expected. The formula (D.28) for two coalescing critical points, however, leads to the expression

$$I(k) \approx e^{jk}\frac{2\pi}{k^{1/3}(1-\epsilon^2)^{1/3}}\mathrm{Ai}\left[-\frac{k^{2/3}\epsilon^2}{(1-\epsilon^2)^{1/3}}\right], \tag{D.48}$$

which is finite when $\epsilon = 0$. In Figure D.4, asymptotic expressions (D.47) and (D.48) are plotted as dashed and dotted lines, respectively, and compared with the exact value of $I(k)$ (solid line) for $k = 100$. It is seen that the solid line agrees well with the dotted line when $0 \le \epsilon < 0.32$ and with the dashed line when $\epsilon > 0.24$, implying that on the interval $0.24 < \epsilon < 0.32$ the local approximations (D.47) and (D.48) flawlessly overlap, thus covering the whole range of values of the parameter ϵ.

The asymptotic formula (D.28) assumes that $A(t)$ is finite at $t = t_0$. If $A(t_0) = 0$, then the next term in the asymptotic expansion of $I(k)$ is required, which turns out to be by a factor $k^{-1/3}$ smaller than (D.28) and proportional to $A'(t_0)$ and $\mathrm{Ai}'(w)$ (see problem D.3).

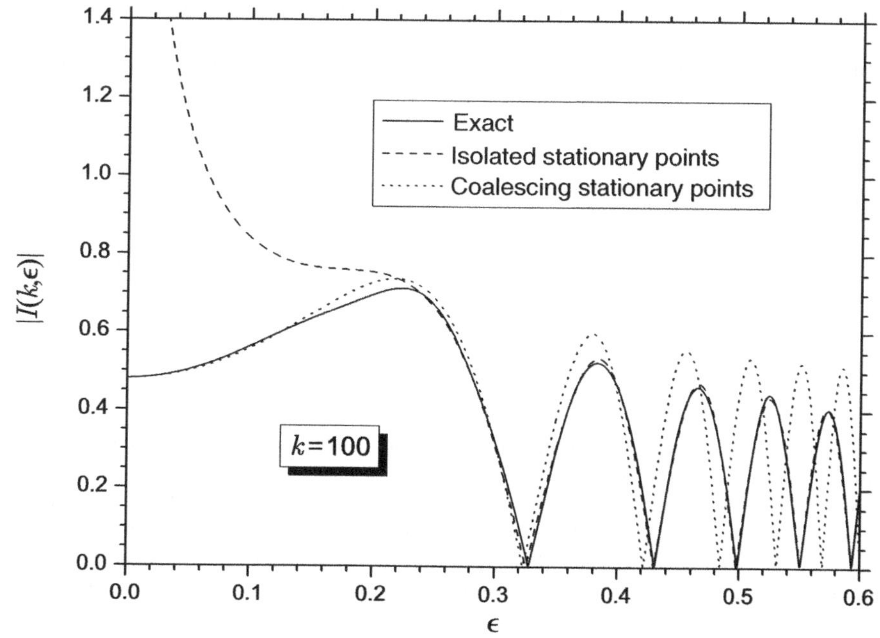

Figure D.4 Comparison of the accuracy of two approximations for an integral with coalescing critical points.

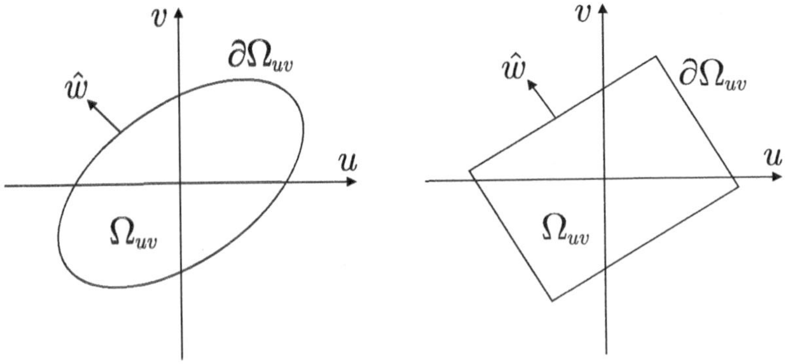

Figure D.5 Integration area Ω_{uv} in the double integral (D.49). The boundary $\partial\Omega_{uv}$ can be smooth (left) or piecewise smooth (right). $\hat{w}$ is the outward unit vector normal to the boundary.

D.3 Two-dimensional Integrals

The PO solutions of scattering problems are given by surface integrals with rapidly oscillating integrand functions, which, on a suitable parametrization of the surface, can be expressed as double integrals. Asymptotic evaluation of such integrals is basically similar to the one-dimensional case but also has its specific features and is more involved.

Consider the integral

$$I(k) = \int_{\Omega_{uv}} A(u, v)\, e^{jkS(u,v)}\, du dv, \tag{D.49}$$

where Ω_{uv} is a finite or infinite domain with a (piecewise) smooth boundary $\partial\Omega_{uv}$ (Figure D.5). We again assume that k is a large positive parameter, the phase function $S(u, v)$ is real-valued, the amplitude function $A(u, v)$ is real- or complex-valued, and both functions have no singularities in the integration area, so that all the integrals occurring in the following analysis exist.

D.3.1 Stationary Point in the Integration Domain

Similarly to the one-dimensional case, the behavior of the integral for $k \to \infty$ depends on the presence of stationary points in the integration domain. Since the phase function is now a function of the two variables u and v, the definition of the stationary phase (D.3) changes to

$$\frac{\partial}{\partial u} S(u_c, v_c) = 0, \quad \frac{\partial}{\partial v} S(u_c, v_c) = 0, \tag{D.50}$$

from which the coordinates $u = u_c$ and $v = v_c$ of the stationary point can be found. The points at which equations (D.50) hold are called critical points of the first kind.

First, let us show that if equations (D.50) have a solution within Ω_{uv}, then $I(k) = O(k^{-1})$. To this end, select a δ neighborhood of the critical point, beyond which the oscillations of the integrand function almost completely cancel each other, and note that since this neighborhood is small when k is large, the amplitude function can be replaced with its value at the critical point $A_c = A(u_c, v_c)$ and the phase function with the first terms of the Taylor expansion

$$S(u, v) \approx S_c + \frac{1}{2}\alpha(u - u_c)^2 + \frac{1}{2}\beta(v - v_c)^2 + \gamma(u - u_c)(v - v_c), \tag{D.51}$$

where

$$S_c = S(u_c, v_c), \quad \alpha = \frac{\partial^2 S}{\partial u^2}(u_c, v_c), \quad \beta = \frac{\partial^2 S}{\partial v^2}(u_c, v_c), \quad \gamma = \frac{\partial^2 S}{\partial u \partial v}(u_c, v_c). \tag{D.52}$$

Here the second derivatives of $S(u, v)$ are assumed continuous in a neighborhood of the stationary point, implying that $\partial^2 S/(\partial u \partial v) = \partial^2 S/(\partial v \partial u)$.

Introducing, for brevity, new variables

$$\xi = u - u_c, \quad \eta = v - v_c \tag{D.53}$$

leads to the following estimate for the contribution from the critical point

$$I_c(k) \approx A_c e^{jkS_c} \int_{-\delta}^{+\delta} \int_{-\delta}^{+\delta} \exp\left[jk\left(\frac{1}{2}\alpha\xi^2 + \frac{1}{2}\beta\eta^2 + \gamma\xi\eta\right)\right] d\xi d\eta. \tag{D.54}$$

The domain of integration in (D.54) can be extended to the full plane by letting δ go to $+\infty$,

$$I_c(k) \approx A_c e^{jkS_c} \int_{-\infty}^{+\infty} \int_{-\infty}^{+\infty} \exp\left[jk\left(\frac{1}{2}\alpha\xi^2 + \frac{1}{2}\beta\eta^2 + \gamma\xi\eta\right)\right] d\xi d\eta, \tag{D.55}$$

which introduces only an asymptotically small error since by the choice of δ the selected neighborhood of the critical point defines almost fully the value of I_c.

The integral in (D.55) can be calculated as an iterated integral, and if the integration with respect to ξ is to be performed first, then the integral

$$\int_{-\infty}^{+\infty} \exp\left[jk\left(\frac{1}{2}\alpha\xi^2 + \gamma\xi\eta\right)\right] d\xi \tag{D.56}$$

can be rewritten as

$$\exp\left(-jk\frac{\gamma^2\eta^2}{2\alpha}\right) \int_{-\infty}^{+\infty} \exp\left[jk\frac{\alpha}{2}\left(\xi + \frac{\gamma}{\alpha}\eta\right)^2\right] d\xi \tag{D.57}$$

and, by the change of the integration variable $\xi' = \xi + \gamma\eta/\alpha$, reduced to a Fresnel-type integral (B.1) with $X \to \infty$, which is evaluated in closed form by using (D.18),

$$\exp\left(-jk\frac{\gamma^2\eta^2}{2\alpha}\right) \int_{-\infty}^{+\infty} \exp\left(jk\frac{\alpha}{2}\xi'^2\right) d\xi' = \exp\left(j\frac{\pi}{4}\operatorname{sgn}\alpha - jk\frac{\gamma^2\eta^2}{2\alpha}\right) \sqrt{\frac{2\pi}{k|\alpha|}}. \tag{D.58}$$

The remaining integral with respect to η is also of the Fresnel type,

$$\int_{-\infty}^{+\infty} \exp\left[j\frac{k}{2}\left(\beta - \frac{\gamma^2}{\alpha}\right)\eta^2\right] d\eta = \exp\left[j\frac{\pi}{4}\operatorname{sgn}\left(\beta - \frac{\gamma^2}{\alpha}\right)\right] \sqrt{\frac{2\pi}{k|\beta - \gamma^2/\alpha|}}, \tag{D.59}$$

and thus

$$I_c(k) \approx \frac{2\pi\sigma_c A_c}{k\,|\alpha\beta - \gamma^2|^{1/2}} e^{jkS_c} \tag{D.60}$$

with

$$\sigma_c = \exp\left\{j\frac{\pi}{4}\left[\operatorname{sgn}\alpha + \operatorname{sgn}\left(\beta - \frac{\gamma^2}{\alpha}\right)\right]\right\} = \begin{cases} j, & \alpha > 0, \quad \alpha\beta - \gamma^2 > 0 \\ 1, & \alpha > 0, \quad \alpha\beta - \gamma^2 < 0 \\ 1, & \alpha < 0, \quad \alpha\beta - \gamma^2 < 0 \\ -j, & \alpha < 0, \quad \alpha\beta - \gamma^2 > 0 \end{cases}. \tag{D.61}$$

Equation (D.60) is sometimes written in terms of the Hessian matrix of the phase function

$$\mathbf{H}(S) = \begin{pmatrix} \dfrac{\partial^2 S}{\partial u^2} & \dfrac{\partial^2 S}{\partial u \partial v} \\ \dfrac{\partial^2 S}{\partial v \partial u} & \dfrac{\partial^2 S}{\partial v^2} \end{pmatrix}. \tag{D.62}$$

Denoting $\mathbf{H}(S)$ at the stationary point by $\mathbf{H}_c$ and using (D.52) give

$$\mathbf{H}_c = \begin{pmatrix} \alpha & \gamma \\ \gamma & \beta \end{pmatrix} \tag{D.63}$$

with $\det \mathbf{H}_c = \alpha\beta - \gamma^2$, and if λ_1 and λ_2 are the eigenvalues of $\mathbf{H}_c$, then (D.60) can be expressed as (problem D.4)

$$I_c(k) \approx \frac{2\pi A_c}{k|\det \mathbf{H}_c|^{1/2}} e^{jkS_c + j\frac{\pi}{4}(\operatorname{sgn}\lambda_1 + \operatorname{sgn}\lambda_2)}. \tag{D.64}$$

The determinant of the Hessian matrix is invariant with respect to rotation of the coordinate system (u, v) used in the integration. The sign of $\det \mathbf{H}_c$ determines the topology of the function $S(u, v)$ at the stationary point. If $\det \mathbf{H}_c > 0$, then $S(u, v)$ has a minimum or maximum and the difference $S(u, v) - S_c$ grows (minimum) or decays (maximum) as the point (u, v) goes away from (u_c, v_c) in any direction. Such stationary points are called elliptic. If $\det \mathbf{H}_c < 0$, then $S(u, v)$ in a vicinity of the stationary point looks like a saddle, implying that $S(u, v) - S_c$ may grow or decay, depending on the direction in which (u, v) goes away from (u_c, v_c). Such stationary points are called hyperbolic. The derived asymptotic formulas, (D.60) or (D.64), are applicable in both cases. The case in which $\det \mathbf{H}_c = 0$ describes the so-called parabolic stationary point; the derived equations are inapplicable in this case and the asymptotic evaluation of the integral requires a special treatment.

If the integration domain includes several isolated critical points, then a sum of the individual contributions from every point gives the asymptotic value of the integral (D.49). The case of coalescing critical points, which means that the points are located in an asymptotically small proximity to each other, requires special treatment. In this case, $\det \mathbf{H}_c \to 0$ and (D.60) and (D.64) are no longer valid.

In problems of scattering, critical points of the first kind are typically responsible for the waves scattered from a surface in the specular direction.

D.3.2 Stationary Point near the Boundary of the Integration Domain

Now assume that a stationary point (critical point of the first kind) is located in an asymptotically small vicinity of the boundary of the integration domain and the boundary is smooth and has no corner points. In this case, only a part of the region essential for integration contributes to the integral, which has to be accounted for in the asymptotic evaluation of (D.49).

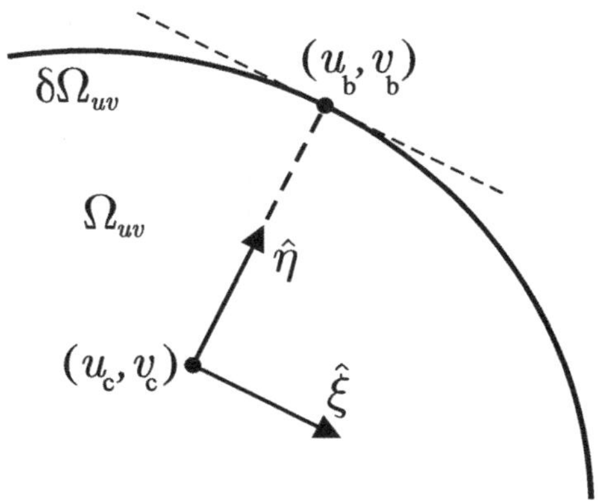

Figure D.6 Stationary point near the boundary of the integration domain.

Similarly to the treatment of an isolated stationary point in Ω_{uv}, we introduce a local coordinate system $(\hat{\xi}, \hat{\eta})$ centered at the stationary point (u_c, v_c) but choose the orientation of the axes so that the axis $\hat{\eta}$ is directed to the point of the boundary nearest to the stationary point, which we denote by (u_b, v_b) (Figure D.6). The coordinates of the boundary point are $\xi = 0$ and $\eta = \eta_b$, and the value of η_b follows from the expansion (D.51) in terms of S_c and $S_b = S(u_b, v_b)$ as

$$\eta_b = \pm \left| \frac{2}{\beta}(S_b - S_c) \right|^{1/2} \tag{D.65}$$

with the upper sign when the stationary point lies in Ω_{uv} and the lower sign otherwise.

Assuming that the curvature radius of $\delta\Omega_{uv}$ at (u_b, v_b) is much greater than the size of the domain essential for integration (which is asymptotically small as k goes to infinity), we can consider the boundary as a straight line tangent to the boundary at (u_b, v_b) to integrate in (D.54) from $-\infty$ to $+\infty$ with respect to ξ. The integration range with respect to η is, however, limited by $\eta = \eta_b$, and instead of (D.55) we have

$$I_c(k) \approx A_c e^{jkS_c} \sqrt{\frac{2\pi}{k|\alpha|}} e^{j\frac{\pi}{4}\operatorname{sgn}\alpha} \int_{-\infty}^{\eta_b} \exp\left[j\frac{k}{2}\left(\beta - \frac{\gamma^2}{\alpha}\right)\eta^2 \right] d\eta. \tag{D.66}$$

When $\beta < \gamma^2/\alpha$, the substitution $s = \eta|(k/2)(\beta - \gamma^2/\alpha)|^{1/2}$ transforms the integral directly to the Fresnel integral (B.1). When $\beta > \gamma^2/\alpha$, this substitution leads to a Fresnel integral with j replaced with $-j$, which can be expressed by complex conjugation of the integral (B.1) since all the involved parameters are real-valued. The result reads

$$I_c(k) \approx \frac{2\pi\sigma_c A_c e^{jkS_c}}{k\,|\alpha\beta - \gamma^2|^{1/2}} \begin{cases} F(X_0) & \text{when} \quad \beta < \gamma^2/\alpha \\ F^*(X_0) & \text{when} \quad \beta > \gamma^2/\alpha \end{cases}, \tag{D.67}$$

where σ_c is as in (D.61) and

$$X_0 = \pm \left| \frac{k}{\alpha\beta}(\alpha\beta - \gamma^2)(S_c - S_b) \right|^{1/2} \tag{D.68}$$

with $X_0 > 0$ when the stationary point lies in Ω_{uv} and $X_0 < 0$ otherwise. Formula (D.67) differs from (D.60) only by the factor $F(X_0)$ or $F^*(X_0)$, and if the stationary point departs from the boundary toward the interior of Ω_{uv} (X_0 is large and positive), then according to (B.11) $F(X_0) \approx 1$ and (D.67) recovers (D.60). When the stationary point moves off in the exterior of Ω_{uv} (X_0 is large and negative), expression (D.67) goes to zero as expected since the integration domain does not contain the stationary point anymore. If the stationary point is exactly on the boundary, then $X_0 = 0$ and $F(0) = 1/2$, implying that its contribution is equal to one half of that for a well-isolated stationary point in Ω_{uv}. The properties of the Fresnel integral are illustrated in Figure B.1.

Equation (D.67) is a uniform asymptotic approximation of (D.49), which describes in a continuous way the transition of a critical point of the first kind across the boundary of the integration domain, and if $I(k)$ represents the PO approximation for the field scattered by a surface, then (D.67) describes the field in the region between the lit and shadowed portions of space.

D.3.3 Contribution from the Boundary of the Integration Domain

Now let us estimate the integral (D.49) in the case that the stationary point equations (D.50) have no solution within the domain of integration. Then, the quantity

$$|\nabla_{uv} S|^2 = \left(\frac{\partial S}{\partial u}\right)^2 + \left(\frac{\partial S}{\partial v}\right)^2, \tag{D.69}$$

where ∇_{uv} denotes the two-dimensional gradient

$$\nabla_{uv} = \hat{u}\frac{\partial}{\partial u} + \hat{v}\frac{\partial}{\partial v} \tag{D.70}$$

with $\hat{u}$ and $\hat{v}$ being the two-dimensional unit vectors of the Cartesian coordinate system on the u–v plane (Figure D.5), does not vanish inside Ω_{uv} and we can integrate by parts in (D.49). Since $I(k)$ is a double integral, a two-dimensional extension of the formula for integration by parts is used

$$\int_{\Omega_{uv}} f\frac{\partial g}{\partial u}\,\mathrm{d}u\mathrm{d}v = \oint_{\partial\Omega_{uv}} fgw_u\,\mathrm{d}l - \int_{\Omega_{uv}} g\frac{\partial f}{\partial u}\,\mathrm{d}u\mathrm{d}v \tag{D.71}$$

$$\int_{\Omega_{uv}} f\frac{\partial g}{\partial v}\,\mathrm{d}u\mathrm{d}v = \oint_{\partial\Omega_{uv}} fgw_v\,\mathrm{d}l - \int_{\Omega_{uv}} g\frac{\partial f}{\partial v}\,\mathrm{d}u\mathrm{d}v, \tag{D.72}$$

where f and g are two continuously differentiable functions, w_u and w_v are the Cartesian coordinates of $\hat{w}$, the outward unit vector normal to $\partial\Omega_{uv}$ (Figure D.5), and $\mathrm{d}l$ is the infinitesimal element of the arc length in the line integrals. By using the identity

$$jke^{jkS(u,v)} = \frac{1}{|\nabla_{uv}S|^2}\left(\frac{\partial S}{\partial u}\frac{\partial}{\partial u} + \frac{\partial S}{\partial v}\frac{\partial}{\partial v}\right)\mathrm{e}^{jkS(u,v)} \tag{D.73}$$

we rewrite (D.49) as

$$I(k) = \frac{1}{jk}\int_{\Omega_{uv}} \frac{A(u,v)}{|\nabla_{uv}S|^2}\left(\frac{\partial S}{\partial u}\frac{\partial}{\partial u} + \frac{\partial S}{\partial v}\frac{\partial}{\partial v}\right)\mathrm{e}^{jkS(u,v)}\,\mathrm{d}u\mathrm{d}v \tag{D.74}$$

and using (D.71) and (D.72) obtain

$$I(k) = \frac{1}{jk}\oint_{\partial\Omega_{uv}} \frac{A(u,v)}{|\nabla_{uv}S|^2}\left(w_u\frac{\partial S}{\partial u} + w_v\frac{\partial S}{\partial v}\right)\mathrm{e}^{jkS(u,v)}\,\mathrm{d}l + \frac{1}{jk}I_1(k), \tag{D.75}$$

where

$$I_1(k) = \int_{\Omega_{uv}} A_1(u,v)\, e^{jkS(u,v)}\, du dv \tag{D.76}$$

with

$$A_1(u,v) = -\frac{\partial}{\partial u}\left(\frac{A(u,v)}{|\nabla_{uv}S|^2}\frac{\partial S}{\partial u}\right) - \frac{\partial}{\partial v}\left(\frac{A(u,v)}{|\nabla_{uv}S|^2}\frac{\partial S}{\partial v}\right). \tag{D.77}$$

By using the two-dimensional gradient (D.70), the involved functions can be expressed in the more compact form

$$w_u\frac{\partial S}{\partial u} + w_v\frac{\partial S}{\partial v} = \hat{w}\cdot\nabla_{uv}S \tag{D.78}$$

$$A_1(u,v) = -\nabla_{uv}\cdot\left[A(u,v)\frac{\nabla_{uv}S}{|\nabla_{uv}S|^2}\right]. \tag{D.79}$$

The integral $I_1(k)$ has the same form as the original integral $I(k)$, and integrating by parts in it will lead us to the estimate $I_1(k) = O(k^{-1})$, implying that the principal contribution to $I(k)$ comes from the line integral

$$I(k) \approx \frac{1}{jk}\oint_{\partial\Omega_{uv}} \frac{A(u,v)}{|\nabla_{uv}S|^2}\left(\hat{w}\cdot\nabla_{uv}S\right) e^{jkS(u,v)}\, dl. \tag{D.80}$$

To evaluate this integral, we transform it to the Riemann integral by using a parametric representation of the closed curve $\partial\Omega_{uv}$

$$u = U(t), \quad v = V(t), \quad 0 \le t \le 2\pi \tag{D.81}$$

with $U(0) = U(2\pi)$ and $V(0) = V(2\pi)$, which gives the formula

$$I(k) \approx \frac{1}{jk}\int_0^{2\pi} \tilde{A}(t)\, e^{jk\tilde{S}(t)}\, dt, \tag{D.82}$$

where

$$\tilde{S}(t) = S(U(t), V(t)) \tag{D.83}$$

$$\tilde{A}(t) = \sqrt{U'^2(t) + V'^2(t)}\left[\frac{A(u,v)}{|\nabla_{uv}S|^2}\left(\hat{w}\cdot\nabla_{uv}S\right)\right]_{u=U(t),v=V(t)}. \tag{D.84}$$

Integral (D.82) is a one-dimensional integral of the type (D.1). Depending on the properties of the boundary $\partial\Omega_{uv}$ and the phase function $\tilde{S}(t)$, there can be three different cases.

Constant Phase Function

If $\tilde{S}(t)$ is independent of t, then the exponential factor is constant on the whole integration interval and can be factored out to give

$$I(k) \approx \frac{1}{jk} e^{jk\tilde{S}} \int_0^{2\pi} \tilde{A}(t)\, dt, \tag{D.85}$$

where the integral in (D.85) is of order unity with respect to k. This situation is encountered in problems of scattering from rotationally symmetric bodies with incidence and scattering directions along the symmetry axis.

Critical Point of the Second Kind

The second case takes place when $\tilde{S}(t)$ has a stationary point as a function of t, that is,

$$\frac{d\tilde{S}}{dt}(t_c) = 0. \tag{D.86}$$

According to (D.19), every such point brings a contribution of the form

$$I_c(k) \approx \tilde{A}_c \sqrt{\frac{2\pi}{k^3 |\tilde{S}''_c|}} \exp\left[jk\tilde{S}_c + j\frac{\pi}{4}\operatorname{sgn}\tilde{S}''_c - j\frac{\pi}{2}\right], \tag{D.87}$$

and the integral (D.82) is given by a sum of these contributions. Note that for any smooth and closed curve $\partial\Omega_{uv}$, the phase function $\tilde{S}(t)$ is 2π periodic and continuous, which implies at least two critical points (at the maximum and the minimum of the phase function) on the integration interval.

Thus, the critical point of the second kind is a stationary point of the phase function on the boundary of the integration domain and its contribution is by a factor of $O(k^{-1/2})$ smaller than that from stationary points of the first kind. Such contributions describe waves scattered at the edges of scattering surfaces.

Critical Points of the Third Kind

If equation (D.86) has no solutions on the interval $0 \leq t \leq 2\pi$, then contributions from the endpoints of the integration interval and singular points of the integrand function become important. A typical situation is that the contour $\partial\Omega_{uv}$ is not smooth at a number of points with coordinates $U_n = U(t_n)$ and $V_n = V(t_n)$ with $0 = t_1 < t_2 < \ldots < t_{N-1} < t_N = 2\pi$ (e.g., Ω_{uv} is a polygon), implying that $U'(t)$ and $V'(t)$, and therefore $\tilde{A}(t)$ and $\tilde{S}(t)$, are discontinuous at $t = t_n$. The integral (D.82) becomes a sum of contributions from the intervals

$$I(k) \approx \frac{1}{jk} \sum_{n=1}^{N-1} \int_{t_n}^{t_{n+1}} \tilde{A}(t)\, e^{jk\tilde{S}(t)}\, dt \tag{D.88}$$

and the integration by parts according to (D.7) in every integral involved in (D.88) gives the asymptotic approximation

$$I(k) \approx \frac{1}{(jk)^2} \left\{ \frac{\tilde{A}}{\tilde{S}'} e^{jk\tilde{S}} \Big|_{t_1}^{t_N} - \sum_{n=2}^{N-1} \frac{\tilde{A}}{\tilde{S}'} e^{jk\tilde{S}} \Big|_{t_n-0}^{t_n+0} \right\}, \tag{D.89}$$

where the notation $t_n - 0$ ($t_n + 0$) indicates that the limiting value of a function at the point t_n is to be taken from the left (right). Formula (D.89) suggests that corner points on the boundary of the integration domain lead to contributions of the order k^{-2} to the integral (D.49). Such contributions are asymptotically small compared to those from the critical points of the first and second kind, which are $O(k^{-1})$ and $O(k^{-3/2})$, respectively.

The PO solution for scattering from polygonal plates can be described entirely in terms of the corner point contributions (section 9.3.3).

D.3.4 Kontorovich's Formula

In the case when the integration domain includes an elliptic stationary point ($\alpha\beta > \gamma^2$), it is still possible to asymptotically reduce the integral (D.49) to a line integral over the boundary of the integration domain and obtain a representation similar to (D.80). We outline the derivation; the details can be found in Kontorovich and Muravév (1952).

The analysis is based on the formula

$$Ae^{jkS} = \frac{j}{k}\left[e^{jkS}\nabla_{uv} \cdot \mathbf{B} - \nabla_{uv} \cdot (e^{jkS}\mathbf{B})\right], \tag{D.90}$$

where

$$\mathbf{B} = \frac{A\mathbf{Q}}{\mathbf{Q} \cdot \nabla_{uv}S} \tag{D.91}$$

and $\mathbf{Q}$ is an arbitrary differentiable vector function. The analysis in section D.3.3 uses $\mathbf{Q} = \nabla_{uv}S$, but with a stationary point in the integration domain this choice would lead to an unacceptable singularity in the integrand. However, choosing $\mathbf{Q}$ such that in a vicinity of the stationary point it behaves as $\mathbf{Q} \approx \hat{\rho}$ with

$$\hat{\rho} = \frac{(u - u_c)\hat{u} + (v - v_c)\hat{v}}{\sqrt{(u - u_c)^2 + (v - v_c)^2}}, \tag{D.92}$$

the two-dimensional integral can be reduced to line integrals. Away from the stationary point, $\mathbf{Q}$ may approach $\nabla_{uv}S$ or any other function since, as we shall see, the choice does not influence the asymptotic value of the line integral.

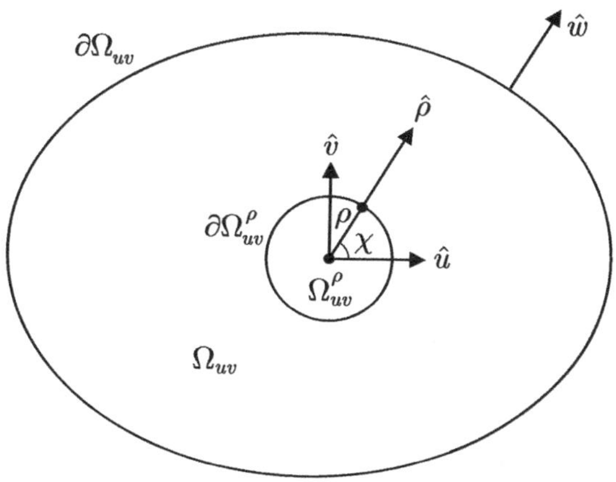

Figure D.7 Integration domains Ω_{uv} and Ω^{ρ}_{uv} and their boundaries $\delta\Omega_{uv}$ and $\delta\Omega^{\rho}_{uv}$. The domain Ω^{ρ}_{uv} is a circle of the radius ρ centered at the critical point. $\hat{\rho}$ is the unit vector normal to $\delta\Omega^{\rho}_{uv}$, and $\hat{w}$ is the unit vector normal to $\delta\Omega_{uv}$.

Assuming this choice of $\mathbf{Q}$, define a circle Ω_{uv}^{ρ} of a small radius ρ centered at the stationary point (Figure D.7). Denoting the rest of the integration domain by $\Omega_{uv} - \Omega_{uv}^{\rho}$ we write

$$I(k) = \int_{\Omega_{uv}^{\rho}} A e^{jkS} \, \mathrm{d}u\mathrm{d}v + \int_{\Omega_{uv}-\Omega_{uv}^{\rho}} A e^{jkS} \, \mathrm{d}u\mathrm{d}v. \tag{D.93}$$

The first integral vanishes when $\rho \to 0$, since its integrand is a bounded function. With any $\rho > 0$ the integration domain in the second integral does not contain the stationary point, and with the help of (D.90) and the two-dimensional divergence theorem (A.28) the integral can be rewritten as

$$\int_{\Omega_{uv}-\Omega_{uv}^{\rho}} A e^{jkS} \, \mathrm{d}u\mathrm{d}v = \frac{j}{k} \left\{ \int_{\Omega_{uv}-\Omega_{uv}^{\rho}} e^{jkS} \nabla_{uv} \cdot \left[\frac{\mathbf{Q}A}{\mathbf{Q} \cdot \nabla_{uv} S} \right] \mathrm{d}u\mathrm{d}v - \oint_{\partial\Omega_{uv}^{\rho}} e^{jkS} \frac{A\mathbf{Q} \cdot \hat{w}}{\mathbf{Q} \cdot \nabla_{uv} S} \mathrm{d}l - \oint_{\partial\Omega_{uv}} e^{jkS} \frac{A\mathbf{Q} \cdot \hat{w}}{\mathbf{Q} \cdot \nabla_{uv} S} \mathrm{d}l \right\}, \tag{D.94}$$

where $\partial\Omega_{uv}^{\rho}$ and $\partial\Omega_{uv}$ are the boundaries of the integration domain and $\hat{w}$ is the unit normal directed into the exterior of the integration domain (Figure D.7).

The two-dimensional integral in (D.94) with the function $\mathbf{Q}$ described above is $O(k^{-1})$ and can be neglected when compared with the line integrals (Kontorovich and Muravév 1952). Let us now evaluate the line integral over $\partial\Omega_{uv}^{\rho}$. Denoting the integral by I_{ρ}, substituting $\mathbf{Q} = \hat{\rho}$ and $\hat{w} = -\hat{\rho}$, and letting ρ go to zero, we obtain

$$I_{\rho} \approx -e^{jkS_c} A_c \oint_{\partial\Omega_{uv}^{\rho}} \left(\frac{\partial S}{\partial \rho} \right)^{-1} \mathrm{d}l, \tag{D.95}$$

where the functions continuous and bounded at $\rho = 0$ are replaced with their limits at the stationary point. Introducing the polar coordinates defined by $u - u_c = \rho \cos\chi$ and $v - v_c = \rho \sin\chi$ and using (D.51), we can write

$$\frac{\partial S}{\partial \rho} \approx \rho(\alpha \cos^2\chi + 2\gamma \cos\chi \sin\chi + \beta \sin^2\chi), \tag{D.96}$$

so that

$$\oint_{\partial\Omega_{uv}^{\rho}} \left(\frac{\partial S}{\partial \rho} \right)^{-1} \mathrm{d}l \approx \int_0^{2\pi} \frac{\mathrm{d}\chi}{\alpha \cos^2\chi + 2\gamma \cos\chi \sin\chi + \beta \sin^2\chi}. \tag{D.97}$$

When $\alpha\beta > \gamma^2$, the integrand has no infinities on the integration interval and the integral is convergent. The substitution $z = \tan\chi$ reduces it to a simpler form

$$\int_0^{2\pi} \frac{\mathrm{d}\chi}{\alpha \cos^2\chi + 2\gamma \cos\chi \sin\chi + \beta \sin^2\chi} = 2 \int_{-\infty}^{+\infty} \frac{\mathrm{d}z}{\alpha + 2\gamma z + \beta z^2} \tag{D.98}$$

with an integrand having complex conjugate poles at

$$z = \frac{1}{\beta} \left(-\gamma \pm j\sqrt{\alpha\beta - \gamma^2} \right) \tag{D.99}$$

on the complex z plane. The integral is given by the residue at any of the poles[3] so that in the limit $\rho \to 0$,

$$I_\rho \to -A_c e^{jkS_c} \frac{2\pi \operatorname{sgn}\beta}{\sqrt{\alpha\beta - \gamma^2}}. \tag{D.100}$$

Combing the results we obtain

$$I(k) = \frac{2\pi j \operatorname{sgn}\beta A_c}{k\sqrt{\alpha\beta - \gamma^2}} e^{jkS_c} - \frac{j}{k} \oint_{\partial\Omega_{uv}} e^{jkS} \frac{A\mathbf{Q}\cdot\hat{w}}{\mathbf{Q}\cdot\nabla_{uv}S}\, dl + O\left(\frac{1}{k^2}\right). \tag{D.101}$$

The first term is equal to the stationary point contribution described by (D.60) and (D.61). The line integral coincides with (D.80) if $\mathbf{Q}$ is chosen to approach $\nabla_{uv}S$ at the boundary of the integration domain. It can be seen, however, that the value of the line integral is independent of $\mathbf{Q}$ when $k \to \infty$. Indeed, it follows from (D.86) that the main contribution to the line integral comes from the point or points at which $\partial S/(\partial l) = 0$, implying that at those points $\nabla_{uv}S$ is perpendicular to the boundary of the integration domain and the ratio $\mathbf{Q}\cdot\hat{w}/(\mathbf{Q}\cdot\nabla_{uv}S)$ is independent of the choice of $\mathbf{Q}$.

Formula (D.101) extends (D.80) to the case in which the integration domain has a stationary point of the elliptic type. Whether or not a hyperbolic stationary point can be included is still an open question. Reduction of the two-dimensional integral (D.49) to a one-dimensional integral means a dramatic saving in computational complexity and significant physical implications if one recalls that in the context of PO, contributions from stationary points represent GO rays and the line integral accounts for the fields diffracted by the rim of an aperture or by an edge of the scattering surface.

D.3.5 Integrand Vanishing on the Boundary

The integral (D.49) is of the order of magnitude k^{-2} with respect to the large parameter k when the amplitude function $A(u, v)$ vanishes at the boundary of the integration domain and the phase function $S(u, v)$ has no critical points in the interior of the integration domain. This case is physically meaningful as it describes scattering of waves from a conical point on a scattering surface (section 9.2), and we consider it briefly.

We limit our analysis to a specific case, in which the integral of interest has the form

$$I(k) = \int_{v_-}^{v_+} \int_0^\infty A(u, v) e^{jkS(u,v)}\, du dv, \tag{D.102}$$

the integration domain is a semi-infinite strip defined by the conditions $0 \le u < \infty$ and $v_- \le v \le v_+$, and at the boundary $u = 0$ the amplitude and phase functions are such that for every value of v

$$A(0, v) = 0, \quad S(0, v) = S_0, \quad \frac{\partial A}{\partial u}(0, v) \ne 0, \quad \frac{\partial S}{\partial u}(0, v) \ne 0, \tag{D.103}$$

where S_0 is a constant independent of v. As shown in section 9.2, such an integral describes the PO solution in the vicinity of a conical point on parametrization of the scattering surface by polar coordinates with variables u and v having the meaning of "radius" and "azimuth", respectively.

Let us estimate (D.102) for large values of k. To this end, we treat $I(k)$ as an iterated integral in which the integration with respect to u is performed first. Since the partial derivative of $S(u, v)$ with respect to u does not vanish in the integration domain, we can asymptotically evaluate the interior integral by integration by parts, and since $A(u, v)$ vanishes at the

[3] The factor sgnβ is missing from a corresponding formula in Kontorovich and Muravév (1952).

endpoint $u = 0$ we have to apply the integration by parts twice. Using the formula (D.9) with $a = 0$ and $b = \infty$, and neglecting the contribution from the point $u = \infty$, we arrive at the estimation

$$I(k) = -\frac{1}{k^2} e^{jkS_0} \int_{v_-}^{v_+} A(0, v) \left[\frac{\partial S}{\partial u}(0, v)\right]^{-2} dv + o\left(\frac{1}{k^2}\right). \tag{D.104}$$

Equation (D.104) solves the problem of asymptotic evaluation of the integral (D.102) since the integral with respect to v does not involve the rapidly oscillating exponent and can be easily evaluated numerically or sometimes even analytically.

D.3.6 Summary of the Two-dimensional Stationary-phase Method

The analysis presented in section D.3 has shown that the principal contributions to the double integral $I(k)$ of the form (D.49) arise from asymptotically small neighborhoods of certain points, called critical points, which are located in the interior or on the boundary of the domain of integration. The analytical form of the contribution depends on the kind of the critical point.

- Critical points of the first kind are points within the domain of integration at which the phase $S(u, v)$ is stationary (D.50). The contribution of the point to the integral (D.49) is given by the formula (D.60) and is $O(k^{-1})$. The case in which the critical point approaches the boundary of the integration domain is described by multiplying the contribution of the critical point with a Fresnel integral (see (D.67)).
- Critical points of the second kind are the points on the boundary of the integration domain at which the tangential derivative of $S(u, v)$ vanishes. If t is the parameter of arc length, measured along the boundary curve, then the critical points are found from the relation (D.86) and their contribution from (D.87). The order of the contribution is $O\left(k^{-3/2}\right)$.
- Critical points of the third kind are corner points on the boundary of the integration domain, that is, points at which the slope of the boundary curve is discontinuous, for example the integration domain is a polygon. The contribution of the points is described by (D.89) and is of the order $O(k^{-2})$. The same asymptotic behavior with respect to k is observed if the amplitude function vanishes at the boundary (D.104). It should be noted that these contributions are of the same order as the correction term in the asymptotic expansion of the contribution from the critical point of the first kind. However, in the absence of the critical points within the integration domain the $O(k^{-2})$ contributions become important.

A special case in which the phase function is constant on the boundary curve $\partial\Omega_{uv}$ leads to the estimate (D.85) with $I(k) = O(k^{-1})$. If the phase function $S(u, v)$ is constant throughout the whole integration region Ω_{uv}, the integrand of (D.49) is no longer rapidly oscillating and $I(k) = O(1)$.

There are further cases (coalescing critical points, integrable infinities in the amplitude function, etc.), in which the asymptotic formulas presented in this section become infinite and therefore inapplicable, requiring special treatment. Asymptotic representations of two-dimensional integrals which remain valid in such cases can be found, for example in Bleistein and Handelsman (1969), Borovikov (1994), Chako (1965), Fedoryuk (1987), Karatygin and Rosov (1972), and McClure and Wong (1991). The problem is that there is a great variety of special cases, with every case needing special treatment as no universal uniform solution seems to be possible. The only truly uniform solution is the original integral.

Direct numerical evaluation of two-dimensional integrals with rapidly oscillating integrand function is time-consuming but, with the progress in computation technologies, may become feasible. Evaluation of $I(k)$ by numerical double integration is described by Ludwig (1968). The integral (D.49) with constant $A(u, v)$ and linear $S(u, v)$ can be exactly reduced to a line integral over the boundary of the integration domain (Gordon 1975) and easily evaluated (section 9.3). Extension to quadratic phase function is addressed in Carluccio *et al.* (2010).

D.4 Problems

1. Derive an asymptotic approximation for the normalized Airy function $v(x)$ with $x \to -\infty$ by applying the method of stationary phase to the integral representation (E.171).

Solution: With $x = -|x|$,

$$v(x) = \frac{1}{\sqrt{\pi}} \int_0^{+\infty} \cos\left(\frac{1}{3}\gamma^3 - |x|\gamma\right) \mathrm{d}\gamma,$$

which can be expressed as $v(x) = \mathrm{Re}\, I$, where

$$I = \frac{1}{\sqrt{\pi}} \int_0^{+\infty} \mathrm{e}^{j\left(\frac{1}{3}\gamma^3 - |x|\gamma\right)} \mathrm{d}\gamma.$$

The change of the integration variable $\gamma = |x|^{1/2} t$ transforms the integral to the form (D.1) with $a = 0$, $b = +\infty$, $k = |x|^{3/2}$, $A(t) = |x|^{1/2}\pi^{-1/2}$, and $S(t) = -t + t^3/3$. Equation (D.3) has two solutions $t_c = \pm 1$, of which only $t_c = 1$ is of interest since it is located within the domain of integration. At the stationary point, $S_c = -2/3$, $S''_c = 2$, and application of the formula (D.19) gives

$$I \approx |x|^{-1/4} \exp\left(-j\frac{2}{3}|x|^{3/2} + \frac{\pi}{4}\right)$$

and therefore

$$v(x) \approx |x|^{-1/4} \cos\left(\frac{2}{3}|x|^{3/2} - \frac{\pi}{4}\right).$$

2. What is the asymptotic behavior of the integral

$$I(k, x) = \int_{-\infty}^{x} \mathrm{e}^{-jkt^2}\, \mathrm{d}t$$

as $k \to +\infty$? Parameter x is real-valued, positive or negative.

Hint: Consider separately two cases: (a) $x < 0$ and (b) $x > 0$. To check the result, express $I(k, x)$ through the Fresnel integral (Appendix B).

Solution: Apply the formula (D.4) in the case (a) and (D.19) in the case (b) to obtain

$$I(k, x) \approx \begin{cases} \frac{j}{2kx}\mathrm{e}^{-jkx^2}, & x < 0 \\ \sqrt{\frac{\pi}{k}}\mathrm{e}^{-j\frac{\pi}{4}}, & x > 0 \end{cases}.$$

The relation to the Fresnel integral is as follows:

$$I(k, x) = \sqrt{\frac{\pi}{k}}\mathrm{e}^{-j\frac{\pi}{4}} F\left(x\sqrt{k}\right).$$

3. Derive an asymptotic approximation for the integral (D.1) with $k \to \infty$ when the phase function $S(t)$ has two stationary points in an asymptotically small vicinity of a point $t = t_0$ and the amplitude function is such that $A(t) \approx A(t_0) + A'(t_0)(t - t_0)$ with $A(t_0) = 0$.

Answer:

$$I(k) \approx -2\pi j \mathrm{e}^{jkS(t_0)} A'(t_0) \mathrm{sgn}\left[S'''(t_0)\right] \left[\frac{2}{k|S'''(t_0)|}\right]^{2/3} \mathrm{Ai}'(q), \tag{D.105}$$

where q is defined in (D.27).

Hint: Follow the derivation of the formula (D.28) and use the relation

$$\int_{-\infty}^{+\infty} \exp\left(jq\tau + j\frac{1}{3}\tau^3\right)\tau\,\mathrm{d}\tau = -2\pi j\mathrm{Ai}'(q). \tag{D.106}$$

4. Prove the formula (D.64) for the stationary point contribution.
Hint: Note that, as apparent from (D.55), the coordinate frame $\hat{\xi} = \hat{u}$ and $\hat{\eta} = \hat{v}$ can be rotated by an arbitrary angle without changing the result of integration. Chose the rotation angle that makes the Hessian matrix (D.62) diagonal.
Solution: First rewrite (D.51) in the vector form

$$S(u, v) - S_c \approx \mathbf{a}^{\mathrm{T}} \cdot \mathbf{H}_c \cdot \mathbf{a}, \tag{D.107}$$

where $\mathbf{a} = (\xi, \eta)^{\mathrm{T}}$ is a column vector, $\mathbf{a}^{\mathrm{T}} = (\xi, \eta)$ is a row vector with $\xi = u - u_c$ and $\eta = v - v_c$, and $\mathbf{H}_c$ is the Hessian matrix (D.62) at the stationary point $u = u_c$ and $v = v_c$. If ξ' and η' are coordinates in the rotated system and $\mathbf{a}' = (\xi', \eta')^{\mathrm{T}}$ and $(\mathbf{a}')^{\mathrm{T}} = (\xi', \eta')$, then the matrix

$$\mathrm{U}(\chi) = \begin{pmatrix} \cos\chi & -\sin\chi \\ \sin\chi & \cos\chi \end{pmatrix} \tag{D.108}$$

relates $\mathbf{a}$ and $\mathbf{a}'$ as

$$\mathbf{a} = \mathbf{U}(\chi) \cdot \mathbf{a}', \quad \mathbf{a}^{\mathrm{T}} = (\mathbf{a}')^{\mathrm{T}} \cdot \mathbf{U}^{\mathrm{T}}(\chi), \tag{D.109}$$

implying that (D.107) can be rewritten as

$$S(u, v) - S_c \approx (\mathbf{a}')^{\mathrm{T}} \cdot \mathbf{H}'_c \cdot \mathbf{a}' \tag{D.110}$$

with

$$\mathbf{H}'_c = \mathbf{U}^{\mathrm{T}}(\chi) \cdot \mathbf{H}_c \cdot \mathbf{U}(\chi). \tag{D.111}$$

Since $\mathbf{H}_c$ is a real symmetric matrix and $\mathbf{U}(\chi)$ is a real orthogonal matrix, $\mathbf{H}'_c$ can be made diagonal by the transformation (D.111) with a suitably chosen $\mathbf{U}(\chi)$ (symmetric Schur decomposition), that is, with a suitable value of the rotation angle χ. In that case, $\alpha = \lambda_1$, $\beta = \lambda_2$, and $\gamma = 0$, where $\lambda_{1,2}$ are eigenvalues of $\mathbf{H}'_c$, and equation (D.61) for the parameter σ_c in (D.60) takes the form

$$\sigma_c = \exp\left[j\frac{\pi}{4}(\mathrm{sgn}\lambda_1 + \mathrm{sgn}\lambda_2)\right]. \tag{D.112}$$

Now, since $\mathbf{U}^{\mathrm{T}}(\chi) = \mathbf{U}^{-1}(\chi)$, (D.111) is a similarity transformation and, therefore, changes neither the eigenvalues nor the determinant of the matrix under transformation, implying the validity of (D.64) at every value of χ.

References

Abramowitz M and Stegun IA (eds) 1965 *Handbook of Mathematical Functions with Formulas, Graphs, and Mathematical Tables*, Dover Publications.
Airy GB 1838 On the intensity of light in the neighborhood of a caustic. *Trans. Cambridge Phil. Soc.*, **6**, 379–402.
Bleistein N and Handelsman RA 1969 Uniform asymptotic expansions of double integrals. *J. Math. Anal. Appl.*, **27**, 434–453.
Borovikov VA 1994 *Uniform Stationary Phase Method*, The Institution of Electrical Engineers, London.

Carluccio G, Albani M, and Pathak PH 2010 Uniform asymptotic evaluation of surface integrals with polygonal integration domains in terms of UTD transition functions. *IEEE Trans. Antennas Propag.*, **58**, 1155–1163.

Chako N 1965 Asymptotic expansions of double and multiple integrals occurring in diffraction theory. *J. Inst. Math. Appl.*, **1**, 372–422.

Cooke JC 1982 Stationary phase in two dimensions. *IMA J. Appl. Math.*, **29**, 25–37.

Fedoryuk MV 1987 *Asymptotics: Integrals and Series*, Nauka, Moscow (in Russian).

Focke J 1954 Asymptotische Entwicklungen mittels der Methode der stationären Phase. *Ber. Verh. Sachs. Ak. Wiss., Math.-Nat. Kl.*, **101**, no. 3, 1–48.

Gordon WB 1975 Far-field approximations to the Kirchhoff–Helmholtz representations of scattered fields. *IEEE Trans. Antennas Propag.*, **23**, 590–592.

James GL 1976 *Geometrical Theory of Diffraction for Electromagnetic Waves*, Peter Peregrinus, London.

Jones DS and Kline M 1958 Asymptotic expansion of multiple integrals and the method of stationary phase. *J. Phys. Maths.*, **37**, 1–28.

Karatygin VA and Rosov VA 1972 Method of stationary phase for double integral with arbitrary located stationary point. *J. Comput. Math. Phys.*, **12**, 1391–1405.

Kelvin 1887 On the waves produced by a simple impulse in water of any depth, or in dispersive medium. *Philos. Mag. Ser. 5*, **23**, 252–255.

Kline M and Kay IW 1965 *Electromagnetic Theory and Geometrical Optics*, John Wiley & Sons.

Kontorovich MI and Muravév YK 1952 Derivation of the laws of reflection of geometrical optics on the basis of an asymptotic treatment of the diffraction problem. *Zhurn. Techn. Fiz.*, **22**, no. 3, 394–407 (in Russian).

Ludwig AC 1968 Computation of radiation patterns involving numerical double integration. *IEEE Trans. Antennas Propag.*, **16**, 767–769.

McClure JP and Wang R 1991 Two-dimensional stationary phase approximation: stationary point at a corner. *SIAM J. Math. Anal.*, **22**, 500–523.

Olver FWJ 1974 *Asymptotics and Special Functions*, Academic Press.

van Kampen NG 1949 An asymptotic treatment of diffraction problems. *Physica*, **14**, 575–589.

Wong R 1989 *Asymptotic Approximations of Integrals*, Academic Press.

E

Asymptotic Approximations of Bessel Functions of Large Argument and Arbitrary Order

E.1 Introduction

Bessel functions (also known as cylindrical functions) play an important role in the electromagnetic theory as they appear in solutions of Maxwell's equations in cylindrical and spherical coordinates. This book uses these functions to describe electromagnetic scattering from wedges (Chapter 5), cylinders (Chapter 6), and spheres (Chapter 7). Bessel functions have been studied in depth and there are multiple books on the subject, presenting their theory, analytical properties, and tables of numerical values, for example Abramowitz and Stegun (1965), Erdélyi *et al.* (1953), Kratzer and Franz (1960), Magnus *et al.* (1966), Ryshik and Gradstein (1957), and Watson (1962). This appendix addresses a less known but important aspect of their theory, namely, asymptotic approximations of the functions for *large* argument z and *arbitrary* order ν, with *complex-valued* z and ν. These approximations are required in the high-frequency analysis of electromagnetic scattering as the arguments of the involved Bessel functions, proportional to the frequency, are large real- or complex- (in lossy media) valued parameters. Furthermore, in the high-frequency case, the series solutions obtained by the separation of variables are numerically inefficient and physically non-transparent, so they are transformed into another form which involves contour integration over a path of infinite extent in the complex ν plane (Watson's transformation, sections 6.4.1 and 7.3.4). The complex integration requires knowledge of analytical properties of the Bessel functions over the whole complex ν plane, including the location of zeros.

In addition to the large-argument approximations, this book refers to a number of further properties of Bessel functions, most notably to their analytical properties on the complex ν plane, relations to Riccati–Bessel functions, and small-argument approximations. The appendix starts with the necessary definitions and properties of the functions and briefly addresses the challenges associated with large-argument approximations of Bessel functions before proceeding with the main sections E.2 and E.3.

E.1.1 Basic Definitions and Properties

This appendix is devoted to the Bessel functions of the first kind, $J_\nu(z)$, and the Bessel functions of the third kind (also known as first and second Hankel functions), $H_\nu^{(1)}(z)$ and $H_\nu^{(2)}(z)$. Only two functions are linearly independent as they are related by the formula

$$2J_\nu(z) = H_\nu^{(1)}(z) + H_\nu^{(2)}(z)\,. \qquad \text{(E.1)}$$

Modern Electromagnetic Scattering Theory with Applications, First Edition. Andrey V. Osipov and Sergei A. Tretyakov.

The argument z and the order ν can be arbitrary complex numbers. The Bessel functions are holomorphic functions of z throughout the z plane cut along the negative real axis, and for fixed z each is an entire function of ν. When ν is any integer number, $J_\nu(z)$ is an entire function of z. When z and ν are real-valued, the Hankel functions are complex conjugates of each other,

$$H_\nu^{(1)}(z) = \left[H_\nu^{(2)}(z)\right]^*, \quad H_\nu^{(2)}(z) = \left[H_\nu^{(1)}(z)\right]^*. \tag{E.2}$$

The Bessel function of the second kind or the Neumann function, $Y_\nu(z)$, because of its properties at $z = 0$ and ∞, only seldom occurs in solutions of scattering problems. It is related to the other Bessel functions by the formulas

$$H_\nu^{(1)}(z) = J_\nu(z) + jY_\nu(z) \tag{E.3}$$

$$H_\nu^{(2)}(z) = J_\nu(z) - jY_\nu(z) \tag{E.4}$$

and

$$Y_\nu(z) = \frac{J_\nu(z)\cos(\nu\pi) - J_{-\nu}(z)}{\sin(\nu\pi)}. \tag{E.5}$$

The functions are linearly independent solutions of the Bessel differential equation

$$\frac{d^2}{dz^2}Z_\nu(z) + \frac{1}{z}\frac{d}{dz}Z_\nu(z) + \left(1 - \frac{\nu^2}{z^2}\right)Z_\nu(z) = 0, \tag{E.6}$$

where $Z_\nu(z)$ denotes any of the Bessel functions, with the Wronskian determinants

$$J_\nu(z)\frac{dY_\nu(z)}{dz} - \frac{dJ_\nu(z)}{dz}Y_\nu(z) = \frac{2}{\pi z} \tag{E.7}$$

$$J_\nu(z)\frac{dH_\nu^{(2)}(z)}{dz} - \frac{dJ_\nu(z)}{dz}H_\nu^{(2)}(z) = \frac{2}{j\pi z} \tag{E.8}$$

$$H_\nu^{(1)}(z)\frac{dH_\nu^{(2)}(z)}{dz} - \frac{dH_\nu^{(1)}(z)}{dz}H_\nu^{(2)}(z) = \frac{4}{j\pi z}. \tag{E.9}$$

For arbitrary values of z and ν,

$$J_\nu(z) = \left(\frac{z}{2}\right)^\nu \sum_{k=0}^{\infty} \frac{(-z^2/4)^k}{k!\,\Gamma(\nu + k + 1)}, \tag{E.10}$$

where Γ is the Gamma function and $k! = 1 \cdot 2 \cdot \ldots \cdot k$ with $0! = 1$. Provided that ν is not an integer, the corresponding series expansion for $Y_\nu(z)$ follows from (E.5) and (E.10); otherwise, when $\nu = n$ with n being a positive integer or zero,

$$Y_n(z) = \frac{2}{\pi}\ln\left(\frac{z}{2}\right)J_n(z) - \left(\frac{z}{2}\right)^{-n}\sum_{k=0}^{n-1}\frac{(n-k-1)!}{\pi k!}\left(\frac{z^2}{4}\right)^k - \left(\frac{z}{2}\right)^n\sum_{k=0}^{\infty}\frac{\psi(k+1)+\psi(n+k+1)}{\pi k!\,(n+k)!}\left(-\frac{z^2}{4}\right)^k, \tag{E.11}$$

where $\psi(z) = \Gamma'(z)/\Gamma(z)$ is the Digamma function such that

$$\psi(1) = -\gamma, \quad \psi(n) = -\gamma + \sum_{k=1}^{n-1}\frac{1}{k}(n \geq 2) \tag{E.12}$$

with $\gamma = 0.57721\ldots$ (Euler's constant). Series expansions of the Hankel functions result from (E.3) and (E.4). The series expansions are particularly useful for describing Bessel functions with small arguments.

Solutions of scattering problems in spherical coordinates (Chapter 7) include the spherical Bessel functions $j_\nu(z)$ and $h_\nu^{(1),(2)}(z)$ or the closely related Riccati–Bessel functions $\psi_\nu(z)$ and $\zeta_\nu^{(1),(2)}(z)$. Both are expressed through the ordinary Bessel functions as

$$j_\nu(z) = \sqrt{\frac{\pi}{2z}} J_{\nu+\frac{1}{2}}(z) = \frac{1}{2}\left[h_\nu^{(1)}(z) + h_\nu^{(2)}(z)\right] \tag{E.13}$$

$$h_\nu^{(1),(2)}(z) = \sqrt{\frac{\pi}{2z}} H_{\nu+\frac{1}{2}}^{(1),(2)}(z) \tag{E.14}$$

$$\psi_\nu(z) = \sqrt{\frac{\pi z}{2}} J_{\nu+\frac{1}{2}}(z) = \frac{1}{2}\left[\zeta_\nu^{(1)}(z) + \zeta_\nu^{(2)}(z)\right] \tag{E.15}$$

$$\zeta_\nu^{(1),(2)}(z) = \sqrt{\frac{\pi z}{2}} H_{\nu+\frac{1}{2}}^{(1),(2)}(z) \tag{E.16}$$

and thus do not need a special treatment.

This appendix describes asymptotic approximations of Bessel functions $J_\nu(z)$, $H_\nu^{(1)}(z)$, and $H_\nu^{(2)}(z)$ on the whole complex ν plane. The approximations of Hankel functions are to be compliant with the general symmetry relations

$$H_{-\nu}^{(1)}(z) = \mathrm{e}^{j\nu\pi} H_\nu^{(1)}(z) \tag{E.17}$$

$$H_{-\nu}^{(2)}(z) = \mathrm{e}^{-j\nu\pi} H_\nu^{(2)}(z)\,. \tag{E.18}$$

The asymptotic formulas are obtained under the assumption that $\mathrm{Re}\, z \geq 0$ but, if necessary, can be extended beyond this limit by using the exact relations

$$H_\nu^{(1)}(z\mathrm{e}^{j\pi}) = -\mathrm{e}^{-j\nu\pi} H_\nu^{(2)}(z) \tag{E.19}$$

$$H_\nu^{(2)}(z\mathrm{e}^{-j\pi}) = -\mathrm{e}^{j\nu\pi} H_\nu^{(1)}(z) \tag{E.20}$$

$$J_\nu(z\mathrm{e}^{\pm j\pi}) = \mathrm{e}^{\pm j\nu\pi} J_\nu(z)\,, \tag{E.21}$$

which can be useful, for example, to describe the modified Bessel functions

$$I_\nu(z) = \begin{cases} \mathrm{e}^{-j\nu\pi/2} J_\nu(z\mathrm{e}^{j\pi/2}), & -\pi < \arg z \leq \pi/2 \\ \mathrm{e}^{3j\nu\pi/2} J_\nu(z\mathrm{e}^{-3j\pi/2}), & \pi/2 < \arg z \leq \pi \end{cases} \tag{E.22}$$

$$K_\nu(z) = \frac{\pi j}{2} \begin{cases} \mathrm{e}^{j\nu\pi/2} H_\nu^{(1)}(z\mathrm{e}^{j\pi/2}), & -\pi < \arg z \leq \pi/2 \\ -\mathrm{e}^{-j\nu\pi/2} H_\nu^{(2)}(z\mathrm{e}^{-j\pi/2}), & -\pi/2 < \arg z \leq \pi \end{cases} \tag{E.23}$$

or solutions of scattering problems involving materials with a negative real part of the refraction index (DNG materials, section 4.2.4).

E.1.2 Large-argument Approximations ($|z| \gg 1$)

The well-known large-argument (Hankel) approximations

$$H_\nu^{(1)}(z) = \sqrt{\frac{2}{\pi z}} \mathrm{e}^{j\left(z - \frac{1}{2}\nu\pi - \frac{1}{4}\pi\right)} \tag{E.24}$$

$$H_\nu^{(2)}(z) = \sqrt{\frac{2}{\pi z}} \mathrm{e}^{-j\left(z - \frac{1}{2}\nu\pi - \frac{1}{4}\pi\right)} \tag{E.25}$$

$$J_\nu(z) = \sqrt{\frac{2}{\pi z}} \cos\left(z - \frac{1}{2}\nu\pi - \frac{1}{4}\pi\right) \tag{E.26}$$

are limited to values of $|\nu|$ much smaller than $|z|$ and cannot be used over the whole complex ν plane. In this book we often use more general Debye's approximations

$$H_\nu^{(1)}(z) = S_\nu^{(1)}(z) \tag{E.27}$$
$$H_\nu^{(2)}(z) = S_\nu^{(2)}(z) \tag{E.28}$$
$$2J_\nu(z) = S_\nu^{(1)}(z) + S_\nu^{(2)}(z), \tag{E.29}$$

where

$$S_\nu^{(1)}(z) = \sqrt{\frac{2}{\pi z}} \frac{1}{\sqrt[4]{1-(\nu/z)^2}} \exp\left[j\Phi(\nu,z) - j\frac{\pi}{4}\right] \tag{E.30}$$

$$S_\nu^{(2)}(z) = \sqrt{\frac{2}{\pi z}} \frac{1}{\sqrt[4]{1-(\nu/z)^2}} \exp\left[-j\Phi(\nu,z) + j\frac{\pi}{4}\right] \tag{E.31}$$

and

$$\Phi(\nu,z) = z\left[\sqrt{1-\left(\frac{\nu}{z}\right)^2} - \frac{\nu}{z}\arccos\frac{\nu}{z}\right] \tag{E.32}$$

with multi-valued functions understood in the arithmetic sense, that is,

$$0 \le \sqrt{1-(\nu/z)^2} \le 1, \quad 0 \le \sqrt[4]{1-(\nu/z)^2} \le 1, \quad \pi \ge \arccos(\nu/z) \ge 0, \tag{E.33}$$

when $-1 \le \nu/z \le 1$. The ratio $\Phi(\nu,z)/z$ is a function of $\xi = \nu/z$ only, and its behavior in the interval $-1 \le \xi \le 1$ is illustrated in Figure E.1.

Approximations (E.27)–(E.29) recover (E.24)–(E.26) when $|\nu/z| \ll 1$ but are valid over a broader range of values of ν. For example, for real-valued, positive and large z, they are applicable when $-z + \delta\sqrt[3]{z} < \nu < z - \delta\sqrt[3]{z}$ with $\delta = O(1)$. However, when ν is outside of the range, for example $\nu \ge z$, or is complex-valued, or when z is complex-valued, approximations (E.27)–(E.29) may not apply. Asymptotic representations of Bessel functions with $|z| \gg 1$, which are applicable to rather arbitrary values of ν, small or large, real- or complex-valued are called Debye's approximations.[1] Debye's approximations of every Bessel function are in general linear combinations of the two basic functions $S_\nu^{(1,2)}(z)$ with coefficients that are constant within specific regions on the complex ν plane but may abruptly change their values at the boundaries between the regions. Approximations (E.27)–(E.29) correspond to a particular case of these.

The Debye approximations fail when ν is close to $\pm z$, in the sense that the parameter

$$\delta = \frac{|z \pm \nu|}{\sqrt[3]{|z|}} \tag{E.34}$$

is comparable to or smaller than unity. In the δ vicinity of the points $\nu = \pm z$, the so-called transition region (or Nicholson-type) asymptotic approximations in terms of the Airy functions are used. The approximation can be constructed so that it is applicable near either $\nu = z$ or $\nu = -z$ but not near both points.

The Debye and Nicholson-type asymptotic approximations have an overlapping area of applicability and thus provide a complete description of the Bessel functions of large argument z on the whole complex ν plane. Figures E.2 and E.3 illustrate the typical features of the Debye and Nicholson-type (the one applicable near $\nu = z$ is plotted) approximations for Bessel functions. The Debye approximation is good almost everywhere, except for the vicinity of the points $\nu = \pm z$

[1] The Debye approximations are sometimes defined as those valid for large values of both ν and z. However, the approximations also include the case of small ν.

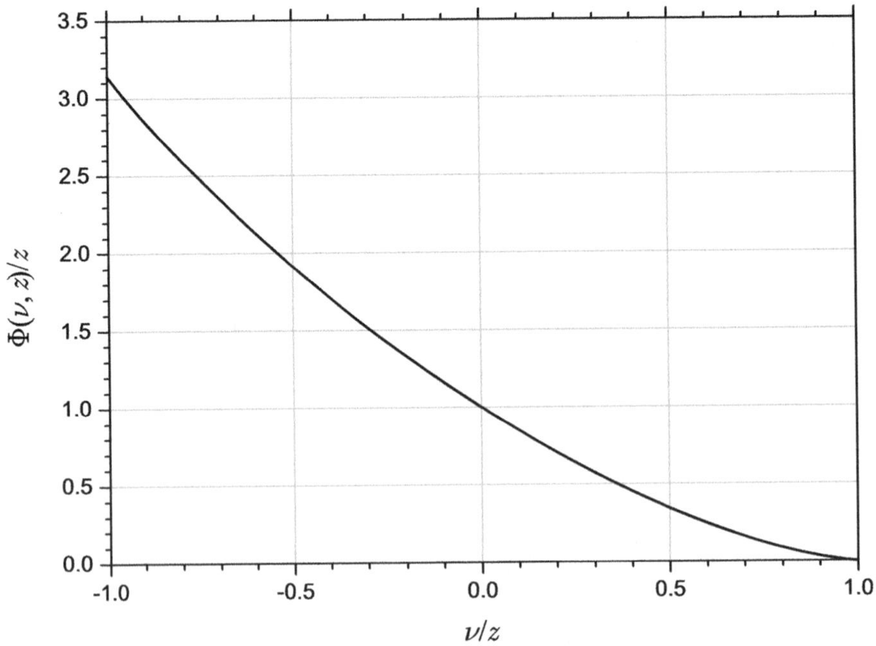

Figure E.1 Behavior of the phase function in the range $-1 \leq \nu/z \leq 1$.

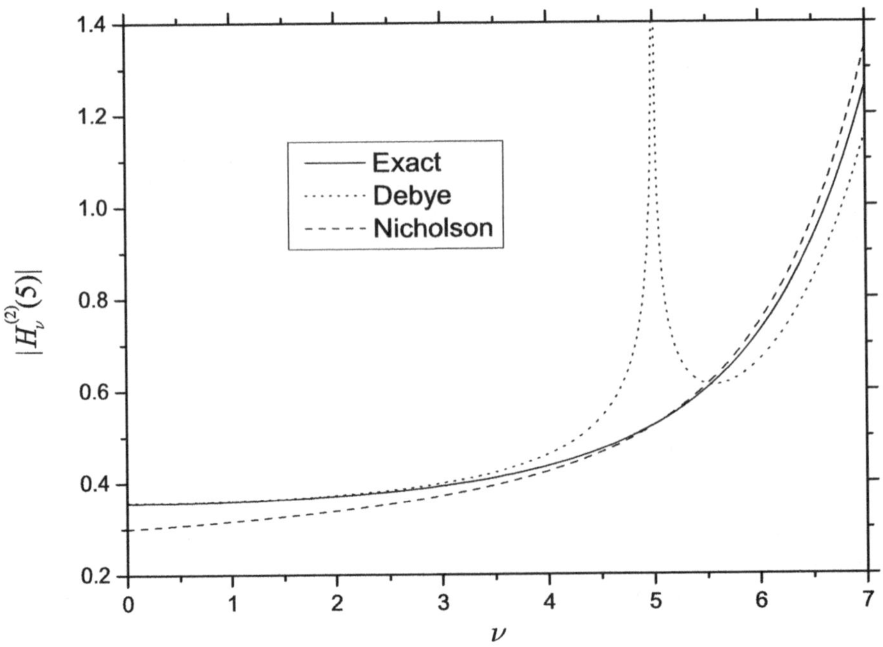

Figure E.2 Absolute value of $H_\nu^{(2)}(z)$ as a function of ν for $z = 5$: exact vs. Debye's and Nicholson-type approximations.

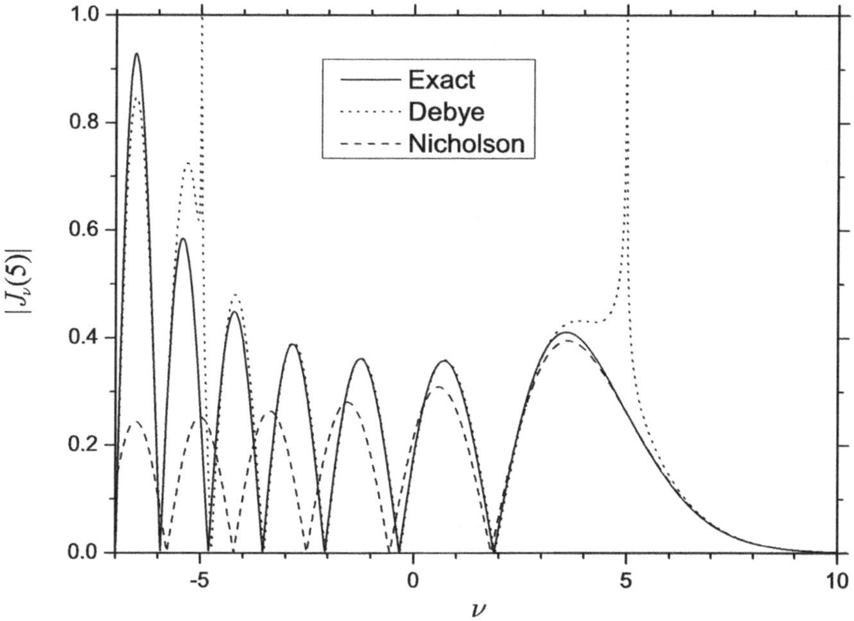

Figure E.3 Absolute value of $J_\nu(z)$ as a function of ν for $z = 5$: exact vs. Debye's and Nicholson-type approximations.

where it is singular. The Nicholson-type approximation provides a good approximation near the singular point $\nu = z$, but as the difference $|\nu - z|$ increases, the approximation quality worsens.

For the sake of completeness it should be noted that uniform representations, which are valid for arbitrary values of ν and contain the Debye and transition region approximations as particular cases, are also available (expressions of the Langer type) (Langer 1931; Olver 1974), but they are more complicated from the analytical viewpoint and therefore less suitable for the analysis of physical problems.

E.1.3 Content of the Appendix

Debye's approximations of the Bessel functions are addressed in detail in Debye (1909c, 1910), Watson (1962), Kratzer and Franz (1960), and Petrashen *et al.* (1953), and compact overviews can be found in Felsen and Marcuvitz (1973) and Bouman *et al.* (1987). The main problem associated with the Debye approximations is the Stokes phenomenon (Stokes 1857), which leads to partitioning of the complex ν plane into regions with different asymptotic approximations of the functions. The results presented in the literature are sometimes inconvenient for practical use because of a missing explicit description of the shape of the regions and indirect asymptotic formulas which express auxiliary functions in terms of combinations of Bessel functions, partly with arguments from other sheets of the complex z plane. As a result, explicit formulas for $J_\nu(z)$ and $H_\nu^{(1,2)}(z)$ are missing in several regions on the complex ν plane.

The goal of this appendix is to present the explicit Debye-type approximations for the Bessel functions of the first and third kind in every region on the complex ν plane. A special emphasis is placed on the asymptotic description of $J_\nu(z)$ and its ν zeros, which is of crucial importance for the analysis of scattering from homogeneous electrically large spheres (sections 7.3.4 and 7.5.5). It is important to note that in the regions where $J_\nu(z)$ is asymptotically small, knowledge of asymptotic approximations for $H_\nu^{(1,2)}(z)$ may be insufficient to derive an asymptotic approximation of $J_\nu(z)$ from (E.1).

The key idea of the following analysis is that the Debye formulas can be interpreted in terms of Liouville–Green approximations to the solutions of the Bessel differential equation extended into the whole complex ν plane by applying connection formulas of the WKB method[2] (Makarov and Osipov 1986, 1987). This approach allows a simple interpretation of the boundaries between the regions as the so-called Stokes lines employed in the WKB method. The results are summarized in Table E.1 and then specialized to each of the Bessel functions. Furthermore, the behavior of the functions for $|\nu| \to \infty$ and the position of zeros on the complex ν plane are described. Knowledge of the WKB method is not necessary for understanding the material, and the reader mostly interested in the final asymptotic formulas can go directly to section E.2.3, which gives the principal information in a condensed form.

Approximations of the Bessel functions in a neighborhood of the points $\nu = \pm z$ (asymptotic expansions in the transition regions) are briefly addressed in section E.3. These formulas, which involve the Airy function $\mathrm{Ai}(z)$ or related functions $u(t)$, $w_1(t)$, and $w_2(t)$, are used in the analysis of electromagnetic fields near caustics and penumbra regions (sections 6.4.3 and 6.4.4).

E.2 Debye's Asymptotic Approximations

The conventional way of constructing asymptotic approximations for Bessel functions when ν and z are arbitrary large numbers consists of asymptotic evaluation of suitable integral representations of the functions by the method of steepest descents (Debye 1909c, 1910). The resulting formulas are called Debye's asymptotic expansions (Watson 1962). This method, however, turns out to be cumbersome and leads to indirect representations of the functions when ν and z are complex variables. In this section we show that the use of WKB method permits an alternative derivation of the asymptotic representations even for complex-valued ν and z (Makarov and Osipov 1986, 1987). The section begins with a short outline of the Debye method. The rest of the section describes application of the WKB method to the construction of Debye's asymptotic approximations for complex-valued ν and z, provided that $|z|$ is a large parameter.

E.2.1 Debye's Method

The various types of Bessel functions can be represented by integrals of the form

$$\int e^{jz\cos\alpha + j\nu\alpha}\,d\alpha \tag{E.35}$$

taken along suitable contours in the complex α plane (e.g., Morse and Feshbach (1953, ch. 20, §2)). For example, when $\mathrm{Re}\, z > 0$,

$$H_\nu^{(1)}(z) = \frac{1}{\pi} e^{-j\nu\pi/2} \int_{C_1} e^{jz\cos\alpha + j\nu\alpha}\,d\alpha \tag{E.36}$$

$$H_\nu^{(2)}(z) = \frac{1}{\pi} e^{-j\nu\pi/2} \int_{C_2} e^{jz\cos\alpha + j\nu\alpha}\,d\alpha \tag{E.37}$$

$$2J_\nu(z) = \frac{1}{\pi} e^{-j\nu\pi/2} \int_{C_3} e^{jz\cos\alpha + j\nu\alpha}\,d\alpha, \tag{E.38}$$

where the integration contours are shown in Figure E.4. The topology of the integration contours ensures the relation (E.1).

When $|z|$ is large, one can attempt to evaluate these integrals by the steepest descent method by deforming the integration contours onto suitable steepest descent paths (SDP) that go through one or two saddle points α_c, which are some solutions

[2] The method is named after physicists Wentzel, Kramers, and Brillouin, who developed it for the Schrödinger equation in 1926. However, a general method of approximating solutions to linear, second-order differential equations, which includes the Schrödinger equation, had been developed in 1923 by Jeffreys (Jeffreys 1925) and that is why other often-used acronyms for the method include JWKB approximation and WKBJ approximation, where the "J" stands for Jeffreys. Another name is the phase-integral method. To find more about the Liouville–Green approximations and the WKB method see, for example, Olver (1974), Felsen and Marcuvitz (1973), and Heading (1962).

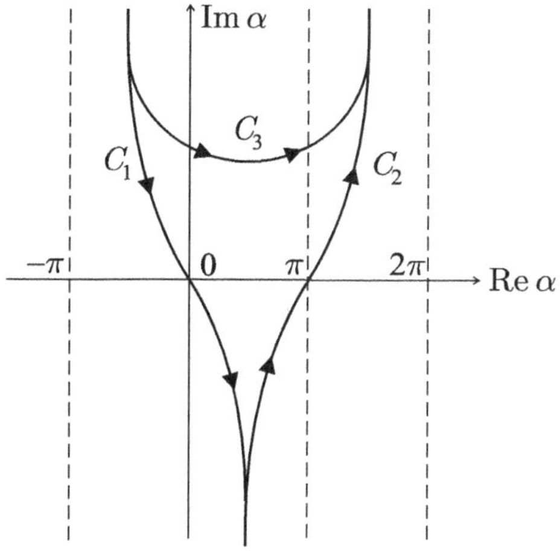

Figure E.4 Integration paths in integral representations of Bessel functions.

of the equation $\sin\alpha_c = \nu/z$. For example, when $|\nu/z| \ll 1$, the contour C_1 is deformed to go through the saddle point $\alpha_c \approx \nu/z$, the contour C_2 through the point $\alpha_c \approx \pi - \nu/z$, and C_3 through both of these, which gives the approximations (E.24)–(E.26). When $|\nu|$ is comparable with or greater than $|z|$, however, the location and the shape of the SDPs on the complex α plane may not allow a deformation of the integration contours in (E.36)–(E.38) onto the SDPs, which is the main complication of the approach.

In the framework of Debye's method, the Bessel functions are expressed in terms of two base functions which are defined as integrals similar to (E.36) and (E.37),

$$S_\nu^{(1)}(z) = \frac{1}{\pi}\mathrm{e}^{-j\nu\pi/2}\int_{L_1} \mathrm{e}^{jz\cos\alpha + j\nu\alpha}\,\mathrm{d}\alpha \tag{E.39}$$

$$S_\nu^{(2)}(z) = \frac{1}{\pi}\mathrm{e}^{-j\nu\pi/2}\int_{L_2} \mathrm{e}^{jz\cos\alpha + j\nu\alpha}\,\mathrm{d}\alpha, \tag{E.40}$$

where L_1 is the SDP that goes through the saddle point with $-\pi/2 < \mathrm{Re}\,\alpha_c < \pi/2$ and L_2 is the SDP that goes through the saddle point with $\pi/2 < \mathrm{Re}\,\alpha_c < 3\pi/2$. If the parameter δ defined in (E.34) is of the order unity or greater, then these integrals can be expanded into the asymptotic series

$$S_\nu^{(1)}(z) = \frac{\exp\left[\nu(\tanh\gamma - \gamma) - \frac{1}{4}j\pi\right]}{\left(-\frac{1}{2}j\nu\pi\tanh\gamma\right)^{\frac{1}{2}}}\sum_{m=0}^{\infty}\frac{\Gamma\left(m+\frac{1}{2}\right)}{\Gamma\left(\frac{1}{2}\right)}\frac{A_m}{\left(\frac{1}{2}\nu\tanh\gamma\right)^m} \tag{E.41}$$

$$S_\nu^{(2)}(z) = \frac{\exp\left[-\nu(\tanh\gamma - \gamma) + \frac{1}{4}j\pi\right]}{\left(-\frac{1}{2}j\nu\pi\tanh\gamma\right)^{\frac{1}{2}}}\sum_{m=0}^{\infty}\frac{\Gamma\left(m+\frac{1}{2}\right)}{\Gamma\left(\frac{1}{2}\right)}\frac{A_m}{\left(-\frac{1}{2}\nu\tanh\gamma\right)^m}, \tag{E.42}$$

where

$$\cosh\gamma = \frac{\nu}{z} \tag{E.43}$$

$$\arg\left(-\frac{1}{2}j\nu\pi\tanh\gamma\right) = \arg z + \arg(-j\sinh\gamma) \tag{E.44}$$

and the value of $\arg(-j\sinh\gamma)$ which lies between $-\pi/2$ and $\pi/2$ is to be taken. The first three expansion coefficients are given by the relations

$$A_0 = 1, \quad A_1 = \frac{1}{8} - \frac{5}{24}\coth^2\gamma, \quad A_2 = \frac{3}{128} - \frac{77}{576}\coth^2\gamma + \frac{385}{3456}\coth^4\gamma. \tag{E.45}$$

The higher-order coefficients can be found in Abramowitz and Stegun (1965, p. 366) and Kratzer and Franz (1960, p. 322).

The key problem of Debye's method is that the SDPs $L_{1,2}$ are not necessarily equivalent to the contours $C_{1,2}$ from the definitions of the Hankel functions, for example depending on the values of ν and z they may end in different strips on the complex α plane, so that $S_\nu^{(1)}(z)$ is not necessarily equal to $H_\nu^{(1)}(z)$ and $S_\nu^{(2)}(z)$ to $H_\nu^{(2)}(z)$. In general, a given Bessel function $Z_\nu(z)$ is a linear combination

$$Z_\nu(z) = c_1 S_\nu^{(1)}(z) + c_2 S_\nu^{(2)}(z) \tag{E.46}$$

with some coefficients $c_{1,2}$. Finding the coefficients is relatively straightforward for real-valued z and ν (Debye 1909c) but becomes difficult for complex-valued z and ν because of the complex topology of the involved steepest descent contours (Debye 1910; Petrashen *et al.* 1953; Watson 1962). The results of the analysis are formulated on the complex ν/z plane divided into several regions with boundaries defined by implicit parametric equations. In every region, $S_\nu^{(1,2)}(z)$ are expressed, in general, as combinations of several Bessel functions, from which an asymptotic representation of a desired Bessel function has to be extracted (Watson 1962, section 8.6). This indirect and non-transparent representation of the asymptotic results complicates their practical use.

E.2.2 WKB Approximation

The difficulties associated with the asymptotic evaluation of the integral representations of Bessel functions with complex-valued z and ν can be alleviated if we notice that functions $S_\nu^{(1,2)}(z)$ given by expansions (E.41) and (E.42) are two linearly independent solutions of the Bessel differential equation in the Liouville–Green approximation. This observation significantly simplifies the asymptotic analysis since it suggests applicability of the WKB method to the extension of the asymptotic approximations to the complex values of z and ν.

We will study the asymptotic approximations on the complex plane of the order ν which is a parameter in the Bessel differential equation.[3] The points $\nu = \pm z$ are the turning points of the Liouville–Green approximation, in the close vicinity of which asymptotic expansions (E.41) and (E.42) are invalid. Outside of these vicinities, asymptotic representations of the Bessel functions are given by linear combinations of the linearly independent functions $S_\nu^{(1,2)}(z)$, as in (E.46). The piecewise behavior and the abrupt changes in the values of c_1 and c_2 with respect to variation of z or ν is then a result of the Stokes phenomenon. The boundaries between the regions with different asymptotic representations of the Bessel functions can be interpreted as the Stokes lines and easily determined. A further advantage of the WKB approach is that the asymptotic representations for Bessel functions can be extended from a single region into the whole complex ν plane by building connection formulas for the turning points $\nu = \pm z$. As the argument z changes, the shape of the regions on the complex ν plane may change, but it does not change the asymptotic representations (i.e., the parameters c_1 and c_2) in the regions.

Solutions of Bessel Differential Equation in Liouville–Green Approximation

The Liouville–Green approximation permits approximate solution of ordinary differential equations of second order (e.g., Olver (1974)). The solutions are in general limited to the values of the argument between the so-called turning points

[3] Note that this approach differs from the standard WKB method, in which the complex plane of the argument is used to extend the Liouville–Green solutions of differential equations around the turning points.

of the differential equation. At the turning points, the character of the solutions of the differential equation changes, for example from oscillating to exponentially decaying or growing. In solutions of diffraction problems these points are associated with caustics and shadow boundaries on smooth surfaces. The WKB method provides a way of connecting these local solutions, defined on isolated intervals, into a global solution valid everywhere.

Let us derive a pair of linearly independent solutions of the Bessel differential equation (E.6) in the Liouville–Green approximation. We will assume that z is large and complex-valued, while the ratio ν/z can be arbitrary. The first step consists of the substitution

$$Z_\nu(z) = \frac{1}{\sqrt{z}} X_\nu(z), \tag{E.47}$$

which transforms (E.6) into

$$\frac{\mathrm{d}^2}{\mathrm{d}z^2} X_\nu(z) + \left[Q_0(z,\nu) + \frac{1}{4z^2} \right] X_\nu(z) = 0 \tag{E.48}$$

with

$$Q_0(z,\nu) = 1 - \left(\frac{\nu}{z}\right)^2. \tag{E.49}$$

The logarithmic derivative of $X_\nu(z)$

$$\alpha(\nu,z) = -j\frac{\mathrm{d}}{\mathrm{d}z} \ln X_\nu(z) \tag{E.50}$$

satisfies the Riccati equation

$$-j\frac{\partial}{\partial z}\alpha(\nu,z) + \alpha(\nu,z)^2 = Q_0(z,\nu) + \frac{1}{4z^2}. \tag{E.51}$$

Once $\alpha(\nu,z)$ is known, the solution of (E.6) is determined from

$$Z_\nu(z) = \frac{1}{\sqrt{z}} \exp\left(j \int^{z} \alpha(\nu,t)\mathrm{d}t \right), \tag{E.52}$$

where the lower end of the integration range can be arbitrary as it determines a multiplicative constant in the solution.

The Liouville–Green approximation can be seen as an approximate solution of equation (E.51) under the assumption that $\alpha(\nu,z)^2$ is much greater than $\partial\alpha(\nu,z)/\partial z$. One searches for the solution in the form of the perturbation series

$$\alpha(\nu,z) = \alpha_{-1}(\nu,z) + \alpha_0(\nu,z) + \alpha_1(\nu,z) + \ldots \tag{E.53}$$

Insertion of (E.53) in (E.51) leads to a subsequent determination of the series members. Since z is a large parameter, the second term on the right-hand side of (E.51) is small and in the lowest order approximation

$$\alpha_{-1}^2(\nu,z) = Q_0(z,\nu), \tag{E.54}$$

implying that

$$\alpha_{-1}(\nu,z) = \sqrt{1 - \left(\frac{\nu}{z}\right)^2}. \tag{E.55}$$

For the next term of (E.53) we have the equation

$$-j\frac{\partial}{\partial z}\alpha_{-1}(\nu,z)+2\alpha_{-1}(\nu,z)\alpha_0(\nu,z)=0, \tag{E.56}$$

where the term $1/(4z^2)$ on the right-hand side of (E.51) is again neglected as it is small compared to

$$\frac{\partial}{\partial z}\alpha_{-1}(\nu,z)=\frac{1}{z}\frac{(\nu/z)^2}{\sqrt{1-(\nu/z)^2}}. \tag{E.57}$$

The solution of (E.56) is given by the formula

$$\alpha_0(\nu,z)=\frac{j}{2z}\frac{(\nu/z)^2}{1-(\nu/z)^2}. \tag{E.58}$$

Comparing with (E.55) suggests that with $|z|\gg 1$, $\alpha_0(\nu,z)$ is asymptotically smaller than $\alpha_{-1}(\nu,z)$, provided that $|1-(\nu/z)^2|$ is not too small.

The third member on the right-hand side of (E.53) results from

$$2\alpha_{-1}(\nu,z)\alpha_1(\nu,z)=\frac{1}{4z^2}+j\frac{\partial}{\partial z}\alpha_0(\nu,z)-\alpha_0^2(\nu,z) \tag{E.59}$$

as

$$\alpha_1(\nu,z)=\frac{1}{8z^2}\frac{1+4(\nu/z)^2}{[1-(\nu/z)^2]^{5/2}}. \tag{E.60}$$

Retaining three terms in (E.53) recovers the first two terms in the Debye expansion of the function $S_\nu^{(1)}(z)$, which are given by (E.41) and (E.45). Indeed, choosing the constant factor and the lower ends of integration in (E.52) according to

$$Z_\nu(z)=\sqrt{\frac{\pi}{2jz}}\exp\left(j\int_\nu^z\alpha_{-1}(\nu,t)\mathrm{d}t+j\int_\infty^z\alpha_0(\nu,t)\mathrm{d}t+j\int_\infty^z\alpha_1(\nu,t)\mathrm{d}t\right) \tag{E.61}$$

and integrating

$$\int_\nu^z\alpha_{-1}(\nu,t)\mathrm{d}t=z\sqrt{1-\left(\frac{\nu}{z}\right)^2}-\nu\arccos\frac{\nu}{z} \tag{E.62}$$

$$\int_\infty^z\alpha_0(\nu,t)\mathrm{d}t=\frac{j}{4}\ln\left[1-\left(\frac{\nu}{z}\right)^2\right] \tag{E.63}$$

$$\int_\infty^z\alpha_1(\nu,t)\mathrm{d}t=-\frac{3+2(\nu/z)^2}{24z[1-(\nu/z)^2]^{3/2}}, \tag{E.64}$$

we obtain the formula

$$Z_\nu(z)=\sqrt{\frac{2}{\pi z}}\frac{\exp\left[j\Phi(\nu,z)-j\frac{\pi}{4}\right]}{\sqrt[4]{1-(\nu/z)^2}}\exp\left\{-j\frac{3+2(\nu/z)^2}{24z[1-(\nu/z)^2]^{3/2}}\right\} \tag{E.65}$$

with $\Phi(\nu, z)$ defined in (E.32). As $|z| \to \infty$, the argument of the outermost exponential factor on the right-hand side of (E.65) is small, and we can approximate the factor as

$$\exp\left\{-j\frac{3+2(\nu/z)^2}{24z\left[1-(\nu/z)^2\right]^{3/2}}\right\} \approx 1 - j\frac{3+2(\nu/z)^2}{24z[1-(\nu/z)^2]^{3/2}}. \tag{E.66}$$

Using (E.66) in (E.65) and replacing ν/z according to (E.43) give an expression that is identical with the first two terms in expansion (E.41),

$$S_\nu^{(1)}(z) = \sqrt{\frac{2}{\pi z}}\frac{A_\nu\left[jz\sqrt{1-(\nu/z)^2}\right]}{\sqrt[4]{1-(\nu/z)^2}}\exp\left[j\Phi(\nu, z) - j\frac{\pi}{4}\right], \tag{E.67}$$

where

$$A_\nu(T) = 1 + \frac{1}{8T}\left(1 - \frac{5\nu^2}{3T^2}\right). \tag{E.68}$$

The second solution of the Bessel differential equation, the function $S_\nu^{(2)}(z)$, which is given by the Debye expansion (E.42), results from a Liouville–Green approximation described above if the second solution of the quadratic equation (E.54)

$$\alpha_{-1}(\nu, z) = -\sqrt{1-\left(\frac{\nu}{z}\right)^2} \tag{E.69}$$

is used to initialize the sequence (E.53),

$$S_\nu^{(2)}(z) = \sqrt{\frac{2}{\pi z}}\frac{A_\nu\left[-jz\sqrt{1-(\nu/z)^2}\right]}{\sqrt[4]{1-(\nu/z)^2}}\exp\left[-j\Phi(\nu, z) + j\frac{\pi}{4}\right]. \tag{E.70}$$

Equation (E.68) defines the first two terms in an asymptotic expansion of $A_\nu(T)$. The higher-order terms can be in principle obtained by retaining higher-order terms in the sequence (E.53). The condition of applicability of the expansion can be obtained by requiring that the second term on the right-hand side of (E.68) be small compared to the first one. It is easily seen that the second term grows as T decreases, and the applicability condition can be written as

$$\left|\frac{5\nu^2}{24T^3}\right| \ll 1 \tag{E.71}$$

or, on replacing $|T|$ with $|z^2 - \nu^2|^{1/2}$,

$$|z^2 - \nu^2| \gg \left(\frac{5|\nu|^2}{24}\right)^{2/3} \approx 0.35|\nu|^{4/3}. \tag{E.72}$$

Thus, the Debye expansions are inapplicable in the vicinity of the points $\nu = \pm z$. The size of this vicinity follows from (E.72) and is estimated by (E.34).

The Debye asymptotic formulas permit differentiation with respect to z and ν. To obtain the leading terms in the asymptotic representations of the derivatives it is sufficient to account only for the quickly varying exponential functions in (E.67) and (E.70). For example,

$$\frac{\partial}{\partial z} S_\nu^{(1)}(z) \approx j\sqrt{1-\left(\frac{\nu}{z}\right)^2}\, S_\nu^{(1)}(z) \tag{E.73}$$

$$\frac{\partial}{\partial z} S_\nu^{(2)}(z) \approx -j\sqrt{1-\left(\frac{\nu}{z}\right)^2}\, S_\nu^{(2)}(z) \tag{E.74}$$

since

$$\frac{\partial}{\partial z}\Phi(\nu,z) = \sqrt{1-\left(\frac{\nu}{z}\right)^2}. \tag{E.75}$$

Derivatives with respect to ν are calculated in a similar manner. Here we just note the following useful formulas:

$$\frac{\partial}{\partial \nu}\Phi(\nu,z) = -\arccos\frac{\nu}{z}, \quad \frac{\partial^2}{\partial \nu^2}\Phi(\nu,z) = \frac{1}{z\sqrt{1-\left(\frac{\nu}{z}\right)^2}}. \tag{E.76}$$

As can be seen from definition (E.32) of $\Phi(\nu,z)$, functions (E.67) and (E.70) are oscillating when $-1 < \nu/z < 1$ and exponentially growing/decaying when $\nu/z > 1$ or $\nu/z < -1$, which means that the points $\nu = \pm z$ are the turning points of the Bessel differential equation (E.6). To extend the approximate solutions of differential equations beyond a turning point, the WKB method uses analytical properties of the solutions as functions of a complex variable, and since we work on the complex ν plane, we will next study the analytical properties of $S_\nu^{(1)}(z)$ and $S_\nu^{(1)}(z)$ as functions of the complex variable ν.

$S_\nu^{(1)}(z)$ and $S_\nu^{(2)}(z)$ on the Complex ν Plane

On the complex ν plane, expressions (E.67) and (E.70) have branch points at $\nu = \pm z$ and thus have to be considered on a Riemann multiple-sheet surface with the sheets connected through branch cuts. We specify the suitable – "working" – sheet of the Riemann surface by conditions (E.33).

It is convenient to describe the analytical properties of the multi-valued functions in terms of the variable

$$\xi = \frac{\nu}{z}. \tag{E.77}$$

If we define the branch cuts by the condition $\mathrm{Re}\sqrt{1-(\nu/z)^2} = 0$, then on the complex ξ plane they have the shape of two semi-infinite lines starting from $\xi = \pm 1$ and going to $\xi = \pm\infty$ (Figure E.5). The opposite sides of the branch cuts

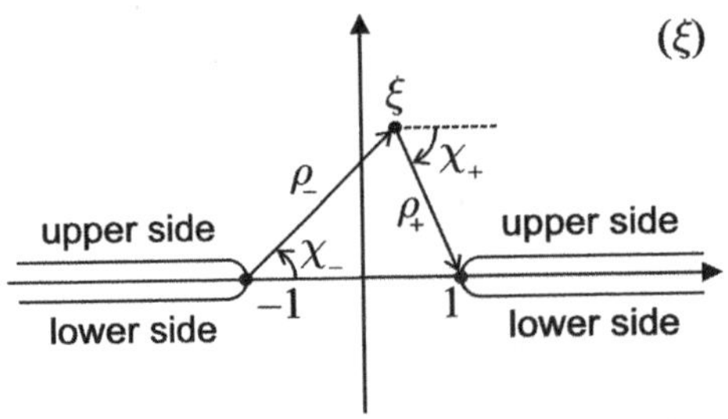

Figure E.5 Branch cuts on the complex ξ plane.

will be referred to as the upper and lower sides as they are seen in the figure. The values of the radical functions on the selected sheet of the Riemann surface are determined by the relations

$$\sqrt{1-\xi^2} = \sqrt{\rho_-\rho_+}\, e^{j(\chi_-+\chi_+)/2} \tag{E.78}$$

$$\sqrt[4]{1-\xi^2} = \sqrt[4]{\rho_-\rho_+}\, e^{j(\chi_-+\chi_+)/4}, \tag{E.79}$$

where $\rho_\pm$ and $\chi_\pm$ are the moduli and the arguments of the complex quantities $1 \mp \xi = \rho_\pm \exp\left(j\chi_\pm\right)$ such that $\rho_\pm \geq 0$ and $-\pi \leq \chi_\pm \leq \pi$. For example, the conditions $\chi_+ = -\pi$ and $\chi_- = 0$ define a point ξ on the upper side of the branch cut $1 \leq \xi < \infty$ where

$$\sqrt{1-\xi^2} = -j\left|1-\xi^2\right|^{1/2}, \quad \sqrt[4]{1-\xi^2} = e^{-j\pi/4}\left|1-\xi^2\right|^{1/4}. \tag{E.80}$$

On the lower side of the branch cut ($\chi_+ = \pi$ and $\chi_- = 0$),

$$\sqrt{1-\xi^2} = j\left|1-\xi^2\right|^{1/2}, \quad \sqrt[4]{1-\xi^2} = e^{j\pi/4}\left|1-\xi^2\right|^{1/4}. \tag{E.81}$$

Functions $\sqrt{1-\xi^2}$ and $\sqrt[4]{1-\xi^2}$ defined by (E.78) and (E.79) are even functions of ξ, provided that the points ξ and $-\xi$ belong to the same sheet of the Riemann surface.

The formula

$$\arccos \xi = -j \ln\left(\xi + j\sqrt{1-\xi^2}\right) \tag{E.82}$$

is a useful way of extending the inverse trigonometric function into the complex ξ plane. For example, on the branch cut $1 \leq \xi < +\infty$ the function becomes purely imaginary,

$$\arccos \xi = \mp j \ln\left(\xi + \left|\xi^2 - 1\right|^{1/2}\right), \tag{E.83}$$

where the upper (lower) sign on the right-hand side of (E.83) corresponds to the upper (lower) side of the branch cut. The values of the function at the points $-\xi$ and ξ, both on the selected sheet of the Riemann surface, are connected by the relation

$$\arccos(-\xi) = \pi - \arccos \xi. \tag{E.84}$$

Thus, on the branch cut $-\infty < \xi \leq -1$,

$$\arccos \xi = \pi \mp j \ln\left(|\xi| + \left|\xi^2 - 1\right|^{1/2}\right) \tag{E.85}$$

and again the upper (lower) sign corresponds to the upper (lower) side of the branch cut. On the interval $-1 < \xi < 1$, equation (E.82) gives $\pi \geq \arccos \xi \geq 0$, as required by the definition of functions (E.67) and (E.70).

Because of the described symmetries of the involved functions on the working sheet of the Riemann surface, the phase function (E.32) satisfies the relation

$$\Phi(-\nu, z) = \Phi(\nu, z) + \nu\pi \tag{E.86}$$

and therefore

$$S^{(1)}_{-\nu}(z) = e^{j\nu\pi} S^{(1)}_{\nu}(z) \tag{E.87}$$

$$S^{(2)}_{-\nu}(z) = e^{-j\nu\pi} S^{(2)}_{\nu}(z), \tag{E.88}$$

which as we shall see a bit later ensures compliance of the Debye asymptotic representations (E.67) and (E.70) with the general symmetry relations (E.17) and (E.18).

The jumps in the values of the multi-valued functions on the branch cuts – equations (E.80), (E.81), (E.83), and (E.85) – lead to the following transition relations

$$S_\nu^{(1)}(z)\big|_{\text{upper side}} = S_\nu^{(2)}(z)\big|_{\text{lower side}} \tag{E.89}$$

$$S_\nu^{(2)}(z)\big|_{\text{upper side}} = -S_\nu^{(1)}(z)\big|_{\text{lower side}} \tag{E.90}$$

on the branch cut that starts at the point $\nu = z$ and

$$S_\nu^{(1)}(z)\big|_{\text{upper side}} = -e^{-2j\nu\pi}\, S_\nu^{(2)}(z)\big|_{\text{lower side}} \tag{E.91}$$

$$S_\nu^{(2)}(z)\big|_{\text{upper side}} = e^{2j\nu\pi}\, S_\nu^{(1)}(z)\big|_{\text{lower side}} \tag{E.92}$$

on the branch cut that starts from $\nu = -z$. In contrast to $S_\nu^{(1,2)}(z)$, the Bessel functions $H_\nu^{(1)}(z)$, $H_\nu^{(2)}(z)$, and $J_\nu(z)$ are single-valued functions of ν. Thus, the asymptotic representations (E.46) must change their analytical form in accordance with the relations (E.89)–(E.92) to be continuous at the branch cuts.

Consider the behavior of functions (E.67) and (E.70) in the limits of small and large orders of the Bessel functions. If the order ν is small so that $|\nu|^2 \ll |z|$, then in the approximation of the phase function

$$\Phi(\nu, z) \approx z\left(1 - \frac{\pi}{2}\xi + \frac{1}{2}\xi^2 + \ldots\right) \tag{E.93}$$

the terms starting from the third one in square brackets can be neglected and the base functions simplify to

$$S_\nu^{(1)}(z) \approx \sqrt{\frac{2}{\pi z}}\, e^{j\left(z - \nu\frac{\pi}{2} - \frac{\pi}{4}\right)}, \quad S_\nu^{(2)}(z) \approx \sqrt{\frac{2}{\pi z}}\, e^{-j\left(z - \nu\frac{\pi}{2} - \frac{\pi}{4}\right)}. \tag{E.94}$$

In the alternative case that $|\nu| \to \infty$, the multi-valued functions can be expressed as

$$\sqrt{1-\xi^2} = \xi e^{-j\sigma\pi/2}\sqrt{1 - \frac{1}{\xi^2}} \approx \xi e^{-j\sigma\pi/2}\left[1 - \frac{1}{2\xi^2} - \frac{1}{8\xi^4} + O\left(\frac{1}{\xi^6}\right)\right] \tag{E.95}$$

$$\sqrt[4]{1-\xi^2} = \sqrt{\xi} e^{-j\sigma\pi/4}\sqrt[4]{1 - \frac{1}{\xi^2}} \approx \sqrt{\xi} e^{-j\sigma\pi/4}\left[1 - \frac{1}{4\xi^2} + O\left(\frac{1}{\xi^4}\right)\right] \tag{E.96}$$

$$\arccos\xi \approx -j\sigma\ln(2\xi) \tag{E.97}$$

with

$$\sigma = \begin{cases} 1, & 0 < \arg\xi < \pi \\ -1, & -\pi < \arg\xi < 0 \end{cases}, \tag{E.98}$$

leading to the approximation

$$\Phi(\nu, z) \approx -j\sigma\nu\left(\ln\frac{e}{2\xi} - \frac{1}{2\xi^2} + \ldots\right). \tag{E.99}$$

If $|\nu| \gg |z|^2$, the second term in the brackets can be neglected to give the large-order approximation of the base functions:

$$S_\nu^{(1)}(z) \approx e^{j\pi(\sigma-1)/4}\sqrt{\frac{2}{\pi\nu}}\left(\frac{2\nu}{ez}\right)^{-\sigma\nu} \tag{E.100}$$

$$S_\nu^{(2)}(z) \approx e^{j\pi(\sigma+1)/4}\sqrt{\frac{2}{\pi\nu}}\left(\frac{2\nu}{ez}\right)^{\sigma\nu}. \tag{E.101}$$

These approximations are valid on the working sheet of the Riemann surface, excluding the branch cuts, so $\arg \nu \neq \arg z$ and $|\arg \nu - \arg z| < \pi$.

Stokes Lines

The Stokes lines are the lines in the complex plane at which asymptotic representations of solutions of ordinary differential equations of second order may abruptly change. Since the asymptotic solutions are typically given by series expansions which satisfy the differential equation up to an arbitrary inverse power of the large parameter, only an exponentially small discontinuity is allowed. At the Stokes lines the two linearly independent solutions of the differential equation become non-oscillating, one exponentially small compared to the other, and it is the coefficient in front of this exponentially small component that may jump (Felsen and Marcuvitz 1973; Olver 1974; Stokes 1857).

Functions $S_\nu^{(1,2)}(z)$ are non-oscillating if

$$\operatorname{Re}\Phi(\nu, z) = 0, \tag{E.102}$$

and this equation defines a group of curves on the complex ν plane, going from the point $\nu = z$ at which the condition (E.102) is always satisfied since $\Phi(z, z) = 0$. Coefficients of the Bessel equation (E.6) are even functions of ν and, therefore, functions $S_{-\nu}^{(1)}(z)$ and $S_{-\nu}^{(2)}(z)$ constitute another pair of linearly independent solutions of the Bessel differential equation, which because of (E.87) and (E.88) are non-oscillating if

$$\operatorname{Re}[\Phi(\nu, z) + \nu\pi] = 0. \tag{E.103}$$

This equation defines the Stokes lines that start at the point $\nu = -z$. Because of the symmetry relation (E.86), the Stokes lines defined by (E.102) and (E.103) are symmetric with respect to the point $\nu = 0$.

A complementary group of lines, which will be referred to as anti-Stokes lines,[4] define the geometric location of the points where both base functions are purely oscillating. These lines are also important as these are often the lines of zeros of the involved functions. In the case of the Bessel differential equation, the anti-Stokes lines are determined by the conditions

$$\operatorname{Im}\Phi(\nu, z) = 0 \tag{E.104}$$

$$\operatorname{Im}[\Phi(\nu, z) + \nu\pi] = 0. \tag{E.105}$$

Let us describe the shape of the lines on the complex ν plane in dependence on $\arg z$. Consider solutions of the equations (E.102) and (E.104) in the vicinity of the point $\nu = z$. If $\nu - z = \epsilon \exp(j\chi)$ with a small positive ϵ, then

$$\Phi(\nu, z) \approx \frac{2\sqrt{2}}{3} z \left(1 - \frac{\nu}{z}\right)^{3/2}, \tag{E.106}$$

so there are three Stokes and three anti-Stokes lines leaving the point $\nu = z$ in the directions $\chi = (\arg z + 2m\pi)/3$ and $\chi = (\arg z + \pi + 2m\pi)/3$, respectively, with $m = -1, 0, 1$.

To determine the position of the ends of the lines ($|\nu| \to \infty$), approximation (E.99) can be used, and equations (E.102) and (E.104) become

$$\operatorname{Re}\nu \approx \frac{\operatorname{Im}\nu}{\arg z - \arg \nu} \ln\left|\frac{2\nu}{ez}\right|, \tag{E.107}$$

$$\operatorname{Im}\nu \approx \frac{\operatorname{Re}\nu}{\arg \nu - \arg z} \ln\left|\frac{2\nu}{ez}\right|, \tag{E.108}$$

[4] This sounds a bit unaesthetic and can be confused with anti-Stokes lines in Raman spectra. That is why some authors refer to the anti-Stokes lines as principal curves (see historical notes at the end of Chapter 13 in Olver (1974)).

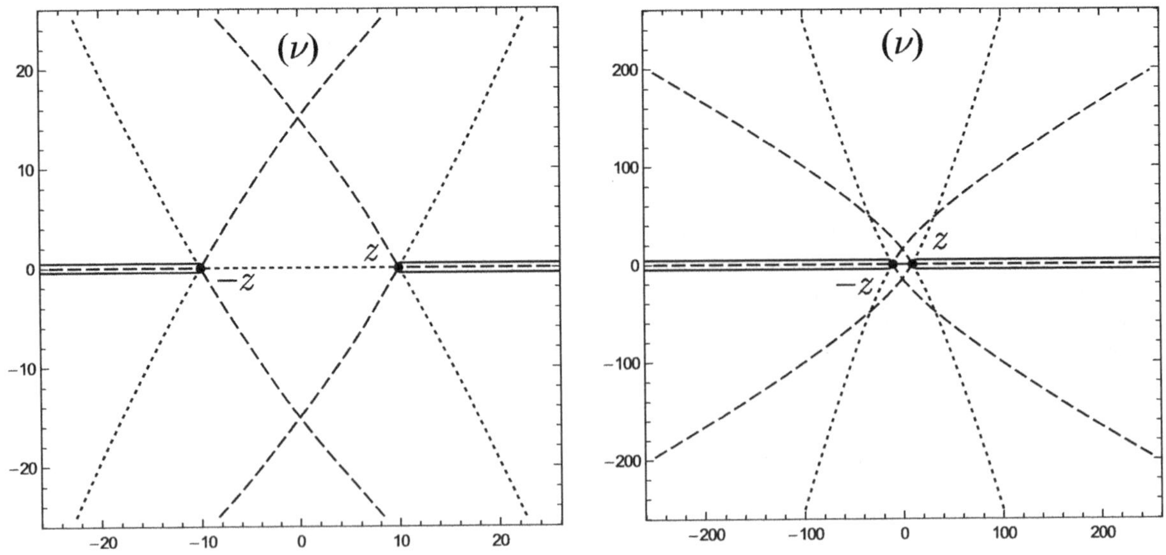

Figure E.6 Computer-generated images of Stokes (dashed) and anti-Stokes (dotted) lines on the complex ν plane when $z = 10$.

respectively. Thus, according to (E.107), at the ends of the Stokes lines $|\text{Re}\,\nu/\text{Im}\,\nu| \to \infty$, implying that $\arg\nu \to 0, \pm\pi$. The anti-Stokes lines go to infinity in such a way that $\arg\nu \to \pm\pi/2$ because $|\text{Re}\,\nu/\text{Im}\,\nu| \to 0$ according to (E.108). In both cases, the position of the ends is independent of the value of arg z.

Figures E.6 and E.7 show the topology of the curves on the complex ν plane for a real- and complex-valued argument z, respectively. Due to the logarithmic nature of the asymptotic estimates (E.107) and (E.108), the ends of the (anti-)

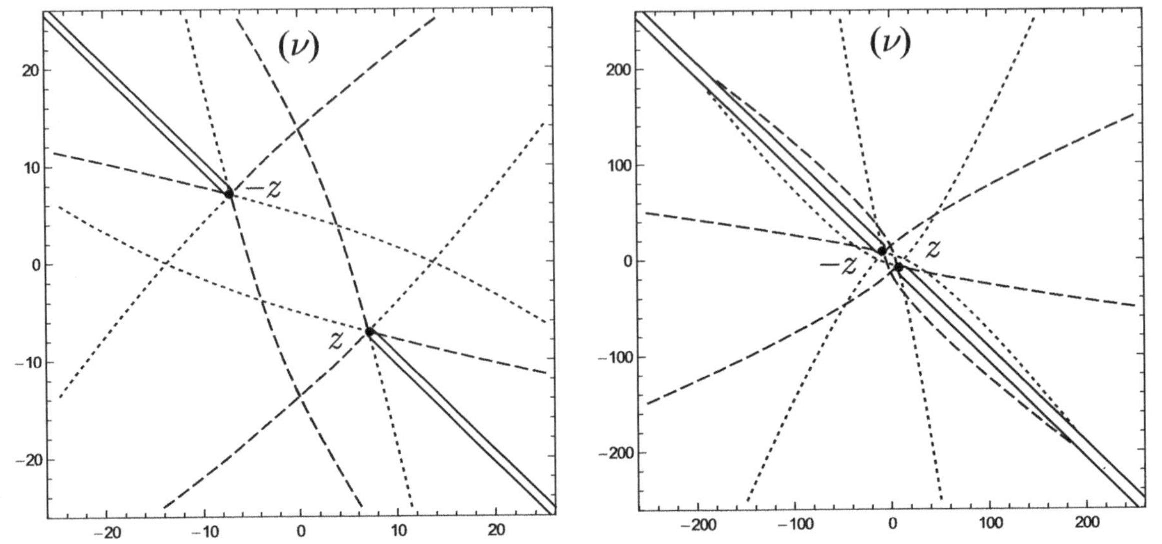

Figure E.7 Stokes (dashed) and anti-Stokes (dotted) lines on the complex ν plane when $z = 10\mathrm{e}^{-j\pi/4}$. On the right, two Stokes and two anti-Stokes lines cross the branch cuts and go into the adjacent sheets of the Riemann surface.

Stokes lines approach the expected dependence only slowly, and to see these, large-scale plots (to the right in Figures E.6 and E.7) are needed. The lines may cross the branch cuts and go to the adjacent sheets of the Riemann surface.

Connection Formulas

Now we are ready to define the Debye asymptotic expansions of Bessel functions on the whole complex ν plane. This is achieved by building the so-called WKB connection formulas that relate asymptotic representations of a solution of an ordinary second-order differential equation in the regions separated by Stokes lines on a complex plane. The main principle is that the analytical form of the asymptotic representation may change only at the Stokes lines and only in the exponentially smaller term to ensure continuity of the asymptotic representation in an asymptotic sense.

In the case of Bessel functions on the complex ν plane, this means that their asymptotic representations have the form of a linear combination of functions $S_\nu^{(1)}(z)$ or $S_\nu^{(2)}(z)$, as in (E.46), with coefficients c_1 and c_2 that are independent of the argument z but are piecewise-continuous functions of the order ν. On the Stokes lines that come from the point $\nu = z$, $S_\nu^{(1)}(z)$ and $S_\nu^{(2)}(z)$ do not oscillate and one of them is exponentially small compared to another. On the Stokes lines coming from the point $\nu = -z$, however, it is functions $S_{-\nu}^{(1)}(z) = \exp(j\nu\pi)S_\nu^{(1)}(z)$ and $S_{-\nu}^{(2)}(z) = \exp(-j\nu\pi)S_\nu^{(2)}(z)$ that do not oscillate, so representation (E.46) should be rewritten as a linear combination of $S_{-\nu}^{(1)}(z)$ and $S_{-\nu}^{(2)}(z)$ before requiring the asymptotic continuity. Furthermore, at the branch cuts the analytical form of the asymptotic representations also change according to the relations (E.89)–(E.92) to compensate for the jumps in the values of the involved multi-valued functions.

Referring to the analysis of the previous section, we note that for every value of $\arg z$ the Stokes lines divide the complex ν plane in no more than nine regions, and as the value of $\arg z$ varies in the interval $-\pi/2 \leq \arg z \leq \pi/2$, one can identify 11 such regions in total. These are labeled with roman numerals from I to XI and shown in Figure E.8 for $\arg z \neq 0, \pm\pi/2$. In the limiting case $\arg z = 0$, two of the Stokes lines coincide with the branch cuts, regions IV, VI, X,

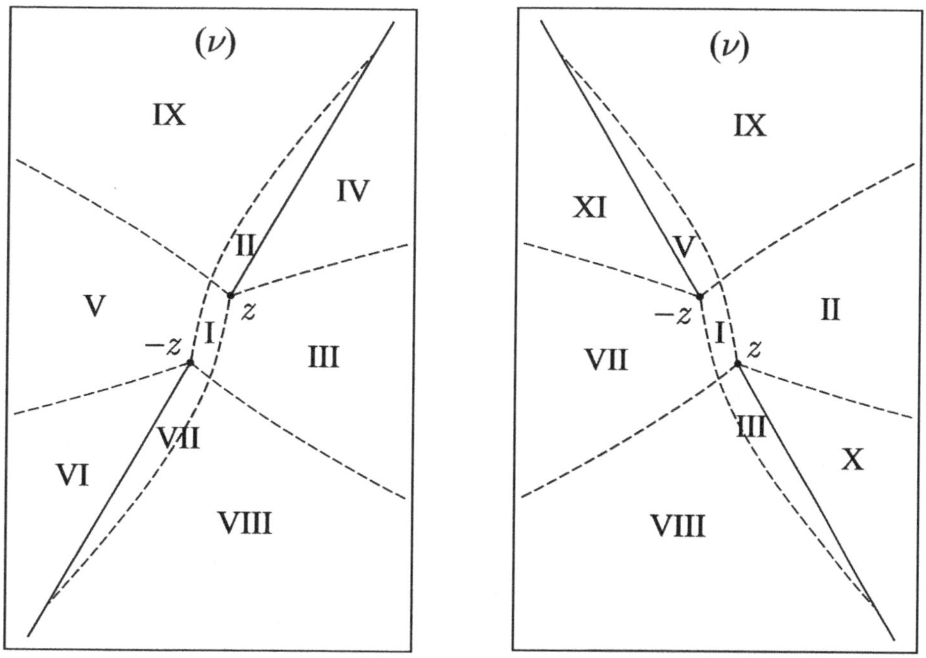

Figure E.8 WKB regions on the complex ν plane when $0 < \arg z < \pi/2$ (left) and $-\pi/2 < \arg z < 0$ (right). The boundaries between the regions are the Stokes lines (dashed) and the branch cuts (solid).

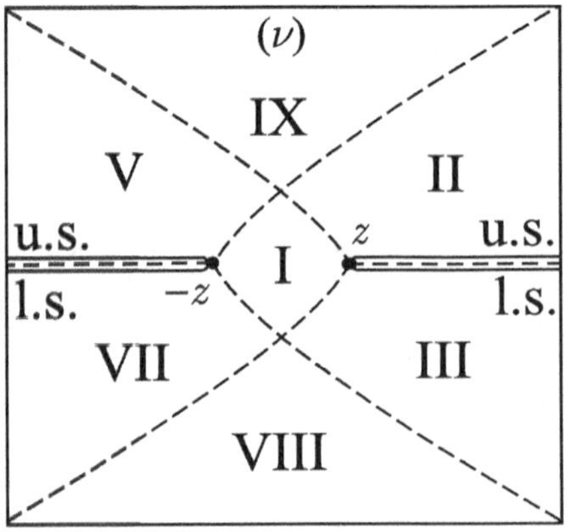

Figure E.9 WKB regions on the complex ν plane when $\arg z = 0$. The lower and upper sides of the branch cuts are labeled l.s. and u.s., respectively.

and XI disappear, and there remain seven regions I, II, III, V, VII, VIII, and IX on the complex ν plane (Figure E.9). If z is purely imaginary, then regions I, II, and VII ($\arg z = \pi/2$) or I, III, and V ($\arg z = -\pi/2$) are no longer present on the working sheet of the Riemann surface (Figure E.10).

We begin with derivation of connection relations for the case where $-\pi/2 < \arg z < 0$ (Figure E.8, right). Let $Z_\nu(z)$ be a solution of the Bessel differential equation that in the region I has the asymptotic representation

$$Z_\nu(z) = c_1 S_\nu^{(1)}(z) + c_2 S_\nu^{(2)}(z), \quad \nu \in \mathrm{I}, \tag{E.109}$$

where coefficients $c_{1,2}$ are known and $S_\nu^{(1)}(z)$ and $S_\nu^{(2)}(z)$ are given by asymptotic expansions (E.67) and (E.70). On the Stokes line between regions I and II, function $S_\nu^{(2)}(z)$ is exponentially small, implying that in region II the coefficient in front of the dominating function $S_\nu^{(1)}(z)$ must be the same as in (E.109). The coefficient in front of the small function $S_\nu^{(2)}(z)$

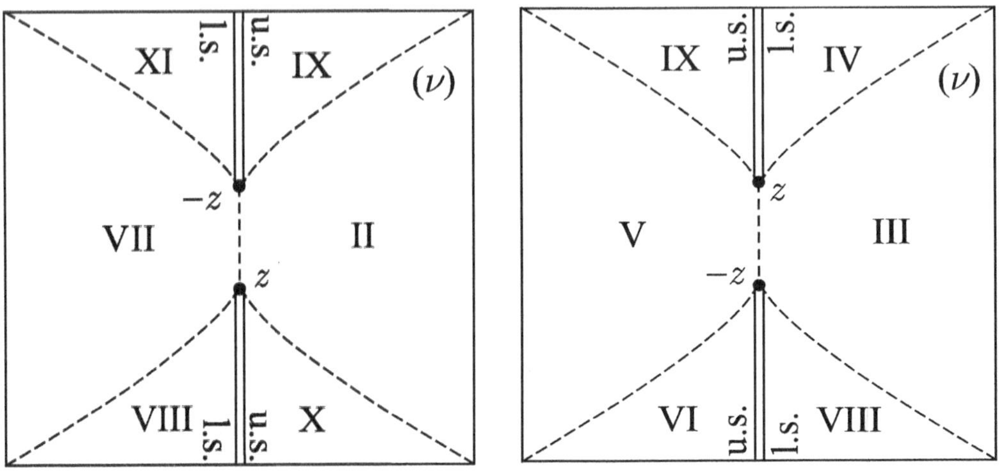

Figure E.10 WKB regions on the complex ν plane when $\arg z = -\pi/2$ (left) and $\arg z = \pi/2$ (right).

may differ from its value in region I, except for the case where $Z_\nu(z)$ does not include the exponentially large function. Thus, in region II we can represent function $Z_\nu(z)$ by

$$Z_\nu(z) = c_1 S_\nu^{(1)}(z) + (c_2 + s_1 c_1) S_\nu^{(2)}(z), \tag{E.110}$$

where s_1 is a parameter that describes the jump at the Stokes line. Note that representation (E.110) ensures the continuity of $Z_\nu(z)$ when $c_1 = 0$.

Representation of $Z_\nu(z)$ in the neighboring region X follows from (E.110) as

$$Z_\nu(z) = [c_1 + s_2(c_2 + s_1 c_1)] S_\nu^{(1)}(z) + (c_2 + s_1 c_1) S_\nu^{(2)}(z) \tag{E.111}$$

on accounting for the fact that function $S_\nu^{(1)}(z)$ is exponentially small on the boundary between regions II and X. Function $S_\nu^{(1)}(z)$ is exponentially small on the boundary between regions I and III as well, so in region III we can write

$$Z_\nu(z) = (c_1 + s_3 c_2) S_\nu^{(1)}(z) + c_2 S_\nu^{(2)}(z). \tag{E.112}$$

Parameters s_1, s_2, and s_3 can be determined by matching representations in regions III and X at the branch cut which separates the regions. By using relations (E.89) and (E.90) and equating the coefficients in front of the linearly independent functions $S_\nu^{(1)}(z)$ and $S_\nu^{(2)}(z)$ we obtain the equations

$$c_1(1 + s_1) + c_2(1 + s_3) = 0 \tag{E.113}$$

$$c_1(1 + s_1 s_2) + c_2(s_2 - 1) = 0 \tag{E.114}$$

which are satisfied with

$$s_1 = -1, \quad s_2 = 1, \quad s_3 = -1. \tag{E.115}$$

Thus, in the regions around the turning point $\nu = z$, the function $Z_\nu(z)$, which is given by (E.109) in region I, has the following representations:

$$Z_\nu(z) = \begin{cases} c_1 S_\nu^{(1)}(z) + (c_2 - c_1) S_\nu^{(2)}(z), & \nu \in \text{II} \\ (c_1 - c_2) S_\nu^{(1)}(z) + c_2 S_\nu^{(2)}(z), & \nu \in \text{III} \\ c_2 S_\nu^{(1)}(z) + (c_2 - c_1) S_\nu^{(2)}(z), & \nu \in \text{X} \end{cases}. \tag{E.116}$$

To determine asymptotic representations of $Z_\nu(z)$ in regions V, VII, and XI around the second turning point $\nu = -z$, we express (E.109) as

$$Z_\nu(z) = \tilde{c}_1 e^{j\nu\pi} S_\nu^{(1)}(z) + \tilde{c}_2 e^{-j\nu\pi} S_\nu^{(2)}(z), \tag{E.117}$$

where

$$\tilde{c}_1 = e^{-j\nu\pi} c_1, \quad \tilde{c}_2 = e^{j\nu\pi} c_2, \tag{E.118}$$

and using the fact that $\exp(j\nu\pi) S_\nu^{(1)}(z)$ is exponentially small at the boundaries between regions I and V and between VII and XI while $\exp(-j\nu\pi) S_\nu^{(2)}(z)$ is exponentially small at the boundary between regions I and VII, obtain the representations

$$Z_\nu(z) = \begin{cases} \left(\tilde{c}_1 + s_4 \tilde{c}_2\right) e^{j\nu\pi} S_\nu^{(1)}(z) + \tilde{c}_2 e^{-j\nu\pi} S_\nu^{(2)}(z), & \nu \in \text{V} \\ \tilde{c}_1 e^{j\nu\pi} S_\nu^{(1)}(z) + \left(s_5 \tilde{c}_1 + \tilde{c}_2\right) e^{-j\nu\pi} S_\nu^{(2)}(z), & \nu \in \text{VII} \\ \left[\tilde{c}_1 + s_6\left(s_5 \tilde{c}_1 + \tilde{c}_2\right)\right] e^{j\nu\pi} S_\nu^{(1)}(z) + \left(s_5 \tilde{c}_1 + \tilde{c}_2\right) e^{-j\nu\pi} S_\nu^{(2)}(z), & \nu \in \text{XI} \end{cases}. \tag{E.119}$$

Imposing the condition of continuity of the representations on the branch cut which separates regions V and XI, and using relations (E.91) and (E.92) between $\exp(j\nu\pi)S_\nu^{(1)}(z)$ and $\exp(-j\nu\pi)S_\nu^{(2)}(z)$ on the opposite sides of the branch cut, we arrive at the equations

$$\tilde{c}_1(1+s_5s_6)+\tilde{c}_2(s_6-1)=0 \tag{E.120}$$

$$\tilde{c}_1(1+s_5)+\tilde{c}_2(1+s_4)=0 \tag{E.121}$$

and determine coefficients s_4, s_5, and s_6 as

$$s_4=-1, \quad s_5=-1, \quad s_6=1. \tag{E.122}$$

Using (E.118) and (E.122) in (E.119) gives representations of $Z_\nu(z)$ which are asymptotically continuous in the regions around the point $\nu=-z$,

$$Z_\nu(z)=\begin{cases}\left(c_1-e^{2j\nu\pi}c_2\right)S_\nu^{(1)}(z)+c_2S_\nu^{(2)}(z), & \nu\in \text{V}\\ c_1S_\nu^{(1)}(z)+\left(c_2-e^{-2j\nu\pi}c_1\right)S_\nu^{(2)}(z), & \nu\in\text{VII}\\ c_2e^{2j\nu\pi}S_\nu^{(1)}(z)+\left(c_2-e^{-2j\nu\pi}c_1\right)S_\nu^{(2)}(z), & \nu\in\text{XI}\end{cases}. \tag{E.123}$$

Regions VIII and IX have no contact with the turning points. However, on their boundaries functions $S_\nu^{(1)}(z)$ and $S_\nu^{(2)}(z)$ or $\exp(j\nu\pi)S_\nu^{(1)}(z)$ and $\exp(-j\nu\pi)S_\nu^{(2)}(z)$ exponentially grow or decay, so we can use the same method as in the regions contacting the turning points to specify the form of asymptotic representations there. Consider, for example, region VIII. Representations of $Z_\nu(z)$ in the neighboring regions III and VII are already known, and any of these can be extended into region VIII. While extending from VII to VIII, we note that the boundary is the Stokes line at which $S_\nu^{(1)}(z)$ is exponentially small with the jump parameter s_3 already determined in (E.115), so that we obtain

$$Z_\nu(z)=\left[c_1(1+e^{-2j\nu\pi})-c_2\right]S_\nu^{(1)}(z)+\left(c_2-e^{-2j\nu\pi}c_1\right)S_\nu^{(2)}(z). \tag{E.124}$$

Alternatively, we could use the known representation of $Z_\nu(z)$ in region III to arrive at the alternative formula

$$Z_\nu(z)=(c_1-c_2)S_\nu^{(1)}(z)+\left[c_2\left(1+e^{-2j\nu\pi}\right)-e^{-2j\nu\pi}c_1\right]S_\nu^{(2)}(z) \tag{E.125}$$

by using the fact that at the boundary between regions III and VIII function $\exp(-j\nu\pi)S_\nu^{(2)}(z)$ is exponentially small and the respective jump parameter s_5 is available from (E.122). Representations (E.124) and (E.125) are asymptotically equivalent as in region VIII the imaginary part of ν is large and negative, implying that $|\exp(-2j\nu\pi)|\ll 1$ and

$$Z_\nu(z)=(c_1-c_2)S_\nu^{(1)}(z)+\left[c_2-e^{-2j\nu\pi}c_1\right]S_\nu^{(2)}(z), \quad \nu\in\text{VIII}. \tag{E.126}$$

In a similar way, one determines the representation of $Z_\nu(z)$ in region IX, which can be done by using the known representations either in region II or V. The result does not depend on the choice of the starting region and on accounting for the fact that in region IX $|\exp(2j\nu\pi)|\ll 1$, it has the form

$$Z_\nu(z)=\left(c_1-e^{2j\nu\pi}c_2\right)S_\nu^{(1)}(z)+(c_2-c_1)S_\nu^{(2)}(z), \quad \nu\in\text{IX}. \tag{E.127}$$

This result completes the extension of the asymptotic representation (E.109) from region I into the whole complex ν plane when $-\pi/2<\arg z<0$. The case of $0<\arg z<\pi/2$ can be analyzed in a similar manner. Note that in this case, regions X and XI disappear and two new regions IV and VI enter the working sheet of the complex ν plane (Figure E.8, left). Building WKB connection relations, we obtain

$$Z_\nu(z) = \begin{cases} (c_1 - c_2)S_\nu^{(1)}(z) + c_1 S_\nu^{(2)}(z), & \nu \in \mathrm{IV} \\ (c_1 - e^{2j\nu\pi}c_2)S_\nu^{(1)}(z) + e^{-2j\nu\pi}c_1 S_\nu^{(2)}(z), & \nu \in \mathrm{VI} \end{cases}. \tag{E.128}$$

Equations (E.116), (E.123), and (E.126)–(E.128) describe asymptotic representations of an arbitrary Bessel function $Z_\nu(z)$ in the whole complex ν plane when $|z| \gg 1$ and $|\arg z| \le \pi/2$, provided that $Z_\nu(z)$ is given by (E.109) in region I.

E.2.3 Bessel Functions on the Complex ν Plane

Summary of Debye's Approximations

Now we are ready to write down Debye-type asymptotic representations of $H_\nu^{(1)}(z)$, $H_\nu^{(2)}(z)$, and $J_\nu(z)$ with arbitrary complex order ν. To apply the results from section E.2.2, we need to know the representations of the functions in region I. Since coefficients c_1 and c_2 do not change as z varies (only the shape of the region changes) and are continuous functions of ν, it is sufficient to know the asymptotic representations, for example, for real-valued z and ν. These are available from (Debye 1909c) and using Debye's results for $0 < \nu < z$ we obtain the formulas

$$H_\nu^{(1)}(z) = S_\nu^{(1)}(z) \tag{E.129}$$

$$H_\nu^{(2)}(z) = S_\nu^{(2)}(z) \tag{E.130}$$

$$2J_\nu(z) = S_\nu^{(1)}(z) + S_\nu^{(2)}(z), \tag{E.131}$$

which are valid in the whole of region I. These equations coincide with (E.27)–(E.29) if $S_\nu^{(1)}(z)$ and $S_\nu^{(2)}(z)$ are represented by (E.67) and (E.70) with $A_\nu(T) = 1$. In the middle of region I, in a vicinity of the point $\nu = 0$ where $|\nu|^2 \ll |z|$, $S_\nu^{(1)}(z)$ and $S_\nu^{(2)}(z)$ can be replaced with large-argument approximations (E.94), and formulas (E.129)–(E.131) recover the Hankel approximations (E.24)–(E.26).

Using the values of coefficients c_1 and c_2 from (E.129)–(E.131) in connection formulas (E.116), (E.123), and (E.126)–(E.128), the Debye asymptotic representations of the Bessel functions $J_\nu(z)$, $H_\nu^{(1)}(z)$, and $H_\nu^{(2)}(z)$ are determined in every region of the complex ν plane. Table E.1 lists the values of the coefficients in front of functions $S_\nu^{(1)}(z)$ and $S_\nu^{(2)}(z)$. The asymptotic approximations with coefficients $c_{1,2}$ from Table E.1 are compliant (up to exponentially small terms) with symmetry properties (E.17) and (E.18) and with the general relation (E.1).

The asymptotic representations can be simplified because aside from an immediate vicinity of anti-Stokes lines, one of the two terms in (E.46) is small and can be neglected. Using this fact we can describe the asymptotic behavior of every Bessel function on a complex ν plane divided into three regions only. The division of the complex plane is shown in

Table E.1 Coefficients c_1 (left columns) and c_2 (right columns) in the Debye asymptotic representations (E.46) of the Bessel functions in different regions on the complex ν plane.

region	$H_\nu^{(1)}(z)$		$H_\nu^{(2)}(z)$		$2J_\nu(z)$	
I	1	0	0	1	1	1
II	1	-1	0	1	1	0
III	1	0	-1	1	0	1
IV	1	1	-1	0	0	1
V	1	0	$-e^{2j\nu\pi}$	1	$1-e^{2j\nu\pi}$	1
VI	1	$e^{-2j\nu\pi}$	$-e^{2j\nu\pi}$	0	$1-e^{2j\nu\pi}$	0
VII	1	$-e^{-2j\nu\pi}$	0	1	1	$1-e^{-2j\nu\pi}$
VIII	1	$-e^{-2j\nu\pi}$	-1	1	0	1
IX	1	-1	$-e^{2j\nu\pi}$	1	1	0
X	0	-1	1	1	1	0
XI	0	$-e^{-2j\nu\pi}$	$e^{2j\nu\pi}$	1	0	$1-e^{-2j\nu\pi}$

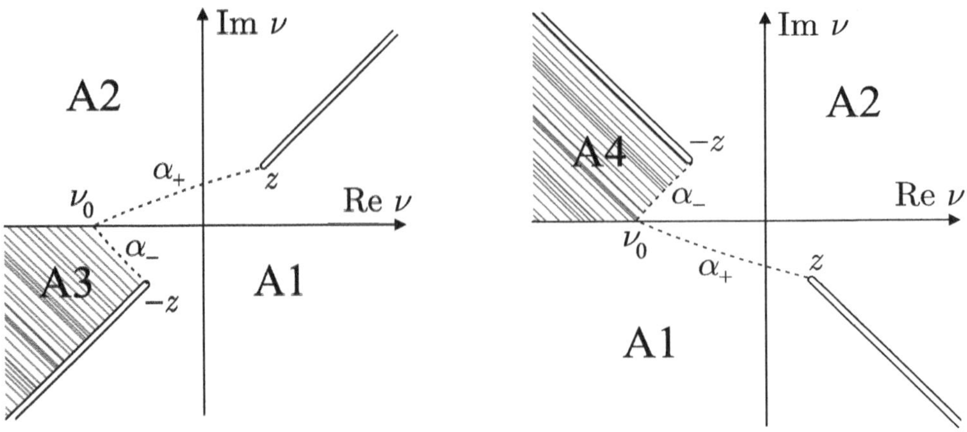

Figure E.11 Regions with different asymptotic representations of $J_\nu(z)$ when $\arg z > 0$ (left) and $\arg z < 0$ (right). Zeros of $J_\nu(z)$ are asymptotically close to the pieces of the anti-Stokes lines $\alpha_\pm$ and to the interval $-\infty < \nu < \nu_0$, where ν_0 is the point of intersection of $\alpha_\pm$ with the real axis.

Figures E.11–E.13. The regions are separated by anti-Stokes lines (dashed) and by the branch cuts. Inside the regions we have the following representations:

$$2J_\nu(z) \approx \begin{cases} S_\nu^{(2)}(z), & \nu \in \mathrm{A1} \\ S_\nu^{(1)}(z), & \nu \in \mathrm{A2} \\ -\mathrm{e}^{2j\nu\pi} S_\nu^{(1)}(z), & \nu \in \mathrm{A3} \\ -\mathrm{e}^{-2j\nu\pi} S_\nu^{(2)}(z), & \nu \in \mathrm{A4} \end{cases} \tag{E.132}$$

$$H_\nu^{(1)}(z) \approx \begin{cases} S_\nu^{(1)}(z), & \nu \in \mathrm{B1} \\ -S_\nu^{(2)}(z), & \nu \in \mathrm{B2} \\ -\mathrm{e}^{-2j\nu\pi} S_\nu^{(2)}(z), & \nu \in \mathrm{B3} \end{cases} \tag{E.133}$$

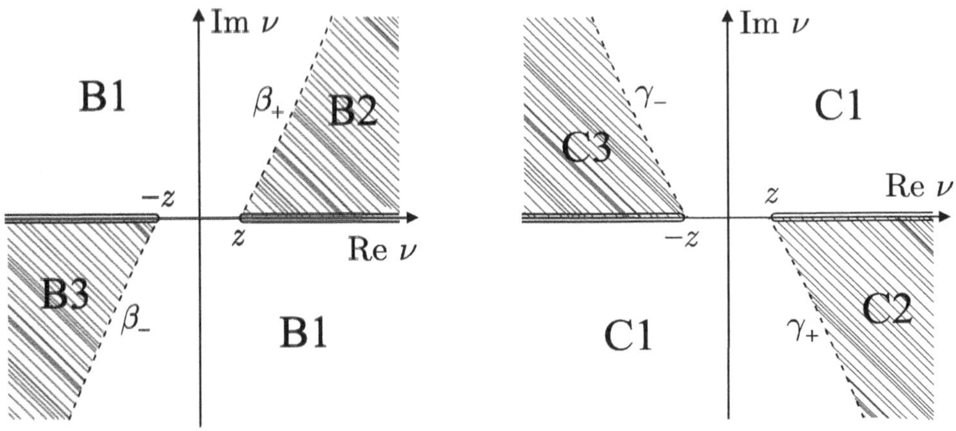

Figure E.12 Regions with different asymptotic representations of $H_\nu^{(1)}(z)$ (left) and $H_\nu^{(2)}(z)$ (right) when $\arg z = 0$. The anti-Stokes lines $\beta_\pm$ and $\gamma_\pm$ show the position of zeros of the Hankel functions.

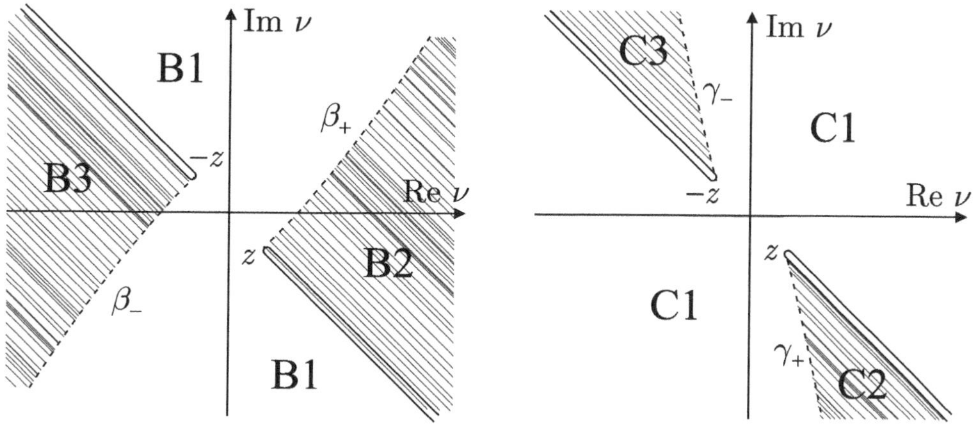

Figure E.13 The same as Figure E.12 but for arg $z < 0$.

$$H_\nu^{(2)}(z) \approx \begin{cases} S_\nu^{(2)}(z), & \nu \in \mathrm{C1} \\ -S_\nu^{(1)}(z), & \nu \in \mathrm{C2} \\ -\mathrm{e}^{2j\nu\pi} S_\nu^{(1)}(z), & \nu \in \mathrm{C3} \end{cases} . \tag{E.134}$$

In the immediate vicinity of the anti-Stokes lines, asymptotic representations from the adjacent regions should be added to each other. In a vicinity of point ν_0 (Figure E.11), asymptotic representations from all three regions are to be summed up.

Zeros of Bessel Functions on the Complex ν Plane

Anti-Stokes lines (E.104) and (E.105) are potential locations of ν zeros of Bessel functions (Kratzer and Franz 1960, pp. 323–332). In order for zeros to appear, however, it is necessary that a function is given by a superposition of the two base functions. Now that we know the asymptotic representations of the functions everywhere on the complex ν plane, we can determine the location of zeros and give simple approximate formulas for them (Makarov and Osipov 1986, 1987, 2000).

Along the anti-Stokes line β_+ separating regions B1 and B2 (Figures E.12 and E.13),

$$H_\nu^{(1)}(z) \approx S_\nu^{(1)}(z) - S_\nu^{(2)}(z) \tag{E.135}$$

and since the base functions $S_\nu^{(1)}(z)$ and $S_\nu^{(2)}(z)$ are comparable to each other and oscillatory there, $H_\nu^{(1)}(z)$ may vanish at the points ν_s with $s = 1, 2, \ldots$, which are approximately determined by the equation

$$\sin\left[\Phi(\nu_s, z) - \frac{\pi}{4}\right] = 0. \tag{E.136}$$

At the anti-Stokes line β_- that separates regions B1 and B3,

$$H_\nu^{(1)}(z) \approx \mathrm{e}^{-j\nu\pi}\left[\mathrm{e}^{j\nu\pi} S_\nu^{(1)}(z) - \mathrm{e}^{-j\nu\pi} S_\nu^{(2)}(z)\right] \tag{E.137}$$

and thus the equation for the zeros reads

$$\sin\left[\Phi(\nu_s, z) + \nu_s\pi - \frac{\pi}{4}\right] = 0. \tag{E.138}$$

Table E.2 Zeros of $H_\nu^{(2)}(z)$ near $\nu = z$ when $z = 100$: ν_s, exact value; $\nu_s^{\rm tr}$, transition region approximation (E.176); $\nu_s^{\rm D}$, Debye approximation (E.136). Parameters t_s, $t_s^{\rm tr}$, and $t_s^{\rm D}$ are defined by (E.163).

s	ν_s	$\nu_s^{\rm tr}$	$\nu_s^{\rm D}$	$\lvert t_s - t_s^{\rm tr}\rvert$	$\lvert t_s - t_s^{\rm D}\rvert$
1	$104.29 - 7.48j$	$104.31 - 7.46j$	$104.26 - 7.42j$	0.007	0.018
2	$107.49 - 13.11j$	$107.53 - 13.04j$	$107.48 - 13.09j$	0.020	0.006
3	$110.10 - 17.73j$	$110.17 - 17.61j$	$110.10 - 17.72j$	0.037	0.003
10	$123.29 - 41.57j$	$123.63 - 40.93j$	$123.29 - 41.57j$	0.198	0.0006

Because of (E.86) the two branches of zeros defined by (E.136) and (E.138) are symmetric to each other with respect to the point $\nu = 0$ and thus conform with the general symmetry property (E.17) of the Hankel function on the complex ν plane.

For the Hankel function $H_\nu^{(2)}(z)$ it follows from (E.134) that

$$H_\nu^{(2)}(z) \approx -S_\nu^{(1)}(z) + S_\nu^{(2)}(z) \tag{E.139}$$

at the anti-Stokes line γ_+ between regions C1 and C2 and

$$H_\nu^{(2)}(z) \approx -e^{j\nu\pi}\left[e^{j\nu\pi}S_\nu^{(1)}(z) - e^{-j\nu\pi}S_\nu^{(2)}(z)\right] \tag{E.140}$$

at the anti-Stokes line γ_- between regions C1 and C3, and therefore the zeros again satisfy equations (E.136) and (E.138) and are grouped in two branches along the respective anti-Stokes lines (Figures E.12 and E.13). Numerical values of several first zeros of $H_\nu^{(2)}(z)$, exact and approximate, obtained with (E.136), are compared in Table E.2.

The structure of ν zeros of $J_\nu(z)$ is more complicated because of the intersection of two anti-Stokes lines at a point ν_0 at the negative real axis (Figure E.11). Along the anti-Stokes line α_+ separating regions A1 and A2

$$2J_\nu(z) \approx S_\nu^{(1)}(z) + S_\nu^{(2)}(z), \tag{E.141}$$

and the zeros between the points $\nu = z$ and $\nu = \nu_0$ can be determined from the equation

$$\cos\left[\Phi(\nu_s, z) - \frac{\pi}{4}\right] = 0. \tag{E.142}$$

At the anti-Stokes line α_- between the points $\nu = -z$ and $\nu = \nu_0$, it follows from equations (E.132), (E.137), and (E.140) that $2J_\nu(z) \approx H_\nu^{(2)}(z)$ ($\arg z > 0$) or $H_\nu^{(1)}(z)$ ($\arg z < 0$) and thus the zeros of $J_\nu(z)$ in this region are asymptotically close to the zeros of the respective Hankel functions defined by the equation (E.138).

The function $J_\nu(z)$ has, however, one more branch of zeros which does not follow any anti-Stokes line. At the semi-infinite portion $-\infty < \nu < \nu_0$ of the real axis, one of the base functions dominates, leading to the single-term asymptotic formula

$$2J_\nu(z) \approx \begin{cases} S_\nu^{(1)}(z)(1 - e^{2j\nu\pi}), & \arg z > 0 \\ S_\nu^{(2)}(z)(1 - e^{-2j\nu\pi}), & \arg z < 0 \end{cases}. \tag{E.143}$$

In this case, it is the coefficient in front of the exponentially large base function that is an oscillating function of ν and leads to zeros. According to (E.143), the zeros are asymptotically close to negative integers.[5] In an asymptotic vicinity of

[5] Note that the zeros cannot coincide with negative integers $-m$ since this assumption, because of the property $J_{-m}(z) = (-1)^m J_m(z)$, would imply zeros at $\nu = m$, which is not true. The zeros are always slightly shifted from $-m$ due to the exponentially small correction terms which are omitted in (E.143).

the point ν_0 the three branches of zeros meet each other. To summarize, the ν zeros of $J_\nu(z)$ fill the semi-infinite interval $-\infty < \nu < \nu_0$ and the portions of the anti-Stokes lines $\alpha_\pm$ between the points $\nu = \pm z$ and ν_0. In the limit of $\arg z = 0$, all zeros of $J_\nu(z)$ reside along the semi-infinite portion $-\infty < \nu < z$ of the real axis.

Zeros of the derivatives $H_\nu^{(1)\prime}(z)$, $H_\nu^{(2)\prime}(z)$ and $J_\nu'(z)$ are located at the same lines in the complex ν plane as the zeros of the respective functions. The corresponding equations can be obtained by taken a derivative of (E.136), (E.138), (E.142), and (E.143) with respect to z, which, in the case of equations (E.136), (E.138), and (E.142), effectively replaces the sine-function with the cosine-function and vice versa.

Bessel Functions of Large Order

The behavior of Bessel functions when $|\nu| \to \infty$ follows from representations (E.132)–(E.134) and the estimates (E.100) and (E.101) for the base functions. Determine, for example, the portions of the complex ν plane where $J_\nu(z)$, $H_\nu^{(1)}(z)$, and $H_\nu^{(1)}(z)$ vanish. In regions A1 and A2, the Bessel function of the first kind is estimated as

$$|J_\nu(z)| \approx \frac{1}{\sqrt{2\pi|\nu|}} \left(\frac{2|\nu|}{\mathrm{e}|z|}\right)^{-|\nu|\cos\psi} \mathrm{e}^{|\nu|\sin\psi(\psi-\arg z)}, \tag{E.144}$$

where $\psi = \arg \nu$ for brevity, and the function decays at infinity as $O\left(|\nu|^{-|\nu|\cos\psi}\right)$ for $-\pi/2 < \psi < \pi/2$. When $\mathrm{Re}\,\nu = a$ is fixed and $\mathrm{Im}\,\nu \to \pm\infty$, then $\psi = \pm\pi/2$ and the estimation (E.144) becomes

$$|J_\nu(z)| \approx \frac{1}{\sqrt{2\pi|\nu|}} \left(\frac{2|\nu|}{\mathrm{e}|z|}\right)^{-a} \mathrm{e}^{|\nu|\left(\frac{\pi}{2}\mp\arg z\right)}, \tag{E.145}$$

which grows exponentially when $|\arg z| < \pi/2$. In the sector $\pi/2 < |\psi| \le \pi$ function $J_\nu(z)$ increases as $O\left(|\nu|^{-|\nu|\cos\psi}\right)$ when $|\nu| \to \infty$.

The Hankel functions grow along arbitrary directions in the complex ν plane except for the directions $\arg \nu = \pm\pi/2$, in which case they are estimated by the relations

$$|H_\nu^{(1)}(z)| \approx \sqrt{\frac{2}{\pi|\nu|}} \left(\frac{2|\nu|}{\mathrm{e}|z|}\right)^{\mp a} \mathrm{e}^{-|\nu|\left(\arg z\mp\frac{\pi}{2}\right)} \tag{E.146}$$

$$|H_\nu^{(2)}(z)| \approx \sqrt{\frac{2}{\pi|\nu|}} \left(\frac{2|\nu|}{\mathrm{e}|z|}\right)^{\pm a} \mathrm{e}^{|\nu|\left(\arg z\mp\frac{\pi}{2}\right)}, \tag{E.147}$$

where $a = \mathrm{Re}\,\nu$, implying that $H_\nu^{(1)}(z)$ vanishes when $\mathrm{Im}\,\nu \to -\infty$, and $H_\nu^{(2)}(z)$ vanishes when $\mathrm{Im}\,\nu \to +\infty$. The functions may vanish for both $\mathrm{Im}\,\nu \to +\infty$ and $\mathrm{Im}\,\nu \to -\infty$ if $\mathrm{Re}\,\nu > -1/2$ and $\arg z = \pi/2$ (function $H_\nu^{(1)}(z)$) or $\arg z = -\pi/2$ (function $H_\nu^{(2)}(z)$).

E.3 Almost Equal Argument and Order

If both ν and z are large but $|\nu \pm z|$ is of the order $|z|^{1/3}$ or smaller, then parameter δ in (E.34) is small and the Debye asymptotic expansions are no longer applicable. Two coalescing saddle points complicate evaluation of the integrals (E.36)–(E.38) with the steepest descent method. The Liouville–Green approximation becomes inapplicable as well since z approaches a turning point of the Bessel differential equation. In these transition regions around the points $\nu = \pm z$ a different kind of approximation, in terms of the Airy functions, is necessary. Because of the symmetry relations (E.17) and (E.18) it is sufficient to describe the Bessel functions in the vicinity of the point $\nu = z$. In this section we transform the results available in the literature (Abramowitz and Stegun 1965; Erdélyi *et al.* 1953; Kratzer and Franz 1960; Magnus *et al.* 1966; Olver 1974; Ryshik and Gradstein 1957; Watson 1962) into the form which is used in the main body of the book. A reader interested in the details of the derivations is referred to a clear description of the approach in Fock (1965).

E.3.1 Approximations in Terms of Airy Functions

In the transition region, the Bessel functions of the first and the second kind can be expressed as

$$J_\nu(\nu + \zeta\nu^{1/3}) = \frac{2^{1/3}}{\nu^{1/3}}\mathrm{Ai}(-2^{1/3}\zeta) + O\left(\frac{1}{|\nu|}\right) \tag{E.148}$$

$$Y_\nu(\nu + \zeta\nu^{1/3}) = -\frac{2^{1/3}}{\nu^{1/3}}\mathrm{Bi}(-2^{1/3}\zeta) + O\left(\frac{1}{|\nu|}\right), \tag{E.149}$$

where Ai(t) and Bi(t) are the Airy functions, ζ is fixed, $|\nu|$ is large, and $|\arg \nu| < \pi/2$ (Abramowitz and Stegun 1965). The relations between the Bessel functions

$$H_\nu^{(1)}(z) = J_\nu(z) + jY_\nu(z) \tag{E.150}$$

$$H_\nu^{(2)}(z) = J_\nu(z) - jY_\nu(z) \tag{E.151}$$

lead to the corresponding approximations for the Hankel functions

$$H_\nu^{(1)}\left(\nu + \zeta\nu^{1/3}\right) = \frac{2^{1/3}}{\nu^{1/3}}[\mathrm{Ai}(-2^{1/3}\zeta) - j\mathrm{Bi}(-2^{1/3}\zeta)] + O\left(\frac{1}{|\nu|}\right) \tag{E.152}$$

$$H_\nu^{(2)}\left(\nu + \zeta\nu^{1/3}\right) = -\frac{2^{1/3}}{\nu^{1/3}}[\mathrm{Ai}(-2^{1/3}\zeta) + j\mathrm{Bi}(-2^{1/3}\zeta)] + O\left(\frac{1}{|\nu|}\right), \tag{E.153}$$

which by using the identities

$$\mathrm{Ai}(t) \mp j\mathrm{Bi}(t) = 2\mathrm{e}^{\mp j\pi/3}\mathrm{Ai}(t\mathrm{e}^{\pm 2j\pi/3}) \tag{E.154}$$

can be rewritten as

$$H_\nu^{(1)}(\nu + \zeta\nu^{1/3}) = \frac{2^{4/3}}{\nu^{1/3}}\mathrm{e}^{-j\pi/3}\mathrm{Ai}(-2^{1/3}\zeta\mathrm{e}^{2j\pi/3}) + O\left(\frac{1}{|\nu|}\right) \tag{E.155}$$

$$H_\nu^{(2)}(\nu + \zeta\nu^{1/3}) = \frac{2^{4/3}}{\nu^{1/3}}\mathrm{e}^{j\pi/3}\mathrm{Ai}(-2^{1/3}\zeta\mathrm{e}^{-2j\pi/3}) + O\left(\frac{1}{|\nu|}\right). \tag{E.156}$$

By denoting $z = \nu + \zeta\nu^{1/3}$ in (E.148), (E.155), and (E.156) we obtain

$$J_\nu(z) = \frac{2^{1/3}}{\nu^{1/3}}\mathrm{Ai}\left[\frac{2^{1/3}}{\nu^{1/3}}(\nu - z)\right] + O\left(\frac{1}{|\nu|}\right) \tag{E.157}$$

$$H_\nu^{(1)}(z) = \frac{2^{4/3}}{\nu^{1/3}}\mathrm{e}^{-j\pi/3}\mathrm{Ai}\left[\mathrm{e}^{2j\pi/3}\frac{2^{1/3}}{\nu^{1/3}}(\nu - z)\right] + O\left(\frac{1}{|\nu|}\right) \tag{E.158}$$

$$H_\nu^{(2)}(z) = \frac{2^{4/3}}{\nu^{1/3}}\mathrm{e}^{j\pi/3}\mathrm{Ai}\left[\mathrm{e}^{-2j\pi/3}\frac{2^{1/3}}{\nu^{1/3}}(\nu - z)\right] + O\left(\frac{1}{|\nu|}\right), \tag{E.159}$$

and since $\nu \approx z$, $\nu^{1/3}$ can be replaced with $z^{1/3}$ to give

$$J_\nu(z) = \frac{2^{1/3}}{z^{1/3}} \mathrm{Ai}(t) + O\left(\frac{1}{|z|}\right) \quad \text{(E.160)}$$

$$H_\nu^{(1)}(z) = \frac{2^{4/3}}{z^{1/3}} \mathrm{e}^{-j\pi/3} \mathrm{Ai}(\mathrm{e}^{2j\pi/3} t) + O\left(\frac{1}{|z|}\right) \quad \text{(E.161)}$$

$$H_\nu^{(2)}(z) = \frac{2^{4/3}}{z^{1/3}} \mathrm{e}^{j\pi/3} \mathrm{Ai}(\mathrm{e}^{-2j\pi/3} t) + O\left(\frac{1}{|z|}\right), \quad \text{(E.162)}$$

where

$$t = \frac{2^{1/3}}{z^{1/3}} (\nu - z). \quad \text{(E.163)}$$

E.3.2 Approximations in Terms of Normalized Airy Functions

By using the Fock notation (Fock 1965) for the Airy functions

$$v(t) = \sqrt{\pi} \mathrm{Ai}(t) \quad \text{(E.164)}$$

$$w_1(t) = \sqrt{\pi} [\mathrm{Bi}(t) + j\mathrm{Ai}(t)] \quad \text{(E.165)}$$

$$w_2(t) = \sqrt{\pi} [\mathrm{Bi}(t) - j\mathrm{Ai}(t)] \quad \text{(E.166)}$$

and the relation (E.154), equations (E.160)–(E.162) can be rewritten as

$$J_\nu(z) = \frac{2^{1/3}}{\sqrt{\pi} z^{1/3}} v(t) + O\left(\frac{1}{|z|}\right) \quad \text{(E.167)}$$

$$H_\nu^{(1)}(z) = -\frac{j2^{1/3}}{\sqrt{\pi} z^{1/3}} w_1(t) + O\left(\frac{1}{|z|}\right) \quad \text{(E.168)}$$

$$H_\nu^{(2)}(z) = \frac{j2^{1/3}}{\sqrt{\pi} z^{1/3}} w_2(t) + O\left(\frac{1}{|z|}\right). \quad \text{(E.169)}$$

Functions $w_{1,2}(t)$ are defined by the integrals

$$w_{1,2}(t) = \frac{1}{\sqrt{\pi}} \int_{\Gamma_{1,2}} \mathrm{e}^{t\gamma - \frac{1}{3}\gamma^3} \mathrm{d}\gamma, \quad \text{(E.170)}$$

where the integration path Γ_1 (Γ_2) runs from infinity to zero along the line $\arg\gamma = -2\pi/3$ ($\arg\gamma = 2\pi/3$) and from zero to infinity along the positive real axis. The integrals (E.170) converge for all complex values of t. When t is real, the function $v(t)$ can be represented by the well-known integral

$$v(t) = \frac{1}{\sqrt{\pi}} \int_0^{+\infty} \cos\left(t\gamma + \frac{1}{3}\gamma^3\right) \mathrm{d}\gamma \quad \text{(E.171)}$$

introduced in Airy's investigations of the intensity of light in the neighborhood of a caustic (Airy 1838).

The derivatives of the Bessel functions with respect to z in the transition region can be obtained by differentiation of equations (E.160)–(E.162) or (E.167)–(E.169). The leading terms are obtained by assuming that the factors in front of the Airy functions are constant and by using the relation

$$\frac{dt}{dz} \approx -\frac{2^{1/3}}{z^{1/3}}. \tag{E.172}$$

For example, the differentiation of equations (E.167)–(E.169) leads to the formulas

$$J_\nu'(z) = -\frac{2^{2/3}}{\sqrt{\pi} z^{2/3}} v'(t) + O\left(\frac{1}{|z|^{4/3}}\right) \tag{E.173}$$

$$H_\nu^{(1)\prime}(z) = \frac{j2^{2/3}}{\sqrt{\pi} z^{2/3}} w_1'(t) + O\left(\frac{1}{|z|^{4/3}}\right) \tag{E.174}$$

$$H_\nu^{(2)\prime}(z) \approx -\frac{j2^{2/3}}{\sqrt{\pi} z^{2/3}} w_2'(t) + O\left(\frac{1}{|z|^{4/3}}\right). \tag{E.175}$$

E.3.3 Zeros in the Neighborhood of the Points $\nu = \pm z$

By using approximations from sections E.3.1 and E.3.2 we can describe the position of zeros of the Bessel functions in the neighborhood of the points $\nu = \pm z$. Functions $\mathrm{Ai}(t)$ and $\mathrm{Ai}'(t)$ have zeros on the negative real axis only. The first three zeros of the functions ($s = 1, 2, 3$) are presented in Table 6.1. A table of the zeros with $s \leq 10$ can be found in Abramowitz and Stegun (1965). If t_s with $s = 1, 2, \ldots$ is a zero, then $t_s = -|t_s|$ and it follows from the transition region approximations (E.160)–(E.162) and the definition (E.163) that the zeros ν_s of the Bessel functions can be expressed as

$$\nu_s \approx z + \sqrt[3]{\frac{z}{2}}\,|t_s|\,e^{j\chi} \tag{E.176}$$

with $\chi = \pi, \pi/3$, and $-\pi/3$ for $J_\nu(z)$, $H_\nu^{(1)}(z)$, and $H_\nu^{(2)}(z)$, respectively. Thus, the zeros are located along the lines leaving the point $\nu = z$ at the angles χ that are exactly the directions at which the anti-Stokes lines α_+, β_+, and γ_+ go from the point $\nu = z$ (see equation (E.106) and Figures E.11–E.13). The zeros of the derivatives of the Bessel functions (E.173)–(E.175) with respect to z are given by the same formula (E.176), in which t_s are now zeros of $\mathrm{Ai}'(t)$.

The zeros in the neighborhood of the point $\nu = -z$ are found by using the symmetry relations (E.17) and (E.18). The procedure is straightforward for the Hankel functions, leading to zeros at points $-\nu_s$ with $s = 1, 2, \ldots$ which are located along the straight lines tangential at the point $\nu = -z$ to the anti-Stokes lines β_- and γ_- shown in Figures E.12 and E.13. To describe the Bessel function of the first kind in a vicinity of the point $\nu = -z$, the relations (E.1), (E.17), and (E.18) can be used to give the approximation

$$J_{-\nu}(z) \approx \frac{2^{1/3}}{z^{1/3}}\left[e^{j\nu\pi - j\pi/3}\mathrm{Ai}(e^{2j\pi/3}t) + e^{-j\nu\pi + j\pi/3}\mathrm{Ai}(e^{-2j\pi/3}t)\right]. \tag{E.177}$$

Since $\nu \approx z$, the terms in the square brackets are of the order $O\left[\exp(\pm\pi \mathrm{Im}\, z)\right]$ and one of them (depending on whether $\mathrm{Im}\, z$ is positive or negative) can be neglected if $|\mathrm{Im}\, z|$ is sufficiently large, resulting in the approximation

$$2J_{-\nu}(z) \approx \begin{cases} H_{-\nu}^{(2)}(z), & \mathrm{Im}\, z > 0 \\ H_{-\nu}^{(1)}(z), & \mathrm{Im}\, z < 0 \end{cases}. \tag{E.178}$$

Thus, the zeros of $J_\nu(z)$ in the vicinity of $\nu = -z$ are close to the zeros of $H_\nu^{(2)}(z)$ if $\mathrm{Im}\, z > 0$ and to the zeros of $H_\nu^{(1)}(z)$ if $\mathrm{Im}\, z < 0$. The zeros are located along the line tangential at the point $\nu = -z$ to the anti-Stokes line α_- (Figure E.11). When $\mathrm{Im}\, z = 0$, both terms in (E.177) should be accounted for.

By construction, the transition region approximation (E.176) is expected to provide the most accurate results for the zeros of the Bessel functions with $s = 1$ that are nearest to the points $\nu = \pm z$. The approximation becomes less accurate as s increases since the zeros with greater values of s are located farther from the points $\nu = \pm z$. In contrast, the accuracy of the Debye approximation improves as s grows. The numerical data presented in Table E.2 illustrate this general tendency by the example of zeros of $H_\nu^{(2)}(z)$. It is seen that the zero ν_1 is better described in the transition region approximation. However, the Debye approximation becomes superior for $s \geq 2$. Another example is given by Table 6.1, which shows that even a simplified Debye approximation with $\Phi(\nu, z)$ approximated by (E.106) leads to sufficiently accurate results once $s \geq 2$.

References

Abramowitz M and Stegun IA (eds) 1965 *Handbook of Mathematical Functions with Formulas, Graphs, and Mathematical Tables*, Dover Publications.

Airy GB 1838 On the intensity of light in the neighborhood of a caustic. *Trans. Cambridge Phil. Soc.*, **6**, 379–402.

Bouman JJ, Senior TBA, and Uslenghi PLE (eds) 1987 *Electromagnetic and Acoustic Scattering by Simple Shapes*, Hemisphere Publishing Corporation.

Debye P 1909 Näherungsformeln für die Zylinderfunktionen für große Werte des Arguments und unbeschränkt veränderliche Werte des Index. *Math. Ann.*, **67**, 535–558.

Debye P 1910 Semikonvergente Entwicklungen für die Zylinderfunktionen und ihre Ausdehnung ins Komplexe. *Sitzungsber. d. Kön. Bay. Ak. Wiss. München, Math.-Phys. Kl.*, **40**, no. 5, 1–29.

Erdélyi A, Magnus W, Oberhettinger F, and Tricomi FG 1953 *Higher Transcendental Functions*, vol. 2, McGraw-Hill.

Felsen LB and Marcuvitz N 1973 *Radiation and Scattering of Waves*, Prentice-Hall.

Fock VA 1965 *Electromagnetic Diffraction and Propagation Problems*, Pergamon Press.

Heading J 1962 *An Introduction to Phase-integral Methods*, John Wiley & Sons.

Jeffreys H 1925 On certain approximate solutions of linear differential equations of the second order. *Proc. London Math. Soc.*, **23**, 428–436.

Kratzer A and Franz W 1960 *Transzendende Funktionen*, Akademische Verlagsgesellschaft Geest & Portig K.-G., Leipzig.

Langer RE 1931 On the asymptotic solutions of ordinary differential equations, with an application to the Bessel functions of large order. *Trans. Amer. Math. Soc.*, **33**, 23–64.

Magnus W, Oberhettinger F, and Soni RP 1966 *Formulas and Theorems for the Special Functions of Mathematical Physics*. Springer-Verlag.

Makarov GI and Osipov AV 1986 To the subject of Debye's asymptotics for cylindrical functions. *Problems of Diffraction and Wave Propagation*, **20**, 3–23, Leningrad State University (in Russian).

Makarov GI and Osipov AV 1987 On Debye's asymptotic representations of cylindrical functions. *Vestnik Leningr. Univ., Ser. 4: Fiz. Khim.*, no. 2, 47–52 (in Russian).

Makarov GI and Osipov AV 2000 Zeros of the Bessel function of the first kind and the role of continuous spectrum in the problem of electromagnetic diffraction by a finitely conducting sphere. *Vestnik Leningr. Univ., Ser. 4: Fiz. Khim.*, no. 1, 29–38 (in Russian).

Morse PM and Feshbach H 1953 *Methods of Theoretical Physics*, McGraw-Hill.

Olver FWJ 1974 *Asymptotics and Special Functions*, Academic Press.

Petrashen GI, Smirnova NS, and Makarov GI 1953 On asymptotic representations of Bessel functions. *Leningrad Gos. Univ. U. Zap.: Ser. Mat. Nauk*, no. 170 (27), 7–95 (in Russian).

Ryshik IM and Gradstein IS 1957 *Tables of Series, Products, and Integrals*, VEB, Berlin.

Stokes GG 1857 On the discontinuity of arbitrary constants which appear in divergent developments. *Trans. Cambridge Phil. Soc.*, **9**, 166–187. Reprinted in *Math. Phys. Papers*, **4**, 77–109, Cambridge University Press, 1904.

Watson GN 1962 *A Treatise on the Theory of Bessel Functions*, Cambridge.

Index

Modern Electromagnetic Scattering Theory with Applications, First Edition. Andrey V. Osipov and Sergei A. Tretyakov.

Printed and bound by CPI Group (UK) Ltd, Croydon, CR0 4YY
16/09/2026
14963389-0008